NanoScience and Technology

NanoScience and Technology

The series NanoScience and Technology is focused on the fascinating nano-world, mesoscopic physics, analysis with atomic resolution, nano and quantum-effect devices, nanomechanics and atomic-scale processes. All the basic aspects and technology-oriented developments in this emerging discipline are covered by comprehensive and timely books. The series constitutes a survey of the relevant special topics, which are presented by leading experts in the field. These books will appeal to researchers, engineers, and advanced students.

Please view available titles in *NanoScience and Technology* on series homepage
http://www.springer.com/series/3705/

Bharat Bhushan
(Editor)

Scanning Probe Microscopy in Nanoscience and Nanotechnology 2

With 453 Figures

Springer

Editor
Professor Bharat Bhushan
Ohio State University
Nanoprobe Laboratory for Bio- and Nanotechnology and Biomimetics (NLB^2)
201, West 19th Avenue
Columbus, OH 43210-1142, USA
E-mail: bhushan.2@osu.edu

NanoScience and Technology ISSN 1434-4904
ISBN 978-3-642-10496-1 e-ISBN 978-3-642-10497-8
DOI 10.1007/978-3-642-10497-8
Springer Heidelberg Dordrecht London New York

Cover design: eStudio Calamar Steinen

Printed on acid-free paper

Springer is part of Springer Science+Business Media (www.spinger.com)

Foreword

Nature is the best example of a system functioning on the nanometer scale, where the involved materials, energy consumption, and data handling are optimized. Opening the doors to the nanoworld, the emergence of the scanning tunneling microscope in 1982 and the atomic force microscope in 1986 led to a shift of paradigm in the understanding and perception of matter at its most fundamental level. As a consequence, new revolutionary concepts stimulated a number of new technologies. The current volume of *Scanning Probe Methods in Nanoscience and Nanotechnology* shows that these methods are still making a tremendous impact on many disciplines that range from fundamental physics and chemistry through information technology, spintronics, quantum computing, and molecular electronics, all the way to life sciences. Indeed, more than 6,000 AFM-related papers were published in 2008 alone, bringing the total to more than 70,000 since its invention, according to the Web of Science, and the STM has inspired a total of 20,000 papers. There are also more than 500 patents related to the various forms of scanning probe microscopes. Commercialization of the technology started at the end of the 1980s, and approximately 12,000 commercial systems have been sold so far to customers in areas as diverse as fundamental research, the car industry, and even the fashion industry. There are also a significant number of home-built systems in operation. Some 60–80 companies are involved in manufacturing SPM and related instruments. Indeed, not even the sky seems to be the limit for AFM technology. The Rosetta mission to comet 67P launched by the European Space Agency in 2004 includes an AFM in its Micro-Imaging Dust Analysis System (MIDAS) instrument. The goal of this mission, which is expected to touch down on 67P in 2014, is to analyze particle size distributions in comet material. NASA's Phoenix mission to Mars in 2008 included an AFM for similar studies (collaboration between the Universities of Neuchâtel and Basel, as well as with Nanosurf GmbH).

What does the future hold? Nanotechnology is still dominated to a certain extent by the top-down approach where miniaturization plays a crucial role. However, there is a worldwide effort of meeting the bottom-up approach of self-assembly and self-organization that has been so successfully implemented in the natural world. Researchers are trying to unravel nature's secrets on a nanometer scale to create a new generation of materials, devices, and systems that will spectacularly outperform those we have today in information technology, medicine and biology, environmental technologies, the energy industry, and beyond. As we better understand how nature is doing "things" on a fundamental level, achievements like clean chemistry or clean processing will emerge along with how to handle waste problems and not polluting the environment. New smart materials, hybrid or heterostructured, as well as carbon nanotubes, a variety of nanowires or graphene could be the ingredients for novel energy-saving devices. In order to understand the whole functionality of a cell, Systems Biology Institutes have been established with the hope of artificially synthesizing a cell in a bottom-up approach. Nanomedicine, including noninvasive diagnostics, will be more and more on the agenda, fighting diseases on the molecular level, e.g., new kinds of drug-delivery systems based on peptides or block copolymer nanocontainers are investigated as possible carriers to target carcinogenic cells. Biology is driven by chemistry; however, the scaffold, the gears, the knots and bolts, e.g., in cell membranes, is nanomechanics, a template that nature has orchestrated during eons of evolution and worthwhile trying to copy and implement in novel nanodevices.

However, to keep this worldwide effort alive, the interdisciplinary structure of nano requires a new breed of scientists educated in all science disciplines with no language barriers, ready to make an impact on all the global challenges ahead where nanotechnology can be applied.[1] Scanning probe microscopy and related methods will still play an important role in many of these investigations, helping to capitalize on this fundamental knowledge, beneficial for future technologies and to mankind.

Prof. Christoph Gerber
(Co-invertor of AFM)
(Formerly at IBM Zurich Research Laboratory)
National Competence Center for Research in Nanoscale Science (NCCR)
Institute for Physics
University of Basel
Switzerland

[1] Nano-curriculum at the University of Basel, http://www.nccr.nano.org/NCCR/study.

Preface

The emergence and proliferation of proximal probes, in particular tip-based microscopies, have found applications in a large number of fields of scientific and industrial interest. These allow investigations down to the atomic scale. The recent focus on nanotechnology has made probe-based methods indispensible. The present editor coedited with Prof. H. Fuchs 13 volumes on applied scanning probe methods (SPM) from 2004 to 2009. These volumes have provided a timely comprehensive overview of SPM applications.

The success of the Springer Series Applied Scanning Probe Methods and the rapidly expanding activities in scanning probe development and applications in nanoscience and nanotechnology worldwide makes it a natural step to collect further specific results in the fields of development of scanning probe microscopy techniques, characterization, and industrial applications, particularly in nanoscience, nanotechnology, and biomimetics. In 2010, the editor launched a series of volumes on Scanning Probe Microscopy in Nanoscience and Nanotechnology. This second volume provides insight into the recent work of leading specialists in their respective fields. The focus in this volume is on the fundamental developments in SPM techniques.

This volume introduces many technical concepts and improvements of existing scanning probe techniques and covers a broad and impressive spectrum of recent SPM development and application in many fields of technology, biology, and medicine. The chapters are given under three major headings: *Scanning Probe Microscopy Techniques*, *Characterization*, and *Industrial Applications*. After introducing new developments in scanning probe microscopy, characterization data in various applications of scientific and technological interest is presented. Next, chapters on various industrial applications are presented. Characterization data and industrial applications include studies of biological materials, nanostructures, and nanotubes.

The chapters are written by leading researchers and application scientists from all over the world and from various industries to provide a broader perspective. The field is progressing so fast that there is a need for a set of volumes every 12–18 months to capture the latest developments.

We gratefully acknowledge the support of all authors representing leading scientists in academia and industry for the highly valuable contribution to this volume.

We also cordially thank the series editor Claus Ascheron and his staff member Adelheid Duhm from Springer for their continued support during the publication process.

We sincerely hope that readers find this volume to be scientifically stimulating and rewarding.

Columbus, OH, USA
September 2010

Bharat Bhushan

Contents

Contributors

Stefano Agnoli Dipartimento di Scienze Chimiche e unità di ricerca, INSTM, Università di Padova, Via Marzolo 1, 35131 Padova, Italy, stefano.agnoli@unipd.it

Johan Angenete Nanofactory Instruments, Sven Hultins gata 9A, 412 88 Gothenburg, Sweden, johan.angenete@nanofactory.com

Arsen Babajanyan Department of Physics and Basic Science Institute for Cell Damage Control, Sogang University, Seoul 121-742, Korea, barsen12@gmail.com

Mirko Ballarini Department of Physics, Politecnico di Torino, 10129 Torino, Italy, mirko.ballarini@polito.it

Igor Beinik Institute of Physics, University of Leoben, 8700, Leoben, Austria, igor.beinik@unileoben.ac.at

Prof. Bharat Bhushan Nanoprobe Laboratory for Bio- & Nanotechnology and Biomimetics (NLB^2), 201 W. 19th Avenue, Ohio State University, Columbus, OH 43210-1142, USA, bhushan.2@osu.edu

Sophie Bistac Université de Haute Alsace, Laboratoire de Chimie Organique, Bio-organique et Macromoléculaire (COBM), Ecole Nationale Supérieure de Chimie de Mulhouse (ENSCMu), 3, rue Alfred Werner, 68093 Mulhouse Cedex, France, sophie.bistac-brogly@uha.fr

Dieter Blaas Max F. Perutz Laboratories, Medical University of Vienna, 1030 Vienna, Austria, dieter.blaas@meduniwien.ac.at

Maurice Brogly Université de Haute Alsace, Laboratoire de Chimie Organique, Bio-organique et Macromoléculaire (COBM), Ecole Nationale Supérieure de Chimie de Mulhouse (ENSCMu), 3, rue Alfred Werner, 68093 Mulhouse Cedex, France, maurice.brogly@uha.fr

Yuguang Cai Department of Chemistry, University of Kentucky, Chemistry-Physics Building 104, Lexington, KY 40506, USA, ycai3@email.uky.edu

J.P. Carmo Department of Industrial Electronics, University of Minho, Campus Azurem, 4800-058 Guimaraes, Portugal, jcarmo@dei.uminho.pt

Emanuele Cavaliere Dipartimento di Matematica e Fisica, Università Cattolica del Sacro Cuore, Via dei Musei 41, KY 25121, Brescia, Italy, e.cavaliere@dmf.unicatt.it

Valeria Chiono Department of Mechanics, Politecnico di Torino, 10129 Torino, Italy, valeria.chiono@polito.it

Prof. Gianluca Ciardelli Department of Mechanics, Politecnico di Torino, 10129 Torino, Italy, gianluca.ciardelli@polito.it

Dr. Robert Cook Material Measurement Laboratory, National Institute of Standards and Technology, 100 Bureau Drive, Gaithersburg, MD 20899, USA, robert.cook@nist.gov

J.H. Correia Department of Industrial Electronics, University of Minho, Campus Azurem, 4800-058 Guimaraes, Portugal, higino.correia@dei.uminho.pt

Prof. Mario D'Acunto Istituto di Scienza e Tecnologia Informatica, ISTI-CNR, Via Moruzzi 1, 56124 Pisa, Italy, mario.dacunto@isti.cnr.it

Sissi de Beer Physics of Complex Fluids, University of Twente, P.O. Box 217, 7500 AE Enschede, The Netherlands, s.j.a.debeer@utwente.nl

Dr. Frank DelRio Material Measurement Laboratory, National Institute of Standards and Technology, 100 Bureau Drive, Gaithersburg, MD 20899, USA, frank.delrio@nist.gov

Emiliano Descrovi Department of Physics, Politecnico di Torino, 10129 Torino, Italy, emiliano.descrovi@polito.it

Daniel Ebeling Physics of Complex Fluids, University of Twente, P.O. Box 217, 7500 AE Enschede, The Netherlands, d.ebeling@utwente.nl

Andreas Ebner Institute of Biophysics, Johannes Kepler University, Altenbergerstrasse 69, 4040 Linz, Austria, andreas.ebner@jku.at

Dr. Robert Eibl Institute of Pathology, Technical University of Munich, Plainburgstrasse 8, D-83457 Bayerisch Gmain, Germany, robeibl@stanfordmedalumni.org

Ahmad Fahs Université de Haute Alsace, Laboratoire de Chimie Organique, Bio-organique et Macromoléculaire (COBM), Ecole Nationale Supérieure de Chimie de Mulhouse (ENSCMu), 3, rue Alfred Werner, 68093 Mulhouse Cedex, France, ahmad.fahs@uha.fr

Prof. Barry A. Friedman Department of Physics, Sam Houston State University, Texas 77341, USA, phy_baf@shsu.edu

J.K. Garleff Department of Applied Physics, COBRA Inter-University Research Institute, Eindhoven University of Technology, PO Box 513, Eindhoven 5600 MB, The Netherlands, j.k.garleff@tue.nl

Luca Gavioli Dipartimento di Matematica e Fisica, Università Cattolica del Sacro Cuore, Via dei Musei 41, KY 25121, Brescia, Italy, luca.gavioli@unicatt.it

Piergiorgio Gentile Department of Mechanics, Politecnico di Torino, 10129 Torino, Italy, piergiorgio.gentile@polito.it

Filippo Giannazzo CNR-IMM Stradale, Primosole, 50, 95121 Catania, Italy, filippo.giannazzo@imm.cnr.it

Fabrizio Giorgis Department of Physics, Politecnico di Torino, 10129 Torino, Italy, fabrizio.giorgis@polito.it

L.M. Gonçalves Department of Industrial Electronics, University of Minho, Campus Azurem, 4800-058 Guimaraes, Portugal, lgoncalves@dei.uminho.pt

Gaetano Granozzi Dipartimento di Scienze Chimiche e unità di ricerca, INSTM, Università di Padova, Via Marzolo 1, 35131 Padova, Italy, gaetano.granozzi@unipd.it

Hermann Gruber University of Linz, Institute for Biophysics, Altenbergerstrasse 69, 4040 Linz, Austria, hermann.gruber@jku.at

Prof. Peter Hinterdorfer University of Linz, Institute for Biophysics, Altenbergerstrasse 69, 4040 Linz, Austria, peter.hinterdorfer@jku.at

Stefan Howorka Department of Chemistry, Institute of Structural Molecular Biology, University College London, London WC1H 0AJ, UK, s.howorka@ucl.ac.uk

Dr. Chun-Sheng Jiang National Renewable Energy Laboratory, 1617 Cole Blvd., Golden, CO 80401, USA, Chun.sheng.jiang@nrel.gov

Markus Kastner Institute of Biophysics, Johannes Kepler University, Altenbergerstrasse 69, 4040 Linz, Austria, markus.kastner@jku.at

Dr. G. Djuidjé Kenmoé Laboratoire de Mécanique, Département de Physique, Faculté des Sciences, Université de Yaoundé I, B.P. 812 Yaoundé, Cameroun, kdjuidje@yahoo.fr

Ferry Kienberger Agilent Technologies Austria GmbH, 4040 Linz, Austria, ferry_kienberger@agilent.com

Timoléon C. Kofané Laboratoire de Mécanique, Département de Physique, Faculté des Sciences, Université de Yaoundé I, B.P. 812 Yaoundé, Cameroun, tckofane@yahoo.com

P.M. Koenraad Department of Applied Physics, COBRA Inter-University Research Institute, Eindhoven University of Technology, PO Box 513, Eindhoven 5600 MB, The Netherlands, p.m.koenraad@tue.nl

Dr. V. Körstgens Physik-Department E13, Lehrstuhl für Funktionale Materialien, Technische Universität München, James-Franck-Str. 1, 85747 Garching, Germany, volker.koerstgens@ph.tum.de

Dr. Kiejin Lee Department of Physics and Basic Science Institute for Cell Damage Control, Sogang University, Seoul 121-742, Korea, klee@sogang.ac.kr

Sy-Hwang Liou Department of Physics and Astronomy, Nebraska Center for Materials and Nanoscience, University of Nebraska, Lincoln, NE 68588-0111, USA, sliou@unl.edu

Anselm Loges Eberhard Karls Universität Tübingen Institute for Geosciences Wilhelmstraße 56, D-72074 Tübingen, Germany, anselm.loges@uni-tuebingen.de

Lingbo Lu Department of Chemistry, University of Kentucky, 505 Rose Street, Lexington, Kentucky 40506, Llu3@uky.edu

Harutyun Melikyan Department of Physics and Basic Science Institute for Cell Damage Control, Sogang University, Seoul 121-742, Korea, harut80@gmail.com

Prof. Peter Müller-Buschbaum Physik-Department E13, Lehrstuhl für Funktionale Materialien, Technische Universität München, James-Franck-Str. 1, 85747 Garching, Germany, muellerb@ph.tum.de

Prof. Frieder Mugele Physics of Complex Fluids, University of Twente, P.O. Box 217, 7500 AE Enschede, Netherlands, f.mugele@utwente.nl

Alexandra Nafari Nanofactory Instruments, Sven Hultins gata 9A, SE-412 88 Gothenburg, Sweden, alexandra.nafari@chalmers.se

Prof. Klaus Georg Nickel Eberhard Karls Universität Tübingen, Institute for Geosciences, Wilhelmstraße 56, D-72074 Tübingen, Germany, klaus.nickel@uni-tuebingen.de

Prof. Michael Nosonovsky Department of Mechanical Engineering, EMS Building, Room E371G, University of Wisconsin-Milwaukee, Milwaukee, WI 53201-0413, USA, nosonovs@uwm.edu

Prof. Håkan Olin Mid Sweden University, SE-851 70, Sundsvall, Sweden, hakan.olin@miun.se

Dr. Guido Paolicelli CNR-Istituto Nanoscienze, Centro S3 Via Campi 213/A 41125 Modena, Italy, guido.paolicelli@unimore.it

Philipp D. Pollheimer Institute of Biophysics, Johannes Kepler University, Altenbergerstrasse 69, 4040 Linz, Austria, philipp.pollheimer@jku.at

Volker Presser Drexel University, Nano Materials Group, PA 134, Philadelphia, USA, drexel@applied-mineralogy.com

Dr. Vito Raineri CNR-IMM Stradale, Primosole, 50, 95121 Catania, Italy, vito.raineri@imm.cnr.it

Christian Rankl Measurement Research Lab, Agilent Technologies Austria GmbH, Aubrunnerweg 11, 4040 Linz, Austria, christian_rankl@agilent.com

Emanuele Rimini CNR-IMM Stradale, Primosole, 50, 95121 Catania, Italy, emanuele.rimini@ct.infn.it

Dr. Massimo Rovatti Dipartimento di Fisica, Università di Modena e Reggio Emilia Via Campi 213/A 41125 Modena, Italy
and
CNR-Istituto Nanoscienze, Centro S3 Via Campi 213/A 41125 Modena, Italy, massimo.rovatti@unimore.it

Anke Sanz-Velasco Chalmers University of Technology, Kemivägen 9, SE-412 96, Gothenburg, Sweden, anke.sanz-velasco@chalmers.se

Ali Fatih Sarioglu Electrical Engineering, Stanford University, E.L. Ginzton Laboratory, Stanford, CA 94305-4088, USA, sarioglu@stanford.edu

Susanna Sartori Department of Mechanics, Politecnico di Torino, 10129 Torino, Italy, susanna.sartori@polito.it

Prof. Olav Solgaard Electrical Engineering, Stanford University, E.L. Ginzton Laboratory, Stanford, CA 94305-4088, USA, solgaard@stanford.edu

Yuki Suzuki Laboratory of Plasma Membrane and Nuclear Signaling, Graduate School of Biostudies, Kyoto University, Yoshida-Konoe-cho, Sakyo-ku, Kyoto 606-8501, Japan, ysuzuki79.m07@lif.kyoto-u.ac.jp

Krister Svensson Karlstad University, Universitetsgatan 2, 651 88 Karlstad, Sweden, krister.s@kau.se

Prof. Kunio Takeyasu Laboratory of Plasma Membrane and Nuclear Signaling, Graduate School of Biostudies, Kyoto University, Yoshida-Konoe-cho, Sakyo-ku, Kyoto 606-8501, Japan, takeyasu@lif.kyoto-u.ac.jp

Prof. Christian Teichert Institute of Physics, University of Leoben, 8700, Leoben, Austria, christian.teichert@unileoben.ac.at

J.M. Ulloa Semiconductors group, Institute for Systems based on Optoelectronics and Microtechnology (ISOM) – Universidad Politécnica de Madrid, Ciudad Universitaria s/n 28040 Madrid, Spain, jmulloa@die.upm.es

Prof. Sergio Valeri Dipartimento di Fisica, Università di Modena e Reggio Emilia Via Campi 213/A 41125 Modena, Italy
and
CNR-Istituto Nanoscienze, Centro S3 Via Campi 213/A 41125 Modena, Italy, sergio.valeri@unimore.it

Dirk van den Ende Physics of Complex Fluids, University of Twente, P.O. Box 217, 7500 AE Enschede, Netherlands, h.t.m.vandenEnde@utwente.nl

Linda Wildling University of Linz, Institute for Biophysics, 4040 Linz, Austria, linda.wildling@jku.at

Seung-Ho Yang National Institute of Standards and Technology, 100 Bureau Dr., Stop 8520, Gaithersburg, MD 20899-8520, USA, seung.yang@nist.gov

Masatoshi Yokokawa Laboratory of Plasma Membrane and Nuclear Signaling, Graduate School of Biostudies, Kyoto University, Yoshida-Konoe-cho, Sakyo-ku, Kyoto 606-8501, Japan, yokokawa@lif.kyoto-u.ac.jp

Shige H. Yoshimura Laboratory of Plasma Membrane and Nuclear Signaling, Graduate School of Biostudies, Kyoto University, Yoshida-Konoe-cho, Sakyo-ku, Kyoto 606-8501, Japan, yoshimura@lif.kyoto-u.ac.jp

Lanping Yue Nebraska Center for Materials and Nanoscience, University of Nebraska, Lincoln, NE 68588-0111, USA, lyue2@unl.edu

Huan Zhang National Institute of Standards and Technology, 100 Bureau Dr., Stop 8520, Gaithersburg, MD 20899-8520, USA, ZhangH@utrc.utc.com

Rong Zhu University of Linz, Institute for Biophysics, 4040 Linz, Austria, rong.zhu@jku.at

Part I
Scanning Probe Microscopy Techniques

Chapter 1
Time-Resolved Tapping-Mode Atomic Force Microscopy

Ali Fatih Sarioglu and Olav Solgaard

Abstract Atomic force microscopy has unprecedented potential for quantitative mapping of material-specific surface properties on the nanoscale. Unfortunately, methods developed for local stiffness measurements suffer from low operational speeds and they require large forces to be applied to the surface, limiting resolution and precluding measurements on soft materials such as polymers and biological samples. On the other hand, tapping-mode AFM, which is well suited to soft materials due to its gentle interaction with the surface, cannot be used to recover information on the tip–sample interaction (and hence, on the material properties) due to limited mechanical bandwidth offered by the resonant AFM probe.

In this chapter, a technique, called Time-resolved Tapping-mode Atomic Force Microscopy, designed for rapid quantitative material characterization on the nanoscale is described. The technique is based on time-resolved measurement of tip–sample interaction forces during tapping-mode AFM imaging by a specially designed micromachined AFM probe. The probe has an integrated high-bandwidth interferometric force sensor that is used to resolve tip–sample interaction forces with high sensitivity and temporal resolution. In the first part of the chapter, the theory, design, and fabrication of the probes are described in detail. Then quantitative force measurements with microsecond time resolution in tapping-mode imaging are presented. Finally, higher harmonic images based on the interaction force measurements are presented for various samples, demonstrating the range of applications of the technique.

1.1 Introduction

Atomic force microscopy (AFM) [1] is one of the main techniques for high-resolution nondestructive measurement of surface topography. Its operation is based on the mechanical interaction of an atomically sharp tip at the end of a flexible cantilever beam with the sample surface. By raster scanning the sharp tip over the sample surface and recording its displacement, AFM provides atomic scale resolution [2] and can be employed in a variety of environments, such as air, vacuum and liquids, and over wide range of temperatures. The high resolution together with such versatility makes this microscopy technique applicable over a wide range of problems in various disciplines from biology to material science.

Beyond topography measurements, there is an increasing interest in the use of AFM for force measurements and quantitative mapping of local elastic and viscoelastic surface properties. In fact, there are several AFM-based techniques already developed for local mechanical measurements, such as nanoindentation [3], force volume imaging [4], ultrasonic force microscopy [5, 6], pulsed-force microscopy [7, 8], and force modulation microscopy [9, 10]. However, these techniques either require excessive forces to be applied on the surface, limiting their use to stiff materials, or rely on the measurement of a parameter whose value is affected by several processes other than the sample elastic properties, complicating the interpretation of the results of these techniques.

Tapping-mode atomic force microscopy (TM-AFM) [11] is a dynamic imaging technique, in which the probe is oscillated close to its resonance frequency and contacts with the sample briefly once in every oscillation. The cantilever oscillation amplitude is kept constant by the feedback loop, and the feedback signal to the piezoelectric actuator is recorded as a topography image. One of the advantages of the TM-AFM imaging is that it eliminates the continuous tip–sample contact that leads to large frictional forces that may damage both the tip and the sample. This makes this mode especially well suited for imaging of weakly immobilized macromolecules, as well as soft samples such as polymer surfaces and biological samples [12]. Consequently, this technique is one of the most widely used AFM imaging modes.

In addition to being a popular technique for topography measurements, TM-AFM can be used for compositional mapping of surfaces by recording the cantilever phase [13, 14]. In fact, TM-AFM offers unique advantages for quantitative material characterization. The dynamic interaction forces between the tip and the sample contain important information on adhesive and elastic properties of the surface. Therefore, measurement of interaction forces is a more direct method for material characterization. Measurement of tip–sample interaction forces have been demonstrated by recording the higher harmonics of the cantilever oscillations [15, 16]. Unfortunately, these measurements suffer from low signal-to-noise ratio mainly due to the limited mechanical bandwidth of the resonant cantilever in TM-AFM. Recently, several techniques have been introduced to increase the SNR in measurements of the higher frequency components of the tip–sample interaction forces. These developments can be summarized as engineering the cantilever frequency response by designing special cantilever geometries [17, 18], driving the cantilever at its multiple eigenmodes simultaneously [19], employing fast and wide-bandwidth membrane-based probes [20, 21] and using specialized cantilevers with off-axis tip to utilize the higher frequency torsional mode [22, 23]. In this chapter, we focus on direct measurement of the nonlinear tip–sample interaction forces in TM-AFM using AFM probes with integrated high-bandwidth interferometric force sensors that offer increased temporal resolution [24, 25]. These probes combine the advantages of TM-AFM imaging and AFM force spectroscopy in a single device. They are unique in the sense that high temporal resolution in force measurements is contributed by the reduced mass of the force sensor. In addition, the integrated differential interferometric sensor enables independent measurement of cantilever motion and interaction force by minimizing crosstalk between cantilever oscillations and tip motion.

The organization of the chapter is as follows: In Sect. 1.2, the tip–sample interaction forces are explained using analytical models (Sect. 1.2.1) and the mechanical response of a typical AFM probe is discussed (Sect. 1.2.2). In Sect. 1.3, we introduce an AFM probe that is developed to measure the interaction forces in TM-AFM with an interferometric high-bandwidth force sensor. First, the dynamics of the probe in TM-AFM is simulated using analytical models (Sect. 1.3.1). The design and operating principles together with the details of microfabrication process is explained in detail (Sect. 1.3.2–1.3.4). Next, the detection techniques, characterization results, and calibration procedures for quantitative force measurements are presented. (Sects. 1.3.5 and 1.3.6) Finally, the section is concluded with the experimental force measurement results (Sect. 1.3.7). In Sect. 1.4, the compositional mapping of various surfaces by utilizing the higher harmonics of the measured tip–sample interaction forces is demonstrated. Specifically, high-bandwidth harmonic imaging results of a checkerboard test sample (Sect. 1.4.1), self-assembled monolayer (SAM) of alkanethiols on gold surface (Sect. 1.4.2) and SBS triblock copolymer film (Sect. 1.4.3) are demonstrated. Finally, the chapter is concluded with a summary and discussion in Sect. 1.5.

1.2 Tip–Sample Interactions in TM-AFM

In the following section, first, the tip–sample interaction forces in TM-AFM are explained using analytical models and the mechanical response of a typical AFM probe under these forces is discussed.

1.2.1 Interaction Forces in TM-AFM

The tip–sample interactions in TM-AFM are dynamic in nature due to the oscillatory motion of the probe [26]. When the tip and sample are close, but not in contact, the tip is pulled towards the sample by long-range attractive forces. These forces originate due to Van der Waals interaction and are relatively weak. On the other hand, as the tip makes contact with the surface, the tip is pushed away by elastic forces that arise due to surface indentation. Therefore, in every cantilever oscillation period, the tip traverses both the attractive and repulsive force regimes.

In this chapter, to model the tip–sample interaction, we use the Dejarguin–Muller–Toporov (DMT) contact model [27,28]. In this model, the AFM tip and the sample surface are modeled as a sphere and a flat half-space, respectively. At large tip–sample distances, the attractive forces between the tip and the sample surface are derived from the Van der Waals energy. For the geometry of an interacting sphere and a flat surface, the Van der Waals force is given by

$$F_{\text{att}} = \frac{-HR}{6z^2}. \tag{1.1}$$

In this expression, H is the Hamaker constant, R is the radius of curvature of the AFM tip, and z is the distance between the tip and the sample surface.

As the tip approaches and contacts with the sample, repulsive forces emerge in addition to attractive forces. These shorter range forces originate from Pauli and ionic interactions between atoms. In the DMT model, Van der Waals forces between the two bodies are assumed to be constant when the tip–sample distance is smaller than a certain parameter referred as the intermolecular distance. In addition, the repulsive forces are modeled as elastic forces occurring between two bodies due to Hertz contact theory [29]. By combining these two effects, the total interaction force in the repulsive regime is given by

$$F_{\text{rep}} = -\frac{HR}{6a_0^2} + \frac{4}{3}E^*\sqrt{R}d^{3/2}. \tag{1.2}$$

In Eq. (1.2), a_0 is the intermolecular distance, d is the sample indentation and E^* is the reduced Young's modulus of the tip–sample contact. The reduced Young's modulus of the tip–sample contact is given by [29]

$$E^* = \left[\frac{1-\upsilon_{\text{t}}^2}{E_{\text{t}}} + \frac{1-\upsilon_{\text{s}}^2}{E_{\text{s}}}\right]^{-1}, \tag{1.3}$$

where E_{t} (E_{s}) and υ_{t} (υ_{s}) are the tip (sample) Young's modulus and Poisson ratio, respectively.

1.2.2 Cantilever Dynamics and Mechanical Bandwidth in TM-AFM

In this section, we analyze the frequency response of an AFM probe when operated in TM-AFM and discuss the limitations imposed by the probe mechanical bandwidth on the measurements of tip–sample interaction forces. As mentioned before, in TM-AFM the tip interacts with the surface repeatedly at a rate close to the cantilever resonance frequency. Therefore, in steady state, the tip–sample interaction force waveform is treated as a periodic signal and can be expressed as a Fourier series.

$$F_{\text{ts}}(t) = \frac{1}{2}a_0 + \sum_{n=1}^{\infty} a_n\cos(n\omega t) + b_n\sin(n\omega t), \tag{1.4}$$

where the Fourier coefficients are given by

$$a_n = \frac{\omega}{\pi} \int_{-\frac{\pi}{\omega}}^{\frac{\pi}{\omega}} F_{\mathrm{ts}}(t) \cos(n\omega t)\mathrm{d}t, \tag{1.5}$$

$$b_n = \frac{\omega}{\pi} \int_{-\frac{\pi}{\omega}}^{\frac{\pi}{\omega}} F_{\mathrm{ts}}(t) \sin(n\omega t)\mathrm{d}t. \tag{1.6}$$

In the Fourier series of the tip–sample interaction signal, the higher harmonics at the integer multiples of the cantilever drive frequency are due to nonlinearity in the tip–sample interaction discussed in the previous section.

It follows from the discussion above that time-resolved measurements of tip–sample interaction forces in TM-AFM requires a mechanical bandwidth that covers the higher harmonics of the interaction force at the integer multiples of the drive frequency. However, the mechanical response of a resonant AFM cantilever is inherently limited beyond its fundamental resonance frequency. This limitation can be understood from an analysis of the cantilever transfer function.

Figure 1.1 shows the frequency response of a typical rectangular AFM cantilever calculated using analysis (FEA). In this plot, the vertical axis is the displacement gain, which is defined as the ratio of tip displacement amplitude to its DC response, whereas the horizontal axis is the frequency. The peaks in the displacement gain are due to fundamental and higher order flexural resonances of the cantilever. From the plot, we see that below its fundamental resonance frequency, the probe has a flat band response, whereas the mechanical response of the cantilever beam is significantly reduced beyond its fundamental resonance frequency except in close vicinity to higher order modes. As a result, higher harmonics of the periodic interaction force signal are lost unless they are in the close vicinity of higher order resonances and tip–sample interaction forces cannot be fully recovered from the cantilever displacement.

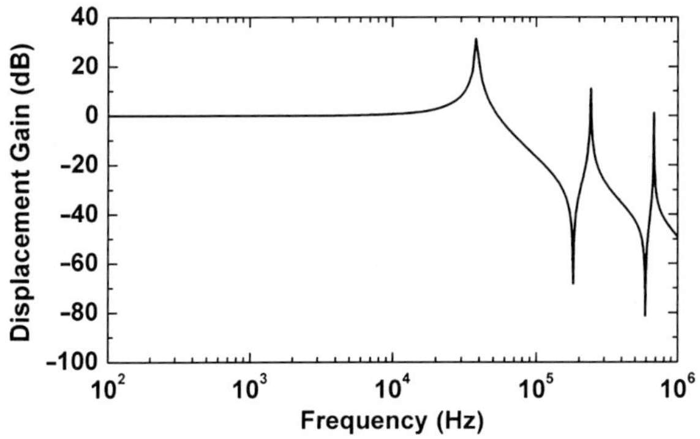

Fig. 1.1 Simulated mechanical response of a typical rectangular AFM cantilever. The cantilever is 300-μm long, 60-μm wide, and 2.5-μm thick and is made of single crystal silicon

Note that the limited bandwidth of TM-AFM is a result of the cantilever being driven on resonance. In fact, forces between an AFM tip and the sample surface can be measured from the displacements of the cantilever when tip–sample interaction force spectrum is band limited within the flat band of the cantilever. Therefore, when the AFM cantilevers are used for force measurement applications, they are either operated statically, or the dynamic interaction rate is kept well below the cantilever resonance frequency. However, these techniques sacrifice the advantages of a resonant probe such as speed, gentle tip–sample interaction, and tip-limited spatial resolution.

1.3 AFM Probes with Integrated Interferometric High Bandwidth Force Sensors

In the previous section, we discussed the potential benefits of measurement of tip–sample interaction forces in TM-AFM. We also pointed out that measurement of these forces remains a difficult challenge due to the limited mechanical bandwidth of conventional AFM probes operated in TM-AFM mode.

In this section, we discuss a new type of AFM probe [24, 25] that can quantitatively measure the interaction forces between the tip and the sample surface during TM-AFM imaging. This probe has an integrated high-bandwidth force sensor that provides the necessary temporal resolution for time-resolved force measurements. To gain insight into the operation of these probes, first the probe is modeled as a coupled damped harmonic oscillator system. Using the tip–sample contact model described in Sect. 1.2.1, the probe dynamics in TM-AFM operation is simulated, and the design principles are explained. In particular, important aspects of mechanical and optical designs are illuminated. In the following sections, the microfabrication process used in realizing these probes is described. The finished probes are characterized and procedures for calibration are explained. Finally, the section is concluded by presenting time-resolved force measurements.

The geometry of the AFM probe with an integrated high-bandwidth force sensor is shown in Fig. 1.2. The probe is mainly composed of a rectangular cantilever beam and an interferometric force sensor. The force sensor is a diffraction grating at the end of the cantilever beam. The sharp tip resides at the very end of the force sensor and is coupled to a stiff and small mechanical resonator that is a part of the diffraction grating. This tip-coupled resonator is much smaller and stiffer than the cantilever beam and therefore has a higher mechanical bandwidth than the soft and massive cantilever beam. As a result, the resonator remains mechanically responsive to the high frequency components of the nonlinear tip–sample interaction forces where the response of the resonant cantilever is limited. Consequently, the tip–sample interaction forces can be obtained through the relative displacement of this resonator with respect to the cantilever beam.

This probe structure combines the advantages of TM-AFM imaging and AFM force spectroscopy in a single device. Specifically, the resonant cantilever

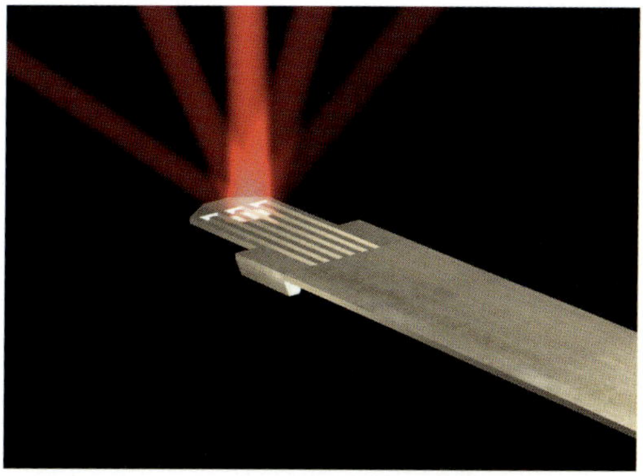

Fig. 1.2 A computer drawing showing the geometry of our AFM probe with an integrated force sensor. The probe is illuminated with a laser beam and the resulting diffraction pattern is also indicated

component provides gentle tip–sample interaction and high resolution as in TM-AFM, whereas the force sensor operated below its resonance can provide quantitative time-resolved measurements of tip–sample interaction forces.

1.3.1 Model

Damped harmonic oscillator models have been widely used to describe probe mechanics in TM-AFM [30–37]. When combined with a tip–sample contact model, these models have proved to be useful in understanding basic characteristics of cantilever motion and tip–sample interactions. In this section, using similar models, we show that with the addition of a high-bandwidth force sensor to an AFM probe, it is possible to measure the tip–sample interaction forces with high temporal resolution.

The model to simulate the dynamics of the probes with high bandwidth force sensors consists of two damped harmonic oscillators coupled to each other. A schematic of the model together with the assumed coordinate system is shown in Fig. 1.3. To simulate TM-AFM operation, the cantilever (the larger mass) is excited with a sinusoidal force at a frequency close to the system's fundamental resonance frequency. On the other hand, the forces due to periodic tip–sample interaction acts on the high-bandwidth force sensor (smaller mass). The equations of motion for the system can be written as:

$$m_c \frac{\partial^2 z_c}{\partial t^2} = -k_c z_c - \gamma_c \frac{\partial z_c}{\partial t} - k_s (z_c - z_s) - \gamma_s \left(\frac{\partial z_c}{\partial t} - \frac{\partial z_s}{\partial t} \right) + F_{dr}, \tag{1.7}$$

$$m_s \frac{\partial^2 z_s}{\partial t^2} = -k_s (z_s - z_c) - \gamma_s \left(\frac{\partial z_s}{\partial t} - \frac{\partial z_c}{\partial t} \right) + F_{ts}, \tag{1.8}$$

$$\gamma_c = m_c \frac{\omega_{0c}}{Q_c}, \tag{1.9}$$

$$\omega_{0c} = \sqrt{\frac{k_c}{m_c}}, \tag{1.10}$$

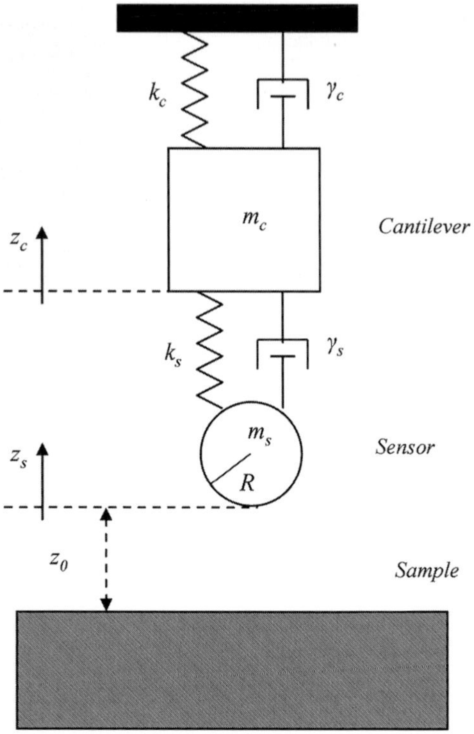

Fig. 1.3 Coupled damped harmonic oscillators used to model the operation of a probe with an integrated high-bandwidth force sensor in TM-AFM imaging

$$\gamma_s = m_s \frac{\omega_{0s}}{Q_s}, \tag{1.11}$$

$$\omega_{0s} = \sqrt{\frac{k_s}{m_s}}. \tag{1.12}$$

In these equations, m_c (m_s), z_c (z_s), γ_c (γ_s), k_c (k_s) and Q_c (Q_s) are the effective mass, instantaneous position, damping coefficient, spring constant, and quality factor of the cantilever beam (force sensor), respectively. In addition, ω is the drive frequency, while ω_{0c} (ω_{0s}) is the fundamental resonance frequency of the cantilever (force sensor). F_{dr} is the driving force applied to the cantilever and F_{ts} is the tip–sample interaction force acting on the force sensor. In the calculations, the tip–sample interaction forces are modeled using the DMT contact model described in Sect. 1.2.1 and is given by

$$F_{ts} = \begin{cases} \frac{-HR}{6(z_0+z_c)^2} & z_0 + z_c \geq a_0 \\ -\frac{HR}{6a_0^2} + \frac{4}{3}E^*\sqrt{R}\,(a_0 - z_0 - z_c)^{3/2} & z_0 + z_c < a_0 \end{cases}, \tag{1.13}$$

By solving (1.7)–(1.13) using numerical methods, the cantilever and force sensor positions are calculated as a function of time. Using these results, the

Table 1.1 Parameters used for TM-AFM calculations for a probe with a high-bandwidth force sensor

Parameter	Value	Parameter	Value
m_c	1×10^{-7} g	F_{dr}	10^{-8} N
k_c	10 N/m	H	5×10^{-20} J
m_s	1×10^{-9} g	z_0	3.5×10^{-8} m
k_s	1,000 N/m	R	10×10^{-9} m
Q_c	100	a_0	1×10^{-10} m
Q_s	100	E_t	1.29×10^{11} Pa
ω_{0c}	1×10^4 rad/s	υ_t	0.28
ω_{0s}	1×10^6 rad/s	E_s	100×10^9 Pa
ω	9.855×10^3 rad/s	υ_s	0.3

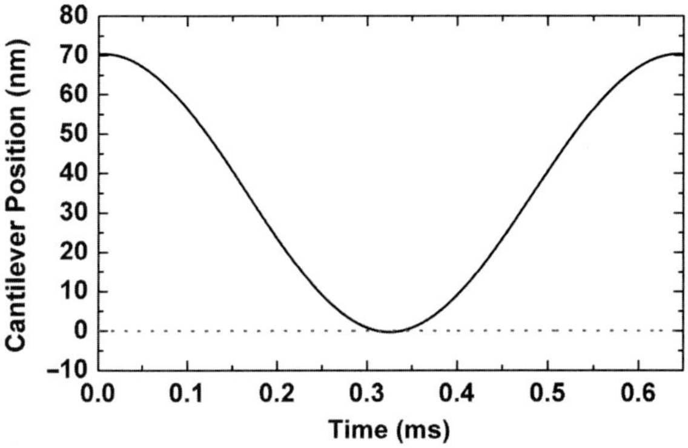

Fig. 1.4 Calculated position of the cantilever with the integrated force sensor. The *red dotted line* indicates the sample surface position

time-dependent interaction forces between the sensor and the sample surface are calculated. The values of the model parameters used for these calculations are listed in Table 1.1. Note from the table that the force sensor mass is assumed to be much smaller than the cantilever mass, while the sensor spring constant is chosen much larger than the cantilever spring constant. This ensures that the resonance frequency of the force sensor is higher than the fundamental resonance frequency of the system and results in a higher force sensor mechanical bandwidth.

Figures 1.4 and 1.5 show the calculated positions of cantilever and force sensor, respectively after the system reaches steady-state for a single oscillation period. The simulation results show that the force sensor closely follows the cantilever motion for most of the oscillation period. This is due to the fact that the drive frequency is well below the resonance frequency of the force sensor, and therefore, the force sensor is in phase with the cantilever motion. However, a close-up of the response of the system in the vicinity of tip–sample contact (shown in Fig. 1.6) shows that this synchronization between cantilever and force sensor oscillations is lost during the tip–sample interaction. This can be explained as follows. The tip–sample

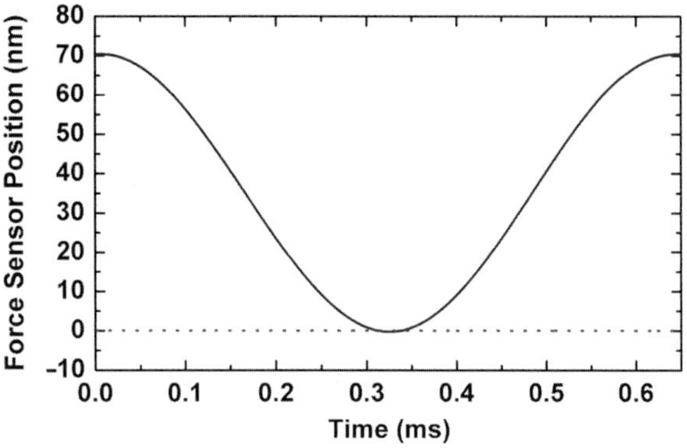

Fig. 1.5 Calculated position of the force sensor position. The *red dotted line* indicates the sample surface position

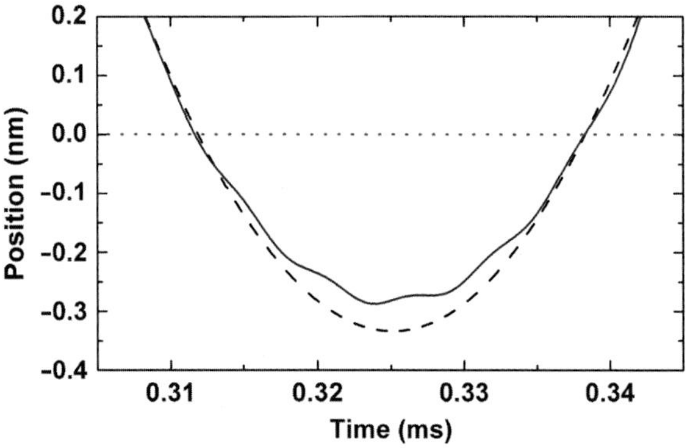

Fig. 1.6 Calculated positions of the cantilever (*black dashed line*) and the force sensor at times close to the tip–sample contact. The *red dotted line* indicates the sample surface position

interaction is nonlinear and therefore introduces higher frequency harmonics into the system as we have discussed in Sect. 1.2. In the simulated model, these forces act directly on the force sensor and are then transferred to the cantilever. Consequently, designed with a higher resonance frequency, the force sensor can capture a number of higher harmonics of the interaction force within its flat band spectrum and displace in response to nonlinear interaction forces. In contrast, the resonantly driven cantilever has a limited response at higher frequencies as we have discussed before, and cannot respond to the interaction forces.

Simulation results show that the relative motion of the force sensor with respect to the cantilever is sensitive to the interaction forces, so the differential motion provides a way to recover the tip–sample force in TM-AFM. Specifically, the interaction forces can be found by simply multiplying the relative displacement of the force sensor ($z_s - z_c$) by the spring constant of the force sensor (k_s). Figure 1.7 shows the simulated tip–sample interaction forces (black dashed line) together with the

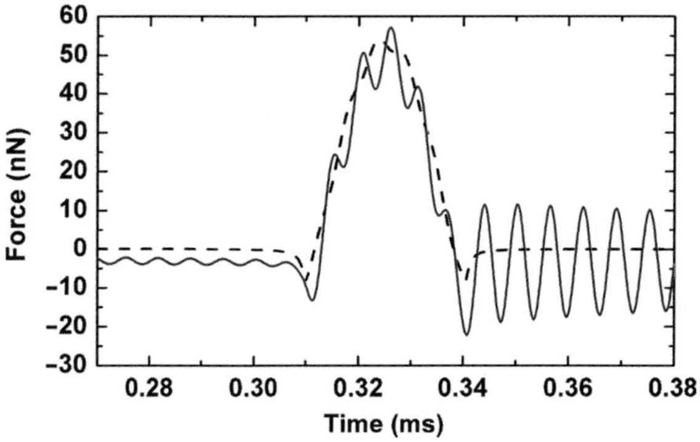

Fig. 1.7 Simulated tip–sample interaction force (*black dashed line*) and tip–sample interaction force, estimated based on the force sensor displacement relative to the cantilever (*blue line*)

estimated force (blue) based on the differential sensor displacement. The estimated force signal fits the actual tip-sample interaction force quite well, except that it adds high frequency oscillations that are due to under-damped nature of the force sensor. The oscillations are triggered by the interaction force high frequency harmonics that are close to sensor resonance frequency. They are more pronounced with the increasing quality factor of the force sensor. This result also indicates that the force sensor acts as a mechanical amplifier for certain harmonics of the interaction force and this can be utilized for higher harmonic imaging as will be discussed in Sect. 1.4. The disadvantage of this ringing effect, however, is that it distorts the force measurements and complicates calibration, so it should be corrected. This can be achieved with the knowledge of the force sensor transfer function, as will be discussed in Sect. 1.3.7.

To summarize, using a simple harmonic oscillator model, we have shown that an integrated high-bandwidth force sensor on the cantilever solves the problem of limited mechanical bandwidth in TM-AFM and provides improved temporal resolution to resolve the interaction forces.

1.3.2 Interferometric Grating Sensor

In the actual probe geometry (shown in Fig. 1.2), the large and small masses correspond to cantilever beam and the tip-coupled force sensor at the end, respectively. In order to measure the differential tip-cantilever displacement, an interferometric grating displacement sensor [38–40] is used. The grating sensor is designed such that the tip is mechanically coupled to alternating grating fingers, while the rest of the grating fingers are isolated from each other and remain free. Therefore, when there is force acting on the tip, the tip-coupled fingers displace relative to the rest of the fingers. To demonstrate the force sensor operation, the sensor response under the

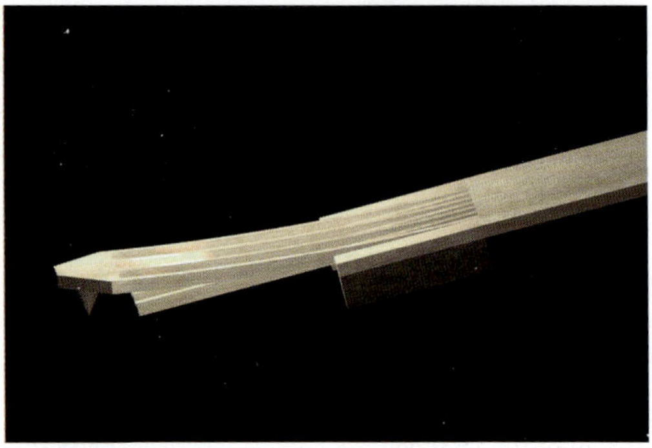

Fig. 1.8 Computer drawing showing the response of the force sensor when the tip interacts with the surface

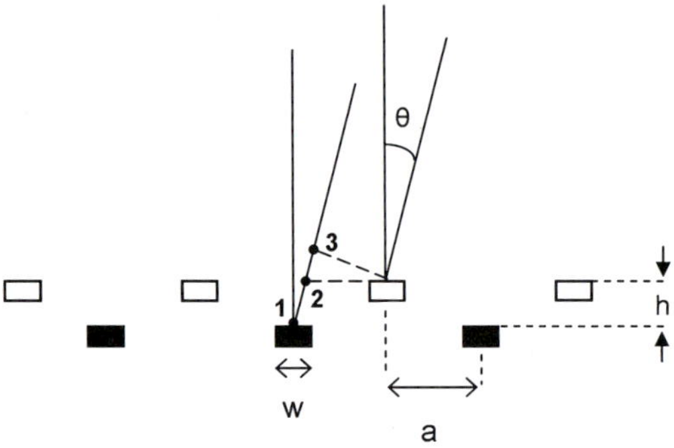

Fig. 1.9 Schematic of the cross section of the differential interferometric grating sensor. The moving (reference) fingers are represented by *bright* (*dark*) features

effect of a tip-sample force is depicted in Fig. 1.8, where the relative displacement is exaggerated to clarify the concept.

In operation, the differential grating sensor is illuminated with a focused laser beam and the reflected light forms a diffraction pattern. Modeling of the sensor operation requires the calculation of the diffracted light intensity pattern as a function of relative vertical displacement between the moving and reference finger sets. For this purpose, the sensor is treated as a two set of mirror surfaces that are vertically displaced as shown in Fig. 1.9. For a grating that is composed of N identical fingers of width w with a period of $2a$, the wave amplitude at the far field is calculated using Fraunhofer integral [41] and is given by

$$\Psi(\theta) \propto \left[\frac{\sin(\pi w \cos\alpha \sin\theta/\lambda)}{\pi w \cos\alpha \sin\theta/\lambda}\right]\left[\frac{1-\exp(-\mathrm{i}4\pi Na \cos\alpha \sin\theta/\lambda)}{1-\exp(-\mathrm{i}4\pi a \cos\alpha \sin\theta/\lambda)}\right]. \quad (1.14)$$

The first part of this expression gives the amplitude of the diffracted wave for a single infinitely long finger of width w as a function of the diffraction angle, whereas the second part is due to the interference of the reflected waves from the periodic grating fingers. Note that, (1.14) also includes the effect of an incident angle α along the grating fingers, since in practical implementation of optical lever, this angle is never zero. The grating sensor can be considered to be formed of two such gratings, namely the moving and reference gratings that are displaced in the vertical direction. Therefore, to calculate the total diffracted field Ψ_t, the diffracted wave amplitudes from both moving fingers and reference fingers should be added. Note that the reflected waves due to moving and reference finger sets are identical except for the phase difference due to height offset between the two finger sets. Therefore, the total diffracted wave amplitude can be expressed as

$$\Psi_t(\theta) = \Psi(\theta) + \Psi(\theta)\,e^{i\Omega(\theta)}. \tag{1.15}$$

The phase difference Ω is a function of the diffraction angle and it is calculated by multiplying the wave number by the optical path difference between the two waves and is expressed as

$$\Omega = \frac{2\pi}{\lambda}\cos\alpha\,[h + \ell_{12} + \ell_{23}]. \tag{1.16}$$

where α is the incident angle along the grating fingers and h is the vertical height difference between the finger sets. ℓ_{12} and ℓ_{23} are the distances between points (1,2) and (2,3) from Fig. 1.9, respectively, and are given by

$$\ell_{12} = \frac{h}{\cos\theta}, \tag{1.17}$$

$$\ell_{23} = [a - h\tan(\theta)]\sin\theta. \tag{1.18}$$

Substituting (1.17) and (1.18) in (1.16), the phase difference between the waves reflected from moving and reference fingers is given by

$$\Omega = \frac{2\pi}{\lambda}h\cos\alpha\,[(1+\cos\theta) + (a/h)\sin\theta]. \tag{1.19}$$

The total diffracted wave amplitude is calculated using (1.14) and (1.15) and is given by

$$\Psi_t(\theta) \propto \left[\frac{\sin(\pi w\cos\alpha\sin\theta/\lambda)}{\pi w\cos\alpha\sin\theta/\lambda}\right]\left[\frac{1-\exp(-i4\pi Na\cos\alpha\sin\theta/\lambda)}{1-\exp(-i4\pi a\cos\alpha\sin\theta/\lambda)}\right][1+\exp(i\Omega)]. \tag{1.20}$$

The diffracted light intensity as a function of the diffraction angle is calculated by the magnitude square of the wave amplitude given in (1.20) and is expressed as

$$I(\theta) \propto \left[\frac{\sin[(\pi w\cos\alpha\sin\theta)/\lambda]}{(\pi w\cos\alpha\sin\theta)/\lambda}\right]^2\left[\frac{\sin[(N\pi 2a\cos\alpha\sin\theta)/\lambda]}{\sin[(\pi 2a\cos\alpha\sin\theta)/\lambda]}\right]^2\cos^2\left(\frac{\Omega}{2}\right). \tag{1.21}$$

In this expression, the first term is the light intensity due to single grating finger, the second term is due to the interference of the reflected waves from multiple grating fingers and the final term is due to the interference of the reflected waves from the moving and reference finger sets. This formulation is compact and simple to evaluate numerically, but the fact that Ω is a function of θ, makes it hard to interpret.

For a formulation that separates diffraction angle and phase shift, and therefore is easier to interpret, see Ref. [42].

Using (1.21), the diffracted light intensities as a function of diffraction angle are calculated for a set of finger displacements. In the calculations, it is assumed that the grating has three finger pairs. Both the moving and reference grating fingers are 3 μm wide and the gap between the adjacent fingers is 3 μm. The grating is illuminated with a plane wave with a wavelength of 670 nm with a 0° incidence angle. In Fig. 1.10, the calculated far-field diffraction patterns for three different values of finger displacement are presented. The light intensities are normalized to the maximum intensity of the zeroth order. From these plots, it is observed that when there is no vertical displacement between the two finger sets, the diffraction pattern is composed of the specular reflection of the fingers as the dominant order (also called zeroth diffraction order) together with higher order modes that are called even modes. As one of the finger sets displaces with respect to another, some of the light intensity is partially transferred to new diffraction orders which are referred as odd orders. As the distance between the two finger sets reaches $\lambda/4$, where λ is the wavelength of laser illumination, the light intensity at the odd orders are maximized whereas the light intensities at the even orders including the zeroth order are minimized.

In order to calculate the light intensity of a particular diffraction order as a function of vertical finger displacement, the diffraction angles that correspond to the even and odd diffraction orders are determined using

$$2a \sin\theta \cos\alpha = m\lambda, \tag{1.22}$$

where m is an even (odd) integer for even (odd) orders. Substituting (1.22) in (1.19) and assuming a small diffraction angle, the phase difference between the lights reflected from the moving and reference fingers corresponding to a particular diffraction order can be expressed as

$$\Omega_m = \frac{2\pi}{\lambda} h \left[2\cos\alpha + (m\lambda/2h)\right]. \tag{1.23}$$

Using (1.23) in (1.21), the relation between the light intensity of a particular diffraction order and vertical displacement between the moving and reference finger sets can be expressed as

$$I_{\text{even}} \propto \cos^2\left(\frac{2\pi}{\lambda} h \cos\alpha\right), \tag{1.24}$$

$$I_{\text{odd}} \propto \sin^2\left(\frac{2\pi}{\lambda} h \cos\alpha\right). \tag{1.25}$$

Figure 1.11 shows the calculated normalized light intensities in the zeroth and first diffraction orders as a function of finger displacement for the grating of Fig. 1.10, when the sensor is illuminated in normal direction. The light intensities in adjacent

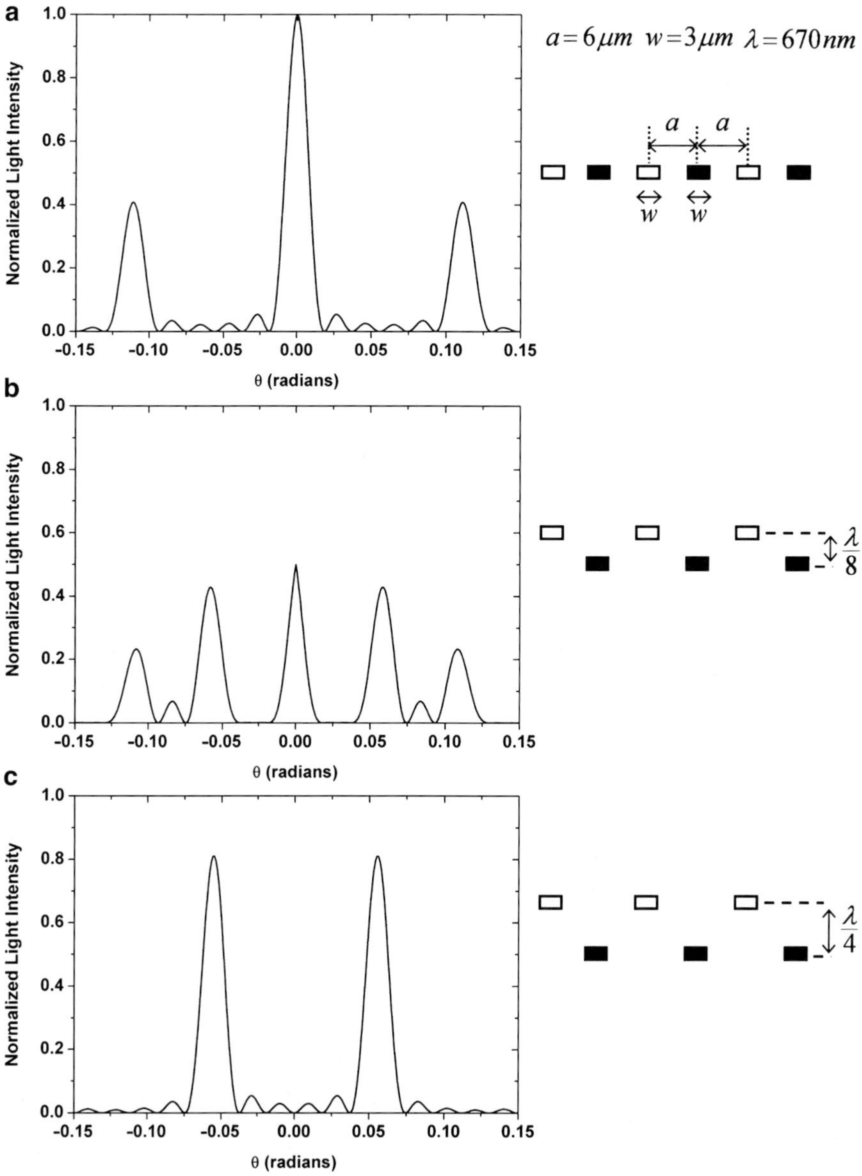

Fig. 1.10 Calculated far field diffracted light intensities as a function of diffraction angle (*left*) for various relative displacements of grating sensor moving and reference finger sets (*right*). Light intensity calculations are performed for three cases with (**a**) no relative displacement, (**b**) a displacement of $\lambda/8$, and (**c**) a displacement of $\lambda/4$, with λ being the wavelength of illumination

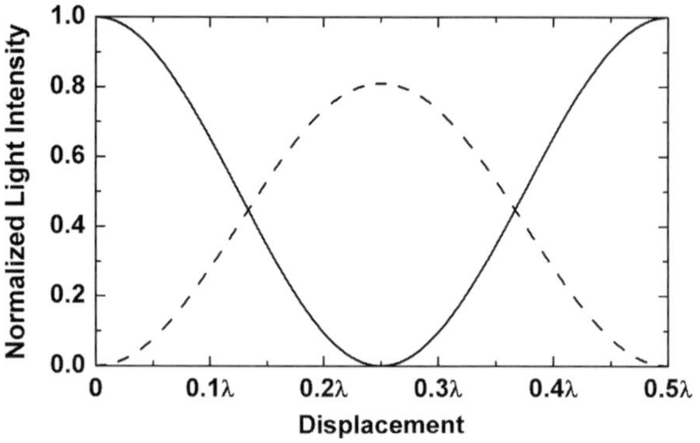

Fig. 1.11 Normalized diffracted light intensities at the zeroth order (*solid line*) and at the first order (*dashed line*) as function of vertical displacement between the grating finger sets. This result is for a grating with 50% duty cycle, i.e., $w/a = 0.5$

diffracted orders change in opposite directions as a function of finger displacement, so differential measurement of zeroth and first diffraction order light intensities, together with the total intensity, enable measurement of relative tip displacement with respect to the cantilever body.

An important design criterion for the differential grating sensor is its sensitivity. In the grating sensor, the finger displacement is measured by recording the light intensity at the diffracted orders using a photodiode. Therefore, the sensitivity of the grating sensor can be defined as $|\partial I/\partial h|$ (i.e., absolute light intensity change at a diffraction order for a unit vertical finger displacement). Using (1.24) and (1.25), the change in the intensity of even and odd diffraction orders are given by

$$\left|\frac{\partial I_m}{\partial h}\right| \propto \sin\left(\frac{4\pi}{\lambda} h \cos\alpha\right), \tag{1.26}$$

From (1.26), the displacement sensitivity of the differential grating sensor varies with the finger displacement. Furthermore, the highest displacement sensitivity is achieved when the fingers are displaced by an odd integer multiple of $\lambda/8$. This height offset also ensures that the sensor operates as linearly as possible since $\left|\partial^2 I_m/\partial h^2\right|$ (i.e., the sensitivity change as a function of displacement) is minimized. Consequently, to achieve force measurements with maximum sensitivity and linearity, the grating sensor should have an initial one-eighth of a wavelength height offset between the reference and moving finger sets. Considering the fact that we use a 670-nm laser diode in our experiments and the cantilever is illuminated with a 10° incidence angle in an AFM system with optical lever detection, we design our interferometric grating sensors such that there is an initial height offset of 85 nm between the reference and moving finger sets.

Another important factor in designing differential grating sensor is the separation of diffraction orders so that the light intensities at each order do not interfere with each other and can be measured individually. In order to ensure this, the separation of the diffracted orders and their optical-beam width on the observation plane should be taken into account. We define the beam width of a diffracted order w_{diff} as the distance between the first two locations with zero light intensity on each side of the maximum intensity point. From (1.21), the beam width w_θ of a diffraction order in angular space can be expressed as

$$w_\theta = \frac{\lambda}{aN\cos\alpha}, \tag{1.27}$$

where N is the number of finger pairs and a is the distance between two adjacent reference and moving fingers. In addition, the angular spacing (i.e., angle difference between the center locations with maximum light intensity) between the zeroth- and first-order diffraction orders is given by

$$s_\theta = \frac{\lambda}{2a\cos\alpha}. \tag{1.28}$$

From (1.27) and (1.28), the ratio of spacing to beam width is given by

$$\frac{s_\theta}{w_\theta} = \frac{\lambda/2a\cos\alpha}{\lambda/aN\cos\alpha} = \frac{N}{2}. \tag{1.29}$$

For the diffracted orders to be well separated, the separation between the orders should be at least as large as the diffracted order beam width. Therefore, from (1.29), it is required that $N \geq 2$. In our probe design, we chose N to be 3 so that the separation of the diffracted orders is ensured. The number of finger pairs is limited due to mechanical considerations and lithography limitations of fabrication process.

1.3.3 Sensor Mechanical Response & Temporal Resolution

The temporal resolution of interaction force measurements with our probes depends on the mechanical bandwidth, and hence, the fundamental resonance frequency of the gating force sensor. In TM-AFM, the tip–sample interaction occurs periodically at a rate close to the cantilever fundamental resonance frequency. Therefore, the ratio of grating sensor resonance frequency to the cantilever resonance frequency is an important design parameter that determines the relative temporal resolution within the probe oscillation period in TM-AFM imaging.

As mentioned before, the tip–sample interaction force waveform strongly depends on the elastic properties of the sample. Specifically, with increasing Young's modulus, the tip–sample contact occurs in a smaller fraction of the cantilever oscillation cycle. [30] Therefore, tip–sample forces on a stiff sample require higher temporal resolution for interaction force measurements than a compliant sample. This means that as the sample gets stiffer, the ratio of force sensor resonance frequency to cantilever resonance frequency should be increased for accurate force measurements [43].

In our devices, the ratio of sensor resonance frequency to cantilever resonance frequency can be tuned by changing the dimensions of the probe. The first-order design rule is to reduce the tip mass as much as possible to get a high finger resonance. In addition, we can tune the ratio of sensor resonance frequency to cantilever resonance frequency by adjusting the ratio of grating force sensor finger length to the cantilever beam length. As this ratio decreases, the ratio of sensor to cantilever resonance frequency increases, and therefore, the relative temporal resolution is increased. Moreover, modifying the length of sensor or cantilever does not complicate the fabrication process and therefore it is the preferred way of tuning the relative temporal resolution of our probes.

Increasing the relative temporal resolution by decreasing the length of the grating fingers, however, comes with a cost of increased stiffness, which reduces the sensitivity of the force measurements. Due to this trade-off, sensor sensitivity and its temporal resolution should be optimized for materials with different elasticity. Therefore, we design our probes with a range of different sensor to cantilever beam length ratios so that we can achieve optimum performance on different materials.

Calibrated measurements with the interferometric grating force sensor require knowledge of its complete mechanical transfer function. Accurate measurements and modeling of the spectral response are necessary for calculation of the tip–sample interaction force from differential sensor signals. Therefore, we must know the responses of both moving and reference fingers to the forces acting on the tip and the cantilever beam. In the ideal case, the sensor grating fingers should move relatively only when there is force acting on the tip. In reality, however, the flexural oscillations of the cantilever couple to the differential grating signal. This mechanical coupling is due to unequal responses of the moving and reference fingers to the flexural cantilever oscillations. Specifically, the tip-coupled moving finger set is longer and has a concentrated mass at the end, and therefore, has a lower flexural resonance frequency than the shorter and lighter grating reference fingers. In order to minimize the resulting differences in the spectral responses of the two, the outer fingers of the tip-coupled moving fingers are shortened by incorporating thick support regions around the grating. This increases the stiffness of the moving part and therefore decreases the difference between the resonance frequencies of the moving and reference grating fingers.

The simulated frequency responses of the differential grating sensor with and without balancing are shown in Fig. 1.12. From these figures, we see that balancing minimizes the effect of mechanical coupling of cantilever oscillations into the differential sensor grating signal. The coupling is only significant around the flexural resonances of the cantilever. Consequently, in the balanced case the spectral response of the differential grating sensor can be approximated as a simple harmonic oscillator. The balancing of the sensor has two advantages. First, it eliminates antiresonances in the mechanical response which might cause information loss if the displacement at that frequency is below the detector noise level. Second, it simplifies the experimental characterization of the sensor response to obtain the

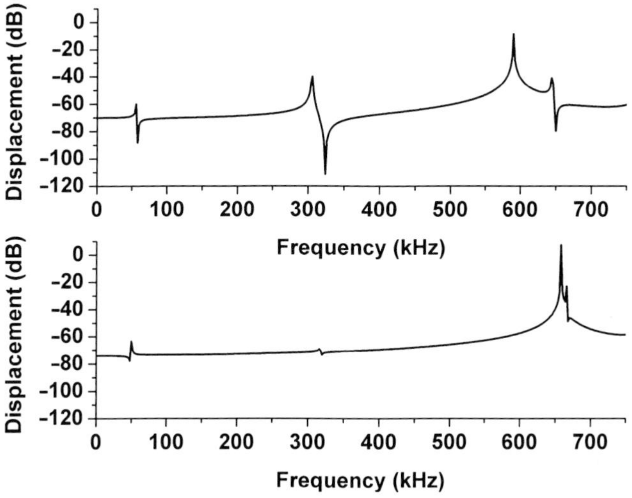

Fig. 1.12 Simulated transfer function of the force sensor without (*upper*) and with (*lower*) the support regions. The support equalizes the resonance frequencies of the moving fingers and the reference fingers and minimizes coupling between the force sensor and the cantilever. Reprinted with permission from [25]. Copyright 2009, IEEE

time-resolved tip–sample interaction force. The entire response can be estimated with the measurement of sensor resonance frequency and quality factor.

1.3.4 Fabrication

The fabrication process is outlined in Fig. 1.13. The probes are fabricated in parallel using optical lithography. The fabrication process starts with a silicon-on-insulator (SOI) wafer (Fig. 1.13a). First, the height offset between the reference and moving grating fingers to ensure maximum force sensitivity is created using Local Oxidation of Silicon (LOCOS). In this process, an 80-nm-thick layer of stoichiometric silicon nitride film is deposited using low pressure chemical vapor deposition (LPCVD) following an oxidation step to grow a 40-nm-thick oxide layer (Fig. 1.13b). This particular combination of film thicknesses ensures that the tensile stress on the silicon nitride film is compensated by the compressive stress in the oxide layer. Next, the silicon nitride layer is etched down to the underlying oxide layer using plasma etching, and then a thermal oxidation at 1,000°C is performed. During this oxidation process, the dense nitride layer acts as a diffusion barrier for oxygen, and therefore, oxidation of the silicon takes place only in the regions that are not covered by the nitride film, i.e., the reference fingers (Fig. 1.13c). The duration of the oxidation step to achieve the correct height offset is determined using the Deal–Grove model [44]. The advantages of using LOCOS instead of plasma etching to pattern the surface

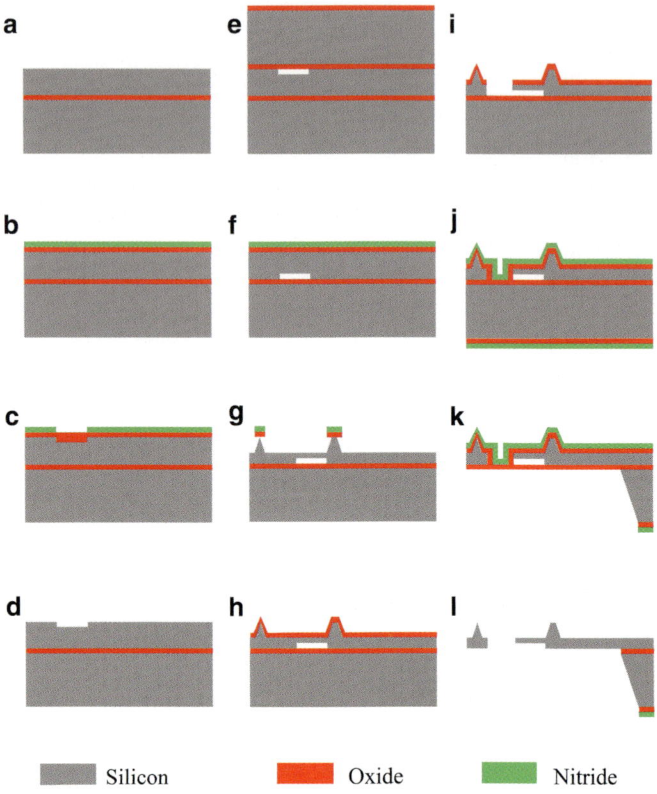

Fig. 1.13 Fabrication process for AFM probes with integrated interferometric high-bandwidth force sensors

are that the LOCOS process provides more accurate etch depth control and also preserves the low surface roughness of the Si wafer, which is crucial for the performance of the interferometric grating sensor. Following the LOCOS step, the silicon nitride film is etched in a phosphoric acid at 150°C and the oxide layer is etched using 6:1 Buffered Oxide Etch (BOE) (Fig. 1.13d).

Next, a double side polished Si wafer with a 1-μm-thick oxide layer is fusion-bonded to the patterned device layer of the SOI wafer. The bonding is performed initially at room temperature and is completed by a wet oxidation at 1,050°C for 2 h (Fig. 1.13e). The oxide layer grown during this step is used as a masking material for the subsequent TMAH etching. First, the oxide layer on the substrate of the original SOI wafer is removed using mechanical grinding and the substrate of the starting SOI wafer is etched in a 20% TMAH solution in water at 95°C. Following the etch, the masking oxide layers are stripped using 6:1 BOE.

Note that at this stage of the process, the height offsets for the grating sensor are buried at the interface between the device layer and buried oxide (BOX) layer of the SOI wafer. Next, a 1-μm-thick oxide film is grown at 1,000°C, followed by the deposition of 100-nm-thick silicon nitride film using LPCVD (Fig. 1.13f). To create a masking layer for the tip and thick support regions around the grating sensor, the nitride film and the underlying oxide films are etched using plasma etching and in a 6:1 BOE solution, respectively. Following this, the tips and the thick support regions are created by SF_6 based isotropic plasma etching until the oxide/nitride tip masks are released (Fig. 1.13g). Next, the tips are oxide-sharpened at 950°C for 2 h (Fig. 1.13h) [45]. The oxide layer grown in the tip sharpening process is used as a masking layer for ion etching the Si device layer to create the cantilever with the grating sensor (Fig. 1.13i). To protect the probe from the subsequent wet etch, the surface is covered with a 1-μm-thick tetraethyl orthosilicate (TEOS) film and then a 300-nm-thick low-stress LPCVD silicon nitride film is deposited as a masking layer for the next KOH etch (Fig. 1.13j). Before KOH etching, the nitride and oxide masking layers on the backside are patterned using plasma etching and 6:1 BOE, respectively. Next, the Si substrate is etched through the backside nitride mask using a 30% KOH solution at 80°C (Fig. 1.13k). Finally, the probes are released by first etching the nitride layer on the front surface using a plasma etch and stripping the remaining oxide masking films in a 6:1 BOE solution (Fig. 1.13l). An SEM micrograph of the finished probe is shown in Fig. 1.14, and a close-up showing details of the interferometric sensor, including the height offset at the base of the reference fingers, is given in Fig. 1.15.

1.3.5 Detection Schemes

Bending of flexible cantilevers can be detected using various techniques, including the optical lever [46,47], optical interferometers [48,49], and piezoresistive sensors [50]. Due to its simplicity, the most popular is the optical lever, in which a collimated laser beam is reflected from the back of the cantilever to a four-quadrant photodiode. The position of the reflected laser spot on the photodiode changes due to flexural or torsional bending of the cantilever beam. Resulting vertical and lateral changes

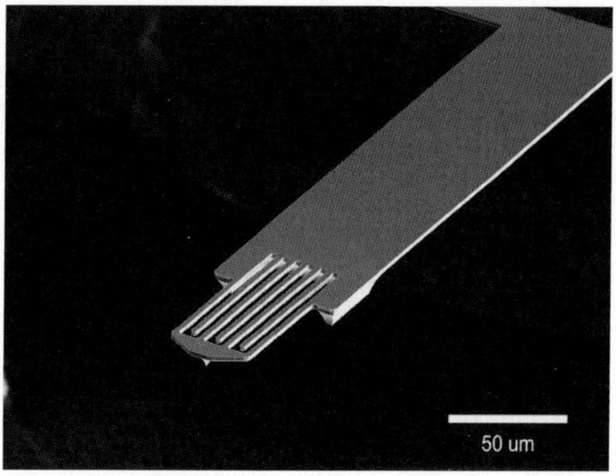

Fig. 1.14 SEM micrograph of an AFM probe with integrated interferometric high-bandwidth force sensor. Reprinted with permission from [25]. Copyright 2009, IEEE

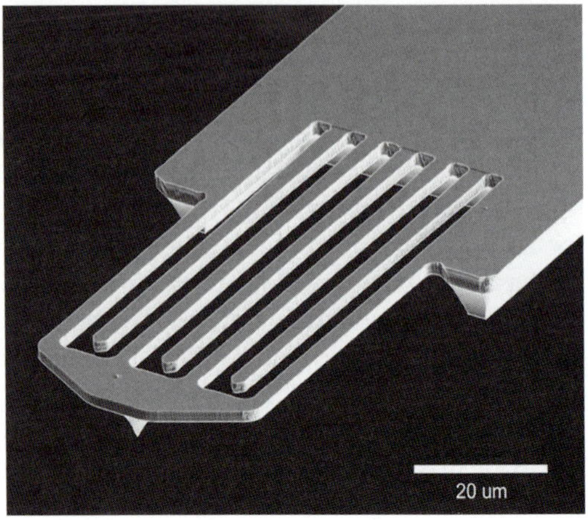

Fig. 1.15 Close-up of the high-bandwidth force sensor. Reprinted with permission from [25]. Copyright 2009, IEEE

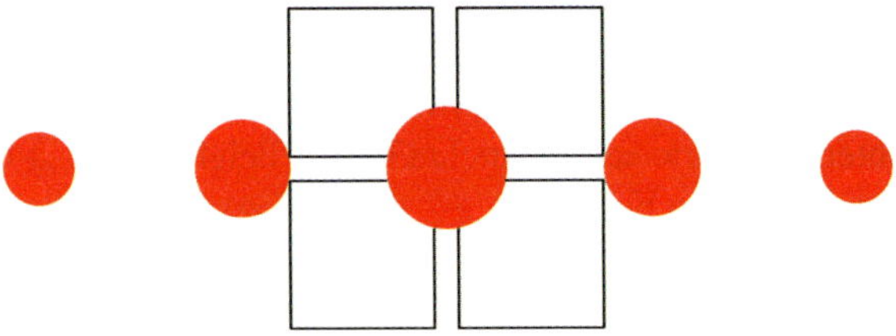

Fig. 1.16 Alignment of diffracted orders (*red circles*) on the built-in AFM quadrant photodiode. In this setting, the vertical position gives the cantilever displacement, and the total light intensity gives the sensor displacement and force

in the laser spot position are obtained by measuring the light intensity difference between the top/bottom and left/right halves of the four-quadrant photodiode.

There are two ways to operate our probes using a single focused laser beam and the built-in four-quadrant photodiode in a conventional AFM system that employs optical lever detection. In the first method, one of the diffracted laser spots is placed at the center of the quadrant photodiode as shown in Fig. 1.16. The other spots can either be left out of the active area if the size of the photodiode is small enough or they can be blocked using an aperture. In this configuration, the flexural oscillations of the cantilever cause the diffracted laser spot to move in the vertical direction and are measured by tracking the position of the diffracted laser spot using the vertical difference signal of the four-quadrant photodiode. For interaction force measurements, the relative tip displacement with respect to the cantilever can be simultaneously measured by changes in the total light intensity of the diffracted spot by adding the outputs of all the individual cells of the four-quadrant diode.

The second approach requires the two adjacent diffracted laser spots to be placed onto the left and right halves of the four-quadrant photodiode as shown in Fig. 1.17. Note that the spots are placed such that they are vertically centered. In this setting, the flexural oscillations of the cantilever can be measured by detecting the

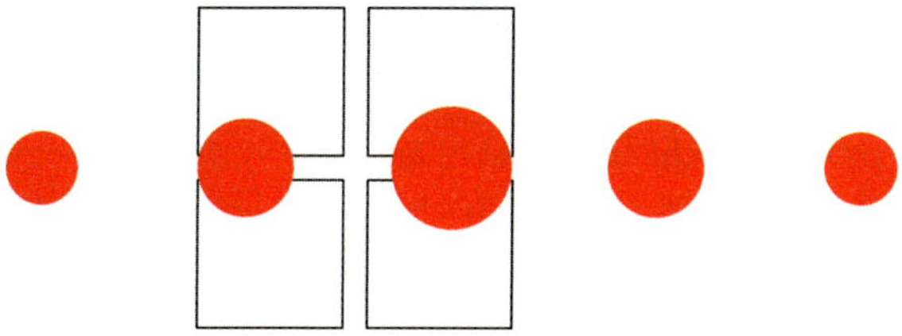

Fig. 1.17 Alignment of diffracted orders (*red circles*) on the built-in AFM quadrant photodiode. In this setting, the vertical position gives the cantilever displacement, and the lateral difference in intensity gives the sensor displacement and force

collective vertical displacement of the two adjacent diffracted spots. Similar to the first method, this is accomplished by using the vertical difference signal of the four-quadrant photodetector. On the other hand, the relative tip displacement with respect to the cantilever is simultaneously detected by measuring the difference in the light intensities of the two adjacent diffracted orders. This is simply achieved by using lateral difference output of the four-quadrant phototodetector.

The power of these detection techniques is that they provide simultaneous measurements of the flexural cantilever displacement and the relative tip–cantilever displacement. Both the detection techniques are combinations of optical lever and interferometry, yet they use AFM's own laser diode, built-in four-quadrant photodiode and already available standard output signals for common AFM operating modes. In addition, both techniques enable independent measurements of the tip and cantilever displacements because the positions and the intensities of the diffracted spots are orthogonal parameters that do not affect each other.

Despite their similarities, the two measurement modes are distinct in terms of their ease of use and noise performance. In particular, the first method that uses a single diffracted laser spot is simpler to implement in practice since it does not impose any extra limitations on the size of the photodiode. However, in this method, the relative tip displacement measurements are affected by laser instabilities since the intensity fluctuations of the laser diode output are directly coupled to the intensity measurements of the spot. This issue is not as significant in the second method since the differential measurement of light intensities of two spots minimizes common mode noise due to laser diode intensity and wavelength instabilities. In addition, a dark zone between the two spatially separated diffraction modes on the photodiode eliminate possible crosstalk between vertical and lateral signals in tapping-mode operation that might occur due to the misalignment between the cantilever and photodiode axes. From practical point of view, the second method requires more specialized design, because the four-quadrant photodiode must be large enough that each lateral half can accommodate a single diffracted spot. This means that the photodiode is designed for a specific probe-to-photodiode distance and cantilever geometry.

1.3.6 Characterization and Calibration

The fabricated probes are characterized in an AFM system by measuring the static and dynamic responses of the grating force sensor. In these measurements, a focused laser beam is positioned on the grating force sensor and the zeroth and first diffracted spots are placed on the four-quadrant photodiode as explained in the previous section.

To characterize the grating sensor response for large displacements and to confirm the initial height offset in the fabricated devices, we use a grating force sensor test structure with 200-μm long interdigitated fingers directly attached to a substrate. By driving the AFM piezoelectric actuator with a triangular waveform, the tip of test structure is pushed in and out of a hard surface at a rate of 100 Hz and the specular reflection intensity is recorded. In this experiment, the dynamic response of the fingers can be neglected since the sensor resonance frequencies are much higher than 100 Hz. Figure 1.18 shows the reflected light intensity modulation recorded in one of the approach/retract cycles of the grating sensor test structure. In this measurement, tip–sample contact occurs at $t = 0.5$ ms and the tip is pushed into the sample until $t = 2$ ms. Then the tip is pulled away and it loses contact with the surface at $t = 3.75$ ms. The asymmetry between the approach and retract phases is due to hysteresis in the tip–sample contact, which is likely a result of a water meniscus between the tip and sample. Specifically, from $t = 3.3$ ms to $t = 3.75$ ms, the tip is pulled by the sample and the fingers are displaced in the opposite direction. This effect is useful for characterization purposes, because it enables observations of the optical response due to attractive forces. In Fig. 1.18, right after the tip-contact at $t = 0.5$ms, the light intensity of the zeroth order decreases in an approximately

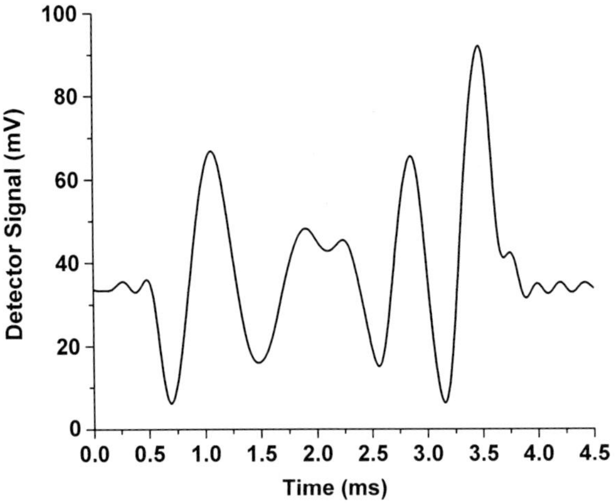

Fig. 1.18 Reflected intensity as the sensor is pushed down and pulled back from surface with a piezoelectric actuator at a rate of 100 Hz. Reprinted with permission from [25]. Copyright 2009, IEEE

linear dependence on the finger displacement. This observation confirms the initial height offset in the fabricated devices, because an unbiased grating sensor would be insensitive to the grating finger displacements at the point of contact [from (1.24)]. Using the retract portion of the results in Fig. 1.18, the initial height offset in the fabricated force sensor is found to be 93.2 nm. The calculated optimum bias that includes the effect of 10° cantilever tilt is 85 nm. Therefore, these results also confirm the initial height offset and show that the fabricated devices are biased close to optimum.

Figure 1.18 also shows that the modulation index of the reflected light is decreasing with the increasing finger displacement. This effect is due to curvature of the grating fingers with the increasing tip displacement. This does not create a problem for our measurements, because under normal TM-AFM imaging conditions, the sensor grating fingers displace less than a nanometer, given that the peak tip–sample forces do not exceed a few of tens of nanonewtons.

To calibrate the photodetector output for force measurements, the probe is engaged on a hard surface and the tip–sample force is modulated by modulating the calibrated AFM piezoelectric actuator with a triangular waveform. The force acting on the tip is calculated by multiplying the probe (cantilever and the force sensor combined) spring constant and the piezoactuator displacement. In this process, the probe spring constant is determined by the thermal tune method [51, 52]. To calibrate the photodetector output for quantitative force measurements, the output signal is measured for a specific tip–sample interaction force and the conversion parameter is calculated. Note that this parameter not only depends on the probe properties but also depends on the specifications of the AFM photodetector, i.e., the photodiode responsivity and transimpedance gain of the photodetector circuit.

Finally, to characterize the dynamic response of the grating sensor, we measure its step response and determine its resonance frequency and quality factor. This is achieved by recording the photodetector signal as the tip goes in and out of contact with a sample. Under ambient conditions, sample surfaces are covered with a thin layer of water, so the tip–sample rupture due to breaking of the capillary neck effectively presents a force step function on the tip. From these measurements, the grating sensor resonance frequency and its quality factor are calculated by measuring the frequency and the decay time of the sensor ringing oscillations, respectively.

1.3.7 Time-Resolved Force Measurements

In this section, we demonstrate the operation of our probes in tapping mode and present force measurement results. In these measurements, the probe is driven close to its fundamental resonance by a piezoactuator from its base and the cantilever oscillations and relative tip motion are measured by observing the location and intensity of the two adjacent diffracted spots on a four-quadrant photodetector as explained before.

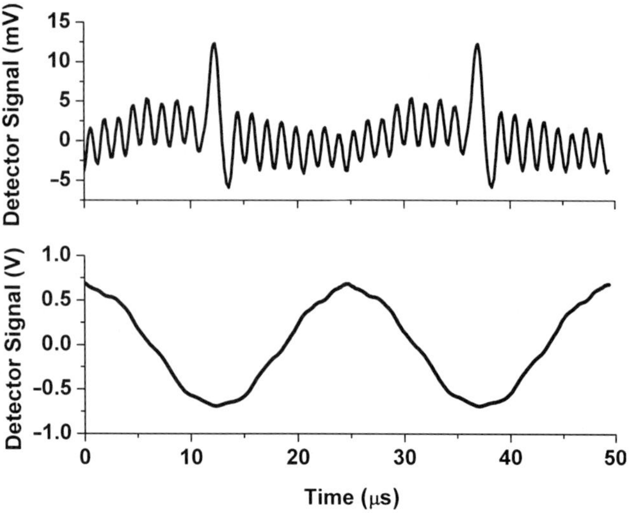

Fig. 1.19 Oscilloscope traces for tip (*upper*) and cantilever (*lower*) displacement. Reprinted with permission from [25]. Copyright 2009, IEEE

Figure 1.19 shows oscilloscope traces of cantilever motion and relative tip displacement signals measured as the probe periodically interacts with the same location on a sample. The measurements clearly show that the tip–sample interactions are time-resolved from the tip-displacement signal. Moreover, the force measurements are in agreement with the cantilever displacement signal, because tip–sample forces are measured only when the cantilever is close to the sample surface as expected. Note that the time-resolved force signals also show hysteresis in the tip–sample interaction due to larger adhesive force during tip retraction. The measurements also show that the cantilever trajectory remains close to sinusoidal, with small distortions due to the higher harmonics introduced by the tip–sample interaction. In the specific probe of Fig. 1.19, the cantilever beam is 250-μm long, 60-μm wide and 2.5-μm thick, and the grating fingers are 70-μm long, 3-μm wide and 2.5-μm thick. This geometry leads to a grating-sensor resonance frequency that is approximately 18 times the cantilever fundamental resonance frequency. The ringing oscillations present in the tip displacement signal are due to the vibrations of the high-bandwidth force sensor with a frequency close to its fundamental resonance frequency. These are triggered by the higher harmonics of the tip–sample interaction. The ringing decays very little between the consecutive tapping events, indicating a high quality factor of the force sensor in air, consistent with the results of our analysis in Sect. 1.3.1.

In order to perform quantitative force measurements, the grating sensor signal should be processed to compensate the frequency response of the grating sensor. For this purpose, the sampled grating sensor signal is filtered using the inverse of the estimated transfer function of the force sensor in the previous section [15,16]. In this process, a cut-off frequency is introduced in the calculations to prevent divergence above the resonance frequency. This inverse filtering cancels the effect of nonuniform mechanical gain of the grating sensor, and therefore removes the ringing due to the resonance. However, after this step the processed waveform may still be distorted due to several nonidealities such as nonlinear photodiode response and coupling of cantilever flexural resonances to the grating signal. To remove these effects, we use the assumption that the tip–sample interaction force should be close to zero except at times close to the tip–sample contact. Therefore, the contributions of the nonidealities are approximated by using a curve fitting algorithm based on least-squares method. For this curve fitting process, the harmonics of the cantilever drive signal are used as the basis set. This procedure is only applied on the parts of the waveform, in which the tip and sample are not expected to be in contact. By subtracting the estimated fit from the original, a corrected waveform is obtained. Finally, the calibration parameters measured with the procedures outlined in the previous section,

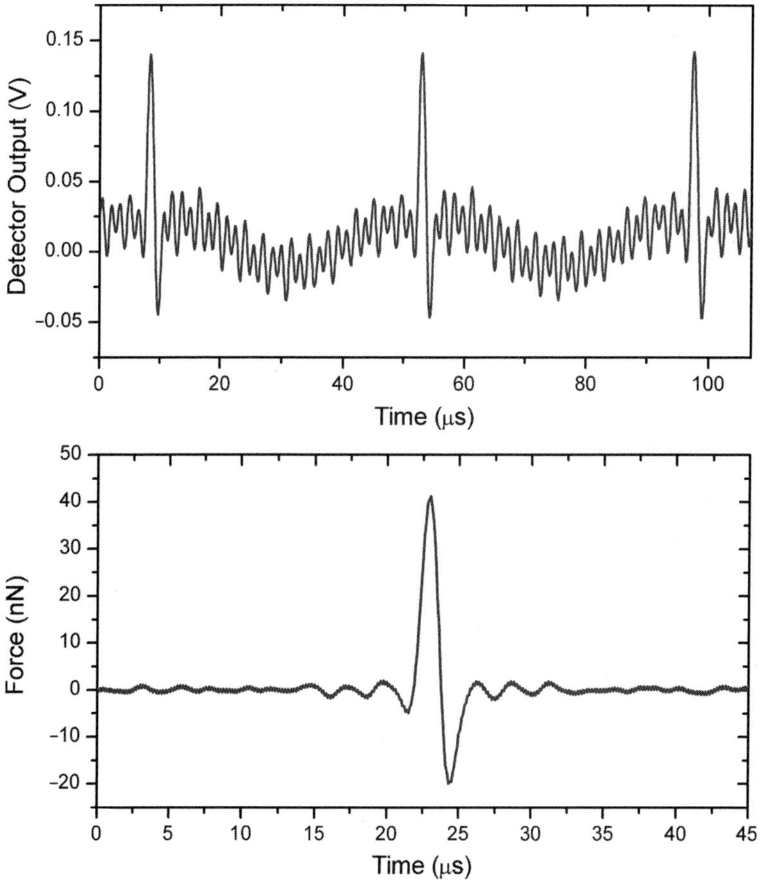

Fig. 1.20 Raw (*top*) and processed (*bottom*) force signals acquired during TM-AFM imaging of a Si sample. Reprinted with permission from [24]. Copyright 2008, American Institute of Physics

are used to convert the photodetector output into force values. Figure 1.20 shows the interaction–force waveform before and after this signal processing has been carried out. These measurements were performed on a Si sample in ambient conditions using a probe with a force sensor that has 30 times higher mechanical bandwidth than the cantilever.

1.4 Imaging Applications

In previous sections, we introduced and analyzed an AFM probe that can resolve tip–sample interaction forces during TM-AFM imaging. We showed that these probes can resolve tip–sample interaction with high temporal resolution and can be used in standard AFM systems without requiring excessive modifications. In this section, we use these probes to form images that are based on time-resolved tip–sample force measurements. First, the imaging technique and experimental setup is explained, and imaging results on composite surfaces, including block copolymers and SAMs of alkanethiols are presented.

As discussed in Sect. 1.2.2, the spectrum of the periodic and nonlinear interaction force waveform in TM-AFM contains harmonics at the integer multiples of the cantilever drive frequency. Therefore, a simple way to create AFM images based on the tip–sample interaction force information is to record the amplitude of one of the higher harmonics of the quasi-periodic force signal [17, 53]. A schematic of our imaging setup is shown in Fig. 1.21. Higher harmonic imaging experiments are done using a commercial AFM system. The AFM's built-in 670 nm laser diode and the four-quadrant photodetector are used for detection of the diffraction spots that

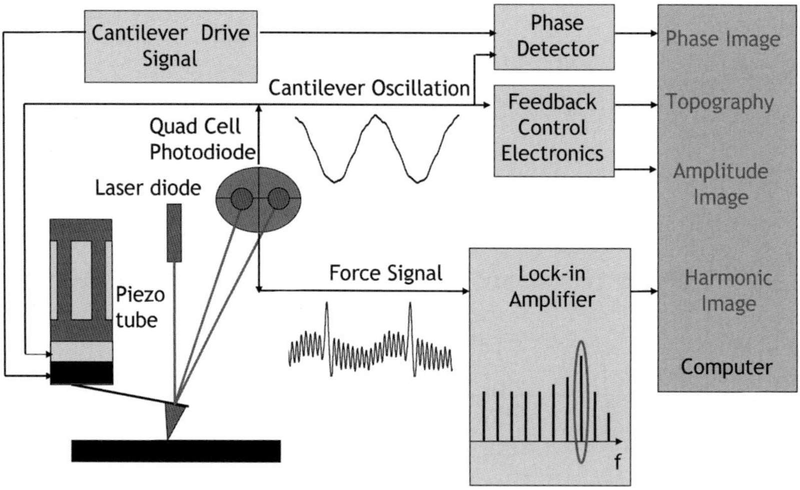

Fig. 1.21 Schematic of experimental setup used for harmonic imaging

are aligned on the photodetector as explained in Sect. 1.3.5. The cantilever oscillation and the force signals are obtained from the vertical and lateral differential channels of the photodetector. In operation, the feedback to the piezoelectric actuator is set to keep a constant cantilever oscillation amplitude, and a lock-in amplifier is used to measure the signal power of one specific higher harmonics of the force signal. To achieve this, two signal generators sharing a common time base are used. The first one is used to drive the cantilever close to its fundamental resonance frequency. The second one is used to generate the reference harmonic signal for the lock-in amplifier. The frequency of this signal is an integer multiple of the cantilever drive frequency. The specific imaging harmonic is simply chosen by adjusting the frequency of the reference signal. In our experiments, we use the harmonic in the vicinity of the grating sensor resonance in order to benefit from the improved sensitivity due to resonance enhancement.

Harmonic imaging using our probes provides advantages over applying the same technique with a regular AFM probe. In the cantilever displacement spectrum, only harmonics that are close to flexural resonances of the cantilever are recovered [17], whereas the force sensor signal can provide a full spectrum harmonics due to larger mechanical bandwidth of the force sensor. Our probes provide freedom to choose a specific harmonic from a range of harmonics to obtain the optimum material contrast and enables simultaneous images of several harmonics.

1.4.1 Nanomechanical Material Mapping

To demonstrate the material mapping capabilities of our probes, we tested our probes on a control sample that we prepared by using focused ion beam-assisted material deposition on a Si surface. In this sample, 150 nm high $1 \times 1\,\mu$m Pt and TEOS islands are deposited in a 5×5 checkerboard pattern. In imaging this sample, the probe is driven close to its resonance frequency at 44.16 kHz and the spectral component at 441.6 kHz is recorded using lock-in detection to create the 10th harmonic image.

Figure 1.22 shows the simultaneously captured topography, amplitude, phase, and 10th harmonic images of the checkerboard test sample. In the 10th harmonic image, the Pt and TEOS islands are clearly distinguishable so that the checkerboard deposition pattern is revealed. The contrast in the harmonic image is due to the fact that the tip–sample interaction waveform in TM-AFM is affected by the local mechanical properties of the sample. From the conventional TM-AFM images in Fig. 1.22, we see that the checkerboard pattern is also apparent in the topography and phase image. However, the contrast mechanism of the topography image is the physical topographical properties of the surface. For this particular sample, the topography image indicates that the Pt islands are higher in topography compared to TEOS islands. The phase image contrast in TM-AFM is created by inelastic tip–sample interactions [54]. Therefore, the phase image of Fig. 1.22 indicates that the average dissipated energy on the two material surfaces is different. This shows that our imaging modality adds information to that obtained by the traditional imaging methods.

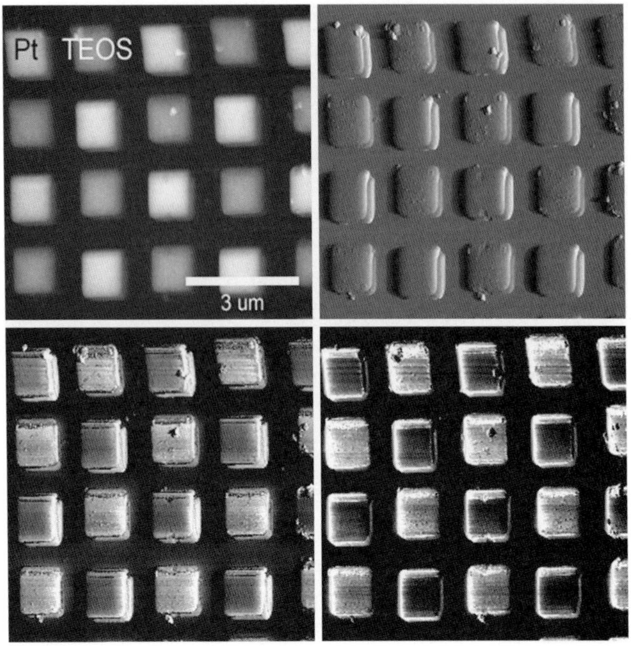

Fig. 1.22 Topography (*upper-left*), amplitude (*upper-right*), phase (*lower-left*) and 10th harmonic (*lower-right*) images of the checkerboard test sample. Reprinted with permission from [25]. Copyright 2009, IEEE

1.4.2 Imaging of Molecular Structures in Self Assembled Monolayers

The nanoscale structure of molecules has strong effects on their mechanical properties. In this section, we demonstrate that our probes can discriminate between two distinct structural forms of the same molecule through changes in its mechanical properties. In this experiment, the sample is a self assembled monolayer (SAM) of C_{18} molecules on a gold-coated mica surface.

Formation of SAM alkanethiols on surfaces has been a subject of several studies [55, 56]. These studies show that during the formation of alkanethiol SAMs, the head groups adsorb on the surface quickly to minimize their free energy. These molecules are initially in a noncrystalline state called the lying-down state. Over time the molecules pack more closely to form a crystalline state called the standing-up state. The ratio of the two structural states on the gold surface depends on the time that the gold surface is exposed to the solution containing $C_{18}SH$ molecules during sample preparation.

During sample preparation, the exposure time is adjusted such that the sample is covered with both lying-down and standing-up states. A 500×500 nm area of this sample is scanned as the probe is driven at 40.5 kHz and amplitude of the 18th harmonic of the force sensor signal is recorded to obtain the harmonic image.

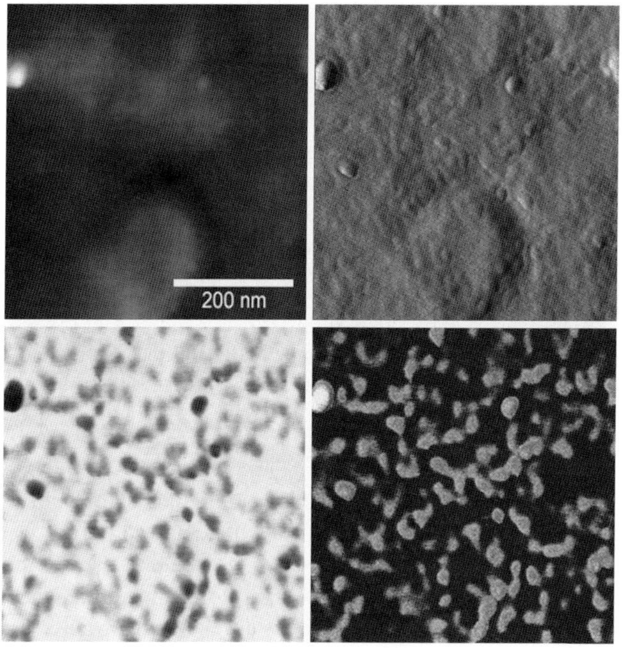

Fig. 1.23 Topography (*upper-left*), amplitude (*upper-right*), phase (*lower-left*) and 18th harmonic (*lower-right*) images of C_{18} self assembled monolayer on gold. Reprinted with permission from [25]. Copyright 2009, IEEE

Figure 1.23 shows the topography, amplitude, phase, and 18th harmonic images of the gold surface covered with both vertical and horizontal C_{18} molecules. The height difference between the two molecular states is too small to be clearly seen in the topography image. The amplitude image, being sensitive to topography gradients, shows the boundaries between the two states. From the harmonic image, we see that the two different states of the C_{18} molecules are clearly distinguished. This is an expected result since the structural properties of the molecules and their packing density are closely related to their mechanical properties. Also note that the results from harmonic image are in qualitative agreement with the phase image, even though the contrast mechanisms are vastly different.

1.4.3 Imaging Microphase Seperation in Triblock Copolymer

We also used thin block copolymer films to demonstrate the nanomechanical mapping capability of our probes. AFM and its related techniques are widely used for studying polymer surfaces [57]. Block copolymers are particularly interesting since they are composed of different polymer subunits and they form nanostructures with varying mechanical properties on the nanoscale [58].

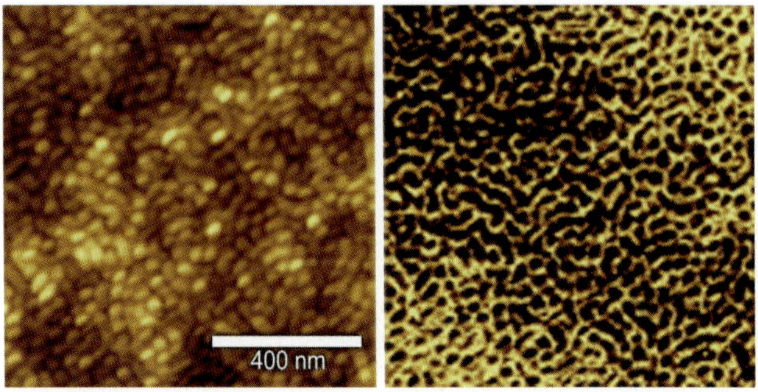

Fig. 1.24 Topography (*left*) and simultaneously acquired 17th harmonic image (*right*) of thin film of SBS triblock copolymer (Scan size: $1 \times 1\,\mu$m). Reprinted with permission from [24]. Copyright 2008, American Institute of Physics

Our sample in this experiment is a triblock copolymer, namely, SBS. This polymer is formed by repeating polystyrene, polybutadiene, and polystyrene polymer blocks, with a periodicity on the order of 40–50 nm [59]. At room temperature, the polybutadiene blocks are above their glass transition temperature, and hence are in rubbery state, whereas the polystyrene blocks are below their glass transition temperature and are in glassy state. This contrast in the elastic properties of the polymer blocks at room temperature makes the SBS polymer an interesting sample for mapping of surface elasticity. Figure 1.24 shows topography and 17th harmonic of the sample. The 17th harmonic image clearly shows the microphase separation between the two subunits. The bright and dark regions in the 17th harmonic image correspond to stiff polystyrene and compliant polybutadiene blocks, respectively.

1.5 Conclusion

In this chapter we present a technique for quantitative nanoscale analysis of materials based on the time-resolved measurement of tip–sample interaction forces in TM-AFM. The technique combines the advantages of TM-AFM imaging, namely nondestructive, high resolution, and fast operation, with the quantitative force-measurement capabilities of static AFM force measurement methods. This development therefore transforms TM-AFM from a technique useful primarily for physical topography measurements to a tool for quantitative material analysis with high sensitivity and resolution, giving it the potential for solving a wide range of measurement problems in material science, biology and medicine.

References

1. G. Binnig, C.F. Quate, C. Gerber, Atomic force microscope. Phys. Rev. Lett. **56**, 930 (1986)
2. F.J. Giessibl, S. Hembacher, H. Bielefeldt, J. Mannhart, Subatomic features on the silicon (111)-(7 × 7) surface observed by atomic force microscopy. Science **289**, 422 (2000)
3. N.A. Burnham, R. J. Colton, Measuring the nanomechanical properties and surface forces of materials using an atomic force microscope. J. Vac. Sci. Technol. A **7**, 2906 (1989)
4. M. Radmacher, J.P. Cleveland, M. Fritz, H.G. Hansma, P.K. Hansma, Mapping interaction forces with the atomic force microscope. Biophys. J. **66**, 2159 (1994)
5. K. Yamanaka, H. Ogiso, O. Kolosov, Ultrasonic force microscopy for nanometer resolution subsurface imaging. Appl. Phys. Lett. **64**, 178 (1994)
6. O.V. Kolosov, M.R. Castell, C.D. Marsh, G.A.D. Briggs, T.I. Kamins, R.S. Williams, Imaging the elastic nanostructure of Ge Islands by ultrasonic force microscopy. Phys. Rev. Lett. **81**, 1046 (1998)
7. A. Rosa-Zeiser, E. Weilandt, S. Hild, O. Marti, The simultaneous measurement of elastic, electrostatic and adhesive properties by scanning force microscopy: pulsed-force mode operation. Meas. Sci. Technol. **8**, 1333 (1997)
8. H. Krotil, T. Stifter, H. Waschipky, K. Weishaupt, S. Hild, O. Marti, Pulsed force mode: a new method for the investigation of surface properties. Surf. Interface Anal. **27**, 336 (1999)
9. P. Maivald, H.J. Butt, S.A.C. Gould, C.B. Prater, B. Drake, J.A. Gurley, V.B. Elings, P.K. Hansma, Using force modulation to image surface elasticities with the atomic force microscope. Nanotechnology **2**, 103 (1991)
10. M. Radmacher, R.W. Tillmann, H.E. Gaub, Imaging viscoelasticity by force modulation with the atomic force microscope. Biophys. J. **64**, 735 (1993)
11. Q. Zhong, D. Inniss, K. Kjoller, V.B. Elings, Fractured polymer/silica fiber surface studied by tapping mode atomic force microscopy. Surf. Sci. **290**, 688 (1993)
12. C. Moller, M. Allen, V. Elings, A. Engel, D.J. Muller, Tapping-mode atomic force microscopy produces faithful high-resolution images of protein surfaces. Biophys. J. **77**(2), 1150–1158 (1999)
13. D.A. Chernoff, *Proceedings of Microscopy and Microanalysis 1995* (Jones and Begell, New York, 1995)
14. S.N. Magonov, V. Elings, V. S. Papkov, AFM study of thermotropic structural transitions in poly(diethylsiloxane). Polymer **38**, 297 (1997)
15. M. Stark, R.W. Stark, W.M. Heckl, and R. Guckenberger, Inverting dynamic force microscopy: From signals to time-resolved interaction forces. PNAS **99**, 8473 (2002)
16. J. Legleiter, M. Park, B. Cusick, T. Kowalewski, Scanning probe acceleration microscopy (SPAM) in fluids: Mapping mechanical properties of surfaces at the nanoscale. PNAS **103**, 4813 2006)
17. O. Sahin, G. Yaralioglu, R. Grow, S.F. Zappe, A. Atalar, C.F. Quate, O. Solgaard, High resolution imaging of elastic properties using harmonic cantilevers. Sens. Actuators A **114**, 183 (2004)
18. S. Sadewasser, G. Villanueva, J.A. Plaza, Special cantilever geometry for the access of higher oscillation modes in atomic force microscopy. Appl. Phys. Lett. **89**, 033106 (2006)
19. R. Proksch, Multifrequency, repulsive-mode amplitude-modulated atomic force microscopy. Appl. Phys. Lett. **89**, 113121 (2006)
20. A.G. Onaran, M. Balantekin, W. Lee, W.L. Hughes, B.A. Buchine, R.O. Guldiken, Z. Parlak, C.F. Quate, F.L. Degertekin, A new atomic force microscope probe with force sensing integrated readout and active tip. Rev. Sci. Instrum. **77**, 023501 (2006)
21. M. Balantekin, A.G. Onaran, F.L. Degertekin, Quantitative mechanical characterization of materials at the nanoscale through direct measurement of time-resolved tip-sample interaction forces. Nanotechnology **19**, 085704 (2008)
22. O. Sahin, S. Magonov, C. Su, C.F. Quate, O. Solgaard, An atomic force microscope tip designed to measure time-varying nanomechanical forces. Nat. Nanotechnol. **2**, 507 (2007)

23. O. Sahin, N. Erina, High-resolution and large dynamic range nanomechanical mapping in tapping-mode atomic force microscopy. Nanotechnology **19**, 445717 (2008)
24. A.F. Sarioglu, O. Solgaard, Cantilevers with integrated sensor for time-resolved force measurement in tapping-mode atomic force microscopy. Appl. Phys. Lett. **93**, 023114 (2008)
25. A.F. Sarioglu, M. Liu, O. Solgaard, Interferometric force sensing AFM probes for nanomechanical mapping of material properties, in *Proceedings of the 15th International Conference on Solid-State Sensors, Actuators and Microsystems – IEEE Transducers*, Denver, CO, USA, 2009, pp. 1634–1637
26. R. Garcia, A. San Paulo, Attractive and repulsive tip-sample interaction regimes in tapping-mode atomic force microscopy. Phys. Rev. B **60**, 4961 (1999)
27. J. Israelachvili, *Intermolecular and Surface Forces* (Academic, London, 2003)
28. B.V. Derjaguin, V.M. Muller, Y.P. Toporov, Effect of contact deformations on the adhesion of particles. J. Colloid Interface Sci. **53**, 314 (1975)
29. L.D. Landau, E.M. Lifshitz, *Theory of Elasticity* (Pergamon, New York, 1986)
30. J. Tamayo, R. Garcia, Deformation, contact time, and phase contrast in tapping mode scanning force microscopy. Langmuir **12**, 4430 (1996)
31. A.S. Paulo, R. Garcia, Unifying theory of tapping mode atomic force microscopy. Phys. Rev. B **66**, 041406 (2002)
32. A.S. Paulo, R. Garcia, Tip-surface forces, amplitude, and energy dissipation in amplitude modulation (tapping mode) force microscopy. Phys. Rev. B. **64**, 193411 (2001)
33. J. Chen, R.K. Workman, D. Sarid, R. Hoper, Numerical simulations of a scanning force microscope with a large-amplitude vibrating cantilever. Nanotechnology **5**, 199 (1994)
34. S.I. Lee, S.W. Howell, A. Raman, R. Reifenberger, Nonlinear dynamics of microcantilevers in tapping mode atomic force microscopy: A comparison between theory and experiment. Phys. Rev. B **66**, 115409 (2002)
35. T.R. Rodriguez, R. Garcia, Tip motion in amplitude modulation (tapping-mode) atomic-force microscopy: Comparison between continuous and point-mass models. Appl. Phys. Lett. **80**, 1646 (2002)
36. O. Sahin, A. Atalar, Analysis of tip-sample interaction in tapping-mode atomic force microscope using an electrical circuit simulator. Appl. Phys. Lett. **78**, 2973 (2001)
37. M. Balantekin, A. Atalar, Power dissipation analysis in tapping-mode atomic force microscopy. Phys. Rev. B **67**, 193404 (2003)
38. O. Solgaard, F.S.A. Sandejas, D.M. Bloom, Deformable grating optical modulator. Opt. Lett. **17**, 688 (1992)
39. S.R. Manalis, S.C. Minne, A. Atalar, C.F. Quate, Interdigital cantilevers for atomic force microscopy. Appl. Phys. Lett. **69**, 3944 (1996)
40. G.G. Yaralioglu, A. Atalar, S.R. Manalis, C.F. Quate, Analysis and design of an interdigital cantilever as a displacement sensor. J. Appl. Phys. **83**, 7405 (1998)
41. R.J. Bell, *Introductory Fourier Transform Spectroscopy* (Academic, New York, 1972)
42. O. Solgaard, *Photonic Microsystems*, Chapter 10.5 (Springer, Heidelberg, 2009)
43. O. Sahin, A. Atalar, C.F. Quate, O. Solgaard, Resonant harmonic response in tapping-mode atomic force microscopy. Phys. Rev. B. **69**, 165416 (2004)
44. B.E. Deal, A.S. Grove, General Relationship for the Thermal Oxidation of Silicon. J. Appl. Phys. **36**, 3770 (1965)
45. T.S. Ravi, R.B. Marcus, D. Liu, Oxidation sharpening of silicon tips. J. Vac. Sci. Technol. B **9**, 2733 (1991)
46. G. Meyer, N.M. Amer, Novel optical approach to atomic force microscopy. Appl. Phys. Lett. **53**, 1045 (1988)
47. S. Alexander, L. Hellemans, O. Marti, J. Schneir, V. Elings, P.K. Hansma, M. Longmire, J. Gurley, An atomic-resolution atomic-force microscope implemented using an optical lever. J. Appl. Phys. **65**, 164 (1989)
48. Y. Martin, C.C. Williams, H.K. Wickramasinghe, Atomic force microscope-force mapping and profiling on a sub 100-Å scale. J. Appl. Phys. **61**, 4723 (1987)
49. D. Rugar, H.J. Mamin, P. Guethner, Improved fiber-optic interferometer for atomic force microscopy. Appl. Phys. Lett. **55**, 2588 (1989)

50. M. Tortonese, R.C. Barrett, C.F. Quate, Atomic resolution with an atomic force microscope using piezoresistive detection. Appl. Phys. Lett. **62**, 834 (1993)
51. J.L. Hutter, J. Bechhoefer, Calibration of atomic-force microscope tips. Rev. Sci. Instrum. **64**, 1868 (1993)
52. B. Ohler, Cantilever spring constant calibration using laser Doppler vibrometry. Rev. Sci. Instrum. **78**, 063701 (2007)
53. R.W. Stark, W.M. Heckl, Higher harmonics imaging in tapping-mode atomic-force microscopy. Rev. Sci. Instrum. **74**, 5111 (2003)
54. J.P. Cleveland, B. Anczykowski, A.E. Schmid, V.B. Elings, Energy dissipation in tapping-mode atomic force microscopy. Appl. Phys. Lett. **72**, 2613 (1998)
55. G.E. Poirier, E.D. Pylant, The self-assembly mechanism of alkanethiols on Au(111). Science **272**, 1145 (1996)
56. M. Liu, N.A. Amro, G. Liu, Nanografting for surface physical chemistry. Annu. Rev. Phys. Chem. **59**, 367 (2008)
57. S.N. Magonov, D.H. Reneker, Characterization of polymer surfaces with atomic force microscopy. Annu. Rev. Mater. Sci. **27**, 175 (1997)
58. L. Leibler, Theory of microphase separation in block copolymers. Macromolecules **13**, 1602 (1980)
59. S.N. Magonov, J. Cleveland, V. Elings, D. Denley, M.-H. Whangbo, Tapping-mode atomic force microscopy study of the near-surface composition of a styrene-butadiene-styrene triblock copolymer film. Surf. Sci. **389**, 201 (1997)

Chapter 2
Small Amplitude Atomic Force Spectroscopy

Sissi de Beer, Dirk van den Ende, Daniel Ebeling, and Frieder Mugele

Abstract Over the years atomic force microscopy has developed from a pure imaging technique to a tool that can be employed for measuring quantitative tip–sample interaction forces. In this chapter we provide an overview of various techniques to extract quantitative tip–sample forces focusing on both amplitude modulation and frequency modulation atomic force spectroscopy (AM-AFS and FM-AFS) in the small amplitude regime. We discuss different actuation methods, such as sample modulation, magnetic actuation, and by far the most widely used acoustic driving. Also the corresponding force-inversion methodology is discussed for – in particular – ambient liquid.

2.1 Introduction

Since its invention by Binnig et al. [1] in the 1980s, the atomic force microscope (AFM) has evolved into one of the most powerful tools for nanotechnology. Nowadays, the AFM is used in numerous research fields varying from biology to solid state physics and in many industries such as the semi-conductor or the automotive industry.

The AFM is most commonly used for topographical imaging of surfaces, where a static or dynamic feedback mechanism on the cantilever (Figure 2.1) is applied to follow the surface in great detail (down to atomic resolution).

In contact mode imaging (as the static feedback is often called), the tip of the cantilever is kept in contact with the surface (Figure 2.1a). Due to the repulsive or adhesive forces between the tip and the surface, the cantilever bends upward or downward as soon as the topography of the surface, respectively increases or decreases in height. While the tip scans the surface (in x and y), the bending of the cantilever is measured by the quadrant photo-detector (or sometimes by interference detection, for a review on detection-techniques see Ref. [2]). The deflection is given by detector signals $(a + b) - (c + d)$ (see Fig. 2.1a) and kept constant by adjusting the z piezo below the sample surface. The change in feedback-voltage applied to the z piezo is the measured height signal and gives the topography of

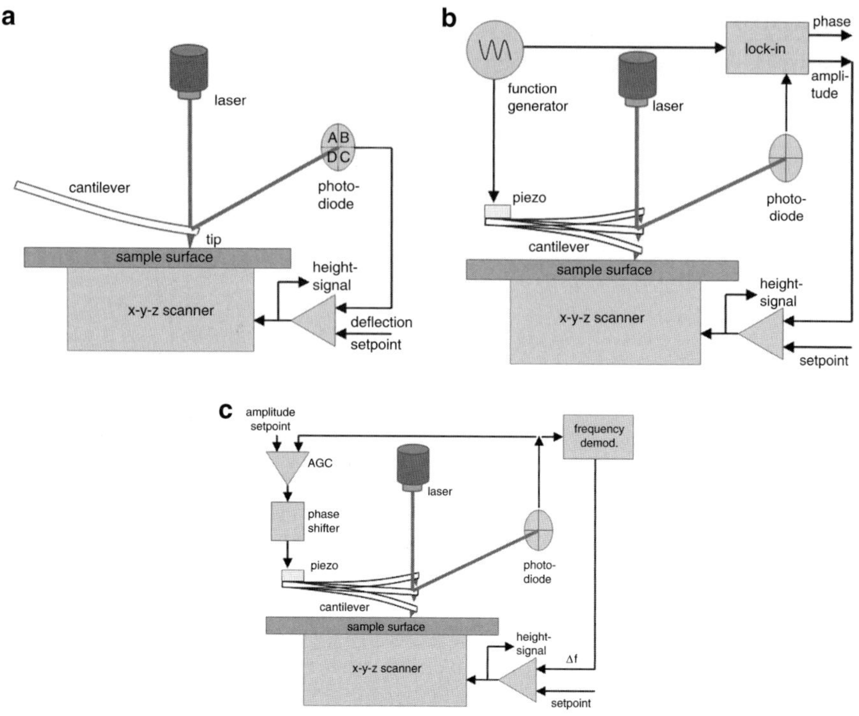

Fig. 2.1 Feedback mechanisms for: (**a**) Contact mode, the deflection is measured by the photodiode and kept constant by adjusting the z piezo. (**b**)Amplitude modulation, via the lock-in amplifier the amplitude and the phase response of the cantilever is measured and the amplitude is kept constant by adjusting the z piezo. (**c**) Frequency modulation, via the frequency demodulator the frequency shift is measured and kept constant by adjusting the z piezo

the scanned surface. Note that in AFM the feedback is on the force and therefore artificial height differences might be observed due to changing surface properties over the sample (e.g. elasticity). Next to a static feedback mechanism (such as contact mode) dynamic methods are also used: Amplitude Modulation (AM-) and Frequency Modulation (FM-) AFM. The advantage of dynamic AFM is that the tip is not necessarily in contact with the surface, and so, during imaging the (lateral, but sometimes also nomal) forces between tip and sample are much less than in contact mode. In both dynamic methods, the cantilever is driven with an amplitude of typically 0.5–100 nm. This can be accomplished by a drive piezo (as in Figure 2.1b, c,) or magnetic actuation. In contrast to simply measuring the static deflection, one now measures the amplitude and phase or frequency shift (with respect to the driving signal) of the cantilever oscillation. As will be explained in more detail below, upon approach of the cantilever toward the surface, the response (amplitude, phase, and resonance frequency) will change due to tip surface interactions. For AM-AFM (Figure 2.1b), the drive frequency and the drive amplitude are kept constant, while the amplitude and phase response of the cantilever are monitored via a lock-in amplifier. Subsequently, the amplitude is compared to the set-point value and kept constant by adjusting the z piezo. Like in contact mode the feedback voltage gives the height-signal. A disadvantage of AM-AFM is the notorious bi-stability due to the nonlinear response of the cantilever caused by nonlinear tip–sample forces [3], which can be overcome using a phase-feedback [4] or FM-AFM [5]. FM-AFM can be applied in two modes, constant amplitude (CA) [6] and constant excitation (CE) mode [7]. In the following, we will focus on the constant amplitude mode, since this technique is most widely used. Figure 2.1c shows a typical driving scheme for FM-AFM. In FM-AFM the cantilever is driven with a fixed phase lag (normally close to resonance, i.e., $-90°$). This can be accomplished by self-excitation ([5], Figure 2.1c) or a phase locked loop (PLL) [8]. In self-excitation, the signal from the photo-detector is amplified, phase-shifted by 90° and then used to drive the cantilever for a guaranteed driving at resonance, while with a PLL the cantilever is driven at the actual measured resonance frequency. At the same time, the measured amplitude response of the cantilever is kept constant with the automatic gain controller (AGC). Due to changes in the height of the scanned surface, the resonance frequency will change. The shift in resonance frequency is compared to the set-point value and kept constant by adjusting the z piezo. Again the z-piezo voltage gives the height signal.

Even though the force on the sample can be significantly reduced using dynamic AFM, for some specific samples (e.g. liquid–air interfaces [9]), and especially in a liquid environment, it can still be too large to image features without deforming them. In this case, a Q-control [10, 11] can be used to reduce the force on the sample even more. For a good review on the different imaging feedback methods see [12].

Although imaging was the number one application of the AFM, over the years, more and more people realized that the AFM can be also used to quantitatively study the surface or sample properties, which can be done via either force spectroscopy or one of the imaging modes.

During imaging in contact mode the friction forces can be studied by monitoring the torsional bending of the cantilever [13], $(a + c) - (b + d)$ in Figure 2.1a, while in AM-AFM the phase signal can, although not straightforward, be related to the energy dissipation [14]. Recently also imaging methods have been developed involving the analysis of higher harmonics [15] and higher oscillation modes [16] of the AFM cantilever. These techniques give access to sample properties such as elasticity and adhesion.

In force distance mode, the topography of the sample surface is no longer studied, but the response of the cantilever is monitored, while it is approached and retracted from the surface. From the response of the cantilever the tip–sample forces are extracted as a function of tip–surface distance. This can be done in contact mode, where the deflection z of the cantilever is translated into a force F using $F = -k_c z$, with k_c the spring constant of the cantilever. Although this method is simple and straightforward to apply, it has some disadvantages, such as the snap-to-contact as soon as the attractive force gradient $\mathrm{d}F/\mathrm{d}z|_{z_c} = k_{\mathrm{int}}$ is larger than the spring constant k_c, which implies that only a small part of the attractive tip–sample force can be measured. This can be overcome using stiffer cantilevers at the expense of sensitivity loss in the force. Nevertheless, with stiff cantilevers, complete high-resolution force profiles can also be extracted [17] in contact mode. This is accomplished by monitoring the Brownian noise instead of the average deflection of the relatively stiff cantilever upon approach and retracts.

Another method to extract the complete force-profile is via dynamic AFM, which has the additional advantage that next to the conservative forces the dissipative forces can also be measured. This can again be done in AM- or FM-AFM mode. Over the years, many techniques have been developed to extract the conservative and dissipative forces, varying from small to large amplitude or even universal methods for both AM-AFM [18–25] and FM-AFM [26–28].

In this chapter, we will discuss the small amplitude force-inversion techniques developed for AM-AFM [22–25] and FM-AFM [28]. The exact force-inversion methodology depends on the driving scheme used in the experiments, as will be discussed in the next sections.

2.2 Small Amplitude Spectroscopy

Small amplitude spectroscopy is a method of dynamic force spectroscopy in which the amplitude response of the cantilever is kept small enough to justify linearization of the conservative and dissipative tip–sample forces. In atomic force spectroscopy the cantilever is usually modeled as a simple harmonic oscillator affected by the tip–sample interactions F_{ts}:

$$m \cdot \ddot{z} + \gamma_{\mathrm{c}} \cdot \dot{z} + k_{\mathrm{c}} \cdot z = F_{\mathrm{dr}} + F_{\mathrm{ts}}. \tag{2.1}$$

Here, $z(t)$ describes the displacement of the cantilever, k_{c} is the cantilevers spring constant, m is the total effective mass (when relevant including the added mass caused by the motion of the surrounding liquid), γ_{c} is the viscous damping around (or of) the cantilever and F_{dr} is the driving force applied to the cantilever and F_{ts} is the tip–sample force.

The system properties k_{c}, m, and γ_{c} can be found using a thermal noise spectrum of the cantilever. Integration of the thermal noise spectrum results in the spring constant k_{c} [29, 30] and fitting the thermal noise spectrum with

$$A = A_{\mathrm{white}} + \frac{A_0 f_0^4}{\left(f^2 - f_0^2\right)^2 + \left(\frac{f f_0}{Q}\right)^2}$$

results in the calibration constants Q and $\omega_0 = 2\pi f_0$. From these calibration constants and the spring constant we can calculate the mass via $m = \omega_0^2 / k_{\mathrm{c}}$ and the damping via $\gamma_{\mathrm{c}} = m\omega_0 / Q$. How the driving force is determined depends on the actuation and detection method and will therefore be explained in the different subsections below. In our measurements we do not know the absolute value of the phase (there are also phase-lags due to, e.g., electronics). So before analyzing the results, the phase (far away, i.e., without interactions) needs to be shifted to the correct value. This phase-shift also depends on the actuation and detection method. The correct value of the phase can be found using the equations for the phase given below (with the interactions set to zero).

In small amplitude spectroscopy the tip–sample force F_{ts} at the (quasi-statically moved) average cantilever position z_c is rewritten using a Taylor expansion:

$$F_{ts}(z_c + z, \dot{z}) = F_{ts}(z_c, 0) - k_{int}z - \gamma_{int}\dot{z} - \frac{1}{2}k'_{int}z^2 - \frac{1}{2}\gamma'_{int}z\dot{z} + \text{H.O.T.}, \tag{2.2}$$

where H.O.T. are the Higher Order Terms, $k_{int} = -\,\mathrm{d}F_{ts}/\mathrm{d}z|_{z_c}$ is the interaction stiffness, γ_{int} is the interaction damping and $F_{ts}(z_c, 0)$ is the equilibrium force on the cantilever. Note that these tip–sample force parameters k_{int} and γ_{int} are independent of the specific measurement technique and should be identical for all data extracted by the different techniques described below. When applying a linearization only the first order terms of (2.2) are taken into account. This approximation is therefore only valid as long as the amplitude $A \ll 2k_{int}/k'_{int}$ and $A \ll 2\gamma_{int}/\gamma'_{int}$ [24]. After substitution of the linearized tip–sample force, (2.1) can be solved for the conservative interaction stiffness and the dissipative interaction damping. The resulting equations we call the force-inversion formulae. The exact form of these formulae depends on the actuation and detection technique. In a liquid (low $Q = m\omega_0\,/\,\gamma_c$, where $\omega_0 = \sqrt{k_c\,/\,m}$) environment the differences between these techniques become more pronounced and can no longer be neglected. Therefore we will explain the different techniques using typical AFM-in-liquid characteristics and (cantilever-) properties (spring constant $k_c = 2\,\mathrm{N/m}$, quality factor $Q = 3$, measured frequency at the amplitude maximum $f_{res} = \frac{\omega_0}{2\pi}\sqrt{1 - \frac{1}{2Q^2}} = 40\,\mathrm{kHz}$).

2.2.1 *Actuation Techniques*

To perform dynamic AFM spectroscopy several driving schemes can be used. The by far most widely used technique is acoustic driving (see e.g. [21, 23, 24, 26]), where a small piezo is used to drive the cantilever at the backside. The downside of this technique is that spurious resonances are produced next to the fundamental resonance [31]. Although, this is not necessarily a problem for imaging, for spectroscopy, a clean cantilever response (resonance curve) is needed for the correct characterization of the system. This problem can be overcome by applying a more direct force to the cantilever, like in sample modulation [25] or a magnetic driving scheme [22, 27]. The exact driving and detection method determines the *measured* amplitude/phase versus frequency response and the shift in resonance frequency of the cantilever [32] and therefore the force-inversion formulae, which will be discussed in more detail below.

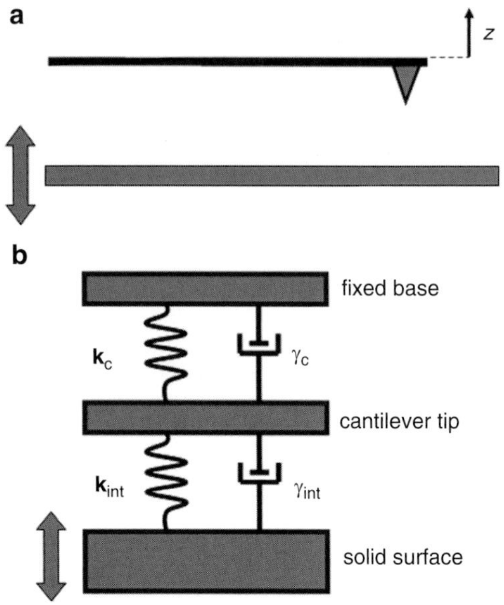

Fig. 2.2 (**a**) In sample modulation the response of the cantilever is monitored while the sample is oscillated. (**b**) Spring-dashpot representation of sample modulation

2.2.1.1 Sample Modulation

In sample modulation [25] the z piezo (on which the sample is mounted) is oscillated with a small amplitude (Fig. 2.2). Due to tip–sample interactions the motion of the cantilever is coupled to the motion of the sample. Consequently the cantilever will only respond close to the surface where the tip–sample forces come into play. For small amplitude spectroscopy the equation describing the dynamics of the cantilever is given by:

$$m \cdot \ddot{z} + \gamma_c \cdot \dot{z} + k_c \cdot z = k_{int}(d - z) + \gamma_{int}(\dot{d} - \dot{z}). \tag{2.3}$$

In which $z(t)$ is the measured response of the cantilever and $d(t)$ is the driving motion of the sample. Using the Ansatz $z = Ae^{i(\omega t + \varphi)}$ and $d = A_d e^{i\omega t}$ (where ω is the drive frequency, A is the measured amplitude, φ is the measured phase, and A_d is the drive amplitude), (2.3) can be solved for the amplitude A and the phase φ:

$$\frac{A}{A_d} = \frac{k_{int}\sqrt{1 + (\omega\gamma_{int}/k_{int})^2}}{\sqrt{(k_{int} + k_c - m\omega^2)^2 + \omega^2(\gamma_{int} + \gamma_c)^2}} \tag{2.4a}$$

and

$$\tan\phi = \frac{-\omega\left(\gamma_{int}\left(m\omega^2 - k_c\right) + \gamma_c k_{int}\right)}{k_{int}\left(k_{int} + k_c - m\omega^2\right) + \omega^2\gamma_{int}(\gamma_{int} + \gamma_c)}. \tag{2.4b}$$

Figure 2.3 shows the amplitude and phase response of the cantilever for a typical sample modulation configuration in liquid. Figure 2.3 shows that for a positive interaction stiffness the resonance frequency of the system shifts up, while for a negative

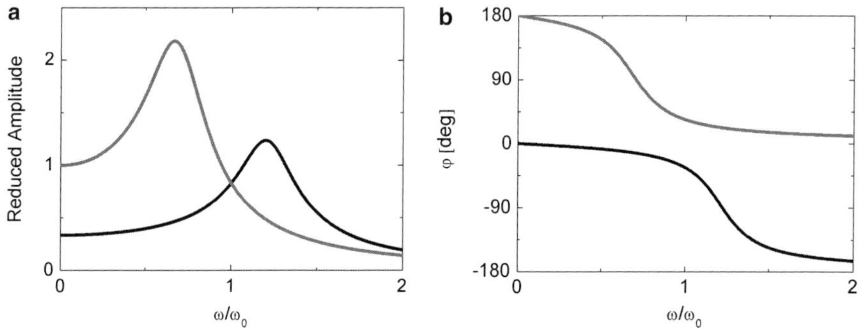

Fig. 2.3 The effect of a positive and a negative interaction stiffness ($k_{\text{int}} = +0.5k_c$ *black* and $-0.5k_c$ *gray*) on the amplitude (**a**) and phase (**b**) spectra of the cantilever in a sample modulation system

interaction stiffness the resonance frequency goes down. A positive interaction stiffness will increase the total stiffness and therefore increase the resonance frequency of the system, while a negative interaction stiffness will decrease the total stiffness and the resonance frequency of the system.

Also note that the phase is negative for a positive interaction stiffness and positive for a negative interaction stiffness. This implies an increased sensitivity in the phase for systems with an oscillatory stiffness (like the oscillatory solvation forces).

Figure 2.4 shows the effect of an increasing interaction damping on the spectra of a typical sample modulation configuration in liquid. For an increasing damping the amplitude response increases, due to the larger coupling with the modulating sample surface. The frequency f_{res} at which we find a maximum in the amplitude is, for a simple harmonic oscillator, expected to go down for an increasing damping coefficient (decreasing Q) with:

$$f_{\text{res}} = \frac{\omega_0}{2\pi}\sqrt{1 - \frac{1}{2Q^2}}.$$

This behavior can also qualitatively be observed in Figure 2.4a. The phase response is 90° for $\omega \to 0$ and decreases to $-90°$ for $\omega \to \infty$. This decrease is more gradual for a higher damping.

Force inversion AM Sample Modulation

In Amplitude Modulation (AM) sample modulation atomic force spectroscopy the sample is driven with a fixed drive amplitude A_d and drive frequency ω. The amplitude A and phase φ response of the cantilever are monitored upon approach and retract towards and from the oscillating sample surface. To extract

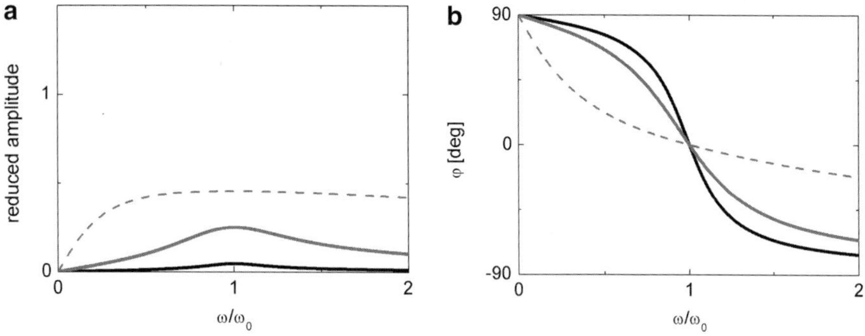

Fig. 2.4 The effect in an increasing interaction damping ($\gamma_{\text{int}} = 0.1\,\gamma_c$ *black*, $1\,\gamma_c$ *gray* and $10\gamma_c$ *dashed*) on the amplitude (**a**) and phase (**b**) spectra of the cantilever in a sample modulation system

the interaction stiffness k_{int} and the damping γ_{int} from the measured amplitude and phase response we need to solve (2.3) for k_{int} and γ_{int}.

After substituting the Ansatz $z = A\mathrm{e}^{\mathrm{i}(\omega t+\varphi)}$ and $d = A_{\mathrm{d}}\mathrm{e}^{\mathrm{i}\omega t}$ and their derivatives in (2.3) we find: (See Appendix for the complete derivation. For the limit $\omega \to 0$ see Ref. [32])

$$k_{\text{int}} = -k_c + m\omega^2 + \frac{\left(k_c - m\omega^2\right)\left(1 - \frac{A}{A_d}\cos\phi\right) - \gamma_c\omega\frac{A}{A_d}\sin\phi}{1 - 2\left(\frac{A}{A_d}\right)\cos\phi + \left(\frac{A}{A_d}\right)^2} \tag{2.5a}$$

and

$$\gamma_{\text{int}} = -\gamma_c + \frac{\gamma_c\omega\left(1 - \frac{A}{A_d}\cos\phi\right) + \left(k_c - m\omega^2\right)\frac{A}{A_d}\sin\phi}{\omega\left(1 - 2\left(\frac{A}{A_d}\right)\cos\phi + \left(\frac{A}{A_d}\right)^2\right)}. \tag{2.5b}$$

From the calibration constants k_c, Q and f_{res}, the damping γ_c and the mass m can be calculated and together with the measured amplitude A and phase φ be used to extract the distance dependent conservative and dissipative interaction forces. The drive amplitude A_d can be determined from the measured amplitude of the cantilever in full contact with the sample surface.

Force Inversion FM Sample Modulation

In FM atomic force spectroscopy we want to extract the interaction stiffness and interaction damping from the measured frequency shift and change in applied drive amplitude to keep the measured amplitude constant. For this we can use the same formulae as derived for AM AFM. In contrast to AM AFM, where the amplitude A and phase φ are monitored while the drive frequency ω and the drive amplitude A_d are kept constant, now ω and A_d are monitored while A and φ are kept constant.

When the phase is locked on $-90°$, (2.5a) and (2.5b) reduce to:

$$k_{\text{int}} = -k_{\text{c}} + m\omega^2 + \frac{\left(k_{\text{c}} - m\omega^2\right) + \gamma_{\text{c}}\omega \frac{A}{A_{\text{d}}}}{1 + \left(\frac{A}{A_{\text{d}}}\right)^2} \quad (2.6a)$$

and

$$\gamma_{\text{int}} = -\gamma_{\text{c}} + \frac{\gamma_{\text{c}}\omega - \left(k_{\text{c}} - m\omega^2\right)\frac{A}{A_{\text{d}}}}{\omega\left(1 + \left(\frac{A}{A_{\text{d}}}\right)^2\right)}. \quad (2.6b)$$

2.2.1.2 Magnetic Driving

In magnetic driving either the cantilever is coated with a magnetic or magneto-restrictive material or a small magnetic particle is attached to the end of the cantilever. With a coil an oscillating magnetic field is applied in order to drive the cantilever (Figure 2.5a). The forcing of the cantilever is direct and with deflection detection (or interferometric detection) the total motion of the cantilever is measured.

The equation of motion for this system is described by:

$$m \cdot \ddot{z} + \gamma_{\text{c}} \cdot \dot{z} + k_{\text{c}} \cdot z = F_0 \text{e}^{\text{i}\omega t} - \gamma_{\text{int}}\dot{z} - k_{\text{int}}z. \quad (2.7)$$

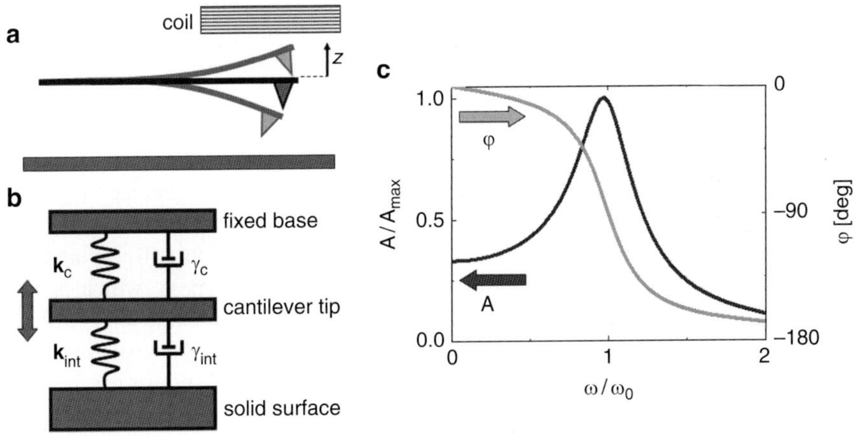

Fig. 2.5 (**a**) In magnetic actuation the response of a magnetic cantilever is monitored while an oscillating magnetic field is applied. (**b**) Spring-dashpot representation of magnetic actuation. (**c**) Amplitude and phase response for magnetic driving with a typical AFM configuration in liquid without interactions

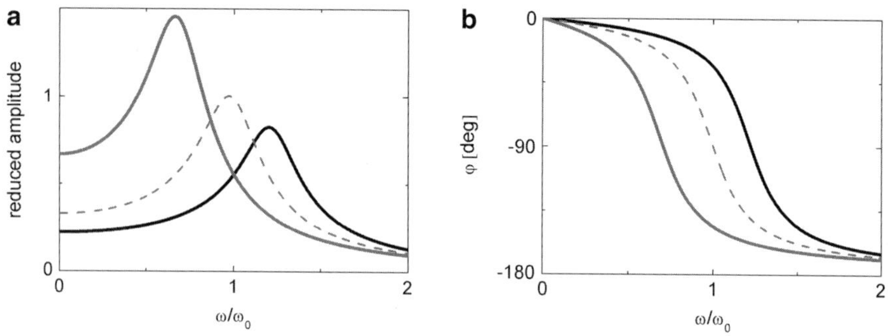

Fig. 2.6 The effect of a positive and a negative interaction stiffness ($k_{\mathrm{int}} = +0.5k_{\mathrm{c}}$ *black* and $-0.5k_{\mathrm{c}}$ *gray*) on the amplitude (**a**) and phase (**b**) spectra of the cantilever using magnetic actuation

Using the Ansatz $z = A\mathrm{e}^{\mathrm{i}(\omega t + \phi)}$ (2.7) can be solved for the amplitude A and phase φ:

$$A = \frac{F_0}{\sqrt{(k_{\mathrm{tot}} - m\omega^2)^2 + (\omega\gamma_{\mathrm{tot}})^2}} \tag{2.8a}$$

and

$$\tan\phi = \frac{-\omega\gamma_{\mathrm{tot}}}{-m\omega^2 + k_{\mathrm{tot}}}. \tag{2.8b}$$

Figure 2.5c shows the amplitude and phase response as a function of frequency for a magnetically driven cantilever in liquid without tip–sample interactions. Inhere we see the well-established typical response of a harmonic oscillator. Due to the damping the amplitude-maximum is at a lower frequency then the resonance frequency $\omega_0 = \sqrt{k_{\mathrm{c}} / m}$ and the phase shows a gradual decrease from 0 to $-180°$. The phase is exactly $-90°$ for ω_0.

Figure 2.6 shows the effect of a positive and a negative interaction stiffness on the amplitude and phase response for a typical magnetic drive configuration in liquid (spring constant $k_{\mathrm{c}} = 2\,\mathrm{N/m}$, quality factor $Q = 3$, amplitude resonance frequency $f_{\mathrm{res}} = 40\,\mathrm{kHz}$).

The resonance frequency increases for a positive interaction stiffness and the resonance frequency decreases for a negative interaction stiffness, which is also expected from its definition: $\omega_0 + \Delta\omega = \sqrt{k_{\mathrm{tot}} / m}$, where k_{tot} is the total stiffness (cantilever stiffness + interaction stiffness). Since the forcing is constant and the static deflection z is given by $z = -F / k_{\mathrm{tot}}$ the amplitude response for $\omega \to 0$ is higher for a negative stiffness. This effect can qualitatively be observed for all frequencies.

Figure 2.7 shows the effect of an increasing interaction damping on the amplitude and phase response of a magnetically driven cantilever in liquid. Since an increasing interaction damping increases the total damping the frequency at which we find the maximum amplitude is expected to go down according to

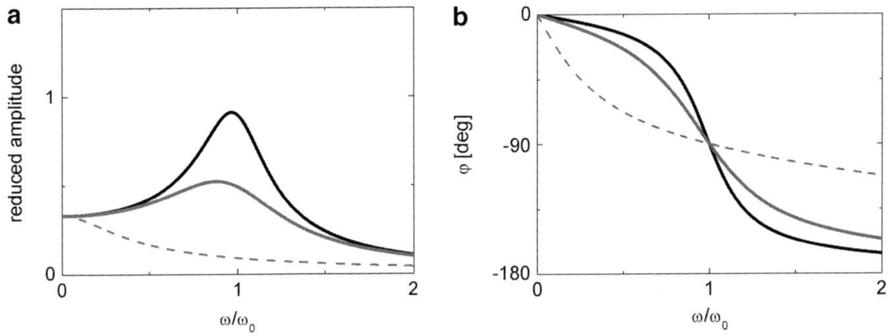

Fig. 2.7 The effect in an increasing interaction damping ($\gamma_{\text{int}} = 0.1\,\gamma_c$ *black*, $1\,\gamma_c$ *gray* and $10\gamma_c$ *dashed*) on the amplitude (**a**) and phase (**b**) spectra of the cantilever using magnetic actuation

$$f_{\text{res}} = \frac{\omega_0}{2\pi}\sqrt{1 - \frac{1}{2Q^2}}.$$

The phase shows for a higher damping a more gradual decrease from 0° at $\omega \to 0$ to $-180°$ at $\omega \to \infty$, but for all values of the interaction damping the phase is $-90°$ at ω_0.

Force Inversion AM Magnetic Driving

In order to extract the physical interaction forces from the calibration constants and the measured amplitude and phase response, we need to solve (2.7) for the interaction stiffness k_{int} and damping γ_{int}, yielding (see, e.g. [22, 33]):

$$k_{\text{int}} = -k_c + m\omega^2 + \frac{F_0}{A}\cos\phi \tag{2.9a}$$

and:

$$\gamma_{\text{int}} = \frac{-F_0}{A\omega}\sin\phi - \gamma_c, \tag{2.9b}$$

where F_0 can be determined from the amplitude far away from the surface (where the interactions are zero) using (2.8a).

Force Inversion FM Magnetic Driving

For the derivation of the force inversion formulae in FM AFM (magnetic drive) we use the fact that the phase is locked at $-90°$. When the phase is $-90°$, the system is driven at resonance. As shown above in Figures 2.6 and 2.7, the shift in the resonance frequency is only determined by the interaction stiffness and not by the interaction damping. So using $\omega_0 + \Delta\omega = \sqrt{k_{tot}/m}$ we can directly calculate the interaction stiffness from the frequency shift:

$$\begin{aligned} k_{int} &= m\,(\omega_0 + \Delta\omega)^2 - k_c \\ &= m\omega_0^2 \left(\frac{2\Delta\omega}{\omega_0} + \left(\frac{\Delta\omega}{\omega_0} \right)^2 \right). \end{aligned} \tag{2.10a}$$

For $k_{int} \ll k_c$ (2.10a) reduces to [28]: $k_{int} = \dfrac{2k_c\Delta\omega}{\omega_0}$

$$\gamma_{int} = \frac{F_0}{A\,(\omega_0 + \Delta\omega)} - \gamma_c. \tag{2.10b}$$

Another way to characterize the dissipation in the system is by determining the dissipated energy per cycle E_{dis}. To do so the amplitude of the oscillating cantilever is kept constant by adjusting the driving force F, and E_{dis} is calculated using $E_{dis} = \int_0^T F v \mathrm{d}t$ resulting in [34]:

$$E_{dis} = \frac{\pi k_c A^2}{Q} \frac{F}{F_0}, \tag{2.10c}$$

where F/F_0 is the relative change in driving force to keep the measured amplitude constant.

2.2.1.3 Acoustic Driving

Acoustic driving is the most widely used actuation technique. In acoustic driving a small drive piezo at the back of the cantilever (Figure 2.8a) is used to oscillate the cantilever. When using deflection detection modeling the system is not as straightforward as for magnetic driving. Since in deflection detection only the motion relative to the driving motion $z_d(t)$ is measured, the actual motion of the cantilever $z(t)$ can be significantly different from the measured signal $d(t)$ (compare Figures 2.8c to 2.5c). This can be circumvented by using an interferometric detection setup, where the total motion of the cantilever is measured. (Although one still needs to be alert for the spurious resonances characteristic for acoustic driving.) When using interferometric detection the forces can be extracted using the equations given in Sect. 2.1.2.

For acoustic driving with deflection detection the equation of motion is given by:

$$m \cdot \ddot{z} + \gamma_{tot} \cdot \dot{z} + k_{tot} \cdot z = k_c \cdot z_d, \tag{2.11}$$

where $k_{tot} = k_c + k_{int}$ and $\gamma_{tot} = \gamma_c + \gamma_{int}$ are the total stiffness and damping, respectively, while the measured deflection will be given by $d(t) = z(t) - z_d(t)$.

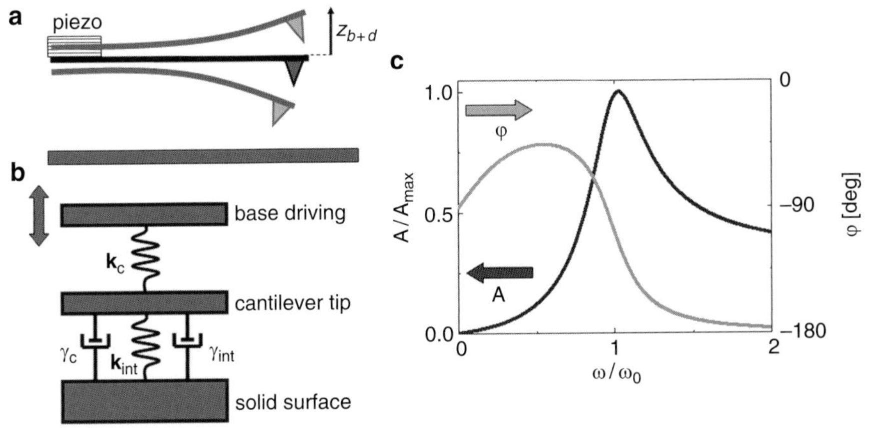

Fig. 2.8 (**a**) In acoustic driving the deflection (for deflection detection) or the total motion (for interference detection) of the cantilever is monitored while it is driven with a small piezo at the backside. (**b**) Spring-dashpot representation of acoustic actuation. (**c**) Amplitude and phase response for acoustic driving and deflection detection for a typical AFM configuration in liquid without interactions

To find the measured amplitude and phase response we solve (2.11) using the Ansatz $z(t) = A_{\text{tot}} \exp(\mathrm{i}(\omega t + \psi_{\text{tot}})) = d(t) + z_{\text{d}}(t) = A \exp(\mathrm{i}(\omega t + \phi)) + A_{\text{d}} \exp(\mathrm{i}\omega t)$, in which ω is the drive frequency, A and φ are the measured deflection amplitude and phase of the deflection $d(t)$ and A_{d} is the amplitude of the base-motion (see e.g., [24]):

$$A = \frac{A_{\text{d}} \sqrt{(k_{\text{c}} - k_{\text{tot}} + m\omega^2)^2 + (\omega\gamma_{\text{tot}})^2}}{\sqrt{(k_{\text{tot}} - m\omega^2)^2 + (\omega\gamma_{\text{tot}})^2}} \tag{2.12a}$$

and

$$\tan\phi = \frac{-k_{\text{c}}\omega\gamma_{\text{tot}}}{k_{\text{c}}\left(-m\omega^2 + k_{\text{tot}}\right) - \left(-m\omega^2 + k_{\text{tot}}\right)^2 - \left(\omega\gamma_{\text{tot}}\right)^2}. \tag{2.12b}$$

Note that the total amplitude A_{tot} of the tip motion, given by $A_{\text{tot}} = \sqrt{(A\sin\phi)^2 + (A\cos\phi + A_{\text{d}})^2}$, can be substantially different from the measured deflection amplitude A in (2.4a). Please be warned: The linearization of (2.2) is only justified when $A_{\text{tot}} \ll 2k_{\text{int}} / k'_{\text{int}}$ and $A_{\text{tot}} \ll 2\gamma_{\text{int}} / \gamma'_{\text{int}}$ [24]. Figure 2.8c shows the amplitude and phase spectra for a typical configuration in liquid without interactions. Note that Figure 2.8c is significantly different from Figure 2.5c. Figure 2.8c shows that the amplitude reduces to zero for $\omega \to 0$ and a finite value for $\omega \to \infty$. Moreover the phase reduces again to $-90°$ for $\omega \to 0$.

Figure 2.9 shows how the amplitude and phase spectra of Figure 2.8c are affected by a positive or negative interaction stiffness. Note that the amplitude for $\omega \to 0$ increases both for a negative and for a positive interaction stiffness. Moreover, the phase becomes extremely sensitive for variations of k_{int} at $\omega \to 0$.

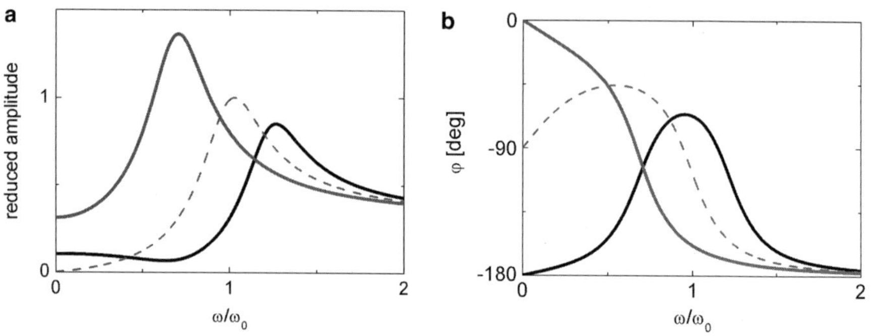

Fig. 2.9 The effect of a positive and a negative interaction stiffness ($k_{\mathrm{int}} = +0.5k_{\mathrm{c}}$ *black* and $-0.5k_{\mathrm{c}}$ *gray*) on the amplitude (**a**) and phase (**b**) spectra of the cantilever for acoustic driving using deflection detection

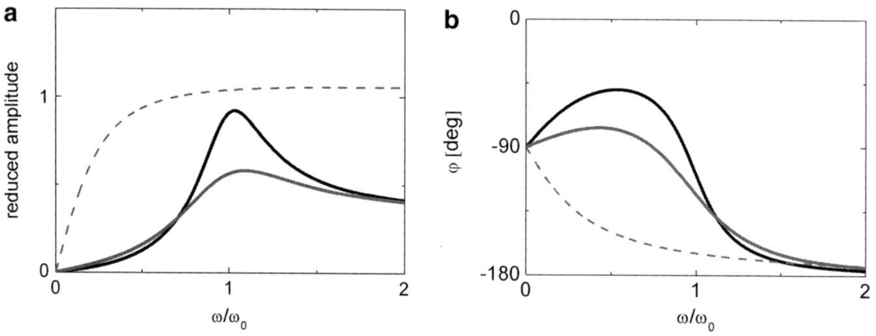

Fig. 2.10 The effect in an increasing interaction damping ($\gamma_{\mathrm{int}} = 0.1\gamma_{\mathrm{c}}$ black, $1\gamma_{\mathrm{c}}$ *gray* and $10\gamma_{\mathrm{c}}$ *dashed*) on the amplitude (**a**) and phase (**b**) spectra of the cantilever for acoustic driving using deflection detection

Figure 2.10 shows the effect of an increasing interaction damping on the spectra. In contrast to the magnetically driven system the amplitude resonance frequency does not go down for an increasing damping. Again, the phase at low frequencies is very sensitive for variations in γ_{int}.

Force Inversion AM Acoustic Driving

In order to extract the physical interaction forces from the calibration constants and the measured amplitude and phase response, we need to solve (2.11) for the interaction stiffness k_{int} and damping γ_{int}, yielding [23, 24]:

$$k_{\mathrm{int}} = -k_{\mathrm{c}} + m\omega^2 + \frac{k_{\mathrm{c}} A_{\mathrm{d}}(A_{\mathrm{d}} + A\cos\phi)}{A_{\mathrm{d}}^2 + A^2 + 2A_{\mathrm{d}}A\cos\phi} \tag{2.13a}$$

and:

$$\gamma_{\text{int}} = \frac{-k_{\text{c}} A_{\text{d}} A \sin\phi}{\omega\left(A_{\text{d}}^2 + A^2 + 2A_{\text{d}} A \cos\phi\right)} - \gamma_{\text{c}}, \tag{2.13b}$$

where A_{d} can be calculated from the measured free amplitude A far away from the surface (at e.g., 10 nm) using (2.12a) with the interactions set to zero.

Force Inversion FM Acoustic Driving

To extract the interaction forces from the frequency shift and change in drive amplitude we can (as described above) use the same formulae as derived for AM AFM. When the phase is locked on $-90°$, (2.13a) and (2.13b) reduce to:

$$k_{\text{int}} = -k_{\text{c}} + m\omega^2 + \frac{k_{\text{c}} A_{\text{d}}^2}{A_{\text{d}}^2 + A^2} \tag{2.13a}$$

and

$$\gamma_{\text{int}} = \frac{k_{\text{c}} A_{\text{d}} A}{\omega\left(A_{\text{d}}^2 + A^2\right)} - \gamma_{\text{c}}. \tag{2.13b}$$

2.2.2 *Effect Frequency Dependent Damping*

In our derivations we assumed that the viscous damping γ_{c} and the total mass m due to the liquid surrounding the cantilever are constant. But is this assumption justified?

The interaction of a solid object oscillating in a viscous fluid does in fact depend on the frequency [35]. Following Sader [36], we describe the hydrodynamic loading on the cantilever by a hydrodynamic function $\Gamma = \Gamma' + \mathrm{i}\Gamma''$, which yields an added mass and a damping given by $m_{\text{added}} = m - m_{\text{cant}} = \rho(\pi/4)w^2 L\Gamma'$ and $\gamma_{\text{c}} = \rho(\pi/4)w^2 L\omega\Gamma''$ [37]. Here, ρ, w, and L denote, respectively, the density of liquid, and the width and the length of the cantilever. Γ depends on the viscous penetration depth $\delta = \sqrt{2\eta / \rho\omega}$ (η: viscosity of the liquid) and the cantilever geometry as $\Gamma' = a_1 + a_2(\delta / w)$ and

$$\Gamma'' = b_1 \frac{\delta}{w} + b_2 \left(\frac{\delta}{w}\right)^2$$

with $a_1 = 1.0533$, $a_2 = 3.7997$, $b_1 = 3.8018$ and $b_2 = 2.736$. With the calibration constants (k, Q, $\omega_{0,\text{air}}$ and $\omega_{0,\text{liquid}}$) applied at resonance, the frequency independent prefactor in both the frequency-dependent added mass and damping can be calculated.

Figure 2.11a shows typical spectra for a *magnetically* driven cantilever in Octamethylcyclotetrasiloxane (spring constant $k_{\text{c}} = 2\,\text{N/m}$, quality factor $Q = 3$, amplitude resonance frequency $f_{\text{res}} = 40\,\text{kHz}$, $\rho = 956\,\text{kg/m}^3$, $\eta = 22\,\text{mPas}$) calculated for a constant damping and added mass and for a frequency dependent damping and added mass (dashed). The amplitude response curves appear

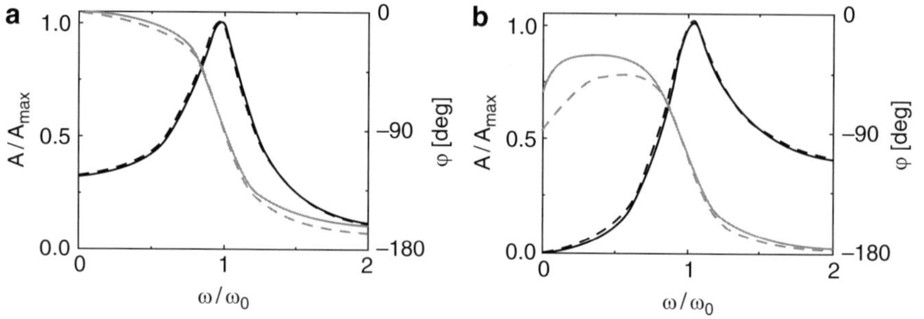

Fig. 2.11 The effect of including a frequency dependent added mass and damping (*dashed*) on the spectra of (**a**) a magnetically driven cantilever and (**b**) an acoustically driven cantilever

similar. But for the phase response above resonance some deviations occur. Since phase-offsets are known to cause cross-coupling between the extracted conservative and dissipative forces [23], these deviations can not be neglected. So, for a cantilever magnetically driven above resonance the frequency dependent added mass and damping need to be taken into account. Figure 2.11b shows the spectra for an *acoustically* driven cantilever (with the same properties as in Figure 2.11a) and again calculated for a constant damping and added mass and for a frequency dependent damping and added mass (dashed). Once more, the amplitude response is the same in both situations. Nevertheless the phase response significantly deviates now below resonance. So, for a cantilever acoustically driven (using deflection detection) below resonance the frequency dependent added mass and damping definitely need to be taken into account. We would like to mark that the conclusions described above are only guidelines. To reduce systematic errors in your analysis it is always recommended to include the frequency dependent added mass and damping in the equations for the amplitude, phase and force-inversion formulae described in the sections above.

2.3 Summary

In this chapter, we have presented and discussed the different driving schemes and the resulting force inversion formulae for AM and FM small amplitude atomic force spectroscopy. We have shown that in liquid (low Q environment) the measured response of the cantilever strongly depends on the actuation and detection technique. Moreover, to avoid phase-errors a frequency dependent damping should be taken into account for the calculation of the calibration constant.

So, in order to extract the correct forces from your dynamic AFM measurements, you have to:

1. Determine your actuation and detection method.
2. Find the accompanying force inversion formulae (depending on the actuation/detection method used).

3. Take into account the frequency dependent added mass and damping.
4. Do a correct calibration (depending on the actuation/detection method used).
5. Shift the phase to the correct value (depending on the actuation/detection method used).

Appendix

To find the interaction stiffness k_{int} and damping γ_{int} for a sample modulation setup we solve the equation of motion:

$$m \cdot \ddot{z} + \gamma_{\text{c}} \cdot \dot{z} + k_{\text{c}} \cdot z = k_{\text{int}}(d - z) + \gamma_{\text{int}}(\dot{d} - \dot{z}). \tag{2.14}$$

We use the ansatz $z_{\text{d}} = A_{\text{d}}\text{e}^{\text{i}\omega t}$, $z = A\text{e}^{\text{i}\varphi}\text{e}^{\text{i}\omega t}$ and obtain:

$$\left(k_{\text{tot}} - m\omega^2 + \text{i}\gamma_{\text{tot}}\omega\right) A(\cos\phi + \text{i}\sin\phi) = (k_{\text{int}} + \text{i}\gamma_{\text{int}}\omega)A_{\text{d}}, \tag{2.15}$$

where $k_{\text{tot}} = k_{\text{c}} + k_{\text{int}}$ and $\gamma_{\text{tot}} = \gamma_{\text{c}} + \gamma_{\text{int}}$ are the total stiffness and damping, respectively.

With:

$$\begin{aligned} K &= k_{\text{tot}} - m\omega^2 \quad & \Gamma &= \gamma_{\text{tot}}\omega, \\ K_{\text{c}} &= k_{\text{c}} - m\omega^2 \quad & \Gamma_{\text{c}} &= \gamma_{\text{c}}\omega. \end{aligned} \tag{2.16}$$

We express k_{int} and γ_{int} as:

$$k_{\text{int}} = K - K_{\text{c}} \quad \gamma_{\text{int}} = (\Gamma - \Gamma_{\text{c}})/\omega. \tag{2.17}$$

To derive equations for k_{int} and γ_{int}, we solve for K and Γ:

$$\frac{z}{z_{\text{d}}} = H(\cos\phi + \text{i}\sin\phi) = \frac{k_{\text{int}} + \text{i}\gamma_{\text{int}}\omega}{k_{\text{tot}} - m + \text{i}\gamma_{\text{tot}}\omega} = \frac{(K - K_{\text{c}}) + \text{i}(\Gamma - \Gamma_{\text{c}})}{K + \text{i}\Gamma} \tag{2.18}$$

With:

$$H = A/A_{\text{d}} \tag{2.19}$$

Hence

$$\begin{aligned} H\cos\phi + \text{i}H\sin\phi &= \frac{((K - K_{\text{c}}) + \text{i}(\Gamma - \Gamma_{\text{c}})) - (K - \text{i}\Gamma)}{K^2 + \Gamma^2}, \\ &= 1 - \frac{KK_{\text{c}} + \Gamma\Gamma_{\text{c}}}{K^2 + \Gamma^2} + \text{i}\frac{\Gamma K_{\text{c}} - K\Gamma_{\text{c}}}{K^2 + \Gamma^2}. \end{aligned} \tag{2.20}$$

or

$$\frac{KK_c + \Gamma\Gamma_c}{K^2 + \Gamma^2} = 1 - H\cos\phi,$$
$$\frac{\Gamma K_c - K\Gamma_c}{K^2 + \Gamma^2} = H\sin\phi. \tag{2.21}$$

The set of equations in (2.21) are quadratic in K and Γ but linear in:

$$\tilde{K} = \frac{K}{K^2 + \Gamma^2},$$
$$\tilde{\Gamma} = \frac{\Gamma}{K^2 + \Gamma^2}. \tag{2.22}$$

So one obtains:

$$\begin{pmatrix} K_c & \Gamma_c \\ -\Gamma_c & K_c \end{pmatrix} \begin{pmatrix} \tilde{K} \\ \tilde{\Gamma} \end{pmatrix} = \begin{pmatrix} 1 - H\cos\phi \\ H\sin\phi \end{pmatrix}. \tag{2.23}$$

Inversion of (2.23) results in:

$$\begin{pmatrix} \tilde{K} \\ \tilde{\Gamma} \end{pmatrix} = \begin{pmatrix} \tilde{K}_c & -\tilde{\Gamma}_c \\ \tilde{\Gamma}_c & \tilde{K}_c \end{pmatrix} \begin{pmatrix} 1 - H\cos\phi \\ H\sin\phi \end{pmatrix}, \tag{2.24}$$

where

$$\tilde{K}_c = \frac{K_c}{K_c^2 + \Gamma_c^2},$$
$$\tilde{\Gamma}_c = \frac{\Gamma_c}{K_c^2 + \Gamma_c^2}. \tag{2.25}$$

From (2.24) one obtains:

$$\tilde{K} = \tilde{K}_c(1 - H\cos\phi) - \tilde{\Gamma}_c H\sin\phi,$$
$$\tilde{\Gamma} = \tilde{\Gamma}_c(1 - H\cos\phi) + \tilde{K}_c H\sin\phi. \tag{2.26}$$

Using (2.22) and (2.25) or the inverse transformations:

$$K = \frac{\tilde{K}}{\tilde{K}^2 + \tilde{\Gamma}^2},$$
$$\Gamma = \frac{\tilde{\Gamma}}{\tilde{K}^2 + \tilde{\Gamma}^2}. \tag{2.27}$$

we rewrite (2.26) into:

$$K = \frac{K_c(1 - H\cos\phi) - \Gamma_c(H\sin\phi)}{(1 - H\cos\phi)^2 + (H\sin\phi)^2},$$
$$\Gamma = \frac{\Gamma_c(1 - H\cos\phi) + K_c(H\sin\phi)}{(1 - H\cos\phi)^2 + (H\sin\phi)^2}. \tag{2.28}$$

Now we know K and Γ, we use (2.17) to eventually obtain:

$$\begin{aligned} k_{\text{int}} &= K - K_c = \frac{K_c(1 - H\cos\phi) - \Gamma_c(H\sin\phi)}{(1 - H\cos\phi)^2 + (H\sin\phi)^2} - K_c, \\ \gamma_{\text{int}} &= (\Gamma - \Gamma_c)/\omega = \left(\frac{\Gamma_c(1 - H\cos\phi) + K_c(H\sin\phi)}{(1 - H\cos\phi)^2 + (H\sin\phi)^2} - \Gamma_c\right)/\omega. \end{aligned} \tag{2.29}$$

Using (2.16) and (2.19) we obtain the results of (2.5a) and (2.5b):

$$k_{\text{int}} = -k_c + m\omega^2 + \frac{\left(k_c - m\omega^2\right)\left(1 - \frac{A}{A_d}\cos\phi\right) - \gamma_c\omega\frac{A}{A_d}\sin\phi}{\left(1 - \frac{A}{A_d}\cos\phi\right)^2 + \left(\frac{A}{A_d}\sin\phi\right)^2}$$

and

$$\gamma_{\text{int}} = -\gamma_c + \frac{\gamma_c\omega\left(1 - \frac{A}{A_d}\cos\phi\right) + \left(k_c - m\omega^2\right)\frac{A}{A_d}\sin\phi}{\omega\left(\left(1 - \frac{A}{A_d}\cos\phi\right)^2 + \left(\frac{A}{A_d}\sin\phi\right)^2\right)}.$$

References

1. G. Binnig, F.C. Quate, Ch. Gerber, Phys. Rev. Lett. **9**, 930 (1986)
2. O. Marti, in *Handbook of Micro/Nanotribology*, 2nd edn. ed. by B. Bhushan (CRC, Boca Raton, FL, 1999), pp. 81–144
3. R. Garcia, R. Perez, Surf. Sci. Reports **47**, 197–301 (2002)
4. Y. Sugawara, N. Kobayashi, M. Kawakami, Y.J. Li, Y. Naitoh, M. Kageshima, Appl. Phys. Lett. **90**, 194104 (2007)
5. H. Hoelscher, A. Schwarz, W. Allers, U.D. Schwarz, R. Wiesendanger, Phys. Rev. B **61**, 12678 (2000)
6. T.R. Albrecht, P. Grutter, D. Horne, D. Rugar, J. Appl. Phys. **69**, 668 (1991)
7. H. Ueyama, Y. Sugawara, S. Morita, Appl. Phys. A **66**, s295 (1998)
8. Ch. Loppacher, M. Bammerlin, F. Battiston, M. Guggisberg, D. Müller, H.R. Hidber, R. Lüthi, E. Meyer, H.J. Güntherodt, Appl. Phys. A **66**, S215–S218 (1998)
9. B.M. Borkent, S. de Beer, F. Mugele, D. Lohse, Langmuir **26**, 260–268 (2010)
10. B. Anczykowski, J.P. Cleveland, D. Krueger, V. Elings, H. Fuchs, Appl. Phys. A: Mater. Sci. Process. **66**, S885 (1998)
11. D. Ebeling, H. Hoelscher, B. Anczykowski, Appl. Phys. Lett. **89**, 203511 (2006)
12. H. Hoelscher, A Schirmeisen, in *Advances in Imaging and Electron Physics*, vol. 135, 41–101 (2005)
13. C.M. Mate, G.M. McClelland, R. Erlandsson, S. Chiang, Phys. Rev. Lett. **59**, 1942–1945 (1987)
14. J.P. Cleveland, B. Anczykowski, A.E. Schmid, V.B. Elings, Appl. Phys. Lett. **72**, 2613–2615 (1998)
15. O. Sahin, S. Magonov, C. Su, C.F. Quate, O. Solgaard, Nat. Nanotechnol. **2**, 507 (2007)
16. X. Xu, J. Melcher, S. Basak, R. Reifenberger, A. Raman, Phys. Rev. Lett. **102**, 060801 (2009)
17. P.D. Ashby, C.M. Lieber, J. Am. Chem. Soc. **126**, 16973 (2004)
18. H. Hoelscher, Appl. Phys. Lett. **89**, 123109 (2006)
19. M.H. Lee, W.H. Jhe, Phys. Rev. Lett. **97**, 036104 (2006)

20. S.Q. Hu, A. Raman, Nanotechnology **19**, 375704 (2008)
21. A.J. Katan, M.H. van Es, T.H. Oosterkamp, Nanotechnology **20**, 165703 (2009)
22. S.J. O'Shea, M.E. Welland, Langmuir **14**, 4186 (1998)
23. C. Jai, T. Cohen-Bouhacina, A. Maali, Appl. Phys. Lett. **90**, 113512 (2007)
24. S. de Beer, D. van den Ende, F. Mugele, nanotechnology **21**, 325703 (2010)
25. R. Lim, S.F.Y. Li, S.J. O'Shea, Langmuir **18**, 6116 (2002)
26. D. Ebeling, H. Hoelscher, J. Appl. Phys. **102**, 114310 (2007)
27. J.E. Sader, S.P. Jarvis, Appl. Phys. Lett. **84**, 1803 (2004)
28. F.J. Giessibl, Rev. Mod. Phys. **75**, 949 (2003)
29. J.L. Hutter, J. Bechhoefer, Rev. Sci. Instrum. **64** 1868 (1993)
30. B. Ohler Veeco Application note: Practical Advice on the Determination of Cantilever Spring Constants, (2007)
31. A. Maali, C. Hurth, T. Cohen-Boxuhacina, Appl. Phys. Lett. **88**, 163504 (2006)
32. N.A. Burnham, G. Gremaud, A.J. Kulik, P.-J. Gallo, F. Oulevay, J. Vac. Sci. Technol. B **14**, 1308–1312 (1996)
33. A. Maali, T. Cohen-Bouhacina, G. Couturier, J.-P. Aimé, Phys. Rev. Lett. 96, 086105 (2006)
34. W. Hofbauer, R.J. Ho, R. Hairulnizam, N.N. Gosvami, S.J. O'Shea, Phys. Rev. B **80**, 134104 (2009)
35. L. Landau, F. Lifshitz, *Fluid Mechanics, Theoretical Physics*, vol. 6, (Mir, Moscow, 1971)
36. J.E. Sader, J. Appl. Phys **84**, 64 (1998)
37. A. Maali, C. Hurth, R. Boisgard, C. Jai, T. Cohen-Bouhacina, J.-P. Aime, J. Appl. Phys. **97**, 074907 (2005)

Chapter 3
Combining Scanning Probe Microscopy and Transmission Electron Microscopy

Alexandra Nafari, Johan Angenete, Krister Svensson, Anke Sanz-Velasco, and Håkan Olin

Abstract This chapter is a review of an in situ method where a scanning probe microscope (SPM) has been combined with a transmission electron microscope (TEM). By inserting a miniaturized SPM inside a TEM, a large set of open problems can be addressed and, perhaps more importantly, one may start to think about experiments in a new kind of laboratory, an in situ TEM probing laboratory, where the TEM is transformed from a microscope for still images to a real-time local probing tool. In this method, called TEMSPM, the TEM is used for imaging and analysis of a sample and SPM tip, while the SPM is used for probing of electrical and mechanical properties or for local manipulation of the sample. This chapter covers both instrumental and applicational aspects of TEMSPM.

Abbreviations

a-C	Amorphous carbon
AFM	Atomic force microscopy/microscope
CBED	Convergent beam electron diffraction
EBID	Electron beam induced deposition
ED	Electron diffraction
EDS	Energy dispersive X-ray spectroscopy
EELS	Electron energy loss spectroscopy
EFTEM	Energy filtered transmission electron microscopy/microscope
FIB	Focused ion beam
HREM	High resolution electron microscopy
HRTEM	High resolution transmission electron microscope
MEMS	Micro-electro-mechanical systems
NEMS	Nano-electro-mechanical systems
REM	Reflection electron microscopy
SAED	Selected area electron diffraction
SEM	Scanning electron microscopy/microscope

SPM	Scanning probe microscopy/microscope
STEM	Scanning transmission electron microscopy/microscope
STM	Scanning tunneling microscopy/microscope
SWNT	Single-wall nanotube
TEM	Transmission electron microscopy/microscope
TEMAFM	Combined TEM and AFM
TEMSPM	Combined TEM and SPM
TEMSTM	Combined TEM and STM

3.1 Introduction

3.1.1 Why Combine SPM and TEM?

The transmission electron microscope (TEM) is the only currently available method that offers sub-nanometer resolution imaging at high frame rates: TV rate or faster. As such, it offers unique possibilities for detailed and real-time observation of the probe–sample interaction in scanning probe microscopy (SPM).

Firstly, in standard SPM, the tip shape and tip–sample distance is usually unknown. By combining SPM with TEM, the imaging capability of the TEM can be used to monitor the positioning of the SPM probe relative to the sample. In this way, the approach can be controlled to avoid unintentional contact between probe and sample. The TEM monitoring is a powerful tool especially in cases when a particular feature of the sample needs to be found or a specific location of the sample should be investigated. In cases where the probe may have an irregular shape (which is the case with many STM tips), it may be of importance to control which part of the tip approaches the sample first. Moreover, because electrostatic and chemical forces between the tip and sample are highly dependent on the radius of the tip and the tip–sample distance, and, as we will see later, using TEM to image the tip and sample during an experiment yields surprising findings. The TEM imaging can also be used to extract the contact area between the tip and sample, which is of importance when studying the fundamentals of friction or the electrical conductance of point contacts.

Besides using the TEM as a monitoring tool for positioning and approach, it can also be used to extract complementary information, not accessible by SPM alone. The TEM can be used to follow changes in the physical properties of the sample and probe as functions of experimental parameters, such as field and current in the STM case, or force in the AFM case. Furthermore, with the imaging capabilities of TEM, morphological or structural information of the sample and/or the probe can be collected, for example, to detect the existence of an oxide layer on the probe or sample, or to observe whether the shape of the tip remains unchanged during an experiment. On a more detailed level, by using electron diffraction (ED) methods or high resolution TEM (HRTEM), it is also possible to investigate the crystallographic properties of the probe and the sample, such as phase changes due to resistive heating or mechanical deformation. The TEM also offers a variety of spectroscopic analysis techniques, of which energy dispersive X-ray spectroscopy (EDS), electron energy loss spectroscopy (EELS), and energy filtered TEM (EFTEM) are the most predominant methods. With these, information about the chemical composition and electronic structure can be extracted. Such chemical information is hard to obtain by SPM methods alone. This additional information is very useful during an SPM experiment for confirmation of the sample phase or to follow phase changes or diffusion and electromigration processes, to give a few examples. A third way to utilize the combination of SPM and TEM is to use the SPM to manipulate a sample inside the TEM. Such manipulation could include resistive heating to clean away an oxide layer, picking up and move nanoobjects with the probe or using the probe to mechanically deform the sample in a controlled way, and analyze the mechanical response.

The TEMSPM has been used for almost two decades, however, there is still no generally accepted nomenclature for this method. There are many in situ TEM techniques such as chemical reactions, heating, cooling, or optical detection that are not combined with scanning probe instruments. Takayanagi and co-workers coined the term TEM–STM [1, 2] for their early in situ experiments and this was used by several groups and later a related technique was called TEM–AFM [3]. But there are also variants such as STM–TEM [4, 5], in situ TEM probing, or TEM–SPM. Therefore, to differentiate truly combined SPM and TEM technology from other in situ TEM techniques and to facilitate literature studies in the field, we propose that the term "TEMSPM" is used when referring to the use of SPM methods combined with TEM, regardless of whether or not the SPM instrument is used in scanning or point probing mode. In more specific cases, the terms "TEMSTM and TEMAFM" can be used, and "TEMSPM" should be viewed as the general family of the techniques. Note that the acronyms now are without hyphen. This is in analogy with naming conventions that give names like HRTEM (High-Resolution TEM) but also, and this is of some importance, it will facilitate literature search. When a hyphen is used, most search machines will return results not only with, for example TEM–STM, but also studies where the two methods are used separately.

In this chapter, work done in the field of combining SPM methods with TEM will be reviewed. There are no other reviews covering the full spectrum of the TEMSPM method, however, Wang et al. have made several reviews on their extensive early TEMSPM work [6–9]. Furthermore, Golberg et al. have published a review on their recent work on characterizing various nanotube materials using the TEMSPM technique [10] and Nelson et al. have reviewed the nanorobotic aspects of carbon nanotubes [11–13]. A special issue of the MRS bulletin reviewed general in situ TEM techniques [14], including a section on TEMSPM. In a review published by Kociak et al. [15], the possibilities of a TEM nanolaboratory were elaborated on and a book on general in situ TEM techniques edited by Banhart can be found in [16]. Although this chapter aims to provide a comprehensive report of methods related to the TEMSPM technique, two related in situ electron microscopy probing methods are not covered: in situ TEM nanoindentation and probing in scanning electron microscopes (SEM). Further information about the in situ TEM nanoindentation is available in reviews by Stach et al. [17–19], and probing experiments inside SEM are, among others, reviewed in [20].

3.2 Some Aspects of TEM Instrumentation

The imaging principle in TEM is based on the transmission of an electron beam through a thin sample. As the electrons interact with the sample during their passage through it, the transmitted beam carries information from the sample. This information can be extracted as images, diffractograms, or spectra and, often after some interpretation, be used for analysis of the sample structure and composition. The electron beam is generated by an electron gun, and the beam is formed by a series of apertures and magnetic lenses before and after the passage through the sample. The TEM sample thickness must be kept small in order to allow electron transmission through it. The optimum sample thickness depends on several factors such as sample composition, mode of analysis, and acceleration voltage of the TEM, and typically, the thickness is in the range of 20–300 nm.

In most TEM instrument designs, the sample is mounted into a specimen holder, which basically consists of a 20–40 cm long rod, as shown in Fig. 3.1. The sample is supported by a device at the end of the rod, and the rod itself is entered through a vacuum port on the side of the TEM instrument so that the sample is oriented perpendicular to the electron beam. Inside standard TEM instruments, space is limited and the most limiting factor for an in situ TEM instrument is the space available between the two pole pieces of the objective lens, where the sample is positioned, which can be seen in Fig. 3.2. In order to decrease the focal length of the lens, this gap is kept small, in many cases it can be in the order of 4–5 mm, but there are instruments with much smaller ($\leq$ 2 mm) and larger ($\geq$ 10 mm) pole gaps.

The most straightforward way to place an SPM instrument inside a TEM is to miniaturize the SPM so that it fits inside the space available, that is, that the SPM is small enough to fit inside the typical dimensions of a conventional TEM specimen

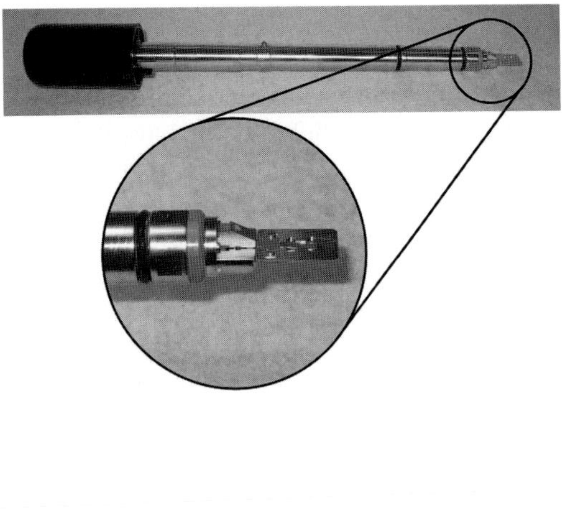

Fig. 3.1 Typical standard TEM specimen holder. In this case, the diameter of the shaft is approximately 18 mm, and the specimen support part is about 1 mm high

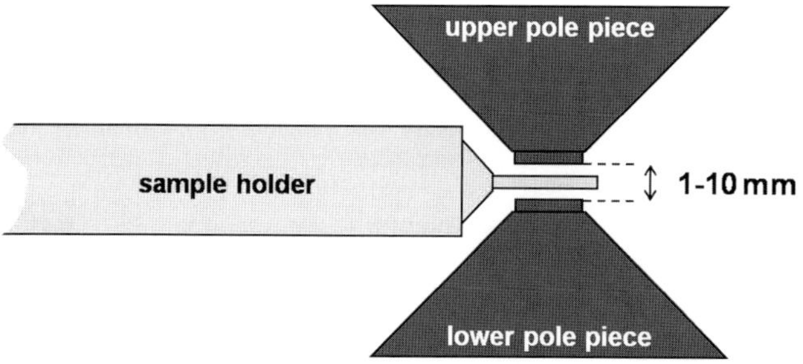

Fig. 3.2 Schematic of the upper and lower pole piece of the objective lens, with a sample holder inserted between the pole pieces

holder. The other approach is to modify the design of the TEM to accommodate a larger SPM device. The first approach has attracted the most interest because redesigning a TEM instrument is usually a complicated task. There are, however, a few examples of rebuilt TEMs in which an STM and an AFM have been incorporated [2, 21]. The different principles by which TEMSPM instruments have been constructed will be presented in this chapter.

3.3 Incorporating an STM Inside a TEM Instrument

The main challenge when incorporating an STM inside a TEM is the limited space available inside the pole-piece gap of the TEM. A traditional STM design is simply too large to be placed in a TEM, and in order to understand this challenge we first take a closer look at how the STM instrument has been developed. It began in the early 1970s when Young and co-workers envisioned an instrument, the "Topografiner," that would image a surface by sweeping a sharp tip over the surface at a constant, noncontacting distance [22]. In a method still practiced today, a tip is scanned over the surface in a raster pattern and by keeping the current constant the tip describes a trajectory that generates the surface topography. The current flowing in the tunneling mode is sensitive to the distance between tip and sample, which is of the order of only 1 nm. This puts very high demands on the mechanical stability of such an instrument. The mechanical stability was improved upon about a decade later by Binnig and Rohrer [23]. They dubbed their instrument a "scanning tunneling microscope" as it was, for the first time, able to acquire images in the tunneling mode and then display atomic resolution on semiconductor surfaces. Since then the mechanical stability and the controlling hardware and software have been improved even further.

To design an STM, a fine motion control of the tip (or sample) is required. This is usually performed by several piezo ceramic parts (or a single tube), giving the tip a movable range of a few micrometers in three dimensions. A coarse approach mechanism is also required to place the sample and tip in close proximity so that

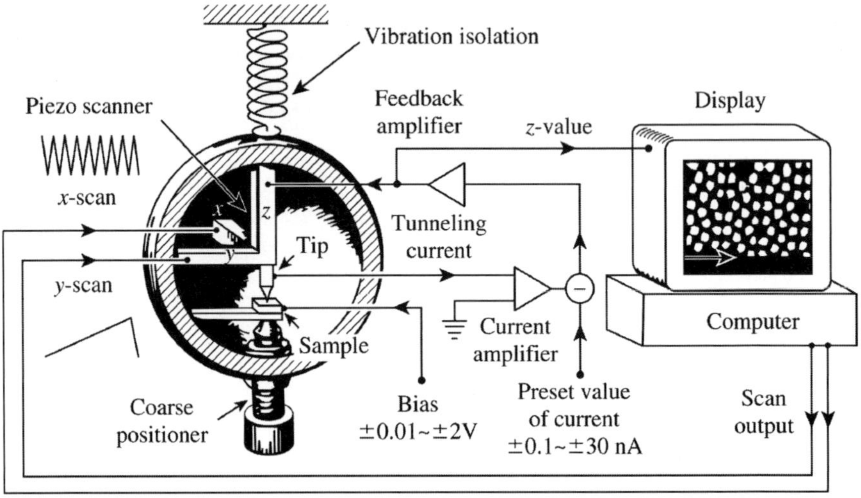

Fig. 3.3 Schematic drawing of the different parts that are needed for a complete STM system. Image is from [24]

the piezo motion is enough for the imaging. These parts should be kept small and sufficiently rigid to minimize the vibrational amplitudes of the tip in relation to the sample. A schematic drawing of the STM construction is shown in Fig. 3.3.

In connection to the development of the STM, there was also a desire to be able to "see" how these new instruments were actually operating. In 1986, Gerber et al. combined an STM with an SEM in order to study the STM in operation [25]. There were also ideas to use the STM and SEM in combination in order to extend the imaging capabilities of the fairly limited SEM instruments of the time 26]. To achieve a better resolution of the electron microscope part, an STM was fitted inside the sample holder of a TEM by Spence [27]. The tip and sample could be observed using the TEM in reflection mode (REM). The first studies were primarily aimed at monitoring the tip–sample interaction and to better understand how the STM operated and imaged a surface [28]. Detailed and valuable data were collected revealing tip changes during deliberate tip alterations as well as regular scanning [2, 29]. After the first STMs were incorporated into TEM instruments, it became apparent that the combined instrument was much more than a mere combination of two imaging methods. Instead, it can be considered to have movable, high precision, electrical probe available inside the imaging TEM. Applying this perspective extends the capabilities of the TEM beyond passive sample observation. The latest developments have indeed gone in that direction, and the modern TEMSTM systems are rarely used for actual STM imaging; instead they are used for sometimes very intricate manipulations and characterizations of complex nanostructured materials (see Sect. 3.3.1).

As mentioned earlier, the main challenge when combining STM and TEM is the limited space available within the pole-piece gap of a TEM. The most common approach to fit an STM in a TEM is to custom make a side-entry holder that includes a coarse motion control and a fine piezo motion, all within the regular size of a sample holder, and thus no modifications to the TEM itself are needed. The first report of using such a configuration was by Spence [27], where a single piezo tube was placed inside a side-entry holder and driven back and forth by a linear motor. For that particular TEM, the pole gap was fairly large (9 mm) and the STM part was kept below 7 mm in diameter. The same design ideas have since then been used by others for more demanding pole pieces. The pole-piece gaps of high resolution instruments can be rather small and require holders with sub-millimeter thickness in the sample region. A one-dimensional coarse motion of the tip towards the sample can be realized by an inch worm [27,28,30], a micrometer screw [1,31], or a stepper motor [32]. In order to obtain a high mechanical stability, the linear drive can be decoupled from the piezo when the coarse motion is not needed.

An alternative approach worth mentioning here is a custom-made micro-STM instrument (μ-STM) presented by several groups [33,34], which consisted of micro-machined silicon structures in order to build up a complete STM instrument in one single chip. A comb structure was used for the linear drive of the tip toward the sample and out-of-plane motion was achieved by torsional motion [34]. These instruments were fairly limited because there was no obvious way to incorporate traditional tip and sample materials. To our knowledge, the development of the μ-STM has not continued past these initial attempts.

More demanding applications have emerged when one wants to study electron transparent materials in electron transmission mode. The samples must be very thin (typically 20–300 nm), and the limited range of the piezo makes ex situ alignments rather impractical as the tip has to effectively find a knife edge inside the TEM. Ideally one should have full three-dimensional coarse motion of the tip (or sample) so that any part of the sample can be reached in situ. The first real attempts at full three-dimensional coarse motion were presented by Olin et al. [32]. In this approach, a coarse motion mechanism was fitted to the end of the piezo scanner tube, thus allowing full three-dimensional coarse and fine motion while maintaining atomic level resolution in both STM and TEM mode [35]. Figure 3.4 shows the TEM sample with this coarse motion incorporated. The coarse motion mechanism utilizes a tip holder connected to the piezo by six springs that clamp around a metal sphere. When the piezo is moved at typical scan speeds the tip holder with its low mass simply follows the motion of the piezo. When much faster, jerking, motions are performed by the piezo, the tip holder will slide against the sphere that is rigidly attached to piezo. In this way, the tip holder and the tip apex can be moved by coarse and fine motion in all three directions. The detailed construction of the springs limits the coarse motion to around 1 mm (in all three directions), which is enough to cover the typical sample area inside a TEM. The friction force is also large enough to allow physical manipulations of nanostructures with the tip, as these are generally in the μN range or lower, while the friction force of the tip holder is in the range of hundreds of mN [35]. This construction can be adapted to any side-entry TEM, and the instrument can be further customized on the fixed side to accommodate special samples, such as those with several electrodes. The fixed side could also contain a

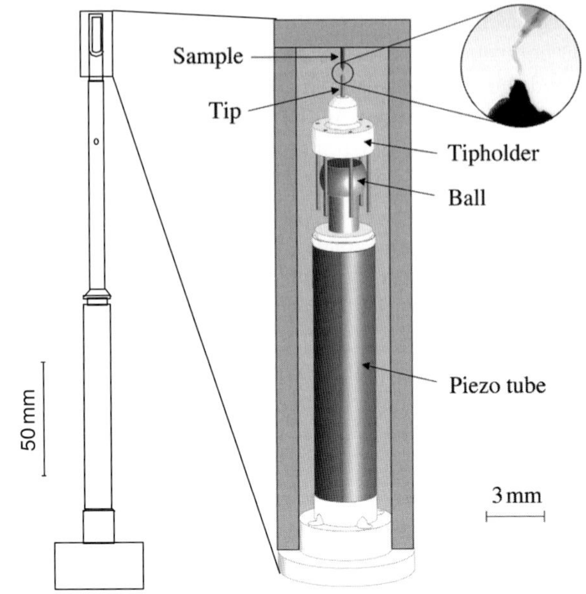

Fig. 3.4 Schematic illustration of the TEMSTM holder with the integrated three-dimensional coarse motion mechanism. The *inset* shows TEM image of sample and tip. Image is from [35]

force sensor, and thereby an AFM has essentially been joined with the TEM. This design forms the basis for commercial TEMSPM instruments [36].

3.3.1 Applications of TEMSTM

The earliest STM instrument inside a TEM by Spence et al. was mainly considered to be a complementary microscope for surface imaging [27, 37, 38]. Other early studies include the work by Takayanagi et al. on the interaction between a STM tip and a sample [2, 29]. Later it was realized that the real power was in the use of an STM tip for local probing of electrical properties and manipulation of a sample.

3.3.1.1 Electron Transport Studies

In one of the first studies when using the TEMSTM as a probe instead of as an additional microscope, Takayanagi et al. used the STM tip both for forming atomic-sized gold wires and measuring the conductance of these wires [39]. By pressing two gold tips together and then retracting them, a nanosized wire was formed before rupture. The entire process of nanowire formation was imaged by TEM and correlated simultaneously with electrical measurements. The thinnest wires were only one atom

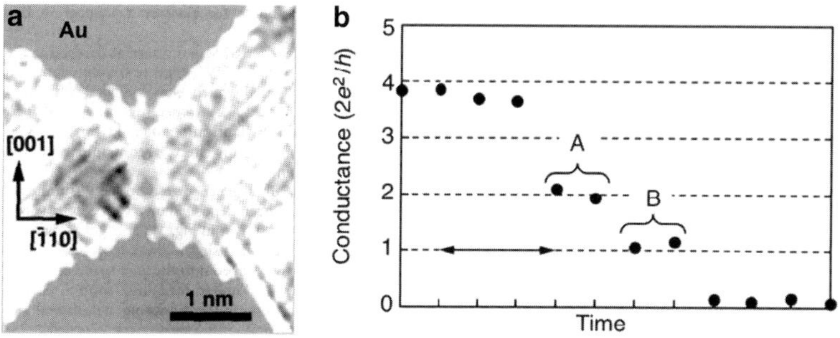

Fig. 3.5 Quantized conductance of atomic-sized gold wires studied by TEMSTM. (**a**) TEM image of a wire consisting of four gold atoms. (**b**) The corresponding conductance measurement, in units of $G_0 = 2e^2/h \sim (13\,\mathrm{k}\Omega)^{-1}$, while withdrawing the tip. Images are from [39]

wide, as can be seen in Fig. 3.5. In such small conductors, the electron transport is ballistic and the conductance is quantized in units of $2e^2/h \sim (13\,\mathrm{k}\Omega)^{-1}$, where e is the electron charge and h the Planck's constant. These studies were done to answer the open question about the microstructure of earlier observations of quantized conductance in similar ex situ STM experiments (for a review see [40]). The work by Takayanagi et al. solved this problem and gave direct evidence of the structure of these atomic sized conductors [41–43]. However, there were also new surprises, such as the anomalously long distance between the atoms in the wire.

Using the same kind of pressing-retraction method, Erts et al. studied larger point contacts of gold nanowires ranging from atomic size to 20 nm in diameter [44]. For the smallest ones, the electron transport was ballistic, but for the larger ones the transport became more and more diffusive. The degree of diffusive electron transport was determined by using the TEM to measure the radius and the STM for contact formation and conductance measurements. The electron mean free path was found to be ten times shorter than the bulk value. This indicates that the earlier calculation of the radius of the point contacts from conductance measurements is misleading by an order of magnitude.

With the increasing interest in carbon nanotubes there have been numerous investigations of their electrical properties [45, 46]. Multiwall carbon nanotubes, concentrically stacked tubular sheets of graphite, were extended in a telescopic manner using an STM probe and their nonlinear electrical resistance change was monitored [47]. In this type of study it is important to have proper electrical contacts, as will be discussed in Sect. 3.5. For example, in an early study by Frank et al. [48], the electron transport of carbon nanotubes was determined by dipping the nanotube in a drop of mercury. With this contacting technique, ballistic conduction was suggested; however, it is not without problems. Recently, it has been demonstrated that the nanowire does not actually enter the mercury droplet due to the high surface tension and a formation of a skin on the mercury surface [49]. This behavior gives a conductivity that appears to be independent of the distance between the electrodes, while in fact the distance remains the same during the dipping process.

Also, other materials such as III–V semiconductors, such as InAs, have been electrically characterized [50] using TEMSTM. The advantage of the TEMSTM method is the correlation between the microstructure determined by TEM imaging and analysis and the direct electrical measurement, but also the simple way to select and contact single nanowires. A challenge for the future is to provide a method to extend this two-probe method to permit four-point measurement in order to avoid the often unknown contact resistance.

3.3.1.2 Field Emission

In situ TEM probing of field emission was first studied by Wang et al. [6, 7, 31] and Kuzumaki et al. [51–53], which revealed the structural change of the electron emitting carbon nanotubes. Cumings et al. [54] performed electron holography of field-emitting carbon nanotubes to determine the magnitude and spatial distribution of the electric field surrounding the tubes. Sveningsson et al. [5] measured the electron emission from carbon nanotubes and observed that for low currents the emission was of a standard Fowler–Nordheim type, while for higher emission currents a nonlinear behavior occurred due to thermally enhanced emission caused by Joule heating of the carbon nanotube. A detailed study of the correlation of the structure of carbon nanotubes and electron emission as well as the structural change during emission was done by Wang et al. [55]. An interesting combination of the resonance method to determine the work function [56, 57] and field emission was done by Xu et al. [58]. In the Fowler–Nordheim model, the field emission current is determined by two factors, the field enhancement factor β and the work function ϕ, and the emission current is

$$I = KF^2/\phi \exp(-B\phi^{3/2}/F), \tag{3.1}$$

where $B = 6.83 \times 10^9$ V eV$^{3/2}$ m^{-1} and K is a constant. The local electric field F is related to the applied voltage V as $F = \beta V/d$, where d is the distance between the electrodes, as seen in Fig. 3.6. The standard way to determine β from measurements is to use a Fowler–Nordheim plot, $\ln(I/V^2)$ vs. $1/V$ (Fig. 3.6), and from the plot the slope $(-B\phi^{3/2}d/\beta)$ is determined. If the work function is known, the field enhancement factor can then be calculated. The work function is sensitive to small changes of the emitting nanotube; however, by using the TEMSTM in the mechanical resonance mode, the work function could be measured in the same setup. In another combined mechanical resonance-field emission study, Xu et al. [59] showed how the field emission current had a periodic oscillation component with a frequency that was twice the mechanical resonance frequency.

3.3.1.3 Electromigration

Svensson et al. studied the electromigration of iron particles inside carbon nanotubes [60]. Above a critical current density of 7×10^6 A cm^{-2} the iron particles started

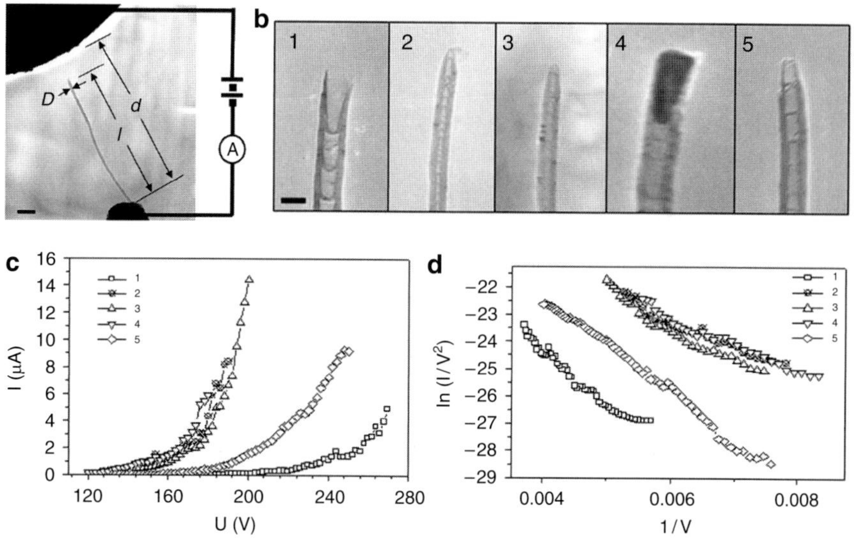

Fig. 3.6 TEMSTM study of field emission from carbon nanotubes. (**a**) TEM image of the nanotube and the electrodes with geometrical definitions. *Scale bar* is 200 nm. (**b**) TEM images of the microstructure of the emitting part of the carbon nanotubes. *Scale bar* is 50 nm. (**c**) Current–voltage curves for the different nanotubes. (**d**) The corresponding Fowler–Nordheim plot. Images are from [58]

to migrate in the same direction as the electron current. The "wind" of the electrons transfers a moment to the particles, which is known as electromigration (see Fig. 3.7h–k). In the same study, these effects were used as a kind of nanopipette to reversibly eject, deposit, and retrieve the iron particles from the end of the carbon nanotube, as shown in Fig. 3.7a–g. Electromigration has been used in several studies; for example, for transport of indium particles on the surface of carbon nanotubes [61], for welding carbon nanotubes [62], for mass transport through the walls of carbon nanotube [63], for transport of CuI particles in carbon nanotubes [64], and in applications for archival memory devices [65].

3.3.1.4 Joule Heating

A high current through a nanowire or nanotube could lead to resistive heating, called Joule heating. This current, I, will inject a heating power, P, of I^2R, where R is the resistance of the nanowire. The heat will lead to an increase in temperature that can easily reach the melting point of the material. Joule heating can be used to anneal samples, evaporate some of the material in a wire, or to decompose compounds. Joule heating is also, for example, used for welding materials together or removing contaminations.

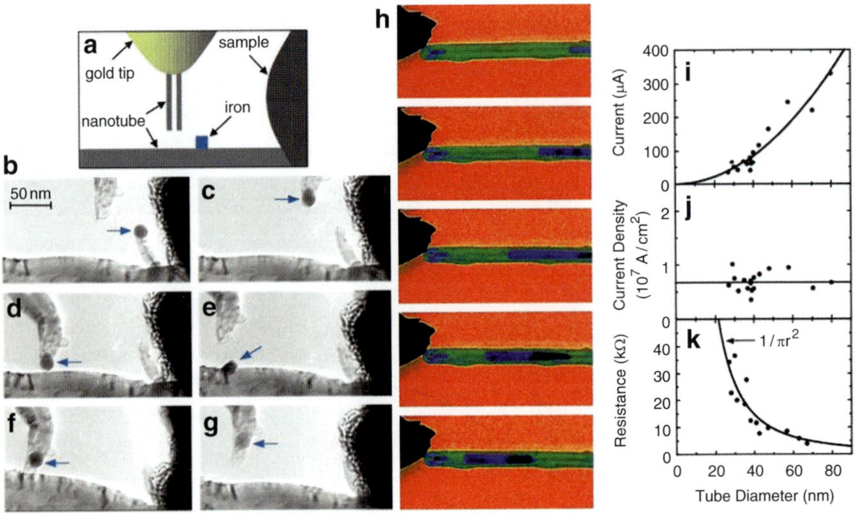

Fig. 3.7 Electromigration experiments using the TEMSTM. (**a–f**) Sequential TEM images demonstrating the nanopipette action. The transferred iron is indicated by an *arrow*. (**a**) Schematic drawing of the setup. (**b**) Iron has been migrated to the end of a fixed nanotube. (**c**) Iron is retrieved using a nanotube attached to the movable tip. (**d**) The nanotube is directed at the side of a large nanotube. (**e**) The iron is deposited. (**f**) The iron particle can be retrieved again. (**g**) The iron particle is ready to be deposited elsewhere. (**h**) Electromigration of an iron particle inside a carbon nanotube. (**i–k**) Electrical data measured on iron-containing carbon nanotubes as a function of diameter. (**j**) The iron particle starts to move above a certain critical current density. (**k**) The resistance measurement also shows that the electron transport is diffusive, because the resistance is inversely proportional to the cross-sectional area. Images are from [60]

Annealing and Evaporation

Annealing by in situ Joule heating on carbon nanostructures has been performed by Huang and co-workers, for example, on the crystallization of amorphous carbon (a-C) into carbon nanotubes [66] with a resulting increase in electrical conductivity [67]. The elevated temperature also allows super plastic deformation of the carbon nanotubes [68], that is, when the material is deformed well beyond its usual breaking point, usually exceeding 100% during tensile deformation. Such a super plastic state is achieved at a high temperature, typically at half the melting temperature. Increasing the electric power beyond that point will instead lead to breakdown of the nanotube [69], unless, as shown by Zettl et al., a controlled application of Joule heating, electromigration, and electron beam irradiation could reduce the diameter of a carbon nanotube, shrinking the tube from several nanometers down to near zero [70]. In situ Joule heating is also used to study other forms of carbon and graphene sublimation, and multilayer edge reconstruction has been observed [71] in such processes.

The structural variation of a-C nanocontacts during voltage application was observed while simultaneously measuring the I–V characteristics of a-C nanocontact

[72]. The maximum current density and resistivity of a-C nanocontacts were 2.8×10^{11} A m^{-2} and 8.5×10^{-3} Ωm, respectively, at a bias voltage of 2.3 V. When the voltage was reduced below 1.4 V, the current was maintained at zero and during this voltage reduction, the a-C nanocontacts transformed into graphitic nanocontacts with a thickness of 2–3 atomic layers. The observation presented in [72] shows that stable graphitic nanocontact is obtained when the first voltage applied.

Decomposition

Decomposition of compound materials has also been studied using in situ TEM Joule heating, for example, in the form of GaN nanowires [73] and BN nanotubes [74, 75]. Both the GaN nanowires and the BN nanotubes show thermal decomposition by leaving nanoparticles (Ga in the first case and B in the second) behind on the surface of the nanowires before failure. In another study of a compound nanowire, both the decomposition and the annealing phases were observed [76]. $Mo_6S_3I_6$ decomposed to metallic Mo below 1,000°C, when the S and I atoms evaporated. The annealing process of the remaining Mo nanowires revealed details about grain growth and the increased conductivity due to a lower degree of scattering, as seen in Fig. 3.8 [76].

Growth

Joule heating can also be used for growing structures inside a TEM. For example, carbon nanotubes [77] and carbon fullerenes have been grown in the gap between two gold electrodes [78]. By depositing gold nanoparticles on carbon nanotubes and using Joule heating of the carbon nanotube, it is possible to grow carbon nanocages using the gold nanoparticles as templates [79]. By increasing the temperature with Joule heating, the gold nanoparticles were evaporated leaving the empty carbon cages behind. In a similar study, a-C was Joule heated, and by using platinum particles as templates, 1–5 graphite layer thick cages were grown [80]. In another study, the TEMSTM and electron beam induced deposition (EBID), a-C nanowires were grown and graphitized mediated by iron particles [81].

Welding

Another application of Joule heating is in the welding of nanostructures, for example, interconnections of nanobuilding blocks, such as carbon nanotubes, nanowires, nanobelts, nanohelixes, and other structures for the assembly of nanoelectronics or nanoelectromechanical systems (NEMS). Compared to other interconnection processes, electrical spot welding has several interesting aspects: (a) a low electron current can induce melting more efficiently than irradiation-based techniques involving high electron beams, focused ion beams (FIB), or lasers; (b) the welding site can readily be selected by using a three-dimensional positioning and

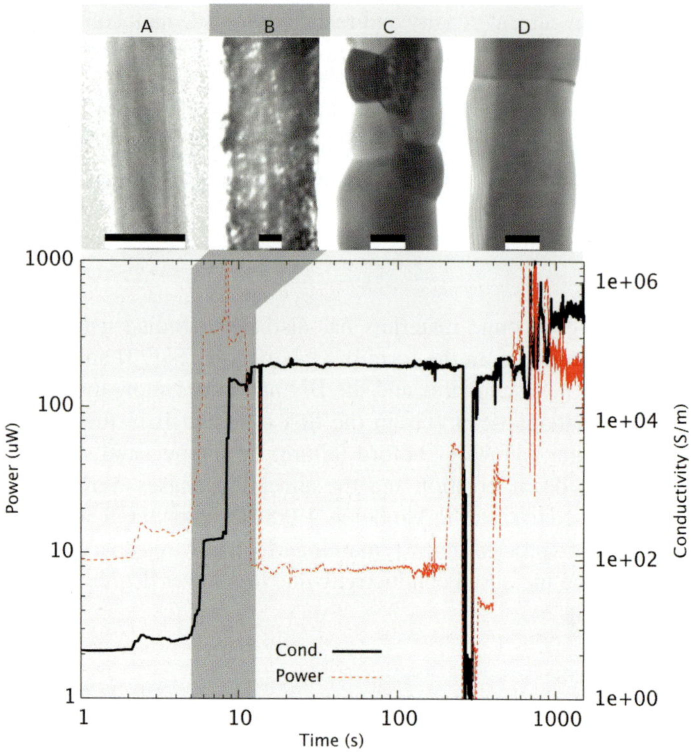

Fig. 3.8 Joule heating of a compound nanowire, $Mo_6S_3I_6$, demonstrating both decomposition and annealing. (**a**) A fresh sample $Mo_6S_3I_6$ is shown. (**b**) The $Mo_6S_3I_6$ nanowire has decomposed and only Mo remains. (**c**) The nanowire is annealed. (**d**) The nanowire wire after further annealing. Note that the TEM images are from different nanowires. The *scale bars* are 100 nm in (**a**) and 50 nm in (**b–d**). Images are from [76]

manipulation probe, which may potentially enable three-dimensional prototyping and assembling; (c) the melting process is very fast, compared with, for example, high intensity electron beam or FIB. Nelson et al. have reviewed several aspects of nanospot welding [11–13].

In the early study by Hirayama et al., bundles of single-wall nanotubes (SWNT) were spot welded together [82]. At a voltage of 2 V and a current of 20 nA, which gives an injected power of 40 nW, the bundles adhered to each other permanently. This was interpreted as spot welding of the tubes by Joule heating.

Spot welding using single crystalline copper filled carbon nanotubes was investigated by Dong et al. [62]. Controlled melting and flowing of copper inside nanotube shells was realized by applying a bias voltage between 1.5 and 2.5 V. The melting was a result of Joule heating, while the flowing was caused by electromigration, where the latter occurred at a critical current density of 3×10^6 A cm^{-2}. The flow rate of the copper was approximately 10 nm s^{-1} at 2.5 V bias, which allowed precise control of the mass transport.

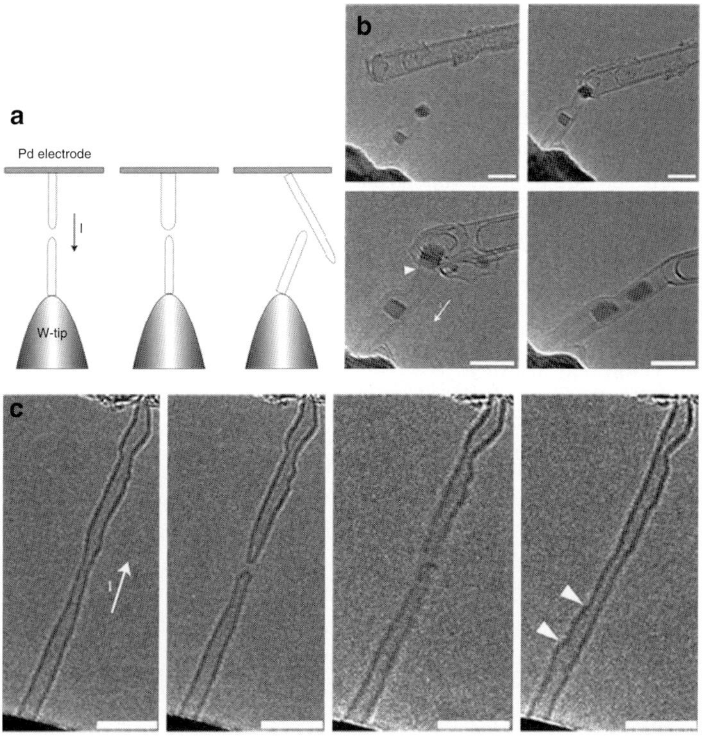

Fig. 3.9 Joining carbon nanotubes with Joule heating. (**a**) A schematic of three typical geometries joined. (**b**) Joining two nanotubes with different diameters using a tungsten particle. The *scale bars* are 5 nm. (**c**) Breaking and joining SWNTs. The *scale bars* are 5 nm. Images are from [83]

Jin et al. [83] reported on a similar welding process of Joule heating to seamlessly join two carbon nanotubes of similar diameter to form new carbon nanotube structures. With the assistance of tungsten particles, carbon nanotubes of different diameters could be joined, as seen in Fig. 3.9b. To join two SWNTs with identical diameters, the nanotubes were first broken into two parts using an electrical breakdown process. The two resulting capped SWNTs could be mechanically recontacted by moving the SWNT mounted on a tungsten wire on the STM probe in close proximity to the second SWNT mounted on a counter electrode. The voltage and current were raised from zero, and by passing threshold voltages, the SWNT joined again into one carbon nanotube. This process of breaking and joining could be repeated up to seven times. Images from these procedures are seen in Fig. 3.9.

Another form of welding, not related to Joule heating, is based on EBID of a-C. This welding process was used by Wang et al. to fix bended carbon nanotubes into a desired morphology [84]. The freestanding nanotubes were bent by an STM probe and then fixed by the deposition of a-C onto the bent area. The mechanical strength of the bent carbon nanotube may be greatly enhanced by increasing the amount of a-C on the deposition area. The electrical conduction of the nanotube has been observed to be independent from the bending deformation and the deposition of a-C. Furthermore, metals such as tungsten, gold, and platinum can be deposited on carbon nanotubes. Using the EBID process, an organometallic compound can be decomposed to fix the bent carbon nanotubes or even connect them to each other [85] and to mount them on a substrate [86]. The disadvantage of using metals is that metal particles can be melted, vaporized, or moved along the carbon nanotubes by Joule heating and electromigration. Also, there is a substantial amount of organic residues left from the EBID process that might be of consequence if a low ohmic contact is needed.

3.3.1.5 Mechanical Studies

Even though the TEMSTM is an electrical probe, there are several ways that it can be used to analyze mechanical properties. For example, Poncharal et al. [7,31,87,88] applied an alternating electric field between an STM tip and a separated carbon nanotube which made the nanotube vibrate. By increasing the frequency of the alternating voltage, the nanotube will reach its resonance frequency as observed by TEM imaging, as shown in Fig. 3.10. The TEM images give information about size, and along with the information about the resonance frequency, the elastic modulus can be calculated from the Euler–Bernoulli beam equation:

$$f_n = \frac{\beta_n^2 D}{8\pi L^2}\sqrt{\frac{E_b}{\rho}}, \tag{3.2}$$

where f is the frequency, β_n are constants for the nth harmonic: $\beta_1 = 1.875$, $\beta_2 = 4.694$, L is the length, D is the diameter, ρ is the density, and E_b is the effective

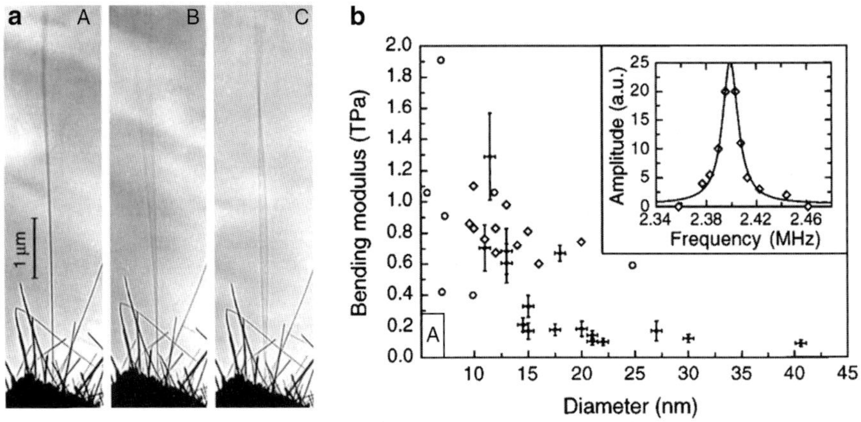

Fig. 3.10 (**a**) The TEM image shows how carbon nanotubes were excited using the TEMSTM. (**b**) The resonance frequency and bending modulus measured. Images are from [87]

bending modulus. From the same kind of measurement, the damping or Q-value can be deduced from the amplitude change around the resonance frequency. High elastic modulus values are expected for carbon nanotubes, for carbon nanotubes with diameters below 5 nm, there have been reports on elastic modulus values up to 1 TPa, comparable to diamond. The mechanical resonance could also be induced by thermal energy and this is visible for long and weak nanowires and nanotubes [89] By comparing the thermal energy, $k_B T$, with the mechanical energy (kA^2, where k is the spring constant and A the amplitude of the vibrating tube), the thermally induced vibrations can be calculated. For a long and thin nanotube or nanowire the spring constant will be low, making this effect readily visible in the TEM as a blur of the nanotube image.

The work function of nanosized materials can also be measured using a variant of the resonance method [56, 57] in a way similar to the Kelvin probe method. In this technique, the resonance frequency is first found as described above, and then by adding a static voltage V_{dc} to the alternating voltage V_{ac}, the amplitude of the oscillating nanowire will change. It can be shown that when V_{dc} is equal to the difference in work function between the nanowire and the counter electrode, the mechanical amplitude will be zero. A false signal could be observed if a high V_{dc} (around several volts) is applied when the distance between the electrodes is short. The oscillating nanowire will be under tension from the electrostatic field, and like tuning the resonance frequency of a spring by applying tension, the resonance frequency will shift away leading to lower amplitudes. This technique was used to determine the work function of carbon nanotubes that were either 0.3 eV below or 0.6 eV above graphite. This measurement was interpreted as the identification of metallic and semiconducting carbon nanotubes [57].

Several other mechanical studies using TEMSTM have been done, including linear bearings in the form of carbon nanotubes sliding inside each other [90], contact formation between gold wires [91], or even NEMS devices in the form of a nanorelay consisting of Ge or Si nanowires [92, 93] or $Mo_6S_3I_6$ nanowires [94].

3.4 Incorporating an AFM Inside a TEM Instrument

The atomic force microscope (AFM) was introduced by Binnig et al. [95] in 1986 as a continuation of the STM. Instead of tunneling currents, the AFM utilizes forces between surfaces to generate an image of the topography; thus, the AFM is not restricted to imaging conductive materials but is also capable of resolving features on an atomic scale for nonconductors. There are two main imaging modes available in the AFM, contact and noncontact mode. In contact mode, the AFM tip is in contact with the sample during imaging and in noncontact mode the AFM tip is excited to its resonance frequency and scanned along the surface at a close proximity.

As the AFM technique was developed, there was a desire to perform advanced mechanical characterizations by utilizing the AFM as a local force probe on, for

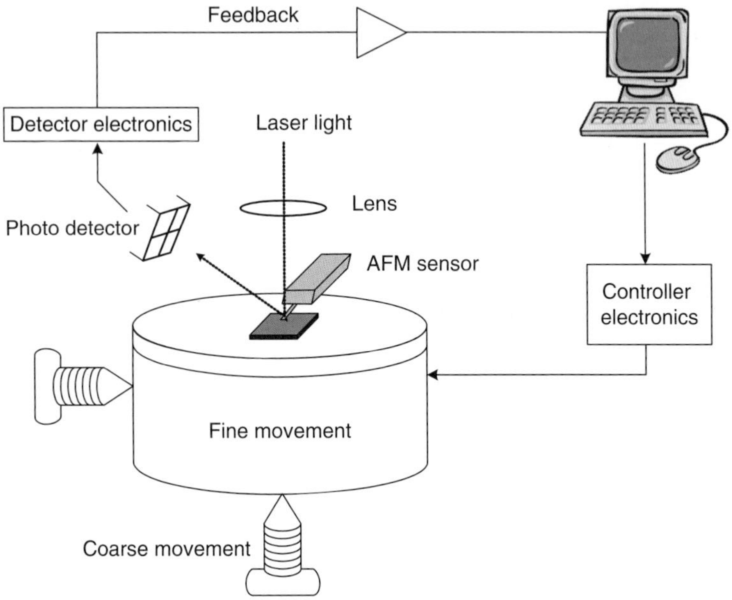

Fig. 3.11 Schematic of a standard atomic force microscope with an optical deflection detection system

example, nanowires. To do this, simultaneous imaging is needed. To enable such experiments, the AFM was first combined with the SEM [96, 97] in 1994 and for higher imaging resolution with the TEM [3, 21] in 2001. The high imaging resolution of the TEMAFM enables measurements on nanostructures such as nanowires and nanoparticles, opening up new possibilities in nanoscale mechanical characterization.

As mentioned previously, the main challenge of a TEMAFM instrument is to fit an AFM inside the TEM pole-piece gap, as illustrated in Fig. 3.1b. A standard AFM system consists of a coarse and fine motion to align the sample and tip, and a system to measure the deflection of the tip in order to be able to control the applied force. In standard AFMs, an optical system for detection of the deflection is the most common configuration. A sketch of a standard AFM system is shown in Fig. 3.11. To construct a TEMAFM instrument, all of these components need to be fitted in the TEM pole-gap.

3.4.1 Optical Force Detection Systems

One of the first TEMAFM systems was presented by Kizuka et al. [21] in 2001. In this work, a HRTEM was rebuilt to incorporate an AFM and an STM. The coarse motion in this system was realized using the positioning stage of the TEM, the

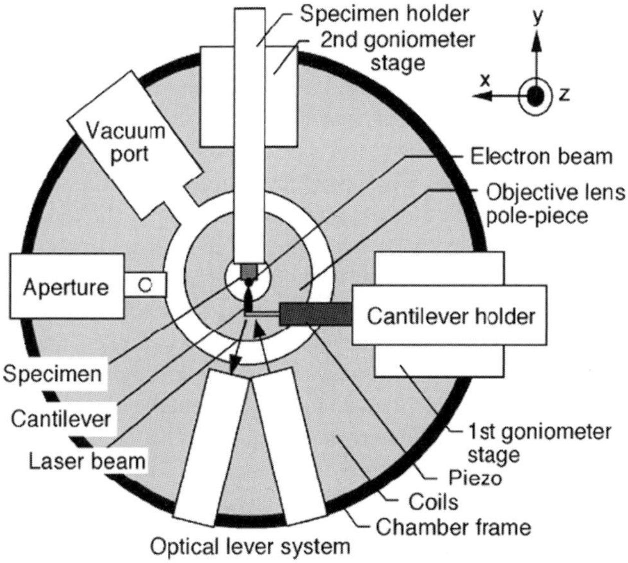

Fig. 3.12 A cross section of the TEM column showing modifications to facilitate an AFM system with laser detection system as presented in [21]. Image is from [21]

goniometer. In [21], the TEM was rebuilt to include two goniometer stages, where the sample is mounted on one stage with an AFM sensor, and a piezoelectric actuator on the other, as shown in Fig. 3.12. To perform an experiment, the two goniometers are used to align the sample and tip, while the piezoelectric actuator (on which the AFM sensor is mounted) is used for high precision movement. The deflection of the cantilever, and consequently the force, is measured using an optical detection system built into the TEM column.

The TEMAFM instrument presented by Kizuka and co-workers is the only known example in which a TEM has been extensively redesigned to fit an AFM. The two main reasons it has not been done previously are the significant cost related to rebuilding a TEM and the extensive collaboration with a TEM manufacturer that is required. Other approaches to implement a TEMAFM for which the TEM does not need to be redesigned have attracted more attention.

3.4.2 Non-optical Force Detection Systems

Simple Spring Approach

The simplest way to measure forces inside a TEM is to use a spring with a known force constant and then image the deflection of the spring using the TEM.

Practically, this is done by placing a standard AFM cantilever at the fixed position in a TEMSTM holder described in Sect. 3.3. Such a configuration was presented by Erts et al. [3] in 2001.This scheme eliminates the extra space needed for an optical deflection detection system, and by placing the AFM sensor on the fixed side, additional inaccuracies in the force caused by the characteristics of the fine movement, that is, the piezo, were avoided. The AFM cantilever was used as a mechanical spring, for which the spring constant is known. The force was extracted using Hooke's law, $F = -kd$, where k is the spring constant of the cantilever, d is the deflection of the cantilever, and F is the force. This configuration has been used to study the mechanical properties of single carbon nanotubes [51–53, 98], welded gold nanowires [99], and the initial contact and adhesion forces in a gold nanocontact [100]. Figure 3.13 shows TEM images of a gold nanocontact experiment and the force measured [100]. The gold nanocontact was created by bringing a gold coated AFM tip into contact with a gold wire and then retracting the tip from the wire.

This technique is straightforward to use when the TEM holder used is equipped with a positioning system, as it does not require any reconstruction other than mounting of an AFM cantilever. It is possible to perform mechanical characterization; however, for more advanced analyses, the technique is limited by two main factors. First, the sampling rate will be low, as seen in Fig. 3.13b. Because the procedure is done manually, every force point is calculated from the cantilever deflection, which in turn is extracted from a comparison of TEM images captured during the experiment. Second, with this technique the force range is limited by the image magnification because the deflection must be visible in the image. As a result of these limitations, a compact force measurement system not reliant on the TEM imaging is desired.

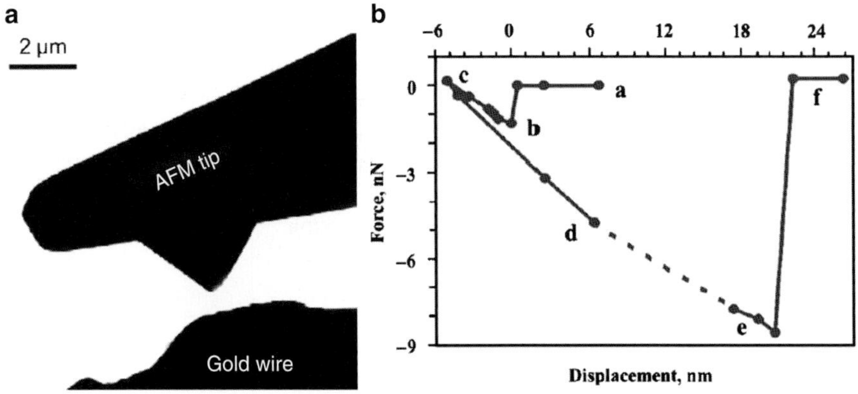

Fig. 3.13 TEMAFM measurement in which two gold surfaces were brought in contact and retracted to study the surface interaction. (**a**) Low magnification TEM image of the gold coated AFM tip and the gold wire. (**b**) The force–distance curve when retracting the gold surfaces from each other. Images are from [100]

Electrical Force Sensors

In order to enable direct force measurement without redesigning the TEM column, a compact electrical AFM sensor in combination with a miniaturized positioning system is needed. Different force detection techniques have been developed for standard AFMs, including current tunneling [95] optical interferometery [101, 102] and electrical techniques such as capacitive [103, 104], piezoelectric [105], and piezoresistive [106] techniques. For in situ TEM instrumentation, capacitive force detection has been used for measuring forces in the μN range utilizing MEMS devices [107,108] and a miniaturized capacitive transducer based on [109]. Thus far, for in situ TEMAFM, which measures forces in the nN range, only piezoresistive detection has been utilized. The detection sensor, presented by Nafari et al. [110], was custom designed to fit inside a Phillips CM200. In that system, the dimensions of the sensor were chosen such that firstly, the sensor fits in the TEM pole-piece gap, and secondly, the sensor tip is centered in the TEM holder aligned to the electron beam. Considering these factors, the final dimensions of the sensor were restricted to 1.3 mm × 1.2 mm × 0.5 mm. The critical dimension is the width of the sensor (1.2 mm) as it will be mounted on the side of the holder so that it will not block the electron beam. The AFM tip is created in the fabrication process as an integral part of the sensor and is made of single crystalline silicon. The arrangement of the sensor and positioning system is shown in Fig. 3.14. This sensor has been integrated in a TEMAFM system, available commercially [36].

The resistors on this sensor chip were assembled in a Wheatstone bridge configuration integrated on the sensor chip as shown in Fig. 3.15. One of the resistors was doped into a reference cantilever (R_2 in Fig. 3.15a) for temperature compensation, which is especially important when operating in vacuum. This type of temperature compensation is described in detail by Thaysen et al. [111].

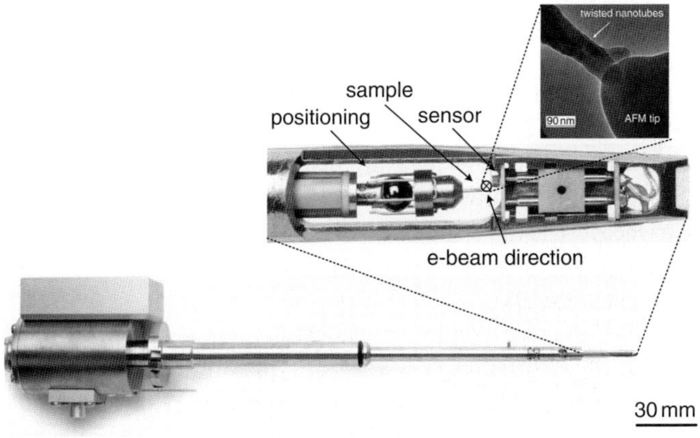

Fig. 3.14 Side view image of the TEMAFM system presented by Nafari et al. [110]

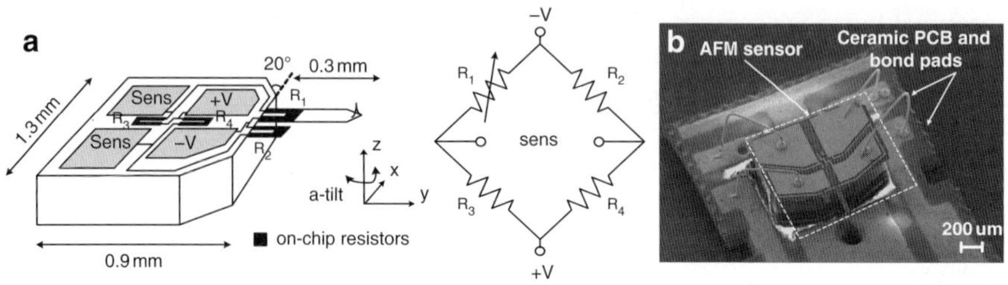

Fig. 3.15 The TEMAFM sensor presented in [110]. (a) The TEMAFM sensor design with the Wheatstone bridge configuration on chip. (b) SEM image of the TEMAFM sensor. Images are from [110]

3.4.3 TEMAFM Applications

While the TEMSTM was first used as an additional microscope, the TEMAFM that was introduced later was directly used as a local probe, with the TEM as a guide for locating suitable samples and for sample–tip characterization. As a local force probe, the TEMAFM has been a useful tool for studying the mechanical properties of individual nanostructures, friction at the nanoscale, and force interactions in nano and atomic sized contacts.

3.4.3.1 Elastic Measurements

A direct application of the TEMAFM is to study the elastic properties of nanostructures, such as nanowires and nanoparticles. There are several important issues related to the elastic properties of nanowires and nanotubes. The record high elastic modulus of carbon nanotubes of up to 1 TPa, as earlier observed by the TEMSTM method, raised questions of the elasticity of other nanostructures. The mechanical properties, including the elastic modulus and damping, are the key parameters in designing NEMS devices. The TEMSTM has been used to address these issues by the resonance method as described in Sect. 3.3.1, but a more direct way is to utilize the TEMAFM with its ability to measure nN forces during bending of a nanowire and to use the TEM for imaging of the nanowire size and shape. These structures are challenging to measure using a standard AFM because the AFM cannot be used for both imaging and point force measurements simultaneously, and the sample requires an elaborate preparation [9].

The TEMAFM characterization of nanostructures is generally performed by first attaching the nanostructures to a wire, typically using electrically conductive glue, and then mounting it in the sample holder. With the sample in place and inserted into the TEM, the nanowire is aligned to the AFM tip and a force can be applied. For example, in a study presented by Golberg et al. [112], the TEMAFM was used to investigate the elastic deformation of multiwalled BN nanotubes using the electrical force sensing TEMAFM system described in Sect. 3.4. Single-BN nanotubes were compressed, and it was found that they were elastic up

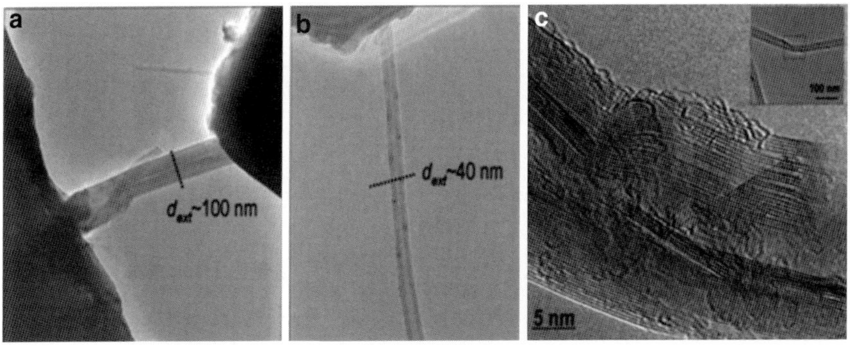

Fig. 3.16 TEMAFM elasticity study of BN nanotube. (**a**) A TEM image of the AFM tip, seen to the *right* in the image, connected to a BN nanotube. (**b**) A TEM image of a thinner BN nanotube connected to an AFM cantilever. (**c**) A high resolution image of the deformed layers in the multiwire structure. The images are from [112]

to a kinking of 90°, and the deformed layers after that point were studied, as shown in Fig. 3.16. By using the dimensions of the structures from TEM imaging and the measured force–displacement curves, a Young's modulus of 0.5–0.6 TPa was estimated for the BN nanotubes. The TEMAFM method has been used for a number of materials such as ZnO nanobelts [113], ZnO nanowires [114], TiO_2 nanotubes [115], carbon nanotubes [116–118], carbon nanocages [119–121], and C60 fullerene nanowhiskers [122, 123]. Golberg et al. have recently published a review of their TEMAFM work, summarizing the possibilities and challenges of the TEMAFM technique [10].

3.4.3.2 Electromechanical Properties

To investigate the electromechanical properties of nanostructures, a dual usage of TEMSTM and TEMAFM has been explored. In an experiment by Wang et al. [124] in 2009, the TEMSTM was used to modify cobalt-filled carbon nanotubes by driving a current through the nanotubes. Depending on the strength of the current, the cobalt particle was used to cut, repair, and interconnect carbon nanotubes with high precision. After the carbon nanotubes had been modified, the TEMAFM system was utilized to assess the mechanical properties. Figure 3.17 shows TEM images from this experiment. This technology opens the possibility to freely engineer advanced nanowire networks. Furthermore, in the work of Lu et al. [99] in 2010, the TEMAFM and the TEMSTM were used, separately, to investigate the properties of welded gold nanowires. With the help of in situ TEM techniques, it was shown that welding at the nanoscale was possible by simply joining two single-crystalline gold nanowires at relatively low pressure. TEM images and in situ probing showed that the welds were nearly perfect, with the same strength and conductivity as the rest of the wire.

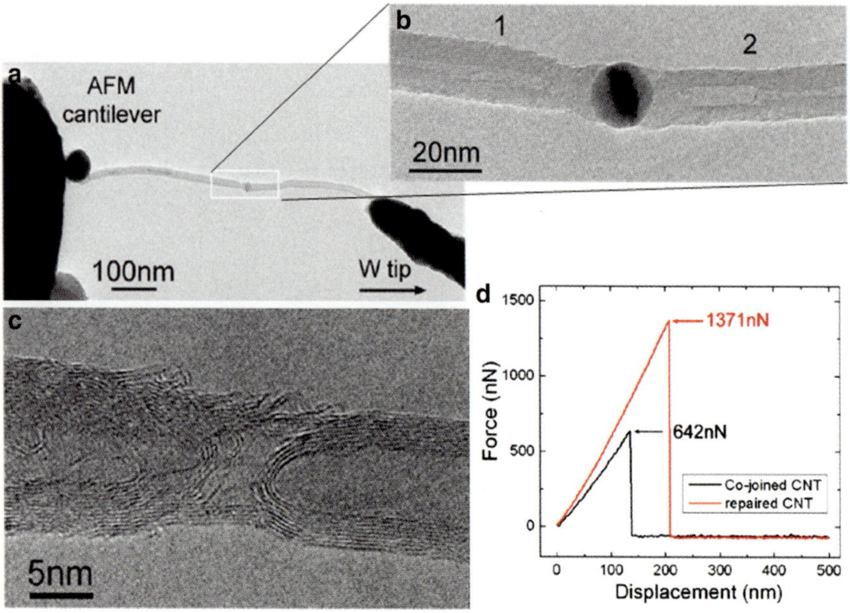

Fig. 3.17 TEMAFM measurement of welding of cobalt filled carbon nanotubes. (**a**) A carbon nanotube suspended by the AFM cantilever and a tungsten tip. (**b**) A zoom-in of the cobalt particle. (**c**) The carbon nanotubes joined without a cobalt particle. (**d**) The mechanical strength of a cobalt joined carbon nanotube and a repaired carbon nanotube. The images are from [124]

In a further development of the TEMAFM technique, an AFM sensor that allows simultaneous current measurements has been presented [36]. This conductive TEMAFM was used in the work of Costa et al. [125] to characterize the electromechanical properties of carbon nanotubes with crystalline fillings. It was shown that the filling dominated the mechanical properties of the carbon nanotube, and through electrical measurements, the semiconducting behavior of the filled carbon nanotube was observed simultaneously.

3.4.3.3 Atomic Scale Wires

The study by Takayangi et al. [39] on the correlation between the structure and the conductance quantization of atomic sized nanowires using the TEMSTM described in the Sect. 3.3.1.1 solved the open question regarding the structure of these small conductors. However, there were surprises in the study made by Takayanagi et al. [39], such as the anomalously long distance between the gold atoms. In addition, both theoretical [126] and experimental works using standard AFM [127] showed a correlation between the conductance and the tensile force; whenever the conductance changed one quantum unit ($2e^2/h$), the force made a jump of about 1 nN (for a review see [40]).

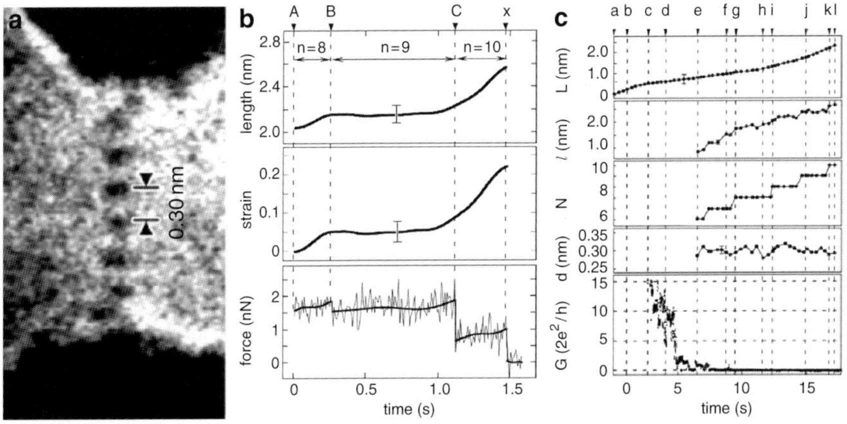

Fig. 3.18 TEMAFM study of single atom wide gold wires. (**a**) A single atomic wire created by separating two gold electrodes is seen. An interatomic distance of 0.3 Å can be measured from the image. (**b**) The force, strain, and length of the wires are seen as the number of atoms in the chain goes from eight to ten. (**c**) Conduction measurement while increasing the separation distance of the tip and plate. *L* is the separation distance, *l* is the total length of the atomic wire, *N* is the number of atoms, *d* is the interatomic distance, and *G* is the conductance of the wire during the tensile deformation. The images are from [124]

The TEMAFM might be a good tool to provide further insight into these problems. In a study by Kizuka, such an experiment was done on atomic sized gold nanowires [128]. Kizuka found several new surprises; for example, for ten atoms long wires a decrease in conductivity was observed, while for wires shorter than five atoms an increase was seen. It was also found that the minimum measured tensile force was in multiples of 3.9 nN. TEM images, force, and conductance measurements from the experiments are shown in Fig. 3.18. This kind of study indicates that the TEMAFM should be an ideal tool for studying atomic scale wires.

Such atomic sized TEMAFM experiments have also been performed for other studies such as on palladium [129] and iridium [130] atomic wide wires. Gold atomic wires have also been studied using a conductance feedback loop to maintain a constant number of electron transport channels [131]. Using basically the same setup, larger contacts for different materials have also been studied. For example, the correlation between stress and electromigration for copper [132, 133] and slip during tensile deformation of palladium [134] and silicon contacts [135, 136] have been investigated.

3.4.3.4 Friction and Adhesion

Friction between sliding bodies is the result of the formation, deformation, and breaking of a large number of contacts between two bodies. There has been a major burst of activity in studies of friction since 1987, when the first standard AFM study of friction was done (for reviews, see [137–139]). With the standard AFM, single contacts are being studied, which may provide insight to the microscopic origin of friction. However, it is the real contact area that is, along with the loading force, of key importance in understanding friction, and, as noted earlier, the contact area information is only indirectly accessible using standard AFM instruments. Thus, the TEMAFM might give a unique opportunity to study friction using the TEM to image the contact area between the AFM tip and sample correlated with the friction and adhesion forces controlled by a loading force.

An early study of adhesion using TEMAFM [100] showed that the adhesive contact between two gold samples could best be described by the Maugis adhesion theory. Another observation in the same study was that the AFM jump-in-contact distance was larger than expected. The observed value of the jump-in-contact distance did not agree with the calculated value. The parameters to calculate the jump distance, the tip–sample distance, and the radius of tip and sample, are all observable by TEM, in contrast to standard AFM where the true tip–sample is unknown, and the tip radius can only be measured before and after the actual experiment using high-resolution microscopes. This larger distance was due to the migration of gold atoms into the gap by the attractive van der Waals forces. This work was done using a simple spring AFM cantilever, and the motion of the AFM tip, as viewed by TEM, gave a measure of the force. However, for extensive friction studies, a complete AFM setup is desired. Such a TEMAFM was used in the friction study by Fujusawa and Kizuka in 2003 [140], which focused on the frictional tip effect during AFM scanning and a small lateral displacement of the AFM tip during imaging. This displacement was shown to have an effect on the force–displacement plots and the scanned image when the scan size was in the same order as the lateral displacement.

3.5 Combined TEM and SPM Sample Preparation

One major difference between conventional TEM and TEMSPM samples is that for TEMSPM experiments both sample and SPM tip have to be placed orthogonally with respect to the electron beam for imaging, all the while keeping the electron transparent part of the TEM sample accessible to the SPM tip. Conventional TEM grids are therefore not practical to use for in situ TEM manipulation. For in situ TEM applications, a wire is normally used as support on which samples and custom-made substrates can be glued onto. Figure 3.19 shows sketches of common types of sample mountings used for in situ TEM manipulation. These sample preparation methods and STM probe preparation will be discussed in this section; however, AFM probes will not be addressed, as they are in most cases purchased and are generally not altered by the researchers themselves. More information on electrical AFM force sensors where a specialized MEMS fabrication scheme may be required can be found in Sect. 3.4.

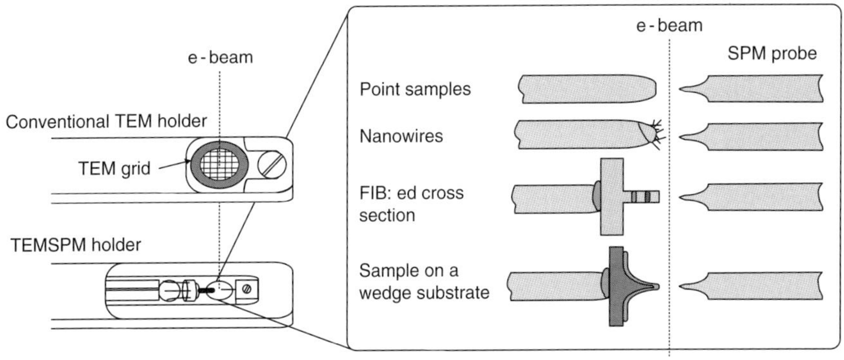

Fig. 3.19 Sketches of a conventional TEM holder, TEMSPM holder, and samples prepared for in situ TEM probing

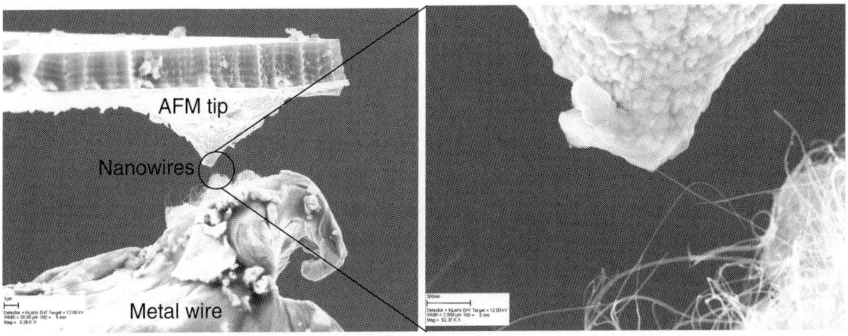

Fig. 3.20 Nanowires glued on a metallic wire ready for TEMSPM experiments

3.5.1 Nanowires and Nanoparticles

As shown throughout this chapter, nanowires are a common sample in TEMSPM experiments. Compared to conventional TEM samples, nanowires already possess the thin cross section required for electron transmission, and thus the sample can be prepared in a matter of minutes. The standard method is to use electrically conductive glue to ensure a reliable electrical and mechanical contact as in [60, 92]. The procedure is to dip the supporting wire, often an inert metal wire cut with a pair of scissors at an angle, into electrically conductive glue and then dip this conducting glue wire into a powder of nanowires or nanotubes [141]. An SEM image of nanowires attached to a support wire can be seen in Fig. 3.20.

Another simple way to prepare nanowire samples is to dip the supporting wire directly in the nanowire powder and simply let the surface adhesion forces keep the nanowires in place [142]. Hirayama et al. [82], for example, used an ethanol solution of single-walled carbon nanotubes to prepare a tip in this way. If a single nanowire is needed, in contrast to the large number of nanowires obtained by method described

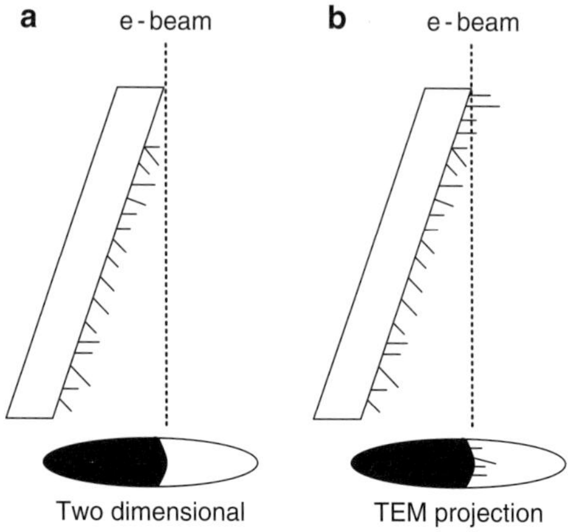

Fig. 3.21 Nanowire grown on a flat substrate. (**a**) Sketch of the TEM image when the nanowires shadow the projection image. (**b**) Sketch of the TEM image when the nanowires are not shadowed by the substrate

above, which might be desired if both contact regions are to be investigated, a more elaborate method is needed. A single nanowire can be of interest when both contact regions are to be investigated. In [143], a SEM-FIB equipped with a nanomanipulator was used to extract a single carbon nanotube, and the FIB was used to weld it to a gold wire.

There have also been reports on nanowires grown directly on a metallic wire as in [92, 144]. This type of sample is ideal for TEMSPM experiments. The wires already have a good mechanical and electrical contact to the wire and are easily accessible by an SPM probe.

For nanowires grown directly on a silicon substrate, the substrate needs to first be broken into smaller parts, typically 2 mm × 2 mm. Here it is important to note that because the substrate will always be mounted at a small angle; only nanowires on the edge of the substrate will be seen in the two-dimensional projection image obtained in the TEM, as illustrated in Fig. 3.21. Consequently, the substrate must be cut in a way that does not damage the nanowires at the edges. Standard diamond dicing is generally not a delicate enough method to use, as there is both cooling water present and chipping on the edges of the sample that can damage the nanowires. An alternative method is to cleave the substrate using a side cutter at the edge of the sample and let the silicon break along its path. Thus, the nanowires along the cleaved edge are not damaged. This sample preparation technique has been used in [50], among others.

To study nanoparticles with TEMSPM the sample can be prepared in several ways. There are examples when nanoparticles have been directly deposited on a wedge type substrate as seen in Fig. 3.27 [145] or they can be dispersed on a flat surface as done in [146]. One simple and common method to prepare the samples is to let a drop of solution of the nanoparticles dry on a TEM grid. This technique is also applicable in TEMSPM, but with the grid replaced by a tip or a half grid.

An alternative method to measure on nanoparticles is to grow them directly on nanowires or nanotubes. Zhang et al. [147] have made a simple, water based, protocol for growing gold nanoparticles on carbon nanotubes. A gold salt is reduced by

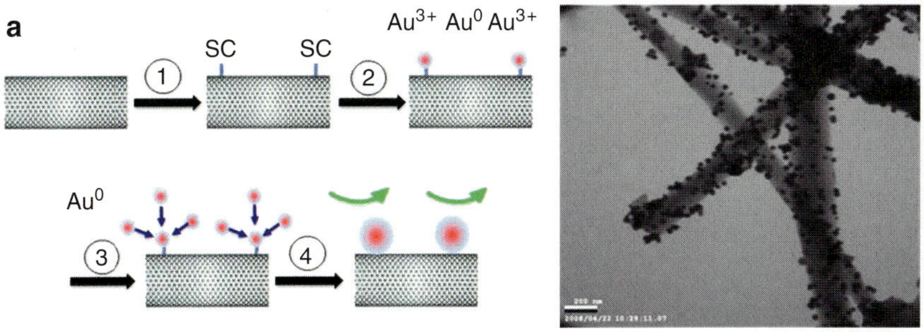

Fig. 3.22 Schematic drawing for synthesizing of gold nanoparticles on carbon nanotubes. (**a**) The procedure of preparing carbon nanotubes with gold particles is shown here. Carbon nanotubes were first premixed with sodium citrate (SC) by ultrasonication (1). $HAuCl_4$ was added into the reaction system causing reduction of Au^{3+} ions to Au^0 (2). The reduced gold atom on the carbon nanotube acts as a seed for self-assembling of other gold atoms (3). Gold nanoparticles are deposited on the carbon nanotube (4). (**b**) A TEM image of 30 nm gold nanoparticles decorating carbon nanotubes. Images are from [147]

sodium citrate and allowed to grow on carbon nanotubes, as seen in Fig. 3.22. The size and density of the nanoparticles could be altered by changing the concentrations of the solvents. The method is not restricted to carbon nanotubes but is generally applicable; for example, nanowires of $Mo_6S_3I_6$ [76] and clay particles [148] can be decorated by gold nanoparticles in this way. Other kinds of nanoparticles, such as platinum, palladium, or silver might be deposited in a similar way [149].

3.5.2 A Proper Electrical Contact for TEMSPM

The best way to make an electrode for electrical probing in a TEMSTM is to mechanically cut a wire from an inert metal such as gold or platinum. A common procedure is to use a pair of scissors or small wire cutters and cut the wire at a 45° angle. This will produce a rough tip with a large number of clean asperities on the nanoscale. Inside the TEM, one of these nanotips closest to the sample is used for probing.

Etched probes are usually more difficult to use and require elaborate cleaning steps. One reason to use an etched tip is that it provides a better geometry in cases where a high aspect ratio tip is required to reach the area of interest. Such etched wires can also be very sharp, with a tip radius as low as 5 nm [150]. In standard STM, the cut tips are the most common, but when operating the STM in UHV, electropolished tungsten is also a common tip. The electropolished tungsten tips are covered with a tungsten oxide layer, but in UHV this layer can be sublimated away at 800°C by heating it with a field emission current [151]. Gold can also be electropolished, but the procedure normally leaves a nanometer contamination layer

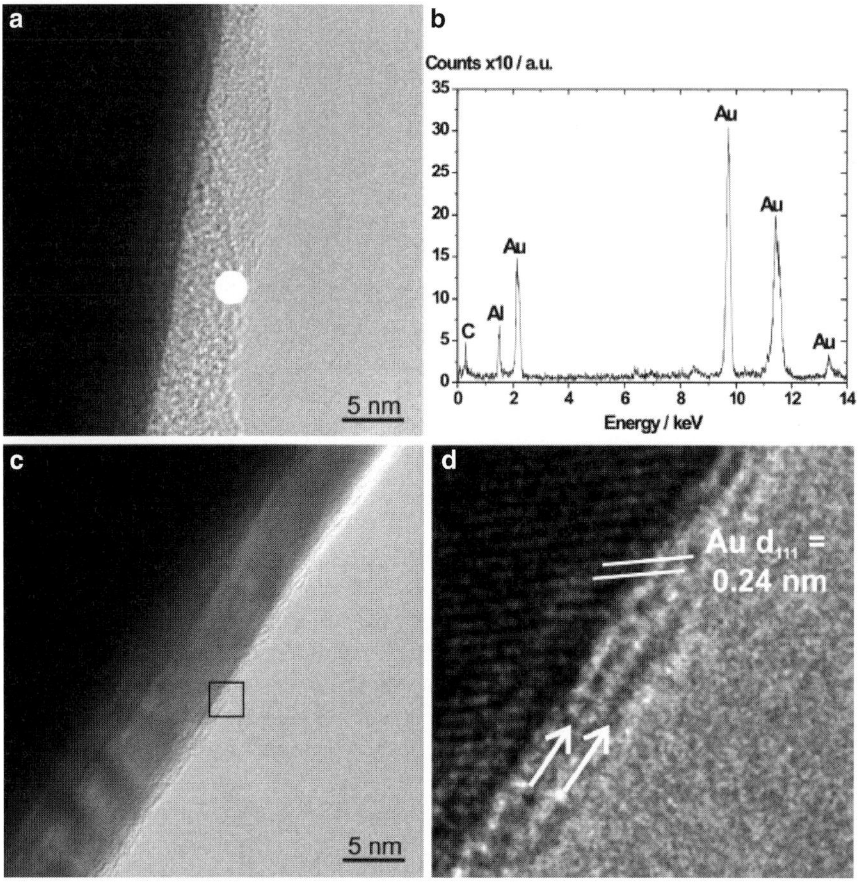

Fig. 3.23 (**a**) An electropolished gold tip with a contamination layer on the surface making it hard to make a proper electrical contact. (**b**) An EDS measurement showing that the contamination layer contains carbon. (**c**) A mechanically cut wire is shown to be more clean. (**d**) Here a carbon layer, 1 nm is seen on the cut gold electrode. Images are from [152]

thick as shown by Costa et al. [152]. Figure 3.23a shows this contamination layer, and the cleanliness of the simple cutting technique is demonstrated in Fig. 3.23c.

An indicator of bad electrical contact between two metals is a nonlinear I–V curve, whereas a proper contact will produce a linear curve, as seen in Fig. 3.24. Such nonlinearity is a signature of a tunneling barrier. The nonlinearity is due to the fact that the applied voltage will change the tunneling barrier. At higher bias voltage, the barrier will be both lower and thinner and thereby increase the tunneling probability and the electron current [153]. This tunneling barrier often originates from contaminations in the form of hydrocarbons present in the air and in the residual gases of the TEM column. This is seen in Fig. 3.23d, where a thin carbon layer is present on the mechanically cut gold wire. One way to remove such contaminations

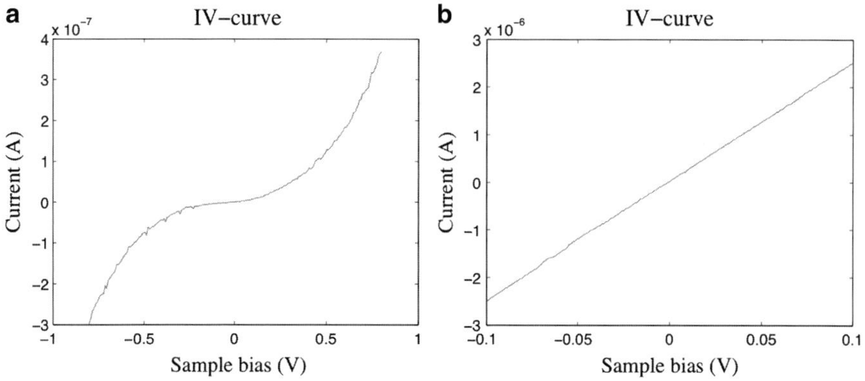

Fig. 3.24 (**a**) Typical nonlinear I–V curve for a bad contact. The curve is symmetric around zero voltage. (**b**) A proper contact. Image is from [143]

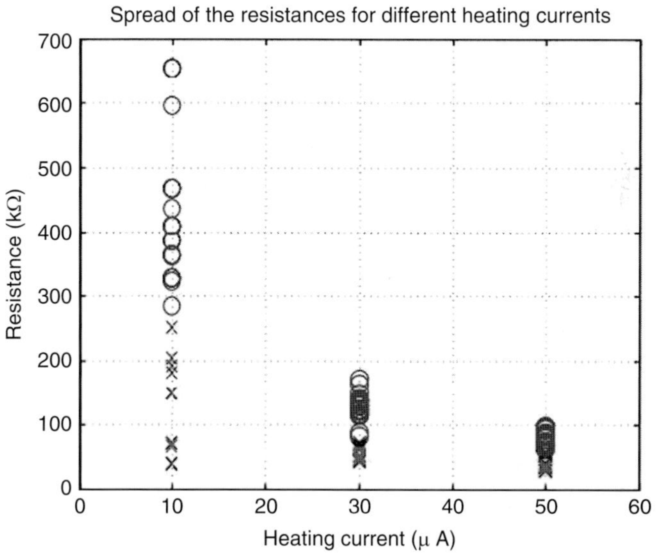

Fig. 3.25 Resistance vs. heating current, along different positions for two separate carbon nanotubes. Image is from [143]

in situ is to apply a high current which leads to Joule heating. This process is self-adjusting because the highest heating power will be dissipated where the highest resistance is present in the circuit ($P = RI^2$), that is, the location of the high ohmic tunneling barrier. Such an effect is shown in Fig. 3.25 where the high-ohmic contact between a carbon nanotube and a gold tip is lowered by increasing the current.

One further consideration regarding electrical contacts in the TEM is the effect of the electron beam on the hydrocarbon contamination layer inside the TEM chamber.

It is well known that when the contamination layer inside the electron microscope is exposed to the electron beam, a growth of a-C will occur there. This phenomenon, EBID, could be used for growing structures but could also result in a bad electrical contact. When the TEM electron beam irradiates a surface, the hydrocarbon will decompose leaving a stable a-C layer behind. More hydrocarbons will quickly cover the a-C structure and if the area is exposed further, the a-C structure will grow in thickness. As shown by Peng et al. [154], the hydrocarbon contamination layer could be evaporated away by Joule heating and thus inhibit the EBID effect. At a temperature of 430°C the EBID process disappeared, pointing at a way to keep a sample clean under high vacuum conditions.

3.5.3 Lamella Samples

For the study of bulk materials or heterogenic sandwich structures with TEMSPM techniques, the sample needs to be in the shape of a thin lamella. For TEMSPM, while it is important that the lamella is thin enough for TEM imaging, it should also be well supported so that it does not bend when a force is applied. If the rear parts of the sample bend, this could change the characteristics of the sample in unintended areas. For example, in a TEMSTM study of magnetic tunnel junctions [155], H-bar shaped electron transparent windows were prepared using FIB. These electron transparent windows were 15 μm across and measured 100 nm in thickness. By using windows, the structure becomes more stable and easily accessible with an STM tip. In another work, a cross section of a $LaAlO_3/SrTiO_3$ thin heterostructure film was measured using TEMSTM [156]. The sample preparation, TEM images, and I–V curves are seen in Fig. 3.26. This sample was prepared by grinding and ion milling. The FIB was not used in this case because the implanted ions from the ion beam could alter the electrical properties. As seen in Fig. 3.26a, the geometry of the sample was prepared such that the STM tip could contact the electron transparent region while being imaged in the same plane.

An alternative way of investigating thin films is to directly deposit a film on a micromachined substrate [18, 157]. An example of such a substrate is seen in Fig. 3.27. By using micromachining, the substrate can be designed to have a steady support and an electron transparent edge. The investigated material can then be deposited directly on the substrate and, if needed, further modified using FIB.

3.5.4 Electron Beam Irradiation Effects

Another issue regarding in situ TEM measurement is the impact of using a high energy electron beam on the sample. The energy of an electron beam used in the TEM is typically 100–300 keV, and this can damage the samples during image

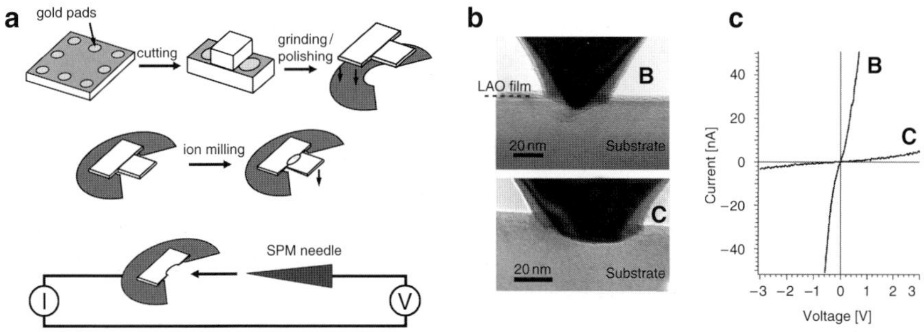

Fig. 3.26 (**a**) The procedure of how a $LaAlO_3/SrTiO_3$ thin film was prepared for TEMSTM measurements. First, a cross section of a macro sample is cut, and then it is thinned down and attached to a support shaped as a half TEM grid. The sample is further thinned to expose an electron transparent area. (**b**) TEM images of the prepared sample and I–V curve taken. Images are from [156]

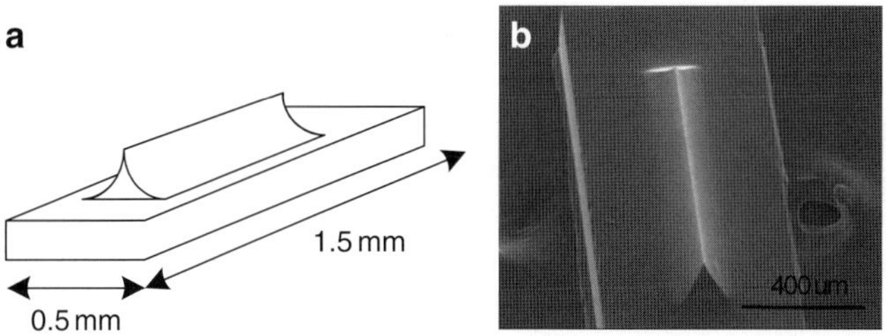

Fig. 3.27 A silicon substrate for TEM sample mounting. (**a**) A sketch of a wedge shaped silicon structure. (**b**) A SEM image of the silicon wedge. Images are from [157]

acquisition. The magnitude of this ever present effect depends on the kinetic energy of the electrons, the current density at the sample, and the accumulated dose [158]. This phenomenon might be useful as in the work of Takayanagi et al. [39] where they thinned down a gold film until only single atom wide wires remained, or as in the studies by Jin et al. [159] where graphene was modified until a single wire of carbon remained. Another example, by Ugarte, is the creation of carbon onions by irradiating a-C [160]. While a high energy electron beam can be useful, it is also more important to be aware of the electron irradiation in order to not alter the specimen under study (for a review, see [161]). For example, carbon nanotubes imaged at 200 kV in a TEM will quickly be destroyed as shown in [161]. One solution is to decrease the acceleration voltage in the TEM; for the case of carbon nanotubes a voltage below 120 kV is recommended as this is the threshold for knock on damage. However, even below that threshold there are still effects from the electron beam, as can be seen in Fig. 3.28. The carbon nanotube in Fig. 3.28 was intentionally exposed a prolonged period to 80 keV.

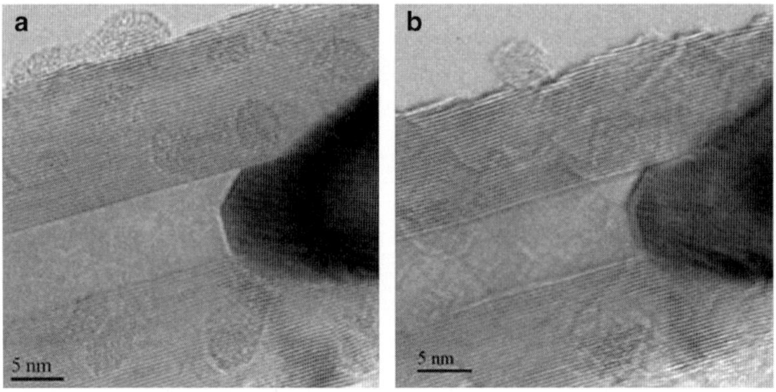

Fig. 3.28 TEM images of iron filled multiwall carbon nanotubes. (**a**) Before prolonged exposure to the electron beam. (**b**) After prolonged electron beam irradiation at 80 keV impact energy

3.6 Conclusion

With the addition of TEM capabilities to the already established STM and AFM techniques, many new possibilities have emerged. As the SPM analysis and manipulation of the sample now can be followed in real time by high resolution imaging in TEM, experimental data that was not previously accessible can be extracted.

In the case of TEMSTM, the use of TEM has increased the fundamental understanding of the tip–sample interaction and has been instrumental in the analysis of materials on the nanoscale. For example, single atom chains have been created and analyzed, electrical properties of single nanoscale structures have been investigated, and details of electromigration phenomena have been revealed. In many cases, the combination of TEM and STM has enabled the use of the STM in completely new ways, such as using the tip for Joule heating of nanowires or welding of carbon nanotubes.

By combining TEM and AFM to TEMAFM, a nanoforce detection system capable of investigating the mechanical properties of individual nanostructures has been created. When correlating the real-time monitoring in the TEM with the force–displacement data from the AFM, details about the deformation mechanisms can be obtained, explaining features in the mechanical data. The analysis of mechanical properties of single nanotubes and nanowires, studies of nanoscale friction, and fracture was for the first time enabled by the TEMAFM.

With TEMSPM techniques, now more than ever before, more detailed data about nanostructures are accessible and the development of such techniques and their applications hold great promises for development of new materials with significantly different properties than previously possible.

References

1. M. Iwatsuki, K. Murooka, S. Kitamura, K. Takayanagi, Y. Harada, Scanning tunneling microscope (STM) for conventional transmission electron microscope (TEM). J. Electron Microsc. **40**, 48–53 (1991)
2. Y. Naitoh, K. Takayanagi, M. Tomitori, Visualization of tip-surface geometry at atomic distance by TEM-STM holder. Surf. Sci. **357–358**, 208–212 (1996)
3. D. Erts, A. Lohmus, R. Lohmus, H. Olin, Instrumentation of STM and AFM combined with transmission electron microscope. Appl. Phys. A Mater. Sci. Process. **72**, S71–S74 (2001)
4. K. Takayanagi, Y. Kondo, H. Ohnishi, Suspended gold nanowires: ballistic transport of electrons. JSAP Int. **3**, 3–8 (2001)
5. M. Sveningsson, K. Hansen, K. Svensson, E. Olsson, E. Campbell, Quantifying temperature-enhanced electron field emission from individual carbon nanotubes. Phys. Rev. B. **72** (2005)
6. Z.L. Wang, P. Poncharal, W.A. de Heer, Nanomeasurements in transmission electron microscopy. Microsc. Microanal. **6**, 224–230 (2000)
7. Z.L. Wang, P. Poncharal, W.A. de Heer, Measuring physical and mechanical properties of individual carbon nanotubes by in situ TEM. J. Phys. Chem. Solids. **61**, 1025–1030 (2000)
8. Z.L. Wang, R.P. Gao, Z.W. Pan, Z.R. Dai, Nano-scale mechanics of nanotubes, nanowires, and nanobelts. Adv. Eng. Mater. **3**, 657 (2001)
9. X. Han, Z. Zang, Z.L. Wang, Experimental nanomechanics of onedimensional nanomaterials by in situ microscopy. Nano: Briefs Rev. **2**, 249–271 (2007)
10. D. Golberg, P.M. Costa, M. Mitome, Y. Bando, Properties and engineering of individual inorganic nanotubes in a transmission electron microscope. J. Mater. Chem. **19**, 909 (2009)
11. L. Dong, A. Subramanian, B.J. Nelson, Carbon nanotubes for nanorobotics. Nano Today. **2**, 12–21 (2007)
12. L. Dong, K. Shou, D.R. Frutiger, A. Subramanian, L. Zhang, B.J. Nelson, X. Tao, X. Zhang, Engineering multiwalled carbon nanotubes inside a transmission electron microscope using nanorobotic manipulation. IEEE Trans. Nanotechnol. **7**, 508–517 (2008)
13. L. Dong, X. Tao, L. Zhang, X. Zhang, B.J. Nelson, Plumbing the depths of the nanometer scale. IEEE Nanotechnol. Mag. **4**, 13–22 (2010)
14. "MRS BULLENTIN," MRS Bull. **33** (2008)
15. M. Kociak, M. Kobylko, S. Mazzucco, R. Bernard, A.Y. Kasumov, C. Colliex, TEM nanolaboratory. Imaging Microsc. **10**, 26–27 (2008)
16. F. Banhart, *In-Situ Electron Microscopy at High Resolution* (World Scientific, Singapore, 2008)
17. E.A. Stach, In-situ TEM – a tool for quantitative observations of deformation behavior in thin films and nano-structured materials. Workshop on new materials science enabled by in situ microscopies, DOE BES, 2001
18. E.A. Stach, T. Freeman, A. Minor, D.K. Owen, J. Cumings, M. Wall, T. Chraska, R. Hull, J. Morris, A. Zettl, U. Dahmen, Development of a nanoindenter for in situ transmission electron microscopy. Microsc. Microanal. **7**, 507–517 (2001)
19. A.M. Minor, S.A. Asif, Z. Shan, E.A. Stach, E. Cyrankowski, T.J. Wyrobek, O.L. Warren, A new view of the onset of plasticity during the nanoindentation of aluminium. Nat. Mater. **5**, 697–702 (2006)
20. R. Gupta, R.E. Stallcup, in *Introduction to In Situ Nanomanipulation for Nanomaterials Engineering*, ed. by W. Zhou, Z.L. Wang. Scanning Microscopy for Nanotechnology – Techniques and Applications (Springer, Heidelberg, 2007), pp. 192–223
21. T. Kizuka, H. Ohmi, T. Sumi, K. Kumazawa, S. Deguchi, M. Naruse, S. Fujusawa, A. Yabe, Y. Enomoto, Simultaneous observation of millisecond dynamics in atomistic structure, force and conductance on the basis of transmission electron microscopy. Jpn. J. Appl. Phys. **40**, 170–173 (2001)
22. R. Young, The topografiner: an instrument for measuring surface microtopography. Rev. Sci. Instrum. **43**, 999 (1972)

23. G. Binnig, H. Rohrer, Surface imaging by scanning tunneling microscopy. Ultramicroscopy. **11**, 157–160 (1983)
24. C.J. Chen, *Introduction to Scanning Tunneling Microscopy* (Oxford University Press, Oxford, 1993)
25. C. Gerber, G. Binnig, H. Fuchs, O. Marti, H. Rohrer, Scanning tunneling microscope combined with a scanning electron microscope. Rev. Sci. Instrum. **57**, 221–224 (1986)
26. T. Ichinokawa, Y. Miyazaki, Y. Koga, Scanning tunneling microscope combined with scanning electron microscope. Ultramicroscopy. **23**, 115–118 (1987)
27. J.C. Spence, A scanning tunneling microscope in a side-entry holder for reflection electron microscopy in the Philips EM400. Ultramicroscopy. **25**, 165–169 (1988)
28. W.K. Lo, J.C. Spence, Investigation of STM image artifacts by in situ reflection electron microscopy. Ultramicroscopy, **48**, 433–444 (1993)
29. H. Ohnishi, Y. Kondo, K. Takayanagi, UHV electron microscope and simultaneous STM observation of gold stepped surfaces. Surf. Sci. **415**, L1061–L1064 (1998)
30. Y. Oshima, K. Mouri, H. Hirayama, K. Takayanagi, Development of a miniature STM holder for study of electronic conductance of metal nanowires in UHV–TEM. Surf. Sci. **531**, 209–216 (2003)
31. Z.L. Wang, P. Poncharal, W.A. de Heer, Nanomeasurements of individual carbon nanotubes by in situ TEM. Pure Appl. Chem. **72**, 209–219 (2000)
32. R. Lohmus, D. Erts, A. Lohmus, K. Svensson, Y. Jompol, H. Olin, STM and AFM instrumentation combined with transmission electron microscope. Phys. Low-Dimensional Struct. **3–4**, 81–89 (2001)
33. M.I. Lutwyche, Y. Wada, Manufacture of micromechanical scanning tunneling microscopes for observation of the tip avex in a transmission electron microscope. Sens. Actuat. A: Phys. **48**, 127–136 (1995)
34. Y. Xu, N.C. Macdonald, S.A. Miller, Integrated micro-scanning tunneling microscope. Appl. Phys. Lett. **67**, 2305 (1995)
35. K. Svensson, Y. Jompol, H. Olin, E. Olsson, Compact design of a transmission electron microscope-scanning tunneling microscope holder with three-dimensional coarse motion. Rev. Sci. Instrum. **74**, 4945 (2003)
36. "Nanofactory Instruments; www.nanofactory.com."
37. J.C. Spence, W. Lo, M. Kuwabara, Observation of the graphite surface by reflection electron microscopy during STM operation. Ultramicroscopy, **33**, 69–82 (1990)
38. W. Lo, J.C. Spence, Investigation of STM image artifacts by in-situ reflection electron microscopy. Ultramicroscopy, **48**, 433–444 (1993)
39. H. Ohnishi, Y. Kondo, K. Takayanagi, Quantized conductance through individual rows of suspended gold atoms. Nature. **395**, 2–5 (1998)
40. N. Agraït, A.L. Yeyati, J.M. van Ruitenbeek, Quantum properties of atomic-sized conductors. Phys. Rep. **377**, 81–279 (2003)
41. Y. Kurui, Y. Oshima, M. Okamoto, K. Takayanagi, Integer conductance quantization of gold atomic sheets. Phys. Rev. B. **77** (2008)
42. Y. Kurui, Y. Oshima, M. Okamaoto, K. Takayanagi, Conductance quantization/dequantization in gold nanowires due to multiple reflection at the interface. Phys. Rev. B. **79**, 165414 (2009)
43. K. Takayanagi, Y. Oshima, Y. Kurui, Conductance quantization of gold nanowires as a ballistic conductor. Phys. Rev. Lett. **2**, 47–50 (2009)
44. D. Erts, H. Olin, L. Ryen, E. Olsson, A. Thölén, Maxwell and Sharvin conductance in gold point contacts investigated using TEM-STM. Phys. Rev. B, **61**, 12725–12727 (2000)
45. Z. Aslam, M. Abraham, R. Brydson, A. Brown, B. Rand, Initial studies using a combined TEM – scanning tunnelling microscopy (STM) side entry sample holder. J. Phys.: Conf. Ser. **26**, 54–58 (2006)
46. Z. Aslam, M. Abraham, A. Brown, B. Rand, R. Brydson, Electronic property investigations of single-walled carbon nanotube bundles in situ within a transmission electron microscope: an evaluation. J. Microsc. **231**, 144–155 (2008)
47. J. Cumings, A. Zettl, Field emission and current-voltage properties of boron nitride nanotubes. Solid State Commun. **129**, 661–664 (2004)

48. S. Frank, P. Poncharal, Z.L. Wang, W.A. de Heer, Carbon nanotube quantum resistors. Science. **280**, 1744–1746 (1998)
49. H.U. Strand, K. Svensson, E. Olsson, Critical aspects of liquid metal immersion methods for characterization of electron transport properties in carbon nanotubes. In preparation
50. M.W. Larsson, L.R. Wallenberg, A.I. Persson, L. Samuelson, Probing of individual semiconductor nanowhiskers by TEM-STM. Microsc. Microanal. **10**, 41–46 (2004)
51. T. Kuzumaki, H. Yasuhiro, T. Kizuka, In-situ atomistic observation of carbon nanotubes during field emission. AIP Conf. Proc. 281–284 (2001)
52. T. Kuzumaki, H. Sawada, H. Ichinose, Y. Horiike, T. Kizuka, Selective processing of individual carbon nanotubes using dual-nanomanipulator installed in transmission electron microscope. Appl. Phys. Lett. **79**, 4580–4582 (2001)
53. T. Kuzumaki, Y. Horiike, T. Kizuka, T. Kona, C. Oshima, Y. Mitsuda, The dynamic observation of the field emission site of electrons on a carbon nanotube tip. Diamond Relat. Mater. **13**, 1907–1913 (2004)
54. J. Cumings, A. Zettl, M.R. McCartney, J.C. Spence, Electron holography of field-emitting carbon nanotubes. Phys. Rev. Lett. **88**, 1–4 (2002)
55. M. Wang, Q. Chen, L. Peng, Grinding a nanotube. Adv. Mater. **20**, 724–728 (2008)
56. R.P. Gao, Z.W. Pan, Z.L. Wang, Work function at the tips of multiwalled carbon nanotubes. Appl. Phys. Lett. **78**, 1757–1759 (2001)
57. X. Bai, E.G. Wang, P. Gao, Z.L. Wang, Measuring the work function at a nanobelt tip and at a nanoparticle surface. Nano Lett. **3**, 1147–1150 (2003)
58. Z. Xu, X.D. Bai, E.G. Wang, Z.L. Wang, Field emission of individual carbon nanotube with in situ tip image and real work function. Appl. Phys. Lett. **87**, 163106 (2005)
59. Z. Xu, X.D. Bai, E.G. Wang, Z.L. Wang, Dynamic in situ field emission of a nanotube at electromechanical resonance. J. Phys.: Condens. Matter. **17**, L507–L512 (2005)
60. K. Svensson, H. Olin, E. Olsson, Nanopipettes for metal transport. Phys. Rev. Lett. **93**, 14590 (2004)
61. B.C. Regan, S. Aloni, R.O. Ritchie, U. Dahmen, A. Zettl, Carbon nanotubes as nanoscale mass conveyors. Nature. **428**, 924–927 (2004)
62. L. Dong, X. Tao, L. Zhang, X. Zhang, B.J. Nelson, Nanorobotic spot welding: controlled metal deposition with attogram precision from copper-filled carbon nanotubes. Nano Lett. **7**, 58–63 (2007)
63. L. Dong, X. Tao, M. Hamdi, L. Zhang, X. Zhang, A. Ferreira, B.J. Nelson, Nanotube fluidic junctions: internanotube attogram mass transport through walls. Nano Lett. **9**, 210–214 (2009)
64. P.M. Costa, D. Golberg, M. Mitome, S. Hampel, A. Leonhardt, B. Buchner, Y. Bando, Stepwise current-driven release of attogram quantities of copper iodide encapsulated in carbon nanotubes. Nano Lett. **8**, 3120–3125 (2008)
65. G.E. Begtrup, W. Gannett, T.D. Yuzvinsky, V.H. Crespi, A. Zettl, Nanoscale reversible mass transport for archival memory. Nano Lett. **9**, 1835–1838 (2009)
66. J.Y. Huang, S. Chen, Z.F. Ren, G. Chen, M.S. Dresselhaus, Real-time observation of tubule formation from amorphous carbon nanowires under high-bias Joule heating. Nano Lett. **6**, 1699–1705 (2006)
67. S. Chen, J.Y. Huang, Z. Wang, K. Kempa, G. Chen, Z.F. Ren, High-bias-induced structure and the corresponding electronic property changes in carbon nanotubes. Appl. Phys. Lett. **87**, 263107 (2005)
68. J. Huang, S. Chen, Z.Q. Wang, K. Kempa, Y.M. Wang, S.H. Jo, G. Chen, M.S. Dresselhaus, Z.F. Ren, Superplastic carbon nanotubes. Nature. **439**, 281 (2006)
69. J.Y. Huang, S. Chen, S.H. Jo, Z. Wang, D.X. Han, G. Chen, M.S. Dresselhaus, Z.F. Ren, Atomic-scale imaging of wall-by-wall breakdown and concurrent transport measurements in multiwall carbon nanotubes. Phys. Rev. Lett. **94**, 236802 (2005)
70. T.D. Yuzvinsky, W. Mickelson, S. Aloni, G.E. Begtrup, A. Kis, A. Zettl, Shrinking a carbon nanotube. Nano Lett. **6**, 2718–2722 (2006)
71. J.Y. Huang, F. Ding, B.I. Yakobson, P. Lu, L. Qi, J. Li, In situ observation of graphene sublimation and multi-layer edge reconstructions. Proc. Natl Acad. Sci. USA. **106**, 10103–10108 (2009)

72. K. Saito, J. Fujii, T. Kizuka, Electric conduction of amorphous carbon and graphitic nanocontacts. Jpn. J. Appl. Phys. **48**, 010218 (2009)
73. T. Westover, R. Jones, J.Y. Huang, G. Wang, E. Lai, A. Talin, Photoluminescence, thermal transport, and breakdown in joule-heated GaN nanowires. Nano Lett. **9**, 257–263 (2009)
74. Z. Xu, D. Golberg, Y. Bando, Electrical field-assisted thermal decomposition of boron nitride nanotube: experiments and first principle calculations. Chem. Phys. Lett. **480**, 110–112 (2009)
75. Z. Xu, D. Golberg, Y. Bando, In situ TEM-STM recorded kinetics of boron nitride nanotube failure under current flow. Nano Lett. **9**, 2251–2254 (2009)
76. M. Hummelgård, R. Zhang, T. Carlberg, D. Vengust, D. Dvorsek, D. Mihailovic, H. Olin, Nanowire transformation and annealing by Joule heating. Nanotechnology. **21**, 165704 (2010)
77. J. Yamashita, H. Hirayama, Y. Ohshima, K. Takayanagi, Growth of a single-wall carbon nanotube in the gap of scanning tunneling microscope. Appl. Phys. Lett. **74**, 2450–2452 (1999)
78. M. Yoshida, Y. Kurui, Y. Oshima, K. Takayanagi, In-Situ observation of the fabrication process of a single shell carbon fullerene nano-contact using transmission electron microscope–scanning tunneling microscope. Jpn. J. Appl. Phys. **46**, L67–L69 (2007)
79. R. Zhang, M. Hummelgård, H. Olin, Carbon nanocages grown by gold templating. Carbon **48**, 424–430 (2010)
80. T. Kizuka, R. Kato, K. Miyazawa, Structure of hollow carbon nanocapsules synthesized by resistive heating. Carbon **47**, 138–144 (2009)
81. C.H. Jin, J.Y. Wang, Q. Chen, L.M. Peng, In situ fabrication and graphitization of amorphous carbon nanowires and their electrical properties. J. Phys. Chem. B. **110**, 5423–5428 (2006)
82. H. Hirayama, Y. Kawamoto, Y. Ohshima, K. Takayanagi, Nanospot welding of carbon nanotubes. Appl. Phys. Lett. **79**, 1169 (2001)
83. C. Jin, K. Suenaga, S. Iijima, Plumbing carbon nanotubes. Nat. Nanotechnol. **3**, 17–21 (2008)
84. M. Wang, L. Peng, J. Wang, Q. Chen, Shaping carbon nanotubes and the effects on their electrical and mechanical properties. Adv. Funct. Mater. **16**, 1462–1468 (2006)
85. D.N. Madsen, K. Mølhave, R. Mateiu, A.M. Rasmussen, M. Brorson, C.J. Jacobsen, P. Bøggild, Soldering of nanotubes onto microelectrodes. Nano Lett. **3**, 47–49 (2003)
86. L. de Knoop, K. Svensson, H. Petterson, E. Olsson, Extraction and local probing of individual carbon nanotubes. AIP Conference Proceedings, Aip, 2005, pp. 118–123
87. P. Poncharal, Electrostatic deflections and electromechanical resonances of carbon nanotubes. Science. **283**, 1513–1516 (1999)
88. R.P. Gao, Z.L. Wang, Z. Bai, W.A. de Heer, L. Dai, M. Gao, Nanomechanics of individual carbon nanotubes from pyrolytically grown arrays. Phys. Rev. Lett. **85**, 622–625 (2000)
89. M.M. Treacy, T.W. Ebbesen, J.M. Gibson, Exceptionally high Young's modulus observed for individual carbon nanotubes. Nature. **381**, 678–680 (1996)
90. J. Cumings, A. Zettl, Low-friction nanoscale linear bearing realized from multiwall carbon nanotubes. Science. **289**, 602–604 (2000)
91. T. Kizuka, Atomistic visualization of deformation in gold. Phys. Rev. B. **57**, 11158–11163 (1998)
92. K.J. Ziegler, D. Lyons, J.D. Holmes, D. Erts, B. Polyakov, H. Olin, K. Svensson, E. Olsson, Bistable nanoelectromechanical devices. Appl. Phys. Lett. **84**, 4074 (2004)
93. J. Andzane, N. Petkov, A.I. Livshits, J.J. Boland, J.D. Holmes, D. Erts, Two-terminal nanoelectromechanical devices based on germanium nanowires. Nano Lett. **9**, 1824–1829 (2009)
94. J. Andzane, J. Prikulis, D. Dvorsek, D. Mihailovic, D. Erts, Two-terminal nanoelectromechanical bistable switches based on molybdenum-sulfur-iodine molecular wire bundles. Nanotechnology. **21**, 125706 (2010)
95. G. Binnig, C.F. Quate, C. Gerber, Atomic force microscope. Phys. Rev. Lett. **56**, 930 (1986)
96. A.V. Ermakop, E.L. Garfunkel, A novel AFM/STM/SEM. Rev. Sci. Instrum. **65**, 2653–2654 (1994)
97. U. Stahl, C.W. Yuan, A.L. de Lozanne, M. Tortonese, Atomic force microscope using piezoresistive cantilevers and combined with a scanning electron microscope. Appl. Phys. Lett. **65**, 2878 (1994)
98. K. Jensen, W. Mickelson, A. Kis, A. Zettl, Buckling and kinking force measurements on individual multiwalled carbon nanotubes. Phys. Rev. B. **76**, 1–5 (2007)

99. Y. Lu, J.Y. Huang, C. Wang, S. Sun, J. Lou, Cold welding of ultrathin gold nanowires. Nat. Nanotechnol. 1–7 (2010)
100. D. Erts, A. Lohmus, R. Lohmus, H. Olin, A.V. Pokropivny, L. Ryen, K. Svensson, Force interactions and adhesion of gold contacts using a combined atomic force microscope and transmission electron microscope. Appl. Surf. Sci. **188**, 460–466 (2002)
101. C.M. Mate, G.M. McClelland, R. Erlandsson, S. Chiang, Atomic-scale friction of a tungsten tip on a graphite surface. Phys. Rev. Lett. **59**, 1942 (1987)
102. Y. Martin, C.C. Williams, H.K. Wickramasinghe, Atomic force microscope-force mapping and profiling on a sub 100-Å scale. J. Appl. Phys. **61**, 4723 (1987)
103. G. Neubauer, S.R. Cohen, G.M. McClelland, D. Horne, C.M. Mate, Force microscopy with a bidirectional capacitance sensor. Rev. Sci. Instrum. **61**, 2296 (1990)
104. T. Go¨ddenhenrich, H. Lemke, U. Hartmann, C. Heiden, Force microscope with capacitive displacement detection. J. Vac. Sci. Technol. A: Vac. **8**, 383 (1990)
105. T. Itoh, T. Suga, Piezoelectric force sensor for scanning force microscopy. Engineering. **43**, 305–310 (1994)
106. M. Tortonese, R.C. Barrett, C.F. Quate, Atomic resolution with an atomic force microscope using piezoresistive detection. Appl. Phys. Lett. **62**, 834 (1993)
107. A. Nafari, A. Danilov, H. Rödjegård, P. Enoksson, H. Olin, A micromachined nanoindentation force sensor. Sens. Actuat. A: Phys. **123–124**, 44–49 (2005)
108. H.D. Espinosa, Y. Zhu, B. Peng, A MEMS device for in situ TEM/AFM/SEM/STM testing of carbon nanotubes and nanowires. 2002 SEM annual conference, Milwaukee, 2002, pp. 1–5
109. S.A. Asif, K.J. Wahl, R.J. Colton, Nanoindentation and contact stiffness measurement using force modulation with a capacitive load-displacement transducer. Rev. Sci. Instrum. **70**, 2408 (1999)
110. A. Nafari, D. Karlen, C. Rusu, K. Svensson, H. Olin, P. Enoksson, MEMS sensor for in situ TEM atomic force microscopy. J. Microelectromech. Syst. **17**, 328–333 (2008)
111. J. Thaysen, A. Boisen, O. Hansen, S. Bouwstra, Atomic force microscopy probe with piezoresistive read-out and a highly symmetrical Wheatstone bridge arrangement. Sens. Actuat. A: Phys. **83**, 47–53 (2000)
112. D. Golberg, P.M. Costa, O. Lourie, M. Mitome, X. Bai, K. Kurashima, C. Zhi, C. Tang, Y. Bando, Direct force measurements and kinking under elastic deformation of individual multiwalled boron nitride nanotubes. Nano Lett. **7**, 2146–2151 (2007)
113. A. Asthana, K. Momeni, A. Prasad, Y.K. Yap, R.S. Yassar, In situ probing of electromechanical properties of an individual ZnO nanobelt. Appl. Phys. Lett. **95**, 172106 (2009)
114. P.M. Costa, D. Golberg, G. Shen, M. Mitome, Y. Bando, ZnO low-dimensional structures: electrical properties measured inside a transmission electron microscope. J. Mater. Sci. **43**, 1460–1470 (2007)
115. T. Shokuhfar, G.K. Arumugam, P.A. Heiden, R.S. Yassar, C. Friedrich, Direct compressive measurements of individual titanium dioxide nanotubes. ACS Nano. **3**, 3098–3102 (2009)
116. T. Kuzumaki, Y. Mitsuda, Nanoscale mechanics of carbon nanotube evaluated by nanoprobe manipulation in transmission electron microscope. Jpn. J. Appl. Phys. **45**, 364–368 (2006)
117. X. Bai, D. Golberg, Y. Bando, C. Zhi, C. Tang, M. Mitome, K. Kurashima, Deformation-driven electrical transport of individual boron nitride nanotubes. Nano Lett. **7**, 632–637 (2007)
118. T. Kizuka, Direct atomistic observation of deformation in multiwalled carbon nanotubes. Phys. Rev. B. **59**, 4646–4649 (1999)
119. K. Asaka, R. Kato, K. Miyazawa, T. Kizuka, Deformation of multiwalled nanometer-sized carbon capsules. Appl. Phys. Lett. **89**, 191914 (2006)
120. H. Ghassemi, Y.K. Yap, R.S. Yassar, In-situ nanomechanical testing of one dimensional material. NSTI-Nanotech 2009, Houston (2009).
121. K. Asaka, K. Miyazawa, T. Kizuka, The toughness of multi-wall carbon nanocapsules. Nanotechnology. **20**, 385705 (2009)
122. R. Kato, K. Asaka, K. Miyazawa, T. Kizuka, In situ high-resolution transmission electron microscopy of elastic deformation and fracture of nanometer-sized fullerene C 60 whiskers. Jpn. J. Appl. Phys. **45**, 8024–8026 (2006)

123. K. Saito, K. Miyazawa, T. Kizuka, Bending process and Young's modulus of fullerene C 60 nanowhiskers. Jpn. J. Appl. Phys. **48**, 010217 (2009)
124. M.S. Wang, Y. Bando, J.A. Rodriguez-Manzo, F. Banhart, D. Golberg, Cobalt nanoparticle-assisted engineering of multiwall carbon nanotubes. ACS Nano. **3**, 2632–2638 (2009)
125. P.M. Costa, U.K. Gautam, M.S. Wang, Y. Bando, D. Golberg, Effect of crystalline filling on the mechanical response of carbon nanotubes. Carbon. **47**, 541–544 (2009)
126. S. Blom, H. Olin, J. Costa-Kramer, N. Garcia, M. Jonson, P. Serena, R. Shekhter, Free-electron model for mesoscopic force fluctuations in nanowires. Phys. Rev. B. **57**, 8830–8833 (1998)
127. N. Agraït, G. Rubio, S. Vieira, Plastic deformation of nanometer-scale gold connective necks. Phys. Rev. Lett. **74**, 3995–3998 (1995)
128. T. Kizuka, Atomic configuration and mechanical and electrical properties of stable gold wires of single-atom width. Phys. Rev. B. **77** (2008)
129. T. Matsuda, T. Kizuka, Palladium wires of single atom width as mechanically controlled switching devices. Jpn. J. Appl. Phys. **45**, L1337–L1339 (2006)
130. M. Ryu, T. Kizuka, Structure, conductance and strength of iridium wires of single atom width. Jpn. J. Appl. Phys. **45**, 8952–8956 (2006)
131. T. Matsuda, T. Kizuka, Structure and tensile force of nanometer- and atomic-sized gold contacts during conductance feedback control. Jpn. J. Appl. Phys. **48**, 125007 (2009)
132. S. Fujisawa, T. Kikkawa, T. Kizuka, Direct observation of electromigration and induced stress in Cu nanowire. Jpn. J. Appl. Phys. **42**, L1433–L1435 (2003)
133. T. Kizuka, H. Aoki, The dynamics of electromigration in copper nanocontacts. Appl. Phys. Express. **2**, 075003 (2009)
134. T. Matsuda, T. Kizuka, Slip sequences during tensile deformation of palladium nanocontacts. Jpn. J. Appl. Phys. **48**, 115003 (2009)
135. T. Kizuka, Y. Takatani, K. Asaka, R. Yoshizaki, Measurements of the atomistic mechanics of single crystalline silicon wires of nanometer width. Phys. Rev. B. **72**, 1–6 (2005)
136. T. Kizuka, Y. Takatani, Growth of silicon nanowires by nanometer-sized tip manipulation. Jpn. J. Appl. Phys. **46**, 5706–5710 (2007)
137. B.N. Persson, Sliding friction. Surf. Sci. Rep. **33**, 83–119 (1999)
138. E. Gnecco, R. Bennewitz, T. Gyalog, E. Meyer, Friction experiments on the nanometre scale. J. Phys.: Condens. Matter. **13**, R619–R642 (2001)
139. B. Bhushan, *Nanotribology and Nanomechanics* (Springer, Heidelberg, 2008)
140. S. Fujisawa, T. Kizuka, Lateral displacement of an AFM tip observed by in-situ TEM/AFM combined microscopy: the effect of the friction in AFM. Tribol. Lett. **15**, 163–168 (2003)
141. "Preparation of sample for in-situ tem-spm operation; www.youtube.com/watch?v=9k9KO5WEyog."
142. P.M. Costa, D. Golberg, M. Mitome, Y. Bando, Nitrogen-doped carbon nanotube structure tailoring and time-resolved transport measurements in a transmission electron microscope. Appl. Phys. Lett. **91**, 223108 (2007)
143. L. de Knoop, Investigation of iron filled multiwalled carbon nanotubes, Chalmers University of Technology (Dept. of Applied Physics), 2005
144. M.M. Yazdanpanah, S.A. Harfenist, A. Safir, R.W. Cohn, Selective self-assembly at room temperature of individual freestanding Ag2Ga alloy nanoneedles. J. Appl. Phys. **98** (2005)
145. J. Deneen, W.M. Mook, A. Minor, W.W. Gerberich, C.B. Carter, In situ deformation of silicon nanospheres. J. Mater. Sci. **41**, 4477–4483 (2006)
146. T. Kizuka, R. Kato, K. Miyazawa, Surface breakdown dynamics of carbon nanocapsules. Nanotechnology. **20**, 105205 (2009)
147. R. Zhang, M. Hummelgård, H. Olin, Simple and efficient gold nanoparticles deposition on carbon nanotubes with controllable particle sizes. Mater. Sci. Eng.: B. **158**, 48–52 (2009)
148. R. Zhang, M. Hummelgård, H. Olin, Simple synthesis of clay-gold nanocomposites with tunable color. Langmuir. **26**, 5823–5828 (2010)
149. K.K. Datta, M. Eswaramoorthy, C.N. Rao, Water-solubilized aminoclay-metal nanoparticle composites and their novel properties. J. Mater. Chem. **17**, 613 (2007)

150. O.L. Guise, J.W. Ahner, M. Jung, P.C. Goughnour, J.T. Yates, Reproducible electrochemical etching of tungsten probe tips. Nano Lett. **2**, 191–193 (2002)
151. I. Ekvall, E. Wahlstrom, D. Claesson, H. Olin, E. Olsson, Preparation and characterization of electrochemically etched W tips for STM. Meas. Sci. Technol. **10**, 11–18 (1999)
152. P.M. Costa, X. Fang, S. Wang, Y. He, Y. Bando, M. Mitome, J. Zou, H. Huang, D. Golberg, Two-probe electrical measurements in transmission electron microscopes–behavioral control of tungsten microwires. Microsc. Res. Tech. **72**, 93–100 (2009)
153. J.G. Simmons, Generalized formula for the electric tunnel effect between similar electrodes separated by a thin insulating film. J. Appl. Phys. **34**, 1793 (1963)
154. X.L. Wei, Y. Liu, Q. Chen, L.M. Peng, Controlling electron-beam-induced carbon deposition on carbon nanotubes by Joule heating. Nanotechnology. **19**, 355304 (2008)
155. A.N. Chiaramonti, L.J. Thompson, W.F. Egelhoff, B.C. Kabius, A.K. Petford-Long, In situ TEM studies of local transport and structure in nanoscale multilayer films. Ultramicroscopy. **108**, 1529–1535 (2008)
156. J. Börjesson, *The Role of Interfacial Microstructure of Perovskite Thin Films: A High Resolution and In Situ Study* (Chalmers University of Technology, Sweden, 2009)
157. A. Nafari, P. Enoksson, H. Olin, Si-wedge for easy TEM sample preparation for in situ probing. Eurosensors, Barcelona, 2005
158. D.B. Williams, B.C. Carter, *Transmission Electron Microscopy* (Plenum, New York, 1996)
159. C. Jin, H. Lan, L. Peng, K. Suenaga, S. Iijima, Deriving carbon atomic chains from graphene. Phys. Rev. Lett. **102**, 1–4 (2009)
160. D. Ugarte, Curling and closure of graphitic networks under electron-beam irradiation. Nature **359**, 707–709 (1992)
161. F. Banhart, Irradiation effects in carbon nanostructures. Rep. Prog. Phys. **62**, 1181–1221 (1999)

Chapter 4
Scanning Probe Microscopy and Grazing-Incidence Small-Angle Scattering as Complementary Tools for the Investigation of Polymer Films and Surfaces

Peter Müller-Buschbaum and Volker Körstgens

Abstract The real-space analysis, based on scanning probe microscopy (SPM) techniques and the advanced scattering technique grazing-incidence small-angle scattering (GISAS), are complementary tools for the structural analysis of nanostructures. GISAS experiments can be performed with X-rays, named grazing-incidence small-angle X-ray scattering (GISAXS) and with neutrons, denoted GISANS, respectively. On selected examples of polymer films and surfaces, these complementarities are illustrated. Master curves of power spectral densities calculated from the SPM data may give an equivalent information as compared with GISAS data as long as the SPM measurement is representative in a statistical meaning for a larger area as it is addressed with GISAS. Moreover, GISAS allows for accessing buried structures that are hardly addressable with SPM.

4.1 Introduction

Nowadays scanning probe microscopy (SPM) techniques offer unique possibilities to investigate local surface properties of material with a high resolution in the nanoscale [1–3]. Many different types of SPM have been established [2]. With the widespread use of commercial instruments, combining easy handling and different operation modes, the imaging of surface structures and the determination of surface properties are possible for material classes such as metals, oxides, ceramics, soft matter, and biological samples. Thus, SPM is a common tool in nanoscience and nanotechnology [4–8], which provides important information on surface topography, surface stiffness, surface magnetization, and many other sample-specific information [9–11]. SPM technologies share the concept of scanning an extremely sharp tip (typically 3–50 nm radius of curvature) across the object surface. This tip is mounted on a flexible cantilever, allowing the tip to follow the surface profile. Operating in real space, direct sample information is gained and frequently displayed in a pictorial way, which allows an easy understanding. The samples do not require a special pretreatment or a partial vacuum but can be observed in air at standard temperature and pressure or under more complex sample environments.

However, despite its successful usage, SPM techniques have limitations. On the one hand, the performance of an SPM instrument is limited by a number of instrumental factors such as the resolution of the mechanical components used to move the tip and measure its position or the sharpness and stability of the probe tip determining the area of contact and the reproducibility of imaging. Moreover, distortion due to tip effects such as geometric effects, point-spread effects, and interaction effects need to be accounted for [12]. On the other hand, SPM measurements are limited to rather small sample surface areas and have difficulties in accessing buried interfaces.

Part of these limitations can be overcome: so far, the major SPM manufacturers have made substantial improvements in mechanical and electronic design. These improvements and advanced electronic calibration routines and ongoing instrumental progress result in a continuous decrease of the limiting instrumental factors. In addition, distortions from tip effects are well accounted by improvements in tip sharpness and the use of deconvolution software.

The remaining limitations are more serious: it is still difficult to achieve information with SPM, for example, about the surface structure, on a larger surface area. An extensive mapping of the sample is either impossible or at least not economical. Buried structures remain inaccessible as long as the nondestructive character of SPM techniques is not sacrificed.

As a consequence, advanced scattering techniques such as grazing-incidence small-angle scattering (GISAS) are a perfect complementary probe for SPM techniques if structures have to be determined [13, 14]. By illuminating large sample areas, with GISAS a statistical relevant structural characterization is possible. Thus, GISAS is well suited to detect a lateral surface or interface roughness, lateral correlations, sizes and shapes of objects such as particles positioned on top of the surface or in a surface near region. Moreover, buried interfaces can be probed in a nondestructive way without any special sample preparation. Thus, GISAS compensates the limitations of SPM techniques. Moreover, GISAS offers additional advantages but has limitations as well.

The main additional advantage is that GISAS measurements can be performed with X-rays (GISAXS method) or with neutrons (GISANS method). Therefore the chosen probe, X-rays or neutrons, determines the contrast in the experiment for a given system or allows for a change in contrast. Moreover, no scanning of the sample position is required, which offers access to fast measurements addressing kinetics during in situ investigations [15–17]. However, as all scattering experiments, GISAXS and GISANS operate in reciprocal space, which prevents the easy interpretation of the measurement in terms of real space pictures. Typically, modeling of the measured signal is necessary to get structural parameters from a GISAXS or GISANS experiment. With this respect, in addition to complementarities, real space and reciprocal space data from the scattering experiment can be used together in a very beneficial approach. The SPM experiment can give a structural input for the modeling of the GISAS data as long as pure surface information is under consideration.

Despite its big success in many different types of sample systems and the increasing interest in GISAS, the complementary use with SPM techniques is still very limited. Therefore, this contribution focuses on possibilities of a complementary use of both techniques, SPM and GISAS, putting emphasis to the less well-known technique, namely, GISAS.

The article is structured as follows: In Sect. 4.2, the analysis of SPM data regarding lateral structures is discussed. In Sect. 4.3, the GISAS technique is first described with X-rays as the light source illuminating the sample surface and the similarities and distinctions concerned with the use of neutrons are discussed. Sections 4.4–4.6 aim on experiments that link together SPM and GISAS measurements. Section 4.3 ends this chapter.

4.2 Statistical Analysis of SPM Data

In many cases, there is not only interest in the qualitative picturing of the local surface structure of a sample but also in the quantitative analysis. Quantitative information can be gained by measuring individual objects from SPM data or from a statistical interpretation of the SPM data [18]. Today, both are easily accessed by the offered SPM software and other available software tools.

For example, the deviation of individual heights from a mean surface can be calculated in terms of a root-mean-square (rms) roughness of the sample surface. This rms roughness is a statistical description of structures perpendicular to the surface. Parallel to the sample surface, one way of statistical analysis is the calculation of a power spectral density (PSD) of the SPM data [19, 20].

To determine the PSD, the Fourier transform of an individual SPM dataset is calculated. With this Fourier transformation, the data are transformed from real into reciprocal space (see Fig. 4.1). In the Fourier transform, it is easier to

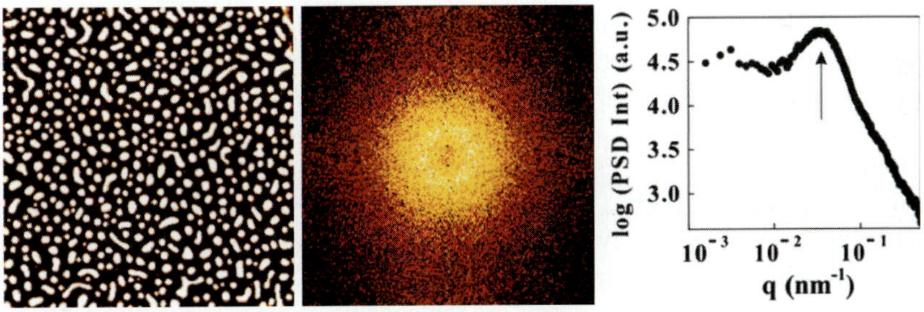

Fig. 4.1 Example of master curve generation from SPM data. *Left*: Topography data of scan size 4 μm × 4 μm measured in noncontact mode presenting the drop-like structures from dewetting of a diblock copolymer film. Brightness represents the structure's height. *Middle*: Calculated two-dimensional Fourier transformed, showing the ring-shaped intensity (sign of an isotropic structure) in Fourier space. *Right*: Master curve formed by combining power spectral density (PSD) functions that resulted from radial averaging of intensity distribution in Fourier space of different scan ranges. The position of the peak shown by the *arrow* relates to the most prominent in-plane length. Reprinted with permission from [21]. Copyright 2007 American Chemical Society

diminish isotropic from anisotropic structures. Isotropic structures have no preferential rotational orientation in real space and the corresponding Fourier transform exhibits a ring-shaped intensity and it is isotropic again. In contrast, an anisotropic structure gives rise to Fourier transform that has an anisotropic intensity distribution. In case of the isotropic intensity distribution, a radial average improves the statistics of the resulting PSD. By using different magnifications in SPM measurements (different scan ranges), individual PSD are calculated and combined into one master curve as shown in Fig. 4.1. With such a master curve a broad range in reciprocal space is addressed and it is validated that the individual SPM images describe the sample surface in a statistically significant manner [22]. In many cases, the master curve will show a peak which is indicating a most prominent length scale in the surface topography. The exact position of this peak may be determined with a fit of a Gaussian line shape to the PSD. From the value of peak position q^* in reciprocal space the dominant length scale ξ of the structure is calculated using $\xi = 1/q^*$. In this type of analysis, the surface structure is characterized only by mean distance ξ.

The intensity distribution of the master curve represents the statistics of length scales of lateral structures in reciprocal space on the overall probed surface area (of the SPM data). Therefore, such master curves are comparable to the GISAS information as described later [23].

To gain a better understanding of the statistical analysis via master curve calculation we look into one particular example: A nanostructure consisting of individual droplets was achieved by spin coating of the diblock copolymer polystyrene-*block*-polyisoprene, denoted P(S-*b*-I), as a thin film on a silicon wafer out of toluene solution without any further sample treatment (see [21] for more details of the sample preparation). A nanostructure of individual droplets is typical for dewetted ultra-thin polymer films [24–26]. In case of spin coating of homopolymers and block copolymers [27–29], isolated droplets are observed as in the final stage of the spin coating process the solvent evaporation out of a highly viscous polymer solution. Below a critical polymer concentration no homogeneous layer of the diblock copolymer is formed [29]. Polymer droplets are formed on the substrate because of dewetting and the insufficient material to form a complete layer. The corresponding statistical analysis of such dewetting morphology with the generation of a master curve is shown in Fig. 4.1.

With the addition of maghemite nanoparticles, the topography of the samples changes [21,30,31] as probed with SPM and shown in Fig. 4.2. The polymer nanostructure appears more densely packed, whereas the nanoparticles are located on top of the diblock copolymer superstructure. At very low nanoparticle concentration (1 and 5%) the droplet-like structures are replaced by worm-like structures. Comparable to the observation in polymer films of larger film thickness [32], the presence of nanoparticles retards the dewetting and thus inhibits the droplet formation, which is typical for the final stage of dewetting [21].

Figure 4.3 shows a plot of the master curves constructed out of SPM data measured with different scan ranges. The calculated master curve of the P(S-*b*-I) sample without added nanoparticles (bottom curve in Fig. 4.3) exhibits a peak at position $q^* = 3.22 \times 10^{-2}\,\mathrm{nm}^{-1}$. This peak corresponds to the most prominent in-plane

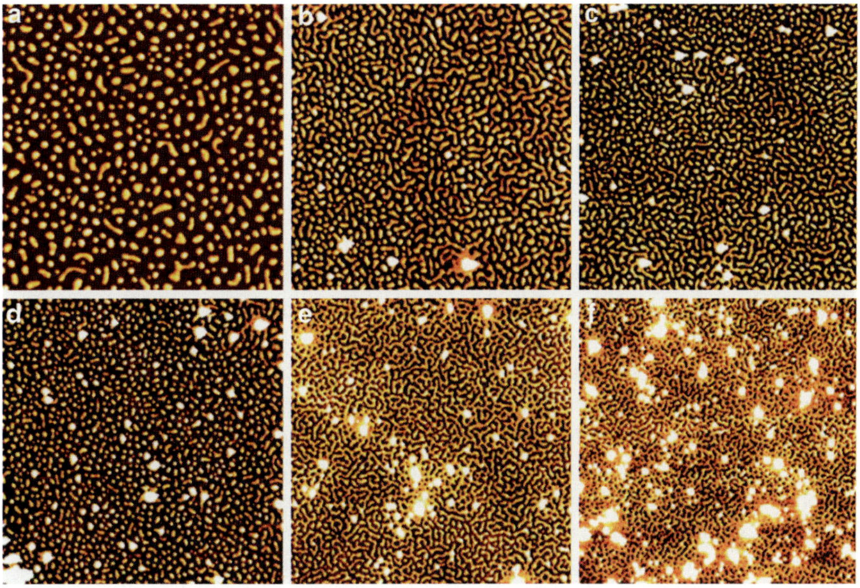

Fig. 4.2 SPM micrographs showing the surface topography of the P(S-*b*-I) nanostructures which result from the addition of different amounts [(**a**) 0%; (**b**) 1%; (**c**) 5%; (**d**) 10%; (**e**) 20%; (**f**) 30% by weight] of nanoparticles and cluster formation at high nanoparticle concentrations. The scan size is 4 μm × 4 μm. Structure height increases with the brightness of the structure in the images. Reprinted with permission from [21]. Copyright 2007 American Chemical Society

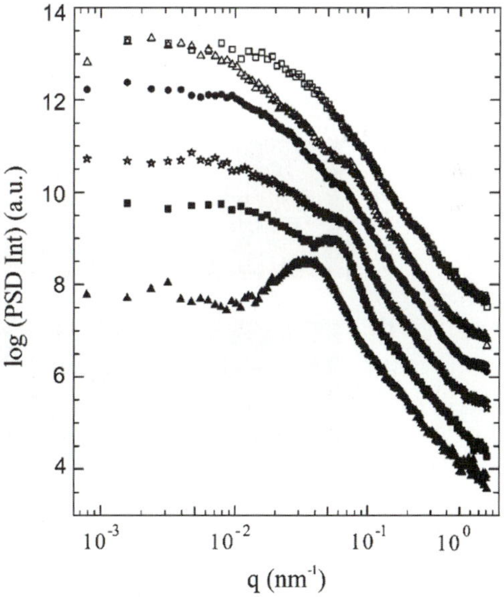

Fig. 4.3 Double-logarithmic plot of PSD master curves for samples containing varying amount of nanoparticles (from bottom to top the nanoparticle concentration increases: 0, 1, 5, 10, 20, and 30% by weight). For a better understanding the master curves are shifted along PSD intensity axis (y-axis). Reprinted with permission from [21]. Copyright 2007 American Chemical Society

length of the polymer structures on the substrate surface. The corresponding lateral length in real space is $d = 2\pi/q^* = 195$ nm. It is given by the distance between two neighboring diblock copolymer droplets. With increasing concentration of nanoparticles (1 and 5%), the related peak in the master curve becomes less pronounced and shifts toward larger q^* values. This shift in reciprocal space indicates a decrease in the lateral distance in real space. At higher nanoparticle concentrations, the peak in the master curve is hardly visible anymore. The presence of nanoparticle clusters strongly dominates the surface structure. Although the clusters are attached to the polymer nanostructure, they remain randomly positioned on top of the polymer part and exhibit no well-defined lateral order. As a consequence, the master curve calculated from the SPM data results in less well-pronounced intensity maxima as compared to simple polymer nanostructures [25, 26].

As a second example for the calculation of master curves from SPM data, we look into the structure of droplets of polystyrene (PS) on silicon substrates again created by dewetting [33]. A confined geometry of the PS is installed by the silicon wafer as solid support and air. Four different substrate surface treatments were carried out to investigate the influence on the dewetting of an initially homogeneous ultra-thin, but metastable PS film. In one of the surface modifications, the silicon substrate was rubbed with a toluene-soaked fuzz-free wipe several times in random direction successively. In the following, this process is referred to as "wiping." After spin coating PS out of toluene solution, samples were annealed at 130°C above the glass transition temperature of PS and quenched to room temperature. The SPM micrographs picturing the surface topography in Fig. 4.4 show the samples pretreated by wiping after different times of annealing. The initially continuous film is destroyed by the appearance of holes, growing in time, connected via channels forming islands and finally spherical droplets. The morphological transitions require movement of the polymer chains near the substrate surface due to the small film thickness under investigation.

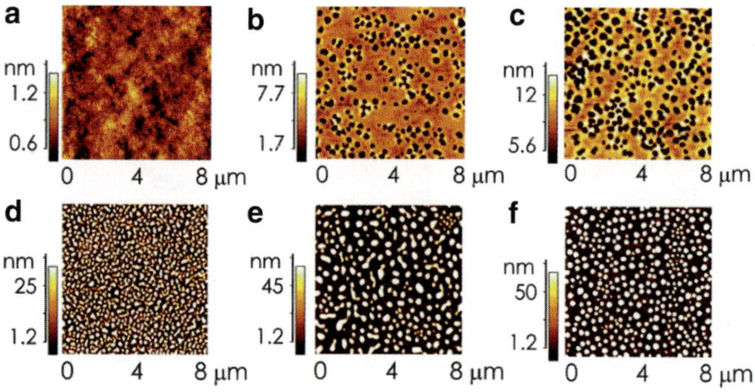

Fig. 4.4 SPM micrographs picturing the surface topography of dewetting of polystyrene after (**a**) 0, (**b**) 10, (**c**) 20, (**d**) 45, (**e**) 90, and (**f**) 180-min heating at 130°C for the samples treated by wiping. Each measurement covers a scan range of 8 μm × 8 μm. The height scale is adjusted to emphasize on the characteristic surface features. Reprinted with permission from [33]. Copyright 2006 American Chemical Society

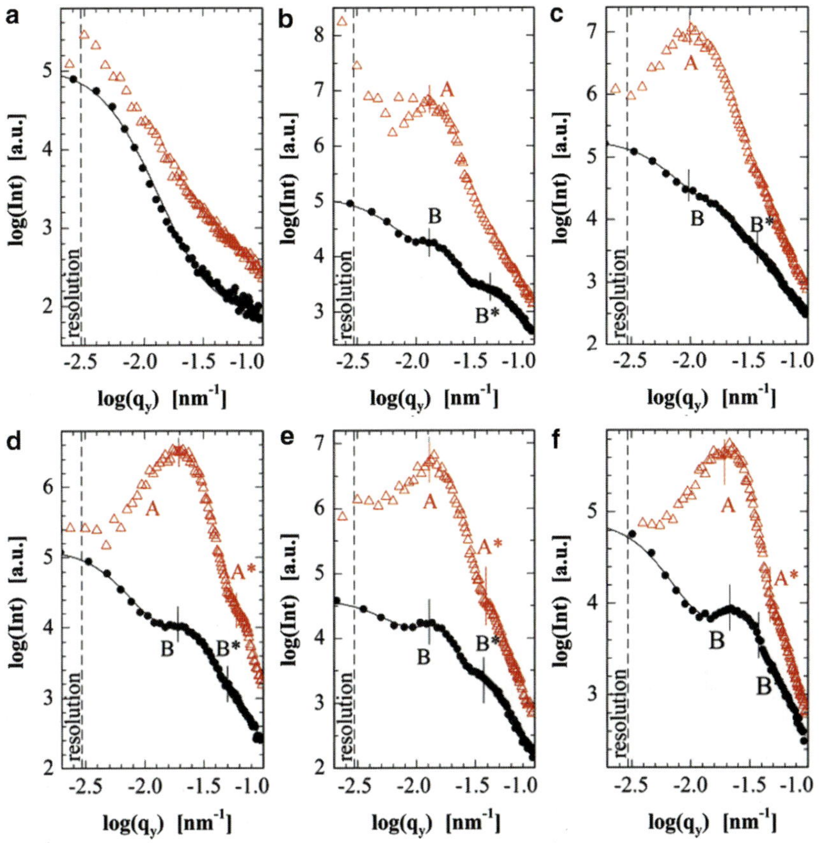

Fig. 4.5 PSD master curves calculated from SPM pictures (*red triangles*) shown in comparison with horizontal line cuts at the critical angle of PS from the two-dimensional GISAXS measurements (*black circles*): In the chosen example the Si was treated by wiping. The shown annealing times are (**a**) 0, (**b**) 10, (**c**) 20, (**d**) 45, (**e**) 90, and (**f**) 180 min at 130°C. The dominant length scales are marked with A, A^*, B, and B^* for the PSD curves and GISAXS cuts, respectively. The resolution limit of the scattering experiment is shown with the *dashed line*. Reprinted with permission from [33]. Copyright 2006 American Chemical Society

The corresponding PSD master curves are shown in Fig. 4.5. In the graphs of Fig. 4.5, not only the master curves and dominant length scales are shown but also GISAXS line cuts giving lateral information are included. These data will be discussed in Sect. 4.4.

So far, all selected examples show isotropic surface structures that have an isotropic Fourier transform. Of course surface structures can be anisotropic as well, which complicates the calculation of the PSD master curve. Radial averaging is no longer possible in case the Fourier transform is anisotropic. We illustrate this with an anisotropic channel structure.

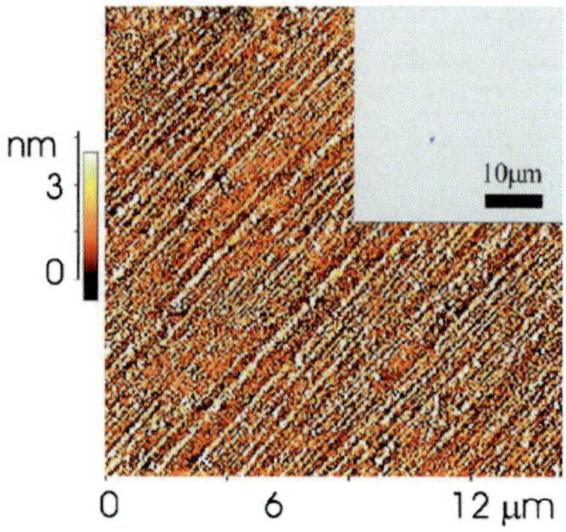

Fig. 4.6 Real-space structural data of the PDMS nanochannels: SPM topography data with $15\,\mu\text{m} \times 15\,\mu\text{m}$ scan range and optical micrograph in the *inset*. Reprinted with permission from [34]

With a simple wiping technique ordered polymeric nanochannel arrays can be produced when the wiping is performed in one spatial direction with high precision [34]. A highly diluted solution of polydimethylsiloxane (PDMS) in isopropanol was deposited onto a standard microscopy glass slide with several successive wipings in parallel to the long axis of the glass slide. The highly diluted PDMS solution dewets the glass surface and with evaporation of the solvent only a small amount of PDMS is transferred onto the glass surface. Figure 4.6 shows an SPM image with a scan range of $15\,\mu\text{m} \times 15\,\mu\text{m}$ of the achieved surface structure.

The sample is tilted with respect to the scanning direction in the SPM measurement to reduce the influence of the scanning or data treatment (flattening) on the determination of the height of the polymer nanochannels. The mean height of the parallel aligned PDMS lines is $H = (3.0 \pm 0.5)$ nm with a mean distance to adjacent lines of $D = (166 \pm 9)$ nm as shown in the line scan of Fig. 4.7c. The Fourier transform of the SPM data is anisotropic (see Fig. 4.7d) but still enables a statistical characterization of the surface pattern. Instead of a radial average selected line cuts in the Fourier transform have to be performed. As compared to the improvement in statistics by radial averaging these line cuts will have poor statistics. Box-shaped cuts can yield an improvement.

The narrow stripes of intensity in Fourier space demonstrate the high degree of perfection of the alignment of PDMS into parallel lines and thus the strong anisotropy of the surface structure. However, only one broad intensity maximum is visible. A perfect surface grating would have resulted in numerous intensity maxima, well separated by the reciprocal grating period. The absence of higher order maxima is a clear indication of a relaxed order of the PDMS lines with respect to the distance between adjacent lines. Therefore, the preparation results rather in nanochannels than in a surface grating. To prove the presence of the polymer nanochannels over a surface area larger than several square millimeters, meaning having prepared a nanochannels array rather than only local nanochannels, SPM is not well suited. In general, for the investigation of surface structures, surface sensitive scattering methods are necessary to get statistical meaningful data of a larger surface area. In the next chapter, GISAS methods will be introduced.

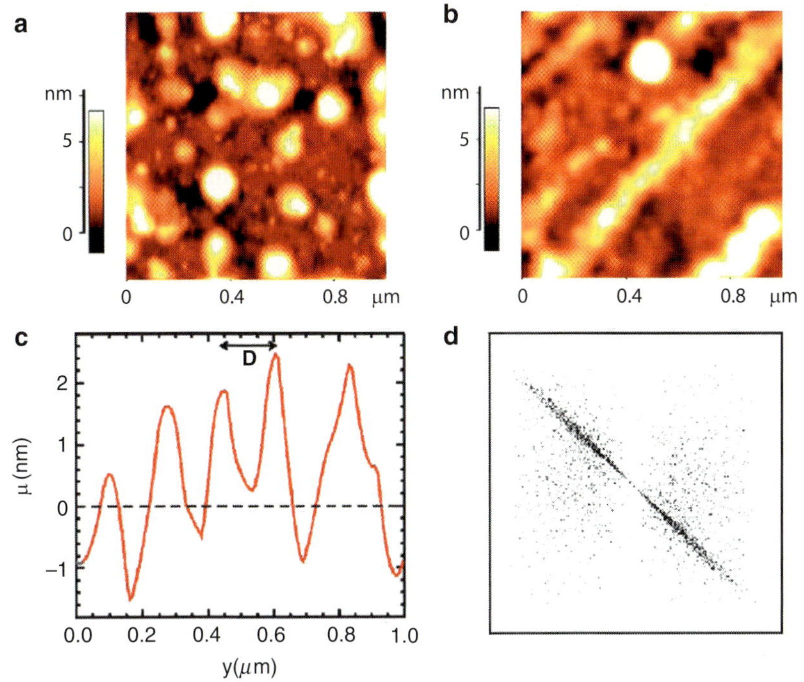

Fig. 4.7 (**a**) For comparison, dewetting structure resulting without wiping and (**b**) zoomed-in SPM data with 1 μm × 1 μm scan range demonstrating imperfections of the PDMS nanochannels. (**c**) Representative line scan ($D = 166\,\mathrm{nm}$), and (**d**) Fourier transformed calculated from the SPM data. Reprinted with permission from [34]

Of course, the analysis using master curves is not limited only to dewetting structures. Surface domains and roughness of polymer gels observed by SPM and analyzed statistically [35], the ordering of self-assembled $Si_{1-x}Ge_x$ islands [36], or self-assembled carbon-induced germanium quantum dots [37] are some among many other examples investigating alternative systems. Moreover it should be mentioned that in addition to the statistical analysis of SPM data, optical microscopy images can be analyzed in a similar way [38]. Instead of height data, the color information is used in the Fourier transformation.

4.3 Introduction to Grazing-Incidence Small-Angle Scattering

Within the last years, grazing-incidence small-angle scattering (GISAS) using X-rays (abbreviated with GISAXS) and neutrons (GISANS) became a flexible and frequently used analysis technique [39]. Both, GISAXS and GISANS are used to probe micro and nano-structured thin films and surfaces in terms of a characterization of micro- and nano-scale density correlations and shape analysis of objects at surfaces or at buried interfaces for various classes of materials such as ceramics, metals, semiconductors, polymers, biopolymers, and soft matter. GISAXS and GISANS are still advanced scattering techniques and as all scattering experiments give results in reciprocal space [25, 26]. In contrast to standard small angle scattering (SAS such as SAXS and SANS), which uses transmission geometry, GISAS is performed in a reflection geometry having a shallow angle of incidence of the beam onto the sample surface. Thus, GISAS has somewhat similarities with SAS and can be understood as a SAS experiment performed in modified scattering geometry (replacing transmission by reflection geometry) [13, 39]. Alternatively, GISAS can be envisaged as the extension of grazing incidence diffraction (GID) [40] to small scattering angles or as a sort of diffuse reflectivity [41]. Consequently, three different X-ray communities, SAS, GID, and diffuse reflectivity are converging through GISAS [14, 39].

In general, GISAXS measurements performed at laboratory X-ray sources are very limited [42, 43]. The full potential of GISAXS is realized with using a synchrotron radiation source (flux, collimation, and choice of wavelength in order to avoid fluorescence or to perform anomalous measurements) [44–46]. GISANS is always restricted to large-scale facilities due to the need for a powerful neutron source [15,47–49]. While there are currently a few dedicated GISAXS and GISANS instruments at large-scale facilities, GISAS capabilities can often be implemented on existing SAS or GID instruments by small modification (addition of precise goniometers and a two-dimensional detector). Due to the increase in interest at GISAXS and GISANS experiments and the widespread applications, new dedicated GISAXS and GISANS instruments have been started recently or are currently under construction. Therefore, in future the access to GISAXS and GISANS experiments will increase and accomplish with the increase in its demand.

Historically, in 1989 the first pioneering GISAXS experiments were performed by Levine and co-workers, investigating the structure of gold nanoparticles on silicon (001) surfaces by using a rotating anode [42, 43]. After several years without being noticed, GISAXS experiments reappeared in the field of hard condensed matter. For example, Salditt and co-workers probed amorphous multilayers on silicon [41, 50] and Naudon and co-workers investigated gold and platinum aggregates in surface regions [51, 52]. By Müller-Buschbaum and co-workers GISAXS was introduced into the field of soft matter and polymers [53, 54]. Whereas the first experiments were focused on static samples, prepared ex-situ and investigated decoupled from preparation, real-time, in-situ measurements as for example performed by Gibaud and co-workers [55] or Ree and coworkers [56] emerged and attracted increasing attention. Thus, GISAXS can be used for the investigation of kinetics as well.

As compared to GISAS measurements with synchrotron radiation, experiments using neutron are still rather rare. First GISANS experiments were performed to investigate thin polymer films that dewetted from the solid support [25, 26]. However, as pointed out the general ideas of GISAXS and GISANS experiments are equal. The experimental setup for both methods includes a completely evacuated pathway, a two-circle goniometer with a z-translation table, and a two-dimensional

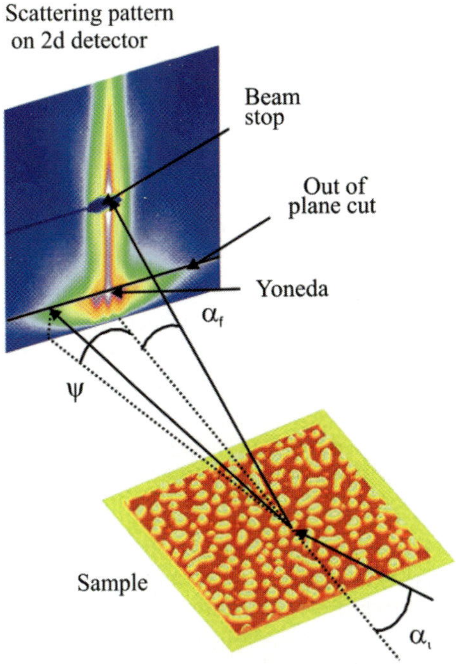

Fig. 4.8 Schematic drawing of the experimental setup in GISAXS or GISANS geometry. The sample surface is placed horizontally. The incident angle is denoted α_i, the exit angle α_f, and the out-of-plane angle ψ. The scattering intensity distribution on the two-dimensional detector (low intensity as dark and high intensity as bright) is presented on a logarithmic scale. The detected scattering pattern shows the diffuse scattering with the split Yoneda peak and specular peak (shielded by a beam stop that protects the detector from a very high intensity of the reflected beam), which are common features of a nanostructured surface. Reprinted with permission from [21]. Copyright 2007 American Chemical Society

detector. Figure 4.8 shows a sketch of a typical GISAS setup with the characteristic angles.

The X-ray or neutron beam impinges onto the sample surface under a shallow angle of incidence α_i and the scattered intensity is probed at an exit angle α_f and an out-of-plane angle ψ. Out of plane refers to the scattering plane defined by α_i and α_f. Typically, a beam stop is needed to block the high intense specular peak on the detector. This specular peak arises from the simple optical law of reflection at $\alpha_i = \alpha_f$. Thus, all the rest of the intensity on the two-dimensional detector is nonspecular or diffuse scattering. Most marked is the so-called Yoneda peak [89], the maximum of the Fresnel transmission function, which appears at the critical angle of exit $\alpha_{f,c}$.

Usually the coordinate system is chosen that way, defining the (x, y)-plane by the sample surface with the x-axis oriented in the direction of the X-ray or neutron beam (with wavelength λ). As a consequence, the z-axis is perpendicular to the sample surface. In this case

$$q_x = 2\pi(\cos\psi\cos\alpha_{\mathrm{f}} - \cos\alpha_{\mathrm{i}})/\lambda$$
$$q_y = 2\pi(\sin\psi\cos\alpha_{\mathrm{f}})/\lambda$$
$$q_z = 2\pi(\sin\alpha_{\mathrm{i}} + \sin\alpha_{\mathrm{f}})/\lambda$$

denote the components of the wave vector **q**. For specular scattering, the wave vector components are $q_x = q_y = 0$ and $q_z \neq 0$, sampling a depth sensitive information only. With off-specular scattering, the lateral component is $q_{||} = (q_x, q_y) \neq 0$ probing the in-plane structure of the sample surface [13].

Isotropic samples cause a two-dimensional scattering pattern that is symmetric with respect to the scattering plane. In general, the full two-dimensional intensity distribution can be modeled with the help of available software such as the program IsGISAXS by Lazzari [57]. As mentioned above, information from SPM measurements can be a helpful input for such simulation of the scattered intensity.

To allow for a direct comparison between SPM information and GISAXS measurement we restrict on a line cut from the two-dimensional GISAXS pattern. This line cut is taken parallel to the sample surface and denoted "out-of-plane cut" (or horizontal cut) as shown in Fig. 4.8. Typically, the Yoneda peak position is selected for such cut to emphasize on a material of choice.

In the framework of the distorted-wave Born approximation (DWBA), the differential cross-section can be approximated by [50]

$$\frac{\mathrm{d}\sigma}{\mathrm{d}\Omega} = \frac{C\pi^2}{\lambda^4}(1-n^2)^2\,|T_{\mathrm{i}}|^2\,|T_{\mathrm{f}}|^2\,F(\vec{q}) \propto F(\vec{q}),$$

where C denotes the illuminated surface area, λ the wavelength used, n the refractive index, $T_{\mathrm{i,f}}$ the Fresnel transmission functions, and $F(q)$ the diffuse scattering factor. Because the incident and exit angle are fixed, the Fresnel transmission functions act only as overall scaling factors and the diffuse scattering factor is directly probed. The intensity is dominated by the Fourier transform of the height–height correlation function of one "effective surface" [58], which gives the name "effective surface approximation" to this approach [14].

For N identical and centro-symmetrical objects with a random orientation, the diffuse scattering factor can be approximated [52]

$$F(\vec{q}) \propto NP(\vec{q})S(\vec{q})$$

to depend on the form factor $P(q)$ of the individual objects and to depend on the structure factor $S(q)$. If the objects are having a well-defined nearest neighbor distance, irrespective of the monodispersity of the individual objects, the structure factor $S(q)$ yields directly the most prominent in-plane length ξ via a simple Bragg law. The form factor $P(q)$ is only the Fourier transform of the shape of the particle. A mathematical description of the form factor depends on the type of object such as cylinder, sphere, or slab. Only in case of highly monodisperse objects, the form factor becomes visible in the GISAS data.

4.4 Comparison of Real and Reciprocal Space Data

The power of SPM in combination with a statistical analysis of lateral surface structures has been shown, for example, for a diblock copolymer nanostructure covered with oxide nanoparticles (see Figs. 4.5 and 4.6) [21, 30, 31]. However, although the examined nanostructures are limited in height, it was not possible to get information about the internal structure with SPM techniques. Also buried structures underneath the nanostructures are not accessible with SPM. Therefore, additional investigation of the diblock copolymer films containing different amounts of nanoparticles used GISAXS [21].

In Fig. 4.9, the two-dimensional GISAXS pattern corresponding to those films investigated with SPM are shown. An incident angle of $\alpha_i = 0.72°$ was chosen

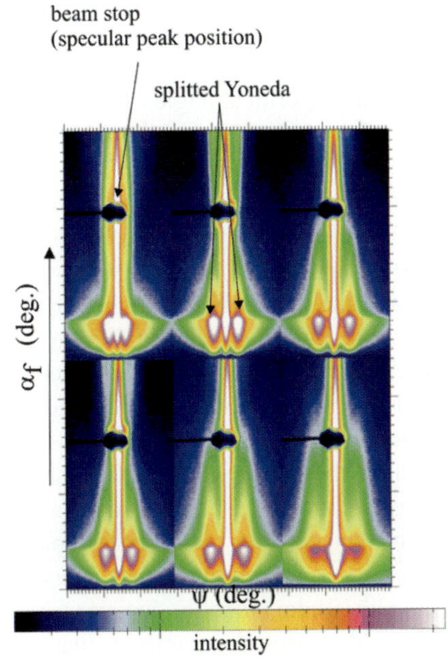

Fig. 4.9 Composite image showing two-dimensional GISAXS scattering patterns of diblock copolymer films with different nanoparticle concentrations. The *top three images* have 0, 1, and 5% (from *left to right*) and the *bottom three* have 10, 20, and 30% (from *left to right*) nanoparticle concentration. In each two-dimensional image, the out-of-plane angle ψ is plotted along the x-axis and the exit angle α_f along the y-axis. The intensity is shown on a logarithmic scale. The color coding was chosen to emphasize the features in the diffuse scattering (*black/blue*), low and *white* high intensities (see *scale bar*). The Yoneda and beam stop shielding the specular peak are marked by *arrows*. Reprinted with permission from [21]. Copyright 2007 American Chemical Society

to ensure a separation of the specular peak (shielded with a beam stop) from the Yoneda peak. The absence of an intensity modulation between the Yoneda peak and the specular peak results from the missing correlation in structures between the substrate surface and the nanostructures [39]. Thus, the nanostructures are individually shaped and positioned on top of the substrate as to be expected in the case of dewetting structures. The common feature in all two-dimensional GISAXS patterns is the splitting of the Yoneda peak in the out-of-plane direction ψ. This splitting is due to the presence of well-defined lateral structures. Only a very limited polydispersity in structural distances allows observing such an intensity distribution [39]. The presence of intensity in the center of the Yoneda peak is caused by the presence of additional large lateral structures which are not resolved in the GISAXS experiment with the used setup. Correspondingly, the intensity maxima due to the splitting of the Yoneda peak occur as side maxima to this central maximum in the intensity. With a change in the nanoparticle concentration, the position of the side maxima shifts along q_y. The change in the allover shape of the Yoneda side maxima as a function of the exit angle α_f reflects the changes in the height of the nanostructures.

However, a simple horizontal cut from an individual two-dimensional GISAXS pattern is sufficient to extract the relevant structural information about the lateral structures. The full treatment of the two-dimensional intensities is not necessary due to the simple scattering patterns. The selected out-of-plane (horizontal) cuts are made at an exit angle equal to the critical angle of the component of one block (in this case PS) present in the diblock copolymer P(S-*b*-I) to become more sensitive to this component.

Figure 4.10 is a plot containing all these cuts from the two-dimensional GISAXS patterns plotted as a function of the nanoparticle concentration. The intensities are plotted as a function of the component of the scattering vector q_y, which is oriented parallel to the sample surface and perpendicular to the X-ray beam. The strong peak in these cuts corresponds to the highly ordered surface structures (marked with I in Fig. 4.10). The position of the peak is determined by a model fit to the data taking into account the form and structure factors as well as the resolution function. The resolution function is defined by the largest detectable in-plane length scales of the sample and is determined by the sample-to-detector distance and the used X-ray optics. Form and structure factors are assumed to have a Lorentzian size distribution. The lateral length introduced by a microphase separation structure was not included into the fit because its contribution is small. Thus, at the expected position (abbreviated with III in Fig. 4.10) the model fit deviates from the data.

Figure 4.11a shows typical structural parameters that are determined from the analysis of the GISAXS data. A comparison between the results obtained from SPM after calculation of a master curve and an out-of-plane cut from the GISAXS measurement is shown for the sample without nanoparticles in Fig. 4.11b. The peak positions (denoted I) in both curves are the same and both peaks have a similar width. Consequently, concerning this large-scale structure, the local observation with SPM agrees nicely with the global information obtained with GISAXS or in other terms the SPM pictures are well representative for the sample surface. The

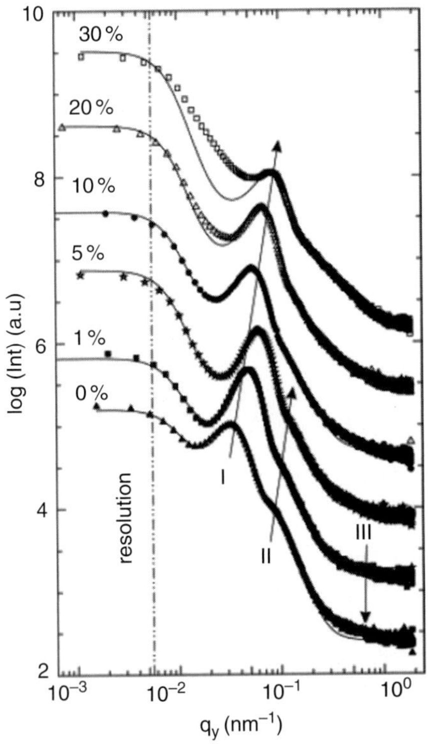

Fig. 4.10 Double-logarithmic plot of out-of-plane cuts (*symbols*) of the two-dimensional intensity as a function of the q_y component of the scattering vector. The *solid lines* are the fit lines for determining the most prominent in-plane length scale. The curves are shifted along the intensity axis for clarity. From the *bottom to the top* the nanoparticle concentration increases as indicated. The *dashed line* indicates the resolution limit of the experiment. The two dominant structures are marked with I and II and position of the expected microphase separation structure with III. The *arrows* display the shift with changing nanoparticle concentration. Reprinted with permission from [21]. Copyright 2007 American Chemical Society

peak at $q^* = 6.35 \times 10^{-2}\,\mathrm{nm}^{-1}$ (denoted II) present in the line cut of the GISAXS data is not present in SPM master curve because it represents the shape of the polymer nanostructures, and SPM delivers only a very limited statistics as compared to the scattering experiment. In addition, the structural length related to the microphase separation structure (denoted III) is only probed with GISAXS because the PSD master curves were constructed from topography data of SPM in noncontact mode. As a result, only the external height changes of the surface are probed in such SPM experiment. The internal phase changes cannot be detected in noncontact condition. Thus SPM gives pure topographic information, whereas GISAXS adds the inner film structure (caused by microphase separation).

In more simplified systems, such as homopolymer islands on solid supports no inner structure can occur due to the absence of a second species. Figure 4.5 shows

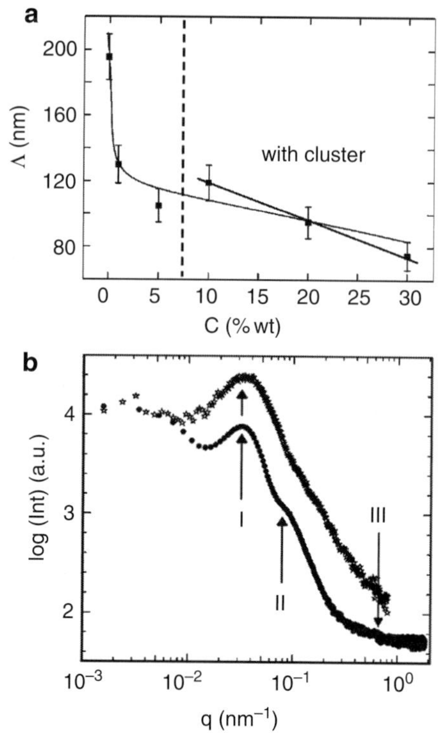

Fig. 4.11 (**a**) The most prominent in-plane lengths Λ of the structure related to peak I in Fig. 4.10 plotted as a function of the nanoparticle concentration C. The *solid lines* are guides to the eye. The *dashed line* indicates the critical concentration for cluster formation. (**b**) Example of a comparison between the out-of-plane cut (*bottom curve*) and the PSD master curve (*top curve*) in the case of the sample containing no nanoparticles. The curves are shifted along y-axis for clarity. The *arrows* show the most prominent in-plane lengths corresponding to structures as explained in the text. Reprinted with permission from [21]. Copyright 2007 American Chemical Society

the comparison of PSD master curves calculated from SPM data with horizontal line cuts from two-dimensional GISAXS patterns for a dewetting PS film after different annealing times. In principle, nice agreement could be expected if the SPM images are fully representative. Therefore, all deviations between GISAXS cut and PSD master curve result in SPM having a too local view only. So it is easy to understand that in the PSD master curves, the peak corresponding to the nearest neighbor droplet distance is better defined. Imperfections on larger scale are not seen in the SPM measurements due to the limited maximal scan range. In contrast, GISAXS has a higher lateral resolution toward large lateral structures. Imperfections cause a smearing of the peak and the peak intensity drops.

4.5 Complementary and In Situ Experiments

Already the example presented above, about diblock copolymer nanostructures with oxide nanoparticles on top, show somewhat complementarities. To make it more clear, in a next example the morphology of dewetted thin polymer-blend films of deuterated polystyrene (dPS) and polyparamethylstyrene (PpMS) on top of silicon surfaces is investigated [59]. The dewetting results from the storage of the samples under toluene vapor atmosphere. After storing the samples for a well-defined time under toluene vapor, the originally homogeneous blend films have dewetted and the substrate surface is covered with individual polymer droplets as shown in Fig. 4.12.

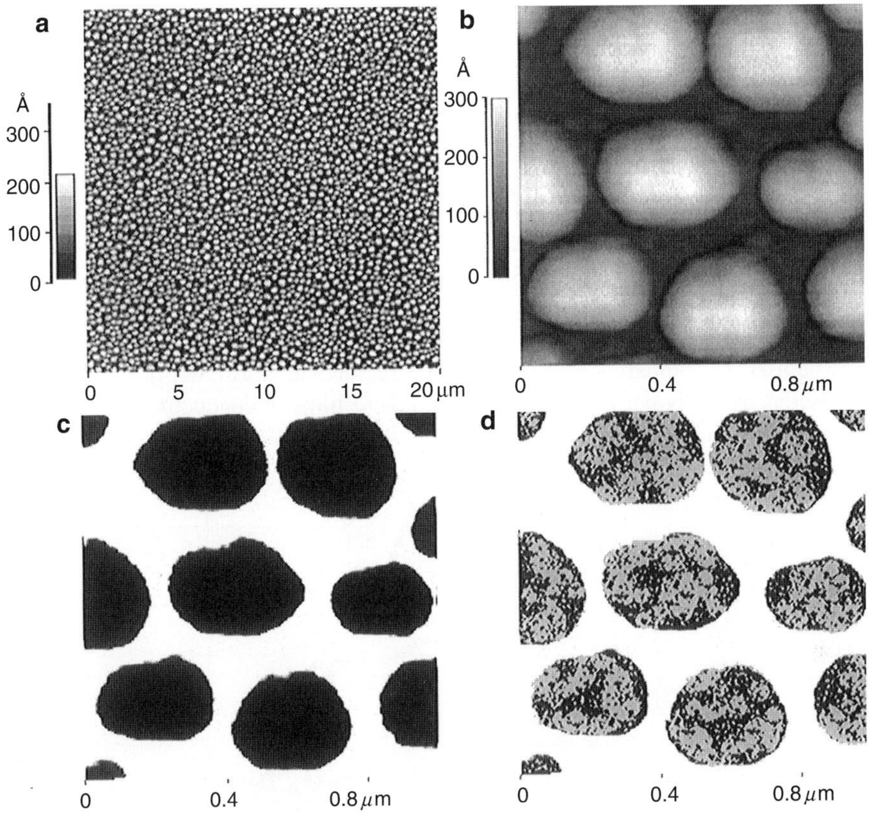

Fig. 4.12 (**a**) SPM picture of the topology of the dewetted confined dPS:PpMS blend film with 31 Å film thickness of the original film. In the shown late stage the topography resulting from a toluene vapor treatment consists of isolated droplets with a regular arrangement. (**b**) Locally zoomed in SPM picture showing isolated droplets. (**c**) Model of the scattering length densities as seen by X-rays. Contrast is only yielded between polymer (*black*) and vacuum (*white*). (**d**) Model of the scattering length densities of dPS (*black*) and PpMS (*gray*), as seen by neutrons. Reprinted with permission from [59]

Figure 4.12a, b shows SPM data on different scan ranges. The observable droplets have a spherical cap shape. With an enhanced resolution it is visible that the droplets are well separated. Note that the elliptical shape as well as the hexagonal arrangement of the droplets is not representative. It is only present in some small-scan ranges as the shown one. This pictures the risk associated with local techniques such as SPM in comparison with averaging techniques like GISAS.

In Fig. 4.13, the PSD master curve from the SPM data is shown in comparison to the out-of-plane scans measured with X-rays (GISAXS – plotted with triangles) and measured with neutrons (GISANS – plotted with crosses). Besides the well-pronounced peak at the position marked with "A," a shoulder at a larger q_y value (marked "B") is present in the SPM data. It can be attributed to the droplet diameter.

The horizontal line cuts from GISAXS and GISANS look different, because in X-ray scattering the blend can be regarded as a homopolymer system without internal contrast while with neutron scattering the main scattering results from the deuterated component dPS only (and there is strong contrast between deuterated and protonated polymer). Therefore, the length scale related to the droplet diameter is more visible in the GISAXS data than in the SPM data (see Fig. 4.13). In the GISANS data, it is only rarely visible due to the changed contrast. Thus from a comparison of both, internal information about the blend composition structure is available. One model, to explain this experimental result, is shown in Fig. 4.12d. The two components of the blend, dPS and PpMS, demixed into a random arrangement inside each droplet.

Of course, the complementary approach is not limited to dewetting structures. It can be used to compare surface and inner structures as illustrated in the example of photoactive polymer blend films [60]. In these films of conducting, photoactive

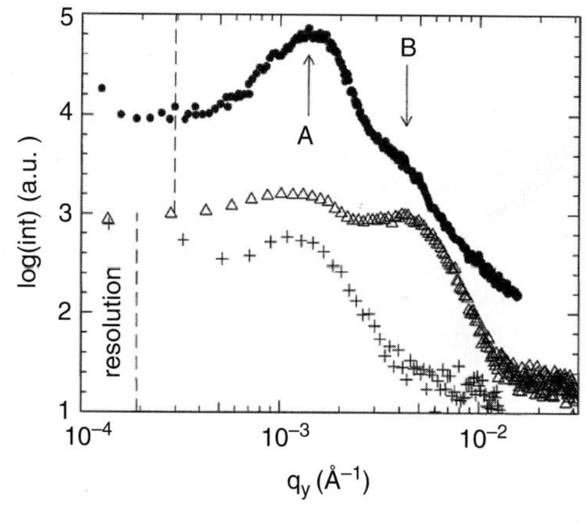

Fig. 4.13 Double logarithmic plot of horizontal slices from the two-dimensional intensity distribution of the dewetted dPS:PpMS sample (film thickness $l = 31$ Å) measured with SPM (*filled circles*), X-rays (*triangles*), and neutrons (*crosses*) (from *top to bottom*). The *dashed line* indicates the resolution limit. The position of characteristic peaks is marked with "A" and "B." The data are shifted for clarity. Reprinted with permission from [59]

polymers (poly[(1-methoxy)-4-(2-ethylhexyloxy)-pphenylene-vinylene] denoted MEH-PPV and poly(3-hexylthiophene-2,5-diyl) denoted P3HT the buried morphology is probed with GISAXS and the surface structure with SPM. In SPM in the topography image, the two polymers cannot be distinguished but in the phase image (compare Fig. 4.14a, b) it is possible to discriminate between hard and soft components on the surface. Hard materials are shown as bright and soft materials as dark areas in the phase image. Thus, the glassy MEH-PPV refers to the bright parts and the soft amorphous P3HT to the dark parts of the phase image. With area analysis of the phase image the ratio of MEH-PPV and P3HT at the surface can be obtained. The PSD function of the investigated film is shown in Fig. 4.14c and compared with the GI(U)SAXS data. While the PSD gives only information about the film surface, the out-of-plane curve represents the lateral structure of the whole film.

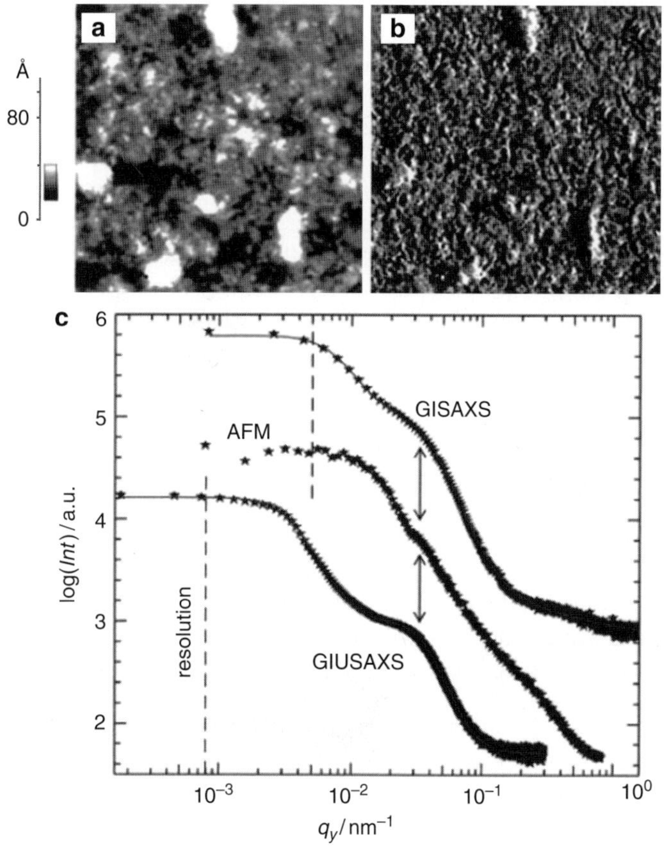

Fig. 4.14 SPM images (1 μm × 1 μm) showing (**a**) the surface topography of a blend film with 10 wt% MEH-PPV and (**b**) the related phase image. In the topography image, the gray scale covers a range from 0 to 8 nm. In the phase image, *dark areas* refer to the softer material (P3HT) and *bright areas* to the harder one (MEH-PPV). (**c**) Comparison of SPM and scattering data in reciprocal space. Reprinted with permission from [60]

The PSD curve shows a sign of a structure peak at the same position as the scattering data. Thus, the same structure size is determined at the film surface and inside the film. From the fitting of the GI(U)SAXS curves the structure size is 320 nm. This size belongs to the distance between the cavities (dark areas) in the topography image obtained by line cuts. Thus, the surface structure extends into the film volume with the same characteristic length scale.

In addition to such complex polymer structures, many other types of morphologies have been probed by combining GISAXS and SPM. Hybrid titanium dioxide (TiO_2) nanostructures with engineered morphologies were produced by Cha and co-workers by a simple synthetic approach based on cooperative sol-gel chemistry and self-assembly of amphiphilic poly(styrene-block-ethylene oxide) (PS-*b*-PEO) block copolymer (BCP) via spin-coating common solutions of BCP and inorganic precursors [61]. TiO_2 nanostructures with two extreme reverse morphologies, i.e., TiO_2 dot-in-PS matrix and PS dot-in-TiO_2 matrix, were obtained and studied with GISAXS.

Chushkin and co-workers reported on the formation of ordered monolayers (two dimensional) and arrays of rods (three dimensional) of magnetic Co nanoparticles in magnetic fields perpendicular to the substrate surface [62]. Rods, about 500 nm in diameter, aligned with the field direction and forming a hexagonal pattern were obtained when higher concentration of colloid and low evaporation rate of the solvent were used. The ordering of nanoparticles in the monolayer was analyzed with GISAXS and described by the local order with hexagonal symmetry. The model of close packing of hard spheres was used for ordering of particles inside the rods. Magnetic features corresponding to the three-dimensional arrays were observed by magnetic force microscopy pointing out that all magnetic moments in the rod are oriented along the field direction.

Further examples on nonpolymeric systems can be found, for example, in the investigation of self-assembled carbon-induced germanium quantum dots studied by GISAXS [37] and the investigation on ordering of self-assembled $Si_{1-x}Ge_x$ islands studied by GISAXS and SPM [36]. Typically, as mentioned before one of the main advantages of the GISAXS technique is that results are obtained with high sampling statistics due to the large illuminated sample area. Where SPM measurements give a real-space image in a reasonable amount of time, it is very beneficial to get information of size distributions with GISAXS.

Because scanning is inherent to SPM techniques, typically high time resolution is difficult to access. In the past, limited available X-ray and neutron flux as well hindered kinetic studies with high time resolution in GISAXS and GISANS. However, with the emerging new instruments dedicated to GISAXS such problems are overcome. Thus in the following, we want to focus on kinetic experiments.

The structural evolution of a single-layer latex film during annealing was studied via GI(U)SAXS and SPM by Hu and co-workers [63]. The GIUSAXS data on latex films annealed at various temperatures ranging from room temperature to 140°C indicated that the structure of the latex thin film beneath the surface changed significantly. The evolution of the out-of-plane scan revealed the surface reconstruction of the latex film. Furthermore, the time-dependent behavior of the structural evolution was followed if the latex film was annealed at a relatively low temperature (60°C). For such kinetic process the restructuring within the film could not be detected with SPM techniques, which probe only surface morphology. The kinetic GIUSAXS complements the information from the SPM measurements.

The kinetic evolution of nanostructures introduced by dewetting can be probed by GISAXS as well. The example of nanostructured polymer films of poly(styrene-block-paramethylstyrene) diblock copolymers, denoted P(Sd-b-pMS), on silicon substrates with a native oxide layer was investigated by Müller-Buschbaum and co-workers [64]. Because the host structure resulting from the destabilization of an initially continuous P(Sd-b-pMS) layer is expected to exhibit characteristic lateral lengths above the resolution limit of a conventional small-angle scattering distance, the GISAXS experiments were performed with a very large sample detector distance.

Figure 4.15 shows six typical two-dimensional GISAXS scattering patterns. Because an incident angle $\alpha_i = 0.5348$ larger than the critical angle of the polymeric material was chosen, the specular and the diffuse scattering contributions are well separated.

With increasing storage time under toluene vapor, the film destabilizes and the intensity of the Yoneda peak increases. In addition, the marked shape of the Yoneda peak changes during ongoing storage. The originally installed continuous P(Sd-bpMS) film of 1.5-nm thickness shows only a slightly increased intensity

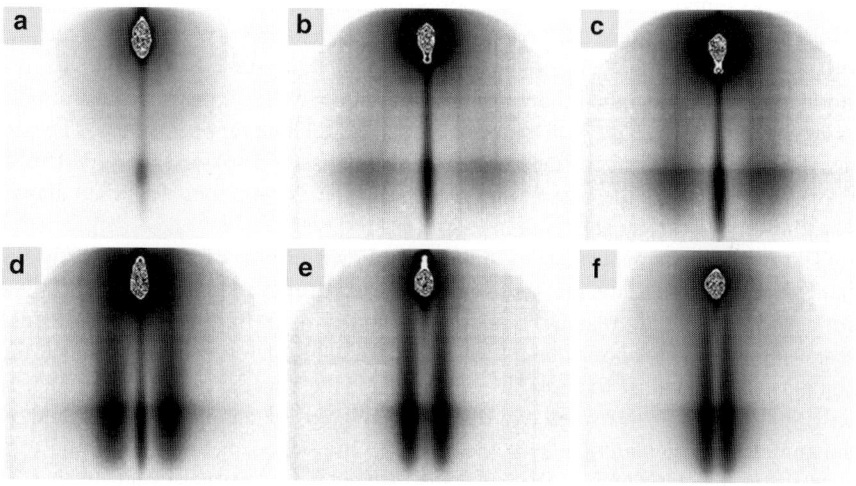

Fig. 4.15 Typical two-dimensional GISAXS scattering patterns measured at a 12.1-m distance between the sample and the detector: (**a**) bare silicon substrate, (**b**) after 0.5 h, (**c**) 1 h, (**d**) 2 h, (**e**) 3 h, and (**f**) 4 h of storage under toluene vapor atmosphere. Each two-dimensional intensity mapping covers a range of $-0.48° < \psi < 0.48°$ in the horizontal direction, and $0.0° < \alpha_f < 0.55°$ in the vertical direction. The intensity is shown on a logarithmic scale. The color coding was chosen to emphasize the features in the diffuse scattering (*white* = low and *black* = high intensity). As a consequence, the specular peak in the top region of each two-dimensional intensity mapping appears with a staggered intensity. Reprinted with permission from [64]

of the diffuse scattering as compared to the bare substrate. During the toluene vapor exposure, the shape of the Yoneda peak is altered significantly. As visible in Fig. 4.15b, the initially central Yoneda peak exhibits two side maxima after 0.5 h of storage. The presence of side maxima is a sign of dominant lateral lengths being present at the surface within the resolution limit of the GISAXS setup [39]. After twice the storage time (see Fig. 4.15c), the position of the side maxima slightly shifts as well as they get more pronounced in the two-dimensional intensity mapping. The shift of the side maxima position towards the plane of reflection (defining $q_y = 0$) can be immediately understood as an increase in the dominant lateral length. After again a doubled storage time, these processes proceed (see Fig. 4.15d), whereas after 3 h of total storage time, the Yoneda intensity distribution again changes markedly (see Fig. 4.15e). The central peak vanishes, and only both side maxima remain. This pattern of a split-up Yoneda peak remains during further storage, as pictured in Fig. 4.15f. The absence of a central Yoneda peak in the two-dimensional intensity mapping is related to the absence of large lateral lengths outside the resolvable length scale regime. It is typically observed in case of supported dominant isolated objects, such as cylinders, pancakes, or spherical caps [36, 57, 59]. It is a fingerprint of a (with respect to polymeric systems) rather high degree of lateral order.

In principle, all the two-dimensional intensity mappings are symmetric to the central vertical axis due to the symmetry imposed by the out-of-plane angle ψ, which covers a range of $\pm 0.48°$.

Because the polymeric material probed within this investigation is placed on top of the solid support, the model of supported particles is well suited. In first approximation, the form factor of the host structure is chosen to obey a cylinder type and the structure factor to behave like a one-dimensional paracrystal. To elucidate the possibilities and limits of the modeling based on IsGISAXS, we restrict to this model assumption. Two examples are focused: 0.5 and 4 h toluene vapor exposure. The corresponding full two-dimensional intensity mappings are displayed in Fig. 4.15b, f. Figure 4.16 displays the best fits obtained with the assumptions described above. The chosen parameters are given in reference [64].

The two-dimensional intensity mappings shown in Fig. 4.16 cover exactly the same angular range as the measured GISAXS data displayed in Fig. 4.15. The color coding was adapted to match the one chosen in Fig. 4.15. Because only the diffuse scattering is modeled, no specular peak is present in Fig. 4.16. As a consequence, the comparison can mainly focus to the region around the Yoneda peaks. On a first view, basic features of the measured two-dimensional intensity mappings are captured by the modeling. In general, the agreement for the late stage is better than for the very early stage. The modeling of the late stage (4-h storage) covers the split-up Yoneda peaks in their positions and shapes. Most deviations are originated from a missing background in the modeled data and affect the regime of large ψ (or q_y) values. The modeling of the early stage (0.5-h storage) agrees in the main Yoneda shape as well, the central peak with two side maxima is reproduced including the position of the side maxima. However, deviations concerning the shape and the intensity are visible.

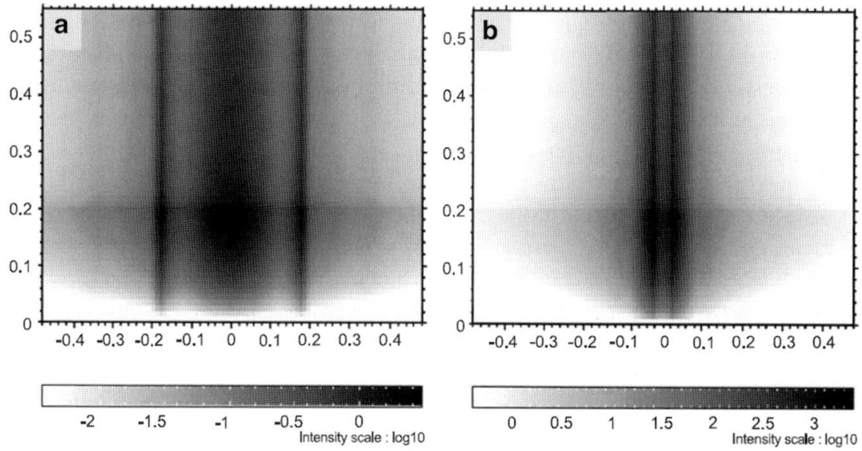

Fig. 4.16 Calculated two-dimensional GISAXS scattering patterns using the program IsGISAXS as described in the text: (**a**) 0.5 and (**b**) 4-h annealing. Each two-dimensional intensity mapping covers a range of $-0.48° < \psi < 0.48°$ in the horizontal direction and $0.0° < \alpha_f < 0.55°$ in the vertical direction. The intensity is shown on a logarithmic scale. The color coding was chosen to be comparable to the measured GISAXS data (*white* = low and *black* = high intensity). Reprinted with permission from [64]

From the good agreement of the master curve extracted from the local SPM data and the statistically significant GISAXS data, we can gain confidence in the SPM data being again statistically representative as well. As a consequence, we can take our SPM data and compare them to the models used in the simulation (see Fig. 4.16).

It became obvious that the assumption of cylindrical-shaped particles, used in the modeling of the GISAXS data, is not justifiable for the early stage. The host structure is ribbon-like with a high degree of interconnection. The surface coverage is still high. During further storage, the host structure coarsens as observed by scattering, and after 4 h, the interconnection gets lost and isolated islands result [88]. Although the island size is rather irregular, the assumption of cylindrical particles fits much better. The characteristic length, as determined with the SPM, matches well with the values used in the fit described above. Thus, the chosen examples picture nicely the limits and possibilities of the modeling for GISAXS data with the available software. In case of the early stage, irrespective of the tried pair correlation function and of the time spent for the modeling, a good agreement between GISAXS data and modeled intensity distribution can never be expected. Although some general features are well described by the modeling, the extracted parameters are wrong due to the chosen type of model. In contrast, in case of the late stage, the used model is well appropriate, and from the modeling, relevant parameters such as object shape, distance, and size as well as the corresponding distributions are extracted.

Thus obviously, modeling of GISAXS (and GISANS) data can be complicated and limitations in the existing software can prevent full data analysis. Here SPM techniques give an important contribution with the real-space picturing of surface structures.

An impressive example how the simulation of two-dimensional GISAXS patterns gives an insight in kinetic structure evolution is given by Kaune and co-workers [65]. They investigated the growth of a thin gold film on a conducting polymer surface in situ with GISAXS with a sputter deposition system installed in the synchrotron instrument. The mobile sputter system was operated in UHV condition [66]. It was possible to follow the growth of gold from nucleation up to a continuous layers of several nm thickness onto a thin film of poly(N-vinylcarbazole) (PVK). Time resolution was achieved by performing the experiment in cycles of gold deposition and subsequently recording the GISAXS data. Figure 4.17 shows four selected GISAXS scattering patterns, taken at gold films with 3.4, 6.5, 9.9, and 16.7 nm thicknesses. The scattering patterns show several prominent features evolving with the proceeding gold deposition. The most prominent feature in the GISAXS patterns is a side peak emerging at larger scattering angles and, with proceeding growth of the gold clusters, shifting towards smaller values. The peak is related to a maximum in the interference function describing the cluster correlation distance. The shifting therefore represents a change in the relative cluster position, i.e., an increasing center-to-center distance. Further well-pronounced features in the GISAXS patterns are the higher order scattering maxima evolving in the vertical direction, indicating a nearly uniform height of the clusters with only small variances. From the scattering maxima in the horizontal direction arise weak streaks bending to the plane of reflection and connecting the lateral maxima with vertical maxima. These streaks originate from a round shape of the clusters, as it is also expected from the nonwetting behavior of gold on polymer surfaces [67].

Again, data analysis was based on the software IsGISAXS dedicated for the simulation of GISAXS scattering patterns [57]. By use of an appropriate model containing all relevant information about the geometry, spatial dimensions, and distribution of an arrangement of nanoparticles, either line cuts or whole two-dimensional scattering patterns can be calculated. For form factor calculation, the DWBA was used, with the assumption of graded interfaces in the perturbated state induced by the particles. Each particle type was described by parameter size, height, orientation, and their respective distributions. For all distributions, Gaussian shapes of the distribution functions were assumed. The interference function was calculated based on the one-dimensional paracrystal model, which is a regular one-dimensional lattice with loss of long-range order induced by a special cumulative disorder. The interference function was described by two parameters, the mean distance of the maxima D and the disorder parameter ω. Again, a Gaussian distribution was assumed. More details are described in [65].

Figure 4.17 shows the two-dimensional simulated scattering patterns together with the corresponding measured data. For modeling, different particle geometries such as spheroids, truncated spheres, cylinders, and parallelepipeds were tried. The best agreement was obtained by use of the parallelepiped geometry, in particular in the description of the intensity distribution of the side peaks. Because the interconnecting streaks in the measured data suggest a round shape of the clusters, a second particle geometry, spheroids, was introduced in the model. The particles of both geometries are arranged statistically with an equal probability of occurrence of 0.5

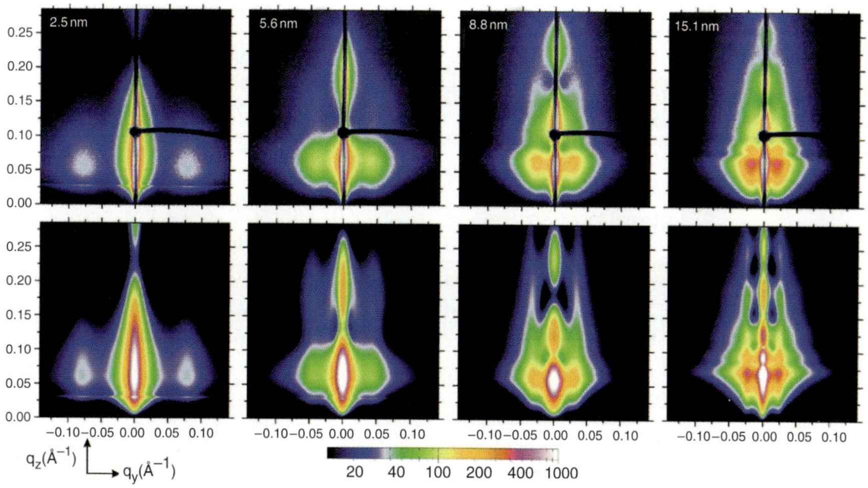

Fig. 4.17 Composite image of four representative measured GISAXS scattering patterns (*upper row*) and corresponding simulations (*lower row*). The images were taken after deposition of gold layers with thicknesses of 2.5, 5.6, 8.8, and 15.1 nm. The *elongated dark line* in the middle of the measured images is the rod-like beam stop and the *single spot* the point-like beam stop to shield the specular beam. The evolving maxima in vertical direction originate from the growth in height, the lateral maximum is related to the center-to-center distance of the clusters. Reprinted with permission from [65]. Copyright 2009 American Chemical Society

for each site. In this case, the form factor is an incoherent sum of the respective particle contributions, and the overall intensity is calculated by use of one interference function describing the assembly of the particles independently of their geometry. Only one average radius value was used in the calculations and assigned to both the parallelepiped and spheroid particles; to account for shape variations, only the radius distribution width was used for fitting. The height of the spheroid particles turned out to be slightly higher than the height of the parallelepiped particles, yielding almost the same volume for the particles of both geometries. With this two-particle-type model, a distinctly better quality of the simulations was obtained. In particular, the arch-like streaks connecting the scattering maxima in horizontal and vertical directions are extremely well described.

The modeling procedure described above was applied to all two-dimensional GISAXS scattering patterns to follow the temporal evolution of the cluster size and spatial distribution, which allowed for the determination of a growth model. This growth model was described by four distinct growth regimes as shown in Fig. 4.18. Three of these four regimes were directly identified by different scaling laws of the geometrical parameters extracted from the fit to the GISAXS data.

The growth process begins with a stage of nucleation, where impinging gold atoms diffuse across the polymer surface and form nuclei when two of them meet. In the second stage, these nuclei grow laterally by capturing further diffusing atoms. As a result, small spherical gold clusters form. At a point where two clusters come

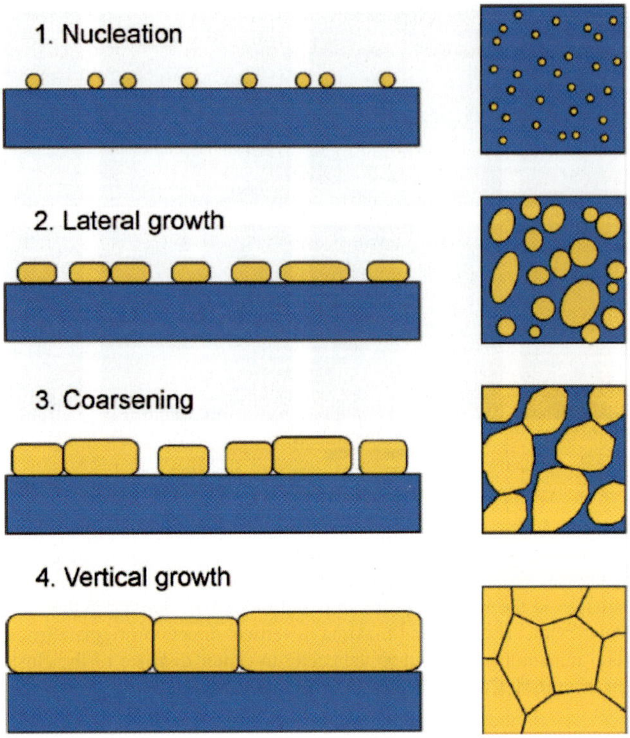

Fig. 4.18 Schematic drawing of gold cluster growth. After the stage of nucleation, cluster growth proceeds mainly in the lateral direction. Coarsening sets in and becomes the dominating process when the surface coverage increases and the clusters get in close contact with each other. With the surface fully covered, adsorption produces only vertical growth and lateral growth is dominated by cluster boundary motion. Reprinted with permission from [65]. Copyright 2009 American Chemical Society

in close contact to each other coalescence occurs and a larger cluster is formed. At a layer thickness of 1.3 nm enhanced coalescence sets in, cluster radius as well cluster distance increase very fast and the clusters start to assume an irregular shape (coarsening stage). In the final stage, the clusters form a continuous layer, which grows vertically in thickness as long as deposition goes on. Simultaneously coarsening continues and a grain structure, as it is well known for vapor-deposited thin films, develops. In addition to the formation of the metal layer on the conducting polymer, metal atoms diffuse into the PVK. An amount of 6.5 wt% gold is incorporated up to a depth of 1.2 nm in the PVK film. Eventually, the gold contact has evolved from isolated islands to a closed layer, and the observation gave us more insight on how a metal film forms on a polymer surface and which processes determine the final film structure.

For comparison, the 25-nm-thick gold layer after complete deposition and a PVK surface without gold deposition were measured with SPM. The image of the gold film shows a rough surface consisting of gold clusters of a round, but not necessarily circular, shape. The clusters are homogeneous in size with a diameter of about 50 nm. Only a few clusters are considerably larger. The film structure is dense, and no interspaces between adjacent clusters with a depth on the order of the cluster height are visible, i.e., the PVK surface is fully covered. However, a direct comparison between SPM and GISAXS is not easy because, on the one hand, SPM was not measured in situ and, on the other hand, with SPM the very final deposition state was probed. As compared to the final state probed with GISAXS, this very final state has a 10-nm-thicker gold layer deposited.

4.6 Combined In Situ GISAXS and SPM Measurements

As the beneficial use of SPM techniques and GISAS as complementary tools has been shown the installation of an SPM instrument into a GISAXS or GISANS instrument at a synchrotron or neutron scattering facility is perhaps the remaining obvious next step. For GISAS experiments, such combination would allow for the surface characterization of the sample at the illuminated sample position. Although many different and complex setups have been implemented into such scattering instruments at large-scale facilities, the integration of a SPM instrument has special problems. One of the difficulties arises from mechanical vibrations which spoil SPM measurements on a level that is tolerated by GISAS measurements [68]. As a consequence, such in situ combinations of SPM and scattering experiments are rare.

The integration of an SPM into a synchrotron radiation beam line is described by Rodrigues et al. [68]. In their approach, a quarz tuning fork is used. According to the authors, the advantages for the use of a quarz tuning fork versus a conventional SPM is that with the natural piezoelectricity used a direct detection of the movement of the oscillator is possible. Therefore, no laser optics is involved and the system is described as robust and easy to install at a beam line. One application, combining SPM and micro-X-ray diffraction was realized at the synchrotron source ESRF by Scheler and co-workers [69]. An SPM instrument with a tuning fork was mounted on the diffractometer of the instrument, thereby allowing for the in situ combination of SPM techniques (both, imaging and indentation) and diffraction. In addition, the use of a micro-focused X-ray beam [70, 71] was necessary to achieve a matching of the illuminated footprint of the X-ray beam on the sample surface and the scanning range of the SPM instrument. From this combination, the investigation of the elastic response of nano-sized objects under external mechanical stress has been made possible. Examples, probed in the unique combination of SPM and diffraction, were so-called semiconductor rolled-up nanotubes and SiGe islands [69, 72]. By these pioneering experiments the door to studies of the influence of nanoscale dimensions on the elastic properties of materials was opened. However, GISAXS experiments were not performed to the knowledge of the authors.

At present, at the new MINAXS instrument at the synchrotron source PETRAIII the authors are setting up a combined instrument which will allow for the investigation of samples by a combination of GISAXS, imaging ellipsometry, and SPM [73].

Alternatively to scattering experiments, there exist other combinations of SPM with synchrotron radiation: For example, the illumination of the tip of a scanning tunneling microscope (STM) with synchrotron radiation may serve as a local probe for photoemission [74–76]. This has been realized in UHV setups and STM images were obtained up to atomic resolution for Si(111) surfaces [77, 78]. Element specific imaging [79] was achieved for Ni dots on a Au-thin film with a resolution of about 20 nm [80] and Ni and Fe checkerboard-patterned samples [81]. In another approach, a scanning capacitance microscope was used to investigate the X-ray absorption fine structure [82].

4.7 Summary and Outlook

The SPM and GISAS have both developed into mighty experimental tools to probe nanoscale structures [83]. In most cases, SPM and GISAS are used solely and not in combination, which might be due to the fact that SPM techniques work in real space and GISAS experiments give results in reciprocal space. Moreover, the techniques are extremely different and thus require very different expertise to make high quality experiments. Due to the possibility to gain complementary information from the use of both techniques, SPM and GISAS, however, the number of experiments that are based on both is increasing over the last years [84–86]. One way to get a direct comparison of both experimental datasets is the transformation of the SPM data into reciprocal space via Fourier transformation. Merged into a so-called master curve the resulting PSD functions can be compared directly with horizontal (or out of plane) line cuts from GISAS data. Alternatively, the SPM structure can be taken as a model for the simulation and fitting of the GISAS data.

On the first route (statistical analysis of SPM data via master curve construction) discrepancies between PSD master curve and GISAS data can be caused by buried structures that contribute to the GISAS signal but not to the SPM data. Thus, complementary information is gained from such comparison and surface and inner structures can be determined together from these techniques. In addition, discrepancies can arise from a missing statistical significance of the SPM data. Either large-scale structures or sample inhomogeneities can be the reason for such effects. Altogether, however, the understanding of the nanoscale structure of the sample is achieved, which makes this route very useful.

On the second route (SPM input into GISAS data modeling) software limitations might be disadvantageous today. For some classes of samples, such as objects on simple planar surfaces without inner structure, the approach is well working. The presence of objects on different hierarchical levels or an inner structure will contribute to complications. As a consequence, SPM information is only rarely used to create models for GISAS data analysis. A fact which, however, might change with improved software capacities in future.

Whereas the advantages of SPM techniques might be more familiar to the reader, we like to summarize the special advantages of GISAS experiments irrespective of based on synchrotron radiation (GISAXS) or neutrons (GISANS):

- Gaining an averaged statistical significant information over all the illuminated sample area (which is for common beam sizes in the range of square centimeters) [53].
- Accessing buried structures, which are located well below the surface and thus are inaccessible to local probe techniques such as SPM [39].
- Measuring depth-dependent structural information using different incident angles [87].
- Probing structures in any kind of complex sample environments ranging from ultra-high vacuum to corrosive gas atmospheres and liquids [15].
- Performing kinetic studies as a function of changes in the external parameters such as temperature, gas pressure, pH value, or ion concentration in situ with a high time resolution [65].

Acknowledgements

The authors thank the Bundesministerium für Bildung und Forschung for financial support (grant 05KS7WO1).

References

1. G. Binnig, C.F. Quate, C. Gerber, Atomic force microscope. Phys. Rev. Lett. **56**, 930–933 (1986)
2. D. Bonnel (ed.), *Scanning Probe Microscopy and Spectroscopy: Theory, Techniques, and Applications*, 2nd edn. (Wiley-VCH, New York, 2001)
3. R. Wiesendanger, *Scanning Probe Microscopy and Spectroscopy: Methods and Applications* (Cambridge University Press, Cambridge, 1998)
4. B. Antic, A. Kremenovic, I. Draganic, P. Colomban, D. Vasiljevic-Racovic, J. Blanusa, M. Tadic, M. Mitric, Effects of O^{2+} ions beam irradiation on crystal structure of rare earth sesquioxides. Appl. Surf. Sci. **255**, 7601–7604 (2009)
5. X. Cui, J.F. Hetke, J.A. Wiler, D.J. Anderson, D.C. Martin, Electrochemical deposition and characterization of conducting polymer polypyrrole/PSS on multichannel neural probes. Sens. Actuat. A **93**, 8–18 (2001)
6. P. Viville, F. Biscarini, J.L. Brédas, R. Lazzaroni, Scaling aspects of the kinetics of thermally induced phase separation in bisphenol-A-polycarbonate/poly(methyl methacrylate) blends. J. Phys. Chem. B **105**, 7499–7507 (2001)
7. H.C. Yang, T.J. Shin, L. Yang, K. Cho, C.Y. Ryo, Z.N. Bao, Effect of mesoscale crystalline structure on the field-effect mobility of regioregular poly(3-hexyl thiophene) in thin-film transistors. Adv. Funct. Mater. **15**, 671–676 (2005)
8. D. Żymierska, J. Auleytner, T. Kobiela, R. Duś, Comparative study of the surface roughness by AFM and GIXR. Phys. Stat. Sol. A **180**, 479–485 (2000)

9. T. Jiang, N. Hall, A. Ho, S. Morin, Quantitative analysis of electrodeposited tin film morphologies by atomic force microscopy. Thin Solid Films **471**, 76–85 (2005)
10. M. Maiti, A.K. Bhowmick, New insights into rubber-clay nanocomposites by AFM imaging. Polymer **47**, 6156–6166 (2006)
11. A. Maksumov, R. Vidu, A. Palazoglu, P. Stroeve, Enhanced feature analysis using wavelets for scanning probe microscopy images of surfaces. J. Colloid Interface Sci. **272**, 365–377 (2004)
12. J. Villarrubia, Morphological estimation of tip geometry for scanned probe microscope. Surf. Sci. **321**, 287–300 (1994)
13. P. Müller-Buschbaum, Structure determination in the thin film geometry using grazing incidence small angle scattering. in *Polymer Surfaces and Interfaces: Characterization, Modification and Applications*, ed. by M. Stamm (Springer, Berlin, 2008), pp. 17–46
14. P. Müller-Buschbaum, A basic introduction to grazing incidence small angle X-ray scattering. In *Special issue of Lecture Notes in Physics on "Applications of Synchrotron Light to Noncrystalline Diffraction in Materials and Life Sciences,"* ed. by T.A. Ezquerra, M. Garcia-Gutierrez, A. Nogales, M. Gomez. vol 776 (Springer, Berlin, 2009), pp. 61–90
15. E. Metwalli, J.-F. Moulin, R. Gebhardt, R. Cubitt, A. Tolkach, U. Kulozik, P. Müller-Buschbaum, Hydration behavior of casein micelles in thin film geometry: a GISANS study. Langmuir **25**, 4124–4131 (2009)
16. W. Wang, E. Metwalli, J. Perlich, C.M. Papadakis, R. Cubitt, P. Müller-Buschbaum, Cyclic switching of water storage in thin block copolymer films containing poly(N-isopropylacrylamide). Macromolecules **42**, 9041–9051 (2009)
17. W. Wang, G. Kaune, J. Perlich, C.M. Papadakis, A.M. Bivigou Koumba, A. Laschewsky, K. Schlage, R. Röhlsberger, S.V. Roth, R. Cubitt, P. Müller-Buschbaum, Swelling and switching kinetics of gold coated end-capped poly(N-isopropylacrylamide) thin films. Macromolecules **43**, 2444–2452 (2010)
18. S.J. Fang, S. Haplepete, W. Chen, C.R. Helms, H. Edwards, Analyzing atomic force microscopy images using spectral methods. J. Appl. Phys. **82**, 5891–5898 (1997)
19. J.S. Bendat, A.G. Piersol, *Random Data: Analysis and Measurement Procedures*. (Wiley-Interscience, New York, 1971)
20. M. Senthilkumar, N.K. Sahoo, S. Thakur, R.B. Tokas, Characterization of microroughness parameters in gadolinium oxide thin films: a study based on extended power spectral density analyses. Appl. Surf. Sci. **252**, 1608–1619 (2005)
21. M.M. Abul Kashem, J. Perlich, L. Schulz, S.V. Roth, W. Petry, P. Müller-Buschbaum, Maghemite nanoparticles on supported diblock copolymer nanostructures. Macromolecules **40**, 5075–5083 (2007)
22. Ph. Dumas, B. Bouffakhreddine, C. Amra, O. Vatel, E. Andre, R. Galindo, F. Salvan, Quantitative microroughness analysis down to the nanometer scale. Europhys. Lett. **22**, 717–722 (1993)
23. J.S. Gutmann, P. Müller-Buschbaum, M. Stamm, Complex pattern formation by phase separation of polymer blends in thin films. Faraday Discuss. **112**, 285–297 (1999)
24. P.F. Green, Wetting and dynamics of structured liquid films. J. Polym. Sci. B **41**, 2219–2235 (2003)
25. P. Müller-Buschbaum, J.S. Gutmann, R. Cubitt, M. Stamm, Probing the in-plane composition of thin polymer films with grazing-incidence small-angle neutron scattering and atomic force microscopy. Colloid Polym. Sci. **277**, 1193–1199 (1999)
26. P. Müller-Buschbaum, J.S. Gutmann, M. Stamm, Dewetting of confined polymer films: an x-ray and neutron scattering study. Phys. Chem. Chem. Phys. **1**, 3857–3863 (1999)
27. P. Müller-Buschbaum, N. Hermsdorf, J.S. Gutmann, M. Stamm, S. Cunis, R. Gehrke, W. Petry, Dewetting of confined diblock copolymer films. J. Macromol. Sci. B **43**, 29–41 (2004)
28. P. Müller-Buschbaum, E. Bauer, O. Wunnicke, M. Stamm, The control of thin morphology by the interplay of dewetting, phase separation and microphase separation. J. Phys.: Condens. Matter **17**, S363–386 (2005)
29. T.G. Stange, R. Mathew, D.F. Evans, W.A. Hendrickson, Scanning tunneling microscopy and atomic force microscopy characterization of polystyrene spin-coated onto silicon surfaces. Langmuir **8**, 920–926 (1992)

30. M.M. Abul Kashem, J. Perlich, L. Schulz, S.V. Roth, P. Müller-Buschbaum, Correlated roughness in polymer films containing maghemite nanoparticles. Macromolecules **41**, 2186–2194 (2008)
31. M.M. Abul Kashem, J. Perlich, A. Diethert, W. Wang, M. Memesa, J.S. Gutmann, E. Majkova, I. Capek, S.V. Roth, W. Petry, P. Müller-Buschbaum, Array of magnetic nanoparticles via particle co-operated self-assembly in block copolymer thin films. Macromolecules **42**, 6202–6208 (2009)
32. K.A. Barnes, A. Karim, J.F. Douglas, A.I. Nakatani, H. Gruell, E.J. Amis, Suppression of dewetting in nanoparticle-filled polymer films. Macromolecules **33**, 4177–4185 (2000)
33. E. Bauer, E. Maurer, T. Mehaddene, S.V. Roth, P. Müller-Buschbaum, Flow in confined geometry introduced by dewetting of ultrathin polystyrene films. Macromolecules **39**, 5087–5094 (2006)
34. P. Müller-Buschbaum, E. Bauer, E. Maurer, K. Schlögl, S.V. Roth, R. Gehrke, Route to create large-area ordered polymeric nanochannel arrays. Appl. Phys. Lett. **88**, 083114 (2006c)
35. A. Suzuki, M. Yamazaki, Y. Kobiki, H. Suzuki, Surface domains and roughness of polymer gels observed by atomic force microscopy. Macromolecules **30**, 2350–2354 (1997)
36. M. Schmidbauer, T. Wiebach, H. Raidt, M. Hanke, R. Köhler, H. Wawra, Ordering of self-assembled $Si_{1-x}Ge_x$ islands studied by grazing incidence small-angle x-ray scattering and atomic force microscopy. Phys. Rev. B **58**, 10523–10531 (1998)
37. J. Stangl, V. Holý, P. Mikulík, G. Bauer, I. Kegel, T.H. Metzger, O.G. Schmidt, C. Lange, K. Eberl, Self-assembled carbon-induced germanium quantum dots studied by grazing-incidence small-angle x-ray scattering. Appl. Phys. Lett. **74**, 3785–3787 (1999)
38. P. Müller-Buschbaum, High-resolution grazing incidence small angle x-ray scattering: investigation of micrometer sized structured polymer films. Prog. Colloid Polym. Sci. **132**, 23–32 (2006)
39. P. Müller-Buschbaum, Grazing incidence small-angle x-ray scattering: an advanced scattering technique for the investigation of nanostructured polymer films. Anal. Bioanal. Chem. **376**, 3–10 (2003)
40. R. Feidenhans'l, Surface structure determination by X-ray diffraction. Surf. Sci. Rep. **10**, 105–188 (1989)
41. T. Salditt, T.H. Metzger, J. Peisl, Kinetic roughness of amorphous multilayers studied by diffuse X-ray scattering. Phys. Rev. Lett. **73**, 2228–2231 (1994)
42. J.R. Levine, J.B. Cohen, Y.W. Chung, P. Georgopoulos, Grazing-incidence small-angle X-ray scattering: new tool for studying thin film growth. J. Appl. Crystallogr. **22**, 528–532 (1989)
43. J.R. Levine, J.B. Cohen, Y.W. Chung, Thin film island growth kinetics: a grazing incidence small angle X-ray scattering study of gold on glass. Surf. Sci. **248**, 215–225 (1991)
44. A. Frömsdorf, R. Capek, S.V. Roth, μ-GISAXS experiment and simulation of a highly ordered model monolayer of PMMA beads. J. Phys. Chem. B **110**, 15166–15171 (2006)
45. N. Hermsdorf, K. Sahre, P. Volodin, M. Stamm, K.J. Eichhorn, S. Cunis, R. Gehrke, P. Panagiotou, T. Titz, P. Müller-Buschbaum, Supported particle track etched polyimide membranes: A grazing incidence small-angle x-ray scattering study. Langmuir **20**, 10303–10310 (2004)
46. E. Metwalli, J.-F. Moulin, J. Perlich, W. Wang, A. Diethert, S.V. Roth, P. Müller-Buschbaum, Polymer-template-assisted growth of gold nanowires using a novel flow-stream technique. Langmuir **25**, 11815–11821 (2009)
47. P. Müller-Buschbaum, M. Wolkenhauer, O. Wunnicke, M. Stamm, R. Cubitt, W. Petry, Structure formation in two-dimensionally confined diblock copolymer films. Langmuir **17**, 5567–5575 (2001)
48. P. Müller-Buschbaum, R. Cubitt, W. Petry, Phase separation of weakly incompatible polymer blends confined in isolated droplets. Appl. Phys. A **74**, S342–S344 (2002)
49. R. Steitz, P. Müller-Buschbaum, S. Schemmel, R. Cubitt, G.H. Findenegg, Lateral structure of a surfactant layer adsorbed at hydrophilic solid/liquid interface. Europhys. Lett. **67**, 962–968 (2004)
50. T. Salditt, T.H. Metzger, J. Peisl, B. Reinker, M. Moske, K. Samwer, Determination of the height-height correlation function of rough surfaces from diffuse X-ray scattering. Europhys. Lett. **32**, 331–335 (1995)

51. A. Naudon, D. Thiaudiere, Grazing-incidence small-angle scattering. Morphology of deposited clusters and nanostructure of thin films. J. Appl. Cryst. **30**, 822–827 (1997)
52. A. Naudon, D. Babonneau, D. Thiaudiere, S. Lequien, Grazing-incidence small-angle X-ray scattering applied to the characterization of aggregates in surface regions. Physica B **283**, 69–74 (2000)
53. P. Müller-Buschbaum, P. Vanhoorne, V. Scheumann, M. Stamm, Observation of nanodewetting structures. Europhys. Lett. **40**, 655–660 (1997)
54. P. Müller-Buschbaum, M. Casagrande, J. Gutmann, T. Kuhlmann, M. Stamm, S. Cunis, G. von Krosigk, U. Lode, R. Gehrke, Determination of micrometer length scales with an X-ray reflection ultra small-angle scattering set-up. Europhys. Lett. **42**, 517–522 (1998)
55. D.A. Doshi, A. Gibaud, N. Liu, D. Sturmayr, A.P. Malanoski, D.R. Dunphy, H. Chen, S. Narayanan, A. MacPhee, J. Wang, S.T. Reed, A.J. Hurd, F. van Swol, C.J. Brinker, In-situ scattering study of continuous silica-surfactant self-assembly during steady-state dip coating. J. Phys. Chem. B **107**, 7683–7688 (2003)
56. B. Lee, J. Yoon, W. Oh, Y. Hwang, K. Heo, K.S. Jin, J. Kim, K.W. Kim, M. Ree, In-situ grazing incidence small-angle X-ray scattering studies on nanopore evolution in low-k organosilicate dielectric thin films. Macromolecules **38**, 3395–3405 (2005)
57. R. Lazzari, IsGISAXS: a program for grazing-incidence small-angle X-ray scattering analysis of supported islands. J. Appl. Cryst. **35**, 406–421 (2002)
58. T. Salditt, T.H. Metzger, J. Peisl, G. Goerigk, Non-specular X-ray scattering from thin films and multilayers with small-angle scattering equipment. J. Phys. D Appl. Phys. **28**, A236–239 (1995)
59. P. Müller-Buschbaum, J.S. Gutmann, M. Stamm, R. Cubitt, S. Cunis, G. von Krosigk, R. Gehrke, W. Petry, Dewetting of thin polymer-blend films examined with GISAS. Physica B **283**, 53–59 (2000)
60. M.A. Ruderer, E. Metwalli, W. Wang, G. Kaune, S.V. Roth, P. Müller-Buschbaum, Thin films of photoactive polymer blends. Chem. Phys. Chem. **10**, 664–671 (2009)
61. M.A. Cha, C. Shin, D. Kannaiyan, Y.H. Jang, S.T. Kochuveedu, D.Y. Ryu, D.H. Kim, A versatile approach to the fabrication of TiO_2 nanostructures with reverse morphology and mesoporous Ag/TiO_2 thin films via cooperative PS-*b*-PEO self-assembly and a sol-gel process. J. Mater. Chem. **19**, 7245–7250 (2009)
62. Y. Chushkin, L. Chitu, Y. Halahovets, S. Luby, E. Majkova, A. Satka, G. Leo, M. Giersig, M. Hilgendorff, V. Holý, O. Konovalov, GISAXS studies of self-assembling of colloidal Co nanoparticles. Mater. Sci. Eng. C **26**, 1136–1140 (2006)
63. S.S. Hu, J. Rieger, S.V. Roth, R. Gehrke, R.J. Leyrer, Y.F. Men, GIUSAXS and SPM studies on surface reconstruction of latex thin films during thermal treatment. Langmuir **25**, 4230–4234 (2009)
64. P. Müller-Buschbaum, N. Hermsdorf, S.V. Roth, J. Wiedersich, S. Cunis, R. Gehrke, Comparative analysis of diblock copolymer films. Spectrochim. Acta B **59**, 1789–1797 (2004)
65. G. Kaune, M.A. Ruderer, E. Metwalli, W. Wang, S. Couet, K. Schlage, R. Röhlsberger, S.V. Roth, P. Müller-Buschbaum, In situ GISAXS study of gold film growth on conducting polymer films. ACS Appl. Mater. Interfaces **1**, 353–360 (2009)
66. S. Couet, T. Diederich, K. Schlage, R. Röhlsberger, A compact UHV deposition system for in situ study of ultrathin films via hard x-ray scattering and spectroscopy. Rev. Sci. Instrum. **79**, 093908 (2008).
67. C. von Bechtolsheim, V. Zaporojtchenko, F. Faupel, Influence of thermal treatment on the morphology and adhesion of gold films on trimethylcyclohexane-polycarbonate. Appl. Surf. Sci. **151**, 119–128 (1999)
68. M.S. Rodrigues, O. Dhez, S. Le Denmat, J. Chevrier, R. Felici, F. Comin, Local detection of X-ray spectroscopies with an in-situ atomic force microscope. J. Inst. **3**, P12004 (2008)
69. T. Scheler, M. Rodriguez, T.W. Cornelius, C. Mocuta, A. Malachias, R. Magahães-Paniago, F. Comin, J. Chevrier, T.H. Metzger, Probing the elastic properties of individual nanostructures by combining in situ atomic force microscopy and micro-x-ray diffraction. Appl. Phys. Lett. **94**, 023109 (2009)

70. P. Müller-Buschbaum, E. Bauer, S. Pfister, S.V. Roth, M. Burghammer, C. Riekel, C. David, U. Thiele, Creation of multi-scale stripe-like patterns in thin polymer blend films. Europhys. Lett. **73**, 35–41 (2006)
71. S.V. Roth, P. Müller-Buschbaum, M. Burghammer, H. Walter, P. Panagiotou, A. Diethert, A. Riekel, Microbeam grazing incidence small angle x-ray scattering – a new method to investigate heterogeneous thin films and multilayers. Spectrochim. Acta B **59**, 1765–1773 (2004)
72. M.S. Rodrigues, T.W. Cornelius, T. Scheler, C. Mocuta, A. Malachias, R. Magahães-Paniago, O. Dhez, F. Comin, T.H. Metzger, J. Chevrier, In situ observation of the elastic deformation of a single epitaxial SiGe crystal by combining atomic force microscopy and micro x-ray diffraction. J. Appl. Phys. **106**, 103525 (2009)
73. V. Körstgens, J. Wiedersich, R. Meier, J. Perlich, S.V. Roth, R. Gehrke, P. Müller-Buschbaum, Combining imaging ellipsometry and grazing incidence small angle x-ray scattering for in situ characterization of polymer nanostructures. Anal. Bioanal. Chem. **396**, 139–149 (2010)
74. J.K. Gimzewski, R. Berndt, R.R. Schittler, Observation of local photoemission using a scanning tunneling microscope. Ultramicroscopy **42–44**, 366–370 (1992)
75. S. Libertino, F. Giannazzo, V. Aiello, A. Scandurra, F. Sinatra, M. Renis, M. Fichera, XPS and AFM characterization of the enzyme glucose oxidase immobilized on SiO_2 surfaces. Langmuir **24**, 1965–1972 (2008)
76. V. Rose, J.W. Freeland, K.E. Gray, S.K. Streiffer, X-ray-excited photoelectron detection using a scanning tunneling microscope. Appl. Phys. Lett. **92**, 193510 (2008)
77. T. Matsushima, T. Okuda, T. Eguchi, M. Ono, A. Harasawa, T. Wakita, A. Kataoka, M. Hamada, A. Kamoshida, Y. Hasegawa, T. Kinoshita, Development and trial measurement of synchrotron-radiation-light-illuminated scanning tunneling microscope. Rev. Sci. Instrum. **75**, 2149–2153 (2004)
78. A. Saito, J. Maruyama, K. Manabe, K. Kitamoto, K. Takahashi, K. Takami, M. Yabashi, Y. Tanaka, D. Miwa, M. Ishii, Y. Takagi, M. Akai-Kasaya, S. Shin, T. Ishikawa, Y. Kuwahara, M. Aono, Development of a scanning tunneling microscope for in situ experiments with a synchrotron radiation hard-X-ray microbeam. J. Synchrotron Radiat. **13**, 216–220 (2006)
79. A. Saito, Y. Takagi, K. Takahashi, H. Hosokawa, K. Hanai, T. Tanaka, M. Akai-Kasaya, Y. Tanaka, S. Shin, T. Ishikawa, Y. Kuwahara, M. Aono, Nanoscale elemental identification by synchrotron-radiation-based scanning tunneling microscopy. Surf. Interface Anal. **40**, 1033–1036 (2008)
80. T. Eguchi, T. Okuda, T. Matsushima, A. Kataoka, A. Harasawa, K. Akiyama, T. Kinoshita, Y. Hasegawa, Element specific imaging by scanning tunneling microscopy combined with synchrotron radiation light. Appl. Phys. Lett. **89**, 243119 (2006)
81. T. Okuda, T. Eguchi, K. Akiyama, A. Harasawa, T. Kinoshita, Y. Hasegawa, M. Kawamori, Y. Haruyama, S. Matsui, Nanoscale chemical imaging by scanning tunneling microscopy assisted by synchrotron radiation. Phys. Rev. Lett. **102**, 105503 (2009)
82. M. Ishii, X-ray absorption fine structure measurement using a scanning capacitance microscope: trial for selective observation of trap centers in the nm region. Jpn. J. Appl. Phys. **41**, 4415–4418 (2002)
83. A.E. Lita, J.E. Sanchez Jr., Characterization of surface structure in sputtered Al films: correlation to microstructure evolution. J. Appl. Phys. **85**, 876–882 (1999)
84. M. Al-Hussein, M.A. Ruderer, E. Metwalli, V. Körstgens, U. Vainio, S.V. Roth, R. Döhrmann, R. Gehrke, R. Gebhardt, M. Burghammer, P. Müller-Buschbaum, Determination of the ordered structure in conjugated-coil diblock copolymers films from a thickness gradient prepared by spin-coated drop technique. Macromolecules **42**, 4230–4236 (2009)
85. G.Y. Liu, P. Fenter, C.E.D. Chidsey, D.F. Ogletree, P. Eisenberger, M. Salmeron, An unexpected packing of fluorinated n-alkane thiols on Au (111) – a combined atomic-force microscopy and x-ray-diffraction study. J. Chem. Phys. **101**, 4301–4306 (1994)
86. F.F. Rossetti, P. Panagiotou, F. Rehfeldt, E. Schneck, M. Dommach, S.S. Funari, A. Timmann, P. Müller-Buschbaum, M. Tanaka, Structures of regenerated cellulose films revealed by grazing incidence small-angle x-ray scattering. Biointerphases **3**, 117–127 (2008)

87. P. Müller-Buschbaum, E. Maurer, E. Bauer, J. Wiedersich, R. Cubitt, Surface versus confinement induced morphology transition in triblock copolymer films: a grazing incidence small angle neutron scattering investigation. Langmuir **22**, 9295–9303 (2006)
88. P. Müller-Buschbaum, R. Cubitt, W. Petry, Nanostructured diblock copolymer films: a grazing incidence small-angle neutron scattering study. Langmuir **19**, 7778–7782 (2003)
89. Y. Yoneda, Anomalous surface reflection of X-rays. Phys. Rev. **131**, 2010–2013 (1963)

Chapter 5
Near-Field Microwave Microscopy for Nanoscience and Nanotechnology

Kiejin Lee, Harutyun Melikyan, Arsen Babajanyan, and Barry Friedman

Abstract We have demonstrated the possibility of near-field microwave imaging of physical structures, such as thin films, bulk material, fluids, etc. by using a near-field microwave microscopy (NFMM). We have developed theoretical models for the microwave reflection coefficient S_{11} and resonant frequency shift $\Delta f/f_0$ dependence on electromagnetic characteristics, in particular, electrical conductivity, dielectric permittivity, magnetic permeability to distinguish the spatial changes of these parameters in materials under various preparation and measurement conditions. The models are based on standard transmission line theory, material perturbation concept as well as finite-element numerical simulation methods. The NFMM is a noncontact, nondestructive and label-free evaluation tool to obtain material properties with high contrast and with high spatial resolution. The smallest detectable change in solvent (glucose) concentration is about 0.5 mg/ml at SNR = 20 dB, the smallest detectable change in permittivity (dielectrics) is about 0.2 at SNR = 30 dB, the smallest detectable change in conductivity (semiconductors and perfect metals) is about 0.01 S/m at SNR = 60 dB, the smallest detectable change in permeability (Permalloy) is about 10 at SNR = 40 dB, and the smallest detectable change in thickness (self-assembled monolayers, SAMs) is 2 nm at SNR = 50 dB. The results clearly show the sensitivity and the usefulness of NFMM for many device applications at microwave frequency such as 3D surface mapping and topography, material characteristics (permittivity, permeability, conductivity, carriers density, etc.) point-by-point distribution, and label-free biosensing (DNA, SAM, aqueous solution of glucose, NaCl, etc.).

5.1 Principles of Microwave Microscope

5.1.1 Introduction

Near-field microwave microscopy (NFMM) is the quantitative way to measure the electrodynamic properties of materials at microwave frequencies [1–7]. This is a direct observation method that allows visualizing electromagnetic properties of thin films and bulk samples. The near-field microwave reflection coefficient (S parameter) image from the sample directly visualizes the dielectric, electric, and magnetic characteristics and their changes This NFMM approach is based on nondestructive probing of a local electromagnetic near-field interaction between the probe-tip and the materials, and can be used in exploring material behaviors [8–13].

There are a number of ways to access the near-field component of electromagnetic fields to gain subwavelength resolution. A technique that uses microwaves, NFMM has a number of interesting features [14–16]. Firstly, there is not much structure in the electromagnetic properties in this frequency range so one is in fact measuring the low frequency properties, which are of considerable practical importance. In addition, "subsurface" information is obtained because microwaves penetrate well into materials, i.e. in good conductors where the skin depth is still of the order of a micron. Finally, in comparison to other frequency regimes, it should be straightforward to obtain quantitative information, again because the electromagnetic properties are relatively simple [17]. Indeed, by consideration of frequency shifts, quantitative measurement of the local dielectric constant and electrical conductivity has been done using numerical modeling based on cavity perturbation theory [18, 19]. Another, perhaps more natural observable in NFMM experiments, is the microwave reflection coefficient S_{11} as a function of frequency (from which, of course, the frequency shift can be obtained) or S_{11} as a function of some sample property for a fixed frequency [20, 21].

Here we review the general concept of near-field interactions between a probe-tip and sample and discuss a quantitative interpretation of near-field microwave images in view of material characteristics such as complex dielectric permittivity (ε), complex magnetic permeability (μ) and electrical conductivity (σ) for dielectrics, metals, semiconductors, alloys, ionic fluids, photovoltaic structures, biological arrays, etc. Recently, several NFMM techniques have been developed for the microwave and millimeter-wave ranges [8–21]. An important ability of the NFMM is noncontact characterization of multilayer structures and single crystals, and thin films. The main advantage of NFMM compared to the usual electrical measurement is that NFMM can directly image the electric properties at interfaces and surfaces with high sensitivity and simplicity, because the NFMM directly evaluates and images the dielectric, conducting and magnetic characteristics and these parameter changes under varying of the external influences.

5.1.2 Near-field Interaction

An important strategy for understanding natural and artificial materials is to study the interaction of the material with electromagnetic fields. The properties of metals, semiconductors, and dielectrics at low frequency have been a fruitful area of investigation. The first experiments were transmission experiments done in the far-field region from source, and typically required that the sample size be of the scale of the wavelength [22–26]. In the second generation some experiments were carried out in resonant cavities, which are on the order of the wavelength in at least one dimension [27–31]. As a result the electrodynamic properties of the sample are still averaged over macroscopic length scales. However, properties of the material can vary on much shorter length scales, even into the nanometer range. Currently, the materials of most interest are complex multicomponent compounds or nanoscale composites and can rarely be made homogeneous on the millimeter or microlength scales required for electrodynamic measurements [32–34]. Finally, the typical dimensions of devices into which some of these functional materials are integrated are orders of magnitude smaller than the wavelength at the operation frequency of the device. This is an issue for dielectric, metal, semiconductor, bulk multilayer, and thin film samples. Electrodynamic measurement where the wavelength is longer than the sample characteristic size has emerged in recent years in condensed matter physics [35,36] and is associated with the concept of near-field interactions between a source and a sample in which evanescent waves are created and interact with the sample. The NFMM is one aspect of near-field interaction.

Electromagnetic waves are created by time-varying currents and charges. Their interactions with materials obey Maxwell's equations supplemented with boundary conditions [37]. Electromagnetic waves can be guided by structures (transmission lines) or propagate in free space. Near-field behavior is most clearly seen surrounding small electric dipoles as shown in Fig. 5.1. The near-field consists of the reactive near-field, also known as the quasi-static near-field, and the radiating near-field also known as the Fresnel zone. In the quasi-static near-field the fields strongly resemble the electrostatic fields of a charge dipole for a dipole antenna (Fig. 5.1a) and the fields of a magnetic dipole for a loop antenna. In large antennas the quasi-static field can be seen near edges.

In the Fresnel zone the waves are clearly not plane waves and may have phase shifts that do not vary linearly with distance from a fictitious phase center. From the near-field to the far-field, electromagnetic radiation changes from spherical waves to plane waves. The far-field is sometimes called the Fraunhoffer region (tip–sample distance is greater than $2d^2/\lambda$, where d is the dipole size and λ is the wavelength). A common feature of all electrically small (less than a wavelength) antennas is

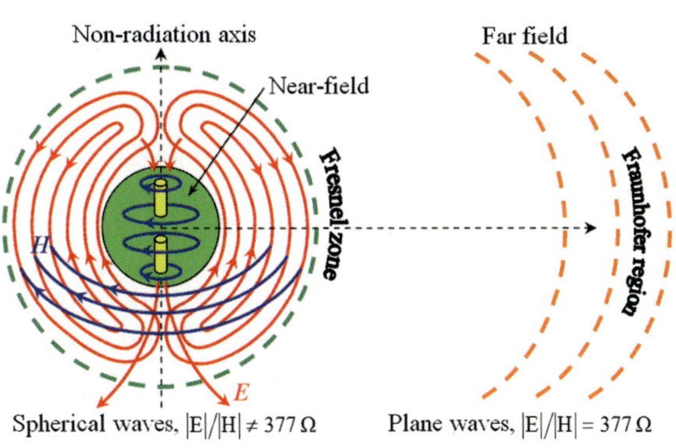

Fig. 5.1 Far-field and near-field interpretation of electromagnetic waves for small electric dipoles as capacitive objects [38]

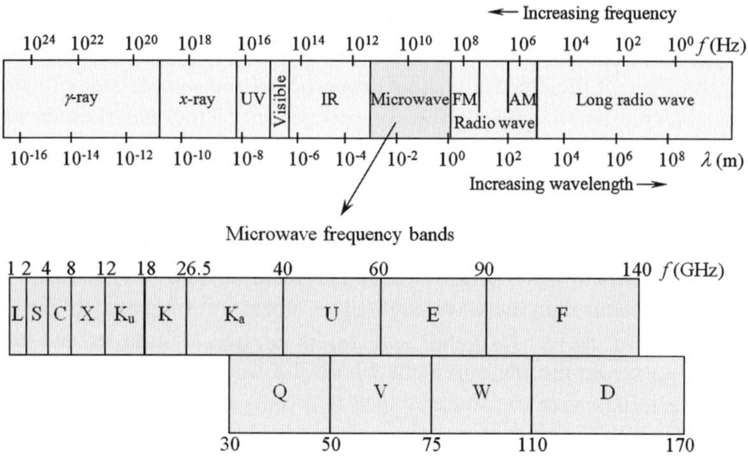

Fig. 5.2 Electromagnetic spectrum

that the near-field excites the environment in which the antenna resides. Far from the source, the spherical electromagnetic waves flatten out and can be treated like plane waves. In the far field, the power density $(E \times H)$ drops as R^{-2}, while in the near-field region it drops as R^{-3}.

5.1.3 Microwave Frequencies

The microwave frequencies are defined as electromagnetic waves between 300 MHz and 300 GHz [39]. Electromagnetic waves below 300 MHz are called very high frequency (VHF) or radio frequency (RF); above 300 GHz submillimeter wave spectrum begins as shown in Fig. 5.2.

The following distinction between millimeter-waves and microwaves is almost universally accepted: frequencies with free-space wavelengths $<$ 1 cm but $>$ 1 mm are referred to as millimeter-waves. Thus, the millimeter-wave spectrum starts at 30 GHz, and runs to 300 GHz, where the wavelength in free-space is $<$ 1 mm. The submillimeter-wave band corresponds to infrared radiation and terahertz frequencies.

Bandwidth (BW) is a measure of the range of the spectrum to which a microwave system can respond. BW is often given in megahertz or gigahertz, calculated from a low frequency f_l to a high frequency f_h, the bandwidth is often given by BW $= f_h - f_l$ and the center frequency $f_c = (f_h + f_l)/2$. The 3-dB BW for a network that has a non-ideal frequency response (which includes all physical networks) is where the transmission coefficient S_{21} falls off from its highest peak by 3 dB.

5.2 Detailed Description of the Near-field Microwave Microscope

5.2.1 *Probe-Tip for NFMM*

Microwaves incident on the sample drive an effective field both inside and outside the sample. To obtain of this effective field, the probe-tip must have dimensions as small as the sample and must be located near the sample, where the effective field is strongly influenced by the sample [40]. In this case the probe-tip and sample form an interacting subsystem. As a result, the effective field of the sample produces microwaves scattered from the probe-tip, which reaches the network analyzer (NA). To extract the microwave near-field interaction during the reflection process, one needs to use a probe-tip (made from special material and with special geometrical shape) that is sensitive only to high spatial frequency components relevant to the local effective field around the specimen. This is why the technique of fabricating fine probes is of primary importance.

Figure 5.3 shows three typical probe-tips and the corresponding three-dimensional (3D) near-field images. The thin tip (a) with apex angle $\theta_1 = 5°$ provides high spatial resolution but low sensitivity for differentiating the signal, while the thick tip (b) with apex angle $\theta_2 = 30°$ provides low spatial resolution but high signal sensitivity. In the case of the hybrid tip (c) with multiangle apex the obtained images have both high spatial resolution and signal sensitivity. The tapered part of thin and thick tips has a simple conical shape and the cone angle θ is clearly defined. The hybrid tip has a multiple tapered shape with three continuously increasing apex angles ($\theta_{31} = 5°$, $\theta_{32} = 10°$, and $\theta_{33} = 30°$) which can be controlled by varying the etching condition. The end-size of the tip makes an important contribution for the determination of the spatial resolution and sensitivity of the microscope.

To achieve the best spatial resolution and sensitivity, we investigated a different type of probe-tip design. In general, high spatial resolution in near-field microscope

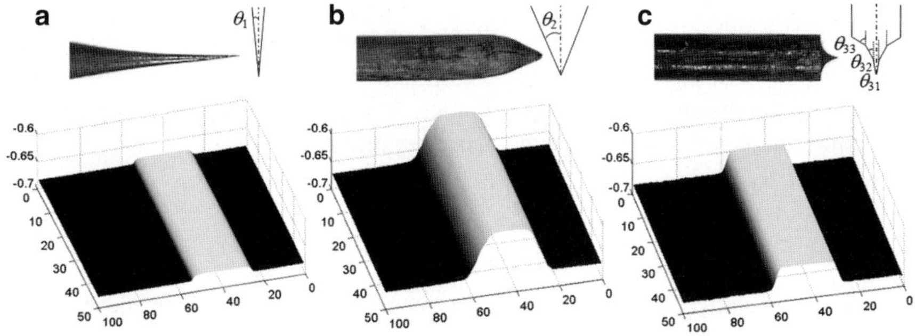

Fig. 5.3 Tip shape and corresponding obtained NFMM images for (**a**) thin tip, (**b**) thick tip, and (**c**) hybrid tip

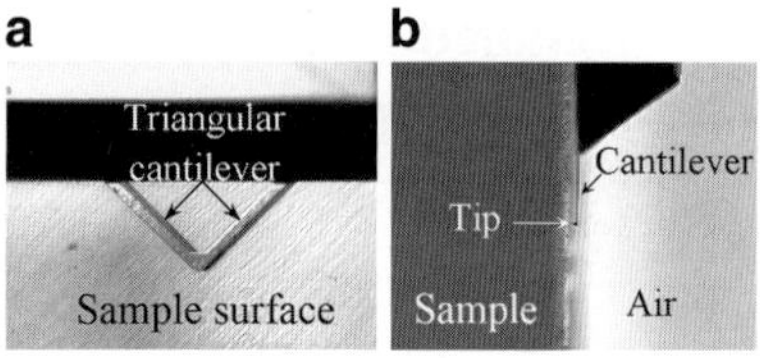

Fig. 5.4 (**a**) The *top* and (**b**) the *side* views of the AFM tip with triangular cantilever close to the sample surface

images is determined primarily by aperture size and high sensitivity can be achieved by the quality factor of the resonator [41]. Thus, we used a high quality dielectric resonator coupled to a commercial AFM cantilever probe-tip (Fig. 5.4), utilizing the distance control sensor provided by a tuning-fork feedback system. The dielectric resonator has a mechanical frequency adjustment and the effective impedance of the probe system, including the AFM tip, can be optimized. This solves the mismatch problem and leads to more sensitive quantitative measurement using a NFMM.

5.2.2 Dipole–Dipole Interaction

As discussed above, the spatial resolution, sensitivity, and contrast in near-field imaging strongly depend on the structure of the probe-tip: the end-size, apex angle, material, etc. Reference [40] gives a simple argument for an optimized structure of a fiber probe for the simultaneous achievement of high spatial resolution, high sensitivity, and high contrast in near-field optical microscopy. This approach also applies for near-field microwave imaging. Since the tip end-size and the sample size are much smaller than the operating wavelength, the argument is based on the theory of two interacting dipoles [42, 43].

The tip and sample are modeled as two dipoles with polarizability α_t and α_s with radii a_t and a_s; they are driven by an external electric field E as shown in Fig. 5.5. The modulation of polarizability due to dipole–dipole coupling is described as

$$\Delta\alpha = \begin{cases} \dfrac{2\alpha_t\alpha_s}{R^3}; & \text{for broadside illumination} \\ -\dfrac{\alpha_t\alpha_s}{R^3}; & \text{for end-on illumination} \end{cases}, \tag{5.1}$$

where R is the distance between the dipoles. The polarizability can be written in terms of the size and their relative dielectric permittivity,

$$\alpha_i = \frac{\varepsilon_i - 1}{\varepsilon_i + 2} a_i^3; \quad i = t, s. \tag{5.2}$$

Note that the relative dielectric permittivity is positive for dielectrics and negative for metals, i.e. for the metallic probe-tip $\varepsilon_t < 0$. In the near-field operation, the electric field detected E field, which is a sum of the unmodulated E_s and the modulated ΔE_s, is expressed as

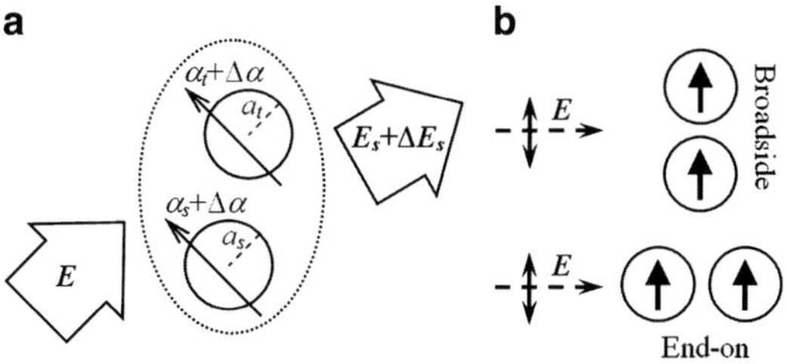

Fig. 5.5 (**a**) Simplified configuration of near-field operation. (**b**) Broadside and end-on illumination of two dipoles [40]

$$|E_s + \Delta E_s|^2 \propto [(\alpha_t + \Delta\alpha) + (\alpha_s + \Delta\alpha)]^2 |E|^2 \approx \left\{(\alpha_t + \alpha_s)^2 |E|^2\right\} + \left\{4(\alpha_t + \alpha_s)\Delta\alpha |E|^2\right\}. \tag{5.3}$$

The first term in (5.3) is background signal and independent of the tip and sample position. The second term is strongly dependent on the separation of two dipoles. It can be inferred that the maximum spatial resolution of near-field microscope is determined by the tip end-size and the other parts of probe contribute to the enhancement of contrast rather than to the achievement of high resolution. For the constant-height operation mode (the distance between tip and substrate is constant) with decreasing of the tip end-size, the image size becomes smaller and the signal intensity decreases (Fig. 5.3). To determine the optimal end-size, the image contrast should be taken into account. The image contrast is the ratio of modulated and unmodulated scattering field intensity:

$$C = \frac{4(\alpha_t + \alpha_s)\Delta\alpha |E|^2}{(\alpha_t + \alpha_s)^2 |E|^2} = \frac{4\Delta\alpha}{\alpha_t + \alpha_s} = \begin{cases} \dfrac{8}{R^3}\dfrac{\alpha_t\alpha_s}{\alpha_t + \alpha_s}; & \text{for broadside illumination} \\ -\dfrac{4}{R^3}\dfrac{\alpha_t\alpha_s}{\alpha_t + \alpha_s}; & \text{for end-on illumination} \end{cases}. \tag{5.4}$$

The modulated intensity saturates with increasing end-size because only the tip end, which is close to the sample, contributes to the polarizability modulation thus the contrast decreases monotonously. On the other hand, the modulated signal decreases with a reduction of the end-size, so the contrast also deteriorates. The conclusion from this analysis is that the optimized end-size is the same as the characteristic size of the sample. Also, the smaller the cone angle, the smaller the undesirable background signal due to the dipole coupling between the sample and the tapered part of the tip [40].

5.2.3 Tip–sample Distance Control in NFMM

One of the strengths of near-field microscopy is that measurements can be made without any physical contact between the probe-tip and the sample or the device being measured. However, the measurement requires that the separation between

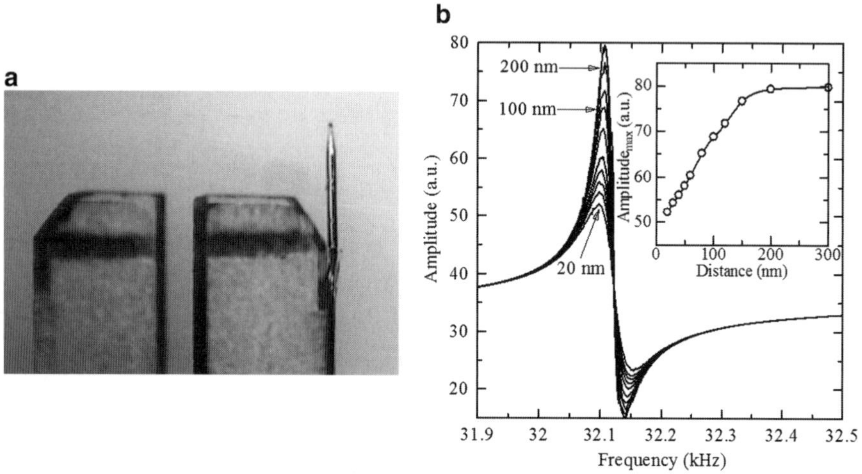

Fig. 5.6 (**a**) Image of a 32 kHz resonant frequency quartz tuning-fork with an attached sharpened tungsten tip. (**b**) The amplitude–frequency characteristic of the quartz tuning-fork resonant system at various tip–sample distances. The *inset* shows the amplitude maximum vs. tip–sample‘ distance for W/Au interface

the probe-tip and the surface of the sample be small compared to the characteristic tip size and it has to be kept constant. The precision of any near-field measurement is directly related to the precision with which the tip–sample distance can be maintained. Although the absolute value of the mean height is not critical, the variance must be less than 1% of the tip size in order to obtain high precision measurements. For example, for a tip with a characteristic end-size of 0.1 μm, the tip–sample separation will have to be maintained at approximately 10 nm, with a precision of 1 nm. This precision can be obtained using a shear-force distance control system [44–46].

The basic idea is that a probe-tip is flexible and can therefore be mounted onto a quartz tuning-fork oscillator with amplitude of a few nanometers. Figure 5.6 shows (a) an image of a 32 kHz resonant frequency quartz tuning-fork and (b) the amplitude–frequency characteristic at various tip–sample distances. The inset shows the amplitude maximum of the quartz tuning-fork vibration vs. tip–sample distance for a tungsten sharpened tip and an Au thin film surface. As the tip the metal wire is brought into close proximity of the sample surface, the amplitude and frequency of the tip oscillations are changed by interactions between the tip and the sample surface. The motion of the probe-tip is detected electrically and a feedback loop allows for precise distance control down to 10 nm. In addition, the height at which this control can be performed is a function of the amplitude of the oscillation. For smaller amplitude, a smaller control distance can be maintained as shown in the inset of Fig. 5.6b. It has been demonstrated that this type of force feedback can be effectively applied to probe structures making it applicable for use in NFMM. The general principal is that the tip–sample separation is modulated with fixed amplitude and frequency, and the modulated signal is used to control the separation.

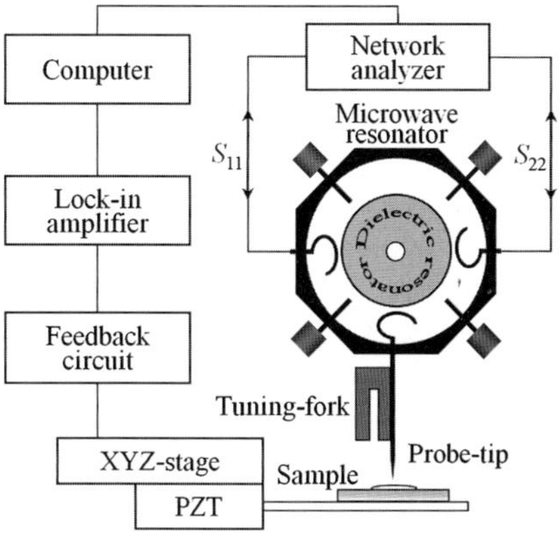

Fig. 5.7 A schematic of NFMM experimental setup

5.2.4 The Basic Experimental Setup of NFMM

To illustrate the basic operation of NFMM we have presented a detailed description. In our design of NFMM experiments, the probe-tip is simply a sharpened metal wire (Fig. 5.3) or metal coated commercial triangular AFM probe-tip (Fig. 5.4) in which voltage oscillations are present at microwave frequencies [5, 47–53]. Figure 5.7 shows the experimental setup of the NFMM. The probe-tip is brought to a distance of 10 nm from the sample surface. Because tip–sample interactions take place over a distance of nanometers, much less than the microwave wavelength (∼1 cm), these interactions occur in the near-field region. If the sample is translated under the tip, changes in tip–sample coupling can be used to obtain images of the sample with submicron resolution. The resolution is governed by tip geometry rather than the diffraction limit. At its other end, the probe is inserted into a microwave resonance cavity, or "resonator," populated by standing microwaves. The resonator microwave fields drive the electronic oscillations in the probe and the microwaves are continually excited through an input wire line connected to an external source. We designed a NFMM with a tuningfork distance control system to keep a constant distance between the sample and the probe-tip [48–51].

The typical conical shape probe-tip is made of tungsten or stainless-steel wire with a diameter of 50 μm with tapered end-size of about 1 μm prepared by a chemical etching method (apex angle is 30°). We also use a commercial metal covered AFM probe-tip cantilever. The length, width, and thickness of the cantilever are 290 μm, 40 μm, and 1 μm, respectively. The apex angle of the tip was 30° and the height is about 20 μm. The tip curvature radius is less than 30 nm. The cantilever probe-tip is coated by Au film (20 nm) on a Cr sublayer (20 nm). An AFM probe-tip exhibited very high input impedance, thus the reflection coefficient measurement using these miniature probes has a very low sensitivity. Thus, in order to boost the sensitivity of the probe, we employed a dielectric resonator and tuned the resonance cavity to match the impedance of 50 Ω. The probe-tip was oriented perpendicular to the sample surface and the other end of the tip was directly connected to a coupling loop in the dielectric resonator. The resonance frequency of a given TE_{011} mode (4–4.5 GHz) was measured with a NA (Agilent 8753ES). To drive the tuning-fork, an AC voltage was applied to one contact on the tuning-fork (Fig. 5.6) at its resonant frequency using the oscillator of a lock-in amplifier (Signal Recovery 7,265). The output from the lock-in amplifier was fed into the feedback system to control the tip–sample distance (10 nm) using a piezoelectric tube (PZT) that supports the sample stage. All NFMM measurements were done at the same sample-tip distance. The sample was mounted onto an x–y–z-translation stage for coarse adjustment which was driven by a computer-controlled microstepping motor with a resolution of 100 nm, whereas fine movement of the sample was controlled by a PZT tube with 10 nm resolution.

5.3 Theory of Near-field Microwave Microscope

5.3.1 Transmission Line Theory

Transmission line theory can be approached from the extension of circuit theory or from Maxwell's equations [37]. A transmission line is a distributed parameter network, where voltages and currents can vary in magnitude and phase over its length. A transmission line is often schematically represented as a two-wire line and is a conducting structure that guides an electromagnetic wave. Most transmission lines use two conductors, where one is considered ground. This includes coax (the outer conductor is ground), microstrip, and strip-line. The transmission lines have two important properties that depend on their geometry: their inductance per unit length, L' and their capacitance per unit length, C'. The characteristic impedance of a system is calculated as the square root of the ratio of these two:

$$Z_c = \sqrt{\frac{L'}{C'}}. \tag{5.5}$$

Note that higher inductance translates to higher impedance, and higher capacitance translates to lower impedance. The exact characteristic impedance of free-space is $Z_0 = 377\,\Omega$. Impedance matching of source and load is important to get maximum effective connection and minimum return losses.

In order to match the impedance of the probe-tip–sample system for near-field imaging, we study the reflection of microwaves (scattered from the probe as a source) from the sample surface and the mismatching condition due to the external influence of the structure of the sample. Here we present the transmission line theory model for the probe-tip–sample system. For simplicity, we consider the two-layer sample: a thin film on the bulk substrate as shown in Fig. 5.8.

An expression of the dependence of reflection coefficient S_{11} depends on material parameters of the sample can be derived by using standard transmission line theory, assuming impedance matching between the probe and the microwave source

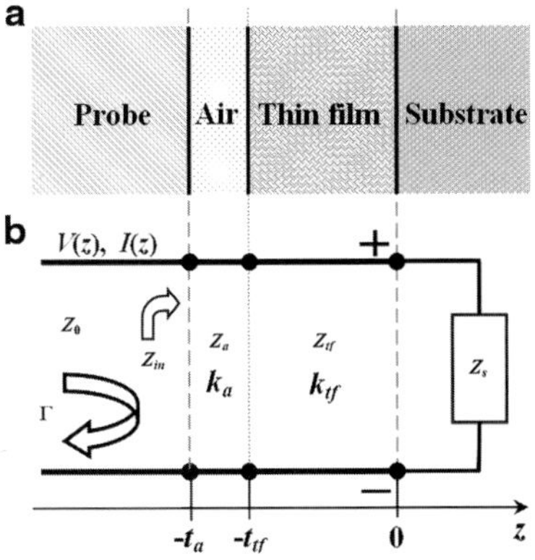

Fig. 5.8 (**a**) Schematic and (**b**) equivalent transmission line circuit of the probe-tip and the two layer sample system

$$S_{11} = 20\log\Gamma = 20\log\frac{Z_{\text{in}} - Z_0}{Z_{\text{in}} + Z_0}, \tag{5.6}$$

where Z_0 is the impedance of probe-tip ($50\,\Omega$) and Z_{in} is the complex input impedance of the sample. The calculation of the input impedance is performed through the transmission line analogy and the impedance transformation [39] for every layer with layer parameters; thickness, wave number, and surface impedance:

$$Z_{\text{in}} = Z_{\text{tf}}\frac{Z_{\text{s}} + \mathrm{j}Z_{\text{tf}}\tan(k_{\text{tf}}t_{\text{tf}})}{Z_{\text{tf}} + \mathrm{j}Z_{\text{s}}\tan(k_{\text{tf}}t_{\text{tf}})}, \tag{5.7}$$

where j ($\mathrm{j}^2 = -1$) is the standard imaginary unit, Z_{s} is the impedance of the substrate, and Z_{tf}, k_{tf}, t_{tf} are the impedance, wave number, and thickness of the thin film, respectively. The impedance and wave number of the thin film interface defined as [39]

$$Z_{\text{tf}} = \begin{cases} \sqrt{\dfrac{\mu_{\text{tf}}\mu_0}{\varepsilon_{\text{tf}}\varepsilon_0}} = Z_a\sqrt{\dfrac{\mu_{\text{tf}}}{\varepsilon_{\text{tf}}}}, & \text{for dielectrics} \\ (1+\mathrm{j})\sqrt{\pi f \mu_{\text{tf}}\mu_0\sigma_{\text{tf}}} = (1+\mathrm{j})\dfrac{1}{\delta_{\text{tf}}}, & \text{for metals} \end{cases}, \tag{5.8}$$

$$k_{\text{tf}} = \begin{cases} \mathrm{j}2\pi f\sqrt{\mu_{\text{tf}}\mu_0\varepsilon_{\text{tf}}\varepsilon_0} = k_a\sqrt{\mu_{\text{tf}}\varepsilon_{\text{tf}}}, & \text{for dielectrics} \\ (1+\mathrm{j})\sqrt{\dfrac{\pi f \mu_{\text{tf}}\mu_0}{\sigma_{\text{tf}}}} = (1+\mathrm{j})\dfrac{1}{\sigma_{\text{tf}}\delta_{\text{tf}}}, & \text{for metals} \end{cases}. \tag{5.9}$$

Here μ_0 and ε_0 are the magnetic permeability and dielectric permittivity of free-space and equal $4\pi \times 10^{-7}\,\text{H/m}$ and $8.854 \times 10^{-12}\,\text{F/m}$, respectively. Physical parameters μ_{tf} and ε_{tf} are the relative magnetic permeability and relative dielectric permittivity, and f is the operating frequency. The electromagnetic wave penetration length defined as $\delta_{\text{in}} = 1/\sqrt{\pi f\, \mu_{\text{tf}}\mu_0\sigma_{\text{tf}}}$ is called the skin depth. Applying (5.7) to the dielectric/metal interface, we obtain the effective surface impedance [54]:

$$Z_{\text{in}} \approx \mathrm{j} Z_{\text{tf}} \tan(k_{\text{tf}} t_{\text{tf}}), \tag{5.10}$$

which does not easily reduce to Z_{tf}; it is immediately apparent that resonance can appear. For the semiconductor thin films the dielectric permittivity is complex with

$$\varepsilon_{\text{tf}} = \varepsilon'_{\text{tf}} + \mathrm{j}\varepsilon''_{\text{tf}} = \varepsilon'_{\text{tf}} + \mathrm{j}\frac{\sigma_{\text{tf}}}{2\pi f \varepsilon_0}, \tag{5.11}$$

where ε'_{tf} and $\varepsilon''_{\text{tf}}$ $\varepsilon''_{\text{tf}}$ are the real and the imaginary parts of dielectric permittivity, and σ_{tf} is the conductivity of the thin film (Fig. 5.8).

The above is the general transmission line solution for reflection of normally incident wave at the interface for lossy materials.

5.3.2 *Perturbation Theory*

In practice cavity resonators are often tuned by making small changes in their shape or by the introduction of small pieces of dielectric or metallic material. For example, the resonant frequency of a cavity can be easily tuned by changing the cavity volume or adding dielectric material or by a metallic screw. This approach allows the determination of material characteristics (permittivity, conductivity, permeability, etc.) by measuring the shift in resonant frequency [55,56]. One useful technique to do this is the perturbation method, which assumes that the actual fields of the cavity with a small material perturbation are not greatly different from those of the unperturbed cavity.

Figure 5.9 shows a cavity perturbed by a change in the permittivity ($\Delta\varepsilon$) and permeability ($\Delta\mu$) of the material of the cavity in the volume (ΔV) of the cavity. If $\vec{E}_0$, $\vec{H}_0$ are the fields of the original cavity, and $\vec{E}$, $\vec{H}$ are the fields of the perturbed cavity, then Maxwell's curl equations can be written for the two cases as

$$\begin{aligned} \nabla \times \vec{E}_0 &= -\mathrm{j}2\pi f_0 \mu \vec{H}_0, \\ \nabla \times \vec{H}_0 &= \mathrm{j}2\pi f_0 \varepsilon \vec{E}_0, \\ \nabla \times \vec{E} &= -\mathrm{j}2\pi f(\mu + \Delta\mu)\vec{H}, \\ \nabla \times \vec{H} &= \mathrm{j}2\pi f(\varepsilon + \Delta\varepsilon)\vec{E}, \end{aligned} \tag{5.12}$$

where f_0 and f are the resonant frequency of the original and perturbed cavity, respectively. For small changes, $\Delta\varepsilon$ and $\Delta\mu$, the perturbed fields can be approximated by the original fields, to give the fractional change in the resonant frequency as [39]

$$\frac{\Delta f}{f_0} = \frac{f - f_0}{f_0} = -\frac{\int_{v_0}\left(\Delta\varepsilon\left|E_0\right|^2 + \Delta\mu\left|H_0\right|^2\right)\mathrm{d}v}{\int_{v_0}\left(\varepsilon\left|E_0\right|^2 + \mu\left|H_0\right|^2\right)\mathrm{d}v}. \tag{5.13}$$

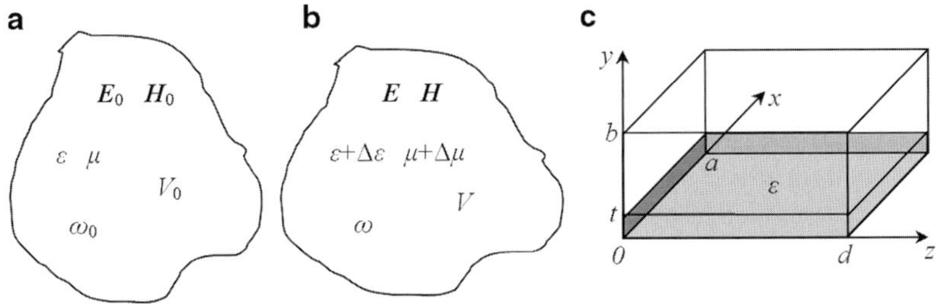

Fig. 5.9 (**a**) Original and (**b**) perturbed resonant cavity perturbed by a change in the permittivity or permeability of the cavity material or by a change of the cavity volume and (**c**) a rectangular cavity operating in the TE_{011} mode.

For a rectangular cavity operating in the TE_{011} mode (Fig. 5.9c), the changes due to material filling can be written as $\Delta V = S_0 \Delta t$, $S_0 = ad$, $\Delta\varepsilon = (\varepsilon - 1)\varepsilon_0$, $\Delta\mu = 0$. Then in the numerator of (5.13) the integral can be evaluated as

$$\int_{v_0} (\Delta\varepsilon |E_0|^2 + \Delta\mu |H_0|^2)\mathrm{d}v = (\varepsilon - 1)\varepsilon_0 \int_{x=0}^{a} \int_{y=0}^{t} \int_{z=0}^{d} |E_y|^2 \mathrm{d}x\, \mathrm{d}y\, \mathrm{d}z$$
$$= \frac{(\varepsilon - 1)\varepsilon_0 |E_0|^2 S_0 t}{4}, \qquad (5.14)$$

where t is the height of the perturbed media. The denominator of (5.13) is proportional to the total energy in the perturbed cavity, thus to the order to which we are working

$$\int_{v_0} (\varepsilon |E_0|^2 + \mu |H_0|^2)\mathrm{d}v = \frac{\varepsilon_0 |E_0|^2 S_0 h}{2}, \qquad (5.15)$$

where h is the height of original media. Finally, the fractional change in the resonant frequency is given by

$$\frac{\Delta f}{f_0} = -\frac{(\varepsilon - 1)t}{2h}. \qquad (5.16)$$

This result shows that any increase in permittivity or permeability at any point in the cavity will decrease the resonant frequency. Thus this decrease in resonant frequency can be related to the increase in stored energy of the perturbed cavity according to (5.13).

By consideration of frequency shifts obtained by NFMM, quantitative measurement of the local electromagnetic characteristics has been done using numerical modeling based on cavity perturbation theory [18, 19].

5.3.3 Finite-Element Model

Another, perhaps more natural observable in NFMM experiments, is the reflection coefficient S_{11} as a function of frequency (from which, of course, the frequency shift can be obtained) or S_{11} as a function of material property for a fixed frequency [20,21]. However, to model S_{11} is not very straightforward in that one cannot work solely with electrostatics. Furthermore, there are distance scales of very different sizes, for example, the cavity of order centimeters and the tip–sample distance of order microns or less. Since analytic methods appear to be difficult, we have applied a numerical approach, the finite-element method (FEM) [57–59], to study S_{11} for a model NFMM.

Very roughly, the experimental NFMM system consists of a cylindrical microwave cavity with three "ports"; two coaxial cables to input and output microwaves and the third port consisting of a hole in the cavity with a thin rod protruding out, the probe-tip (Fig. 5.7). Inside the cavity is a dielectric resonator with a high dielectric constant and which is very stable to temperature variation. The probe-tip scans the sample at close distance (less than a micron); the effect of the sample is observed by measuring S_{11} in the plane where the microwaves are input.

Due to the two coaxial cables and protruding rod, the system described above is fully 3D, even though the cylindrical cavity has axial symmetry. Of course, from the standpoint of numerical calculation, it is much easier to work in two dimensions or in three dimensions with axial symmetry. Therefore, to simplify the calculations, we have considered a modification of the experimental system that maintains axial symmetry. The cavity is taken to be a cylinder with coaxial cables along the axis of symmetry, one inputting the microwaves, while the second provides a route for the microwaves to exit the cavity. The central conductor of the exit coaxial cable is extended and held in close proximity to the sample. The FEM is designed to provide physical insights, and is not intended to be directly compared with experiment [60]. By using a commercial finite-element program, COMSOL 3.3, we shall see that this model system is able to reproduce a number of the features of the experimental system.

The FEM consists of a cylindrical air-filled cavity of height 2 cm and diameter 4 cm with two coaxial cables along the axis of symmetry (Fig. 5.10a). The top

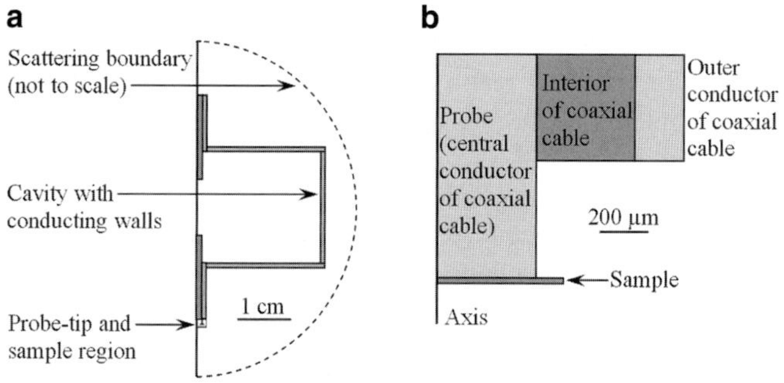

Fig. 5.10 Scale drawing (**a**) of the physical model (the scattering boundary is not to scale) and (**b**) of probe-tip of the NFMM. The *light gray* regions are electric conductors, and the *dark gray* regions are dielectrics with variable values. The *nonshaded* region is a vacuum

coaxial cable is the input and the bottom cable is the output. The measurement of S_{11} is made on the plane located at the input.

Each coaxial cable has a 0.8 mm diameter inner conductor surrounded by a 0.4 mm thick insulator and a 0.2 mm thick outer conductor. Each inner conductor extends 5 mm into the cavity and the end of the output coaxial cable (probe-tip) lies a small distance above the sample (Fig. 5.10b). The lengths of the central conductors of the coaxial cables (extending into the cylinder) are similar to that in the experimental system. Both the probe-tip–sample distance and the sample thickness are varied in this study.

In a NFMM, the cavity serves as a resonator with resonant frequencies given for TM modes by

$$\omega_{mnp} = c\sqrt{\frac{{x_{mn}}^2}{R^2} + \frac{\pi^2 p^2}{d^2}} \quad (m, p = 0, 1, 2, \ldots; n = 1, 2, 3, \ldots), \tag{5.17}$$

where x_{mn} are the nth roots of the equation $J_m(x) = 0$ [50]. For this cavity, the lowest resonant frequency $f_{010} = \omega_{010}/2\pi = 5.74\,\text{GHz}$ is the lowest TM mode.

The focus of this study was to explore the effects of various samples located near the probe-tip. In order for our physical model to give realistic results, care must be taken in choosing the boundary conditions. The natural choice for the boundary condition along the symmetry axis is "axial symmetry." To minimize the effects of the outer boundary, chosen to be spherical, a "scattering" boundary condition is used: $\vec{n} \times (\nabla \times H_\phi) - jkH_\phi = 0$. All of the conductors in the device (central conductor and outer conductor in the coaxial cable, walls of resonant cavity) are treated as "perfect electric conductors" ($\vec{n} \times \vec{E} = 0$). Boundaries between insulators, air, and the sample are treated using the "continuity" boundary condition: $\vec{n} \times (\vec{E}_1 - \vec{E}_2) = 0$. Finally, the electromagnetic wave is fed into the input coaxial cable using a "port" boundary condition.

Another important component of a FEM is the grid. The grid elements in this model are triangular, with the minimum element size defined by the smallest geometry in the model (the 1 μm gap between the sample and the tip). Note that these grid elements are much smaller than the wavelength of 5 cm corresponding to 5.7 GHz microwaves. Since the elements farther from the tip are less important, the size of the grid elements increases to a scale set by the largest objects in the model. To ensure an accurate result, the grid is refined once, adaptively based on the results of the first calculation. Because the geometry of each model is different, there are slight differences in the number of grid elements from model to model. On average, an adaptively refined model will contain 20,000 grid elements, with the majority of those elements concentrated in the sample and the gap between the probe-tip and the sample.

An important consideration when modeling a NFMM is that the system must be carefully tuned to achieve a sharp resonance [5]. This is consistent with the experimental observation that impedance matching is a crucial ingredient when constructing a NFMM with high sensitivity and subwavelength spatial resolution. Generally, tuning is accomplished by making changes to "convenient" parameters in the system, such as the cavity dimensions, or less realistically from the standpoint of experiment, the permeability of the cavity or of the coaxial cables. In the experimental systems of interest, impedance matching is achieved by a tuning screw that controls a metal plate inside the cavity. However, computationally, this method is rather difficult to implement. Instead, we have achieved a matched system by varying the magnetic permeability μ in the region between the inner and the outer conductors of the coaxes. By carefully choosing μ through "trial and error," a value of $S_{11} = -40$ dB was realized at the resonant frequency ($\sim$5.8 GHz) in the computational model when no sample was present. Once this tuning was achieved, the value of μ was fixed as various samples were introduced.

Finally, in designing the FEM for NFMM, the length l, which the central conductor extends out the bottom coaxial cable, is a critical parameter. We have chosen the value of l to be 0.5 mm. Larger values of l give smaller frequency shifts and changes in S_{11} which makes the device less sensitive. A natural solution is make l as small as possible, $l = 0$, having the central conductor end flush with the sides. Unfortunately, when one does this, it becomes increasingly difficult to match the impedance of the coaxes to the cavity through the variation of μ. Hence $l = 0.5$ mm is a compromise between the need of having a good match and high sensitivity.

The results shown in Fig. 5.11A are for cylindrical of radius 0.5 mm and thickness 0.1 mm. The samples were chosen to be "pure" dielectrics (conductivity $\sigma = 0$) with relative permittivity ε_r varying from 1 (air) to 240 ($SrTiO_3$) (Fig. 5.11A). The distance from the tip to the sample was fixed to be 1 μm. One sees the resonance of the air "sample" occurs at a frequency of 5.8 GHz close to the lowest resonance of the cylinder, 5.74 GHz, with the difference being due to the presence of the leads. To obtain the good match between the coaxial cables and cavity (i.e. $S_{11} \approx -40$ dB for air at resonance) a value of $\mu = 4.137$ was used inside the cables. As discussed in the previous section, this choice was not meant to be realistic, but was picked for computational simplicity. From the inset of Fig. 5.11A one sees changes in magnitude of S_{11} are similar between the model and the experiment (Fig. 5.17A). On the other hand, the frequency shifts in Fig. 5.11A are at least 30 times larger than those of the experiment, and in the wrong direction. When compared to a device that has a more similar geometry [19] ($\lambda/4$ resonator), the shifts calculated in the model are in the right direction (toward lower frequencies) and roughly the same magnitude. One of the most convenient quantities to measure in the experiment is S_{11} at a fixed frequency. The inset of Fig. 5.11A shows the dependences of the microwave reflection coefficient S_{11} (left axis) and resonant frequency shifts $\Delta f / f_0$ (right axis) on the sample relative permittivity at 5.8 GHz. Solid line is the guideline for eyes. The variation of S_{11} with ε at a fixed frequency is a result of two factors [6]. Firstly, the S_{11} curves shift with frequency. Hence for a fixed frequency a different part of the S_{11} curve is "sampled" for different ε. The second factor is that the shape of the S_{11} curve vs. frequency varies with ε. That is, for larger values of ε, the peak becomes shallower and more rounded. We see from inset of Fig. 5.11A, the computational NFMM is quite sensitive for ε in the range 1–10, with sensitivity being rather poor for ε greater than 100.

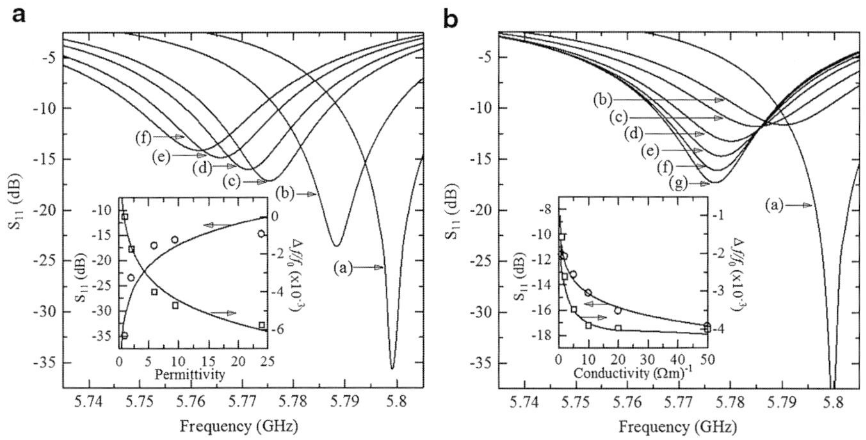

Fig. 5.11 Results of the FEM: (**A**) S_{11} vs. frequency for various dielectrics. The values of the relative dielectric permittivity are: (a) 1, (b) 2.1, (c) 6, (d) 9.5, (e) 24, and (f) 240. (**B**) S_{11} vs. frequency for various conductors. The conductivities in units of $(\Omega\ \mathrm{m})^{-1}$ are: (a) 0, (b) 1, (c) 2, (d) 5, (e) 10, (f) 20, and (g) 50. In all cases the relative permittivity is 1. The *insets* show the dependences of the microwave reflection coefficient S_{11} (*left axis*) and resonant frequency shifts $\Delta f/f_0$ (*right axis*) on the sample permittivity (**A**) and conductivity (**B**) at 5.8 GHz. *Solid line* is the guideline for eyes

Figure 5.11B shows S_{11} vs. frequency is plotted for "metals" with various conductivities σ (these parameter values were selected for illustrative purposes and do not correspond to any real materials). The inset of Fig. 5.11B shows the dependences of the microwave reflection coefficient S_{11} (left axis) and resonant frequency shifts $\Delta f/f_0$ (right axis) on the sample conductivity at 5.8 GHz. Solid line is the guideline for eyes. A 0.1 mm $\times$ 0.5 mm sample with a tip–sample distance of 10 μm is used for six metal samples of conductivity 1, 2, 5, 10, 20, and 50 $(\Omega\mathrm{m})^{-1}$. One observes conductivities can be resolved up to $\sigma = 50\,(\Omega\mathrm{m})^{-1}$; for larger σ the S_{11} curves fall atop each other. Thus, our calculation would give for real metals, i.e. gold, a curve identical to the $\sigma = 50\,(\Omega\mathrm{m})^{-1}$ result. This is not in accord with experimental NFMM measurements (Fig. 5.17B), which can distinguish between good metals $\sigma = 10^7\,(\Omega\mathrm{m})^{-1}$ [61]. This is consistent with our previously mentioned poor sensitivity for large values of permittivity. In addition, for the samples with finite conductivity, the frequency shifts can be to higher frequency rather shifts to lower frequency indicated by Fig. 5.11B. Another feature not in agreement with the experiment [61] is the depth of the resonance in the model increases with increasing conductivity.

There are three major differences between the model and the experiment. The first two are unavoidable because of the cylindrical symmetry of the two-dimensional (2D) model: the device has leads entering the cavity radially, while the second model has leads entering the cavity axially. The experiment makes use of square samples, while the model requires cylindrical samples. The third difference is that the device has a tuning disk that can be raised or lowered to optimize the response of the cavity and a dielectric resonator at the center of the cavity. The FEM is tuned by adjusting the relative permittivity of the material inside the coaxial lead. However, it should be noted that introducing the dielectric resonator and optimizing the size of the model results in better agreement between the model and the device [58].

5.4 Electromagnetic Field Distribution

5.4.1 Probe-tip–fluid Interaction

Electromagnetic field distribution and concentration of electromagnetic energy are important aspects for the description of changes of the material properties probed by NFMM. A common and powerful tool for the determination of the electromagnetic field distribution in complex geometries with variable shapes and inhomogeneous media is numerical simulation using FEM [57–59] in the CST Microwave Studio and Ansoft/HFSS programming codes, which are standard software packages for numerical modeling and simulation of high frequency electromagnetic signals.

To visualize the electromagnetic field distribution during measurement, the simulated electromagnetic field interaction between the cylindrical probe-tip and a glucose solution was obtained using CST Microwave Studio software. The results are presented in Fig. 5.12. From the simulated electromagnetic energy distribution, it is seen that as the glucose concentration increases from 1 to 500 mg/ml (50 wt%), the relative permittivity of glucose solution increases and the electric energy density increases from 20 to 30 mJ/m^3 at the maximum. The impedance of probe-tip–sample system becomes less mismatched with impedance of the dielectric resonator, thus the amplitude of the reflection coefficient S_{11} was decreased. At

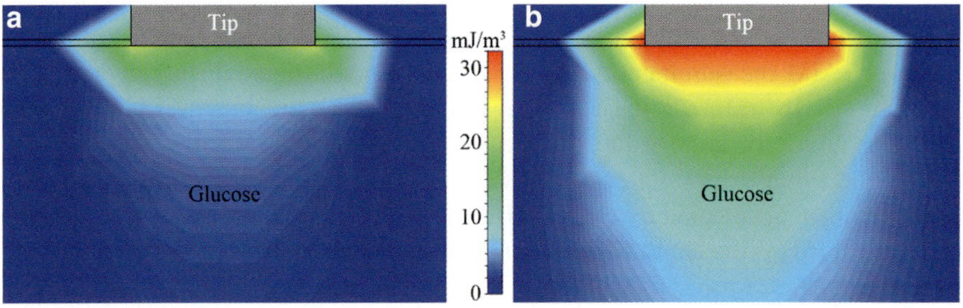

Fig. 5.12 The simulated images for the electric energy densities of the electromagnetic interaction between the cylindrical probe-tip and glucose solution with the glucose concentration of (**a**) 1 mg/ml and (**b**) 500 mg/ml. The glucose solution temperature was 37°C, and the operation frequency was 4.6 GHz

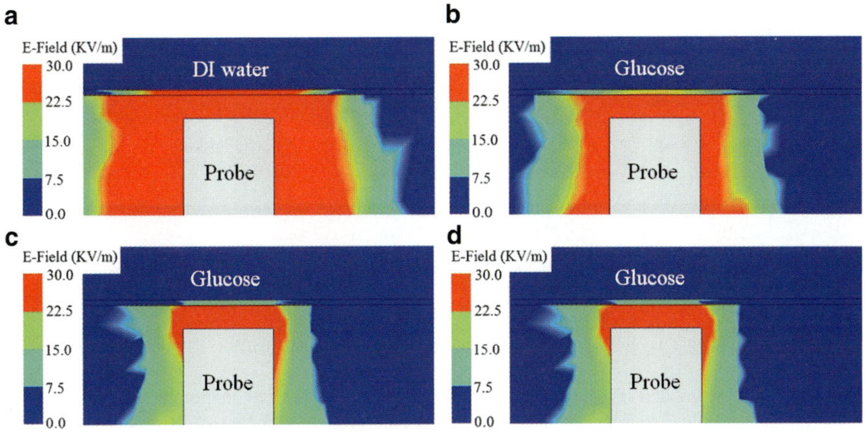

Fig. 5.13 The simulated images of the electromagnetic near-field interaction between a $\lambda/2$ dielectric resonator and (**a**) DI water, (**b**) 100 mg/ml, (**c**) 200 mg/ml, and (**d**) 300 mg/ml glucose aqueous solutions. The glucose solution temperature was 25°C and the operation frequency was 1.7 GHz

the same time the magnetic energy density was still almost the same (not shown here). Note that the electric energy density exceeds the magnetic energy density by about 3 times at the same position.

Figure 5.13 shows the visualized electromagnetic field distribution during measurement. The near-field interaction between the aqueous glucose solution [from de-ionized (DI) water to 300 mg/ml glucose concentration] and the $\lambda/2$ dielectric resonator sensor (microstrip type) was obtained by using Ansoft/HFSS simulation package [62]. As the glucose concentration increases, the relative permittivity of the solution increases and the electromagnetic interaction between the $\lambda/2$ dielectric resonator and the sample decreases. The impedance of probe-tip–sample system becomes less mismatched with impedance of $\lambda/2$ dielectric resonator, thus the amplitude of the electrical field is decreased.

5.4.2 *Probe-tip–photosensitive Heterojunction Interaction*

To visualize the electromagnetic field distribution at heterojunction, we have imaged the near-field interaction between the probe-tip and silicon solar cell under illumination by using Ansoft/HFSS. Figure 5.14 shows the (a), (b) electric and (c), (d) magnetic field distribution between the conical stainless-steel probe-tip and silicon solar cells (a), (c) in dark condition and (b), (d) under illumination with intensity of 160 $\mathrm{mW/cm^2}$ [63]. As the incident light intensity increases, the conductivity of p-type silicon layer increases and the electrical component of the electromagnetic field increases in the solar cell as shown in Fig. 5.14a, b. At the same time

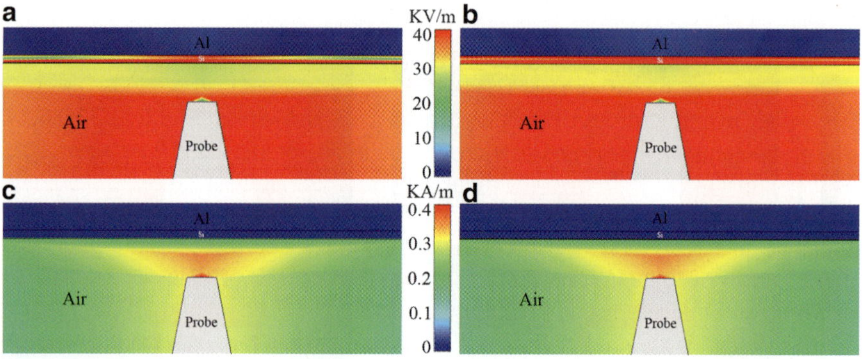

Fig. 5.14 The simulated images for the (**a**), (**b**) electric and (**c**), (**d**) magnetic field distribution between the pyramidal stainless-steel probe-tip and silicon solar cells (**a**), (**c**) in dark condition and (**b**), (**d**) under illumination with intensity of $160\,\mathrm{mW/cm^2}$. The operation frequency was 4.1 GHz

the magnetic component of the electromagnetic field is not changed as shown in Fig. 5.14c, d.

5.4.3 Probe-Tip–Ferromagnetic Thin Film, Magnetic Domain Interaction

To visualize the electromagnetic field distribution for magnetic domains, we have imaged the near-field interaction between the probe-tip and CoCrPtTa soft magnetic layer of hard disk (HD) by using CST Microwave Studio, as shown in Fig. 5.15.

As the external magnetic field intensity increases, the relative permeability of CoCrPtTa soft magnetic layer increases and the magnetic component of the electromagnetic field between the probe-tip and HD decreases in the magnetic layer and increases in air as shown in Fig. 5.15a, b. At the same time the electric component of the electromagnetic field is not changed as shown in Fig. 5.15c, d [53].

Figure 5.16 shows the visualized near-field electromagnetic field distribution between the Py/Si(100) film and the tungsten conical probe-tip under external in-plane magnetic field obtained by a CST Microwave Studio software. The results are shown in Fig. 5.16 where only the probe-tip–sample part is shown in the z-plane under (a), (c) 0 A/m and (b), (d) 650 A/m applied external magnetic field. From the simulated electromagnetic energy distribution, it is seen that as the intensity of the applied external magnetic field intensity is increased from 0 to 650 A/m, the relative permeability of Py decreases [the relative permeability of Si(100) is not changed] from 1.9 to 1.6 mH/m (decreasing about 17.5%) and the magnetic energy density decreases from 2.8 to $2.3\,\mathrm{mJ/m^3}$ at the maximum (decreasing about 17.8%) as shown in Fig. 5.16a, b. The impedance of probe-tip–sample system becomes more mismatched with impedance of the dielectric resonator, thus the amplitude of

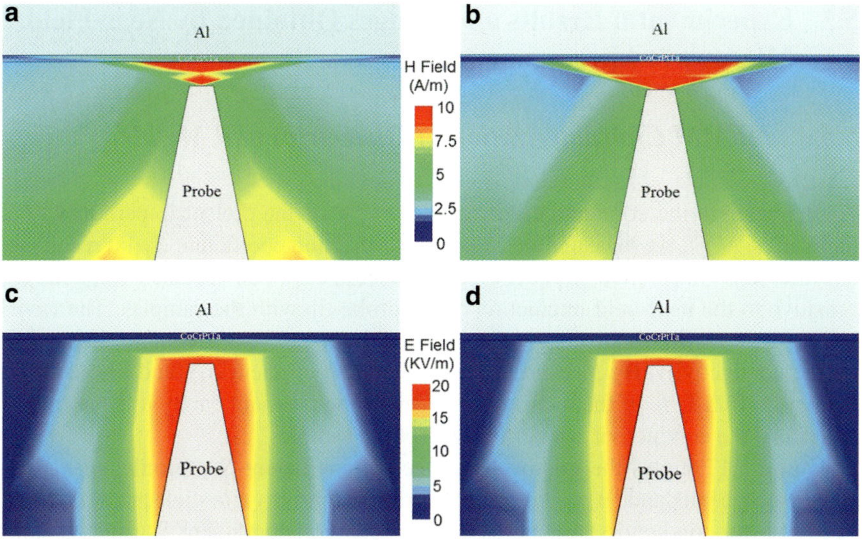

Fig. 5.15 The simulated images for (**a**), (**b**) the magnetic field and (**c**), (**d**) the electric field of the electromagnetic near-field interaction between a conical tungsten probe-tip and the CoCrPtTa soft magnetic layer of HD under external magnetic field intensities (**a**), (**c**) 0 Oe and (**b**), (**d**) 100 Oe

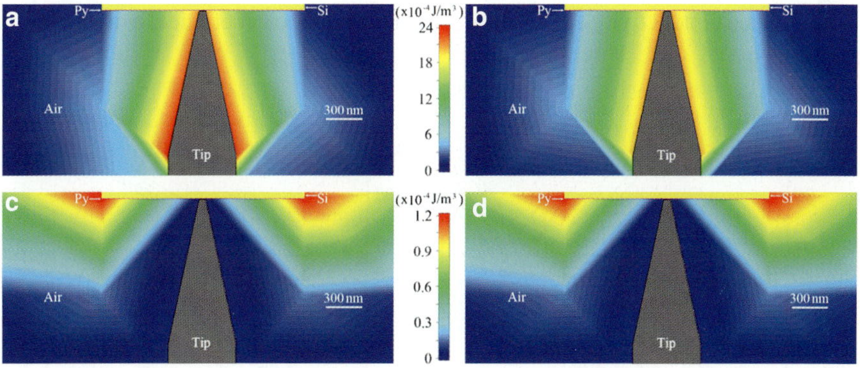

Fig. 5.16 The simulated images for (**a**), (**b**) the magnetic and (**c**), (**d**) the electric energy densities of the electromagnetic near-field interaction between the tungsten conical probe-tip and patterned Py thin film under applied external in-plane magnetic field with intensities of (**a**), (**c**) 0 A/m, and (**b**), (**d**) 650 A/m

the magnetic field decreases. At the same time the electric energy density is almost the same due to the independence of electrical characteristics (dielectric permittivity and conductivity) on the applied external magnetic field as shown in Fig. 5.16c, d. Note that the magnetic energy density exceeds the electric energy density by about 20 times at the same position.

5.5 Experimental Results and Images Obtained by Near-Field Microwave Microscope

5.5.1 NFMM Characterization of Dielectrics and Metals

To characterize the conductivity of metals, $\sigma(f)$, or the dielectric permittivity of dielectrics, $\varepsilon(f)$, we measure the changes of reflection coefficient S_{11} at resonance and the shift of resonant frequency position [58, 65]. The resonant frequency is sensitive to the near-field interaction of the probe-tip with the samples. The cavity and resonator store electromagnetic energy with the resonant frequency depending on the conductivity. By measuring the reflection coefficient S_{11} and resonant frequency shift $\Delta f/f_0$ of the microwave sweep of the resonant cavity it is possible to determine the conductivity or permittivity of the samples.

Figure 5.17 shows the microwave reflection coefficient S_{11} profiles of various (A) dielectric bulk and (B) metal sheets. As the permittivity (for dielectrics) and conductivity (for conductors) are increased, the reflection coefficient S_{11} and resonant frequency shift $\Delta f/f_0$ increases from air to $LaAlO_3$ and from tungsten to silver, respectively, as shown in the insets of Fig. 5.17A, B. Note that resonant frequency shifts to the right on the frequency scale (decreasing of the resonant frequency due to perturbation of the material characteristics: ε and σ). Clearly the change of the permittivity and conductivity affects the microwave reflection coefficient at the resonant frequency.

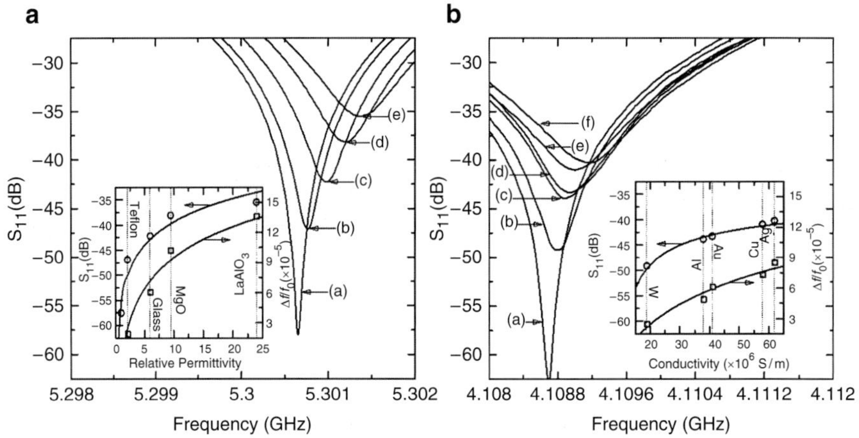

Fig. 5.17 (**A**) Microwave reflection coefficient S_{11} vs. frequency for various dielectrics: (a) air ($\varepsilon_r = 1$), (b) Teflon ($\varepsilon_r = 2.1$), (c) glass ($\varepsilon_r = 6$), (d) MgO ($\varepsilon_r = 9.5$), and (e) $LaAlO_3$ ($\varepsilon_r = 24$). (**B**) Microwave reflection coefficient S_{11} vs. frequency for air (without sample) and for various metals: (b) tungsten ($\sigma = 1.85 \times 10^7$ S/m), (c) aluminum ($\sigma = 3.82 \times 10^7$ S/m), (d) gold ($\sigma = 4.10 \times 10^7$ S/m), (e) copper ($\sigma = 5.81 \times 10^7$ S/m), and (f) silver ($\sigma = 6.17 \times 10^7$ S/m). The *insets* show the dependence of the microwave reflection coefficient S_{11} (*left axis*) and resonant frequency shifts $\Delta f/f_0$ (*right axis*) on the sample permittivity (**A**) and conductivity (**B**) near 5.3 and 4.1 GHz (at the S_{11} minimum), respectively. The *solid line* is a guide for the eyes

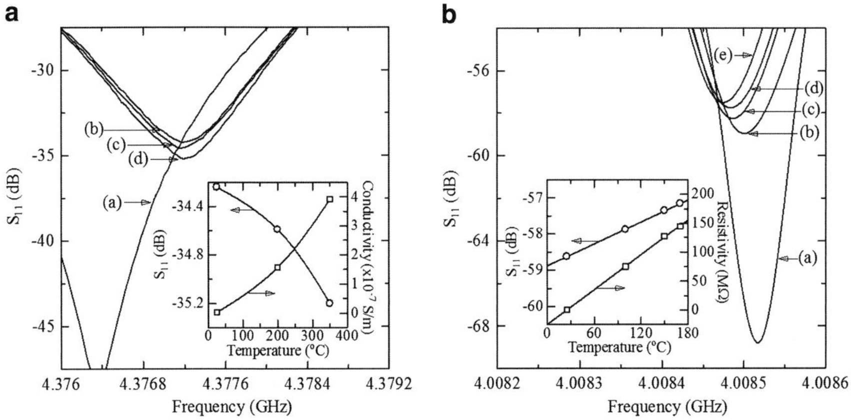

Fig. 5.18 (**A**) The microwave reflection coefficient S_{11} profile for CuPc thin film prepared at (b) RT and annealed at (c) 200°C and (d) 350°C temperatures [curve (a) corresponds to measurement without sample, i.e. in air]. The *inset* shows the microwave reflection coefficient S_{11} (*left axis*) and intrinsic conductivity (*right axis*) dependences of the annealing temperature for the CuPc thin films near 4.37 GHz. The *solid line* is a guide for the eyes. (**B**) Measured reflection coefficients S_{11} for (a) air and for Alq3 thin films with 100 nm thicknesses and with substrate heating temperature of (b) RT, (c) 100°C, (d) 150°C, and (e) 170°C. The *inset* shows the microwave reflection coefficient S_{11} (*left axis*) and sheet resistance (*right axis*) dependence on substrate heating temperature. The *solid line* is a guide for the eyes

5.5.2 NFMM Characterization of Semiconductor Thin Films

To characterize the complex dielectric permittivity of semiconductors under various external influences we measure the changes of reflection coefficient S_{11} at the resonance frequency for CuPc and Alq_3 thin films prepared under various temperature and treatment regimes [64, 65].

Figure 5.18A shows the measured microwave reflection coefficient S_{11} profile for (a) air (without sample) and for CuPc thin films (the thickness is 100 nm) annealed at (b) RT, (c) 200°C, and (d) 350°C for 1 h. The matched resonant curve of the air has a minimum level of −52.09 dB, which is the reference level of S_{11} of our measurements. As the annealing temperature is increased up to 350°C, the minimum reflection coefficient S_{11} increases from −34.24 to −35.21 dB, as shown in the inset of Fig. 5.18A (left axis). The change of reflection coefficient depends on the conductivity of CuPc thin films. Note, that 100 nm CuPc thin film conductivity is about 10^{-10} S/m for as-growth thin films [18]. Compared to RT, the conductivity of CuPc thin film increases about 83 times for 200°C and 100 times at the 350°C annealing as shown in the inset of Fig. 5.18A (right axis).

Figure 5.18B shows the measured microwave reflection coefficient S_{11} for Alq_3 thin film (the thickness is 100 nm) samples with different substrate heating temperatures. The change of the reflection coefficient S_{11} shows direct dependence on the heating temperature. The matched resonant curve for air (a) has a minimum of −68.8 dB at about 4 GHz, which is the reference level of S_{11} of measurements. Substrate heating temperature is increased as (b) RT, (c) 100°C, (d) 150°C, and (e) 170°C. As the heating temperature increases from RT to 170°C, the sample reflection coefficient minimum S_{11} linearly increases from −58.6 to −57.2 dB and the sheet resistance of Alq3 increases from 0.54 to 144.5 MΩ. For the investigated range the empirical dependences of the reflection coefficient on heating temperature and on the sheet resistance of the Alq3 thin films were both approximately linearly with slopes of $\Delta S_{11}/\Delta t = 0.01\,(\mathrm{dB}/°\mathrm{C})$ and $\Delta S_{11}/\Delta R_s = 0.01\,(\mathrm{dB}/\mathrm{M}\Omega)$, respectively. So the empirical dependence of the sheet resistance on the heating temperature is linear with the slope $\Delta R_s/\Delta t = 1.0007\,(\mathrm{M}\Omega/°\mathrm{C})$.

5.5.3 NFMM Characterization of DNA Array, SAMs, and Mixture Fluids

In bioassay applications, DNA detection is often accomplished in an array format. For investigating this type of sample a prototype array is prepared consisting of an immobilized sequence of DNA, P2 (CGT TGT AAA ACG ACG GCC AG) [12,66]. Figure 5.19 shows NFMM images of the (a) as-prepared single-DNA array and (b) after dehybridization to sequences complementary to P2 (5′ CGT TGT AAA ACG ACG GCC AG-$(CH_2)_3$-SH 3′) spots (the image of the hybridized DNA not shown here). Sequence-specific hybridization is the dominant change identified. From the data, deactivation of a fraction of capture strands (e.g. through crosslinking at internal sites within the hybridization sequence) would be expected to decrease the change in S_{11} realizable in response to hybridization of immobilized sequence

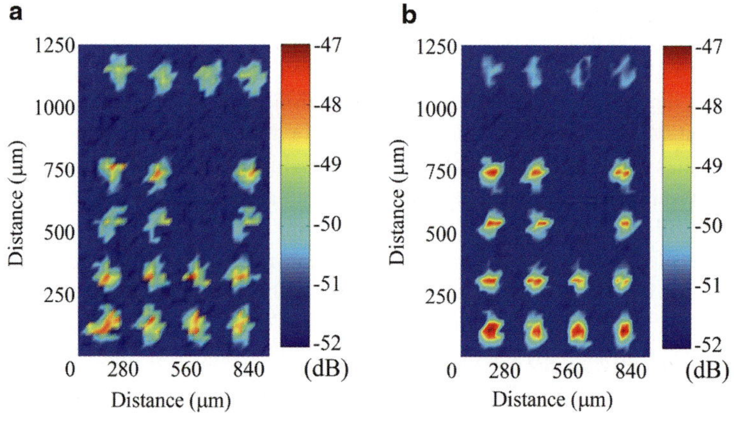

Fig. 5.19 The NFMM images confirm layout of the immobilized P2 sequences: (**a**) the as-prepared single DNA array, (**b**) dehybridization to sequences complementary to P2

P2. Moreover, target molecules can absorb through sequence-nonspecific physical interactions, in this instance increasing the measured change.

These various contributions to S_{11}, representing important but difficult-to-track aspects of surface modification with DNA molecules and hybridization protocols, merit a more detailed study separate from the current focus on the near-field microwave imaging technique. Hybridization between target (free) and capture (immobilized) sequences leads to changes in the microwave reflection coefficient (S_{11}) which are measured by the NFMM. These changes are caused by hybridization-induced modification of the dielectric constant profile of the DNA film. The response curve of log $S_{11}(f_R{}^{ref})$ with nucleotide ($\upsilon\tau$) coverage $\sigma_{\upsilon\tau}$ using a DNA-free area as reference was calibrated. The data can be empirically fit to a linear relationship with dlog S_{11}/dlog$\sigma_{\upsilon\tau}$ $= 1.97 \pm 0.27$ [12]. For an SNR of 9, and at log $S_{11} = -2.0$ when root-means-square noise in log S_{11} is 7.7×10^{-9}. This response allows detection of changes in $\sigma_{\upsilon\tau}$ of 3×10^6 nt/cm^2, corresponding to 1.4×10^5 21mer strands per cm^2, which compares favorably with fluorescence-based scanners widely used in microarray applications. Our NFMM system was reported capable of detecting coverages down to 1×10^9 fluorophores per cm^2 or about 300 zmol (10^{-21}) in a 150 μm diameter microarray spot.

We demonstrate that near-field microwave probing technique can achieve the noncontact detection of the thickness of self-assembled monolayers (SAMs) by measuring the microwave reflection coefficient S_{11} at an operating frequency near 5.3 GHz. The formation and structure of SAMs based on metal bonding have been well-characterized as a model system for understanding organic monolayer interfaces [67–69]. SAMs on a gold (Au) interface are of significant interest and have been widely investigated since the adsorption of di-n-alkyl disulfides on gold has been reported [70, 71]. SAMs of n-alkanethiol have been investigated and characterized by various methods, which have revealed a densely packed, highly ordered, and oriented monolayer structure on gold.

Figure 5.20A shows frequency-sweep data for the reflection coefficient S_{11} measured from a series of n-alkanethiol SAMs at a tip–sample separation of about 10 nm. Alkanethiol SAMs thicknesses were increased systematically from about 0.7 nm to over 1.4 nm by adjustment of the alkyl chain length. The SAMs were prepared on gold-coated glass slides with the thickness of the gold equal to the skin depth 1 μm of the microwave radiation so as to approximate a semi-infinite medium. As the frequency of the microwave source is swept, each trace in Fig. 5.20A exhibits a minimum corresponding to the standing wave mode for that particular resonator-sample combination. The minimum is most pronounced for the hexanethiol (C6) specimen with 0.72 nm thickness since this sample was used as the reference level. The reflection coefficient change ΔS_{11} is directly related to ΔZ, which is directly related to the thickness change Δt of SAMs films. Clear changes are evident in response to differences in SAM thickness. Note that, as the SAMs thickness increased, the impedance of SAMs increased and the intensity of the reflection coefficient S_{11} increased according to transmission line theory model (5.6) and (5.7) [52]. Thus, the thickness (chain length) of the SAMs can be inferred by the intensity of the reflection coefficient at the resonance frequency. The inset of Fig. 5.20A

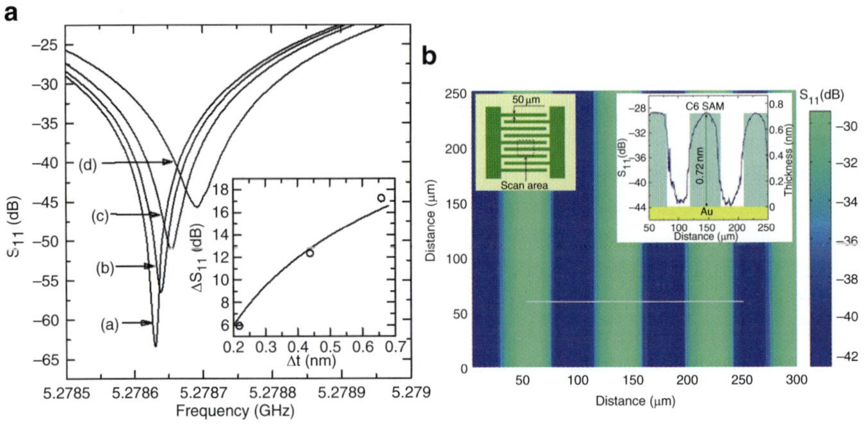

Fig. 5.20 (**A**) Microwave reflection coefficients S_{11} profiles for a series of n-alkanethiol SAMs: (a) hexanethiol (C6), (b) octanethiol (C8), (c) decanethiol (C10), and (d) dodecanethiol (C12). The *inset* shows the reflection coefficient changes ΔS_{11} of the series of n-alkanethiol SAMs, at the reflection minimum, plotted against the changes of thickness of samples. A *solid line* shows a fit to the transmission line theory model. (**B**) 2D NFMM image of the microwave reflection coefficient changes of patterned SAMs on Au substrates. The *left inset* shows SAMs pattern as a periodic structure with 50 μm stripe width. The *right inset* shows the cross-sectional profile indicated by *white line* (left axis) and schematic topography (*right axis*) of C6 SAMs sample pattern with 0.72 nm thickness

plots the measured increase in reflection coefficient variation, ΔS_{11} from C6 to C8, C10, and C12, at the reflection minimum vs. thickness changes, Δt of SAMs. Film thickness was assumed to increase by $\Delta t = 0.22$ nm per two methylene units [72]. Clearly the change of the thickness of SAMs affected the change of the microwave reflection coefficient S_{11}. To demonstrate the ability of NFMM to characterize SAMs properties, we imaged the patterned SAMs (see the left inset of Fig. 5.20B).

Figure 5.20B shows 2D microwave reflection coefficient S_{11} image of hexanethiol (C6) SAMs with the thickness of 0.72 nm. The microwave images showed clear contrast between the SAMs (upper level) and the Au substrate (lower level). The right inset of Fig. 5.20B shows a C6 SAMs microwave reflection coefficient S_{11} line scan and the topography of the SAMs sample pattern. The possibility of determining different thiol terminations is determined by physical properties differences in the microwave range.

5.5.4 *Biosensing of Fluids by a NFMM*

The real-time detection and quantification of concentration of glucose, NaCl and other compounds in fluids with high sensitivity, selectivity, and accuracy is required in many different areas. Accurate and rapid measurement of the glucose concentration is very important in biological analysis and clinical monitoring and in the food processing industry [73–76]. Aqueous NaCl solutions play a fundamental role in many chemical processes in a variety of chemical and biological systems [77, 78]. Here, we monitor the glucose concentration using a microwave dielectric resonator technique. To determine whether the device is capable of differentiating between glucose and other chemicals in solution, we measured the reflection coefficient S_{11} of the mixed solution of glucose with NaCl [48, 53, 79].

Figure 5.21A shows the microwave reflection coefficient S_{11} profiles of aqueous glucose for concentrations ranging from DI water to 50 mg/ml glucose concentration with the interval of 10 mg/ml. As the glucose concentration increased, the dielectric permittivity of the glucose solution increased and the intensity of the reflection coefficient S_{11} decreased. The inset of Fig. 5.21A shows the microwave reflection coefficient S_{11} as a function of concentration. The theoretical curves were derived by using (5.6)–(5.9). As the concentration increased up to 50 mg/ml, the dielectric permittivity of the glucose solution increased and the intensity of the reflection coefficient S_{11} decreased as shown in the inset of Fig. 5.21A. Here, the permittivity of the aqueous glucose solution could be estimated by the intensity of the reflection coefficient, because the dependence of relative permittivity on solute glucose concentration is approximately linear and is often expressed as the molar increment of the glucose solution. We also found a good agreement with theory for higher glucose concentration (0–300 mg/ml) for the variation of the reflection coefficient S_{11} with glucose concentration [62]. From the linear relationship as

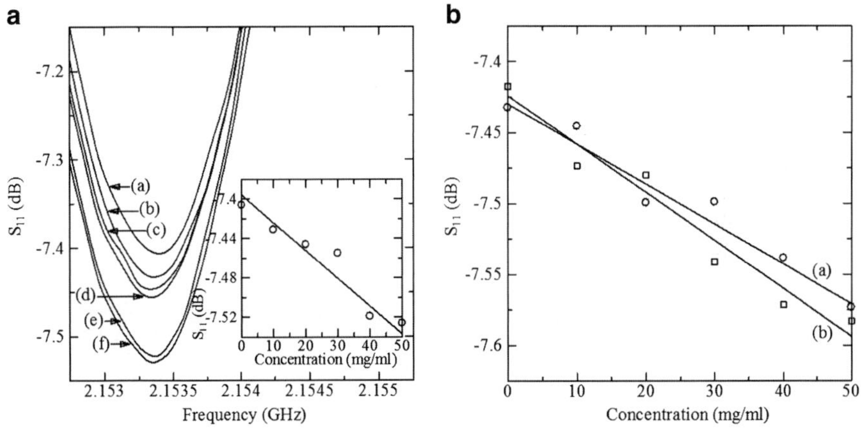

Fig. 5.21 (**A**) Measured microwave reflection coefficient S_{11} for (a) DI water and for different glucose concentrations: (b) 10, (c) 20, (d) 30, (e) 40, and (f) 50 mg/ml with glucose solution volume of 1 ml. The *inset* shows microwave reflection coefficient S_{11} plotted as a function of the glucose concentration. A *solid line* shows a fit to (5.6) with $Z_0 = 7.6 + j\ 3.2\,\Omega$. (**B**) Measured microwave reflection coefficient minimum S_{11} of the mixture solution of glucose and NaCl plotted as a function of (a) glucose and (b) NaCl concentration with fixed (a) NaCl and (b) glucose concentration at 10 mg/ml. The solution volume was 1 ml. A *solid line* shows a fit to (5.6) with $Z_0 = 7.6 + j\ 3.2\,\Omega$

a function of glucose concentration, as observed at lower glucose concentrations, $\Delta S_{11}/\Delta c = 0.0028\,\text{dB}/(\text{mg/ml})$. The measured increase in relative permittivity as the solute concentration of glucose is raised by 1 unit was $\gamma' = 0.0577\,(\text{mg/ml})^{-1}$ and $\gamma'' = 0.0015\,(\text{mg/ml})^{-1}$.

We measured the reflection coefficient S_{11} of the mixed solution of glucose with NaCl. For this experiment the concentration of NaCl in solution was kept constant (10 mg/ml) and only glucose concentration was changed or the concentration of glucose in solution was kept constant (10 mg/ml) and only NaCl concentration was changed. Figure 5.21B shows the microwave reflection coefficient S_{11} for glucose and NaCl mixture solution with (a) the glucose concentration and (b) the NaCl concentration variation. As the NaCl concentration increased, the relative permittivity of the mixture solution increased as expected and the intensity of the reflection coefficient S_{11} decreased. The measured increase in relative permittivity as the solute concentration of NaCl is raised by 1 unit was $\gamma' = 0.0695\,(\text{mg/ml})^{-1}$ and $\gamma'' = 0.0019\,(\text{mg/ml})^{-1}$. The sensitivity of the microwave reflection coefficient vs. the NaCl concentration was $\Delta S_{11}/\Delta c = 0.0034\,\text{dB}/(\text{mg/ml})$. Note that we could determine the concentrations of glucose (or NaCl) in solution through the separate measurement of the loss tangent $\tan\delta$ and the reflection coefficient S_{11} [79].

5.5.5 NFMM Characterization of Solar Cells

Solar cells work using a semiconductor that has been doped to produce two different regions, an n-doped region and a p-doped region. Across this heterojunction, the two types of charge carrier, electrons and holes, are able to cross. In doing so, they deplete the region from which they came and transfer their charges to the new region. These migrations of charges result in a potential gradient or electrical slope, which charge carrier tends to slide down as they approach the junction. In particular, our solar cells consisted of five layers: the metallic back contact, the p-type semiconductor, the n-type semiconductor, the antireflection coating, and the transparent adhesive. How the reflection coefficient S_{11} depends on the photovoltaic conducting properties of solar cells can be derived by using standard transmission line theory [39].

Figure 5.22A shows the measured microwave reflection coefficient S_{11} of solar cells for (a) the dark regime and the white incident light intensities of (b) 122 mW/cm^2, (c) 146 mW/cm^2, (d) 163 mW/cm^2, and (e) 166 mW/cm^2. As the frequency of the microwave source is swept, each trace in Fig. 5.22A exhibited a minimum corresponding to the standing wave mode for that particular resonator-sample combination. The matched resonance curve for the (a) dark regime has a minimum level of −40.7 dB, which is the reference level for the microwave reflection coefficient S_{11} of solar cells in measurements. As the light intensity increased, the photoconductivity of solar cell increased and the reflection coefficient S_{11} increased as shown in the inset of Fig. 5.22A. We found a good agreement with theory between the

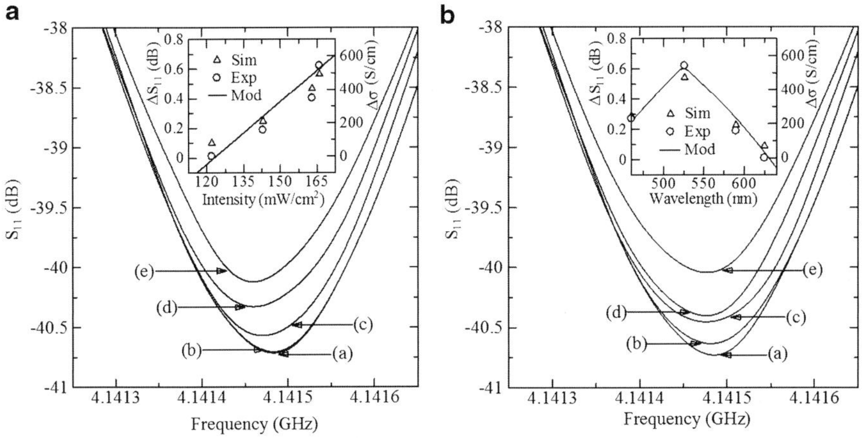

Fig. 5.22 (**A**) Measured microwave reflection coefficient, S_{11} of the solar cells for (a) the dark condition and for the various white light intensities. The *inset* shows the change in microwave reflection coefficient minimum, ΔS_{11} (*left axis*) and the change in photoconductivity (*right axis*), $\Delta\sigma$, plotted as a function of the incident light intensity. The *solid line* shows a fit to the microwave reflection coefficient S_{11} with $Z_0 = 50\,\Omega$ and a S_{11} offset of -42.38 dB, the *open-circle* symbols show the measured data, and the *open-triangle* symbols show the results calculated by a computer simulation. (**B**) Measured microwave reflection coefficient, S_{11}, of the solar cells for (a) the dark condition and for the various incident light wavelengths. The *inset* shows the change in microwave reflection coefficient minimum, ΔS_{11} (*left axis*) and the change in photoconductivity (*right axis*), $\Delta\sigma$, plotted as a function of the incident light wavelength. The *solid line* shows a fit to the microwave reflection coefficient S_{11} with $Z_0 = 50\,\Omega$ and a S_{11} offset of -45.73 dB, the *open-circle* symbols show the measured data, and the *open-triangle* symbols show the results calculated by a computer simulation [63]

variation of reflection coefficient S_{11} and the incident light intensities with a fit to the S_{11} formula with an offset of $S_{11} = -42.38$ dB.

Such an offset is necessary, because our simple theory is phenomenological; the near-field effect is not properly taken into account. On the other hand, the results obtained by using a computer simulation program Ansoft/HFSS (open-triangle symbols) also confirm the experimental results (open-circle symbols) as shown in the inset of Fig. 5.22. From the simulated electromagnetic field distribution (Fig. 5.14), it is seen that as the incident light intensity increased, the photoconductivity of the n-type silicon layer increased and the electromagnetic field between the probe-tip and solar cell was decreased in the solar cell and increased in air. The impedance of the probe-tip and solar cells system becomes more mismatched with the impedance of the dielectric resonator, thus the amplitude of the reflection coefficient S_{11} was increased. We assumed the photoconductivity of n-type silicon layer increased as the incident light intensity increased due to the carrier motion through the p–n junction. The conductivity is directly proportional to the carrier mobility and the incident light intensity. The lifetime of the charge carriers decreased as the incident light intensity increased and the product $\mu_n \times \tau_n = 3.75\,\text{cm}^2/\text{V}$ appeared almost constant independent of the incident light intensity. Thus, the change of dynamic conductivity depends only on the incident light intensity. The obtained maximum short circuit current, I_{sc}, open circuit voltage, V_{oc}, and output power, P_{max}, were 0.29 mA, 0.62 V, and 0.53 mW, respectively. The estimated fill factor, FF, was $FF = P_{max}/(I_{sc} \times V_{oc}) = 0.58$.

Figure 5.22B shows the measured microwave reflection coefficient S_{11} of solar cells for (a) the dark regime and the incident light wavelengths of (b) 625 nm (Red), (c) 590 nm (Yellow), (d) 460 nm (Blue), and (e) 526 nm (Green) with the fixed light intensities of 166 mW/cm^2. The light wavelength variation changed the photoconductivity of the solar cell which changed the reflection coefficient S_{11} as shown in the inset of Fig. 5.22B. We found a good agreement with theory between the variation of reflection coefficient S_{11} and the incident light wavelengths with a fit to the microwave reflection coefficient S_{11} with offset of −45.73 dB. The results obtained by using a computer simulation program Ansoft/HFSS (open-triangle symbols) confirm the experimental results (open-circle symbols) as shown in the inset of Fig. 5.22B. From the simulated electromagnetic field distribution (not presented here), it is seen that the photoconductivity of n-type silicon layer has a maximum at 526 nm, and the electromagnetic field between the probe-tip and solar cell is smaller in the solar cell and bigger in air at the same wavelength. The impedance of the probe-tip and solar cells system becomes more mismatched with the impedance of the dielectric resonator at wavelength of 526 nm, thus the amplitude of the reflection coefficient S_{11} was increased. The solar cell is sensitive to the incident light photon energy due to different conversion efficiency for the incident light wavelength. Note that the maximum intensity was observable at 526 nm with the conversion efficiency $\eta = 0.36$. The efficiency coefficients of wavelengths of 460 nm, 590 nm, and 625 nm were 0.31, 0.25, and 0.19, respectively. The obtained maximum short circuit current, I_{sc}, open circuit voltage, V_{oc}, and output power, P_{max}, were 0.26 mA, 0.49 V, and 0.36 mW, respectively, at a wavelength of 526 nm.

To visualize the photoconductivity of solar cells under external light sources, we directly imaged the reflection coefficient changes ΔS_{11} of solar cells at 4.1 GHz. Figure 5.23 shows the 3D NFMM images of the microwave reflection coefficient changes $\Delta S_{11} = \left|S_{11}^{i} - S_{11}^{0}\right|$ of solar cells dependence on (A) the incident light intensities and (B) on the incident light wavelengths, where S_{11}^{0} is the reflection coefficient for dark condition and $i = 1, 2, 3$ indicates the incident light intensities of (a) 122 mW/cm^2, (b) 146 mW/cm^2, (c) 166 mW/cm^2 for intensity variation and (d) 625 nm (Red), (e) 526 nm (Green), (f) 460 nm (Blue) with the light intensity of 166 mW/cm^2 for wavelength variation, respectively. The lower baseline of the 3D images is the minimum level of the Ag electrode. In order to evaluate the pure photoconductivity effect at the solar cells, we subtracted the level of the dark condition from the level of the light condition. Note that the reflection coefficient changes originate from the changes of the photoconductivity of solar cells. The changes of reflection coefficient ΔS_{11} are significantly increased when the incident light intensity is bigger than 122 mW/cm^2, which is the threshold electron injection. In addition, the changes of reflection coefficient ΔS_{11} show the maximum at 526 nm incident light wavelength. The above NFMM image directly visualizes how the reflection coefficient changes ΔS_{11} of solar cells depend on the external light

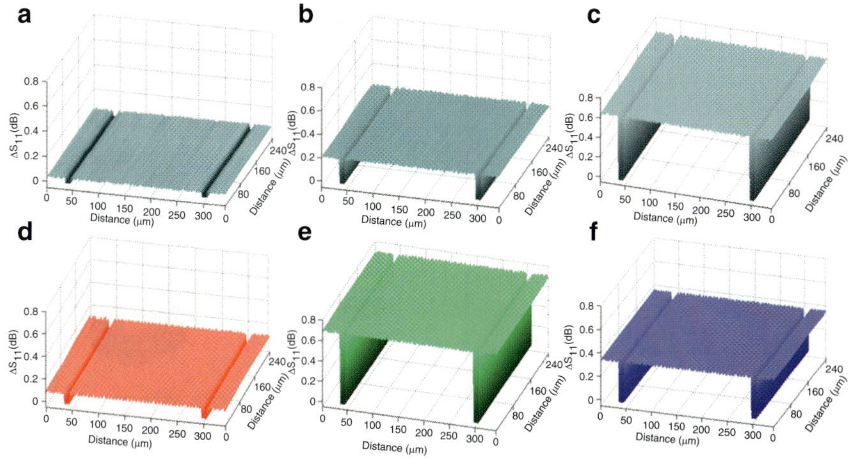

Fig. 5.23 3D NFMM images of microwave reflection coefficient changes of solar cells dependence on the incident light intensity and wavelength with the scan area of 340 × 240 μm^2 at 4.1 GHz. Here $\Delta S_{11} = S_{11}^{i} - S_{11}^{0}$, where S_{11}^{0} is the reflection coefficient for dark condition and $i = 1, 2, 3$ indicates the incident white light intensities of (**a**) 122 mW/cm^2, (**b**) 146 mW/cm^2, and (**c**) 166 mW/cm^2 and the incident light wavelengths of (**d**) 625 nm (*Red*), (**e**) 526 nm (*Green*), and (**f**) 460 nm (*Blue*) with fixed intensity of 166 mW/cm^2

intensity and wavelength of the solar cells. So, the NFMM could directly image the photoconductivity changes inside the solar cells by measuring the change of reflection coefficient S_{11} from the cell.

5.5.6 NFMM Characterization of Organic FET

Direct probing of carrier motion in an organic thin film can be done using a NFMM. This direct observation method allows visualization of conductivity profiles in organic thin films. The near-field microwave reflection coefficient S_{11} image from the organic thin film directly visualizes the conductivity dependence on bias voltages and can be used in exploring carrier behaviors in devices such as OTFTs. The results obtained with NFMM are confirmed by the complementally visualized profile of the electric field distribution which is obtained by using an electric field induced optical second harmonic generation (EFISHG) method [80–85].

The NFMM probes the conductivity $\sigma(x)$ directly in the microwave frequency region. The change of microwave reflection coefficient traces the channel conductivity distribution. The conductivity along the pentacene channel is caused by accumulated charge and intrinsic carriers, and is given in the transmission line model by the expression

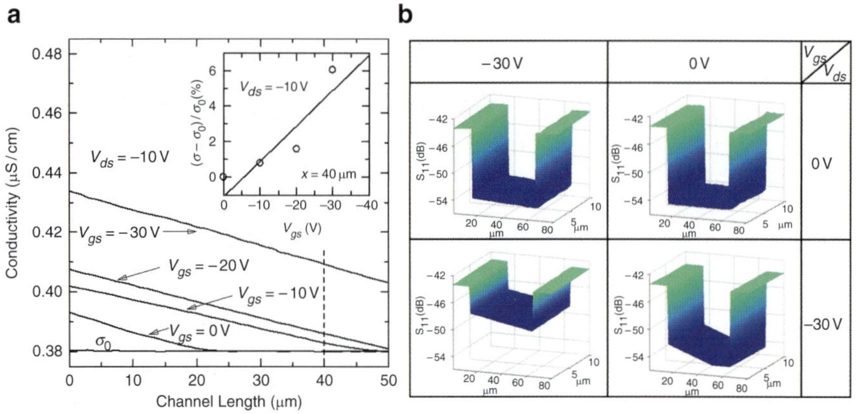

Fig. 5.24 (**a**) The estimated conductivity distribution of the pentacene FET source–drain channel at the various negative gate biases. The *inset* shows the relative change of estimated conductivity (measured by NFMM) $(\sigma - \sigma_0)/\sigma_0$ vs. applied gate bias at a channel position of $x = 40\,\mu\text{m}$ and $V_{\text{ds}} = -10\,\text{V}$. The *solid line* is a fit to (5.5) with $C_{\text{g}} = 7 \times 10^{-9}\,\text{F/cm}^2$ and $\mu_{\text{p}} = 5 \times 10^{-4}\,\text{cm}^2/(\text{V s})$. (**b**) 3D NFMM image of the pentacene FET source–drain channel at the various negative gate biases

$$\sigma = \sigma_0 + \mu C_{\text{g}} \left[V_{\text{ds}} \left(1 - \sqrt{\frac{L-x}{L}} \right) - V_{\text{gs}} \right]; \tag{5.18}$$

in the region $0 \leq x \leq x_p$, whereas in the region $x_p \leq x$, $\sigma = \sigma_0$. Here μ is the mobility of the carriers, σ_0 is the intrinsic conductivity of the organic thin film (for example, for the pentacene $\sigma_0 = 0.38 \times 10^{-6}\,\text{S/cm}$ [81]), and x_p represents the maximum position where holes injected from the source are allowed to accumulate [82].

Figure 5.24a shows the estimated conductivity distribution in the pentacene FET source–drain channel at $V_{\text{ds}} = -10\,\text{V}$. As the magnitude of the negative gate bias increased the conductivity of the channel increased and a nonuniform conductivity distribution, with low density near the drain, was obtained. The inset of Fig. 5.24a shows the measured conductivity (from NFMM) vs. (5.18). Based on this good agreement, we conclude that the injection charges dominate in the conductivity. On the other hand, at $V_{\text{gs}} = 0\,\text{V}$, a convex point x_p, corresponding to the boundary between charge accumulated and exhausted region, is seen. Equation (5.18) accounts well for this result taking into account a threshold voltage $V_{\text{th}} = -10\,\text{V}$, that is by replacing V_{gs} with $V_{\text{gs}} - V_{\text{th}}$. Figure 5.24b shows the 3D NFMM images from the channel of a pentacene FET at various negative source–drain and gate bias voltages. Voltages are applied to the drain and gate electrodes with the source electrode grounded. The change of microwave reflection coefficient traces the channel conductivity distribution.

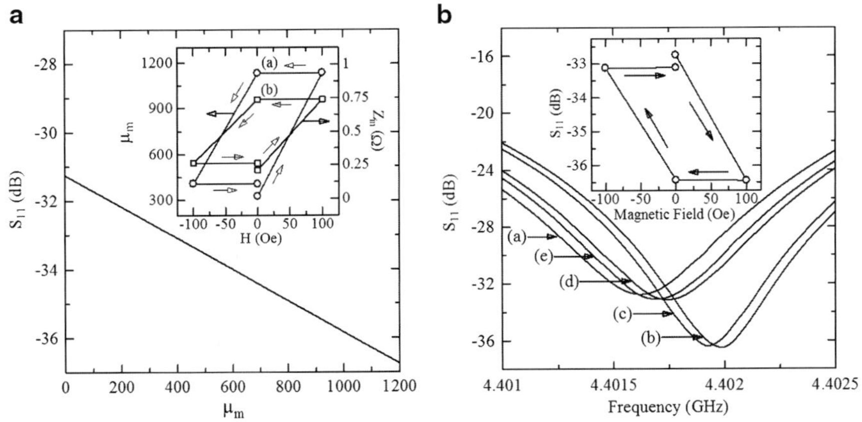

Fig. 5.25 (**A**) Microwave reflection coefficient S_{11} vs. relative permeability μ_m of the CoCrPtTa soft magnetic layer. The *inset* shows (a) the relative permeability μ_m (left axis) and (b) system input impedance Z_{in} (right axis) as a function of applied external magnetic field. The *arrows* indicate the sweep direction. (**B**) Microwave reflection coefficient S_{11} profiles under different external magnetic field intensities: (a) 0 Oe, (b) +100 Oe, (c) 0 Oe, (d) −100 Oe, and (e) 0 Oe. The *inset* shows the measured microwave reflection coefficient S_{11} of the hard disk as a function of the external magnetic field at the resonant frequency. The *arrows* indicate the sweep direction

5.5.7 NFMM Characterization of Magnetic Domains

We demonstrate the measurement of electromagnetic properties of magnetic layers in a HD platter by using a NSMM. As the magnetization changed, the intensity of the reflection coefficient S_{11} varied. The electromagnetic properties of a hard disk were estimated by measuring the microwave reflection coefficient S_{11}.

Figure 5.25A shows the estimated microwave reflection coefficient dependence on the magnetic relative permeability of the CoCrPtTa soft magnetic layer of the HD plater. The inset shows (a) the relative permeability μ_m and (b) system input impedance Z_{in} changes due to variation of the external magnetic field. The transmission line theory [39] predicts that the microwave reflection coefficient S_{11} decreases and the input impedance increases as the magnetic field increases due to the increased relative permeability μ_m. Figure 5.25B shows the measured microwave reflection coefficient S_{11} profile of a 1.2-Gb HD platter under external magnetic intensities of (a) 0 Oe, (b) +100 Oe, (c) 0 Oe, (d) −100 Oe, and (e) 0 Oe. The magnetic field was applied parallel to the HD surface. As the frequency of the microwave source is swept, each trace in Fig. 5.25B exhibited a minimum corresponding to the standing wave mode for that particular resonator-sample combination. As the magnetic field intensity increased up to +100 Oe, the reflection coefficient minimum decreased. As the magnetic field intensity further decreased to 0 Oe, then −100 Oe and finally 0 Oe, the reflection coefficient minimum increased again and demonstrated hysteresis behavior. These changes of the microwave reflection coefficient

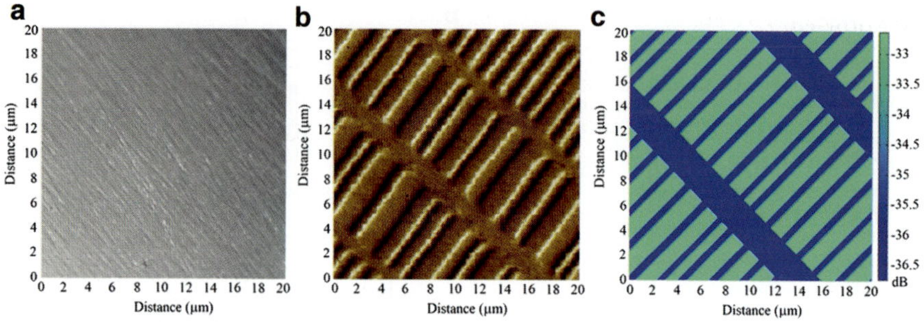

Fig. 5.26 2D (**a**) topographic, (**b**) MFM, and (**c**) NFMM images of the hard disk surface. Scanned surface area was 20 μm × 20 μm

S_{11} were caused by changes of the magnetic permeability. The inset of Fig. 5.25B shows the measured magnetic hysteresis loop of the hard disk sample by using a NFMM system. The arrow indicates the sweep direction. Applied external magnetic field was parallel to the sample surface. The measured reflection coefficient S_{11} is directly related to the magnetization. By measuring the quantitative change of the reflection coefficient S_{11} due to the magnetic moment change, we could find the magnetization change of the hard disk due to the variation of the external magnetic fields.

Here, we also report a NFMM system incorporating an AFM cantilever probe-tip at an operating frequency of $f = 4.4$ GHz [53] (Fig. 5.4). We demonstrate improved sensitivity and spatial resolution better than 50 nm for the magnetic domain images of a 1.2 Gb magnetic HD platter under an external magnetic field. To demonstrate local microwave characterization of magnetic domains by NFMM we imaged domains and directly compared with images received by MFM. The dependence of the magnetic domains on the external magnetic field could be imaged by measuring the microwave reflection coefficient S_{11} and interpreted by transmission line theory [39]. When an external magnetic field is applied, the domains align such that all the domains reorient themselves in the direction of the external field. To visualize the magnetic domains of a hard disk under the external magnetic field, we directly imaged the microwave reflection coefficient S_{11} of the HD at 4.4 GHz.

Figure 5.26 shows the 2D (a) topographic, (b) MFM, and (c) NFMM images of a hard disk surface with surface area of 20 μm × 20 μm under 100 Oe external magnetic field applied parallel to the surface. Note that the reflection coefficient changes originate from the changes of the magnetic permeability of the CoCrPtTa magnetic layer due to its change under external magnetic field. The changes of reflection coefficient S_{11} are significantly increased until the external magnetic field intensity reaches 50 Oe with further increase of magnetization up to 100 Oe. This nanoscale measurement of the magnetic domains has a great potential for investigating the magnetic profile with high sensitivity.

Acknowledgements

This work was supported by Sogang University Special Research Grant (2009, 2010), Seoul Research and Business Development Program (10816), Korea Research Foundation Grant (KRF-2008-615-C00001), and National Research Foundation of Korea Grant (2009-0073557, 2009-0093822, and 2010-0012037).

The author would like to thank Prof. Brian Oetiker and Prof. Khachatur Nerkararyan for careful reading and comments of the manuscript.

References

1. B. Michel, W. Mizutani, R. Schierle, A. Jarosch, W. Knop, H. Benedickter, W. Bachtold, H. Rohrer, Rev. Sci. Instrum. **63**, 4080 (1992)
2. D.W. van der Weide, Appl. Phys. Lett. **70**, 677 (1997)
3. B. Knoll, F. Keilmann, A. Kramer, R. Guckenberger, Appl. Phys. Lett. **70**, 2667 (1997)
4. C. Gao X.-D. Xiang, Rev. Sci. Instrum. **69**, 3846 (1998)
5. J. Kim, M.S. Kim, K. Lee, J. Lee, D. Cha, B. Friedman Meas. Sci. Technol. **14**, 7 (2003)
6. M. Tabib-Azar, Y. Wang, IEEE Trans. Microwave Theory Tech. **52**, 971 (2004)
7. A. Imtiaz, T. Baldwin, H.T. Nembach, T.M. Wallis, P. Kabos, Appl. Phys. Lett. **90**, 243105 (2007)
8. D.E. Steinhauer, C.P. Vlahacos, S.K. Dutta, F.C. Wellstood, S.M. Anlage, Appl. Phys. Lett. **71**, 1736 (1998)
9. M. Tabib-Azar, P.S. Pathak, G. Ponchak, S. LeClair, Rev. Sci. Instrum. **70**, 2783 (1999)
10. M.E. Rodrıguez, A. Mandelis, G. Pan, J.A. Garcia, J. Appl. Phys. **87**, 8113 (2000)
11. M. Kunst, F. Wunsch, D. Jokisch, Mater. Sci. Eng. **B102**, 173 (2003)
12. B. Friedman, M.A. Gaspar, S. Kalachikov, K. Lee, R. Levicky, G. Shen, H. Yoo, J. Am. Chem. Soc. **127**, 9666 (2005)
13. D.I. Mircea T.W. Clinton, Appl. Phys. Lett. **90**, 142504 (2007)
14. M. Tabib-Azar, N.S. Shoemaker, S. Harris, Meas. Sci. Technol. **4**, 583 (1993)
15. M.S. Kim, S. Kim, J. Kim, K. Lee, B. Friedman, J. Kim, J. Lee, Rev. Sci. Instrum. **74**, 3675 (2003)
16. M. Golosovsky, D. Davidov, Appl. Phys. Lett. **68**, 1579 (1996)
17. S.V. Kalinin, A. Gruverman (eds), *Scanning Probe Microscopy: Electrical and Electromechanical Phenomena at the Nanoscale* (Springer, New York, 2006)
18. D.E. Steinhauer, C.P. Vlahacos, F.C. Wellstood, S.M. Anlage, C. Canedy, R. Ramesh, A. Stanishevsky, J. Melngailis, Rev. Sci. Instrum. **72**, 2073 (2001)
19. C. Gao, B. Hu, I. Takeuchi, K.S. Chang, X.-D. Xiang, G. Wang, Meas. Sci. Technol. **16**, 248 (2005)
20. M. Abu-Teir, M. Golosovsky, D. Davidov, A. Frenkel, H. Goldberger, Rev. Sci. Instrum. **72**, 2073 (2001)
21. B. Friedman, M.A. Gaspar, S. Kalachikov, K. Lee, R. Levicky, G. Shen, H. Yoo, J. Am. Chem. Soc. **127**, 9666 (2005)
22. E.H. Synge, Phil. Mag. **13**, 297 (1932)
23. D.A. Holmes, D.L. Feucht, H. Jacobs, Solid-Stat. Electron. **7**, 267 (1964)
24. P.C. Kothari, S. Prasad, Physica **B+C81**, 125 (1976)
25. M.Ya. Azbel, Phys. Lett. **A70**, 455 (1979)
26. M. Bramanti, J. Biomed. Eng. **5**, 240 (1983)
27. F. Buchy, B. Lemaire, Phys. Lett. **A26**, 170 (1968)
28. W. Schilz, B. Schiek, Adv. Electron. Electron Phys. **55**, 309 (1981)
29. E.T. Thostenson, T.-W. Chou, Composites **A30**, 1055 (1999)

30. J.T. Donohue, J. Gardelle, T. Lefevre, J.L. Rullier, C. Vermare, S.M. Lidia, Y. Meurdesoif, Nucl. Instrum. Methods Phys. Res. **A445**, 307 (2000)
31. D. Luo, C. Yu, L. He, C. Lu, D. Gao, Cryobiology **53**, 288 (2006)
32. M. Tabib-Azar, N.S. Shoemaker, S. Harris, Meas. Sci. Technol **4**, 583 (1993)
33. M. Golosovsky, D. Davidov, Appl. Phys. Lett. **68**, 1579 (1996)
34. X.-D. Xiang, C. Gao, Mat. Character. **48**, 117 (2002)
35. A. Imtiaz, S.M. Anlage, J.D. Barry, J. Melngailis, Appl. Phys. Lett. **90**, 143106 (2007)
36. V.V. Talanov, A. Scherz, R.L. Moreland, A.R. Schwartz, Appl. Phys. Lett. **88**, 134106 (2006)
37. D.J.Griffiths, *Introduction to Electrodynamics*, 3rd edn. (Prentice, New York, 1999)
38. M. Bass (ed.), *Handbook of Optics: Geometrical and Physical Optics, Polarized Light, Components and Instruments*, vol. 1, 3rd edn. (McGraw Hill, New York, 2010)
39. D. Pozar, *Microwave Engineering* (Addison-Wesley, New York, 1990)
40. M. Ohatsu (ed.), *Near-Field Nano/Ato Optics and Technology* (Springer, Tokyo, 1998)
41. M. Tabib-Azar, Y. Wang, IEEE Trans. Microwave Theory Tech. **52**, 971 (2004)
42. J.D. Jackson, *Classical Electrodynamics* (Wiley, New York, 1975)
43. F. Zenhausern, Y. Martin, H.K. Wickramasinghe, Science **269**, 1083 (1995)
44. D.A. Lapshin, E.E. Kobylkin, V.S. Letokhov, Ultramicroscopy **83**, 17 (2000)
45. D. Haefliger, M. Muenchinger, G. Schitter, A. Stemmer, Sens. Actuators **A103**, 353 (2003)
46. G.Y. Shang, W.H. Qiao, F.H. Leia, J.-F. Angiboust, M. Troyon, M. Manfait, Ultramicroscopy **105**, 324 (2005)
47. S. Kim, H. You, K. Lee, B. Friedman, M. Gaspar, R. Levicky, Appl. Phys. Lett. **86**, 153506 (2005)
48. A. Babajanyan, J. Kim, S. Kim, K. Lee, B. Friedman, Appl. Phys. Lett. **89**, 183504 (2006)
49. A. Babajanyan, K. Lee, E. Lim, T. Manaka, M. Iwamoto, B. Friedman, Appl. Phys. Lett. **90**, 182104 (2007)
50. E. Lim, T. Manaka, M. Iwamoto, B. Friedman, A. Babajanyan, S. Kim, Y. Yoon, S. Kim, K. Lee, Thin Solid Films **516**, 2573 (2008)
51. H. Melikyan, A. Hovsepyan, T. Sargsyan, Y. Yoon, H. Yoo, A. Babajanyan, K. Lee, Ultramicroscopy **108**, 1030 (2008)
52. A. Babajanyan, H. Melikyan, T. Sargsyan, S. Kim, J. Kim, K. Lee, B. Friedman, Thin Solid Films **517**, 5597 (2009)
53. H. Melikyan, T. Sargsyan, A. Babajanyan, S. Kim, J. Kim, K. Lee, B. Friedman, J. Magn. Magn. Mater. **321**, 2483 (2009)
54. E. Silva, M. Lanucara, R. Marcon, Supercond. Sci. Technol. **9**, 934 (1996)
55. J. Kim, A. Babajanyan, A. Hovsepyan, K. Lee, B. Friedman, Rev. Sci. Instrum. **79**, 089107 (2008)
56. T. Sargsyan, A. Hovsepyan, H. Melikyan, Y. Yoon, H. Lee, A. Babajanyan, M. Kim, D. Cha, K. Lee, Ultramicroscopy **108**, 1062 (2008)
57. R. Coccioli, T. Itoh, G. Pelosi, P.P. Silvester, IEEE Anten. Propagat. Mag. **38**, 37 (1996)
58. B. Friedman, B. Oetiker, K. Lee, J. Kor. Phys. Soc. **52**, 588 (2008)
59. L. Liua, S. Yi, L.S. Ong, K.S. Chian, Thin Solid Films **462–463**, 436 (2004)
60. J. Kim, K. Lee, B. Friedman, D. Cha, Appl. Phys. Lett. **83**, 1032 (2003)
61. S. Yun, S. Na, A. Babajanyan, H. Kim, B. Friedman, K. Lee, Thin Solid Films **515**, 1354 (2006)
62. S. Kim, J. Kim, A. Babajanyan, K. Lee, B. Friedman, Curr. Appl. Phys. **9**, 856 (2009)
63. A. Hovsepyan, A. Babajanyan, T. Sargsyan, H. Melikyan, S. Kim, J. Kim, K. Lee, B. Friedman, J. Appl. Phys. **106**, 114901 (2009)
64. A. Hovsepyan, H. Lee, T. Sargsyan, H. Melikyan, Y. Yoon, A. Babajanyan, B. Friedman, K. Lee, Ultramicroscopy **108**, 1058 (2008)
65. L. Enkhtur, R. Galbadrakh, T. Sargsyan, H. Melikyan, A. Babajanyan, B. Friedman, K. Lee, MNUS J.: Physics **309**, 23 (2009)
66. S. Kim, Y. Jang, S. Kim, T.-D. Kim, H. Melikyan, A. Babajanyan, K. Lee, B. Friedman, J. Nanosci. Nanotechnol. **11**, 1 (2011)
67. M. Tominaga, Sh. Maetsu, A. Kubo, I. Taniguchi, J. Electroanal. Chem. **603**, 203 (2007)
68. Zh. Cao, N. Gu, F. Gong, Mater. Sci. Eng. **C27**, 773 (2007)

69. A. Widge, M. Jeffries-El, X. Cui, C. Lagenaur, Y. Matsuoka, Biosens. Bioelectron. **22**, 1723 (2007)
70. Y. Hou, Y. Chen, N. Amro, K. Wadu-Mesthrige, P. Andreana, G. Liu, P. Wang, Chem. Commun. **19**, 1831 (2000)
71. P. Laibinis, C. Bain, R. Nuzzoand, G. Whitesides, J. Phys. Chem. **99**, 7663 (1995)
72. M.D. Porter, T.B. Bright, D.L. Allara, C.E.D. Chiddey, J. Am. Chem. Soc. **109**, 3559 (1987)
73. A. Heller, Annu. Rev. Biomed. Eng. **1**, 153 (1999)
74. J. Homola, Anal. Bioanal. Chem. **377**, 528 (2003)
75. N. Forrow, S. Bayliff, Biosens. Bioelectron. **21**, 581 (2005)
76. L. Mayor, R. Moreira, F. Chenlo, A.M. Sereno, J. Food. Eng. **81**, 595 (2003)
77. K. Rahmouni, M. Keddam, A. Srhiri, H. Takenouti, Corros. Sci. **47**, 3249 (2005)
78. K. Lee, A. Babajanyan, C. Kim, S. Kim, B. Friedman, Sens. Actuators **A148**, 28 (2008)
79. T. Manaka, E. Lim, R. Tamura, M. Iwamoto, Appl. Phys. Lett. **87**, 222107 (2005)
80. T. Manaka, E. Lim, R. Tamura, D. Yamada, M. Iwamoto, Appl. Phys. Lett. **89**, 072113 (2006)
81. P. Parisse, M. Passacantando, S. Picozzi, L. Ottaviano, Org. Electron. **7**, 403 (2006)
82. R. Tamura, E. Lim, T. Manaka, M. Iwamoto, J. Appl. Phys. **100**, 114515 (2006)
83. J.D. Jackson, *Classical Electrodynamics*, 2nd edn. (Wiley, New York, 1975)
84. H. Yoo, J. Kim, A. Babajanyan, S. Kim, K. Lee, Key Eng. Mater. **321–323**, 1457 (2006)
85. J.H. Jiang, D.L. Wu, Atmos. Sci. Let. **5**, 146 (2004)

Chapter 6
Single Cluster AFM Manipulation: a Specialized Tool to Explore and Control Nanotribology Effects

Guido Paolicelli, Massimo Rovatti, and Sergio Valeri

Abstract A new experimental approach dedicated to the analysis of friction effects on individual nano-contacts has been recently developed. Controlled nano-contacts are obtained depositing well-prepared nano-clusters on clean surfaces. Then clusters are moved in a controlled way using the atomic force microscope (AFM) set-up. Tribology information is obtained analyzing, at the onset or during the cluster translation, the torsional bending of the cantilever or the energy dissipation of the tip in amplitude modulation feedback (AM-AFM). Among these two possible AFM modes, we will describe in detail the analysis of the energy dissipation signal. As an example of this technique, recent results on the contact area dependence of frictional forces on antimony or gold clusters on HOPG will be presented.

6.1 Introduction

The atomic force microscope (AFM), initially developed to achieve ultimate surface imaging and immediately recognized as a key instrument to perform nanoscale tribology [1], nowadays can be considered an invaluable characterization technique in surface science. Using special tips or dedicated controlling methods, the AFM is able to map local magnetic fields, work function variations and surface hardness, to mention just few examples, with a lateral resolution that easily reaches the nanoscale range. The AFM is also exploited as a versatile tool in the field of nanotechnology, to locally modify surfaces at the nanoscale [2], to build up artificial nanostructures [3,4] and to control and precisely manipulate surface nano-objects [5–10]. In the latter case, small movements are usually obtained pushing a nano-object (equivalently nano-cluster in the following) with the AFM tip from the side while a complete and controlled displacement is obtained composing a pathway with a sequence of single small movements. In the following, we will refer to a complete process as a *manipulation procedure*.

A vast literature exists which describes the optimization of different *manipulation procedures* ([11–13] and references therein) but the topic of this contribution, even if it is connected to the general problem of controlling movements at the

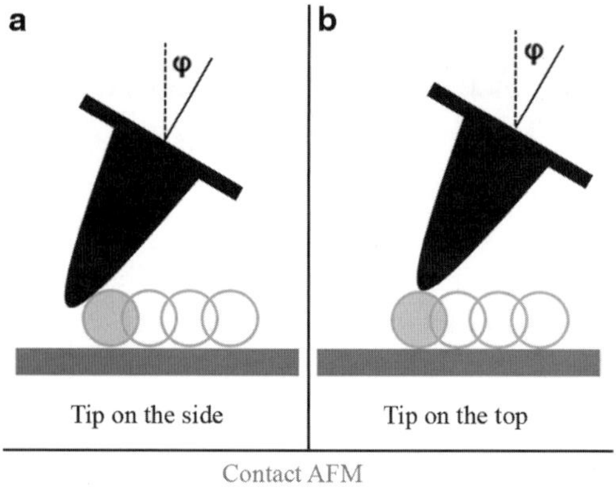

Fig. 6.1 AFM used in contact mode to push the cluster by the side (**a**) or to drag it with the tip placed on top of the cluster (**b**). Friction forces are measured from the torsional bending of the cantilever φ like in a standard friction force microscopy (FFM). Tip and cluster dimensions are not on scale. Usually this method is applied to clusters having lateral size much larger with respect to the curvature radius of the tip [14–16]

nanoscale, is different. In the following, we will describe how clusters manipulation by AFM could be exploited to extract tribology information on isolated, well-defined nano-contacts.

The subject has been explored systematically only in the last few years after the pioneering work of Lüthi et al. on C60 molecules in 1994 [6]. Two different approaches have been developed to manipulate nano-clusters and to obtain tribology information from these measurements. The first one is evident, the AFM is used in contact mode to push the cluster by the side (Fig. 6.1a) or to drag it with the tip placed on top of the cluster (Fig. 6.1b). In that case friction forces are measured from the torsional bending of the cantilever, e.g., in standard friction force microscope (FFM) [14–17]. The underlying hypothesis is that tip–cluster interaction gives a contribution to the overall bending that can be either properly subtracted or neglected with respect to the cluster/surface contribution. The second approach relies on the use of the AFM in dynamic mode with amplitude modulation feedback (AM-AFM or tapping mode, Fig. 6.2). In this method, depinning and movements are induced by the intermittent tip pushing on the cluster and this effect is correlated to the energy dissipation signal reconstructed from phase shift measurements [19, 22].

The direct evaluation of a friction force corresponding to the induced cluster movement is the main advantage of measuring in contact mode. Nevertheless, particles having an overall dimension comparable or smaller than the typical tip apex cannot be easily addressed because aligning the sliding path of the cluster to the torsional movement of the cantilever may result in a very critical issue. The opposite situation characterizes the dynamic manipulation mode. The periodic tip–cluster interaction, typical of AM-AFM, has been proved to induce depinning events with particles having comparable dimensions with respect to tip apex and to produce smooth cluster movements along a precise direction [20, 23, 24]. Nevertheless, the

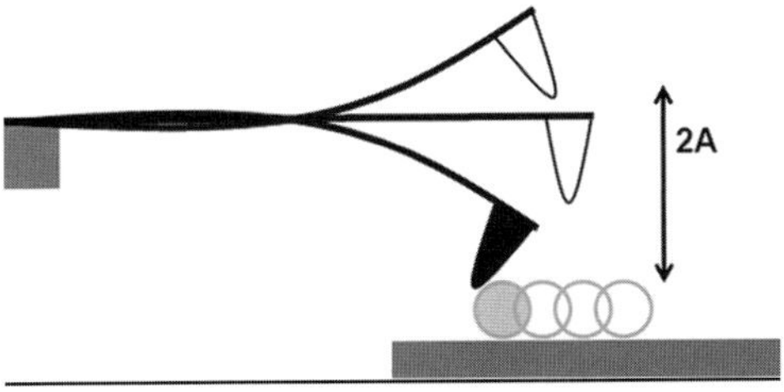

Fig. 6.2 AFM used with amplitude modulation feedback. The tip oscillates with a fixed peak to peak amplitude $2A$. The periodic interactions between tip and cluster induce depinning events on particles having comparable dimensions with respect to tip apex and produce smooth cluster movements [18–20]. Here depinning and induced movements may be correlated to the energy dissipation signal reconstructed from phase shift measurements [17, 21]

connection of the corresponding energy dissipation signal to the friction forces acting on the system cluster/surface requires an elaborate modeling.

Independent of the particular mode adopted to induce clusters movement and to extract friction information, the main advantage of this method as compared to FFM is on the control of the sliding interface (Fig. 6.3). The contact region between the cluster and the substrate now represents the interface we are measuring on, while the AFM tip is part of our apparatus and it is ideally disentangled from the physical system in exam. Using well-established deposition techniques, clean and controlled nano-interfaces can be obtained by preparing first the surface and then depositing the overlayer in the form of nano-clusters. In comparison, during FFM measurements only the substrate can be properly cleaned and characterized while the tip, which is the sliding counterpart, is usually much more difficult to prepare and to maintain in stable conditions. Moreover, the use of nano-clusters allows a better control on shape and size of the interface. The contact area can be varied by simply choosing a cluster with a proper size and its shape can be monitored before and after the measurement.

In the following, we will concentrate on cluster manipulation by dynamic AFM. In the final part of this presentation we will show how this class of experiments can shed light into the fundamental problem of contact area dependence of frictional forces at the nanoscale.

6.2 Manipulation and Friction Effects Explored by Dynamic AFM

6.2.1 Experimental Evidences

The amplitude modulation is the dynamic AFM mode regularly used for imaging of soft and hard materials both in air and liquid environments [25]. In this operational

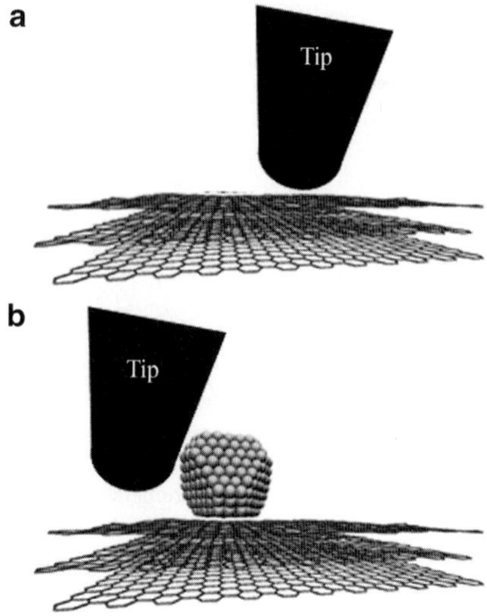

Fig. 6.3 Schematic representation of the analysed interface during FFM measurements (**a**) and cluster manipulation (**b**). In the first case usually the substrate only can be properly cleaned and characterized while the tip is amorphous and difficult to maintain in stable conditions. In the second case the contact region can be controlled on either side using well-established deposition or growing techniques for both the cluster and the surface. The AFM tip is ideally disentangled from the physical system in exam

mode the cantilever is driven by an oscillating piezo placed at the base of the beam. The frequency is held fixed at or close to the cantilever proper resonance; so the tip oscillation amplitude becomes relatively large and linearly depends on the amplitude of driver movement. Approaching the substrate, tip oscillations are damped because of the interaction with surface and the damping effect depends on the separation distance. By introducing a feedback loop, which acts on the separation between cantilever base and surface, and keeping reduced but constant the tip oscillation amplitude while scanning over the sample, a topographic map is obtained. The method is usually called *tapping mode* because the feedback is set in such a way that tip at the bottom of each oscillation cycle interacts with the substrate both in the attractive and eventually in the repulsive regime. Even if this method is widely used on delicate samples also, during the imaging of loosely bonded nano-objects, unintended particle translations are often obtained because feedback activation always requires some interaction between the tip and the substrate. These movements are uncontrolled because the signal optimization is focused to obtain high-quality images. Nevertheless, if the interaction strength is carefully tuned, selected detachments and controlled movements of nano-clusters may be achieved.

In Fig. 6.4 we show the effectiveness of this method. Panel (a) is the initial topographic image of a gold spherical nano-cluster (nominal diameter 42 nm) deposited on freshly cleaved HOPG graphite. In panel (b) the oscillation amplitude of the driver piezo, which is easily controlled even line by line during a tapping mode

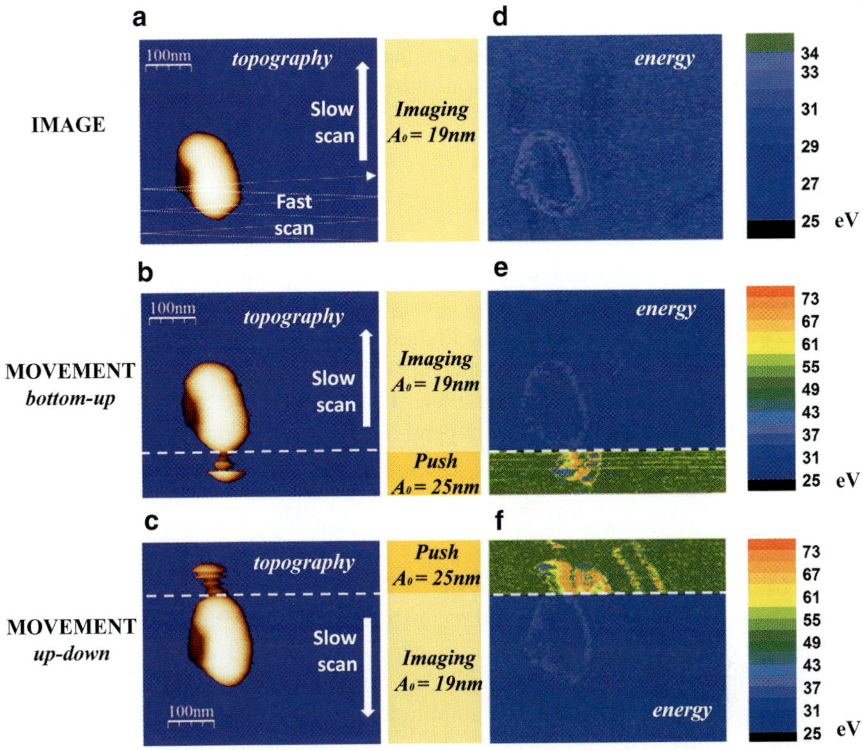

Fig. 6.4 Panel (**a**) Topographic image of a gold nano-cluster deposited on a freshly cleaved HOPG graphite (the blue plane is a 10-nm cut-off introduced for graphical reason). The measured height is in agreement with the nominal diameter of the deposited clusters while the shape strongly depends on tip dimensions. The corresponding energy dissipation signal is shown in panel (**d**). Panel (**b**): The tip oscillation amplitude A_0 is increased at the beginning of the scan (*bottom*) by increasing the piezo amplitude oscillations (the amplitude set point is held fixed along the whole scan). As a consequence the energy dissipation signal increases [panel (**e**)], and the particle detachment is induced. After few scan lines A_0 is reduced to the optimized value for imaging and the particle movement is stopped, the scan continues as a normal image acquisition. In panels (**c**) and (**f**) the same procedure is repeated while scanning the image from the top. The topographic images which correspond to panels (**a**)–(**c**), correctly do not reveal the variation of tip oscillation amplitude while the simultaneous acquisition of the phase shift signal, transformed into energy dissipation per cycle by (6.2), makes evident the energy dissipated into the substrate by the tip–cantilever system [panels (**d**)–(**f**)]

image, is increased at the beginning of the scan (bottom) to induce particle detachment and suddenly reduced to the optimized value for imaging to stop the particle movement. In panel (c) we repeat the same procedure scanning the image from the top. Correctly, the topographic images which correspond to panels (a–c) do not reveal the variation of driver oscillation amplitude because the amplitude set-point is kept fixed, while the simultaneous acquisition of the phase shift signal makes evident the effect of increasing the piezo oscillation amplitude. In particular, the use of the phase shift signal in AM-AFM mode allows to control and to evaluate the energy released to the substrate by the tip–cantilever system. The dissipated energy per cycle corresponds to phase variations through the relation [26, 27]:

$$E_{\rm tip} = \pi k A \left(A_{\rm d} \sin(\phi) - \frac{A}{Q} \right), \tag{6.1}$$

where $A_{\rm d}$ is the oscillation amplitude of the piezo and A is the oscillation amplitude of the tip near the surface, usually called amplitude set-point. Q is the quality factor of the resonance curve, k is the elastic constant of the cantilever and φ is the phase shift between the external driving oscillation and the tip response. It is useful to replace the piezo oscillation amplitude $A_{\rm d}$ with A_0, the tip oscillation amplitude far from the surface, because this parameter can be easily measured and calibrated during the experiment and it linearly depends on $A_{\rm d}$ at the resonance ($A_0 = Q\ A_{\rm d}$):

$$E_{\rm tip} = \frac{\pi k A}{Q} (A_0 \sin(\phi) - A). \tag{6.2}$$

Panels (d–f) of Fig. 6.4 represent the signal $E_{\rm tip}$ corresponding to the topographic images presented on panels (a–c) and reconstructed from the phase data by (6.2).

The analysis of these images puts in evidence the typical effect that characterizes the interaction between the AFM oscillating tip and loosely bonded nano-objects. On one hand, a large tip oscillation amplitude, obtained by increasing the forcing oscillator amplitude $A_{\rm d}$, produces an increase of energy dissipation during tip–substrate interaction and, above a given threshold, it induces the detachment of nano-clusters. On the other hand, above the threshold, a smooth and controlled movement of the nano-cluster is usually obtained most likely as a result of many subsequent detachments and small movements.

In both cases the overall dynamical behavior of these nano-objects seems to be remarkably similar to that of macroscopic systems subjected to an external mechanical excitation. First to begin a cluster movement, a given threshold force should be applied, then the object moves but the displacement is damped by sliding friction. In our experimental conditions clusters may be repeatedly tapped and eventually detached and moved by the tip which continuously oscillates and scans over the surface. So the combination of these small displacements with AFM tip movements generates a visible track that we will call *trajectory* in the following. According to this model, dynamic cluster manipulation can help understanding nanoscale tribology effects both for what concern the static as well as the sliding friction behavior. In the first case the energy dissipation signal can be conveniently used to characterize the onset of sliding, while in the other case the sliding distance or the overall trajectory behavior are the interesting parameters to possibly extract tribology information on kinetic friction.

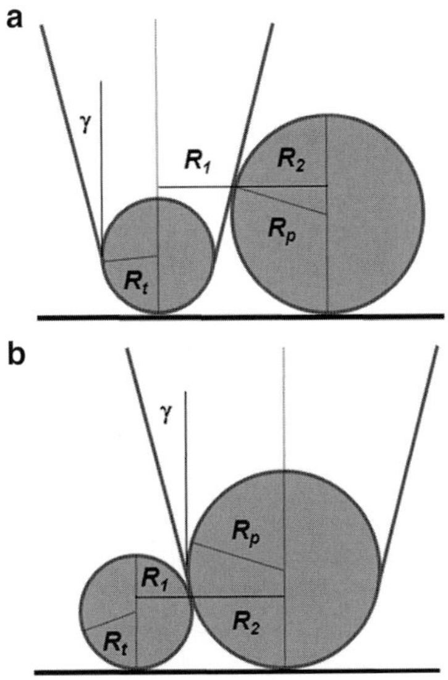

Fig. 6.5 Schematic representation of shape and relative positions of the tip and cluster. The relevant dimension parameter is the effective radius $R = R_1 + R_2$. Panel (**a**): $R_p > R_t[1-\sin(\gamma)]$, the cluster interacts with the conical part of the tip and we have: $R_1 = R_t[1 + \sin(\gamma)]\tan(\gamma) + R_t[1 - \sin(\gamma)]/[\cos(\gamma)]$, $R_2 = R_p\cos(\gamma)$. Panel (**b**) $R_p < R_t[1 - \sin(\gamma)]$, the cluster interacts with the spherical part of the tip and we have: $R_1 = (2\sqrt{R_1 R_p})/(1 + R_p/R_t)$, $R_2 = (2\sqrt{R_1 R_p})/(1 + R_t/R_p)$

6.2.2 *Controlled Movements*

A detailed analysis of the cluster movements induced by the interaction with an AFM oscillating tip has been developed by A. Rao and colleagues [23] from the University of Basel. The tip is approximated with a truncated cone terminated with a hemispherical apex while the cluster is described by a complete sphere. The two possible geometrical configurations are described in Fig. 6.5. The tip–particle interaction is modeled as a collision between two hard bodies so that an elementary cluster displacement occurs along the direction connecting the centers of tip apex and cluster (Fig. 6.6). Moreover, the elementary clusters movements on the surface are supposed to be strongly damped by friction effects. Following these hypotheses, the overall trajectory is calculated as the sum of different small displacements due to the subsequent tip–cluster interactions and it turns out to depend on the AFM scan parameters as well as on tip and particle size. The authors distinguish two possible AFM scan movements, the "raster path "and the "zig–zag path" (Fig. 6.6a, b) and describe the trajectory direction as a function of the reduced radius R (see Fig. 6.5) and the spacing b between consecutive scan lines. The analytical relation and the corresponding conclusions are different in the two cases, in particular for "raster path" the trajectory results in a straight line which forms an angle α with respect to the fast scan direction and the direction α depends only on the parameters R, b.

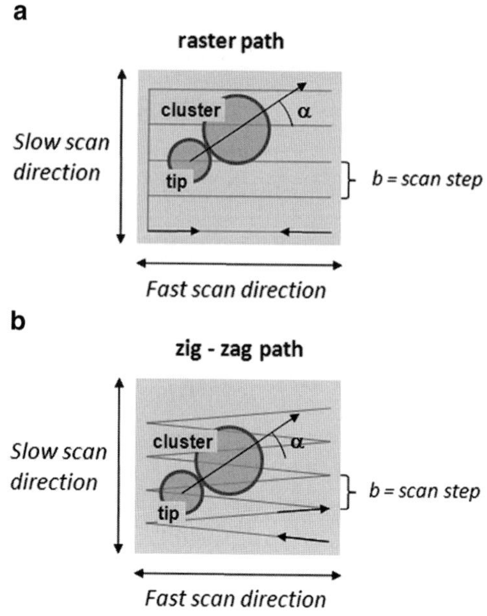

Fig. 6.6 Typical scan movements of the AFM tip over the sample. Panel (**a**): raster scan; panel (**b**): zig–zag scan. The direction of cluster movements is sketched on both panels. Note that we always refer the movements to the fast scan direction

In the other case (zig–zag path) no preferred direction exists because the angle α depends also on the initial position of the moving cluster. In the limit of $b/R \rightarrow 0$ the angle α always approaches the value of 90°, that is, the clusters are moved in a direction normal to the fast scan. The analytical formulas describing the trajectory deflection have been compared with manipulation experiments performed in air and ambient conditions on gold spherical clusters (nominal diameter 25 nm) deposited on silicon. The results show a good qualitative correspondence of the trajectory deflection α as a function of the scan step parameter b in the range 2–20 nm ([23], Fig. 6.5). The parameters utilized during the acquisition of Fig. 6.4 correspond to the limit $b/R \rightarrow 0$. In this case, even if our system performs a zig–zag scan, the overall movement results normal to the fast scan direction.

Recently Rao et al. [24] have improved their analysis simulating cluster trajectories as a function of the sliding distance d. The results have been compared to experimental test on gold clusters (nominal diameter 50 nm) deposited on silicon.

The simulated behavior shows that the larger the friction force between the particle and the surface, corresponding to $d \ll b$, the more regular are the trajectories. The regularity is measured from the deviation of each trajectory steps from the ideal one, and this quantity can also be extracted from the measured cluster pathways. From the direct comparison of the observed discontinuities with the simulated ones, the authors may conclude that the typical sliding distance d in this particular case is quite large, about 20 nm, with respect to the lattice constant of the substrate. This is not surprising as the actual silicon surface is covered by an amorphous native oxide layer. The authors also pointed out that it is difficult to extract information on the friction forces which damp the cluster sliding motion, because the initial velocity acquired by the particle during the interaction with the AFM tip remains unknown. To overcome this drawback, the collision process between the tip and the particle should be better understood both from experimental and theoretical points of view.

6.2.3 Depinning and Energy Dissipation

We have shown in Fig. 6.4 that forced and controlled movements of weakly bonded nano-clusters can be obtained by increasing the dither piezo oscillation amplitude while working in AM-AFM. We have also shown that this action corresponds to an increasing energy dissipation during the tip–substrate interaction. Actually this method has been exploited by different groups to force and characterize the sliding onset of nano-clusters in dedicated experiments [18–24]. As an example, Ritter et al. [21] have analyzed the contact area dependence of antimony clusters depinning on HOPG and Mougin et al. [19] have studied the depinning of gold clusters on silicon as a function of temperature. The energy dissipation signal per cycle introduced by (6.2), or equivalently the power dissipation per cycle, is the parameter that uniquely describes all these different experiments. Thus it seems reasonable assuming the energy dissipation being directly linked to the static friction force that cluster has to overcome to initiate the sliding.

In the following of this section we would like to analyze in detail this assumption in order to assess the limit of the measurement procedure, while in section 6.3 the main experimental result obtained with this technique will be reviewed.

Extracting force characteristics from dynamic AFM measurements is a difficult task because tip experiences a non-linear force field during its periodical movement towards and away from the surface. Nevertheless, the description of tip dynamics in AM-AFM has reached, in the last years, an high level of accuracy and several analytical approaches have been developed which allow to overcome this difficulty [28–30]. The tip–cantilever system interacting with a surface is described as an externally driven ($F_{\text{ext}} = A_{\text{d}}k\cos(\omega_0 t)$), harmonic oscillator with damping, governed by the standard single variable differential equation of a point-mass oscillator. The equation of motion is solved by neglecting high harmonic contributions and concentrating on steady-state solution: $z(t) = d + A\cos(\omega_0 t + \phi)$. The correctness of this model has been positively tested comparing the analytical behavior of force and energy dissipation curves as a function of tip–sample separation with dedicated experimental tests and with numerical results obtained with standard force model [31–33]. Nowadays this approach is widely used to obtain quantitative information by dynamic AFM such as, for example, the force distribution maps [34].

Particularly interesting to our purpose is the analysis performed by Hölscher [31] on the energy dissipation behavior as a function of tip–sample separation in AM-AFM. Firstly, the general significance of (6.2) introduced by Anczykowski et al. [27]

on the basis of energy conservation flow is confirmed within this model. Secondly, the authors show that energy dissipation curve with respect to tip–sample separation can be completely reconstructed from phase data even in the repulsive regime. This analysis gives support to the transformation of phase signal into energy signal we have used to obtain the panels (d–f) in Fig. 6.4 [35, 36].

The fundamental guess of neglecting high harmonic contributions has been tested by Ritter et al. [21] during the manipulation procedure in AM-AFM and found to be largely satisfied because the ratio of the higher harmonics to the resonant frequency does not exceed about 1% even during strong tip–sample interaction.

To summarize, frequency, amplitude, and phase shifts are the measurable quantities associated with damping effects in the model but, using the microscope with amplitude feedback at fixed frequency, that is, the AM-AFM mode, we force the system to show all the dissipation effects in the phase shift. This result, under the hypothesis that steady-state solution can be approximated by a single harmonic oscillator function, is condensed in (6.2), which besides contains only parameters that can be controlled and measured during a standard AM-AFM measurement.

The following step is to link the energy dissipation signal to microscopic dissipative interactions and in particular to the mechanisms acting at the onset or during the cluster sliding. This step is one of the most delicate for the correct interpretation of clusters detachment and movement with AM-AFM.

To better understand the possible mechanisms it is useful to analyze Fig. 6.7 which is similar to Fig. 6.4 but shows the complete *manipulation sequence* developed by our group to manipulate and analyze clusters.

This sequence is composed by an alternation of *imaging* and *manipulation scan*. We select an area and perform a first image with optimized AM-AFM parameters (Fig. 6.7a). In the following scan, Fig. 6.7b, we intentionally increase only the free amplitude oscillation A_0, keeping it fixed along the complete scan. This step, called *manipulation scan*, is repeated slowly increasing A_0 until detachment events start to occur (Fig. 6.7c). Combining this behavior with those presented in Fig. 6.4 and taking into account the result of Rao et al. [24] we can describe the situation as follows:

Initially, a small increase of A_0 corresponds to a dissipation localized on the cluster but the cluster itself remains stable. Further by increasing A_0 the dissipation also increases and detachment events followed by small displacements may occur. Moreover, by reducing A_0 stable imaging can be retrieved, while by maintaining a large A_0 a series of pinning and depinning events is obtained.

Looking in detail to the *trajectory* on Fig. 6.8 three phases can be distinguished. Initially the cluster is stable and the energy dissipation has an almost uniform distribution in the range 100–110 eV. Then the first depinning occurs and the cluster moves up and right with a displacement of about 40 nm. Afterward it continues to move with small displacements for about 300 nm up to the next large movement to the right. Note that during the smooth translation phase, the energy dissipation profile remains almost uniform and centered on values equal to those measured on the initial stable cluster position. This behavior indicates that the energy dissipation signal cannot capture the cluster depinning event, which indeed is strictly a

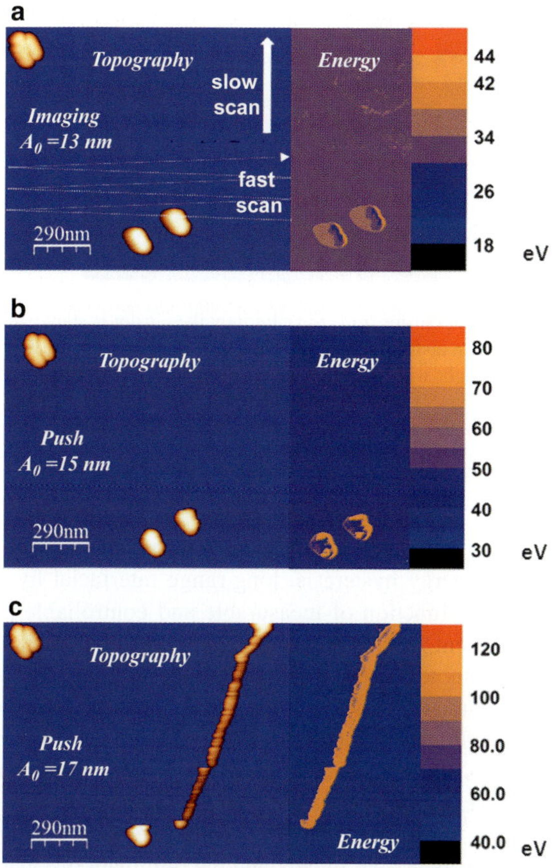

Fig. 6.7 On the *left* of each panel the height signal is presented (blue background is a 10-nm cut-off plane introduced for graphical reasons). On the *right* a zoom of the corresponding energy dissipation signal obtained from the simultaneous acquisition of the phase signal is shown (with the appropriate color scale). The collection of these panels corresponds to a *manipulation sequence*. Panel (**a**): Topographic image with optimized AM-AFM parameters. Panel (**b**): Manipulation scan, only the free amplitude oscillation A_0 has been intentionally increased, keeping it fixed along the complete scan. Panel (**c**): the free amplitude oscillation A_0 has been further increased and the first detachment events occur

non-adiabatic process, but it registers the dissipation when the cluster gets pinned on the substrate, typically at the end of each single translation. This behavior is in agreement with the model consisting of a succession of pinning and depinning events.

On the other hand, it is also evident that the dissipation on particles is a clear indicator of the energy threshold necessary to obtain the detachment, because cluster movement responds directly to energy dissipation variation (see Fig. 6.4).

A detailed analysis of the possible dissipation mechanisms between an oscillating tip and a flat surface in the absence of plastic deformation has been performed

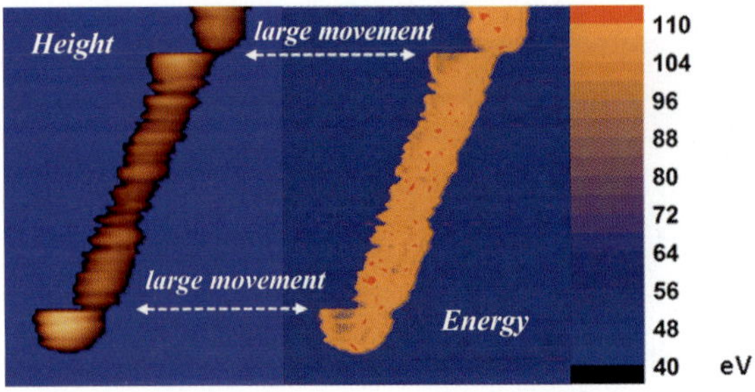

Fig. 6.8 Zoom into a typical cluster *trajectory*. Initially the cluster is stable and the energy dissipation has an almost uniform distribution in the range 100–110 eV. Then the first depinning occurs, the cluster moves up and right with a displacement of about 40 nm and it continues to move with small displacements for about 300 nm up to the next large movement to the right. During this large and smooth translation the energy dissipation profile remains uniform and centered on values equal to measured one on the initial stable cluster position

by Garcia et al. [26, 33]. The authors have calculated the energy dissipation per cycle due to surface energy hysteresis, long-range interfacial hysteresis or visco elasticity damping as a function of measurable and controllable parameters such as the amplitude set point A and the free amplitude oscillation A_0 ([26], Fig. 6.1). Moving from weak interaction ($A \approx A_0$) to strong damping ($A \ll A_0$) the calculated curves predict the existence of a maximum, but the three mechanisms give rise to different behaviors as a function of the ratio A/A_0. In particular the two hysteresis mechanisms show a broad or even flat region of maximum dissipation while the profile corresponding to visco elastic damping has a pronounced maximum at a $A/A_0 = 0.5$. Making a comparison between experimental results and these theoretical curves may help understand the dominant dissipation channel for a particular physical system. As an example, the authors measured in air the dissipation between a silicon tip and a Si(100) surface passivated with native silicon dioxide finding a remarkable agreement with the curves calculated considering both surface energy and long-range interfacial hysteresis ([26], Fig. 6.2b). They also measured the response of the same tip on a polystyrene and polyboutadiene blend fiding a complete agreement with a viscous damping dissipation mechanism ([26], Fig. 6.2c).

In agreement with these results we found, on the system composed by gold nanocluster on HOPG graphite, that energy dissipation on clusters initially increases by decreasing A with fixed A_0. Then the process is limited by the occurrence of cluster detachment and movement which indeed behaves as a new dissipation channel. A typical cluster manipulation obtained by varying A with fixed A_0 is presented in Fig. 6.9. Starting from the relative weak interaction regime characteristic of the imaging scan (Fig. 6.9a), depinning and movements are induced when A/A_0 approaches the value 0.5 (Fig. 6.9b, bottom). Then the movement can be easily stopped by restoring the A parameter value characteristic of the imaging scan (Fig. 6.9b, up). This behavior is completely equivalent to that one presented

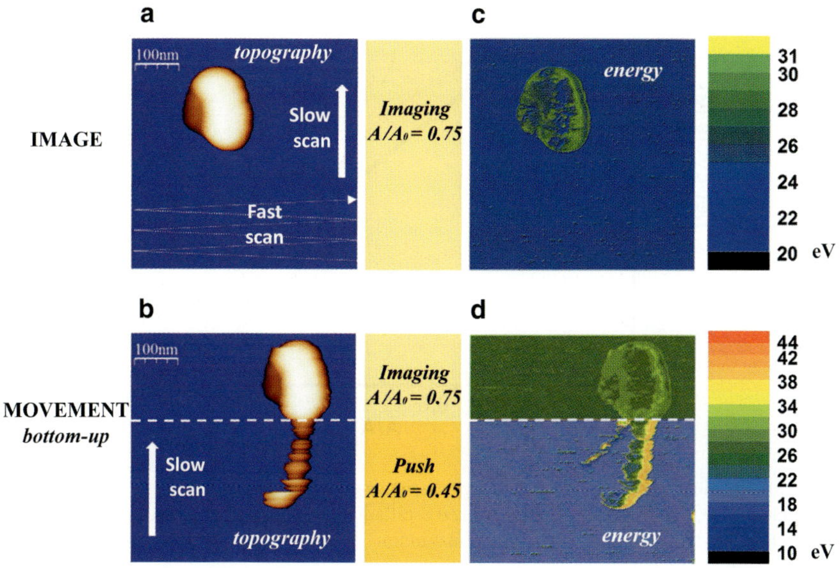

Fig. 6.9 Typical cluster *manipulation* obtained by varying A with fixed A_0. Panel (**a**): Topographic image of a gold nano-cluster deposited on a freshly cleaved HOPG graphite. The corresponding energy dissipation signal is shown in panel (**c**). Panels (**b**, **d**): The amplitude set-point A is reduced at the beginning of the scan (*bottom*) while maintaining the oscillation amplitude A_0 fixed $(A/A_0 = 0.5)$. In this condition the particle detachment is induced, then after few scan lines A is increased to the optimized value for imaging and the particle movement is stopped. The scan continues as a normal image acquisition

in Figs. 6.4 and 6.7 and obtained by increasing A_0. In Fig. 6.10 we report a typical profile of the experimental energy dissipation increase measured on cluster to compare this behavior with the curves predicted by Garcia et al. A direct and accurate comparison is difficult because the accessible range of A/A_0 is limited by the cluster depinning event, nevertheless the absence of an extended maximum plateau suggests that surface energy hysteresis associated with sample deformation gives a negligible contribution.

The energy dissipation signal obtained from the phase shift measurement cannot directly capture the energy released to the cluster to initiate sliding, as we have already pointed out, rather it measures the dissipation on the system composed by cluster bonded on the surface until it remains stable. Then the analysis of the energy dissipation signal as a function of A or A_0 may not only help understand the initial damping mechanism but it also gives clear evidence of the existence of a threshold value for the energy dissipated by the tip on the cluster above which the sliding event becomes an active dissipation mechanism.

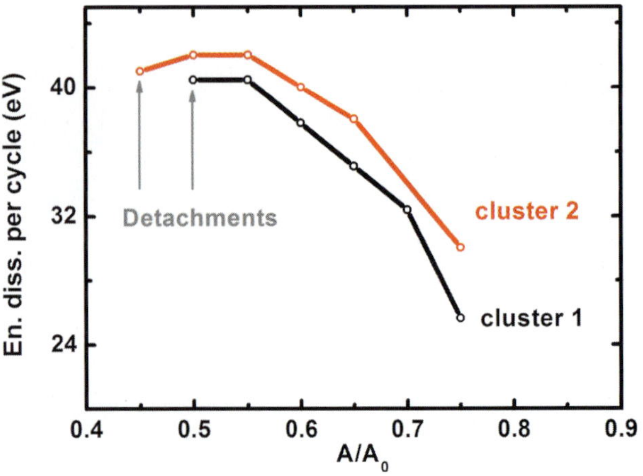

Fig. 6.10 Profile of the experimental energy dissipation increase measured on two different clusters reducing the amplitude set-point A at fixed oscillation amplitude A_0. Usually depinning events occur when the dissipation has reached the maximum

6.3 The Problem of Contact Area in Nanotribology Explored by AFM Cluster Manipulation

The contact area dependence of frictional forces in the nanometer regime is a topic of strong scientific interest and an argument of debate since the advent of the friction force microscope (FFM). The problem is whether or not the original argument by Bowden and Tabor [37], which distinguishes between the apparent contact area and the real contact area and assumes that there is a proportionality between the friction force and the actual contact area, is valid at the nanoscale.

Unfortunately when the friction force F_f between a nanometric tip and a substrate is measured as a function of the externally applied loading force F_l, the variation of contact area as a function of F_l plays a significant role. So the FFM data need to be interpreted with the help of specific contact mechanics models which, in particular at these length scales, may hamper the correct interpretation of the data.

Only recently, with the help of dedicated AFM cluster manipulation experiments, an evaluation of contact area dependence of frictional forces independent of contact mechanics models has been obtained.

The first systematic attempt to solve this problem is reported on the work of Ritter and collegues [21]. Antimony nanoparticles were grown on HOPG graphite and molybdenum disulfide (MoS_2) in ultra-high-vacuum (UHV) conditions. Then controlled translation of nanoparticles was induced in ambient conditions ([21], Fig. 6.2), using the method we have discussed in the previous chapter, that is, slowly increasing the amplitude oscillation A_0 while working in AM-AFM mode. During manipulation, the power dissipated by tip–sample interaction was measured and the threshold value which marks the transition to sliding was recorded as a function of cluster dimensions.

Antimony nanoparticles grown both on HOPG and molybdenum disulfide have ramified and relatively flat shape. In this case, particle height ranges from 20 to 35 nm and lateral dimension from 120 to 400 nm. The corresponding contact areas are set between 10^4 and $11 \times 10^4\,\text{nm}^2$. Being the lateral dimension of the particle bigger than typical tip radius (10 nm) one can approximate the effective contact area with the measured shape of the particle.

The result unambiguously shows that the threshold value of the power dissipation necessary for translation of antimony nanoparticles for both HOPG and molybdenum disulfide surfaces depends linearly on the contact area ([21], Fig. 6.4). Nevertheless, the comparison of this result with FFM experiment requires to model the relationship between dissipated power and frictional forces. This step is nothing but trivial, as we have discussed in section 6.2. Different models have been discussed by Aruliah et al. [38] in connection with this experiment showing that a linear relation may also exist but the hypothesis leading to that conclusion seems to be in contradiction with recent results of Rao et al. [24].

A significant step for the understanding of the system composed by antimony nanoparticles on HOPG has been recently obtained by Dietzel and colleagues. Using AFM cluster manipulation in contact mode they found clear evidence of the linear dependence between contact area and sliding friction force in the nanoscale regime [15].

The authors present about 50 displacement events obtained on particles with contact areas between 22,000 nm^2 and 3,10,000 nm^2. They have used two different cantilevers and different, but identically prepared, samples. The main data were obtained performing both the deposition and the manipulation measurements in UHV conditions. Displacement events have been divided into two distinct regimes ([15], Fig. 6.3a, b). The majority featured substantial frictional resistance but about one-quarter of the events showed almost no detectable friction (these events were only found for islands smaller than 90,000 nm^2). In the first case, the results suggest a linear dependence on contact area and so a constant shear stress $\tau = F_l/A = (1.04 \pm 0.06)\,\text{MPa}$. Since the normal force experienced by the particles is due to adhesion, which scales linearly with area, an area-independent friction coefficient follows, reinforcing Amontons' law also at the nanoscale.

The coexistence of "normal" and superlubric behavior is explained by the authors by the competition of structural superlubricity between crystalline clusters and HOPG surface and the counter effect due to the presence of small amounts of mobile molecules such as hydrocarbon or water molecules [39]. These particles can be trapped on the HOPG surface during the initial cleaning procedure and indeed, decreasing the accuracy of cleaning procedure, the ratio of superlubric to normal translations also decreases.

A complementary set of data, performed in ambient conditions, show a similar linear behavior with an increased shear stress $\tau = F_l/A = (40 \pm 1)\,\text{MPa}$ ([15], Fig. 6.3c) and few superlubric transitions (about 8%).

Starting from the observation that frictionless translations were found only on "small" particles, our group has performed a set of experiments with the aim of reducing the analyzed contact areas of about one order of magnitude. In this case, particles may have comparable dimensions with respect to the tip apex so the analysis of the experimental procedure is an essential aspect for the correct interpretation of the measurements.

Contact mode cluster translation gives a direct evaluation of the sliding resistance of the nanoparticle if the cantilever torsional movement remains aligned with respect to the sliding path of the cluster. This aspect has proved particularly important for particle sizes below 50 nm and drove us to use the dynamic AFMs method discussed in detail in section 6.2 even knowing that the dissipated energy signal gives only indirect indications on the dynamic behavior of the cluster.

Clusters obtained by thermal evaporation are usually characterized by a broad distribution with respect to size and this effect may help to easily explore a wide range of contact areas. Clearly, to fully exploit this possibility each single cluster before being manipulated has to be imaged in order to characterize its size. Unfortunately the direct evaluation of nano-cluster lateral dimension is strongly affected by the tip shape if the particle and the tip apex have comparable size. To overcome this problem we use gold nano-clusters obtained by chemical synthesis (see [19] and reference therein for details on their production process). They can be prepared with specific shape and well definite size distribution and then conveniently deposited on the surface in ambient condition without altering these characteristics. Moreover, the distribution of a large number of similar clusters mimics a set of identical nano-contacts, so that repeating the measurement on different clusters is equivalent to accumulating data on a single, well-controlled nano-contact.

We decide to use gold clusters not only because they can be easily chemically synthesized but mostly because they represent, when deposited on HOPG surface, a prototypical system to the understanding of friction effects at the nanoscale. Specifically, clusters may show adhesive and frictional properties that vary dramatically with temperature and real contact area [40–42].

Gold spherical clusters characterized by two different sizes, diameters of 24 ± 3 and 42 ± 4 nm, are separately deposited from an aqueous solution on freshly cleaved HOPG graphite in air. Their size distribution is checked by transmission electron microscopy (TEM) before deposition and then confirmed by scanning electron microscopy (SEM) and by AFM heights analysis after the deposition confirming that clusters have nearly spherical shape. Manipulation experiments have been performed under ambient conditions ($T \approx 25$°C, relative humidity 40%) using a commercial AFM microscope (mod. Enviroscope by VEECO) with cantilevers characterized by nominal spring constants between 5 and 50 N/m.

To evaluate the size dependence of detachment energy we follow a detailed measurement protocol composed by an alternation of imaging and manipulation scan, see for example Fig. 6.7. We perform a first image with optimized AM-AFM parameters on a suitable area, typically 1 μm^2. In Fig. 6.11 (manipulation) we intentionally increase the free amplitude oscillation A_0, keeping it fixed along the complete scan. Within a manipulation scan we evaluate the maximum dissipated

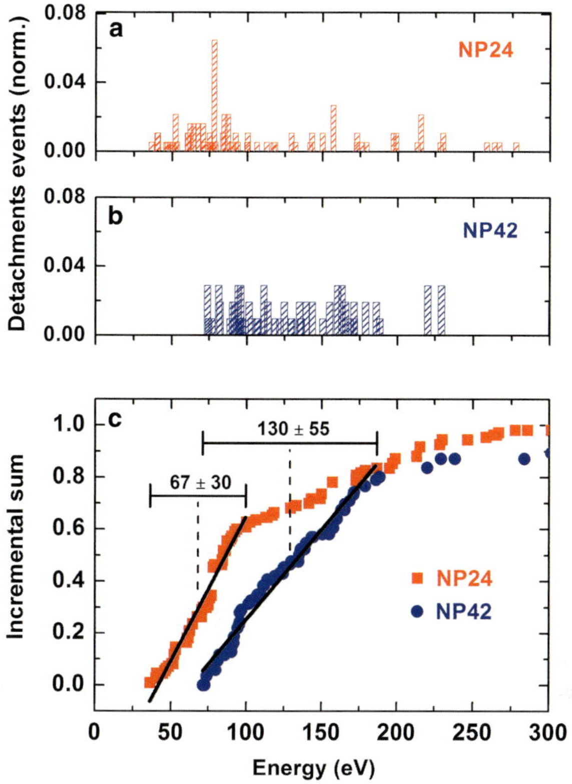

Fig. 6.11 Panels (**a**), (**b**): Referring to 24 and 42 nm diameter clusters respectively, and showing the number of detachment events (normalized to the total number of single clusters) and their associated detachment energy. Data are sorted by increasing energy and represent the collection of a number of manipulation sequences containing about 140 single clusters each, covering an equivalent area of 10 μm^2. Panel (**c**) contains the same data but presented as the incremental sum of detachments events. The linear increase of incremental sum represents regions where detachment occurrence has random uniform probability. Detachment energy thresholds are located in the center of these regions, while their width corresponds to the maximum associated errors ($E_{24} = 67 \pm 30$ eV and $E_{42} = 130 \pm 55$ eV). Reprinted with permission from [20]. Copyright 2009, American Institute of Physics

energy on each clusters eventually being detached and we record the number of detachment events by comparison with the subsequent imaging step. The procedure is repeated until any movement appears during two subsequent manipulation steps.

The experimental data in Fig. 6.11 summarize the size dependence of detachment energy measurements for both cluster sizes. Panels a and b show the raw data representing detachments events sorted by increasing energy. They correspond to the collection of a number of manipulation sequences containing about 140 single clusters each and covering an equivalent area of 10 μm^2. Panel c contains the same data but presented as the incremental sum of detachments events. Curves are normalized to the total number of single clusters.

The overall behavior is similar for both cluster families but the energy regions where detachments occur are clearly different and well identifiable. The curve representing the incremental sum consists of a region where detachments occurrence increases linearly followed by an almost flat plateau indicating that all the detachment events have already taken place. The linear increase typically represents a random uniform probability so the detachment energy threshold can be conveniently located in the centre of this region, while its width corresponds to the maximum associated error. The detachment energy thresholds result in $E_{24} = 67 \pm 30\,\mathrm{eV}$ for 24 nm particles and $E_{42} = 130 \pm 55\,\mathrm{eV}$ for 42-nm particles.

The comparison of our results with Dietzel measurements is difficult because the different techniques used highlight two separate aspects of the friction process. In our case, sensitivity is on the threshold of depinning while in the other case measurement is sensitive to sliding friction. On the contrary, a useful comparison can be established with measurements of Ritter because their results are obtained with the same technique, in ambient condition although on a slightly different physical system.

We take as a reference the angular coefficient of the linear fit of antimony cluster detachment on HOPG ([21], Fig. 6.4), which has the dimension of a power divided by a surface area. Then we translate power into energy according to the oscillation frequency used in the experiment and compare these results with our findings in Table 6.1.

The value corresponding to 24-nm particles is in perfect agreement with previous finding while the value corresponding to 42-nm particles overlaps within the error bars. This result is consistent with the behavior of antimony and gold clusters on HOPG which exhibits a large and similar thermal diffusion coefficient at room temperature. But, more importantly, the comparison of these values suggests a linear decrease of detachment energy with respect to the contact area for particles having a lateral dimensions in the tenth nanometer scale. These data confirm, and at the same time extend by more than one order of magnitude with respect to the contact area, the results obtained on antimony nano-clusters.

Finally, we must highlight, in accordance with the results of Dietzel, that superlubric translations have also been observed with a frequency of about 5%. These events, as shown in Fig. 6.12, usually follow a "normal" movement phase and suggest that this behavior is due to random and momentary realization of a clean interface between an amorphous surface (cluster) and an ordered surface (graphite).

Table 6.1 Comparison between the detachment energy thresholds per unit area evaluated on the 24 and 42 nm particle and the value corresponding to the linear fit of results by Ritter et al. [21]

NP24	NP42	Ritter et al. [21]
$0.15 \pm 0.07\ (\mathrm{eV/nm^2})$	$0.09 \pm 0.04\ (\mathrm{eV/nm^2})$	$0.14 \pm 0.01\ (\mathrm{eV/nm^2})$

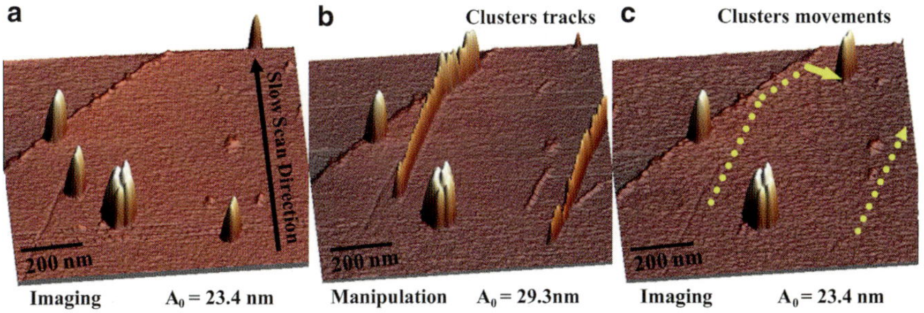

Fig. 6.12 Manipulation sequence relative to 24 nm gold clusters. Panel (**a**): Initial topographic image. Panel (**b**): Manipulation scan: two single clusters are detached and pushed along the slow scan direction; panel (**c**): final topographic image: cluster on the right has left the scan area, cluster on the centre now is sitting on the right upper corner. Note the large jump at the end of its trajectory (120 nm long) which reveals a frictionless movement preceded by a "normal" trajectory and followed by a pinning event

6.4 Conclusion

In conclusion we have presented a new experimental approach to study the friction behavior of single nano-contacts obtained by depositing well-prepared nano-clusters on clean surfaces. The clusters are moved in a controlled way using the AFM set-up and the information is obtained analyzing the torsional bending of the cantilever (FFM) or the energy dissipation of the tip (through the phase shift signal) at the onset or during the cluster translation. Among these two possible AFM modes we concentrate on the analysis of the energy dissipation signal which is obtained by operating the system in AM-AFM.

We show that the periodic tip–cluster interaction, typical of AM-AFM, induces depinning events and produces smooth cluster movements of particles having comparable or smaller dimensions with respect to tip apex. Nevertheless the connection of the corresponding energy dissipation signal to the friction forces acting on the system cluster/surface is only indirect and it requires a specific model.

Using the AFM in contact mode to push the cluster gives a direct evaluation of the friction force corresponding to the induced cluster movement. Nevertheless, aligning the torsional bending of the cantilever to the sliding path of cluster having an overall dimension comparable or smaller than the typical tip apex may result in a very critical issue.

The main advantage of using cluster manipulation as compared to FFM is on the control of the sliding interface. The AFM tip is ideally disentangled from the physical system in exam because the contact region between the cluster and the substrate now represents the interface we are measuring on. So a wide range of well-established cleaning and deposition techniques can be used to prepare first the surface, and then depositing the overlayer in the form of nano-clusters. In comparison, during FFM measurements, the tip, which is one of the interface side, is much more difficult to prepare and to maintain in stable conditions. Usually only the substrate can be properly cleaned and characterized before and after a FFM measurement.

As an example of using cluster manipulation to study friction behavior at the nanoscale we have presented recent measurements on the contact area dependence of frictional forces. Using cluster manipulation with both methods on comparable systems (antimony or gold cluster on HOPG) a similar behavior is obtained. Friction force during sliding decreases linearly with the contact area with the exception of few events where movements without apparent friction have occurred. Experiments performed in UHV conditions and in ambient environment produced similar results but an increase of frictionless translations has been noticed while working in UHV and very clean conditions. This observation supports the hypothesis that at these small contact areas the linear dependence of frictional force is due to the presence of small amounts of mobile molecules trapped at the interface. Also the energy dissipation at the onset of movement decreases linearly as a function of contact area for antimony island featuring contact between 10^4 and 11×10^4 nm^2. The same analysis on smaller gold clusters indicates a reduction consistent with a linear decrease but new and detailed experiments will be performed to confirm this behavior.

Acknowledgements

We are grateful to A. Vanossi (SISSA, Trieste and CNR, Democritos, Trieste, Italy) and G. Mistura (Department of Physics, University of Padova, Italy) for critical and fruitful discussions on this subject. We thank E. Gnecco (NCCR Nanoscale Science and Institute of Physics, University of Basel, Switzerland) and K. Mougin (CNRS-ICSI Mulhouse, France) for scientific discussions. The authors also wish to acknowledge the financial support from PRRIITT Regione Emilia Romagna, Intermech Project and from the European Science Foundation, Program FANAS through ACOF and AFRI projects.

References

1. C.M. Mate, G.M. McClelland, R. Erlandsson, S. Chiang, Atomic-scale friction of a tungsten tip on a graphite surface. Phys. Rev. Lett. **59**, 1942 (1987)
2. R. Garcia, R.V. Martinez, J. Martinez, Nano-chemistry and scanning probe nanolithographies. Chem. Soc. Rev. **35**, 29 (2005)
3. T. Junno, S.B. Carlsson, H. Xu, L. Montelius, L. Samuelson, Fabrication of quantum devices by Angstrom-level manipulation of nanoparticles with an atomic force microscope. Appl. Phys. Lett. **72**, 548 (1998)
4. M. Sitti, Survey of nanomanipulation system. Proc. IEEE–NANO **01**, 75 (2001)
5. T. Junno, K. Deppert, L. Montelius, L. Samuelson, Controlled manipulation of nanoparticles with an atomic force microscope. Appl. Phys. Lett. **66**, 3621 (1995)
6. R. Lüthi, E. Meyer, H. Haefke, L. Howald, W. Gutmannsbauer, H.J. Güntherodt, Sled-type motion on the nanometer scale: determination of dissipation and cohesive energies of C_{60}. Science **266**, 1979 (1994)

7. M.R. Falvo, R.M. Taylor, A. Helser, V. Chi, F.P. Brooks Jr, S. Washburn, R. Superfine, Nanometre-scale rolling and sliding of carbon nanotubes. Nature **397**, 236 (1999)
8. Z. Liu, Y. Yang, Y. Qu, Z. Dong, W.J. Li, Y. Wang, Vibration-mode based real-time nanoimaging and nanomanipulation. Proceedings of the 2007 International Conference on Nanotechnology (NANO '07), vol. 515. (IEEE, Hong Kong, 2007)
9. B. Mokaberi, J. Yun, M. Wang, A.A.G. Requicha, Automated nanomanipulation with atomic force microscopes. Proceedings of the 2007 International Conference on Robotics and Automation (ICRA'07), vol. 1406 (IEEE, Rome, 2007)
10. H. Xie, D.S. Haliyo, S. Regnier, Parallel imaging/manipulation force microscopy. Appl. Phys. Lett. **94**, 153106 (2009)
11. G. Li, N. Xi, H. Chen, M. Yu, A. Saeed, Assembly of nanostructure using AFM based nanomanipulation system. Proceedings of the 2007 International Conference on Robotics and Automation (ICRA '04) (New Orleans, 2007), pp. 428–433
12. A.A.G. Requicha, *Nanomanipulation with the Atomic Force Microscope*, ed. by R. Waser, Nanotechnology, Volume 3: Information Technology. Wiley-VCH, Weinheim, 2008), pp. 239–273.
13. A.A.G. Requicha, Nanorobots, NEMS and nanoassembly. Proc. IEEE, special issue on nanoelectronics and nanoscale processing **91**(11), 1922–1933 (2003)
14. D. Dietzel, T. Mönninghoff, L. Jansen, H. Fuchs, C. Ritter, U.D. Schwarz, A. Schirmeisen, Interfacial friction obtained by lateral manipulation of nanoparticles using atomic force microscopy techniques. J. Appl. Phys. **102**, 084306 (2007).
15. D. Dietzel, C. Ritter, T. Mönninghoff, H. Fuchs, A. Schirmeisen, U.D. Schwarz, Frictional duality observed during nanoparticle sliding. Phys. Rev. Lett. **101**, 125505 (2008)
16. M. Palacio, B. Bhushan, A nonscale friction investigation during the manipulation of nanoparticles in controlled environments. Nanotechnology **19**, 315710 (2008)
17. D. Dietzel, M. Feldmann, H. Fuchs, U.D. Schwarz, A. Schirmeisen, Transition from static to kinetic friction of metallic nanoparticles. Appl. Phys. Lett. **95**, 053104 (2009)
18. G. Paolicelli, K. Mougin, A. Vanossi, S. Valeri, Adhesion detachment and movement of gold nanoclusters induced by dynamic atomic force microscopy. J. Phys.: Condens. Matter **20**, 354011 (2008)
19. K. Mougin, E. Gnecco, A. Rao, M.T. Cuberes, S. Jayaraman, E.W. McFarland, H. Haidara, E. Meyer, Manipulation of gold nanoparticles: influence of surface chemistry,temperature, and environment (vacuum versus ambient atmosphere). Langmuir **24**, 1577 (2008)
20. G. Paolicelli, M. Rovatti, A. Vanossi, S. Valeri, Controlling single cluster dynamics at the nanoscale. Appl. Phys. Lett. **95**, 143121 (2009)
21. C. Ritter, M. Heyde, B. Stegemann, K. Rademann, U.D. Schwarz, Contact-area dependence of frictional forces: moving adsorbed antimony nanoparticles. Phys. Rev. B **71**, 085405 (2005)
22. C. Ritter, M. Heyde, U.D. Schwarz, K. Rademann, Controlled translational manipulation of small latex spheres by dynamic force microscopy. Langmuir **18**, 7798 (2002)
23. A. Rao, E. Gnecco, D. Marchetto, K. Mougin, M. Schönenberger, S. Valeri, E. Meyer, The analytical relations between particles and probe trajectories in atomic force microscope nanomanipulation. Nanotechnology **20** (2009)
24. A. Rao, M.L. Wille, E. Gnecco, K. Mougin, E. Meyer, Trajectory fluctuations accompanying the manipulation of spherical nanoparticles. Phys. Rev. B **80**, 193405 (2009)
25. B. Bhushan (ed.), *Handbook of nanotechnology, Part C and Part D*, 2nd edn. (Springer, Berlin, 2007)
26. R. Garcia, C.J. Gomez, N.F. Martinez, S. Patil, C. Dietz, R. Magerle, Identification of nanoscale dissipation processes by dynamic atomic force microscopy. Phys. Rev. Lett. **97**, 016103 (2006)
27. B. Anczykowski, B. Gotsmann, H. Fuchs, J.P. Cleveland, V.B. Elings, How to measure energy dissipation in dynamic mode atomic force microscopy. Appl. Surf. Sci. **140**, 376 (1999)
28. A. San Paulo, R. García, Tip-surface forces, amplitude, and energy dissipation in amplitude modulation force microscopy. Phys. Rev. B **64**, 193411 (2001)
29. M. Lee, W. Jhe, General theory of amplitude-modulation atomic force microscopy. Phys. Rev. Lett. **97**, 036104 (2006)

30. T.R. Rodríguez, R. García, Tip motion in amplitude modulation tapping-mode atomic-force microscopy: comparison between continuous and point-mass models. Appl. Phys. Lett. **80**, 1646 (2002)
31. H. Hölscher, Quantitative measurement of tip–sample interactions in amplitude modulation AFM. Appl. Phys. Lett. **89**, 123109 (2006).
32. O. Sahin, Time-varying tip–sample force measurements and steady-state dynamics in tapping-mode atomic force microscopy. Phys. Rev. B **77**, 115405 (2008)
33. N.F. Martinez, R. Garcia, Measuring phase shifts and energy dissipation with amplitude modulation atomic force microscopy. Nanotechnology **17**, S167 (2006)
34. B.J. Albers, T.C. Schwendemann, M.Z. Baykara, N. Pilet, M. Liebmann, E.I. Altman, U.D. Schwarz, Three-dimensional imaging of short-range chemical forces with picometre resolution. Nat Nanotechnol. **4**, 307 (2009)
35. K. Schroeter, A. Petzold, T. Henze, T. Thurn-Albrecht, Quantitative analysis of scanning force microscopy data using harmonic models. Macromolecules **42**, 1114 (2009)
36. O. Sahin, N. Erina, High-resolution and large dynamic range nanomechanical mapping in tapping-mode atomic force microscopy. Nanotechnology **19**, 445717 (2008)
37. F.P. Bowden, D. Tabor, *The Friction and Lubrication of Solids. Oxford Classic Texts in the Physical Sciences*, Paperback edn. (2001)
38. D.A. Aruliah, M.H. Muser, U.D. Schwarz, Calculation of the threshold force and threshold power to move adsorbed nanoparticles. Phys. Rev. B **71**, 085406 (2005)
39. M.H. Müser, L. Wenning, M.O. Robbins, Simple microscopic theory of amontons's laws for static friction. Phys. Rev. Lett. **86**, 1295 (2001)
40. L. Bardotti, P. Jensen, A. Hoareau, M. Treilleux, B. Cabaud, A. Perez, F. Cadete Santos Aires, Diffusion and aggregation of large antimony and gold clusters deposited on graphite. Surf. Sci. **367**, 276 (1996)
41. S. Pisov, E. Tosatti, U. Tartaglino, A. Vanossi, Gold clusters sliding on graphite: a possible quartz crystal microbalance experiment? J. Phys.: Condens. Matter **19**, 305015 (2007)
42. W.D. Luedtke, U. Landman: Slip diffusion and lévy flights of an adsorbed gold nanocluster. Phys. Rev. Lett. **82**, 3835 (1999)

Part II
Characterization

Chapter 7
Cell Adhesion Receptors Studied by AFM-Based Single-Molecule Force Spectroscopy

Robert H. Eibl

Abstract Cell adhesion receptors are expressed on the surface of cells and can mediate specific binding to other cells and to the extracellular matrix. Atomic force microscopy (AFM) can be used to study the unbinding force of specific receptor–ligand interactions down to the single-molecule level and, at the same time, between two living cells. This review provides an overview on AFM force measurements related to cell adhesion in biology, immunology and medicine. This includes the force measurement between two isolated molecules, as well as studies of adhesion receptors in their natural environment of an intact living cell. In biological systems the typical force ranges from 20 to 200 pN (pico-Newton) to rupture a single bond of a receptor–ligand pair. For illustration of the method, the interaction of an integrin receptor, very late antigen 4 (VLA-4) with its ligand vascular cell adhesion molecule 1 (VCAM-1), is described. White blood cells, as well as metastasizing melanoma cells, can use VLA-4 receptors on their surface to recognize the receptor VCAM-1 on the endothelial cell lining of blood vessels and to attach rapidly to the endothelial surface, thereby resisting the blood stream, so that the trafficking cell can extravasate. This example demonstrates how the complex process of rapid and physiological activation of integrins on lymphocytes can be resolved by modern applications of nanotechnology. AFM measurements of cell adhesion receptors at the single-molecule level and, at the same time, on an intact living cell can be used to reconstruct the important process of cell migration. The multiple steps involved in lymphocyte homing can be compared to similar, but postulated steps in organ-specific tumour metastasis. Such a combination of addressing current questions in biomedical research with innovative nanotechnology will soon lead to further applications in pharmacology, immunology and cancer research.

Abbreviations

$\alpha_4\beta_1$	Alpha(4)beta(1) (integrin VLA-4)
$\alpha_L\beta_2$	Alpha(L)beta(2) (integrin LFA-1)
AFM	Atomic force microscope

Anti-VLA-4	Antibody against VLA-4
ConA	Concanavalin A
CXCL12	CXC chemokine receptor ligand 12 (SDF-1)
CXCR4	CXC chemokine receptor 4; (receptor for SDF-1)
F_c	Crystallisable fragment (of an antibody)
G*RGD*SP	Amino-acid sequence, RGD-containing peptide (binding to, for example, integrin $\alpha_5\beta_1$)
ICAM-1	Intercellular cell adhesion antigen 1
LFA-1	Lymphocyte function-associated antigen 1 (integrin $\alpha_L\beta_2$)
pN	Pico-Newton
RGD	Arginine–glycine–aspartic acid (aminoacid sequence found in several integrin ligands)
SDF-1	Stromal-derived factor 1 (CXCL12)
SFM	Scanning force microscope
SLeX ag	Sialylated Lewis X antigen
SPM	Scanning probe microscope
STM	Scanning tunneling microscope
VCAM-1	Vascular cell adhesion molecule 1
VLA-4	Very late antigen 4 (integrin $\alpha_4\beta_1$)

7.1 Introduction

In 1986, Binnig, Quate and Gerber introduced the first atomic force microscope (AFM), also known as scanning force microscope (SFM) [1]. This AFM was invented as an adaptation from the scanning tunneling microscope (STM) developed by Binnig and Rohrer in 1981. Both researchers won the Nobel Prize in Physics in 1986. Both scanning probe microscopes (SPM), the AFM and the STM, are as mechanic microscopes not affected by the optical diffraction limit, a well-known limiting factor for resolution in conventional light microscopy [2, 3]. Meanwhile, the AFM technique has evolved into a wide range of applications from basic science into numerous industrial applications (Figs. 7.1–7.3), and more recently, into the study of live cell–cell interactions (Figs. 7.4 and 7.5). At the imaging level, the AFM allows for a resolution down to single atoms and can reveal subatomic features on a surface, i.e. an improvement of several orders of magnitudes over the capacity of conventional light microscopy. Lower resolutions allow for AFM-imaging of objects and their surfaces at resolutions comparable to applications in light microscopy. Another widely used application for the AFM is the detection of adhesive or repellent forces measured as force–distance curves. This force-spectroscopic mode of operation has been used to study even weak forces in the range of pico-Newton (pN) and less. Such forces are necessary to break atomic bonds, interfere with van-der-Waals forces, or are required for to stretch biological molecules and even allow for the rupture of such bonds.

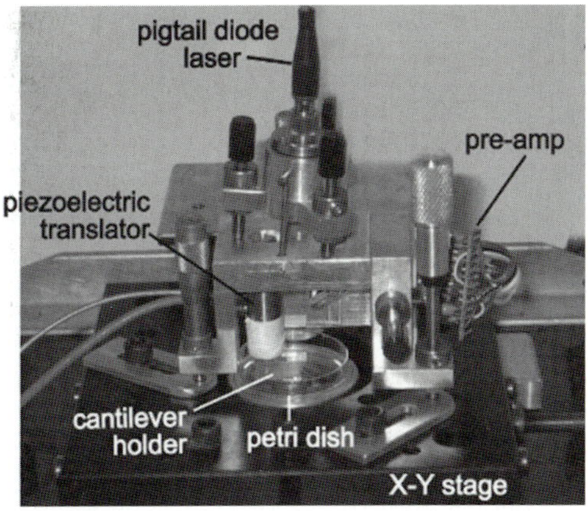

Fig. 7.1 Photograph of a custom-made AFM force spectroscope to measure adhesion forces of single receptor–ligand bonds on living cells. The piezoelectric translator moves the cellular probe up and down onto another cell or substrate within the Petri dish. A laser beam is adjusted onto the back of the cantilever and reflected into a segmented photodiode. The signal is sent to a computer and later translated into the unbinding force. An inverted light microscope is attached from underneath the X–Y stage to place the cell onto the cantilever and to adjust the position of the laser onto the back of the cantilever

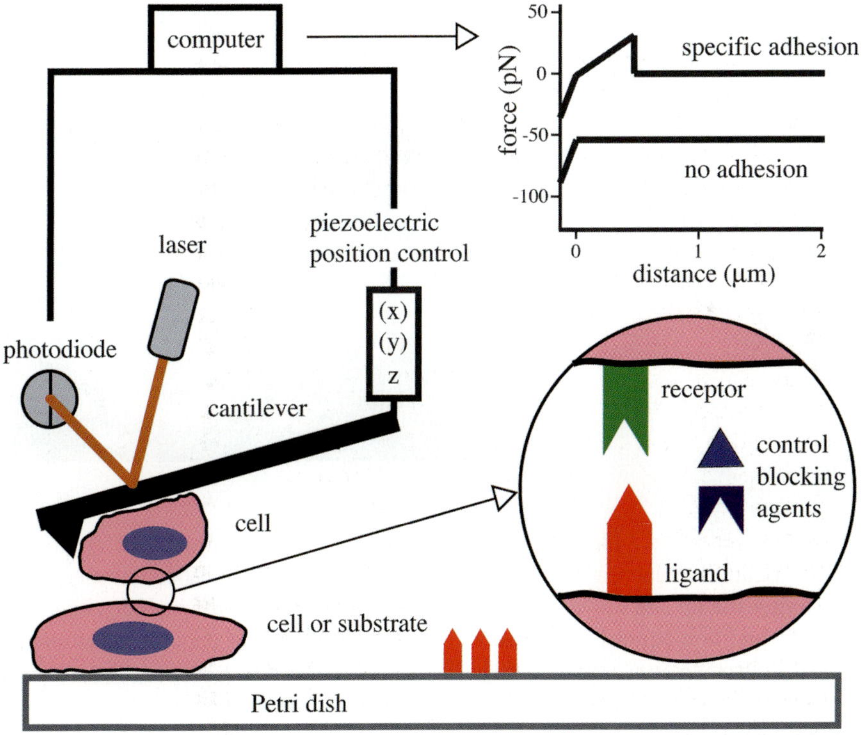

Fig. 7.2 Schematic setup of an AFM force spectroscope to measure unbinding forces of single receptor–ligand bonds on living cells

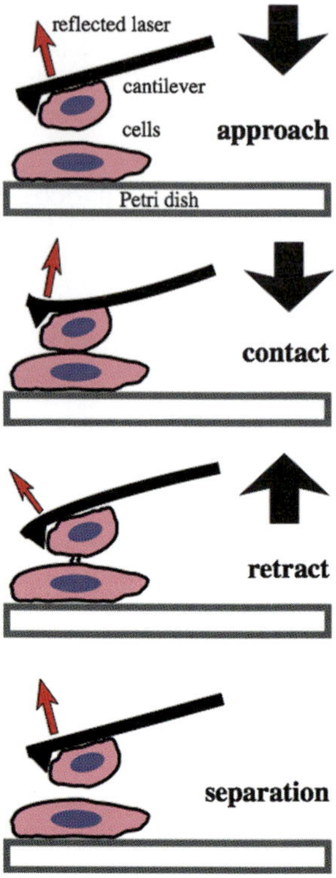

Fig. 7.3 Scanning steps in AFM based force spectroscopy for cell–cell adhesion measurements

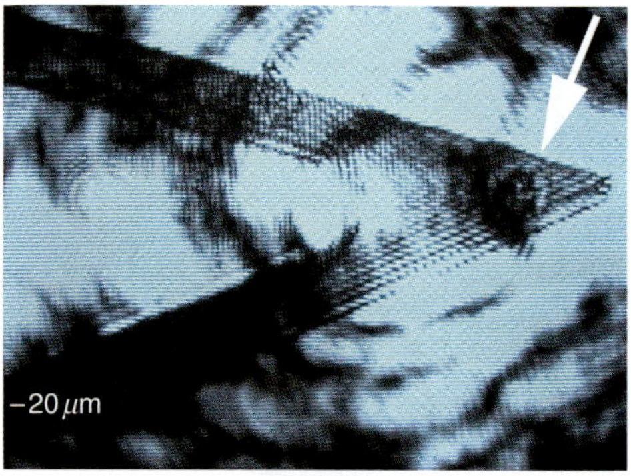

Fig. 7.4 Placing a cell on the cantilever. Photomicrograph of a living cell attached to the tip of an AFM cantilever previously functionalized with conA

In contrast to the STM, the AFM does not require the sample to be composed of electrically conducting material, because the AFM uses a reflected laser beam to monitor the up and down movements of a cantilever tip. This includes its use in aqueous and non-aquous solutions, as long as they are clear enough for an undisturbed laser reflection and detection. Therefore, the AFM appears to be an ideal instrument for direct measurements of biological samples, a unique advantage over the STM.

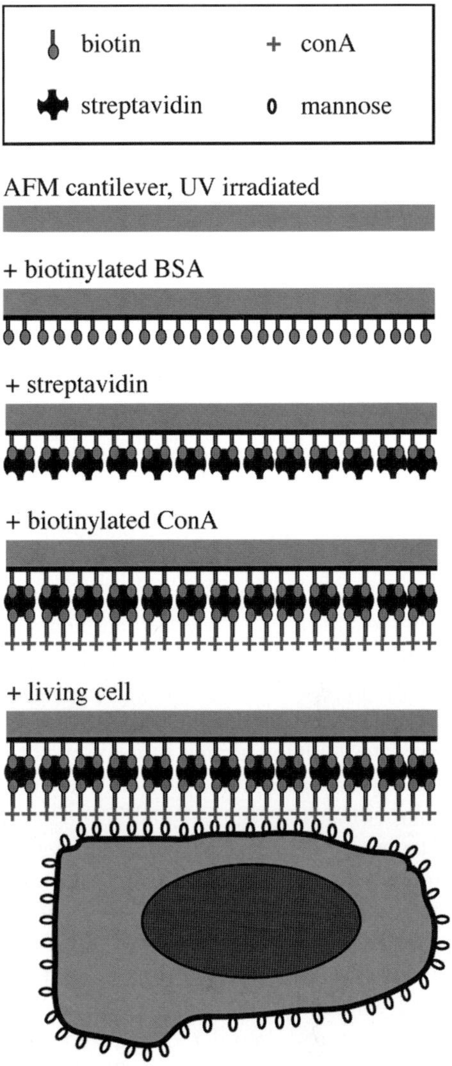

Fig. 7.5 Scheme of immobilizing an intact living cell to the AFM cantilever

The enormous imaging resolution power set aside, already, the power of an AFM to directly measure cell adhesion forces brings the AFM force spectroscope into the focus of cell adhesion research (Fig. 7.2), as there are immunologists, cell biologists and cancer or metastasis scientists, who are in general not very familiar with this technology of nanoscience. In addition, physicists may use the AFM on purified and sometimes non-functional or not fully functional molecules. Therefore, these molecules are studied in vitro, in isolation from the complexity of an intact living cell and not as highly motile integrant of an active cell membrane. At a first glance, it may appear challenging to measure the discrete unbinding force of a single pair of two adhering distinct adhesion receptors, regardless of whether these receptors were isolated and immobilized to a substrate or reside within a cell membrane of a living cell. Despite this unique potential of the method that allows for measurements at a resolution that detects single pairs of cell adhesion receptors and the extent of modulation of the receptor interaction on intact cells, the method slowly enters the fields of physics, cell biology, immunology and pharmacology [4–16]. Nevertheless, the more and more manufacturers provide AFM apparatuses adjusted for measurements on living cells. If necessary, there is often a choice to combine these Bio-AFMs with other state-of-the-art microscopes. The aim of this chapter is to provide an insight for both beginner scientists as well as the experienced principal investigator who are interested to use AFM techniques for answering their questions arising from interdisciplinary studies.

7.2 AFM-Based Single-Molecule Force Spectroscopy

In addition to the scenario depicted in Figs. 7.2 and 7.3, it is not a necessity for single-molecule force spectroscopy to use living cells or to perform such experiments in aqueous solutions. Instead, the tip of the cantilever can be functionalized with the substance of interest and then mechanically probed against a surface with another substrate. The reflection of the laser is collected in a photodiode and the resulting force–distance curve (Fig. 7.6) is collected in a computer for further analysis. A sudden step in the backward scan of such a force–distance cycle directly correlates with the rupture force of one or more adhesive bonds (Figs. 7.2 and 7.6). The up and down movements of the cantilever tip are induced by a piezoelectric control element (Figs. 7.1 and 7.2). For force spectroscopy experiments one laser beam and one piezoelectric control element in z-axis are necessary and sufficient. Additional laser beams with piezoelectric control elements in x- and y-axis allow the correlation of the detected force with the exact position of the probe. Although manufacturers provide AFMs with three lasers, several laboratories prefer custom-made AFMs with an optimized design for high-resolution in the force spectroscopy mode which requires only one laser in z-axis but may have an elongated size of the mechanical parts in order to reduce the background and to increase the resolution of weak forces of much less than 20 pN. In addition, Benoit was able to significantly increase the resolution of his custom-made AFM system by breaking away one of the two arms of the triangular cantilever [17, 18]. Others may prefer the limitations of an undisturbed cantilever and try to reduce the background noise in order to improve the signal-to-background ratio in their experimental setup. In force–distance curves and after statistical analysis of usually hundreds of individual force scans, the average binding or repellent force can be evaluated in force histograms (see below). For stretching experiments of single molecules, the substrate is completely bound between the tip of the cantilever and the opposing surface. In such an approach, Rief and colleagues were able to characterize the consecutive opening of several loops of a single protein molecule, titin, and relate this to the function of the muscle protein [14].

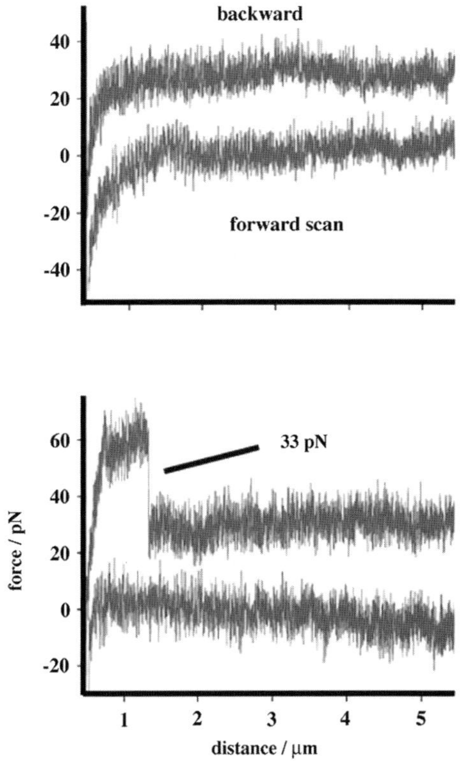

Fig. 7.6 AFM force scans of cell–cell interactions. Force–distance diagram of forward and backward scans. In the case of no adhesion after short contact of a B16 melanoma cell to a bEnd.3 endothelial cell, both the backward and forward scans look similar and show the indentation force (40 pN) superimposed by the thermal noise, whereas an adhesion event of about 33 pN can be easily detected by the step in the backward force scan

7.3 Receptor–Ligand Interactions

Although not related to cell adhesion, the avidin–biotin complex can serve in many disciplines as a reference model for ligand–receptor interactions [19–24]. The technique of AFM force spectroscopy was introduced to characterize the dynamic response of individual avidin–biotin complexes to a pulling force. In 1994, Florin and colleagues quantified the unbinding forces of individual members of the avidin–biotin complex. The avidin was functionalized onto the tip of the AFM cantilever, and the biotin, iminobiotin, or desbiotin molecules, respectively, were immobilized on agarose beads. The average rupture forces ranged from 85 to 160 pN per single bond, depending on which member of the biotin family was probed against the avidin. As a further derivation of these findings, Moy and colleagues a few months later presented a model to establish the correlation among binding forces, intermolecular potential and molecular function. In 2010, AFM measurements have been used to characterize a new variant of streptavidin, named traptavidin, which can bind more strongly to its biotin ligand (Table 7.1), pointing to an advantage over conventional avidin or streptavidin in laboratory applications.

An overview of AFM studies on the avidin–biotin complex at the single-molecule level is shown in Table 7.1 and compared to other receptor–ligand interactions in biology [25–47]. Most of these reports include measurements performed under different conditions, as they are allowing variable off-rates by changing the velocity of the approaching and retracting probes. As many of these reports describe, the experimental setup did allow for ambient temperature and pH changes. For many cell adhesion molecules, both parameters, temperature and pH, are critical. Such unphysiologic conditions may not prevent the adhesion of cells and still allow the measurements of single bond ruptures. One could conclude that these conditions are not critical or may support the measurements, as the cell becomes less sticky with a reduced number of receptors or a temperature-dependent slower binding rate. AFM measurements at room temperature are much easier to perform than under higher temperatures of physiologic conditions (37°C), since the cantilever itself is sensitive to changes in temperature and to faster moving water molecules which can result in thermal drifts and higher background noise, respectively.

7.4 Cell Adhesion Interactions on Living Cells

It appears that not only all cells of multi-cellular organisms but also lower organisms depend on specific interactions with their environment. The adhesion between cells or of cells to substrate also appears to be strongly regulated, e.g. white blood cells within the blood stream should not adhere very much to each other or to the endothelial cells at the vessel wall. In 2004 Butcher and Springer received the Crafoord Prize for their independent work on lymphocyte trafficking [49–55]. In the 1990s, they had revealed distinct adhesive steps during the trafficking and attachment of white blood cells on the endothelial wall of blood vessels. They showed that lymphocytes use specific adhesion receptors to extravasate into lymph nodes or at sites of inflammation. The distinct steps involved include initial tethering, rolling and sticking (Fig. 7.7). Basically, all selectins but also CD44 and very few other receptors are known today to mediate initial tethering and rolling of the cell, whereas the integrins lymphocyte function-associated antigen 1 (LFA-1, integrin $\alpha_L\beta_2$), very late antigen 4 (VLA-4, integrin $\alpha_4\beta_1$) and integrin $\alpha_4\beta_7$ usually mediate the later steps of firm sticking and extravasation. Interestingly, these three integrin receptors can mediate the rolling part as well, but the amount of this rolling function varies and depends on the cell type, the presence, amount and combination of receptors and corresponding signalling molecules, as well as the local environment and the velocity or shear force within the blood stream. It is known for lymphocytes, that an activating chemokine like stroma derived factor - 1 (SDF-1) can immediately trigger the rapid activation of an integrin receptor, e.g. VLA-4, from a low affinity state to a high affinity state, resulting in a rapid arrest of a rolling lymphocyte.

Table 7.1 Unbinding forces of individual pairs of cell adhesion receptors

Receptor	Ligand	Interacting cell or substrate	Force (pN)	References
		Cell–cell		
E-cadherin	E-cadherin	L-M(TK-)-L-M(TK-)	73	Panorchan et al. [25]
N-cadherin	N-cadherin	L-M(TK-)-L-M(TK-)	30	Panorchan et al. [25]
Integrin $\alpha_4\beta_1$ (VLA-4)	Integrin $\alpha_4\beta_1$ (VLA-4)	B16 melanoma cell -B16 melanoma cell	20–40	Eibl and Benoit (unpubl.)
Integrin $\alpha_4\beta_1$ (VLA-4)	VCAM-1	B16 melanoma cell -bEnd.3 endothelial cell	33	Eibl and Benoit [11]
CsA	CsA	Dictyostelium/amoeboid cell -dictyostelium/amoeboid cell	23	Benoit et al. [17]
		Cell–substrate		
Integrin $\alpha_2\beta_1$	Collagen I	CHO cell-protein ligand	65	Taubenberger et al. [26]
Integrin $\alpha_4\beta_1$ (VLA-4)	VCAM-1	B16 melanoma cell -VCAM -1/F_c fusion protein	21–45	Eibl and Benoit (unpubl.)
Integrin $\alpha_4\beta_1$ (VLA-4)	VCAM-1	Lymphocyte -VCAM-1/F_c fusion protein	30–70	Eibl [4]
Integrin $\alpha_4\beta_1$ (VLA-4)	VCAM-1	U937 monocytic cell -VCAM-1/F_c fusion protein	25–170	Zhang et al. [27, 28]
Integrin $\alpha_5\beta_1$	Fibronectin	K562 myeloid leukemic cell -protein	40–160	Li et al. [29]
Integrin $\alpha_L\beta_2$ (LFA-1)	ICAM-1	3A9 lymphoma cell -ICAM-1/F_c fusion protein	100–300 ???	Zhang et al. [30]
Integrin $\alpha_L\beta_2$ (LFA-1)	ICAM-1	3A9 lymphoma cell -ICAM-1/F_c fusion protein	35–70 ???	Zhang et al. [31]
Integrin $\alpha_L\beta_2$ (LFA-1)	ICAM-1	3A9 lymphoma cell -ICAM-1/F_c fusion protein	100–300	Zhang et al. [30]
Integrin $\alpha_L\beta_2$ (LFA-1)	ICAM-2	Jurkat -ICAM-2/F_c fusion protein	40	Wojcikiewicz et al. (2006) [32]
Integrin $\alpha_M\beta_2$ (Mac-1)	ICAM-1	CHO -ICAM-2/F_c fusion protein	60	Yang et al. (2007) [33]
Carbohydrates	ConA	NIH3T3 fibroblast -lectin protein	86	Chen and Moy [34]
Integrin $\alpha_V\beta_3$	Osteopontin	Osteoclast-protein ligand	50	Lehenkari and Horton [35]
Integrin $\alpha_V\beta_3$	Echistatin	Osteoclast-protein ligand	97	Lehenkari and Horton [35]
E-selectin	PSGL-1	PMN cell-protein	140	Hanley et al. (2004) [36]
L-selectin	PSGL-1	PMN cell-protein	80	Hanley et al. (2004) [36]
P-selectin	PSGL-1	PMN cell-protein	130	Hanley et al. (2003) [37]
Integrin $\alpha_V\beta_3$	G*RGD*SP	Osteoclast-RGD peptide	42	Lehenkari and Horton [35]
Saccharides from blood types A	Helix pomatia lectin	Red blood cell -lectin protein	65	Grandbois et al. (2000) [38]

(*Continued*)

Table 7.1 (Continued)

Receptor	Ligand	Interacting cell or substrate	Force (pN)	References
		Substrate–substrate (related to cell adhesion)		
Integrin $\alpha_V\beta_3$	G*RGD*SP	Protein-RGD peptide	32	Kokkoli et al. [39]
P-selectin	PSGL-1	Protein-protein	115–165	Fritz et al. [40]
P-selectin	sLeX	F_c fusion protein -carbohydrate complex	40–200	Zhang et al. [27, 28]
VE-cadherin	VE-cadherin	F_c fusion protein -F_c fusion protein	15–150	Baumgartner et al. [41]
Carbohydrate	ConA	Carbohydrate -lectin protein	35–65	Dettmann et al. [42]
Proteoglycan	Proteoglycan	Proteoglycan -proteoglycan (from marine sponge)	400	Dammer et al. [12]
		Substrate–substrate (unrelated to cell adhesion)		
HSA	Antibody	Protein -antibody	244	Hinterdorfer et al. [43]
Avidin	Biotin	Protein-organic molecule	115–170	Yuan et al. [20]
Avidin	Biotin	Protein-organic molecule	5–170	Merkel et al. [23]
Avidin	Biotin	Protein-organic molecule	200	Wong et al. [44]
Avidin	Biotin	Protein-organic molecule	160	Florin et al. [22]
Avidin	Desthiobiotin	Protein-organic molecule	94	Moy et al. [21]
Avidin	Iminobiotin	Protein-organic molecule	85	Moy et al. [21]
Strepavidin	Biotin	Protein-organic molecule	257	Moy et al. [21]
Strepavidin	Biotin	Protein-organic molecule	63	Chivers et al. [24]
Traptavidin	Biotin	Protein-organic molecule	71	Chivers et al. [21]
Strepavidin	Iminobiotin	Protein-organic molecule	135	Moy et al. [21]

Rupture forces of single adhesion receptor–ligand interactions revealed by AFM force spectroscopy on living cells (cell–cell or cell–substrate interactions) and between isolated adhesion receptors (substrate–substrate interaction). The (strept)avidin–biotin system is included as the classical reference for one of the strongest single receptor–ligand interactions in biology, although not directly related to cellular adhesion

Interestingly, VLA-4, was found to mediate similar effects of rolling and arrest of metastasizing melanoma cells, although addition of SDF-1 showed no effect, that is, no better adhesion (Fig. 7.8). In order to compare cancer progression and organ-specific metastasis with lymphocyte trafficking, it appeared to be tempting to transfer both rolling experiments, with melanoma cells and with lymphocytes, respectively, into AFM measurements, especially at the single molecule level.

It took several years from characterizing the first single-molecule interactions of the avidin–biotin system to study cell adhesion receptors between living cells and at the single-molecule level. Over the last decades, cell adhesion receptors were studied with all suitable methods of biochemistry, molecular genetics, histology, immunohistochemistry, and the functional and structural characterization of the proteins. In many cases, even the time-consuming generation of transgenic or knock-out mouse strains was used to overexpress or mutate adhesion receptors.

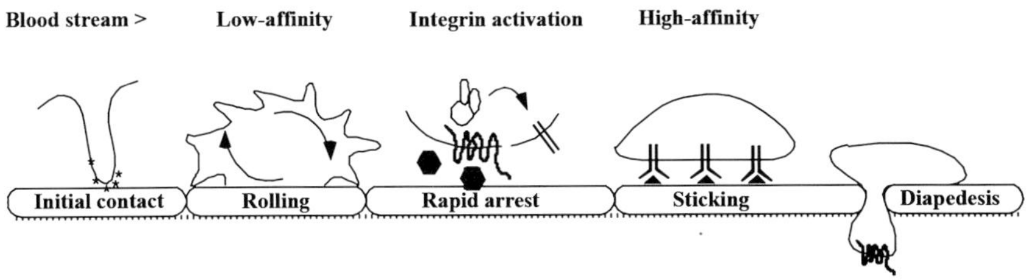

Fig. 7.7 Scheme of lymphocyte adhesion under shear (modified; courtesy of E.C. Butcher)

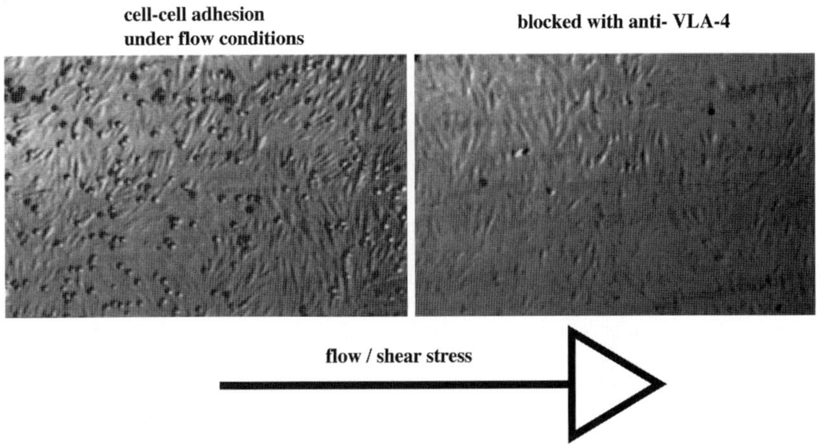

Fig. 7.8 Photomicrographs of rolling melanoma cells attaching on a monolayer of endothelial cells (*left*). A function-blocking antibody against the adhesion receptor VLA-4 blocks the cell adhesion

These approaches helped to elucidate many surprising functions of adhesion receptors and their regulation in health and disease. As an example, the long known lymphocyte homing and putative metastasis and stem cell receptor CD44 could be identified as a weak and usually masked mediator of rolling (Fig. 7.7) after crossing of several mice strains, which did not express the known major rolling receptors, but still showed the phenomenon of rolling in intravital microscopy experiments.

It took again several years from the first proof of concept, i.e. that AFM can be used to study VLA-4 interactions on living cells (Figs. 7.9 and 7.10) [11], to the specific modulation of such VLA-4 receptors before and after physiologic activation by a chemokine like SDF-1 (Figs. 7.11–7.13) [4]. In fact, this became the first measurement at the single-molecule level of any physiologic integrin activation by any chemokine on any cell type.

In principle, AFM force spectroscopy could be used to study multi-bond ruptures on a living cell (Fig. 7.10). Typical force scans reveal several steps, each representing at least one, but sometimes several bond ruptures at the same time. The integral area below the force curve is often referred to as work force [43]. Manipulation of a

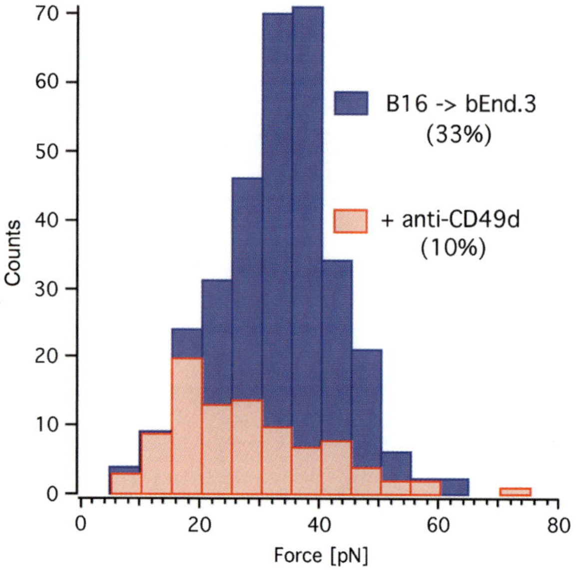

Fig. 7.9 Histogram of two typical AFM force spectra between melanoma and endothelial cells at the resolution of single receptor–ligand bonds. Addition of a function-blocking antibody against VLA-4 on the melanoma cells blocks the binding to VCAM-1 on the endothelial cell (lower columns)

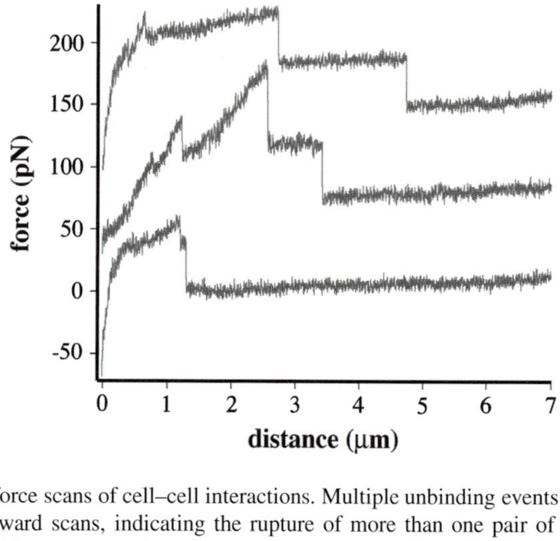

Fig. 7.10 AFM force scans of cell–cell interactions. Multiple unbinding events are shown in each of the three backward scans, indicating the rupture of more than one pair of VLA-4–VCAM-1 receptors, each of them expressed on the surface of a B16 melanoma cell or a bEnd.3 endothelial cell, respectively

living cell with different agents can also result in changes of the adhesive properties, e.g. phorbol ester can augment the adhesive force of a cell, whereas Mg+ ions or removal by EDTA may inhibit or reduce calcium-dependent cell adhesion. With a

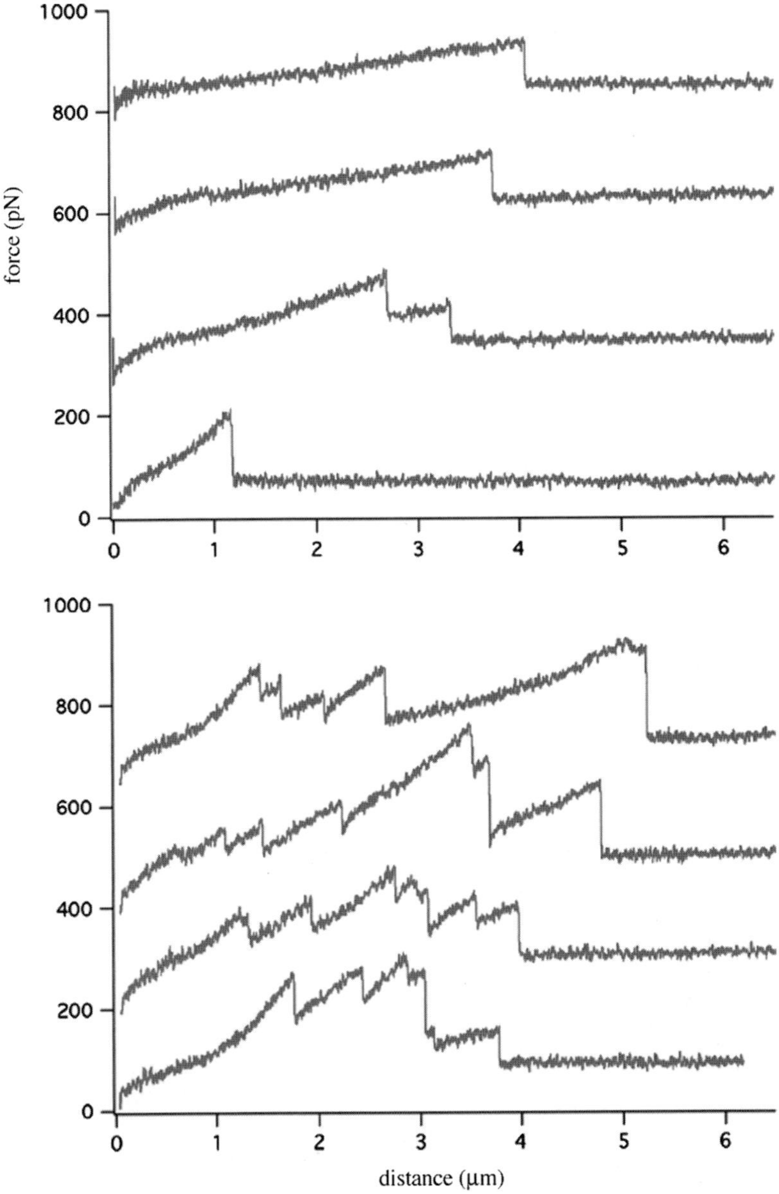

Fig. 7.11 Effects of chemokine SDF-1 on AFM force scans of VLA-4–VCAM-1 cell–substrate interactions. Before activation (above), and after activation with the chemokine SDF-1

good resolution of the AFM, reduction of background noise and choosing the best conditions of rare adhesion events, there is a good chance to measure even single ruptures on living cells.

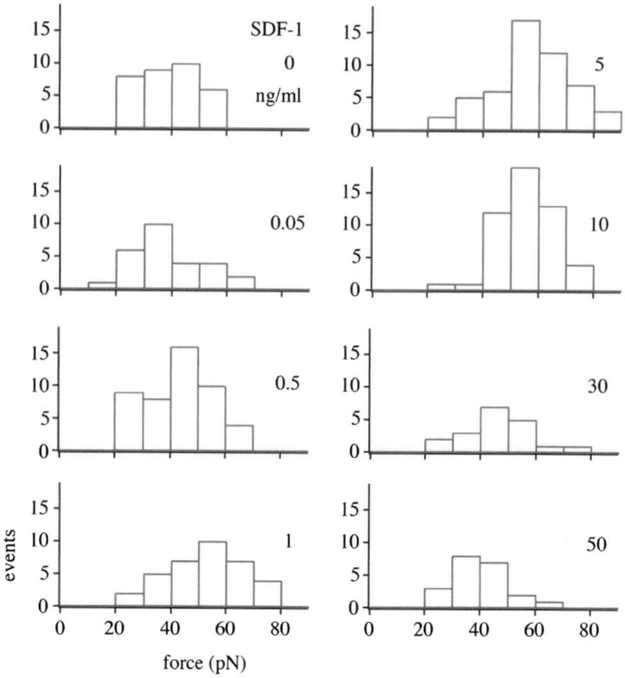

Fig. 7.12 Affinity modulation of single VLA-4 receptors on a living cell by SDF-1. AFM force histograms of a lymphoma cell expressing integrin VLA-4 probed on immobilized VCAM-1 fusion protein. Increasing SDF-1 concentrations show a shift of the rupture forces to higher forces, then shift back to normal

Although it may appear more suitable today to use one immobilized ligand and to probe it with one living cell, it is in fact possible to bring two living cells into short and weak contact with each other in order to measure one single rupture of a bond of a specific pair of adhesion receptors between the two living cells. Despite the possibility of measuring more than just one bond rupture (Fig. 7.10), the major question will then be to prove the specificity of the postulated interaction. In 1998, Thie and colleagues were able to detect single bond ruptures with AFM between mammalian cells, but then could only speculate on the nature of the putative pairs of receptors, which included several integrins as a possibility [55]. Usually, mammalian cells carry several different types of functional adhesion receptors at the same time on their surface. In order to measure just one pair of receptors it is important to choose a suitable combination of cells.

In 2000, Benoit and colleagues (Table 7.1) presented AFM measurements of the force necessary to unbind the single adhesion bonds of receptors between two living amoeboid *Dictyostelium* cells at a resolution of 23 pN per average rupture [17]. In their system the specificity was obtained through the use of mutant cell lines that could either adhere to all other cells of the same subline or did not adhere at all. The

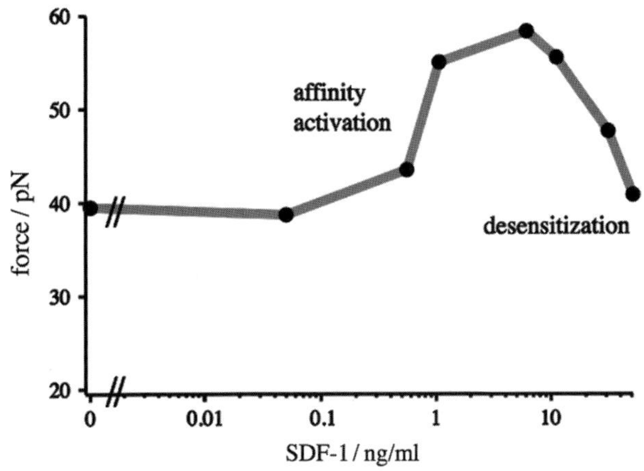

Fig. 7.13 Dose–response effects of chemokine SDF-1 revealed by AFM force spectroscopy. The dose–response curve shows the mean value of forces from the histograms shown in Fig. 7.12. The peak of unbinding single VLA-4–VCAM-1 adhesion bonds at physiologic conditions indicates a physiologic switch to a higher affinity state of the integrin VLA-4

sensitivity of measuring single bonds at high probability was achieved by reduction of the adhesion rate. With an adhesion rate of 30% or less the probability of measuring the rupture of single pairs of receptors increased to 85% or more.

Interestingly, at higher adhesion rates, force peaks of two-, three- and fourfold of the 23 pN per single bond appeared with mathematical approaches in their force histograms, which was consistent with the idea of simultaneous ruptures of two, three or four bonds.

In 2004, Eibl and Benoit presented the first single-molecule AFM measurement of the integrin VLA4 and VCAM-1 between two living mammalian cells (Fig. 7.9) [11]. The extent of the specificity of the interaction was expected from previous rolling experiments (Fig. 7.8). In both types of experiments, the rolling assay as well as the AFM force measurements, function-blocking antibodies against each type of receptor, respectively, blocked the adhesion formation, whereas control antibodies showed no effect. Later, Eibl replaced the endothelial cell with a recombinant VCAM-1 fusion protein that was immobilized on the Petri dish and obtained similar results. These results were interpreted as proof of concept, i.e. the AFM can be used to mimic the rolling of cells in vivo and to measure specific VLA-4/VCAM-1 interactions on living cells and at the same time at the single-molecule level. This setup then allowed the use of another VLA4-expressing cell type instead of melanoma cells. In melanoma cells, the VLA-4 receptor appears to exist in just one binding state. For lymphocytes expressing VLA-4, at least two different activation states were postulated from many types of experiments, including rolling experiments in flow chambers and intravital microscopy in lymph nodes of a mouse using fluorescent labelling. Using such lymphoid cells, AFM measurements could detect an activation of the integrin VLA-4 after exposure to the chemokine SDF-1 on the single-molecule level as well as for multiple-rupture events (Figs. 7.11–7.13). Further studies and comparisons to CD44 may elucidate the sometimes paradox functions of VLA-4 in tumour metastasis and lymphocyte rolling [refs-refs]. VLA-4 receptors may support the arrest and metastasis of melanoma cells via the ligand VCAM-1 on endothelial cells, whereas VLA-4 binding to VLA-4 on an adjacent tumour cell may prevent the migration into the blood stream [45, 56–63].

7.5 Limitations of the AFM Method

An interdisciplinary approach to measure the nanoscale forces within the biological samples of living cells can be very time consuming and may represent a major challenge for any laboratory not yet familiar with AFM. The construction or adaptation of a suitable AFM to measure cell adhesion interactions under physiologic conditions of temperature can take several months or even longer, reducing within the same time the resources and dampening the enthusiasm of students and researchers accordingly. For members of a physics laboratory, who are not yet familiar with biological techniques, it is a major challenge to properly culture cells and maintain optimal growth conditions. The experienced biochemist knows how to work with highly active proteins, like chemokines, that are often extremely sensitive to repeated freeze–thaw cycles, but such work can hardly be performed successfully by a physics student not trained in biology or immunology. Nevertheless, after having solved all problems arisen during the setup of such AFM measurements, the experiment itself remains still rather time-consuming and will still fall behind any standard requirement in a modern high-throughput screen. Therefore, the technique does not seem to be attractive for the pharmaceutical industry to be applied in tests of thousands of substances in order to identify inhibitors, activators or modifiers of any cell adhesion pathway. Nevertheless, the technique could be adapted to reassess and compare just some hundreds of candidate substances of which some may be used in further tests and, perhaps, in clinical trials. The preparation of the tips of the cantilever can be time- and labour consuming. The final protocol has to adequately address the research question but has to be feasible at the same time. One possible drawback, especially for immunologists performing studies on lymphocytes, is that many substances which can be used to attach a cell to the cantilever may activate the attached lymphocyte thereby interfering with the real experiment. Even well-chosen alternatives to, for example, ConA have the potential to activate one or the other signalling pathway within the lymphocyte and interfere with the scientific question. On one hand, the possibility to obtain such artificially produced results may deter many immunologists from applying the method. On the other hand, it often appears to be necessary to manipulate adhesion receptors of cells growing as monolayers strongly attached to the Petri dish and to their neighbouring cells. It has been proved useful to test several techniques to singularize the attached cells, including mild trypsinization which involves partly digestion of cell adhesion molecules, or short-term removal of calcium ions by EDTA, or both. In addition, using otherwise firmly attaching cell cultures, passaged just every few hours up to every day before the experiment, may increase the chances to get single cells of good quality after mild trypsinization or calcium removal.

For statistical analysis, many force–distance curves have to be carefully acquired, usually for several different cells of the same type, and analysed. In some experiments the conditions of measurement, especially the temperature and pH have to be controlled. Otherwise, e.g. a highly pH-sensitive cell adhesion receptor is measured under changing conditions of pH. By far, not all cell–cell interactions can be measured with AFM force spectroscopy. A major restriction is the contact time. So far, only short contacts can be measured successfully, since a longer contact time may result in tight adhesion of the cell to another cell or to the substrate. The optimal contact time depends often on the receptor and the cell type used. Receptors with a slow binding regime may be masked in many experiments. Of course, the ideal cell type may be a cell with only one specific adhesion receptor and without any further specific or unspecific binding activities. Unspecific binding, e.g. as a result of attracting charges on protein and carbohydrate molecules cannot be avoided. Therefore, the right choice of cells with the right adhesion receptor is crucial for such experiments. Since the technology of measuring adhesion of living cells at the single-molecule level is still in its beginning, future developments may overcome current limitations.

Acknowledgements

The author gratefully acknowledges Hermann Gaub, Martin Benoit, Vincent Moy, Xiaohui (Frank) Zhang, Olga Vinogradova and Wolfang Knoll for their expert advice in the field of AFM and nanotechnology; Irving L. Weissman, Eugene C. Butcher, Marcus Hubbe, Bernhard Holzmann, Uwe Gosslar, James Campbell, Albert Zlotnik, Anja Müller, Horst Kessler for their support and material in the field of rolling, chemokines and tumour stem cell biology; Timothy A. Springer and Ronen Alon for discussion; Heinz Höfler who supported the technology transfer of rolling experiments to AFM nanotechnology from its beginning; Stephan Bärtsch and Matthias Maiwald for critical comments on the manuscript and Markus Schneemann for reading. Part of this work was supported by stipends from the German research council (DFG, Bonn), the German cancer research center (DKFZ, Heidelberg), major travelling grants from the Max Planck Society and the Technical University of Munich as well as a Dean's fellowship from Stanford University and tuition sponsoring by Irv Weissman.

References

1. G. Binnig, C. Quate, C. Gerber, Phys. Rev. Lett. **56**, 930 (1986)
2. F. Giessibl, S. Hembacher, H. Bielefeldt, J. Mannhart, Science **289**, 422 (2000)
3. F.J. Giessibl, Science **267**, 68 (1995)

4. R.H. Eibl, in *Applied Scanning Probe Methods*, vol.12, ed by B. Bhushan, H. Fuchs (Springer, Berlin, 2009) p. 1
5. R.H. Eibl, in *Advances in Single Molecule Research for Biology and Nanoscience*, ed by P. Hinterdorfer, G. Schuetz, P. Pohl (Trauner, Linz, 2007), p. 40
6. R.H. Eibl, in *Immunology 2004*, vol. 1 (Medimond, Bologna, 2004) p. 115
7. R.H. Eibl, I.L. Weissman, H.E. Gaub, M. Benoit, Clin. Invest. Med. Suppl. **27**, 4 (2004)
8. R.H. Eibl, V.T. Moy, Methods Mol. Biol. **305**, 439 (2005)
9. R.H. Eibl, V.T. Moy, in *Recent Research Development in Biophysics*, ed by S.G. Pandalai (TRN, Trivandrum, 2004) p. 235
10. R.H. Eibl, V.T. Moy, in *Protein-Ligand Interactions*, ed by G.U. Nienhaus (Humana Press, Totowa, NJ, 2005) p. 450
11. R.H. Eibl, M. Benoit, IEE Proc. Nanobiotechnol. **151**, 128 (2004)
12. U. Dammer, O. Popescu, P. Wagner, D. Anselmetti, H.J. Güntherodt, G.N. Misevic, Science **267**, 1173 (1995)
13. P. Hinterdorfer, Y.F. Dufrêne, Nat. Methods **3**, 347 (2006)
14. M. Rief, H. Clausen-Schaumann, H.E. Gaub, Nat. Struct. Biol. **6**, 346 (1999)
15. M. Rief, M. Gautel, F. Oesterhelt, J.M. Fernandez, H.E. Gaub, Science **276**, 1109 (1997)
16. H.W. Wu, T. Kuhn, V.T. Moy, Scanning **20**, 389 (1998)
17. M. Benoit, D. Gabriel, G. Gerisch, H.E. Gaub Nat. Cell Biol. **2**, 313 (2000)
18. M. Benoit, H.E. Gaub, Cells Tissues Organs **172**, 174 (2002)
19. J. Wong, A. Chilkoti, V.T. Moy, Biomol. Eng. **16**, 45 (1999)
20. C. Yuan, A. Chen, P. Kolb, V.T. Moy, Biochemistry **39**, 10219 (2000)
21. V.T. Moy, E.L. Florin, H.E. Gaub, Science **266**, 257 (1994)
22. E.L. Florin, V.T. Moy, H.E. Gaub, Science **264**, 415 (1994)
23. R. Merkel, P. Nassoy, A. Leung, K. Ritchie, E. Evans, Nature **397**, 50 (1999)
24. C.E. Chivers, E. Crozat, C. Chu, V.T. Moy, D.J. Sherratt, M. Howarth, Nat. Methods **5**, 391 (2010)
25. P. Panorchan, M.S. Thompson, K.J. Davis, Y. Tseng, K. Konstantopoulos, D Wirtz, J. Cell Sci. **119**, 66 (2006)
26. A. Taubenberger, D.A. Cisneros, J. Friedrichs, P.H. Puech, D.J. Muller, C.M. Franz Mol. Biol. Cell **18**, 1634 (2007)
27. X. Zhang, A. Chen, D. De Leon, H. Li, E. Noiri, V.T. Moy, M.S. Goligorsky, Am. J. Physiol. Heart Circ. Physiol. **286**, H359 (2004)
28. X. Zhang, D.F. Bogorin, V.T. Moy, Chemphyschem **5**, 175 (2004)
29. F. Li, S.D. Redick, H.P. Erickson, V.T. Moy, Biophys. J. **84**, 1252 (2003)
30. X. Zhang, E. Wojcikiewicz, V.T. Moy, Biophys. J. **83**, 2270 (2002)
31. X. Zhang, E.P. Wojcikiewicz, V.T. Moy, Exp. Biol. Med. **231**, 1306 (2006)
32. E.P. Wojcikiewicz, M.H. Abdulreda, X. Zhang, V.T. Moy, Biomacromolecules **7**, 3188 (2006)
33. H. Yang, J. Yu, G. Fu, X. Shi, L. Xiao, Y. Chen, X. Fang, C. He, Exp. Cell Res. **313**, 3497 (2007)
34. A. Chen, V.T. Moy, Biophys. J. **78**, 2814 (2000)
35. P.P. Lehenkari, M.A. Horton, Biochem. Biophys. Res. Commun. **259**, 645 (1999)
36. W.D. Hanley, D. Wirtz, K. Konstantopoulos, J. Cell Sci. **117**, 2503 (2004)
37. W. Hanley, O. McCarty, S. Jadhav, Y. Tseng, D. Wirtz, K. Konstantopoulos, J. Biol. Chem. **278**, 10556 (2003)
38. M. Grandbois, W. Dettmann, M. Benoit, H.E. Gaub, J. Histochem. Cytochem. **48**, 719 (2000)
39. E. Kokkoli, S.E. Ochsenhirt, M. Tirrell, Langmuir **20**, 2397 (2004)
40. J. Fritz, A.G. Katopodis, F. Kolbinger, D. Anselmetti, Proc. Natl. Acad. Sci. U S A **95**, 12283 (1998)
41. W. Baumgartner, P. Hinterdorfer, W. Ness, A. Raab, D. Vestweber, H. Schindler, D. Drenckhahn, Proc. Natl. Acad. Sci. U S A **97**, 4005 (2000)
42. W. Dettmann, M. Grandbois, S. André, M. Benoit, A.K. Wehle, H. Kaltner, H.J. Gabius, H.E. Gaub, Arch. Biochem. Biophys. **383**, 157 (2000)
43. P. Hinterdorfer, W. Baumgartner, H.J. Gruber, K. Schilcher, H. Schindler, Proc. Natl. Acad. Sci. U S A **93**, 3477 (1996)

44. S.S. Wong, E. Joselevich, A.T. Woolley, C.L. Cheung, C.M. Lieber, Nature **394**, 52 (1998)
45. M. Thie, R. Röspel, W. Dettmann, M. Benoit, M. Ludwig, H.E. Gaub, H.W. Denker, Hum. Reprod. **13**, 3211 (1998)
46. W. Baumgartner, N. Golenhofen, N. Grundhöfer, J. Wiegand, D. Drenckhahn, J. Neurosci. **23**, 11008 (2003)
47. A. Chen, V.T. Moy, Methods Cell Biol. **68**, 301 (2002)
48. E.C. Butcher, Cell **67**, 1033 (1991)
49. T.A. Springer, Cell **76**, 301 (1994)
50. C. Berlin, R.F. Bargatze, J.J. Campbell, U.H. von Andrian, M.C. Szabo, S.R. Hasslen, R.D. Nelson, E.L. Berg, S.L. Erlandsen, E.C. Butcher, Cell **80**, 413 (1995)
51. J.J. Campbell, J. Hedrick, A. Zlotnik, M.A. Siani, D.A. Thompson, E.C. Butcher, Science **279**, 381 (1998)
52. J.J. Campbell, G. Haraldsen, J. Pan, J. Rottman, S. Qin, P. Ponath, D.P. Andrew, R. Warnke, N. Ruffing, N. Kassam, et al. Nature **400**, 776 (1999)
53. P. Altevogt, M. Hubbe, M. Ruppert, J. Lohr, P. von Hoegen, M. Sammar, D.P. Andrew, L. McEvoy, M.J. Humphries, E.C. Butcher, J. Exp. Med. **182**, 345 (1995)
54. R.A. Warnock, S. Askari, E.C. Butcher, U.H. von Andrian, J. Exp. Med. **187**, 205 (1998)
55. E.P. Wojcikiewicz, X. Zhang, A. Chen, V.T. Moy, J. Cell Sci. **116**, 2531 (2003)
56. E.C. Butcher, L.J. Picker, Science **272**, 60 (1996)
57. H.C. DeGrendele, P. Estess, L.J. Picker, M.H. Siegelman, J. Exp. Med. **183**, 1119 (1996)
58. H.C. DeGrendele, P. Estess, M.H. Siegelman, Science **278**, 672 (1997)
59. W.M. Gallatin, I.L. Weissman, E.C. Butcher, Nature **304**, 30 (1983)
60. I. Weissman, Nature **215**, 315 (1967)
61. F. Qian, D.L. Vaux, I.L. Weissman, Cell **77**, 335 (1994)
62. F. Qian, D. Hanahan, I.L. Weissman, Proc. Natl. Acad. Sci. U S A **98**, 3976 (2001)
63. U. Gosslar, P. Jonas, A. Luz, A. Lifka, D. Naor, A. Hamann, B. Holzmann, Proc. Natl. Acad. Sci. U S A **93**, 4821 (1996)
64. R.H. Eibl, P. Kleihues, P.S. Jat, O.D. Wiestler, Am. J. Pathol. **144**, 556 (1994)
65. U. Gunthert, M. Hofmann, W. Rudy, S. Reber, M. Zoller, I. Haussmann, S. Matzku, A. Wenzel, H. Ponta, P. Herrlich, Cell **65**, 13 (1991)
66. R.H. Eibl, T. Pietsch, J. Moll, P. Skroch-Angel, K.H. Heider, K. von Ammon, O.D. Wiestler, H. Ponta, P. Kleihues, P. Henrrlich, J. Neurooncol. **26**, 165 (1995)
67. H.B.J. Stamper, J.J. Woodruff, J. Exp. Med. **144**, 828 (1976)
68. I. Stamenkovic, M. Amiot, J.M. Pesando, B. Seed, Cell **56**, 1057 (1989)

Chapter 8
Biological Application of Fast-Scanning Atomic Force Microscopy

Yuki Suzuki, Masatoshi Yokokawa, Shige H. Yoshimura, and Kunio Takeyasu

Abstract Recent developments in microscopy technology are remarkable; improvements in temporal and spatial resolution allow us to detect transient states of molecules and biomolecular complexes at the single-molecule level. Atomic force microscopy (AFM) has become an important contributor to physiological and biological cell research. It is a powerful tool in studying the structural organization and dynamics of various biological macromolecules and their assemblies. Current AFM techniques facilitate qualitative and quantitative measurements of diverse cellular processes, ranging from the forces generated by protein–membrane interaction to the mechanism of enzyme reaction at the single-molecule level. In this chapter, we shall focus on the most-recently established technique, fast-scanning AFM. A significant feature of this technique is its capability of directly analyzing conformational changes of biological macromolecules and intermolecular interactions at a single-molecule level on a subsecond time scale under near-physiological conditions. Atomic resolution can be achieved by X-ray crystallography and high resolution electron microscopy (EM). On the other hand, real-time movement of proteins in cells can be monitored under fluorescence microscopy. However, the movement detected under fluorescence microscopy does not uncover the dynamics of the protein shape itself. The applications outlined here, regarding fast-scanning AFM applications, fill the gap between X-ray crystallography/EM and fluorescence microscopy.

8.1 Introduction

Structural changes of biological macromolecules are tightly related to the dynamics of their functions. Static structures of a number of proteins have been determined by X-ray crystallography and high-resolution electron microscopy (EM), while the functions of biological macromolecules have been the target of biochemists, physiologists, biophysicists, as well as cell biologists. Since a simple assembly of static structural information does not provide an answer to how protein functions at the molecular level, different types of approaches are required to overcome this dilemma. Fortunately, single-molecule techniques that can track and measure individual biological macromolecules at work are now available, and therefore, have been employed to obtain clear information on the structure–function relationship of biological macromolecules [1–5].

Recent developments in fluorescence microscopy allow us to analyze the translocation or rotational motion of a fluorescent spot emitted from a fluorophore attached to molecules of interest [4, 6]. Nevertheless, the observed behavior of the fluorescent spot does not provide direct information on how the labeled molecule actually behaves; information obtained is not sufficient to fully understand how such biological macromolecules manifest their functions related to their structures. Therefore, direct visualization of individual molecules in action at the nanometer spatial and millisecond temporal resolutions has been desired.

Atomic force microscopy (AFM) [7] is a powerful tool for imaging individual biological macromolecules [8–10]. Its capability to directly visualize unfixed or unstained biological specimens at nanometer resolution has been suitable for the morphological analyses of various biological macromolecules ranging from a single protein (~nm) to a large protein complex (~100 nm). The development of tapping mode AFM in liquid has made it possible to study the structural and mechanical properties of biological macromolecules under "physiological conditions" [11–16]. For example, in the late 1990s, using AFM, the movement of RNA polymerases [13, 14], the reaction of type I restriction enzymes [12], and the behavior of DNase [11] were examined by time-lapse imaging in solution. In these studies, maximum scan rates varied between 0.5 and 2 frames min^{-1}, which limited the time resolution of the reactions requiring a subsecond time scale. Unfortunately, the scanning rates of commercially available AFMs (several seconds to minutes per frame) are too slow to monitor many physiological processes. Therefore, improvement of the time resolution has been an important issue in the biological applications of AFM.

An extraordinary improvement in the device was made by Ando's group at Kanazawa University in 2001 [17]. This newly developed fast-scanning AFM has a miniaturized cantilever and a scanning stage to reduce mechanical response time of the cantilever and to prevent the onset of resonant motion during high-speed scanning. The fast temporal resolution of 1–3 frames per second (fps) allows the dynamics of biological macromolecules to be analyzed more closely in the subsecond time scale. Applications of this device have included kinetic analyses of conformational changes of proteins and DNA [18–20], reaction mechanisms of enzymes [18, 19, 21], and dynamic behavior of motor proteins [17, 22] and nucleosomes [23]. Both the "subsecond time frame" and "nm scale" single-molecule observations of functional biological macromolecules cannot be achieved by other techniques.

8.2 Principles of Biological Fast-Scanning AFM

Initial attempts toward m-second time resolution date back to the early 1990s. In 1994, Hansma's group first tried to detect structural dynamics of biological molecules in liquid with m-second time resolution by a conventional (slow-speed) AFM [24, 25]. The cantilever position was fixed on a single molecule without two-dimensional scanning, and the conformational changes of the protein were detected as height fluctuations (height changes). This approach dramatically improved the temporal resolution and demonstrated the potential of AFM to analyze various kinds of biological processes in real time scale. However, their system still had limitations, that is, the approach completely sacrificed the two-dimensional information in the x–y directions, and the scan speed was still limited by the piezo scanner, the cantilever response, and so on. To overcome these limitations, the AFM with much faster scanning and sampling speed to x–y–z directions was required.

Between 1999 and 2003, mainly three research groups, Hansma's, Miles', and Ando's groups, have reported the instrumentations of fast-scanning AFM with unique ideas and features. The basic construction of such fast-scanning AFM is quite similar to that of conventional AFM, however various devices have been optimized for high-speed scanning.

8.2.1 Hansma's Fast-Scanning AFM

In 1999, Hansma's group first reported a small cantilever system for fast imaging and fast force spectroscopy [26, 27]. As mentioned above, the response (resonance frequency) of an AFM cantilever is one of the main limitations in achieving high-speed AFM. Their miniaturized cantilevers (14–24 μm long, 5 μm wide, and 102 nm thick) made it possible to break the limitation. Decreasing cantilever dimensions gave a much higher resonance frequency than larger cantilevers while simultaneously providing a similar spring constant [28]. Using these cantilevers, they clearly observed individual molecules of the co-chaperonin protein, GroES, binding to and dissociating from individual chaperonin GroEL in tapping mode with a time resolution of 20 Hz (20 lines s^{-1}) [29]. It made a great impact in the field of chaperonin research because the results indicated a new reaction model of the GroEL–GroES system, and the hypothesis was immediately confirmed by other researchers using single-fluorescence observation technique [30] (see detail in Sect. 8.4.1.1).

8.2.2 Miles' Fast-Scanning AFM

Miles' group developed an ultra-high-speed AFM [31–33]. The most unique aspect of their AFM is in the scanner system, with the combination of a normal piezo scanner, flexure stage, and a special tuning fork scanner [34] as the x-scanner. In this system, the fast line scan (x-axis) was produced by a tuning fork oscillating at its resonance frequency of about 32 kHz and the slow line scan (y-axis) was produced by the flexure stage, which enables ultra high-speed scanning (1,300 fps). However, the scanning speed is too fast to trace sample surfaces using the previous conventional AFM feedback system. Therefore, "constant height mode" [34] was developed based on the contact-mode imaging method [35, 36], in which the probe's z-position is maintained constant using a normal piezo scanner during the scan in the x- and y-directions with a constant force averaged over the timescale of one frame (about 1 ms). Topographic information is obtained through evaluation from deflection of the cantilever [36, 37]. The simplicity of this instrumental design and the principles of image capturing make it easy to speed up the scanning rate. Using this system, movies of soft biological samples, for example, collagen fibers with a time resolution of 1,280 fps in air and liquid environments [32], and human chromosomes with 20 fps (with scan scale of about a square micrometer) in liquid condition [33] have been obtained.

8.2.3 Ando's Fast-Scanning AFM

Ando's group, in collaboration with Olympus Corp., developed the most successful fast-scanning AFM for biological application [17, 18, 20, 22, 38, 39]. There were a number of improvements in their AFM system, for example, extremely small cantilevers (similar to Hansma's group), optical beam deflection detection system for its small cantilever, high-speed piezo scanner and electronics. The small cantilever, made of silicon nitride (spring constant of 0.2 N m^{-1}), had dimensions of 6 μm long, 2 μm wide, and 100 nm thick, which results in a resonance frequency of 3.5 MHz in air and 1.2 MHz in water [40]. In order to utilize this extremely small cantilever, which is too small to see and to obtain a strong enough deflection signal from, they combined it with an optical microscope to focus an incident laser beam onto the cantilever through a 20× objective lens with a long working distance (CFI Plan Fluor ELWS 20xC, Nikon). They also developed the special high-speed scanner with a small sample stage (about 1 mm^2).

The advantages of their AFM over the other high-speed AFMs are the following: (a) the tapping mode minimizes sample damage [41–43], and (b) the high speed and well controllable piezo scanner enables image capturing with a time resolution of 40 fps and high spatial resolution in liquid conditions. This is a quite straightforward and logical evolution of fast-scanning AFM for biological applications.

In the following sections, the word "fast-scanning AFM" shall be used to indicate the tapping-mode AFM in solution based on the Ando's and Hansma's systems.

8.3 Effects of a Scanning Probe and Mica Surface on Biological Specimens

In AFM, the molecules floating in solution are completely invisible. Therefore, adsorption of molecules to a substrate surface is a prerequisite for AFM imaging. Preparing appropriate substrate conditions is a key to successful imaging. In addition, evaluation of the effects of a scanning probe on the movement of the molecule is critical for reliable imaging, because interaction between the scanning probe and the sample is inevitable in tapping mode AFM. This section describes several examples of the approaches to the development of sample preparation procedures and the evaluation of the effects of the tip motion onto DNA and protein samples.

8.3.1 Experimental Conditions Required for Fast-Scanning AFM Imaging

Success in time-lapse analysis by AFM requires conditions that satisfy two conflicting requirements: (a) attachment of the molecules of interest onto the substrate surface, and at the same time (b) assuring free movement of the molecules on the substrate surface. To image the dynamics of biological macromolecules, the molecules must retain enough mobility to allow their functional/structural changes. At the same time, the molecules must be stably adsorbed to the substrate surface and remain within the scanning area during imaging.

Mica has been frequently used as a substrate in the imaging of biological samples because it has a flat surface at the atomic level over a wide area. Because the mica surface is negatively charged in solution, the technique of chemisorptions has often been used for the adsorption of negatively charged samples, such as nucleic acids and various protein–DNA complexes. This method has made a large contribution in the successful static and time-lapse imaging of biological samples [10, 12–14, 44]. However, in high-speed imaging of protein–protein interactions, unmodified straight mica may not be a suitable substrate. Because interactions between samples and the mica surface are achieved by electrostatic interaction, it is sometimes difficult to control the orientation of the molecule on the substrate surface. Therefore, the mica surface needs to be modified depending on the specimens.

Site-specific attachment of the molecules to the substrate surface was achieved by using two-dimensional crystals of streptavidin formed on biotin-containing planar lipid layers [45]. Characterization of streptavidin crystals showed its appropriate properties as a substrate for high-speed imaging; minimal roughness, and resistance to non-specific binding to protein. The applicability of these surfaces was tested by observing GroEL–GroES interaction and actin polymerization [45]. GroEL carrying a biotin moiety at the outer surface of the equatorial domain was immobilized on the streptavidin crystals. Using this sample, binding of GroES to GroEL was successfully visualized by fast-scanning AFM. Actin filaments formed by copolymerization of biotinylated G-actin with nonbiotinylated G-actin were stably immobilized on the streptavidin layer. Polymerization process at the filament end in the presence of non-biotinylated G-actin was visualized without non-specific binding of G-actin onto the substrate surface. Applications of streptavidin two-dimensional crystals will provide a new choice of substrate for fast-scanning AFM imaging, although we have to select appropriate two-dimensional crystal structures in each experiment and optimize a buffer condition which allows both the stable formation of two-dimensional crystals and the interaction between molecules of interest.

8.3.2 Effects of High-Speed Scanning on the Behavior of DNA in Solution

Effects of the scanning probe on the movement of DNA on the substrate have been evaluated by analyzing the movement of linear DNA or circular DNA at different scanning rates [23, 46]. A comparison of the motion of the free end of the linear DNA in 1 s at different scan rates revealed that the direction of the motion was apparently random [46]. The innate circular topology of plasmid DNA allows the analysis of fluctuating motion of DNA in any direction. Consecutive images of a plasmid in solution showed swinging motions of DNA strands without a breakage of the strands (Fig. 8.1a). Fluctuating motions of DNA strands in both x- and y-directions on the mica surface appeared to be equal, even though the scanning velocity was much greater in the x-axis (fast-scanning direction) than in the y-axis (slow-scanning direction) (Fig. 8.1b). This indicates that the scanning probe has little effect on the DNA motion.

The influence of the scanning probe onto these fluctuating motions could be determined by calculating elastic bending energy, which is one of the parameters to feature DNA conformations. The elastic bending energy can be estimated by analyzing the bending motion of the DNA strand. The elastic bending energy of DNA is calculated as follows (Fig. 8.1c),

$$E_{\text{bend}} = \sum \kappa(1 - \cos\,\theta_i) = \sum \kappa(1 - \vec{r}_i \cdot \vec{r}_{i+1} / \left|\vec{r}_i\right| \cdot \left|\vec{r}_{i+1}\right|) \qquad (8.1)$$

where $\kappa / k_{\text{B}}T \approx l_{\text{p}} \approx 50$ and l_{p} is persistence length (50 nm).

The probability distribution of the bending energy obtained at different scan rates (1–3 fps) showed similar distribution patterns (Fig. 8.1d), indicating that thermal energy is essentially constant. In other words, there is no injection of thermal energy that originates from the scanning motion of the AFM probe, and therefore, the observed motion of DNA is caused mainly by thermal fluctuation.

8.3.3 Effects of High-Speed Scanning on Protein Movement

Whether the movement of protein particles is caused by thermal diffusion on the mica surface or by the mechanical force from the scanning probe is also a critical

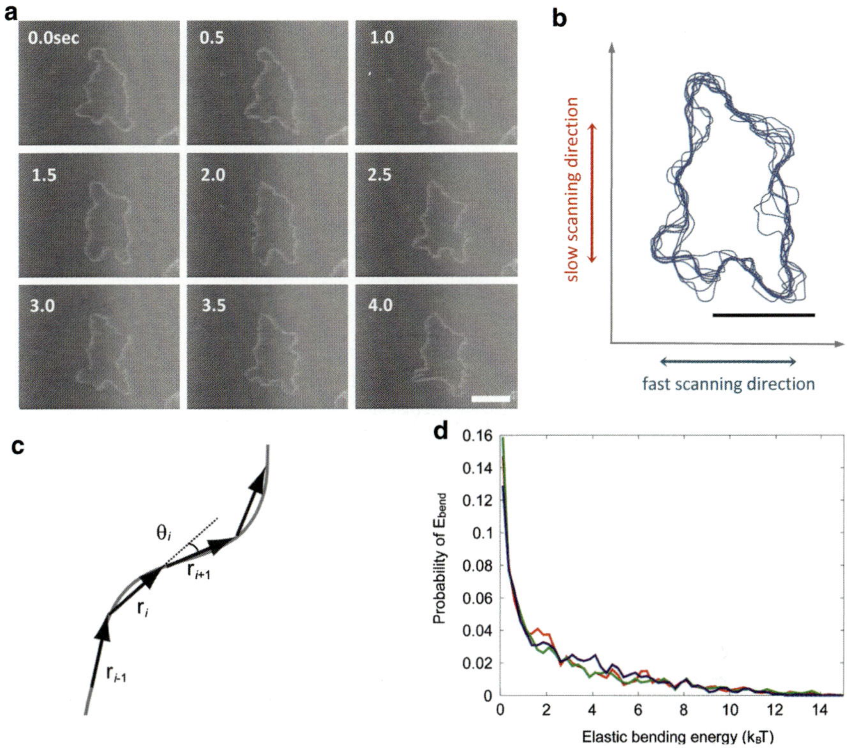

Fig. 8.1 Analysis of the movement of plasmid DNA. (**a**) Successive time-lapse images of a pGEMEX-1 obtained at 2 frames per second (fps). The elapsed time is shown in each image. *Scale bar*: 200 nm. (**b**) DNA strands from the ten successive images were overlaid to show DNA movement. *Scale bar*: 200 nm. The fast-scanning (parallel to x-axis) and slow-scanning directions (parallel to y-axis) are shown by the *blue and red arrows*, respectively. (**c**) Analytical image: DNA was fitted by the vectors. The angle of θ was calculated as the inner product of the vectors. (**d**) Probability distributions of bending energy; *red, green, and blue lines* are 3, 2, and 1 fps, respectively

question. If there is no effect from the scanning probe, protein particles should show random movements on the substrate surface. These movements could be determined by the two-dimensional diffusion coefficients (D_{2D}). The D_{2D} of a molecule is given by the following equation:

$$D_{2D} = <\Delta r_{2D}{}^2> /4\Delta t \tag{8.2}$$

where $<\Delta r_{2D}{}^2>$ is the two-dimensional mean-square displacement of a molecule during the time interval Δt.

Figure 8.2 represents a sequential record of a restriction endonuclease, *Apa*I (see Sect. 8.4.2.2), movement on the mica surface. The standard deviations of the diffusion coefficients were large (Table 8.1), possibly due to the heterogeneity of

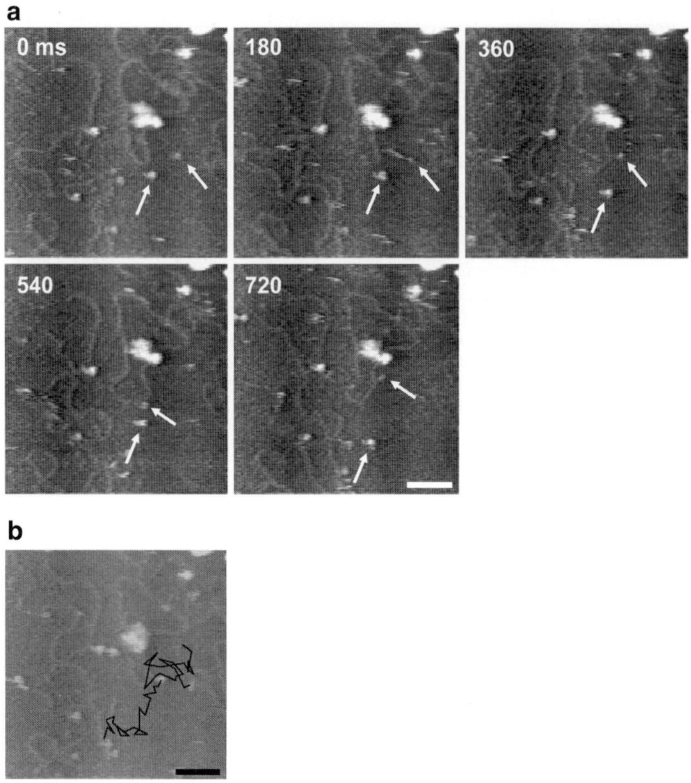

Table 1: Two-dimensional diffusion coefficients at each scan rate

Scan rate Δt, ms	n	$\langle \Delta r^2_{2D} \rangle$, × 10^3 nm^2 (mean ± SE)	$\langle \Delta r^2_{2D} \rangle$,× 10^{11}cm^{2}s^{-1} (mean ± SE)
180	34	1.46 ± 0.21	2.02 ± 0.30
360	16	2.79 ± 0.54	1.94 ± 0.38
540	11	4.43 ± 1.42	2.05 ± 0.66
640	53	5.84 ± 1.04	2.28 ± 0.41

Fig. 8.2 Dynamic movement of *Apa*I on a mica surface. (**a**) Successive time-lapse images of a *Apa*I–DNA reaction obtained at 5.6 fps. Movement of *Apa*I particles (*white arrows*) can be seen. *Scale bar*: 100 nm. (**b**) Sequential record of *Apa*I movement on the mica surface. *Apa*I molecules were traced in 30 successive images. *Scale bar*: 100 nm

the surface property of mica or unequal adsorption of ions on the surface, which strongly affects the binding force between the molecule and mica surface. However, the mean square displacement of the molecules, $< \Delta r_{2D}{}^2 >$, linearly increased with time, and the diffusion coefficients at different scan rates were almost in the same order (Table 8.1). These results indicate that the movement of the molecules on the mica surface is mainly caused by Brownian motion, and not influenced by

Table 8.1 Two-dimensional diffusion coefficients at each scan rate

Scan rate Δt (ms)	n	$<\Delta r^2{}_{2D}>$, $\times 10^3$ nm^2 (mean $\pm$ SE)	$<\Delta r^2{}_{2D}>$, $\times 10^{11}$ cm^2 s^{-1} (mean $\pm$ SE)
180	34	1.46 ± 0.21	2.02 ± 0.30
360	16	2.79 ± 0.54	1.94 ± 0.38
540	11	4.43 ± 1.42	2.05 ± 0.66
640	53	5.84 ± 1.04	2.28 ± 0.41

the scanning probe, although the fast-scanning probe occasionally causes apparent artifacts with irregular movement of the molecules.

8.4 Application to Biological Macromolecule Interactions

8.4.1 Application to Protein–Protein Interaction

Most biological macromolecules interact with each other in a highly specific and regulated manner, especially protein–protein interactions, as they play a key role in all fundamental regulatory systems in cells [47–50]. Since AFM can directly visualize unfixed or unstained biological specimens at nanometer resolution, this technique, together with the newly developed fast-scanning AFM, allows us to visualize and analyze the structural dynamics of protein–protein interaction at the single-molecule level.

8.4.1.1 Single-Molecule Kinetics Analyses of Chaperonin Reaction

The efficient folding of most newly synthesized proteins requires the assistance by molecular chaperones that prevent protein misfolding and aggregation in the crowded cell environment [51–53]. One of the best-characterized chaperone systems is the *Escherichia coli* chaperonin GroEL–GroES system [51, 54, 55]. The detailed three-dimensional structures of GroEL, GroES, and the GroEL/GroES complex have been resolved by X-ray crystallography [56–60], EM [61–63] and AFM [64] (Fig. 8.3a, b). GroEL consists of two homo-heptameric rings that are stacked back-to-back, whereas GroES is a single homo-heptameric ring. The chaperonin reaction cycle begins with the binding of a polypeptide (unfolded protein) and GroES to one side of GroEL, which results in a formation of the GroEL/polypeptide/GroES complex in the presence of nucleotides. Once the GroEL/polypeptide/GroES complex is formed, the polypeptide chain is (re-)folded into its functional conformation in a large central cavity in the GroEL ring. Then, the GroEL/GroES complex decays by the dissociation of GroES, followed by a release of the folded protein from the cavity [65–69].

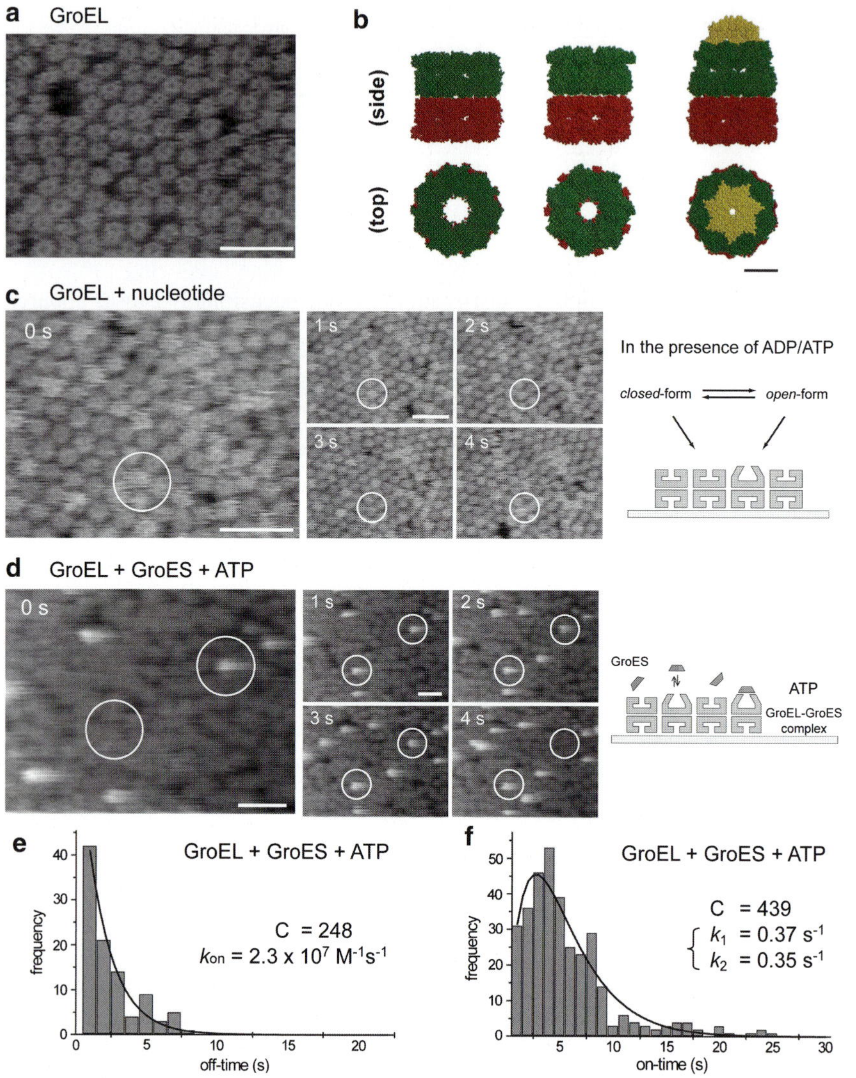

Fig. 8.3 Single-molecule imaging of chaperonin GroEL–GroES dynamics. (**a**) Topographic high-resolution image of GroEL by fast-scanning AFM. The GroEL molecules were lightly absorbed on a mica surface in buffer solution. In this image, the central channels of the GroEL molecules are visible. The image (192 × 144 pixels) was taken with a scan rate of 2 fps. *Scale bar*: 50 nm. (**b**) Three-dimensional representation of GroEL structures. *Left*: The crystal structure of GroEL (pdb 1GRL). *Center*: The cryo-EM map of the GroEL(D398A)-ATP complex (pdb 2C7E). *Right*: The crystal structure of the GroEL–GroES–ADP complex (pdb 1AON). The side views (*upper*) and the top views (*lower*) of van der Waals space-filled models of each structure, highlighting the individual ring structures of GroEL (lower GroEL ring, *red*; upper GroEL ring, *green*) and GroES (*yellow*), are shown. *Scale bar*: 5.0 nm. (**c**) Time-lapse sequence AFM images of GroEL on a mica surface in buffer solution were taken in the presence of 2 μM ATP and 500 nM GroES. GroES molecules were repeatedly found to bind to and dissociate from GroEL. ATP concentration was kept low (unsaturated condition) in order to observe and analyze single molecule events.

The chaperonin reaction was first visualized at a single-molecule level by Hansma's group in 2000 using a small cantilever AFM [29]. They measured the turnover rate of ATP-dependent dissociation events of the GroEL/GroES complexes using a single-line two-dimensional scan method. The method, in which an AFM probe repeatedly scanned a single line, provided a higher time resolution than the normal three-dimensional observation technique. Interestingly, their preliminary results of the distribution of the complex's lifetime did not show a single exponential decay, which indicates that the dissociation reaction of GroES from GroEL is not governed by a single rate-determining step, but rather obeys two sequential transitions. This result was immediately supported by the following research using other single-molecule observation techniques and traditional biochemical experiments [30, 70].

Concomitant with the advancement of the fast-scanning AFM, the reaction mechanism of GroEL/GroES has been deeply analyzed. Takeyasu's group clearly visualized and kinetically analyzed an ATP-dependent association–dissociation event of the GroEL/GroES complex and an ATP/ADP-dependent *open-closed* transition of GroEL without GroES with 10 m-second time resolution [20]. GroES molecules repeatedly associated with and dissociated from GroEL in the presence of both GroES and ATP (Fig. 8.3d). The duration time of GroES-unbound form of GroEL ("off time") was measured (Fig. 8.3e). The histogram of the "off-time" was well fitted to a single exponential curve, indicating that the GroES binding was a stochastic process. The association rate constant of GroES to GroEL (k_{on}) ($2.3 \times 10^7\ M^{-1}s^{-1}$) was calculated by curve fitting, which is half the value measured by previous bulk phase experiments ($4.6 \times 10^7\ M^{-1}\ s^{-1}$) [30, 68, 71]. This is reasonable because one side of the GroEL ring is inaccessible to GroES in this experimental condition. Thus, the semi-immobilized GroEL sample does not lose the activity for GroES binding and the effect of probe scanning to the GroES binding reaction is low enough to be ignored.

Fig. 8.3 (continued)
In the AFM images at 0 s, for example, a 4-nm taller structure (GroEL–GroES complex) can be seen in the *right white circle*. (**e**, **f**) Statistical analyses of GroES binding and dissociation events of GroEL in the presence of ATP. Histogram in (**e**) shows the distribution of the duration time between the GroES dissociation from and the next GroES binding to individual GroEL molecule in the presence of 25 nM GroES and 100 μM ATP. Some GroELs that do not undergo conformational changes at all over the period of observation in the presence of ATP was excluded from the analyses. The histogram was fitted with a single-exponential function as the following equation: $H(t) = C_3 k_3 \exp(-k_3 t)$, where $H(t)$ is the number of GroEL that had an off time t, C_3 is the number of the total events, and k_3 is the rate constant. The rate constant (k_3) was obtained by the nonlinear least-square curve fitting method. Then, the association rate constant for GroES binding (k_{on}) was calculated. Histogram in (**f**) shows the distribution of the lifetime of GroEL–GroES complex. Individual GroES molecules that attached to and dissociated from GroEL molecules repeatedly were used for the analyses over the observation time of ~10 min. The *solid line* was obtained by the least-square fitting to the equation: $C_1 k_2 k_2'[\exp(-k_2 t) - \exp(-k_2' t)]/(k_2 - k_2')$ derived from the two-step reaction of scheme (8.1)

The duration time of GroES-bound GroEL ("on time") in the presence of 500-nM GroES and 2-μM ATP was also measured and summarized in Fig. 8.3f. This histogram had a maximum peak at 4 s (the average lifetime is 6 s) and did not show a single exponential decay, indicating that the dissociation reaction of GroES from GroEL is probably not governed by a single rate-determining step, but rather obeys two sequential transitions (two timer model), as previously proposed by Taguchi et al. [30, 70].

$$[\text{GroEL/GroES}] \underset{k_2}{\rightarrow} [\text{GroEL/GroES}]^* \underset{k_2}{\rightarrow} [\text{GroEL}] + [\text{GroES}] \qquad (8.3)$$

The histogram data was well-fitted to the equation deduced from the scheme (8.3): $G(t) = C_2 k_2 k_2{}' \, [\exp(-k_2 t) - \exp(-k_2{}' t)]/(k_2{}' - k_2)$, where $G(t)$ is the number of GroEL/GroES complexes that had a lifetime t, $k_2 = 0.37\,\text{s}^{-1}$, $k_2{}' = 0.35\,\text{s}^{-1}$, and $C_2 = 295$ (the number of the total events). They also analyzed the ATP concentration dependency on GroES-retained time. The time constants of the lifetimes of two independent states were 3 and 2–3 s, respectively, and the sum was 5–6 s, which well agrees with the previous value (7–15 s) estimated in bulk-phase kinetics [30, 65, 67–69].

ATP/ADP-bound GroEL can change its conformation "from *closed* to *open*" without additional ATP hydrolysis (Fig. 8.3c). In order to monitor the conformational changes of GroEL with a higher time resolution, a single-line two-dimensional scan method [29], in which a single line was repeatedly scanned at a rate of 250–1,000 Hz, was used. This approach clearly demonstrated that both the "*closed* to *open*" and the "*open* to *closed*" conformational changes were observed as a 1-nm height difference, and completed within at least 10 ms in most GroEL molecules (more than 95%). After this rapid "*closed* to *open*" conformational change, GroEL kept an *open* conformation for a few seconds. These properties of the *open-closed* conformational changes were also observed in the presence of ADP, ATP, and ATP analogues. Furthermore, the lifetime of the *open* conformation in the presence of ADP (∼1.0 s) was apparently shorter than those of ATP and ATP analogues (2–3 s), meaning that ADP-bound *open*-form exists and is structurally less stable than ATP-bound form. Those results indicate that GroEL has at least two distinct *open*-conformations in the presence of nucleotides: ATP-bound pre-hydrolysis*open*-form and ADP-bound *open*-form, and the ATP hydrolysis in *open*-form destabilizes its *open*-conformation and induces the "*open* to *closed*" conformational change of GroEL.

8.4.1.2 Single-Molecule Morphological Analyses of Motor Proteins

Myosin V is a cargo-protein processive motor that takes steps along actin filaments [72–74]. Like many other processive motors, such as kinesin [75–77] and dynein [78,79], it has two heads held together by a coiled-coil stalk. Each head of myosin V contains a catalytic domain responsible for actin binding, ATP hydrolysis, and a light chain binding domain that likely acts as a lever arm to amplify nucleotide dependent conformational changes in the catalytic domain. About ten years ago, it was unclear how the two heads of myosin V coordinated to produce steps. Since then, two mechanisms have been proposed to account for this movement based on previous biochemical and biophysical studies. One was a hand-over-hand "walking" model [73, 80–82] and the other was the so-called inchworm model [83], in which one myosin V head always leads, moving forward before the trailing head catches up. Therefore, in order to conclude the molecular mechanism of the motor proteins, a new technique to analyze "motion" of a single molecule was strongly required.

In 2001, Ando and his colleagues reported the motion of single myosin V on mica (not on actin filaments) by his high-speed AFM, which could capture a 100 × 100 pixels image within 80 ms in aqueous solution [17]. They showed the changes in the locations of the myosin heads, neck, and tail regions on mica surface in the presence of ATP. Although they could not address the question of the molecular mechanism of motor function, they clearly demonstrated the possibility of using fast-scanning AFM for reaction analyses on a single biological macromolecule. We should note that they recently succeeded in visualizing the "walking" reaction of single myosin V molecules on actin filaments by their fast-scanning AFM [22].

8.4.2 Application to DNA–Protein Interaction

Genomic DNA interacts with a number of proteins in cells. Numerous biological processes (including recombination, transcription, and repair) rely on such DNA–protein interactions. The AFM has allowed us to analyze the site of protein binding, the angle of the DNA bending upon the protein binding, the size of protein-mediated DNA loop(s), the degree of oligomerization of protein on DNA, and others [84]. Examination of the protein–DNA dynamics with the fast-scanning AFM provides clues that elucidate the principles of the protein–DNA interaction mechanism.

8.4.2.1 Dynamics of DNA-Targeted Enzyme Reaction

Restriction endonucleases serve as useful models in the study of how intracellular protein machinery functions on DNA. The restriction endonucleases are classified into three main groups (types I, II, and III) according to their subunit composition, cofactor requirement, and mode of action [85, 86]. Type II restriction endonucleases do not require ATP for their activity, whereas types I and III absolutely do. More than 3,500 type II restriction endonucleases have been characterized [87] and subdivided into at least 11 distinct subclasses [85, 86]. There has been much interest in how site-specific proteins search for their recognition site on DNA. The reaction between DNA and restriction endonucleases has been thought to begin with the random binding of the enzyme to DNA. However, it has been shown that proteins are able to find their recognition site 100–1,000 times faster than what would be expected from random diffusion [88]. Therefore, the enzymes should have certain mechanisms to facilitate the finding of the specific sequences. Although several models have been proposed, such as a one-dimensional diffusion (sliding) model

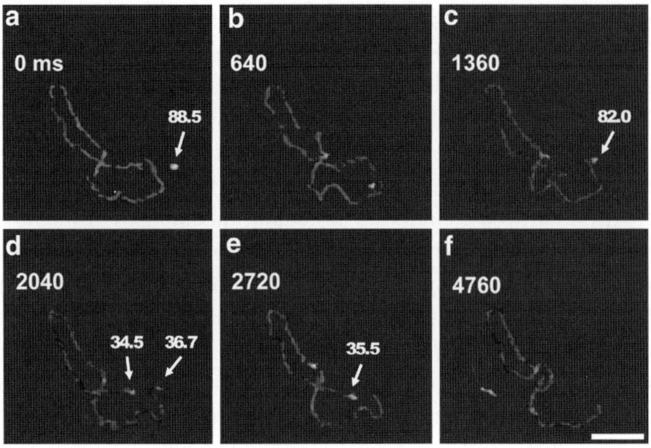

Fig. 8.4 Interaction between *Apa*I and DNA. Successive time-lapse images of a *Apa*I–DNA reaction obtained at 1.6 fps. The elapsed time is shown in each image. During the period of observation (time 0–4,760 ms), several intermediate states of the DNA cleaving process by *Apa*I were caught and traced. A volume analysis of the real-time AFM images showed that *Apa*I bound to DNA as a dimer (Fig. 8.5a, c, *white arrow*). Then, the *Apa*I dimer digested DNA within 640 ms (Fig. 8.5c, d). Immediately after the digestion, the *Apa*I dimer was separated into two monomers, each of which remained on each cleaved end (Fig. 8.5d). Finally, one of the *Apa*I monomers disappeared (Fig. 8.5e). *Scale bar*: 100 nm

[89, 90] and a microscopic dissociation–reassociation (hopping) model [91, 92], the details of the reaction mechanism are still unclear.

One of the best-characterized subclass is type IIP, which includes *Eco*RI, *Eco*RV, *Sma*I, and *Apa*I [93], and is frequently used for recombinant DNA technology. The type IIP restriction endonucleases recognize short palindromic DNA sequences (4–8 bp long), and, in the presence of Mg^{2+} ion as a cofactor, cleave the DNA at a given position within or close to their recognition sequences [85]. Fast-scanning AFM was applied to investigate the interaction between *Apa*I and DNA substrates (Fig. 8.4). A series of images revealed that *Apa*I bound to DNA as a dimer and slid along DNA in a one-dimensional diffusion manner [21]. When it encountered a specific DNA sequence, the enzyme halted for a moment to digest the DNA. Immediately after digestion, the *Apa*I dimer separated into two monomers, each of which remained on the DNA end and then eventually dissociated from the DNA end.

In the previous biochemical and microscopic studies, some type II restriction endonucleases, such as *Eco*RI, *Eco*RV, and *Sma*I, are assumed to follow the successive cleavage model in which the enzymes remain associated with DNA after the cleavage, and continue to scan for additional recognition sites [90, 94]. However, the fast-scanning AFM analyses suggest that the successive cleavage model cannot be applied to *Apa*I. The *Apa*I dimer should have a longer binding time than the monomer because the dimers of type IIP restriction endonucleases structurally enclose DNA. Therefore, the separation of the *Apa*I dimer might have significance in accelerating the dissociation of *Apa*I from the cleaved DNA after the digestion. This quick dissociation of *Apa*I would directly enhance the chance of *Apa*I to meet another DNA.

8.4.2.2 Dynamics of More Complex Protein–DNA Interaction

The formation of protein–DNA complexes, including several distinct sites on DNA, can be seen in various biological processes, such as recombination, replication, transcriptional regulation, and repair. Restriction endonucleases, which bind two or more cognate sites for their activity, are widely spread [95]. Such a multi-site mechanism suggests that restriction endonucleases serve as evolutionary precursors to many DNA regulatory factors [96, 97]. How spatially separated sites come into contact with each other is one of the most interesting subjects in understanding the fundamental process in gene functions.

The restriction endonuclease *Eco*RII belongs to the type IIE enzyme group that binds two DNA recognition sites to cleave one of the sites [86]. Binding and translocation dynamics of the *Eco*RII protein to and along DNA were visualized by fast-scanning AFM [19]. The analyses showed that the protein can translocate along DNA to search for the specific binding site (Fig. 8.5a). Once it specifically binds to one of the recognition sites, the protein is capable of searching for the second specific binding site while it stays on the first site (Fig. 8.5b). This is the first direct evidence describing the mechanism of searching for the second cognate site. The obtained results also suggested that the translocation along the DNA strand occurs either by sliding or hopping mechanisms when it searches for the initial and the second binding sites (Fig. 8.5c).

A restriction endonuclease, EcoP15I, is a type III enzyme that requires the specific asymmetric recognition sites where two of these enzymes are held together in a head-to-head orientation for DNA cleavage [98, 99]. EcoP15I was previously thought to translocate along the DNA strand by ATP-dependent manner and form a loop between the recognition sites. However, several studies on the type III enzymes reached a conflicting conclusion that no loops were formed [100]. Time-lapse imaging by fast-scanning AFM has demonstrated that Ecop15I bound to its cognate recognition site can make another contact with nonspecific DNA region and the intervening DNA loop in the complex can be translocated unidirectionally in an ATP-dependent manner [18].

It is interesting to consider how fast restriction enzymes translocate on DNA. Analyses of sequential image sets gave the translocation rates of $30\,\mathrm{bp\,s^{-1}}$ for *Eco*RII and $80\,\mathrm{bp\,s^{-1}}$ for EcoP15I. These values are probably underestimated because of the surface friction present in AFM experiments. Nevertheless, these values are higher than the expected rates of other cellular events, such as RNA transcription, which occurs with the rate of $20\,\mathrm{bp\,s^{-1}}$ [101]. A higher velocity of translocation suggests that the translocation mechanism might be essential for the facilitation in finding the specific sequences. Currently, total internal reflection fluorescence (TIRF) is the most common technique for studying translocation of proteins along DNA. In such an experiment, DNA is stretched over the glass surface and movements of fluorescently labeled proteins along the DNA are visualized. However, the measurement of looped translocations cannot be analyzed with such a technique because the DNA is stretched prior to the experiments. Thus, the ability of fast-scanning AFM to image multisite protein–DNA complexes with DNA

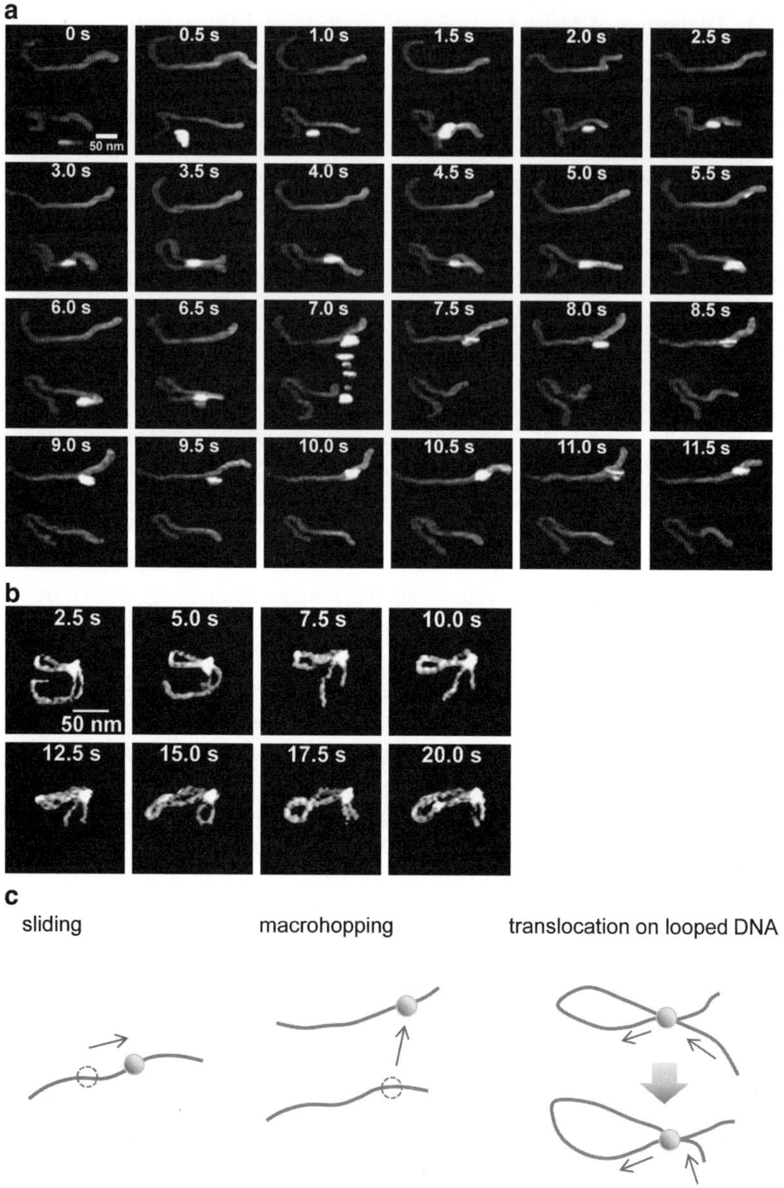

Fig. 8.5 Dynamic *Eco*RII–DNA Interactions. Successive time-lapse images of *Eco*RII–DNA interaction obtained at 2 fps. The elapsed time is shown in each image. (**a**) *Eco*RII translocation and macrohopping. The protein is seen to bind and form loops. In addition, the protein is seen to bind apparent nonspecific regions of the DNA and translocate down the DNA (from 5.0 to 6.0 s and from 7.5 to 11.5 s). The protein is seen to "come away" from and rebind to the protein (8.0 and 9.5 s). Macrohopping to an adjacent DNA strand is seen as well (7.0 s). The smearing effect is due to the raster scan pattern of the microscope in which the tip is tracking the protein as it moves. (**b**) *Eco*RII translocation on looped DNA. Individual frames are shown every 2.5 s. As the long arm gradually becomes shorter, the loop length increases. (**c**) Schematic illustration of sliding, macrohopping, and translocation on looped DNA

in a relaxed conformation is not amenable to other methods such as TIRF imaging techniques.

8.4.2.3 Nucleosome Dynamics: Sliding and Disruption

Eukaryotic genomic DNA interacts with a number of proteins and is folded into higher-order chromatin fibers through several hierarchical packaging mechanisms. The most fundamental structural unit of chromatin is the nucleosome, which is composed of about 146 bp of DNA wrapping 1.75 turns around a histone octamer [102–104]. The structural properties of the nucleosome and its stability has been one of the major topics in chromatin studies, because the exposure from or occlusion into chromatin of genomic DNA is closely related to the regulation of gene functions. It has been biochemically demonstrated that nucleosomes spontaneously reposition themselves along DNA when the temperature or the ionic strength of the solution increases [105–107]. A combination of AFM and molecular simulations has demonstrated that this sliding motion is caused by a manifestation of Brownian motion [108]. These investigations indicate that thermal fluctuation is possibly responsible for nucleosome dynamics.

By taking advantage of fast-scanning AFM, the dynamics of nucleosomes in solution were analyzed. In the sequential images of poly-nucleosomal arrays, the sliding of the nucleosome core particle and the dissociation of histone particle were visualized [23]. The sliding showed limited fluctuation within ~ 50 nm along the DNA strand (Fig. 8.6a). It was also demonstrated that the histone dissociation occurs in at least two distinct ways; (a) all of the core histones (2A, 2B, 3, and 4) are released from the DNA at once under the current limited-time resolution (0.5 s frame^{-1}), (b) histone subunits (possibly 2A and 2B, and 3 and 4, in pairs) sequentially dissociate from the nucleosome. Based on these observations, the second dissociation in the latter case occurs within a few seconds after the first dissociation, whereas the first step rarely occurs even within a hundred seconds (Fig. 8.6b). That is to say, it is likely that the second step of the disassembly occurs much faster than the first step, and that the intermediate state of the disruption (between the first and the second steps) is very unstable and easily proceeds to the completely dissociated state. These observations help us to develop the molecular mechanisms of nucleosome repositioning involved in chromatin dynamics.

8.5 Mechanisms of Signal Transduction at the Single-Molecule Level

Structural dynamics of the cell membrane and membrane proteins are also attractive targets of fast-scanning AFM. Cellular membranes have various important functions in living cells, such as determination and maintenance of cell shape, energy generation and transformation, signal transduction, cell motility, formation of focal adhesions, and so on [109–111]. Protein compositions of the membrane have been

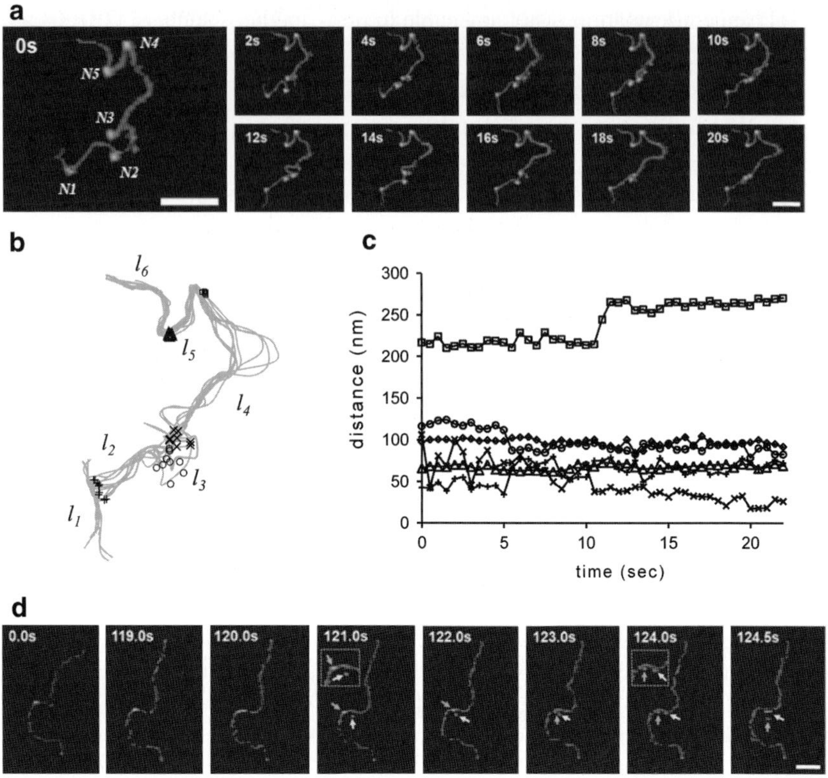

Fig. 8.6 Analysis of nucleosome dynamics. (**a**) Time-lapse images of a reconstituted nucleosomal array obtained at 2 fps. Images at every 2.0 s are selected and sorted (the elapsed time is shown in each image). The images were trimmed from the original scan size of 800 × 600 nm. *Scale bar*: 100 nm. (**b**) Movement of nucleosomes on mica. The DNA chain of the nucleosomal array was traced at every 2.0 s and overlaid. The center of five nucleosomes [*N1* (*plus*), *N2* (*open circle*), *N3* (*times*), *N4* (*empty square*), *N5* (*up triangle*)] were indicated. (**c**) The changes in the DNA length between adjacent nucleosomes (or nucleosome and the closer end of DNA) over a time period of 22 s. l_1 (between *N1* and the closer end of DNA), l_2 (between *N1* and *N2*), l_3 (between *N2* and *N3*), l_4 (between *N3* and *N4*), l_5 (between *N4* and *N5*), and l_6 (between *N5* and the closer end of DNA) are indicated by (*plus*), (*open circle*), (*times*), (*empty square*), (*up triangle*) and (*dot*), respectively. (**d**) Sequential dissociation of histone subunits from the nucleosome array. Time-lapse images of reconstituted nucleosomal array were obtained at 2 fps. Dissociated particles were indicated by *arrows*. *Scale bar*: 100 nm

greatly characterized and are ready for structural and functional analyses at the single-molecule level. In fact, various membrane proteins, such as trimeric bacteriorhodopsin in purple membranes [112], F-type ATPase [113–115], Rho in native disc membranes [116], the cyanobacterial photosystem [117], and S-layer protein [118], have been visualized by conventional AFM. Fast-scanning AFM will be even more powerful for the structure–function analyses of such membrane proteins.

8.5.1 Conformational Changes of Ligand-Gated Ion Channels

P2X receptors ($P2X_nRs$) are cell-surface ATP-gated cation channels [119]. A single functional P2X receptor channel is composed of three subunits, each of which is predicted to have a large extracellular domain, two transmenbrane domains, and N and C termini intracellular domains. $P2X_nRs$ have different permeability states that were originally discovered by Cockcroft and Gomperts [120].

For example, the permeability of $P2X_4Rs$ is greatly affected by extracellular Ca^{2+} levels [121]. In the presence of Ca^{2+}, $P2X_4R$ acts as a small cation-permeable channel pore, but in the absence of Ca^{2+} $P2X_4R$ is supposed to form a larger pore that allows larger molecules including *N*-methyl-D-glucamine $(NMDG)^+$, propidium iodide, and ethidium bromide to pass. Although there must be a relationship between the permeability and structural changes in the extracellular domain, how the activated architecture of the $P2X_4Rs$ in the extracellular domain is related to the response to ATP binding is poorly understood.

Shinozaki et al. successfully visualized the ATP-induced structural changes in $P2X_4Rs$ by fast-scanning AFM [122]. The image analyses revealed that $P2X_4Rs$ have an upward orientation on a mica surface. Time-lapse imaging of the ATP-induced structural changes in $P2X_4Rs$ revealed two different structural changes: (a) from one circular structure to a trimeric structure and (b) the subsequent moving away of each lobe (pore dilation-like change). In order to consider the functional importance of this pore dilation of $P2X_4R$, Ca^{2+} and dye uptake imaging of membrane reconstituted $P2X_4R$ was also performed using fluorescence microscopy. In the absence of Ca^{2+}, both pore dilation-like structural changes in the extracellular domains and ethidium bromide uptake were detected. At 2 mM Ca^{2+}, no ethidium bromide uptake was observed but there was a Ca^{2+} flow via $P2X_4Rs$. Under this condition, the pore dilation-like structure of $P2X_4Rs$ was not observed, but $P2X_4Rs$ exhibited a disengaged trimer similar to the structure seen immediately after ATP binding in Ca^{2+} starved conditions. These results provide direct evidence of structural changes in the extracellular domains of $P2X_4Rs$ that are involved in permeability dynamics.

8.5.2 Conformational Changes of G-protein Coupled Receptors

Bacteriorhodopsin [123, 124] is an archaeal *Halobacterium salinarum*'s protein, which is the first membrane protein whose two-dimentional crystal structure has been determined at a 7 Å resolution by EM in 1975 [125]. Its structure and function have been analyzed in great detail using a multitude of different experimental techniques including AFM [112, 126, 127] and has become the best-understood example of vectorial catalysis. Bacteriorhodopsin acts as a proton pump, i.e., it captures the energy of "green" light (500–650 nm, max 568 nm) and uses it to make an electrochemical proton gradient across the cell membrane. The resulting proton gradient is subsequently converted into chemical energy. Bacteriorhodopsin contains seven transmembrane α-helices (termed A–G) enclosing chromophore retinal [128–130]. The light-induced conformational changes in bacteriorhodopsin have been investigated by various methods such as X-ray diffraction and electron crystallography [129, 131–140], leading to a consensus that the proton channel at cytoplasmic surface is opened by the tilting of helix F away from the protein centre [134, 137, 139]. However, dynamic molecular processes of bacteriorhodopsin in response to photoactivation are not fully understood.

Using fast-scanning AFM, the light-induced conformational changes in the D96N bacteriorhodopsin mutant, which has a longer photocycle ($\sim$10 s) than the WT ($\sim$10 ms) but retains its proton pumping activity, were monitored [38]. Upon illumination with green light, a cytoplasmic portion of each bacteriorhodopsin monomer was brought into contact with adjacent trimers and returned to the unphotolysed state in a few seconds after light-off. This dynamical change of bacteriorhodopsin was reproducible in repeated dark-illumination cycles. The mass center analysis of the individual bacteriorhodopsin monomers imaged during the cycles showed that all the activated bacteriorhodopsin monomers exhibited displacements of their center of mass, with an average displacement of 0.69 ± 0.15 nm. Traces of the mass center positions on the AFM images taken under dark-illumination cycles demonstrated that the photo-induced movement of bacteriorhodopsin includes counterclockwise rotation (7.4 ± 2.2) around the trimer center.

8.5.3 Direct Visualization of Albers–Post Scheme of P-Type ATPases

After the first discovery of the sodium pump by Jens Skou at 1957 [141], a number of enzymes with similar properties have subsequently been identified and studied [142]. Now, it is well known that ion pumps, together with ion channels, form ubiquitous and fundamental ion regulatory systems in cells. The P-type ATPases, e.g., Ca^{2+}-ATPase, Na^+/K^+-ATPase, and H^+-ATPase, form a superfamily of membrane-spanning cation transporters (ion pumps) which transport specific cations across the membrane by utilizing the energy obtained from ATP hydrolysis.

One of the best-characterized P-type ATPases is the *rabbit* Sarco(endo)plasmic reticulum (SR) Ca^{2+}-ATPase (SERCA) [142, 143], which plays a major role in muscle relaxation [144, 145]. Ca^{2+} is released into the cytoplasm from the SR through the Ca^{2+}-release channel and removes the troponin inhibition of the actin–myosin interaction and induces muscle contraction. When the muscle relaxes, Ca^{2+} needs to be removed from the cytoplasm by the SERCA, which accumulates Ca^{2+} inside the SR against its concentration gradient using energy obtained from ATP hydrolysis. The importance of the SR Ca^{2+} pump was realized in the early 1960s by Ebashi and Lipmann [146, 147]. Since then, most of the molecular components in the regulation of skeletal muscle contraction have been identified, crystallized, and their genes cloned [144, 145, 148]. Their molecular mechanisms have been well studied by traditional biochemical experiments. The functional cycle of P-type ATPases,

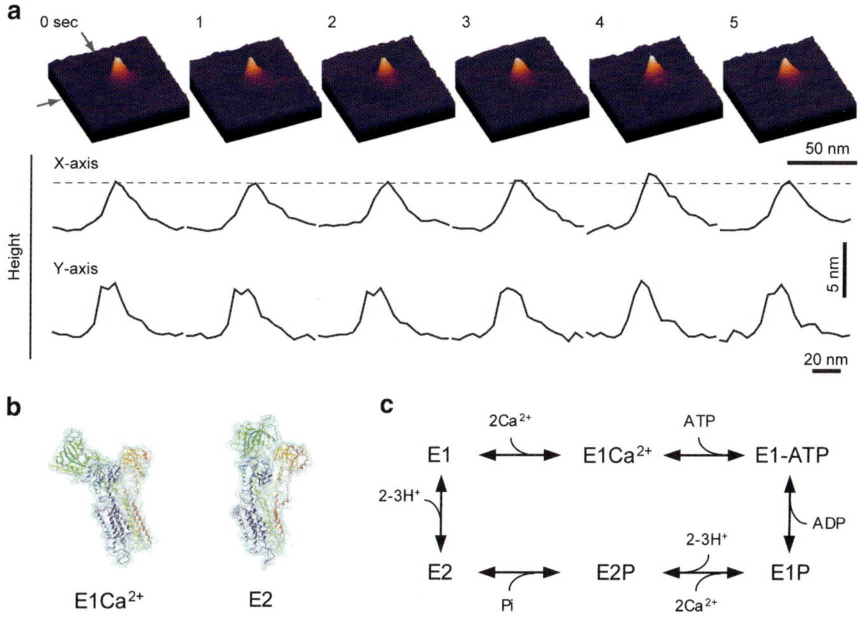

Fig. 8.7 Single-molecule imaging of SERCA dynamics in the presence of nucleotides and calcium ions. (**a**) Time-lapse sequence AFM images of the SERCA in the SR membrane on a mica surface in buffer solution were taken in the presence of 10 nM ATP and 100 μM free Ca^{2+} with 192 × 144 pixels at a rate of 1 fps. The images (40 × 40 pixels) presented here were cut from the original data without any modification. *Scale bars*: 20 nm. The Z scale is 20 nm (*Lookup table: gem*). The resulting profiles are shown in the corresponding *lower panel*. The *broken line* indicates a height of 5.5 nm from the membrane surface (**b**) three-dimensional structures of SERCA: $E1Ca^{2+}$ form (*left*) and E2 form (*right*). (**c**) Simplified Albers–Post scheme for SERCA

including SERCA, is frequently described by the Albers–Post scheme ($E1/E2$ theory) (Fig. 8.7c). In the reversible catalytic cycle, the binding of two Ca^{2+} to SERCA with high affinity from the cytoplasmic side of the membrane ($E1$Ca form) triggers autophosphorylation at Asp^{351} by Mg-ATP ($E1$P form). Then, after a series of conformational changes, SERCA goes to the de-phosphorylated (*E2*) and the initial Ca^{2+} unbound form ($E1$) in the scheme. Until now, the detailed three-dimensional structures of SERCA in many states in the theoretical cycle have been resolved by X-ray crystallography [149–153]. These crystallographic studies clearly show that SERCA undergoes dynamical conformational changes during the ion transporting reaction.

The structural dynamics of the SERCA in the rabbit SR membrane was also successfully captured by using fast-scanning AFM. Takeyasu's group visualized the conformational changes of the SERCA with a 1-ms time resolution. The analyses of individual SERCA molecules in the presence of both ATP and free Ca^{2+} revealed *up*-and-*down* motions and shape changes between taller (compacted) and shorter (open and Y-shape) forms of SERCA molecules (Fig. 8.7a), corresponding to the conformational changes of SERCA during ATP-mediated ion transporting reactions (Fig. 8.7b, c). Interestingly, the number of "*up* and *down*" conformational changes of SERCA per unit time was dependent on ATP concentration, and the motion was strongly inhibited by adding thapsigargin [154, 155] to the buffer solution. Considering the crystallography data and the fact that, under normal buffer conditions, the SERCA reaction usually goes in one direction (catalytic direction) in the Albers–Post scheme (Fig. 8.7c) [142, 156, 157], one *up*-and-*down* motion corresponds with one catalytic cycle (Ca^{2+}-binding shorter conformation → ATP hydrolysis-mediated elevated conformations → Ca^{2+}-binding shorter conformation), and the number of peaks must correspond to the velocity of the catalytic cycle of SERCA. Actually, the tendencies of their results were quite similar to those from previous biochemical data (ATPase activity and Ca^{2+} uptake) [158, 159].

Interestingly, under physiological ATP conditions, the Ca^{2+} pump did not exhibit *up*-and-*down* motion, and showed the up-configuration throughout. These results indicate the possibility that, in physiological ATP concentration (mM order), SERCA does not transit the $E1Ca^{2+}$ state [Fig. 8.7b (left)] [149], in which SERCA has the shortest structure, and has a catalytic pathway different from the ordinary Albers–Post scheme. This hypothesis is further supported by previous X-ray crystallography research [160, 161]; the SERCA assumes its compact (taller) structure during the whole reaction cycle in the physiological working condition. It is also notable that many biochemical experiments have shown that ATP exerts an additional stimulatory effect on the reaction cycle under higher [ATP] conditions ($>100\,\mu M$) [162].

8.6 Conclusion

We have summarized advantages, disadvantages, and unique features of fast-scanning AFM. Applications of fast-scanning AFM to protein–DNA complexes have provided new insights into the fundamental protein-dependent molecular mechanisms of DNA function. Applications of fast-scanning AFM to the molecular mechanisms of signal transduction via membrane proteins have enabled the initial trial of nano-scale analyses on the conformational changes of the first signal reception systems at the single-molecule level. Thus, we are sure that the qualitative and quantitative (kinetics) analyses of biological phenomenon by fast-scanning AFM will enhance our understanding of the important biological processes with high temporal and spatial resolution.

References

1. W.J. Greenleaf, M.T. Woodside, S.M. Block, High-resolution, single-molecule measurements of biomolecular motion. Annu. Rev. Biophys. Biomol. Struct. **36**, 171–190 (2007)
2. K.C. Neuman, A. Nagy, Single-molecule force spectroscopy: optical tweezers, magnetic tweezers and atomic force microscopy. Nat. Meth. **5**(6), 491–505 (2008)

3. R. Roy, S. Hohng, T. Ha, A practical guide to single-molecule FRET. Nat. Meth. **5**(6), 507–516 (2008)
4. Y. Sako, T. Yanagida, Single-molecule visualization in cell biology. Nat. Rev. Mol. Cell Biol. Suppl, SS1–SS5 (2003)
5. N.G. Walter, C.Y. Huang, A.J. Manzo, M.A. Sobhy, Do-it-yourself guide: how to use the modern single-molecule toolkit. Nat. Meth. **5**(6), 475–489 (2008)
6. Y. Sako, Imaging single molecules in living cells for systems biology. Mol. Syst. Biol. **2**, 56 (2006)
7. G. Binnig, C.F. Quate, C. Gerber, Atomic force microscope. Phys. Rev. Lett. **56**(9), 930–933 (1986).
8. A. Engel, D.J. Muller, Observing single biomolecules at work with the atomic force microscope. Nat. Struct. Biol. **7**(9), 715–718 (2000)
9. D. Fotiadis, S. Scheuring, S.A. Muller, A. Engel, D.J. Muller, Imaging and manipulation of biological structures with the AFM. Micron. **33**(4), 385–397 (2002)
10. Y. Hirano, H. Takahashi, M. Kumeta, K. Hizume, Y. Hirai, S. Otsuka, S.H. Yoshimura, K. Takeyasu, Nuclear architecture and chromatin dynamics revealed by atomic force microscopy in combination with biochemistry and cell biology. Pflugers Arch. **456**(1), 139–153 (2008)
11. H.G. Abdelhady, S. Allen, M.C. Davies, C.J. Roberts, S.J. Tendler, P.M. Williams, Direct real-time molecular scale visualisation of the degradation of condensed DNA complexes exposed to DNase I. Nucleic Acids Res. **31**(14), 4001–4005 (2003)
12. D.J. Ellis, D.T. Dryden, T. Berge, J.M. Edwardson, R.M. Henderson, Direct observation of DNA translocation and cleavage by the EcoKI endonuclease using atomic force microscopy. Nat. Struct. Biol. **6**(1), 15–17 (1999)
13. M. Guthold, X. Zhu, C. Rivetti, G. Yang, N.H. Thomson, S. Kasas, H.G. Hansma, B. Smith, P.K. Hansma, C. Bustamante, Direct observation of one-dimensional diffusion and transcription by *Escherichia coli* RNA polymerase. Biophys. J. **77**(4), 2284–2294 (1999)
14. S. Kasas, N.H. Thomson, B.L. Smith, H.G. Hansma, X. Zhu, M. Guthold, C. Bustamante, E.T. Kool, M. Kashlev, P.K. Hansma, *Escherichia coli* RNA polymerase activity observed using atomic force microscopy. Biochemistry. **36**(3), 461–468 (1997)
15. S.J. van Noort, K.O. van der Werf, A.P. Eker, C. Wyman, B.G. de Grooth, N.F. van Hulst, J. Greve, Direct visualization of dynamic protein-DNA interactions with a dedicated atomic force microscope. Biophys. J. **74**(6), 2840–2849 (1998)
16. M. Bezanilla, B. Drake, E. Nudler, M. Kashlev, P.K. Hansma, H.G. Hansma, Motion and enzymatic degradation of DNA in the atomic force microscope. Biophys. J. **67**(6), 2454–2459 (1994)
17. T. Ando, N. Kodera, E. Takai, D. Maruyama, K. Saito, A. Toda, A high-speed atomic force microscope for studying biological macromolecules. Proc. Natl. Acad. Sci. USA. **98**(22), 12468–12472 (2001)
18. N. Crampton, M. Yokokawa, D.T. Dryden, J.M. Edwardson, D.N. Rao, K. Takeyasu, S.H. Yoshimura, R.M. Henderson, Fast-scan atomic force microscopy reveals that the type III restriction enzyme EcoP15I is capable of DNA translocation and looping. Proc. Natl. Acad. Sci. USA. **104**(31), 12755–12760 (2007)
19. J.L. Gilmore, Y. Suzuki, G. Tamulaitis, V. Siksnys, K. Takeyasu, Y.L. Lyubchenko, Single-molecule dynamics of the DNA-EcoRII protein complexes revealed with high-speed atomic force microscopy. Biochemistry. **48**(44), 10492–10498 (2009)
20. M. Yokokawa, C. Wada, T. Ando, N. Sakai, A. Yagi, S.H. Yoshimura, K. Takeyasu, Fast-scanning atomic force microscopy reveals the ATP/ADP-dependent conformational changes of GroEL. EMBO J. **25**(19), 4567–4576 (2006)
21. M. Yokokawa, S.H. Yoshimura, Y. Naito, T. Ando, A. Yagi, N. Sakai, K. Takeyasu, Fast-scanning atomic force microscopy reveals the molecular mechanism of DNA cleavage by ApaI endonuclease. IEE Proc. Nanobiotehnol. **153**(4), 60–66 (2006)
22. T. Ando, T. Uchihashi, N. Kodera, D. Yamamoto, A. Miyagi, M. Taniguchi, H. Yamashita, High-speed AFM and nano-visualization of biomolecular processes. Pflugers Arch. Eur. J. Physiol. **456**(1), 211–225 (2008)

23. Y. Suzuki Y. Higuchi, K. Hizume, M. Yokokawa, S.H. Yoshimura, K. Yoshikawa, K. Takeyasu, Molecular dynamics of DNA and nucleosomes in solution studied by fast-scanning atomic force microscopy. Ultramicroscopy. **110**(6), 682–688 (2010)
24. N.H. Thomson, M. Fritz, M. Radmacher, J.P. Cleveland, C.F. Schmidt, P.K. Hansma, Protein tracking and detection of protein motion using atomic force microscopy. Biophys. J. **70**(5), 2421–2431 (1996)
25. M. Radmacher, M. Fritz, H.G. Hansma, P.K. Hansma, Direct observation of enzyme activity with the atomic force microscope. Science. **265**(5178), 1577–1579 (1994)
26. M.B.Viani, T.E. Schaffer, A. Chand, M. Rief, H.E. Gaub, P.K. Hansma, Small cantilevers for force spectroscopy of single molecules. J. Appl. Phys. **86**(4), 2258–2262 (1999)
27. M.B.Viani, T.E. Schaffer, G.T. Paloczi, L.I. Pietrasanta, B.L. Smith, J.B. Thompson, M. Richter, M. Rief, H.E. Gaub, K.W. Plaxco, A.N. Cleland, H.G. Hansma, P.K. Hansma, Fast imaging and fast force spectroscopy of single biopolymers with a new atomic force microscope designed for small cantilevers. Rev. Sci. Instrum. **70**(11), 4300–4303 (1999)
28. D.A.Walters, J.P. Cleveland, N.H. Thomson, P.K. Hansma, M.A. Wendman, G. Gurley, V. Elings, Short cantilevers for atomic force microscopy. Rev. Sci. Instrum. **67**(10), 3583–3590 (1996)
29. M.B.Viani, L.I. Pietrasanta, J.B. Thompson, A. Chand, I.C. Gebeshuber, J.H. Kindt, M. Richter, H.G. Hansma, P.K. Hansma, Probing protein-protein interactions in real time. Nat. Struct. Biol. **7**(8), 644–647 (2000)
30. H. Taguchi, T. Ueno, H. Tadakuma, M. Yoshida, T. Funatsu, Single-molecule observation of protein-protein interactions in the chaperonin system. Nat. Biotechnol. **19**(9), 861–865 (2001)
31. A.D.L. Humphris, M.J. Miles, J.K. Hobbs, A mechanical microscope: high-speed atomic force microscopy. Appl. Phys. Lett. **86**(3) (2005)
32. L.M. Picco, L. Bozec, A. Ulcinas, D.J. Engledew, M. Antognozzi, M.A. Horton, M.J. Miles, Breaking the speed limit with atomic force microscopy. Nanotechnology. **18**(4) (2007)
33. L.M. Picco, P.G. Dunton, A. Ulcinas, D.J. Engledew, O. Hoshi, T. Ushiki, M.J. Miles, High-speed AFM of human chromosomes in liquid. Nanotechnology. **19**(38) (2008)
34. A.D.L. Humphris, J.K. Hobbs, M.J. Miles, Ultrahigh-speed scanning near-field optical microscopy capable of over 100 frames per second. Appl. Phys. Lett. **83**(1), 6–8 (2003)
35. G. Meyer, N.M. Amer, Novel optical approach to atomic force microscopy. Appl. Phys. Lett. **53**(12), 1045–1047 (1988)
36. S. Alexander, L. Hellemans, O. Marti, J. Schneir, V. Elings, P.K. Hansma, M. Longmire, J. Gurley, An atomic-resolution atomic-force microscope implemented using an optical-lever. J. Appl. Phys. **65**(1), 164–167 (1989).
37. G. Meyer, N.M. Amer, Simultaneous measurement of lateral and normal forces with an optical-beam-deflection atomic force microscope. Appl. Phys. Lett. **57**(20), 2089–2091 (1990).
38. M. Shibata, H. Yamashita, T. Uchihashi, H. Kandori, T. Ando, High-speed atomic force microscopy shows dynamic molecular processes in photoactivated bacteriorhodopsin. Nat. Nanotechnol. **5**(3), 208–212 (2010).
39. T. Ando, T. Uchihashi, N. Kodera, D. Yamamoto, M. Taniguchi, A. Miyagi, H. Yamashita, High-speed atomic force microscopy for observing dynamic biomolecular processes. J. Mol. Recognit. **20**(6), 448–458 (2007).
40. M. Kitazawa, K. Shiotani, A. Toda, Batch fabrication of sharpened silicon nitride tips. Jpn. J. Appl. Phys. Part 1: Regular Papers Short Notes Rev. Papers. **42**(7B), 4844–4847 (2003).
41. Q. Zhong, D. Inniss, K. Kjoller, V.B. Elings, Fractured polymer silica fiber surface studied by tapping mode atomic-force microscopy. Surf. Sci. **290**(1–2), L688–L692 (1993)
42. P.K. Hansma, J.P. Cleveland, M. Radmacher, D.A. Walters, P.E. Hillner, M. Bezanilla, M. Fritz, D. Vie, H.G. Hansma, C.B. Prater, J. Massie, L. Fukunaga, J. Gurley, V. Elings, Tapping mode atomic-force microscopy in liquids. Appl. Phys. Lett. **64**(13), 1738–1740 (1994)
43. C.A.J. Putman, K.O. Vanderwerf, B.G. Degrooth, N.F. Vanhulst, J. Greve, Tapping mode atomic-force microscopy in liquid. Appl. Phys. Lett. **64**(18), 2454–2456 (2000)
44. Y.L. Lyubchenko, L.S. Shlyakhtenko, AFM for analysis of structure and dynamics of DNA and protein-DNA complexes. Methods. **47**(3), 206–213 (2009)

45. D. Yamamoto, N. Nagura, S. Omote, M. Taniguchi, T. Ando, Streptavidin 2D crystal substrates for visualizing biomolecular processes by atomic force microscopy. Biophys. J. **97**(8), 2358–2367 (2009)
46. M. Kobayashi, K. Sumitomo, K. Torimitsu, Real-time imaging of DNA-streptavidin complex formation in solution using a high-speed atomic force microscope. Ultramicroscopy. **107** (2–3), 184–190 (2007)
47. L. Giot, J.S. Bader, C. Brouwer, A. Chaudhuri, B. Kuang, Y. Li, Y.L. Hao, C.E. Ooi, B. Godwin, E. Vitols, G. Vijayadamodar, P. Pochart, H. Machineni, M. Welsh, Y. Kong, B. Zerhusen, R. Malcolm, Z. Varrone, A. Collis, M. Minto, S. Burgess, L. McDaniel, E. Stimpson, F. Spriggs, J. Williams, K. Neurath, N. Ioime, M. Agee, E. Voss, K. Furtak, R. Renzulli, N. Aanensen, S. Carrolla, E. Bickelhaupt, Y. Lazovatsky, A. DaSilva, J. Zhong, C.A. Stanyon, R.L. Finley, Jr., K.P. White, M. Braverman, T. Jarvie, S. Gold, M. Leach, J. Knight, R.A. Shimkets, M.P. McKenna, J. Chant, J.M. Rothberg, A protein interaction map of *Drosophila melanogaster*. Science. **302**(5651), 1727–1736 (2003)
48. S. Li, C.M. Armstrong, N. Bertin, H. Ge, S. Milstein, M. Boxem, P.O. Vidalain, J.D. Han, A. Chesneau, T. Hao, D.S. Goldberg, N. Li, M. Martinez, J.F. Rual, P. Lamesch, L. Xu, M. Tewari, S.L. Wong, L.V. Zhang, G.F. Berriz, L. Jacotot, P. Vaglio, J. Reboul, T. Hirozane-Kishikawa, Q. Li, H.W. Gabel, A. Elewa, B. Baumgartner, D.J. Rose, H. Yu, S. Bosak, R. Sequerra, A. Fraser, S.E. Mango, W.M. Saxton, S. Strome, S. Van Den Heuvel, F. Piano, J. Vandenhaute, C. Sardet, M. Gerstein, L. Doucette-Stamm, K.C. Gunsalus, J.W. Harper, M.E. Cusick, F.P. Roth, D.E. Hill, M. Vidal, A map of the interactome network of the metazoan C. elegans. Science. **303**(5657), 540–543 (2004)
49. N.J. Krogan, G. Cagney, H. Yu, G. Zhong, X. Guo, A. Ignatchenko, J. Li, S. Pu, N. Datta, A.P. Tikuisis, T. Punna, J.M. Peregrin-Alvarez, M. Shales, X. Zhang, M. Davey, M.D. Robinson, A. Paccanaro, J.E. Bray, A. Sheung, B. Beattie, D.P. Richards, V. Canadien, A. Lalev, F. Mena, P. Wong, A. Starostine, M.M. Canete, J. Vlasblom, S. Wu, C. Orsi, S.R. Collins, S. Chandran, R. Haw, J.J. Rilstone, K. Gandi, N.J. Thompson, G. Musso, P. St Onge, S. Ghanny, M.H. Lam, G. Butland, A.M. Altaf-Ul, S. Kanaya, A. Shilatifard, E. O'Shea, J.S. Weissman, C.J. Ingles, T.R. Hughes, J. Parkinson, M. Gerstein, S.J. Wodak, A. Emili, J.F. Greenblatt, Global landscape of protein complexes in the yeast Saccharomyces cerevisiae. Nature. **440**(7084), 637–643 (2006)
50. A.C. Gavin, P. Aloy, P. Grandi, R. Krause, M. Boesche, M. Marzioch, C. Rau, L.J. Jensen, S. Bastuck, B. Dumpelfeld, A. Edelmann, M.A. Heurtier, V. Hoffman, C. Hoefert, K. Klein, M. Hudak, A.M. Michon, M. Schelder, M. Schirle, M. Remor, T. Rudi, S. Hooper, A. Bauer, T. Bouwmeester, G. Casari, G. Drewes, G. Neubauer, J.M. Rick, B. Kuster, P. Bork, R.B. Russell, G. Superti-Furga, Proteome survey reveals modularity of the yeast cell machinery. Nature. **440**(7084), 631–636 (2006)
51. B. Bukau, A.L. Horwich, The Hsp70 and Hsp60 chaperone machines. Cell. **92**(3), 351–366 (1998)
52. L.E. Hightower, Heat shock, stress proteins, chaperones, and proteotoxicity. Cell. **66**(2), 191–197 (1991).
53. F.U. Hartl, M. Hayer-Hartl, Molecular chaperones in the cytosol: from nascent chain to folded protein. Science. **295**(5561), 1852–1858 (2002)
54. H. Grallert, J. Buchner, Review: a structural view of the GroE chaperone cycle. J. Struct. Biol. **135**(2), 95–103 (2001)
55. N.A. Ranson, H.E. White, H.R. Saibil, Chaperonins. Biochem. J. **333(**Pt 2**)**, 233–242 (1998)
56. K. Braig, Z. Otwinowski, R. Hegde, D.C. Boisvert, A. Joachimiak, A.L. Horwich, P.B. Sigler, The crystal structure of the bacterial chaperonin GroEL at 2.8 A. Nature. **371**(6498), 578–586 (1994)
57. C. Chaudhry, G.W. Farr, M.J. Todd, H.S. Rye, A.T. Brunger, P.D. Adams, A.L. Horwich, P.B. Sigler, Role of the gamma-phosphate of ATP in triggering protein folding by GroEL-GroES: function, structure and energetics. EMBO J. **22**(19), 4877–4887 (2003)
58. J.F. Hunt, A.J. Weaver, S.J. Landry, L. Gierasch, J. Deisenhofer, The crystal structure of the GroES co-chaperonin at 2.8 A resolution. Nature. **379**(6560), 37–45 (1996)

59. S.C. Mande, V. Mehra, B.R. Bloom, W.G. Hol, Structure of the heat shock protein chaperonin-10 of *Mycobacterium leprae*. Science. **271**(5246), 203–207 (1996)
60. Z. Xu, A.L. Horwich, P.B. Sigler, The crystal structure of the asymmetric GroEL-GroES-(ADP)7 chaperonin complex. Nature. **388**(6644), 741–750 (1997)
61. S. Chen, A.M. Roseman, A.S. Hunter, S.P. Wood, S.G. Burston, N.A. Ranson, A.R. Clarke, H.R. Saibil, Location of a folding protein and shape changes in GroEL-GroES complexes imaged by cryo-electron microscopy. Nature. **371**(6494), 261–264 (1994)
62. N.A. Ranson, D.K. Clare, G.W. Farr, D. Houldershaw, A.L. Horwich, H.R. Saibil, Allosteric signaling of ATP hydrolysis in GroEL-GroES complexes. Nat. Struct. Mol. Biol. **13**(2), 147–152 (2006)
63. M. Schmidt, K. Rutkat, R. Rachel, G. Pfeifer, R. Jaenicke, P. Viitanen, G. Lorimer, J. Buchner, Symmetric complexes of GroE chaperonins as part of the functional cycle. Science. **265**(5172), 656–659 (1994)
64. J. Mou, S. Sheng, R. Ho, Z. Shao, Chaperonins GroEL and GroES: views from atomic force microscopy. Biophys. J. **71**(4), 2213–2221 (1996)
65. S.G. Burston, N.A. Ranson, A.R. Clarke, The origins and consequences of asymmetry in the chaperonin reaction cycle. J. Mol. Biol. **249**(1), 138–152 (1995)
66. M.K. Hayer-Hartl, J. Martin, F.U. Hartl, Asymmetrical interaction of GroEL and GroES in the ATPase cycle of assisted protein folding. Science. **269**(5225), 836–841 (1995)
67. H.S. Rye, S.G. Burston, W.A. Fenton, J.M. Beechem, Z. Xu, P.B. Sigler, A.L. Horwich, Distinct actions of cis and trans ATP within the double ring of the chaperonin GroEL. Nature. **388**(6644), 792–798 (1997)
68. H.S. Rye, A.M. Roseman, S. Chen, K. Furtak, W.A. Fenton, H.R. Saibil, A.L. Horwich, GroEL-GroES cycling: ATP and nonnative polypeptide direct alternation of folding-active rings. Cell. **97**(3), 325–338 (1999)
69. J.S. Weissman, H.S. Rye, W.A. Fenton, J.M. Beechem, A.L. Horwich, Characterization of the active intermediate of a GroEL-GroES-mediated protein folding reaction. Cell. **84**(3), 481–490 (1996)
70. T. Ueno, H. Taguchi, H. Tadakuma, M. Yoshida, T. Funatsu, GroEL mediates protein folding with a two successive timer mechanism. Mol. Cell. **14**(4), 423–434 (2004)
71. M. Taniguchi, T. Yoshimi, K. Hongo, T. Mizobata, Y. Kawata, Stopped-flow fluorescence analysis of the conformational changes in the GroEL apical domain: relationships between movements in the apical domain and the quaternary structure of GroEL. J. Biol. Chem. **279**(16), 16368–16376 (2004)
72. A.D. Mehta, R.S. Rock, M. Rief, J.A. Spudich, M.S. Mooseker, R.E. Cheney, Myosin-V is a processive actin-based motor. Nature. **400**(6744), 590–593 (1999)
73. A. Mehta, Myosin learns to walk. J. Cell Sci. **114**(Pt 11), 1981–1998 (2001)
74. C. Veigel, F. Wang, M.L. Bartoo, J.R. Sellers, J.E. Molloy, The gated gait of the processive molecular motor, myosin V. Nat. Cell Biol. **4**(1), 59–65 (2002)
75. F.J. Kull, E.P. Sablin, R. Lau, R.J. Fletterick, R.D. Vale, Crystal structure of the kinesin motor domain reveals a structural similarity to myosin. Nature. **380**(6574), 550–555 (1996)
76. G.S. Bloomn, S.A. Endow, Motor proteins. 1: kinesins. Protein Profile. **1**(10), 1059–1116 (1994)
77. L.S. Goldstein, With apologies to scheherazade: tails of 1001 kinesin motors. Annu. Rev. Genet. **27**, 319–351 (1993)
78. S.A. Burgess, M.L. Walker, H. Sakakibara, P.J. Knight, K. Oiwa, Dynein structure and power stroke. Nature. **421**(6924), 715–718 (2003)
79. I.R. Gibbons, Dynein family of motor proteins: present status and future questions. Cell Motil. Cytoskeleton. **32**(2), 136–144 (1995)
80. J.N. Forkey, M.E. Quinlan, M.A. Shaw, J.E. Corrie, Y.E. Goldman, Three-dimensional structural dynamics of myosin V by single-molecule fluorescence polarization. Nature. **422**(6930), 399–404 (2003)
81. E.M. De La Cruz, A.L. Wells, S.S. Rosenfeld, E.M. Ostap, H.L. Sweeney, The kinetic mechanism of myosin V. Proc. Natl. Acad. Sci. USA. **96**(24), 13726–13731 (1999)

82. A. Yildiz J.N. Forkey, S.A. McKinney, T. Ha, Y.E. Goldman, P.R. Selvin, Myosin V walks hand-over-hand: single fluorophore imaging with 1.5-nm localization. Science. **300**(5628), 2061–2065 (2003)
83. W. Hua, J. Chung, J. Gelles, Distinguishing inchworm and hand-over-hand processive kinesin movement by neck rotation measurements. Science. **295**(5556), 844–848 (2002)
84. K. Takeyasu, H. Maruyama, Y. Suzuki, K. Hizume, S.H. Yoshimura, Modern atomic force microscopy and its application to the study of genome architecture, ed. by B. Bhushan. In *Scanning Probe Microscopy in Nanoscience and Nanotechnology* (Springer, Berlin, 2010), 723.
85. A. Pingoud, A. Jeltsch, Structure and function of type II restriction endonucleases. Nucleic Acids Res. **29**(18), 3705–3727 (2001)
86. R.J. Roberts, M. Belfort, T. Bestor, A.S. Bhagwat, T.A. Bickle, J. Bitinaite, R.M. Blumenthal, S. Degtyarev, D.T. Dryden, K. Dybvig, K. Firman, E.S. Gromova, R.I. Gumport, S.E. Halford, S. Hattman, J. Heitman, D.P. Hornby, A. Janulaitis, A. Jeltsch, J. Josephsen, A. Kiss, T.R. Klaenhammer, I. Kobayashi, H. Kong, D.H. Kruger, S. Lacks, M.G. Marinus, M. Miyahara, R.D. Morgan, N.E. Murray, V. Nagaraja, A. Piekarowicz, A. Pingoud, E. Raleigh, D.N. Rao, N. Reich, V.E. Repin, E.U. Selker, P.C. Shaw, D.C. Stein, B.L. Stoddard, W. Szybalski, T.A. Trautner, J.L. Van Etten, J.M. Vitor, G.G. Wilson, S.Y. Xu, A nomenclature for restriction enzymes, DNA methyltransferases, homing endonucleases and their genes. Nucleic Acids Res. **31**(7), 1805–1812 (2003)
87. R.J. Roberts, T. Vincze, J. Posfai, D. Macelis, REBASE: restriction enzymes and methyltransferases. Nucleic Acids Res. **31**(1), 418–420 (2003)
88. A.D. Riggs, S. Bourgeois, M. Cohn, The lac repressor-operator interaction. 3. Kinetic studies. J. Mol. Biol. **53**(3), 401–417 (1970)
89. H.J. Ehbrecht, A. Pingoud, C. Urbanke, G. Maass, C. Gualerzi, Linear diffusion of restriction endonucleases on DNA. J. Biol. Chem. **260**(10), 6160–6166 (1985)
90. B.J. Terry, W.E. Jack, P. Modrich, Facilitated diffusion during catalysis by EcoRI endonuclease. Nonspecific interactions in EcoRI catalysis. J. Biol. Chem. **260**(24), 13130–13137 (1985)
91. D.M. Gowers, S.E. Halford, Protein motion from non-specific to specific DNA by three-dimensional routes aided by supercoiling. EMBO J. **22**(6), 1410–1418 (2003)
92. N.P. Stanford, M.D. Szczelkun, J.F. Marko, S.E. Halford, One- and three-dimensional pathways for proteins to reach specific DNA sites. EMBO J. **19**(23), 6546–6557 (2000)
93. R.J. Roberts, Restriction and modification enzymes and their recognition sequences. Nucleic Acids Res. **10**(5), r117–144 (1982)
94. B. Schafer, H. Gemeinhardt, K.O. Greulich, Direct microscopic observation of the time course of single-molecule dna restriction reactions. We thank Dr. Werner Wolf for helpful discussions and the VW-foundation, Grant No. I/75099 for supporting this work. Angew. Chem. Int. Ed. Engl. **40**(24), 4663–4666 (2001)
95. S.E. Halford, A.J. Welsh, M.D. Szczelkun, Enzyme-mediated DNA looping. Annu. Rev. Biophys. Biomol. Struct. **33**, 1–24 (2004)
96. M. Mucke, G. Grelle, J. Behlke, R. Kraft, D.H. Kruger, M. Reuter, EcoRII: a restriction enzyme evolving recombination functions? EMBO J. **21**(19), 5262–5268 (2002)
97. M. Mucke, D.H. Kruger, M. Reuter, Diversity of type II restriction endonucleases that require two DNA recognition sites. Nucleic Acids Res. **31**(21), 6079–6084 (2003)
98. P. Janscak, U. Sandmeier, M.D. Szczelkun, T.A. Bickle, Subunit assembly and mode of DNA cleavage of the type III restriction endonucleases EcoP1I and EcoP15I. J. Mol. Biol. **306**(3), 417–431 (2001)
99. A. Meisel, P. Mackeldanz, T.A. Bickle, D.H. Kruger, C. Schroeder, Type III restriction endonucleases translocate DNA in a reaction driven by recognition site-specific ATP hydrolysis. EMBO J. **14**(12), 2958–2966 (1995)
100. L.J. Peakman, M.D. Szczelkun, DNA communications by Type III restriction endonucleases—confirmation of 1D translocation over 3D looping. Nucleic Acids Res. **32**(14), 4166–4174 (2004)
101. A. Kornberg, T.A. Baker, *DNA Replication* (Freeman, New York, 1999), 246.

102. R.D. Kornberg, Chromatin structure: a repeating unit of histones and DNA. Science. **184**(139), 868–871 (1974)
103. K. Luger, A.W. Mader, R.K. Richmond, D.F. Sargent, T.J. Richmond, Crystal structure of the nucleosome core particle at 2.8 A resolution. Nature. **389**(6648), 251–260 (1974)
104. P. Oudet, M. Gross-Bellard, P. Chambon, Electron microscopic and biochemical evidence that chromatin structure is a repeating unit. Cell. **4**(4), 281–300 (1975)
105. G. Meersseman, S. Pennings, E.M. Bradbury, Mobile nucleosomes – a general behavior. EMBO J. **11**(8), 2951–2959 (1992)
106. S. Pennings, G. Meersseman, E.M. Bradbury, Mobility of positioned nucleosomes on 5 S rDNA. J. Mol. Biol. **220**(1), 101–110 (1991)
107. S. Pennings, G. Meersseman, E.M. Bradbury, Linker histones H1 and H5 prevent the mobility of positioned nucleosomes. Proc. Natl. Acad. Sci. USA. **91**(22), 10275–10279 (1994)
108. T. Sakaue, K. Yoshikawa, S.H. Yoshimura, K. Takeyasu, Histone core slips along DNA and prefers positioning at the chain end. Phys. Rev. Lett. **87**(7), 078105 (2001)
109. M.T. Cabeen, C. Jacobs-Wagner, Bacterial cell shape. Nat. Rev. Microbiol. **3**(8), 601–610 (2005)
110. S.S. Branda, S. Vik, L. Friedman, R. Kolter, Biofilms: the matrix revisited. Trends Microbiol. **13**(1), 20–26 (2005)
111. D.M. Engelman, Membranes are more mosaic than fluid. Nature. **438**(7068), 578–580 (2005)
112. D.J. Muller, J.B. Heymann, F. Oesterhelt, C. Moller, H. Gaub, G. Buldt, A. Engel, Atomic force microscopy of native purple membrane. Biochim. Biophys. Acta. **1460**(1), 27–38 (2000)
113. H. Stahlberg, D.J. Muller, K. Suda, D. Fotiadis, A. Engel, T. Meier, U. Matthey, P. Dimroth, Bacterial Na(+)-ATP synthase has an undecameric rotor. EMBO Rep. **2**(3), 229–233 (2001)
114. D. Pogoryelov, J. Yu, T. Meier, J. Vonck, P. Dimroth, D.J. Muller, The c15 ring of the Spirulina platensis F-ATP synthase: F1/F0 symmetry mismatch is not obligatory. EMBO Rep. **6**(11), 1040–1044 (2005)
115. K. Takeyasu, H. Omote, S. Nettikadan, F. Tokumasu, A. Iwamoto-Kihara, M. Futai, Molecular imaging of *Escherichia coli* F0F1-ATPase in reconstituted membranes using atomic force microscopy. FEBS Lett. **392**(2), 110–113 (1996)
116. D. Fotiadis, Y. Liang, S. Filipek, D.A. Saperstein, A. Engel, K. Palczewski, Atomic-force microscopy: rhodopsin dimers in native disc membranes. Nature. **421**(6919), 127–128 (2003)
117. D. Fotiadis, D.J. Muller, G. Tsiotis, L. Hasler, P. Tittmann, T. Mini, P. Jeno, H. Gross, A. Engel, Surface analysis of the photosystem I complex by electron and atomic force microscopy. J. Mol. Biol. **283**(1), 83–94 (1998)
118. D.J. Muller, W. Baumeister, A. Engel, Controlled unzipping of a bacterial surface layer with atomic force microscopy. Proc. Natl. Acad. Sci. USA. **96**(23), 13170–13174 (1999)
119. B.S. Khakh, R.A. North, P2X receptors as cell-surface ATP sensors in health and disease. Nature. **442**(7102), 527–532 (2006)
120. S. Cockcroft, B.D. Gomperts, ATP induces nucleotide permeability in rat mast cells. Nature. **279**(5713), 541–542 (1979)
121. B.S. Khakh, X.R. Bao, C. Labarca, H.A. Lester, Neuronal P2X transmitter-gated cation channels change their ion selectivity in seconds. Nat. Neurosci. **2**(4), 322–330 (1999)
122. Y. Shinozaki, K. Sumitomo, M. Tsuda, S. Koizumi, K. Inoue, K. Torimitsu, Direct observation of ATP-induced conformational changes in single P2X4 receptors. PLoS Biol. **7**(5), e103 (2009)
123. J.K. Lanyi, Bacteriorhodopsin. Annu. Rev. Physiol. **66**, 665–688 (2004)
124. U. Haupts, J. Tittor, D. Oesterhelt, Closing in on bacteriorhodopsin: progress in understanding the molecule. Annu. Rev. Biophys. Biomol. Struct. **28**, 367–399 (1999)
125. R. Henderson, P.N. Unwin, Three-dimensional model of purple membrane obtained by electron microscopy. Nature. **257**(5521), 28–32 (1975)
126. A. Engel, H.E. Gaub, Structure and mechanics of membrane proteins. Annu. Rev. Biochem. **77**, 127–148 (2008)
127. D.J. Muller, G. Buldt, A. Engel, Force-induced conformational change of bacteriorhodopsin. J. Mol. Biol. **249**(2), 239–243 (1995)

128. Y. Kimura, D.G. Vassylyev, A. Miyazawa, A. Kidera, M. Matsushima, K. Mitsuoka, K. Murata, T. Hirai, Y. Fujiyoshi, Surface of bacteriorhodopsin revealed by high-resolution electron crystallography. Nature. **389**(6647), 206–211 (1997)
129. H. Luecke, B. Schobert, H.T. Richter, J.P. Cartailler, J.K. Lanyi, Structural changes in bacteriorhodopsin during ion transport at 2 angstrom resolution. Science. **286**(5438), 255–261 (1999)
130. H. Luecke, B. Schobert, H.T. Richter, J.P. Cartailler, J.K. Lanyi, Structure of bacteriorhodopsin at 1.55 A resolution. J. Mol. Biol. **291**(4), 899–911 (1999)
131. L.S.Brown, R. Needleman, J.K. Lanyi, Conformational change of the E-F interhelical loop in the M photointermediate of bacteriorhodopsin. J. Mol. Biol. **317**, 471–478 (2002)
132. N.A. Dencher, D. Dresselhaus, G. Zaccai, G. Buldt, Structural changes in bacteriorhodopsin during proton translocation revealed by neutron diffraction. Proc. Natl. Acad. Sci. USA. **86**, 7876–7879 (1989)
133. H. Kamikubo, Structure of the N intermediate of bacteriorhodopsin revealed by X-ray diffraction. Proc. Natl. Acad. Sci. USA. **93**, 1386–1390 (1996)
134. H.J. Sass, Structural alterations for proton translocation in the M state of wild-type bacteriorhodopsin. Nature. **406**, 649–653 (2000)
135. M. Shibata, H. Kandori, FTIR studies of internal water molecules in the Schiff base region of bacteriorhodopsin. Biochemistry. **44**, 7406–7413 (2005)
136. S. Subramaniam, M. Gerstein, D. Oesterhelt, R. Henderson, Electron diffraction analysis of structural changes in the photocycle of bacteriorhodopsin. EMBO J. **12**, 1–8 (1993)
137. S. Subramaniam, R. Henderson, Molecular mechanism of vectorial proton translocation by bacteriorhodopsin. Nature. **406**, 653–657 (2000)
138. T.E. Thorgeirsson, Transient channel-opening in bacteriorhodopsin: an EPR study. J. Mol. Biol. **273**, 951–957 (1997)
139. J. Vonck, Structure of the bacteriorhodopsin mutant F219 L N intermediate revealed by electron crystallography. EMBO J. **19**, 2152–2160 (2000)
140. W. Xiao, L.S. Brown, R. Needleman, J.K. Lanyi, Y.K. Shin, Light-induced rotation of a transmembrane alpha-helix in bacteriorhodopsin. J. Mol. Biol. **304**, 715–721 (2000)
141. J.C. Skou, The influence of some cations on an adenosine triphosphatase from peripheral nerves. Biochim. Biophys. Acta. **23**(2), 394–401 (1957)
142. W. Kuhlbrandt, Biology, structure and mechanism of P-type ATPases. Nat. Rev. Mol. Cell Biol. **5**(4), 282–295 (2004)
143. W. Hasselbach, M. Makinose, The calcium pump of the "relaxing granules" of muscle and its dependence on ATP-splitting. Biochem. Z. **333**, 518–528 (1961)
144. I. Ohtsuki, S. Morimoto, Troponin: regulatory function and disorders. Biochem. Biophys. Res. Commun. **369**(1), 62–73 (2008)
145. M.W. Berchtold, H. Brinkmeier, M. Muntener, Calcium ion in skeletal muscle: its crucial role for muscle function, plasticity, and disease. Physiol. Rev. **80**(3), 1215–1265 (2000)
146. S. Ebashi, F. Lipmann, Adenosine triphosphate-linked concentration of calcium ions in a particulate fraction of rabbit muscle. J. Cell Biol. **14**(3), 389 (1962)
147. S. Ebashi, Third component participating in the superprecipitation of 'natural actomyosin.' Nature. **200**, 1010 (1963)
148. S. Ebashi, A. Kodama, A new protein factor promoting aggregation of tropomyosin. J. Biochem. **58**(1), 107–108 (1965)
149. C. Toyoshima, M. Nakasako, H. Nomura, H. Ogawa, Crystal structure of the calcium pump of sarcoplasmic reticulum at 2.6 A resolution. Nature. **405**(6787), 647–655 (2000)
150. C. Toyoshima, H. Nomura, Structural changes in the calcium pump accompanying the dissociation of calcium. Nature. **418**(6898), 605–611 (2002)
151. C. Toyoshima, T. Mizutani, Crystal structure of the calcium pump with a bound ATP analogue. Nature. **430**(6999), 529–535 (2004)
152. C. Toyoshima, H. Nomura, T. Tsuda, Lumenal gating mechanism revealed in calcium pump crystal structures with phosphate analogues. Nature. **432**(7015), 361–368 (2004)
153. T.L. Sorensen, J.V. Moller, P. Nissen, Phosphoryl transfer and calcium ion occlusion in the calcium pump. Science. **304**(5677), 1672–1675 (2004)

154. Y. Sagara, G. Inesi, Inhibition of the sarcoplasmic reticulum Ca2+ transport ATPase by thapsigargin at subnanomolar concentrations. J. Biol. Chem. **266**(21), 13503–13506 (1991)
155. G. Inesi, Y. Sagara, Thapsigargin, a high affinity and global inhibitor of intracellular Ca2+ transport ATPases. Arch. Biochem. Biophys. **298**(2), 313–317 (1992)
156. J.V. Moller, B. Juul, M. le Maire, Structural organization, ion transport, and energy transduction of P-type ATPases. Biochim. Biophys. Acta. **1286**(1), 1–51 (1996)
157. L. de Meis, A.L. Vianna, Energy interconversion by the Ca2+-dependent ATPase of the sarcoplasmic reticulum. Annu. Rev. Biochem. **48**, 275–292 (1979)
158. S. Verjovski-Almeida, G. Inesi, Fast-kinetic evidence for an activating effect of ATP on the Ca2+ transport of sarcoplasmic reticulum ATPase. J. Biol. Chem. **254**(1), 18–21 (1979)
159. L. Dode, B. Vilsen, K. Van Baelen, F. Wuytack, J.D. Clausen, J.P. Andersen, Dissection of the functional differences between sarco(endo)plasmic reticulum Ca2+-ATPase (SERCA) 1 and 3 isoforms by steady-state and transient kinetic analyses. J. Biol. Chem. **277**(47), 45579–45591 (2002)
160. C. Olesen, M. Picard, A.M. Winther, C. Gyrup, J.P. Morth, C. Oxvig, J.V. Moller, P. Nissen, The structural basis of calcium transport by the calcium pump. Nature. **450**(7172), 1036–1042 (2007)
161. A.M. Jensen, T.L. Sorensen, C. Olesen, J.V. Moller, P. Nissen, Modulatory and catalytic modes of ATP binding by the calcium pump. EMBO J. **25**(11), 2305–2314 (2006)
162. R.L. Post, C. Hegyvary, S. Kume, Activation by adenosine triphosphate in the phosphorylation kinetics of sodium and potassium ion transport adenosine triphosphatase. J. Biol. Chem. **247**(20), 6530–6540 (1972)

Chapter 9
Transport Properties of Graphene with Nanoscale Lateral Resolution

Filippo Giannazzo, Vito Raineri, and Emanuele Rimini

Abstract Graphene, a two-dimensional sheet of C atoms, is the object of many scientific studies, especially due to the outstanding electronic properties. The behavior of graphene based nano-devices or mesoscopic devices is ultimately influenced by the lateral homogeneity of the carrier density, electron mean free path, and mobility on the layer. This chapter reviews the main applications of scanning probe methods to the local (nanoscale) characterization of the electronic properties of graphene.

Abbreviations

AFM	Atomic force microscopy
C–V	Capacitance–voltage
I–V	Current–voltage
DG	Deposited graphene
EG	Epitaxial graphene
FLG	Few layers of graphene
MOS	Metal-oxide-semiconductor
MOSFET	Metal-oxide-semiconductor field-effect transistor
nanoMIS	Nanometric metal–insulator–semiconductor
SCM	Scanning capacitance microscopy
SCS	Scanning capacitance spectroscopy
SPM	Scanning probe microscopy
STM	Scanning tunneling microscopy
SBH	Schottky barrier height
SET	Single electron transistor
SLG	Single layer of graphene
2D	Two-dimensional
2DEG	Two-dimensional electron gas

9.1 Introduction

Graphene [1, 2], a two-dimensional (2D) layer of sp^2 hybridized carbon atoms arranged in a honeycomb lattice (see Fig. 9.1a), is currently the object of large scientific interests, due to many unique electrical, magnetic [3–5], optical [6], thermal [7, 8], thermoelectric [9] and mechanical properties [10–12]. Some of these properties are summarized in Table 9.1.

Single layers of graphene were experimentally separated for the first time in 2004 [16], although monoatomic carbon layers were seen in previous studies on ultrathin graphitic films [17]. Before 2004, it was commonly believed that 2D crystals could not exist at temperatures above 0 K, but should bend and crumple to form a three-dimensional (3D) structure, due to thermal fluctuation of the lattice atoms [18].

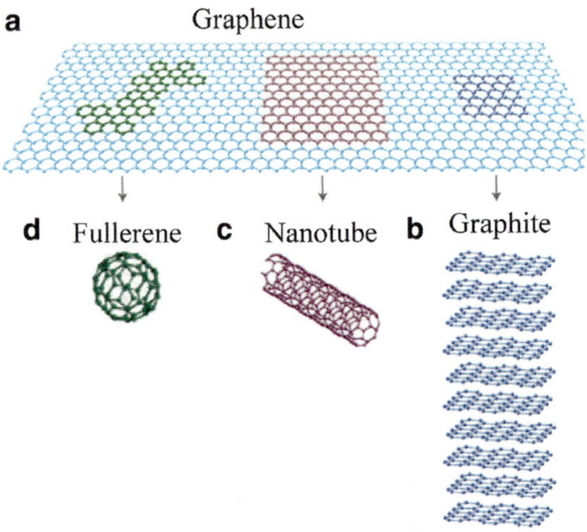

Fig. 9.1 Representation of the four allotropic forms of sp^2 carbon: (**a**) graphene (2D), (**b**) graphite (3D), (**c**) carbon nanotubes (1D), (**d**) fullerene (0D)

Table 9.1 Main physical properties of graphene and possible applications

Physical properties		Possible applications
Electronic	Giant "intrinsic" mobility of graphene 2DEG ($\sim 2 \times 10^5$ cm^2 V^{-1} s^{-1}) [13, 14]	High frequency (GHz–THz) devices
	Large "intrinsic" electron mean free path (~ 1 μm) [13]	"Zero energy loss" devices
	High specific capacitance (100 F g^{-1}) [15]	Highly efficient passive components (ultracapacitors)
Thermal	Very high thermal conductivity (~ 50 W cm^{-1} K^{-1}) [7]	"Zero energy loss" devices
Magnetic	Electrical spin-current injection and detection up to 300 K [3–5]	Spintronics
Mechanical	Young's modulus 0.5 TPa [10–12]	MEMS, NEMS
Optical	High transparency ($\sim 97.7\%$) both at visible and near-infrared wavelengths [6].	Transparent electrode for solar cells

Nevertheless, graphene has been investigated from a theoretical point of view for more than 60 years, since the electronic structure of this 2D allotrope of sp^2 carbon was the basis for the understanding of the electronic properties of the other allotropic forms. In fact, graphite was modeled with stacks of graphene layers weakly coupled by van der Waals forces (see Fig. 9.1b). Carbon nanotubes, the one-dimensional (1D) allotropic form, were modeled by rolling graphene along a given direction and reconnecting the carbon bonds (see Fig. 9.1c). Finally, fullerenes, molecules where carbon atoms are arranged spherically [zero-dimensional (0D) objects with discrete energy states], can be thought as wrapped-up graphene (see Fig. 9.1d).

Although one atomic layer thick, graphene exhibits a breaking strength 200 times greater than steel, its spring constant is in the range 1–5 $\mathrm{N\,m^{-1}}$ and the Young's modulus is 0.5 TPa. The near-room temperature thermal conductivity of graphene was recently measured to be between $(4.84 \pm 0.44) \times 10^3$ and $(5.30 \pm 0.48) \times 10^3\,\mathrm{W\,m^{-1}\,K^{-1}}$.

The most attractive properties of this novel material are certainly the electronic properties. Low energy electronic excitations in graphene exhibit a linear E–k dispersion relation, formally similar to that of phonons and Dirac-fermions [19]. Hence, charge carriers propagate in the 2D lattice as quasi particles with zero effective mass and their behavior has to described by a Dirac-like equation (rather than the usual Schrödinger equation). These amazing similarities between graphene electrons and relativistic particles resulted especially exciting for the theoreticians, who conceived the possibility of benchtop experiments on graphene to investigate complex relativistic quantum phenomena [20]. From the practical point of view, graphene is especially interesting for the outstanding transport properties of the two-dimensional electron gas (2DEG): electrons in ideal graphene exhibit a giant intrinsic mobility ($>2 \times 10^5\,\mathrm{cm^2\,V^{-1}\,s^{-1}}$) and can travel for micrometers without scattering even at room temperature. Furthermore, graphene can sustain current densities [21] six orders of magnitude higher than copper without suffering of electromigration problems, and exhibits excellent thermal conductivity, allowing very efficient heat dissipation. Due to the high specific capacitance ($\sim$100 $\mathrm{F\,g^{-1}}$) graphene-based composites are interesting candidates for the fabrication of "ultracapacitors" for very efficient charge and energy storage [15]. In Table 9.2, the electronic properties of graphene 2DEG are compared with those of electrons in common bulk semiconductors, and with those of the 2DEG present at the heterointerface AlGaN/GaN.

Since the early studies on graphene, scanning probe microscopy (SPM) methods played an important role in the investigation of the physical properties of this material on the nanometer scale. This is clearly due to the peculiar structure of graphene (one atom thick layer weakly interacting with a substrate or even freestanding), which makes it directly accessible to a scanning probe.

Atomic force microscopy (AFM) is routinely used in combination with optical microscopy (OM) [22,23] and micro-Raman (μR) [24] spectroscopy as nondestructive method to measure the number of layers in samples composed by few layers of graphene (FLG) deposited on a substrate. As an example, Fig. 9.2 reports the OM image (a) and AFM scan in tapping mode (b) of FLG obtained by mechanical

Table 9.2 Electronic properties of graphene compared with those of common bulk semiconductors and with those of the 2DEG at AlGaN/GaN heterointerface

	Si	Ge	GaAS	4H-SiC	GaN	AlGaN/GaN 2DEG	Graphene
Energy band gap (eV) @ 300 K	1.1	0.67	1.43	3.3	3.4	3.4	$\sim$0
Electron effective mass (m^*/m_e)	1.08	0.55	0.067	0.3	0.19	0.19	$\sim$0
Electron mobility ($cm^2\,V^{-1}\,s^{-1}$) @300 K	1,350	3,900	4,600	800	1,300	1,500–2,000	2×10^5
Saturated electron drift velocity v_s (10^7 cm s^{-1})	1	0.6	2	2	3	3	>5
Carrier concentration (cm^{-3})	10^{15}	10^{15}	10^{15}	10^{15}	10^{15}	10^{19}–10^{20}	10^{19}–10^{20}

exfoliation of highly oriented pirolitic graphite (HOPG) and deposited on top of a SiO_2 substrate [25]. In Fig. 9.2c, a line profile extracted from the AFM scan in the inset is reported also. While the height difference between the first and second and between the second and third graphene layers corresponds to the interplane distance between graphene layers in HOPG ($\sim$0.35 nm), the height difference between the substrate and the first layer ($\sim$0.72 nm) is typically higher, that is, it is affected by an offset. In Fig. 9.2d, the measured height is reported versus. the number of layers n (the data are collected on a large number of graphene samples). These data have been fitted with the linear relation, $h = t_0 + nt_{gr}$, where the interlayer spacing t_{gr} and the offset t_0 are the fitting parameters. The best fit has been obtained for the values of $t_{gr} = 0.36 \pm 0.01$ nm and $t_0 = 0.28 \pm 0.06$ nm. This calibration curve is used to determine the number of layers from an AFM image. However, the calibration procedure has to be repeated if the substrate material is changed, being the offset t_0 depending on the specific substrate.

The value of this offset is partially related to the roughness of the substrate on which graphene is deposited, but can also depend on the working principle of the AFM measurement itself [25,26]. In tapping mode AFM, when the tip is close to the sample surface, the oscillation amplitude of the cantilever is strongly dependent on the local force gradient felt by the tip. During a scan, the tip–sample force gradient is kept constant by a feedback loop which changes the Z piezo extension (i.e., the tip–sample distance) to maintain the oscillation amplitude fixed. The AFM map is obtained as a 2D map of the Z piezo elongation for each tip positions. This map corresponds to the "actual" surface morphology only if the scanned surface is uniformly coated by the same material, while it can be influenced by the different force gradient between the Si tip and different materials on the surface. In particular, when the AFM tip is scanned across the step between a single layer and SiO_2, if the force gradient is larger on graphene than on SiO_2, the tip will move closer to

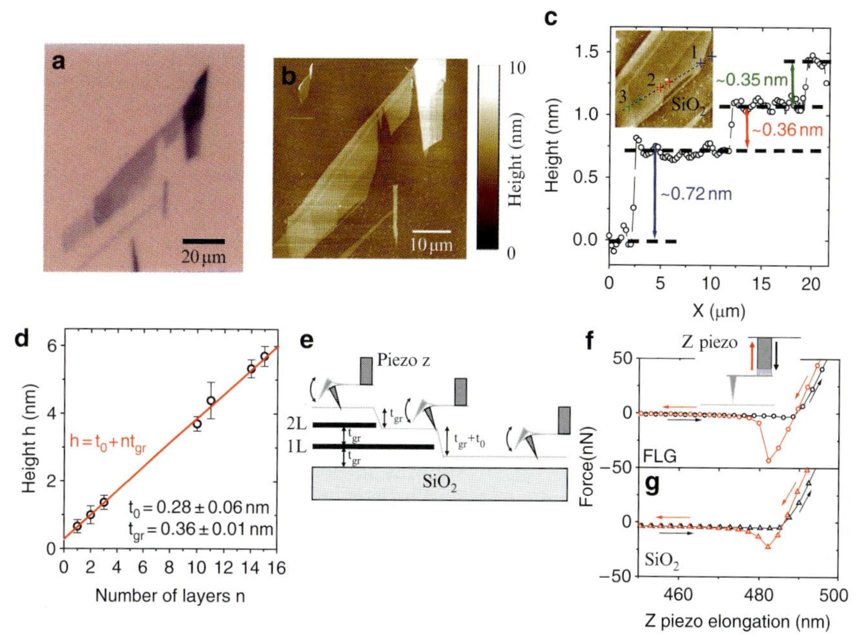

Fig. 9.2 Optical microscopy (**a**) and tapping mode AFM image (**b**) of a sample with few layers of graphene, obtained by mechanical exfoliation of HOPG and deposited on SiO2. In (**c**) a line profile extracted from the AFM scan in the *inset* is reported. In (**d**) calibration curve reporting the measured height vs. the number of layers. Illustration of the interaction between the oscillating AFM tip and surface, when the tip is scanned across the step between a single layer of graphene and SiO_2 (**e**). The force–distance curves measured on FLG (**f**) and on SiO_2 (**g**), respectively. The black curves are collected with the tip approaching to the surface, while the red ones with the curves retracting from the surface. Data from [25]

SiO_2 than to graphene to maintain the oscillation amplitude fixed. This picture is illustrated in Fig. 9.2e.

Local force-spectroscopy measurements were also carried out with the tip on SiO_2 and on FLG. The tip–sample force F as a function of tip–sample distance d was measured in static conditions, i.e., no oscillation was imposed to the cantilever. The distance d was changed changing the Z piezo elongation, while the force F was measured by measuring the cantilever deflection for each value of d. Cantilevers with elastic constant $k = 20\,\mathrm{N\,m^{-1}}$ were used for these measurements. The force–distance curves measured on FLG and on SiO_2 are reported in Fig. 9.2f, g, respectively. The black curves are collected with the tip approaching to the surface, while the red ones with the curves retracting from the surface. The different shape between the red and black curve is due to the adhesion force contribution, which tends to keep the tip attached to the sample surface during the Z piezo retraction. As it is evident the adhesion force between the tip and graphene is higher than between the tip and SiO_2. Hence, during tapping mode measurements, the oscillating tip felt in an oscillation cycle a higher force gradient when it is on graphene than when it is on SiO_2. This confirms the picture in Fig. 9.2e.

Furthermore, high resolution AFM and scanning tunneling microscopy (STM) provides information on the lattice structure [27], as well as on the peculiar mesoscopic corrugations (ripples) of these membranes [28]. AFM has been also used to characterize the mechanical properties of suspended graphene membranes [29].

In the past five years, several methods have been demonstrated to synthesize graphene, i.e. (a) epitaxial growth on SiC by thermal decomposition at high temperatures [30–34], (b) chemical vapor deposition on metal catalysts (Ni, Cu, etc.) [35–39], and (c) chemical exfoliation of HOPG [40]. In all of these cases, scanning probe methods have been extensively applied to understand the growth mechanisms of SLG or FLG on the different substrates and the relation between the graphene lattice structure and that of the substrate.

Due to the "explosion" of interest on graphene physics and technology, the literature on the SPM on graphene has grown accordingly. The aim of this work will be to review the application of scanning probe methods to local (nanoscale) characterization of the electronic transport properties of graphene.

The bandstructure, the density of states, and the transport properties (electron mean free path, mobility) of "ideal" graphene will be first introduced. Therefore, some of the commonly observed discrepancies between the theoretical behavior and the experimentally measured transport properties on graphene samples will be discussed. Although a detailed explanation of these discrepancies has not been furnished, it is common opinion that the interaction of graphene with the external environment (adsorbed impurities, interaction with the substrate, etc.) as well as lattice defects in graphene itself strongly affects the transport properties. The use of different scanning probe methods allowed to determine the lateral variations of the electronic and transport properties due to these inhomogeneities.

9.2 Transport Properties of Graphene

9.2.1 Electronic Bandstructure and Dispersion Relation

Most of the unusual electronic properties of graphene derive from its peculiar lattice structure. Graphene is made out of carbon atoms arranged on a planar honeycomb structure composed of hexagons. This is a consequence of the sp^2 hybridization between one s orbital and two p orbitals of carbon, which leads to a trigonal planar structure with the formation of a σ bond between atoms separated by 1.42 Å. The σ bond is responsible for the robustness of the graphene lattice structure. The unaffected p orbital, which is perpendicular to the planar structure, can bind covalently with neighboring carbon atoms, leading to the formation of a delocalized π bond.

P.R. Wallace in 1946 calculated the band structure of graphene [41] and showed its semimetallic behavior and the peculiar linear dispersion $E = E(k)$ of low energy electronic excitations. The semimetallic behavior is a consequence of the Pauli principle: the σ band has a filled shell and, hence, forms a deep valence

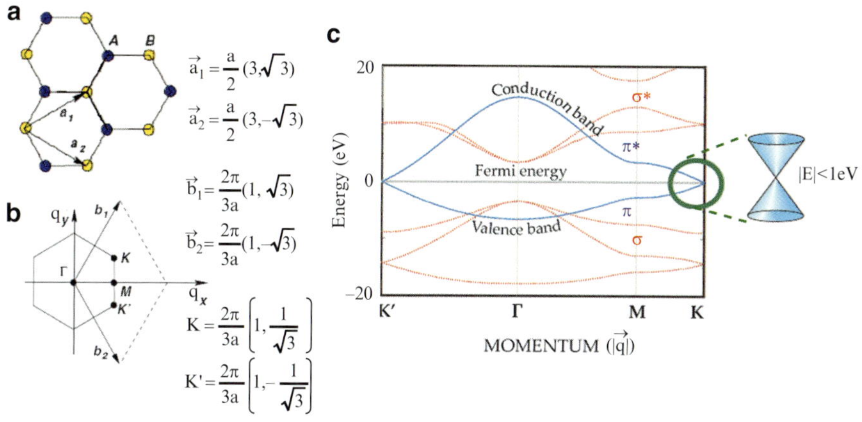

Fig. 9.3 Representation of the graphene direct lattice (**a**) and reciprocal lattice (**b**) In (**c**), the full band structure of graphene is shown

band, whereas each p orbital is occupied by one electron and the π band is then half filled. The detailed band structure was calculated by solving the Schrodinger equation for electrons in the graphene lattice using the tight-binding approximation. The direct-lattice, formed by two atoms per unit cell, is represented in Fig. 9.3a, where $a \approx 1.42$ Å is the carbon–carbon distance and $\mathbf{a}_1$ and $\mathbf{a}_2$ are the two basis vectors of the direct lattice. The first Brillouin zone of the reciprocal lattice is shown in Fig. 9.3b, where $\mathbf{b}_1$ and $\mathbf{b}_2$ are the two basis vectors of the reciprocal lattice. Of particular importance for the physics of graphene are the two points K and K' at the corners of the Brillouin zone, which are named Dirac points.

In Fig. 9.3c, the full band structure of graphene is shown. The red lines represent the σ bands, whereas the blue lines represent the π bands. In particular, the π bands have the form:

$$E_{\pm}\left(\vec{q}\right) = \pm t\sqrt{3 + f\left(\vec{q}\right)}, \tag{9.1}$$

where $\mathbf{q}$ is the momentum measured from the origin (Γ), the parameter $t \approx 3.1$ eV is the nearest-neighbor hopping energy and

$$f\left(\vec{q}\right) = 2\cos\left(\sqrt{3}q_y a\right) + 4\cos\left(\frac{\sqrt{3}}{2}q_y a\right)\cos\left(\frac{3}{2}q_x a\right). \tag{9.2}$$

Interestingly, the dispersion relation close to the Dirac points is linear (see inset of Fig. 9.3c). It can be obtained from the full band structure, expressing $\mathbf{q}$ as $\mathbf{q} = \mathbf{q}_K + \mathbf{k}$ (where $\mathbf{q}_K$ is the momentum of the Dirac point K measured from the origin and $\mathbf{k}$ is the momentum measured relatively to the point K) and expanding (9.1) for small values of $|\mathbf{k}|$. The resulting expression is

$$E_{\pm}\left(\vec{k}\right) = \pm\hbar v_{\mathrm{F}}\left|\vec{k}\right|, \tag{9.3}$$

where v_F is the Fermi velocity, given by $v_F = 3ta/2\hbar$, with a value $v_F \approx 1 \times 10^6\,\mathrm{m\,s^{-1}}$. The maximum of the valence band and the minimum of the conduction band are coincident, i.e., the energy band gap is null at the Dirac point. Furthermore, in neutral graphene, the chemical potential (i.e., the Fermi energy) crosses exactly the Dirac point. Under this point of view, graphene is considered as a zero bandgap semiconductor or a semimetal. However, the linear dispersion relation is quite different than the parabolic relation typical of common semiconductors. As shown in Fig. 9.3c, this linear dispersion is valid only at electron energies $|E| < 1\,\mathrm{eV}$, whereas at higher energies deviations from linearity have to be considered. Noteworthy, this energy range is wide enough to observe many peculiar electronic properties even at room temperature (*RT*). In particular, the energy spectrum of graphene electrons mimics that of massless Dirac fermions or photons, but they move with a velocity v_F, which is 300 times smaller than the velocity of light c. Hence, many of the unusual properties of quantum electrodynamics (QED) can show up in graphene but at much smaller velocities.

The linear dispersion relation has been directly measured on epitaxially grown graphene on SiC by angle resolved photoemission spectroscopy (ARPES) [42, 43]. In this technique, a beam of monochromatic light with energy $\hbar\omega$ is incident perpendicular to the sample surface and electrons inside the sample are emitted with kinetic energy E_{kin}. The energy conservation law requires $E_{\mathrm{kin}} = \hbar\omega - \phi - E$, where ϕ is the work function of the sample and E the electrons binding energy. The conservation of the momentum parallel to the surface also holds for the case of a 2D system. Because the momentum of photon is negligible compared to that of the electron in a solid, the momentum of the outgoing electron is approximately equal to that of the electron inside the solid. By detecting both the energy and the emission angle of the outgoing electron, the in-plane momentum is obtained. In an ARPES map, the intensity of the photoemitted electrons is reported as a function of the energy E and of the momentum k.

Epitaxial graphene (EG) is obtained from high-temperature thermal treatments performed on SiC either in ultrahigh vacuum or in standard secondary vacuum or even at atmospheric pressure in an inert gas atmosphere. Both the (0001) Si and (000–1) C faces of hexagonal SiC polytypes (4H and 6H) have been used but large differences have been found in the structural and electronic properties of EG grown on the two faces [44, 45]. As an example, graphene growth on the Si face at fixed temperature is a self-limited process and single or FLG can be obtained by properly setting the process conditions. In contrast, the number of graphene layers grown on the C face is typically more difficult to be controlled. Both experimental and theoretical studies have shown that EG synthesis on the (0001) face of SiC occurs through a series of complex surface reconstructions [46–50]. This starts from an initial (Si-rich) $(\sqrt{3} \times \sqrt{3})R30°$ phase which rapidly converts into a second (C-rich) $(6\sqrt{3} \times 6\sqrt{3})R30°$ reconstruction when the temperature increases. This layer may be more or less defective with more or less dangling bonds at the interface with the Si face (see schematic in Fig. 9.4a). The ARPES measurements in Fig. 9.4a1 indicate that its electronic structure is not that of graphene single layer, i.e., no linear dispersion is observed. This layer is simply an intermediate (buried) buffer

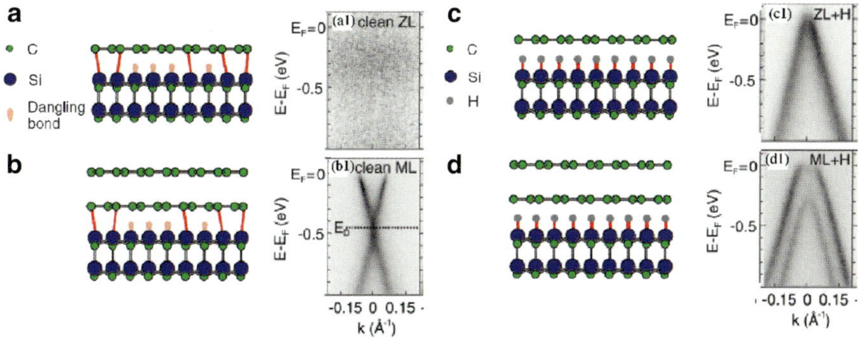

Fig. 9.4 Schematic representation of the interface between the zero layer (ZL) and the Si face of 4H-SiC (**a**). In (**a1**) the angle resolved photoemission spectroscopy (ARPES) of the system in (**a**) is reported. Representation of the system formed by ZL and the first graphene layer (**b**). In (**b1**) the ARPES spectra of the system in (**b**) is reported, showing a linear dispersion relation [51]. Representation of the interface between the ZL and the Si face of 4H-SiC after passivation of the dangling bonds by a thermal treatment in hydrogen (**c**). In (**c1**) the ARPES spectra of the system in (**c**) is reported, showing that the energy spectrum of the ZL converts to that of a "neutral" graphene single layer after that treatment. Representation of the system formed by ZL and the first graphene layer after the same treatment (**d**). In (**d1**) the ARPES spectra of the system in (**d**) is reported, showing that the energy spectrum of graphene monolayer on top of the ZL is converted to that of a bilayer. Data from [51]

layer (zero layer, ZL) with a large percentage of sp^2 hybridization. As shown in Fig. 9.4b1, the first graphene layer which forms on the ZL exhibits, instead, a linear dispersion relation [51], in agreement with the schematic in Fig. 9.4b. However, it is worth noting that the Fermi level does not correspond to the Dirac point, as expected for neutral graphene, but it is shifted $\sim$0.5 eV in the conduction band. This shift is explained in terms of a positive charge associated to the dangling bonds between the ZL and the bulk substrate. It has been shown that, after passivation of the dangling bonds by proper thermal treatments in hydrogen, the energy spectrum of the ZL converts to that of a "neutral" graphene single layer [51], i.e., a linear dispersion relation is recovered, with the Fermi energy coincident with the Dirac point (see Fig. 9.4c, c1). Furthermore, after the same treatment, the energy spectrum of the graphene monolayer on top of the ZL is converted to that of a bilayer (see Fig. 9.4d, d1). These results clarify the role of the interface on the electronic properties of EG on (0001) SiC. The graphene areas sampled by ARPES are in the order of mm^2 and, hence the measured dispersion relation is an average on those areas. As a matter of fact, lateral variations in the charge density at the interface can cause local changes in the electronic properties of EG, but cannot be evidenced by this technique. For this reason, these have been recently investigated by SPM, as discussed in Sect. 9.3.4.

9.2.2 Density of States

A direct consequence of the linear dispersion relation (for $|E| < 1\,\text{eV}$) is that the density of states $D(E)$ in graphene depends linearly on the energy as

$$D(E) = \frac{g_s g_v}{2\pi(\hbar v_F)^2}|E| = \frac{2}{\pi(\hbar v_F)^2}|E| \tag{9.4}$$

and vanishes at the Dirac point; $g_s = 2$ and $g_v = 2$ are the spin and valley (K and K' points) degeneracies, respectively. This behavior is very different than in common 2D electron systems obtained at semiconductor heterointerfaces, where the dispersion relation is parabolic and the density of states is step-like with respect to the energy.

In Fig. 9.5a, the calculated $D(E)$ is reported, showing that the linear dependence (red line) is strictly valid for energies $|E| < 1\,\text{eV}$, whereas for larger energies deviations from the linear behavior occur.

Theoretical works have demonstrated that "disorder" causes strong modifications of the density of states at the Dirac point, whereas far from it $D(E)$ remains linearly dependent of the energy [52]. This is illustrated in Fig. 9.5b, where the calculated density of states in a graphene lattice with a vacancy (i.e., a missing C atom) is reported.

9.2.3 Carrier Density

The carrier density depends on the temperature T and on the chemical potential (i.e., on the Fermi level position E_F with respect to the Dirac point), according to

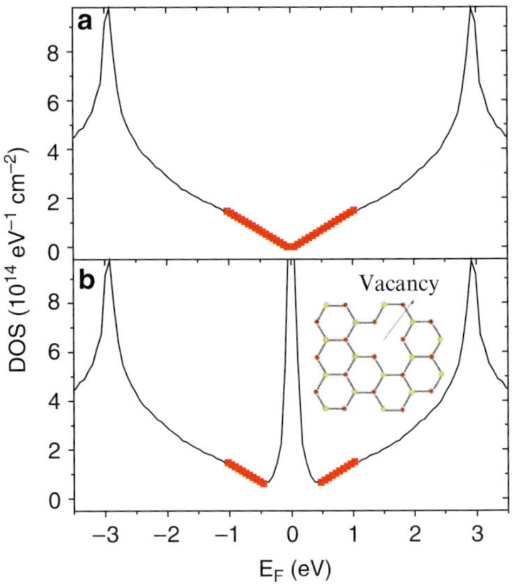

Fig. 9.5 Calculated density of states vs. energy $D(E)$ for ideal graphene (**a**) and for a graphene lattice with vacancies, i.e., missing C atoms (**b**). Data from [52]

$$n(E_F, T) = \int_0^\infty dE \cdot D(E) \cdot f(E, E_F, T), \tag{9.5}$$

where $f(E, E_F, T)$ is the Fermi–Dirac distribution function:

$$f(E, E_F, T) = \frac{1}{1 + \exp\left[\frac{E - E_F}{k_B T}\right]}. \tag{9.6}$$

In neutral graphene, the Fermi level coincides with the Dirac point, whereas it shifts due to charged impurities present on the layer and/or at the interface with the substrate. A shift of the Fermi level can be intentionally induced by an external electric field and, hence, the carrier density can be modulated by modulating the intensity of this field. The electric field is typically applied by a bias applied to a conductive gate, which is separated from graphene by a dielectric layer. Figure 9.6 shows the calculated carrier density n at temperature $T = 300$ K as a function of the Fermi level position. It is worth noting the parabolic dependence of n on E_F. In particular, for a negative gate bias, E_F is shifted in the valence band and the hole density is increased (see inset (i) in Fig. 9.6), whereas, for a positive gate bias, E_F is shifted in the conduction band and the electron density is increased (see inset (iii) in Fig. 9.6). This important property, demonstrated in the first experiments on graphene exfoliated from HOPG, opened the way to the fabrication of field effect devices in graphene. Typical values of the carrier density range from $\sim 10^{11}$ to $\sim 10^{13}$ cm^{-2} for E_F ranging from $\sim$10 to $\sim$400 meV.

From (9.5) and (9.6), in neutral graphene (where $E_F = 0$), the intrinsic carrier density is

$$n_i = p_i = \frac{\pi}{6} \left(\frac{k_B T}{\hbar v_F} \right)^2. \tag{9.7}$$

In particular, at temperatures approaching to 0 K a vanishing carrier density is theoretically expected in neutral graphene, whereas at room temperature a uniform carrier density $n_i \approx 9 \times 10^{10}$ cm^{-2} is expected.

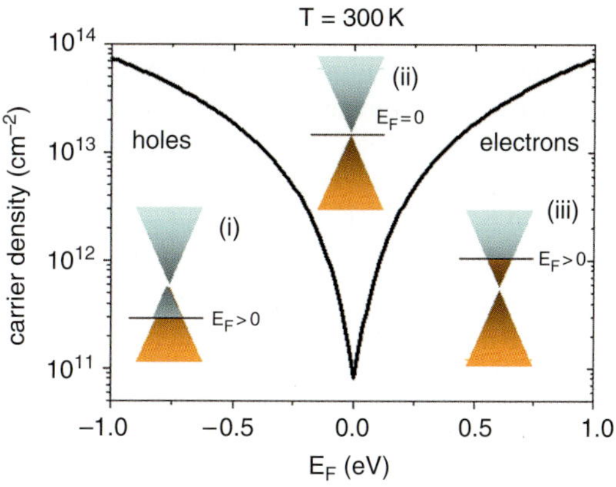

Fig. 9.6 Calculated carrier density as a function of the Fermi level position at 300 k

9.2.4 Quantum Capacitance

A basic physical property of graphene, related to the carrier density and to the density of states, is the quantum capacitance C_{q}, defined as

$$C_{\mathrm{q}} = \frac{\mathrm{d}Q}{\mathrm{d}E_{\mathrm{F}}} = Aq\frac{\mathrm{d}(n-p)}{\mathrm{d}E_{\mathrm{F}}}, \tag{9.8}$$

being $\mathrm{d}Q$ and $\mathrm{d}E_{\mathrm{F}}$ the variations of the charge and of the Fermi level, respectively, induced by a gate bias, and A the biased area of the graphene sheet. The following expression for the quantum capacitance per unit area $C_{\mathrm{q}}{}'$ of a perfect graphene sheet has been derived starting from the linear dispersion relation for the 2D free-electron gas [53]:

$$C_{\mathrm{q}}{}' = \frac{2q^2 k_{\mathrm{B}}T}{\pi\,(\hbar v_{\mathrm{F}})^2}\ln\left[2\left(1+\cosh\left(\frac{E_{\mathrm{F}}}{k_{\mathrm{B}}T}\right)\right)\right]. \tag{9.9}$$

When $E_{\mathrm{F}} \gg k_{\mathrm{B}}T$, i.e., far from the Dirac point, (9.9) reduces to

$$C_{\mathrm{q}}{}' \approx q^2\frac{2}{\pi}\frac{E_{\mathrm{F}}}{(\hbar v_{\mathrm{F}})^2} = \frac{2q^2}{\hbar v_{\mathrm{F}}\sqrt{\pi}}\sqrt{n}, \tag{9.10}$$

where the Fermi energy is related to the carrier density as $E_{\mathrm{F}} = \hbar v_{\mathrm{F}}(\pi n)^{1/2}$. The theory provides a quantitative description of graphene quantum capacitance in terms of Fermi velocity, carrier density, temperature, and fundamental physical quantities. The theoretical behavior of $C_{\mathrm{q}}{}'$ vs. E_{F} at $T = 300\,\mathrm{K}$ is reported in Fig. 9.7 (blue line). It is worth noting that $C_{\mathrm{q}}{}'$ has a minimum value $C_{\mathrm{q,min}} = 2q^2 k_{\mathrm{B}}T/\pi(\hbar v_{\mathrm{F}})^2\ln(4)$ at the Dirac point. Furthermore, it increases linearly with E_{F} with a slope of $23\,\mu\mathrm{F\,cm^{-2}\,V^{-1}}$ and it is symmetric with respect to the Dirac point. Comparing (9.4) with (9.10), it is clear that

$$C_{\mathrm{q}}{}' \approx q^2 D(E_F). \tag{9.11}$$

The experimental determination of the quantum capacitance of graphene on a dielectric substrate (SiO_2, semi-insulating SiC, etc.) is difficult. In fact, the dielectric capacitance contribution (which is typically several orders of magnitude lower than the quantum capacitance) adds in series to C_q and completely masks it. Recently, a measurement of the quantum capacitance as a function of gate potential has been performed on graphene immersed in an ionic liquid electrolyte inside an electrochemical cell [54]. The ionic liquid is chemically inert and stable. More relevant, owing to the high ionic concentration ($2.9 \times 10^{21}\,\mathrm{cm^{-3}}$) of the used ionic liquid (BMIM-PF6), the Debye length at the interface with graphene is very small ($t \approx 0.3\,\mathrm{nm}$) and yields a capacitance contribution $C_{\mathrm{i}}' = \varepsilon_0\varepsilon/t \approx 21\,\mu\mathrm{F\,cm^{-2}}$, where $\varepsilon_0 = 8.85 \times 10^{-12}\,\mathrm{F\,m^{-1}}$ and $\varepsilon \approx 7$ is the dielectric constant of the ionic liquid. This value is comparable with the quantum capacitance and it then easier to extract $C_{\mathrm{q}}{}'$ from the total capacitance.

In Fig. 9.7, the measured per unit area quantum capacitance is reported (open circles) and it differs from the theoretical curve, both for symmetry and minimum value, $C_{q,min}'$. The theoretical model assumes a pure and perfect graphene. In reality, various impurities and defects exist. It has been reported that charged impurities in substrates cause local potential fluctuations and electron/hole puddles in graphene.

In the absence of microscopic capacitance measurements sensitive to the local potential fluctuations, these local fluctuations were taken into account by the authors, as a first approximation, considering an additional effective carrier density n^* induced by the impurities. These impurities induce a shift in the position of the capacitance minimum. Furthermore, the expression of the quantum capacitance per unit area is modified as

$$C_q' = \frac{2q^2}{\hbar v_F \sqrt{\pi}} \sqrt{|n_G| + |n^*|}, \tag{9.12}$$

where n_G is the carrier density caused by the gate potential:

$$n_G = \left(\frac{2E_F}{q\hbar v_F \sqrt{\pi}} \right)^2. \tag{9.13}$$

In Fig. 9.7, the fit of the experimental C_q' using (9.12) is also reported (red line). The fit gives an effective carrier density induced by the impurities $n^* = 8 \times 10^{11}\ \mathrm{cm}^{-2}$.

In Sect. 9.3.2, local measurements of the graphene quantum capacitance by SPM will be discussed.

9.2.5 *Transport Properties: Mobility, Electron Mean Free Path*

Graphene is especially interesting for its outstanding transport properties, that is, the electron mean free path l and the carrier mobility μ. These properties are typically extracted from the measured conductivity σ or resistivity $\rho = 1/\sigma$. In fact,

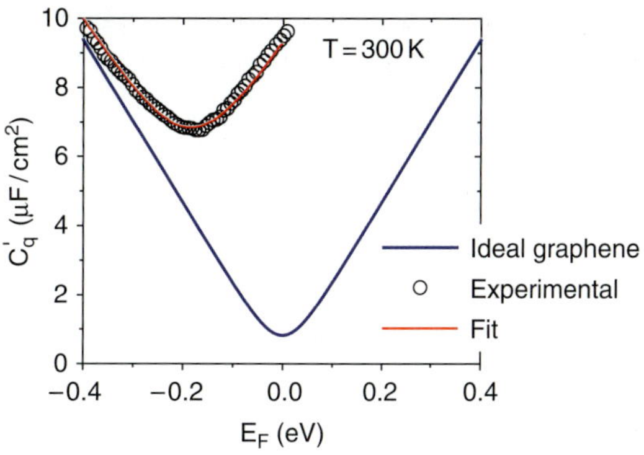

Fig. 9.7 Calculated quantum capacitance per unit area for ideal graphene (no defects or charged impurities). Measured quantum capacitance compared with the theoretical prediction for graphene with an effective density of charged impurities $n^* = 8 \times 10^{11}\ \mathrm{cm}^{-2}$. Data from [54]

conductivity is related to the carrier density and to the mobility as

$$\sigma = qn\mu. \tag{9.14}$$

From a theoretical point of view, a semiclassical approach based on the Boltzmann transport theory and on the relaxation time approximation can be used to find an expression of the conductivity [55, 56].

The relaxation time $\tau(E_F)$ is related to the electron mean free path l as

$$l = v_F\tau(E_F). \tag{9.15}$$

According to [55], for a fixed value of the Fermi energy, σ can be expressed in terms of the relaxation time as

$$\sigma = \frac{q^2 v_F^2}{2} D(E_F)\tau(E_F). \tag{9.16}$$

From (9.14):

$$\mu = \frac{q v_F^2}{2n} D(E_F)\tau(E_F). \tag{9.17}$$

From (9.4), the following expression for the dependence of μ on the carrier density is obtained:

$$\mu = \frac{q v_F \tau}{\hbar\sqrt{\pi n}} = \frac{ql}{\hbar\sqrt{\pi n}}. \tag{9.18}$$

The relaxation time depends on the peculiar scattering mechanism limiting the electron mean free path in graphene.

9.2.5.1 Intrinsic Transport Properties

In an ideal graphene sheet, that is, a freestanding planar sheet, free from contaminations or lattice defects and at a finite temperature T, the electron mean free path is limited by scattering with phonons. At low energy electronic excitations, the scattering cross-section with longitudinal acoustic (LA) phonons is higher with respect to other phonon modes. According to [55], the electron mean free path limited by scattering with LA phonons (l_{LAphon}) depends on the electron density and on the temperature T as

$$l_{\mathrm{LAphon}}(n) \approx \frac{\hbar^2 \rho_{\mathrm{gr}} v_S^2 v_F^2}{\sqrt{\pi} D_A^2 k_B T} n^{-1/2}, \tag{9.19}$$

where ρ_{gr} is the graphene mass density ($\rho_{\mathrm{gr}} \approx 7.6 \times 10^{-7}\,\mathrm{kg\,m^{-2}}$) [14], D_A is the acoustic deformation potential ($D_A \approx 18\,\mathrm{eV}$) [14], v_s is the acoustic phonons group velocity ($v_s \approx 2.1 \times 10^4\,\mathrm{m\,s^{-1}}$), and k_B is the Boltzmann constant.

According to (9.18), the corresponding drift mobility μ_{LAphon} is proportional to n^{-1} and to T^{-1}:

$$\mu_{\mathrm{LAphon}}(n) = \frac{q\hbar \rho_{\mathrm{gr}} v_S^2 v_F^2}{\pi D_A^2 k_B T}\frac{1}{n}. \tag{9.20}$$

Finally, the intrinsic graphene conductivity should be independent of n:

$$\sigma_{\mathrm{LAphon}} = \frac{q^2 \hbar \rho_{\mathrm{gr}} v_{\mathrm{S}}^2 v_{\mathrm{F}}^2}{\pi D_{\mathrm{A}}^2 k_{\mathrm{B}} T}. \tag{9.21}$$

In Fig. 9.8a–c, the calculated μ_{LAphon}, l_{LAphon}, and σ_{LAphon} (normalized to $4q^2/\pi h$) at three different temperatures ($T = 10$, 100, and 300 K) are reported vs. the carrier density n.

According to the considered semiclassical model, both the mean free path and mobility of electrons should diverge for T and n approaching zero [see (9.19) and (9.20)]. In this case (9.16) is not any more appropriate to describe the transport behavior in graphene, but the Landauer formalism for ballistic transport has to be applied to evaluate the carrier conductivity [57, 58]. More generally, the latter formalism is appropriate to describe the current transport through a sample of width W

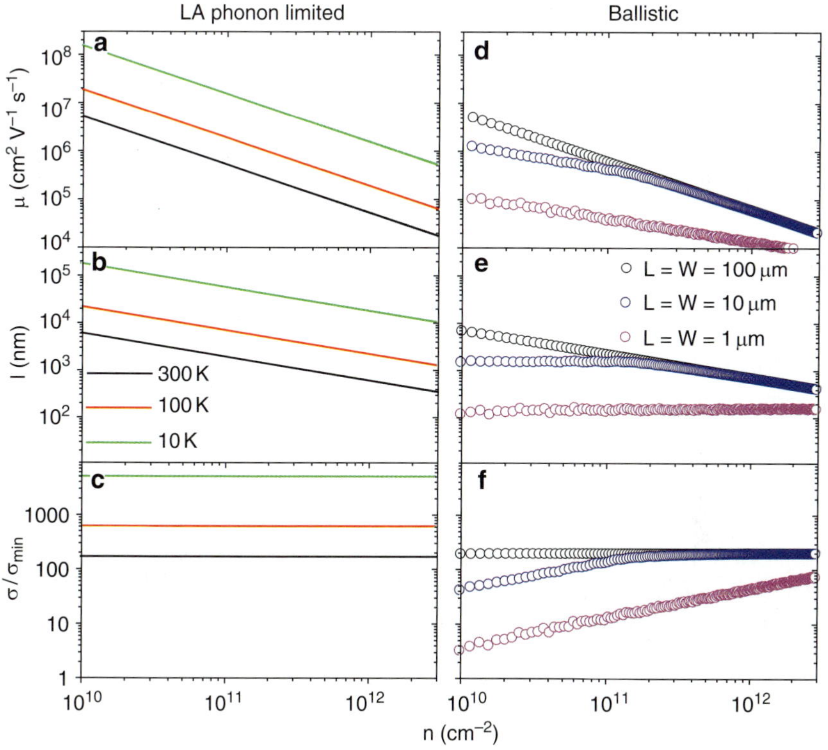

Fig. 9.8 Calculated electron mobility μ_{Laphon} (**a**), electron mean free path l_{Laphon} (**b**), and conductivity σ_{LAphon} (normalized to $4q^2/\pi h$) (**c**) limited by scattering with LA phonons for three different temperatures ($T = 10$, 100, and 300 K) vs. the carrier density n. Calculated mobility μ_{bal} (**d**), electron mean free path l_{bal} (**e**), and conductivity σ_{bal} (normalized to $4q^2/\pi h$) (**f**) vs. n in the ballistic limit for graphene sheets of different widths W, contacted by leads with separation L ($L = W = 100$, 10, and 1 μm)

between two leads with separation L, whenever the electron mean free path is comparable to the lead separation. In this case, the ballistic conductivity is expressed by the Landauer formula:

$$\sigma_{\mathrm{bal}} = \frac{L}{W}\frac{4q^2}{h}\sum_{n=1}^{\infty} T_n, \tag{9.22}$$

where the summation is over all available longitudinal transport channels and T_n is the transmission probability in the nth channel, which depends on sample width W, lead separation L, and chemical potential (Fermi energy E_{F}), as [59]:

$$T_n = \left| \frac{k_n}{k_n \cos(k_n L) + \mathrm{i}\left(\frac{E_{\mathrm{F}}}{\hbar v_{\mathrm{F}}}\right)\sin(k_n L)} \right|^2 \tag{9.23}$$

being

$$k_n = \sqrt{\left(\frac{E_{\mathrm{F}}}{\hbar v_{\mathrm{F}}}\right)^2 - \frac{\pi^2}{W^2}\left(n + \frac{1}{2}\right)^2}. \tag{9.24}$$

In Fig. 9.8f, the calculated conductivity σ_{bal} (normalized to $4q^2/\pi h$) for $L = W = 100$, 10, and 1 μm is reported vs. the carrier density. It is worth noting that, according to this formalism, a decrease of the conductivity with decreasing the carrier density is expected for graphene samples with smaller dimensions (10 and 1 μm).

In the framework of this formalism, the minimum value of the conductivity has been theoretically predicted at the Dirac point for ballistic transport mediated by evanescent modes:

$$\sigma_{\min} = \frac{4q^2}{\pi h} = 4.92 \times 10^{-5}\,\Omega. \tag{9.25}$$

Furthermore, in Fig. 9.8d, e, the mobility and the electron mean free path obtained from σ_{bal} through (9.14) and (9.18) are reported also.

9.2.5.2 Transport Properties Limited by Extrinsic Scattering Mechanisms

The previous predictions refer to ideal graphene, that is, an infinite perfectly planar sheet of C atoms, free from lattice defects and minimally affected by interactions with the environment. However, graphene sheets are not exactly planar (they exhibit intrinsic ripples), are not infinite (present edges), lie on a substrate (though the interaction between the layer and the substrate material is typically weak, i.e., van der Waals like), and a certain amount of contaminations are present on their surface or at the interface with the substrate. Furthermore, a non-null density of lattice defects is present in the sheet. All of these sources act as local (nanoscale) perturbations and alter, to some extent, the electronic properties of a perfect graphene sheet. In fact, defects or impurities introduce local potentials fluctuations that have a double effect:

1. To introduce spatial inhomogeneities in the carrier density
2. To act as scattering sources causing the reduction of the electron mean free path and mobility

Both effects play a role in determining the overall electronic behavior of macroscopic graphene devices. However, the first one is the dominating effect when the Fermi level is close to the Dirac point (i.e., for very low average carrier density), whereas the second one is the dominating effect when the Fermi level is far from the Dirac point.

In particular, the identification of the scattering mechanisms limiting transport properties is one of the hot topics in the research on graphene. This problem has been addressed by combining experimental investigations and theoretical modeling. From a theoretical point of view, two types of scattering sources are typically considered: (a) the long-range scatterers, like charges impurities, which interact with graphene electrons through a screened Coulomb potential; (b) the short-range scatterers, like point defects in the lattice, which interact with electrons through a short-range potential. From the experimental point of view, transport properties are typically determined from the electrical characterization (by four point probe and/or Hall effect measurements) on macroscopic (micrometer size) graphene devices. The experimental results are compared with the theoretical expectations and the discrepancies are explained with models considering the role played by the different scattering mechanisms. A complementary approach, based on the use of scanning probe methods, is used to directly image on the nanoscale the lateral inhomogeneities in the electronic and transport properties. This will be the object of the Sect. 9.3.

9.2.5.3 Electronic Transport Close to the Dirac Point

The effect of defects and impurities is especially deleterious near the Dirac point, where the screening of the potential fluctuations by the 2DEG is weak, due to the low carrier density.

In Fig. 9.9 (black circles), the conductivity of graphene deposited on a SiO_2/Si substrate is reported vs. the potential V_{gr} of the graphene sheet [13]. This potential (related to the Fermi energy as $V_{gr} = E_F/q$) is a function of the bias applied to the n^+–Si backgate, while the conductivity is determined by four-point probe measurements (see schematic *i* in Fig. 9.9). The measurements were carried out at $T = 100\,K$. The conductivity increases with the potential for V_{gr} higher than $V_{gr0} = 31\,mV$, whereas it is independent of the potential for $V_{gr} < V_{gr0}$ and saturates to a minimum value $\sigma_0 = 5\sigma_{min}$. This is in contrast with the predictions for ideal graphene, where the conductivity is expected to decrease continuously with the potential to the minimum value σ_{min}. This discrepancy was explained in terms of the random potential fluctuations due to charges trapped in the SiO_2 substrate or located at the graphene/SiO_2 interface or on the graphene surface. These fluctuations

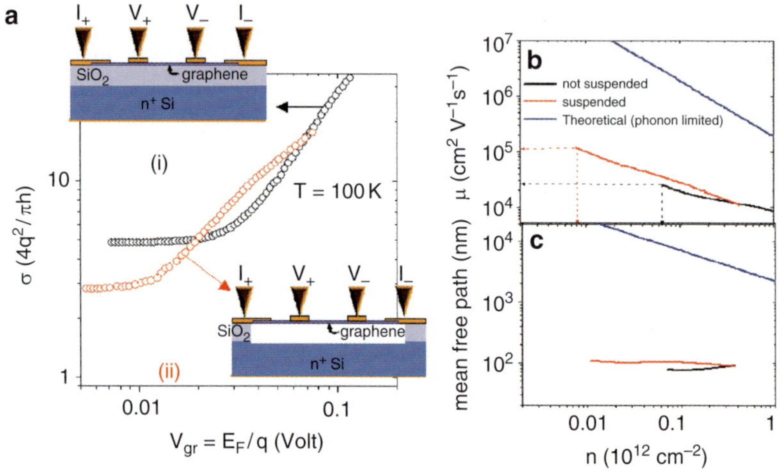

Fig. 9.9 (**a**) Conductivity measured by four point probes of deposited graphene (DG) (*black line*) on a SiO_2/Si substrate [see schematic i] and of suspended graphene (*red line*) after SiO_2 etching [see schematic i] vs. the potential $V_{gr} = E_F/q$ (with E_F the Fermi energy). Mobility (**b**) and electron mean free path (**c**) vs. n extracted from the measured conductivity for graphene on SiO_2 (*black line*) and for suspended graphene (*red line*). The calculated (LA phonon limited) mobility and electron mean free path are also reported (*blue lines*). Data from [13]

superimpose to the average potential V_{gr} induced by external gating. It was supposed that, for $0 < V_{gr} < V_{gr0}$, the potential fluctuations cause the splitting of the charge distribution in the graphene sheet into "electron–hole puddles." Hence, within that energy range, the effect of gating is limited to a redistribution of carriers between the electrons and holes puddles without significantly changing the total carrier density. This picture has been confirmed by nanoscale measurements of the charge distribution close to the Dirac point by scanning probe methods, as discussed in Sect. 9.3.1.

The role played by the interface with the SiO_2 substrate was demonstrated by performing conductivity measurements on free-standing graphene samples, where the SiO_2 layer was locally etched (in buffered HF) under the graphene sheet, which remained suspended and attached to the metal leads (see schematic *ii* in Fig. 9.9). The conductivity measured at 100 K on a suspended sample is reported for comparison in Fig. 9.9a (red circles). In this case, the minimum conductivity σ_0 and the potential V_{gr0} are reduced to $\sigma_0 = 3\sigma_{min}$ and $V_{gr0} = 15\,\text{mV}$.

9.2.5.4 Transport Far from the Dirac Point

Measurements of carrier mobility and mean free path in graphene are only meaningful far from the Dirac point, that is, outside the puddle regime, where the carrier density is controlled by the gate. For instance, Fig. 9.9b shows the density dependence of the mobility calculated from the measured conductivity ($\mu = \sigma/en$)

for graphene on SiO_2 (black line) and suspended graphene (red line). At $T = 100\,\mathrm{K}$, just outside the puddle regime (i.e., for $V_{gr} = V_{gr0} \approx 32\,\mathrm{mV}$) the carrier density of graphene on SiO_2 is $n_0 \approx 6 \times 10^{10}\,\mathrm{cm}^{-2}$ and the mobility $\mu_0 \sim 25{,}000\,\mathrm{cm}^2\,\mathrm{V}^{-1}\,\mathrm{s}^{-1}$. In the case of suspended graphene, $V_{gr0} \approx 15\,\mathrm{mV}$, $n_0 \approx 7 \times 10^9\,\mathrm{cm}^{-2}$, and $\mu_0 \sim 120{,}000\,\mathrm{cm}^2\,\mathrm{V}^{-1}\,\mathrm{s}^{-1}$. The mobility decreases with increasing the carrier densities and becomes similar in the two samples for $n > 6 \times 10^{11}\,\mathrm{cm}^{-2}$. In Fig. 9.9e the corresponding mean free paths, related to the mobility according to (9.18), are reported.

The experimental values of mobility and electron mean free path are much lower if compared with the theoretical expectations of μ_{LAphon} and l_{LAphon}, showing that transport properties are limited by extrinsic scattering mechanisms. Three main extrinsic scattering mechanisms have been identified: (a) scattering due to the Coulomb potential generated by charged impurities; (b) scattering due to the short-range potential generated by point defects (vacancies) in the graphene lattice; (c) scattering due to the long-range potential associated with the surface polar phonons (SPP) of the substrate.

According to [55], the electron mean free path l_{Coul} limited by scattering with charged impurities on graphene surface or at the interface with the substrate depends on the electron density as

$$l_{\mathrm{Coul}}(n) = \frac{16\varepsilon_0^2\varepsilon^2\hbar^2 v_{\mathrm{F}}^2}{Z^2 q^4 N_{\mathrm{Coul}}}\left(1 + \frac{q^2}{\pi\hbar v_{\mathrm{F}}\varepsilon_0\varepsilon}\right)^2 \sqrt{\pi n}, \tag{9.26}$$

where N_{Coul} is the impurity density, $\hbar$ the Planck's reduced constant, Z the impurities charge number, q the electron charge, ε_0 the vacuum absolute dielectric constant, and ε the average between SiO_2 and vacuum relative dielectric constants. According to (9.18), the electron mobility limited by scattering with charged impurities is

$$\mu_{\mathrm{Coul}} = \frac{16\varepsilon_0^2\varepsilon^2\hbar v_{\mathrm{F}}^2}{Z^2 q^3 N_{\mathrm{Coul}}}\left(1 + \frac{q^2}{\pi\hbar v_{\mathrm{F}}\varepsilon_0\varepsilon}\right)^2, \tag{9.27}$$

i.e., it is independent of the carrier density n and on the temperature T. The mobility μ_{Coul} is inversely proportional to the charged impurity density N_{Coul}. Furthermore, it depends on the dielectric constant of the substrate and of the overlayer.

Finally, the conductivity limited by charged impurity scattering depends linearly on n:

$$\sigma_{\mathrm{Coul}} = \frac{16\varepsilon_0^2\varepsilon^2\hbar v_{\mathrm{F}}^2}{Z^2 q^2 N_{\mathrm{Coul}}}\left(1 + \frac{q^2}{\pi\hbar v_{\mathrm{F}}\varepsilon_0\varepsilon}\right)^2 n. \tag{9.28}$$

The electron mean free path limited by scattering by vacancies (l_{vac}) can be expressed as a function of the carrier density as

$$l_{\mathrm{vac}}(n) = \frac{1}{\pi^2 N_{\mathrm{vac}}}\sqrt{n}\left[\ln\left(\sqrt{n}R_0\right)\right]^2. \tag{9.29}$$

The drift mobility limited by vacancies is only slightly dependent on the carrier density:

$$\mu_{\mathrm{vac}}(n) = \frac{q}{\pi^2 \hbar N_{\mathrm{vac}}} \left[\ln\left(\sqrt{\pi n} R_0\right)\right]^2. \tag{9.30}$$

Finally, the conductivity limited by vacancies depends almost linearly on the carrier density:

$$\sigma_{\mathrm{vac}}(n) = \frac{q^2}{\pi^2 \hbar N_{\mathrm{vac}}} \left[\ln\left(\sqrt{\pi n} R_0\right)\right]^2 n. \tag{9.31}$$

When graphene is deposited on a polar substrate, the SPP of the substrate produce an electric field that couples to the electrons on the nearby graphene. The following relation has been proposed to express the dependence of the resistivity limited by SPP, i.e., ρ_{SPP}, on the carrier density n and on the temperature.

$$\rho_{\mathrm{SPP}}(n,T) = \frac{B}{n^\alpha} \sum_i \left[\frac{1}{\exp\left(\frac{\hbar\omega_i}{k_B T}\right) - 1}\right]. \tag{9.32}$$

In (9.32), B is a fitting parameter, $\alpha \approx 1.04$, and ω_{i} are the characteristic resonant frequencies of the surface phonons of the substrate.

Recently, the mechanisms of electron scattering in graphene deposited on SiO_2/Si have been experimentally investigated by measuring the temperature and carrier density dependent four-probe resistivity $\rho(n,T)$ [59]. Figure 9.10a shows ρ vs. T (in the range 16–485 K) at different carrier densities (induced by the gate voltage V_g applied to the Si substrate). At low T (< 100 K) the $\rho(T)$ curves are linear in temperature, whereas, at high T, ρ is highly nonlinear in T. The data were fitted by three terms:

$$\rho(n,T) = \rho_0(n) + \rho_1(T) + \rho_{\mathrm{h}T}(n,T). \tag{9.33}$$

The first two terms were determined by linear fitting of the low temperature curves. The term $\rho_0(n)$, obtained as the intercept at $T = 0$, was found to depend on the carrier density as n^{-1}, as shown in Fig. 9.10b, black open circles.

Due to this peculiar dependence, it was associated to the charged impurities and/or vacancies limited transport, i.e., $\rho_0(n) = \rho_{\mathrm{Coul}}(n) + \rho_{\mathrm{vac}}(n)$. The term $\rho_1(T)$ was determined as the first-order term in the linear fit. Its slope, $(4.0 \pm 0.5) \times 10^{-6} h/q^2$, is independent of the carrier density. It was associated to the acoustic phonon scattering, i.e., to $\rho_{\mathrm{LAphon}} = (\pi D_A{}^2 k_B)/(4q^2\hbar\rho_{\mathrm{gr}} v_s{}^2 v_F{}^2)T = (3.8 \times 10^{-6} h/q^2)T$.

The third term $\rho_{\mathrm{h}T}$ in (9.33) has a strong (activated) temperature dependence and it depends roughly on n^{-1} (see Fig. 9.10b, red circles). It was associated to scattering by SPP modes of SiO_2:

$$\rho_{\mathrm{SPP}}(n,T) = \frac{B}{n^\alpha}\left[\frac{1}{\exp\left(\frac{59\,\mathrm{meV}}{k_B T}\right) - 1} + \frac{6.5}{\exp\left(\frac{155\,\mathrm{meV}}{k_B T}\right) - 1}\right]. \tag{9.34}$$

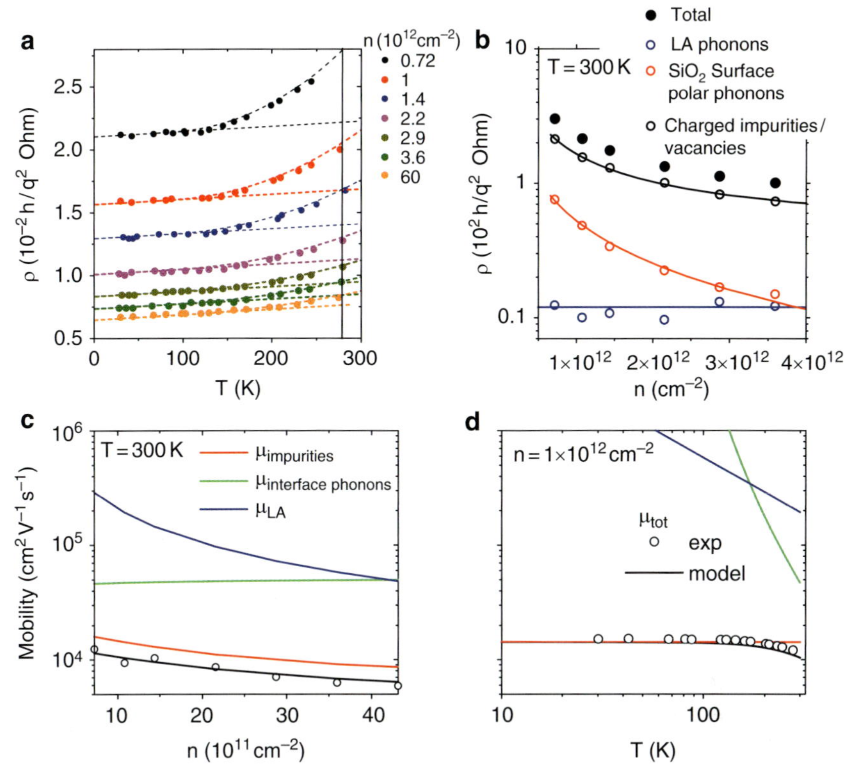

Fig. 9.10 (**a**) Measured resistivity ρ of graphene deposited on SiO_2/Si vs. temperature T (from 16 to 485 K) at different carrier densities (changed by the gate voltage V_g applied to the Si substrate). Measured data are fitted with theoretical expressions of resistivity limited by scattering with charged impurities/vacancies ρ_0, LA phonons (ρ_{Laphon}), and SPP of the SiO_2 substrate (ρ_{SPP}). (**b**) Carrier density dependence of the three components of the resistivity (ρ_0, ρ_{Laphon}, and ρ_{SPP}) at room temperature. (**c**) Electron mobility $\mu = 1/nq\rho$ at room temperature for the different resistivity contributions as a function of the carrier density. (**d**) Temperature dependence of the mobility at $n = 1 \times 10^{12}\,\mathrm{cm}^{-2}$, as well as the limits due to scattering by LA phonons, polar optical phonons of the SiO_2 substrate, and impurities/vacancies. Data from [14].

The two strongest surface optical phonon modes in SiO_2 have $E_1 = 59\,\mathrm{meV}$ and $E_2 = 155\,\mathrm{meV}$, with a ratio of coupling to the electrons of 1:6.5. The SPP results in a long-range potential, which gives rise to a density-dependent resistivity in graphene, similar to charged impurity scattering. Figure 9.10b shows the dependence of the three components of the resistivity (ρ_0, ρ_{LAphon}, and ρ_{SPP}) on the carrier density at room temperature.

Figure 9.10c shows the corresponding room-temperature mobility $\mu = 1/nq\rho$ for the different resistivity contribution as a function of the carrier density. Figure 9.10d shows the temperature dependence of the mobility at $n = 1 \times 10^{12}\,\mathrm{cm}^{-2}$, as well as the limits due to scattering by LA phonons, polar optical

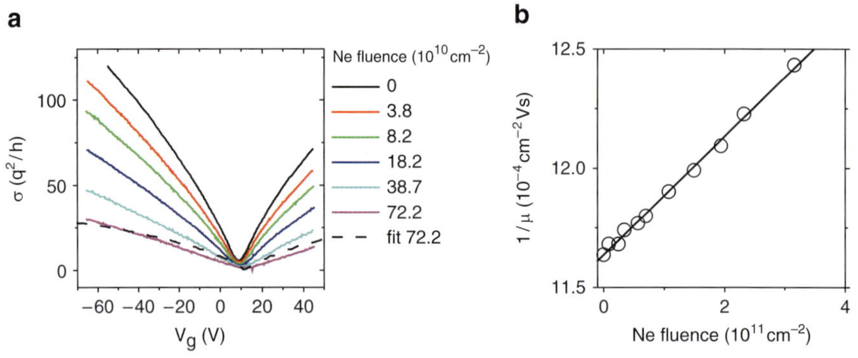

Fig. 9.11 (**a**) Conductivity vs. gate bias for pristine graphene flakes exfoliates from HOPG and deposited on a SiO_2/Si substrate and for graphene flakes irradiated with 500 eV Ne^+ ions at different fluences. (**b**) Reciprocal of the mobility $1/\mu$ vs. the irradiated fluence (*symbols*). The line is the fit with the relation $1/\mu = 1/\mu_0 + \alpha N_{vac}$, where μ_0 is the mobility of pristine graphene and $\alpha = [\pi/\ln(p^{1/2} < n >^{1/2} R_0)]^2\hbar/q$, with $< n >$ the average carried density and $R_0 = 0.23$–0.29 nm the "vacancy radius." It was assumed that a vacancy is generated in graphene by each incident Ne^+ ion. Data from [60]

phonons of the SiO_2 substrate, and impurities/vacancies. As shown in Fig. 9.10d, impurity/vacancies scattering is still the dominant factor limiting the mobility for $T = 300$ K.

According to (9.28) and (9.30), the conductivity limited by scattering with charged impurities and with point defects (vacancies) exhibits a similar dependence on the carrier density n and on the density of scattering centers (N_{Coul} and N_{vac}, respectively), i.e., σ_{Coul} and σ_{vac} increase linearly with n and decrease as $1/N_{Coul}$ or $1/N_{vac}$. In order to disentangle the effect of these two different scattering sources on the carrier mobility, conductivity measurements have been recently performed on graphene samples subjected to plasma treatment or ion irradiation, in order to intentionally introduce defects (vacancies) in the ultrathin C sheets. In particular, irradiation with ions at fixed energy allows a good control of the density of defects through control of the irradiated fluence.

As an example, in Fig. 9.11a, the conductivity vs. gate bias is reported for pristine graphene flakes exfoliates from HOPG and deposited on a SiO_2/Si substrate and for graphene flakes irradiated with 500 eV Ne^+ ions at different fluences [60]. A decrease of the conductivity with increasing the irradiated fluence is evident. The mobility μ extracted from the measured conductivity was found to be almost constant with the carrier density. In Fig. 9.11b, it is shown that $1/\mu$ grows linearly with the fluence. Assuming that a vacancy was generated in graphene by each incident Ne^+ ion, the data were fitted with the relation $1/\mu = 1/\mu_0 + \alpha N_{vac}$ [derived from (9.30)], where μ_0 is the mobility of pristine graphene and $\alpha = [\pi^2/\ln(\pi^{1/2} < n >^{1/2} R_0)]^2\hbar/q$, with $< n >$ the average carried density and $R_0 = 0.23$–0.29 nm the "vacancy radius."

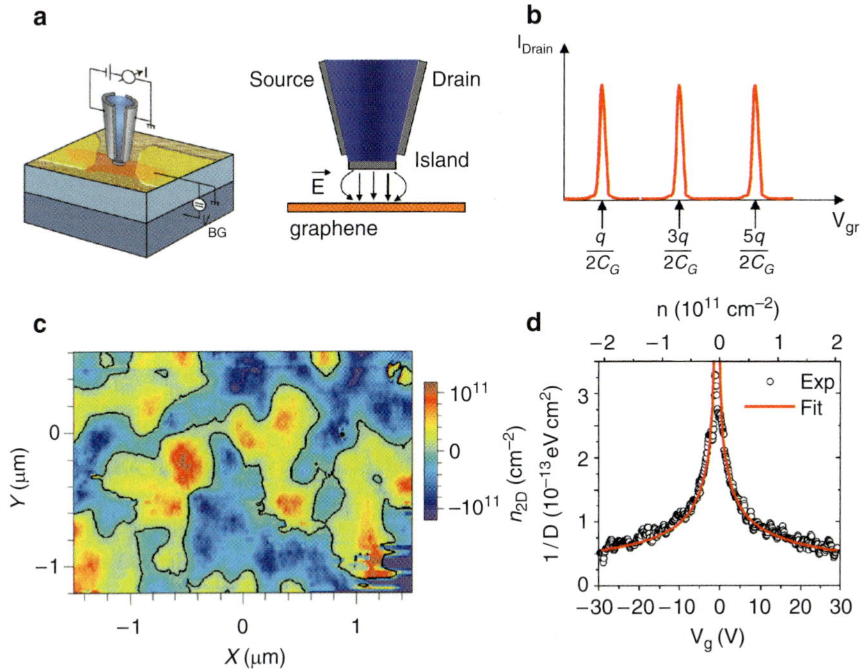

Fig. 9.12 (**a**) Schematic diagram of the experimental setup of single electron transistor microscopy and (**b**) representation of the behavior of the drain current as a function of the graphene potential. (**c**) Map of the carrier density distribution n at the Dirac point. (**d**) Inverse density of states for a fixed location of the SET as a function of the back-gate voltage. Data from [62]

9.3 Local Transport Properties of Graphene by Scanning Probe Methods

9.3.1 Lateral Inhomogeneity in the Carrier Density and in the Density of States

9.3.1.1 Scanning Single Electron Transistor Microscopy

Maps of the carrier density lateral distribution in graphene with submicrometer resolution were determined for the first time using a scanning probe method named scanning single electron transistor (SET) microscopy [61].

A schematic diagram of the experimental setup is shown in Fig. 9.12a. The probe is represented by a SET fabricated at the end of a sharp glass tip: the source and drain are two electrodes at the sides of the probe, which are separated from a very small metal island at the apex by two tunnel junctions. The diameter of the SET is about 100 nm and the probe is scanned in close proximity (roughly 50 nm) to the sample surface.

The charge flow between source and drain through the island is ruled by the Coulomb blockade effect. The current tunnels through the junctions at a rate determined by the island's electrostatic potential with respect to the source and drain. This potential is in turn controlled by the electric field that the island experiences from external sources, such as nearby fixed charges. At low temperature and proper drain bias, the current flowing through the SET fluctuates periodically as the external electric field increases. In fact, the current passes through a full period each time the electric field lines terminating on the island induce a charge of exactly one additional electron (see Fig. 9.12b). Hence, monitoring of the current through the SET as it is scanned over the sample provides a means of mapping the electric field emanating from the sample. This system is capable of measuring the local electrostatic potential Φ with microvolt sensitivity and to detect local variations of the carrier density $\sim$1% of an electron charge (i.e., $0.01q$). The spatial resolution is close to the island size (100 nm).

Local electrostatic potential measurements were carried out (at low temperature 0.3 K) on single layers of graphene deposited on a SiO_2/Si substrate [62]. A potential V_g applied to Si backside allowed also to change the global Fermi level in graphene, thus changing the average carrier density. In the absence of current, any change in the local electrostatic potential Φ measured by SET is equal in magnitude and opposite in sign to the changes in the local chemical potential (i.e., the difference between E_F and the "local" Dirac energy E_D) of graphene. The lateral fluctuations of $E_D(x, y)$ can be converted in lateral variations of the carrier density $n(x, y)$, according to the relation

$$n(x, y) = \frac{|E_F - E_D(x, y)|^2}{\pi(\hbar v_F)^2}. \tag{9.35}$$

In Fig. 9.12c, a map of $n(x, y)$ collected for $V_g = 0$ is reported [62]. This was the first experimental demonstration that, close to the Dirac point, the carrier landscape is composed by spatially separated electron/hole puddles.

The SET allows also local measurements of the inverse of the density of states in graphene. This was obtained as the derivative of the local electrostatic potential as a function of the average carrier density ($d\Phi/dn$). In fact, in the absence of any transport current in graphene,

$$\left|\frac{d\Phi}{dn}\right| = \left|\frac{dE_F}{dn}\right| = \frac{1}{D(E_F)}. \tag{9.36}$$

The variation dn was induced by a small signal modulating bias applied to the backgate and the $d\Phi$ values were measured by a lock-in amplifier. Figure 9.12d shows the inverse density of states for a fixed location of the SET as a function of the backgate voltage [62]. The maximum in the experimental $1/D$ trace clearly identifies the position of the Dirac point at this particular location. The absence of a singularity at the Dirac point in the experiment is ascribed to disorder broadening and experimental resolution. The data are fitted with the dependence of the inverse density of states on the carrier density $1/D = \hbar v_F(4\pi/|n|)^{1/2}$ and the fit parameter is the Fermi velocity. The red line in Fig. 9.12d shows the result of such a fit and yields $v_F = 1.1 \times 10^6 \pm 0.1 \times 10^6\,\mathrm{m\,s^{-1}}$.

9.3.1.2 Scanning Tunneling Microscopy and Spectroscopy

Scanning tunneling microscopy and spectroscopy were used to probe charge carriers inhomogeneity in graphene with a spatial resolution two orders of magnitude higher than previous scanning SET microscopy measurements.

Zhang et al. [63] reported measurements on graphene monolayers on SiO_2, as illustrated in the schematic of Fig. 9.13a. When the STM tip is placed at a distance z from the graphene surface, and a potential V_{tip} is applied between the tip and graphene, the tunneling current I is a function of z, V_{tip}, and of the local density of states (LDOS) at the Fermi level, $D(E_F)$, in graphene, i.e.,

$$I \propto V_{tip} D(E_F) \exp(-kz), \tag{9.37}$$

where

$$k = \frac{\sqrt{2m(\Delta - qV_{tip})}}{\hbar} \tag{9.38}$$

being m the free electron mass and Δ the tunnel barrier height. For low biases ($V_{tip} \ll \Delta/q$) typically applied to the tip, the current I reduces to

$$I \propto V_{tip} D(E_F) \exp\left(-\frac{\sqrt{2m\Delta}}{\hbar} z\right). \tag{9.39}$$

If the tip is scanned on the sample surface at fixed tunneling bias V_{tip}, and the distance z is adjusted (point by point) in order to maintain a constant current, the resulting map of the z values is mainly related to the surface topography (although, to some extent, it is affected also by the lateral variations in the LDOS). Figure 9.13b shows the STM topography of a typical $30 \times 30\,\mathrm{nm}^2$ area of a graphene monolayer on SiO_2. Random corrugations with lateral dimension of a few nanometres and a vertical dimension of $\sim$1.5 Å (RMS) can be observed, due to roughness in the underlying SiO_2 surface and/or intrinsic ripples of the graphene sheet. An atomic resolution map of the graphene lattice is also reported in the inset of Fig. 9.13b.

Scanning tunneling spectroscopy is performed maintaining a fixed distance $z=z_0$ of the tip from the sample surface and collecting the derivative of tunnel current with respect to tip bias $\mathrm{d}I/\mathrm{d}V_{tip}$ [the current variation $\mathrm{d}I$ is induced by a small amplitude ($\mathrm{d}V_{tip}$) modulating bias and $\mathrm{d}I/\mathrm{d}V_{tip}$ is measured through a lock-in system]. The derivative

$$\frac{\mathrm{d}I}{\mathrm{d}V_{tip}} \propto D(E_F) \exp\left(-\frac{\sqrt{2m\Gamma}}{\hbar} z_0\right) \tag{9.40}$$

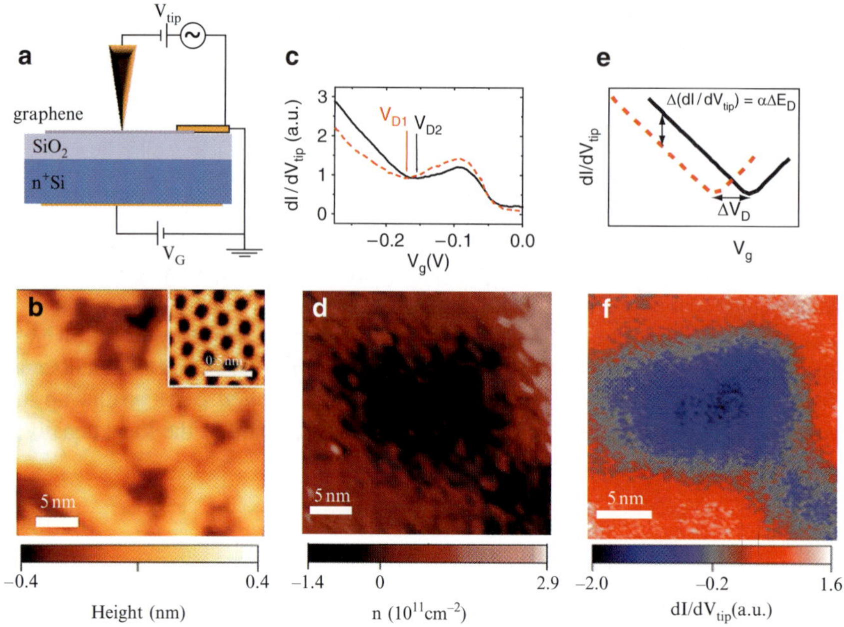

Fig. 9.13 (**a**) Schematic diagram of the experimental setup of scanning tunneling microscopy (STM). (**b**) STM topography of a typical 30 × 30 nm^2 area of a graphene monolayer on SiO_2. (**c**) Two $\mathrm{d}I/\mathrm{d}V_{\mathrm{tip}}$ vs. the gate bias (V_g) spectra taken with the tip fixed on two different points in (**b**). (**d**) Carrrier-density map, $n(x, y)$, for the same area as shown in (**b**) at an applied gate voltage of $V_g = 15$ V. (**e**) Illustration of the proportionality between $\mathrm{d}I/\mathrm{d}V$ and the local density of states of graphene in the vicinity of the Dirac point. (**f**) $\mathrm{d}I/\mathrm{d}V$ maps taken at a fixed bias ($V_{\mathrm{tip}} \sim -0.25$ V, $V_g = 15$ V), which shows the same charge puddle obtained from direct E_D mapping in (**d**). Data from [63]

is mainly related to the graphene LDOS at the Fermi level. It is worth noting that the LDOS determined by STS is different than that measured by SET microscopy, being the latter one defined on a mesoscopic scale (∼100 nm).

Figure 9.13c shows two $\mathrm{d}I/\mathrm{d}V_{\mathrm{tip}}$ vs. the gate bias (V_g) spectra taken with the tip fixed on two different points in Fig. 9.13b. The graphene LDOS at the Dirac point shows a local minimum, which is reflected by a dip in the tunneling spectra of graphene. The energy position of the dip, qV_D, marks the Dirac-point energy, E_D, offset by a constant ∼ 63 meV shift that arises from the loss of energy experienced by electrons as they inelastically tunnel into graphene by generating a phonon. Spatial variation in the measured value of E_D reflects the spatial profile of charge inhomogeneity in graphene. Charge puddles can thus be mapped by measuring the tunnel spectrum at every pixel over a given area and identifying E_D at each point. A Dirac point map, $E_D(x, y)$, can be converted into a carrrier-density map, $n(x, y)$, through the relation $n(x, y) = E_D{}^2(x, y)/\pi(\hbar v_F)^2$. Figure 9.13d shows such a map of E_D for the same area as shown in Fig. 9.13b at an applied gate voltage of $V_g = 15$ V. Fluctuations in the Dirac-point energy up to ∼30 meV are resolved, corresponding to charge–density fluctuations of $\sim 4 \times 10^{11}\,\text{cm}^{-2}$. A single puddle of electrons with a width of ~20 nm can be seen over this area. The inhomogeneous graphene charge density has been determined by mapping the Dirac point in the LDOS for each tip position.

Charge puddles can also be probed by spatially mapping the tunneling differential conductance, dI/dV_{tip}, for a fixed sample bias V_g held slightly below V_D. The basis for using this second technique to measure charge puddles is illustrated in Fig. 9.13e, where it is shown that, in the vicinity of the Dirac point (i.e., V_g slightly below V_D), lateral variations in dI/dV_{tip} are proportional to lateral variations in the Dirac point position. Hence, dI/dV maps taken at a fixed V_g close to V_D can produce a map of E_D, up to a multiplicative factor. This is shown in Fig. 9.13f, where a dI/dV_{tip} map acquired at fixed bias ($V_g = 15\,\text{V}$) in the same area as in Fig. 9.13c shows a charge puddle of the same shape as obtained from direct E_D mapping. Clearly, this technique reduces data acquisition time and is particularly suited for measuring large graphene areas containing multiple charge puddles.

9.3.2 *Nanoscale Measurements of Graphene Quantum Capacitance*

Local capacitance measurements on graphene monolayers deposited on a SiO_2/n^+Si substrate have been recently performed by scanning capacitance spectroscopy (SCS) [64,65]. A schematic representation of the measuring setup is reported in Fig. 9.14a.

The graphene/insulator/semiconductor backgate system can be described as a metal–insulator–semiconductor (MIS) capacitor. As already discussed in Sect. 9.2.4, under electric fields, graphene does not behave as a "classical" metal film, but manifests itself as a capacitor, whose capacitance (the quantum capacitance C_q) adds in series to the insulator and semiconductor capacitance contributions.

Local measurements were carried out by using Pt coated n^+-Si tips as probes. A modulating bias

$$\Delta V = \frac{V_g}{2} + \frac{V_g}{2}\sin(\omega t)$$

with amplitude V_g and frequency $\omega = 100\,\text{kHz}$ was applied between the Si n^+ backgate and the probe. An ultra-high-sensitive ($10^{-21}\,\text{F}\,\text{Hz}^{-1/2}$) capacitance sensor connected to the conductive AFM tip allowed to measure, through a lock-in system, the capacitance variation $\Delta C(V_g)$ induced by the modulating bias.

The AFM tip in contact with bare SiO_2 forms a MOS capacitor with capacitance $A_{tip} \times C'_{MOS}$, where C'_{MOS} is the per unit area capacitance of the MOS system and $A_{tip} = \pi r_{tip}^2$ is the tip contact area with tip radius r_{tip}. When the tip is in contact with graphene, a positive gate bias causes the accumulation of electrons in graphene to screen the positively charged uncompensated ions in the depletion layer at the SiO_2/Si interface. In contrast, a negative gate bias induces the accumulation of holes in graphene to screen the electrons at the SiO_2/Si interface. In both cases, the induced carriers in graphene are distributed over an effective area A_{eff}. This is schematically illustrated in Fig. 9.14b. The total capacitance, when the tip is on

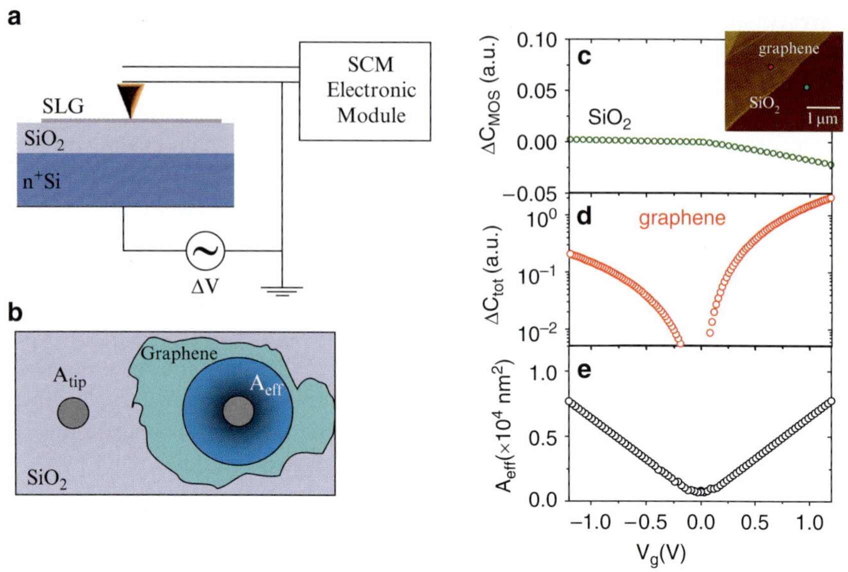

Fig. 9.14 (**a**) Schematic representation of the measuring setup for scanning capacitance spectroscopy and (**b**) illustration of the effective area A_{eff} in graphene. Two representative $\Delta C - V_g$ characteristics measured with the tip in contact with bare SiO_2 (**c**) and with the tip on graphene (**d**). The tip positions are indicated in the AFM image in the inset of (**c**). (**e**) Graphene capacitor effective area A_{eff} vs. the gate bias, obtained from the ratio of the curves in (**d**) and (**c**). Data from [64, 65]

graphene, is $C_{tot} = A_{eff} \times C'_{tot}$, where C'_{tot} is the series combination of C'_{MOS} and C'_q, the per unit area quantum capacitance of graphene. Since $C'_q \gg C'_{MOS}$ for the typical thicknesses ($\sim$ 100 or $\sim$ 300 nm) of the adopted SiO_2 layers, C_{tot} can be expressed as $C_{tot} = A_{eff} \times C'_{MOS}$. Hence, $C_{tot}/C_{MOS} = A_{eff}/A_{tip}$.

Figure 9.14c, d shows two representative $\Delta C - V_g$ characteristics measured with the tip in contact on two distinct positions on a sample with graphene deposited on 100-nm SiO_2. The curve in Fig. 9.14c was measured with the tip on bare SiO_2 and exhibits the typical behavior observed for a MIS capacitor on a n-type semiconductor. The curve in Fig. 9.14d, measured with the tip on graphene, has a minimum value close to $V_g = 0$ and increases both for negative and positive V_g values. The experimental dependence of the effective area on the gate bias can be obtained from the ratio of the curves in Fig. 9.14c, d, i.e.,

$$A_{eff} = A_{tip} \frac{|\Delta C_{tot}|}{|\Delta C_{MOS}|}. \tag{9.41}$$

A_{eff} exhibits a minimum at $V_g \sim 0$ and increases linearly with V_g in a symmetric way for both bias polarities. $A_{eff} = \pi L_{eff}{}^2$ represents the effective area of the graphene capacitor responding to the AC bias, i.e., L_{eff} is the distance from the tip contact where the induced screening charges (electrons/holes) spread in graphene. A_{eff} is much smaller than the area of the entire graphene sheet ($\sim 100\,\mu\text{m}^2$ in this specific case). It will be shown in Sect. 9.3.3 that the effective length $L_{\text{eff}} = (A_{\text{eff}}/\pi)^{1/2}$ corresponds to the "local" electron (hole) mean free path in graphene.

The measurement of the effectively biased graphene area around the tip allowed to determine the corresponding local quantum capacitance in graphene. It has been shown in [64, 65] that the local C_q depends on V_g and on the effective area as

$$C_q(V_g) = \frac{A_{\text{eff}}}{\sqrt{2\pi}\hbar v_F f(A_{\text{eff}})} \sqrt{\frac{q^3 \varepsilon_0 \varepsilon_{\text{ox}} V_g}{t_{\text{ox}}}},$$

where $f(A_{\text{eff}})$ is a function of A_{eff}.

By measuring the $\Delta C - V_g$ characteristics on an array point on the graphene sheet, the lateral variations of the quantum capacitance have been determined. Figure 9.15a shows the C_q vs. V_g curves obtained on a matrix of 5×5 different tip positions (with a separation of $1\,\mu\text{m} \times 1\,\mu\text{m}$) on a sample with a graphene monolayer deposited on 300-nm SiO_2. The histogram of the C_q values measured at fixed V_g on the different positions is also reported in the inset of Fig. 9.15a. Finally, since the quantum capacitance per unit area is related to the density of states by (9.11), the distribution of the LDOS has been extracted from the local quantum capacitance and is reported in Fig. 9.15b. It is worth noting that, similarly to the LDOS determined by SET microscopy [62], also the LDOS obtained by capacitance measurements is a quantity defined on the mesoscopic scale, and it differs from the LDOS determined by STM [63].

9.3.3 *Local Electron Mean Free Path and Mobility in Graphene*

As discussed in Sect. 9.2.5.2, one of the main issues in the research on transport properties of graphene is to understand how the various scattering sources (charged impurities, lattice defects, and lattice disorder) limit the electron mobility. The random distribution of these scattering sources causes local inhomogeneities in the transport properties, which adversely influence the reproducible operation of graphene nanodevices. Under this point of view, nanoscale resolution methods are required to probe the lateral homogeneity of the transport properties in graphene sheets. As a matter of fact, electron mobility is commonly measured on test patterns fabricated on micrometer size samples and is, therefore, an average value over that area.

Recently, SCS on graphene was applied to determine the "local" electron mean free path l (defined as the average of the distances traveled by electrons between "few" subsequent scattering events) and mobility μ [66].

As discussed in Sect. 9.3.2, in SCS measurements the high frequency modulating bias induces a local accumulation of electrons (holes) in graphene under the

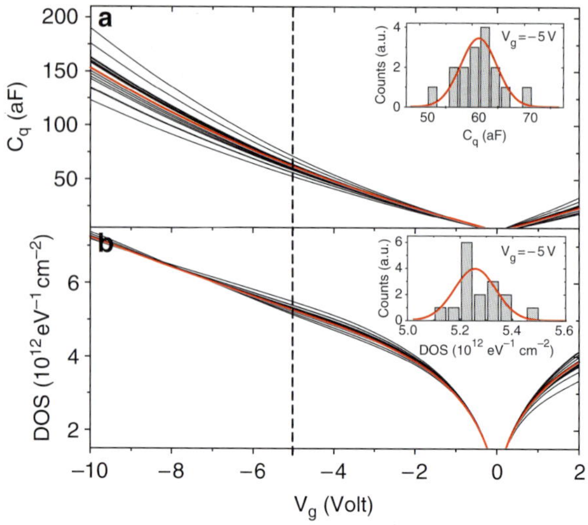

Fig. 9.15 (**a**) Quantum capacitance C_q vs. V_g curves obtained on a matrix of 5 × 5 different tip positions (with a separation of 1 μm × 1 μm) on a sample with a graphene monolayer deposited on 300-nm SiO_2. The histogram of the C_q values measured at fixed V_g on the different positions is also reported in the inset (**a**). Density of states vs. V_g extracted form C_q vs. V_g curves in (**a**). The histogram of the DOS at fixed V_g on the different positions is also reported in the inset (**b**). Data from [64]

tip contact. The induced charge carriers diffuse over a length L_{eff} from the tip contact, traveling at velocity v_F. Their diffusivity in graphene can be expressed as $D = v_F L_{\mathrm{eff}}/2$. On the other hand, the diffusivity D can be related to electron mobility μ by the generalized Einstein relation $D/\mu = n/(q\partial n/\partial E_F)$, being q the electron charge, n the electron density, and E_F the Fermi energy in graphene. As discussed in Sect. 9.2.5.4, the concepts of mobility and carrier density in graphene are meaningful when the Fermi level is far from the Dirac point but is still in the linear region of the dispersion relation. In these conditions, the density of states is linearly dependent on E_F. As a consequence, the carrier density can be expressed as $n = E_F{}^2/(\pi\hbar^2 v_F{}^2)$, and $D/\mu = E_F/2q$. Furthermore, far from the Dirac point, Boltzmann transport theory can be used to describe the electronic transport in graphene and μ can be expressed in terms of l as $\mu = qv_F l/E_F$. As a consequence, L_{eff} measured by SCS corresponds to l. This continuum treatment, using the concepts of diffusivity and mobility and the Einstein relation, is justified on the "mesoscopic" length scale investigated by SCS (from 10 to 100 nm). A quantum mechanical treatment of electron scattering phenomena, based on the interference of incident and reflected electron wavefunctions by a single scattering center, is required on a length scale of ∼ 1 nm, as recently demonstrated in some investigations by STM [67].

In Fig. 9.16a, L_{eff} is reported versus. $n^{1/2}$. L_{eff} is extracted from the A_{eff} vs. V_g curve in Fig. 9.14e, and n is related to V_g as $n = C_{ox}{}' V_g/q$, with $C_{ox}{}'$ the oxide

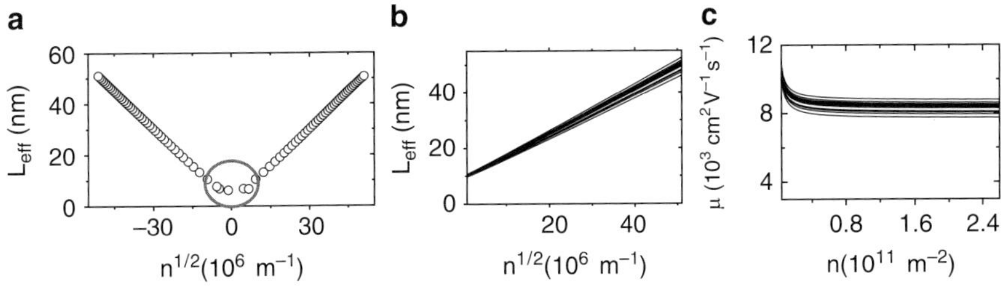

Fig. 9.16 (**a**) L_{eff} vs. $n^{1/2}$. L_{eff} is extracted from the A_{eff} vs. V_{g} curve in Fig. 9.14e, and n is related to V_{g} as $n = C_{\text{ox}}{}' V_{\text{g}}/q$, with $C_{\text{ox}}{}'$ the oxide capacitance per unit area ($C_{\text{ox}}{}' = \varepsilon_{\text{ox}}\varepsilon_0/t_{\text{ox}}$, being ε_0 the vacuum permittivity, ε_{ox} and t_{ox} the relative permittivity and the thickness of the SiO_2 film). (**b**) Local electron mean free path l vs. $n^{1/2}$ curves extracted from the $|\Delta C| - V_{\text{g}}$ characteristics on an array of 100 tip positions on graphene. The observed spread in the local values of l can be explained in terms of the random spatial distribution of the charged impurities and/or vacancies. (**c**) Local mobility μ vs. n. Data from [66]

capacitance per unit area ($C_{\text{ox}}{}' = \varepsilon_{\text{ox}}\varepsilon_0/t_{\text{ox}}$, being ε_0 the vacuum permittivity, ε_{ox} and t_{ox} the relative permittivity and the thickness of the SiO_2 film) [66].

As discussed in Sect. 9.2.5.4, the electron mean free path limited by charged impurities and vacancies scattering is expected to increase as $n^{1/2}$, instead that limited by scattering with phonons is proportional to $n^{-1/2}$. The obtained results clearly show an increase of l with $n^{1/2}$, indicating that the main scattering sources are charged impurities and/or lattice vacancies. The impurities are either adsorbed on graphene surface or located at the interface with the SiO_2 substrate.

Figure 9.16b reports the local l vs. $n^{1/2}$ curves extracted from the $|\Delta C| - V_{\text{g}}$ characteristics measured on an array of 100 tip positions. The observed spread in the local values of l can be explained in terms of the random spatial distribution of the charged impurities and/or vacancies. Finally, from (9.18), the local mobility, μ, was obtained from the local mean free path and its trend vs. n is reported in Fig. 9.16c. It is worth noting that, except at very low carrier densities, μ is almost independent of n. The spread in the measured curves is again associated with the distribution of charged impurities/vacancies on graphene.

In order to understand the effect of a distribution of defects (vacancies) on the local transport properties, as-deposited graphene (DG) flakes on SiO_2 (100 nm)/Si were also irradiated with C^+ ions at 500 keV [66, 68]. The projected range of the implanted ions is ~1 μm, quite deep into the Si substrate. This minimizes the damage both in the SiO_2 layer and at its interface with Si [69]. Ion fluences ranging from 1×10^{13} to $1 \times 10^{14}\ \text{cm}^{-2}$ were used.

In Fig. 9.17, the local mobility measured by SCS on an array of 100 tip positions is reported vs. V_{g} and n for irradiated graphene with the lowest (a) and the highest fluence (b). A larger spread with respect to pristine graphene is clearly observed. To better quantify this spread, the histograms of μ at $V_{\text{g}} = 1\ \text{V}$ on pristine and irradiated graphene are also reported in Fig. 9.17c–e. The corresponding l values are also indicated (upper scale). For pristine graphene, the distribution exhibits a single peak at $\mu_0 \approx 8{,}200\ \text{cm}^2\ \text{V}^{-1}\ \text{s}^{-1}$ with FWHM $\approx 180\ \text{cm}^2\ \text{V}^{-1}\ \text{s}^{-1}$.

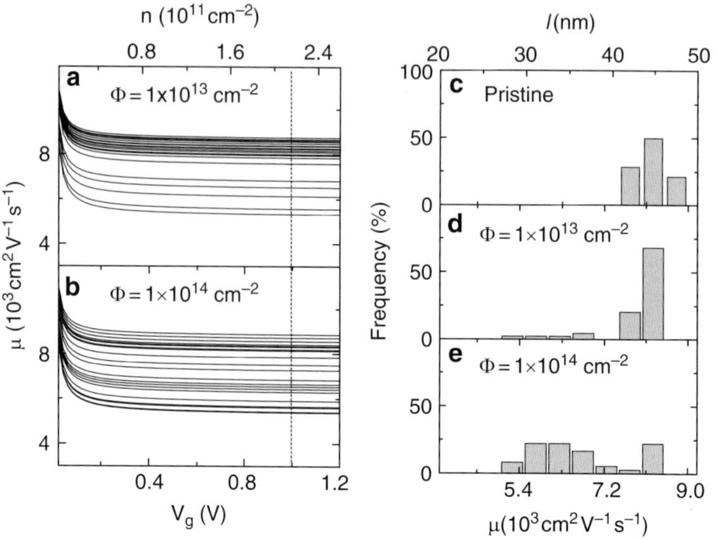

Fig. 9.17 Local mobility vs. V_g and n measured on an array of tip positions by SCS for irradiated graphene with 1×10^{13} (**a**) and 1×10^{14} C^+ ions cm^{-2} (**b**). Histograms of μ at $V_g = 1$ V on pristine (**c**) and irradiated graphene with the lowest (**d**) and highest dose (**e**). The corresponding l values are also indicated (*upper scale*). Data from [66].

This value is comparable to the typically reported mobility values at room temperature ($\sim 10^4\,\mathrm{cm^2\,V^{-1}\,s^{-1}}$), determined by sheet resistance and/or Hall effect measurements on graphene exfoliated from HOPG and placed on SiO_2 [14] or on other substrates [70]. In irradiated graphene, in addition to a narrow peak at $\mu_0 \approx 8{,}200\,\mathrm{cm^2\,V^{-1}\,s^{-1}}$, a broader distribution extending from ~5,000 to ~7,500 and centered around $\mu \approx 6{,}200\,\mathrm{cm^2\,V^{-1}\,s^{-1}}$ is obtained, which is associated to the locally damaged regions. It is worth noting that the number of counts under this part of the distribution increases with the fluence.

9.3.4 Local Electronic Properties of Epitaxial Graphene/4H-SiC (0001) Interface

As discussed in Sect. 9.2.1, the interface between EG and 4H-SiC (0001) strongly affects the overall electronic and transport properties of the 2DEG in EG. In particular, the presence of a buffer layer (ZL), which may be more or less defective with more or less Si dangling bonds at the interface with the Si face, makes EG n-type doped and causes an overall degradation of the carrier mobility.

Recently, the properties of the interface between EG and ZL have been investigated with nanoscale lateral resolution and compared with those of graphene mechanically exfoliated from HOPG and deposited on 4H-SiC (0001) [71, 72]. In fact, in the case of DG no ZL is present at the interface. The EG was grown on n^+-doped (0001) 4H-SiC substrate, 8° off axis, with a weakly doped n^--epitaxial layer on top, in an inductively heated reactor at a temperature of 2,000°C and under 1 atm. Ar pressure [32]. A piece of the same 4H-SiC wafer (not subjected to annealing) was used as the substrate for DG. This local investigation has been performed by measuring the current flowing across the interface using torsion resonance conductive atomic force microscopy (TR-CAFM) and scanning current spectroscopy (SCS).

The TR-CAFM (see schematic in Fig. 9.18a) is a dynamic scanning probe method that allows nondestructive electrical measurements from a conductive tip oscillating in a torsional or twisting mode in close proximity to the sample surface (0.3–3 nm). When a bias is applied between the tip and the sample backside, a map of the current flowing from the tip to sample surface is acquired. This noncontact method has a distinct advantage over the conventional conductive AFM performed in contact mode because of the absence of shear forces that can damage the graphene sheets. Commercially available Si n^+-doped probes with platinum (Pt) coating were used, with a typical radius of the apex of 10–20 nm. The TR-CAFM was used to measure, simultaneously, topographic and current maps on the sample surface. Figure 9.18b shows a TR-CAFM current map taken on DG at tip bias of 1 V. The brighter portions in the current map (region carrying a higher current) correspond to graphene in the morphological map. After identifying the SiC regions coated with DG, local I–V measurements were carried out by SCS. In SCS the conductive tip is placed in contact on a discrete array of positions on the sample surface noninvasively, lifting the tip by 20 nm at every interval. A swift transition between TR-CAFM and SCS is possible without altering the sample position. The representative positions of the tip during the SCS scan are depicted in Fig. 9.18b and the recorded I–V curves are shown in Fig. 9.18c. All I–V curves exhibit a rectifying behavior, with a low leakage current under reverse bias and a sudden increase in the current for positive bias higher than a threshold voltage. It is worth noting that the acquired curves can be divided into two groups, which correspond to the tip positions on DG and bare SiC substrate. It is clear that DG forms a Schottky contact on top of the 4H-SiC layer.

Based on these results on DG, TR-CAFM was used to get local information on the graphene coverage on EG. Figure 9.18e shows a typical TR-CAFM current map along with the morphology collected on EG (Fig. 9.18d). The surface is corrugated with a measured roughness of 16 nm, which is significantly higher than the 0.3-nm value of the pristine 4H-SiC substrate. This comes from the large step-bunching phenomenon commonly observed on off-axis SiC substrates after high-temperature annealing ($T > 1{,}400$°C). Graphene grows on the SiC surface following these steps. By comparing the morphology and current maps, brighter regions in the current map appear evident, corresponding to areas covered by a few MLs of graphene. The I–V characteristics collected by SCS in the same sample region are reported in Fig. 9.18f.

Similarly to the sample with DG, the I–V curves on the sample with EG can be divided into two distinct families: the curves acquired on an area covered with

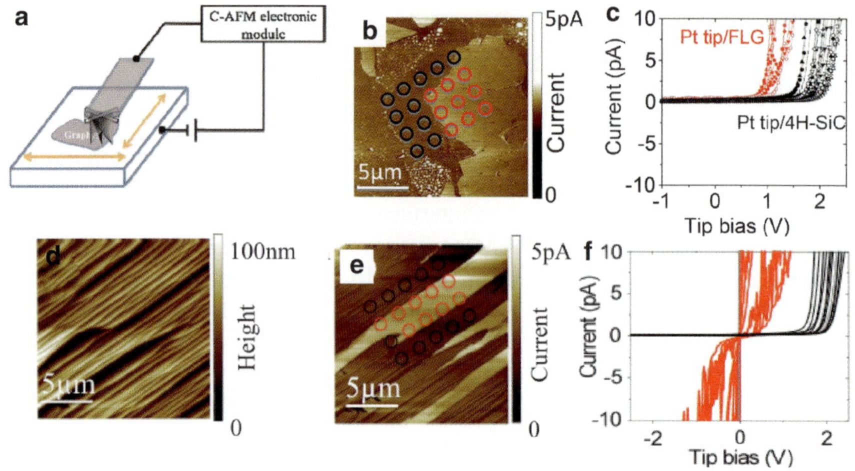

Fig. 9.18 (**a**) Schematic representation of the TR-CAFM setup. (**b**) TR-CAFM current map taken on DG on 4H-SiC (0001). The brighter portions in the current map (region carrying a higher current) correspond to graphene coated regions. (**c**) Local I–V curves measured at different tip positions (depicted in b). (**d**) Morphology and (**e**) TRCAFM current map on epitaxial graphene grown on 4H-SiC (0001). The *brighter regions* in the current map correspond to areas covered by a few MLs of graphene. (**f**) I–V curves measured by SCS in the sample positions indicated in (**e**). Data from [72]

graphene and those acquired on an area free of graphene. The I–V characteristics on the graphene-free regions remained similar to those measured on the bare 4H-SiC surface in the sample not subjected to thermal treatment (see Fig. 9.18c). The curves obtained on the regions coated with EG show completely different I–V characteristics. The leakage current is now very large with, in some cases, a clear Ohmic behavior. This is a direct evidence that the Schottky barrier height (SBH) at the EG/4H-SiC interface has been reduced with respect to DG/4H-SiC with, in some cases, a possibility of direct tunnel injection.

These SBH values have been evaluated for each of the I–V curves in Fig. 9.18c, f, using the procedure of Giannazzo et al. [73] and [74]. The histograms of the corresponding values have been reported in Fig. 9.19a, b, respectively. While the SBHs for Pt-tip/4H-SiC remain similar for both samples, the average SBH value for EG on 4H-SiC appears very much reduced ($\Phi_{\mathrm{EG}} = 0.36 \pm 0.1\,\mathrm{eV}$) with respect to DG ($\Phi_{\mathrm{DG}} = 0.85 \pm 0.06\,\mathrm{eV}$). This fact is especially interesting considering that the SBHs on positions uncovered with graphene remain similar. The SBH difference must then be explained in terms of the band diagrams schematically drawn in Fig. 9.19c, d. Here, E_{C} is the energy of the conduction band edge, E_{F} is the Fermi energy at the bulk 4H-SiC, $E_{\mathrm{F,DG}}$ is the Fermi energy of DG, and $E_{\mathrm{F,EG}}$ the Fermi energy of EG. Due to the weak (van der Waals-like) interaction with the substrate, DG can be considered neutral (or close to neutral) in the absence of external bias. In this case, $E_{\mathrm{F,gr}}$ should (more or less) coincide with the Dirac point energy (E_{Dirac}). Hence, in the first approximation, the SBH between DG and the n-type 4H-SiC

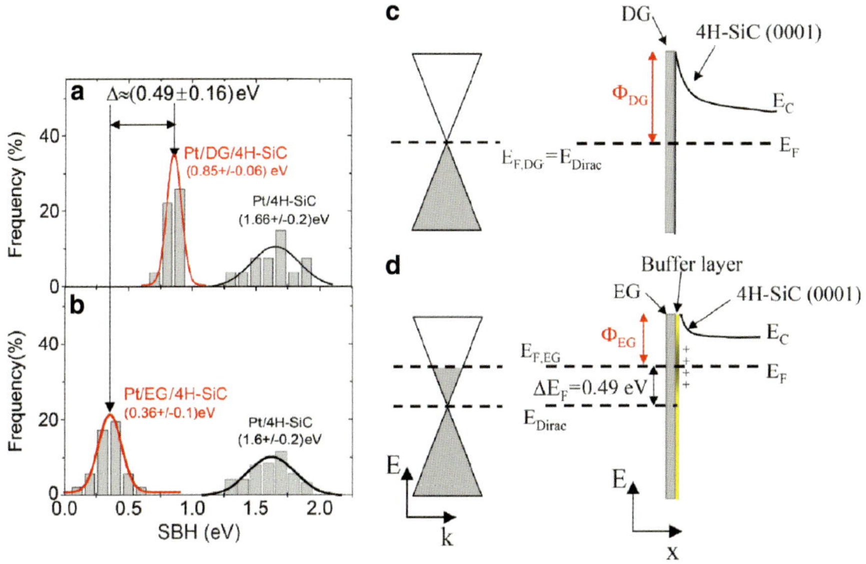

Fig. 9.19 Histograms of the local Schottky barrier height (SBH) values of graphene deposited (**a**) and grown on 4H-SiC (0001) (**b**). Schematic representation of the band diagrams for graphene deposited (**c**) and grown on 4H-SiC (0001) (**d**). Data from [72].

epilayer should be given by the difference between the neutral graphene work function ($W_{gr} \approx 4.5\,\text{eV}$) and the 4H-SiC electron affinity ($\chi_{4H\text{-}SiC} \approx 3.7\,\text{eV}$). The value of Φ_{DG} is in good agreement with the value predicted by the simple model, i.e., 0.8 eV. The SBH in the case of EG can be explained in terms of a pinning of $E_{F,EG}$ at an average value $\Delta E_F \approx 0.49\,\text{eV}$ higher than $E_{F,DG}$, as shown in Fig. 9.19d. This pinning is a direct evidence of the presence of positively charged states located at the interface between the Si face of 4H-SiC and the ZL. The spread of SBH values for EG/4H-SiC is an evidence of a nonuniform density of the interface states. Based on these data, an average value of the interface charge density of $1.8 \times 10^{13}\,\text{cm}^{-2}$ has been estimated, but the local values were found to vary from about 1×10^{12} to $5 \times 10^{13}\,\text{cm}^{-2}$ at a few μ m distance.

9.4 Conclusion

The main applications of SPM to the local characterization of graphene electronic properties have been discussed. Scanning SET microscopy and STM were applied to demonstrate the local inhomogeneity of the carrier density close to the Dirac point and to measure the LDOS. The lateral variations of the graphene quantum capacitance were measured by SCS. A method based on SCS was employed to determine the local electron mean free path and mobility on graphene. Finally, the electronic properties of the interface between graphene epitaxially grown on 4H-SiC (0001) and the substrate were investigated by SCS. The results of nanoscale measurements allowed to explain some peculiar aspects of the electrical behavior of macroscopic devices fabricated on graphene.

Acknowledgements

F. Roccaforte, R. Lo Nigro, A. La Magna, S. Sonde, and P. Fiorenza from CNR-IMM, Catania, are acknowledged for useful discussions. This work has been supported, in part, by the European Science Foundation (ESF) under the EUROCORE program EuroGRAPHENE, with GRAPHIC-RF coordinated project.

References

1. A.K. Geim, K.S. Novoselov, The rise of grapheme. Nat. Mater. **6**, 183 (2007)
2. A.K. Geim, Graphene: status and prospects. Science **324**, 1530 (2009)
3. N. Tombros, C. Jozsa, M. Popinciuc, H.T. Jonkman, B.J. van Wees, Electronic spin transport and spin precession in single graphene layers at room temperature. Nature **448**, 571 (2007)
4. S. Cho, Y.-F. Chen, M. S. Fuhrer, Gate-tunable graphene spin valve. Appl. Phys. Lett. **91**, 123105 (2007)
5. H. Min, G. Borghi, M. Polini, A.H. MacDonald, Pseudospin magnetism in grapheme. Phys. Rev. B. **77**, 041407(R) (2008)
6. R.R. Nair, P. Blake, A.N. Grigorenko, K.S. Novoselov, T.J. Booth, T. Stauber, N.M.R. Peres, A.K. Geim, Fine structure constant defines visual transparency of graphene. Science. **320**, 1308 (2008)
7. A.A. Balandin, S. Ghosh, W. Bao, I. Calizo, D. Teweldebrahn, F. Miao, C.N. Lau, Superior thermal conductivity of single-layer graphene. Nano Lett. **8**, 902 (2008)
8. K. Saito, J. Nakamura, A. Natori, Ballistic thermal conductance of a graphene sheet. Phys. Rev. B. **76**, 115409 (2007)
9. Y.M. Zuev, W. Chang, P. Kim, Thermoelectric and magnetothermoelectric transport measurements of graphene. Phys. Rev. Lett. **102**, 096807 (2009)
10. C. Lee, X. Wei, J.W. Kysar, J. Hone, Measurement of the elastic properties and intrinsic strength of monolayer graphene. Science. **321**, 385 (2008)
11. I.W. Frank, D.M. Tanenbaum, A.M. Van Der Zande, P.L. McEuen, Mechanical properties of suspended graphene sheets. J. Vac. Sci. Technol. B. **25**, 2558 (2007)
12. T.J. Booth, P. Blake, R.R. Nair, D. Jiang, E.W. Hill, U. Bangert, A. Bleloch, M. Gass, K.S. Novoselov, M.I. Katsnelson, A.K. Geim, Macroscopic graphene membranes and their extraordinary stiffness. Nano Lett. **8**, 2442 (2008)
13. D. Xu, I. Skachko, A. Barker, E.Y. Aandrei, Approaching ballistic transport in suspended grapheme. Nat. Nanotechnol. **3**, 491 (2008)
14. J.H. Chen, C. Jang, S. Xiao, M. Ishigami, M.S. Fuhrer, Intrinsic and extrinsic performance limits of graphene devices on SiO_2. Nat Nanotechnol **3**, 206 (2008)
15. M.D. Stoller, S. Park, Y. Zhu, J. An, R.S. Ruoff, Graphene-based ultracapacitors. Nano Lett. **8**, 3498 (2008)
16. K.S. Novoselov, et al., Electric field effect in atomically thin carbon films. Science. **306**, 666 (2004)
17. H.P. Boehm, A. Clauss, G. Fischer, U. Hofmann, in *Surface Properties of Extremely Thin Graphite Lamellae*. Proceedings of the Fifth Conference on Carbon (Pergamon Press, Oxford, 1962)
18. I.M. Lifshitz, J. Exp. Theor. Phys. **22**, 475 (1952)

19. A. Castro Neto, et al. The electronic properties of grapheme. Rev. Mod. Phys. **81**, 109 (2009)
20. M.I. Katsnelson, K.S. Novoselov, A.K. Geim, Chiral tunnelling and the Klein paradox in grapheme. Nat. Phys. **2**, 620 (2006)
21. R. Murali, Y. Yang, K. Brenner, T. Beck, J.D. Meindl, Breakdown current density of graphene nanoribbons. Appl. Phys. Lett. **94**, 243114 (2009)
22. S. Roddaro, P. Pingue, V. Piazza, V. Pellegrini, F. Beltram, The optical visibility of graphene: interference colors of ultrathin graphite on SiO_2. Nano Lett. **7**, 2707 (2007)
23. Z.H. Ni, H.M. Wang, J. Kasim, H.M. Fan, T. Yu, Y.H. Wu, Y.P. Feng, Z.X. Shen, Graphene thickness determination using reflection and contrast spectroscopy. Nano Lett. **7**, 2758 (2007)
24. A.C. Ferrari, J.C. Meyer, V. Scardaci, C. Casiraghi, M. Lazzeri, F. Mauri, S. Piscanec, D. Jiang, K.S. Novoselov, S. Roth, A.K. Geim, Raman spectrum of graphene and graphene layers. Phys. Rev. Lett. **97**, 187401 (2006)
25. F. Giannazzo, S. Sonde, V. Raineri, G. Patanè, G. Compagnini, F. Aliotta, R. Ponterio, E. Rimini, Optical, morphological and spectroscopic characterization of graphene on SiO_2. Phys. Status Solidi C. **7**, 1251 (2010)
26. P. Nemes-Incze, Z. Osvath, K. Kamaras, L.P. Biro, Anomalies in thickness measurements of graphene and few layer graphite crystals by tapping mode atomic force microscopy. Carbon. **46**, 1435 (2008)
27. M. Ishigami, et al., Atomic structure of graphene on SiO_2. Nano Lett. **7**, 1643 (2007)
28. A. Fasolino, J.H. Los, M.I. Katsnelson, Intrinsic ripples in grapheme. Nat. Mater. **6**, 858 (2007)
29. D. Garcia-Sanchez, A.M. van der Zande, A. San Paulo, B. Lassagne, P.L. McEuen, A. Bachtold, Imaging mechanical vibrations in suspended graphene sheets. Nano Lett. **8**, 1399 (2008)
30. C. Berger, Z. Song, T. Li, X. Li, A.Y. Ogbazghi, R. Feng, Z. Dai, A.N. Marchenkov, E.H. Conrad, P.N. First, W.A. de Heer, Ultrathin epitaxial graphite: 2D electron gas properties and a route toward graphene-based nanoelectronics. J. Phys. Chem. B. **108**, 19912 (2004)
31. C. Berger, Z. Song, X. Li, X. Wu, N. Brown, C. Naud, D. Mayou, T. Li, J. Hass, A.N. Marchenkov, E.H. Conrad, P.N. First, W.A. de Heer, Electronic confinement and coherence in patterned epitaxial graphene. Science. **312**, 1191 (2006)
32. C. Virojanadara, M. Syvajarvi, R. Yakimova, L.I. Johansson, A.A. Zakharov, T. Balasubramanian, Homogeneous large-area graphene layer growth on 6H -SiC(0001). Phys. Rev. B. **78**, 245403 (2008)
33. K.V. Emtsev, A. Bostwick, K. Horn, J. Jobst, G.L. Kellogg, L. Ley, J.L. McChesney, T. Ohta, S.A. Reshanov, J. Rohrl, E. Rotenberg, A.K. Schmid, D. Waldmann, H.B. Weber, T. Seyller, Towards wafer-size graphene layers by atmospheric pressure graphitization of SiC(0001). Nat. Mater. **8**, 203 (2009)
34. Y.Q. Wu, P.D. Ye, M.A. Capano, Y. Xuan, Y. Sui, M. Qi, J.A. Cooper, T. Shen, D. Pandey, G. Prakash, R. Reifenberger, Top-gated graphene field-effect-transistors formed by decomposition of SiC. Appl. Phys. Lett. **92**, 092102 (2008)
35. K.S. Kim, Y. Zhao, H. Jang, S.Y. Lee, J.M. Kim, K.S. Kim, J.-H. Ahn, P. Kim, J.-Y. Choi, B. H. Hong, Large-scale pattern growth of graphene films for stretchable transparent electrodes. Nature. **457**, 706 (2009)
36. X. Li, et al., Large-area synthesis of high-quality and uniform graphene films on copper foils. Science. **324**, 1312 (2009)
37. X. Li, W. Cai, L. Colombo, R.S. Ruoff, Evolution of graphene growth on Ni and Cu by carbon isotope labeling. Nano Lett. **9**, 4268 (2009)
38. Q. Yu, J. Lian, S. Siriponglert, H. Li, Y.P. Chen, S.-S. Pei, Graphene segregated on Ni surfaces and transferred to insulators. Appl. Phys. Lett. **93**, 113103 (2008)
39. Y. Lee, S. Bae, H. Jang, S. Jang, S.-E. Zhu, S.H. Sim, Y. Il Song, B.H. Hong, J.-H. Ahn, Wafer-scale synthesis and transfer of graphene films. Nano Lett. **10**, 490 (2010)
40. M. Choucair, P. Thordarson, J.A. Stride, Gram-scale production of graphene based on solvothermal synthesis and sonication. Nat. Nanotechnol. **4**, 30 (2008)
41. P.R. Wallace, The band theory of graphite. Phys. Rev. **71**, 622 (1947)
42. T. Ohta, A. Bostwick, T. Seyller, K. Horn, E. Rotenberg, Controlling the electronic structure of bilayer graphene. Science. **313**, 951 (2006)

43. M. Sprinkle, D. Siegel, Y. Hu, J. Hicks, A. Tejeda, A. Taleb-Ibrahimi, P. Le Fevre, F. Bertran, S. Vizzini, H. Enriquez, S. Chiang, P. Soukiassian, C. Berger, W.A. de Heer, A. Lanzara, E.H. Conrad, First direct observation of a nearly ideal graphene band structure. Phys. Rev. Lett. **103**, 226803 (2009)
44. F. Varchon, R. Feng, J. Hass, X. Li, B. Ngoc Nguyen, C. Naud, P. Mallet, J.-Y. Veuillen, C. Berger, E.H. Conrad, L. Magaud, Electronic structure of epitaxial graphene layers on SiC: effect of the substrate. Phys. Rev. Lett. **99**, 126805 (2007)
45. N. Camara, J.-R. Huntzinger, G. Rius, A. Tiberj, N. Mestres, F. Perez-Murano, P. Godignon, J. Camassel, Anisotropic growth of long isolated graphene ribbons on the C face of graphite-capped 6H-SiC. Phys. Rev. B. **80**, 125410 (2009)
46. K.V. Emtsev, F. Speck, Th. Seyller, L. Ley, J.D. Riley, Interaction, growth, and ordering of epitaxial graphene on SiC{0001} surfaces: A comparative photoelectron spectroscopy study. Phys. Rev. B. **77**, 155303 (2008)
47. J. Borysiuk, R. Bożek, W. Strupiński, A. Wysmołek, K. Grodecki, R. StJpniewski, J.M. Baranowski, Transmission electron microscopy and scanning tunneling microscopy investigations of graphene on 4H-SiC (0001). J. Appl. Phys. **105**, 023503 (2009)
48. A. Mattausch, O. Pankratov, Ab initio study of graphene on SiC. Phys. Rev. Lett. **99**, 076802 (2007)
49. S. Kim, J. Ihm, H.J. Choi, Y.-W. Son, Origin of anomalous electronic structures of epitaxial graphene on silicon carbide. Phys. Rev. Lett. **100**, 176802 (2008)
50. S.Y. Zhou, G.-H. Gweon, A.V. Fedorov, P.N. First, W.A. De Heer, D.-H. Lee, F. Guinea, A.H. Castro Neto, A. Lanzara, Substrate-induced bandgap opening in epitaxial grapheme. Nat. Mater. **6**, 770 (2007)
51. C. Riedl, C. Coletti, T. Iwasaki, A.A. Zakharov, U. Starke, Quasi-free-standing epitaxial graphene on SiC obtained by hydrogen intercalation. Phys. Rev. Lett. **103**, 246804 (2009)
52. V.M. Pereira, J.M.B. Lopes dos Santos, A.H. Castro Neto, Modeling disorder in grapheme. Phys. Rev. B. **77**, 115109 (2008)
53. T. Fang, A. Konar, H.L. Xing, D. Jena, Carrier statistics and quantum capacitance of graphene sheets and ribbons. Appl. Phys. Lett. **91**, 092109 (2007)
54. J. Xia, F. Chen, J. Li, N. Tao, Measurement of the quantum capacitance of grapheme. Nat. Nanotechnol. **4**, 505 (2009)
55. T. Stauber, N.M.R. Peres, F. Guinea, Electronic transport in graphene: a semiclassical approach including midgap states. Phys. Rev. B. **76**, 205423 (2007)
56. E.H. Hwang, S. Adam, S. DasSarma, Carrier transport in two-dimensional graphene layers. Phys. Rev. Lett. **98**, 186806 (2007)
57. N.M.R. Peres, A.H. Castro Neto, F. Guinea, Conductance quantization in mesoscopic grapheme. Phys. Rev. B. **73**, 195411 (2006).
58. F. Miao, S. Wijeratne, Y. Zhang, U.C. Coskun, W. Bao, C.N. Lau, Phase-coherent transport in graphene quantum billiards. Science. **317**, 1530 (2007)
59. J. Tworzydło, B. Trauzettel, M. Titov, A. Rycerz, C.W.J. Beenakker, Sub-Poissonian shot noise in graphene. Phys. Rev. Lett., **96** 246802, (2006)
60. J.-H. Chen, W.G. Cullen, C. Jang, M.S. Fuhrer, E.D. Williams, Defect scattering in grapheme. Phys. Rev. Lett. **102**, 236805 (2009)
61. M.J. Yoo, T.A. Fulton, H.F. Hess, R.L. Willett, L.N. Dunkleberger, R.J. Chichester, L.N. Pfeiffer, K.W. West, Scanning single-electron transistor microscopy: imaging individual charges. Science. **276**, 579 (1997)
62. J. Martin, N. Akerman, G. Ulbricht, T. Lohamann, J.H. Smet, K. Von Klitzing, A. Yacobi, Observation of electron–hole puddles in graphene using a scanning single-electron transistor. Nat. Phys. **4**, 144 (2008)
63. Y. Zhang, V.W. Brar, C. Girit, A. Zettl, M.F. Crommie, Origin of spatial charge inhomogeneity in grapheme. Nat. Phys. **5**, 822 (2009)
64. F. Giannazzo, S. Sonde, V. Raineri, E. Rimini, Screening length and quantum capacitance in graphene by scanning probe microscopy. Nano Lett. **9**, 23 (2009)
65. S. Sonde, F. Giannazzo, V. Raineri, E. Rimini, Dielectric thickness dependence of capacitive behavior in graphene deposited on silicon dioxide. J. Vac. Sci. Technol. B. **27**, 868 (2009)

66. F. Giannazzo, S. Sonde, V. Raineri, E. Rimini, Irradiation damage in graphene on SiO_2 probed by local mobility measurements. Appl. Phys. Lett. **95**, 263109 (2009)
67. G.M. Rutter, J.N. Crain, N.P. Guisinger, T. Li, P.N. First, J.A. Stroscio, Scattering and interference in epitaxial graphene. Science. **317**, 219 (2007)
68. G. Compagnini, F. Giannazzo, S. Sonde, V. Raineri, E. Rimini, Ion irradiation and defect formation in single layer grapheme. Carbon **47**, 3201 (2009)
69. S. Sonde, F. Giannazzo, V. Raineri, E. Rimini, Nanoscale capacitive behaviour of ion irradiated graphene on silicon oxide substrate. Phys. Status Solidi B. **247**, 907 (2010)
70. L.A. Ponomarenko, R. Yang, T.M. Mohiuddin, M.I. Katsnelson, K.S. Novoselov, S.V. Morozov, A.A. Zhukov, F. Schedin, E.W. Hill, A.K. Geim, Effect of a high-κ environment on charge carrier mobility in graphene. Phys. Rev. Lett. **102**, 206603 (2009)
71. S. Sonde, F. Giannazzo, V. Raineri, R. Yakimova, J.-R. Huntzinger, A. Tiberj, J. Camassel, Electrical properties of the graphene/4H-SiC (0001) interface probed by scanning current spectroscopy. Phys. Rev. B. **80**, 241406(R) (2009)
72. S. Sonde, F. Giannazzo, V. Raineri, E. Rimini, Investigation of graphene–SiC interface by nanoscale electrical characterization. Phys. Status Solidi B. **247**, 912 (2010)
73. F. Giannazzo, F. Roccaforte, V. Raineri, S.F. Liotta, EPL. **74**, 686 (2006)
74. F. Giannazzo, F. Roccaforte, F. Iucolano, V. Raineri, F. Ruffino, M.G. Grimaldi, Nanoscale current transport through Schottky contacts on wide bandgap semiconductors. J. Vac. Sci. Technol. B. **27**, 789 (2009)

Chapter 10
Magnetic Force Microscopy Studies of Magnetic Features and Nanostructures

Lanping Yue and Sy-Hwang Liou

Abstract The study of small magnetic features and nanostructures has attracted much attention due to interest in both technological applications and fundamental research in micromagnetism. For their characterization, a visualization technique with high lateral resolution is required. Among the wealth of techniques, magnetic force microscopy (MFM) has become a powerful tool for visualizing submicron-sized domain structures. This is mainly due to its ease of use without any specific sample preparation and the high lateral resolution of a few 10 nm. MFM is a sensitive and useful technique for direct observation of magnetic domains and their magnetic behavior, which can help elucidate properties of magnetic films and nanostructures. This chapter reviews MFM techniques and applications that demonstrate the achievement of MFM in magnetic materials and nanoscience research. The review focuses on the current MFM study involved with magnetic features and nanostructures, including magnetic interactions in nanostructured thin films and nanomagnetic patterns with special emphasis on the recent research in micromagnetism acquiring from our SPM laboratory.

10.1 Magnetic Force Microscopy

10.1.1 Introduction

Magnetic force microscopy (MFM) is one of the most successfully used magnetic imaging techniques for studying and analyzing both scientific and technologically important magnetic phenomena in surface structures at micro- and nanometer scales. Since the development of MFM in 1987 [1, 2], which is derived from atomic force microscopy (AFM) [3], it has been widely used for imaging magnetic domain structures on a variety of magnetic materials, such as high-density recording media, ultrathin films, nanoparticles, patterned elements, template-mediated assembly of nanoclusters, as well as other magnetic features and nanostructures [4–47].

Nanostructured magnetic materials have attracted intense research interest over recent years due to their significant applications in magnetic recording media

[48], magnetic random access memory (MRAM) [49], and spintronics [50, 51] or magnetoelectronics [52]. With the structural sizes on the nanometer scale, it is challenging to characterize magnetic properties of such nanostructures with small signals. For their characterization, a visualization technique with high lateral resolution is required. The magnetic force microscope with high spatial resolution MFM probes is a powerful tool to locally observe the magnetization distribution or magnetization reversal process of a single nanoparticle. Also, the sample preparation for MFM measurement is relatively simple. Furthermore, the external magnetic field can be applied easily to the sample in order to study the magnetic switching process in situ or observe the magnetic domain structure changes directly by varying the field. MFM plays a key role in nanomagnetism research that is concerned with the underlying physics of the fabrication and analysis of nano-scaled structures [38].

In this chapter, we start with a short introduction to the basic principles of MFM and review this technique briefly. Then we describe a few improvements in the fabrication of high-resolution MFM tips in our laboratory. The main attention in this chapter is focused on showing a gallery of nanostructure MFM images, with examples from recent investigations that have demonstrated the high spatial resolution or have the potential to improve it in the future. We hope to provide an insight into MFM imaging by giving a reliable interpretation of MFM images.

10.1.2 MFM Basic Principles

A magnetic force microscope is a scanning probe microscope. The principle of MFM is very much like that of AFM [3, 8]. A flexible MFM cantilever probe oscillates near its resonant frequency and scans laterally at a constant height above the sample surface. The tip interacts with the magnetic stray field emanating from the sample, which changes the resonance frequency of the cantilever. The shifts in resonance frequency is in the range of 1–50 Hz for cantilevers having a resonance frequency $f_0 \sim 100\,\text{kHz}$ [53]. The frequency shift (Δf) of the cantilever is proportional to the vertical gradient of the magnetic force on the tip, which can be detected in one of the following ways: phase detection, which measures the cantilever's phase of oscillation relative to the piezo drive; amplitude detection, which tracks variation in oscillation amplitude; and frequency modulation, which detects shift in resonant frequency directly. We use phase detection more often due to its ease of use, greater signal-to-noise ratio, and reduced artifacts content. The variations in the magnetic interaction force between the cantilever tip and the sample surface of interest can be measured by a position-sensitive optical deflection detector, which is used to detect the cantilever motion and hence to measure the force gradient due to the magnetostatic interactions between the magnetic sample and the MFM tip. The force gradient, as small as $10^{-6}\,\text{N/m}$, can be easily detected [1].

The MFM images do not correspond directly to the sample magnetization in a simple way. The magnetostatic force between the magnetic tip and magnetic sample is relatively long-ranged (10–100 nm) and the signal contains a contribution from regions far from the tip. In general, the magnetostatic force gradient is detected in MFM [21]. Magnetic interaction between a sample and a magnetic tip is extremely complicated because of the long range of magnetic forces and their complex sources, as well as the potential distortion effect of tip–sample interaction [32, 38, 47, 54, 55]. However, in many cases such an influence can be neglected. MFM uses the lift mode of interleave scanning to obtain images. The lift scan height between tip and sample surface is typically 10–100 nm. Operating MFM in the tapping/lift mode allows the imaging of relatively weak but long-range magnetic interactions while minimizing the influence of topography, which can efficiently separate the topographic contrast and magnetic signal [16].

Due to the shape of the tip, the tip moment is stronger and more stable for axial magnetization, tips are magnetized along their axes, in general. In other words, tip fields are well-defined localized, and the cantilever deflection senses the force mainly in one direction [38]. To simplify, assuming that the force and the direction of the cantilever are aligned in the z direction, i.e., the tip's magnetic dipole moment is along the z-axis and magnetization of x and y components of the tip are zero; and also the magnetostatic forces arise purely from the magnetic dipoles in the tip interacting with dipoles in the sample. In this case, the magnetic force on the tip is the gradient of the magnetostatic interaction energy, and the force gradient is proportional to the second spatial derivative of the individual components of the magnetic field emanating from the sample [56]:

$$F'_{z,z} = \mu_0 \int_{V'} M_z \frac{\partial^2 H_z}{\partial z^2} \mathrm{d}V'$$

where M_z is the magnetic moment of the tip, H the stray field from the sample, and V is the volume of magnetic material on the tip. The permeability μ_0 has a value of $4\pi \times 10^{-7}$ for free space. This simplified equation has been used for the qualitative interpretation of the interaction between the tip and the sample reasonably.

10.1.3 MFM Image Contrast

MFM uses a magnetic tip to measure the magnetic force gradient distribution on a sample's surface by oscillating the cantilever normal to the surface at its resonant frequency. In practice, the MFM tips are most often magnetized along the z-axis direction of the tip which is perpendicular to the sample surface. Just the z component of magnetization in the magnetic sample can be detected. In MFM images, the magnetic transitions appear as either dark or bright, corresponding to attractive or repulsive force gradient, respectively.

The cantilever's resonant frequency changes in response to a force gradient. The attractive force with negative force gradient makes the cantilever effectively "softer" thereby reducing the resonant frequency of the cantilever (usually displayed as a dark signal). Conversely, the effect of repulsive forces makes the cantilever

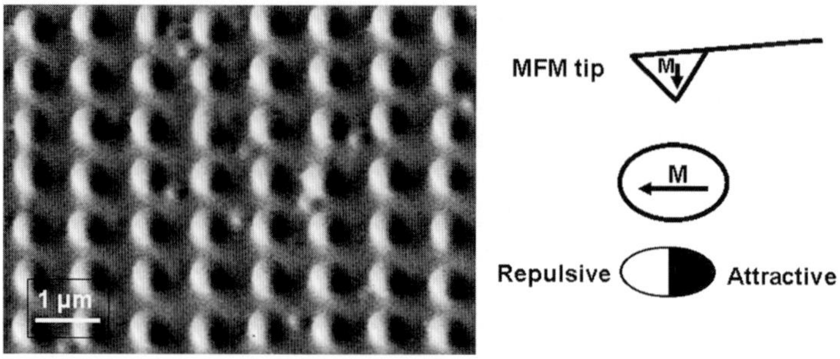

Fig. 10.1 Magnetic image of FePt nanodot arrays (*left*) and schematic drawing (*right*) illustrate the contrast mechanism of MFM. The magnetization of dot points from *dark* to *bright* under an in-plane magnetic field

effectively "stiffer" thereby increasing the resonant frequency of the cantilever, and this is coded as a bright signal by the NanoScope software.

Figure 10.1 shows the magnetic image of FePt nanodots produced by laser interference lithography along with a schematic illustration, which give us an intuitive way to understand the MFM contrast mechanism. For example, the magnetization of the tip is perpendicular to the sample surface, while the in-plane magnetic field was applied to the dots (the magnetization of dot points from dark to bright). The tip field interacts with the stray field coming from the edge of the nanodots, thus forming bright and dark contrast for repulsive and attractive force gradient, respectively. In this case, the bright spot indicates the tip is repulsive to the dot, while the black spot indicates it is attractive.

10.1.4 Magnetic Imaging Resolution

Magnetic imaging resolution is affected by the combination of many factors, such as the properties of the tip including mechanical geometry and magnetic structure, tip–sample distance (lift height), instrument sensitivity, and the background noise level, etc. In general, the system and environmental conditions, including electrical, mechanical, and acoustic noises, are related to the vertical resolution issues; while the tip shape and tip–sample distance affect the lateral resolution issues mostly.

High resolution usually means that it is capable of separating the minimum spacing between two closely spaced objects. If we image an array with a number of equally closely spaced lines, the definition of resolution can quite naturally be transferred to the spatial frequency domain as the minimum spatial wavelength that can still be observed [57, 58]. In general, the resolution of MFM with NanoScope III operated in ambient conditions is routinely obtained in the range of 30–50 nm; it can provide images with lateral resolution down to around 20 nm when magnetically coated tips remain in best condition.

Tip sharpness is the most important factor which determines the ultimate limit of MFM resolution. Smaller tips will provide higher spatial resolution. We will discuss this in the next section of this chapter.

The lift scan height is also an important parameter affecting image resolution and signal sensitivity. A larger lift scan height will decrease tip stray fields, hence also the sample perturbation, but leads to lower signal and degraded resolution. A smaller lift scan height will give better resolution to resolve fine magnetic features, as long as it is not much smaller than the surface roughness. Otherwise, the tip will strike the sample surface on the low point of its swing in the image. On the one hand, the best resolution is achieved with lift scan height at the smallest value possible that does not make the tip crash into the sample surface. On the other hand, the lift scan height of tip has to be at least 10 nm to keep away from the van der Waal's attractive force and capillary force, and also to avoid the effect of mapping topography on the magnetic signal. To improve the MFM image resolution in a simple way, we can try reducing the lift scan height for relatively smooth samples by using the high-quality MFM tips with improved tip shape and size.

In practice, we should comprehensively manage all the variables, such as: (1) tip selection, (2) sample preparation, and (3) optimizing the imaging parameters including choosing the right scan size and scan speed, optimizing the setpoint, and optimizing the feedback gains, etc., in order to achieve high-resolution imaging.

10.2 High-Resolution MFM Tips

Higher resolution MFM is required for studying small domain structures due to the decrease of the size of nanostructures and the increase of the density of recording media. In order to improve the spatial resolution of magnetic images, a further development of advanced MFM tips is of central importance and remains the challenge. An MFM probe is a sharp tip on a flexible cantilever, coated by a thin magnetic film with a typical thickness of 10–50 nm. Tips are usually magnetized with the field aligned along the tip axis (perpendicular to the sample surface). The MFM then senses force gradient due to the perpendicular component of the sample's stray field. The basic requirements for high-resolution MFM tips are: well-defined magnetization direction, low stray magnetic field, and very sharp tip ends.

A well-defined magnetization direction is achieved by using tips coated with high coercivity magnetic particles. High coercivity makes the tip resistant to reversal by the stray field from hard magnetic samples (such as recording medium or permanent magnets). This assures that the magnetization direction of the tip remains undisturbed by high stray field samples, which also makes it useful for the domain imaging under an applied magnetic field.

Low stray magnetic field of the MFM tips can be achieved by reducing the magnetic volume of the tip which can substantially reduce the emitted magnetic stray field from the MFM tip, and thus the influence of magnetic tip on the sample is negligible. Low stray magnetic field MFM tips with low magnetic moment show

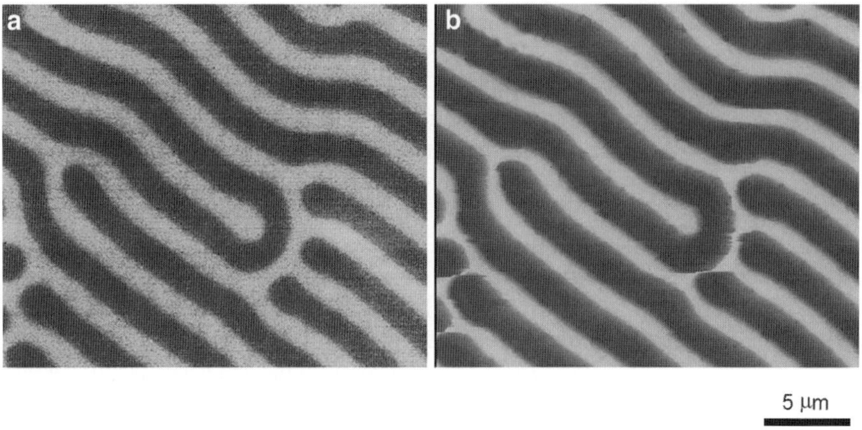

Fig. 10.2 Magnetic domain images of a garnet in the same area using (**a**) a high coercivity and low stray magnetic field FePt MFM tip, (**b**) a commercial CoCr MFM tip, respectively

great promise for high-resolution domain images of magnetic samples, especially for imaging of soft magnetic materials (low magnetic coercive field materials such as garnets) due to the fact that the domain structures of soft magnetic materials are easily disturbed by the high stray field MFM tips [54, 55]. Figure 10.2a, b is the MFM image of a garnet scanned in the same area using a high coercivity and low stray magnetic field FePt MFM tip [59] and a commercial CoCr MFM tip, respectively. As shown in Fig. 10.2a, the magnetic domains of a garnet film are clearly observed. There is no observable interference of the domain pattern of the garnet by the FePt MFM tip. While in Fig. 10.2b, the domain width appears disturbed by the commercial CoCr MFM tip.

The size of the tip end radius obviously determines the ultimate resolution of the MFM imaging. A magnetic tip that is suitable for high-resolution MFM imaging should have lateral dimensions in the nanometer regime near 30 nm or lower. The dimensions of a typical commercially available MFM probe used for MFM imaging are: cantilever width $w = 25$–$35\,\mu$m, cantilever length $l = 100$–$300\,\mu$m, tip height $h = 15\,\mu$m and tip radius $r = 30$–50 nm. As it is obvious from these dimensions, the lateral resolution is dominated by the tip radius.

There are many ways to develop and improve the magnetic tips for higher resolution MFM measurements [59–69]. One effective method used in our laboratory to enhance the resolution of domain images is through the development of advanced MFM probes with high coercivity and low stray field [59]. The commercial AFM Si tips were first coated with a thin magnetic CoPt layer and annealed at 650°C for about 10 h to form a face-centered-tetragonal (fct) structure ($L1_0$ phase), and then milled away the unnecessary materials around the tip end by the focused ion beam (FIB) technique. $L1_0$ phase CoPt exhibits high coercivity ($H_c = 15\,$kOe) and high saturation magnetization ($M_s = 800\,\mathrm{emu/cm^3}$) [70]. The high H_c ensures that the magnetization of the CoPt-coated MFM probe is stable during scanning a

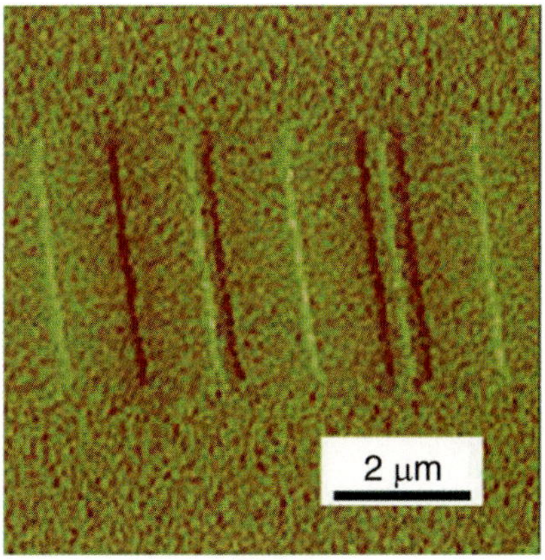

Fig. 10.3 MFM image of a hard disk film (a reference sample from NIST)

strong magnetic sample and/or under an applied external magnetic field; the high M_s can ensure that even a relatively small volume tip has enough magnetic moment to obtain strong signals. As shown in Fig. 10.3, the magnetic transitions and magnetic grains in a recording disk are well distinguishable by using a high H_c CoPt MFM tip with better defined magnetization direction perpendicular to the sample surface. Another example, as shown in Fig. 10.4, demonstrated that these high H_c and high M_s CoPt MFM tips can be used in the study of magnetic domain images of soft magnetic materials in the presence of an external magnetic field. A systematic domain evolution under different magnetic fields can clearly be observed by MFM in a garnet film sample. In this case, the magnetic field was applied along the easy axis (perpendicular to the film surface). The magnetization curve was measured by a SQUID magnetometer, as shown in Fig. 10.5. The labels are corresponding to those MFM images from (a) to (t) in Fig. 10.4, which shows that the magnetic domain patterns can be correlated directly with the magnetization curve.

Figure 10.6a shows a typical MFM tip with high H_c CoPt particles on the apex of the tip. The phase map reconstructed from the hologram of the same tip is shown in Fig. 10.6b [71]. The alternating black and white lines are the magnetic flux lines that emerge from the MFM tip. The magnetic stray field that originates from the CoPt particles on the tip can be estimated from the analysis of the hologram data. We use FIB milling to remove the magnetic material around the tip end as shown in Fig. 10.6c. The precise FIB milling is very important to reduce the amount of magnetic materials deposited on the tip but leave only a tiny bit of magnetic particle at the tip apex. In this way, the magnetic stray field generated by the tip end area was reduced, and thus improved the spatial resolution of the MFM images. Ultimate lateral resolution of MFM images near 10 nm has been reported by using the FIB-modified CoPt MFM tips [67].

The fabricated high H_c CoPt MFM probe with post-FIB modification was evaluated by MFM imaging under ambient conditions by using a Digital Instrument Dimension 3100 SPM (VEECO Instruments) operated in tapping/lift mode with the lift height of 10 nm. We tested a high-density recording medium with a fine

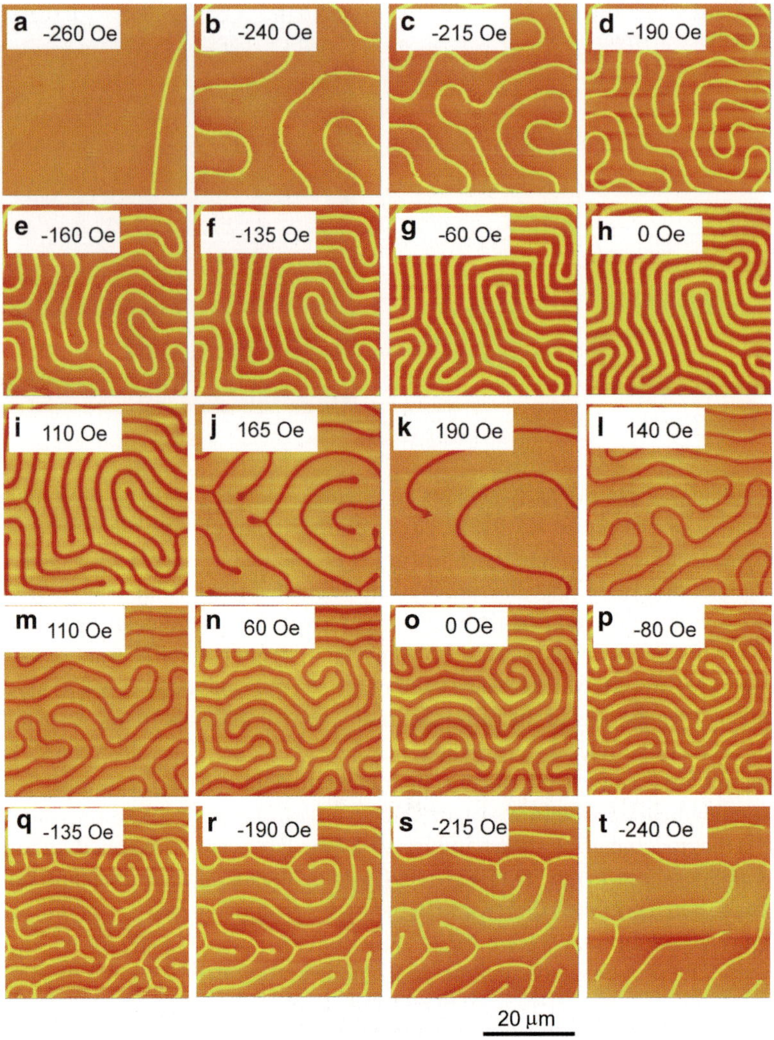

Fig. 10.4 Magnetic domain images of a garnet film under different magnetic fields

bit pattern from 800 to 1,000 kfci (kilo flux change per inch) as shown in Fig. 10.7 [67]. Tracks of 800, 900, and 1,000 kfci correspond to bit sizes of 32, 28, and 25 nm, respectively. Note that a single magnetic bit with fine magnetic features is visible in Fig. 10.7 and recording bits with a density as high as 900 kfci are well resolved. It is demonstrated that the high coercivity and low stray field MFM tip ensures a less invasive and higher resolution imaging. We characterized the image resolution by performing a Fourier transformation to define a cutoff frequency which corresponds to the real space resolution, a spatial resolution as good as 11 nm under ambient conditions with a commercial magnetic force microscope is achieved [67].

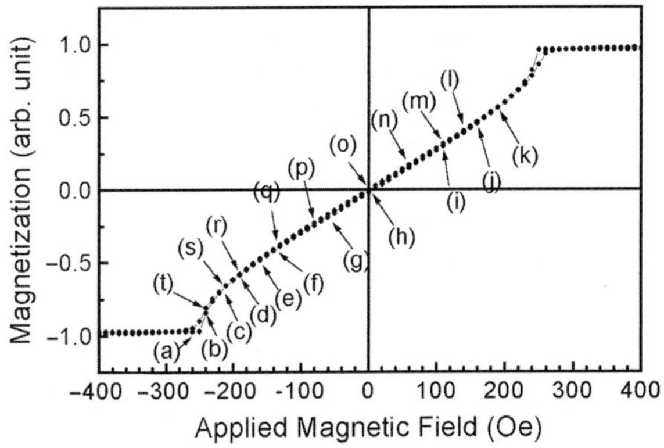

Fig. 10.5 The magnetization curve of a garnet film measured with the field perpendicular to the film surface. The labels correspond to the MFM images in Fig. 10.4

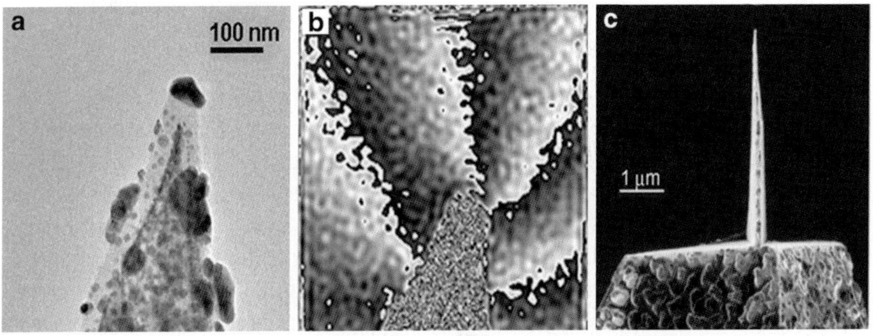

Fig. 10.6 (**a**) High-magnification SEM image of the very end of the tip coated with CoPt particles; (**b**) electro-holography graph of the same tip; and (**c**) an FIB-modified MFM tip with high coercivity and low stray field

High coercivity FePt-coated carbon nanocones (CNCs) have been fabricated recently for high-resolution MFM imaging by direct-current plasma-enhanced chemical vapor deposition growth of nanocones on tipless cantilevers followed by sputtering and annealing processes [68, 69]. The CNC probe tip has high aspect ratio and sharper probe geometry, which can reduce the magnetic volume on the tip, hence increasing the spatial resolution of the MFM images, and is also needed to acquire topographical and MFM images inside deep structures. From the analysis of the power spectrum of the track profiles for the images obtained by using $L1_0$ phase FePt-coated CNC probes, MFM images with 20 nm lateral resolution have been achieved. Compared to the conventional MFM tips (50 nm resolution routinely, and 30 nm or better when in good condition), the CNC probe provides much sharper images with a significantly improved resolution. As we know, the major portion of the magnetic moment of the probe mostly comes from the tip apex. However, the magnetic dipole interaction is a long-range interaction that includes not only the

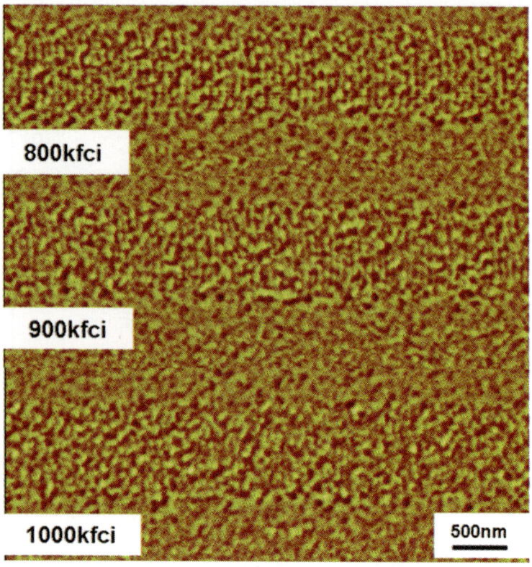

Fig. 10.7 High-resolution MFM image of a CoCrPt-SiO_2/Ru perpendicular recording media tracks with linear recording densities from 800 to 1,000 kfci [67]

contribution between the sample and the magnetic materials near the tip, but also that from extended areas of the tip. Therefore, such an improved image resolution is believed to be the result of a small cone angle of the CNC tip, which reduces the lateral magnetic interaction volume on the sidewall of the CNC tip. Another significant advantage worth noting is that shorter tip–sample spacing is allowed in the case of the FePt-coated CNC probe without degrading the resolution, which can enhance the magnetic contrast as well as the ability to trace the surface topography of the sample. This is primarily due to the high aspect ratio and sharp tip geometry of the CNC probes in general [69].

These advanced MFM probes have shown excellent performance for observation of magnetic nanostructures with lateral resolution of less than 20 nm. These appear to be promising for many future MFM applications, including high-aspect-ratio patterns and nanoscale magnetic bits, due to their slim shape configuration. Much better resolution could be obtained by further optimizing the tip radius and thickness of the magnetic film coating and selecting cantilevers with moderate resonant frequency in order to obtain an optimum compromise between image contrast and resolution. Research work to further improve the process parameters is underway.

10.3 Magnetic Domains

Since H'amos et al. [72, 73] first observed domains experimentally in 1931, Bloch [74] described the concept of domain walls in 1932 and later Landau and Lifshitz [75] introduced the modern domain theory in 1935. Since then, the experimental and general aspects of magnetic domains and domain walls have been a fascinating and technologically important research subject [76–80].

The domain structure consists of uniformly magnetized domains separated by narrow boundaries. This means that the individual magnetic spins of the atoms are aligned with one another and point in the same direction in each domain. However, different domains are magnetized in different directions and a ferromagnetic material generally divides into many tiny magnetic domains to reduce their magnetostatic energy. The narrow boundaries of rapidly changing spin direction between neighboring domains are called domain walls, where the magnetization rotates coherently from the direction in one domain to that in the next domain. The formation of domain walls is energetically favored. In general, magnetic domain walls are regions with large stray fields as opposed to domains, in which the spins are usually parallel to the sides of the crystals to minimize stray fields.

The width of a domain wall depends on the exchange energy (A) and the magnetocrystalline anisotropy energy (K_1) of the material. The domain-wall width δ in uniaxial bulk magnets is giving by $\delta = \pi\sqrt{A/K}$ (Bloch-wall width) [74, 76]. So we see that larger anisotropy energy favors narrow domain walls. On the contrary, if the exchange energy is larger than the anisotropy energy, the domain wall will be thicker. The calculated domain-wall widths are typical 5 nm for hard magnetic materials and 100 nm for soft materials, respectively [80]. As a measure of the domain wall width (W) seen by MFM, we measure the full width at half maximum (FWHM) of the MFM signal across a domain boundary by using the section profile analysis. The domain wall energy per unit area is $E = 2\pi\sqrt{AK}$, with range from about 0.1 $\mathrm{mJ/m^2}$ for soft materials to about 50 $\mathrm{mJ/m^3}$ in very hard materials. A detailed description, however, would go beyond the scope of this article and can be found elsewhere [76, 80].

Magnetic domains and domain walls can take many configurations which strongly depend on anisotropy and sample geometry. We illustrate some forms in Fig. 10.8 [80], such as uniformly magnetized single domain (a, h), two domains with 180°

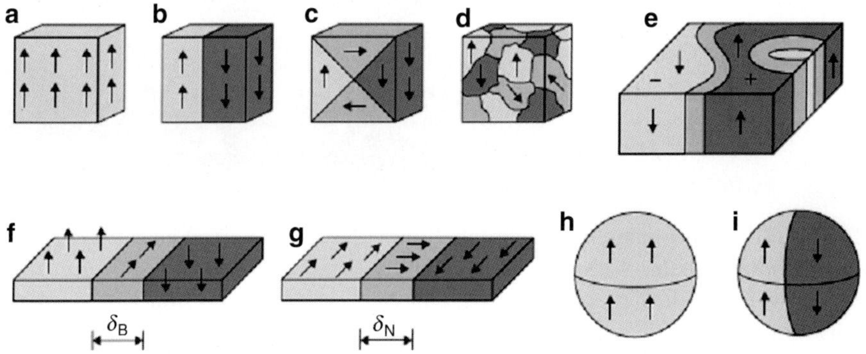

Fig. 10.8 Schematic diagram of magnetic domain configurations for (**a**) single domain (uniformly magnetized); (**b**) two domains with 180° Bloch wall; (**c**) four domains (flux-closure) with 90° walls; (**d**) multi-domains separated by domain walls; (**e**) complicated domain structure in a polycrystalline magnet; (**f**) Bloch wall in a thin film with perpendicular anisotropy; (**g**) Néel wall in a thin film with in-plane anisotropy; (**h**) very small particle with single domain, and (**i**) two-domain configuration in fairly small particle with uniaxial anisotropy [80]

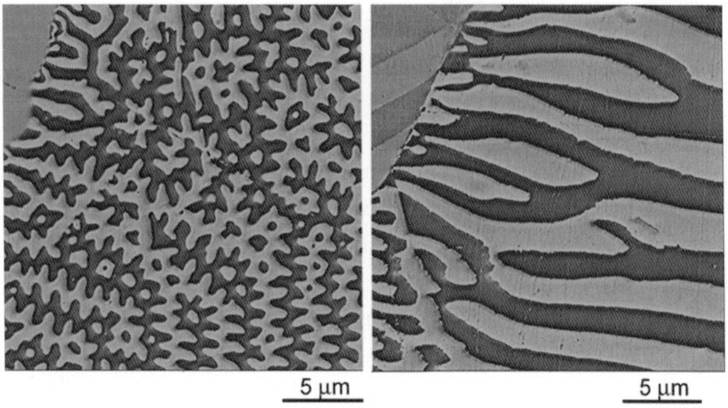

Fig. 10.9 MFM images of an $Nd_2Fe_{14}B$ magnet with the *top view* surface scan (*left*) and the *side view* surface scan (*right*), respectively

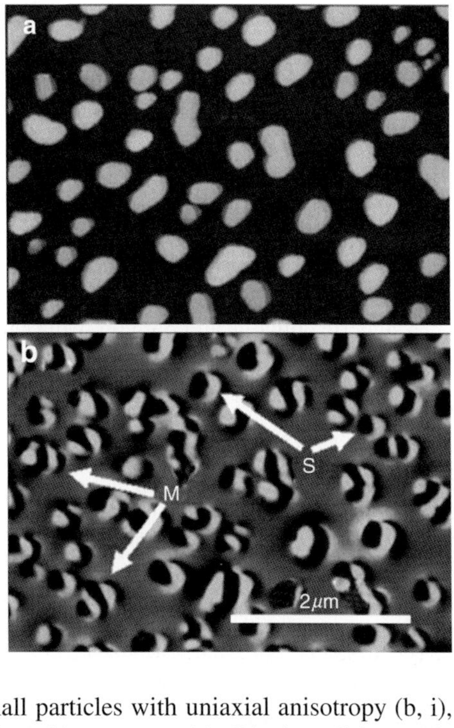

Fig. 10.10 (**a**) AFM and (**b**) MFM images of CoPt particles. *M* indicates multi-domain crystallites; *S* indicates single-domain crystallites [70]

Bloch wall as encountered in fairly small particles with uniaxial anisotropy (b, i), flux-closure domains with 90° walls in cubic magnets (c), Bloch wall in a thin film with perpendicular anisotropy (f), and Néel wall in a thin film with in-plane anisotropy (g).

MFM is a promising imaging technique for measuring magnetic domain features, due to its high sensitivity and resolution. Figure 10.9 shows an example of MFM images of a $Nd_2Fe_{14}B$ magnet with the top view surface scan (left) and the side view surface scan (right), respectively. The spike domain structure corresponds to the schematic domain structure of Fig. 10.8e. Figure 10.10 shows topographic and magnetic images of CoPt particles, from a 5-nm CoPt thin film annealed at

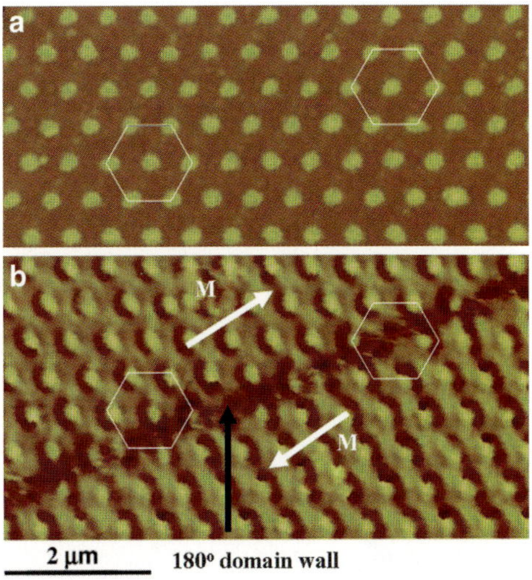

Fig. 10.11 (**a**) AFM topographic image of Co-C nanodots with a diameter of about 250 nm; (**b**) 180° domain wall in Co-C nanodot arrays, as probed by MFM [29]

750°C for 3 h. The domain structures of these separated crystalline grains are single-domain or multi-domain as indicated by "*S*" and "*M*" in Fig. 10.10b, respectively. The size of a single-domain crystallite is between 100 and 200 nm [70]. Figure 10.11 shows AFM (a) and MFM (b) images of Co-C nanodot arrays prepared by laser interference lithography in the remanence state (after an ac demagnetization field) [29]. The reversed contrast of the dot arrays in the MFM image (indicated by two white arrows) shows the existence of 180° domain walls (as indicated by the black arrows). As we can see by comparing AFM and MFM images, the domain wall is located not only between the dots, but also on the dot site, as shown in hexagons in Fig. 10.11.

Figure 10.12 shows a number of interesting domain features in an elliptical magnetic sample: (1) there is one vertex structure close to the center and another vertex close to the edge; (2) there are two cross-tie domains close to the edge of the sample; (3) there are "ripple" domains coming from the center vertex to the edge of the sample. The origin of the "ripple" domains is due to the edge defects. The edge roughness is also responsible for the variety of domain structures appeared in this sample. The cross-tie domain structures are frequently observed in a large-sized magnetic feature with a few micrometers.

A vortex core with perpendicular magnetization is expected to appear in submicron-sized magnetic dots with appropriate shape and thickness [24]. Figure 10.13 is an MFM image showing the domain structure of 50-nm-thick permalloy circular dots with diameter of about 1 μm. The image shows two types of vortex core with up- and down-magnitization (a clearly contrasted spot) at the center of each dot. Each dot is in a counterclockwise (white center) or clockwise (black center) vortex state (a curling magnetic structure). The orientation of magnetic spins in the center of each dot is perpendicular to the plane of the dot, which is pointing out of the plane for the white center, while the black center points into the plane.

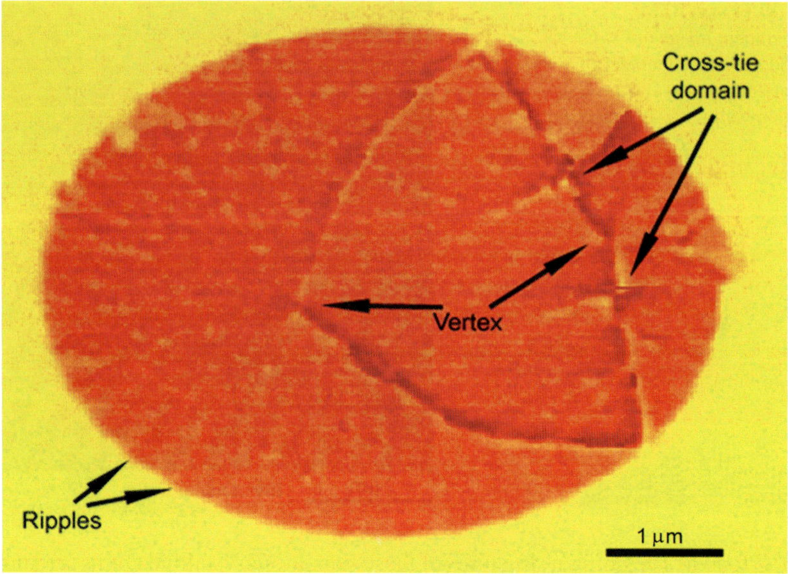

Fig. 10.12 MFM image of an elliptical sample with the size of about 4.5 μm × 6 μm

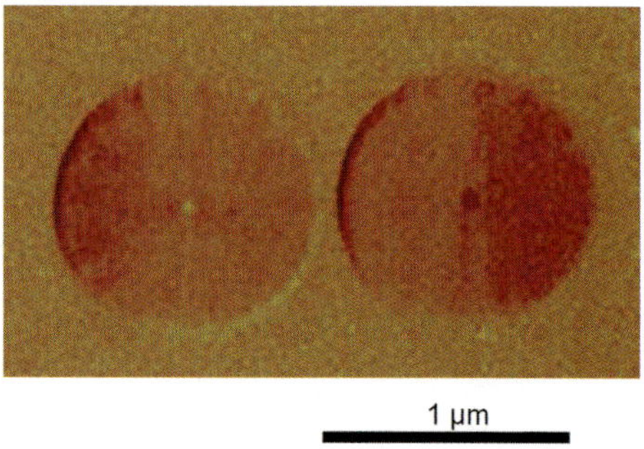

Fig. 10.13 MFM image of magnetic dots showing the vortex core structure

The domain walls are usually pinned by imperfections. These imperfections include dislocations, grain boundaries, lattice distortions, and even stresses within the crystal. These defects inhibit domain wall propagation through the medium. Usually, a greater applied magnetic field is required for domain wall motion. An applied magnetic field tends to align the magnetization of spins parallel to the field, which leads to domain wall motion and magnetization rotation. The domain-wall motion will depend on the demagnetizing energy and applied field, while the magnetization orientations are determined mainly by the anisotropy, which resists the rotation of the magnetization. The domain-wall motion becomes irreversible after removing a large enough applied field. The main reason for this is that the domain wall energy is not a constant but varies in an irregular manner because of imperfections in the specimen [39]. These crystallographic defects may prevent the domain walls from returning to their original unmagnetized configurations.

10.4 Patterned Nanomagnetic Films

There is widely increasing interest in the study of patterned magnetic nanoelements, nanodots, and nanostructures due to their fundamental issues and potential or realized technological applications in ultra-high-density recording media, MRAM, magnetic logic, and spinelectronic devices [81–85]. The unique magnetic properties of patterned nanostructures made from thin films can be tailored by controlling the shape, size, and spacing of elements using nanofabrication technologies such as FIB milling [35,36], electron-beam lithography [27,81], and interferometric annealing [42, 82]. In this section, we demonstrate that MFM is a very sensitive tool for imaging such small magnetic patterns from the submicron scale down to nanometer dimensions that can be seen from the following examples.

10.4.1 FIB Milled Patterns

10.4.1.1 FIB Milling

Focused ion beam milling is a convenient and efficient single-step high-resolution nanofabrication technique for directly patterning nanostructures by removing or milling material from films. The milling spot size of the FIB (FEI Strata FIB 201 work station) is varied from 8 to 500 nm, which makes it possible to achieve 20 nm resolution in patterning. Using FIB, a variation of shapes and dimensions of nanometer-scale magnetic features can be obtained easily and precisely as a one-step process. Therefore, it is a very flexible method for the fabrication of patterned nanostructures of arbitrary shapes and sizes for research.

10.4.1.2 Magnetic Interactions of $Ni_{80}Fe_{20}$ Arrays

Magnetization reversal behavior of nanostructures has been studied extensively from both the scientific and technical points of view. The magnetic interaction of submicrometer-sized arrays is an important topic for micromagnetism for studying these arrays' behavior. One of the most prominent examples for soft magnetic materials is $Ni_{80}Fe_{20}$, in which the magnetic structures are dominated by magnetostatic energy or demagnetization energy (in contrast to hard magnetic materials in which the anisotropy energy dominates).

The submicrometer-sized single and paired $Ni_{80}Fe_{20}$ bars patterned by FIB milling were characterized by using MFM and micromagnetic simulation to study the magnetic reversal behavior of these bars by Gao et al. [36]. Figure 10.14 shows the field-dependent magnetic domain images of a single and a set of paired $Ni_{80}Fe_{20}$ bars to illustrate the magnetic reversal behavior of the bars. All the MFM images were obtained under different magnetic fields using tapping/lift mode with CoPt MFM tips at a lift height of 20 nm. The 300 nm $\times$ 1.5 μm $\times$ 30 nm single bars and

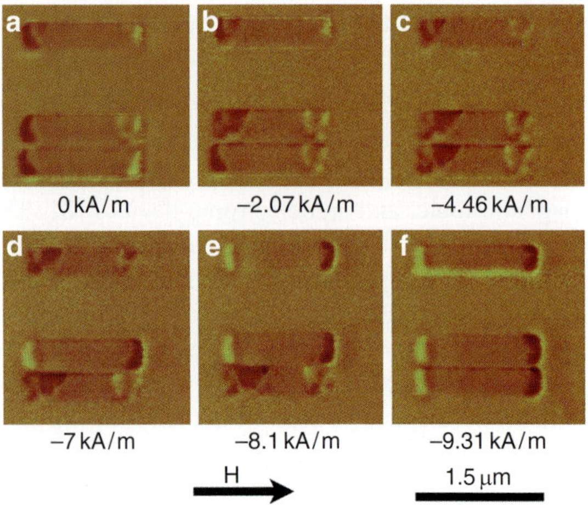

Fig. 10.14 Field-dependent magnetic domain structures of single $Ni_{80}Fe_{20}$ bars (300 nm × 1.5 μm × 30 nm) and same-size paired bars with a gap of 40 nm. The sample was saturated with a field 79.6 kA/m along the long axis of the bars and then decreased to (**a**) 0 kA/m, (**b**) −2.07 kA/m, (**c**) −4.46 kA/m, (**d**) −7 kA/m, (**e**) −8.1 kA/m, and (**f**) −9.31 kA/m [36] (1 kA/m = 12.56 Oe)

paired bars with a 40-nm gap are observed simultaneously. The sample was saturated with a field of 79.6 kA/m (= 1 kOe) along the long axis of the bars. At zero field, the magnetization direction of the bars did not change and was still along the direction of the applied magnetic field (Fig. 10.14a). As the field switched to negative, end domains and vortices were formed first (Fig. 10.14b), then domain walls propagated along the bars (Fig. 10.14c) and were stopped by the pinning sites. The pinning sites could be produced by surface or edge roughness, defects, or disorder in the film (such as random anisotropy). The magnetic reversal process at low magnetic field for the bars is due to the domain wall motion instead of coherent rotation. At a field of −7 kA/m, only one of the paired bars is reversed (Fig. 10.14d). As field changed to −8.1 kA/m, the single bar reversed (Fig. 10.14e), but the magnetization direction of paired bars is still antiparallel to each other. This means that the external magnetic field can switch the paired bars individually. At larger negative field, all the bars switched (Fig. 10.14f). The fact that the switching field of single bars is larger than the reversing field of only one of the paired bars and less than that of both paired bars indicates that magnetostatic interaction (long-range dipolar interaction) exists between the closely paired bars. The interaction makes the antiparallel magnetization directions of paired bars a stable state. This is very important for application because these magnetic interactions will influence the H_c and switching field distribution of the $Ni_{80}Fe_{20}$ arrays. MFM can directly determine the individual segment differences and switching-field distributions of patterned magnetic structures.

Micromagnetic simulations are of great help to understand the magnetization reversal behavior by using the OOMMF software package developed by NIST [86]. The micromagnetic simulation of the paired $Ni_{80}Fe_{20}$ bars is shown in Fig. 10.15 [36]. After saturation, paired bars form a symmetrically inverted C-state (C-state is named because the domain structure resembles like the letter C) at the remanent field state (Fig. 10.15a). With smaller negative field applied, the magnetization reversal proceeds through the formation of two vortices (end domains) as seen in

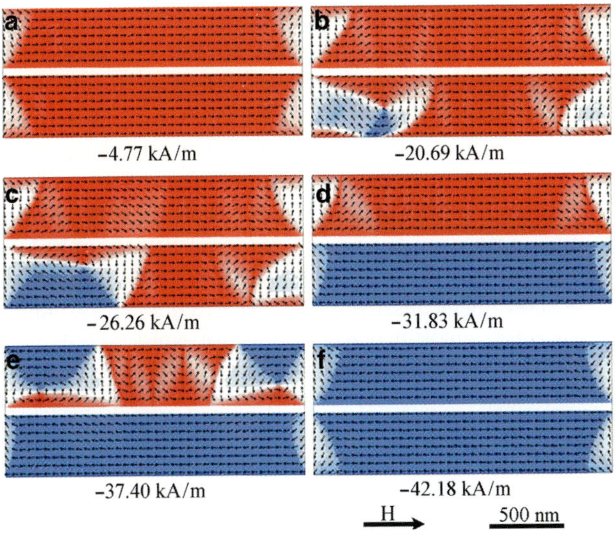

Fig. 10.15 Micromagnetic simulations on the paired $Ni_{80}Fe_{20}$ bars (300 nm × 1.5 μm × 30 nm) with a gap of 40 nm. The sample was saturated along the long axis of the bars and then decreased to (**a**) −4.77 kA/m, (b) −20.69 kA/m, (**c**) −26.26 kA/m, (**d**) −31.83 kA/m, (**e**) −37.40 kA/m and (**f**) −42.18 kA/m [36]

Fig. 10.15b, c. Due to the broken symmetry (random anisotropy), the reversal of one of the bars happens first (Fig. 10.15d). Although the general picture of the reversal of this bar is similar to the single-bar case, there are some deviations because of the magnetostatic interactions with the neighboring bar. Then the system creates a closure for the magnetic field (magnetization of paired bars is aligned in opposite directions). It is a stable state for the paired bars system. At somewhat larger applied field, the system is forced out of this potential well and the second bar also switched (Fig. 10.15f). The simulation results are consistent with the MFM measurement (Fig. 10.14), which clarified the magnetization reversal behavior of the paired bars.

10.4.2 Arrays of Magnetic Dots by Direct Laser Patterning

10.4.2.1 Direct Laser Interference Patterning

Direct laser interference patterning (DLIP) has been demonstrated as a relatively simple and efficient method for fabrication of a large area of periodic magnetic dot arrays or patterns [42,87–90]. In this technique, interference between several coherent laser beams results in intensity maxima which can direct and locally modify both the film topography and the material properties. In contrast to most nanofabrication techniques, DLIP allows thin films to be patterned directly, without the need for an intermediary photoresist mask or hard mask and other subsequent treatments. Specifically, the DLIP technique uses interference between two or more counter-propagating laser beams of a given wavelength to pattern a film through local "annealing" or ablation. The standing wave interference pattern produces alternating light and dark regions with a spacing determined by the wavelength and the angles at which the beams intersect. The minimum periodicity achievable by DLIP is $[\lambda/(2 \sin \theta)]$, where λ is the wavelength and θ is the angle between the incidence light and the sample surface normal. For a Nd:YAG laser operating at its second harmonic wavelength (532 nm), the periods of line and dot patterns down to 300 nm can be fabricated. By choosing the laser wavelength and adjusting θ, smaller periods can be achieved down to 100 nm. With one or more pulse exposures, many patterns, such as lines, elongated lines, dots, anti-dots, and even more complex nanostructures can be formed and fabricated [90].

Figure 10.16 shows MFM (left side) and AFM (right side) two-dimensional images of Co-C (100 nm) film patterned by two-pulse "crossed" two-beam interference patterning [90]. The "crossed beam" patterning leads to square arrays of dots or antidots, depending on the laser pulse energy. Note in particular that there is little magnetic contrast in the dots themselves, but that there is considerable domain structure in the partially ablated regions between the dots. It appears that the high temperatures during the ablation process have annealed the remaining film, forming Co or Co-rich clusters, and resulting in significant magnetization. In contrast, the local heating of the dots (in the less intense regions of the interference pattern) was not sufficient to cause such local atomic rearrangements. The MFM image in Fig. 10.16 (left side) shows considerable domain structure in the regions between the dots.

The DLIP technique has been explained in detail by Polushkin et al. [87–90]. Aktag provided a comprehensive review of this technique [90] and applied it to produce an "anisotropy" lattice in Co/Pt multilayers [42]. As will be seen in the following section, DLIP is a powerful emerging tool to form interesting and potentially

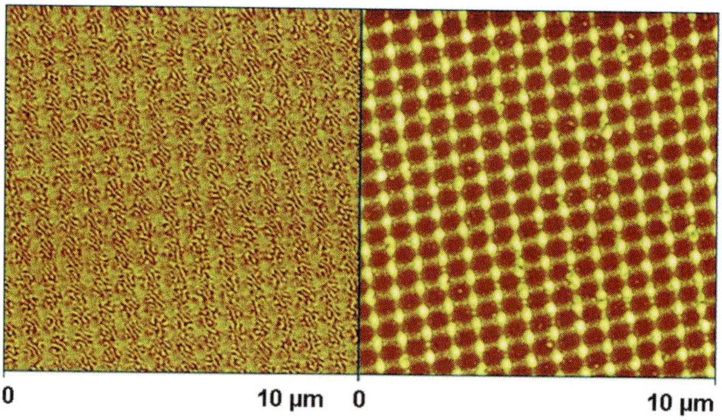

Fig. 10.16 MFM (*left*) and AFM (*right*) images of Co-C film patterned by two-pulse "crossed" two-beam interference patterning

useful magnetic structures; and MFM is a unique imaging technique for studying these structures at micro- and nanoscales, both because of its high resolution and its ability to operate in situ under applied magnetic fields.

10.4.2.2 In Situ MFM Imaging Under Applied Magnetic Fields

In-situ MFM imaging under a magnetic field can track the movement and change of domain walls, providing a glimpse of the magnetic mechanisms leading to the reversal behavior and other properties. Applications of in-situ magnetic fields on Co/Pt : SiO_x multilayers prepared by DLIP were carried out in the presence of both in-plane and perpendicular applied fields. A Digital Instrument Dimension D3100 SPM with homemade high-resolution and high coercivity CoPt MFM tips was used to obtain the entire magnetic domain mapping here. The Co/Pt multilayer film was chosen because it not only exhibits desirable magnetic properties of much interest but also is an attractive candidate for ultrahigh-density information storage applications and other technologies [42,90–94]. The initial Co/Pt multilayer thin films used here consist of several Co (0.4 nm)/Pt (1 nm) bilayers with 20-nm SiO_2 overcoat, grown on silicon substrate by magnetron sputtering. These multilayer films with 7–11 bilayers have perpendicular magnetic anisotropy and square hysteresis loops with a coercivity in the range of 100–200 Oe before the patterning [42]. Figure 10.17 shows MFM (a) and AFM (b) images of a [Co (0.4 nm)/Pt (1 nm)] × 7 multilayer film after DLIP using three beams, with incidence angles of 25° for both x and y directions. The MFM image (left side) shows that the domain of dots has elliptical shape with major axis of 790 nm and minor axis of 490 nm approximately. Each dot has a "dark" region (attractive force) at one end and a "light" region (repulsive force) at the other end. It seems that the moments behave as if they have in-plane anisotropy within a dot; whereas the moments in the region between the dots behave as having perpendicular anisotropy. This kind of "anisotropy" lattice in Co/Pt thin film multilayers holds promise for the study of model nanoscale magnetic systems. In the

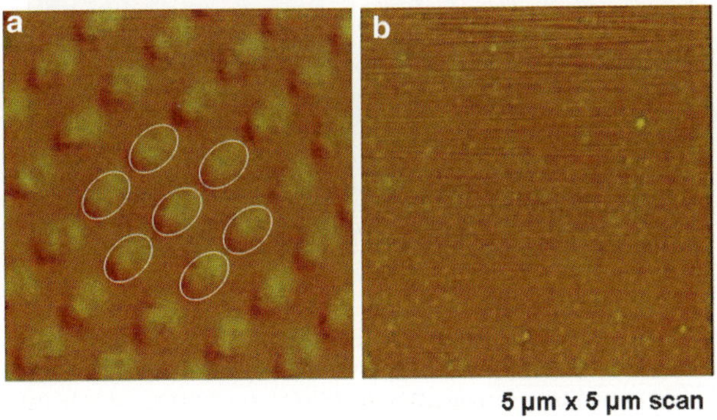

Fig. 10.17 (**a**) MFM and (**b**) AFM images of Co/Pt multilayer film after laser interference patterning using three beams [42]

AFM image shown in Fig. 10.17b, it was observed that the topographic change on the Co/Pt : SiO_x multilayers is quite small or absent. The measurement results indicated that the magnetic properties of Co/Pt film were modified by the intense light, but not the film topography. Thus it is confirmed that the DLIP method can locally change the intrinsic magnetic properties, with virtually no changes in topography of the sample surface.

Figure 10.18 shows a series of MFM images in various applied fields perpendicular to the sample surface. The film had been previously saturated with a negative magnetic field of 9 kOe applied into the film surface using an external electromagnet. The first MFM image in Fig. 10.18a shows a complex domain configuration,

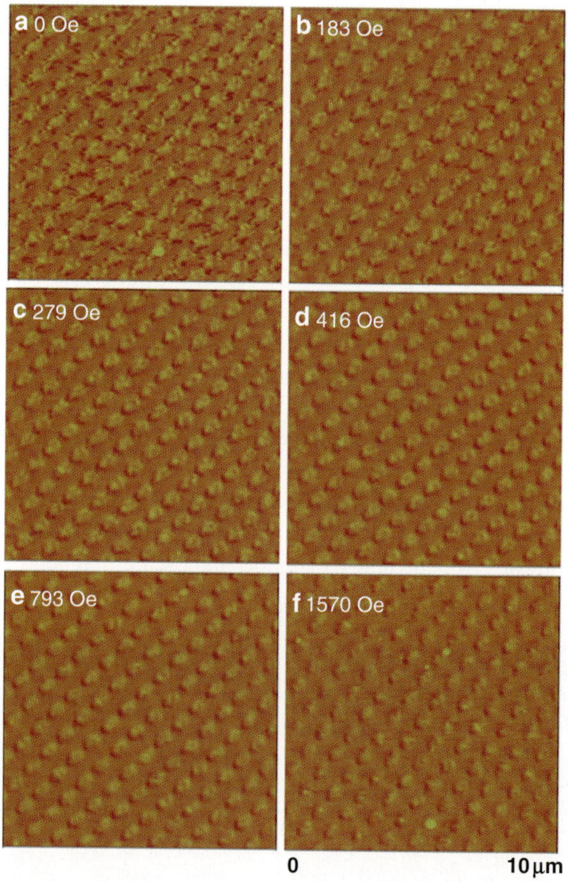

Fig. 10.18 Field-dependent MFM images of magnetic dots in Co/Pt multilayers with applied perpendicular fields. The image (**a**) was taken in the remanent state after saturation in a perpendicular field of -9 kOe, and then increased to (**b**) 183 Oe, (**c**) 279 Oe, (**d**) 416 Oe, (**e**) 793 Oe, and (**f**) 1,570 Oe. Each image is $10 \times 10\,\mu m^2$ [42]

which was taken in the remanence state. Subsequently an increasing in-situ positive field, range from 183 to 1,570 Oe, was applied using a home-built set of rotating pole permanent magnets which provide an adjustable field perpendicular to the sample plane. As the applied field is increased, a recognizable dot domain pattern starts to appear with increasing clarity, which demonstrates that the film has a regular array of "dot" domain structure with in-plane magnetization in a background matrix of perpendicular magnetization. When a fairly modest field is applied in the perpendicular direction such as 183 Oe in Fig. 10.18b, the moments outside the dots will align with the field immediately, since the coercivity of the unpatterned film is only 105 Oe [42]. Thus the perpendicular component (after patterning) reverse direction quite easily as the field is increased. But the in-plane moments (in the dots) were much harder to pull out of plane until high enough field was applied. At the applied field of about 1,570 Oe, the magnetic contrast between the dots and the background diminishes considerably. This indicates that the in-plane moments of the dots are pulled out-of-plane. Therefore, the contrast of dot domains in the last MFM image is significantly reduced at high applied perpendicular field.

Figure 10.19 shows a series of MFM images taken with in-plane applied fields. The sample was first saturated by a negative in-plane field of 3.6 kOe along the dots direction. Then positive fields ranging from 85 to 3,010 Oe were applied parallel to the line of dots from left to right of the sample plane, as labeled in Fig. 10.19. The MFM image in Fig. 10.19a was taken in the remanent state. This image shows a relatively uniform chainlike line of dots in the horizontal direction, which suggests that the moments in the regions to the left and right of each dot are quite strongly coupled to the moments in the dots. The evolution of the MFM images indicates that the magnetic dots themselves showed relatively little contrast, but the regions between dots (where there is perpendicular magnetic anisotropy) show a rich domain structure (the similar phenomenon is also observed in Fig. 10.16). As mentioned before, one expects a complicated interaction between the in-plain moments of the dots and the perpendicular moments of the dots spacing regions, and this can lead to complex domain structures. When the in-plane field is increased, the rich domain structure appearing in areas between dots diminishes gradually because the moments that are pointing perpendicular in regions between dots start to align with the field slowly due to the strong perpendicular anisotropy. There is very little magnetic contrast in the last MFM image after an applied field of 3,010 Oe, as shown in Fig. 10.19f, which means that the magnetization in the film is almost entirely pulled in plane under this field. This result is consistent with the observed in-plane magnetic hysteresis loops [42].

In summary, the well-defined domain structures and their evolution with an externally applied field have been analyzed. Note that the domain structures at remanence state seem quite complicated for both perpendicular and in-plain fields. This is expected, because the dot and background region moments are interacting with each other to minimize the overall energy. The observed magnetic domain patterns indicate the existence of inter-dot exchange interactions, mediated by the magnetic

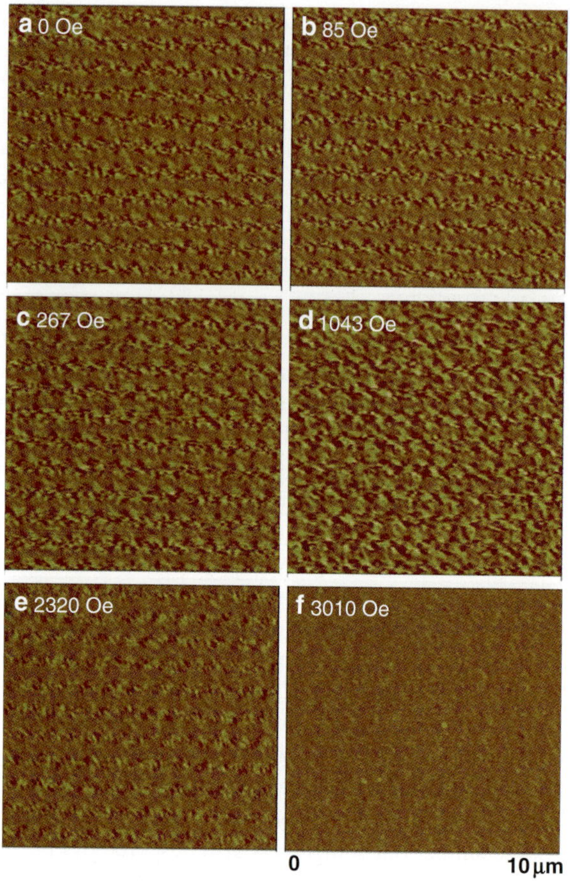

Fig. 10.19 Field-dependent MFM images of magnetic dots in Co/Pt multilayers with applied in-plane fields. The image (**a**) was taken in the remanent state after saturation in an in-plane field of −3.6 kOe, and then increased to (**b**) 85 Oe, (**c**) 267 Oe, (**d**) 1,034 Oe, (**e**) 2,320 Oe, and (**f**) 3,010 Oe. Each image is $10 \times 10\,\mu m^2$ [42]

matrix. This result suggests that the removal of the exchange between the patterned dots is very important for magnetic devices. Direct laser patterning does have the significant advantage that both local material properties and film topography can be modified quite easily in some materials. A wide variety of predicted patterns could be produced by varying the angles of incidence, the polarizations of the beams, and the wavelength and intensity of the beams. Perhaps the biggest drawback of DLIP is that array spacing is limited to dimensions greater than $\lambda/2$. Approaching the significantly smaller periods down to sub-100 nm will require much shorter wavelength coherent beams; it might be possible with synchrotron sources or X-ray radiation.

10.5 Template-Mediated Assembly of FePt Nanoclusters

The ordered fct structure ($L1_0$ phase) FePt nanoclusters have large magnetocrystalline anisotropy, high magnetization, and chemical stability. They are promising as building blocks for the fabrication of future high-density recording media, highly sensitive magnetic sensors, and high-performance permanent magnets [95–98]. Fabrication of $L1_0$ ordered clusters with narrow size distribution, the control of cluster location, magnetic orientation, and their interactions are important research issues both for studying nanomagnetism and for technological applications. Assembly of FePt clusters in highly ordered nanoporous alumina templates under magnetic field to obtain oriented nanoarrays with high perpendicular anisotropy were reported [37]. There are three main procedures for the sample preparation: First, synthesis of $L1_0$ FePt clusters with an average size of about 11 nm by the hydrogen reduction method; next, fabrication of an ordered porous alumina template by two-step anodization in a 0.3-M aqueous solution of oxalic acid; and then assembly of FePt clusters into nanoporous alumina templates in an external magnetic field up to 10 kOe to obtain perpendicular anisotropy. The template-mediated assembly of FePt nanoclusters under a magnetic field is a complex process. The capillary effect from the pores to cluster suspension, the cluster alignment driven by magnetic field, the magnetic dipolar interaction among clusters, interactions of the FePt clusters with alumina and surfactants will all contribute to the assembly process.

Hydrogen reduction is a very efficient way to produce FePt $L1_0$-ordered nanoparticles with high coercivity [37]. When the selected FePt clusters with average size of 11 nm were assembled into highly ordered nanoporous alumina arrays, the presence of an external magnetic field aligned their easy axis which produced magnetic films with anisotropic behavior. The coercivities of a template-mediated assembly of $L1_0$-structured FePt nanoclusters measured parallel and perpendicular to the pores are 13.4 and 10.2 kOe, respectively. A fabricated hexagonal ordered array of porous aluminum has pore size of about 60 nm, and the center-to-center distance of about 110 nm. It is interesting to note that essentially all of the clusters enter the pores after assembly, which means that an ordered pore array will create an ordered array of clusters, as shown in Fig. 10.20. The ordered magnetic nanoarrays were measured using two types of MFM tips from the bottom of the pores after removing the aluminum matrix. For a CoPt probe with coercivity as high as 15 kOe, the MFM image of nanoclusters assembly in Fig. 10.20a shows bright contrast spots (left), which indicates a repulsive force gradient. The strong MFM signal demonstrates that all clusters reached the bottom of the pores and formed an ordered nanopattern with a well-defined magnetization orientation. For the CoCr probe, with smaller coercivity field of $\sim$400 Oe, the magnetic contrast difference is visible as shown in Fig. 10.20b, which means that the magnetization direction of the tip may be disturbed by high stray field of the sample. The interaction between the MFM tip stray field and the magnetically hard FePt clusters is responsible for the contrast in the MFM images. To study magnetic images, using more than one type of MFM tip may help in the qualitative and/or quantitative analysis of MFM data.

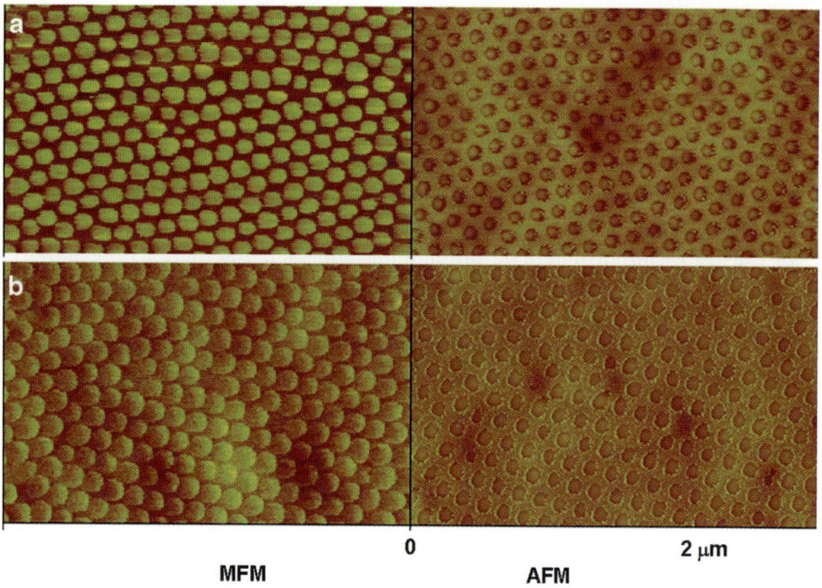

Fig. 10.20 MFM (*left*) and AFM topography (*right*) images at the bottom of the pores after removing the aluminum matrix of FePt nanoparticles assembly by using (**a**) CoPt probe with high H_c of 15 kOe and (**b**) CoCr probe with smaller H_c of 400 Oe, respectively

In summary, template-mediated assembly of nanoclusters is a process using preformed monodisperse magnetic nanoparticles as building blocks for fabrication of uniform patterned magnetic nanostructures smaller than sub-100 nm over a relatively large area. The $L1_0$-ordered FePt clusters assembled in nanoporous template formed ordered magnetic arrays with well-defined magnetization orientation, which exhibits distinct perpendicular anisotropy behavior caused by the external magnetic field. Grow monodispersed FePt particles in nanoporous alumina and assemble in magnetic field to obtain perpendicular anisotropy is a novel and simple approach to tailor the location of nanoclusters and their magnetic orientation, which may find applications in high-density magnetic recording media, nanodevices, and other related nanotechnology.

10.6 Interlayer Exchange-Coupled Nanocomposite Thin Films

Interlayer exchange-coupled ferromagnetic (FM)/antiferromagnetic (AF) layered thin films are essential components for the design of magnetic storage and logic devices, such as spin valves, magnetic recording media, memory devices, and magnetoresistive sensors. They are expected to have many advantages over single layers, including improved density and stability, enhanced switching speed, and higher reliability of the switching process [99–101]. Since the first observation of an AF coupling between FM layers through a nonmagnetic spacer [102], interlayer exchange coupling (IEC) has been the subject of numerous studies. It has been shown that the coupling can be FM or AF depending on the spacer layer thickness [34, 41, 103]. In this section, we will concentrate on two kind of model systems, i.e., interlayer coupling between FM layers with perpendicular anisotropy [(Co/Pt)/NiO/(CoPt) multilayer] and with in-plane anisotropy (Co/Ru/Co trilayers) studied by MFM.

10.6.1 (Co/Pt)/NiO/(CoPt) Multilayers with Perpendicular Anisotropy

10.6.1.1 Introduction

In the (Co/Pt)/NiO/(CoPt) multilayer system, the CoPt multilayers have perpendicular easy magnetization axis, while the NiO spins lie in plane. It is observed that there is an oscillatory coupling between Co/Pt multilayers on either side of NiO spacer layer [34, 41]. The period of oscillation corresponds to the AF ordering period of the NiO, suggesting a correlation of the coupling with the AF order. The theoretical explanation [104] assumes there is an exchange interaction at the NiO/Co interfaces as well as the AF super-exchange interaction between Ni moments within the NiO interlayer. The canting of NiO spins must occur in order to allow the IEC to propagate across the NiO film, leading to either FM or AF coupling. As described in Ref. [104], NiO is an antiferromagnet with antiparallel-aligned magnetic moments in the adjacent Ni (111) planes. If a NiO (111) film is placed between two FM layers, their magnetic moments tend to align parallel (antiparallel) to the magnetic moments of the top and bottom Ni monolayers (MLs) in the NiO due to a FM (AF) exchange coupling at the interface. Hence, the magnetic moments of the two FM layers tend to align parallel for the odd number of NiO MLs and antiparallel for the even number of NiO MLs.

10.6.1.2 MFM Images of Varying NiO Thickness

The behavior of magnetic domains in these (Co/Pt)/NiO/(CoPt) multilayers studied by MFM is presented in Refs. [34], [40] and [41]. It has been shown that interlayer thickness determines strongly the exchange coupling in the multilayer system. An example is shown in Fig. 10.21 [41]. The domain images of FM- and AF-coupled samples with different NiO thicknesses show quite different domain configurations. For NiO thickness of 8 and 12 Å samples (corresponding to AF coupling), only closed domain walls are observed. The only contrast appears in the domain wall regions, as can be seen in Fig. 10.21a,d, since the Co/Pt layers order

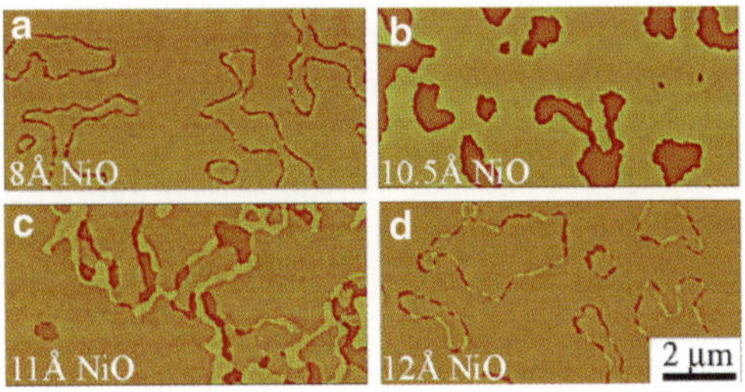

Fig. 10.21 MFM images of Co/Pt multilayers with 8, 10.5, 11, and 12 Å NiO interlayer thicknesses. There are domain overlap regions (closed domain walls) observed in (**a**) and (**d**) with antiferromagnetic interlayer coupling, (**b**) up and down domains (ferromagnetic interlayer coupling), (**c**) domain overlap (slightly antiferromagnetic coupled) [41]

antiferromagnetically domain by domain [34]. For FM-coupled samples with 10.5 Å NiO, the up and down domains are observed clearly in Fig. 10.21b. In the 11 Å NiO sample, very weak AF coupling is visible in Fig. 10.21c. In this sample, with very weak interlayer exchange energy, magnetostatic effects play a large role, leading to domain overlaps that are a significant fraction of the domain size.

In FM-coupled samples with up and down domain features, the domain-by-domain coupling implies that an up (down) domain in the upper Co/Pt layer is in perfect alignment with an up (down) domain in the lower Co/Pt layer, leading to a net upward (downward) magnetization. Whereas in AF-coupled samples, an up (down) domain in the upper Co/Pt layer is in alignment with a down (up) domain in the lower Co/Pt layer, leading to a zero net magnetization. Hence in AF-coupled samples, the only contrast is seen in the regions of the domain walls where the magnetization changes in the FM domain overlap region. The orientation flips from up to down in the overlap region to minimize magnetostatic energy.

10.6.1.3 Domain Overlap

It was found there is a relative shift between the domains of the two magnetic layers originating from a competition between the magnetostatic interaction and the AF IEC [41, 105]. Baruth et al. [40] developed a simple model that provides a quantitative explanation of the formation of these overlap regions that occur in antiferromagnetically coupled (Co/Pt)/NiO/(CoPt) multilayer samples. The relationship between the domain overlap width δ and the coupling strength J_{IEC} can be described as $\delta \sim 1/J_{\mathrm{IEC}}$, i.e., the domain overlap grows with decreasing exchange coupling strength. The calculated width of domain overlap is consistent with the experimental data obtained from MFM images by measuring the FWHM of the average line scans. Their results are important for application of magnetic layered structures with perpendicular anisotropy in advanced magnetoresistive devices.

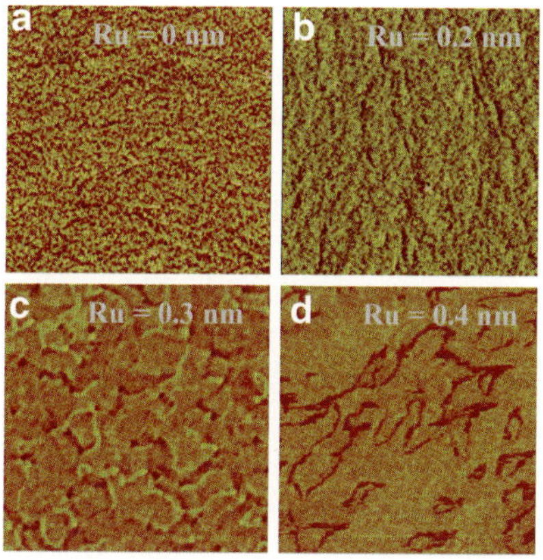

Fig. 10.22 The evolution of MFM images of the magnetic domain and domain wall structures of as-deposited Co (5 nm)/Ru (x)/Co (5 nm) thin films with Ru thickness of (**a**) 0 nm, (**b**) 0.2 nm, (**c**) 0.3 nm, and (**d**) 0.4 nm. Each image is $8 \times 8\,\mu m^2$ in size [46]

10.6.2 Co/Ru/Co Trilayers with In-Plane Anisotropy

MFM has proven to be an effective tool for detecting FM/AF interlayer coupling and exploring magnetic domain structures in exchange-coupled layered thin films. Ultrathin Co/Ru/Co trilayers consisting of two Co layers of equal thickness (5 nm) with in-plane anisotropy, exchange-coupled through a nonmagnetic Ru spacer layer of variable thickness was studied by MFM [46]. The effect of the variable Ru interlayer thickness on the exchange coupling and thus the magnetic domain structure during the FM/AF coupling transition in Co/Ru/Co films is demonstrated in Fig. 10.22. The sign and magnitude of the interlayer exchange are well tuned by the thickness of the Ru interlayer. When the Ru spacer $x = 0$, Co thin film has a general FM coarse-grained stripe domain structure. The bright and dark contrast of domains corresponds to high concentrations of positive and negative poles. For $x = 0.2$ nm, the image shows a complicated ripple domain pattern. For $x = 0.3$ nm, the bright and dark overlap domain walls are observed, which means that very weakly AF coupling is formed. For $x = 0.4$ nm, the closed-loop domain walls appear as dominantly black bands in antiparallel "no contrast" domains. In other words, large AF domains are separated by high-contrast dark domain walls. Increasing the thickness of Ru interlayer ($x \geq 0.5$ nm) will decrease the AF coupling strength between the two magnetic layers gradually, which results in the formation of complicated stripe domain structures. This series of MFM images of the Co/Ru/Co trilayer as a function of interlayer thickness show clearly that the exchange coupling strength and thus the domain configuration are very sensitive to the thickness of the Ru interlayer.

Depending on the thickness of Ru, the Co/Ru/Co thin films have two types of IEC: FM and AF coupling. For FM domains, the magnetization changes from one stripe domain to the next, corresponding to the areas with the contrast of light and dark, respectively. For AF domains, the antiparallel arrays of magnetization in two Co layers occupy most of the area in the image and form the low contrast domains, while the parallel channels at the boundary of the antiparallel arrays form strong dark-contrast closed-loop domain walls, suggestive of AF interlayer coupling between the top and bottom Co layers [46,105,106]. Clearly, the interlayer thickness is a large factor during the FM–AF transition in Co/Ru/Co films. For Ru thickness x less than 0.2 nm, the trilayer appears to be FM coupled; when $x \sim 0.3$ nm, the AF coupling through Ru becomes comparable to the FM coupling; for the films with Ru thickness of 0.4 nm, the MFM image displays a distinct signature of AF coupling. The variation of domains and domain-wall formation with Ru thickness provides insight into the FM–AF transition behavior, which enables us to see how the relative orientation of Co magnetization vectors in a FM/AF/FM structure varies gradually from parallel to antiparallel orientation as a function of Ru thickness varying from 0 to 0.4 nm. Again, the results clearly indicate that MFM offers a reliable method for visualization of magnetic domains with high spatial resolution and sensitivity in exchange-coupled layered thin films.

10.7 Conclusion (Outlook)

Magnetic domain images with very high lateral resolution (dependent mainly on the end radius of the tip) are easily obtained by MFM measurements that are non-destructive for the sample. As nanoscience and nanotechnology require nanometer-scale magnetic features for fundamental studies and applications in the fields of magnetic recording and magneto-electronics, MFM combined with other measurement tools can help us in understanding the unique magnetic properties of small features down to nanometer scale. MFM is well suited to study patterned nanomagnetic films and elements. MFM is also suitable to identify the magnetic exchange coupling of layered thin films. Notice the importance of MFM not only because of its capability for imaging the magnetic domain structures, but also for its usefulness as a powerful tool to characterize the magnetization reversal behavior of the nanostructures. The magnetization switching and switching-field variability can be studied by MFM measurements in-situ in the presence of an external magnetic field. These are the primary reasons for an increasing use of scanning probe microscopy-based methods for characterizing the local magnetization reversal behavior of magnetic materials. High-resolution visualization of magnetic nanostructures with MFM continues to attract the attention of researchers and there is promising for further improvement of image resolution. We expect that there are various applications of MFM in the future.

Acknowledgements

The authors are grateful to all those contributors involved in the work presented here but like to mention specially the cooperation with Drs. Lan Gao, Lu Yuan,

Aliekber Aktag, Yuchen Sui, Andrew Baruth, Zhongyuan Liu, and Zhen Li; and Professors Roger Kirby, Shireen Adenwalla, Nikolay Polushkin, Ralph Skomski, and David Sellmyer for their excellent contributions and many helpful discussions which enable us to present the recent research achievements in MFM field. We also would like to thank Prof. Sitaram Jaswal and Prof. Edger Pearlstein for their careful proofreading of this chapter, and for their critical comments and helpful suggestions. Financial supports from NSF-MRSEC Award DMR-0820521, ARO W911NF-08-1-0311, ARL W911NF-09-2-0039, and the Nebraska Center for Materials and Nanoscience are gratefully acknowledged.

References

1. Y. Martin, H. Wickramasinghe, Magnetic imaging by "force microscopy" with 1000 Å resolution. Appl. Phys. Lett. **50**, 1455–1457 (1987)
2. J. Saenz, N. Garcia, P. Grutter, E. Meyer, H. Heinzelmann, R. Wiesendanger, L. Rosenthaler, H. Hidber, H. Guntherodt, Magnetic domain structure by measuring magnetic forces. J. Appl. Phys. **62**, 4293–4295 (1987)
3. G. Binnig, C. Quate, C.h. Gerber, Atomic force microscope. Phys. Rev. Lett. **56**, 930 (1986)
4. H. Mamin, D. Rugar, J. Stern, B. Teris, S. Lambert, Force microscopy of magnetization patterns in longitudinal recording media. Appl. Phys. Lett. **53**, 1563 (1988)
5. H. Mamin, D. Rugar, J. Stern, R. Fontana, P. Kasiraj, Magnetic force microscopy of thin permalloy films. Appl. Phys. Lett. **55**, 318 (1989)
6. P. Grutter, A. Wadas, E. Meyer, H. Hidber, H. Guntherodt, Magnetic force microscopy of CoCr thin film. J. Appl. Phys. **66**, 6001 (1989)
7. A. Wadas, P. Grutter, H. Guntherodt, Analysis of magnetic bit pattern by magnetic force microscopy. J. Vac. Sci. Technol. **A8**(1), 416 (1990)
8. D. Rugar, H. Mamin, P. Guethner, S. Lambert, J. Stern, I. McFadyen, T. Yogi, Magnetic force microscopy: general principles and application to longitudinal recording media. J. Appl. Phys. **68**, 1169 (1990)
9. U. Hartmann, Magnetic force microscopy and its application to longitudinal thin films. Adv. Mater. **2**(11), 550 (1990)
10. C. Schoenenberger, S. Alvarado, S. Lambert, I. Sanders, Magnetic force microscopy and its applications to longitudinal thin films. J. Magn. Magn. Mater. **93**, 123 (1991)
11. S. Mueller-Pfiffer, M. Schneider, W. Zinn, Imaging of magnetic domain walls in iron with a magnetic force microscope: a numerical study. Phys. Rev. B. **49**(22), 15745 (1994)
12. M. Lederman, S. Schultz, M. Ozaki, Measurement of the dynamics of the magnetization reversal in individual single-domain ferromagnetic particles. Phys. Rev. Lett. **73**, 1986 (1994)
13. S. Gider, J. Shi, D. Awschalom, P. Hopkins, K. Campman, A. Gossard, A. Kent, S. Molnar, Imaging and magnetometry of switching in nanometer-scale iron particles. Appl. Phys. Lett. **69**, 3269 (1996)
14. S. Foss, R. Proksch, E. Dahlberg, B. Moskowitz, B. Walsh, Localized micromagnetic perturbation of domain walls in magnetite using a magnetic force microscopy. Appl. Phys. Lett. **69**, 4326 (1996)
15. H. Hug, B. Stiefel, A. Moser, I. Parashikov, A. Klicznik, D. Lipp, H. Guntherodt, G. Bochi, D. Paul, R. O'Handley, Magnetic domain structure in ultrathin Cu/Ni/Cu/Si(001) films. J. Appl. Phys. **79**, 5609 (1996)
16. Digital Instruments Support Note No. 229. Rev. B: Magnetic force microscopy. Digital Instruments, Inc. Santa Barbara, CA. 1–17, (1996)
17. L. Billiard, J. Miltat, A. Thiaville, S. Dubois, J. Duvail, L. Piraux, Observing magnetic nanowires by means of magnetic force microscopy. J. Magn. Magn. Mater. **190**, 1 (1998)

18. A. Fernandez, M. Gibbons, M. Wall, C. Cerjan, Magnetic domain structure and magnetization reversal in submicron-scale Co dots. J. Magn. Magn. Mater. **190**, 71 (1998)
19. U. Memmert, P. Leinenbach, J. Losch, H. Hartmann, Ultrahigh vacuum magnetic force microscopy: domain imaging on in situ grown Fe(100) thin films. J. Magn. Magn. Mater. **190**, 124 (1998)
20. T. Aign, P. Meyer, S. Lemerle, J. Jamet, J. Ferré, V. Mathet, C. Chappert, J. Gierak, C. Vieu, F. Rousseaux, H. Launois, H. Bernas, Magnetization reversal in arrays of perpendicularly magnetized ultrathin dots coupled by dipolar interaction. Phys. Rev. Lett. **81**, 5656 (1998)
21. S. Porthun, L. Abelmann, C. Lodder, Magnetic force microscopy of thin film media for high density magnetic recording. J. Magn. Magn. Mater. **182**, 238 (1998)
22. J. Wittborn, K. Rao, R. Priksch, I. Revenko, E. Dahlberg, D. Bazylinski, Magnitization reversal observation and manipulation of chains of nanoscale magnetic particles using the magnetic force microscope. Nanostruct. Mater. **12**(5–8), 1149 (1999)
23. R. Gomez, T. Luu, A. Pak, K. Kirk, J. Chapman, Domain configurations of nanostructured Permalloy elements. J. Appl. Phys. **85**, 6163 (1999)
24. T. Shinjo, T. Okuno, R. Hassdorf, K. Shigeto, T. Ono, Magnetic vortex core observation in circular dots of permalloy. Science **289**, 930 (2000)
25. A. Asenjo, D. Garcia, J. Garcia, C. Prados, M. Vazquez, Magnetic force microscopy study of dense stripe domains in Fe-B/Co-Si-B multilayers and the evolution under an external applied field. Phys. Rev. B **62**, 6538 (2000)
26. J. lohau, A. Carl, S. Kirsch, E. Wassermann, Magnetization reversal and coercivity of a single-domain Co/Pt dot measured with a calibrated magnetic force microscope tip. Appl. Phys. Lett. **78**, 2020 (2001)
27. Xiaobin-Zhu, P. Grutter, V. Metlushko, B. Ilic, Magnetic force microscopy study of electron-beam-patterned soft permalloy particles: technique and magnetization behavior. Phys. Rev. B **66**(2), 024423 (2002)
28. J. Garcia, A. Thiaville, J. Miltat, MFM imaging of nanowires and elongated patterned elements. J. Magn. Magn. Mater. **249**, 163 (2002)
29. L. Gao, S. Liou, M. Zheng, R. Skomski, M. Yan, D. Sellmyer, Magnetic force microscopy observations of the magnetic behavior in Co-C nanodot arrays. J. Appl. Phys. **91**, 7311 (2002)
30. Xiaobin-Zhu, P. Grutter, Magnetic force microscopy studies of patterned magnetic structures. IEEE Trans. Magn. **39**, 3420 (2003)
31. O. Hellwig, T. Kirk, J. Kortright, A. Berger, E. Fullerton, A new phase diagram for layered antiferromagnetic films. Nat. Mater. **2**, 112 (2003)
32. E. Meyer, H.J. Hug, R. Bennewitz, *Scanning Probe Microscopy* (Springer-Verlag, Berlin/Heidelberg, 2004)
33. K. Sorge, A. Kashyap, R. Skomski, L. Yue, L. Gao, R. Kirby, S.H. Liou, D.J. Sellmyer, Interaction and switching behavior of anisotropic magnetic dots. J. Appl. Phys. **95**(11), 7414–7416 (2004)
34. Z.Y. Liu, L.Yue, D.J. Keavney, S. Adenwalla, Oscillations of interlayer exchange coupling in [Pt/Co]n/NiO/[CoPt]n multilayers with perpendicular anisotropy: dependence on NiO and Pt thicknesses. Phys. Rev. B **70**, 224423 (2004)
35. L. Gao, L. Yuan, L. Nicholl, R. Sabiryanov, Z.Y. Liu, S. Adenwalla, S.H. Liou, Domain structure and magnetoresistance in $Ni_{81}Fe_{19}$ zigzag wires. J. Magn. Magn. Mater. **272–276**, 1301 (2004)
36. L. Gao, D.Q. Feng, L. Yuan, T. Yokota, R. Sabirianov, S.H. Liou, M.D. Chabot, D. Porpora, J. Moreland, A study of magnetic interactions of $Ni_{80}Fe_{20}$ arrays using ultra-sensitive microcantilever torque magnetometry. J. Appl. Phys. **95**, 7010 (2004)
37. Y.C. Sui, W. Liu, L.P. Yue, X.Z. Li, R. Skomski, D.J. Sellmyer, Template-mediated assembly of FePt $L1_0$ cluster under external magnetic field. J. Appl. Phys. **97**, 10J304 (2005)
38. H. Hopster, H.P. Oepen, *Magnetic Microscopy of Nanostructures* (Springer-Verlag, Berlin/Heidelberg, 2005)
39. L. Gao, Magnetic interactions in nanostructured films, Ph.D. thesis, University of Nebraska, USA, 2005

40. A. Baruth, L. Yuan, J. Burton, K. Janicka, E. Tsymbal, S. Liou, S. Adenwalla, Domain overlap in antiferromagnetically coupled [Co/Pt]/NiO/[Co/Pt] multilayers. Appl. Phys. Lett. **89**, 202505 (2006)
41. A. Baruth, D. Keavney, J. Burton, K. Janicka, E. Tsymbal, L. Yuan, S. Liou, S. Adenwalla, Origin of the interlayer exchange coupling in [Co/Pt]/Nio/[CoPt] multilayers studied with XAS, XMCD, and micromagnetic modeling. Phys. Rev. B. **74**, 054419 (2006)
42. A. Aktag, S. Michalski, L. Yue, R. Kirby, S. Liou, Formation of an anisotropy lattice in Co/Pt multilayers by direct laser interference patterning. J. Appl. Phys. **99**, 093901 (2006)
43. N. Polushkin, S. Michalski, L. Yue, R. Kirby, Evidence of long-wavelength collective excitations in magnetic superlattices. Phys. Rev. Lett. **97**, 256401 (2006)
44. X. Rui, J.E. Shield, Z. Sun, L. Yue, Y. Xu, D.J. Sellmyer, Z. Liu, D.J. Miller, High energy product exchange-spring FePt/Fe cluster nanocomposite permanent magnets. J. Magn. Magn. Mater. **305**, 76–82 (2006)
45. A. Asenjo, M. Jaafar, E. Gonzalez, J. Martin, M. Vazquez, J. Vicent, Imaging magnetic domains in Ni nanostructures. J. Magn. Magn. Mater. **310**, e936 (2007)
46. L. Yue, Z. Li, R. Kirby, D. Sellmyer, MFM study of interlayer exchange coupled Co/Ru/Co films: effect of Ru layer thickness. Ultramicroscopy, **109**, 1040 (2009)
47. V. Mironov, B. Gribkoc, S. Vdovichev, S. Gusev, A. Fraerman, O. Ermolaeva, A. Shubin, A. Alexeev, P. Zhdan, C. Binns, Magnetic force microscope tip-induced remagnetization of CoPt nanodisks with perpendicular anisotropy. J. Appl. Phys. **106**, 053911 (2009)
48. K. Stoev, F. Liu, Y. Chen, X. Dang, P. Luo, J. Chen, J. Wang, K. Kung, M. Lederman, M. Re, G. Choe, J.N. Zhou, M. Yu, Demonstration and characterization of $130\,\mathrm{Gb/in}^2$ magnetic recording systems. J. Appl. Phys. **93**, 6552 (2003)
49. M. Johnson, Hybrid ferromagnet-semiconductor devices. J. Vac. Sci. Technol. A **16**, 1806 (1998)
50. S. Mangin, D. Ravelosona, J. Katine, M. Carey, B. Terris, E. Fullerton, Current-induced magnetization reversal in nanopillars with perpendicular anisotropy. Nat. Mater. **5**, 210 (2006)
51. D. Andrew, Kent: Spintronics: a nanomagnet oscillator. Nat. Mater. **6**, 399 (2007)
52. G.A. Prinz, Magnetoelectronics. Science **282**(5394), 1660 (1998)
53. Digital Instruments, Dimension 3100 Series Scanning Probe Microscope Instruction Manual, Chapter 13, Digital Instruments, Inc. Santa Barbara, CA. (1998)
54. S. Tomlinson, A. Farley, S. Hoon, M. Valera, Interactions between soft magnetic samples and MFM tips. J. Magn. Magn. Mater. **157–158**, 557 (1996)
55. A. Thiaville, L. Belliard, D. Majer, E. Zeldov, J. Miltat, Measurement of the stray field emanating from magnetic force microscope tips by Hall Effect microsensors. J. Appl. Phys. **82**, 3182 (1997)
56. P. Grutter, H. Mamin, D. Rugar, in *Magnetic Force Microscope*, ed. R. Wiesendanger, H.-J. Gunterodt. Scanning Tunneling Microscopy, vol. 28 (Springer-Verlag, Berlin, 1992), pp. 151–207.
57. L. Abelmann, A. van den Bos, C. Lodder, in *Magnetic Force Microscopy – Towards Higher Resolution*, ed. H. Hopster, H.P. Oepen. Magnetic Microscopy of Nanostructures (Springer-Verlag, Berlin Heidelberg, 2005) pp. 253–283
58. L. Abelmann, S. Porthun, M. Haast, C. Lodder, A. Moser, M. Best, P. Vanschendel, B. Stiefel, H. Hug, G. Heydon, A. Farley, S. Hoon, T. Pfaffelhuber, R. Proksch, K. Babcock, Comparing the resolution of magnetic force microscopes using the CAMST reference samples. J. Magn. Magn. Mater. **190**, 135 (1998)
59. S.H. Liou, Advanced magnetic force microscopy tips for imaging domains, *in* Handbook of Advanced Magnetic Materials, ed. by Y. Liu, D.J. Sellmyer, D. Shindo, vol. 2 (Springer, New York, 2006), pp. 1–23
60. P. Grutter, D. Rugar, H. Mamin, G. Castillo, S. Lambert, C. Lin, R. Valletta, O. Wolter, Batch fabricated sensors for magnetic force microscopy. Appl. Phys. Lett. **57**, 1820 (1990)
61. P. Hopkins, J. Moreland, S. Malhotra, S.H. Liou, Superparamagnetic magnetic force microscopy tips. J. Appl. Phys. **79**, 6448 (1996)
62. S.H. Liou, Y.D. Yao, Development of high coercivity magnetic force microscopy tips. J. Magn. Magn. Mater. **190**, 130 (1998)

63. J. Lohau, S. Kirsch, A. Carl, G. Dumpich, E. Wassermann, Quantitative determination of effective dipole and monopole moments of magnetic force microscopy tips. J. Appl. Phys. **86**, 3410 (1999)
64. L. Folks, M. Best, P. Rice, B. Terris, D. Weller, J. Chapman, Perforated tips for high-resolution in-plane magnetic force microscopy. Appl. Phys. Lett. **76**, 909 (2000)
65. Z. Liu, Y. Dan, Q. Jinjun, Y. Wu, Magnetic force microscopy using focused ion beam sharpened tip with deposited antiferro–ferromagnetic multiple layers. J. Appl. Phys. **91**, 8843 (2002)
66. G. Phillips, M. Siekman, L. Abelmann, J. Lodder, High resolution magnetic force microscopy using focused ion beam modified tips. Appl. Phys. Lett. **81**, 865 (2002)
67. L. Gao, L.P. Yue, T. Yokota, R. Skiomski, S.H. Liou, H. Takahoshi, H. Saito, S. Ishio, Focused ion beam milled CoPt magnetic force microscopy tips for high resolution domain images. IEEE Trans. Magn, **40**, 2194–2196 (2004)
68. I.-C. Chen, L.-H. Chen, A. Gapin, S. Jin, L. Yuan, S.-H. Liou, Iron–platinum-coated carbon nanocone probes on tipless cantilevers for high resolution magnetic force imaging. Nanotechnology **19**, 075501 (2008)
69. I.-C. Chen, L.-H. Chen, X.-R. Ye, C. Daraio, S. Jin, C. Orme, A. Quist, R. Lal, Extremely sharp carbon nanocone probes for atomic force microscopy imaging. Appl. Phys. Lett. **88**, 153102 (2006)
70. S.H. Liou, S. Huang, E. Klimek, R. Kirby, Y. Yao, Enhancement of coercivity in nanometer-size CoPt crystallites. J. Appl. Phys. **85**, 4334 (1999)
71. S. Signoretti, C. Beeli, S.H. Liou, Electron holography quantitative measurements on magnetic force microscopy probes. J. Magn. Magn. Mater. **272**, 2167 (2004)
72. L. Hámos, P. Thiessen, Uber die Sichtbarmachung von Bezirken verschiedenen ferromagnetischen Zustandes fester Körper. Z. Phys. **71**, 442–444 (1931)
73. F. Bitter, On inhomogeneities in the magnetization of ferromagnetic materials. Phys. Rev. **38**, 1903–1905 (1931)
74. F. Bloch, Zur Theorie des Austauschproblems und der Remanenzerscheinung der Ferromagnetika. Z. Phys. **74**, 295 (1932)
75. L. Landau, E. Lifshitz, On the theory of the dispersion of magnetic permeability in ferromagnetic bodies. Phys. Z. Sowjetunion **8**, 153 (1935)
76. A. Hubert, R. Schafer, *Magnetic Domains: The Analysis of Magnetic Microstructures* (Springer-Verlag, Berlin, Heidelberg, New-York, 1998)
77. D. Atkinson, D. Allwood, G. Xiong, M. Cooke, C. Faulkner, R. Cowburn, Magnetic domain-wall dynamics in a submicrometre ferromagnetic structure. Nat. Mater. **2**, 85 (2003)
78. M. Redjdal, J. Giusti, M. Ruane, F. Humphrey, Structure dependent stray fields from domain walls in permalloy films. IEEE Trans. Magn, **39**, 1267 (2003)
79. D. Allwood, G. Xiong, C. Faulkner, D. Atkinson, D. Petit, R. Cowburn, Magnetic domain-wall logic. Science **309**, 1688 (2005)
80. R. Skomski, *Simple Models of Magnetism* (University Press, Oxford, 2008), pp. 123–128
81. S. Stephen, Y. Chou, Patterned magnetic nanostructures and quantized magnetic disks. Proc. IEEE **85**, 652 (1997)
82. C. Ross, H. Smith, T. Savas, M. Schattenburg, M. Farhoud, M. Hwang, M. Walsh, R. Ram, Fabrication of patterned media for high density magnetic storage. J. Vac. Sci. Technol. B **17**, 3168–3174 (1999)
83. S. Wolf, D. Awschalom, R. Buhrman, J. Daughton, S. Molnar, M. Roukes, A. Chtchelkanova, D. Treger, Spintronics: a spin-based electronics vision for the future. Science **294**, 1488 (2001)
84. V. Novosad, M. Grimsditch, J. Darrouzet, J. Pearson, S. Bader, V. Metlushko, K. Guslienko, Y. Otani, H. Shima, K. Fukamichi, Shape effect on magnetization reversal in chains of interacting ferromagnetic elements. Appl. Phys. Lett. **82**, 3716 (2003)
85. J. Lodder, Methods for preparing patterned media for high-density recording. J. Magn. Magn. Mater. **272–276**, 1692 (2004)
86. J. Donahue, R.D. McMichael, Exchange energy representations in computational micromagnetics. Physica B **233**, 272 (1997)

87. N. Polushkin, S. Gusev, M. Drozdov, Y. Verevkin, V. Petryakov, Arrays of magnetic wires created in phase-separating Fe-containing alloys by interference laser irradiation. J. Appl. Phys. **81**, 5478 (1997)
88. N. Polushkin, J. Wittborn, C. Canalias, K. Rao, A. Alexeev, A. Popkov, Local magnetostrictive response of small magnetic entities in artificial Fe − Cr composites. J. Appl. Phys. **92**, 2779 (2002)
89. M. Zheng, M. Yu, Y. Liu, R. Skomski, S. Liou, D. Sellmyer, V. Petryakov, Y. Verevkin, N. Polushkin, N. Salashchenko, Magnetic nanodot arrays produced by direct laser interference lithography. Appl. Phys. Lett. **79**, 2606 (2001)
90. A. Aktag, Direct laser interference pattering of magnetic thin films, Ph.D. thesis, University of Nebraska, USA, 2004
91. Z. Li, P. Carcia, Microstructural dependence of magnetic properties of Pt/Co multilayer thin films. J. Appl. Phys. **71**, 842 (1992)
92. J. Du, S. Wang, C. Ryby, A. Khapikov, J. Liu, A. Barnard, W. Harrell, Magnetic and structural properties of annealed CoPt/C multilayers. J. Magn. Magn. Mater. **231**, 231 (2001)
93. G. Hu, T. Thomson, M. Albrecht, M. Best, B. Terris, C. Rettner, S. Raoux, G. McClelland, M. Hart, Magnetic and recording properties of Co/Pd islands on prepatterned substrates. J. Appl. Phys. **95**, 7013 (2004)
94. K. Pandey, J. Chen, G. Chow, J. Hu, $L1_0$ CoPt − Ta_2O_5 exchange coupled multilayer media for magnetic recording. Appl. Phys. Lett. **94**, 232502 (2009)
95. S. Sun, C. Murray, D. Weller, L. Folks, A. Moser, Monodisperse FePt nanoparticles and ferromagnetic FePt nanocrystal superlattices. Science 287, 1989 (2000)
96. H. Zeng, J. Li, J. Liu, Z. Wang, S. Sun, Exchange-coupled nanocomposite magnets by nanoparticle self-assembly. Nature **420**, 395 (2002)
97. D. Sellmyer, Applied physics: strong magnets by self-assembly. Nature **420**, 374 (2002)
98. Y. Sui, L. Yue, R. Skomski, X. Li, J. Zhou, D. Sellmyer, CoPt hard magnetic nano- particle films synthesized by high temperature chemical reduction. J. Appl. Phys. **93**, 7571 (2003)
99. E. Abarra, A. Inomata, H. Sato, I. Okamoto, Y. Mizoshita, Longitudinal magnetic recording media with thermal stabilization layers. Appl. Phys. Lett. **77**, 2581 (2000)
100. E. Fullerton, D. Margulies, M.E. Schabes, M. Carey, B. Gurney, A. Moser, M. Best, G. Zeltzer, K. Rubin, H. Rosen, M. Doerner, Antiferromagnetically coupled magnetic media layers for thermally stable high-density recording. Appl. Phys. Lett. **77**, 3806 (2000)
101. S. Mangin, D. Ravelosona, J. Katine, M. Carey, B. Terris, E. Fullerton, Current-induced magnetization reversal in nanopillars with perpendicular anisotropy. Nat. Mater. **5**, 210 (2006)
102. P. Grunberg, R. Schreiber, Y. Pang, M. Brodsky, H. Sowers, Layered magnetic structures: evidence for antiferromagnetic coupling of Fe layers across Cr interlayers. Phys. Rev. Lett. **57**, 2442 (1986)
103. S. Parkin, N. More, P. Roche, Oscillations in exchange coupling and magneto- resistance in metallic superlattice structures: Co/Ru, Co/Cr, and Fe/Cr. Phys. Rev. Lett. **64**, 2304 (1990)
104. M. Zhuravlev, E. Tsymbal, S. Jaswal, Exchange model for oscillatory interlayer coupling and induced unidirectional anisotropy in $[Pt/Co]_3/NiO/[Pt/Co]_3$ multilayers. Phys. Rev. Lett. **92**, 219703 (2004)
105. O. Hellwig, A. Berger, E. Fullerton, Domain walls in antiferromagnetically coupled multilayer films. Phys. Rev. Lett. **91**, 197203 (2003)
106. Z. Li, R. Skomski, S. Michalski, L.P. Yue, R. Kirby, Magnetic antiphase domains in Co/Ru/Co trilayers. J. Appl. Phys. **107**, 09D303 (2010)

Chapter 11
Semiconductors Studied by Cross-sectional Scanning Tunneling Microscopy

J.K. Garleff, J.M. Ulloa, and P.M. Koenraad

11.1 Introduction

Scanning tunneling microscopy (STM) has been extremely successful due to the fact that it allowed visualizing atomic scale details of a clean surface layer in real space. In the field of semiconductor materials, the observation of the 7×7 reconstruction of the annealed Si(111) surface [5] immediately showed its enormous potential by resolving the long-standing discussion of the actual surface reconstruction of this silicon surface. Also in the field of semiconductor growth it has been extremely useful that STM can image the growth surface in atomic detail. Already quickly after the invention of the STM technique it was realized that cleavage is promising way to create a clean semiconductor surface [28]. It was straightforward to look "inside" semiconductor materials by cleaving an epitaxial grown semiconductor multilayer structure and imaging the cleaved facet with an STM [2]. The fact that STM is only sensitive to the topmost layer of the cleaved structure allows to resolve atomic details such as alloy fluctuations with atomic scale precision something which is impossible even with atomically resolved transmission electron microscopy (TEM) due the fact the information shown in TEM images is obtained by electrons that travel through many atomic layers.

In this chapter, we focus on cross-sectional scanning tunneling microscopy (X-STM) applied to cleaved semiconductor structures. The cleavage process conserves the atomic structure of the cleaved semiconductor nanostructures and the original lattice position of impurity atoms. X-STM is therefore well suited to study the composition of ternary and quaternary compound semiconductor nanostructures in their undistorted configuration. Especially the spatial distribution of the different components can be explored on the atomic scale with X-STM. It is thus not only a very powerful tool to investigate atomic scale properties of semiconductor nanostructures but also the properties of individual impurities in a semiconductor material.

11.2 Cleaving Methods and Geometries

In general the X-STM technique is based on cleaving a sample by brittle facture, as a convenient way to create a clean surface. In contrast to repeated cycles of thermal treatment and sputter erosion applied to the sample in other commonly used techniques to prepare a clean, atomically flat surface, cleaving is a "cold" preparation process. Thermal cleaning steps may result in diffusion of the atoms close to the surface and for compound materials such as III–V semiconductors temperatures above ~500 K even lead to decomposition of the material rather than to cleaning the surface. The constituent atoms including impurity atoms buried in the crystal do not undergo a random diffusion in case of preparing a surface via cleavage, because no elevated temperatures are involved. The resulting conservation of the atomic arrangement by the crack passing through the material has been deduced from the investigation of atomic terraces, steps, and kinks on GaAs(110) [86, 87] and Si(111) [40]. Neither the observed distances between the step edges nor the density of kinks agree with the theoretical predictions for a surface in thermal equilibrium configuration. The direct comparison between a Si(001) surface (prepared by thermal cleaning [57]) scanned at 520 K (shown in Fig. 11.1a) and a cleaved Si(111) surface scanned at 300 K (shown in Fig. 11.1b) proves this conclusion. In thermal equilibrium, the step edges meander much more than on the cleaved surface. The different terrace width distribution between the equilibrium configuration and the experimental results is clearly seen in Fig. 11.2 for the furrow configuration of GaAs(110) [86,87], proving once more the nonequilibrium character of a cleaved surface.

Brittle fracture approaches the ideal situation of simply removing the upper part of the crystal. However, brittle fracture is a highly nonlinear process that is still far

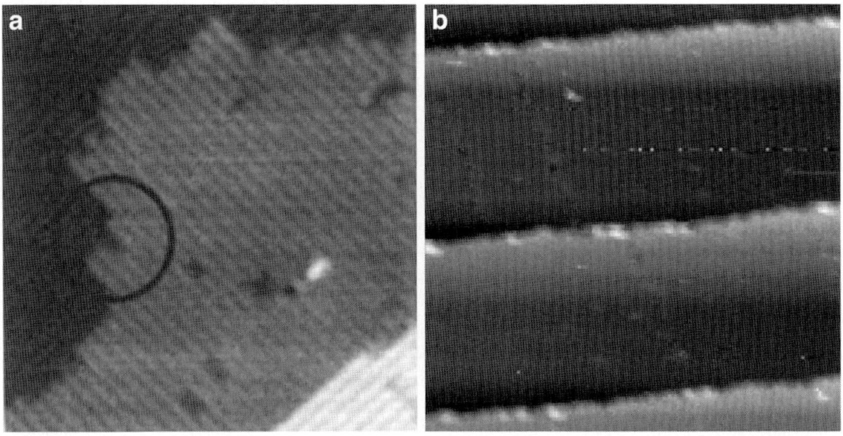

Fig. 11.1 Constant current topography images of a Si(001) surface in thermal equilibrium at 520 K (**a**, from Ref. [57]) and a freshly cleaved Si(111) surface at 300 K (**b** from Ref. [40])

MÖHRE HER....ODER ICH FÖHNE !!

Dez.
05
iker
htsmarkt -

0 bis 16.00 Uhr geöffnet !

Fahrrad-Dratz :
selbst zusammen und lassen
schenken !

ere Schneebar :
- Stollen - Lebkuchen

eits eingetroffen ...

mberg, Tel: 0951-12428

d ein gutes Neues Jahr !

Fürs Kind

Schlichtes 3/4-lg. Komm.-Kleid (extra weit), Gr. 152, 50 €, **Schuhe,** Gr. 36, 10 €, **2 Fahrr.-Helme,** rosa, je 3 € ☎09503/505787 ab 18 h

Vk. Kombi-Ki.-Wag., neu, VB 120 € ☎0173/2119376

Elektr. Babyschaukel + Motor, Badewannenständ., Babykorb tragbar, Kleidung ab Gr. 56, u.v.m., günst. ☎09547/60286

Kindergruppe St. Stephan hat Platz - ab 1 1/2 J. ☎16910

Gesundheit

Die neue Kräuterkapsel –
der geniale Weg zur **guten Figur!**
Jutta Lintner ☎ 0 95 05 / 78 88

MassageRaum
Touch*Life*-Hot Stone und Klangschalenmassage auch zum Verschenken
Claudia Maria und Jürgen Rattinger
96103 Hallstadt • Tel. 0951 / 72585

Weihnachtsangebote !!!
Med. Massage 20 min, 11,--€
Hot-Stone-Massage 80 min, 40,--€
Aromamassage 80 min, 36,--€
Fußreflexzonenmassage 40 min, 24,--€
Nadine Schlott, Physiotherapeutin
Stufenburgstr. 18, 96169 Lauter
☎ 0 95 44 - 98 50 49

Florija's
Aroma - Oel - Massagen
Fußpflege,
Kornstraße 6
(kostenl. Parken vor der Praxis
Nähe ALDI Münchner Ring
Termine: ☎ 993 12 85

Weihnachtsgeschenkideen: Ayurveda Ganzkörperöl- massagen & Qi Gong z. Kennenlernen in Bbg. Frisan Claudia ☎0951/1890891 od. 0176/ 20442764

Hausrat

Tausche Gefrierschrank geg. E-Herd ☎0951/3027176

Bauknecht Gefrierschrank, Maße: 95*60, Preis: VB 50 €

Braut Vertmode
Memmelsdorf/Schmerldorf

Neuw. Brautkleid, Gr. 36/38 inkl. Schleier u. Handschuhe, VB 150 € ☎0173/5133458

2 He-Trachtenanzüge dkl.-grün/ braun, sehr gut erh., Gr. 50/52, Pr. VS ☎09547/1535

2 Skianzüge, Gr. 140 u. 176 zu verk. ☎0951/61811

Schlichtes Komm.-Kleid, Gr. 134, m. Zubeh., 60 € ☎0951/ 1337298

Komm.-Kleid, Gr. 146,5, extra weit, m. Zub., Schuhe, Gr. 34 ☎09502/8268

Vk. Kleider, Gr. 36/38, wenig getrag., mehr. **Schuhe/Stiefel,** Gr. 40 ☎0951/1868125 od. 0176/22019388 abends

Vk. neuw. Pelzjacke (Nerz), Gr. 40-42, VB 350 € ☎0173/ 8032421 ab 18 h

Hübsche Pelzjacke, Gr. 42, 100 € ☎0951/15252

Beilagenhinweis

Bitte beachten Sie die Prospekte in dieser Ausgabe!

Vollbelegung:
Marktkauf, Bamberg

Teilbelegung:
Ertl Spielwarengigant, Hallstadt
dm-Drogerie, Bamberg
KIK-Märkte

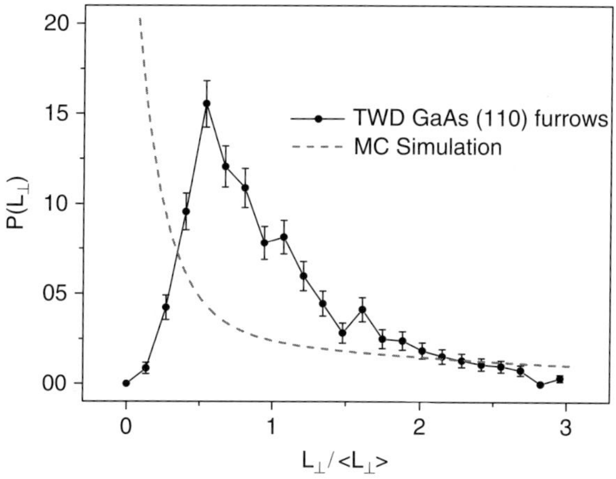

Fig. 11.2 Terrace width distribution (TWD) observed on GaAs(110) in X-STM compared with a MC simulation (from Ref. [88])

from being fully understood [1, 18, 36–38, 40, 48, 49, 71, 86, 87, 89, 91]. The large scale of dimensions involved in the cleavage process – close to the crack tip single atomic bonds have to be considered, whereas the distribution of stress, strain, and torque spread over the whole sample clearly exceeding the mm scale – is challenging for numerical simulations. Despite the attempts that have been made, first principle calculations still cannot handle the quantum mechanical description of the large numbers of atoms involved in brittle fracture of realistic samples. Molecular dynamics (MD) calculations are in a somewhat better position as they simplify the quantum mechanical inter-atomic interactions by a (semi) empirical potential. However, the nonlinear Lennard-Jones or Morse potentials in combination with $\sim 10^{14}$ atoms in 2D, or even $\sim 10^{20}$ atoms in a 3D approach, are at the limits of available computing power. Furthermore finding a well-suited potential is challenging, because brittle fracture depends on details of the assumed potential at large inter-atomic distance. This long-range part is very difficult to verify independently, because only melting or evaporating the material drive the atoms to comparably large distances. Therefore, MD calculations depend on the choice of the potential.

Nevertheless, clever combinations of atomistic MD methods [56] close to the crack tip with less demanding descriptions by continuum mechanics [92] for the rest of the sample are employed to tackle this problem. However, the strong nonlinearity leads to chaotic behavior of the crack traveling through the material. Experimentally it is still a delicate issue to achieve flat terraces on the micron scale, especially if strained samples such as a nonlattice-matched heterostructure are cleaved. Therefore, we briefly present a few commonly used techniques to successfully cleave semiconductors. A number of surfaces can exclusively be prepared by brittle fracture or this method is the most convenient way to access them. Examples of surfaces that cannot be prepared by other means than brittle fracture are $Si(111)-2\times1$ and $Ge(111)-2\times1$. We also include studies of the cleavage behavior of (110) surfaces of III–V and II–VI materials, because most X-STM investigations are conducted on these materials.

There are several crucial conditions for the preparation of a surface by cleaving. The method is restricted to brittle materials that offer a natural cleavage plane, as e.g., the (110) surface for III–V materials and the (111) surface for Si and Ge [6]. The cleavage plane has to be charge neutral, and its surface energy has to be low. Therefore the cleaving direction cannot be chosen at will, and comparison of different surfaces of the same material system is not possible. The technical setup to cleave the samples has to be implemented using manipulators in a vacuum chamber – otherwise the surface would oxidize or catch up contamination quickly. On top of the facilities to cleave the sample, the STM has to be equipped with sufficient coarse movement of the tip relative to the sample and elaborated optical approach to find the cleaved surface with the tip. This is mainly due to the fact that good cleavage with flat surfaces on the atomic scale in many cases calls for rather thin samples on the order of 100 μm thickness.

As already stated, X-STM is based on the possibility to cleave a sample in a way that a smooth surface is formed. Brittle fracture depends nonlinearly on the load applied to the material on length scales from the individual atoms at the crack tip to the macroscopic dimensions of the sample where the external force is applied. Because these parameters are not fully controlled, the result of cleaving a sample cannot be predicted. Nevertheless, techniques have been developed that allow cleaving samples with a high success rate. Each group working on X-STM has its own recipe in order to cleave their samples. Normally, the recipes have to be implemented inside an existing UHV system, so not all desirable operations can easily be executed. Furthermore, the cleaved surface has to be produced within the range of the coarse approach of the STM system. Especially the height at which the sample is cleaved is crucial in case of using a beetle-type scanner, because this design allows a rather limited range of coarse movement in z direction (=removing or approaching the tip) of ∼1.5 mm (e.g. Ref. 88). In the following, a few general remarks about brittle fracture will be discussed and some ways of how samples are cleaved in practice will be presented.

Three principal modes of applying load to a sample are distinguished in engineering sciences as shown in Fig. 11.3. The load modes are classified by the orientation of the pair of forces relative to the opening cleaved facets and the crack front. Mode I

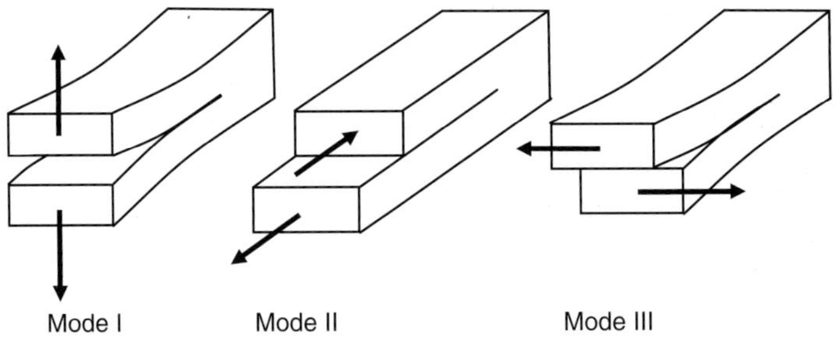

Fig. 11.3 Principal load modes defined by their orientation relative to the cleavage plane and to the crack front line

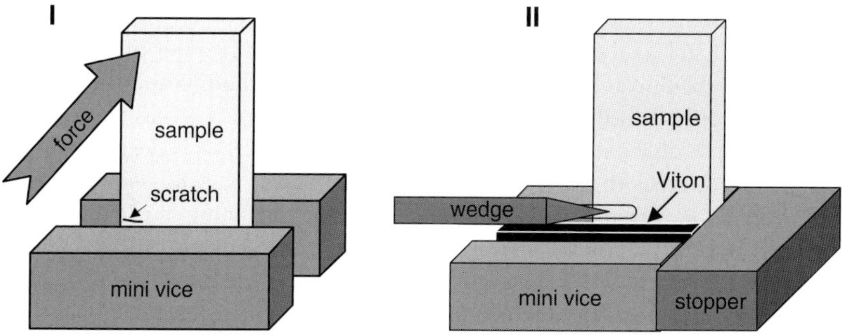

Fig. 11.4 Geometries and methods used in X-STM: I – breaking a scratched sample, II –cleaving a sample by pressing a wedge into a notch (from Ref. [40])

consists of strain applied perpendicular to the facets and to the crack front, mode II consists of a shear force parallel to the facets and perpendicular to the crack front, whereas in mode III a shear force is applied parallel to the facets and to the crack front. Neglecting binding angles, only mode I will break bonds on the atomic scale. However, cleaving procedures used for X-STM are mostly restricted by the technical possibilities in the UHV chamber such that a combination of different load modes is applied to the sample. Some very brittle materials like InAs or cuprate high-T_c superconductors [107] expose an atomically flat facet when simply breaking a rod that has been mounted in upright position on a sample holder by bending it. Achieving flat facets on many other materials especially containing heterostructures requires more sophisticated methods.

Figure. 11.4 schematically shows two commonly used cleavage geometries. Typical sample dimensions are 10 mm long, 3 mm wide, and 0.1–0.2 mm thick. Most X-STM work nowadays is based on cleaving III–V and II–VI crystals cleaved by using method I. A scratch is prepared close to the edge of the sample before it is clamped in a mini vice. In order to cleave the sample, force is applied at the edge of the sample plate as depicted by the arrow.

The crack always has to travel along the cleavage plane of the crystal. Because this cleavage plane cannot be perfectly aligned with the vice, the crack has to travel apart from the clamping blocks. Otherwise, the different orientations of the cleavage plane of the crystal and of the vice will disturb the crack propagation. This results in a rough surface. Clamping the sample only at the edge below the scratch is a possibility to further reduce the distortion of the crack propagation by the vice.

The main goal of cleaving samples for X-STM is to expose a flat surface. Therefore, the generation of step edges and other surface defects has to be reduced. An atomically flat surface has the lowest energy because surface defects like vacancies, adatoms, and step edges cost energy. During cleavage, this energy is released from the elastic energy that is deposited on the sample by the applied external load. In order to achieve a flat surface, the amount of elastic energy has to be reduced. This calls for a good starting point for the crack in order to cleave the sample at low load.

At a given external load, the local load depends on the geometry of the crack tip. External and local load are connected by the inverse square root of radius of the crack tip. The local load has to exceed the bond strength in order to make the crack propagate. In order to reduce the elastic energy in the sample this has to be achieved at the lowest possible external load. So a very sharp pre-crack is desired as a starting point for the crack to travel though the sample. Such a starting point is prepared experimentally in different ways: cutting a notch or scratching the sample. Initiating the crack propagation by a scratch results in much flatter surfaces than using a notch for III–V and II–VI materials. Scratches are made either with a diamond pen or with a tungsten–carbide needle, and can be applied manually. In order to scratch in a reproducible manner, automatic systems were implemented that control the geometry and applied force during scratching. No comparing studies with significant statistics have been done yet to find out which exact clamping position and which method to scratch the sample is superior. Different combinations of parameters were found that allow to prepare atomically flat terraces with only a few steps per square micron for many III–V and II–VI materials and heterostructures with a good success rate.

Achieving a flat surface by cleaving other materials can be difficult. In Si and Ge, the (111) and the (110) surface compete as natural cleavage planes. In general, the cleaved facets are rougher on group IV semiconductors than on the III–V materials. Si has been cleaved exhibiting the (111) surface by method I [101]. However, the facets were rather rough. The success rate depends on unknown details of the procedure, and was not reproduced in our group. Therefore we developed method II (see Fig. 11.4), which is based on cutting a slot into one side of the sample. The sample is clamped in a vice with a stopper at the opposite side facet, and a layer of viton (UHV capable rubber) in order to clamp it more tightly. A wedge is pressed into the slot until the sample cleaves. This geometry guarantees cleavage at the desired height and yields reasonably flat surfaces for the investigation of the surface reconstruction and of individual impurities in homogeneously doped samples. It has also been used to cleave Si along the (110) plane, which has a 15% larger surface energy than the Si(111) surface. Preliminary experiments outside UHV achieved much larger flat terraces, which was attributed to the stopper supporting the complete sample. Furthermore, the wedge must slide smoothly into the slot in order to expand it – what applies a mode I load, as depicted in Fig. 11.3, at the bottom of the slot. Therefore, the wedge has to consist of sufficiently hard material. No flat cleavage is achieved if the side facets of the wedge are rough and thus stick at the entrance of the slot. This confirms the impact of the detailed load mode on brittle fracture.

Optical microscopy images of cleaved GaAs(110) and Si(111) surfaces are shown in Fig. 11.5. The characteristic pattern on the surface is illustrated by the schematic representation next to it. Each pattern clearly depends on the cleaved material and on the applied geometry. Six different areas appear when cleaving GaAs along (110) using geometry I [89]. Similarly four different areas are observed in case of Si cleaved along (111) using the same method, whereas a rather different structure consisting of only two different areas is formed in case of Si cleaved along (111) in geometry II. The comparison of optical microscopy images with the results

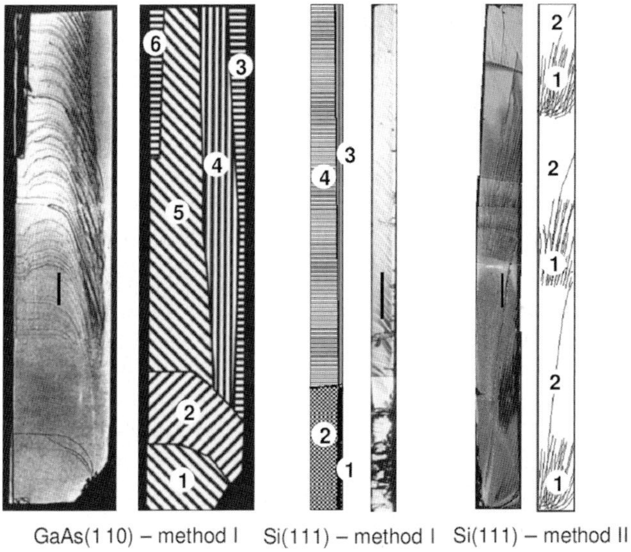

Fig. 11.5 Characteristic crack patterns exhibited on GaAs(110) and Si(111) surfaces after cleavage using method I or II. In case of using method I, the scratch was located at the bottom right corner of the samples, the notch. In case of using method II, the notch was at the bottom of the sample (from Refs. [40, 89])

of atomic force microscopy (AFM) and STM [40, 89] proves that cleavage-induced steps are parallel on all length scales. Unfortunately, the microscopic propagation of the crack inside the material is not known, which hinders a direct connection of the orientation of the structures found on the surface with the direction of crack propagation. In case of geometry II, the crack macroscopically travels from the notch toward the back facet of the sample. The steps are mainly parallel to this direction.

11.3 Properties of Cleaved Surfaces

11.3.1 The (111) Surface of Silicon and Germanium

Cleaved surfaces can be terminated by a metastable surface reconstruction. X-STM is the only way to investigate these reconstructions on the atomic scale in real space if they are exclusively formed by brittle fracture and destroyed by thermal treatment. A prominent example for a cleavage-induced reconstruction is the 2×1 structure on the (111) surface of Si and Ge. It is formed by cleavage at room temperature, whereas thermal treatment results in the 7×7 structure on Si(111), and the $c2 \times 8$ structure on Ge(111).

Scanning tunneling microscopy was used to study the temperature-dependent transformation of the Si(111) surface [30]. The 2×1-reconstructed surface was prepared by cleavage and subsequently annealed. The transition from the 2×1 to the 7×7 structure involves two intermediate configurations: at lower temperatures a disordered 1×1 surface covered with adatoms appeared, and at elevated temperature a well-ordered 5×5 structure was found. The kinetics of the transformation rate was measured as a function of time for temperatures in the range 550–700 K. Below 600 K, the 2×1 surface transforms mainly into disordered adatom arrangements. The transformed areas grow nonlinearly with time, because the number of excess adatoms on the surface is limited. Above 600 K, domains of the 5×5 and 7×7 structure are formed on the disordered part of the surface. Even on the remaining 2×1 surface, the 5×5 domains start to grow. The transformation rates for these processes are linear with time. The temperature dependence of the growth rates leads to an activation energy of 2.1 ± 0.5 eV for the transformation process.

The 2×1 reconstruction is especially interesting because the surface atoms form π-bonded chains [81]. The electronic states connected to the chains are highly anisotropic with a strong coupling between atoms along the chain and minor hopping perpendicular to the chains. As expected, the Ge(111) surface is quite similar to Si(111). They have been investigated and compared by X-STM and tunneling spectroscopy by Feenstra [31,33]. Ge(111) as well as Si(111) form the 2×1 reconstruction on the cleaved surface, even though the stable configurations after thermal cleaning are clearly different on both materials: the 7×7 structure on Si(111) and a $c2 \times 8$ reconstruction on Ge(111). The geometry and energetic properties have been studied theoretically by DFT calculations [96] to investigate the origin of the different configurations of the (111) surface of group IV semiconductors (C, Si, and Ge).

The electronic structure of Si(111) $- 2 \times 1$ was studied with STM and scanning tunneling spectroscopy (STS) in more detail on the sub-Å scale at low temperatures. The results were compared with ab initio calculations of surface unit cell [41]. Due to the sub-Å resolution in the lateral dimension, these measurements resolved the voltage-dependent electronic structure of the 2×1 reconstruction within the surface unit cell. The dominant effects are a characteristic LDOS shift between the up and the down atom when crossing the surface band gap, a similar shift inside the empty surface state band with a transition region around $E - E_F = +0.4$ eV, and a transition to dominating [2–11]-orientated corrugation close to the band edges. All three observations fully agree with the theoretical results based on the π-bonded chain model.

The anisotropic electronic band structure of the surface states leads to a quasi 1D electronic system that has been proven by the detection of a Coulomb gap due to the small capacitance of this low-dimensional system [42]. The gap width depends linearly on the inverse length of the addressed chain segment between neighboring atomic steps and agrees quantitatively with the classical approach of a Coulomb blockade resulting from the capacitance of a 1D object of finite length.

11.3.2 The (110) Surface of Silicon

Several reconstructions are also induced by the different ways of preparing the Si(110) surface. The annealed Si(110) surface [4, 80] has been investigated in much more detail than the cleaved surface. In STM measurements 5×1 and 16×2 reconstructed domains were found, whereas scattering techniques like LEED and RHEED additionally reported the following structures: 2×1, 7×1, 9×1, and 4×5 [117]. Although a Si crystal can indeed be cleaved macroscopically exposing a flat (110) surface [18, 67], the surface appears rough on the sub-nm scale in AFM as well as in STM experiments [40]. Atomic force microscopy revealed terraces with a height difference roughly resembling monatomic steps. Lines along the [011] direction on the terraces seem to reflect a reconstruction on the surface. However, the lateral resolution of the instrument on the order of tenths of nanometers does not allow clear-cut evidence. STM scans on the Si(110) cleavage surface in UHV indeed provide sufficient resolution, but they nevertheless did not unravel the surface structure of Si(110). An "atomically rough" surface was found with stripes roughly along the ⟨110⟩ direction in the surface. The peak-to-peak amplitude of the stripe pattern significantly exceeds the height of monatomic steps. No convincing periodicity is found in the pattern, and STM did not confirm a reconstruction consisting of enhanced lines parallel to the ⟨110⟩ direction. It remains an open question why an "atomically rough" seems to be favorable on cleaved Si(110) facets. The large number of surface defects should lead to a strongly increased surface energy with respect to the perfectly flat configuration. This might be connected to the large energy at the crack tip traveling through the crystal.

11.3.3 The (110) Surface of III–V Semiconductors

Quite similar to the cleavage-induced reconstruction on the elementary semiconductors Si and Ge, also on compound semiconductors cleavage exposes a clean and flat surface that is hardly available by other techniques. The atoms on/in the cleaved GaAs(110) surface relax due their reduced coordination. Since the unit cell has the same size at the surface as in the bulk, the rearranged configuration is rather called a relaxation than a reconstruction. It modifies the electronic properties at the surface less dramatically than the reconstructions on Si(111), because the electronic surface states are located in the projected bulk bands. Nevertheless, four different surface states are observed in X-STM measurements [23, 83] by their specific corrugation that is known from ab-initio calculations [15]. The absence of surface states in the band gap furthermore leaves the Fermi level unpinned at the surface. The electric field of the metallic tip therefore bends the bands in the first few nanometers below the surface, and the externally applied voltages no longer directly correspond to the energy scale of the band structure. The so-called tip-induced band bending (TIBB) [29] can be calculated similarly as the band bending in the semiconducting side of a Schottky contact. However, a number of crucial parameters (e.g., tip radius and tip work function) can only be estimated. The comparison of the external voltages at which the surface states are observed experimentally with their predicted energetic positions in the band structure served to quantify the TIBB [84].

Recently also the nonpolar (1–100) surface of GaN as a III–V semiconductor with a hexagonal crystal structure has been studied with X-STM [7]. As predicted by theory, no cleavage-induced reconstruction is observed. At clean surface areas with defect concentrations below $2 \times 10^{12}\,\mathrm{cm}^{-2}$, the Fermi energy is not pinned, whereas a defect concentration of $3 \times 10^{13}\,\mathrm{cm}^{-2}$ in other areas is sufficient to pin the Fermi level at the surface.

11.3.4 The (110) Surface of II–VI Semiconductors

In X-STM experiments performed on the II–VI materials ZnTe and CdZnTe, dark lines were found at negative polarity parallel to the [1–10]-direction [111]. In high resolution these lines turn out to consist of a row of vacancies. Charge neutrality suggests that along these lines both the Te and the Zn atoms are missing. Since the lines are found in higher concentration in areas with increased density of cleavage-induced steps, they are addressed as a special kind of cleavage-induced surface defect. The lines are typically found on II–VI, but never on III–V material although the (110) surfaces of both material systems are similar in theory. The difference seems to stem from the different hardness and ionic character of II–VI in comparison with III–V materials [3, 119].

11.4 Semiconductor Bulk Properties

11.4.1 Ordering in Semiconductor Alloys

Most of the III–V compound semiconductors form the zinc-blende crystal structure, which consists of two fcc sublattices with the group III element on one, and with the group V element on the other sublattices [e.g., Ref. 6]. Different chemical elements from group III and from group V can be mixed to build up ternary, quaternary, or even more complicated semiconductor alloys as long as the lattice mismatch is sufficiently small to grow single crystals of these alloys. Well-known examples are $Ga_xAl_{1-x}As$, $GaAs_xP_{1-x}$, $In_xGa_{1-x}As_yP_{1-y}$, where x and y can be varied over a wide range. In contrast to dual materials like GaAs or InP that are perfectly ordered with one element on each sublattice, the introduction of a second group III or group V material opens up an extra ordering parameter. The new component can either form a sublattice or it can be located on random lattice positions in the material. A variety of ordered structures has been observed in III–V semiconductors, e.g., the CuAu structure in (AlGa)As [60] and CuPtB in

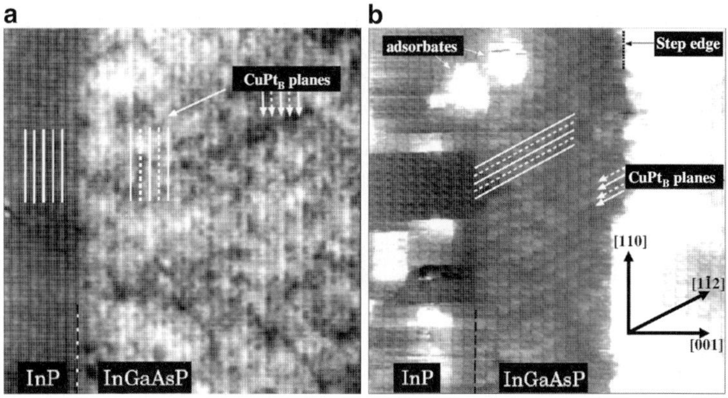

Fig. 11.6 Constant current images of InGaAsP (1–10) (**a**) and (110) surface (**b**) (from Ref. [54])

(GaIn)P [46]. An overview is given, e.g., by Stringfellow and Chen [98]. Nonlocal approaches to measure the ordering or random distribution of semiconductor alloys as LEED, RHEED, or detecting changes in the band gap or the lattice phonon modes always give an average over rather big volumes. This can be a main disadvantage because the ordering domains can be quite small – down to typical length scales of even smaller than 5 nm. The effects of opposite domains might cancel out, and macroscopic approaches thus underestimate the degree of ordering within the domains.

Macroscopic ordering depends on several parameters during the growth of the material [109]: growth rate, growth temperature, the ratio between element III and element V during growth, as well as the incident angle in molecular beam epitaxy (MBE) and the doping of the substrate. On top of this, the measured ordering might differ between these macroscopic approaches and the atomic configuration on the atomic scale within a domain. X-STM can detect the positions of individual atoms in the cleaved surface. Therefore, it determines the microscopic ordering. This has been done for ternary (GaIn)P [51] as well as for quaternary $In_{0.75}Ga_{0.25}As_{0.5}P_{0.5}$ [54]. As expected for $CuPt_B$ ordering – that means a sequence of alternating 1–11 layers which are rich of one species, a qualitative difference was observed between the (110) and the 1–10 surface. On 1–10, the ordering planes intersect the cleaved surface parallel to the (110) direction, which is also followed by the atomic rows of the relaxed surface atoms, whereas the intersection between the (110) cleavage surface with the 1–11 ordering planes runs along the 1–12 direction. This is shown in Fig. 11.6.

As the different species can be distinguished in topographic images in the STM, the ordering has been determined by directly counting the atoms of both species in each row. This can be done within the domains, because the domain walls and anti-phase boundaries are visible in the STM images. The result is an unexpected high-ordering parameter of 0.73(±3). That means 85% of the As and P atoms are found on the positions where they are supposed to be located in perfect CuPtB. In

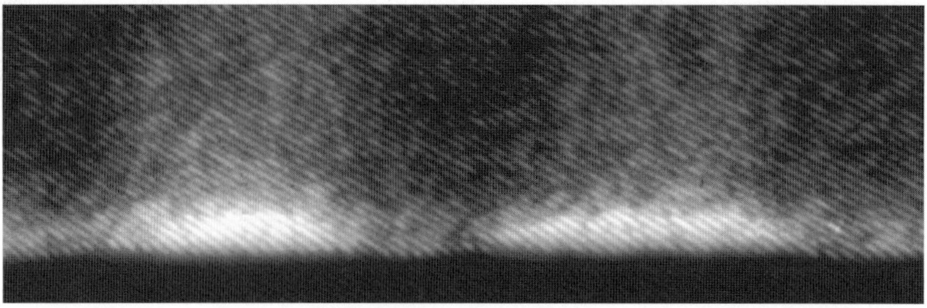

Fig. 11.7 Filled-state topography image of a layer of InAs quantum dots grown on (311B) InP and capped with InGaAs The capping layer shows a clear inhomogeneity due to composition modulation in the InGaAs capping layer. Image size 92 nm by 20 nm and the growth direction is along the vertical direction of the image (from Ref. [13])

contrast to the high ordering on the group V sublattice, no significant ordering was found for the group III sublattice [54]. Fluctuations of the contrast on a length scale of a few nm give rise to the idea of 3D clustering of the both types of group III atoms on their sublattice [82].

11.4.2 Phase Separation Effects

3D clustering or composition modulation is often observed in semiconductor alloys. The phase separation process can be induced by the presence of a strain field, as is the case, e.g., when capping quantum dots (QDs) with ternary alloys [13, 104]. On the other hand, composition modulation can also simply be the result of a strong miscibility gap between the two binary components. Figure. 11.7 shows InAs QDs grown on (311B) InP and capped with InGaAs. An inhomogeneous contrast showing clearly brighter regions is observed due to composition modulation in the InGaAs capping layer. Since in high voltage images the relaxation is proportional to the strain (see Fig. 11.11), the relaxation is in this case proportional to the In content: the brighter regions are In-rich regions and the darker ones In-poor regions. Strong composition modulation in GaAsSb which is likely due to the large miscibility gap of GaAsSb has also been reported [70]. By comparing the outward relaxation profile between Sb-rich and Sb-poor regions, strong fluctuations of 12% in the Sb content have been found in high Sb content layers [106].

11.5 Low-Dimensional Semiconductor Nanostructures

Low-dimensional semiconductor structures like quantum wells (QW) and self-assembled QDs have attracted much attention in the last decades [8, 93]. Since the first observation of confinement-induced quantum effects in the form of a step-like absorption spectra in a GaAs/AlGaAs QW in 1974 [21], the QW-based technology experienced a very fast development [99, 102, 123], and QWs are currently used in many optoelectronic devices at a commercial level. These nanostructures are typically grown pseudomorphically on a substrate with a different lattice constant and are therefore strained. QWs, as well as QDs, are typically grown by techniques like MBE or metal-organic vapor phase epitaxy that allow a precise control of the interfaces at the monolayer scale.

QDs are more recent nanostructures. They are very interesting from a scientific point of view because they form nearly ideal zero-dimensional systems in which quantum confinement effects become very important. These unique properties make them also very interesting from the technological point of view. For example, QDs are employed in QD lasers [27, 52], single electron transistors [118], mid-infrared detectors [95, 108], single-photon sources [73, 122], etc. QDs are commonly created by Stranski-Krastanov growth mode when a certain material (i.e., InAs) is deposited on a substrate with a bigger lattice constant (i.e., GaAs or InP) [22]. Above a certain critical thickness, 3D islands are spontaneously formed on top of a wetting layer (WL) to reduce the strain energy. Once created, the QDs are subsequently capped, a step which is required for any device application.

For any application based on QWs or QDs, an accurate control of their electronic properties is required. The electron and hole states in the nanostructures are very sensitive to its size, shape and composition (as well as to the surrounding material), and therefore, an accurate control of the QW or QD structural properties is necessary for applications. This explains, for example, the strong effort dedicated in the last years to the structural characterization of QDs [94], which is essential in order to have a deep understanding of the QD formation process. Cross-sectional STM is an especially useful and unique technique for this purpose, because it allows to access the structure of the buried nanostructures at the atomic scale. Since the first X-STM studies of semiconductor nanostructures [26, 62], this technique has provided crucial information which has strongly helped to understand the formation process of semiconductor low-dimensional structures. It has also a strong impact on the field of theory and modelling of nanostructures allowing to make calculations with realistic size, shape, composition, and strain parameters. In the following, a brief summary of the capabilities of X-STM applied to semiconductor nanostructures is shown through some particular examples. All the measurements shown in this section were taken in constant current mode.

11.5.1 Quantum Wells

The composition of a QW can be determined by X-STM by directly counting individual atoms. As mentioned in the previous section, depending on the polarity of the applied bias between tip and sample, either the empty states of the type III elements ($V_{\mathrm{sample}} > 0$) or the filled states of the type V elements ($V_{\mathrm{sample}} < 0$) are imaged. Since impurity-related features can be observed down to several monolayers below

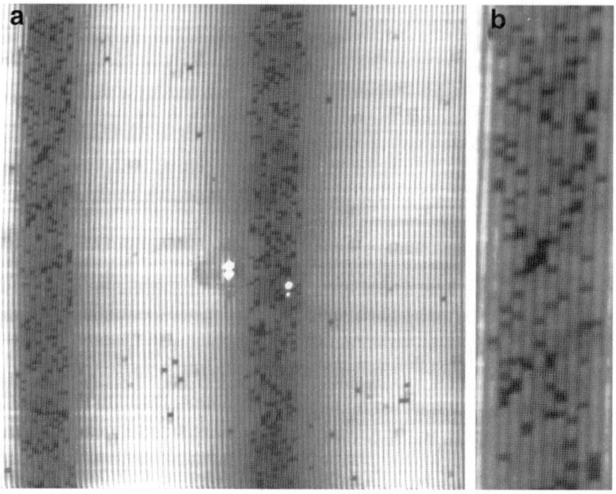

Fig. 11.8 Filled-state topography image (**a**) of 60 nm by 60 nm of the (110) cleavage plane showing two ZnSeTe/ZnTe QWs and (**b**) magnified part of the ZnSeTe QW region. The dark spots are individual Se atoms (from Ref. [111])

the cleaved surface, the first step toward composition mapping is the identification of the atoms in the different planes. This can be done by characterizing and comparing the different observed features, which, as shown above for different impurities, would to some extent change depending on the position below the surface.

Two ZnSeTe/ZnTe QWs can be observed in the filled states X-STM image in Fig. 11.8 [111]. The samples were grown by MBE on a *p*-type phosphorous doped (100) ZnTe substrate. It must be noticed that, while X-STM has been commonly applied to structures based on III–V semiconductors, much less work has been done regarding the II–VI family. The QWs appear as dark layers in which the individual Se atoms can clearly be resolved as dark spots. Due to the different size and bonding configuration, Se atoms appear darker than Te atoms. The dark contrast in the QWs represents the surface relaxation of the tensile-strained ZnSeTe epilayers, which will be discussed later. It can be seen that the QW interfaces are abrupt and smooth, which means that no segregation was present. The thickness of the QWs is found to be 5 nm, which is in agreement with the nominal value.

Furthermore, Se atoms are observed in the ZnTe material in between the QWs. These Se atoms were most likely residual elements in the MBE chamber that have been incorporated in the material during growth of the ZnTe spacers. Since the X-STM image is atomically resolved, the concentration of Se atoms within the QWs can be deduced by simply counting the dark spots. This was done for both QWs and the resulting Se concentration profile as a function of the distance along the growth direction z can be seen in Fig. 11.9. Again, the abrupt QW interfaces can be traced back from this profile. The average concentration of Se atoms within both QWs was deduced to be 17.1 and 17.7%, respectively.

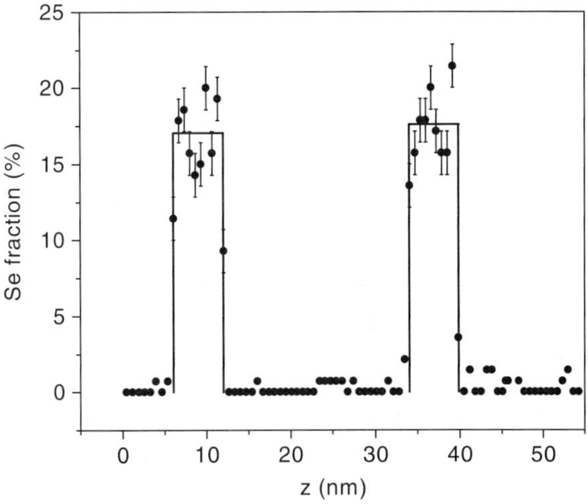

Fig. 11.9 Profile of the Se concentration within two quantum wells as a function of position along the growth direction z. The Se concentration is measured by counting the individual atoms (from Ref. [111])

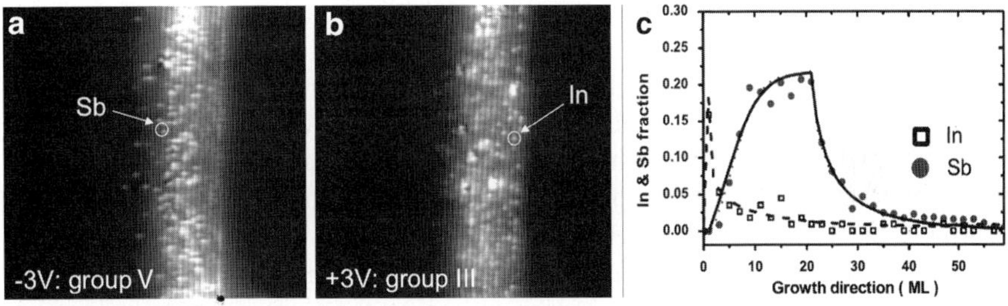

Fig. 11.10 In(Ga)As WL-GaAsSb capping layer system scanned at negative (**a**) and positive (**b**) voltages. Image size 38 nm by 36 nm. Sb and In profiles obtained by counting the In and Sb atoms in the X-STM images and the fitted distribution using a three-layer model for the segregation of In and Sb (**c**, from Ref. [50])

The same analysis can be applied to other 2D structures, like WLs [78]. In the case of InAs/GaAs WLs, an exponential decay with segregation coefficients of $R \cong$ 0.8–0.9 is typically obtained for the In concentration, and the amount of material in the WL is found to depend on the capping material and procedure [78, 105]. Indeed, more complex 2D structures like WL-thin capping layer systems can be characterized at the individual atom level by X-STM. An In(Ga)As WL grown on GaAs and capped with a 5-nm GaAsSb strain-reducing layer is shown in Fig. 11.10. Left and right X-STM images show the same area scanned at negative and positive voltage, respectively. The group V and group III atoms can be distinguished in this way, allowing an independent measurement of the In and Sb profiles by counting the individual atoms. The obtained profiles are also shown in Fig. 11.10. An average Sb content of 20% is found in the capping layer. Furthermore significant Sb segregation is found. Contrary to what one could expect, there is almost no intermixing between the In(Ga)As WL and the GaAsSb capping layer. As shown in Fig. 11.10c, this behavior can be well explained using a three-layer model to simulate the segregation processes of the In and Sb atoms during growth [50].

An alternative method to deduce the composition of a QW is by analyzing the strain-induced surface outward relaxation. The electronic contribution to the apparent height in the STM image can be suppressed to a large extent by scanning at high voltages [10, 34]. Since the tunneling probability decays exponentially with the height of the effective tunneling barrier, the states with the highest energy contribute most to the tunneling current. If a sufficiently large voltage is applied during empty-state imaging of a heterostructure, the states contributing most to the tunneling current have energies far above the conduction band edge and the contribution of the lower lying states is negligible. In this way, the electronic contrast induced by the different band gap energies becomes negligible. In the case of filled states tunneling (electron tunneling from sample to tip), the valence band offset in a heterostructure is smaller with respect to the total barrier height than the conduction band offset, since the effective barrier height for tunneling is increased by the band gap. Therefore, the resulting electronic contrast is smaller than in the case of empty states tunneling (electron tunneling from tip to sample). Thus, as demonstrated in Refs. [10, 34], at high (preferably negative) sample voltages, the electronic contribution to the height contrast is minimized, enabling the acquisition of the true topography of the structure, such as the outward relaxation of the cleaved surface.

When two materials with a different lattice constant are used in a QW or QD, the system will accumulate strain. Stranski-Krastanov grown QDs are the ultimate example of such strained heterostructures, but also QWs have built-in strain. When such a structure is cleaved, it reduces its built-in tensile or compressive strain by deforming the cleaved surface. Regions under compressive strain bulge outward while tensile strain depresses the cleaved surface (Fig. 11.11). The strain can be measured by X-STM since the distortion normal to the surface can be measured using (filled states) topography imaging, as described before. The measured outward displacement can be used to determine the composition of a strained nanostructure by comparing the experimental data with the calculated relaxation using elasticity theory [19], since the outward relaxation is directly proportional to the strain, which depends on the composition of the QW.

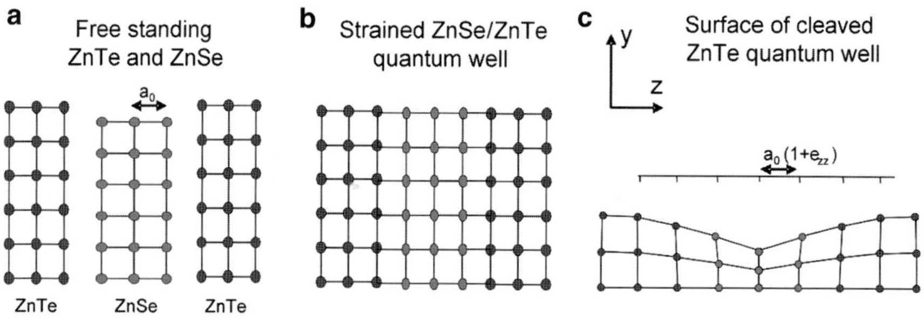

Fig. 11.11 Strain relaxation at the cleaved surface of a strained quantum well. (**a**) Natural lattices of free-standing ZnTe and ZnSe. (**b**) Due to the lattice mismatch, the ZnSe layer is tensile-strained in ZnTe. (**c**) The strain is released at the cleaved surface by outward relaxation

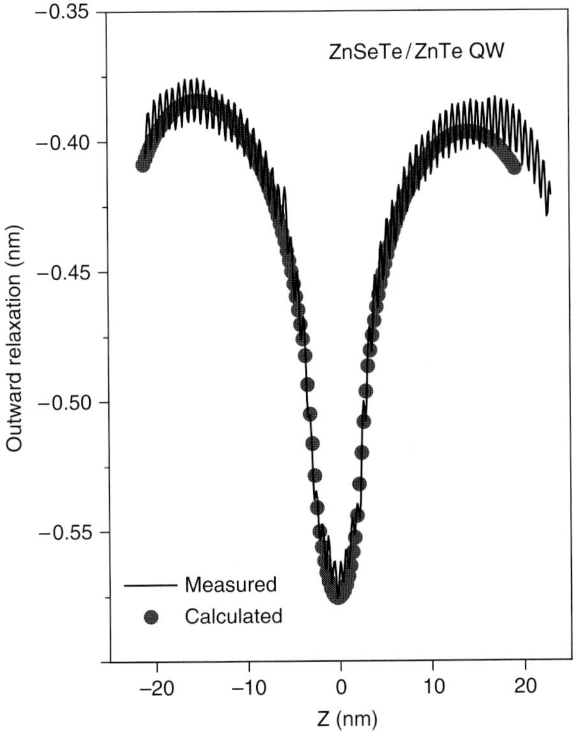

Fig. 11.12 Measured (*line*) and calculated (*dots*) outward relaxation profiles across one of the QWs in Fig. 11.8. The shown calculated profile is obtained when a Se content of 17% is considered (from Ref. [111])

Therefore, the Se concentration in the previously shown ZnSeTe/ZnTe QWs can be determined by using a finite element calculation package to simulate the surface relaxation of the QWs. For this method, continuum elastic theory is used and an isotropic material is assumed. Furthermore, the measured thickness of the QWs (5 nm) is used as an input for the simulation and the Se content is changed in order to obtain the best fit to the outward surface relaxation that is directly obtained from the X-STM images. Both measured and simulated outward relaxation profiles are displayed in Fig. 11.12. The shown fit was found for a Se concentration of 17%, which is in very good agreement with the value that was found by directly counting the Se atoms. This agreement confirms the high accuracy of the outward relaxation analysis technique for determining the composition of a strained QW.

11.5.2 Quantum Dots

The analysis of QDs is more complicated than that of the QWs. First of all, the QDs will be cleaved at a random position. Therefore, the real size and shape of the

Fig. 11.13 Pyramidal GaAs QD in AlGaAs grown by droplet epitaxy measured by X-STM. Image size 40 by 34 nm. The dark area at the top of the QD corresponds to an Al-rich region (phase separation) (from Ref. [55])

QDs can only be known after scanning a large number of dots. The size and shape of the QDs can be deduced by a statistical analysis of the relationship between the height and the base length in a large number of QDs. Pyramidal shape of QDs as shown in Fig. 11.13 for a GaAs dot in AlGaAs grown by droplet epitaxy, or truncated pyramidal shape has been often observed [11, 47, 76, 77, 103, 104]. Only after scanning a large number of cleaved dots, the maximum base length of the cross-section of the dot can be determined, which is indicative for a cleavage near the dot center. The possible size/shape inhomogeneities in the QD layer have to be neglected in this analysis.

On the other hand, QDs are not always truncated pyramids [9], and the shape of QDs made of the same material system can be strongly dependent on the growth conditions. Figure. 11.14a shows the height–base diameter distribution of high temperature grown InAs/GaAs QDs. The biggest dot in the distribution is shown in Fig. 11.14b after applying a local mean equalization filter to the image (which eliminates the background contrast due to strain relaxation and enhances the contrast of individual atomic rows). Therefore, this QD is considered to be cleaved through the middle and represents the real size of the QDs: a height of 8.1 ± 0.4 nm and a diameter of 27 ± 4 nm. Remarkably, the statistics of the height–base diameter relationship strongly deviates from a pyramidal shape (dotted line) typically found by X-STM [11, 47, 76, 77, 103, 104]. Better agreement is found using a lens-shape (dashed line) or ellipsoidal (solid line) QD, which agrees well with the topography of the dot outlined in Fig. 11.14b.

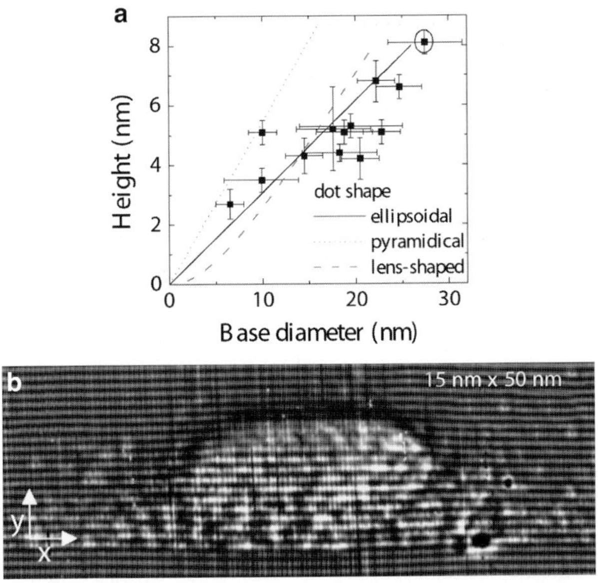

Fig. 11.14 (**a**) Height vs. base diameter for the measured high temperature grown InAs/GaAs QDs (squares). Theoretical curves for ellipsoidal (*solid line*, height $h = 8\,\text{nm}$, diameter $d = 26\,\text{nm}$), pyramidal (*dotted line*, $h = 8.7\,\text{nm}$, base $b = 29\,\text{nm}$) and lens-shaped QDs (*dashed line*, $h = 8\,\text{nm}$, $d = 22\,\text{nm}$). (**b**) Fourier filtered topography image of a representative QD assumed to be cleaved through its middle plane. This dot corresponds to the encircled data point in (**a**) (from Ref. [9])

It was previously shown how in the case of QWs the composition can in general be obtained directly by counting the individual atoms. However, in the case of QDs it becomes more difficult to distinguish individual atoms, and the main way to determine its composition is by fitting to the measured outward relaxation or change in lattice constant. Despite that, in some cases a good estimate of the composition can also be obtained by analyzing the presence of a small amount of impurity atoms. This is for example the case in Stransky-Krastanov InAs QDs when the In content is very close to 100% (Fig. 11.15) [106] or in droplet epitaxy grown GaAs/AlAs QDs with almost 100% Ga content [55]. In the first case, the presence of a few Ga atoms inside the QD can be detected even in filled-state images through the distortion that they induce in their surrounding As atoms (see the dark spots in Fig. 11.15). These dark spots reflect the presence of a small amount of Ga atoms inside the QDs. The amount of Ga is quite small and these MBE grown In(Ga)As/InP QDs are close to 100% InAs [106]. In the second case, a few Al atoms are also clearly observed inside the QD, but the total Al content is negligible and the QDs are almost pure GaAs [55].

The composition of the QDs is more often deduced by the analysis of the outward relaxation in high negative voltage images (topographic images) in a similar way that was done for QWs before. To calculate the outward relaxation of a cleaved QD, we use the package COMSOL. It is based on continuum elasticity theory and

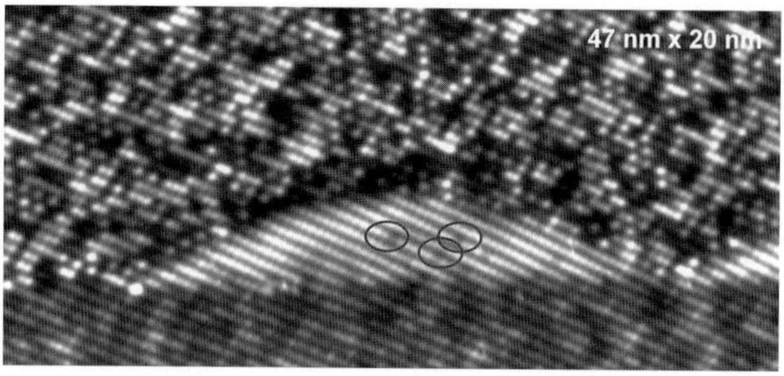

Fig. 11.15 High-resolution local mean equalization filtered image of an GaAsSb-capped In(Ga)As QD grown on a (311)B InP substrate. The *dark spots* inside the QDs marked by the *circles* correspond to individual Ga atoms (from Ref. [106])

performs a finite element calculation to solve the 3D problem, in which an isotropic material is considered. The measured size and shape of the QDs is introduced in the model and the composition is varied until the calculated profile fits to the experimental one. For better accuracy, the measured outward relaxation at different lateral positions in the cleavage plane (Fig. 11.16b) is compared with simulations (Fig. 11.16a). In these calculations, the QD consist of an InGaAs alloy, in which the internal indium concentration profile is described by an expression that uses the indium end concentrations at different points inside the dot and using linear interpolation in between [9, 77, 103]. The dot is assumed to be surrounded by a pure GaAs matrix and is on top of a seven monolayer thick $In_{0.225}Ga_{0.775}As$ WL. The thickness of the WL has been determined experimentally by X-STM at a position far from a QD and agrees well with a total indium content of 1.7 ML. Comparing the measured and modeled outward relaxation, we obtain excellent agreement with an ellipsoidal dot, whose shape is described by the equation $(x/13)^2 + (y/8)^2 = 1$ (dimensions in nm) and with the indium profile depicted in Fig. 11.16c. The indium concentration varies significantly over the dot. The fraction of InAs increases vertically from 70% at the base center to 100% at the top center. In addition, horizontally the fraction InAs decreases to 25% at the base edge, evolving in a 22.5% 7-ML thick WL. From X-STM and photocurrent experiments, it has been shown that low-growth-rate InAs/GaAs QDs have an increasing indium concentration in the growth direction [11, 12, 39]. However, other groups have reported InGaAs QDs with laterally nonuniform indium compositions showing an inverted-triangle, trumpet or truncated reversed-cone shape [61, 63, 77, 114]. In this case, we find that the indium distribution of the studied InAs/GaAs QDs mostly resembles the trumpet shape proposed in [74] and found by X-STM in Refs. [12, 77, 103]. It is important to note that this level of agreement between simulation and data could not be obtained with a lens-shaped dot or using an ellipsoidal dot without WL. The present simulations thus represent the simplest dot shape and composition that matches the data. More complicated faceted dots [17] cannot be disregarded but additional minor changes in the dot shape would not greatly influence the results we have found. These relatively tall dots with large indium gradients differ significantly from some

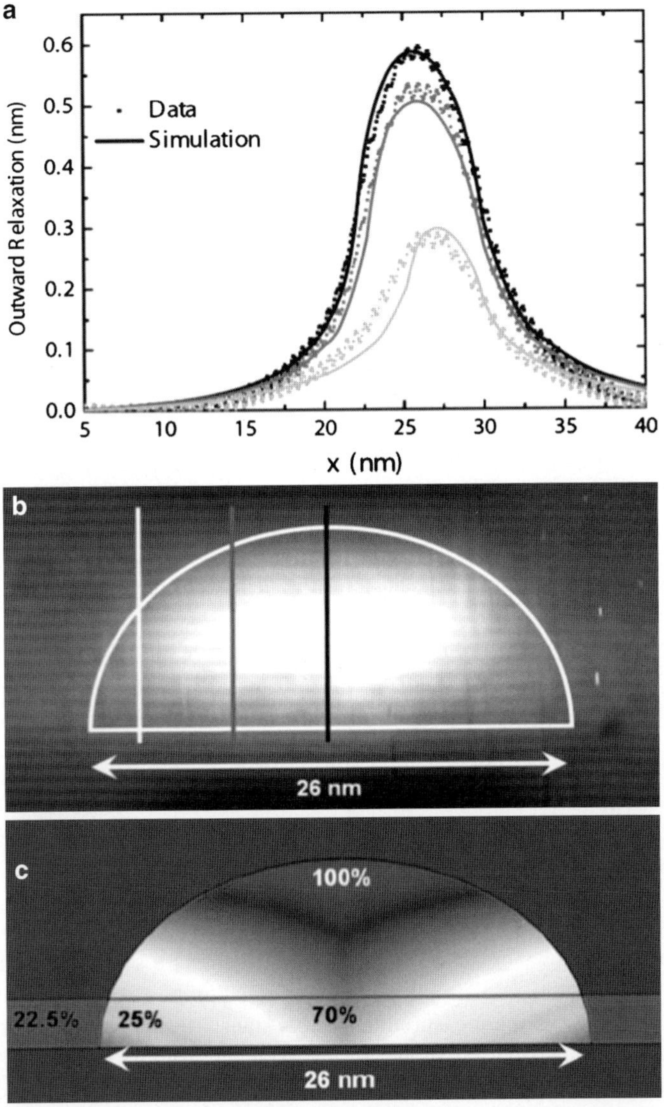

Fig. 11.16 (**a**) Measured (*symbols*) and simulated (*solid lines*) outward relaxation at three different positions with respect to the middle of an InAs QD. (**b**) Topography X-STM image of the tallest QD measured. The z-range of this image is 0.7 nm. The *solid lines* indicate the positions at which the outward relaxation is plotted in (**a**). (**c**) Simulated indium content of an InAs QD. The color scale indicates the percentage of indium arsenide composition and the given numbers correspond to the end values at the top center, base center, base edge, and WL (from Ref. [9])

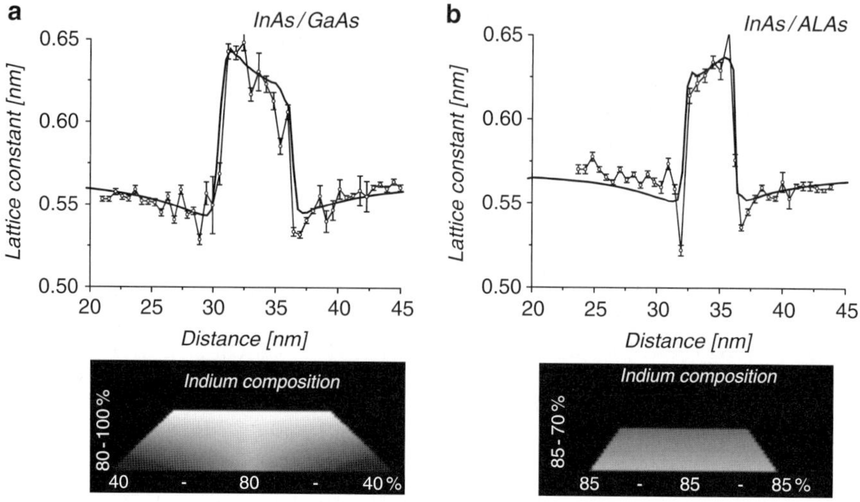

Fig. 11.17 Measured (*blue circles*) and calculated (*black line*) lattice constant profiles through the middle plane in the growth direction of an InAs/GaAs QD (*left*) and an InAs/AlAs QD (*right*). Below: In distribution inside the QD used in the calculation to reproduce the experimental lattice constant profile (from Ref. [77])

of the dots previously used in calculations, where usually a fixed InAs fraction is assumed.

Further confirmation of the QD inhomogeneous composition can be obtained by analyzing the lattice constant inside the QD. The lattice constant is determined experimentally in the middle of the dot by taking line profiles across the cleaved surface and as plotted in Fig. 11.17 measuring the distances between the atomic rows, using the Fourier filtered topography image of Fig. 11.14b. Since the (110) and (1–10) surface planes contain only half of the (001) layers within the zinc-blende crystal, the atomic rows observed in the X-STM image are separated by a bilayer distance in the growth [001] direction, which is equal to the lattice constant a_0.

Figure. 11.17 shows the measured lattice constant, together with the one calculated using the In distribution shown in the sketches, for InAs QDs grown on a GaAs buffer layer (left) and on an AlAs buffer layer (right) [77]. From the change in lattice constant, which is determined by the strain distribution in and around the QDs, it can be seen that there is compressive strain above and below the QDs. For the InAs/GaAs QD, there is a clear increase in lattice constant toward the top of the QD, which indicates an increasing indium concentration. The indium distribution mostly resembles again the trumpet shape proposed by [74] which we describe with a linear gradient in both the growth direction and lateral direction (see Fig. 11.17, left). For the InAs/AlAs QDs, the situation is quite different and the lattice constant decreases toward the top of the dot, indicating a decreasing In content. Best-fit results were obtained by an indium gradient of 85 to 70% decreasing from base to

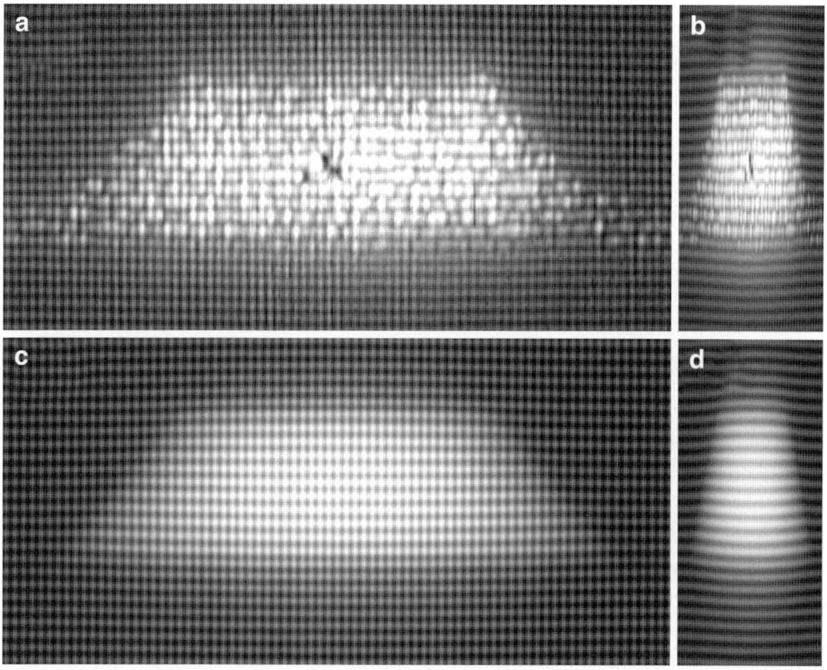

Fig. 11.18 (**a**) Fourier filtered X-STM image of a QD showing the bending of the atomic rows of the cleaved (110) surface due to compressive strain. (**c**) Simulated image based on the calculated strain and outward relaxation of the cleaved surface (a homogeneous In distribution of 70% is considered in the calculation). The image is compressed in the lateral directions in (**b**) and (**d**) to show the bending of the atomic rows more clearly (from Ref. [77])

top of the QD (Fig. 11.17, right). In this case, there is no indication for a lateral gradient in the indium composition.

From the analysis shown above, it is clear that measuring the QD lattice constant can also provide detailed information about the strain and composition of the QD. A mapping of the lattice constant of the QD could be performed by measuring the lattice spacing not only in the middle, but also in different positions along the QD. The result would give an experimental picture of the strain distribution of the QD. Moreover, the strain field of the QD can be calculated with the continuum elasticity model described above, and the strain data can be used to determine the shift of the atomic rows inside and outside the QD. Figure. 11.18 shows a Fourier filtered X-STM image of a low temperature capped InAs/GaAs QD (Fig. 11.18a), together with a simulation of the atomic rows in which the strain-induced shift was included (Fig. 11.18c) [79]. From the good agreement between the calculated and the measured in-plane displacement of the atomic rows, which can be better appreciated in the squeezed images on the left, it can be concluded that this is a real topographic effect due to the strain field of the QD, which can be measured by X-STM.

As shown above, X-STM can give a good approximation to the composition of QWs and QDs and in the case of QDs also the gradients inside the QD can be obtained. Nevertheless, there is a limit to the precision of the composition mapping. Recently we have shown that in order to explain optical spectra of exciton complexes, a simple linear gradient of the indium concentration is not sufficient to explain the observed ordering of the neutral and charged exciton transitions [75]. With X-STM, it is not yet possible to determine such fine details in the indium distribution; however, the dot size and the surface relaxation of a QD as measured by X-STM put severe constraints on the model calculations and thus is still of great value.

11.6 Impurities in Semiconductors

Impurities in semiconductors keep attracting intensive scientific interest, which is strongly stimulated by the crucial role that impurities play for the application of semiconductor materials in optoelectronic devices and since recently magnetic semiconductors. Impurities with a low binding energy easily release free carriers into the valence or conduction band of and serve to dope the semiconductor crystal. Prominent examples for doping atoms with a low binding energy are phosphorus and boron in silicon crystals [64], and carbon, beryllium, zinc, and silicon in III–V semiconductors [68]. Despite its rather high binding energy, the manganese acceptor in gallium arsenide has been extensively studied because of its magnetic moment that allows for the creation of ferromagnetic semiconductor materials [20,25]. Other defects with a high binding energy are sometimes desired because they act as traps to reduce the free carrier. In view of the ever-shrinking dimensions of electro-optical devices, the properties of individual impurities and their mutual interaction are of increasing crucial importance. Excitingly, recently several approaches have been developed and are under intensive study to manipulate just a single impurity in a device.

X-STM is a very suitable tool that allows studying impurities in real space at the atomic scale. Impurities in semiconductor material are not only observed in the topmost layer but can also be detected up to ten monolayers below the cleaved surface. When studying impurities in a semiconductor however, one has to be aware of possible surface-induced modifications of the impurity states compared with its bulk impurity properties. The STM images reflect the wave functions of impurities in or nearby the surface, which might significantly differ from those in the bulk due to the nearby presence of the surface. The latter seems to play a smaller but still appreciable role for impurities below the (110) surface of III–V materials as compared to impurities below the Si(111) surface where the surface properties dominate the contrast of the impurity state [43]. Striking modifications were also found to be induced to the properties of doping atoms in GaAs where the binding energy of Si [112] and Mn [45,97] were shown to be enhanced, and the symmetry of acceptor wave functions is broken [14] for impurities in the first few layers below the surface.

11.6.1 Impurity Atoms in Silicon

Donors (phosphorus) as well as acceptors (boron) were mapped at room temperature below the 2×1 reconstructed Silicon surface [90, 101]. Studying a defect, the first task is normally to identify its specific contrast in topographic STM images. This might seem straightforward, but the chemical contrast in STM is by far not as well understood as, e.g., the analysis of element-specific emission and absorption series in X-ray and electron spectroscopy. By comparison of the observed numbers of defects in the surface with the nominal doping concentration (between 5.5×10^{16} and $6 \times 10^{19}\,\text{cm}^{-3}$) for samples, the contrast pattern of P and B in Si(111) $- 2 \times 1$ was identified. In both cases, it strongly reflects the highly anisotropic electronic structure of the electronic surface states. Due to the enhanced stability of STM at low temperature, the resolution was improved and four slightly different patterns related to the P atoms in the 2×1 reconstructed surface were found [43], as shown in Fig. 11.19. The patterns (a)–(c) were identified as P atoms on distinguishable sites of the reconstructed surface. The lack of P atoms located on one of the four possible sites, and the unidentified contrast pattern (d) seemed to originate from the formation of the reconstructed surface during cleavage. Statistics on the pair distance between the P (respectively B) impurities revealed the random distribution of the dopant atoms in the surface. The conservation of the embedded positions in the cleavage process allowed proving that the doping atoms are indeed statistically distributed in the bulk of Czochralski grown sample material.

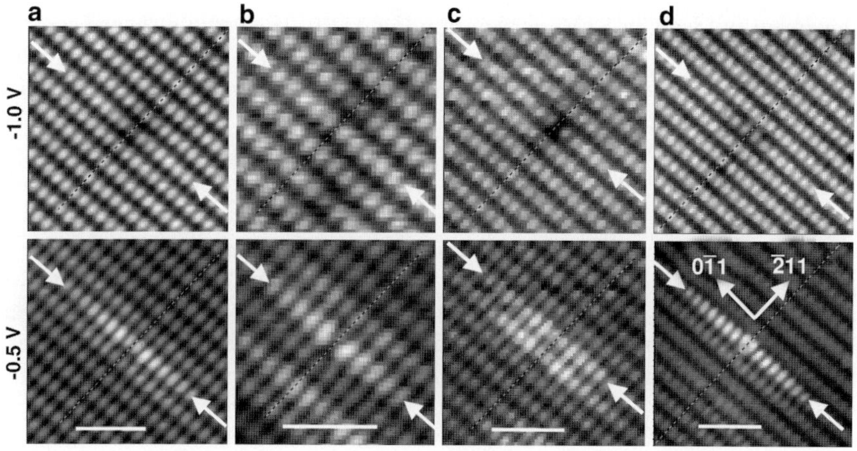

Fig. 11.19 P-induced defects on different lattice site in the 2×1 reconstruction of the Si(111) $- 2 \times 1$ surface. (**a–c**) Substitutional P atoms in the π-bonded chain, (**d**) unidentified P-induced contrast (from Ref. [43])

11.6.2 Impurity Atoms in III–V and II–VI Semiconductors

The first observation of an impurity in a semiconductor by cross-sectional STM was made on Zn-doped GaAs [121]. Quickly after this other impurities have been identified. An early review of the first STM studies of impurities in III/V semiconductor has been given by Ebert [24]. By now the following doping atoms are among those that have been studied below the GaAs cleavage surface: Mn [115], Zn [53, 120], C [113], Be [69], Cd [59], Si [110, 121], vacancies and antisites [24]. Furthermore, Mn has been studied by X-STM in and below the InAs(110) surface [72]. To date, the observation of N atoms in ZnTe [111] is the only observation of impurities in II/VI materials. The GaAs(110) cleavage surface has furthermore been used as a template to incorporate Mn atoms by STM in specific configurations to study their mutual interaction [58]. Finally, the metal–insulator transition by formation of an impurity band has been investigated by X-STM in real space on the cleaved surface of highly doped GaMnAs [85].

The unpinned Fermi level at the GaAs(110) surface opens up the unique possibility to modify the sample potential by the tip voltage and allows one to scan the neutral as well as charged state of an impurity. Figures 11.20 and 11.21 show the contrasts of Mn and Si atoms as examples for donors and acceptors in the GaAs(110) surface in both charge states. In the ionized state, the STM contrast is due to the symmetric Coulomb potential, whereas in the neutral state the wave function of the bound electron or hole is imaged. This has been proven for Si and Mn in GaAs by comparison with theoretical predictions on the shape of the wave function [35, 115]. The s-like donor state of Si in GaAs fits well to the circular symmetric contrast of this impurity found by STM, as shown in Fig. 11.20. The bow-tie-shaped contrast of the Mn acceptor below the GaAs(110) surface shown in Fig. 11.21 reflects the symmetry of the valence band states of GaAs in good agreement with both effective mass and tight binding model calculations.

The experimentally observed contrast of shallow acceptors like Zn and C in GaAs does however not follow the expected symmetry stemming from the valence band states. A triangular contrast was firstly observed for Zn in GaAs [120]. Later it was observed similarly for all other shallow acceptors in GaAs. A nice example is shown in Fig. 11.22. Tunneling into excited states [69] and effects of the complex band structure in the band gap [66] have been suggested to explain the triangular contrast. Very recently it has been shown that the strain that is induced by the buckling in the surface gives the main contribution to the observed triangular patterns

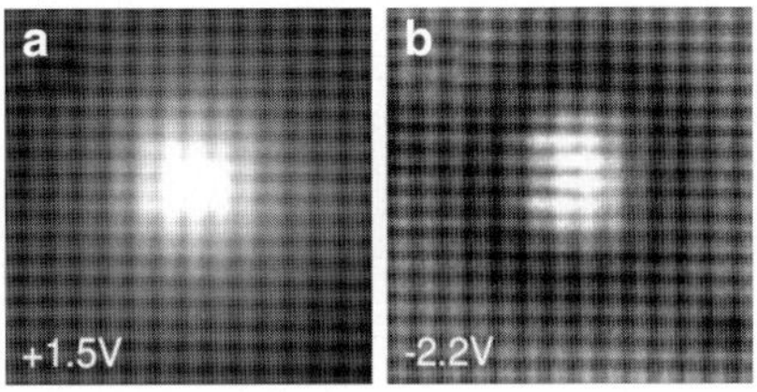

Fig. 11.20 Contrast of the Si donor in GaAs at positive and negative polarity. At positive polarity, the Si atom is in the ionized state (d^+) whereas at negative polarity it is neutral (d^0) (from Ref. [35])

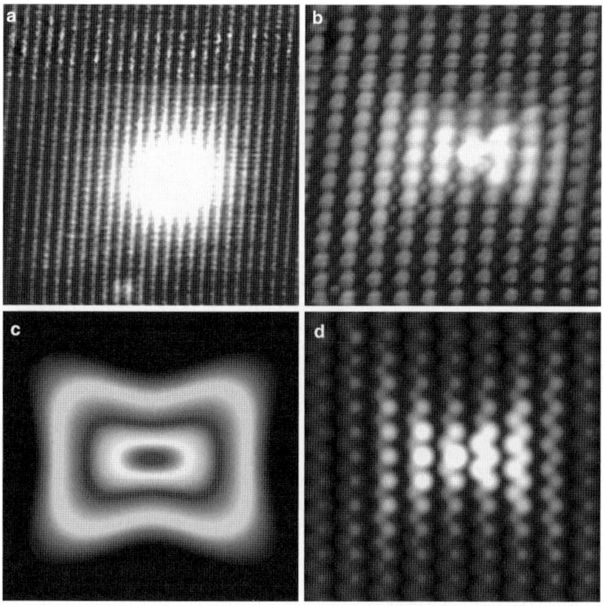

Fig. 11.21 Contrast of a Mn acceptor in the ionized state (**a**) and the neutral state (**b**). The calculated wave function obtained by an effective mass approach (**c**) and by a tight binding model (**d**) agree well with the observed contrast (from Ref. [115])

and also explains the asymmetry of the Mn contrast [14]. Depending on the doping concentration and the noise level, STM measurements typically detect impurities in the first ~10 layers of the crystal. Comparison of chemically identical impurities in different layers below the surface allows studying the surface-induced modifications on its topographic contrast. This has been done, e.g., for Si donors [32] and Mn acceptors [44]. Figure. 11.23 shows the contrast pattern as observed in STM topographies on Mn impurities in GaAs in the ten topmost layers of the cleaved crystal. The depth assignment is based on statistical analysis of the patterns counted in STM images. For deeper layers, more sophisticated methods based on the analysis of the symmetry have been employed and resulted in successful depth assignment. In some cases, even more quantitative details can be extracted from STM topography and especially from STS. Combining the analysis of nanostructures with the investigation of individual impurities allowed for a direct analysis of strain affecting the wave function of Mn acceptors [116].

As mentioned before the unpinned Fermi level at the GaAs(110) surface opens up the unique possibility to modify the sample potential by the tip voltage and allows one to scan the neutral as well as charged state of an impurity. A quantitative analysis of Si impurities in GaAs(110) was based on X-STM measurements. Their charge state was manipulated with the STM tip in a fully controlled way [100]. The ionization occurs when the TIBB, depicted by the cloud below the tip in Fig. 11.24b, is strong enough to lift the donor state above the onset of the conduction band

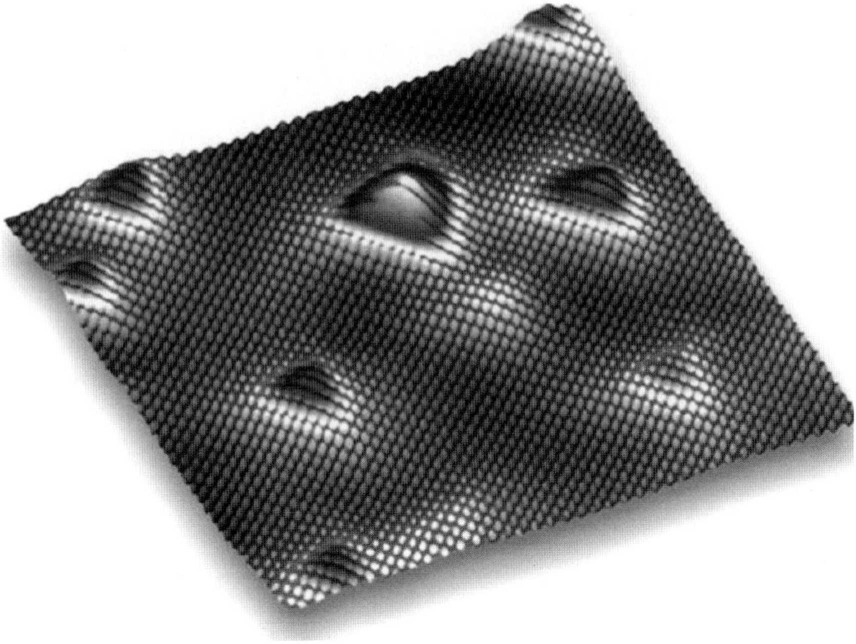

Fig. 11.22 Triangular contrast of shallow Zn acceptors in different layers below the GaAs surface in a high-resolution 3D image (from Ref. [65])

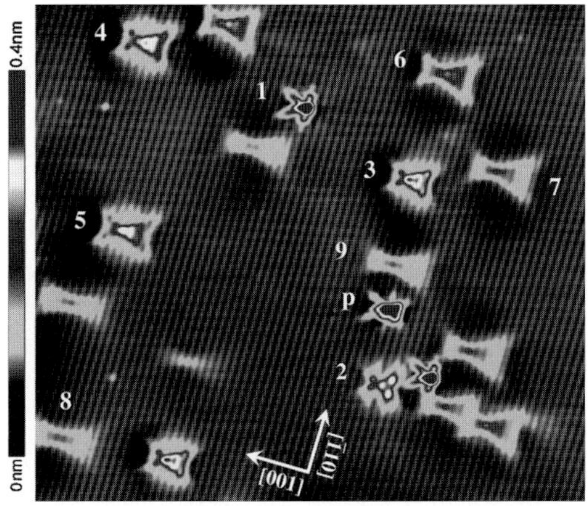

Fig. 11.23 Contrast patterns induced by Mn atoms in different layers below the GaAs(110) surface. The number next to the Mn contrast indicates its depth below the cleaved surface (from Ref. [14])

in the bulk. Due to the radial symmetry of the TIBB, this requires a specific distance between the tip and the donor at a given bias voltage. In topography and in spectroscopy images, a ring occurs around the donor as shown in Fig. 11.24c. The rings are only visible with ultra-sharp tips with a radius of only a few nanometers (see Fig. 11.24a). The voltage dependence of the ring diameter as a function of the

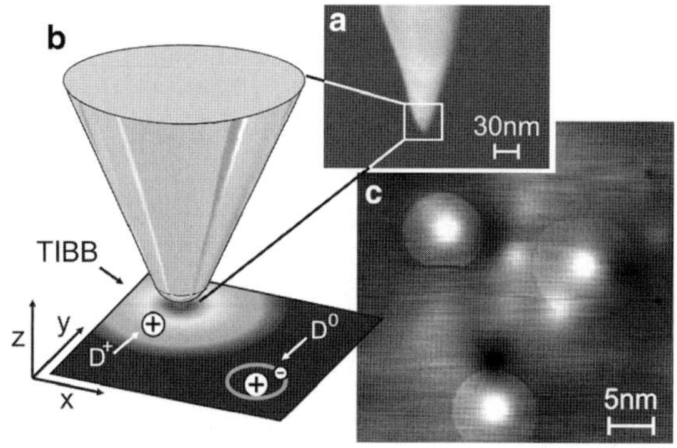

Fig. 11.24 Scanning tunneling spectroscopy of Si donors in different depths below the GaAs(110) surface. The ionization of the donor results in a ring (**c**), with a diameter similar to the radius of the tip apex (**a**). The dependence of the ring diameter on the tunneling parameters is quantitatively described by the tip-induced band bending (**b**). The depth dependence of the rings proves their enhanced binding energy qualitatively

distance between the tip and the surface is quantitatively explained by the calculated TIBB. As the TIBB decays toward the bulk of the sample, one expects the donors closer to surface to ionize at lower voltage than donors in a deeper layer.

Experimentally the opposite is observed. This proves that the donor binding energy depends on the depth of the individual donor below the surface [112]. Translating the TIBB that is required in order to ionize the donor into the binding energy involves assumptions about the Si donor wave function and the electrostatic screening in the surface. A gradual increase of the binding energy is observed from a few meV for the donors 1.5 nm below the surface to $\sim$40 meV in a depth of 0.2 nm. A very similarly increased binding energy has also been reported for Mn acceptors below the GaAs(110) surface [45].

References

1. F.F. Abraham, D. Brodbeck, R.A. Rafey, W.E. Rudge, Phys. Rev. Lett. **73**(2), 272 (1994)
2. O. Albrektsen, D.J. Arend, H.P. Meier, H.W.M. Salemink, Appl. Phys. Lett. **57**, 31 (1990)
3. J.L.A. Alves, Solid State Commun. **87**(11), 1001 (1993)
4. T. An, M. Yoshimura, I. Ono, K. Ueda, Phys. Rev. B. **61**(4) 3006 (2000)
5. G. Binnig, H. Rohrer, C. Gerber, E. Weibel, Phys. Rev. Lett. **50**, 120 (1984)
6. F. Bechstedt, R. Enderlein: *Semiconductor Surfaces and Interfaces* (Akademie Verlag, Berlin, 1988)
7. M. Bertelli, P. Löptien, M. Wenderoth, A. Rizzi, R.G. Ulbrich, M.C. Righi, A. Ferretti, L. Martin-Samos, C.M. Bertoni, A. Catellani, Phys. Rev. B **80**, 115324 (2009)
8. D. Bimberg, M. Grundmann, N.N. Ledentsov, *Quantum Dot Heterostructures* (Wiley, Chichester, UK, 1999)
9. J.H. Blokland, M. Bozkurt, J.M. Ulloa, D. Reuter, A.D. Wieck, P.M. Koenraad, P.C.M. Christianen, J.C. Maan, Appl. Phys. Lett. **94**, 023107 (2009)
10. D.M. Bruls, P.M. Koenraad, M. Hopkinson, J.H. Wolter, H.W.M. Salemink, Appl. Surf. Sci. **190**, 258 (2002)

11. D.M. Bruls, J.W.A.M. Vugs, P.M. Koenraad, H.W.M. Salemink, J.H. Wolter, M. Hopkinson, M.S. Skolnick, F. Long, S.P.A. Gill, Appl. Phys. Lett. **81**, 1708 (2002)
12. D.M. Bruls, P.M. Koenraad, H.W.M. Salemink, J.H. Wolter, M. Hopkinson, M.S. Skolnick, Appl. Phys. Lett. **82**, 3758 (2003)
13. C. Çelebi, J.M. Ulloa, P.M. Koenraad, A. Simon, A. Letoublon, N. Bertru, Appl. Phys. Lett. **89**, 023119 (2006)
14. C. Celebi, J.K. Garleff, A. Yu. Silov, A.M. Yakunin, P.M. Koenraad, W. Van Roy, J.-M. Tang, M.E. Flatté, Phys. Rev. Lett. **104**, 086404 (2010)
15. J.R. Chelikowsky, M.L. Cohen, Phys. Rev. B. **20**, 4150 (1979)
16. C.J. Chen, *Introduction to Scanning Tunneling Microscopy* (Oxford University Press, Oxford, 1993)
17. G. Costantini, A. Rastelli, C. Manzano, P. Acosta-Diaz, G. Katsaros, R. Songmuang, O. Schmidt, H.v. Kanel, K. Kern, J. Cryst. Growth **278**, 38 (2005)
18. T. Cramer, A. Wanner, P. Gumbsch, Phys. Rev. Lett. **85**(4) 788 (2000)
19. J.H. Davies, D.M. Bruls, J.W.A.M. Vugs, P.M. Koenraad, J. Appl. Phys **91**, 4171 (2002)
20. T. Dietl, F. Matsukura, H. Ohno, Phys. Rev. B **66**, 033203 (2002)
21. R. Dingle, W. Wiegmann, C.H. Henry, Phys. Rev. Lett. **33**, 827 (1974)
22. D.J. Eaglesham, M. Cerullo, Phys. Rev. Lett. **64**, 1943 (1990)
23. P. Ebert, B. Engels, P. Richard, K. Schroeder, S. Blügel, C. Domke, M. Heinrich, K. Urban, Phys. Rev. Lett. **77** 2997 (1996)
24. Ph. Ebert, Surf. Sci. Rep. **33**, 121–303 (1999)
25. K.W. Edmonds, N.R.S. Farley, T.K. Johal, G. van der Laan, R.P. Campion, B.L. Gallagher, C.T. Foxon, Phys. Rev. B **71**, 064418 (2005)
26. H. Eisele, O. Flebbe, T. Kalka, C. Preinesberger, F. Heinrichsdorff, A. Krost, D. Bimberg, M. Dähne-Prietsch, J. Appl. Phys. **75**, 106 (1999)
27. S. Fafard, K. Hinzer, S. Raymond, M. Dion, J. McCaffrey, Y. Feng, S. Charbonneau, Science **274**, 1350 (1996)
28. R.M. Feenstra, A.P. Fein, IBM J. Res. Dev. **30**, 466 (1986)
29. R.M. Feenstra, J.A. Stroscio, J. Vac. Sci. Technol. B **5**, 923 (1987)
30. R.M. Feenstra, Surf. Sci. **243**, 151–165 (1991)
31. R.M. Feenstra, Phys. Rev. B. **44**, 13791 (1991)
32. R.M. Feenstra, J.M. Woodall, G.D. Pettit, Phys. Rev. Lett. **71**, 1176 (1993)
33. R.M. Feenstra, Phys. Rev. B. **60**, 4478 (1999)
34. R.M. Feenstra, Physica B **273**, 796 (1999)
35. R.M. Feenstra, G. Meyer, F. Moresco, K.H. Rieder, Phys. Rev. B **66**, 165204 (2002)
36. J.E. Fields, Cont. Phys. **12**(1), 1–31 (1971)
37. J. Fineberg, S.P. Gross, M. Marder, H.L. Swinney, Phys. Rev. Lett. **67**, 457 (1991)
38. L.B. Freund, *Dynamic Fracture Mechanics* (Cambridge University Press, New York, 1990)
39. P.W. Fry, I.E. Itskevich, D.J. Mowbray, M.S. Skolnick, J.J. Finley, J.A. Barker, E.P. O'Reilly, L.R. Wilson, I.A. Larkin, P.A. Maksym, M. Hopkinson, M. Al-Khafaji, J.P.R. David, A.G. Cullis, G. Hill, J.C. Clark, Phys. Rev. Lett. **84**, 733 (2000)
40. J.K. Garleff, Master Thesis, University of Göttingen, 2001
41. J.K. Garleff, M. Wenderoth, K. Sauthoff, R.G. Ulbrich, M. Rohlfing, Phys. Rev. B **70** 245424 (2004)
42. J.K. Garleff, M. Wenderoth, R.G. Ulbrich, C. Sürgers, H.v. Löhneysen, Phys. Rev. B **72** 073406 (2005)
43. J.K. Garleff, M. Wenderoth, R.G. Ulbrich C. Sürgers, H.v. Löhneysen, M. Rohlfing, Phys. Rev. B. **76**, 125322 (2007)
44. J.K. Garleff, C. Celebi, W. Van Roy, J.-M. Tang, M.E. Flatte, P.M. Koenraad, Phys. Rev. B. **78**, 075313 (2008)
45. J.K. Garleff, A.P. Wijnheijmer, A. Yu. Silov, J. van Bree, W. Van Roy, J.-M. Tang, M.E. Flatté, P.M. Koenraad, Phys. Rev. B 82, 035303 (2010)
46. A. Gomyo, T. Suzuki, K. Kobayashi, S. Kawata, I. Hino, T. Yuasa, Appl. Phys. Lett. **50**, 673 (1987)

47. Q. Gong, P. Offermans, R. Nötzel, P.M. Koenraad, J.H. Wolter, Appl. Phys. Lett. **85**, 5697 (2004)
48. S.P. Gross, J. Fineberg, M. Marder, W.D. McCormick, H.L. Swinney, Phys. Rev. Lett. **71**(19) 3162 (1993)
49. P. Gumbsch, S.J. Zhou, B.L. Holian, Phys. Rev. B **55**(6), 3445 (1997)
50. V. Haxha, I. Drouzas, J.M. Ulloa, M. Bozkurt, P.M. Koenraad, D.J. Mowbray, H.Y. Liu, M.J. Steer, M. Hopkinson, M.A. Migliorato, Phys. Rev. B **80**, 165334 (2009)
51. A.J. Heinrich, M. Wenderoth, M.A. Rosentreter, K. Engel, M.A. Schneider, R.G. Ulbrich, E.R. Weber, K. Uchida, Appl. Phys. A **66**, 959 (1998)
52. F. Heinrichsdorff, M.-H. Mao, N. Kirstaedter, A. Krost, D. Bimberg, A.O. Kosogov, P. Werner, Appl. Phys. Lett. **71**, 22 (1997)
53. M.B. Johnson, O. Albrektsen, R.M. Feenstra, H.W.M. Salemink, Appl. Phys. Lett. **63**, 2923 (1993)
54. S. Kagel, Master Thesis, University of Göttingen, 2003
55. J.G. Keizer, J. Bocquel, P.M. Koenraad, T. Mano, T. Noda, K. Sakoda, Appl. Phys. Lett. **96**, 062101 (2010)
56. J.R. Kermode, T. Albaret, D. Sherman, N. Bernstein, P. Gumbsch, M.C. Payne, G. Csanyi, A. De Vita, Nature **455**, 1224 (2008)
57. N. Kitamura, B.S. Swartzentruber, M.G. Lagally, M.B. Webb, Phys. Rev. B, **48**(8) 5704 (1993)
58. D. Kitchen, A. Richardella, J.-M. Tang, M.E. Flatté, A. Yazdani, Nature (London) **442**, 436 (2006)
59. R. de Kort, W. Kets, H. van Kempen, Surf. Sci. 482–485, 495–500 (2001)
60. T.S. Kuan, T.F. Kuech, W.I. Wang, E.L. Wilkie, Phys. Rev. Lett. **54**, 201 (1985)
61. A. Lenz, R. Timm, H. Eisele, C. Hennig, S.K. Becker, R.L. Sellin, U.W. Pohl, D. Bimberg, M. Dhne, Appl. Phys. Lett. **81**, 5150 (2002)
62. B. Lita, R.S. Goldman, J.D. Phillips, P.K. Bhattarchaya, Appl. Phys. Lett. **75**, 2797 (1999)
63. N. Liu, J. Tersoff, O. Baklenov, A.L. holmes Jr., C.K. Shih, Phys. Rev. Lett, **84**, 334 (2000)
64. H.v. Löhneysen, Philos. Trans. R. Soc. Lond. A **356,** 139 (1998)
65. S. Loth, M. Wenderoth, R.G. Ulbrich, S. Malzer, G.H. Döhler, Phys. Rev. B **76**, 235318 (2007)
66. S. Loth, M. Wenderoth, L. Winking, R.G. Ulbrich, S. Malzer, G.H. Döhler, Phys. Rev. Lett. **96**, 066403 (2006)
67. M.A. Lutz, R.M. Feenstra, J.O. Chu, Surf. Sci. **328**, 215 (1995)
68. O. Madelung, *Semiconductors – Basic Data*, 3rd edn. (Springer, New York, 2004)
69. G. Mahieu, B. Grandidier, D. Deresmes, J.P. Nys, D. Stiévenard, Ph. Ebert, Phys. Rev. Lett. **94**, 026407 (2005)
70. H. Mani, A. Joullie, F. Karouta, C. Schiller, J. Appl. Phys. **59**, 2728 (1986)
71. M. Marder, X. Liu, Phys. Rev. Lett. **71** (15) 2417 (1993)
72. F. Marczinowski, J. Wiebe, J.-M. Tang, M.E. Flatté, F. Meier, M. Morgenstern, R. Wiesendanger, Phys. Rev. Lett. **99**, 157202 (2007)
73. P. Michler, A. Kiraz, C. Becher, W.V. Schoenfeld, P.M. Petroff, L. Zhang, E. Hu, A. Imamogùlu1, Science **290**, 2282 (2000)
74. M.A. Migliorato, A.G. Cullis, M. Fearn, J.H. Jefferson, Phys. Rev. B **65**, 115316 (2002)
75. V. Mlinar, M. Bozkurt, J.M. Ulloa, M. Ediger, G. Bester, A. Badolato, P.M. Koenraad, R.J. Warburton, A. Zunger, Phys. Rev. B **80**, 165425 (2009)
76. P. Offermans, P. Koenraad, J. Wolter, K. Pierz, M. Roy, P. Maksym, Physica E **26**, 236 (2005)
77. P. Offermans, P.M. Koenraad, J.H. Wolter, K. Pierz, M. Roy, P.A. Maksym, Phys. Rev. B **72**, 165332 (2005)
78. P. Offermans, P.M. Koenraad, R. Nötzel, J.H. Wolter, K. Pierz, Appl. Phys. Lett. **87**, 111903 (2005)
79. P. Offermans, PhD Thesis, Eindhoven University of Technology 2005
80. W.E. Packard, J.D. Dow, Phys. Rev. B **55**(23) 15643 (1997)
81. K.C. Pandey, Phys. Rev. Lett. **47**(26) 1913 (1981)
82. M. Pfister, M.B. Johnson, S.F. Alvarado, H.W.M. Salemink, U. Marti, D. Martin, F. Morier-Genoud, F.K. Reinhart, Appl. Phys. Lett. **67**, 1459 (1995)

83. G.J. de Raad, D.M. Bruls, P.M. Koenraad, J.H. Wolter, Phys. Rev. B **64**, 075314 (2001)
84. G.J. de Raad, D.M. Bruls, P.M. Koenraad, J.H. Wolter, Phys. Rev. B **66**, 195306 (2002)
85. A. Richardella, P. Roushan, S. Mack, B. Zhou, D.A. Huse, D.D. Awschalom, A. Yazdani, Science **327**, 665 (2010)
86. M.A. Rosentreter, M. Wenderoth, N.H. Theuerkauf, A.J. Heinrich, M.A. Schneider, R.G. Ulbrich, Europhys. Lett. **38**(9) 675–680 (1997)
87. M.A. Rosentreter, M. Wenderoth, N.H. Theuerkrauf, A.J. Heinrich, M.A. Schneider, R.G. Ulbrich, Phys. Rev. B, **56**(16) 10538 (1997)
88. M.A. Rosentreter, PhD Thesis, University of Göttingen, 1997
89. K. Sauthoff M. Wenderoth, A.J. Heinrich, M.A. Rosentreter, K.J. Engel, T.C.G. Reusch, R.G. Ulbrich, Phys. Rev. B **60**, 4789 (1999)
90. M. Schoeck, C. Sürgers, H.v. Löhneysen, Phys. Rev. B **61**, 7622 (2000)
91. E. Sharon, S.P. Gross, J. Fineberg, Phys. Rev. Lett. **74**(25) 5096 (1995)
92. E. Sharon, J. Fineberg, Nature **397**, 333 (1999)
93. M.S. Skolnick, D.J. Mowbray, Annu. Rev. Mater. Res. **34**, 181 (2004)
94. J. Stangl, V. Holý, G. Bauer, Rev. Mod. Phys. **76**, 725 (2004)
95. A.D. Stiff, S. Krishna, P. Bhattacharya, S.P. Kennerly, Appl. Phys. Lett. **79**, 421 (2001)
96. A.A. Stekolnikov, J. Furthmüller, F. Bechstedt, Phys. Rev. B **65**, 115318 (2002)
97. T.O. Strandberg, C.M. Canali, A.H. MacDonald, Phys. Rev. B **80**, 024425 (2009)
98. G.B. Stringfellow, G.S. Chen, J. Vac. Sci. Technol. B **9**, 2182 (1991)
99. N. Tansu, L.J. Mawst, IEEE Photon. Technol. Lett. **14**, 444 (2002)
100. K. Teichmann, M. Wenderoth, S. Loth, R.G. Ulbrich, J.K. Garleff, A.P. Wijnheijmer, P.M. Koenraad, Phys. Rev. Lett. **101**, 076103 (2008)
101. T. Trappmann, et. al., Europhys. Lett. **38**, 177 (1997)
102. W.T. Tsang, Appl. Phys. Lett. **40**, 217 (1982)
103. J.M. Ulloa, I.W.D. Drouzas, P.M. Koenraad, D.J. Mowbray, M.J. Steer, H.Y. Liu, M. Hopkinson, Appl. Phys. Lett. **90**, 213105 (2007)
104. J.M. Ulloa, C. Celebi, P.M. Koenraad, A. Simon, E. Gapihan, A. Letoublon, N. Bertru, I. Drouzas, D.J. Mowbray, M.J. Steer, M. Hopkinson, J. Appl. Phys. **101**, 081707 (2007)
105. J.M. Ulloa, P.M. Koenraad, M. Bonnet-Eymard, A. Létoublon, N. Bertru, J. Appl. Phys. **107**, 074309 (2010)
106. J.M. Ulloa, R. Gargallo-Caballero, M. Bozkurt, M. del Moral, P.M. Koenraad, A. Guzmán, A. Hierro, Phys. Rev. B **81**, 165305 (2010)
107. M. Vershinin, S. Misra, S. Ono, Y. Abe, Y. Ando, A. Yazdani, Science **303** 1995 (2004)
108. S.Y. Wang, S.D. Lin, H.W. Wu, C.P. Lee, Appl. Phys. Lett. **78**, 1023 (2001)
109. S.-H. Wei, D.B. Laks, A. Zunger, Appl. Phys. Lett. **62**, 1937 (1993)
110. M.C.M.M. van der Wielen, A.J.A. van Roij, H. van Kempen, Phys. Rev. Lett. **76**, 1075 (1996)
111. A. Wierts, J.M. Ulloa, C. Çelebi, P.M. Koenraad, H. Boukari, L. Maingault, R. André, H. Mariette, Appl. Phys. Lett. **91**, 161907 (2007)
112. A.P. Wijnheijmer, J.K. Garleff, P.M. Koenraad, K. Teichmann, M. Wenderoth, S. Loth, R.G. Ulbrich, P.A. Maksym, M. Roy, Phys. Rev. Lett. **102**, 166101 (2009)
113. L. Winking, M. Wenderoth, T.C.G. Reusch, R.G. Ulbrich, P.-J. Wilbrandt, R. Kirchheim, S. Malzer, G. Döhler, J. Vac. Sci. Technol. B **23**, 267 (2005)
114. O. Wolst, M. Kahl, M. Schardt, S. Malzer, G.H. Döhler, Physica E **17**, 554 (2003)
115. A.M. Yakunin, A. Yu. Silov, P.M. Koenraad, J.H. Wolter, W. Van Roy, J. De Boeck, J.-M. Tang, M.E. Flatté, Phys. Rev. Lett. **92**, 216806 (2004)
116. A.M. Yakunin, A. Yu. Silov, P.M. Koenraad, J.-M. Tang, M.E. Flatté, J.-L. Primus, W. Van Roy, J. De Boeck, A.M. Monakhov, K.S. Romanov, I.E. Panaiotti, N.S. Averkiev, Nat. Mater. **6**, 512 (2007)
117. Y. Yamamoto, Surf. Sci. **313**, 155 (1994)
118. K. Yano, T. Ishii, T. Hashimoto, T. Kobayashi, F. Murai, K. Seki, IEEE Trans. Electron Devices **41**, 1628 (1994)
119. I. Yonenaga, Philos. Mag. Lett. **82**(10), 535(2002)
120. J.F. Zheng, M.B. Salmeron, E.R. Weber, Appl. Phys. Lett. **64**, 1836 (1994)

121. J.F. Zheng, X. Liu, N. Newman, E.R. Weber, D.F. Ogletree, M. Salmeron, Phys. Rev. Lett. **72**, 1490 (1994)
122. Zhiliang Yuan, B.E. Kardynal, R.M. Stevenson, A.J. Shields, Ch.J. Lobo, K. Cooper, N.S. Beattie, D.A. Ritchie, M. Pepper, Science **295**, 102 (2002)
123. J.P. van der Ziel, R. Dingle, R.C. Miller, W. Wiegmann, W.A. Nordland Jr., Appl. Phys. Lett. **26**, 463 (1975)

Chapter 12
A Novel Approach for Oxide Scale Growth Characterization: Combining Etching with Atomic Force Microscopy

V. Presser, A. Loges, and K.G. Nickel

Abstract The combination of atomic force microscopy (AFM) analysis of oxide scales before and after chemical etching is presented as a fast, powerful method to gain information on oxide scale growth kinetics. While not limited to the field of ceramics, we chose the thermal oxidation of silicon carbide at high temperatures (1,400°C) as an example for the potential of the AFM/etching method. SiC is a promising semiconductor material with many high-temperature applications during which not only simple oxidation, but also crystallization of the initially vitreous silica scale occurs. We demonstrate how AFM/etching analysis of crystalline areas (radialites) yields valuable information on the growth rate of the crystalline and amorphous silica scale. This can be directly translated into a statement on the oxidation passivation potential of a certain oxide scale state/morphology with important consequences for the actual high-temperature application. Also, the influence of impurities is addressed as this is an essential aspect for real-world application of silicon carbide (both, as a refractory and ceramic material).

Nomenclature

A/B	Linear rate constant (k_l)
B	Parabolic rate constant (k_p)
c	Concentration
d	Scale thickness
D	Diffusion coefficient (=effective diffusion coefficient)
g	Gas phase
m	Order of interference color
n	Refraction index
N	Gas molecules per unit volume
P	Pressure
s	Solid
t	Time
x	Radial distance from crystallization center

θ Angle of refracted ray
λ Wavelength
ν Crystallization rate
ρ Density
τ Correction term

12.1 Introduction

An important criterion for modern technical materials is their chemical performance which should not contribute to device degradation under the chosen working conditions. The latter may be extreme: many ceramic materials are exposed to temperatures of more than 1,200°C or even 1,500°C. In oxygen containing atmospheres, oxidation reactions cannot be excluded – indeed, the formation of an oxide layer may prove useful as long as a dense, protective scale is formed, which slows further oxidation. This phenomenon is well known for metals, but can also be found for a wide range of ceramic materials. Binary carbides are a group of such materials: while the metal or metaloid ion is oxidized and contributes to the formation of a dense oxide scale, the carbon is oxidized and diffuses out of the growing oxide.

An ideal model system to study the growth kinetics of oxides grown from the oxidation of carbides is silicon carbide (SiC). For over a century it was used almost exclusively for abrasive and high-temperature applications for which the oxidation behavior played an important role. In the last decade, SiC has also turned into a promising semiconductor material as high-quality silicon carbide single-crystals have become available [1–3]. These two fields (high-temperature application and the semiconductor sector) explain why SiC oxidation has so intensively been studied: in both cases, the oxide scale plays an important role.

For high-temperature applications, the state of the oxide scale growth limits the working conditions of the ceramic material. At high temperatures ($\gg$1,000°C), crystallization occurs and is accompanied by crack formation when cooling below 300°C because of a phase transition of the crystalline silica. This has proven to be a problem for the cyclic oxidation of silicon carbide [4]. Also, the oxide growth kinetics and the active–passive transition temperature have to be considered for the reliability of a ceramic device and the maximum operating conditions in terms of temperature and time.

For the use as a semiconductor, simple oxidation is used to manufacture metal oxide semiconductor field-effect transistors (MOSFET) for electronic application [5]. Here, exact knowledge of the growth kinetics is important so that a defined scale thickness can be achieved. Also, the oxidation routine (dry oxygen, water vapor, etc.) strongly influences the electronic properties at the carbide–oxide interface [6–10]. With the electrical interface properties still being inferior to, for example, the Si–SiO_2 interface, much effort is spent to increase the understanding of the carbide–oxide interface [11, 12].

12.2 Oxidation of Silicon Carbide

Below 900°C, no significant oxidative reaction of silicon carbide can be found when exposed to oxygen containing atmospheres [13]. It is this beneficial oxidation resistance that makes SiC a suitable candidate for high-temperature applications. In this chapter, we adhere to the oxidation of SiC in pure oxygen but the same applies for CO/CO_2 atmospheres or other gas mixtures containing an oxidizing species [11, 14–16]. In this context, water vapor has proven to be an effective oxidizing agent [17–19].

It should be noted that even a short-term exposure of silicon carbide to atmospheric oxygen results in surface reconstruction and the formation of chemically bounded oxygen in the form of monolayer(s), which can be readily detected using, for example, X-ray photoelectron spectroscopy [20, 21] or low energy electron diffraction (LEED) [22–24]. Special treatment (etching, ion bombarding, thermal annealing under ultra-high vacuum conditions) must be applied to obtain a fully oxygen-free SiC surface on an atomic level [25].

A limiting factor for the working conditions of silicon carbide is the active–passive transition temperature which strongly depends on the oxygen partial pressure [26–32]. In atmospheres with low oxygen partial pressures, no silica scale can be formed with gaseous SiO and CO being the stable oxidation products. This is referred to as active oxidation which is accompanied with a bulk mass loss of the sample material. Typical transition temperatures are around 1,400°C for $P_{(O_2)} \approx 10^{-4}$ bar or 1,550°C for $P_{(O_2)} \approx 10^{-3}$ bar [11].

The formation of silica is referred to as passive oxidation and is accompanied with an increase in both bulk mass and net volume. With a growing oxide scale, the underlying silicon carbide becomes more and more protected from the oxidizing gas species: the thicker the oxide scale, the longer the distance for oxygen transportation toward the $SiC–SiO_2$ interface. With a constant diffusion coefficient for the transport of oxygen through the amorphous silica network, the growth rate slows as the scale grows and a parabolic time law can be observed over longer times (cf. Chap. 6).

From thermodynamic calculations there exist several oxidation reactions that led to the formation of silica scale (12.1)–(12.3) [33]:

$$\mathrm{SiC}^{(s)} + 1\mathrm{O}_2^{(g)} \leftrightarrow 1\mathrm{SiO}_2^{(s)} + 1\mathrm{C}^{(s)} \tag{12.1}$$

$$\mathrm{SiC}^{(s)} + 3/2\,\mathrm{O}_2^{(g)} \leftrightarrow 1\mathrm{SiO}_2^{(s)} + 1\mathrm{CO}^{(g)} \tag{12.2}$$

$$\mathrm{SiC}^{(s)} + 2\mathrm{O}_2^{(g)} \leftrightarrow 1\mathrm{SiO}_2^{(s)} + 1\mathrm{CO}_2^{(g)} \tag{12.3}$$

Only (12.2) was observed for hot-gas oxidation (g = gas, s = solid phase) of silicon carbide [13,34–36] but for hydrothermal corrosion also the formation of both carbon and silica was reported [37–39].

Stoichiometric silica [40] is the only stable oxide for thermal oxidation of SiC above 900°C [12, 13]. This was found to be true also for other thermal oxidation reactions of SiC (e.g., in $CO–CO_2$ gas mixtures) [26, 33, 41–45]. At low temperatures oxycarbides have been reported [46,47]. The latter (or at least carbon clusters within a disordered SiO_2 network) seem to play an important role in understanding the electrical properties at the SiO_2–SiC interface [48–51].

As mentioned before, passive oxidation of silicon carbide is characterized by an increase in both bulk mass and volume. The first is a result of (12.2) where 1 mol carbon is replaced by 2 mol oxygen. This translates to a net mass gain of 19.9881 g per formula unit. Also, amorphous silica ($\rho \approx 2.2\,\mathrm{g/cm^3}$; compare to $\rho_{(\alpha-\mathrm{cristobalite})} = 2.3\,\mathrm{g/cm^3}$) [12] is less dense than silicon carbide ($\rho = 3.2\,\mathrm{g/cm^3}$) [31], which results in an increase in volume.

Silicon carbide oxidation is influenced by many factors, among which temperature and oxygen partial pressure (or the gas pressure of the oxidizing gas species, respectively) are the two most important ones [11, 12]. Oxide scale growth is also strongly affected by impurities which, for example, can readily be found when using an alumina tube within the oxidizing furnace [52, 53]. Alternatively, impurities may originate from SiC when using sintered ceramics where sintering aids are present [54].

There are many similarities between the silica formation during the oxidation of silicon carbide and the well-studied silicon oxidation. The latter has attracted much attention with silicon (still) being the most important semiconductor material. In both cases, similar kinetics can be observed while the oxidation rates are different with SiC being much more oxidation resistant. At 1,200°C, a temperature at which both Si and SiC oxidation can be studied, a difference of factor 2.5 between the parabolic rate constant of Si (faster) and SiC (slower) oxidation was reported [55].

There are, however, several significant differences. In case of SiC oxidation, we have to consider the transport of oxygen through the silica scale towards the SiC–SiO_2 interface and the out-diffusion of the gaseous reaction product CO [56]. The latter is of no concern for Si oxidation. Also, there is the much debated phenomenon of SiC oxidation anisotropy: depending on the crystallographic orientation, we find very different oxidation rates for SiC [57, 58]. This phenomenon was discussed in terms of chemical bonds [59, 60] and different reaction products by the proposed interlayer model after Hornetz et al. [48, 49].

12.3 Silica: Growth and Crystallization

Depending on the oxidation temperature, the initially vitreous silica devitrifies and crystalline silica is formed once nuclei of a critical size have been formed over time (Fig. 12.1). In general, we find an Arrhenius equation for the characterization of the correlation between the nucleation rate and the applied oxidation temperature. Any discontinuity on the substrate (e.g., scratches) will act as an area of preferential nucleation (Fig. 12.1a) [61]. The growth kinetics is not affected. The situation is different for impurities or the presence of water vapor which result in both an increased nucleation rate and accelerated crystal growth [62–66].

Crystallization will start on the silica–atmosphere interface as provoked by adsorbed gas molecules causing metastable crystallization centers [67, 68]. Significant

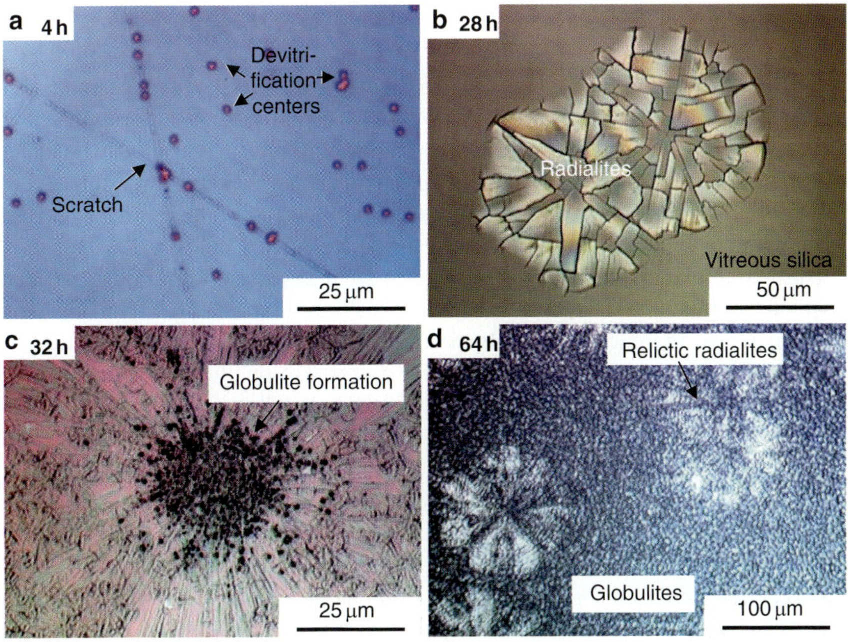

Fig. 12.1 Optical microphotographs of SiO_2 thermally grown on SiC (1,400°C, alumina tube furnace). The different interference colors are a function of oxide scale thickness. (**a**) Surface inhomogeneities locally trigger crystallization. (**b**) Intergrown radialites (R1 type). (**c**) Early stage of globulite formation (R2 type radialite). (**d**) Late stage of globulite formation: almost complete consumption of the radialite lawn by globulites

crystallization can be expected only for temperatures higher than 1,000°C [11]. Depending on the temperature, crystallization occurs after a nucleation phase which, in case of 1,000°C, has been reported to be as long as 100 h [69]. At this temperature, the formation of β-cristobalite has been reported forming rosette-like devitrification centers [70–73] which here are referred to as radialites adapting the nomenclature of [12] (Fig. 12.1b). This type of crystalline structure is typical for β-cristobalite crystallizing within an amorphous silica matrix [73–77]. In case of SiC oxidation, this silica scale is rather thin (hundreds of nanometers) with the crystallization speed being much faster than the oxide growth itself. This is why the resulting crystallization centers are disk-like (e.g., diameter 150μm, thickness 1.2μm; [72]) rather than three-dimensional spheres. The latter would be referred to as spherulite [78].

Due to the sluggishness of the crystalline transformation, cristobalite does not undergo a phase transition into another silica modification even after long annealing times at room temperature ([79] no transformation after 2 years). For lower temperatures and long oxidation periods, other SiO_2 modifications have been identified using transmission electron microscopy (TEM) (700°C quartz/800°C tridymite) [79]. In addition, β-tridymite was found to be a significant phase for some experiments but its occurrence on high-purity single crystals usually indicates the presence of impurities (e.g., sodium) [52].

While nucleation is an exponential process, crystal growth of stoichiometric silica follows a linear time law [80, 81]. Opila [82] reported a crystal growth rate of $0.1\mu m/h$ at 1,300°C and at higher temperatures (1,400°C), a crystallization rate of $2.6 \pm 0.1\mu m/h$ for atmospheres containing 0.6–2.0% water vapor was found [80]. The trend to higher crystallization rates at higher water vapor pressures is evident (e.g., $> 15\mu m/h$ in 0.1 MPa water vapor at 1,426°C) [81, 82].

During the oxidation, each radialite will grow and once a critical size is reached, it will develop smaller subgrains (crystal plates) as seen in Fig. 12.1b. This is a result of stress and energy minimization. Please note that the pronounced cracks between the crystal plates are a result of the β–α transition of cristobalite at around 300°C which is accompanied by a decrease in the bulk volume of approximately 5% [83–86]. This is also the reason why most radialites are easily distinguished from the amorphous silica.

At the beginning of the growth phase, the radialites show a hexagonal shape because of the crystal growth of β-cristobalite. TEM studies showed that the cubic β-cristobalite phase grows with the $\langle 111 \rangle$ axis (almost) normal to radial directions creating a fine lineation coinciding with $\{111\}$ faces [87]. This preferred orientation of β-cristobalite radialites was also corroborated via X-ray microdiffraction [12, 71].

With the radialites growing, one will observe intergrowth of radialites (Fig. 12.1b) depending on the original distribution of critical nuclei and this happens sooner or later. Over time, a homogenous lawn of crystal plates will evolve. For high-purity systems, this is the last stage of the microstructural evolution of silica thermally grown on silicon carbide.

In the presence of impurities, radialites are gradually consumed by $1–2\mu m$ small crystalline spheres which are referred to as globulites (Fig. 12.1c, d). They also consist of β-cristobalite but result from a recrystallization caused by superficial melt formation. Impurities from the furnace atmosphere accumulate on the silica surface over time. As long as the amorphous phase is present, a high amount of impurities like sodium can be incorporated into the vitreous network. Crystalline silica, however, only has a very limited solubility and, in consequence, impurities will accumulate at the crystalline–amorphous boundary [71, 88].

Once a critical impurity concentration is reached, a process of dissolution and reprecipitation gradually transforms the large radialite plates into smaller globulites. This process indeed is a recrystallization as we see a complete loss of the preferred orientation of the globulite lawn compared to areas covered by radialites [71]. Globulite areas are also characterized by significant open pore porosity as seen in Fig. 12.2. It is clear that a globulite lawn will provide inferior oxidation protection compared to a dense crystalline or a dense amorphous silica scale.

The strong influence of impurities becomes evident when we compare samples that were subject to impurities (e.g., NaCl as a source for sodium) with SiC after thermal oxidation in completely impurity free environments (fused quartz tube and dehumidified oxygen). In Fig. 12.3, we see that in the absence of impurities no globulite formation can be observed. Also, the radialites from these high-purity

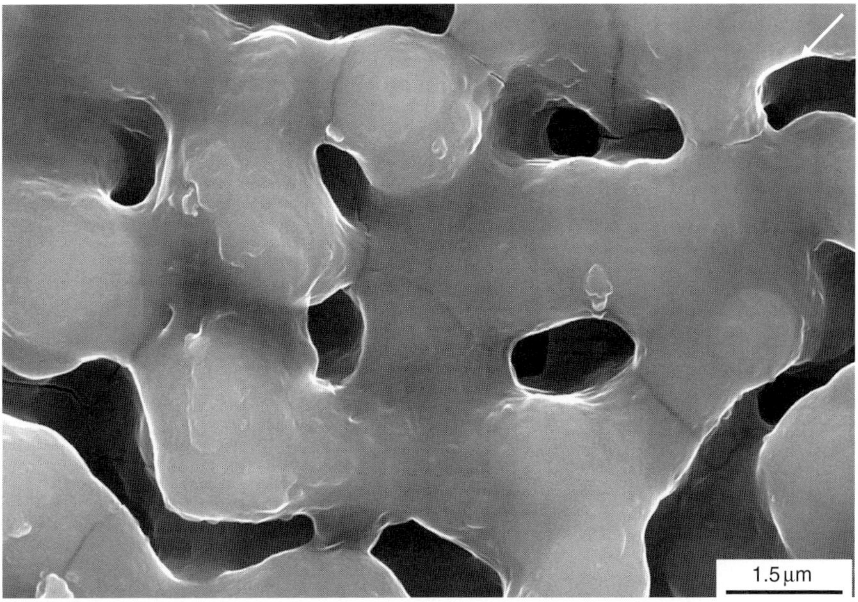

Fig. 12.2 SEM image of the globulites after 64-h thermal oxidation at 1,400°C (alumina tube furnace). Note the extensive pore volume comprised in-between the globulites

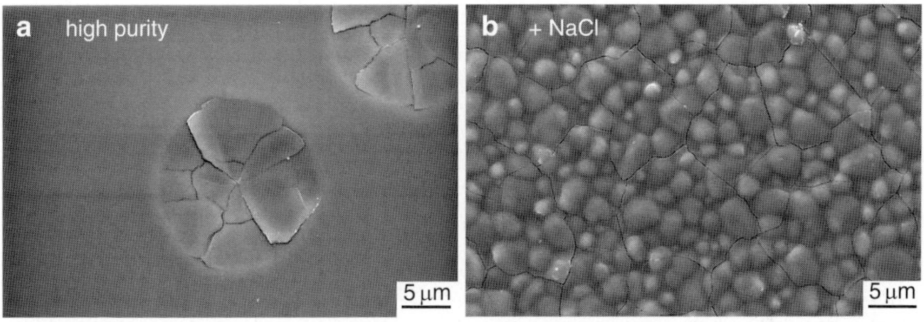

Fig. 12.3 SEM images of the oxide scale obtained after 28 h of thermal oxidation at 1,400°C using a high-purity furnace (fused quartz tube + dehumidified, high purity oxygen). (**a**) In high-purity environments ideal radialite formation can be observed. (**b**) In the presence of intentionally introduced NaCl, extensive globulite formation and accelerated crystallization are observed. Cracks are due to the β–α phase transition below 300°C

environments are significantly smaller in diameter (28 h,1,400°C) compared to experiments carried out in an alumina tube furnace (Fig. 12.1b). Assuming a similar crystallization rate for both cases, the reason for this difference can be found to be the smaller nucleation rate in the absence of impurities. In fact, no radialites were observed for high-purity experiments after 10 h whereas significant radialite formation was observed readily after 4 h when carrying out thermal oxidation inside and alumina tube. It should be noted, however, that there are some differences between the crystallization rate of high-purity silica and impurity-enriched SiO_2 with a higher rate in the latter case [66].

12.4 Etching

Surface morphology tells us a lot about the oxide scale: SiC covered by a thinner oxide scale has seen better oxidation protection than areas covered by a thick oxide layer. Also, we can differentiate between radialites and globulites considering their characteristic shape and dimension. However, we have also seen that crack formation due to the β–α phase transition of cristobalite (and, to a minor extent, because of the mismatch of the thermal expansion coefficient [12]), the surface of silica thermally grown on silicon carbide can locally be very rough. This circumstance makes high resolution AFM extremely difficult.

As most of the outer surface information is affected by the change in morphology during cooling, a more consistent image of the actual scale morphology can be gained from the analysis of the SiC–SiO_2 interface. With a scale thickness of only a few micrometers/several hundred nanometers and a surface morphology that has to be studied one or two orders of magnitude below that high-resolution X-ray tomography is not an adequate tool [89]. Techniques like atom probe tomography [90] or electron tomography [91] are more suitable, but very elaborate in terms of technical equipment and sample preparation. Also, the net area accessible via high resolution methods like these is limited to a few microns (or less).

An easy way to achieve the goal of surface analysis of the covered carbide–oxide interface is AFM subsequently applied after etching. With AFM, easy-to-select areas of the surface can be analyzed. The entire silica scale can be removed using hydrofluoric acid (HF) [92]. Although extreme care has to be taken when operating with this dangerous acid, a completely silica free SiC surface can be obtained. Depending on scale thickness and HF concentration, the exposure time needed for the removal of the oxide scale from the silicon carbide surface ranges from minutes to hours. While silica can be completely removed, silicon carbide will not be affected by HF etching [92–94]. In a recent study, we corroborated that the surface roughness of the HF etched SiC wafer did not change because of the etching [72].

Silicon carbide can be etched using molten salts like KOH [95] or plasma [96,97]. SiC etching is widely used to study defects in SiC like micropipes (=hollow-core dislocation) as the defect-rich areas show preferential etching (so-called decoration) with higher etching rates [98–101]. The reason for this is the residual stress in the surrounding area of, for example, a screw dislocation.

12.5 Scale and Interface Morphology

As mentioned before, the former carbide–oxide interface provides information on the morphology of the oxide scale. For a complete characterization of the oxide scale, however, we need to apply the following steps:

1. *Surface characterization of the silica-atmosphere interface* (*outer surface*). This can be done using AFM or other methods (e.g., laser scanning microscopy, white-light interferometry, etc.).
2. *Determination of the scale thickness.* For this, methods such as ellipsometry or scanning electron microscopy (SEM) on a cross-section sample can be applied. Note that amorphous and crystalline areas show different scale thickness values.
3. *Complete removal of the oxide scale without damaging the carbide.* This can be achieved via HF etching.
4. *Surface characterization of the former carbide–oxide interface* (inner surface) analogously to the previous characterization of the outer surface.

Thanks to the very characteristic morphology and shape of the individual radialites, it is easy to correlate AFM images from the outer and inner surfaces of the oxide scale. The examples shown in the following sections are taken from a recent study [72]. As sample material, high-quality nitrogen doped 6H-SiC (SiCrystal AG, Erlangen, Germany) was chosen and cut to 10.0 mm × 5.0 mm × 0.25 mm samples. Twenty-eight hours of dry thermal oxidation in an alumina tube furnace at 1,400°C in high-purity nondehumidified oxygen yielded a 1.24-μm-thick oxide scale as determined by SEM on a cross-section sample and as corroborated by spectral ellipsometry. During oxidation, the sample rested on two 250-μm-thick vertically aligned 6H-SiC:N plates to reduce the contact area with the support plate, to avoid contamination, and to allow free access for the gas from all directions. These conditions were chosen so that the influence of impurities typical for most furnace atmospheres could be investigated and both radialite and globulite formation would occur. Complete removal of the oxide scale was accomplished with a 40% HF solution at room temperature (24 h).

The chosen oxidation temperature and oxygen exposure yielded samples which were covered to approximately $^1/_4$ by radialites and $^3/_4$ by the amorphous silica scale (Fig. 12.4a). This way many radialites were obtained which had not undergone intergrowth with another radialite nearby. Yet, the mean diameter ranged between 80 and 90μm making identification and analysis of the radialites possible by optical microscopy.

We can distinguish between R1 (Fig. 12.4b, c) and R2 type (Fig. 12.4d, e) radialites. The first (R1 = ideal radialites) shows no sign of globulite formation, yet, while the second type (R2) has already undergone the structural transition towards globulites in the central part of the radialite.

First, we look at R1 radialites. Most of them are easy to distinguish from the amorphous silica scale in which they have grown because of crack formation corresponding to the β–α phase transition of the original devitrification product β-cristobalite (Fig. 12.4b). Some R1 radialites, however, have preserved their *in situ*

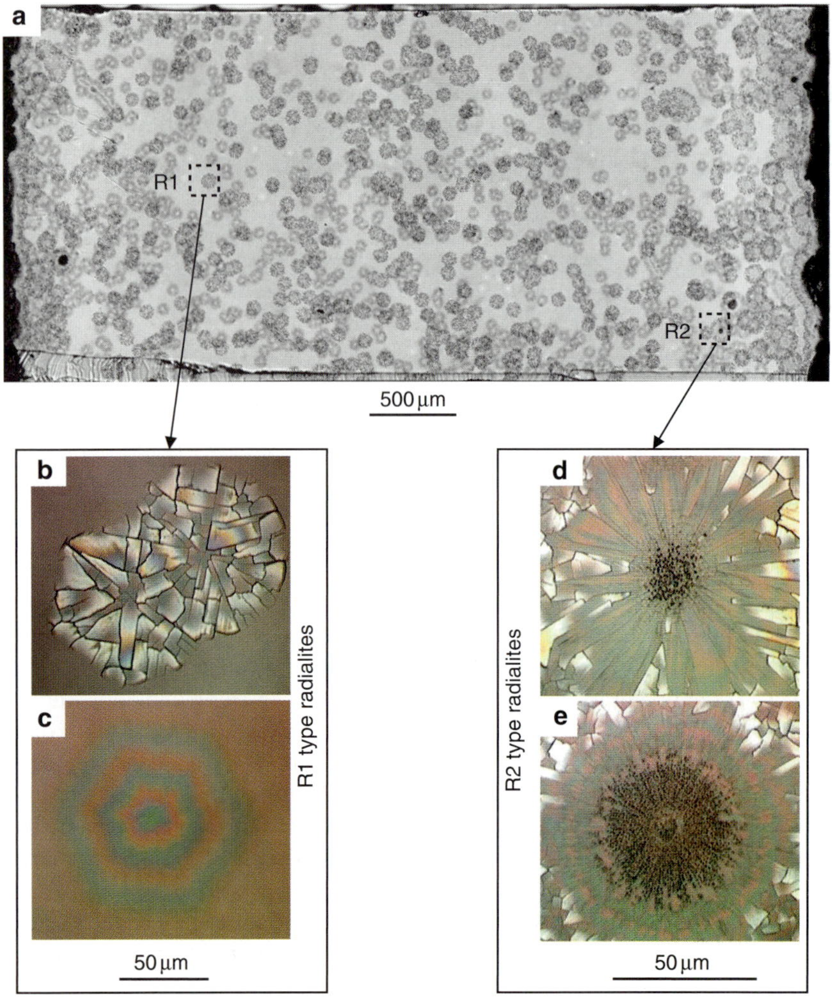

Fig. 12.4 Optical microphotographs of the oxide scale after 28 h of thermal oxidation at 1,400°C. R1 and R2 denote two radialites that will be subject to AFM/HF-etching analysis and which are two typical radialite types [(**b**, **c**): R1 type, (**d**, **e**): R2 type]. After the experiment the radialites are often highly differentiated into several crystal plates (**b**) because of the β–α phase transition of cristobalite. Locally, this differentiation has not yet occurred (**c**). Globulite consumption of the radialites starts at the radialite center (**d**, **e**)

morphology and have not yet undergone a structural differentiation from the surrounding amorphous scale (Fig. 12.4c). Analysis of the interference colors provides valuable information about the scale thickness and morphology: the colors indicate that the radialite must be thinnest in the center with increasing thickness towards its outer edges (Fig. 12.5). Note that the analysis of interference colors has proven

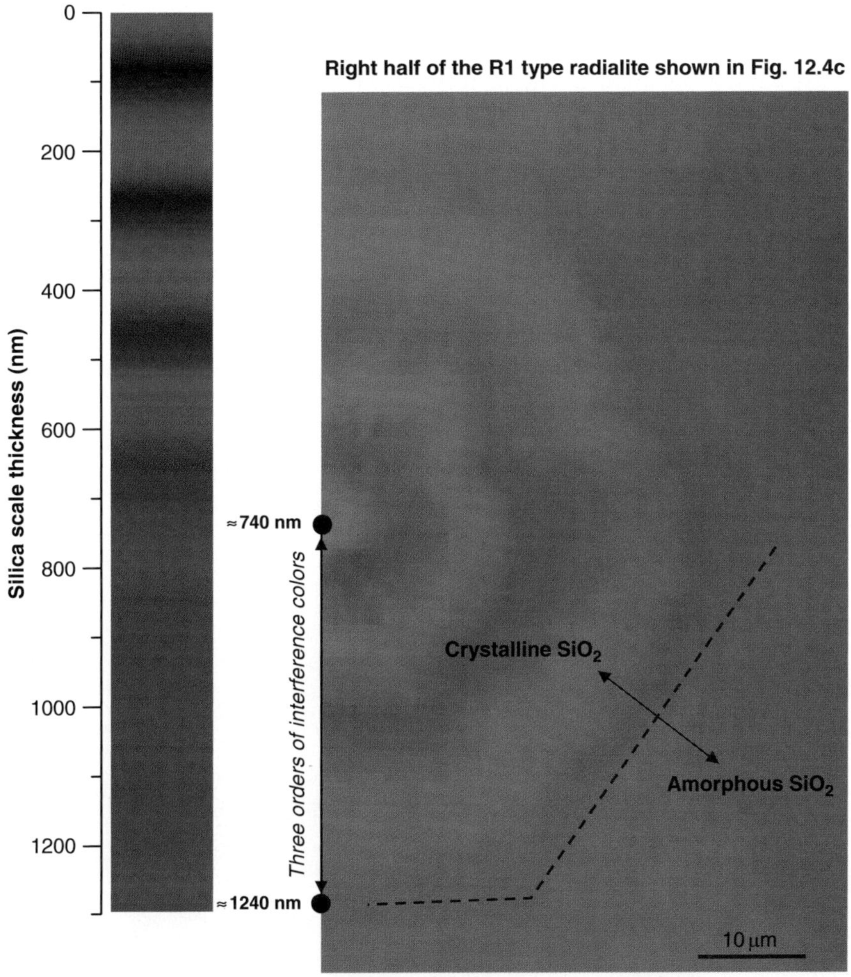

Fig. 12.5 Calculated and observed interference colors of an undifferentiated R1 type radialite showing the different oxide scale thickness

to be a fast and very powerful method for both qualitative phase determination (when all phases have the same thickness) and thickness measurement in the field of geosciences [102, 103].

It is also possible to translate the observed interference colors into discrete thickness values for a known refraction index ($n_{\alpha\text{-cristobalite}} = 1.48$; α-cristobalite is the present phase at ambient temperature). The wavelength of the thickness-dependent (d) interference color can be calculated via (12.4) [72, 102, 104]:

$$\lambda = \frac{2nd\cos(\theta)}{m - \frac{1}{2}}. \tag{12.4}$$

In (12.4), m denotes the interference color order and θ is the angle of the refracted ray (perpendicular incident light, $\theta = 42.51$).

As a result, we see that with three orders of interference colors, the difference between the crystalline oxide scale thickness in the center of the radialite and the outmost edges is 503 ± 15 nm. This is very close to the corresponding value we obtain via combined AFM/HF-etching (532 nm).

Figure 12.6 depicts AFM scans of the outer and inner surface of the oxide scale of an R1 type radiate which has undergone differentiation into several crystal plates during cooling. While the outer surface shows the differentiation of the radialite into several separated crystal plates, the inner surface is smooth in comparison. The inner part of the radialite was found to be thinner than the outer parts which are in agreement with the surface scans: here, we see an elevation on the inner and a

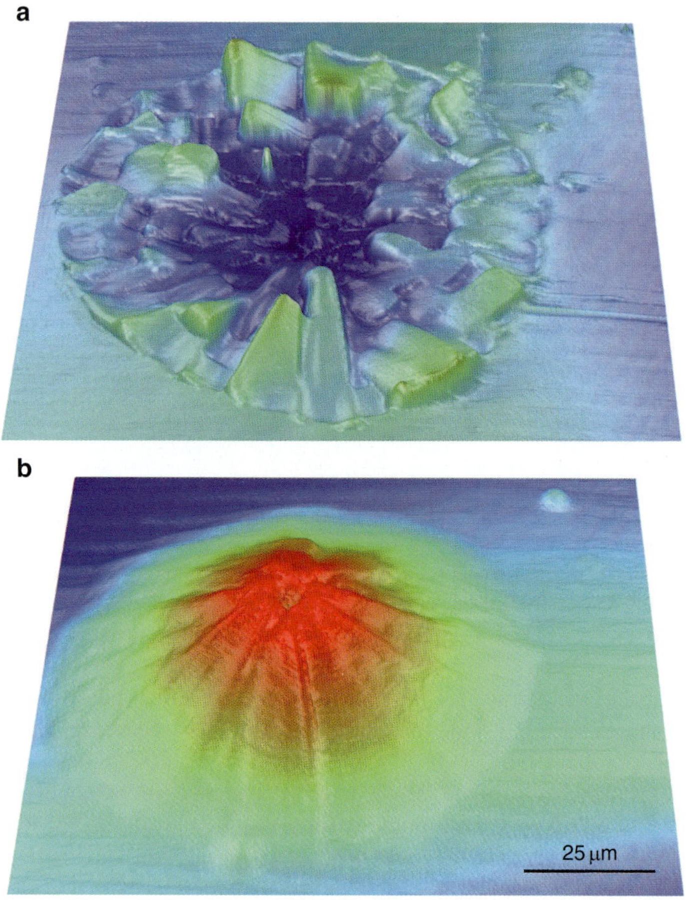

Fig. 12.6 False-color AFM scans of R1 as an example for a typical R1 type radialite. (**a**) Outer surface of the oxide scale. (**b**) Inner surface (= former carbide–oxide interface). Note the vertical superelevation and the perspective distortion

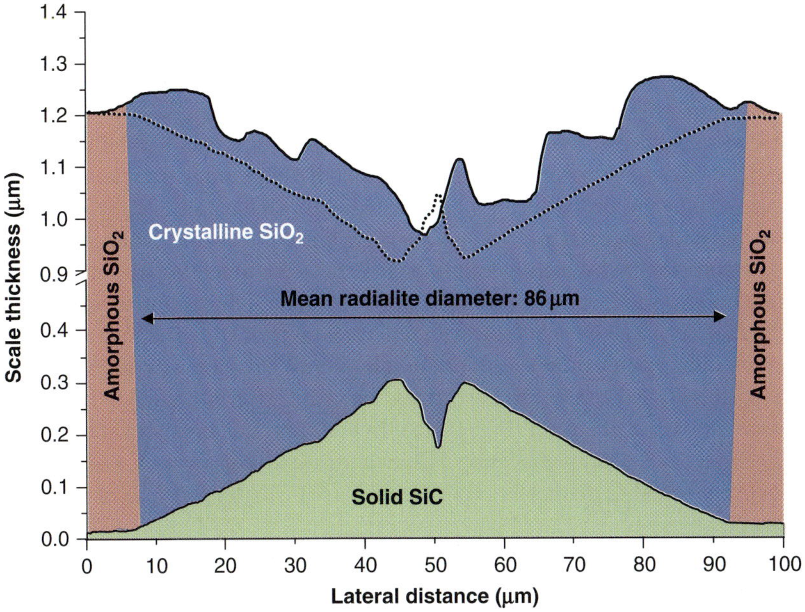

Fig. 12.7 Line scans of the R1 type radialite shown in Fig. 12.6. The *dashed line* is the profile of the carbide–oxide interface horizontally flipped to illustrate how the original radialite shape would have looked like under *in situ* conditions

corresponding depression on the outer surface. This becomes even more evident when we look at a profile scan as provided in Fig. 12.7. To see what the radialite looked like under *in situ* conditions, we have to look at the carbide–oxide interface. Considering this, the upper oxide scale is just a mirror image of the inner oxide surface. In Fig. 12.7, this z-profile is horizontally flipped and compared with actual morphology of the outer surface. The actual scale is higher than expected based on the SiC surface because of uplifting of the individual crystal plates, which also contribute to the uneven shape.

The central part of the radialite is characterized by an especially thick oxide scale. This indicates a more progressive oxidative consumption of the underlying carbide substrate. An explanation is given by the shape of the radialites: in the central part, the strain of the originally uncracked inclined crystal plates is largest and grain boundaries will be the unavoidable consequence. Grain boundaries, on the other side, accelerate the oxidation process as the reactant gas is transported faster towards the carbide–oxide interface.

On the inner surface of the oxide scale, a fine radial lineation is visible that corresponds with the crystal plates as seen from outer surface. This has important consequences for the evolution of radialites as this lineation (both elevations and depression) must have their origin during the thermal oxidation and cannot be a relic from the cooling process. Obviously, grain defragmentation occurs *in situ* as a process of energy minimization while the radialite grows into the surrounding amorphous matrix. This must not be mistaken with the pronounced crack formation we see after cooling the samples from 1,400°C to ambient temperature: such extensive cracks would have rendered the inner surface of oxide scale highly uneven and much less uniform.

We can expand the analysis of the radialite growth to the crystallization rate ν (μm/h). Although it is difficult to assess at which time exactly a nucleus of critical size was formed and crystallization had started, we still can tell that the crystallization rate must have been at least 1.5 μm/h. This value is obtained when dividing the radius of the radialite by the oxidation time. Any larger value for ν is also possible when we consider that crystallization will most likely not have happened at the very moment that oxidation began. However, when we consider areas with scratches and surface inhomogeneity at the edges of the sample, we also find a crystallization rate of approximately 1.5 μm/h. In these areas, crystallization is known to start during the initial oxidation phase (cf. Fig. 12.1a). In addition, this crystallization speed is in agreement with a previous study [12] where we found $\nu = 1.4 \pm 0.2\,\mu\text{m/h}$ from average radialite diameters of 11 samples (6H-SiC:N) and oxidation times of up to 64 h.

Close examination of the radialite diameter furthermore reveals that the radialite is larger on the outer surface (87 μm) compared to its expansion on the carbide–oxide interface (85 μm). As discussed before, superficially adsorbed gas molecules or impurities may act as unstable or metastable crystallization centers because of their contribution to the total energy (adsorption energy) causing the crystallization to start at the oxide–atmosphere interface. The small difference in the observed radialite diameters (i.e., $d_{\text{outer surface}} \approx d_{\text{inner surface}}$) substantiate that crystallization must have started during an early oxidation stage.

This AFM result is corroborated by the use of TEM. Figure 12.8 clearly shows the difference of the upper and lower surface expansion of the crystalline radialite into the amorphous silica matrix. We also see that along the amorphous–crystalline phase boundary, accumulation of gas bubbles and especially near the carbide–oxide interface, exsolved impurities can be observed (Fig. 12.8b). The latter can be understood in terms of the introduced globulite formation mechanism [71] that involves impurity enrichment followed by melt-accelerated recrystallization of the energetically disadvantageous radialite plates. Gas bubbles indicate lower gas diffusion in the crystalline region which will be discussed in more detail in the next section when outlining the kinetics of oxide scale and radialite growth.

Apart from locally accumulated impurities, the carbide–oxide interface is extremely smooth and abrupt on a nanometer level as known from previous studies [105, 106]. In Fig. 12.8c, a high-resolution TEM image is given that shows the abruptness of the carbide–oxide interface. Note that the gray contrast on the SiC side of the carbide–oxide interface corresponds to the stacking order of 6H-SiC.

In summary, we find the typical R1 type radialite morphology (= ideal radialite) to be as follows (Fig. 12.9): with a thinner inner and thicker outer part, the oxide scale of the radialite transients in thickness towards the surrounding vitreous silica matrix. Radialite differentiation into several subplates appears during *in situ* conditions and affects the morphology of the inner surface of the oxide scale. Grain

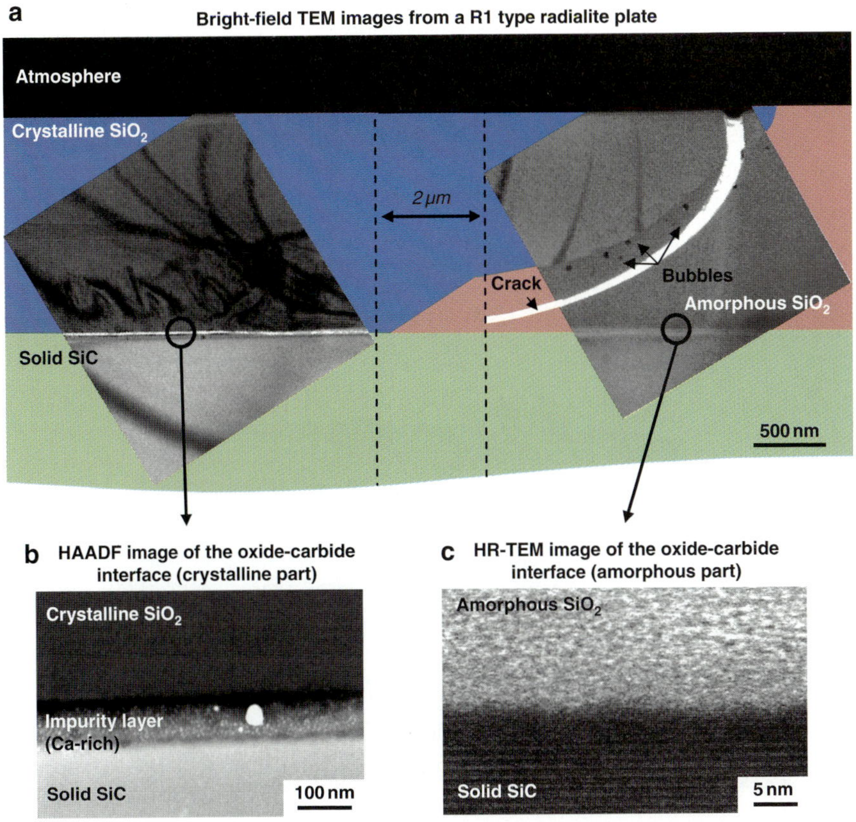

Fig. 12.8 (**a**) Carbide–oxide interface showing the transition between crystalline and amorphous silica. Crack formation was not present under *in situ* conditions and is because of thermal expansion mismatch between crystalline and amorphous silica. (**b**) Exsolved impurities can be found near the carbide–oxide interface underneath crystalline silica. (**c**) The carbide–oxide interface underneath amorphous silica is smooth on a nanometer scale and not dominated by accumulated impurities

boundaries act as weak spots in the oxide scale in terms of oxidation passivation. Much more significant in terms of plate separation, however, is the β–α phase transition of cristobalite during cooling after the actual oxidation took place. The latter causes drastic crack formation because of the increase in bulk crystalline volume. Also, the upper diameter of radialites is larger than the lower one indicating that crystallization must have started at the oxide–atmosphere interface.

In the following paragraph, the scale morphology of an R2 type radialite is discussed. With globulite formation having started (Figs. 12.10 and 12.11), the oxide scale is significantly thicker in the center of the radialite. As expected from the scale morphology depicted in Fig. 12.2, the partially porous globulite scale cannot provide the same oxidation protection for the silicon carbide body as the crystal plates of the

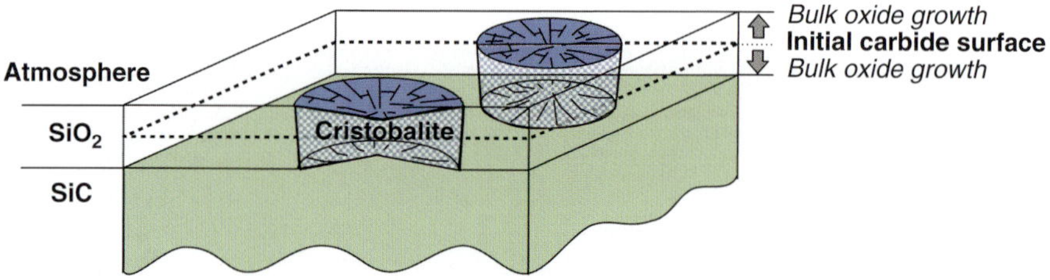

Fig. 12.9 Schematic illustration of R1 type radialites within the amorphous silica matrix thermally grown on silicon carbide. Note the expansion of the vertical scale with radialites being pronouncedly two-dimensional bodies (1–2 μm height vs. 50–200 μm diameter)

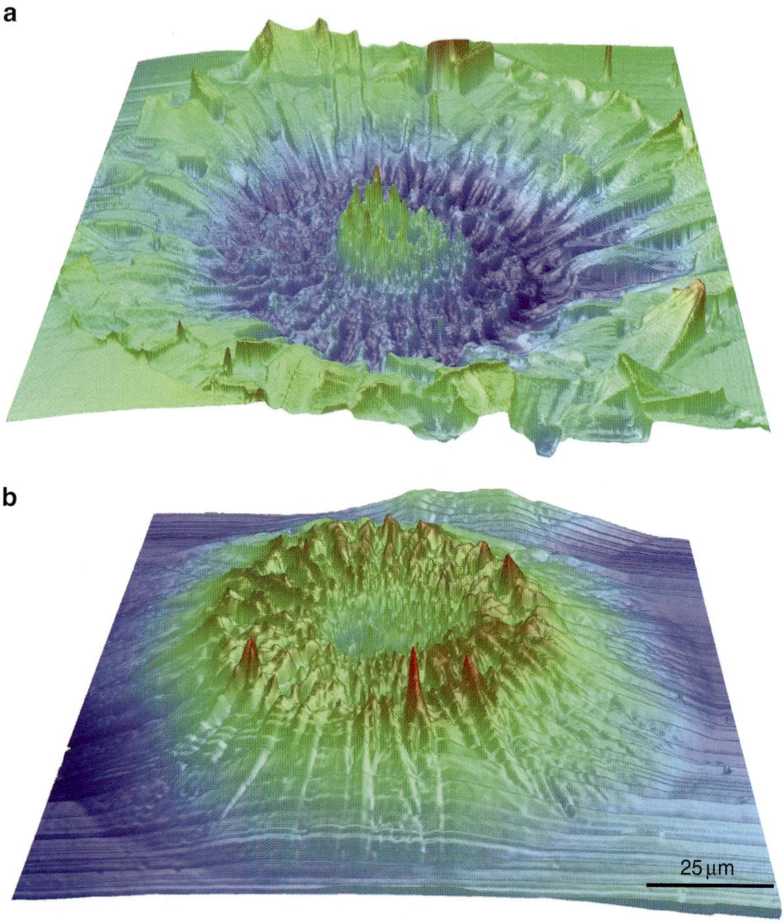

Fig. 12.10 False-color AFM scans of R2 as an example for a typical R2 type radialite. (**a**) Outer surface of the oxide scale. (**b**) Inner surface (= former carbide–oxide interface). Note the expansion of the vertical scale and the perspective distortion. The straight horizontal ripples in (**b**) are analytical artifacts

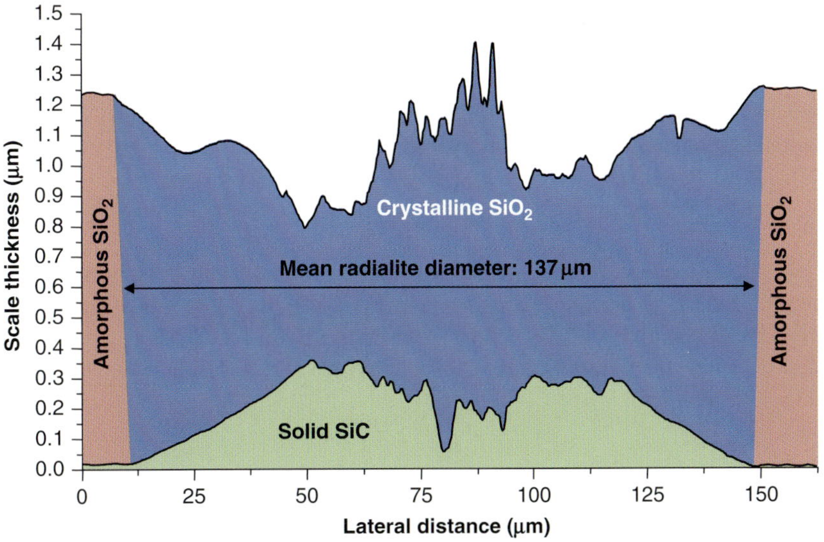

Fig. 12.11 Line scans of the R2 type radialite shown in Fig. 12.10. Note the significant increase in oxide scale thickness in the globulite area near the center of the radialite

radialite. In fact, the inner part shows a scale thickness larger than the surrounding amorphous silica layer (1.40 μm instead of 1.24 μm).

The outer part of the inner surface of the oxide scale shows a smooth transition between carbide and oxide with an almost linear slope. When directly comparing the R1 radialite (Fig. 12.7) with the R2 type radialite in Fig. 12.11, the larger diameter of the R2 crystalline disc is evident (137 μm). When we assume that both R1 and R2 nucleated at the same time during initial oxidation, then the crystallization rate of R2 must have been much larger (2.4 μm/h instead of 1.5 μm/h). With the evident correlation between globulite formation and the presence of impurities, an impurity-related increase in the crystallization rate is likely and in agreement with the literature [66].

12.6 Kinetics: Details and Overall Model

The basic factors that could be rate limiting for silicon carbide oxidation are (a) diffusion of oxygen within the atmosphere, (b) in-diffusion of oxygen through the growing oxide scale, (c) oxidation reaction at the SiC/SiO_2 interface, (d) out-diffusion of the gaseous oxidation product(s) through the silica layer, and (e) out-diffusion of gas phases from the SiO_2 surface back into the atmosphere [107, 108]. Experimental studies show, that with the exception of ultra-thin oxide scales, linear-parabolic kinetics can be observed for the dry thermal oxidation of silicon carbide. This indicates a diffusion-controlled mechanism [11, 108–110]. Literature agrees that the in-diffusion of oxygen is the rate-determining process limiting SiC oxidation [111–114].

The widely accepted Deal and Grove model [115] which was originally developed to describe the oxidation kinetics of pure metallic silicon can be equally adapted for the oxidation of silicon carbide (12.5):

$$d^2 + Ad = B(t + \tau). \qquad (12.5)$$

In (12.5), d denotes the oxide scale thickness, t denotes the oxidation time, and τ compensates for an initial oxide scale present at $t = 0$. The linear rate constant (B/A) is often abbreviated as k_l and is proportional to the reaction rate of the slowest reaction, while the parabolic rate constant (B, k_p) is proportional to the diffusion coefficient D (12.6):

$$B = \frac{2Dc}{N}. \qquad (12.6)$$

D denotes the diffusion coefficient within the oxide scale, c denotes the concentration of the gas species being transported, and N labels the number of gas molecules/ions per one volume unit [115, 116]. Both rate constants are thermally activated parameters following an Arrhenius law with characteristic activation energies [115, 117].

Using the introduced nomenclature (12.5) can be expressed as (12.7):

$$\frac{d^2}{k_p} + \frac{d}{k_l} - \tau = t. \qquad (12.7)$$

Over long oxidation periods (usually several hours), we find a simple parabolic time-law (12.8):

$$d^2 = Bt. \qquad (12.8)$$

From the literature we know that with increasing crystallization the oxidation rate is supposed to shift towards lower rate constants. The reason for this is that the diffusion coefficient of oxygen through crystalline silica is smaller compared to amorphous silica at the same temperature [68, 107, 118]. Differences as large as a factor of 30 between B_{vitreous} and $B_{\text{cristobalite}}$ were reported [119]. This was not corroborated by another study that reported differences with a factor of only ≈ 3 of the corresponding diffusion coefficients (D) at 1,400°C [118]. The differences in the diffusion coefficient explain the different thickness of the amorphous and crystalline oxide scale covering silicon carbide: SiC is consumed faster by oxygen in areas covered by vitreous SiO_2[119]. The exact difference between B_{vitreous} and $B_{\text{cristobalite}}$, however, is still a matter of current dispute [78].

Valuable information on the kinetics of the crystallization rate and oxide scale growth can be gained from detailed analysis of the slope at the former carbide–oxide interface. On first glance, one might assume the latter to be perfectly linear (Fig. 12.12a). Fitting a linear equation, however, yields significant residuals indicating that another regression function must be used (cf. Fig. 12.12). From the growth

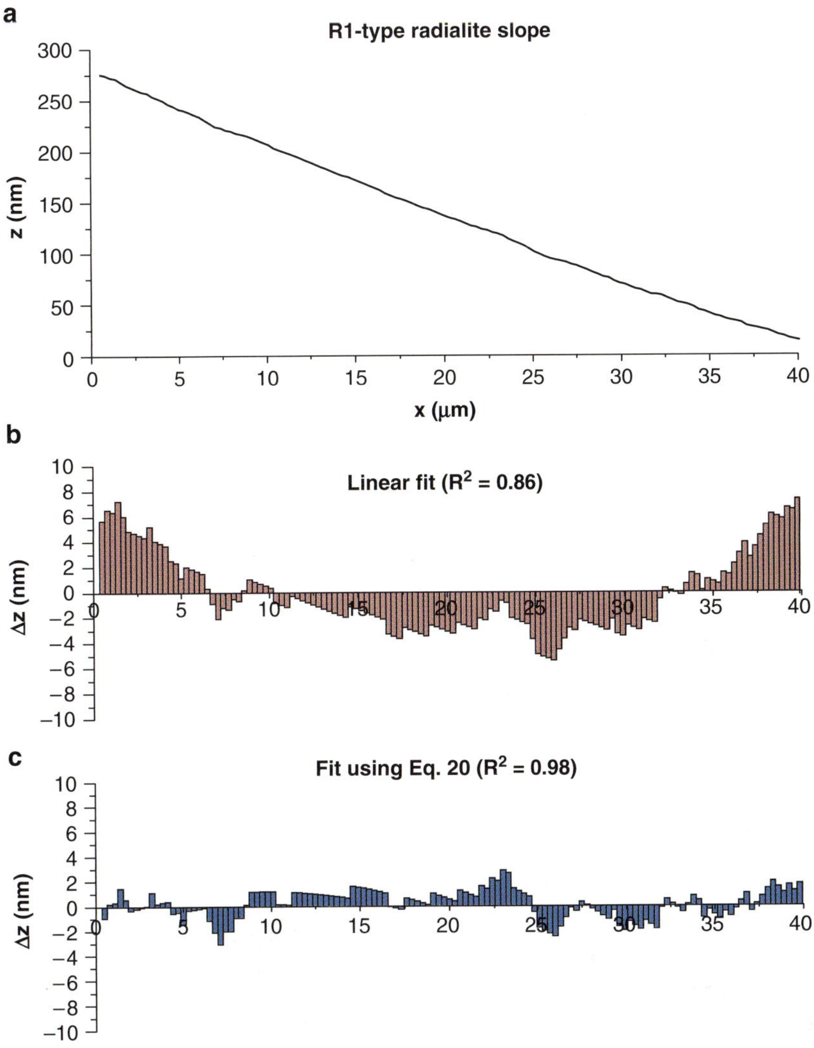

Fig. 12.12 Line profile of the slope of radialite R1 (**a**) and the fitting results using a linear fit (**b**) and (12.20) (**c**). Note the large residuals in case of a single linear fit

kinetics, it is clear that a simple linear time law is not sufficient to describe the slope underneath areas which have seen parabolic scale growth of the amorphous and (later) devitrified scale.

First, we consider scale growth of the ideal vitreous silica scale and assume the growth kinetics to be ideal parabolic with the rate constant B_{vitreous} (12.9) cf. (12.8). Over the time of amorphous silica growth ($t_{\text{vitreous growth}}$), a vitreous SiO_2 scale with thickness d_{vitreous} will form.

$$d_{\text{vitreous}} = \sqrt{B_{\text{vitreous}}\, t_{\text{vitreous growth}}}. \tag{12.9}$$

In the center of the radialite, we assume crystallization to start during initial oxidation. The crystallization front will reach a point at distance x from the crystallization center after a certain amount of time which only depends on the crystallization rate ν. As mentioned before, we know that for stoichiometric silica, ν follows a linear time law [12, 78, 80, 120]. Consequently, $t_{\text{vitreous growth}}$ can be expressed as the quotient of the distance x from the center of the radialite and the crystallization rate ν:

$$t_{\text{vitreous growth}} = \frac{x}{\nu}. \tag{12.10}$$

Having established these basic correlations, we can now calculate the total oxide scale thickness at any point of a specific radialite as a function of the parabolic rate constant from the amorphous growth phase (B_{vitreous}) and the time during which the crystallized oxide scale continued to grow ($B_{\text{cristobalite}}$).

The total oxide scale thickness cannot be calculated by simply adding two parabolic time laws. Instead, we must modify the original Deal and Grove equation (12.5) by using the factor τ. Latter compensates for a finite oxide scale thickness as present when the oxidation starts. In this case, τ can be used to compensate for the vitreous scale growth before the crystalline layer continues bulk growth. For a simple parabolic time law, (12.5) can be modified to (12.11):

$$d = \sqrt{B(t + \tau)}. \tag{12.11}$$

The factor τ has units of time, not scale thickness, so we have to calculate the time that is needed for the formation of the amorphous scale with thickness d_{vitreous} which was present at the moment the radial expanding crystallization front reached that point (12.12). Although there is a difference in the density of vitreous silica and the crystallization product β-cristobalite, that difference is small ($< 5\%$) and can be neglected for the sake of simplicity.

$$\tau_{\text{crystallization}} = \frac{(d_{\text{vitreous}})^2}{B_{\text{cristobalite}}}. \tag{12.12}$$

Using (12.9) for the thickness of the vitreous oxide scale and insert that expression in (12.12), we obtain (12.13):

$$\tau_{\text{crystalliation}} = \frac{B_{\text{vitreous}} t_{\text{vitreous growth}}}{B_{\text{cristobalite}}}. \tag{12.13}$$

The total oxide scale thickness d_{total} (encompassing the vitreous and crystalline growth phase) can be expressed as (12.14):

$$d_{\text{total}} = \sqrt{B_{\text{cristobalite}}(t_{\text{cristobalite growth}} + \tau_{\text{crystallization}})}. \tag{12.14}$$

Inserting (12.13) in (12.14) yields the following expression (12.15):

$$d_{\text{total}} = \sqrt{B_{\text{cristobalite}}(t_{\text{cristobalite growth}} + \frac{B_{\text{vitreous}} t_{\text{vitreous growth}}}{B_{\text{cristobalite}}})}. \tag{12.15}$$

The total oxidation period can be divided into two growth phases (12.16):

$$t_{\text{cristobalite growth}} + t_{\text{vitreous growth}} = t_{\text{total}}. \tag{12.16}$$

Furthermore, $t_{\text{vitreous growth}}$ is equal to the quotient x/ν (12.10). This allows us to reformulate (12.15) and (12.17):

$$d_{\text{total}} = \sqrt{B_{\text{cristobalite}}\left(t_{\text{cristobalite growth}} + \frac{B_{\text{vitreous}}(\frac{x}{\nu})}{B_{\text{cristobalite}}}\right)}. \tag{12.17}$$

Now, we replace $t_{\text{cristobalite growth}}$ (12.18):

$$t_{\text{cristobalite growth}} = t_{\text{total}} - t_{\text{vitreous growth}} = t_{\text{total}} - \frac{x}{\nu}. \tag{12.18}$$

Combining (12.17) and (12.18), we obtain (12.19):

$$d_{\text{total}} = \sqrt{B_{\text{cristobalite}}\left(\left(t_{\text{total}} - \frac{x}{\nu}\right) + \frac{B_{\text{vitreous}}\left(\frac{x}{\nu}\right)}{B_{\text{cristobalite}}}\right)}. \tag{12.19}$$

This equation can be rearranged (12.20):

$$d_{\text{total}} = \sqrt{B_{\text{cristobalite}}\left(t_{\text{total}} - \frac{x}{\nu}\right) + B_{\text{vitreous}}\left(\frac{x}{\nu}\right)}. \tag{12.20}$$

With (12.20), we now have an equation that gives a correlation between the scale thickness at any point on a radialite with distance x from the center of the radialite, the crystallization speed and both parabolic rate constants (B_{vitreous}, $B_{\text{cristobalite}}$). For situations in which the beginning of the crystallization cannot be assumed to be close to the initial oxidation phase, (12.20) must be corrected for that offset period.

As seen from (12.20), the crystallization rate and the B values cannot be calculated independently from each other. A convenient way to determine the crystallization rate was discussed before by simply dividing the radius of an R1 type radialite by the total oxidation time. For this, we have to be sure about the beginning of crystallization, which requires either several experiments at long and short oxidation times and/or *in situ* monitoring of the oxidation process. With oxidation temperatures $\gg$1,200°C, this is technically challenging but the best way for a comprehensive understanding of the crystallization process yielding the most reliable crystallization rate values. Alternatively, we can calculate B_{vitreous} precisely as we have exact knowledge of the thickness of the amorphous silica scale and the oxidation time. Having this parameter fixed, the other two ($B_{\mathrm{cristobalite}}$ and ν) can be obtained by least-square regression.

The quantitative results for the parabolic rate constants as obtained from fitting (12.20) are listed in Table 12.1. With a thickness of 1.24 μm for the surrounding amorphous silica scale, we find for B_{vitreous} a value of $1.34 \times 10^{-17} \mathrm{m}^2/\mathrm{s}$. This value is in agreement with the parabolic rate constant derived from the oxidation study of 11 samples and oxidation times up to 64 h, which were between 1.28×10^{-17} and $1.34 \times 10^{-17} \mathrm{m}^2/\mathrm{s}$ at 1,400°C [12]. A direct comparison between B_{vitreous} and $B_{\mathrm{cristobalite}}$ shows that the parabolic rate constant during vitreous scale growth is larger than the corresponding value for $B_{\mathrm{cristobalite}}$ by a factor of 4.2. This is in excellent agreement with a previous study by Rodriguez-Viejo et al. [118] who reported a ratio of 2.8 for the diffusion coefficients ($\approx B$) of amorphous and crystalline silica at 1,400°C. Larger values as reported by Ogbuji [119] could not be corroborated and are likely a result of different experimental conditions and setup [72]. Also, for the crystallization rate, a value of $\nu = 1.5 \pm 0.1$ μm/h results from fitting of (12.20) what is in agreement with the aforementioned crystallization rate of 1.5 μm/h.

It is important to note that the oxide scale thickness should be calculated based on the slope on the carbide–oxide interface because of the discussed inhomogeneities and discontinuities on the outer surface of the oxide scale. For this we neglect the small curvature of the radialite as a result of $d_{\mathrm{outer\ surface}} > d_{\mathrm{inner\ surface}}$ for the sake of simplicity. A positive side effect of this approach is that it is not necessary to correct the d values for the β–α phase transition of cristobalite.

From the values found for the parabolic rate constant for the vitreous and crystalline growth phase we can conclude that only a few micrometers away from the center of the radialite, more than 50% of the total oxide scale thickness was created during the time in which the oxide scale was still amorphous (Fig. 12.13). That means, of course, that oxide scale growth can drastically be slowed down by rendering the vitreous oxide scale into the crystalline state as long as the latter does not show globulite or crack formation.

When we compare experiments conducted under high-purity and impurity-containing conditions, we see the drastic morphological difference shown in Fig. 12.3. Bearing this in mind, we have to carefully address the influence of impurities on the data presented above. Sodium is potent crystallization trigger. Also, the viscosity of silica is lowered so that the sodium-doped amorphous silica is less protective against the oxidation of the underlying silicon carbide matrix because of the increased diffusion coefficient D [52, 53]. So we have on the one side the

Table 12.1 Results of fitting (12.20) to five radialite slopes (using a constant value for B_{vitreous})

Fixed	$B_{\mathrm{vitreous}} = 1.34 \times 10^{-17}\ \mathrm{m}^2/\mathrm{s}$	
Refined	$B_{\mathrm{cristobalite}}$, ν	
Results	$B_{\mathrm{cristobalite}} = 3.20 \times 10^{-18} \pm 4.87 \times 10^{-20} \mathrm{m}^2/\mathrm{s}$	$\nu = 1.5 \pm 0.1 \mu\mathrm{m/h}$
	$B_{\mathrm{vitreous}} : B_{\mathrm{cristobalite}} = 4.2 : 1$	
	$R^2 = 0.975$	

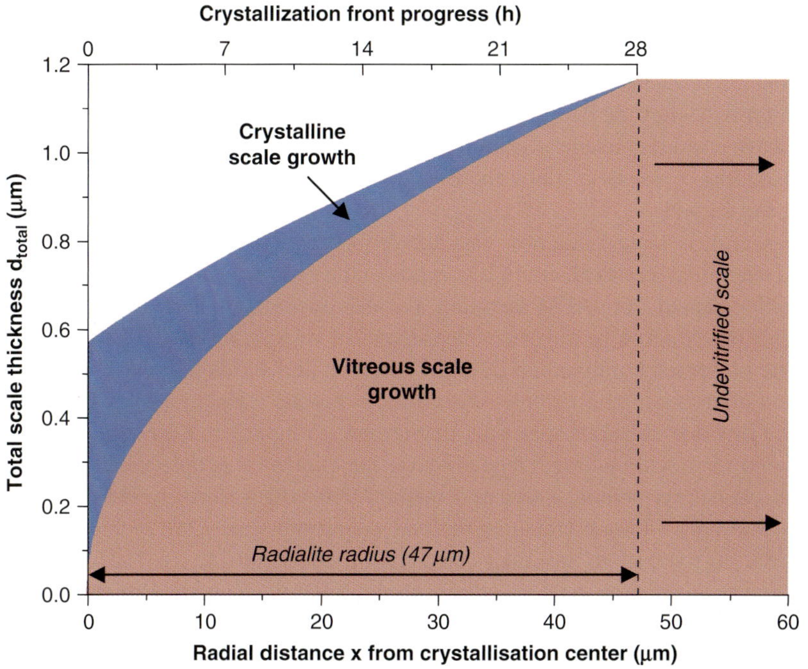

Fig. 12.13 Contribution from the vitreous and crystalline scale growth period to the total oxide layer thickness d_{total}. As seen, most of the total thickness is created during the vitreous scale growth phase because of the faster oxygen diffusion in amorphous silica

negative effects of a less protective, impurity-doped vitreous oxide scale along with globulite formation over longer oxidation periods and on the other side the positive effect of early devitrification providing superior carbide passivation compared to an amorphous oxide scale.

12.7 Conclusion and Outlook

As demonstrated in the example of thermally oxidized silicon carbide, valuable information can be gathered from AFM analysis of the outer and inner surface of the oxide scale. Simple etching is a fast and easy way to study the carbide–oxide interface which is usually covered by silica and not accessible by other methods. Also, we are able to expand a morphology-based analysis from a qualitative level (good/bad oxidation protection) to a quantitative level. The latter allowed us to calculate the total oxide scale thickness as a function of the rate constant of the vitreous and crystalline growth phase and the crystallization rate. While the amorphous scale growth rate can be calculated also by measuring the thickness of the vitreous scale and dividing it by the square root of the oxidation time, this cannot be done for the areas of silicon carbide which are covered by crystalline silica due to its lower diffusion rate compared to amorphous silica. When we look at a typical radialite, we see that the most central part (potentially bearing the information on pure crystalline scale growth) must be disregarded because of the locally increased scale thickness. The latter is a direct result of grain boundary formation as a stress-relief and energy-minimization processes. Along these oxide scale weak spots, accelerated oxidation can be observed.

In fact, it is remarkable how easy the difference between the oxidation rates of SiC covered by crystalline and amorphous silica can be calculated just based on z-profiles gained from AFM scanning. For this information, it usually is necessary to apply oxygen isotope diffusion experiments or very elaborate gravimetric studies.

The presented combination of AFM analyses and etching, however, is not limited to silica on silicon carbide. The same technique equally applies for silica on silicon or other substrate materials that are not to be affected by removal of their oxide by chemical etching. Depending on the choice of sample material, one can easily choose an etching agent and method that will selectively remove the thin film scale only without damaging the underlying substrate. One such example can be found in the case of salt-melt corrosion of silicon nitride (Si_3N_4 [121]). Here, important information on surface morphology based on optical microscopy was gained by comparing the sample before and after removal of the water-soluble part of the sodium glasses that formed during the corrosive attack. This way, the formation of local isles with a chemical composition different from the rest of the sodium-glass layer was identified. Similar results could be obtained from analysis of metals yielding a tomography-like thickness information with a high level of spatial resolution.

Acknowledgements

V.P. gratefully acknowledges the financial support of the Alexander-von-Humboldt-Foundation (Feodor-Lynen scholarship). Dr. Richard Wirth is thanked for his kind help with the HR-TEM analyses. Dipl.-Min. Yannick Hemberger kindly assisted with the AFM measurements.

References

1. S.E. Saddow, A. Agarwal, *Advances in Silicon Carbide Processing and Applications* (Artech House Inc., Norwood, 2004), pp. 212
2. G.L. Harris, *Properties of Silicon Carbide*, vol. 13, 1st edn. (INSPEC, the Institution of Electrical Engineers, London, 1995), pp. 282
3. K.A. Schwetz, in *Handbook of Ceramic Hard Materials,* ed. by R. Riedel. Silicon Carbide Based Hard Materials, vol. 2.(Wiley-VCH, Weilheim, 2000), pp. 683–748
4. E.J. Opila, D.S. Fox, C.A. Barrett, Cyclic oxidation of monolithic silicon carbide and silicon nitride materials. Ceram. Forum Int. **14**(7–8), 367–374 (1993)
5. S. Onda, R. Kumar, K. Hara, SiC integrated MOSFETs. Phys. Status Solidi A. **162**(1), 369–388 (1997)

6. D. Alok, P.K. McLarty, B.J. Baliga, Electrical properties of thermal oxide grown using dry oxidation on p-type 6H-silicon carbide. Appl. Phys. Lett. **65**(17), 2177–2178 (1994)
7. M.R. Baklanov, M. van Hove, G. Mannaert, S. Vanhaelemeersch, H. Bender, T. Conard, K. Maex, Low temperature oxidation and selective etching of chemical vapor deposition a-SiC:H films. J. Vac. Sci. Technol. B. **18**(3), 1281–1287 (2000)
8. P. Friedrichs, E. P. Burte, R. Schörner, Dielectric strength of thermal oxides on 6H-SiC and 4H-SiC. Appl. Phys. Lett. **65**(13), 1665–1667 (1994)
9. J.B. Petit, P.G. Neudeck, L.G. Matus, J.A. Powell, Thermal oxidation of single-crystal silicon carbide: kinetic, electrical, and chemical studies. Springer Proc. Phys. **71**(1), 190–196 (1992)
10. S. Zaima, K. Onoda, Y. Koide, Y. Yasuda, Effects of oxidation conditions on electrical properties of SiC–SiO_2 interfaces. J. Appl. Phys. **68**(12), 6304–6308 (1990)
11. T. Narushima, A. Goto, T. Hirai, Y. Iguchi, High-temperature oxidation of silicon carbide and silicon nitride. Mater. Trans. **38**(10), 821–835 (1997)
12. V. Presser, K.G. Nickel, Silica on silicon carbide. Crit. Rev. Solid State Mater. Sci. **33**(1), 1–99 (2008)
13. V. Haase, G. Kirschstein, H. List, S. Ruprecht, R. Sangster, F. Schröder, W. Töpper, H. Vanecek, W. Heit, J. Schlichting, *Si. Silicon. Supplemental Volume B3* (Springer, Berlin, 1986), pp. 545
14. J.E. Antill, J.B. Warburton, Active to passive transition in the oxidation of SiC. Corros. Sci. **11**(1), 337–342 (1971)
15. M. Balat, R. Berjoan, G. Pichelin, D. Rochman, High-temperature oxidation of sintered silicon carbide under pure CO_2 at low pressure: active–passive transition. Appl. Surf. Sci. **133**(1–2), 115–123 (1998)
16. A. Goto, H. Homma, T. Hirai, Effect of oxygen partial pressure on the high-temperature oxidation of CVD SiC. Corros. Sci. **44**(1), 359–370 (2002)
17. P.J. Jorgensen, M.E. Wadsworth, I.B. Cutler, Effects of water vapor on oxidation of silicon carbide. J. Am. Ceram. Soc. **44**(6), 258–261 (1961)
18. Q.N. Nguyen, E.J. Opila, R.C. Robinson, Oxidation of ultrahigh temperature ceramics in water vapor. J. Electrochem. Soc. **151**(10), B558–B562 (2004)
19. E.J. Opila, R.E. Hann Jr., Paralinear oxidation of CVD SiC in water vapor. J. Am. Ceram. Soc. **80**(1), 197–205 (1997)
20. L. Simon, L. Kubler, A. Ermolieff, T. Billon, Oxidation of 6H-SiC(0001). Microelectron. Eng. **48**(1), 261–264 (1999)
21. M. Sreemany, T.B. Ghosh, B.C. Pai, M. Chakraborty, XPS studies on the oxidation behavior of SiC particles. Mater. Res. Bull. **33**(2), 189–199 (1998)
22. N. Sieber, T. Seyller, R. Graupner, L. Ley, R. Mikalo, P. Hoffmann, D.R. Batchelor, D. Schmeißer, PES and LEED study of hydrogen- and oxygen-terminated 6H-SiC(0001) and (000–1) surfaces. Appl. Surf. Sci. **184**, 278–283 (2001)
23. M. Hollering, J. Bernhardt, J. Schardt, A. Ziegler, R. Graupner, B. Mattern, A.P.J. Stampfl, U. Starke, K. Heinz, L. Ley, Electronic and atomic structures of the 6H-SiC (000–1) surface studied by ARPES, LEED, and XPS. Phys. Rev. B. **58**(8), 4992–5000 (1998)
24. U. Starke, J. Bernhardt, M. Franke, J. Schardt, K. Heinz, Structure and morphology of SiC surfaces studied by LEED, AES, HREELS and STM. Diam. Relat. Mater. **6**(1), 1349–1352 (1997)
25. A.J. van Bommel, J.E. Crombeen, A. van Tooren, LEED and Auger electron observations of the SiC(0001) surface. Surf. Sci. **48**(1), 463–472 (1975)
26. T. Goto, H. Homma, High-temperature active/passive oxidation and bubble formation of CVD SiC in O_2 and CO_2 atmospheres. J. Eur. Ceram. Soc. **22**(14–15), 2749–2756 (2002)
27. E.A. Gulbransen, K.F. Andrew, F.A. Brassart, The oxidation of silicon carbide at 1150° to 1400°C and at 9×10^{-3} to 5×10^{-1} torr oxygen pressure. J. Electrochem. Soc. **113**(12), 1311–1314 (1966)
28. J.W. Hinze, H.C. Graham, The active oxidation of Si and SiC in the viscous gas-flow regime. J. Electrochem. Soc. **123**(7), 1066–1073 (1976)

29. T. Narushima, A. Goto, Y. Iguchi, T. Hirai, High-temperature active oxidation of chemically vapor-deposited silicon carbide in an Ar–O_2 atmosphere. J. Am. Ceram. Soc. **74**(10), 2583–2586 (1991)
30. K.G. Nickel, The role of condensed silicon monoxide in the active-to-passive oxidation transition of silicon carbide. J. Eur. Ceram. Soc. **9**(1), 3–8 (1992)
31. K.G. Nickel, Y.G. Gogotsi, in *Corrosion of Hard Materials*, ed. by R. Riedel. Handbook of Ceramic Hard Materials, vol. 1 (Wiley-VCH, Weinheim, 2000), pp. 140–154
32. B. Schneider, A. Guette, R. Naslain, M. Cataldi, A. Costecalde, A theoretical and experimental approach to the active-to-passive transition in the oxidation of silicon carbide. J. Mater. Sci. **33**, 535–547 (1998)
33. K.G. Nickel, P. Quirmbach, in *Gaskorrosion nichtoxidischer keramischer Werkstoffe,* ed. by J. Kriegesmann. Technische Keramische Werkstoffe (Deutscher Wirtschaftsdienst, Köln, 1991), pp. Kapitel 5.4.1.1, p. 1–76
34. G. Wiebke, Die Oxydation von Siliziumkarbid. Ber. Dtsch. Keram. Ges. **37**(5), 219–226 (1960)
35. J. Schlichting, Siliciumcarbid als Oxidationsbeständiger Hochtemperaturwerkstoff. Oxidations-und Heißkorrosionsverhalten. I. Ber. Dtsch. Keram. Ges. **56**(8), 196–200 (1979)
36. J. Schlichting, Siliciumcarbid als Oxidationsbeständiger Hochtemperaturwerkstoff. Oxidations-und Heißkorrosionsverhalten. II. Ber. Dtsch. Keram. Ges. **56**(8), 156–261 (1979)
37. K.G. Nickel, Y.G. Gogotsi, Hydrothermal degradation of chemical vapor deposited SiC fibres. J. Mater. Sci. **33**, 4357–4364 (1998)
38. T. Kraft, K.G. Nickel, in *Hydrothermal Carbon Coatings of Alpha-SiC Crystals*, ed. by H. Dimigen. Surface Engineering – EUROMAT 99, vol. 11 (Wiley-VCH, Weilheim, 1999), pp. 306–311
39. T. Kraft, K.G. Nickel, Carbon formed by hydrothermal treatment of α-SiC crystals. J. Mater. Chem. **10**(3), 671–680 (2000)
40. E.G. Stein von Kamienski, F. Portheine, J. Stein, A. Gölz, H. Kurz, Charge trapping in dry and wet oxides on N-type 6H-SiC studied by Fowler-Nordheim charge in-jection. J. Appl. Phys. **79**(5), 2529–2534 (1996)
41. N.S. Jacobson, A.J. Eckel, A.K. Misra, D.L. Humphrey, Reactions of SiC with $H_2/H_2O/Ar$ Mixtures at 1300°C. J. Am. Ceram. Soc. **73**(8), 2330–2332 (1990)
42. A.H. Heuer, V.L.K. Lou, Volatility diagrams for silica silicon nitride and silicon carbide and their application to high-temperature decomposition and oxidation. J. Am. Ceram. Soc. **73**(10), 2785–3128 (1990)
43. E.J. Opila, Q.N. Nguyen, Oxidation of chemically-vapor-deposited silicon carbide in carbon dioxide. J. Am. Ceram. Soc. **81**(7), 1949–1952 (1998)
44. J.E. Antill, J.B. Warburton, Oxidation of silicon and silicon carbide in gaseous atmospheres at 1000°C–1300°C. In *Proceedings of the AGARD Conference Proceedings No. 52*, 1970, pp. 10
45. E. Fitzer, R. Ebi, in *Kinetic Studies on the Oxidation of Silicon Carbide*, ed. by K.L. Marshall, T.W. Faust, C.E. Ryan. Silicon Carbide (South Carolina University Press, Columbia, SC, 1973), pp. 320–328
46. C. Önneby, C.G. Pantano, Silicon oxycarbide on SiC surfaces and at the SiC/SiO_2 interface. Vac. Sci. Technol. A. **15**(3), 1597–1602 (1997)
47. R. Pampuch, W. Ptak, S. Jonas, J. Stoch, Formation of ternary Si-O-C phase(s) during oxidation of SiC. Mater. Sci. Monogr. **6**(1), 435–448 (1980)
48. B. Hornetz, H.-J. Michel, J. Halbritter, ARXPS studies of SiO_2–SiC interfaces and oxidation of 6H SiC single crystal Si-(001) and C-(00–1) surfaces. J. Mater. Res. **6**(12), 3088–3094 (1994)
49. B. Hornetz, H.-J. Michel, J. Halbritter, Oxidation and 6H-SiC–SiO_2 interfaces. J. Vac.Sci. Technol. A. **13**(3), 767–771 (1995)

50. G. Lucovsky, H. Niimi, A. Gölz, H. Kurz, Differences between silicon oxycarbide regions at SiC–SiO_2 prepared by plasma-assisted oxidation and thermal oxidations. Appl. Surf. Sci. **123/124**(1), 435–439 (1998)
51. G. Lucovsky, H. Niimi, Remote plasma-assited oxidation of SiC: a low temperature process for SiC–SiO_2 interface formation that eliminates interfacial Si oxycaribde transition regions. J. Phys. Condens. Matter. **16**(1), S1815–S1837 (2004)
52. E. Opila, Influence of alumina reaction tube impurities on the oxidation of chemically-vapor-deposited silicon carbide. J. Am. Ceram. Soc. **78**(4), 1107–1110 (1995)
53. E.J. Opila, D.L. Myers, Alumina volatility in water vapor at elevated temperatures. J. Am. Ceram. Soc. **87**(9), 1701–1705 (2004)
54. K.A. Schwetz, H. Werheit, E. Nold, Sintered and monocrystalline black and green silicon carbide – chemical composition and optical properties. Ber. Dtsch. Keram. Ges. **80**(12), E37–E44 (2003)
55. L.U.J.T. Ogbuji, E.J. Opila, A comparison of the oxidation kinetics of SiC and Si_3N_4. J. Electrochem. Soc. **142**(3), 925–930 (1995)
56. K. Motzfeld, On the rates of oxidation of silicon and of silicon carbide in oxygen and correlation with permeability of silica glass. Acta Chem. Scand. **18**(7), 1595–1606 (1964)
57. C.E. Ramberg, G. Cruiciani, K.E. Spear, R.E. Tressler, C.F. Ramberg Jr., Passive-oxidation kinetics of high-purity silicon carbide from 800° to 1100°C. J. Am. Ceram. Soc. **79**(11), 2897–2911 (1996)
58. K. Christiansen, R. Helbig, Anisotropic oxidation of 6H-SiC. J. Appl. Phys. **79**(6), 3276–3281 (1996)
59. N.G. Wright, C.M. Johnson, A.G. O'Neill, Mechanistic model for oxidation of SiC. Mater. Sci. Eng. B. **B61–62**(1), 468–471 (1999)
60. N.G. Wright, C.M. Johnson, A.G. O'Neill, Oxidation modelling for SiC. Mater. Res. Soc. Symp. Proc. **572**(1), 135–140 (1999)
61. D. Lipinski, H.E. Schwiete, Die Bildung des Cristobalits aus amorphem Siliziumdioxid unter verschiedenen Gasatmosphären. Tonindustrie-Zeitung Keram. Rundschau. **88**(7/8; 10; 11/12), 145–153; 217–225; 258–262 (1964)
62. N.G. Ainslie, C.R. Morelock, D. Turnbull, in *Devitrification Kinetics of Fused SiO_2*, ed. by M.K. Reser, G. Smith, H. Insley. Proceedings of the Symposium on Nucleation and Crystallization in Glasses and Melts, Columbus, 1962, pp. 97–107
63. K.L. More, P.F. Tortorelli, M.K. Ferber, J.R. Keiser, Observations of accelerated silicon carbide ression by oxidation at high water-vapor pressures. J. Am. Ceram. Soc. **83**(1), 211–213 (2000)
64. J.H. Oehler, Hydrothermal crystallization of silica gel. Geol. Soc. Am. Bull. **87**(1), 1143–1152 (1976)
65. F.E. Wagstaff, S.D. Brown, I.B. Cutler, The influence of H_2O and O_2 atmospheres on the crystallization of vitreous silica. Phys. Chem. Glasses. **5**(3), 76–81 (1964)
66. P. Donnadieu, Influence of impurities on plastic flow of silica glass. J. Non-Cryst. Solids. **99**(1), 113–117 (1988)
67. W. Göpel, C. Ziegler, *Einführung in die Materialwissenschaften: Physikalisch-chemische Grundlagen und Anwendungen* (Teubner, Stuttgart, 1996), pp. 490
68. R. Ebi, Hochtemperaturoxidation von Siliziumkarbid und Siliziumnitrid in technischen Ofenatmosphären. In *Fakultät für Chemiewesen*, PhD (Universität Karlsruhe, Karlsruhe, 1973), pp. 123
69. E. Gugel, H.W. Hennicke, P. Schuster, Zur Bildung der SiO_2-Schicht auf SiC. Ber. Dtsch. Keram. Ges. **46**(9), 481–485 (1969)
70. A.H. Heuer, L.U.J.T. Ogbuji, T.E. Mitchell, The microstructure of oxide scales on oxidized Si and SiC single crystals. J. Am. Ceram. Soc. **63**(5–6), 354–355 (1980)
71. V. Presser, A. Loges, R. Wirth, K.G. Nickel, Microstructural evolution of silica on single crystal silicon carbide. Part II: Influence of impurities and defects. J. Am. Ceram. Soc. **92**(8), 1796–1805 (2009)

72. V. Presser, A. Loges, Y. Hemberger, K.G. Nickel, Microstructural evolution of silica on single-crystal silicon carbide. Part I: Devitrification and oxidation rates. J. Am. Ceram. Soc. **92**(3), 724–731 (2009)
73. L. Filipuzzi, R. Naslain, C. Jaussaud, Oxidation kinetics of SiC deposited from CH_3SiCl_3/H_2 under CVI conditions. J. Mater. Sci. **27**(12), 3330–3334 (1992)
74. H.D. Keith, F.J. Padden Jr., A phenomenological theory of spherulitic crystallization. J. Appl. Phys. **34**(8), 2409–2421 (1963)
75. S.D. Pantsukrin, D.V. Kalinin, Kinetics and mechanism of formation of alpha-cristobalite spherulites. Seriya Khim. Nauk. **1**, 27–31 (1988)
76. S.W. Freiman, G.Y. Onoda Jr., A.G. Pincus, Spherulitic crystallization in glasses. In *Proceedings of the Symposium of the Glass Division of the American Ceramic Society*, Ohio, 1971, pp. 141–150
77. M.H. Lewis, J. Metcalf-Johansen, P.S. Bell, Crystallization mechanisms in glass-ceramics. J. Am. Ceram. Soc. **62**(5–6), 278–288 (1979)
78. R.H. Doremus, *Glass Science*, 2nd edn. (Wiley, New York, 1994), pp. 339
79. M.J.-F. Guinel, M.G. Norton, Oxidation of silicon caribde and the formation of silica polymorphs. J. Mater. Res. **21**(10), 2550–2563 (2006)
80. H. Li, M. Tomozawa, V.K. Lou, Effects of nitrogen and carbon ion implantation on devitrification of silica glasses. J. Non-Cryst. Solids. **168**(1), 56–63 (1993)
81. F.E. Wagstaff, K.J. Richards, Kinetics of crystallization of stoichiometric SiO_2 glass in H_2O atmospheres. J. Am. Ceram. Soc. **49**(3), 118–121 (1966)
82. E.J. Opila, Variation of the oxidation rate of silicon carbide with water-vapor pressure. J. Am. Ceram. Soc. **82**(3), 625–636 (1999)
83. O.W. Flörke, Die Modifikationen von SiO_2. Fortschr. Mineral. **44**(2), 181–230 (1967)
84. H. Strunz, E.H. Nickel, *Strunz Mineralogical Tables*, 9th edn. (Schweizerbart'sche Verlagsbuchhandlung, Stuttgart, 2001), pp. 870
85. N.R. Keskar, J.R. Chelikowsky, Structural properties of nine silica polymorphs. Phys. Rev. B. **46**(1), 1–13 (1992)
86. D.R. Peacor, High-temperature single-crystal study of the cristobalite inversion. Z. Kristallogr. **138**(1), 274–298 (1973)
87. L.U.J.T. Ogbuji, Development of oxide scale microstructure on single-crystal SiC. J. Mater. Sci. **16**(1), 2753–2759 (1981)
88. Z. Zheng, R.E. Tressler, K.E. Spear, The effect of sodium contamination on the oxidation of single crystal silicon carbide. Corros. Sci. **33**(4), 545–556 (1992)
89. A. King, G. Johnson, W. Ludwig, Diffraction contrast tomography – a synchrotron radiation technique for mapping polycrystalline microstructures in 3D. Mater. Sci. Forum. **571–572**(1), 207–212 (2008)
90. K. Thompson, P.L. Flaitz, P. Ronsheim, D.J. Larson, T.F. Kelly, Imaging of arsenic cottrell atmospheres around silicon defects by three-dimensional atom probe tomography. Science **317**(5843), 1370–1374 (2007)
91. I. Arslan, E.A. Marquis, M. Homer, M.A. Hekmaty, N.C. Bartelt, Towards better 3-D reconstructions by combining electron tomography and atom-probe tomography. Ultramicroscopy **108**(12), 1579–1585 (2008)
92. M.B. Johnson, M.E. Zvanut, O. Richardson, HF chemical etching of SiO_2 on 4H and 6H SiC. J. Electron. Mater. **29**(3), 368–371 (2000)
93. S. Dhar, O. Seitz, M.D. Halls, S. Choi, Y.J. Chabal, L.C. Feldman, Chemical properties of oxidized silicon carbide surfaces upon etching in hydrofluoric acid. J. Am. Chem. Soc. **131**(46), 16808–16813 (2009)
94. S.J. Schoell, M. Hoeb, I.D. Sharp, W. Steins, M. Eickhoff, M. Stutzmann, M.S. Brandt, Functionalization of 6H-SiC surfaces with organosilanes. Appl. Phys. Lett. **92**(15), 153301–153303 (2008)
95. D. Siche, D. Klimm, T. Hölzel, A. Wohlfahrt, Reproducible defect etching of SiC single crystals. J. Cryst. Growth. **270**(1), 1–6 (2004)
96. S. Dohmae, K. Shibahara, S. Nishino, H. Matsunami, Plasma etching of CVD grown cubic SiC single crystals. Jpn. J. Appl. Phys. **24**(11), L873–L875 (1985)

97. B. Kim, S. Kim, B.T. Lee, Low pressure plasma etching of silicon carbide. Appl. Phys. A. **81**, 793–797 (2005)
98. M. Katsuno, N. Ohtani, J. Takahashi, H. Yashiro, M. Kanaya, Mechanism of molten KOH etching of SiC single crystals: comparative study with thermal oxidation. Jpn. J. Appl. Phys. **38**(1), 4661–4665 (1999)
99. W.C. Say, J.K. Wu, W.L. Chen, Hot corrosion of α-SiC ceramics by V_2O_5 melt. J. Mater. Sci. **25**(3), 1614–1617 (1990)
100. N.W. Thibault, Morphological and structural crystallography and optical properties of silicon carbide. Part I: Morphological crystallography and etching figures. Am. Mineral. **29**(7–8), 249–278 (1944)
101. F. Yang, P. Fei, Indentation behavior of silicon carbide etched by KOH. Electrochem. Solid-State Lett. **8**(2), G51–G53 (2005)
102. W.D. Nesse, *Introduction to Optical Mineralogy* (Oxford University Press, Oxford, 2005), pp. 362
103. H.-R. Wenk, A. Bulakh, *Minerals: Their Constitution and Origin* (Cambridge University Press, Cambridge, 2004), pp. 646
104. E. Hecht, *Optics* (Pearson Education Inc., San Francisco, 2002), pp. 698
105. F. Devynck, F. Giustino, A. Pasquarello, Abrupt model interface for the 4H(1000)SiC-SiO_2 interface. Microelectron. Eng. **80**(1), 38–41 (2005)
106. T. Ohnuma, H. Tsuchida, T. Jikimoto, Interface states in abrupt SiO2/4H- and 6H-SiC(0001) from first-principles: effects of Si dangling bonds, C dangling bonds and C clusters. Mater. Sci. Forum. **457–460**(1), 1297–1300 (2004)
107. R. Ebi, E. Fitzer, K.J. Hüttinger, Oxidationskinetik von SiC-Pulvern bei Temperaturen zwischen 1200°C und 1600°C. High Temp. High Press. **4**(1), 21–25 (1972)
108. K.L. Luthra, Some new perspectives on oxidation of silicon carbide and silicon nitride. J. Am. Ceram. Soc. **74**(5), 1095–1103 (1991)
109. R.F. Adamsky, Oxidation of silicon carbide in the temperature range 1200 to 1500°C. J. Phys. Chem. **63**(2), 305–307 (1959)
110. E.J. Opila, Q.G.N. Nguyen, Oxidation of chemically-vapor-deposited silicon carbide in carbon dioxide. J. Am. Ceram. Soc. **81**(7), 1949–1952 (1998)
111. I.C. Vickridge, D. Tromson, I. Trimaille, J.-J. Ganem, E. Szilagyi, G. Battistig, Oxygen isotopic exchange occuring during dry thermal oxidation of 6H SiC. Nucl. Instrum. Methods Phys. Res. B. **190**(1), 574–578 (2002)
112. C.D. Fung, J.J. Kopanski, Thermal oxidation of 3C silicon carbide single-crystal layers on silicon. Appl. Phys. Lett. **45**(7), 757–759 (1984)
113. V.V. Afanas'ev, M. Bassler, G. Pensl, M. Schulz, Intrinsic SiC/SiO_2 interface states. Phys. Status Solidi A. **162**(1), 321–337 (1997)
114. J.-J. Ganem, I. Trimaille, P. Andre, S. Rigo, Diffusion of near surface defects during the thermal oxidation of silicon. J. Appl. Phys. **81**(12), 8109–8111 (1997)
115. B.E. Deal, A.S. Grove, General relationship for the thermal oxidation of silicon. J. Appl. Phys. **36**(12), 3770–3778 (1965)
116. D. Schmeißer, D.R. Batchelor, R.P. Mikalo, P. Hoffmann, A. Lloyd-Spetz, Oxide growth on SiC(0001) surfaces. Appl. Surf. Sci. **184**(1–4), 340–345 (2001)
117. C. Raynaud, Silica films on silicon carbide: a review of electrical properties and device applications. J. Non-Cryst. Solids. **280**(1), 1–31 (2001)
118. J. Rodriguez-Viejo, F. Sibieude, M.T. Clavaguera-Mora, C. Monty, ^{18}O diffusion through amorphous SiO_2 and cristobalite. Appl. Phys. Lett. **63**(14), 1906–1908 (1993)
119. L.U.J.T. Ogbuji, Effect of oxide devitrification on oxidation kinetics of SiC. J Am Ceram Soc, 80(6), 1544–1550 (1997)
120. F.E. Wagstaff, Crystallization and melting kinetics of cristobalite. J. Am. Ceram. Soc. **52**(12), 650–654 (1969)
121. C. Berthold, *Degradation von Si3N4 als Ventilwerkstoff durch Oxidation und Salzschmelzkorrosion bei Ensatztemperaturen* (Shaker Verlag, Aachen, 1998), pp. 145

Chapter 13
The Scanning Probe-Based Deep Oxidation Lithography and Its Application in Studying the Spreading of Liquid *n*-Alkane

Yuguang Cai and Lingbo Lu

Abstract We have developed a novel chemical patterning method, the scanning probe-based deep oxidation lithography, which enables us to fabricate a carboxylic acid-terminated chemical pattern on a hydrophobic silane film. This lithography method employs a conducting atomic force microscope (AFM) probe to oxidize the silane film under the probe in a humid environment. When a positive bias voltage is applied to the silane film, the area under the probe is oxidized to the partially degraded silane (OTSpd), which is 10 Å lower than the background. The oxidation time and the path of the AFM probe during the oxidation determine the size and shape of the OTSpd chemical pattern. We can use the OTSpd pattern to control shape and size of a droplet of liquid *n*-alkane on a surface because the OTSpd pattern is lyophilic (liquid attractive). This novel chemical pattern method enables us to study the *n*-alkane structures on the surface and investigates how a liquid alkane droplet spreads over patterned surfaces at molecular scale. We found that the alkane formed seaweed-shaped monolayer on the OTSpd surface, where the alkane molecules may tilt at certain angles. Furthermore, we investigated the role of vapor during the spreading of liquid alkane droplet. We found that the vapor of the alkane droplet adsorbed on the nearby surface before the liquid spread. Hence, the liquid actually always spreads on a layer of the adsorbate. It is the surface properties of this adsorbate determine the spreading and wetting of the liquid alkane.

13.1 Introduction

Understanding the behaviors of liquid hydrocarbons at nanometer scale is important for applications like painting, coating, high-resolution inkjet printing, and high-performance lubrication [1]. However, the relevant knowledge is rather limited because of the technical difficulty in studying liquid at nanometer scale. At present, the primarily method for studying liquids is visual observation, which is employed in the contact angle measurement, spreading dynamics studies, and surface tension measurement [2–5]. When the size of liquid under investigation is at nanometer level, we can no longer observe the liquid directly due to the diffraction limit of visible light, nor can we observe the liquid via the electron microscopy, since the liquid would evaporate in the vacuum chamber of the electron microscope. Furthermore, at nanometer scale, it is difficult to manipulate the liquids, such as positioning a droplet at a designated place. The lack of proper methods for studying liquids at nanometer scale leads to the knowledge gap about the behavior of liquids at the nanometer scale, which undermines the development of many relevant areas such as the micro-lubrication in MEMS and design of micro-fluidic devices. Chemical patterns can be used to control the position, shape, and size of liquids at nanometer scale [6–8]. Such capabilities enable us to use the chemical pattern to study the liquid spreading, especially the spreading of hydrocarbons, at nanometer scale. We have developed a scanning probe-based two-step surface patterning method, which can be used to fabricate chemical patterns on flat self-assembled monolayer surfaces [9]. This new method makes the study of the spreading of liquid hydrocarbons at nanometer scale possible. Recently, we have used this method to investigate the spreading and the adsorption structures of long-chain alkanes. A series of discoveries were made by employing this new method. In this chapter, we provide a brief account on this progress in the studies of liquids at nanometer scale.

13.2 Part 1. The Chemical Patterning Method for Alkane Spreading Study

13.2.1 Octadecyltrichlorosilane as the Substrate for Pattern Fabrication

We use the octadecyltrichlorosilane (OTS) self-assembled monolayer on an ultra-flat Silicon(100) wafer as the surface for chemical pattern fabrication and the subsequent liquid spreading experiments. The spreading of liquids is very sensitive to the surface topography and chemistry. The OTS film on the ultra-flat silicon wafer is a topographically and chemically homogenous surface, which is ideal as the substrate for studying liquids. The silicon wafer (1–40Ω cm resistivity, nitrogen doped) was polished to the ultra-flat level with a root mean square (RMS) roughness smaller than 5 Å. The wafer was boiled in the piranha solution (2 parts of 98% sulfuric acid and 1 part of 30% hydrogen peroxide) at 125°C for 15 min to remove all organic contaminants on the surface, generating a hydroxyl-terminated SiO_2 layer on the Si wafer surface. We dipped the wafer in a 5-mM OTS toluene solution for 8 h to grow the OTS silane monolayer, as shown in Scheme 13.1. The precursor molecule, OTS (Scheme 13.1a), reacted with the trace amount of water in the toluene to form silanols (Scheme 13.1b), followed by the cross-linking and condensation reactions between the hydroxyl groups. As a result, the silane molecules cross-linked together and anchored to the SiO_2 surface (Scheme 13.1c) via the Si–O–Si bonds, forming the OTS film (Scheme 13.1d). Grazing incidence X-ray diffraction and X-ray reflectivity studies showed that the C18 chains of the silane molecules were

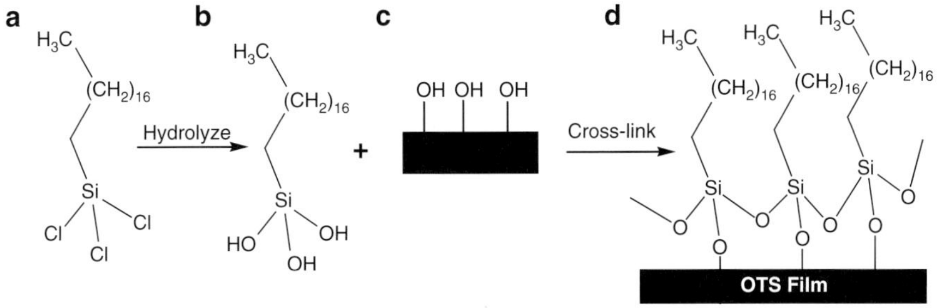

Scheme 13.1 The formation of OTS film on Si(100) wafer

close-packed, and stood perpendicularly on the surface [10]. The OTS film was homogenous, methyl-terminated, and with a thickness of 2.6 nm [10]. Nevertheless, pin-holes still existed in the *as-prepared* OTS film. To obtain a pinhole-free OTS film, the *as-prepared* OTS film was annealed in a sealed vial at 40°C, 100% relative humidity (RH) for 8 h. Then the OTS-coated wafer was dipped in the OTS toluene solution again for another 8 h. After three such cycles, the OTS film is a pinhole-free film, which is called "stabilized OTS" [11–14]. The quality of OTS film is monitored by atomic force microscopy (AFM), Brewster angle IR spectroscopy, contact angle measurement, and ellipsometry. The RMS roughness of an OTS film measured from AFM image is typically smaller than 5 Å and the height variation between any two points of 15 μm apart should be less than 5 Å. The IR characterization also reveals the structure and thickness of the OTS film. Figure 13.1 shows a typical Mid-IR spectrum of the OTS film, which was acquired using the hydroxyl-terminated Si(100) wafer as background. The area of the peaks between 2,800 and 3,000 cm^{-1} should have a typical value of the 54 mA, which corresponds to carbon density of one monolayer of perpendicular and close-packed C18 chains. The peak positions for the systematic and asymmetric stretching of CH_2 are the indication of the order and tightness of the molecular packing inside the film [15]. Higher peak wave number means poorer order of the film. In Fig. 13.1, the CH_2 symmetric stretching $\nu_s(CH_2)$ and CH_2 asymmetric stretching $\nu_a(CH_2)$ peaks are at 2,851 and 2,920 cm^{-1}, which correspond to an ordered in-plane C18 chain structure and tight packing. The blue circle in Fig. 13.1 highlights the broad negative peak around 3,600 cm^{-1}, which is the position of the –OH stretching. This negative peak is an indication that the surface hydroxyl group on the background wafer sample disappeared after the formation of OTS on the wafer. Hence, the disappearance of the surface hydroxyl group confirmed that OTS molecules anchored on the surface through the Si–O–C bond. The thickness of the OTS film measured by ellipsometry has a typical value of 2.6 ± 0.3 nm. The advancing contact angle (water) is 112°, as expected for a good OTS film. Table 13.1 summarizes these values, which were used to monitor the quality of the OTS film throughout our study.

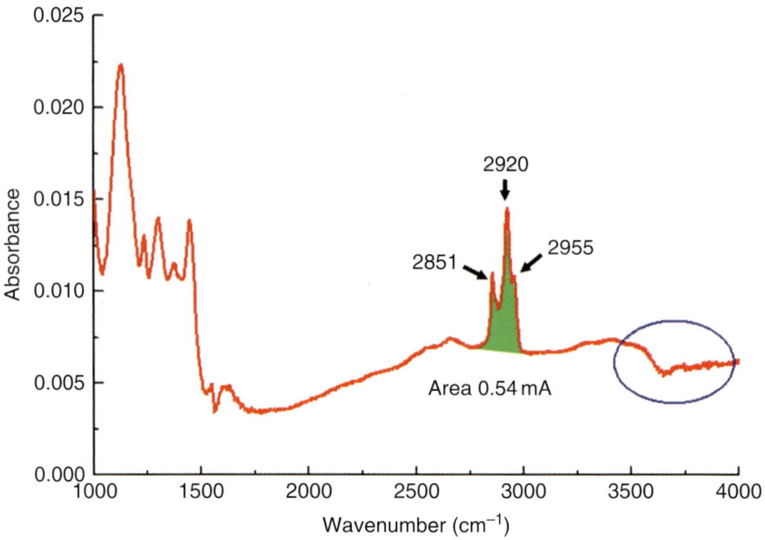

Fig. 13.1 The Mid-IR spectrum of OTS

Table 13.1 The typical parameters of a good OTS film

Properties	Typical value
RMS roughness	< 5Å
Height variation within 15 μm	< 5Å
Peak area between 2,800 and 3,000 cm^{-1} in IR	54 mA
Peak position CH_2 symmetric stretching $\nu_s(CH_2)$	2,851 cm^{-1}
Peak position CH_2 asymmetric stretching $\nu_a(CH_2)$	2,921 cm^{-1}
Advancing contact angle	112°
Thickness by ellipsometry	2.6 ± 0.3 nm

13.2.2 Fabricating Hydrophilic Chemical Patterns on OTS: The Scanning Probe Deep Oxidation Lithography

13.2.2.1 The Experimental Setup

The fabrication of hydrophilic chemical patterns on an OTS surface was conducted using an Agilent PicoPlus 3000 environmental AFM in the contact mode, as shown in Fig. 13.2. The silicon wafer was placed in the environmental chamber at 100% RH and 25°C. A Pt–Ti-coated conducting AFM probe was connected to the virtual ground of the current pre-amplifier. During the pattern fabrication process, the conducting tip contacted the OTS film at first. Due to the capillary condensation in the humid environment, a water meniscus formed between the tip and the surface. Then a +10-V DC bias voltage was applied to the sample, an electrochemical cell was formed. In the cell, the Pt–Ti-coated tip acted as the cathode and the substrate

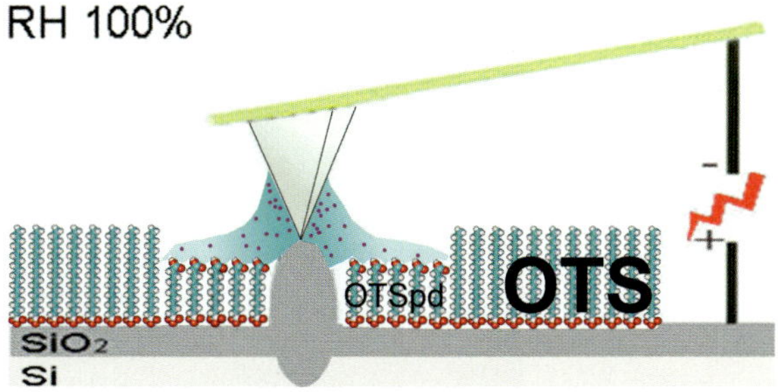

Fig. 13.2 The experimental setup for the deep oxidation lithography on OTS film

acted as the anode. When a bias was applied, the water was electrolyzed, generating hydrogen on the tip surface and oxygen on the substrate. Since the voltage between the anode and the cathode (5–10 V) is much higher than the water electrolysis reaction needs (1.23 V), the over potential generated active oxygen species such as ozone, atomic oxygen, hydroxyl radical (OH) on the substrate, which subsequently oxidized the chemically inert alkane chain. As a result, after the voltage was applied, the OTS surface directly under the conducting AFM probe was oxidized. The oxidized OTS is hydrophilic, partially degraded silane with the carboxylic acid (–COOH) terminus, which is designated as OTSpd.

The setup is the same as the silane oxidation method invented by Sagiv group [11, 12, 16]. However, we chose to use Pt–Ti-coated tip and conduct the oxidation under 100% RH, which caused a much higher degree of oxidation on the OTS film than the original Sagiv method. As a result, the oxidized area is slightly trimmed down (partially degraded) and is lower than the OTS. In contrast, the Sagiv method generated an oxidized pattern with the same height as OTS. A unique characteristic of our method is that the size of OTSpd pattern is not just limited to the tip–surface contact area. We found that size of OTSpd pattern was a function of the dwell time of the bias voltage. The longer the bias lasted, the bigger the pattern was, which is due to the diffusion of the active oxygen species. Once generated under the tip, these species oxidized the OTS molecules directly under the tip, changing the surface from hydrophobic OTS to hydrophilic OTSpd. In addition, they can diffuse out and oxidize silane molecules nearby. The reason is that water immediately adsorbed on the newly formed hydrophilic surface of the OTSpd in the 100% RH environment. Therefore, a thin film of water covered the surface of OTSpd. As a result, the active oxygen species can laterally diffuse outward toward the border of OTSpd/OTS, where they oxidize additional OTS to make the OTSpd pattern bigger. If the voltage bias remained, the active oxygen species would be continuously generated and diffuse outward, which is driven by the radical concentration gradient. Eventually, the OTSpd pattern grows bigger and bigger. The growth kinetics is a typical example of the two-dimensional moving boundary diffusion [17]. Because the OTS surface is homogenous, the diffusion is isotropic. So during oxidation, a stationary AFM probe would generate a round disk, whereas a moving AFM probe

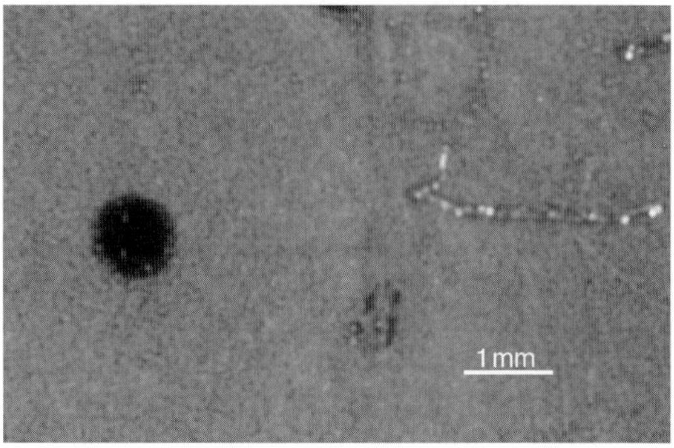

Fig. 13.3 A large OTSpd disk (*dark disk*) fabricate by a single stationary AFM probe

would generate a pattern that followed the probe's path. In Fig. 13.3, we demonstrate that if a 10-V bias was applied and held for 1 h, a stationary AFM probe generated a 0.2-mm diameter OTSpd disk. This unique relation between pattern size and bias dwell time enables us to use a single AFM probe to fabricate patterns ranging in size from ~25 nm to sub-millimeter. This technique sets the stage for us to study how small liquid droplets (25–250 nm, for example) spread over a large area (micrometer or sub-millimeter scale), employing the OTSpd chemical patterns to control the spreading.

13.2.3 The Structure and Chemistry of the OTSpd Pattern

Upon the electrochemical oxidation of the OTS film, two types of reactions occurred at the same time. The first type is that the active oxygen species oxidized the alkyl chains of the OTS molecules and diffused outward laterally, converting the alkyl chain into carboxylic group and chopping the methyl and methylene units on the chain. This reaction partially damaged the OTS film, making the oxidized silane film thinner. The second type is the growth of SiO_2 between the electrodes [18]. Because the SiO_2 has a lower density than silicon, the Si to SiO_2 conversion led to the expansion of volume. Thus, the formation of SiO_2 underneath OTS film causes the apparent height of the oxidized pattern to increase [19]. Overall, the height of the OTSpd pattern is determined by the equilibrium of these two types of reactions. Several factors affect this equilibrium. A higher water vapor pressure leads to a thicker water film on OTSpd. The thick water film facilitates the lateral diffusion of active oxygen species. Hence, increasing the water vapor pressure shifts the equilibrium toward the silane degradation. In contrast, if the substrate is the heavily doped silicon wafer (resistivity $<1\ \Omega$ cm), the current between the wafer and the tip would be much higher, causing the equilibrium to shift toward silicon oxide growth. Thus the OTSpd will not be generated.

To facilitate the generation of OTSpd, we conduct the electrochemical oxidation on wafers with a resistivity between 1 and 40 Ω cm and in a 100% RH environment.

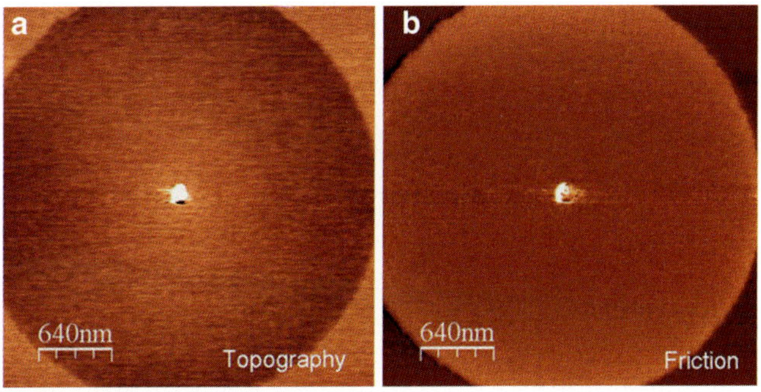

Fig. 13.4 Representative deeply oxidized OTSpd patterns. (**a**) Topography image of an OTSpd disk, which was fabricated by a stationary tip. The *bright spot* in the disk center is the oxide, which can be etched into a deep hole after HF treatment. (**b**) The corresponding friction image

Figure 13.4 shows a representative OTSpd disk fabricated using a stationary tip under the above controlled conditions. The AFM image was acquired using a new tip under the contact mode. Figure 13.4a, b is the topography image and the corresponding phase image. Figure 13.4a shows that there is a peak in the center of the disk. This peak is the SiO_2, which was formed directly under the tip under the intense electric field. After the sample was dipped in 1% HF solution for 2 s, the center peak changed into a deep hole whereas disks remained intact. This HF etching test confirmed that the center peak is made of bare SiO_2.

In Fig. 13.4b, the OTSpd disk has a higher friction. When a tip with a hydrophilic surface moved across a surface, the friction between hydrophilic–hydrophilic surfaces is stronger than the friction between the hydrophilic–hydrophobic surfaces. The image was acquired using a SiO_2 tip (hydrophilic surface) scanning over the surface. In the friction image, the friction over the OTSpd is higher than that of over OTS. Since the OTS surface is hydrophobic, so the OTSpd surface is hydrophilic.

13.2.4 The Depth of the OTSpd Pattern

The accurate measurement of the height/depth of the OTSpd pattern using AFM is not a trivial business. For patterns that have small height difference and large frictional contrast such as OTSpd, the direct height measurement from the contact mode is not reliable. This is because of the friction-topography cross-talks [20]. If the tip scans over an area with a high frictional contrast, the high dragging force will cause the cantilever to twist. The twisting of the cantilever leads to the laser spot to move horizontally on the photodiode detectors of AFM. As shown in Fig. 13.5, when the cantilever is not perfectly aligned vertically to the fast scan direction, or if the lateral torsion of cantilever is not perfectly symmetrical, the laser spot will move not only horizontally, but also up and down. Since the vertical position of the laser spot on the photo detector represents the height in AFM topography, such artificial up and down

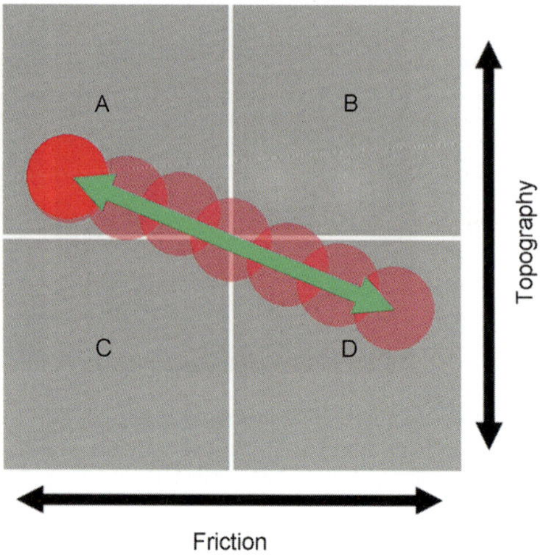

Fig. 13.5 The movement of the laser spot on the four-segment photo detector of AFM when the friction-topography cross-talk occurs

movements of the laser spot create artificial topography signals. The phenomenon is called the friction-topography cross-talks, where the signal in the friction channel pollutes the topography signal. A characteristic of the friction-topography cross-talk is that the apparent heights in the topography channels are opposite for the trace and re-trace scan directions, which is an artifact frequently seen when we characterize OTSpd pattern in contact mode.

We used two different approaches to measure the real depth of the OTSpd pattern. The first approach is to use the tapping mode to directly detect the depth of the OTSpd pattern. In the tapping mode, the tapping motion is always perpendicular to the surface. The cantilever did not twist during scanning, thus avoids the friction-topography cross-talks. After the OTSpd fabrication, we incubated the sample in triethylamine, 1% ethylenediaminetetraacetic acid solution, 10% HCl, and chloroform, respectively, to remove any potential charge, inorganic, and organic contaminants on the OTSpd pattern. Then the sample was immediately installed on the AFM in the environmental chamber filled with ultra-pure nitrogen. Subsequent tapping mode AFM scans indicate that the OTSpd is $\sim$10 Å lower than the OTS background. Figure 13.6.1a shows a representative tapping mode topography image of an OTSpd disk array. The corresponding histogram (Fig. 13.6.1b) indicates the OTSpd disks in this array have the depth of 10.7 ± 1.4Å.

In the second approach, we used a molecule with known length (alkyltrichlorosilane) as a ruler to measure the real depth of the OTSpd. Alkyltrichlorosilane molecules can form an additional monolayer on top of a hydrophilic pattern, such as OTSpd [16]. Based on this mechanism, we incubated an OTSpd pattern in a 5-mM OTS toluene solution. After 8 h, the sample was rinsed and imaged using AFM. The OTS molecules formed a second layer on top of the OTSpd. Figure 13.6.2 shows the contact mode image of the OTSpdOTS bilayer pattern on OTS surface. In this pattern, since the OTSpdOTS is also terminated with methyl group, the frictions of the OTSpdOTS and the OTS background are the same. Therefore, when the

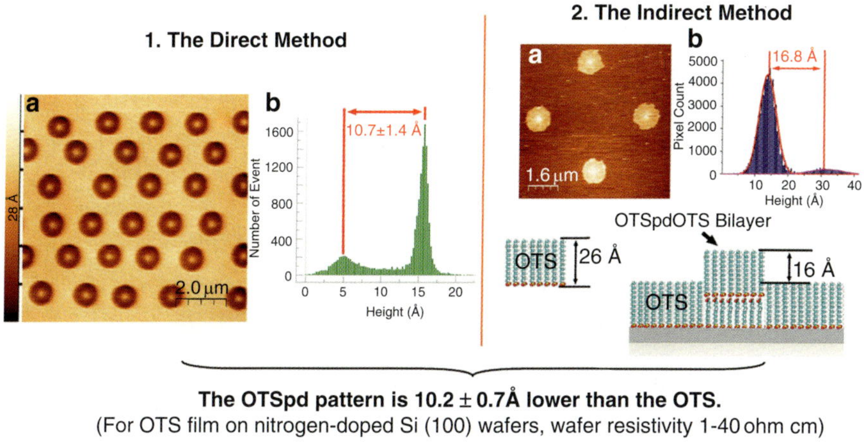

Fig. 13.6 Measuring the depth of OTSpd pattern

tip scanned across the pattern in contact mode, the cantilever did not twist. So the friction-topography cross-talk is eliminated. Under this condition, height obtained from the corresponding topography image is the real height. Here, the histogram (Fig. 13.6.2b) corresponding to the topography image of the OTSpdOTS bilayer (Fig. 13.6.2a) shows that the OTSpdOTS is 16 Å above the OTS. As shown in the model in Fig. 13.6.2c, since the thickness of one OTS layer is 26 Å, the OTSpd is calculated to be 10 Å lower than the OTS background.

In summary, using the direct and indirect methods, we measured the real depth of OTSpd pattern for hundreds of sample. The overall statistical result from the measurements shows that the OTSpd pattern is 10.2 ± 0.7Å lower than the OTS, as long as the OTSpd pattern is fabricated according to our experimental conditions (using the nitrogen-doped silicon wafer with a resistivity between 1 and 40 Ω cm, and conducting the electrochemical oxidation at 100% RH at 25°C).

13.2.5 OTSpd Is Terminated with Carboxylic Acid Group

We used a small molecule fluorescent probe 1-pyrenyldiazomethane (PDAM) to investigate the surface chemistry of the OTSpd pattern. Our result indicates that the OTSpd surface is carboxylic acid (–COOH)-terminated. PDAM is a fluorescent probe that specifically reacts with the –COOH group. It has been widely used to detect the existence of the carboxylic acid group [21]. The reaction is illustrated in Scheme 13.2. In our experiment, after an array of OTSpd disks was fabricated on an OTS-coated wafer, we incubated the sample in PDAM methanol solution for 4 h. After rinsed in methanol and cleaned with the super-critical carbon dioxide snow jet cleaner, we imaged the patterned area using AFM and a Nikon ME 600L optical fluorescent microscope. Figure 13.7a, b shows AFM topography image of the same

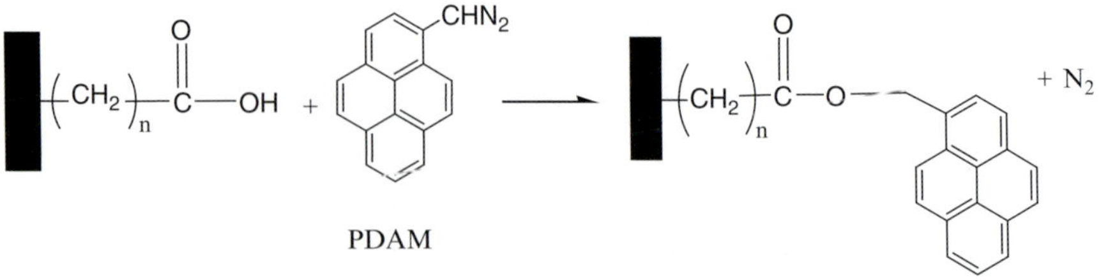

Scheme 13.2 The reaction of PDAM and carboxylic acid-terminated surface

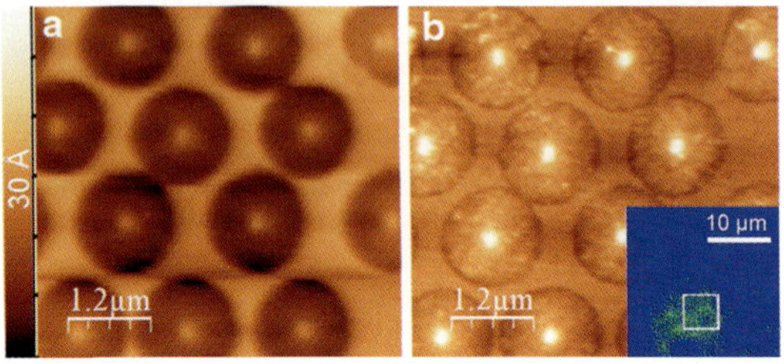

Fig. 13.7 The PDAM reaction on the OTSpd pattern. (**a**) Part of an OTSpd disk array. (Topography image, tapping mode, $6 \times 6\,\mu m^2$). The *bright spot* in the disk center is the SiO_2. (**b**) The exactly same area after the sample reacted with PDAM and was wiped in chloroform. (Topography image acquired in contact mode, $6 \times 6\,\mu m^2$). Both (**a**) and (**b**) are rendered in the same 30 Å height scale. The PDAM molecules coupled with the carboxylic groups of the OTSpd surface, which led to the increase of the height of the pattern. *Inset* in (**b**). The fluorescent image of the pattern after the sample reacted with PDAM and was wiped in chloroform. The PDAM is a fluorescent molecule, which emits fluorescent photons upon excitation. The *white box* corresponds to the exact area shown in (**a**) and (**b**). The *fluorescent signal* (*green*) confirms that the species on the OTSpd pattern is PDAM

region on the sample before and after PDAM incubation, respectively. The OTSpd disks were lower than the OTS before the PDAM incubation. After the incubation, the height of the disk increased to the same as the OTS, indicating that the PDAM reacted with the carboxylic group on the OTSpd surface and was grafted on the OTSpd disks. In the corresponding control test, the same region was also imaged with a very stiff tip with ~40 nN force in contact mode. The disk patterns were still the same height as OTS. The 40 nN force is not strong enough to break a bond, yet the tip can shave off any physisorbed contaminants on the OTSpd surface. The additional materials on the OTSpd disks survived the shaving, indicating they are not physisorbed contaminants. This control test shows that the PDAM on the OTSpd surface is covalently bonded, rather than weakly physisorbed on the OTSpd surface.

The same sample was then imaged with optical fluorescent microscope. The inset of Fig. 13.7b is the fluorescent image of the PDAM-bound OTSpd array. The white box in the inset is the same area shown in Fig. 13.7a, b. The fluorescent microscopy shows that the additional material grafted on the OTSpd was indeed the PDAM, which emitted fluorescence at the expected spectral region. In summary, both the AFM and the fluorescent microscopy show that the PDAM selectively bonded to the OTSpd surface. Since PDAM specifically bonded to carboxylic group, we conclude that the OTSpd is terminated with the carboxylic group.

13.2.6 *The Two-Step Patterning Method for Liquid Spreading Studies*

The OTSpd pattern is a hydrophilic, carboxylic acid-terminated pattern, which is 10.2 ± 0.7Å lower than the OTS. The OTSpd is topographically and chemically homogenous. Due to its carboxylic acid terminus, the pattern has a high surface energy and is lyophilic (liquid attractive). When the sample with the OTSpd pattern is dip-coated with liquid alkane, the liquid alkane will selectively stay on the OTSpd surface. In contrast, no liquid alkane will stay on the methyl-terminated, lyophobic (liquid repelling) OTS surface. So, the OTS pattern can determine the shape and size of the liquid alkane remaining on the surface. This approach has been demonstrated by Sagiv group and known as the "Wetting-Driven Self-Assembly" [22]. On the other aspect, because the OTSpd is lyophilic, the liquid alkane can spread over the OTSpd surface. Furthermore, because the OTSpd is flat, topographically and chemical homogenous, it is the ideal surface for us to study the spreading. Based on these advantages of OTSpd surface, we developed a novel "two-step" patterning method to study the liquid alkane spreading [9]. In the first step, we used the OTSpd pattern to hold a small liquid alkane droplet on surface. In the second step, we fabricated a large OTSpd pattern that surrounds the liquid alkane droplet and then made the liquid alkane to spread within the large OTSpd pattern. Thus, for the first time, we can study how the liquid can spread from a nanometer-sized droplet to sub-micron scale surfaces. Scheme 13.3 illustrates the procedure of the two-step patterning method in detail.

13.2.7 *The Validity of the Two-Step Patterning Approach*

In order to let the liquid alkane spread, the OTSpd fabricated in the first step and the OTSpd pattern fabricated in the second step must be continuous. No OTS shall separate these two regions. We conducted a control test to confirm this. Figure 13.8

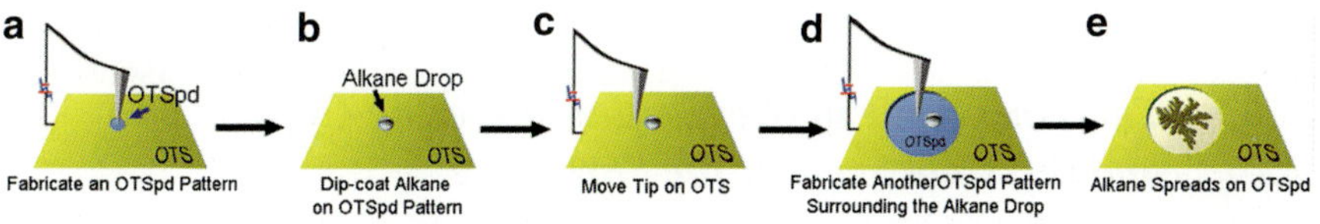

Scheme 13.3 The two-step patterning method developed by our group. (**a**) An OTSpd disk pattern fabricated by local probe deep oxidation. (**b**) Dip-coating the C36 on the OTSpd pattern. By briefly dipping the OTSpd pattern in the liquid alkane, the alkane drop adsorbed and solidified on the OTSpd pattern. (**c**) Fabrication of a new OTSpd area [*blue disk* in (**d**)] surrounding the coated alkane by local probe deep oxidation again. (**d**) At room temperature, the coated alkane is a solid. There is no spreading. (**e**) Upon heating, the alkane drop melts and spreads over the OTSpd surface. Followed by cooling, the alkane liquid freezes, at which point the alkane structure can be characterized by AFM

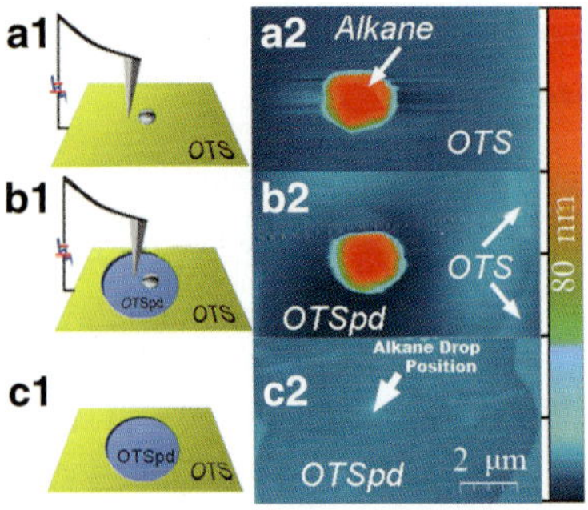

Fig. 13.8 The two-step patterning. (**a**) Alkane molecules (*red dot*) assembled on the OTSpd pattern. (**b**) Fabrication of a new OTSpd region surrounding the alkane drop. (**c**) The surface after the removal of the alkane drop. a1, b1 and c1 are the schematic views. a2, b2, and c2 are the actual AFM scans over the same region corresponding to the schematic views. The AFM images in a2 and b2 show that the two-step patterning does not affect the integrity of the alkane drop. Image c2 shows the OTSpd pattern fabricated surrounding the alkane drop and the OTSpd pattern fabricated in the first step are continuous. No residual OTS separates them

shows the results of our experiment. In the first step, we fabricated a small OTSpd disk and then dip-coated the small OTSpd disk using molten 1-hexatriacontane (C36). When the sample was cooled to room temperature, the C36 drop froze. Figure 13.8a2 shows the frozen alkane drop (red). Figure 13.8a1 is the corresponding schematic model. In the second step, we fabricated a big OTSpd disk that surrounded the frozen C36 drop. Figure 13.8b2 shows the frozen C36 drop and the OTSpd, which appeared as dark blue in the image. Then, we ultrasonicated the sample in chloroform and wiped the sample with chloroform subsequently. This operation completely removed C36 and any other potential organic contaminants from the surface. Therefore, OTSpd patterns fabricated at two different steps were exposed. Next, we imaged the same region, which is shown in Fig. 13.8c2. The OTSpd patterns fabricated in the first step and in the second step fused together, no unreacted OTS barrier was observed. Therefore, during spreading, the liquid alkane can freely spread within the fused OTSpd pattern.

13.2.8 The Time Scale of the Heating–Freezing Cycle and the Time Scale of the Spreading

13.2.8.1 Feasibility of the "Heat–Freeze" Approach to Capture Snapshots of Spreading

Results from the Riegler group showed that it took ~500 s for C30 to spread over 70 μm [23]. In the "heat–freeze" approach, the sample temperature can be reduced from 90°C to 70°C within 10 s through rapid cooling using a peltier heater–cooler.

Because the time scale of freezing is substantially shorter than the time scale of spreading, it is possible to capture snapshots of spreading through rapid cooling.

13.2.8.2 Interference by Surface Freezing

Surface freezing occurs to C14–C50 n-alkane drops. When the drop temperature is decreased to 0–3°C above the alkane bulk melting temperature, the surface layer of the alkane drop forms a crystalline monolayer in which the alkane chain is normal to the surface [24,25]. In our experimental approach, however, the alkane drop spreads at temperatures higher than the surface freezing temperature. After the spreading is complete, the drop is frozen through rapid cooling. Surface freezing requires a slow crystallization process. Rapid cooling, as employed in these studies, suppresses surface freezing. We have conducted spreading tests of C24 (subject to surface freezing when cooled slowly) and naphthalene (not subject to surface freezing). Both of these species spread without surface freezing under rapid cooling conditions. This comparison demonstrated that surface freezing would not occur and would not influence our measurements on long-chain n-alkanes during the proposed studies.

13.3 Part 2. Structures of Long-Chain Alkanes on Surface

The spreading, wetting, capillary action, and adsorption of liquid alkanes are relevant to many important applications like petroleum tertiary recovery and lubrications [26–29]. These liquid behaviors have been well characterized in macroscopic realm. Nevertheless, understanding these behaviors at molecular scale is far from complete. An intrinsic difficulty is that the typical liquid behaviors such as spreading are dynamic. The classic means to characterize a dynamic behavior is to obtain a spatial–temporal record using a movie or a sequence of snapshots. To characterize a motion of nanometer-sized droplet, the spatial–temporal record needs to be fast enough and retains the molecular resolution, which is difficult to meet simultaneously. One of the pioneering solutions is to stop the motion by freezing and characterize the liquid at molecular scale. For instance, Salmeron group had obtained the molecular level STM image of water adsorbed on metal surface at 4 K [30–32], which revealed how wetting occurred at the earliest stage. Our study on liquid alkane follows the same rationale. We aim to understand the behaviors of liquid alkane at nanoscale by studying the structures of solidified alkanes that was frozen during the spreading. From the position and the structures of solidified alkanes adsorbed on the surface, we can infer how the liquid alkane spread.

Structures of solid long-chain alkanes on surface have been studied using a variety of methods such as X-ray diffraction [33–35], neutron diffraction [36–38], ellipsometry [39], optical microscopy [23,40], and AFM [41–43]. Taub group has established the phase diagram of dotriacontane (C32) adsorbed on SiO_2 surface [41]. The thermodynamically stable phase is the parallel phase, in which the alkane molecules lay flat on the surface. Other phases such as the surface bulk phase, surface rotator phase, and the monolayer phase also existed, which adsorbed on top of the parallel layer. The monolayer phase is the metastable phase at the room temperature, which spontaneously forms only in a narrow temperature range (3–5°C) below the alkane's melting point. The monolayer phase layer forms through the cooling of liquid alkane on the surface or through drop-casting solutions of alkane. Basu and Satija have demonstrated that the monolayers formed from these two methods had the same structure [35]. Typically, the layer has the shape of seaweed, which suggests that the formation of the monolayer phase is a diffusion-limited aggregation process. When the liquid is cooled down rapidly, the super-cooled droplet of liquid may also change into the glassy state, which does not have crystalline order but can be observed by AFM. These alkane structural data provide us a solid foundation for studying the liquid's behaviors at nanometer scale. However, these previous results are mainly obtained on the SiO_2 surface, on which alkanes had been heated and cooled for several cycles to generate the thermodynamically stable surface [40, 44]. Nevertheless, the spreading of liquid usually occurs when liquid first contacts the surface. Therefore, to study how liquid alkane spread, we need to focus on how liquid alkane interacts with the substrate when they first contact.

The spreading at nanometer scale is a dynamic and local event. Our two-step patterning method is suitable to characterize such event. We can fabricate a spreadable surface around the solidified alkane drop, and subsequently initiate the spreading by flash-heating the sample, which converts the solidified alkane droplet to liquid. Therefore, when liquid droplet spreads, it is the first time that the liquid contacts with the surface. The two-step pattern method enables us to study the liquid spreading under the first-time contact condition, which is never studied before. In our method, the alkane droplet spreads over the OTSpd pattern. Different stages of spreading can be characterized through sudden freezing at different times, which provides a series of snapshots of the spreading. The chemical patterns, with well-defined depth and chemistry, serve as references to compare the changes during the spreading.

In the following section, we demonstrate a few of new findings we discovered using the OTSpd pattern to study the adsorption structures of the alkanes on the surface.

13.3.1 Alkane Structures on Hydrophilic Surfaces and on Hydrophobic Surfaces

13.3.1.1 The Alkane Tilting in the Seaweed-Shaped Alkane Layers

The monolayer phase alkane usually has the shape of a seaweed, which is frequently observed when the liquid alkane droplet is frozen. Previously, people thought that the seaweed-shaped monolayer consisted of alkane molecules perpendicularly standing on the surface [39, 41]. The thickness of the seaweed monolayer was the

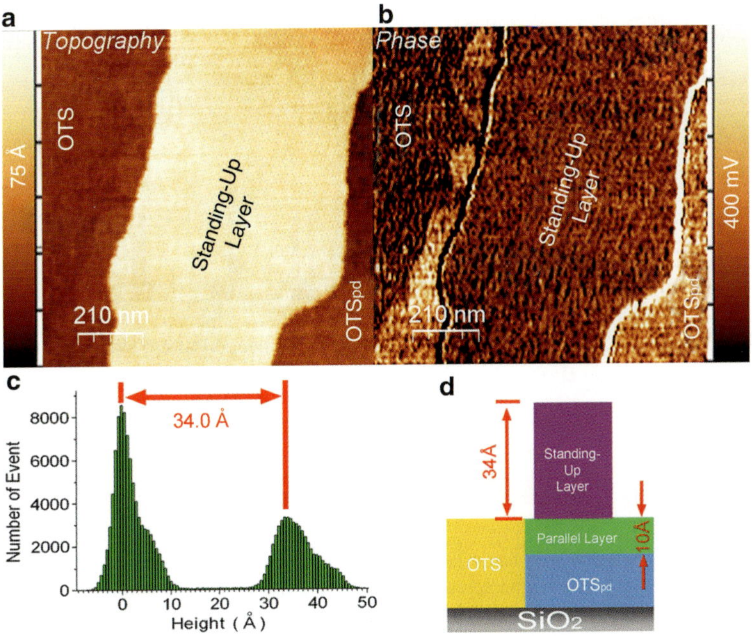

Fig. 13.9 A representative standing-up layer adsorbed near the OTSpd/OTS pattern edge. (**a**) Topography image. (**b**) The corresponding phase image. (**c**) The height histogram of the topography image (**a**). From the *histogram*, the height of the standing-up layer is calculated to be 34 Å. (**d**) The structural model corresponding to (**a**)

same as the length of the alkane molecule. We found this was not the case – the alkane molecules inside the seaweed-shaped monolayer might tilt as well.

Figure 13.9a, b shows a seaweed-shaped C36 monolayer on an OTSpd pattern in the topography and phase channels, respectively. In the topography image (Fig. 13.9a), the brightest part is monolayer phase C36 sitting on the parallel layer. The same region in phase image has the same phase contrast as that of OTS surface. Because the phase signal represents the surface properties and the OTS surface is methyl-terminated, the same phase values over the seaweed layer and the OTS surface indicate that the seaweed layer is also methyl-terminated. The brighter region in Fig. 13.9b is the parallel layer over the OTSpd pattern, which has the same height as the OTS in the topography image. Since the OTSpd pattern is 10.2 Å lower than the OTS surface, the thickness of the parallel layer in the OTSpd pattern should be 10.2 Å. Figure 13.9c is the histogram of the topography image, indicating that the thickness of the standing-up alkane monolayer is 34 Å. Because the full length of C36 molecule is 50.6 Å, and the C36 standing-up monolayer is methyl-terminated, the measured 34 Å thickness indicates that the C36 molecules in the monolayer phase in Fig. 13.9 are tilted [45].

We characterized hundreds of seaweed-shaped monolayer phase alkane on OTSpd surface. We found that the thickness of the standing-up alkane layer was

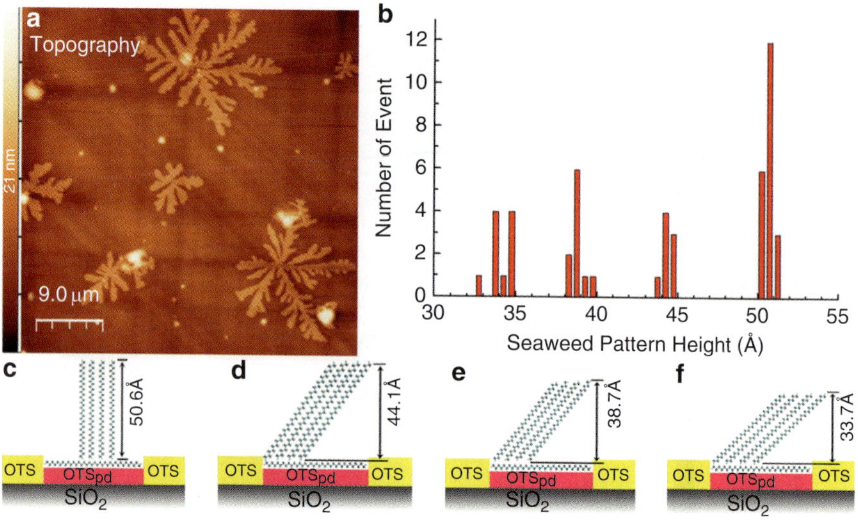

Fig. 13.10 The seaweed-shaped hexatriacontane pattern on the parallel alkane layer. (**a**) Representative hexatriacontane patterns after spreading over the OTSpd surface (Tapping mode, topography image, 45 × 45 μm^2). The OTSpd disk is bigger than the scanned area. Therefore, the OTSpd/OTS border is out of field of view in this figure. (**b**) The height distribution of seaweed patterns. (**c**–**f**) The proposed alkane structural models correspond to different alkane layer thickness. The thickness shown in the figure is the calculated value assuming the alkane full length is 50.6 Å

not always the value corresponding to the perpendicular orientation of the alkane. Instead, the seaweed-shaped layers have a series of quantized thickness values [9]. Figure 13.10a shows a representative topography image of multiple seaweed-shaped standing-up alkane layers. The background is the parallel alkane layer adsorbed on the OTSpd surface. We plotted the thickness distribution of the seaweed-shaped standing-up layers in Fig. 13.10b. The histogram shows that the observed thickness values distribute around four discrete positions: 50.6 Å, 44.3 Å, 38.8 Å, and 34.1 Å. Based on knowledge that alkane molecules may tilt in the standing-up layer (as discussed in the previous paragraph), and considering these typical thickness values, we proposed that the alkane molecules packed according to the models in Fig. 13.10c–f, in which the alkane molecules packed side by side and shifted 0, 2, 3, and 4 methylene unit (1.26 Å) along the chain direction, respectively. Therefore, if the full length of C36 is 50.6 Å, then according to the models in Fig. 13.10c–f, the calculated thickness would be 50.6 Å, 44.1 Å, 38.7 Å, and 33.7 Å, which match the observed data very well. From the statistics of the measured thickness of the standing-up layer, we showed that the monolayer phase alkane had different structures. The alkane molecules may tilt at specific angles.

13.3.2 The Multiple Domains Within a Seaweed-Shaped Layer

The monolayer phase alkane usually has the shape like a seaweed, which is a characteristic fractal structure. The diffusion-limited aggregation (DLA) [46–49] during the monolayer formation was speculated as the mechanism responsible for the seaweed shape. According to the DLA mechanism, mobile alkane molecules randomly walk on the surface when the surface temperature is 1–3°C below the alkane's bulk-phase melting point. When the temperature drops further, an immobile seed molecule forms, which has the standing-up orientation. One mobile alkane molecule will be captured by the seed alkane molecule (the nucleation center) when they encounter. These two molecules, therefore, are packed together, forming a two-molecule assembly with the same orientation. Subsequently, the third mobile alkane molecule will be captured by the two-molecule assembly and is aligned with the seed molecules, thus the nucleation center grows in size. Such a cycle will repeat until all mobile alkane molecules on the surface are captured, forming a seaweed-shaped layer. Under the DLA theory, the mobile molecule is captured at the moment it encounters the seed. It is almost impossible for the alkane molecule to be captured by two different nuclei at the same time. As a direct consequence, two branches of the seaweed will never meet and no holes exist within a fractal. Furthermore, if a seaweed-shaped layer is grown from a single seed, then all molecules within the seaweed layer shall be aligned the same as the seed molecule. Thus, all alkane molecules within the seaweed-shaped layer shall have the same height. However, our data do not agree with these conclusions, indicating the single-seed DLA mechanism does not fully describe the formation process of the seaweed-shaped alkane layer.

We characterized >150 seaweed-shaped alkane layers on the OTSpd surface. A majority of these seaweed-shaped layers have the uniform height (typical values are: 50.6 Å, 44.3 Å, 38.8 Å, and 34.1 Å), which has been discussed in the previous section. However, we found some seaweed-shaped layers are composed of different domains, which had different heights [45]. Holes were also observed in these layers, which cannot be explained by the single-seed DLA mechanism. Figure 13.11 shows an example. The topography image in Fig. 13.11a shows that a branch of the seaweed-shaped layer is composed of two domains – domain A and domain B. The height cross-section across the branch is plotted in Fig. 13.11c, showing that the height of domain A is 43.9 ± 0.9 Å, while the height of domain B is 4.2 ± 1.5 Å lower than domain A. In the corresponding phase image (Fig. 13.11b), a black line separating these two domains is observed, which is the domain boundary. Similar domain boundaries can be identified in the phase image as well, which have been highlighted by the white circles. For example, in the upper right part of the image, a horizontal black line is visible, separating domain C and domain D. However, domain C and D have the same height. It is difficult to distinguish them in the topography image. In addition, holes are observed within the seaweed-shaped layer right on the domain boundary line. These holes are highlighted by green arrows in Fig. 13.11.

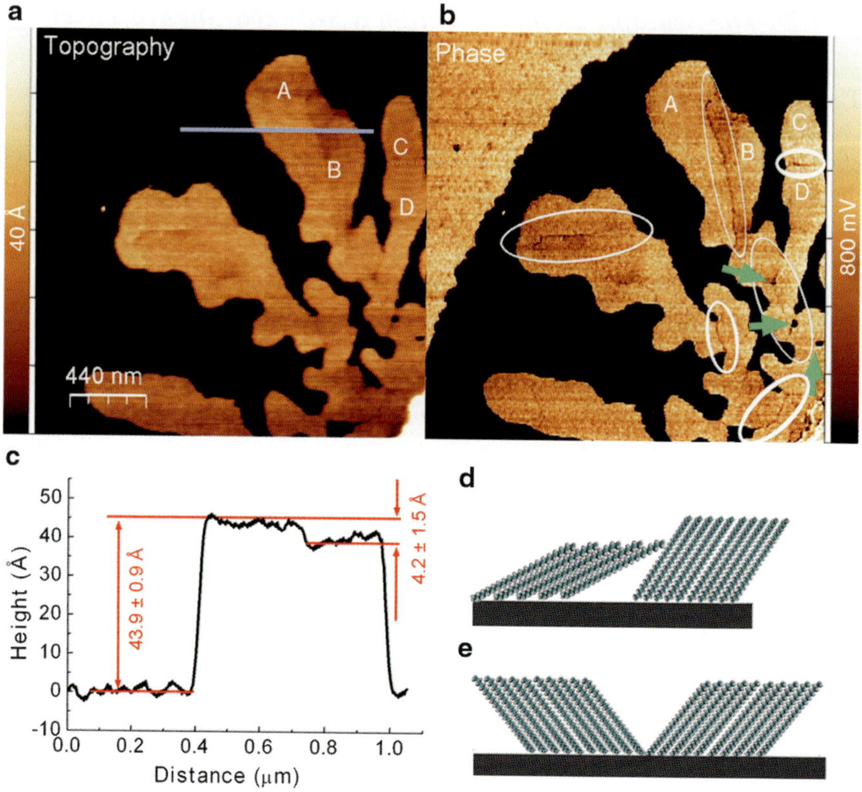

Fig. 13.11 The multiple domain structure within the standing-up layer. (**a**) Topography image of a representative standing-up layer. AC mode, $2.2 \times 2.2\,\mu m^2$. (**b**) The corresponding phase image. The *white circles* highlight the domain boundaries. The *green arrows* point out the holes inside the standing-up layer. The holes are always on the domain boundary line, indicating the coalescence process of two neighboring domains. (**c**) The height cross-sectional plot corresponding to the *blue line* in (**a**). Domain A and B are a pair of domains fused together. (**d**) A structural model shows that different alkane domains may have different heights [such as domains A and B in (**a**)] because the alkane molecules in these two domains tilt in different angles. (**e**) A structural model shows that alkane domains may have the same height [such as domains C and D in (**a**)] because the alkane molecules in these two domains tilt in same angle but in different directions

As discussed in the previous section, the measured heights of domain A and B indicate that the C36 molecules in these domains are tilted. Furthermore, molecules in domain A and in domain B tilt at different angles; molecules in domain C and D have the same tilting angle, but they might tilt at different directions. The tiling structures of domain A, B, C, and D are illustrated in the models in Fig. 13.11d. The multiple domain structure within the seaweed-shaped layer shows that the layer is formed by the fusion of different C36 domains, which have different tilting angles and tilting directions. This conclusion can be further confirmed by the fact that all holes are right on the domain boundaries, because the holes originate from the imperfect matching of the domain boundary during the coalescence of domains.

In summary, we found that the DLA mechanism alone cannot explain the formation process of the seaweed-shaped alkane layer. Coalescence also plays a role in the formation of the seaweed-shaped layer. The seaweed-shaped alkane layer does not always have a uniform height [45].

13.4 Part 3. The Role of Vapor During the Spreading of Liquid Alkane

The science of spreading is established on observations of how a droplet spreads at macroscopic scale. Since Young and Laplace, the approach is to obtain the X, Y, Z dimensions of the droplet as a function of time, and measure the energy dissipations during the spreading in the first step. Then the total energy (the energy dissipation during spreading and the kinetic energy of the liquid, which are functions of surface tension and viscosity) and the droplet's dimension are correlated as a function of time following the classical Newtonian laws. By solving the fluid mechanical equations, the spreading properties of a liquid can be predicted based on its physical properties such a surface tension and viscosity [2,27,50–53]. A major breakthrough in the science of spreading is the discovery of the precursor film surrounding a droplet by Hardy in 1919 [54], which revealed that the contribution of the precursor film in energy dissipation should be considered as well. In his seminal article, de Gennes pointed out that the total energy dissipation during the spreading can be expressed as the form of $T\sum_{\mathrm{w}} + T\sum_{\mathrm{f}} + T\sum_{\mathrm{l}}$, where $T\sum_{\mathrm{w}}$, $T\sum_{\mathrm{f}}$, $T\sum_{\mathrm{l}}$ are the energy dissipations from the spreading of the bulk drop, the spreading of the precursor film, and the adsorption of the liquid molecule on the substrate, respectively [1]. de Gennes also pointed out that the remaining challenging issue in the field is that the role of vapor during spreading had not been considered. This is because the contributions from the *via-liquid* transport and the *via-vapor* transport are difficult to be studied separately, which leaves this challenging issue a mystery until now. For example, as shown in Fig. 13.12, for a specific molecule within the precursor film (the green bar in the figure), it is difficult to tell whether the molecule is transported within the liquid or from the vapor adsorption. The properties and structures of the liquid–solid interface determine the spreading dynamics. Nevertheless, adsorbate molecules will change the structure and properties of the interface. How the molecules in the vapor adsorb on surfaces is unknown at the molecular level. As a consequence, most of the current studies on spreading of the liquids do not consider the effect of the adsorption on the interface (the $T\sum_{\mathrm{l}}$ term) when solving the fluidic mechanics equation. In the case of the spreading of liquid n-alkane, such simplification might not correctly depict what happens on the surface. Liquid n-alkanes have a non-trivial vapor pressure, as a result the concurrent vapor adsorption accompanying the spreading could be significant. Furthermore, the n-alkane molecules usually have large adsorption energy and an ordered adsorption structures on the surface. When considering these properties of the n-alkane, we can speculate that the vapor adsorption during the spreading of n-alkane should not be ignored.

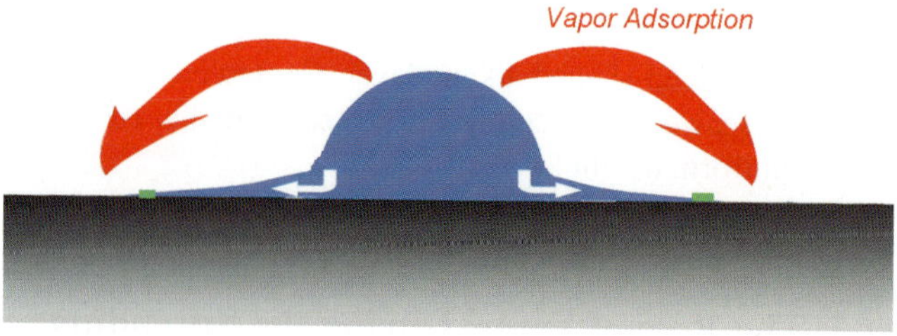

Fig. 13.12 The via-liquid transport and via-vapor transport occur at the same time

In the practical side, understanding the role of vapor during the spreading of liquid alkane is helpful for the development of novel lubricants. Long-chain *n*-alkanes are extensively used in lubricants as the base oil. The functional components in the lubricants are the polar additives, which are called "friction modifiers" [55]. They are supposed to adsorb on the solid surface and reduce the friction of the surface. When two solid surfaces slide, due to the surface roughness, in places that two surfaces have intermittent contact, the adsorbed friction modifiers will be removed. Subsequently, the lubricant elsewhere will slowly replenish the scratched zone and the friction modifier molecules will re-adsorb on the scratched surface. However, if the role of vapor adsorption is considered, the alkane molecules from the vapor of the base oil will adsorb on the scratched surface first, while the friction modifier might not be on the surface at all. To avoid this kind of problems in lubrication, we need the knowledge of the vapor adsorption during the spreading of liquid alkane. This demand motivates us to investigate the role of vapor during the spreading of liquid alkane, using our newly developed two-step patterning method, which enables us to study the vapor adsorption and the spreading separately.

Our scheme for studying the concurrent vapor adsorption is illustrated in Scheme 13.4 [56]. We first coated a C36 drop on a small OTSpd pattern. At room temperature, the C36 drop is frozen as a solid. Next, we used the scanning probe deep oxidation lithography to fabricate a series of OTSpd disks that were 10–50 μm away from the C36 drop. These disks are called "satellite disks." Then we moved the tip close to the frozen C36 drop and applied a pulse, which generated a large OTSpd disk (blue disk in the figure). This OTSpd disk is called the main disk. After the pattern fabrication, we raised the sample temperature to melt the C36 drop and made it to spread over the main disk. Since the main disk and the satellite disks are separated by OTS, the liquid alkane would not spread from the main disk to the satellite disk. During the spreading, since the vapor of C36 was also present, C36 in the vapor could adsorb on clean OTSpd surface of both main disks and the satellite disks. Therefore, studying the C36 molecules on the satellite disks would reveal the role of vapor phase adsorption during the spreading, which avoids the interference from the liquid transport.

Figure 13.13 shows the patterns after the spreading. In the center is the alkane layers spread on the OTSpd main disk, where the standing-up layer is on top of the parallel layer. In the upper left of the image, a small satellite disk can be seen. The

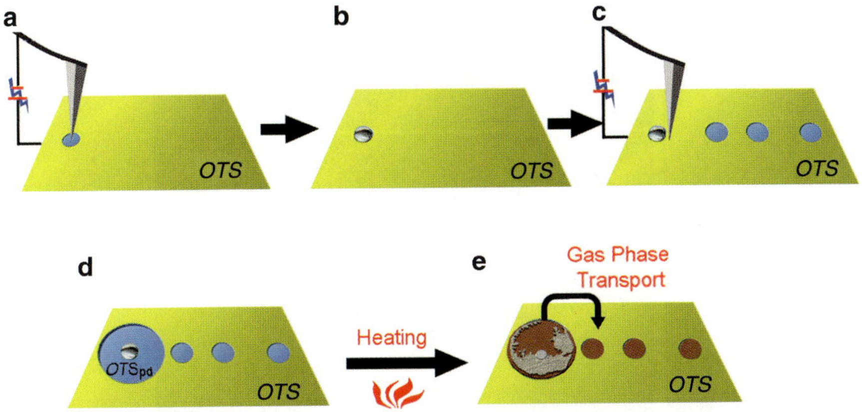

Scheme 13.4 (**a**) An OTSpd disk pattern fabricated by the local probe deep oxidation. (**b**) Dip-coating the n-hexatriacontane on the OTSpd pattern. By briefly dipping the OTSpd pattern in the liquid n-hexatriacontane, the alkane drop adsorbed and solidified on the OTSpd pattern. (**c**) A few satellite OTSpd disks were fabricated on the OTS surface, which were 2–20 μm apart from the alkane drop. Then we moved the conducting AFM tip to the proximity of the alkane drop and gave a voltage pulse to the tip. Thus, a new OTSpd disk (the main disk, *blue disk in the figure*) was fabricated surrounding the coated n-hexatriacontane by local probe deep oxidation. The main disk did not connect to those satellite disks. (**d**) At room temperature, the coated n-hexatriacontane was a solid. There was no spreading. (**e**) When heated, the alkane drop melted. The liquid alkane spread over the OTSpd main disk, whereas the alkane molecules were deposited on the satellite OTSpd disks through the vapor phase transport. Upon cooling, the alkane liquid froze. Then the alkane structures can be characterized by AFM

satellite disk in the green box is amplified and shown in Fig. 13.13c. Similarly, other satellite disks (which are further away from the main disk and out of the field of view in the AFM image) are shown in the panels in Fig. 13.13b, d, e. The AFM image shows that after the spreading, the apparent height of the satellite disk changed from 10 Å below the OTS surface to 10–15 Å higher than the surface. Obviously, there were new materials adsorbed on the OTSpd surface of the satellite disk during spreading. Our control test has shown that a clean OTSpd patterns stay unchanged when the sample is placed in the ultra-pure nitrogen-purged AFM environmental chamber for at least 10 min. On the other hand, the spreading experiment illustrated in Scheme 13.4 (step d to e) was completed within 10 min after the clean OTSpd patterns were fabricated (step c) and the experiment was conducted in the ultra-pure nitrogen-purged AFM environmental chamber as well. Thus, the potential contaminants from the environment (under ultra-pure nitrogen) are not responsible for the additional materials on the OTSpd. Since the vapor was present during the spreading of the alkane, we conclude that the new materials on the satellite OTSpd disks are from the adsorption of the liquid alkane's vapor. Next, we compare the phase signal over the satellite disks and the phase signal of the parallel layer of the main disk. The phase signal represents the AFM tip's energy dissipation during the scanning, which reflects the surface properties. If two patterns have the same phase signal, they have the same surface. Usually the phase signal over samples cannot be compared

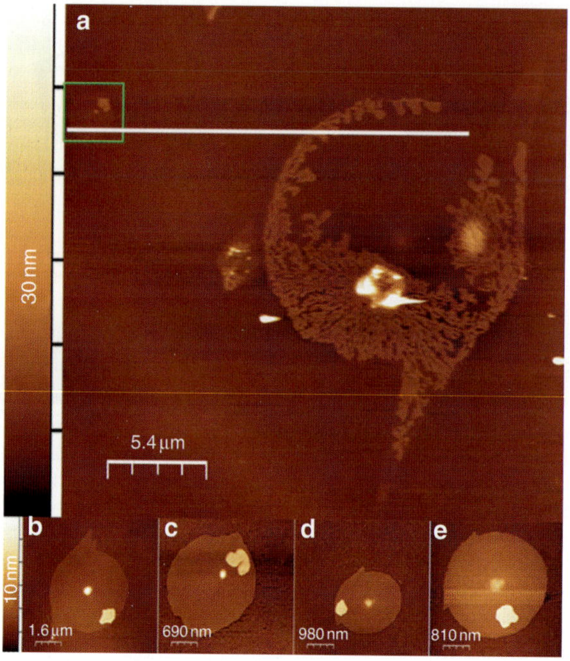

Fig. 13.13 The main OTSpd disk and the satellite OTSpd disks after the spreading. AC mode AFM topography images. (**a**) The main OTSpd disk and a nearby satellite OTSpd disk (in the *green box*). (**b–e**) Satellite OTSpd disks. (**c**) The same OTSpd disk as in the *green box* of (**a**). (**b**, **d**, **e**) Satellite OTSpd disks outside the view of (**a**)

directly, because the value of the phase signal is a function of the tip state and scanning parameters. However, in our case we were able to compare the two disks as they were imaged during the same scan. In Fig. 13.13, we profiled the phase signal along the horizontal white line, which crosses both the satellite disk and the parallel layer on the main disk. Since the AFM tip scanning direction is also horizontal, the phase signals over the satellite disk and over the parallel layer of the main disk were acquired at the same time using the same tip and under the same set of scanning parameters, the phase signals over the satellite disk and over the parallel layer on the main disk can be compared. We find that they are the same. Therefore, we conclude that the new materials on the satellite disk is the alkane parallel layer as well, which is from the vapor adsorption.

The data in Fig. 13.13 only show that the parallel layer on the satellite OTSpd disk originates from the vapor adsorption. Is the parallel layer on any OTSpd surface (include the main OTSpd disk and the satellite OTSpd disks) also from the vapor adsorption? To address this question, we conducted another spreading experiment (from the state in Scheme 13.4d, e) at 70°C, which is 5°C below the melting temperature of C36. After the sample was cooled down to room temperature, we imaged the pattern and show it in Fig. 13.14. The solidified C36 drop in the center is

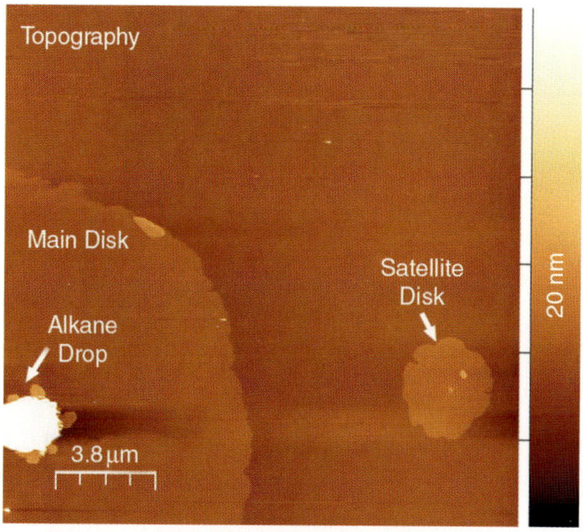

Fig. 13.14 Hexatriacontane covered both the OTSpd main disk and the satellite disks after the sample was heat at 5°C below the melting point of hexatriacontane. (AC mode topography image, $19 \times 19\,\mu m^2$)

still intact, no spreading occurred because the drop never melted. However, the clean OTSpd surface changed to higher than the OTS surface, indicating that new parallel alkane layers adsorbed on the clean OTSpd surface in the main disk at 5°C below the melting point of C36. Since at this temperature, the C36 still has a significant vapor pressure, we conclude from Fig. 13.14 that all the parallel layers on all OTSpd surfaces originate from the vapor adsorption. Furthermore, since the standing-up layer is always on top of the parallel layer, and the standing-up layer forms from the cooling of liquid alkane, we conclude that during the spreading, alkane molecules in vapor adsorbed on the OTSpd surface, forming the parallel layer first, then the liquid arrived on the parallel layer and gradually formed standing-up layer upon cooling [56].

13.4.1 The Stability of the Parallel Layer During the Spreading

The stability of the parallel layer during the spreading is an important question because it determines on what surface the droplet actually spreads. If the parallel layer is stable (i.e., stationary) during the spreading, the droplet spreads on the parallel layer. In contrast, if the interaction between the parallel layer and the substrate is weak, the parallel layer also diffuses outward together with the droplet, the liquid spreads on the substrate. However, because the parallel layer usually is a featureless and complete layer, it is difficult to track the motion of the parallel layer during the spreading.

We found the standing-up layer left a residue on the parallel layer after it was annealed at the temperature above the melting point [45]. The residue can be used as a landmark to track the motion of the parallel layer during the spreading. Figure 13.15a shows a seaweed-shaped standing-up layer of C36 on the parallel

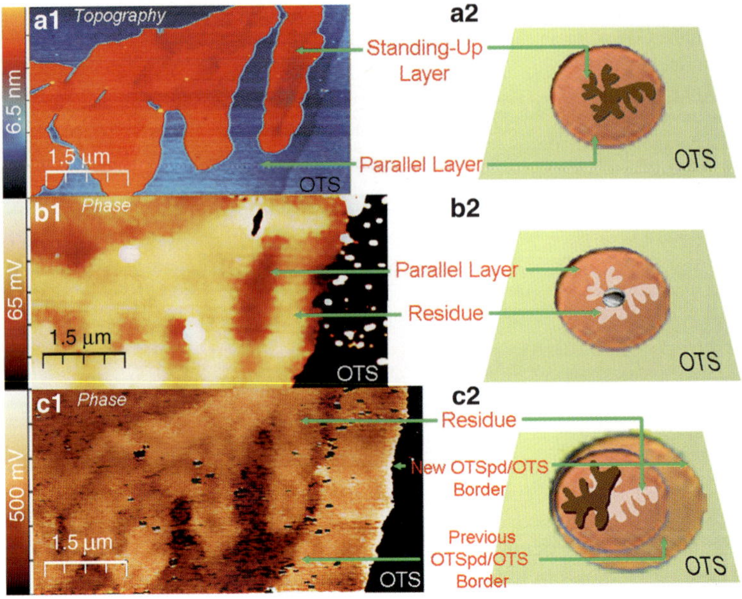

Fig. 13.15 Comparison of the shape of the residue layer after annealing with the shape of the standing-up layer before annealing. (a1) The seaweed-shaped standing-up layer. Topography. AC mode. (a2) The structural model of (a1). (b1) The phase image of the residue after the sample was annealed. (b2) The structural model of (b1). (c1) The phase image of the same region after the second spreading. Because of the third oxidation, the new OTSpd/OTS border moved to the right. After the second spreading, a new parallel layer formed in the region between the old OTSpd/OTS border and the new OTSpd/OTS border (the *blue crescent-shaped* area in Scheme 13.1h). The old OTSpd/OTS border is still recognizable in the phase image. (c2) The structural model of (c1). (a1), (b1), and (c1) are rendered in the same scale

layer. Both the standing-up layer and the parallel layer are inside an OTSpd disk. Figure 13.15b shows the same region on the surface after the sample was annealed. The majority of C36 molecules turned into liquid and flowed to elsewhere. A small fraction of the C36 molecules in the liquids was pinned to the same place of the standing-up layer. Upon cooling, these remaining C36 molecules formed a residue, which can be identified through AFM phase image in Fig. 13.15b. The residue is at the same location and has the same shape as the standing-up layer. The residue has a slightly higher phase contrast than the parallel layer and has a uneven height of 1–4 Å above the parallel layer. We speculate that the residue is composed of loosely packed C36 molecules laying flat on the surface – the same as the parallel layer but with a lower packing density.

To use the residue as a landmark for tracking the motion of parallel layer, we design an experimental scheme, which is illustrated in Scheme 13.5. In the scheme, steps from (a) to (e) are the same procedures as we used to conduct the alkane spreading over the OTSpd surface, which has been illustrated in Scheme 13.3. Then in step (f), we anneal the sample, thus a residue identical to the shape of standing-up

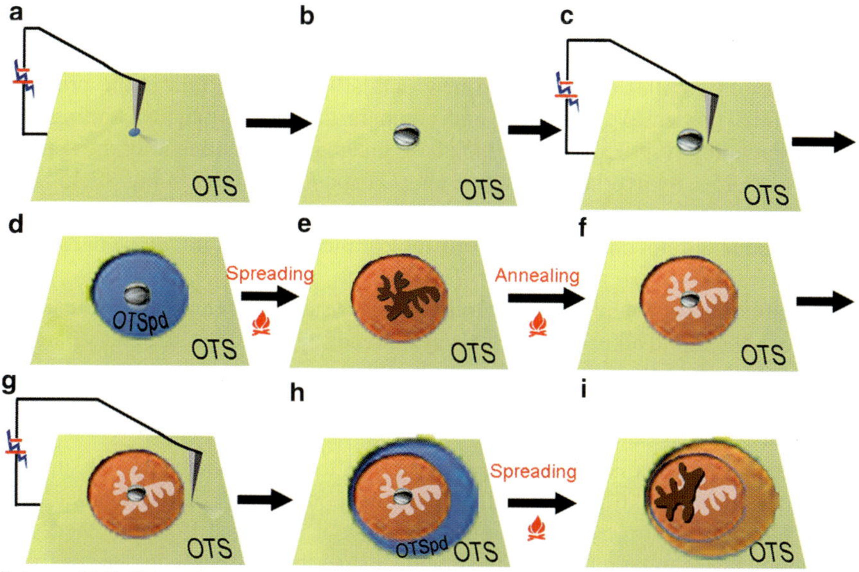

Scheme 13.5 Using the residue as a landmark to track the movement of the parallel layer during the spreading of liquid alkane. (**a**) The first scanning probe oxidation. We fabricated an OTSpd disk (*blue*) on the OTS film (*yellow*). (**b**) Molten C36 was dip-coated on the OTSpd disk at 90°. When the sample was cooled, C36 solidified. (**c**) The second scanning probe oxidation. (**d**) A large OTSpd disk was fabricated, which enclosed the solidified C36 drop. (**e**) The first spreading. The sample was heated to 90°C. The C36 melted and spread on the OTSpd disk. The C36 molecules in the vapor phase adsorbed on OTSpd surface and formed the parallel layer (*tangerine* disk). Additional C36 molecules formed the seaweed-shaped standing-up layer (*dark brown*) on the parallel layer. The sample was then cooled to room temperature and was characterized by AFM. (**f**) Annealing. The sample was heated again and the sample temperature was maintained at 90°C for 2 min. And then the sample was rapidly cooled (<20 s) to room temperature. C36 molecules in the standing-up layer (*dark brown*) receded and left a residue (pale white mark on the tangerine background). After cooling, the liquid alkane solidified again. (**g**) The third scanning probe oxidation. (**h**) A crescent-shaped OTSpd (*blue*) was fabricated after the oxidation. (**i**) The second spreading. The sample was heated to 90°C for 25 s and then rapidly cooled to room temperature. The liquid alkane spread out again. The crescent-shaped OTSpd was also covered with parallel layer (*light tangerine* color). Using the residue as a landmark, we can compare whether the parallel layer in (**h**) moved during the second spreading

layer is fabricated. Then, we move the conducting AFM tip to the OTS surface near the existing pattern and apply another pulse. Thus, in step (g), we use the scanning probe deep oxidation lithography for the third time to create a new OTSpd region, which is illustrated as the blue crescent-shaped region in the scheme. Finally, when we raise the sample temperature above the alkane's melting point, the liquid alkane will spread again over the newly fabricated crescent-shaped OTSpd region. After the sample is cooled to room temperature, we image the sample again and compare the position of the residue before and after the spreading. Therefore, we can tell whether the parallel layer moved during spreading.

Figure 13.15c shows the same residue after the spreading, which corresponds to the step (h) in Scheme 13.5. After the second spreading, molecules in alkane's vapor also adsorbed on the newly fabricated crescent-shaped OTSpd region, forming a new parallel layer. In Fig. 13.15c, we can clearly see that a line in the phase image separates the old parallel layer and the new parallel layer. The line is the previous OTSpd/OTS border. We can align and compare the phase image of the residue before spreading (Fig. 13.15b) and the phase image of the residue after the spreading (Fig. 13.15c) by overlaying the previous OTSpd/OTS border in Fig. 13.15c and the OTSpd/OTS border line in Fig. 13.15b. We found that the shape and position of the residue did not change. Therefore, the parallel layer did not move during the spreading.

In summary, we found that the parallel layer is stationary during the spreading. Therefore, the liquid alkane actually always spreads on top of the stationary parallel layer instead of on the substrate surface. It is the surface properties of the parallel layer that influence how liquid alkane wets and spreads.

The energy dissipation during the spreading has the form of $T\sum_{\mathrm{w}} + T\sum_{\mathrm{f}} + T\sum_{\mathrm{l}}$, the $T\sum_{\mathrm{l}}$ term is the adsorption energy of the liquid molecules on surface. Based on our finding, this term shall be the adsorption energy on the parallel layer rather than on the substrate. The spreading dynamics derived from the energy of dissipation needs to be refined. In the practical side, this finding motivates us to rethink the design of additives in the lubricant. The alkane molecules in the base oil can adsorb on the surface and form stable adsorbate film on the surface. The friction modifier molecules need to have adsorption strong enough to replace the alkane film on the surface in order to reduce the friction.

13.5 Conclusion

The scanning deep oxidation lithography can fabricate surface patterns with defined chemistry and topography, which can be used as references to compare and measure dynamical processes on the surface. The patterning method offers a convenient approach to create a surface pattern ranging from 25 nm to sub-millimeter in size using the same AFM tip. It is an ideal tool to study the nanometer scale liquid behavior and observe how the liquid spreads over a large range. We have used this patterning method to study the long-chain *n*-alkane structures and spreading over surfaces. We found the tilting structure and the multi-center nucleation growth mechanism for the standing-up layer of alkane adsorbed on the surface. We also investigated the role of vapor during the spreading of liquid alkane droplet, which has been an unsolved problem since the time of Young and Laplace. We found that the vapor of the alkane droplet adsorbed on the surrounding surface before the liquid spread. Thus, the liquid actually always spreads on a layer of the adsorbate. It is the surface properties of this adsorbate determine the spreading and wetting of the liquid alkane.

References

1. P.G. de Gennes, Wetting: statics and dynamics. Rev. Mod. Phys. **57**, 827 (1985)
2. L. Leger, J.F. Joanny, Liquid spreading. Rep. Prog. Phys. **55**, 431 (1992)
3. B. He, N.A. Patankar, J. Lee, Multiple equilibrium droplet shapes and design criterion for rough hydrophobic surfaces. Langmuir **19**, 4999 (2003)
4. J. Drelich, J.L. Wilbur, J.D. Miller, G.M. Whitesides, Contact angles for liquid drops at a model heterogeneous surface consisting of alternating and parallel hydrophobic hydrophilic strips. Langmuir **12**, 1913 (1996)
5. J. de Coninck, M.J. de Ruijter, M. Voue, Dynamics of wetting. Curr. Opin. Colloid Interf. Sci. **6**, 49 (2001)
6. A. Checco, Y. Cai, O. Gang, B.M. Ocko, High resolution non-contact AFM imaging of liquids condensed onto chemically nanopatterned surfaces. Ultramicroscopy **106**, 703 (2006)
7. A. Checco, O. Gang, B.M. Ocko, Liquid nanostripes. Phys. Rev. Lett. **96**, 056104 (2006)
8. A. Checco, Liquid spreading under nanoscale confinement. Phys. Rev. Lett. **102**, 106103 (2009)
9. Y.G. Cai, The partially degraded hydrophilic silane pattern and its application in studying the structures of long chain alkane films. Langmuir **25**, 5594 (2009)
10. S.R. Wasserman, G.M. Whitesides, I.M. Tidswell, B.M. Ocko, P.S. Pershan, J.D. Axe, The structure of self-assembled monolayers of alkylsiloxanes on silicon – a comparison of results from ellipsometry and low-angle x-ray reflectivity. J. Am. Chem. Soc. **111**, 5852 (1989)
11. R. Maoz, S.R. Cohen, J. Sagiv, Nanoelectrochemical patterning of monolayer surfaces: toward spatially defined self-assembly of nanostructures. Adv. Mater. **11**, 55 (1999)
12. R. Maoz, E. Frydman, S.R. Cohen, J. Sagiv, "Constructive nanolithography": inert monolayers as patternable templates for in-situ nanofabrication of metal-semiconductor-organic surface structures – a generic approach. Adv. Mater. **12**, 725 (2000)
13. S. Hoeppener, R. Maoz, S.R. Cohen, L.F. Chi, H. Fuchs, J. Sagiv, Metal nanoparticles, nanowires, and contact electrodes self-assembled on patterned monolayer templates – a bottom-up chemical approach. Adv. Mater. **14**, 1036 (2002)
14. S.T. Liu, R. Maoz, G. Schmid, J. Sagiv, Template guided self-assembly of [$Au_5(_5)$] clusters on nanolithographically defined monolayer patterns. Nano Lett. **2**, 1055 (2002)
15. L. Bertilsson, B. Liedberg, Infrared study of thiol monolayer assemblies on gold preparation, characterization, and functionalization of mixed monolayers. Langmuir **9**, 141 (1993)
16. A. Zeira, D. Chowdhury, S. Hoeppene, S. Liu, J. Berson, S.R. Cohen, R. Maoz, J. Sagiv, Patterned organosilane monolayers as lyophobic–lyophilic guiding templates in surface self-assembly: monolayer self-assembly versus wetting driven self-assembly. Langmuir **25**, 13984 (2009)
17. J.L. Duda, M.F. Malone, R.H. Notter, Analysis of 2-dimensional diffusion-controlled moving boundary problems. Int. J. Heat Mass Transf. **18**, 901 (1975)
18. H.C. Day, D.R. Alleea, Selective area oxidation of silicon with a scanning force microscope. Appl. Phys. Lett. **62**, 2691 (1993)
19. R. Garcia, M. Tello, Size and shape controlled growth of molecular nanostructures on silicon oxide templates. Nano Lett. **4**, 1115 (2004)
20. A. Hoffmann, T. Jungk, E. Soergel, Cross-talk correction in atomic force microscopy. Rev. Sci. Instrum. **78**, 016101 (2007)
21. B. Yan, L. Liu, C.A. Astor, Q. Tang, Determination of the absolute amount of resin bound hydroxyl or carboxyl groups for the optimization of solid phase combinatorial and parallel organic synthesis. Anal. Chem. **71**, 4564 (1999)
22. D. Chowdhury, R. Maoz, J. Sagiv, Wetting driven self-assembly as a new approach to template-guided fabrication of metal nanopatterns. Nano Lett. **7**, 1770 (2007)
23. P. Lazar, H. Schollmeyer, H. Riegler, Spreading and two-dimensional mobility of long-chain alkanes at solid/gas interfaces. Phys. Rev. Lett. **94**, 116101 (2005)
24. X.Z. Wu, B.M. Ocko, E.B. Sirota, S.K. Sinha, M. Deutsch, A new surface phase in liquid normal-alkanes. Physica A **200**, 751 (1993)

25. B.M. Ocko, X.Z. Wu, E.B. Sirota, S.K. Sinha, O. Gang, M. Deutsch, Surface freezing in chain molecules: normal alkanes. Phys. Rev. E **55**, 3164 (1997)
26. A.J. Gellman, Vapor lubricant transport in MEMS devices. Tribol. Lett. **17**, 455 (2004)
27. D.T. Wasan, A.D. Nikolov, Spreading of nanofluids on solids. Nature **423**, 156 (2003)
28. G.V. Dedkov, Experimental and theoretical aspects of the modern nanotribology. Phys. Status Solidi A Appl. Res. **179**, 3 (2000)
29. B. Bhushan, Tribology on the macroscale to nanoscale of microelectromechanical system materials: a review. Proc. Inst. Mech. Eng. J J. Eng. Tribol. **215**, 1 (2001)
30. L. Xu, A. Lio, J. Hu, D.F. Ogletree, M. Salmeron, Wetting and capillary phenomena of water on mica. J. Phys. Chem. B **102**, 540 (1998)
31. J. Hu, X.D. Xiao, D.F. Ogletree, M. Salmeron, Imaging the condensation and evaporation of molecularly thin-films of water with nanometer resolution. Science **268**, 267 (1995)
32. T. Mitsui, M.K. Rose, E. Fomin, D.F. Ogletree, M. Salmeron, Water diffusion and clustering on Pd(111). Science **297**, 1850 (2002)
33. Z. Wu, S.N. Ehrlich, B. Matthies, K.W. Herwig, P. Dai, U.G. Volkmann, F.Y. Hansen, H. Taub, Growth of n-alkane films on a single-crystal substrate. Chem. Phys. Lett. **348**, 168 (2001)
34. H. Schollmeyer, B. Struth, H. Riegler, Long chain n-alkanes at SiO_2/air interfaces: molecular ordering, annealing, and surface freezing of triacontane in the case of excess and submonolayer coverage. Langmuir **19**, 5042 (2003)
35. S. Basu, S.K. Satija, In-situ X-ray reflectivity study of alkane films grown from the vapor phase. Langmuir **23**, 8331 (2007)
36. H. Taub, K.W. Herwig, B. Matthies, F.Y. Hansen, Neutron diffraction studies of the structure, growth, and melting of intermediate-length n-alkane films adsorbed on graphite. Inorg. Mater. **35**, 847 (1999)
37. K.W. Herwig, B. Matthies, H. Taub, Solvent effects on the monolayer structure of long n-alkane molecules adsorbed on graphite. Phys. Rev. Lett. **75**, 3154 (1995)
38. A. Diama, B. Matthies, K.W. Herwig, F.Y. Hansen, L. Criswell, H. Mo, M. Bai, H. Taub, Structure and phase transitions of monolayers of intermediate-length n-alkanes on graphite studied by neutron diffraction and molecular dynamics simulation. J. Chem. Phys. **131**, 10 (2009)
39. U.G. Volkmann, M. Pino, L.A. Altamirano, H. Taub, F.Y. Hansen, High-resolution ellipsometric study of an n-alkane film, dotriacontane, adsorbed on a SiO_2 surface. J. Chem. Phys. **116**, 2107 (2002)
40. R. Kohler, P. Lazar, H. Riegler, Optical imaging of thin films with molecular depth resolution. Appl. Phys. Lett. **89**, 241906 (2006)
41. M. Bai, K. Knorr, M.J. Simpson, S. Trogisch, H. Taub, S.N. Ehrlich, H. Mo, U.G. Volkmann, F.Y. Hansen, Nanoscale observation of delayering in alkane films. Europhys. Lett. **79**, 26003 (2007)
42. S. Trogisch, M.J. Simpson, H. Taub, U.G. Volkmann, M. Pino, F.Y. Hansen, Atomic force microscopy measurements of topography and friction on dotriacontane films adsorbed on a SiO_2 surface. J. Chem. Phys. **123**, 154703 (2005)
43. L.P. Van, V. Kyrylyuk, J. Polesel-Maris, F. Thoyer, C. Lubin, J. Cousty, Experimental three-dimensional description of the liquid hexadecane/graphite interface. Langmuir **25**, 639 (2009)
44. H. Riegler, R. Kohler, How pre-melting on surrounding interfaces broadens solid-liquid phase transitions. Nat. Phys. **3**, 890 (2007)
45. L.B. Lu, K.J. Zander, Y.G. Cai, The stability of parallel layer during the alkane spreading and the domain structures of the standing-up layer. Langmuir **26**, 5624 (2010)
46. T.A. Witten, L.M. Sander, Diffusion-limited aggregation, a kinetic critical phenomenon. Phys. Rev. Lett. **47**, 1400 (1981)
47. T.A. Witten, L.M. Sander, Diffusion-limited aggregation. Phys. Rev. B **27**, 5686 (1983)
48. A.J. Hurd, D.W. Schaefer, Diffusion-limited aggregation in 2 dimensions. Phys. Rev. Lett. **54**, 1043 (1985)
49. P. Meakin, Formation of fractal clusters and networks by irreversible diffusion-limited aggregation. Phys. Rev. Lett. **51**, 1119 (1983)

50. M.J.D. Ruijter, M. Charlot, M. Voue', J.L.D. Coninck, Experimental evidence of several time scales in drop spreading. Langmuir **16**, 2363 (2000)
51. H. Xu, S.S. Sheiko, D. Shirvanyants, M. Rubinstein, K.L. Beers, K. Matyjaszewski, Flow-enhanced epitaxial ordering of brush-like macromolecules on graphite. Langmuir **22**, 1254 (2006)
52. M.K. Chaudhury, A. Chaudhury, Super spreading of oil by condensed water drops. Soft Matter **1**, 431 (2005)
53. J.D. Miller, Spreading kinetics for low viscosity n-alkanes on a water surface as recorded by the high-speed video system. Ann. Univ. Mariae Curie Skodowska Lublin Polonia **7**, 105 (1999)
54. W.B. Hardy, The spreading of fluids on glass. Philos. Mag. **38**, 49 (1919)
55. P.A. Willermet, J.M. Pieprzak, D.P. Dailey, R.O. Carter, N.E. Lindsay, L.P. Haack, J.E. Devries, The composition of surface-layers formed in a lubricated cam tappet contact. J. Tribol. Trans. ASME **113**, 38 (1991)
56. L.B. Lu, Y.G. Cai, The role of vapor phase mass transport during the spreading of a long chain alkane drop. Langmuir **25**, 13914 (2009)

Chapter 14
Self-assembled Transition Metal Nanoparticles on Oxide Nanotemplates

Emanuele Cavaliere, Stefano Agnoli, Gaetano Granozzi, and Luca Gavioli

Abstract In this work we present some significant paradigmatic examples of directed self-assembling of transition metals deposited on ultrathin oxide layers grown on single crystal metal surfaces. Two major driving factors are considered in the description of the self-assembling process on these systems, giving rise to ordered arrays of metal nanoparticles (NPs): the presence of defects in the oxide layer and the metal–oxide interaction. We first summarize the structure of different oxide layers, and we discuss the role of geometric and electronic factors in the formation of ordered Au and Fe NP arrays on the selected systems. The discussion suggests that the defective oxide template can drive the self-organization of NPs, if the defects are accessible through the diffusion of metal adatoms on the surface, which in turn, is determined by the metal–oxide interaction.

14.1 Introduction

Metal–oxides interfaces have been in the focus of applied and fundamental research for many years due to their relevance in many different fields such as material science, catalysis, sensors, and microelectronics [1]. In the specific field of nanoscience, controlling the position of both the metal and the oxide at an atomic level is a relevant issue to achieve higher performances and new functionalities. Thence, the interest towards the understanding of the fundamental aspects of metal–oxide heterojunctions has further risen, and effective strategies for a controlled design and fabrication of these systems are under development. In addition, the new interesting chemical properties associated with transition metal nanoparticles (NPs) have stimulated work aiming at preparing metal NP arrays, possibly presenting a spatial ordering, which are stable with respect to the spontaneous coarsening processes driven by the reduction of their surface energy (highly dependent on their sizes). Oxide surfaces quickly appeared to be good candidates to support and stabilize metal NP arrays and most of the real catalysts are formed by an oxide powder where metal NPs are dispersed. Such a type of support presents the advantage of being thermodynamically very stable and, in most cases, chemically inert, so that the catalytic properties of the metal NPs can be exploited.

To optimize the catalyst performance, a detailed understanding of the metal NP–oxide interactions is needed, and this task is best accomplished by using planar model catalyst. The metal–oxide interaction could be in principle studied by examining two rather different systems, that is, metal films/particles on oxide surfaces or oxide films/particles on metal surfaces. Because of their relevance in real catalysts, the former type of systems has been first studied and a well known phenomenon, often referenced as the strong metal–support interaction (SMSI) has been documented in detail [2–9]. The SMSI acronym was first proposed by Tauster et al. in 1978 [2] to describe the suppression of both H_2 and CO chemisorption capacity for noble transition metal (group VIIIA) particles, when adsorbed on TiO_2 or other reducible oxides [2, 10]. Subsequently, the SMSI, later observed in other metal–oxide catalytic systems [2, 5, 8–10], has been examined and explained by the action of two different factors: an electronic and a geometric factor. The first is determined by a perturbation of the electronic structure of the metal catalyst, deriving from charge transfer between the metal and the oxide. The second results from an ultrathin (UT) layer of the reduced oxide support that is physically covering the metal particles (called the encapsulation model), and actively blocks catalytic sites at the metal surface [11, 12]. Unfortunately, the investigation of properties of metal–/oxide interfaces on classical catalytic systems, that is, metals on oxide surfaces, is often limited by the large band gap typical of oxide surfaces, hampering the use of modern surface science tools such as scanning tunneling microscopy (STM). However, the SMSI effect can also be investigated using the second type of model systems, often called as the inverse catalyst model, that is, oxide films/particles on metal surfaces.

Actually, it is now well recognized that the best way to achieve detailed information on the NP–oxide interactions is to carry out surface science experiments on a composite system, where the metal NPs are deposited on UT oxide films epitaxially grown on metal single crystals [13]. Such methodology presents experimental advantages, due to the intrinsic conductivity of the UT film, and also allows to prepare substrates with a well-defined defectivity, where the role of defects can be studied in detail. In addition, with respect to the bulk oxide surfaces, such systems allow an effective control over the size and spatial arrangement of metal NPs because such films can act as a two-dimensional (2D) template [14]. Templating the NP growth by adsorbate-induced surface reconstructions and by employing vicinal metal surfaces has been already largely explored in the past [15]. However, using metal-based 2D templates has scarce applicative perspectives, due to the intrinsic instability of the template itself under realistic conditions. On the contrary, 2D templates based on UT film oxides, thanks to the higher thermal stability, are expected to be more promising for some applications.

In particular, the way the UT oxide layers act as effective 2D templates is associated to the presence of defects, which can operate as preferential nucleation sites for the self-assembling process of the metal NPs. Due to the lattice mismatch between the oxide UT film and the metal substrate, ordered arrays of defects (e.g., ordered arrays of holes or dislocation lines) that can form the motif of the 2D template are often present. As discussed in the following, also electronic effects can direct the NPs self-assembly process. In addition, the affinity with oxygen of the deposited metal NPs can also be an important issue to be taken into consideration, since the 2D template itself can be partially destroyed by the deposited metal, and so the effective NPs ordering can be influenced [16].

In this chapter, we report a comparison between the templating 2D capabilities of three different oxide UT films, that is, with an ordered array of defects (TiO_x), with defects and double-layer structure (Al_2O_3) and with no defects (FeO). Moreover, the effects due to the different nature of the metal NPs, i.e., a noble metal (Au) and a transition metal (Fe), are discussed. In a previous review article the same triad of model systems has been briefly discussed [14], whereas the templating behavior of the Al_2O_3 films with respect to several different metals has been discussed in detail in another review [17].

14.2 The Structure of the UT Oxide Layers

In recent years, much attention has been focused on oxide model systems in the form of UT epitaxial films grown on single-crystal metal surfaces [13, 18–26]. In the following, by the term UT we mean films characterized by thickness in the range up to a few monolayers (MLs), i.e., thickness $\leq 1 - 2$ nm. While the results to date have demonstrated that films with thickness of the order of several MLs are adequate to simulate bulk-terminated oxide surfaces, more and more examples are reported for UT films with radically new structures, stoichiometries and properties. Actually, UT oxide films represent a potentially new class of materials where innovative properties with respect to bulk phases are obtained, due to the interaction with the substrate (metal proximity effects) and/or to the confinement of atomic-scale structures in at least one dimension [27–29]. For example, to optimize the interaction with the substrate, the film can adjust its structure, producing a strain which is maintained until a critical thickness is reached, over which the film recovers the thermodynamically stable bulk structure. Indeed, most of the recent focus on UT films is related to the possibility of stabilizing phases and structures which are not obtainable in bulk form due to thermodynamic and/or kinetic constraints.

A number of review articles have appeared in the literature in the field of UT oxide films; they have been mainly focused on the epitaxial growth procedures of oxide UT films [20, 26, 30] or on their relevance in catalysis [31, 32]. In many cases, the metal–oxide interactions result in interface-stabilized reduced oxide phases [3, 20, 33–39]. The body of this work has shown that the compact (111) metallic surfaces are best suited to grow UT oxide phases. In particular, a common feature observed is the presence of a wide variety of reconstructions with respect to the bulk phases, depending on different factors such as the lattice mismatch, the metal overlayer–oxygen ratio during deposition, and the preparation procedure [20, 34, 38, 40].

Concerning the actual oxide–metal interaction, the contact between a metal and an oxide results in charge redistribution at the interface at the local range and/or the long range. The electron redistribution is driven by principles of energy minimization of the system and of continuity of electric potential in the solid. The close proximity of the oxide to a highly polarizable medium, i.e., the metal support, can modify substantially the physico-chemical properties of the oxide itself. The charged atoms in the oxide induce image charges in the substrate, resulting in strong polarization interactions that can modify the work function of the system, as well as the fundamental electronic properties in the UT film, such as the on-site coulomb repulsion or the charge transfer energy [41]. Strictly connected to the formation of the aforementioned interfacial dipole is the bending of the oxide bands at the interface leading to the formation of a Schottky-type heterojunction [13].

Another interesting phenomenon is the possibility of a direct electronic tunnelling via the UT oxide film from substrate to the adsorbed NP. This can happen if the Fermi level (E_F) of the metal single crystal lies above the empty states of the adsorbate [42]. The modified charging state determines a profound change on the electronic properties and even the geometry of the supported NPs [43]. Therefore the nature of the metal–oxide interface (relative position of the metal E_F with respect to the band structure of the oxide, interfacial hybridization, possibility to undergo an easy polaronic distortion etc), type of oxide material (degree of covalence/ionicity), thickness of the oxide layer, can be actively used as tools to control the shape, arrangement and chemical properties of the metal NPs.

In the following we present a short survey of the most suitable systems for templating metal NPs, i.e., $TiO_x/Pt(111)$, $Al_2O_3/NiAl(111)$ and FeO/Pt(111). They represent different examples where the templating effect is sustained by a different mechanism. For a detailed description of each oxide system we refer the reader to the specific references.

14.2.1 TiOx/Pt(111)

A comprehensive report on the preparation procedures and structure of the several different TiO_x UT films grown on Pt(111) is part of a recent review article [26]. They can be prepared by depositing Ti atoms in oxygen reactive atmosphere on the Pt(111) substrate, held at room temperature under initial UHV conditions. A successive heat treatment of the surface at suitably high temperature, performed in controlled oxidizing or reducing environment (from $P_{O_2} = 5\times10^{-6}$ mbar to UHV), yields the desired surface structure. A more detailed description can be found in Refs. [38, 39, 44].

The detailed structure at an atomic level has been derived from combined low energy electron diffraction (LEED), STM, X-ray photoemission spectroscopy (XPS) and photoelectron diffraction (XPD) measurements [38] and by density functional theory (DFT) simulations of the structure arrangement [39]. The main morphological differences among these phases are to be traced back to the different stacking of titanium and oxygen atoms in the film, and to the presence of ordered arrays of defects. In the present paper we are most concerned with the two phases, called z' - TiO_x and w' - TiO_x, whose STM images are shown in Fig. 14.1a, b. Both of them are formed by a Ti–O double layer (Fig. 14.1c), where the Ti atoms are

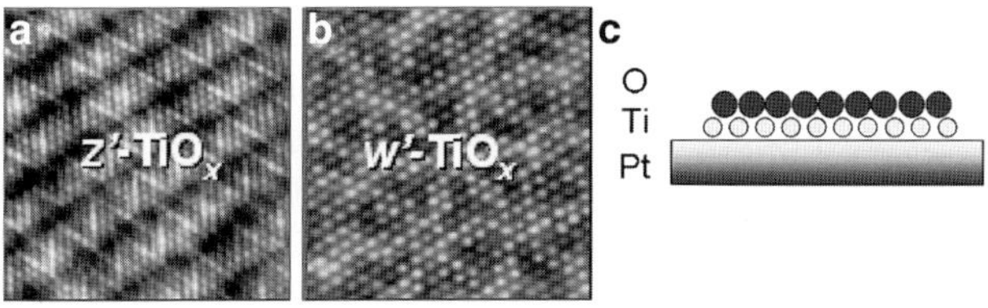

Fig. 14.1 STM images of the two different TiO_x/Pt(111) phases discussed in the present work. (**a**) z' phase, $9 \times 9\,nm^2$, $V = 0.8\,V$, $I = 1.5\,nA$. (**b**) w' phase, $6 \times 6\,nm^2$, $V = 0.2\,V$, $I = 1.0\,nA$. (**c**) Schematic drawing of the Pt/Ti/O stacking sequence. Adapted with permission from Ref. [38] Copyright 2005 by the American Chemical Society and Ref. [62] Copyright 2007 by the American Chemical Society

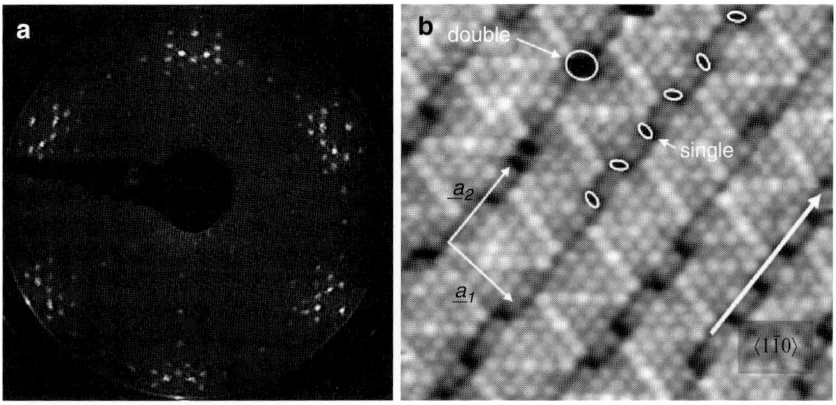

Fig. 14.2 (**a**) LEED pattern of z'-TiO_x/Pt(111) with its three domains. (**b**) z'-TiOx lattice and high-resolution STM image showing the characteristic *zigzag-like* contrast and orientation with respect to substrate lattice. The single and double *picoholes* (see text) are indicated by white ellipsis. Adapted with permission from Ref. [38] Copyright 2005 American Chemical Society

at the interface with the substrate, and the oxygen atoms form the topmost layer. Among the two films, the most interesting with respect to the template effect is the z'-TiO_x one [44] because it presents an ordered array of defects (see below). On the contrary, the w'-TiO_x, while it has a close related stoichiometry, does not present an ordered array of defects, and its structure has been described in terms of a Moiré-like structure [45].

In Fig. 14.2, where we report the LEED pattern and a representative STM image of the z' - TiO_x/Pt(111) system. The STM image reported in Fig. 14.2b shows that the z' - TiO_x/Pt(111) layer is constituted by bright stripes clearly separated by darker rows. On the stripes, one can observe a *zigzag-like* motif, while the darker rows, parallel to the direction of Pt(111), are present at both positive and negative bias, suggesting that these are true topological features, named *troughs*. The arrangement of the Ti layer would continue the fcc stacking of Pt, but the mismatch between the TiO_x lattice parameter (ranging around 0.3 nm) and the Pt one (a = 0.277 nm) induces the creation of denser stripes separated by lower density regions, thus originating the *troughs*. The characteristic STM pattern can be explained in terms of different coordination of Ti atoms: fourfold coordinated Ti atoms are separated by dislocation lines from defective Ti threefold coordinated atoms, arranged in triangles near the stripe edge. Fourfold coordinated Ti atoms are more positively charged, so with higher density of empty states, and are imaged as brighter protrusions in STM images acquired with positive bias [38, 46].

A comparison between STM and DFT [44] calculations reveals that different types of defects are present along the *troughs*, depending on the preparation procedures. Wide troughs are characterized by the presence of titanium vacancy defects, defined as *picoholes* (single in Fig. 14.2b), that can merge to form larger *picoholes* (double in Fig. 14.2b), while narrow *troughs* do not show titanium vacancy defects [44].

Such data show that this oxide layer is potentially very useful to act as nanotemplate for organization of metal NP, since the *picoholes* are the best candidate as nucleation sites during the self-assembling of the deposited metal atoms.

14.2.2 $Al_2O_3/Ni_3Al(111)$

Aluminum oxide UT films can also be grown on several different metallic substrates, and they have been recently reviewed [26]. One of the best substrate to grow alumina films is $Ni_3Al(111)$, exposing the clean surface to 40 L of oxygen with the substrate at 1,000 K (oxygen pressure 3×10^{-8} mbar), followed by annealing to 1,050 K for 5 min [47, 48]. To obtain closed and long-range ordered UT films, the oxidation processes are performed twice [48]. The structure of the aluminium oxide film is rather complicate, and it is commensurate with the Ni_3Al substrate [48, 49].

The real space topography of the alumina superstructure is strongly dependent on the bias voltage, as shown in Fig. 14.3a, b. At 3.2 V bias voltage, one can observe the so-called "network" structure, which can be described as a hexagonal arrangement of dark holes, which are surrounded by smaller hexagonal rings of bright dots (Fig. 14.3a). Note that not all the holes present the same apparent depth. At 2.0 V bias voltage, the STM image shown in Fig. 14.3b, taken at the same spot as in Fig. 14.3a, shows a contrast reversal, with bright protrusions that are defined as the "dot" structure. The profiles taken at the two different biases are compared in Fig. 14.3c, indicating that the shallow holes in Fig. 14.4a can be correlated to the dots at a bias of 2.0 V. Taking these findings into account the superstructure unit cell present in both images is identical. Its lattice constant is $b_{dot} = 4.07$ nm, in good agreement with spin polarized LEED (SPA-LEED) measurements [48]. The described contrast reversal is apparently not only due to an electronic state in the oxide band gap [50], but must be related to a different topography of the alumina film at these sites. The comparison with SPA-LEED data indicates that the "dot" structure is the only true superstructure of the alumina film, and the two different superstructures are schematized in Fig. 14.3d.

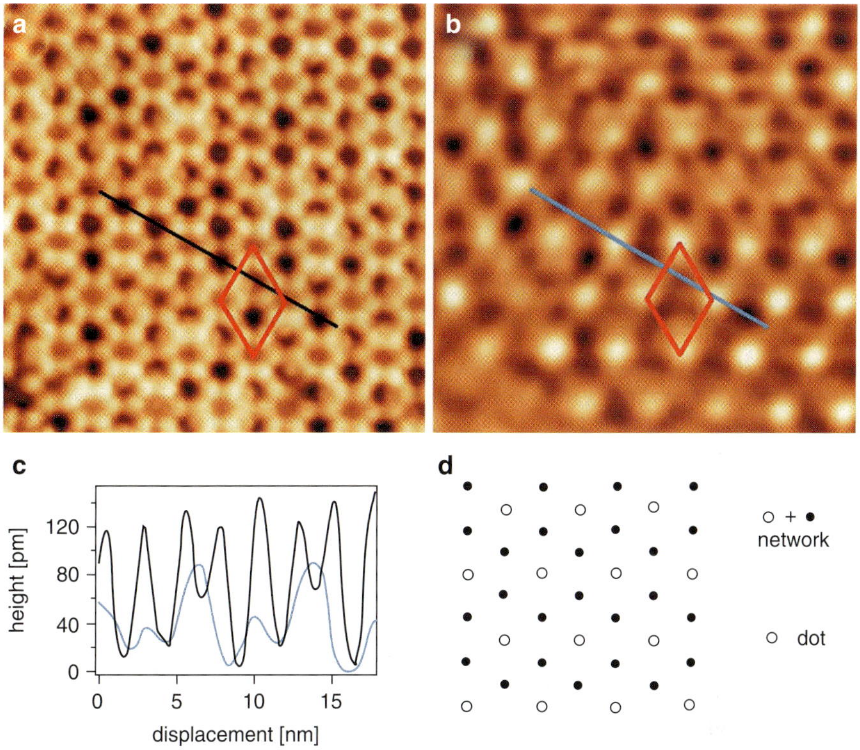

Fig. 14.3 STM topography (image size 27.8 × 27.8 nm^2) of the $Al_2O_3/Ni_3Al(111)$ "network" structure (**a**) taken at a bias voltage of +3.2 V and of the "dot" structure (**b**) taken at a bias voltage of 2.0 V. The *red rhombus* represents the unit cell. In (**c**) the line profile taken from the *black* (**a**) and *blue* (**b**) line are plotted. (**d**) Schematic representation of the "network" and "dot" structure. The *open circles* correspond to the *white spots* of the STM image in (**b**), while the *black circles* correspond to the *dark holes* observed in (**a**). Reprinted from Ref. [48] Copyright 2005 with permission from Elsevier and from Ref. [49]. Reproduced by permission of The Royal Society of Chemistry

Recent STM and DFT studies of this surface have finally revealed the correct atomic geometry of the oxide layer [51]. In Fig. 14.4a the STM data taken on the same system at very low bias voltage are shown, in which the yellow triangles correspond to the black dots of the scheme shown in Fig. 14.3d, while the black hexagon corresponds to the open circles shown in Fig. 14.3d. The white protrusions with triangular arrangement observed in the STM image represent oxygen atoms, and the ball model obtained from comparison with DFT calculations is shown in Fig. 14.4b. Here, the lattice rotation of 23.1° with respect to the substrate, and the stacking sequence of the UT oxide (Al–O–Al–O) is schematized, with the two surface layers (named Al_s and O_s in Fig. 14.4c) being almost coplanar [51]. The comparison with DFT calculations reveals that the hexagon structure with the brightest spot is an actual geometric hole in the Al_2O_3 layer, reaching down the $Ni_3Al(111)$ substrate (*picoholes*) [51].

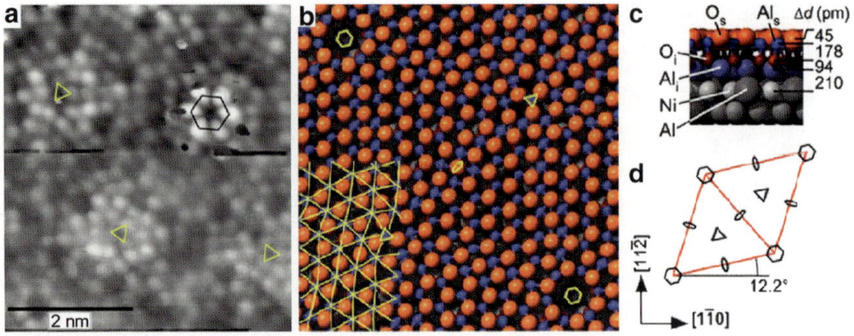

Fig. 14.4 (**a**) STM image of the $Al_2O_3/Ni_3Al(111)$ surface taken at a bias voltage of 58 mV, with *yellow triangles* corresponding to the *black dots* of the scheme in (**d**) and the *black hexagon* corresponding to the *open circles* of (**d**). (**b**) Top view of the ball model of the system. *Red* indicates oxygen atoms, while *blue* reveal Al atoms. (**c**) Side view of the ball model obtained from DFT calculations. (**d**) Unit cell of the oxide layer. Adapted from Ref. [51] Copyright 2007 by the American Physical Society

With respect to the z' - $TiO_x/Pt(111)$ film, the aluminum oxide UT film presents only a single type of defects, and a different arrangement and symmetry of the defects on the surface.

14.2.3 FeO/Pt(111)

The FeO film is usually prepared by evaporating Fe onto the Pt(111) substrate, and subsequently heating the sample to 1,030 K in an oxygen pressure of 10^{-5} torr for ~30 s. [52]. The resulting system, shown in Fig. 14.5a, is a FeO bilayer with an internal periodicity of 0.309 nm, slightly larger than that of the bulk FeO(111) plane (0.304 nm) [52–54]. XPD data show that the O atoms occupy hollow sites with respect to the iron layer underneath [53]. The incommensurability between the oxide layer (0.31 nm periodicity) and the Pt(111) substrate (0.277 nm periodicity), with the misalignment by 0.6°, gives rise to the formation of a Moiré pattern, shown on a larger scale in Fig. 14.5b, and results in a very large unit cell (2.6 nm), schematized in Fig. 14.5c. The origin of the Moiré pattern is explained considering few simple high-symmetry Fe/Pt configurations with a different registry: in some cases the Fe atoms are on top of the Pt atoms, in other cases they are in bridge sites and, still in other regions, in fcc or hcp sites (Fig. 14.5b) [52, 54, 55]. STM images, taken in the near-field emission regime, suggest that the modulated Fe–Pt stack is responsible for corresponding modulations of the surface potential [56].

Note that the corrugation of this oxide layer is very low (below 0.01 nm) and no defects are present inside the domains forming the oxide layer. Hence this UT film is, from the geometric point of view, very different from both the titanium oxide and aluminium oxide counterparts.

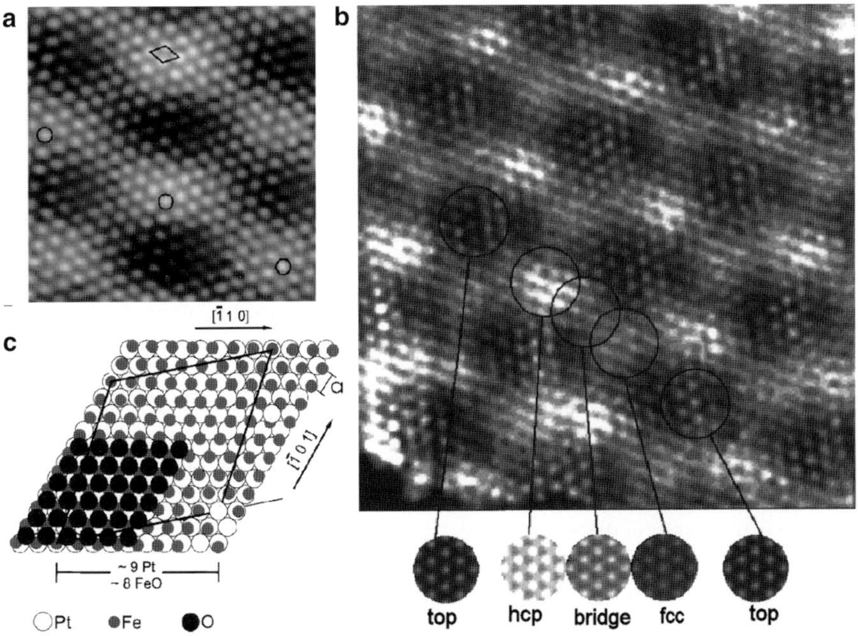

Fig. 14.5 (**a**) A $5.5 \times 5.5\,\text{nm}^2$ STM image of the FeO/Pt(111) system. An atomic periodicity of 0.31 nm is modulated by a large 2.6 nm periodicity creating a Moiré superstructure. The direction of this superstructure indicated by the marked atoms is rotated by about 11° against the small FeO(111)$-(1\times1)$ unit cell that is also indicated. ($V = 0.9\,\text{V}$, $I = 0.3\,\text{nA}$). (**b**) A $6\times6\,\text{nm}^2$ STM image showing the Moiré structure. The small circular images at the *bottom* are calculated images using the ESQC (electron scattering quantum chemistry) method for the various high-symmetry epitaxial configurations and associated with the *circled* regions in the image. (**c**) Model of the FeO(111) bilayer on Pt(111). Adapted from Ref. [52] Copyright 1996 by the American Physical Society and Ref. [54] Copyright 1998 by the American Physical Society

14.3 The Oxide Layers as Nanotemplates for Metal NPs

Before the discussion about specific systems, it is convenient to point out the main factors directing the growth of metal adsorbates on UT oxide films. The defects in the structure of the oxide film (i.e., domain boundaries, steps, kinks, dislocations and point defects) could drive the nucleation of the adsorbed metals. Such kind of templating effect could be referenced as *topographical directed assembly*. As an example, on highly regular UT films such as MgO/Ag(100), colour centre defects were induced by electron bombardment and probed to be effective trapping sites for further deposited Au adatoms [57]. However these kinds of artificially induced defects are randomly distributed on the film surface. To obtain more interesting properties, extended NPs arrays are needed so that a necessary step is obtaining a large density of defects on the template surface. A further important goal is having a periodic and regular array of defects, so that different NP–NP distances and geometries could be tested to put in evidence the possible role of inter-talk between the NPs.

Beside the *topographically directed assembly*, also an *electronically directed assembly* can be envisioned. Actually, in the UT oxide film we can also imagine to have a local modulation of the electronic structure of the film. As an example, band bending at the metal–oxide interface and modification of the work function (WF) [13] of the oxide-on-metal system can be considered. In the case of the monolayer-thick $SiO_2/Mo(112)$ oxide, a 0.42 eV increase in the WF is predicted by DFT calculations [58] and an even higher value (0.8 eV) is experimentally observed [59], while 1 ml of MgO on Mo(100) significantly lowers the WF. We discuss the role of this factor in the specific case of polar oxide films, such as TiO_x or FeO, where the surface dipole layer associated with the reduced and oxygen terminated film is expected to lower the work function of the Pt(111) substrate. The metal NP deposited on the oxide UT film will generate another interface with the consequent charge transfer or polarization phenomena.

Another interesting phenomenon is the possibility of a direct electronic tunnelling *via* the UT oxide film from the metal substrate to the adsorbed metal NPs (already mentioned in Sect. 14.2). The modified charge states of the NPs may have a strong effect both on the electronic properties and the geometry of the single NP, but also on their self-assembly due to possible repulsions between them [42, 60, 61]. Moreover, the local modification of the diffusion barrier landscape within the UT film can play a role during the self-assembly of the NPs. At last, but not at least, the oxygen affinity of the adsorbed metals critically affects the templating effect, as it will be put in evidence in the case of the z' - $TiO_x/Pt(111)$ UT film.

14.3.1 Au and Fe on z' - TiO_x - $Pt(111)$

Our group has investigated in detail the templating effect of the z' - TiO_x UT film with respect to two different metals, i.e., Au and Fe; they have been deposited at room temperature, and the STM images were acquired at the same temperature. Fig. 14.6a shows that starting from the very onset of nucleation, no gold adatoms or clusters are found on the oxide stripes, and all Au NPs are pinned inside the *troughs* with a narrow size dispersion. At higher coverage regime (Fig. 14.6b) the behavior is not modified, and the fast Fourier transform (FFT) reported in the insets shows the preserved alignment of the NPs, that are also almost monodispersed [62]. The broader view of the same region shown in Fig. 14.6c confirms the alignment of Au NPs in the oxide *troughs*, which reflects the symmetry of the three equivalent [110] directions of the Pt substrate. One can also note that on uncovered Pt(111) areas, Au is forming 2D wetting islands, typical of Au/Pt(111) growth mode [63].

Thermodynamic data suggest a low oxygen affinity of Au with oxygen and DFT simulations confirm that Au adatoms experience a weak adsorption energy (0.1 eV) and flat diffusion landscapes on the oxygen-terminated stripes, with a negligible distortion of the oxide film [64]. According to the DFT numbers, at RT Au adatoms quickly reach the *picoholes* in the *troughs* where the adsorption energy is the highest (1.7 eV) [64].

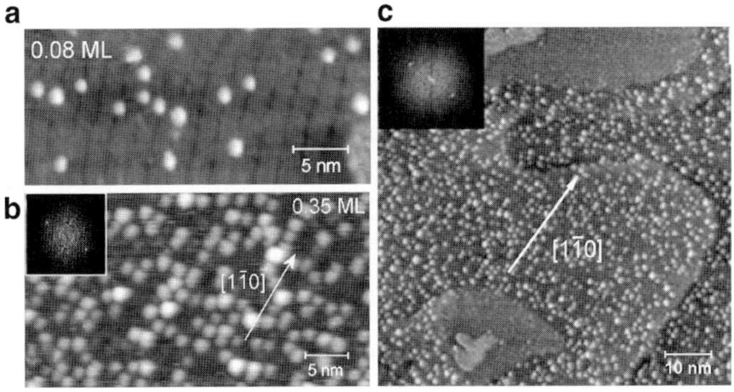

Fig. 14.6 Au NPs on z'-TiO_x/Pt(111) UT film: (**a**) at low coverage ($\theta = 0.08$ ml) ($V = +0.6$ V, $I = 0.6$ nA) all NPs are nucleated in the *troughs*; (**b**) at $\theta = 0.35$ ml, the NPs arrangement into linear arrays following the oxide *troughs* becomes more clear, as shown in the FFT transform of the image in the *upper left inset* frame; (**c**) extended view of the region of frame. The orientation of the z'-TiO_x film and its *troughs* is given with respect to the main Pt(111) crystal direction [1$\bar{1}$0]. Adapted with permission from Ref. [62] Copyright 2007 by the American Chemical Society

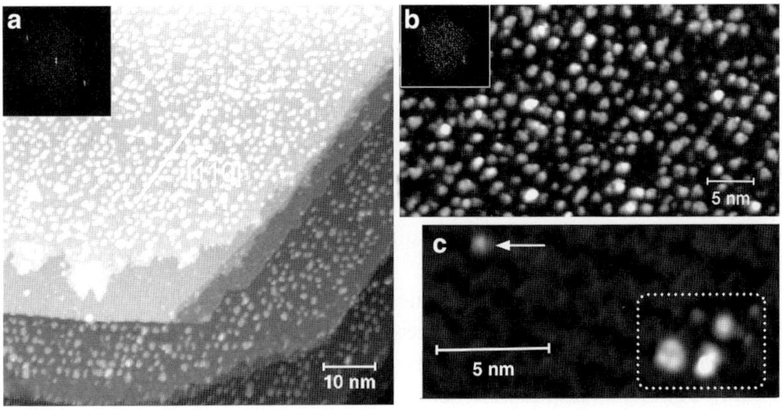

Fig. 14.7 Fe NPs on z'-TiO_x/Pt(111): (**a**) ($\theta = 0.3$ ml, $V = -1.61$ V, $I = 0.25$ nA) [16]; (**b**) a magnified region of frame (**a**) shows less aligned Fe NPs with more irregular shapes; (**c**) at the very low coverage $\theta = 0.03$ ml ($V = +0.89$ V, $I = 0.48$ nA) one can see granularity of the NPs in the *troughs* (*dotted rectangle*) and tiny aggregates pinned on the oxygen terminated stripes (*white arrow*). Adapted from Ref. [16]. Reproduced by permission of the PCCP Owner Societies

The deposition of a more oxygen affine metal, i.e., iron ($\Delta H_{fFe-O} = 250$–300 kJ/mol) yields a different picture. At first sight a large-scale image (Fig. 14.7a) shows again nm-sized Fe NPs aligned and pinned in the oxide *troughs* [16]. Portions of Pt(111) substrate not decorated by the oxide film show the quasi 2D growth of Fe islands [65].

However, a more careful investigation highlights that the alignment of the NPs with respect to the *troughs* centerlines (Δx) is quite effective for Au NPs but much fuzzier for Fe (Fig. 14.7b). The Au NPs appear as a diffuse glow due to the overlapping of featureless 6s orbitals of Au atoms, while the Fe NPs exhibit a granular contrast and irregular shape. In the very low coverage regime ($\theta = 0.03$ ml, (Fig. 14.7c) some larger Fe NPs are pinned in the *troughs*, but also smaller particles (adatoms or dimers outlined in the figure) are adsorbed on the stripes [16]. The presence of adsorbate adatoms or tiny aggregates on the oxide stripes is coherent with the high oxygen affinity of Fe. DFT calculations yield substantially higher diffusion barriers for Fe adatoms and stronger adsorption, from 0.74 to 2.0 eV, depending on oxygen sites considered along the stripe [66]. Adsorption energies on the stripes are as well higher than for Au, ranging from 2.2 to 3.5 eV. Nevertheless, similarly to the Au case, the *picoholes* are the most favored adsorption sites (4.66 eV) also for the Fe aggregates. The higher adhesion of Fe on the oxygen landscape, terminating the oxide surface, ends up in a possible oxidation of adsorbed Fe. Within this hypothesis 4s valence electrons of Fe would be transferred to the oxygen. The depletion of electron density from the Fe 4s orbitals means that there are no s electrons to "glue" one metal center to a neighboring one, and so this is a possible explanation for the apparent granularity of Fe NPs found both in experimental [16] and simulated STM images [66].

Despite these differences, the average NP lateral size and size dispersion are very narrow for both metals: 1.3 ± 0.4 nm for Au and 1.1 ± 0.4 nm for Fe [16]. Kinetic Monte Carlo simulations of metal NPs growth at RT [67] using the values of adsorption energies and diffusion energy barriers derived above produce a narrow size distribution in fair agreement with experimental data for both Au and Fe NPs, confirming the templating effect of the z' - TiO_x film. They also rationalize the irregular, fractal-like shape of the Fe NPs at higher coverage versus the homogeneous shape of the Au NPs as a result of kinetic trapping into non-equilibrium configurations [68].

In conclusion, the adsorption and nucleation of metals on the z' - $TiO_x/Pt(111)$ film seem to be substantially conditioned by the interplay between the oxygen diffusion barrier landscape and defectivity of the oxide film. Such a conclusion has been put in light by examining the behavior of two different metals, i.e., Au and Fe, which are respectively poorly and strongly oxygen affine.

The comparison with the adsorption behavior of the same metals on the flat, not defective, w' - $TiO_x/Pt(111)$ UT film clearly put in evidence the essential role of the *picoholes* in the ordered nucleation of metal NPs observed on the defective z' - TiO_x film. Actually, in the w' - $TiO_x/Pt(111)$ case, Au NPs randomly spread on the substrate are observed: they are quite dispersed and have a definitely larger size (3.0 nm average versus 1.3 nm) (Fig. 14.8a, b, respectively). On the contrary, Fe NPs, though randomly nucleated on the oxide surface, show a size distribution similar to the one previously found on z' - TiO_x film, in agreement with the expected lower mobility of Fe on w' - TiO_x (Fig. 14.8c, d, respectively) [62].

Thence, we can infer that the ordered array of picoholes is an effective template for metals with a low oxygen affinity, where weak adsorption and flat diffusion barriers do not hamper the path of impinging adatoms towards the *troughs*. The growth of metals more reactive towards oxygen is much more limited by kinetics,

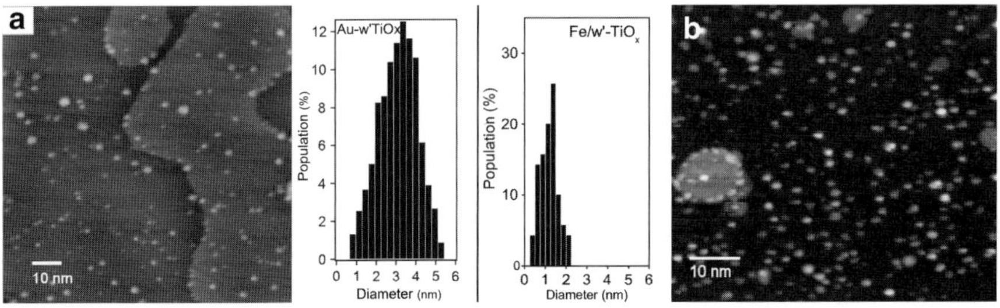

Fig. 14.8 (**a**) Au NPs ($\theta = 0.2$ ml, $V = 1.5$ V, $I = 0.2$ nA) and (**b**) Fe NPs ($\theta = 0.35$ ml, V $=$ 1.6 V, $I = 0.2$ nA) on w' - TiO_x/Pt(111). In the centre frame, histograms of size dispersion of Au (*left*) and Fe (*right*) NPs are shown. Adapted with permission from Ref. [62] Copyright 2007 American Chemical Society

due to strong interactions and high diffusion barriers on the oxygen landscape on these UT films.

14.3.2 Metals on $Al_2O_3/Ni_3Al(111)$

This UT film presents two types of templating sites: the hexagonal pattern of the "network" structure with a 2.6 nm pitch, and the "dot" structure, having a 4.07 nm pitch, and corresponding to an ordered network of *picoholes* reaching down the Ni_3Al(111) substrate (Sect. 14.2.2).

Becker et al. [69] explored the behavior of several adsorbate metals on this alumina film, from noble metals like Ag, Au and Cu, through Mn toward the most reactive (oxygen affine) V. STM images of the very early growth of Au and Ag show a preferential pinning of the metal NPs into the "dot" structure. These data suggest that preferential nucleation occurs in the *picoholes*, like on the z' - TiO_x film above discussed. However, when larger amounts of Au or Ag are deposited, coalescence of NPs occurs and there is no further evidence of ordered spacing of the particles related to the oxide substrate "dot" structure (Fig. 14.9a). Also tip-induced mobility of Ag NPs was observed, suggesting weak metal–oxide film interactions. These observations are again in tune with the low oxygen affinity of the noble metals.

When Cu is deposited on the alumina film, initial nucleation on the "dot" structure is found, as in the case of Ag and Au, confirming that the growth is indeed governed by the topographic features of the "dot" structure. However, in contrast to the Au and Ag growth mode, at higher coverages more ordered and size-controlled Cu NPs are grown, following the periodicity of the "dot" structure (Fig. 14.9b), while the coalescence into larger NPs is observed. The observed behavior suggests a correlation between the metal–oxide interaction and the degree of ordering of the NP array. In fact, when metals with even higher oxygen affinity, like Mn, Fe and V, are deposited on the Al_2O_3 oxide layer, the observed behavior confirms this trend

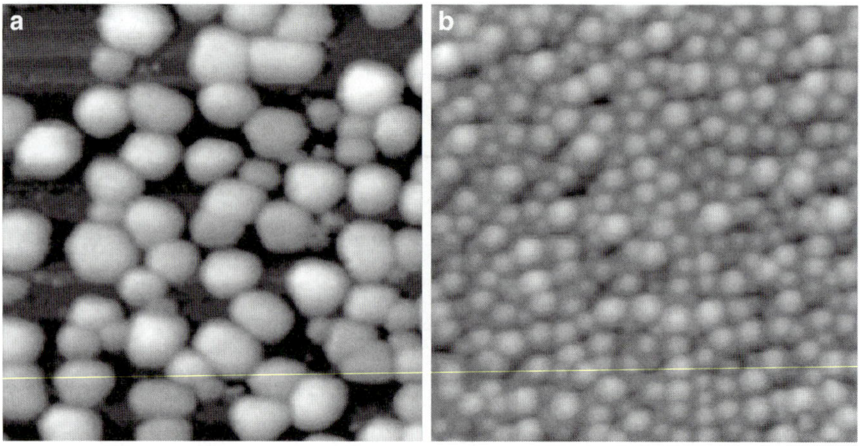

Fig. 14.9 Au (**a**) and Cu (**b**) NPs on $Al_2O_3/Ni_3Al(111)$. Size of the frames is $60 \times 60\,nm^2$ and $72 \times 72\,nm^2$, respectively. Adapted from Ref. [69], New Journal Of Physics, 2002 IOP Publishing

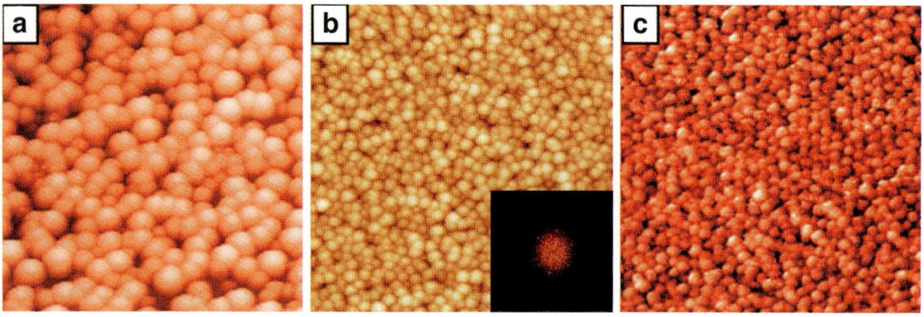

Fig. 14.10 The $68 \times 68\,nm^2$ STM images of (**a**) Mn ($V = 0.7\,V, I = 0.11\,nA$), (**b**) Fe ($V = 0.7\,V, I = 0.078\,nA$) and (**c**) V ($V = 0.7\,V, I = 0.33\,nA$) NPs on $Al_2O_3/Ni_3Al(111)$. Inset of frame (**b**) is the FFT of the image, outlining the honeycomb-like arrangement of the NPs, with a 2.4-nm pitch. Adapted from Ref. [69], New Journal Of Physics, 2002 IOP Publishing and from Ref. [71] Copyright 2006 with permission from Elsevier

[69–71]: a parallel, better order and size control in the NPs pattern is associated with increasing metal–oxide interaction (see Fig. 14.10 from left to right).

The adsorption of more reactive metals outlines also a new feature, deduced from the analysis of the radial distribution function of the nearest-neighbour NP, especially for the most reactive V metal: the appearance of another pattern in the NP spacing, an hexagonal mesh with a 2.6 nm periodicity. This pitch indicates a clear relation to the "network" structure of the underlying alumina, i.e., the true geometrical holes with higher concentration of oxygen atoms (Sect. 14.2.2), hence confirming the major role of the metal NP–oxide interaction in the self-assembling process.

In the case of Fe adsorption, in spite of the apparently disordered nucleation (Fig. 14.10b), the pattern related to the "network" structure is still observed in the FFT of the STM data (inset of Fig. 14.10b) [71]. This observation is explained by

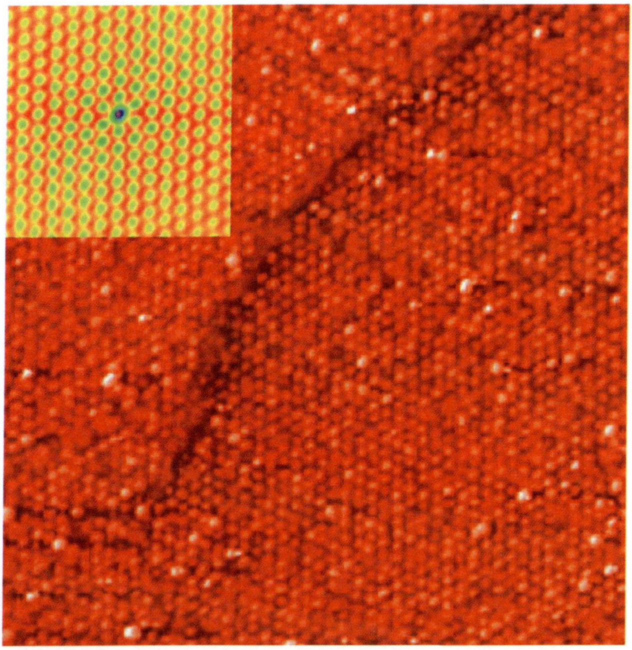

Fig. 14.11 (**a**) STM image ($V = 0.7\,V, I = 100\,pA$) of V NPs on $Al_2O_3/Ni_3Al(111)$ grown at a substrate temperature of Tsub $= 550\,K$. The image size is $125 \times 125\,nm^2$. Inset: autocorrelation function of the image. Adapted from Ref. [69], New Journal Of Physics, 2002 IOP Publishing

the presence of random defects, acting as trapping centres for Fe, located close to the threefold sites of the bare alumina film (triangles shown in Fig. 14.4a, d), and observed in the STM images at positive biases like bright protrusions near threefold oxide sites. Therefore the Fe cluster arrangement is only affected in terms of higher background in the FFT, while the hexagonal structure is still visible. This effect is much less pronounced for Co that nucleates perhaps only in the "network" structure sites with a more ordered arrangement [71].

The case of vanadium is particularly interesting, since the deposition of this metal at 550 K results in an almost perfect arrangement of the NP on the "network" structure sites, with a degree of order much higher than that obtained at room temperature (Fig. 14.11). This observation is clearly related to the role played by kinetics in determining the actual structure of the metal film, through the different diffusion barriers that the NPs are experiencing on the oxide layer.

The deposition of Pd on the $Al_2O_3/Ni_3Al(111)$ film has also been thoroughly investigated. In this case it is possible to obtain a hexagonal array of size selected NPs, simply because the heterogeneous nucleation of metal nuclei on the "dot" structure is energetically very favoured. It has been demonstrated that a careful tuning of temperature and flux of incoming metal atoms is of pivotal importance to obtain a good order of the metal NPs [49]. Taking the cue from basic knowledge gained in these studies, even more complex systems like array of bimetallic NPs have been tackled. Hamm et al. used a two-step procedure to grow ordered PdAu bimetallic NPs [72]: in the first step, Pd is deposited on the "dot" structure so that an arrangement of order stable nuclei is obtained; while in the second step, Au is evaporated on top of this seeding layer. The incoming atoms are then captured by the NPs formed during the first deposition, before they have time to nucleate any pure single metal NP [72]. This same procedure based on sequential deposition has been adopted to grow Fe–Pd [51] and Co–Pd bimetallic NPs [73].

These studies supported by extensive DFT calculations were able to give a microscopic explanation of the reasons at the basis of a good templating effect: the key factor is associated with the possibility of single atoms to have an easy access to the holes that constitute the "dot" structure. A kind of Ehrlich–Scwoebel barrier prevents atoms like Fe or Co to jump into the hole, which is energetically the most favourable nucleation site; accordingly they nucleate on less favourable sites of the oxides which are structurally not well ordered because of the presence of frequent random defects (Fig. 14.12a). On the contrary, in the case of Pd the diffusion barrier almost vanishes and indeed this hole can exert a quite large capture radius (0.4 nm) (Fig. 14.12b).

In summary, the examples discussed in this section indicate that a good templating substrate should indeed posses nucleation sites that are geometrical in nature, but the data also show that these sites are effective in templating metal NPs if the energy landscape for surface diffusion of metal, partly determined by the oxygen affinity, allows the diffusion towards the most energetically favourable adsorption sites.

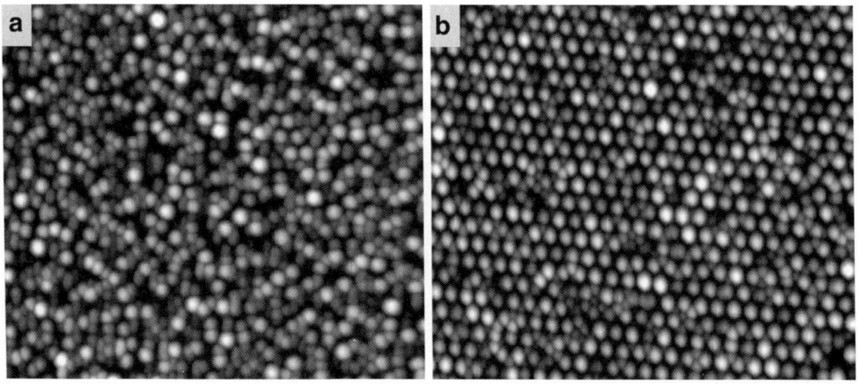

Fig. 14.12 STM images (80-nm wide) after deposition of 1 Å of Fe (**a**) without and (**b**) with predeposition of a 0.03 Å Pd seed layer. Adapted from Ref. [51]. Copyright 2007 by the American Physical Society

14.3.3 Au on FeO/Pt(111)

The adsorption of Au on the FeO/Pt(111) system was investigated with low temperature STM (5 K) and in a very low coverage regime; we should remark that, in clear contrast to the described experiments so far, at this temperature the adsorbed metal diffusion coefficients are extremely low, and only Au adatoms or perhaps small aggregates are expected. However, these working conditions are useful to rule out the role of diffusion barriers in the NP growth.

The FeO UT film shares some structural characters with the above discussed TiO_x films: they are both polar oxygen terminated bilayers where the metal (Fe or Ti) is at the Pt interface and, as for the w' - TiO_x case, the oxide–Pt lattice mismatch gives rise to a Moiré pattern observed in the STM images of the bare oxide surfaces (Sect. 14.2.3). A hard-sphere model of the film predicts that also the surface dipole within the FeO Moiré cell will be modulated by the varying Fe–O distance: the largest dipole is expected for Fe atoms bound to Pt hcp hollow sites and O on Pt on-top sites (largest Pt–O distance), while the smallest dipole is expected for Fe on-top and O fcc hollow arrangement. This model of the surface dipole modulations inside the Moiré cell is also confirmed through STM barrier-height and field emission resonance images of the bare FeO film [56].

Differential conductance $\mathrm{d}I/\mathrm{d}V$ scanning tunnelling spectra (STS) [74] reveal a fingerprint of Au adatoms and aggregates adsorbed on hcp domains, in the form of a resonance located at $V = +0.5\,\mathrm{V}$ bias for adatoms and a broader feature over $V = +0.75\,\mathrm{V}$ bias for NP (Fig. 14.13b). Consequently, conductance images acquired at $V = +0.55\,\mathrm{V}$ in the region of Fig. 14.13a outline an enhanced contrast of Au adatoms adsorbed on hcp regions versus a depressed apparent height of adatoms adsorbed, for example, in fcc sites on the Moiré cell (Fig. 14.13c), whereas differential conductance images acquired at $V = +0.35\,\mathrm{V}$ do not show any contrast difference between the adsorbed adatoms (Fig. 14.13d) (due to the very similar $\mathrm{d}I/\mathrm{d}V$ spectral features of all objects adsorbed at biases out of the resonance region). Moreover, small NPs are identified at higher biases, like in Fig. 14.14d.

When the nominal coverage reaches roughly 1 Au adatom per Moiré cell ($\theta = 0.012\,\mathrm{ml}$), STM images (Fig. 14.14a) show that adsorbate is self-organized

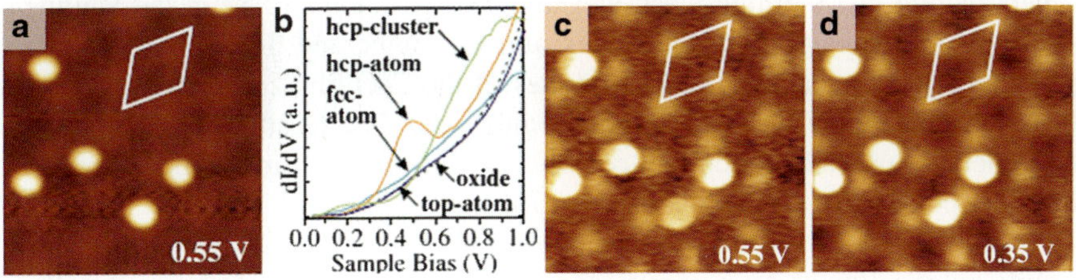

Fig. 14.13 (**a**) STM topographic image of Au adatoms on FeO/Pt(111) taken at 0.55 V and 0.1 nA. (**b**) Conductance spectra of bare FeO and Au adatoms bound to different domains of the Moiré cell. (**c**) Conductance images of the same area as in (**a**) taken at 0.55 V and (**d**) 0.35 V. All images are $130 \times 130\,\text{Å}^2$ in size. Adapted from Ref. [74] Copyright 2005 by the American Physical Society

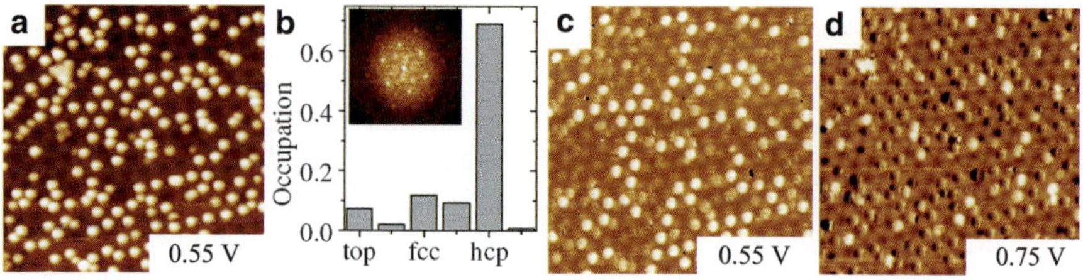

Fig. 14.14 (**a**) STM topographic image of Au adatoms on FeO/Pt(111) taken at 0.55 V and 0.1 nA. (**b**) Histogram of the occupation probability of different domains in the FeO Moiré cell determined from roughly 700 adatom positions. The inset shows a power spectrum of (**a**). (**c**) Conductance images of the same area as in (**a**) taken at 0.55 V and (**d**) 0.75 V. The current was set to 0.1 nA. All images are 430 × 430 Å in size. Adapted from Ref. [74]. Copyright 2005 by the American Physical Society

into a hexagonal network. The pitch of this pattern (0.26 nm) just corresponds to the period of the Moiré superstructure of the FeO film. The FFT transform of the STM image, shown in the inset frame of Fig. 14.14b, clearly outlines this long-range ordered pattern of adatoms, even up to the second nearest neighbours. The authors also performed a statistical analysis aimed at finding not only a long-range ordering of the adsorbate monomers and NP, but also the preferred adsorption sites along the Moiré cell of the FeO film.

The histogram of occupation frequency of adatoms versus cell sites (Fig. 14.14b) depicts a neat preference for the adsorption of Au atoms on hcp sites (70% probability), in contrast to the largely lower 10% or less value for all other kinds of adsorption sites. Quite unexpectedly, the differential conductance images shown in Fig. 14.14c, d that enhance the contrast of adatoms and aggregates, respectively, show a limited coalescence of Au adatoms into larger particles, even at the coverage corresponding to one adatom per Moiré cell. This apparent blocking of the Au aggregation on the FeO film is in sharp contrast to the nucleation behavior observed on nonpolar oxide surfaces, suggesting an effective repulsion between the adatoms [19, 75, 76].

The self-organization of Au adatoms preferentially adsorbed on hcp domains of the FeO film have been explained with the interplay of the following factors: (1) periodic modulations of the surface potential and dipole in the cell of the Moiré superstructure, (2) the high electronegativity of Au atoms (2.54 in the Pauling's scale) to be compared with the values of 1.83 for Fe and 3.44 for O [77]. The adsorption of Au adatoms on the hcp sites of the FeO film can partially heal the larger surface dipole on these specific regions and then locally lower the surface free energy. Such adsorption-induced stabilization mechanism has been observed for other polar oxide surfaces before [69, 78, 79].

Polarization mechanisms can be invoked to explain the adsorption: though a simple polarization of Au adatoms cannot explain the resonance observed at $V = +0.5$ V in the empty state region of the differential conductance STS curves. The last evidence is more in tune with a charge transfer to adatoms from the oxide film and hybridization of valence Au 6s electrons with the electronic states of FeO. This kind of hybridization is expected to generate a new bonding and antibonding state. An analogous antibonding state has already been mapped as a resonance above E_F in the STS spectra of Au adatoms and dimers adsorbed on the NiAl(110) substrate [80]. The electrostatic repulsion between charged Au atoms explains also the long-range ordering of Au monomers and the suppressed aggregation into NPs. The adatom interaction on the FeO film is then attributed to either dipole–dipole repulsion in the case that polarization interaction dominates the Au–FeO bond, or repulsion between negatively charged Au atoms assuming partial electron transfer upon adsorption. This mechanism is not active only on polar UT oxide films. A similar Coulomb repulsion between charged Au adatoms, upon electron tunnelling from the metal substrate through the UT oxide film has been predicted by DFT calculations [42,60] and experimentally observed [61] for Au atoms absorbed on a nonpolar 3-ml thick MgO(100)-like film grown on Ag(100) or Mo(100) substrates. Such a phenomenon is in any case peculiar to the UT oxide layers: charging of deposited neutral Au atoms is neither predicted nor observed on the bulk-terminated MgO(001) oxide surface, though a polarization of electronegative Au adatoms occurs [81].

The self-organization of the adsorbate driven by the Coulomb repulsion is already known for alkali atoms on metal [82] or semiconductor surfaces [83], where adsorption is accompanied by electron transfer from alkali atom to the substrate. As previously mentioned in Sect. 14.2, another driving force for charge transfer is the formation of heterojunctions at the interfaces, both between the oxide film and the metal substrate, and between the oxide film and adsorbed metal adatoms. The electric neutrality of the system has to be conserved, so the Fermi energies of the bare UT oxide-on-metal and metal adatoms or NP have to be aligned at the interface. The electrical contact implies several mechanisms like band bending, the rise of a Schottky-like junction at NP–oxide interface [1], possibly stimulating polarization and charge transfer from or to the adatoms. Also the surface work function of the metal–oxide–metal system will change due to the surface charge redistribution [60].

Berdunov and coworkers investigated also the nucleation at small coverage of Fe and V on a FeO/Pt(111) film [84]: the nominal coverage referred to a (110) plane of bcc Fe and V are respectively $\theta_{Fe} = 0.099$ ml and $\theta_V = 0.023$ ml. LEED patterns acquired upon deposition of this amount of Fe show that the structure of the FeO film is not disrupted (Fig. 14.15c). Fe NPs look quite size controlled and some kind of order can be seen in the NPs arrangement along the FeO film (inset of Fig. 14.15a), following the 2.5–2.6 nm Moiré superstructure. Though the temperature of the substrate during the deposition is not reported, the degree of the spatial order of Au NPs or monomers adsorbed on the FeO film is much sharper than for the Fe adsorbate, as one can deduce from the comparison of the FFT transform of the STM images in Figs. 14.14b and 14.15a)

In Fig. 14.15b the onset of V nucleation is shown. From STM simulations [52] the bright protrusions highlighted by hollow arrows are identified with the hcp sites on the Moiré cell of the FeO film. The unit cell of the superstructure of the film is also outlined (black rhomb). V clusters, pointed by the solid arrows, are pinned where the FeO layer forms fcc stacking sequence with the Pt surface, i.e., the darker regions between the hcp sites. From DFT simulations only a small difference, about

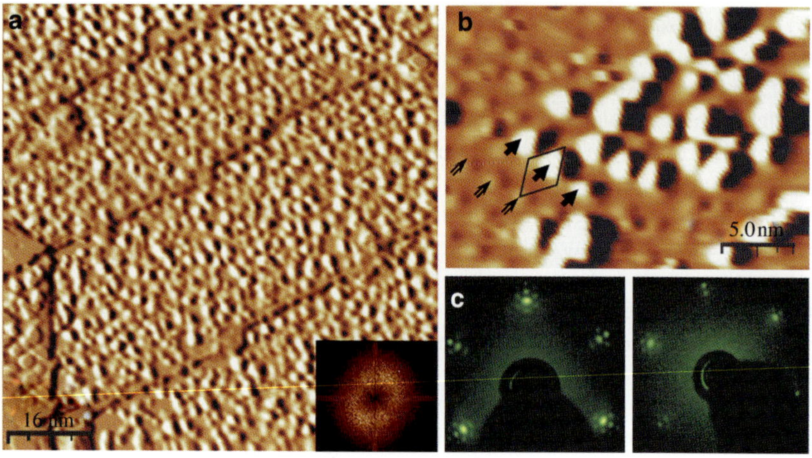

Fig. 14.15 (**a**) STM image of 0.2 Å Fe film grown on FeO/Pt(111) template. NPs of 1-ml thickness nucleate following the long-range surface superstructure of 2.6 nm periodicity. The image is presented in differentiated contrast. The 2D Fourier transformation image shows preferred NP's periodicity of 2.6 ± 0.6 nm. (**b**) Demonstration of "site selective" nucleation of vanadium NPs (*white spots*) after depositing a 0.05 Å vanadium film. The smaller corrugation seen on the image is the surface superstructure. Three *hollow arrows* point out the hcp sites of FeO/Pt(111) and the *solid arrows* indicate the vanadium NPs nucleated at fcc sites. Unit cell of FeO/Pt(111) superstructure is indicated. (**c**) LEED pattern of Au (*left*) and Fe (*right*) NP deposited on FeO/Pt(111). Reprinted from Ref. [85] Copyright 2006, with permission from Elsevier

0.1 eV, is predicted for the adsorption sites with hcp and fcc stacking, so the reason of the 'site selective' nucleation of V has to be traced back over the corrugations of the surface potential.

In the previous discussion the role of kinetics and oxygen affinity were not mentioned. The experiments of the adsorption of Au adatoms on FeO [74] were conducted with low temperature STM at 5 K, so the diffusion of the adsorbate is expected to be very marginal. Nilius et coworkers evaluated also the surface potential landscape felt by Au adatoms and found an adsorption energy on hcp sites of FeO of 10 ± 2 meV higher than elsewhere. In addition, they observed that the hexagonal arrangement of Au nano-objects and blocked coalescence of NPs persists up to 50 K. At room temperature, where the thermal energy is $k_B T = 25$ meV, so higher than the calculated diffusion barriers for Au adatoms on the FeO landscape, a very facile diffusion and nucleation of Au into larger NPs along the surface should be reasonably expected. This hypothesis is confirmed by STM studies of Lemire [85] that show the 3D Volmer–Weber growth mode of Au NPs on the FeO/Pt(111) film, in fine agreement with our experimental findings about the nucleation at room temperature of Au on the $TiO_x/Pt(111)$ films, especially the w' - TiO_x films, and there related to the low oxygen affinity of Au.

As a final consideration, we want to make a comparison with the results obtained on the z' - TiO_x UT polar film; in the previous discussion in Sect. 14.3.1, the charge state of Au and Fe NPs was not mentioned. Actually DFT calculations show a possible adsorption of an Au adatom on a Ti on-top site (0.8 eV); the effects are a local

Ti–O bond breaking, where the underlying Ti atom is lifted up to the oxygen plane and Au atom negatively charged [64]. Such an adsorption energy is however compatible with a finite Au mobility along the stripe at room temperature. Finally, even weaker adsorption on O on-top sites and flat diffusion barriers for Au do not allow to observe Au NPs nucleated on the stripes, at least at room temperature, and Au NPs are pinned only in the *troughs*. Moreover, DFT calculations show that Au NP nucleated in the *picoholes* are neutral [86]. The theoretical prediction is quite different for Fe adatoms, which maximize the coordination with the oxygen, with a preferential adsorption on the O-hollow coordinated sites along the titania stripes. The Ti on-top sites are on the contrary saddles in the diffusion barrier landscape and, due to oxidation, Fe adatoms and particles pinned on the stripes are positively charged [16].

14.4 Conclusions

In this work we have discussed some significant examples of directed self-assembling of transition metals on UT oxide layers grown on single-crystal metal surfaces. The ordered array of *picoholes* of the z'-$TiO_x/Pt(111)$ oxide layer phase is an effective template for low oxygen affinity metals, while for more reactive metals the strong interactions and high diffusion barriers on the oxygen landscape are limiting the template effectiveness. The presence of a double templating structure in the $Al_2O_3/NiAl(111)$ ("dot" and "network" structure) results in different behavior towards the arrangement of deposited metals, that is determined not only by the geometric nature of the oxide layer defects, but also by the metal/oxygen affinity, which is driving the surface diffusion of adatoms. In the last example discussed, where no ordered geometrical defects are present within the oxide layer mesh, the data show that this system is an effective template only at low temperatures, i.e., when surface diffusion is strongly limited, and the templating sites are dictated by electronic effects induced by the lattice mismatch with the supporting substrate. At room temperature, where the diffusion barriers are lowered, nucleation of large NPs in 3D growth mode is observed, indicating that diffusion kinetics has yet another primary role in the effectiveness of the templating substrates.

References

1. Q. Fu, T. Wagner, Surf. Sci. Rep. **62**, 431 (2007)
2. S.J. Tauster, S.C. Fung, J. Garten, J. Am. Chem. Soc. **100**, 170 (1978)
3. G.M. Schwab, *Electronics of Supported Catalysts, Advances in Catalysis*, vol. 27 (Academic Press, London, 1978) p. 1
4. G.L. Haller, D.E. Resasco, Adv. Catal. **36**, 173 (1989)
5. O. Dulub, W. Hebenstreit, U. Diebold, Phys. Rev. Lett. **84**, 3646 (2000)
6. G.L. Haller, J. Catal. **216**, 12 (1003)
7. C.G. Vayenas, S. Brosda, C. Pliangos, J. Catal. **216**, 487 (2003)

8. D.W. Goodman, Catal. Lett. **99**, 1 (2005)
9. Q. Fu, T. Wagner, S. Olliges, H.D. Carstanjen, J. Phys. Chem. B **109**, 944 (2005)
10. S.J. Tauster, S.C. Fung, R.T.K. Baker, J.A. Horsley, Science **211**, 1121 (1981)
11. R.A. Bennett, C.L. Pang, N. Perkins, R.D. Smith, P. Morrall, R.I. Kvon, M. Bowker, J. Phys. Chem. B **106**, 4688 (2002)
12. S. Bernal, J.J. Calvino, M.A. Cauqui, J.M. Gatica, C.L. Cartes, J.A.P. Omil, J.M. Pintado, Catal. Today **77**, 385 (2003)
13. H.J. Freund, G. Pacchioni, Chem. Soc. Rev. **37**, 2224 (2008)
14. C. Becker, K. Wandelt, Top Curr. Chem. **287**, 45 (2009)
15. H. Brune, Surf. Sci. Rep. **31**, 121 (1998)
16. E. Cavaliere, I. Kholmanov, L. Gavioli, F. Sedona, S. Agnoli, G. Granozzi, G. Barcaro, A. Fortunelli, Phys. Chem. Chem. Phys. **11**, 11305 (2009)
17. Q.H. Wu, Curr. Nanosci. **5**, 58 (2009)
18. R.M. Lambert, G. Pacchioni, *Chemisorption and Reactivity on Supported Clusters and Thin Films, NATO ASI Series, Series E: Applied Sciences*, vol. 331 (Kluwer, Dordrecht, 1997)
19. M. Baumer, H.J. Freund, Prog. Surf. Sci. **61**, 127 (1999)
20. S.A. Chambers, Surf. Sci. Rep. **39**, 105 (2000)
21. S. Surnev, M.G. Ramsey, F.P. Netzer, Prog. Surf. Sci. **73**, 117 (2003)
22. S. Schintke, W.D. Schneider, J. Phys. Condens. Matter **16**, R49 (2004)
23. M.S. Chen, D.W. Goodman, J. Phys. Condens. Mattter **20**, 264013 (2008)
24. H.J. Freund, Surf. Sci. **601**, 1438 (2007)
25. J. Goniakowski, F. Finocchi, C. Noguera, Rep. Prog. Phys. **71**, 016501 (2008)
26. Q.H. Wua, A. Fortunelli, G. Granozzi, Int. Rev. Phys. Chem. **28**, 517 (2009)
27. D.K. Fork, J.M. Phillips, R. Ramesh, R.W. Wolf, Mater. Res. Soc. Symp. Proc. **341**, (1994)
28. J.S. Speck, D.K. Fork, R.W. Wolf, T. Shiosaki, Mater. Res. Soc. Symp. Proc. **401**, (1995)
29. E-MRS Spring Conference Proceedings, in *Thin Solid Films*, vol. 400, ed. by G. Granozzi, H.J. Freund, S. Chambers (2001) p. 1
30. R. Franchy, Surf. Sci. Rep. **38**, 195 (2000)
31. J. Schoiswohl, S. Surnev, F.P. Netzer, Top Catal. **36**, 91 (2005)
32. V. Schoiswohl, M. Sock, Q. Chen, G. Thornton, G. Kresse, M.G. Ramsey, S. Surnev, F.P. Netzer, Top Catal. **46**, 137 (2007)
33. S. Surnev, L. Vitali, M.G. Ramsey, F.P. Netzer, G. Kresse, J. Hafner, Phys. Rev. B **61**, 13945 (2000)
34. J. Schoiswohl, G. Kresse, S. Surnev, M. Sock, M.G. Ramsey, F.P. Netzer, Phys. Rev. Lett. **92**, 206103 (2004)
35. F.P. Netzer, Surf. Rev. Lett. **9**, 1553 (2002)
36. J. Schoiswohl, M. Sock, S. Eck, S. Surnev, M.G. Ramsey, F.P. Netzer, G. Kresse, Phys. Rev. B **69**, 155403 (2004)
37. J. Schoiswohl, S. Surnev, M. Sock, S. Eck, M.G. Ramsey, F.P. Netzer, G. Kresse, Phys. Rev. B **71**, 165437 (2005)
38. F. Sedona, G.A. Rizzi, S. Agnoli, S.I. Llabre', F.X. Xamena, A. Papageorgiou, D. Ostermann, M. Sambi, P. Finetti, K. Schierbaum, G. Granozzi, J. Phys. Chem. B **109**, 24411 (2005)
39. G. Barcaro, S. Agnoli, F. Sedona, G.A. Rizzi, A. Fortunelli, G. Granozzi, J. Phys. Chem. C **113**, 5721 (2009)
40. F. Sedona, M. Eusebio, G.A. Rizzi, G. Granozzi, D. Ostermann, K. Schierbaum, Phys. Chem. Chem. Phys. **7**, 697 (2005)
41. S. Altieri, L.H. Tjeng, G.A. Sawatzky, Thin Solid Films **400**, 9 (2001)
42. G. Pacchioni, L. Giordano, M. Baistrocchi, Phys. Rev. Lett. **94**, 226104 (2005)
43. D. Ricci, A. Bongiorno, G. Pacchioni, U. Landman, Phys. Rev. Lett. **97**, 036106 (2006)
44. F. Sedona, G. Granozzi, G. Barcaro, A. Fortunelli, Phys. Rev. B **77**, 115417 (2008)
45. F. Sedona, S. Agnoli, G. Granozzi, J. Phys. Chem. B **110**, 15359 (2006)
46. P. Finetti, F. Sedona, G.A. Rizzi, U. Mick, F. Sutara, M. Svec, V. Matolin, K. Schierbaum, G. Granozzi, J. Phys. Chem. C **111**, 869 (2007)
47. U. Bardi, A. Atrei, G. Rovida, Surf. Sci. **239**, L511 (1990)

48. S. Degen, A. Krupski, M. Kralj, A. Langner, C. Becker, M. Sokolowski, K. Wandelt, Surf. Sci. **576**, L57 (2005)
49. S. Degen, C. Becker, K. Wandelt, Faraday Discuss. **125**, 343 (2004)
50. T. Maroutian, S. Degen, C. Becker, K. Wandelt, R. Berndt, Phys. Rev. B **68**, 155414 (2003)
51. M. Schmid, G. Kresse, A. Buchsbaum, E. Napetschnig, S. Gritschneder, M. Reichling, P. Varga, Phys. Rev. Lett. **99**, 96104 (2007)
52. H.C. Galloway, P. Sautet, M. Salmeron, Phys. Rev. B **54**, R11145 (1996)
53. Y.J. Kim, C. Westphal, R.X. Ynzunza, Z. Wang, H.C. Galloway, M. Salmeron, M.A. Van Hove, C.S. Fadley, Surf. Sci. **416**, 68 (1998)
54. M. Ritter, W. Ranke, W. Weiss, Phys. Rev. B **57**, 7240 (1998)
55. Y.J. Kim, C. Westphal, R.X. Ynzunza, H.C. Galloway, M. Salmeron, M.A. VanHove, C.S. Fadley, Phys. Rev. B **55**, R13448 (1997)
56. E. Rienks, N. Nilius, H.P. Rust, H.J. Freund, Phys. Rev. B **71**, 241404 (2005)
57. M. Sterrer, M. Yulikov, E. Fischbach, M. Heyde, H. Rust, G. Pacchioni, T. Risse, H.J. Freund, Angew. Chem. Int. Ed. **45**, 2630 (2006)
58. L. Giordano, D. Ricci, G. Pacchioni, P. Ugliengo, Surf. Sci. **584**, 225 (2005)
59. S. Wendt, E. Ozensoy, T. Wei, M. Frerichs, Y. Cai, M.S. Chen, D.W. Goodman, Phys. Rev. B **72**, 115409 (2005)
60. L. Giordano, G. Pacchioni, Phys. Chem. Chem. Phys. **8**, 3335 (2006)
61. M. Sterrer, T. Risse, M. Heyde, H.P. Rust, H.J. Freund, Phys. Rev. Lett. **98**, 206103 (2007)
62. F. Sedona, S. Agnoli, M. Fanetti, I. Kholmanov, E. Cavaliere, L. Gavioli, G. Granozzi, J. Phys. Chem. C **111**, 8024 (2007)
63. S. Ogura, K. Fukutani, M. Matsumoto, T. Okano, M. Okada, T. Kawamura, Phys. Rev. B **73**, 125442 (2006)
64. G. Barcaro, A. Fortunelli, G. Granozzi, Phys. Chem. Chem. Phys. **10**, 1876 (2008)
65. D. Repetto, T.Y. Lee, S. Rusponi, J. Honolka, K. Kuhnke, V. Sessi, U. Starke, H. Brune, P. Gambardella, C. Carbone, A. Enders, K. Kern, Phys. Rev. B **74**, 054408 (2006)
66. G. Barcaro, A. Fortunelli, J. Phys. Chem. A **113**, 14860 (2009)
67. G. Barcaro, A. Fortunelli, F. Nita, R. Ferrando, Phys. Rev. Lett. **95**, 246103 (2005)
68. A.L. Barabasi, H.E. Stanley, *Fractal Concepts in Surface Growth* (University of Cambridge, New York, NY, USA, 1995)
69. C. Becker, A. Rosenhahn, A. Wiltner, K. von Bergmann, J. Schneider, P. Pervan, M. Milun, M. Kralj, K. Wandelt, New J. Phys. **4**, 75 (2002)
70. C. Becker, K. von Bergmann, A. Rosenhahn, J. Schneider, K. Wandelt, Surf. Sci. **486**, L443 (2001)
71. A. Lehnert, A. Krupski, S. Degen, K. Franke, R. Decker, S. Rusponi, M. Kralj, C. Becker, H. Brune, K. Wandelt, Surf. Sci. **600**, 1804 (2006)
72. G. Hamm, C. Becker, C.R. Henry, Nanotechnology **17**, 1943 (2006)
73. A. Buchsbaum, M. De Santis, H.C.N. Tolentino, M. Schmid, P. Varga, Phys. Rev. B **81**, 115420 (2010)
74. N. Nilius, E.D.L. Rienks, H.P. Rust, H.J. Freund, Phys. Rev. Lett. **95**, 66101 (2005)
75. S.C. Street, C. Xu, D.W. Goodman, Ann. Rev. Phys. Chem. **48**, 43 (1997)
76. G. Haas, A. Menck, H. Brune, J.V. Barth, J.A. Venables, K. Kern, Phys. Rev. B. **61**, 11105 (2000)
77. *CRC Handbook of Chemistry and Physics*, 81st edn. (Chemical Rubber Company Press, Boca Raton, FL, 2001)
78. A. Barbier, C. Mocuta, G. Renaud, Phys. Rev. B **62**, 16056 (2000)
79. O. Dulub, U. Diebold, G. Kresse, Phys. Rev. Lett. **90**, 016102 (2003)
80. N. Nilius, T.M. Wallis, M. Persson, W. Ho, Phys. Rev. Lett. **90**, 196103 (2003)
81. M. Yulikov, M. Sterrer, M. Heyde, H.P. Rust, T. Risse, H.J. Freund, G. Pacchioni, A. Scagnelli, Phys. Rev. Lett. **96**, 146804 (2006)
82. J. Kliewer, R. Berndt, Surf. Sci. **477**, 250 (2001)
83. L. Gavioli, M.G. Betti, V. Corradini, M. Sancrotti, Phys. Rev. B **70**, 125317 (2004)
84. N. Berdunov, G. Mariotto, K. Balakrishnan, S. Murphy, IV Shvets, Surf. Sci. **600**, L287 (2006)
85. C. Lemire, R. Meyer, S.h.K. Shaikhutdinov, H.J. Freund, Surf. Sci. **552**, 27 (2004)
86. G.A. Rizzi, F. Sedona, L. Artiglia, S. Agnoli, G. Barcaro, A. Fortunelli, E. Cavaliere, L. Gavioli, G. Granozzi, Phys. Chem. Chem. Phys. **11**, 2177 (2009)

Chapter 15
Mechanical and Electrical Properties of Alkanethiol Self-Assembled Monolayers: A Conducting-Probe Atomic Force Microscopy Study

Frank W. DelRio and Robert F. Cook

Abstract The structure–property relationships for methyl-terminated alkanethiol [$CH_3(CH_2)_{n-1}SH$, where n is the number of carbons in the molecular chain] self-assembled monolayers (SAMs) on gold substrates are considered, with a particular emphasis on using conducting-probe atomic force microscopy (CPAFM) to assess the mechanical and electrical properties and near-edge X-ray absorption fine structure (NEXAFS) spectroscopy to determine the molecular structure. CPAFM measurements are analyzed with an elastic adhesive contact model, modified by a first-order elastic perturbation method to account for substrate effects, and a parabolic tunneling model, appropriate for a metal–insulator–metal contact in which the insulator is extremely thin. NEXAFS carbon K-edge spectra are used to compute the dichroic ratio for each film, which provides a quantitative measure of the molecular structure. The combination of the two measurement methods provides clear property trends for this system over a wide range of n and a general methodology for the optimization of SAMs for micro- and nano-electromechanical systems, magnetic storage devices, and other applications.

15.1 Introduction

Self-assembly of surfactant molecules on solid surfaces to form organic monolayer films has proven to be a convenient method of tailoring the physical and chemical properties of metal, oxide, and semiconductor surfaces. A critical factor in the formation of such self-assembled monolayers (SAMs) is that the molecules have one end group that interacts with the surface in a specific and binding manner such that the orientation of the molecules on the surface, at least from an end group perspective, is identical. If molecular density on the surface is great enough, steric and chemical interactions between the usually long-chain molecules can lead to two-dimensional organization of the molecules into stable, ordered structures. The stability of these, often crystalline, structures, leads to reproducibility in SAM properties and thus the ability to use SAMs in surface engineering. The formation and structure of SAMs has been reviewed by Ulman [1] and the kinetics of SAM formation by

Schwartz [2]. In both reviews, attention is focused on two common SAM systems, alkanethiol molecules on gold surfaces and alkylsilane molecules on oxide surfaces.

Thiol-terminated molecules have shown potential for myriad applications when formed into SAMs on metal surfaces: protective layers to slow oxidation (of copper) [3] or control the adsorption of proteins (on gold and silver) [4–6]; photoresistive layers for lithography (of copper and gold) [7, 8]; receptor sites for DNA detection (using gold and silver surfaces) [9]; electrically switchable layers to control wettability [10]; charge injection barriers in organic field-effect transistors [11]; and molecular boundary lubricants to control adhesion [12]. This last example outlined the many potential benefits of SAMs in microelectromechanical systems (MEMS) and magnetic storage devices. In such devices, contacting surfaces in relative motion are integral to device operation and SAMs offer great potential for reductions in adhesion and enhancement in lubrication [13, 14].

The lubrication study also highlighted an aspect of SAM behavior that is critical in the selection of a particular SAM for a given application – the coupling of one SAM property to another. In the study of Patton et al. [12], which focused on the behavior of a MEMS switch, the mechanical and electrical properties were shown to be related, and both changed as a function of the number of switch contact cycles. Underlying the mechanical–electrical relationship are, of course, the separate relationships between both properties and the SAM structure. However, the mechanical properties of the SAM determine the change in the SAM structure on loading, in this case during contact. As the SAM electrical properties will change with changes in structure, a mechanical–electrical coupling will result. This coupling is the focus of this chapter, in particular for contacts that fall in the "middle ground" between the micro-scale, millinewton-load, metal–metal contacts of MEMS devices [15–17], and the pico-scale, sub-nanonewton-load atom–atom contacts of atomic-break junctions [18, 19]. These mid-range, nano-scale, multi-nanonewton contacts are expected to be predominant in nanoelectromechanical systems (NEMS), and, to have effective performance and reliability, will need to be surface engineered by SAMs to achieve the required mechanical and electrical behavior. Measurement of mechanical and electrical properties is thus critical to SAM selection.

The coupling of mechanical and electrical properties of SAMs requires the simultaneous measurement of mechanical and electrical behavior during contact, under load. In this chapter, a conducting-probe atomic force microscopy (CPAFM) approach is demonstrated for simultaneous measurement of SAM mechanical and electrical properties during nano-scale contact loading. The methodology has broad applicability for the optimization of SAMs for MEMS, NEMS, magnetic storage devices, and other applications, and is demonstrated here on a set of methyl-terminated alkanethiols [$CH_3(CH_2)_{n-1}SH$, denoted by C_n, where n is the number of carbons in the molecular chain] on gold surfaces, and structure–property relations for this system are developed over a range of n. Section 15.2 provides a brief background to the structure of alkanethiol SAMs and presents synchrotron X-ray measurements of the structure of the SAMs studied here. Section 15.3 describes CPAFM and presents measurements of SAM mechanical and electrical behavior. Section 15.4 provides the mechanical and electrical theoretical framework required to analyze the experimental measurements. Such analyses are provided in Sect. 15.5 and 15.6 for the mechanical and electrical measurements, respectively. Section 15.7 concludes the chapter and suggests some future research directions.

15.2 Order, Orientation, and Surface Coverage

One of the first studies on the ordered structure of an alkanethiol SAM on a metal surface was by Strong and Whitesides, who used electron diffraction to determine that long-chain alkanethiols ($n = 22$) formed a hexagonal c(7×7) lattice incommensurate with the Au(111) lattice [20]. In reexamining the diffraction patterns, however, Chidsey and Loiacono concluded that the long-chain alkanethiols ($10 \leq n \leq 22$) actually adopt the commensurate $(\sqrt{3}\times\sqrt{3})R30°$ lattice as shown in Fig. 15.1a, which places molecules at the next-nearest-neighbor sites on the Au(111) surface and is a 30° rotation of the incommensurate c(7×7) lattice [21]. In later studies, researchers further refined the model for molecular packing and demonstrated the existence of a larger unit cell. Nuzzo et al. observed band splitting in the infrared (IR) spectra for $n = 22$ at low temperatures, which suggested that the lowest energy structure consisted of a unit cell with *two* chains [22]. Molecular dynamics simulation results also pointed to a unit cell with two chains, with molecules tilted along the equivalent nearest-neighbor direction [23]. While the two-chain model was sufficient to explain the results from these two studies, it could not account for the $3 \times 2\sqrt{3}$ periodicity observed in later helium diffraction [24], grazing incidence X-ray diffraction [25], and scanning tunneling microscopy (STM) [26] measurements. The $3 \times 2\sqrt{3}$ primitive unit cell contains *four* hydrocarbon chains and can also be described as a c(4×2) superlattice of the $(\sqrt{3} \times \sqrt{3})R30°$ hexagonal lattice as demonstrated in Fig. 15.1b. The height corrugations and lattice spacings in the Au nearest-neighbor and next-nearest-neighbor directions are demonstrated in Fig. 15.1c. The larger unit cell has been attributed to both (a) a patterned arrangement of the all-*trans* hydrocarbon chains, as shown in Fig. 15.1a [22–24, 26] and (b) a disulfide head group structure, bought about by a *gauche* defect at the S–C bond [27]. Furthermore, IR spectroscopy and optical ellipsometry data indicate that long-chain alkanethiols form densely packed, crystalline-like films with alkyl chains tilted 20–30° from the surface normal [28, 29], as shown in Fig. 15.2a. The SAMs are thermally stable up to 325 K, but undergo structural changes from oxidation and desorption of the alkanethiol molecules above 350 K [30–32].

There are a number of factors that can affect the structure of SAM films, including the substrate type [33, 34] and crystallographic orientation [20, 35], temperature [30–32], the type of solvent from which the SAM molecules were precipitated [34, 36], ripening time for the formation of the SAM ordered structure [37, 38], and chain length [26, 28, 29, 35]. Here, we focus on the effects of chain length. Electron diffraction measurements have shown that long-chain ($n \geq 9$) alkanethiols form densely packed, crystalline-like monolayers with $(\sqrt{3} \times \sqrt{3})R30°$ molecular order [20, 35], while short-chain alkanethiols result in loosely packed, liquid-like films

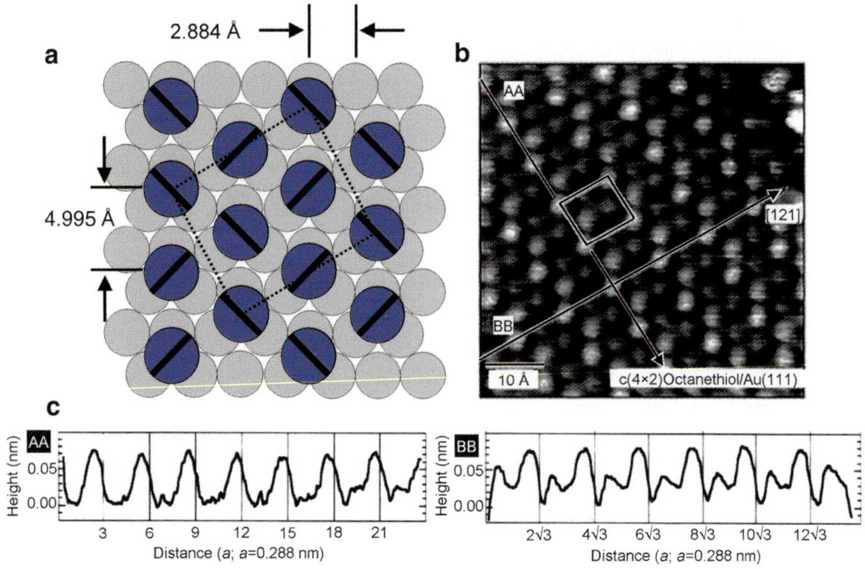

Fig. 15.1 (**a**) Schematic representation of an alkanethiol SAM on Au(111). The *light gray circles* are Au atoms and the *blue circles* are thiol groups. The *diagonal line* in each *blue circle* represents the projection of the plane of the all-*trans* hydrocarbon backbone. The $3 \times 2\sqrt{3}$ primitive unit cell, which can also be described as a $c(4 \times 2)$ superlattice of the $(\sqrt{3} \times \sqrt{3})R30°$ hexagonal lattice, is outlined. (**b**) STM image of octanethiol ($n = 8$) on Au(111). As in (**a**), the $3 \times 2\sqrt{3}$ primitive unit cell is marked. (**c**) Cross-sections of (**b**). *Section AA* shows the height corrugations and repeat distances along the Au nearest-neighbor direction. *Section BB* illustrates the same information in the Au next-nearest-neighbor direction (reprinted with permission from Poirier and Tarlov [26]. Copyright 1994 American Chemical Society)

with $(m\sqrt{3} \times \sqrt{3})R30°$ surface structure (where m varies from 1 to 6 depending on n) [35]. Also, as n decreases, the orientation and surface coverage change as a result of *gauche* defects and thermal desorption as shown schematically in the transition from Fig. 15.2a to Fig. 15.2b. Porter et al. utilized IR spectroscopy and optical ellipsometry to show that the structure becomes increasingly disordered with smaller packing density and surface coverage as n decreases; the trend changes in the film thickness data and the CH_2 and CH_3 peak intensities at $n \approx 9$ suggest that the short-chain molecules adopt a more tilted conformation than the longer chains [28]. Bain et al. later noted a similar trend in contact angle data; the increased disorder in the short-chain films allowed water to interact with the underlying gold surface and more of the CH_2 groups in the alkanethiol chains [29].

These structural changes have also been demonstrated using near edge X-ray absorption fine structure (NEXAFS) spectroscopy, a synchrotron technique that enables the identification of chemical bonds and a determination of their relative number density within a sample [39, 40]. NEXAFS partial electron yield (PEY) spectra at the carbon K-edge for $n = 5, 8, 12$, and 18 at $\theta = 20°$, 44°, and 70° are shown in Fig. 15.2c, where θ is the angle between the sample surface and the photon

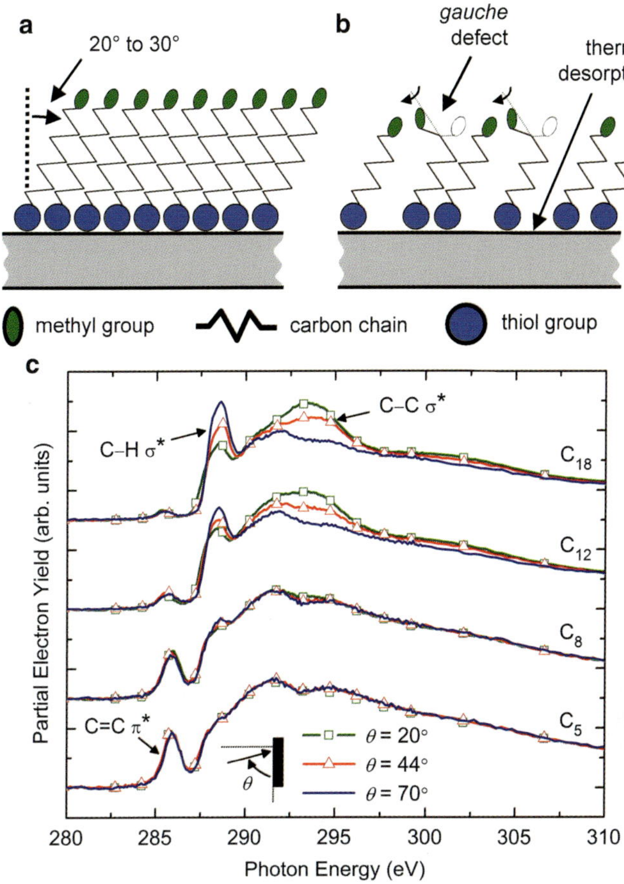

Fig. 15.2 Schematic representation of a (**a**) long-chain and (**b**) short-chain alkanethiol. (**c**) NEXAFS carbon K-edge PEY spectra for $n = 5$, 8, 12, and 18 at various θ (offset for clarity). For $n = 5$ and 8, the spectra are identical. However, $n = 12$ and 18 reveal a strong angular dependence, with C–H σ^* resonance decreasing and C–C σ^* resonance increasing as θ decreases (reprinted with permission from DelRio et al. [40]. Copyright 2010 American Chemical Society)

beam. All spectra exhibit the same characteristic hydrocarbon resonance peaks: the C=C π^* peak at 285.5 eV, the C–H σ^* peak at 288.6 eV, and the C–C σ^* peak at 293.6 eV. For $n = 5$ and 8, the spectra are nearly identical. However, $n = 12$ and 18 reveal a strong angular dependence, with C–H σ^* resonance decreasing and C–C σ^* resonance increasing as θ decreases. Also, as n decreases, the intensity of the C=C π^* resonance peak increases, likely the result of small surface coverage and the subsequent deposition of adventitious hydrocarbons on the Au surface for the short-chain thiols. The peak intensities were quantified by fitting Gaussian peaks to the spectra and integrating the peak areas [41]. The variations in peak intensity with θ are related to the molecular orientation through the dichroic ratio $R_{\mathrm{I}} = (I_{90^\circ} - I_{0^\circ})/(I_{90^\circ} + I_{0^\circ})$, where I_{90° is the peak intensity at $\theta = 90^\circ$ and

$I_{0°}$ is the extrapolated peak intensity at $\theta = 0°$. R_I can vary from -1.0 to $+0.75$, with a more positive value for the C–H σ^* peak corresponding to greater surface normality ($R_I = 0$ may also represent a random distribution of chain orientations) [42]. For $n = 18$, $R_I = 0.50$, indicative of a densely packed, crystalline-like film. R_I decreases to 0.26 for $n = 12$, 0.02 for $n = 8$, and 0.04 for $n = 5$; by definition, this indicates an increase in *average* tilt angle and change in phase from crystalline to amorphous.

Several studies have shown that the exact details regarding the mechanical and electrical behavior of the films are dependent on the molecular structure (e.g., densely packed, crystalline-like vs. loosely packed, liquid-like). As an example, Lio et al. studied the frictional properties of alkanethiol and alkylsilane SAMs as a function of n [43]. For $n > 11$, the thiols and silanes displayed similar frictional behavior, independent of chain length and head group linkage, due to the densely packed, invariant structure of the films. For $n < 11$, however, the frictional trends were dependent on the chain length and head group, with larger values reported for the silanes and decreasing n, presumably from increased disorder from numerous defects and energy-dissipating modes. Holmlin et al. examined the charge transport properties of alkanethiols and oligophenylene thiols as a function of n [44]. At a voltage of 0.5 V, the decay coefficients were $0.87\,\text{Å}^{-1}$ for the alkanethiols and ranged from 0.61 to $0.67\,\text{Å}^{-1}$ for the oligophenylene thiols. In addition, the height of the tunneling barrier varied between 2.0 and 2.9 eV for the alkanethiols and ranged from 1.7 to 2.2 eV for the oligophenylene thiols. From these results, the authors concluded that charge transport occurred via "superexchange" tunneling, or tunneling from interactions between donor and acceptor electrons in the metal electrodes and unoccupied orbitals in the molecular film.

A summary of reported mechanical [45–57] and electrical [44,58–72] properties of alkanethiol SAMs is shown in Tables 15.1 and 15.2, respectively. From this summary, it is clear that the mechanical and electrical properties are not solely dependent on the SAM structure, but also on the specifics of the device fabrication or measurement technique, as evident by the wide range of results for the work of adhesion w, Young's modulus E_{film}, tunneling decay coefficient β, and the tunneling barrier height $\phi_0 - E_F$ for similar values of n. Consequently, it is necessary to use innovative methods to measure *simultaneously* the coupled mechanical and electrical properties of molecular films. Such techniques are critical as the application of a contact load can alter the layer structure (e.g., by introducing *gauche* defects) and therefore changes both the mechanical and electrical properties together.

15.3 Conducting-Probe Atomic Force Microscopy

CPAFM is well suited to make simultaneous mechanical and electrical measurements, as it enables the mechanical load between a probe and a surface to be controlled, while sweeping the voltage applied to the surface, allowing both force–displacement and current–voltage data to be collected simultaneously. The contact

Table 15.1 Summary of reported mechanical properties of alkanethiol SAMs

n	$w(\mathrm{mJ\,m^{-2}})$	E_{film} (GPa)	Technique	Reference
18		0.18, 0.60	AFM	[45]
18		36	AFM	[46]
16	61.4		AFM	[47]
10, 18		1.07, 1.02	AFM	[48]
10, 16		0.003–0.06	Liquid AFM	[49]
12, 22		2.0	CPAFM	[50]
6, 8, 10, 12		54, 52, 48, 75	CPAFM	[51]
9, 10, 11	60		IFM	[52]
12	50	8	IFM	[53]
16	37	5	IFM	[54]
18	122.5	1.5	IFM	[55]
10, 18, 28		14.7, 15.0, 15.5	Theory	[56]
8, 15	119.7	20	Theory	[57]

n is the chain length, w is the work of adhesion, and E_{film} is the Young's modulus of the film
AFM atomic force microscopy, *CPAFM* conducting-probe atomic force microscopy, *IFM* interfacial force microscopy

Table 15.2 Summary of reported electrical properties of alkanethiol SAMs

n	β ($\mathrm{Å^{-1}}$)	$\phi_0 - E_F$ (eV)	Technique	Reference
7, 8, 9, 10, 15	0.85		Electrochemical	[58]
10, 12, 14, 16, 18	0.91		Electrochemical	[59]
8, 10, 12, 14, 16	0.87	2.7, 2.0, 2.3, 2.9, 2.4	Hg junction	[44]
8, 12, 16	0.83	1.83, 1.42, 1.40	Nanopore	[60, 61]
10		2.35	STM	[62]
10, 12	1.2		STM	[63]
16	1.37	1.8	Tuning-fork AFM	[64]
10	1.1	2.2	CPAFM	[65]
8, 10	0.80	2.3	CPAFM	[66]
6, 8, 10	0.88	1.6, 2.1, 2.2	CPAFM	[67]
8, 10, 12, 16	0.98	1.13	CPAFM	[68]
16	0.57	2.05	CPAFM	[69]
6, 8, 10, 12		1.24, 1.28, 1.27, 1.21[a]	CPAFM	[70, 71]
12, 16, 18		1.20, 1.15, 1.26[a]	Crossed wire	[70, 71]
4, 5, 6, 7, 8	0.76	1.3, 3.4	Theory	[72]

n is the chain length, β is the decay coefficient, and $\phi_0 - E_F$ is the barrier height
STM scanning tunneling microscopy, *AFM* atomic force microscopy, *CPAFM* conducting-probe atomic force microscopy
[a] Values represent the transition voltage V_{trans} from direct tunneling to field emission, which are related, but not equivalent, to the barrier height $\phi_0 - E_F$

area of the resulting metal–molecule–metal junctions is small, typically between 20 and 200 nm^2, allowing the deformation and charge transport to take place over a small number of molecules, about 100 for a 50-nm tip [65]. Moreover, CPAFM alleviates the need for special micro- or nano-fabrication processes that are sometimes used to create solid-state metal–molecule–metal heterostructures

[73–75]. In creating these structures, the second electrode is deposited onto the thin molecular film, often at elevated temperatures, which may produce structural defects in the film or bring about penetration of metal atoms into the film. In addition, CPAFM provides a direct measure of the transport properties as a function of applied load, as opposed to STM, where the position of the probe and the conductivity of the film are tied together [62, 63, 76, 77]. In other words, the STM feedback mechanism is junction current, and as the transport properties of the film are unknown, the position of the probe relative to the film is uncertain. As a result, the measured properties are determined by the molecules *and* the vacuum gap if the tip is not in contact or represent only a portion of the molecules when the tip is in contact [65]. The CPAFM avoids this problem by utilizing force as the feedback signal, which allows the position of the probe to be controlled without a priori knowledge of the film properties.

Figure 15.3 shows a schematic representation of a typical CPAFM experimental setup. As with traditional atomic force microscopy (AFM), CPAFM uses a cantilever with a sharp probe to examine a sample surface. At one end, the cantilever beam is attached to a piezoelectric displacement actuator operated by the CPAFM controller. At the other end, the cantilever probe tip interacts with the surface. Displacement d of the cantilever towards the sample surface leads to unstable "snap-on" of the tip to the surface on approach followed by compressive contact to a peak load. Retraction of the cantilever leads to an *almost* reversible decrease of the compressive load F, followed by a stable, increasing tensile load as the tip adheres to the surface, and finally unstable "pull-off" of the tip from the surface. In response to the attractive and repulsive forces encountered during approach and retraction, the cantilever deflects, and this deflection is measured using a laser that

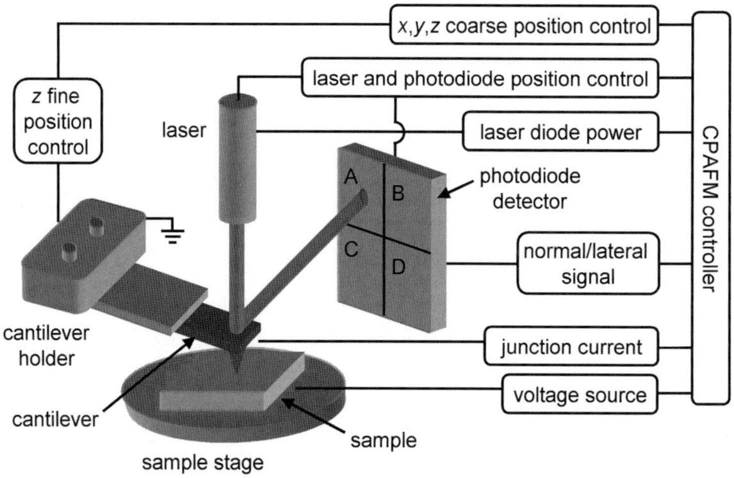

Fig. 15.3 Schematic representation of the conducting-probe atomic force microscopy experimental setup

is reflected from a mirrored surface on the back side of the cantilever onto a split photodiode. The differential output V_{diff} of the split photodiode (i.e., the difference in intensity between the upper and lower sections) is recorded as a function of d. To convert the resulting V_{diff}–d trace to F–d data, it is necessary to first consider the compressive region of the V_{diff}–d trace; V_{diff} is a direct measure of the cantilever deflection in compressive contact because the deformation of the surface is small relative to d. Thus, the slope of the V_{diff}–d trace in this region provides a ratio of the cantilever deflection to V_{diff}, which is known as the inverse optical lever sensitivity S (in units of nm V^{-1}). F can then be found by $F = kSV_{\text{diff}}$, where k is the cantilever spring constant in the normal direction. Usually, k is found by the thermal fluctuation method [78], a nondestructive method based on a cantilever's response to thermal noise (e.g., thermal fluctuations are $\approx 3\,\text{Å}$ for a cantilever spring constant of $0.05\,\text{N m}^{-1}$). Other methods to measure k are based on added mass [79], the unloaded resonant frequency, density, and dimensions of the cantilever [80] or a reference cantilever array [81].

Unlike conventional AFM, however, CPAFM utilizes a conductive probe, which enables current I to flow from the substrate, through the molecular film, and to the probe when a voltage V is applied. The conductive probe tips are formed by either micromachining monolithic metal probes [82] or coating standard rectangular or triangular Si or Si_3N_4 probes with Pt, Ti–Pt, Cr–Au, Co–Cr, and other conductive thin films. In the latter approach, the underlying layer is typically used to enhance the adhesion, hardness, or electromigration properties of the outer film. The film thicknesses are tens of nanometers, which results in probe radii of tens of nanometers. In this study, for example, rectangular Si cantilevers with 60 nm of Co and 20 nm of Cr with an average tip radius of $\approx$20 nm were used. The sample voltage was swept from -2 to $+2$ V and the current recorded at the probe through a current preamplifier. The upper bound on the current measurement was 100 nA, set by the saturation of the current preamplifier.

The CPAFM F–d data for $n = 12$ are shown in Fig. 15.4a. The F–d data include both the extension (loading) of the cantilever towards the sample surface to peak load and the retraction (unloading) of the cantilever away from the sample surface. At various points in the extension and retraction cycle, cantilever displacement was halted and I–V measurements performed as shown in Fig. 15.4b. The hysteresis in the F–d data is most likely from the "reverse-path" effect, which is due to hysteresis and creep in the piezoactuator used to control the displacement of the CPAFM cantilever [83]. This nonideal behavior causes the retraction curve to pass over the extension curve. In other words, at a given d, F is greater on unloading than on loading. This phenomenon, however, would not explain the hysteresis in the I–V data, because the curves were taken at the same values of F (not d) on loading and unloading. Other mechanisms that would explain the I–V hysteresis include adhesion hysteresis, viscoelastic effects, and plastic deformation. A change in the adhesion energy during loading and unloading might be responsible for the observed hysteresis. Schneider et al. noted hysteresis in the adhesion energy of glycine amphiphile bilayers by surface-force apparatus experiments [84]. The hysteresis was attributed to a spatially varying adhesion gradient in the contact

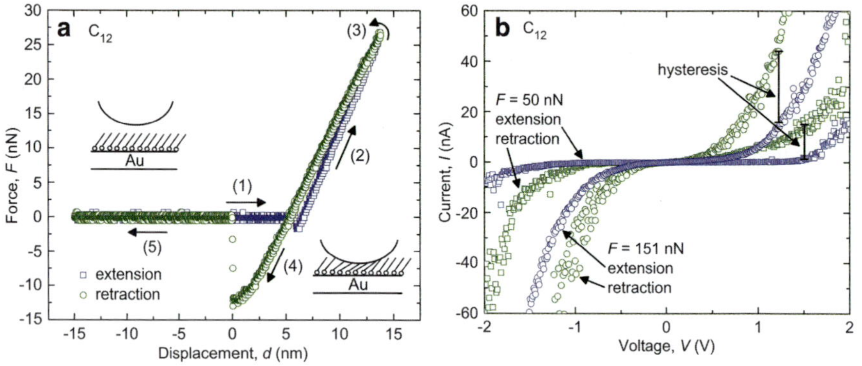

Fig. 15.4 (**a**) F–d experimental data for $n = 12$. The F–d data includes both the extension of the cantilever towards the sample surface (1, 2) to a peak load (3) and the retraction of the cantilever away from the sample surface (4, 5). The schematic diagrams illustrate the deformation state of the monolayer at various stages. (**b**) I–V traces during extension and retraction of the cantilever at $F = 50$ nN and $F = 151$ nN (reprinted with permission from DelRio et al. [40]. Copyright 2010 American Chemical Society)

zone due to pH-dependent surface rearrangement from the snap-on of the tip to the surface. At high pH, both the adhesion and adhesion hysteresis disappeared, which provided further evidence for the spatially varying adhesion gradient model and ensured the deformation was purely elastic. Similar experiments with alkanethiol SAMs, however, tell a different story [51]. In the absence of adhesion, which was achieved via measurements in solution, the hysteresis was still observed and therefore not due to adhesive effects. Another possible mechanism that might explain the hysteresis involves viscoelastic behavior of the alkanethiol SAMs. As the name suggests, a viscoelastic material exhibits both elastic and viscous properties under an applied load; it is the viscous component that gives rise to a hysteresis due to energy dissipation in the material. The amount of hysteresis (i.e., the size of the hysteresis loop) is dependent on the ratio of the relaxation time constant to the loading time [85]. For loading rates in which the data acquisition is on the order of the relaxation time constant, the hysteresis loop would be large. Joyce et al. used an interfacial force microscope to examine the mechanical relaxation of a long-chain ($n = 16$) alkanethiol on gold [85], and in doing so, extracted an effective relaxation time constant of 0.08 s. In this study, the cantilever displacement was halted for several seconds as the I–V measurements were performed, meaning that the alkanethiol SAMs had sufficient time to respond to the applied load. However, it is well known that polymers can exhibit relaxation on multiple time scales, and as a result, it is not possible to completely rule out viscoelasticity as a possible source of the observed hysteresis. Yet another possible reason for the hysteresis is plastic deformation in the alkanethiol SAMs. In this case, the monolayer exhibits both elastic and plastic deformation during loading, but only recovers elastic displacements on unloading, resulting in larger conductance ($G = I/V$) values on unloading. Some studies provide evidence for permanent deformation on unloading [51], while others illustrate that sharp tips can in fact *reversibly* displace alkanethiols [86]. In light of these arguments and to simplify the subsequent analysis, we only consider the unloading portion of each F–d curve, as it can be analyzed by elastic contact theory without regard to irreversibility and hysteresis.

15.4 Theoretical Framework

15.4.1 Elastic Adhesive Contact

Johnson, Kendall, and Roberts (JKR) developed a model to predict the deformed shape of two contacting spheres (radii R_1 and R_2) that interact via attractive surface forces based on a balance between elastic energy and surface energy [87]. To simplify the analysis, it is convenient to consider the case of a rigid sphere in contact with an elastic half-space. Under an applied load F, the rigid sphere deforms the flat surface a distance δ to form a contact over a circle of radius a, given by

$$F = \frac{4E^*a^3}{3R} - 2\sqrt{2\pi E^* w a^3} \tag{15.1}$$

and

$$\delta = \frac{a^2}{R} - \sqrt{\frac{2\pi a w}{E^*}}, \tag{15.2}$$

where w is the work of adhesion, E^* is the reduced modulus of the contact, and R is the radius of the rigid sphere. Here, the reduced modulus is the plane strain modulus of the linear-elastic, isotropic half-space, $E^* = E/(1-\nu^2)$, where E and ν are the Young's modulus and Poisson's ratio, respectively. From (15.1) and (15.2), we see that the contact area $A = \pi a^2$ is larger than that predicted by Hertzian theory [88], and remains finite until a critical pull-off force is reached, at which point the surfaces separate abruptly. In a load-controlled system, such as a compliant CPAFM cantilever, the pull-off force is reached at $-3\pi Rw/2$. At pull-off, the sphere does not separate from the flat surface at the same position in which it came into contact. In fact, the contact is elongated to a pull-off distance of $(3\pi^2 w^2 R/64E^{*2})^{1/3}$.

Derjaguin, Muller, and Toporov (DMT) used a completely different approach to consider the molecular forces in an annular zone around the area of contact [89,90]. Unlike the JKR model, the profile of the contact is assumed to be Hertzian. Originally, using a "thermodynamic" approach, Derjaguin et al. found that at initial contact, the attractive force between the sphere and flat surface is $2\pi Rw$, but as the deformation increases, this force decreases rapidly to πRw [89]. However, as the contact deforms, the adhesive force should become larger, not smaller, than $2\pi Rw$. Muller et al. eventually developed a model consistent with this idea using a more correct "force" method [90]. In the revised model, the attractive force is again $2\pi Rw$

at contact, but this time increases as the sphere and flat are pressed closer together according to

$$F = \frac{4E^*a^3}{3R} - 2\pi Rw \tag{15.3}$$

and

$$\delta = \frac{a^2}{R}. \tag{15.4}$$

At $\delta = 0$, the sphere and flat surface separate at the initial point of contact, and (15.3) reduces to the Bradley solution for a rigid sphere in contact with a rigid flat surface [91].

The JKR and DMT models seem to contradict each other: the pull-off force in the former case is $-3\pi Rw/2$, while the pull-off force in the latter case is $-2\pi Rw$. Tabor compared the two models and briefly outlined some of the problems associated with each of the solutions [92]. In the JKR theory, the adhesive forces outside the area of contact are neglected. These forces, however, must be taken into account when the height h of the "neck" around the contact, which is given by $(Rw^2/E^{*2})^{1/3}$, approaches the interatomic separation z_0. The DMT model includes the adhesive forces along the periphery, but assumes the contact profile is Hertzian in nature (i.e., the surface forces are small enough that their effect on the deformation can be neglected). Muller et al. performed a numerical calculation based on the Lennard–Jones potential and determined the models actually represent the extreme ends of a spectrum defined by the parameter [93]

$$\mu = \left(\frac{Rw^2}{E^{*2}z_0^3}\right)^{\frac{1}{3}}. \tag{15.5}$$

which represents the ratio of elastic deformation to the effective range of surface forces. Thus, the JKR theory is suitable for compliant materials with large, short-range surface forces ($\mu \gg 1$), while the DMT theory is appropriate for stiff contacts with small, long-range attractive forces ($\mu \ll 1$). The F–δ relationships for the JKR and DMT contact models are shown in Fig. 15.5. As $w \to 0$, both models converge to the Hertz theory.

15.4.2 *Effective Elastic Modulus of a Film–Substrate System*

As previously mentioned, for a rigid sphere in contact with a linear-elastic, isotropic flat surface, $E^* = E/(1-\nu^2)$. On the other hand, when the flat surface is a thin film of thickness t on a substrate, as shown in Fig. 15.6, E^* is a function of the Young's moduli and Poisson's ratios of the film (E_{film}, ν_{film}) and substrate (E_{sub}, ν_{sub}). Therefore, to find the "true" film properties, it becomes necessary to either limit δ to avoid substrate effects or fit the experimental data to a film–substrate model. In the former approach, a common rule of thumb to measure *hardness* is to limit the deformation

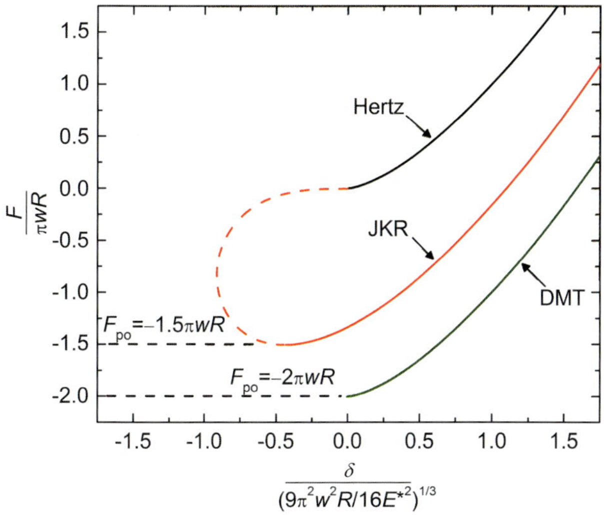

Fig. 15.5 Normalized F–δ curves for the JKR and DMT elastic adhesive contact models

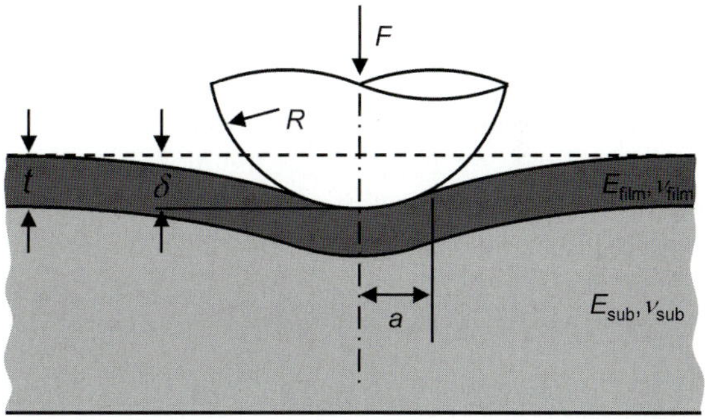

Fig. 15.6 Schematic representation of a rigid sphere in contact with a coated elastic half space. (reprinted with permission from DelRio et al. [40]. Copyright 2010 American Chemical Society)

to $\delta = 0.10t$, because beyond this depth, the plastic zone beneath the interface reaches the substrate and affects the measured film properties [94]. This is not a comprehensive law, however, as the size of the plastic zone at a given δ depends on the film/substrate hardness ratio and the indenter geometry [95, 96]. In determining *elastic* film properties, the extracted values are even more sensitive to the substrate due to the long-range nature of the elastic field under the point of contact, with substrate-dependent properties starting at $\delta = 0.05t$ [96]. For very thin films, it is clear that these guidelines can severely limit the experimental measurement range. As a consequence, it is useful to consider the latter option, which involves fitting the experimental data to an empirical [97, 98] or analytical [99–101] model. The model by Gao et al. is particularly useful due to its simple, closed-form solution [99]. The analysis, which is based on a first-order elastic perturbation method, is in good agreement with finite element modeling for films and substrates with similar elastic properties. However, in the rigid substrate limit (i.e., $E_{\mathrm{sub}} \rightarrow \infty$), the model predicts that E^* depends solely on E_{sub} and not E_{film}, which is clearly incorrect and leads to large errors when $E_{\text{sub}} \gg E_{\text{film}}$. Xu and Pharr improved the model for this case via a simple modification to the first-order perturbation approximation [102]. In the revised model, the effective elastic modulus is given by

$$\frac{1}{E^*} = \frac{1}{2}[1 - \nu_{\text{sub}} + (\nu_{\text{sub}} - \nu_{\text{film}})I_1]\left[\frac{2(1+\nu_{\text{sub}})}{E_{\text{sub}}}(1 - I_0) + \frac{2(1+\nu_{\text{film}})}{E_{\text{film}}}I_0\right], \tag{15.6}$$

where

$$I_0(\xi) = \frac{2}{\pi}\arctan\xi + \frac{1}{2\pi(1-\nu_{\text{sub}})}\left[(1 - 2\nu_{\text{sub}})\xi\ln\frac{1+\xi^2}{\xi^2} - \frac{\xi}{1+\xi^2}\right] \tag{15.7}$$

and

$$I_1(\xi) = \frac{2}{\pi}\arctan\xi + \frac{\xi}{\pi}\ln\frac{1+\xi^2}{\xi^2} \tag{15.8}$$

are weighting functions that account for shear modulus mismatch and Poisson's ratio effects, respectively, and depend on the normalized film thickness, $\xi = t/a$. For $E_{\text{sub}} > E_{\text{film}}$, (15.6) is found to yield results within 10% of those obtained with finite element models over a wide range of $E_{\text{film}}/E_{\text{sub}}$. It is important to point out that the modification to the perturbation method leads to inaccuracies in the rigid film limit. As $E_{\text{film}} \to \infty$, the model predicts a nonzero value for E^*. This should not be an issue in this particular study, however, as the alkanethiol SAMs are expected to be much more compliant than the Au substrate.

15.4.3 Electron Tunneling Through Thin Insulating Films

The link between mechanical and electrical behavior of a metal–metal junction is through the contact radius a. I–V curves are measured at different values of F and the measured variation of G with a are compared with predictions from contact mechanics models for the variation of a with F. The relationship between G and a can take on various forms depending on the principle conduction mechanism, which can be defined by comparing a to both the mean free path l and the Fermi wavelength λ_F of the conducting electrons. For a macroscopic contact ($a > l$), the conductance is due to diffusive transport and is described by the Maxwell conductance $G_M = 2a\sigma$, where σ is the conductivity of the two metals in contact [103]. For smaller contacts ($a < l$), the conductance is due to ballistic transport and is explained by the Sharvin conductance $G_S = (\pi a/\lambda_F)^2 G_0$, where the quantum unit of conductance $G_0 = 2e^2/h = 77\,\mu\text{S}$ is described in terms of electron charge e and Planck's constant h [104]. When $a \approx l$, the transition from G_M to G_S can be described in terms of an interpolation function [105]. As $a \to \lambda_F$, $G \to G_0$. For alkaline and noble metals, the conductance is quantized around integer values of

G_0, whereas for other metals, the conductance is not quantized [106]. Regardless of the metals in question, when they are in intimate contact, it is clear that G cannot be less than G_0.

On the other hand, when the metal–metal junction is separated by a thin insulating layer, such as a deliberately formed organic monolayer [44, 58–72] or an adventitious hydrocarbon [107–110], the conductance values are significantly less than G_0. Wang et al. demonstrated that the dominant conduction mechanism through alkanethiols is in fact direct, nonresonant tunneling when the applied voltage is less than the barrier height [60]. A number of analytical expressions have been developed to describe the I–V characteristics of metal–insulator–metal tunneling junctions [111–113]. The derivations are commonly based on the Wentzel–Kramers–Brillouin (WKB) approximation, which is a semiclassical method for obtaining an approximate solution to the Schrödinger equation. However, the WKB approach is valid only when t is much greater than λ_F, which ranges from 0.5 to 1 nm for most metals [114]. As t approaches λ_F, which is often the case for SAMs, the validity of the WKB approximation comes into question.

Hansen and Brandbyge developed a different approach to describe the I–V characteristics of metal–insulator–metal tunneling junctions [115]. The method assumes a parabolic shape for the tunneling barrier; an energy diagram for a parabolic tunneling barrier is shown in Fig. 15.7. In the unbiased state, $V = 0$, the Fermi energies E_F of the two electrodes are at the same energy level E, whereas in the biased state, the Fermi levels are offset by eV, which enables electrons to travel from left to right through the tunneling barrier (as drawn). The general formula for current due to tunneling is given by

$$I = G_S T_m(V) V = \left(\frac{\pi a}{\lambda_F}\right)^2 G_0 T_m(V) V, \tag{15.9}$$

where $T_m(V)$ is the mean transmission probability averaged over all electrons. For very thin barriers and at small voltages, $T_m(V) \approx 1$. Thus, at this extreme, the I–V behavior is described solely in terms of the Sharvin conductance, or by $I = G_S V$. The transition to the Sharvin behavior is appropriate, as the contact radius at each load is much less than the mean free path, which is on the order of 10 nm for metals at room temperature [114]. In the presence of a realistic barrier, however, $T_m(V) \ll 1$. With the Sommerfeld expansion [114], $T_m(V) \sim T_0(V) + \Delta T(V)$, where $T_0(V)$ is the zero temperature mean transmission and $\Delta T(V)$ is the temperature correction. Using the extended parabolic barrier model [115]

$$T_0(V) = \frac{1}{\gamma^2 E_F eV}\left(\mathrm{Li}_2\{-\exp[-\gamma(eV + \phi_V - E_F)]\} - \mathrm{Li}_2\{-\exp[-\gamma(\phi_V - E_F)]\}\right) \tag{15.10}$$

and

$$\Delta T(V) = \frac{\pi^2}{6}\frac{(k_B T)^2}{E_F eV}\left\{\frac{1}{1+\exp[\gamma(\phi_V - E_F)]} - \frac{1}{1+\exp[\gamma(eV + \phi_V - E_F)]}\right\}, \tag{15.11}$$

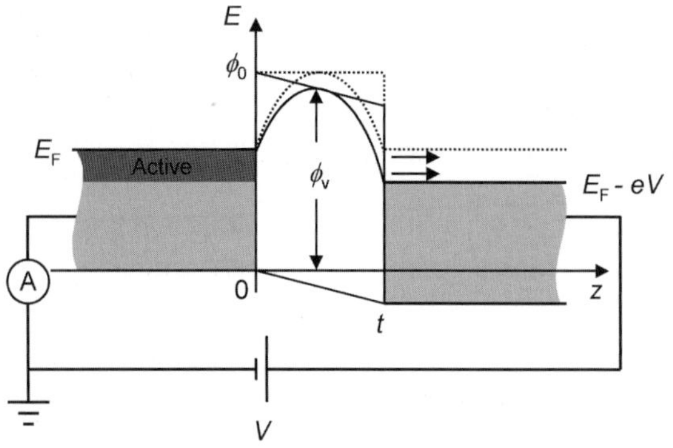

Fig. 15.7 Schematic representation of the electron energy levels in a metal–insulator–metal structure in which the insulator forms a parabolic barrier for electron tunneling between the metals. The *dotted and solid lines* indicate the levels in the unbiased and biased state, respectively (reprinted with permission from DelRio et al. [40]. Copyright 2010 American Chemical Society)

where $\mathrm{Li}_2(z)$ is the dilogarithm function, k_B is Boltzmann's constant, T is the temperature, $\gamma = (\pi^2 t/h)(2m/\phi_0)^{1/2}, \phi_\mathrm{V} = \phi_0(1 - eV/4\phi_0)^2$ is the voltage-dependent maximum barrier height, m is the mass of an electron, and ϕ_0 is the zero bias barrier height. In a metal–molecule–metal junction, $\phi_0 - E_\mathrm{F}$ corresponds to the energy difference between the Fermi level of the metal and the nearest molecular orbital (i.e., the tunneling barrier height), while t corresponds to the thickness of the molecule. Engelkes et al. showed that the nearest molecular orbital for alkanethiol SAMs is the highest occupied molecular orbital (HOMO), as opposed to the lowest unoccupied molecular orbital (LUMO) [67]. In other words, $\phi_0 - E_\mathrm{F}$ could be rewritten as $E_F - E_\mathrm{HOMO}$, where E_HOMO is the energy level of the HOMO. However, to stay true to the original nomenclature, the barrier height will be denoted as $\phi_0 - E_\mathrm{F}$.

15.5 Mechanical Properties

The CPAFM F–δ data (unloading curve only) for $n = 12$ are shown in Fig. 15.8. F–δ data were derived from the raw F–d curves by subtracting the cantilever deflection, F/k. In principle, the work of adhesion w can be found by integrating the F–δ curve, normalized by A, from z_0 to infinity [116, 117]. In most CPAFM experiments, however, the snap-on and pull-off instabilities described above preclude the full F–δ curve from being collected. Thus, it becomes necessary to use the pull-off force F_po, in conjunction with a particular contact theory, to extract values for w. In the JKR model (15.1) and (15.2), w is related to F_po by $w = -F_\mathrm{po}/(3/2\pi R)$, while for the DMT model (15.3) and (15.4), $w = -F_\mathrm{po}/2\pi R$. To find Young's modulus of the film E_film, the JKR and DMT models were fit to the same set of F–δ data as shown in Fig. 15.8, using (15.6)–(15.8) for the reduced modulus E^* and E_film as the

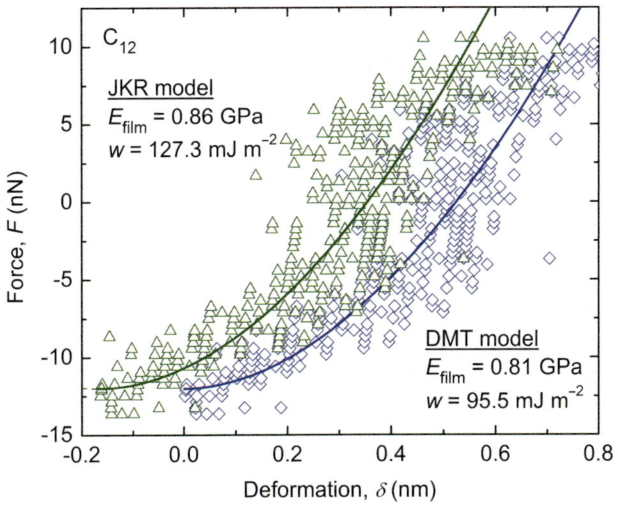

Fig. 15.8 F–δ experimental data (*symbols*) and theoretical fits (*solid lines*) for $n = 12$. The theoretical fits are based on the extended JKR and DMT contact models, using the modulus of the film E_{film} as the fitting parameter. The experimental data are the same in each case, shifted along the horizontal axis to fit the appropriate contact model. Both the JKR and DMT models fit the F–δ data well, but with different values of w and E_{film} (reprinted with permission from DelRio et al. [40]. Copyright 2010 American Chemical Society)

fitting parameters. $E_{\mathrm{sub}} = 77\,\mathrm{GPa}$ and $\nu_{\mathrm{sub}} = 0.42$ were used for the Au substrate [118], and $\nu_{\mathrm{film}} = 0.44$ was used for the monolayer film [55]. Both the JKR and DMT models fit the F–δ data well, but with different values of w and E_{film}. From the JKR model, we find $w = 127.3\,\mathrm{mJ\,m^{-2}}$, $E_{\mathrm{film}} = 0.86\,\mathrm{GPa}(E^* = 7.4\,\mathrm{GPa}$ at $F = 0)$, and $z_0 = 0.35\,\mathrm{nm}$, which yields a Tabor parameter of $\mu = 0.52$ from (15.5). On the other hand, with the DMT model, $w = 95.5\,\mathrm{mJ\,m^{-2}}$, $E_{\mathrm{film}} = 0.81\,\mathrm{GPa}$ $(E^* = 5.3\,\mathrm{GPa}$ at $F = 0)$, and $z_0 = 0.52\,\mathrm{nm}$, which results in $\mu = 0.36$. Thus, even at modest values of F, the DMT theory is a good approximation. As F increases, E^* increases (due to substrate effects) and μ decreases, pushing the system further into the DMT zone. As a result, the DMT model was fit to the CPAFM F–δ data (unloading curves only) for $n = 5$, 8, 12, and 18 as shown in Fig. 15.9 [39,40]. The extracted values for w and E_{film} from several measurements are given in Table 15.3.

For $n = 18$, we find that $w = 82.8\,\mathrm{mJ\,m^{-2}}$, which is in good agreement with results from bulk contact angle studies [119], 46–62 $\mathrm{mJ\,m^{-2}}$, surface force apparatus experiments [120], 56 $\mathrm{mJ\,m^{-2}}$, atomic force microscopy [47], 61.4 $\mathrm{mJ\,m^{-2}}$, interfacial force microscopy [52–55], 37–122.5 $\mathrm{mJ\,m^{-2}}$, and theoretical simulations [57], 119.7 $\mathrm{mJ\,m^{-2}}$, of similar low-energy hydrocarbon molecules as shown in Table 15.1. The value is also consistent with results from Lifshitz theory, which indicates that the adhesion can be attributed mostly to van der Waals forces [52,121]. However, as n decreases, w increases. Berger et al. noted a similar trend with liquid-condensed (LC) and liquid-expanded (LE) domains in a phase-separated lipid

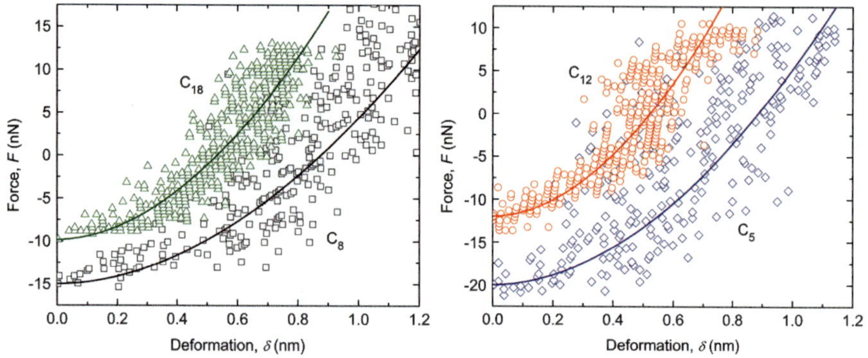

Fig. 15.9 F–δ experimental data (*symbols*) and theoretical fits (*solid lines*) for $n = 5, 8, 12$, and 18. The theoretical fits are based on the extended DMT contact model, using the modulus of the film E_{film} as the fitting parameter. The DMT theory is a good approximation even at modest values of F according to the calculated values for μ (reprinted with permission from DelRio et al. [40]. Copyright 2010 American Chemical Society)

Table 15.3 Mechanical and electrical properties of alkanethiol SAMs determined in this study

n	$w(\mathrm{mJ\,m^{-2}})$[a]	$E_{\mathrm{film}}(\mathrm{GPa})$[a]	$\phi_0 - E_F(\mathrm{eV})$[a]	$\delta_p(\mathrm{nm})$[b]
5	168.3 ± 17.6	0.15 ± 0.04	2.10 ± 0.26	0.38 ± 0.01
8	125.4 ± 9.0	0.28 ± 0.08	2.17 ± 0.30	0.40 ± 0.01
12	104.1 ± 9.8	0.86 ± 0.14	2.25 ± 0.15	0.70 ± 0.04
18	82.8 ± 9.0	1.00 ± 0.18	2.39 ± 0.35	0.40 ± 0.04

[a]Uncertainty values represent two standard deviations from at least five measurements
[b]95% confidence level in the fit

monolayer [122]. The change was attributed to differences in molecular orientation and order: in the LC domains, the molecules were densely packed and the probe tip interacted only with the CH_3 end groups, whereas in the LE domains, the films were liquid like and the adhesion was mainly due to CH_2 groups along the hydrocarbon chain. With $\gamma_{\mathrm{CH2}} = 31\,\mathrm{mJ\,m^{-2}}$ and $\gamma_{\mathrm{CH3}} = 23\,\mathrm{mJ\,m^{-2}}$, $w_{\mathrm{CH2}}/w_{\mathrm{CH3}} = (\gamma_{\mathrm{CH2}}/\gamma_{\mathrm{CH3}})^{1/2} = 1.2$, where γ_a and w_a are the surface energy and work of adhesion for molecule a, respectively. As the average tilt angle increases as n decreases, as shown by the trends in R_{I}, a similar increase in work of adhesion should be expected here due to the additional CH_2 groups at the surface. However, the increase in adhesion goes beyond a simple CH_3 versus CH_2 argument, as evident from $w_{\mathrm{C5}}/w_{\mathrm{C18}} = 2.0$. Koleske et al. showed that the elastic modulus of a film may also affect tip–surface interactions [123]. From Table 15.3, it is clear that E_{film} decreases from 1.0 to 0.15 GPa as n decreases from 18 to 5. As E_{film} decreases, δ at a given F increases, exposing the tip to additional functional groups in the film or to the substrate. Thus, for the short-chain molecules, we would expect an increase in w, as the tip is closer to the substrate and the Hamaker constant for metals is an order of magnitude greater than that for hydrocarbons [121].

The variation of E_{film} with n can be divided into three regions, each the result of a different variation mechanism, but which all derive from the variation of the attractive van der Waals interactions between alkyl chains with n [124]. As n increases for densely packed chains, the stabilization energy provided by additional CH_2 groups decreases, and eventually saturates at $n \approx 10$. Hence, the small decrease in E_{film} as n decreased from 18 to 12 arises from a small decrease in molecular stability. This decrease in stability gives rise to an increase in the concentration of *gauche* defects in the topmost methyl unit [23]. The defects increase the *average* tilt angle of the alkyl chains, which is evident from the decrease in R_{I}. For n decreasing from 12 to 8, E_{film} decreased by an additional factor of three. In this range, the intermolecular forces are no longer invariant with n, but decrease drastically as n decreases, resulting in a greater number of *gauche* defects. As a consequence, there is a change in phase from crystalline to amorphous ($R_{\text{I}} \rightarrow 0$), which reduces the film's resistance to elastic deformation. For $n < 8$, a lack of cohesive energy leads to thermal desorption of the alkyl chains [125] and the loss of surface coverage leads to the decrease in E_{film} over this range. The variation mechanisms above are consistent with the NEXAFS data in Fig. 15.2, which also point to an increase in molecular tilt relative to the normal followed by a change in phase and a loss in surface coverage for decreasing n.

Overall, the results for E_{film} are in agreement with a number of previous studies for both bulk polyethylene and alkanethiol SAMs. For instance, the values for the short-chain ($n = 5$ and 8) and long-chain ($n = 12$ and 18) molecules are consistent with results for low-density (0.17–0.28 GPa) and high-density (1.08 GPa) polyethylene, respectively [118]. Tensile testing of oriented high-density polyethylene revealed a Young's modulus of 0.80 GPa for as-fabricated samples and moduli as high as 3.05 GPa for samples with a significant amount of plastic deformation, with the increase presumably due to strain-induced crystallization (i.e., an increase in overall crystallinity) [126]. While the initial value for the elastic modulus is consistent with our results, an increase in modulus from permanent deformation would not be expected here, as alkanethiol SAMs become disordered, not ordered, at pressures of $\approx$0.8 GPa [86]. A number of scanning probe microscopy experiments have also suggested that E_{film} for long-chain ($n > 10$) alkanethiol SAMs is between 0.1 and 8 GPa [45, 48, 50, 53, 54]. In addition, a hybrid continuum-molecular dynamics analysis of interfacial force microscopy experiments on similar organic monolayers demonstrated an initial tangent modulus of 1.5 GPa [55].

Other studies, however, report much greater moduli for organic monolayers, from 15 to 75 GPa. For instance, Engelkes and Frisbie conducted low-bias conductance measurements of various alkanethiol SAMs as a function of applied load [51]. The conductance data were analyzed with the DMT contact model and tunneling equations to extract the reduced film modulus. The results showed a decrease in modulus (54–48 GPa) with increasing chain length for $n = 6$, 8, and 10, followed by an increase in modulus (75 GPa) for $n = 12$. In the analysis, the moduli were extracted based on the assumption that the measured properties were free from substrate effects. However, as the contact depths were often on the order of the film thicknesses, the underlying substrate most likely played a role, and therefore must be considered to determine the true film properties. To illustrate the role of the substrate, we again fit the DMT model to the CPAFM F–δ data in Fig. 15.9, this time assuming $E^* = E_{\text{film}}/(1-\nu^2_{\text{film}})$ instead of using (15.6)–(15.8). In this revised approach, the extracted values for E_{film} varied from 3.9 to 5.5 GPa, with no observable trend between E_{film} and n. Consequently, by neglecting the substrate, the calculated properties are almost an order of magnitude greater than the true film properties. In another study, Henda et al. utilized a static energy minimization technique to examine the elastic response of alkanethiol SAMs under compression [56]. The calculations showed that the initial tangent moduli of the SAMs were between 10 and 18 GPa, with the values nearly independent of n. The moduli were extracted based on the assumption that the SAMs were in uniaxial compression. However, as periodic boundary conditions were used, the periphery of the film was constrained and the stress state was in fact triaxial [55]. As a result, the reported values for E_{film} were inflated by a factor of $(1-\nu_{\text{film}})/(1-2\nu_{\text{film}})(1+\nu_{\text{film}})$.

15.6 Electrical Properties

The CPAFM I–V traces (recorded during the unloading portions of the F–δ curves only) for $n = 5$, 8, 12, and 18 are shown in Fig. 15.10 [40]. For all values of n, I–V behavior was strongly dependent on contact load, with increased F leading to increased G. In addition, the I–V responses were non-Ohmic, symmetric about zero bias, with conductance values smaller than G_0, which suggests that charge transport involved electron tunneling through an insulating layer. Thus, it becomes necessary to use an electron tunneling model appropriate for ultra-thin insulating layers (15.9)–(15.11) to relate the I–V behavior to a, in conjunction with the DMT contact theory (15.3) and (15.4) to describe the variation of a with F, to correlate the I–V data to F. As shown in Fig. 15.10, the parabolic tunneling model (solid lines) was fit to the experimental data (symbols) with good agreement, using $\phi_0 - E_{\text{F}}$ and t as the fitting parameters. The resulting values for $\phi_0 - E_{\text{F}}$ and t are shown in Fig. 15.11 [40].

For all n, we find that the extracted results for $\phi_0 - E_{\text{F}}$ are effectively independent of F, as there are no clear trends and the average values (represented by dashed lines in Fig. 15.11 and numerically in Table 15.3) typically fall within the uncertainty of measured values. The results for $\phi_0 - E_{\text{F}}$ are in good agreement with previously reported values for alkanethiols, which range from 1.1 to 2.9 eV [44, 60–62, 64–69, 72]. From Table 15.3, it appears as if the barrier height might be chain length dependent; as n decreases from 18 to 5, $\phi_0 - E_{\text{F}}$ decreases from 2.4 to 2.1 eV. In fact, using Student's t-test, we find that the difference for $n = 18$ and 12 is significant at the 95% level, while the differences for $n = 12$ and 8 and $n = 8$ and 5 are significant at the 85% level. There are two possible reasons for a variation in $\phi_0 - E_{\text{F}}$: (a) a change in E_{F} or (b) a change in the HOMO–LUMO energy gap. On the first point, the Fermi energy is expected to remain constant as a function of n, as the same CPAFM tip and substrate were used in all of the experiments. On

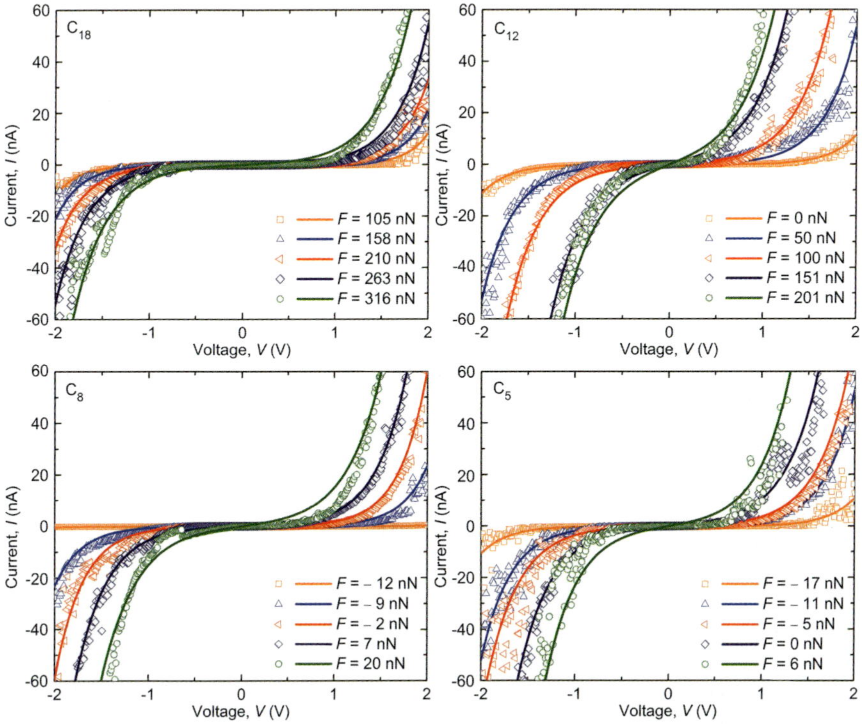

Fig. 15.10 *I–V* responses for $n = 5, 8, 12$, and 18. The *symbols* represent measurements for a single voltage sweep; the *solid lines* represent best fits to the data using a parabolic-barrier tunneling model, with $\phi_0 - E_F$ and t as the fitting parameters (reprinted with permission from DelRio et al. [40]. Copyright 2010 American Chemical Society)

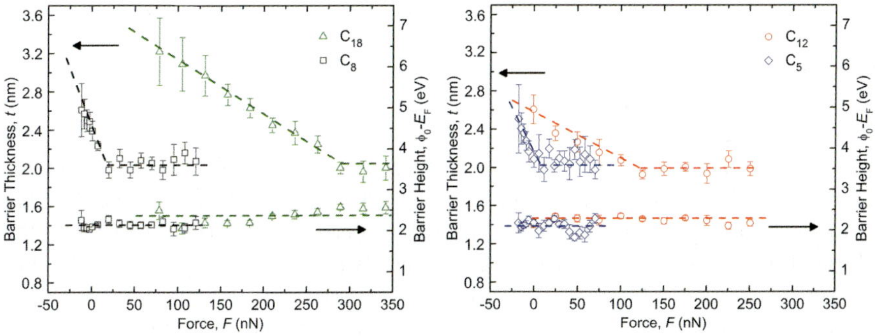

Fig. 15.11 $\phi_0 - E_F$ (*right*) and t (*left*) as a function of F for $n = 5, 8, 12$, and 18. For all n, the extracted results for $\phi_0 - E_F$ are effectively independent of F. The variations in t with F can be divided into two regions. One possible explanation for the change in behavior involves the presence of *two* insulating layers: an oxide layer on the Co–Cr tip and the alkanethiol monolayer on the Au surface (reprinted with permission from DelRio et al. [40]. Copyright 2010 American Chemical Society)

the second point, Boulas et al. conducted photoconductivity experiments at variable excitation wavelengths to extract the HOMO–LUMO energy gaps for several long-chain ($n = 12$, 16, and 18) alkyl-based SAMs [127]. The HOMO–LUMO gap was found to be independent of n, with values ranging from 9.2 to 9.9 eV. As a result, for densely packed, crystalline-like films, $\phi_0 - E_F$ should be independent of n. However, by changing the deposition conditions, the authors were also able to examine some of the same films in a liquid-like state with only partial surface coverage. The resulting photocurrent curves were shifted towards smaller barrier heights, which yields a large decrease (>1 eV) in the HOMO–LUMO energy gap. Thus, it was shown that a *qualitative* relationship exists between the molecular orientation, order, and coverage and the barrier height, presumably due to changes in the free volume between the chains. Here, we arrive at the same conclusion, and provide *quantitative* evidence for the link between SAM structure and transport properties; as n decreases, both R_I and $\phi_0 - E_F$ decrease. This effect has been attributed to image charges generated in the metal electrodes by trapped charges in the molecule, which leads to a localization of charge at the metal–molecule interface and a significant change in the intrinsic energy levels [128]. However, other studies suggest that $\phi_0 - E_F$ is constant for alkanethiols, regardless of the molecular orientation and surface coverage. Beebe et al. used CPAFM and crossed-wire tunnel junctions to investigate the effect of n on the voltage required to induce a transition from direct tunneling to field emission, denoted by V_{trans} [70, 71]. V_{trans} was found to scale linearly with $\phi_0 - E_F$ for π-conjugated thiol SAMs using ultraviolet photoelectron spectroscopy [70]. For both short-chain and long-chain alkanethiols, V_{trans} was constant [71], which led to the conclusion that $\phi_0 - E_F$ was independent of n.

The variations in t with F can be divided into two regions for all n; at small F, t decreases monotonically, while at large F, t remains constant. One possible explanation for the change in behavior involves the presence of *two* insulating layers: an oxide layer (thickness t_{oxide}) on the Co–Cr tip and the alkanethiol monolayer (thickness t_{film}) on the Au surface, as shown in Fig. 15.12. In this scenario, the probe tip compresses the compliant alkanethiol monolayer during the initial stages of contact. Eventually, the alkanethiol layer is compressed between the contacts, and the insulating layer consists mainly of the oxide layer on the Cr surface, which is more difficult to deform. Chromium thin films are known to readily form native oxide layers under ambient conditions; the standard enthalpy of formation for Cr_2O_3, for example, is $\Delta H_f^0 = -1,140\,\mathrm{kJ\,mol^{-1}}$ [129]. In one study, the thickness of the oxide layer varied from 2 to 6 nm, with values depending on the thickness of the chromium film and the underlying substrate material [130]. To estimate t_{oxide} on our CPAFM cantilever, the cantilever surface was sputtered and XPS high resolution region scans were performed initially every 10 s and then at 20 s intervals throughout the sputter etch. Figure 15.13 shows the Cr 2p region for sputtering times of 0, 10, and 220 s. After 10 s, the adventitious C (not shown here) was reduced to 35% of its initial value, and the signals for elemental Cr and Cr_2O_3 increased by factors of 3.7 and 2.4, respectively. In subsequent sputter steps, the elemental Cr peak monotonically increased, while the Cr_2O_3 peak increased another 5% after 20 s and then decreased steadily until it was no longer detectable after 220 s. Subsequent

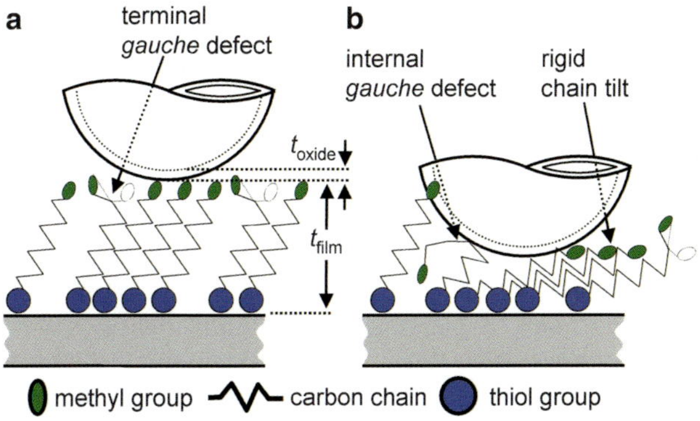

Fig. 15.12 Schematic representation of the Co–Cr tip in contact with the alkanethiol monolayer. At small F, the alkanethiol film and oxide layer both act as insulators, $t = t_{film} + t_{oxide}$. At large F, however, the alkanethiol film is compressed and the oxide layer is the sole interlayer, $t = t_{oxide}$ (reprinted with permission from DelRio et al. [40]. Copyright 2010 American Chemical Society)

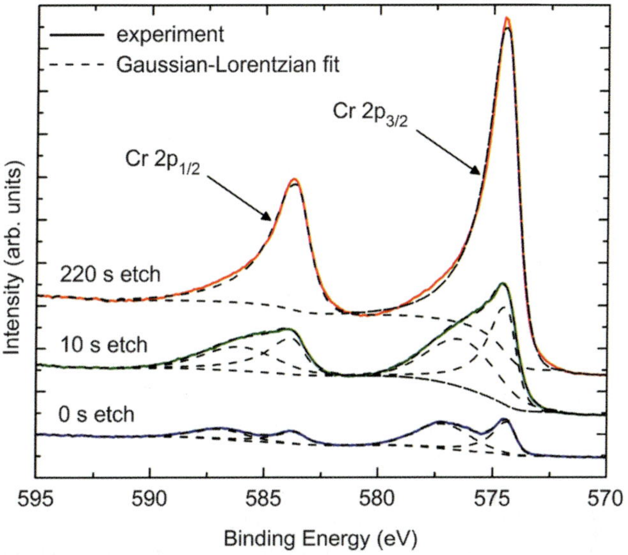

Fig. 15.13 XPS Cr 2p spectra for sputter times of 0, 10, and 220 s (offset for clarity). The elemental Cr peaks are at ≈574.4 and ≈583.6 eV, while the Cr_2O_3 peaks are at ≈576.5 and ≈586.3 eV. After 10 s, the adventitious C (not shown here) was reduced to 35% of its initial value, and the signals for elemental Cr and Cr_2O_3 increased by factors of 3.7 and 2.4, respectively. After 220 s, the Cr_2O_3 peaks were no longer detectable (reprinted with permission from DelRio et al. [40]. Copyright 2010 American Chemical Society)

sputtering did not result in increased signal for the elemental Cr peak. The elemental Cr $2p_{3/2}$ peaks were fit with 30% Gaussian–70% Lorentzian peaks with an exponential blend parameter of 0.8, while the Cr $2p_{3/2}$ peaks for Cr_2O_3 were fit with 70% Gaussian–30% Lorentzian peaks. With these spectra, t_{oxide} was calculated using a standard uniform overlayer model [131] $I_{\mathrm{Cr}} = I^0_{\mathrm{Cr}} \exp(-t_{\mathrm{oxide}}/L_{\mathrm{Cr}})$, where the initial value of Cr, I_{Cr}, was the peak area for the elemental Cr $2p_{3/2}$ of the unsputtered sample, and the final base value, I^0_{Cr}, was an average of the elemental Cr $2p_{3/2}$ peak areas for 260, 280, and 300 s sputter times. NIST Standard Reference Database 82 was used to calculate the practical effective attenuation length [132] L_{Cr} for elemental Cr electrons passing through a Cr_2O_3 overlayer. The calculations were based on the kinetic energy of the base electrons KE = 912.5 eV, asymmetry parameter $\beta_a = 1.45$, and the following Cr_2O_3 parameters: band gap energy [133] of 3.7 eV and density [134] of 5,250 kg m^{-3}. This calculation resulted in a Cr_2O_3 thickness of 3.9 nm. However, as the adventitious C on the surface also contributed to the attenuation, 3.9 nm can be viewed as an upper bound on the Cr_2O_3 thickness. To determine a lower bound for t_{oxide}, the Cr_2O_3 thickness was recalculated with I_{Cr} as the elemental Cr $2p_{3/2}$ peak area after 10 s of sputtering. At this depth, most of the C was etched from the surface, in addition to some of the Cr_2O_3 layer, as evident by the increase in the Cr/Cr_2O_3 signal ratio. This second estimate resulted in a Cr_2O_3 thickness of 1.9 nm. Hence, the XPS results indicate that t_{oxide} was between 1.9 and 3.9 nm. From this analysis, it seems reasonable to assume that the constant values for t at large F, which are all around 2.2 nm as shown in Fig. 15.11, are due to an oxide layer on the Co–Cr tip.

As the tip was retracted from the surface, elastic displacements were eventually recovered from the alkanethiol monolayer, and the extracted values for $t = t_{\mathrm{film}} + t_{\mathrm{oxide}}$, as shown in Fig. 15.12. This observation raises a key point about the mechanical and electrical behavior of the oxide layer on the Co–Cr tip. From a mechanical standpoint, the modulus of the oxide layer E_{oxide} is much greater than E_{film}, and as a result, the deformation of the oxide layer δ_{oxide} will represent only a small portion of δ. Thus, t in (15.6)–(15.8) is just the monolayer thickness, i.e., $t = t_{\mathrm{film}}$, and does not depend on the oxide layer. With an applied voltage, however, the alkanethiol SAM and the oxide layer both act as insulators, and as mentioned above, t is a function of both t_{film} and t_{oxide}. Therefore, to get t_{film} as a function of F in Fig. 15.11, t_{oxide} must be subtracted according to $t_{\mathrm{film}} = t - t_{\mathrm{oxide}}$. For all n, (a) t_{film} increases monotonically as F decreases and (b) t_{film} at $F = 0$ is less than the original film thickness, due to elastic recovery and viscoelastic–plastic deformation, respectively. The average values for the viscoelastic–plastic deformation δ_p, or the difference between t_{film} at $F = 0$ and the original film thickness, are shown in Table 15.3. While δ_p is fairly constant for all n except $n = 12$, it is difficult to draw any definite conclusions about the relationship between δ_p and n at a given F, because the maximum F in each experiment in Fig. 15.10 was different. Salmeron noted that the viscoelastic–plastic deformation occurs due to the three different types of defects shown in Fig. 15.12: terminal *gauche* defects, internal *gauche* defects, and rigid chain tilts [135].

It is possible, however, to "normalize" the data in Fig. 15.11 by plotting $\phi_0 - E_F$ and t as a function of a characteristic strain $\varepsilon = F/AE^*$ as shown in Fig. 15.14 [40]. In doing so, two distinct features become clear, both of which are independent of n. First, $\phi_0 - E_F$ is strain dependent; for positive (tensile) ε, $\phi_0 - E_F$ is constant at $\approx$2.1 eV, but as ε decreases into negative (compressive) values, $\phi_0 - E_F$ increases to

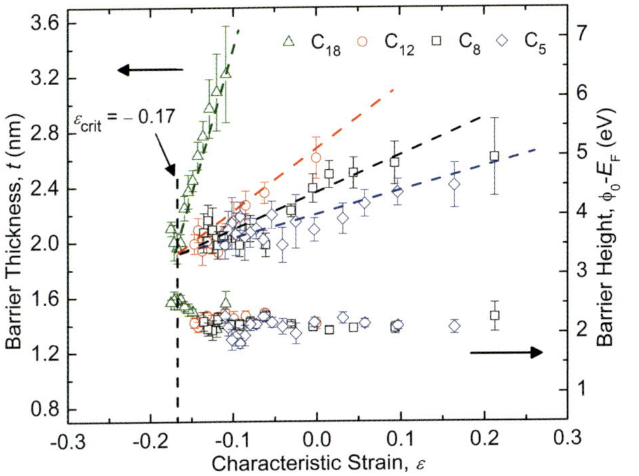

Fig. 15.14 $\phi_0 - E_F$ (*right*) and t (*left*) as a function of ε for $n = 5$, 8, 12, and 18. For positive $\varepsilon, \phi_0 - E_F$ is constant at 2.1 eV, but as ε decreases, $\phi_0 - E_F$ increases to 2.6 eV. The critical strain ε_{crit}, or the strain required to displace the film, is shown to be $\varepsilon_{crit} = -0.17$ (reprinted with permission from DelRio et al. [40]. Copyright 2010 American Chemical Society)

almost 2.6 eV. Slowinski et al. [59] and later Song et al. [136] showed that the dominant charge transport mechanism for alkanethiols transitions from through-bond tunneling, where electrons flow through the alkyl chains via overlapping σ-bonds, to chain-to-chain coupling, which involves charge transfer between adjacent hydrocarbon chains, as the tilt angle increases. In fitting the data to the two-pathway model, the authors demonstrated that the tunneling decay coefficient β, defined as the change in junction resistance at low-bias with n, for chain-to-chain tunneling is greater than that for through-bond tunneling. In other words, chain-to-chain coupling is less efficient for charge transport, but becomes significant at large tilt angles, which is consistent with the increase in $\phi_0 - E_F$ at negative ε in Fig. 15.14. Second, while the rate at which t changes with ε is dependent on n (due to the variations in E_{film} as shown in Table 15.3), the strain at which the film is completely displaced from under the tip is not. From Fig. 15.14, this critical strain ε_{crit} is shown to be $\varepsilon_{crit} = -0.17$, which might represent the strain required to change the sulfur-gold binding structure (e.g., shift a sulfur atom from a hollow site to a bridge site on the gold) to accommodate large tilt angles [69].

15.7 Conclusions and Future Directions

The CPAFM has been demonstrated here as an effective technique for simultaneous measurement of mechanical and electrical properties of SAMs under load. By controlling the contact force at which electrical conduction measurements were performed, a full picture of the coupling between the mechanical and electrical properties of alkanethiol SAMs was obtained under both tensile adhesive loading and compressive indentation loading. A critical feature in analyzing both the mechanical and electrical measurements was fully accounting for the small thickness of the SAM relative to the contact deformation and the Fermi wavelength of the conducting electrons. In the mechanical analysis, it was necessary to use a contact model that accounted explicitly for substrate effects in stiffening the reduced modulus, and in the electrical analysis, it was necessary to use a tunneling model that accounted for bias effects in reducing the tunneling barrier height. Analyzing the measurements in this way enabled the work of adhesion, Young's modulus, barrier thickness, and barrier height to be determined; the latter three exhibiting clear load dependence as the SAM structure was altered during loading. The properties determined were in agreement with independent measurements; in particular, the Young's modulus was in the gigapascal range appropriate to alkane SAMs, and not artificially inflated by substrate effects. The measurements and analyses enabled clear property trends to be established: adhesion decreased, modulus increased, and barrier height increased with increased chain length. Although the ordering and structure of the SAMs altered with chain length, becoming more ordered as the chain length increased, normalizing the SAM deformation by the Young's modulus under load enabled a critical strain to be identified for collapse of the SAM tunneling barrier thickness. A key conclusion from this study is that these insights regarding SAM structure–property relations were only possible through the use of self-consistent models of mechanical and electrical behavior.

As well as revealing properties of SAMs, this study has revealed a property of nano-scale metal–metal electrical contacts that distinguishes them from micro- and pico-scale contacts – nano-scale contacts will typically have an insulating barrier separating the metal electrodes. At the micro-scale, contact loads are great enough to break through any intervening insulating layers and establish metal–metal interfaces and at the pico-scale contacts consist of metallic atom chains. In these latter two cases, the conductance, whether Maxwell, Sharvin, or quantized, is determined predominantly by the contact geometry and is only weakly dependent on the contact load. Nano-scale contacts are not loaded sufficiently to break down intervening insulating layers and do not have the deliberately engineered purity of atomic break junctions. As a consequence, designing nano-scale contacts for MEMS and NEMS applications requires explicit measurement of the load dependence of the contact conductance as altering the load alters the structure and thus the electrical tunneling properties of the remnant intervening insulating layer. SAMs are good model materials for study and formation of such nano-scale contacts, as their reproducible structures that can be controlled by chain length and chemical composition will lead to layers with reproducible mechanical and electrical properties. The combination of CPAFM with SAMs is thus a good vehicle for MEMS and NEMS development.

Additional future directions for CPAFM application include extending the technique to other SAM-surface combinations, including the well-studied alkylsilane-oxide systems, and nonchain and branched-chain SAMs. A principal component of MEMS devices is silicon and hence CPAFM will need to be extended to measure the properties of SAMs when separating semiconducting electrodes, for which the electrical properties are expected to not be symmetric with bias and for which the adhesive properties are expected to be less dependent on chain length relative to metal electrode systems. As was noted here, many systems in both MEMS and microelectronics applications will have thin oxide layers on the surfaces of electrodes. Study of the tunneling properties of these layers will require extension of CPAFM to contacts much stiffer than those associated with SAM-coated surfaces, with much greater electrical tunneling barriers. However, as was performed here, study of such oxides with well-characterized SAM overlayers allows the oxide tunneling properties to be determined by extrapolation of SAM thickness by controlling chain length and contact load. Finally, not all insulating layers separating electrodes will be intentionally engineered, but formed by adventitious organic and inorganic deposition. The CPAFM studies of SAMs with controlled order will provide a firm foundation for study of these less ordered or disordered systems.

References

1. A. Ulman, Formation and structure of self-assembled monolayers. Chem. Rev. **96**, 1533–1554 (1996)
2. D.K. Schwartz, Mechanisms and kinetics of self-assembled monolayer formation. Annu. Rev. Phys. Chem. **52**, 107–137 (2001)
3. P.E. Laibinis, G.M. Whitesides, Self-assembled monolayers of n-alkanethiolates on copper are barrier films that protect the metal against oxidation by air. J. Am. Chem. Soc. **114**, 9022–9028 (1992)
4. K.L. Prime, G.M. Whitesides, Adsorption of proteins onto surfaces containing end-attached oligo(ethylene oxide): a model system using self-assembled monolayers. J. Am. Chem. Soc. **115**, 10714–10721 (1993)
5. S. Herrwerth, W. Eck, S. Reinhardt, M. Grunze, Factors that determine the protein resistance of oligoether self-assembled monolayers – internal hydrophilicity, terminal hydrophilicity, and lateral packing density. J. Am. Chem. Soc. **125**, 9359–9366 (2003)
6. F. Zhang, M.W.A. Skoda, R.M.J. Jacobs, S. Zorn, R.A. Martin, C.M. Martin, G.F. Clark, G. Goerigk, F. Schreiber, Gold nanoparticles decorated with oligo(ethylene glycol) thiols: protein resistance and colloidal stability. J. Phys. Chem. A. **111**, 12229–12237 (2007)
7. M. Geissler, H. Schmid, A. Bietsch, B. Michel, E. Delamarche, Defect-tolerant and directional wet-etch systems for using monolayers as resists. Langmuir. **18**, 2374–2377 (2002)
8. V.C. Sundar, J. Aizerberg, Narrow features in metals at the interfaces between different etch resists. Appl. Phys. Lett. **83**, 2259–2261 (2003)
9. M. Su, S. Li, V.P. Dravid, Microcantilever resonance-based DNA detection with nanoparticle probes. Appl. Phys. Lett. **82**, 3562–3564 (2003)
10. J. Lahann, S. Mitragotri, T.–N. Tran, H. Kaido, J. Sundaram, I.S. Choi, S. Hoffer, G.A. Somorjai, R. Langer, A reversibly switching surface. Science. **299**, 371–374 (2003)
11. K. Asadi, F. Gholamrezaie, E.C.P. Smits, P.W.M. Blom, B. de Boer, Manipulation of charge carrier injection into organic field-effect transistors by self-assembled monolayers of alkanethiols. J. Mater. Chem. **17**, 1947–1953 (2007)
12. S.T. Patton, K.C. Eapen, J.S. Zabinski, J.H. Sanders, A.A. Voevodin, Lubrication of microelectromechanical systems radio frequency switch contacts using self-assembled monolayers. J. Appl. Phys. **102**, 024903 (2007)
13. A.M. Homola, Lubrication issues in magnetic disk storage devices. IEEE Trans. Magn. **32**, 1812–1818 (1996)

14. R. Maboudian, R.T. Howe, Critical review: adhesion in surface micromechanical structures. J. Vac. Sci. Technol. B. **15**, 1–20 (1997)
15. S. Majumder, N.E. McGruer, G.G. Adams, P.M. Zavracky, R.H. Morrison, J. Krim, Study of contacts in an electrostatically actuated microswitch. Sens. Actuators, A. **93**, 19–26 (2001)
16. S.T. Patton, J.S. Zabinski, Fundamental studies of Au contacts in MEMS RF switches. Trib. Lett. **18**, 215–230 (2005)
17. G. Gregori, D.R. Clarke, The interrelation between adhesion, contact creep, and roughness on the life of gold contacts in radio-frequency microswitches. J. Appl. Phys. **100**, 094904 (2006)
18. J.I. Pascual, J. Mendez, J. Gomez-Herrero, A.M. Baro, N. Garcia, U. Landman, W.D. Luedtke, E.N. Bogachek, H.-P. Cheng, Electrical and mechanical properties of metallic nanowires: conductance quantization and localization. J. Vac. Sci. Technol. B. **13**, 1280–1284 (1995)
19. K. Hansen, S.K. Nielsen, M. Brandbyge, E. Laegsgaard, I. Stensgaard, F. Besenbacher, Current-voltage curves of gold quantum point contacts revisited. Appl. Phys. Lett. **77**, 708–710 (2000)
20. L. Strong, G.M. Whitesides, Structures of self-assembled monolayer films of organosulfur compounds adsorbed on gold single crystals: electron diffraction studies. Langmuir. **4**, 546–558 (1988)
21. C.E.D. Chidsey, D.N. Loiacono, Chemical functionality in self-assembled monolayers: structural and electrochemical properties. Langmuir. **6**, 682–691 (1990)
22. R.G. Nuzzo, E.M. Korenic, L.H. Dubois, Studies of the temperature-dependent phase behavior of long chain n-alkyl thiol monolayers on gold. J. Chem. Phys. **93**, 767–773 (1990)
23. J. Hautman, M.L. Klein, Molecular dynamics simulation of the effects of temperature on a dense monolayer of long-chain molecules. J. Chem. Phys. **93**, 7483–7492 (1989)
24. N. Camillone, C.E.D. Chidsey, G.-Y. Liu, G. Scoles, Superlattice structure at the surface of a monolayer of octadecanethiol self-assembled on Au(111). J. Chem. Phys. **98**, 3503–3511 (1993)
25. P. Fenter, P. Eisenberger, K.S. Liang, Chain-length dependence of the structures and phases of $CH_3(CH_2)_{n-1}SH$ self-assembled on Au(111). Phys. Rev. Lett. **70**, 2447–2450 (1993)
26. G.E. Poirier, M.J. Tarlov, The c(4×2) superlattice of n-alkanethiol monolayers self-assembled on Au(111). Langmuir. **10**, 2853–2856 (1994)
27. P. Fenter, A. Eberhardt, P. Eisenberger, Self-assembly of n-alkyl thiols as disulfides on Au (111). Science. **266**, 1216–1218 (1994)
28. M.D. Porter, T.B. Bright, D.L. Allara, C.E.D. Chidsey, Spontaneously organized molecular assemblies: structural characterization of n-alkyl thiol monolayers on gold by optical ellipsometry, infrared spectroscopy, and electrochemistry. J. Am. Chem. Soc. **109**, 3559–3568 (1987)
29. C.D. Bain, E.B. Troughton, Y.-T. Tao, J. Evall, G.M. Whitesides, R.G. Nuzzo, Formation of monolayer films by the spontaneous assembly of organic thiols from solution onto gold. J. Am. Chem. Soc. **111**, 321–335 (1989)
30. E. Delamarche, B. Michel, H. Kang, Ch. Gerber, Thermal stability of self-assembled monolayers. Langmuir. **10**, 4103–4108 (1994)
31. N. Camillone, P. Eisenberger, T.Y. B. Leung, P. Schwartz, G. Scoles, G.E. Poirier, M.J. Tarlov, New monolayer phases of n-alkane thiols self-assembled on Au(111): preparation, surface characterization, and imaging. J. Chem. Phys. **101**, 11031–11036 (1994)
32. D.-H. Tsai, R.A. Zangmeister, L.F. Pease, M.J. Tarlov, M.R. Zachariah, Gas-phase ion-mobility characterization of SAM-functionalized Au nanoparticles. Langmuir. **24**, 8483–8490 (2008)
33. P.E. Laibinis, G.M. Whitesides, D.L. Allara, Y.-T. Tao, A.N. Parikh, R.G. Nuzzo, Comparison of the structures and wetting properties of self-assembled monolayers of *n*-alkanethiols on the coinage metal surfaces, Cu, Ag, Au. J. Am. Chem. Soc. **113**, 7152–7167 (1991)
34. J.B. Schlenoff, M. Li, H. Ly, Stability and self-exchange in alkanethiol monolayers, J. Am. Chem. Soc. **117**, 12528–12536 (1995)
35. L.H. Dubois, B.R. Zegarski, R.G. Nuzzo, Molecular ordering of organosulfur compounds on Au (111) and Au (100): adsorption from solution and in ultrahigh vacuum. J. Chem. Phys. **98**, 678–688 (1993)

36. K.A. Peterlinz, R. Georgiadia, In situ kinetics of self-assembly by surface plasmon resonance spectroscopy, Langmuir. **12**, 4731–4740 (1996)
37. E. Barrena, C. Ocal, M. Salmeron, Evolution of the structure and mechanical stability of self-assembled alkanethiol islands on Au(111) due to diffusion and ripening. J. Chem. Phys. **111**, 9797–9802 (1999)
38. E. Barrena, C. Ocal, M. Salmeron, Molecular packing changes of alkanethiol monolayers on Au (111) under applied pressure. J. Chem. Phys. **113**, 2413–2418 (2000)
39. F.W. DelRio, C. Jaye, D.A. Fischer, R.F. Cook, Elastic and adhesive properties of alkanethiol self-assembled monolayers on gold. Appl. Phys. Lett. **94**, 131909 (2009)
40. F.W. DelRio, K.L. Steffens, C. Jaye, D.A. Fischer, R.F. Cook, Elastic, adhesive, and charge transport properties of a metal-molecule-metal junction: the role of molecular orientation, order, and coverage. Langmuir. **26**, 1688–1699 (2010)
41. D.A. Outka, J. Stöhr, J.P. Rabe, J.D. Swalen, The orientation of Langmuir-Blodgett monolayer using NEXAFS. J. Chem. Phys. **88**, 4076–4087 (1988)
42. S.Sambasivan, S. Hsieh, D.A. Fischer, S.M. Hsu, Effect of self-assembled monolayer film order on nanofriction. J. Vac. Sci. Technol. A. **24**, 1484–1488 (2006)
43. A. Lio, D.H. Charych, M. Salmeron, Comparative atomic force microscopy study of the chain length dependence of frictional properties of alkanethiols on gold and alkylsilanes on mica. J. Phys. Chem. B. **101**, 3800–3805 (1997)
44. R.E. Holmlin, R. Haag, M.L. Chabinyc, R.F. Ismagilov, A.E. Cohen, A. Terfort, A. Rampi, G.M. Whitesides, Electron transport through thin organic films in metal-insulator-metal junctions based on self-assembled monolayers. J. Am. Chem. Soc. **123**, 5075–5085 (2001)
45. W. Kiridena, V. Jain, P.K. Kuo, G. Liu, Nanometer-scale elasticity measurements on organic monolayers using scanning force microscopy. Surf. Interface Anal. **25**, 383–389 (1997)
46. A. Lio, C. Morant, D.F. Ogletree, M. Salmeron, Atomic force microscopy study of the pressure-dependent structural and frictional properties of n-alkanethiols on gold. J. Phys. Chem. B. **101**, 4767–4773 (1997)
47. B. Bhushan, H. Liu, Nanotribological properties and mechanisms of alkylthiol and biphenyl thiol self-assembled monolayers studied by AFM. Phys. Rev. B. **63**, 245412 (2001)
48. W.J. Price, S.A. Leigh, S.M. Hsu, T.E. Patten, G. Liu, Measuring the size dependence of Young's modulus using force modulation atomic force microscopy. J. Phys. Chem. A. **110**, 1382–1388 (2006)
49. G. Oncins, C. Vericat, F. Sanz, Mechanical properties of alkanethiol monolayers studied by force spectroscopy. J. Chem. Phys. **128**, 044701 (2008)
50. M. Salmeron, G. Neubauer, A. Folch, M. Tomitori, D.F. Ogletree, P. Sautet, Viscoelastic and electrical properties of self-assembled monolayers on Au(111) films. Langmuir. **9**, 3600–3611 (1993)
51. V.B. Engelkes, C.D. Frisbie, Simultaneous nanoindentation and electron tunneling through alkanethiol self-assemled monolayers. J. Phys. Chem. B. **110**, 10011–10020 (2006)
52. R.C. Thomas, J.E. Houston, R.M. Crooks, T. Kim, T.A. Michalske, Probing adhesion forces at the molecular scale. J. Am. Chem. Soc. **117**, 3830–3834 (1995)
53. A.R. Burns, J.E. Houston, R.W. Carpick, T.A. Michalske, Friction and molecular deformation in the tensile regime. Phys. Rev. Lett. **82**, 1181–1184 (1999)
54. J.E. Houston, H.I. Kim, Adhesion, friction, and mechanical properties of functionalized alkanethiol self-assembled monolayers. Acc. Chem. Res. **35**, 547–553 (2002)
55. M. Wang, K.M. Liechti, V. Srinivasan, J.M. White, P.J. Rossky, M.T. Stone, A hybrid continuum-molecular analysis of interfacial force microscope experiments on a self-assembled monolayer. J. Appl. Mech. **73**, 769–777 (2006)
56. R. Henda, M. Grunze, A.J. Pertsin, Static energy calculations of stress-strain behavior of self-assembled monolayers. Trib. Lett. **5**, 191–195 (1998)
57. Y. Leng, S. Jiang, Atomic indentation and friction of self-assembled monolayers by hybrid molecular simulations. J. Chem. Phys. **113**, 8800–8806 (2000)
58. K. Weber, L. Hockett, S. Creager, Long-range electronic coupling between ferrocene and gold in alkanethiolate-based monolayers on electrodes. J. Phys. Chem. B. **101**, 8286–8291 (1997)

59. K. Slowinski, R.V. Chamberlain, C.J. Miller, M. Majda, Through-bond and chain-to-chain coupling: two pathways in electron tunneling through liquid alkanethiol monolayers on mercury electrodes. J. Am. Chem. Soc. **119**, 11910–11919 (1997)
60. W. Wang, T. Lee, M.A. Reed, Mechanism of electron conduction in self-assembled alkanethiol monolayer devices. Phys. Rev. B. **68**, 035416 (2003)
61. T. Lee, W. Wang, J.F. Klemic, J.J. Zhang, J. Su, M.A. Reed, Comparison of electronic transport characterization methods for alkanethiol self-assembled monolayers. J. Phys. Chem. B. **108**, 8742–8750 (2004)
62. A. Dhirani, P.-H. Lin, P. Guyot-Sionnest, R.W. Zehner, L.R. Sita, Self-assembled molecular rectifiers. J. Chem. Phys. **106**, 5249–5253 (1997)
63. L.A. Bumm, J.J. Arnold, T.D. Dunbar, D.L. Allara, P.S. Weiss, Electron transfer through organic molecules. J. Phys. Chem. B. **103**, 8122–8127 (1999)
64. F.-R.F. Fan, J. Yang, L. Cai, D.W. Price, S.M. Dirk, D.V. Kosynkin, Y. Yao, A.M. Rawlett, J.M. Tour, A.J. Bard, Charge transport through self-assembled monolayers of compounds of interest in molecular electronics. J. Am. Chem. Soc. **124**, 5550–5560 (2002)
65. D.J. Wold, C.D. Frisbie, Fabrication and characterization of metal-molecule-metal junctions by conducting probe atomic force microscopy. J. Am. Chem. Soc. **123**, 5549–5556 (2001)
66. X.D. Cui, X. Zarate, J. Tomfohr, O.F. Sankey, A. Primak, A.L. Moore, T.A. Moore, D. Gust, G. Harris, S.M. Lindsay, Making electrical contacts to molecular monolayers. Nanotechnology. **13**, 5–14 (2002)
67. V.B. Engelkes, J.M. Beebe, C.D. Frisbie, Length-dependent transport in molecular junctions based on SAMs of alkanethiols and alkanedithiols: effect of metal work function and applied bias on tunneling efficiency and contact resistance. J. Am. Chem. Soc. **126**, 14287–14296 (2004)
68. H. Song, C. Lee, Y. Kang, T. Lee, Electronic transport and tip-loading force effect in self-assembled monolayer studies by conducting atomic force microscopy. Colloids Surf. A: Physicochem. Eng. Aspects. **284–285**, 583–588 (2006)
69. Y. Qi, I. Ratera, J.Y. Park, P.D. Ashby, S.Y. Quek, J.B. Neaton, M. Salmeron, Mechanical and charge transport properties of alkanethiol self-assembled monolayers on a Au(111) surface: the role of molecular tilt. Langmuir. **24**, 2219–2223 (2008)
70. J.M. Beebe, B. Kim, J.W. Gadzuk, C.D. Frisbie, J.G. Kushmerick, Transition from direct tunneling to field emission in metal-molecule-metal junctions. Phys. Rev. Lett. **97**, 026801 (2006)
71. J.M. Beebe, B. Kim, C.D. Frisbie, J.G. Kushmerick, Measuring relative barrier heights in molecular electronic junctions with transition voltage spectroscopy. ACS Nano. **2**, 827–832 (2008)
72. C.-C. Kaun, H. Guo, Resistance of alkanethiol molecular wires. Nano Lett. **3**, 1521–1525 (2003)
73. D. Vuillaume, C. Boulas, J. Collet, J.V. Davidovits, F. Rondelez, Organic insulating films of nanometer thicknesses. Appl. Phys. Lett. **69**, 1646–1648 (1996)
74. C. Zhou, M.R. Deshpande, M.A. Reed, L. Jones, J.M. Tour, Nanoscale metal/self-assembled monolayer/metal heterostructures. Appl. Phys. Lett. **71**, 611–613 (1997)
75. J. Collet, S. Lenfant, D. Vuillaume, O. Bouloussa, F. Rondelez, J.M. Gay, K. Kham, C. Chevrot, High anisotropic conductivity in organic insulator/semiconductor monolayer heterostructure. Appl. Phys. Lett. **76**, 1339–1341 (2000)
76. L.A. Bumm, J.J. Arnold, M.T. Cygan, T.D. Dunbar, T.P. Burgin, L. Jones, D.L. Allara, J.M. Tour, P.S. Weiss, Are single molecular wires conducting? Science. **271**, 1705–1707 (1996)
77. S. Datta, W. Tian, S. Hong, R. Reifenberger, J.I. Henderson, C.P. Kubiak, Current-voltage characteristics of self-assembled monolayers by scanning tunneling microscopy. Phys. Rev. Lett. **79**, 2530–2533 (1997)
78. J.L. Hutter, J. Bechhoefer, Calibration of atomic-force microscope tips. Rev. Sci. Instrum. **64**, 1868–1873 (1993)
79. J.P. Cleveland, S. Manne, D. Bocek, P.K. Hansma, A nondestructive method for determining the spring constant of cantilevers for scanning force microscopy. Rev. Sci. Instrum. **64**, 403–405 (1993)

80. J.E. Sader, J.W.M. Chon, P. Mulvaney, Calibration of rectangular atomic force microscope cantilevers. Rev. Sci. Instrum. **70**, 3967–3969 (1999)
81. R.S. Gates, M.G. Reitsma, Precise atomic force microscope cantilever spring constant calibration using a reference cantilever array. Rev. Sci. Instrum. **78**, 086101 (2007)
82. H.J. Kim, S.K. Choi, S.H. Kang, K.H. Oh, Structural phase transitions of $Ge_2Sb_2Te_5$ cells with TiN electrodes using a homemade W heater tip, Appl. Phys. Lett. **90**, 083103 (2007)
83. S.M. Hues, C.F. Draper, K.P. Lee, R.J. Colton, Effect of PZT and PMN actuator hysteresis and creep on nanoindentation measurements using force microscopy. Rev. Sci. Instrum. **65**, 1561–1565 (1994)
84. J. Schneider, Y. Dori, K. Haverstick, M. Tirrell, R. Sharma, Force titration of Langmuir-Blodgett bilayers of glucine amphiphiles: JKR-type measurements using the surface-force apparatus. Langmuir. **18**, 2702–2709 (2002)
85. S.A. Joyce, R.C. Thomas, J.E. Houston, T.A. Michalske, R.M. Crooks, Mechanical relaxation of organic monolayer films measured by force microscopy. Phys. Rev. Lett. **68**, 2790–2793 (1992)
86. G. Liu, M. Salmeron, Reversible displacement of chemisorbed *n*-alkanethiol molecules on Au(111) surface: an atomic force microscopy study. Langmuir. **10**, 367–370 (1994)
87. K.L. Johnson, K. Kendall, A.D. Roberts, Surface energy and the contact of elastic solids. Proc. Roy. Soc. Lond. A. **324**, 301–313 (1971)
88. H. Hertz, The contact of elastic solids. J. Reine Angew. Math. **92**, 156–171 (1881)
89. B.V. Derjaguin, V.M. Muller, Y.P. Toporov, Effect of contact deformations on the adhesion of particles. J. Colloid Interface Sci. **53**, 314–326 (1975)
90. V.M. Muller, B.V. Derjaguin, Y.P. Toporov, On two methods of calculation of the force of sticking of an elastic sphere to a rigid plane. Colloids Surf. **7**, 251–259 (1983)
91. R.S. Bradley, The cohesive force between solid surfaces and the surface energy of solids. Philos. Mag. **13**, 853–862 (1932)
92. D. Tabor, Surface forces and surface interactions. J. Colloid Interface Sci. **58**, 2–13 (1977)
93. V.M. Muller, V.S. Yushenko, B.V. Derjaguin, On the influence of molecular forces on the deformation of an elastic sphere and its sticking to a rigid plane. J. Colloid Interface Sci. **77**, 91–101 (1980)
94. H. Buckle, *The Science of Hardness Testing and Its Research Applications*, ed. by J.W. Westbrook, H. Conrad (American Society for Materials, Ohio, 1973)
95. D. Lebouvier, P. Gilormini, E. Felder, A kinematic model for plastic indentation of a bilayer. Thin Solid Films. **172**, 227–239 (1989)
96. R. Saha, W.D. Nix, Effects of the substrate on the determination of thin film mechanical properties by nanoindentation. Acta Mater. **50**, 23–38 (2002)
97. M.F. Doerner, W.D. Nix, A method for interpreting the data from depth-sensing indentation instruments. J. Mater. Res. **1**, 601–609 (1986)
98. R.B. King, Elastic analysis of some punch problems for a layered medium. Int. J. Solids Struct. **23**, 1657–1664 (1988)
99. H. Gao, C–H. Chiu, J. Lee, Elastic contact versus indentation modeling of multi-layered materials. Int. J. Solids Struct. **29**, 2471–2492 (1992)
100. E.H. Yoffe, The elastic compliance of a surface film on a substrate. Philos. Mag. Lett. **77**, 69–78 (1998)
101. A. Perriot, E. Barthel, Elastic contact to a coated half-space: effective elastic modulus and real penetration. J. Mater. Res. **19**, 600–608 (2004)
102. H. Xu, G.M. Pharr, An improved relation for the effective elastic compliance of a film/substrate system during indentation by a flat cylindrical punch. Scripta Mater. **55**, 315–318 (2006)
103. J.C. Maxwell, *A Treatise on Electricity and Magnetism* (Oxford University Press, London, 1892)
104. Y.V. Sharvin, A possible method for studying Fermi surfaces. Zh. Eksp. Teor. Fiz. **48**, 984–985 (1965) [Sov. Phys. JETP 21, 655–656 (1965)]
105. B. Nikolic, P.B. Allen, Electron transport through a circular constriction. Phys. Rev. B. **60**, 3963–3969 (1999)

106. B. Ludoph, J.M. van Ruitenbeek, Conductance fluctuations as a tool for investigating the quantum modes in atomic-size metallic contacts. Phys. Rev. B. **61**, 2273–2285 (2000)
107. J. Beale, F. Pease, Limits of high-density, low-force pressure contacts. IEEE Trans. Compon. Packag. Manufact. Technol. A. **17**, 257–262 (1994)
108. J.W. Tringe, T.A. Uhlman, A.C. Oliver, J.E. Houston, A single asperity study of Au/Au electrical contacts. J. Appl. Phys. **93**, 4661–4669 (2003)
109. M. Enachescu, R.W. Carpick, D.F. Ogletree, M. Salmeron, The role of contaminants in the variation of adhesion, friction, and electrical conduction properties of carbide-coated scanning probe tips and Pt(111) in ultrahigh vacuum. J. Appl. Phys. **95**, 7694–7700 (2004)
110. D.-I. Kim, N. Pradeep, F.W. DelRio, R.F. Cook, Mechanical and electrical coupling at metal-insulator-metal nanoscale contacts. Appl. Phys. Lett. **93**, 203102 (2008)
111. R. Stratton, Volt-current characteristics for tunneling through insulating films. J. Phys. Chem. Solids. **23**, 1177–1190 (1962)
112. J.G. Simmons, Generalized formula for the electric tunnel effect between similar electrodes separated by a thin insulating film. J. Appl. Phys. **34**, 1793–1803 (1963)
113. J.G. Simmons, Electric tunnel effect between dissimilar electrodes separated by a thin insulating film. J. Appl. Phys. **34**, 2581–2590 (1963)
114. N.W. Ashcroft, N.D. Mermin, *Solid State Physics* (Holt, Rinehart and Winston, New York, 1976)
115. K. Hansen, M. Brandbyge, Current-voltage relation for thin tunnel barriers: parabolic barrier model. J. Appl. Phys. **95**, 3582–3586 (2004)
116. F.W. DelRio, M.P. de Boer, J.A. Knapp, E.D. Reedy, P.J. Clews, M.L. Dunn, The role of van der Waals force in adhesion of micromachined surfaces. Nat Mater. **4**, 629–634 (2005)
117. F.W. DelRio, M.L. Dunn, M.P. de Boer, Capillary adhesion model for contacting micromachined surfaces. Scripta Mater. **59**, 916–920 (2008)
118. W.D. Callister, *Materials Science and Engineering: An Introduction* (Wiley, New York, 2000)
119. H.W. Fox, W.A. Zisman, The spreading of liquids on low-energy surfaces: hydrocarbon surfaces. J. Colloid Sci. **7**, 428–442 (1952)
120. J.N. Israelachvili, Interfacial forces. J. Vac. Sci. Technol. A. **10**, 2961–2971 (1992)
121. J. Israelachvili, *Intermolecular and Surface Forces* (Academic, London, 1992)
122. C.E.H. Berger, K.O. van der Werf, R.P.H. Kooyman, B.G. de Grooth, J. Greve, Functional group imaging by adhesion AFM applied to lipid monolayers. Langmuir. **11**, 4188–4192 (1995)
123. D.D. Koleske, W.R. Barger, G.U. Lee, R. Colton, Scanning probe microscope study of mixed chain-length phase segregated Langmuir-Blodgett monolayers. J. Mat. Res. Soc. Proc. **464**, 377–383 (1997)
124. X. Xiao, J. Hu, D.H. Charych, M. Salmeron, Chain length dependence of the frictional properties of alkylsilane molecules self-assembled on mica studied by atomic force microscopy. Langmuir. **12**, 235–237 (1996)
125. R.G. Nuzzo, B.R. Zegarski, L.H. Dubois, Fundamental studies of the chemisorption of organosulfur compounds on Au(111). Implications for molecular self-assembly on gold surfaces. J. Am. Chem. Soc. **109**, 733–740 (1987)
126. Z. Bartczak, J. Morawiec, A. Galeski, Deformation of high-density polyethylene produced by rolling with side constraints: mechanical properties of oriented bars. J. Appl. Polym. Sci. **86**, 1405–1412 (2002)
127. C. Boulas, J.V. Davidovits, F. Rondelez, D. Vuillaume, Suppression of charge carrier tunneling through organic self-assembled monolayers. Phys. Rev. Lett. **76**, 4797–4800 (1996)
128. S. Kubatkin, A. Danilov, M. Hjort, J. Cornil, J. –L. Brédas, N. Stuhr-Hansen, P. Hedegard, T. Bjørnholm, Single-electron transistor of a single organic molecule with access to several redox states. Nature. **425**, 698–701 (2003)
129. G.H. Aylward, T.J.V. Findlay, *SI Chemical Data* (Wiley, Sydney, 1974)
130. R.J. Matyi, M.S. Hatzistergos, E. Lifshin, X-ray reflectometry analyses of chromium thin films. Thin Solid Films. **515**, 1286–1293 (2006)
131. D.Y. Petrovykh, H. Kimura-Suda, M.J. Tarlov, L.J. Whitman, Quantitative characterization of DNA films by x-ray photoelectron spectroscopy. Langmuir. **20**, 429–440 (2004)

132. C.J. Powell, A. Jablonski, *NIST Electron Effective-Attenuation-Length Database – Version 1.1* (National Institute of Standards and Technology: Gaithersburg, MD, 2003)
133. H. Tsuchiya, S. Fujimoto, O. Chihara, T. Shibata, Semiconductive behavior of passive films formed on pure Cr and Fe-Cr alloys in sulfuric acid solution. Electrochim. Acta. **47**, 4357–4366 (2002)
134. M. Gasgnier, L. Nevot, Analysis and crystallographic structures of chromium thin films. Phys. Status Solidi A. **66**, 525–540 (1981)
135. M. Salmeron, Generation of defects in model lubricant monolayers and their contribution to energy dissipation in friction. Trib. Lett. **10**, 69–79 (2001)
136. H. Song, H. Lee, T. Lee, Intermolecular chain-to-chain tunneling in metal-alkanethiol-metal junctions. J. Am. Chem. Soc. **129**, 3806–3807 (2007)

Chapter 16
Assessment of Nanoadhesion and Nanofriction Properties of Formulated Cellulose-Based Biopolymers by AFM

Maurice Brogly, Ahmad Fahs, and Sophie Bistac

Abstract Hydroxypropyl methylcellulose (HPMC) constitutes one of the most dedicated polymers used in the production of film coatings for pharmaceutical applications (capsules, tablets, etc.). In order to control the surface properties of HPMC films, additives are frequently incorporated during film formulation: these are in most cases hydrophobic lubricant (like fatty acids) or hydrophilic plasticizer (like polyethylene glycol). The main objective is to present the surface characteristics in terms of structuration, surface morphologies, surface phase separation, surface energy, nanoadhesion, and nanofriction properties of HPMC-formulated films. At nanoscale, atomic force microscopy in contact mode and in friction mode is a powerful tool for studying nanoadhesion and nanofriction. The presented chapter underlines the strong dependence of film surface properties on additive nature, concentration, and/or water sensitivity. It also shows that first the HPMC-additive compatibility seems to be an interesting factor behind the variation of surface properties of HPMC-formulated films, and second that formulation is an effective way to tune surface properties of HPMC biopolymer films.

16.1 Introduction

Hard two-piece capsules have been used for almost a century in the pharmaceutical field, and gelatin constitutes almost exclusively the main material of these capsules due to its excellent characteristic as film forming, availability, and edible compound. Gelatin is not universally compatible with all active ingredients because its high moisture content renders it unsuitable for the encapsulation of hygroscopic drugs or excipients. Furthermore, the use of animal-derived products is subject to strong restrictions owing to the fear of transmissible spongiform encephalopathy (TSE) to the human body. In the last years, there were growing interests and works to examine green-based materials to replace the gelatin-based capsules. Cellulose derivatives are often used to modify the release of drugs in tablet and capsule formulations [1, 2]. Hydroxypropyl methylcellulose (HPMC) appears to be a successful biomaterial alternative for film coating formulations. Indeed HPMC is characterized by its ease of use, availability, water solubility, and nontoxicity. It is primarily used as controlled release matrix in formulated pharmaceutical films [3]. Drug release from these systems is controlled by the hydration of HPMC, which forms a gelatinous layer at the surface of the matrix through which the included drug diffuses. HPMC films must be optimized to achieve the desired properties (in terms of moisture content, barrier properties, mechanical and surface characteristics, etc.). Incorporation of additives such as plasticizers [4], surfactants [5–7], colorants or pigments [8], etc., in film coating formulations constitutes the best solution to improve some specific properties of HPMC film. Several works have focused on the influence of additives on physicochemical properties of HPMC films. It is known that lipid compounds such as waxes, triglycerides (e.g., tristearin), fatty acids (e.g., stearic and palmitic acid), frequently incorporated into HPMC films, lead to a decrease in the water affinity and moisture transfer due to their high hydrophobic properties [9]. Nevertheless, very few studies concern surface properties of HPMCformulated films [10] The present review focuses on the understanding of the effects of additives on the surface characteristics of the formulated films. Surface characteristics of new green-based film coatings will be explored in terms of structuration, surface morphologies, surface phase separation, surface energy, adhesion, and sliding properties. In order to understand how the hydrophobicity/hydrophilicity of the surface can be tuned, the effects of additive concentration on wettability, adhesion, and friction behaviors are discussed. The HPMC-formulated films are analyzed by using contact angle to evaluate film's wettability. Atomic force microscopy (AFM) in tapping mode (TM-AFM) was used to explore surface morphology. Adhesion and friction measurements at nanoscale were performed by chemical force microscopy (CFM) and friction force microscopy (FFM), respectively.

16.2 Application of Cellulose-Based Biopolymers in Pharmaceutical Formulations

Cellulose ethers are extremely versatile materials, which are widely used in pharmaceutical products. They are manufactured by reacting cellulose in alkaline solution and their properties are largely dependent on the nature and extent of substitution. Polymers used for film coating purpose (capsules, tablets) can be classified into three groups [11]:

- The first group includes watersoluble polymers such as methylcellulose (MC), hydroxyethylcellulose (HEC), HPMC, watersoluble vinyl derivatives (polyvinyl alcohol and polyvinyl pyrrolidone), and some acrylic copolymers and high molecular weight polyglycols.
- The second group includes the water-insoluble polymers such as ethyl cellulose (EC), polyvinyl acetate and some acrylates, which are mainly used for drug release control.

- The third group includes polymers which are soluble in a limited range of PH such as HPMC phthalate, cellulose acetate phthalate, or those that dissolve following enzymatic degradation.

Obviously HPMC appears to be one of the most interesting biopolymers. In addition to its main role in film formulation of capsule shell, HPMC polymer plays many other functions in pharmaceutical field. HPMC is used as film coating of tablets. The main reason for the extensive use of HPMC as film coating polymer for capsules and tablets is that it is soluble in some organic solvents and also in water over the entire biological pH range [12]. At low concentrations (2–6%), HPMC is used as binders in pharmaceutical formulation to produce hard tablets and improve the compaction properties during the tableting process (binders are used in the formulation of solid oral dosage forms to hold the pharmaceutical ingredient together in a cohesive mix) [13]. At concentration (15–80%) high-viscosity grade HPMC plays an important role as a controlled release matrix in solid dosage forms [14, 15].

The solid dosage forms must be coated with a "nonfunctional" or "unreactive" polymeric film coating in order to protect a drug from degradation, mask the taste of an unpleasant drug or excipient, or improve the visual elegance of the formulation [16]. The film coating containing cellulose ethers like HPMC extends the drug release by rapidly forming a protective viscous layer around the tablet surface when exposed to aqueous fluids. This hydrated viscous layer controls water penetration into the central dry core of the tablet and prevents disintegration [17].

In liquid dosage forms, HPMC is used as a suspending and thickening agent in pharmaceutical formulations. Suspending agents are used to prevent sedimentation by influencing the rheological behaviors of a suspension. The nontoxicity, viscous property and stability over a wide pH range make it a good suspending agent. At low concentration (0.45–1%) it could be added as a thickening agent in formulations for ophthalmic use [18].

16.3 General Composition of Pharmaceutical Film Coatings

Cellulose ethers are the preferred materials in the formation of film coatings, but films and coatings should satisfy many important criteria such as uniform thickness, free from cracks, and stable on storage and handling. In fact, the formulation of films and coatings needs to be optimized and other film constituents may be added to enhance the final characteristics. We would mainly consider two mainly used constituents here: plasticizers and surfactants.

16.3.1 Plasticizers

Plasticizers are used in film coating formulations to increase the flexibility and reduce the brittleness of the film. They generally account for 20–30% of the formulation. The amount and choice of the plasticizer help to determine the hardness of the final product [19]. They are selected on the basis of their compatibility with the polymer matrix. Plasticizer effects are due to a decrease in intermolecular forces between polymer chains, by insertion and "swelling" mechanisms. This fact lowers the glass transition temperature T_g [20, 21]. The most common plasticizers used for pharmaceutical film coatings are classified into three groups and include:

- Polyols such as glycerol, propylene glycol and polyethylene glycols (PEG) with molecular weights ranging from 200 to 6,000 g mol^{-1}.
- Organic esters such as triacetin, phthalate esters (diethyl, dibutyl) and citrate esters.
- Oils and glycerides such as castor oil, acetylated monoglycerides and fractionated coconut oil.

Solubility and compatibility of the plasticizer with the polymer are of major importance for effective plasticization. The effects of plasticizer can be evaluated by several techniques: tensile test (mechanical properties), determination of the glass transition temperature (thermal properties by using DSC, DMTA), etc.

16.3.2 Surfactants and Lubricants

Surface-active agents (surfactants) are substances which, at low concentrations, adsorb onto the surfaces or interfaces of a system and decrease their surface energy. Surfactants have a characteristic structure, possessing both polar (hydrophilic) and nonpolar (hydrophobic) behaviors in the same molecule. They have been widely used in film coating formulations to improve wetting and spreadability of the coating material. In many pharmaceutical application domains, surfactants improve the solubility or stability of a drug in a liquid preparation, stabilize and modify the texture of a semi-solid preparation, in addition to its important role in the formulation processing of solid dosage forms [22]. They are included in tablet coatings and capsules. Four main groups of surfactants, classified according to the nature of the hydrophilic group, are defined as follows:

- Anionic surfactants, where the hydrophilic group carries a negative charge. Examples of pharmaceutical importance include potassium laurate $CH_3(CH_2)_{10}COO^-K^+$ and sodium lauryl sulfate $CH_3(CH_2)_{11}SO_4^-Na^+$.
- Cationic surfactants, where the hydrophilic group carries a positive charge such as alkylbenzyl dimethylammonium chlorides $C_6H_5CH_2N^+(CH_3)_2RCl^-$, where R represents the alkyls from C_8H_{17} to $C_{18}H_{37}$.
- Amphoteric surfactants (zwitterionic surfactants), where the molecule contains a negative and a positive charge, e.g., *N*-dodecyl-*N,N*-dimethylbetaine $C_{12}H_{25}N^+(CH_3)_2CH_2COO^-$
- Nonionic surfactants, where the molecule carries no charge such as polyoxyethylated glycolmonoethers, sorbitan esters (Spans®) and polysorbates (Tweens®).

Moreover, lubricants are used in film coating formulations to reduce the friction arising at the interface of tablet coatings and die walls during compression and ejection (by interposing an intermediate film at the interface between the tablet and the die wall). They possess anti-adherent characteristics and are also used for manufacture of hard capsules. They may be classified as water-soluble or water-insoluble. The water-insoluble lubricants are generally more effective than water-soluble lubricants and can be used at a lower concentration. Common water-insoluble lubricants include magnesium stearate, calcium stearate, sodium stearate, and stearic acid. Water-soluble lubricants include sodium lauryl sulfate and magnesium lauryl sulfate. For example, sodium lauryl sulfate is used in the production of hard gelatin capsules where it is added to the gelatin solution during the preparation stage. The stainless steel molds are lubricated prior to dipping into the gelatin solution and sodium lauryl sulfate is added to reduce the surface tension of the mixture and causes the mold pins to wet more uniformly [23].

16.4 Structure and Bulk Properties of HPMC Biopolymers

16.4.1 Chemical Structure of HPMC

Cellulose ethers are a class of semi-synthetic polymers obtained by chemical reaction of the hydroxyl groups at positions 2, 3, and/or 6 of the anhydroglucose residues of cellulose, which is made of D-glucopyranose units of conformation chair, bonded through $\beta(1 \rightarrow 4)$ glycosidic linkages. HPMC, one of the cellulose ether, contains two types of substituents: the methoxy group (OCH_3) and the hydroxypropyl group (OC_3H_6OH). The chemical structure of HPMC is shown in Fig. 16.1

The manufacture of HPMC and cellulose ethers is obtained by mixing the cellulose with aqueous sodium hydroxide solution (35–60% w/v) for a predetermined time under controlled temperature and pressure to ensure complete reaction. Treating the cellulose with caustic alkaline solution disrupts hydrogen bonding within the polymeric chains, destroys the crystalline organization and makes the majority of the hydroxyl groups available for chemical modification [24].

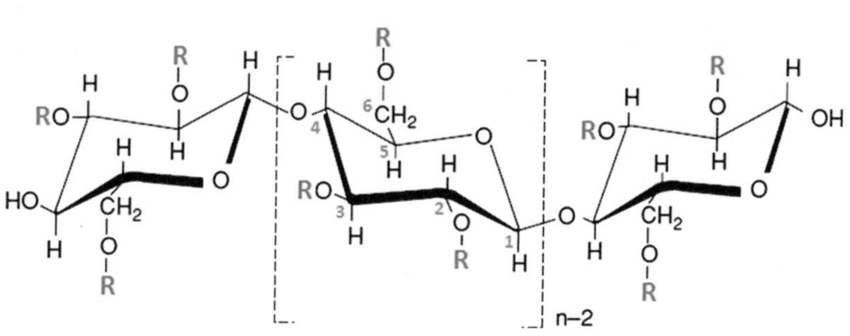

Fig. 16.1 Chemical structure of hydroxypropyl methylcellulose (HPMC), R = H, $-CH_3$ or $-(OCH_2CHCH_3)_xOH$

The preparation of alkali cellulose proceeds as follows:

$$R_{cell} + NaOH \leftrightarrows R_{cell}ONa + H_2O \tag{16.1}$$

To obtain the MC, alkali cellulose reacts with methyl chloride to give MC and sodium chloride according to:

$$R_{cell}OH + CH_3Cl \xrightarrow{NaOH} R_{cell}OCH_3 + NaCl + H_2O \tag{16.2}$$

Hydroxyalkylation is carried out by treating alkali cellulose with an epoxide. HPMC is obtained when MC reacts with propylene oxide. The ring of epoxide opens at the primary carbon leading to a product containing a secondary alcohol function which remains available for further reaction.

$$R_{cell}OH + CH_3CH(O)CH_2 \xrightarrow{NaOH} R_{cell}OCH_2CH(OH)CH_3 + NaOH \tag{16.3}$$

The purification of HPMC is carried out by hot water washing and filtration procedures.

One of the fundamental points to obtain reproducible, quantitative, and reliable data is the calibration procedure, which should be rigorous and systematic for all measurements.

16.4.2 Physicochemical Properties

The physicochemical properties of HPMC polymers are strongly affected by (1) the methoxy group content; (2) the hydroxypropoxy group content; and (3) the molecular weight [25]. However, these products are described by the degree of substitution (DS) and the molar substitution (MS). Each anhydroglucose unit in the cellulose chain has three hydroxyl groups available for modification. Thus, if all three available positions on each unit are substituted, the DS is designated as 3, if average of two on each ring is reacted, the DS would be 2 (Fig. 16.2). The term DS is reserved for substituents that block reactive hydroxyl groups (methoxy groups).

Derivatization of a reactive hydroxyl group with propylene oxide generates a replacement hydroxyl site for further reaction, and then the formation of internal and external hydroxypropoxy groups. Thus, the substitution is described by the MS i.e., the number of moles of hydroxypropyl group per mole of anhydroglucose in the chain (Fig. 16.3)

The United States Pharmacopeia (USP) distinguishes different types of HPMC, classified according to their relative methoxy and hydroxypropoxy contents: HPMC 1828, HPMC 2208, HPMC 2906 and HPMC 2910. The first two numbers indicate the weight percent of methoxy groups, the last two numbers the weight percent of hydroxypropoxy groups, determined after drying at 105°C for 2 h.

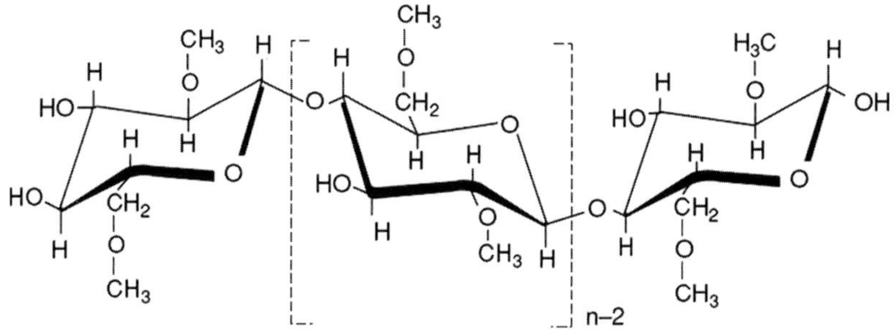

Fig. 16.2 Methylcellulose with a DS of 2.0

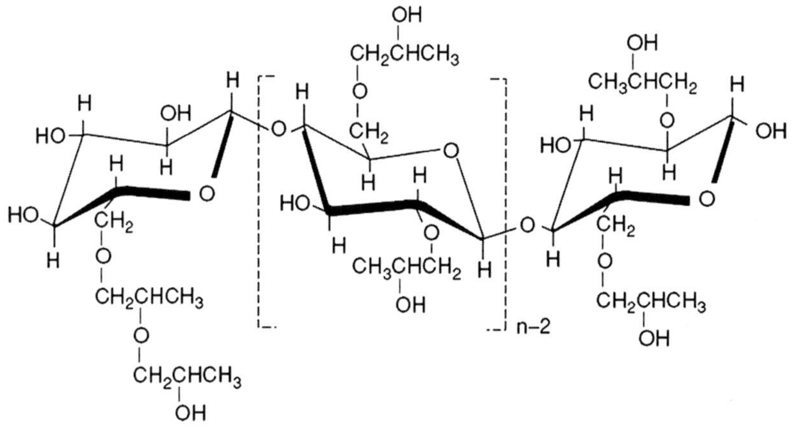

Fig. 16.3 Hydroxypropyl cellulose with an MS of 2.0

Another important parameter of HPMC polymer is its molecular weight. Commercially, HPMC polymers are available with different molecular weights and thus give different solution viscosity. The viscosity designation is defined as the nominal viscosity of a 2% w/w aqueous solution at 20°C. Since viscosity of the HPMC solution is directly correlated with its molecular weight, viscosity measurements are used for quality control as an alternative way of molecular weight determination. The viscosity of the aqueous solution increases with increasing the molecular weight of the polymer or the concentration. HPMC is soluble in cold water and form a viscous solution, but is not soluble in hot water. Additionally, it can be dissolved in most polar organic solvents or binary systems of methylene chloride or chloroform and alcohol. It is practically insoluble in pure chloroform, ethanol, or ether [26] HPMC 2910 has the best solubility in organic solvents, and so it has often been used for aqueous and organic solvent-based films.

Aqueous solution of HPMC presents a gelpoint characteristic at specific temperature which is affected by the nature and the quantity of substitution on anhydroglucose ring (i.e., gel point of HPMC 2910 in aqueous solution [2% w/v] is 56°C and for HPMC 2208 is 70°C). The gel is completely reversible in that it is formed upon

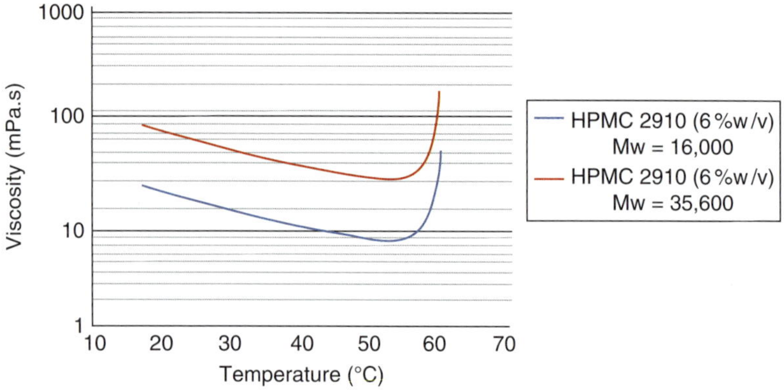

Fig. 16.4 Effect of temperature on the viscosity of aqueous solutions of HPMC. Dramatic increases in viscosity is observed near 60°C, which indicates the occurrence of gelatin [27]

heating yet will liquefy upon cooling. At lower temperature, molecules are hydrated and little polymer–polymer interaction remains (Fig. 16.4) [27].

As the temperature increases, the molecules gradually lose their hydration water as reflected by a decrease in viscosity. When a sufficient dehydration of polymer occurs, a polymer–polymer association takes place as reflected by a dramatic increase in viscosity and the flocculation temperature is reached. Flocculation results from the weakening of hydrogen bonding between polymer and water molecules and the strengthening of interactions between polymer chains [28] Along the cellulose backbone, methyl substitutes constitute hydrophobic zones whereas hydroxypropyl groups are more hydrophilic. The introduction of these groups allows HPMC to behave as a surfactant. Thus HPMCs are adsorbed at fluid interfaces lowering the surface tension. With an increase in the degree of substitution, the polymer hydrophobicity and then the interfacial activity increases [29–31]. The usefulness of these nonionic cellulose ethers such as HPMC is essentially based on four key attributes: efficient thickening, surface activity, film-forming ability and the ability to form thermal gels that melt upon cooling [32].

The thermal behaviors of HPMC polymers depend on their types and/or grades. The substitution of hydroxyl OH (hydrophilic group) on anhydroglucose units by methoxy OCH_3 (hydrophobic group) decreases the hydrophilic character of HPMC polymer, and consequently the amount of adsorbed water, which could be determined by thermal analysis (DSC, TGA). DSC analysis shows that HPMC polymers are purely amorphous [33]. Concerning the glass transition temperature, T_g, increase in the substitution degrees on cellulose chains leads to decrease in the T_g. Thus the T_g of HPMC 2910 is 137°C whereas the T_g of HPMC 2208 is 153°C. Intramolecular and intermolecular interactions can occur between unsubstituted hydroxyl groups and ether oxygen groups introduced by the substituent groups (hydroxypropoxy). The fact that hydrogen bonds decrease as DS increases especially by introducing methoxy substituents, is the essential cause of T_g reduction and also depends thus on the content of the alkyl substituents present on the chains [34]. Moreover the glass transition temperature of HPMCs increases linearly (log scale) with the average molecular weight of the cellulosic chain.

16.5 Physicochemical Properties of HPMC-Formulated Films

Cellulose-based films for pharmaceutical uses must exhibit good mechanical properties, but need to avoid ruptures and cracks in the film. As a consequence, they are formulated with plasticizers. Several studies on plasticization of HPMC films indicate that the PEG has a good effectiveness. The HPMC films plasticized with PEGs became softer and more viscoelastic with increase in plasticizer content. Plasticization efficiency increases as PEG molecular weight decreases. Increasing the concentration of PEGs progressively lowers the tensile strength (TS), Young's modulus, and the glass transition temperature of cellulose ether films [4, 11]. The addition of plasticizers into films also affects the surface properties of the films [35]. Hydrophobic additives are frequently added to control physicochemical properties of HPMC films. Fatty acids frequently incorporated into HPMC films lead to a decrease in the water affinity and moisture transfer due to their hydrophobic properties [36, 37]. The surface-active agents or surfactants are added to coating solutions and films to reduce the surface tension and modify the wetting properties by increasing the spreadability of the coating material [38]. But sometimes, the addition of surfactants to film coating formulations does not only affect the surface properties but also affects the mechanical or other properties of the films [39].

16.5.1 Materials

In this review, we will consider HPMC 2910 (Methocel E3 Premium Low Viscosity LV; Dow Chemical Company, Colorcon, France) as film matrix. This grade has a nominal viscosity of 3 mPa s (cP) at 2% aqueous solution at 20°C. The degree of substitution of methoxy groups (DS) and the mass molar substitution of hydroxypropoxy groups (MS) of HPMC used are 1.87 and 0.25, respectively (by weight percent, the amount of OCH_3 groups is 28.3% [w/w] and $OCH_2(CHOH)CH_3$ groups is 9.1% [w/w]).

Stearic acid and polyethylene glycol 400 (Sigma-Aldrich Chemical Company) were used as additives for film-forming process. PEG 400 is infinitely soluble in water, its density is 1.128 g cm^{-3}. The purity of the used additives is greater than 99%. Complementary characteristics of the two additives are gathered in Table 16.1.

Polyethylene glycol, otherwise known as polyoxyethylene or polyethylene oxide (PEO), is a synthetic polyether that is readily available in a wide range of molecular weights. Compounds with $M_w < 100{,}000$ are usually called PEGs, while higher molecular weight polymers are classified as PEOs. PEG has been found to be nontoxic and used as excipient in different pharmaceutical formulations, foods, and

Table 16.1 Physicochemical characteristics of stearic acid and polyethylene glycol 400

Characteristic	Stearic acid	Polyethylene glycol 400
Chemical structure	$C_{18}H_{36}O_2$	$(C_2H_4O)_mH_2O$
Molecular weight	$284.48\ g\ mol^{-1}$	$380–420\ g\ mol^{-1}$
Structural formula	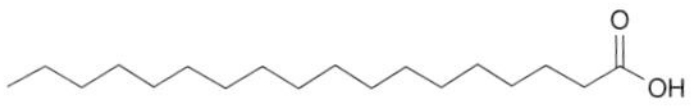	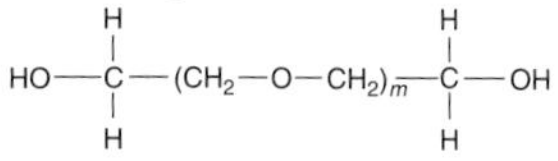
Melting point	69–72°C	4–6°C
Solubility	Ethanol, chloroform, ether, and practically insoluble in water	Soluble and miscible in water, alcohols, acetone
State	White crystalline powder	Clear viscous liquid, colorless or slightly yellow-colored
Stability	Stable and antioxidant material	Stable and hygroscopic

cosmetic fields [40]. Stearic acid is derived from edible fat. It is widely used in pharmaceutical formulations and mainly used in oral formulations as tablets and capsules lubricant [41].

16.5.2 Pure HPMC Film Formation

Pure HPMC films were obtained by preparing the HPMC solutions (6% w/v – HPMC was dissolved by hydration and cooling to 25±2°C) and casting onto a glass plate (Petri dish), then drying at ambient conditions. To obtain 100-μm-thick films, solutions were poured in a uniform layer of 1.6 mm thickness onto a glass plate. The dried films were conditioned at 25 ± 2°C and 30 ± 5% RH before experiments.

16.5.3 Formulation of HPMC–Stearic Acid Films and HPMC–PEG Films

HPMC-formulated films were prepared by incorporation of additives in HPMC solution. The amount of additives was expressed as a weight percentage of HPMC powder. For HPMC–stearic acid (SA) films, fatty acid was firstly dissolved in 10 ml of absolute ethanol, and then added (0.05%, 0.1%, 0.5%, and 1% w/w) to HPMC solution under magnetic agitation. The PEG 400 has been added (10%, 20%, 30%, and 40% w/w) without solubilization in ethanol because it is water soluble and it can be homogenized with HPMC solution. Film-forming solutions were spread onto a glass plate to form uniform layers (1.6 mm of thickness), and dried under natural conditions to obtain 100 μm-thick films. All formulated films were preserved at 25 ± 2°C and 30 ± 5% RH before experiments.

16.5.4 Thermomechanical Properties of HPMC–PEG Films

Dynamic mechanical analysis (DMA) experiments are conducted to measure the storage and loss modulus of formulated films as a function of temperature. The storage modulus G', loss modulus G'' (dissipated energy), and damping coefficient tan δ (ratio of loss modulus to storage modulus) are determined [42]. Cooperative segmental mobility representative of the glass transition phenomenon was considered as the maximum of tan δ. Thermo-mechanical properties of plasticized HPMC films were studied between 0°C and 225°C by DMA. Figure 16.5a, b shows, respectively, the storage modulus G' and tan δ curves of HPMC–PEG films as a function of temperature.

For pure HPMC films, the storage modulus decreases sharply above 130°C and tan δ maximum is recorded at 160°C corresponding to the glass transition phenomenon. The addition of plasticizer shows that only one transition is present indicating the thermodynamic miscibility PEG with HPMC matrix. Addition of PEG leads to decrease both G' and the intensity of the maximum of tan δ. The tan δ peak becomes broader for plasticized HPMC films indicating the distribution of H bond sites in the blend. The position of tan δ peak was also changed according to the PEG content (Fig. 16.6).

DMA analysis showed high HPMC matrix–PEG plasticizer interactions, resulting in a good film plasticizing effect and a decrease of the glass transition of initial HPMC polymer.

The effect of PEG on mechanical properties of HPMC films was also determined by TS test. Figure 16.7 shows the TS and the ultimate elongation (UE) of HPMC films as a function of concentration of PEG (% w/w HPMC). Films were tested at a velocity of 1 mm min^{-1} after a suitable storage period of 7 days at 25 ± 2°C and 55 ± 5% RH.

The TS and UE of pure HPMC film are, respectively, estimated to be 12 MPa and 2%. The introduction of PEG into HPMC films changes significantly the TS and UE of HPMC films. The plasticized films become more flexible and less brittle. Addition of 40% of PEG decreases six times the TS and increases four times the UE. The plasticizer may gradually diffuse into the polymer chains, reduces the intermolecular forces of attraction along the polymer chains, allows the polymer chains to move more easily and, consequently, lowers the elastic modulus and the glass transition temperature [43]. The free space around the HPMC polymer chains increases with introduction of PEG, this fact provides a greater volume for movement of the polymer chains. The mechanical properties are directly related to the structure characteristics of the polymer and glass transition, which decrease with addition of PEG plasticizer [44, 45].

16.5.5 Thermo-Mechanical Properties of HPMC–SA Films

Dynamic mechanical measurements indicate no significant changes of tan δ with incorporation of stearic acid into HPMC films. Tan δ peak is located in the same

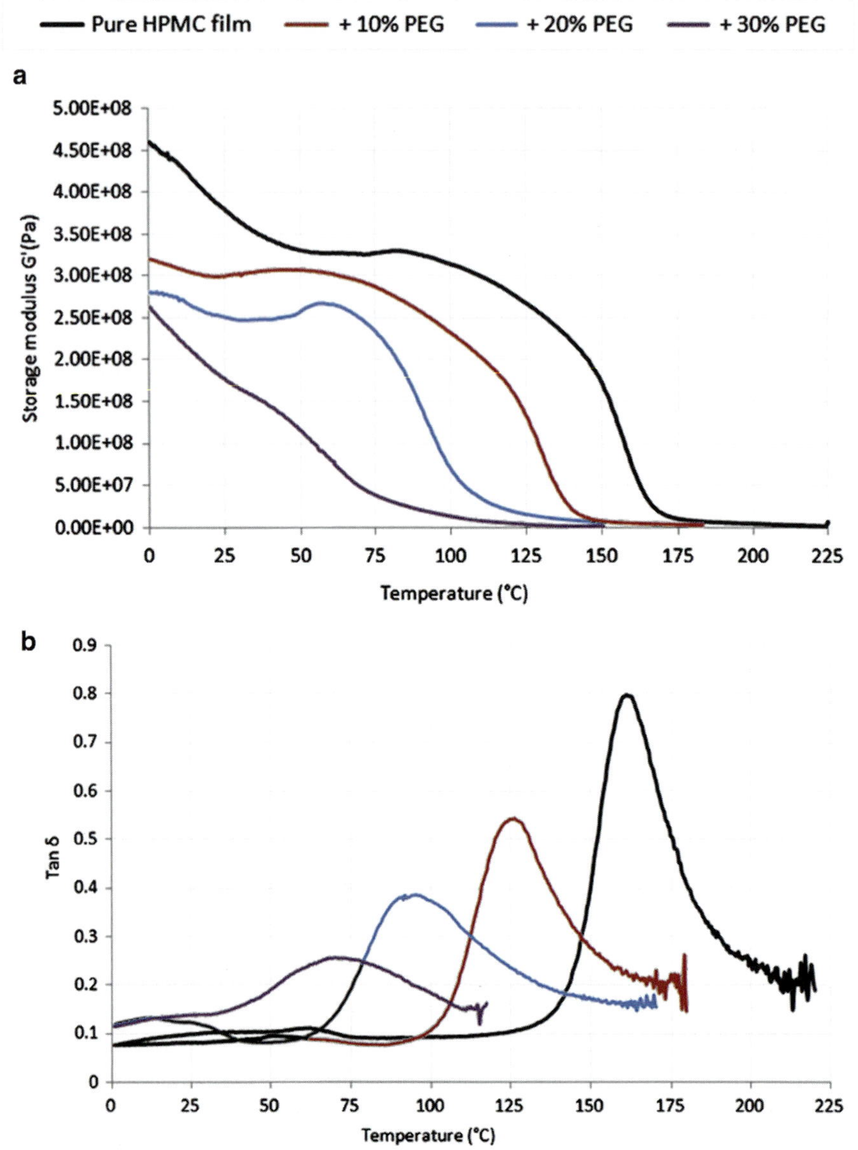

Fig. 16.5 Storage modulus G' (**a**) and tan δ (**b**) curves of HPMC–PEG films measured by DMA (10 Hz)

thermal range for the films containing stearic acid (stearic acid content is ranging from 0 to 1% w/w). In addition, tan δ maximum and tan δ integrated intensity remain quite unchanged according to the stearic acid content. This result also supports the hypothesis of a strong possibility of phase separation between the HPMC matrix and the stearic acid. On the other hand, one must be careful about the

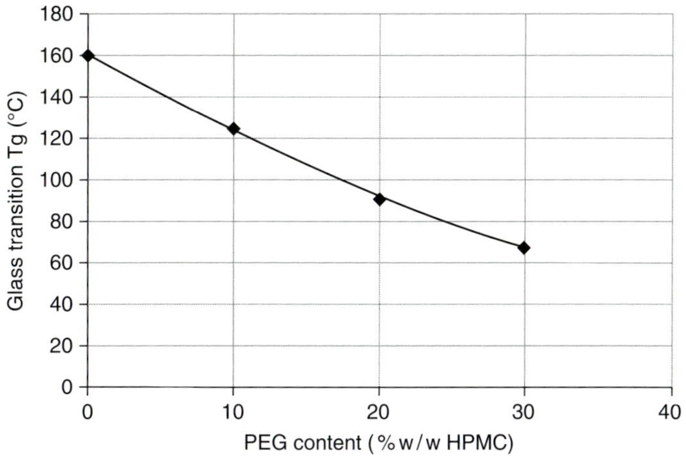

Fig. 16.6 Variation of glass transition tan δ (DMA) according to the PEG content

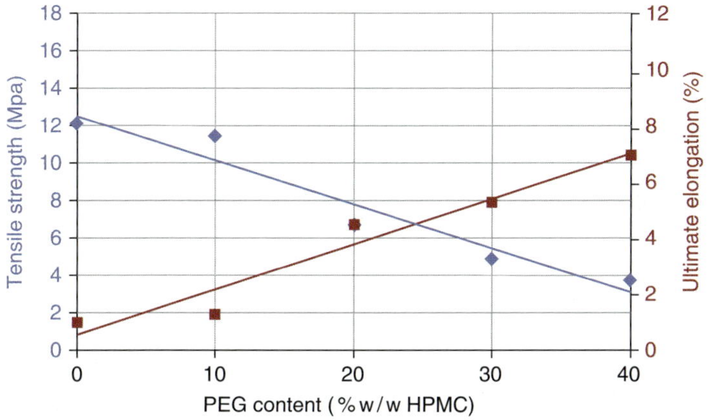

Fig. 16.7 Effect of PEG content on tensile strength (TS) and ultimate elongation (UE) of HPMC films

experimental sensibility of the technique. An amount of 1% additive may not be sufficient to detect the occurred changes. Nevertheless, films appearance (increase of opacity observed for HPMC–SA films) and results obtained by DMA support this hypothesis of phase separation.

The uniaxial tensile experiments show that, even with low content of fatty acid, the mechanical properties of HPMC films are affected. TS and UE values decrease sharply and the test becomes not sensitive to the variation of TS and UE. Incorporation of stearic acid makes the HPMC films more breakable. These observations support the results of literature [46, 47], which reported that the mechanical resistance of film decreases with fatty acid addition. The negative effect of hydrophobic stearic acid on mechanical properties is due to phase separation with HPMC film matrix, resulting in discontinuities within HPMC polymer network or creation of low cohesion domains which decrease mechanical properties.

To summarize, the use of PEG in formulation has a significant effect on the bulk properties of HPMC films. If this plasticizer may provide a solution for film flexibility and facilitates its industrial applications, it may have an important influence on film hydrophilicity. However, the effectiveness of stearic acid on film hydrophobicity may appear with introduction of very low content (1% w/w) due to phase separation, but in this case, it was possible to alter the film softness. As functional properties of HPMCformulated films are directly related to the composition and structure of its components, the influence of nature and content of additives on film surface properties will be investigated in the next section.

16.6 Surface Properties of HPMC-Formulated Films Adhesion

16.6.1 Surface Topography and Morphologies by AFM

The molecular structure of HPMC allows to behave as surfactant, because the molecules preferentially adsorb at the liquid–air interface and lower the solution surface tension. The methyl groups on the cellulose chains constitute the hydrophobic zones whereas the hydroxypropyl groups are more hydrophilic. The surface activity, film-forming ability, thermal gelling, and efficient thickening are the essential characteristics which encourage the usefulness of HPMC [48]. Nevertheless, very few studies concern surface properties of HPMC-formulated films. The present chapter focuses on understanding the effects of additives on the surface characteristics of HPMC-formulated films. Surface characteristics will be explored in terms of structuration, surface morphologies, surface phase separation, surface energy, adhesion, and sliding properties.

The aim of this section is to investigate the surface properties of HPMC-formulated films. In order to control the hydrophobicity/hydrophilicity of the surface, the effects of additive nature and concentration on adhesion and friction behaviors are discussed. The HPMC-formulated films are analyzed by TM-AFM to explore surface morphology whereas adhesion and friction measurements at nanoscale were performed by CFM and FFM, respectively.

16.6.1.1 Surface Imaging of Pure HPMC Film

Surface topographies and morphologies of pure HPMC films were firstly explored using a Nanoscope IIIa AFM (Digital Instruments, Santa Barbara, CA, USA). Silicon probes (Veeco probes), with a spring constant of $5\,N\,m^{-1}$, a resonance frequency of 150 kHz and tip radius <10 nm were used. All images were collected with

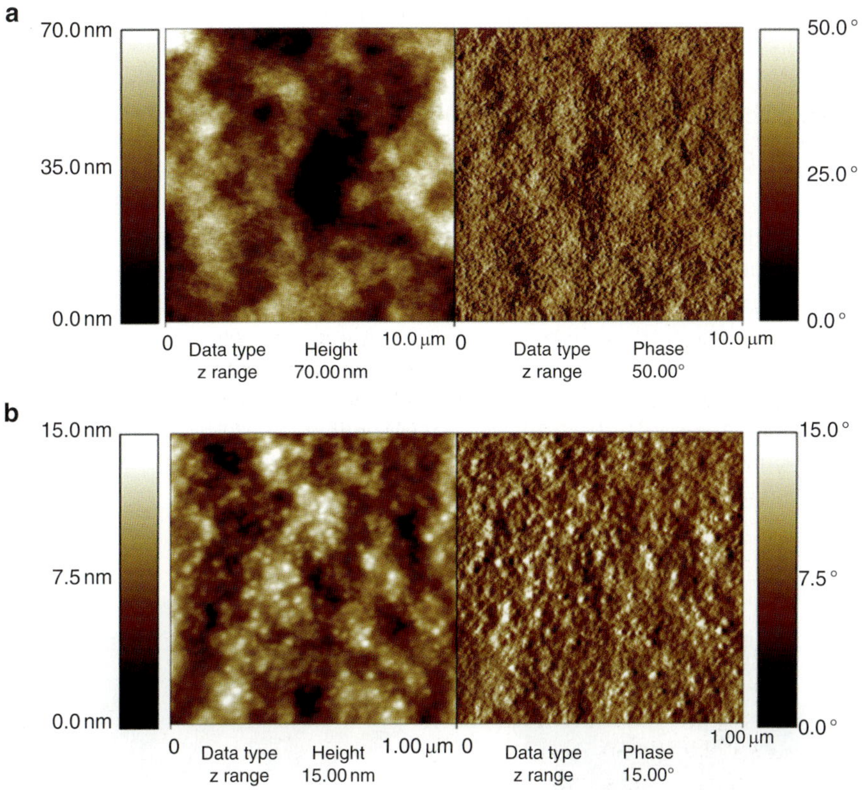

Fig. 16.8 AFM topographic and phase contrast images at two scales: (**a**) ×10 μm and (**b**) 1×1 μm of pure HPMC film

a resolution of 512×512 pixels and scan rate of 0.8 Hz. The surface topography was carried out by TM-AFM.

Figure 16.8 represents two AFM images at two different scales of pure HPMC films, 10 × 10 μm and 1 × 1 μm, respectively.

Phase contrast images represent the phase lag between the piezo excitation oscillation and the tip oscillation. Therefore, such images reveal differences in terms of mechanical and damping properties of the surface. The phase contrast images at large scale (10 × 10 μm) reveal the homogeneity of mechanical/chemical properties of the surface. At smaller scale (1 × 1 μm), phase lag is very low (<10°), indicating even locally the homogeneity of the film surface. On the other hand, the corresponding topographic image (1 × 1 μm) shows the presence of nanodomains, of spherical form, usually called clusters in the literature with a characteristic size 20–30 nm. HPMC nanodomains are due to the existence of aggregates or clusters of HPMC chains in the film-forming solution. The hydrophobic interactions between the hydrophobic substituents of HPMC polymer induce the formation of these clusters [49].

16.6.1.2 Surface Imaging of HPMC–PEG Films

Surface aspect of HPMC–PEG films was also studied by TM-AFM, where the oscillating cantilever lightly taps the films surface. One remembers that this mode gives two major information: the topographical data and phase contrast images. For all the formulated films, nanodomains are always present on topographic images and the phase contrast images obtained present very low phase shift (10–15°). This low phase shift measured is due to high HPMC–PEG compatibility and homogenous film formation, which was previously revealed by thermal and mechanical analysis of these films (see Sect. 16.5). Figure 16.9 shows typical AFM topographic images of pure and HPMC–PEG films.

The scan size of images was $1 \times 1\,\mu m$ and the Z-range was 25 nm. As mentioned above, HPMC film surfaces are formed by spherical nanodomains or clusters having a nanometric size (20–30 nm for pure HPMC films). However, a variation of the characteristic size of the clusters is observed after incorporation of 20% PEG or

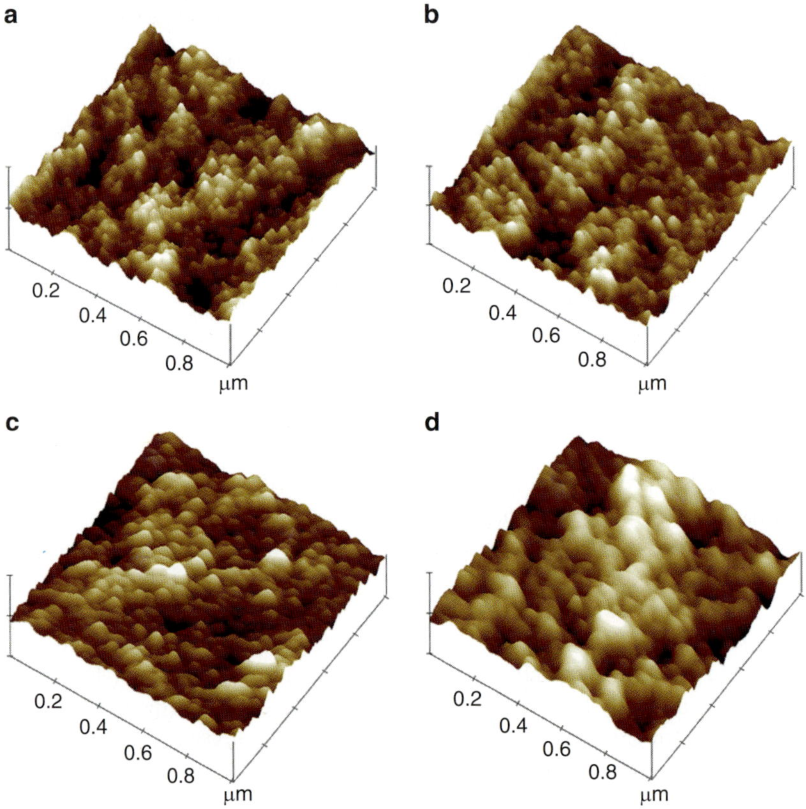

Fig. 16.9 Topographical AFM images ($1 \times 1\,\mu m$) of (**a**) pure HPMC film and HPMC–PEG films with (**b**) 20%, (**c**) 30%, and (**d**) 40% w/w PEG content, respectively

Table 16.2 Values of average roughness and clusters size measured by AFM on HMPC and HPMC–PEG films

Sample	Average roughness (nm)	Clusters size diameter (nm)
Pure HPMC	2.4 ± 0.2	25 ± 5
+20% PEG	2.2 ± 0.2	35 ± 5
+30% PEG	2.6 ± 0.3	50 ± 7
+40% PEG	2.9 ± 0.3	70 ± 10

more. The characteristics of the topographical images obtained are summarized in Table 16.2.

The addition of larger amount into HPMC films increases significantly the cluster's size and an average value of 70 nm is measured for 40% of PEG. Cross-section analysis indicates that the nanodomain's volume is widened, especially in the case of high amount of PEG. No significant difference on surface roughness is observed. On the basis of plasticizing theory, which involves the increase in intermolecular spaces in the polymer and increase in free volume between the polymer chains [50], the variation of cluster's characteristic size observed is proposed to be due to insertion of PEG molecules within HPMC-associated chains. As a consequence, the size of HPMC nanodomains increases.

16.6.1.3 Surface Imaging of HPMC–SA Films

Tapping mode AFM analysis of four HPMC film samples with different stearic acid contents (0%, 0.1%, 0.5%, and 1% w/w) indicates a variation of surface topography as represented in Fig. 16.10. No contrast was observed on the phase images.

The topographic images and the section analysis show clear variation of surface aspect. The scan size of each image is $1 \times 1\,\mu$m and the Z-range is 15 nm. Figure 16.10a concerns pure HPMC film. Figure 16.10b–d is observed for formulated films having stearic acid content equal to 0.1, 0.5, and 1%, respectively. Average surface roughness, R_a, expressed in nm, is equal to 2.3, 1.8, 1.2, and 0.8, respectively. R_a values decrease with stearic acid content from 2.3 nm for pure HPMC films to 0.8 nm for films having 1% stearic acid. The "cluster" morphology, characteristic of pure HPMC, disappears progressively when stearic acid content increases. This effect is proposed to be due to migration of stearic acid molecules at the film surface [46, 51]. For high stearic acid content, the formation of homogeneous layer is suspected. In this case, the migration of stearic acid at the surface is possible and results from a phase separation between the hydrophobic additive and the HPMC polymer matrix. To confirm this hypothesis, nanoadhesion and nanofriction measurements are necessary to characterize the nanoscale properties of the film surface.

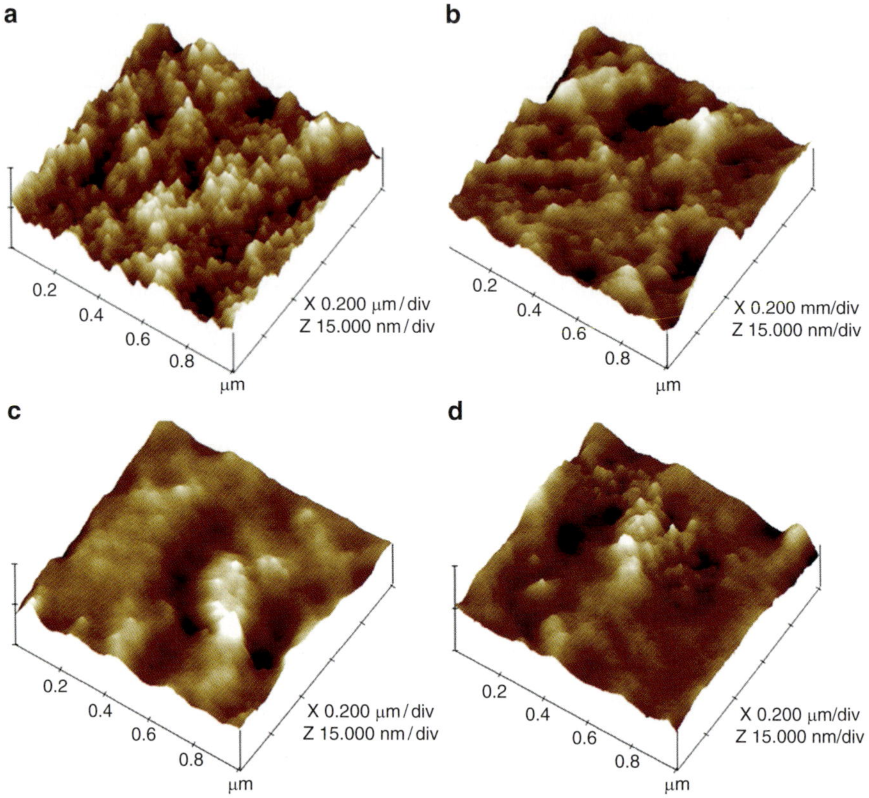

Fig. 16.10 TM-AFM topographic images (1 × 1 μm) of HPMC–stearic acid (SA) films: (**a**) pure HPMC film, (**b**) HPMC + 0.1% SA, (**c**) HPMC + 0.5% SA, and (**d**) HPMC + 1% SA

16.6.2 AFM Force–Distance Experiments

To measure adhesion between two solid surfaces, one of them is suspended on a spring and the adhesion or "pull-of" force needed to separate the two surfaces is deduced knowing the deflection of the spring. In the case of AFM where the first surface is the AFM tip and the second is the substrate, interaction forces between them are calculated through the application of Hooke's law [52]: $F = k \cdot \Delta z$, where k is the cantilever spring constant ($N\,m^{-1}$) and Δz is the cantilever deflection (nm). The measurement of adhesion force at the nanoscale between tip and HPMC-formulated films was performed in contact mode AFM by recording force curve. Silicon nitride (Si_3N_4) cantilevers of 115 μm length, 57-kHz resonance frequency with pyramidal sharpened tips were used. The spring constant was $0.3\,N\,m^{-1}$ and the radius of curvature of the tip was 45 nm. Calibration of the spring constant of the cantilevers is of major importance to calculate adhesion force and to obtain quantitative and reliable data [53].

Force measurements with AFM, in contact mode, consist in detecting the deflection of a spring (or cantilever) bearing a silicon nitride tip at its end, when interacting with the sample surface. The deflection of the cantilever is detected by an optical device (four quadrant photodiodes) while the tip is vertically moved forward and backward thanks to a piezoelectric ceramic (or actuator). Thus, provided that the spring constant of the cantilever is known, one can obtain a deflection–distance (DD) curve and then a force–distance (FD) curve, by using Hooke's law. A schematic representation of a DD curve obtained when probing a hard surface is reported in Fig. 16.11.

In zone A, the cantilever is far from the surface and stays in a state of equilibrium (no interaction with the surface). The cantilever deflection is zero. During the approach to (or withdrawal from) the surface, the tip interacts with the sample and a jump in (or jump-off) contact occurs (zones B [for loading] and E [for unloading]). These instabilities take place because the cantilever becomes mechanically unstable. Usually, for undeformable surfaces, because of mechanical instabilities, jump-in contact is not significant to determine attractive van der Waal's (VDW) forces. When in contact, the cantilever deflection is equal to the piezoelectric ceramic displacement provided no indentation of the substrate occurs (zones C [for loading] and D [for unloading]). An undeformable reference sample (cleaned silicon wafer) is used to scale the DD curve in deflection by fixing to unity the slope value of the contact line. Due to adhesion forces during contact, the jump-off is greater than the jump-in and occurs in position D. Considering the cantilever like a spring, knowing its spring constant, one can obtain the adhesion force between the tip and the sample by using Hooke's law. The reported force values are the averages of 30 measurements at different locations on the HPMC film sample. All measurements were made using the same AFM tip. In order to carry out a quantitative analysis, different experimental points should be taken into consideration such as: cantilever spring

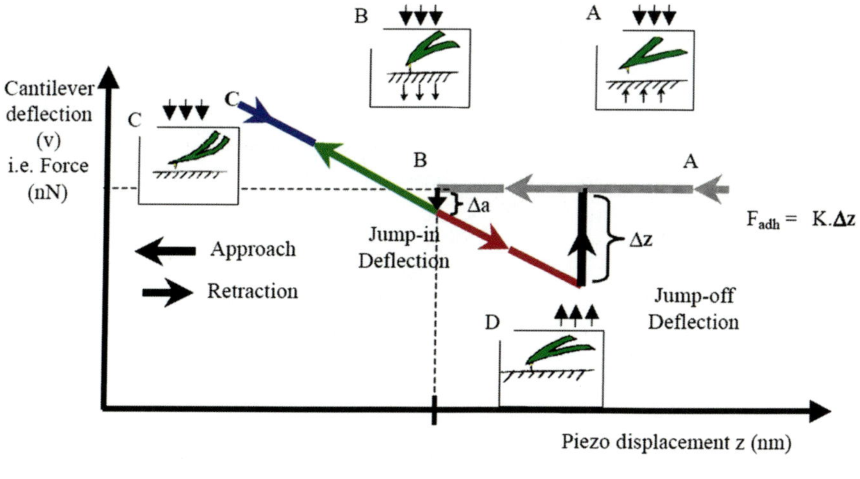

Fig. 16.11 Representation of an AFM deflexion–distance (DD) curve

constant, tip radius, linearity of the photodiodes, piezo driver hysteresis, cantilever and piezo thermal stabilities, and tip contamination [54].

16.6.2.1 Nanoadhesion Force

Determination of Nanoadhesion Properties of HPMC–PEG Films

Using the AFM in contact mode, we have studied how adhesion of HPMC films at nanoscale is influenced by the introduction of PEG. Experimental force curves obtained on HPMC films surface with different content of PEG are shown in Fig. 16.12.

Force curve experiments show that before tip pull-off from film surface, the slope of the DD curve is slightly reduced with PEG incorporation (0.98 for pure film to 0.94 for film containing 40% PEG). In the case of materials having high plastic or viscoelastic properties, this slope is lower than unity [55]. During the separation step, an increase of the pull-off deflexion (Δz) with hydrophilic character of HPMC films surface is observed, and consequently an increase of nanoadhesion force is observed. Adhesion force between AFM tip and samples is calculated by multiplying the cantilever deflexion (nm) by its spring constant ($0.33\,\mathrm{N\,m^{-1}}$). Table 16.3 gathers the slope of DD curves, cantilever deflexion, and the corresponding values of adhesion force (nN).

As shown in Table 16.3 the surface properties of HPMC films depend strongly on PEG content. The increase of PEG content into film makes a continuous increase of

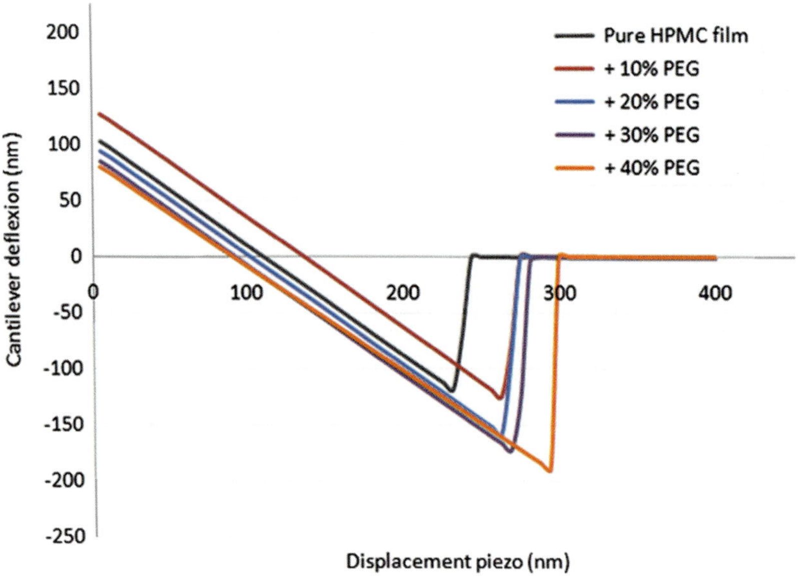

Fig. 16.12 Experimental deflexion–distance curves (retraction) of different HPMC–PEG films

Table 16.3 Experimental values of slopes of DD curves, cantilever deflexion and nano-adhesion force of HPMC–PEG films

Sample	Δz (nm)	Slope	F_{ADH}(nN)
Pure film	115	0.98	37.9
+10% PEG	128	0.98	42.2
+20% PEG	159	0.98	52.5
+30% PEG	183	0.97	60.4
+40% PEG	197	0.94	65.0

the nanoadhesion force. The adhesion force at nanoscale has a linear dependence on PEG content, which can be explained by the good compatibility between the HPMC matrix and the PEG plasticizer. The increase of nanoadhesion properties is due to increase of surface force components.

Determination of Nanoadhesion Properties of HPMC–SA Films

To evaluate nanoadhesion properties, FD curves have been obtained according to the method previously described. The maximum cantilever deflection Δz during tip–substrate separation is directly proportional to the adhesion force. The most important factors which can influence the adhesion forces are the tip–surface interactions and the tip–surface contact area. Since the measurements were made with the same AFM tip and the same tip indentation before retraction, only the intensities of the tip–film surface interactions differ between different samples. Knowing the spring constant of the cantilever, the adhesion force was determined by application of Hook's law as a function of stearic acid content. Figure 16.13 shows the influence of stearic acid content (% w/w HPMC) on nanoadhesion force for different HPMC-formulated films.

The result clearly shows that adhesion force at the nanoscale decreases when stearic acid content increases. With 0.1% stearic acid, a sharp decrease is observed. This result supports, in a quantitative way, the hypothesis of a surface enrichment in hydrophobic species when stearic acid content increases in the HPMC-formulated films. In this case, adhesion force is the result of dispersive interactions (London) between the AFM tip and film surface.

16.6.2.2 Capillary Contribution to Nanoadhesion Force

The origin of the measured adhesive force between AFM tip and HPMC film surface is VDW and capillary forces. This later is due to water meniscus formation [56–58]. The water present on film surfaces is due to the condensation of water vapor from environment. Then, the total adhesion force is given by: $F_{ADH} = F_{VDW} + F_{Cap}$, where F_{VDW} and F_{Cap} represent the respective contribution of VDW and capillary forces.

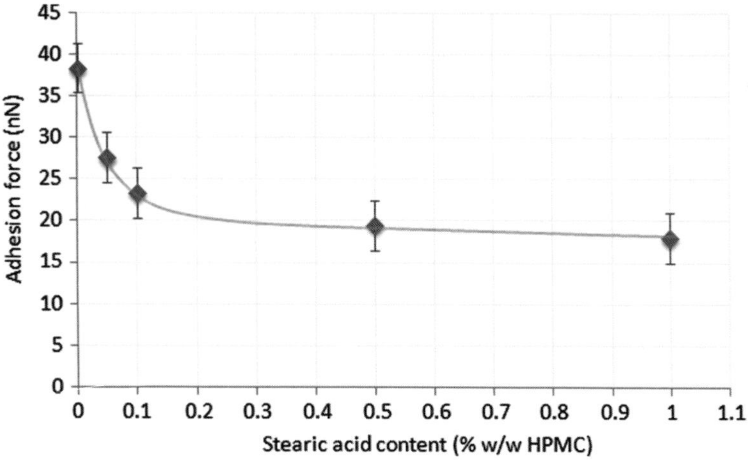

Fig. 16.13 Adhesion force (nN) versus stearic acid content into HPMC films

The contribution of both forces depends on many factors such as distance between AFM tip–sample surface, surface roughness, hydrophobicity or hydrophobicity, and the relative humidity [59]. In our case, the last factor is constant, but the surface roughness and the surface hydrophobic/hydrophilic character of HPMC films depend on PEG or stearic acid content. As experiences were conducted with the same probe, variation of tip radius is avoided.

To a first approximation, since HPMC-formulated films have a T_g much higher than room temperature one can consider that film surfaces are not deformable and purely elastic at room temperature. Thus, the adhesion force between a non-deformable spherical particle (AFM tip) of radius R and a flat surface (HPMC and HPMC-formulated films) in an inert atmosphere is given by [57]: $F_{\mathrm{VDW}} = 4\pi R\gamma_{\mathrm{SV}}$, where R is the tip radius and γ_{SV} is the surface energy of HPMC or HPMC-formulated film.

Contribution of Capillary Force to Nanoadhesion for HPMC–PEG Films

As mentioned above, the origin of the adhesive force measured is VDW attractive and capillary forces. In the present case, it is considered that the factors responsible for adhesion are the distance between AFM tip and sample surface, and the hydrophilic/hydrophobic character of surface. The capillary force depends on a meniscus formed between the tip and the film surface. Although HPMC–PEG films become softer with addition of PEG, the DD curves show only slightly changes of the slope before the tip pull-off. So it appears reasonable to consider that the AFM tip–film surface contact follows the DMT model. In this model, a non-deformable spherical particle (AFM tip) of radius R is in contact with a flat surface. The capillary force contributes to the adhesion force. Figure 16.14 summarizes the variation

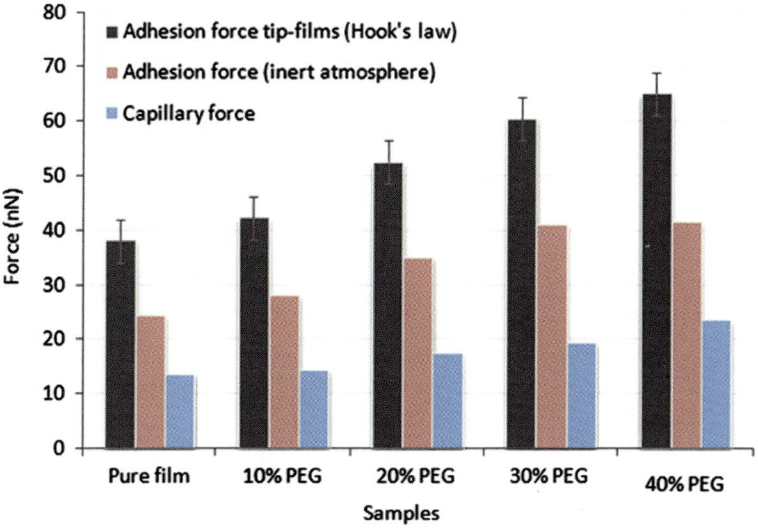

Fig. 16.14 VDW and capillary forces contribution to adhesion force between AFM tip and HPMC–PEG films

of VDW and capillary forces which are deduced by subtracting the interactions in inert atmosphere (represents the VDW force) from adhesion force obtained by AFM FD experiments at ambient.

The increase of adhesion force at nanoscale with PEG content is due to a simultaneous increase of VDW and capillary forces. Increasing of surface free energy has a significant influence on AFM tip–film surface interactions. It is also observed that capillary force increases with hydrophilic film character and no saturation stage is observed with PEG insertion. This result can be explained by the formation of a meniscus bridge between the AFM tip and the film surface leading to extend the adhesive force. Since the surface roughness and the tip radius are the same during experiments, we can consider that the PEG content is the only cause of forces variation for HPMC films. The simultaneous capillary condensation of water on films surface with the addition of PEG increases the contact zone between the AFM probe and film surface. Incorporation of PEG ($\gamma_{\mathrm{PEG}} \approx 48\,\mathrm{mJ\,m^{-2}}$) which has a surface energy higher than HPMC increases logically the surface force components. The thermodynamic miscibility of HPMC–PEG system (modification of glass transition temperature and thermo-mechanical properties) favors a continuous increase of nanoadhesion force.

Contribution of Capillary Force to Nanoadhesion for HPMC–SA Films

Figure 16.15 represents, as a function of stearic acid content (% w/w), the contribution of both forces (VDW and capillary) to the nanoadhesion force between AFM

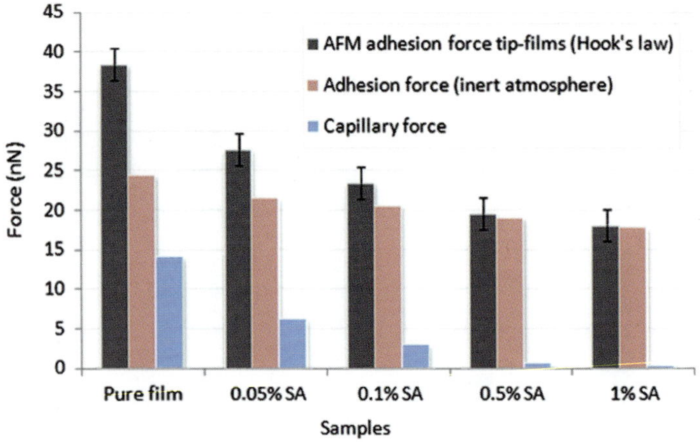

Fig. 16.15 VDW and capillary forces contribution to adhesion force between AFM tip and HPMC–stearic acid films

tip and HPMC films. The values of capillary force were deduced by subtracting the adhesion force in inert atmosphere to the adhesion force calculated on the basis of AFM FD experiments.

Figure 16.15 shows clearly that the decrease of the capillary force is mainly responsible for the decrease of the adhesion force at nanoscale and follows almost the same evolution with addition of stearic acid. The strong decrease of capillary force with 0.1% of stearic acid is remarkable. The presence of stearic acid on HPMC films surface generates a hydrophobic surface and as a consequence a reduction of the condensation of water (Laplace pressure) inside AFM tip–HPMC film contact area. This effect induces a decrease of the capillary force F_{Cap}, resulting thereafter a decrease of the adhesion force F_{ADH} between AFM tip and HPMC–SA films. These results seem in agreement with the tendency observed [60] for the surface free energy decrease and more particularly its polar component, which reflects the absence of dipoles having important polarity. The surface free energy, affected by the presence of hydrophobic stearic acid molecules, has an immediate effect on the adhesion at nanoscale between AFM tip and the surface of HPMC-formulated films.

16.6.3 LFM Nanofriction Experiments

Friction force microscopy is a powerful research tool for the surface characterization of materials and helps to explain friction phenomena at the nanoscale. Under a given applied load, the sample is scanned back and forth (producing a friction loop) in a direction perpendicular to the long axis of the cantilever and the output of the horizontal two quadrants of the photodiode detector is measured. The friction force between the films and the tip produces a twisting of the cantilever. The differential signal between the left and right detectors is denoted as FFM signal $[(L - R)/(L + R)]$. This signal can be related to the degree of twisting, hence the magnitude of friction force can also be influenced by adhesive force at the tip–sample interface. Nanoscale friction is quantified by TMR (trace minus retrace, in Volts, equal to the difference between lateral forces scanning left-to-right and right-to-left). In this case, to calculate friction coefficient, friction force (F_{TMR}) should be divided by the sum of applied load (F_N) and intrinsic adhesive force (F_{ADH}). In the presence of capillary force, the coefficient of friction is given by [52]: $\mu_{nano} = F_{TMR}/(F_N + F_{ADH})$.

A higher TMR value corresponds to a larger friction force. Friction forces (in Volts) can be converted to force units (in Newtons) and absolute friction forces can be obtained using heavy calibration methods [61]. The lateral spring constant depends critically on the tip length and the tip should be centered at the free end. In our study, we will not realize the calibration method in FFM, but the study will be comparative between samples by using the same cantilever for both series of HPMC-additive films. Nanofriction measurements were made using square-pyramidal shaped tips. The Si_3N_4 probe has the same properties than probes used in nanoadhesion, and the same normal spring constant ($0.3\,N\,m^{-1}$). TMR values have been measured for different applied normal loads (for 0.1–3 V). The friction properties of different HPMC films were compared using TMR values. The nanotribological tests were carried out in the air atmosphere. For a given applied load, the TMR value during the sliding of the AFM probe on films surface is constant. The friction force (TMR value) increases when the applied load increases. Friction force is the average of TMR values at different locations on samples surface.

16.6.3.1 Nano Friction Force

Nanofriction of HPMC–PEG Films

Figure 16.16 shows the effect of applied load on frictional force (TMR in mV) of HPMC–PEG films. The evolution of friction force at nanoscale depends on both PEG content and applied load on AFM probe.

Linear relationships are observed for all samples between the friction force (F_{TMR}) and the applied load (F_N) obeying a modified form of Amonton's law:

$$F_{TMR} = F_T^0 + \mu_{nano} F_N$$

where F_T^0 is the friction force at zero applied load, due to adhesion component and μ_{nano} represents the relative friction coefficient (slope of lines) [62]. One can observe from Fig. 16.16 that both μ_{nano} and F_T^0 increase with PEG content and that friction force was always higher when the films become more hydrophilic (increase of line slope with PEG content).

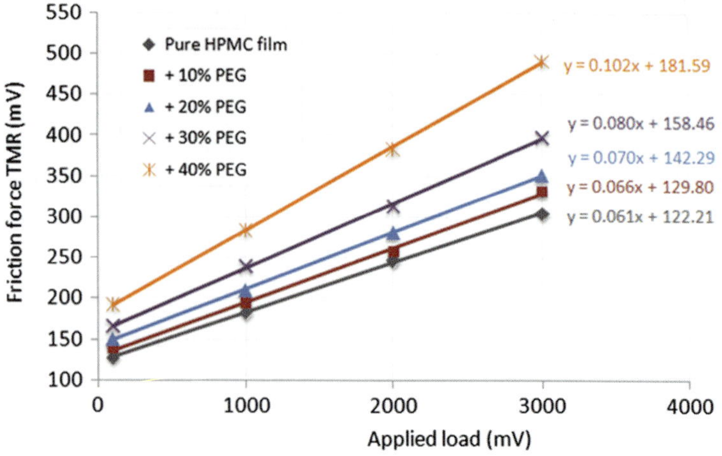

Fig. 16.16 Friction force (mV) between HPMC–PEG films and AFM tip as a function of applied normal load

Table 16.4 Friction force (TMR values in mV) of HPMC–stearic acid films for different applied loads (from 0.1 to 3 V). The error of TMR is ±10 mV

Sample	Friction force (mV) for different applied loads			
	0.1 V	1 V	2 V	3 V
Pure film	132	183	247	309
+0.05% SA	111	157	213	269
+0.01% SA	93	128	183	233
+0.5% SA	82	118	169	210
+1% SA	79	107	156	188

Nanofriction of HPMC–SA Films

To study the nanofriction of HPMC–SA films, we have summarized in Table 16.4 the TMR values obtained for all samples at different normal loads (Volts). Each TMR value is an average of different locations on sample surfaces.

For all samples, the data obtained show that the increase of stearic acid content causes a decrease of the TMR value and thus of the friction force whatever the applied normal load. Pure HPMC film presents the higher friction force for all applied normal load. This tendency was previously observed for nanoadhesion force which was greater for pure HPMC film surface. Friction seems to be correlated to nanoadhesion. In order to study the effect of hydrophobic additive on friction properties of HPMC films, we have estimated the friction coefficient. Data gathered in Table 16.4 indicate that the friction force has a linear dependence on applied load, which can be described by the previously detailed modified form of Amonton's law. The slope of the friction-versus-load curve can be used as the relative friction coefficient because it is supposed to be proportional to the real friction coefficient

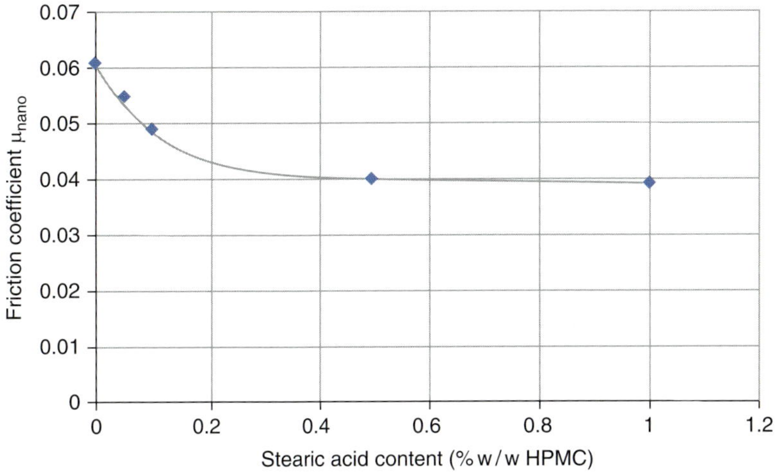

Fig. 16.17 Evolution of the relative friction coefficient of HPMC films versus SA content

[63]. Figure 16.17 shows the variation of relative friction coefficients of various HPMC–SAformulated films.

It can be seen that the pure film has the highest friction coefficient value (0.061). Addition of 1% of stearic acid within HPMC film decreases 36% the nanofriction coefficient. One can suspect that the addition of stearic acid into HPMC films increases the contact area between AFM tip and surface, because the surface roughness becomes smoother. Therefore, the contribution of VDW interactions would be greater, but in contrast a strong decrease of capillary forces also occurs when the amount of stearic acid increases. Friction force lowering with stearic acid shows that capillary contribution is the dominant factor. In other words, the effect of hydrophobic chains leads to decrease the tip–HPMC film surface interactions and thus leads to decrease adhesion and friction forces. The surface migration of stearic acid molecules induces the formation of an easy sliding layer at the HPMC film surface. These observations indicate that capillary condensation is an important factor in addition to the frictional properties of HPMC-formulated films under ambient conditions, which agrees with the results obtained for cellulosic films [64].

16.6.3.2 Interplay Between Nanoadhesion and Nanofriction

Nanoadhesion Interplay in Nanofriction of HPMC–PEG Films

The dependence of capillary force on friction properties of HPMC is shown in Fig. 16.18.

A simultaneous increase of friction force with capillary forces on films surface is observed. This result indicates that capillary condensation could be an important

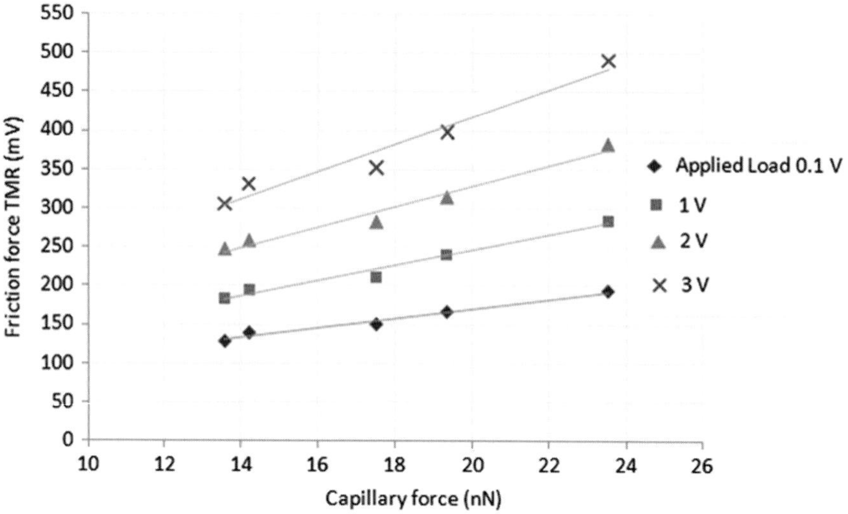

Fig. 16.18 Variation of friction force (TMR in mV) as a function of capillary forces

factor and influences the frictional properties of HPMC films under ambient conditions. These results are in agreement with those of Garoff and Zauscher [64], which reported that the increase of relative humidity leads to the formation of water multilayers on surface of cellulosic films and increases the friction force. Nevertheless the friction force is not only related to the applied load (Amonton's law), but it also depends on adhesion force which is proportional to the area of contact and to the adhesion shear strength. Figure 16.19 shows the friction force at zero applied loads versus the nanoadhesion force of HPMC–PEG films.

Figure 16.19 displays the interplay of nanoadhesion force in friction measurements. The contribution of nanoadhesion strength constitutes a supplementary factor that contributes to increase the friction properties of HPMC films. This result also shows the validity of Amonton's modified law in nanofriction experiments.

Nanoadhesion Interplay in Nanofriction of HPMC–SA Films

The increase of applied normal load induces a higher friction force (in agreement with modified Amonton's law). The evolutions of TMR versus applied load are linear for both HPMC and HPMC–SA-formulated films (cf. Table 16.4. The intercept of the friction force versus normal load line at a zero volt value of applied force corresponds to the adhesion force. TMR values obtained at zero value of normal force are represented in Fig. 16.20 as a function of stearic acid content. A constant decrease of TMR (at zero volt) is observed.

The evolution of TMR at zero load versus stearic acid content follows the same tendency than previously observed for adhesion force at nanoscale. As a

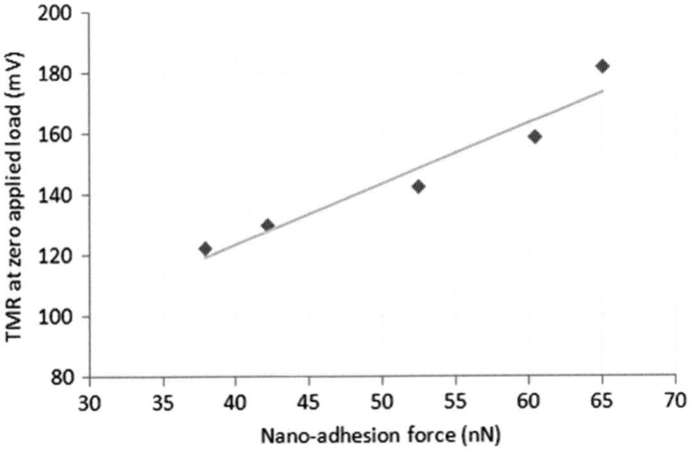

Fig. 16.19 Variation of TMR (at $F_N = 0$ V) versus nanoadhesion force determined by AFM force–distance measurements, for HPMC–PEG systems

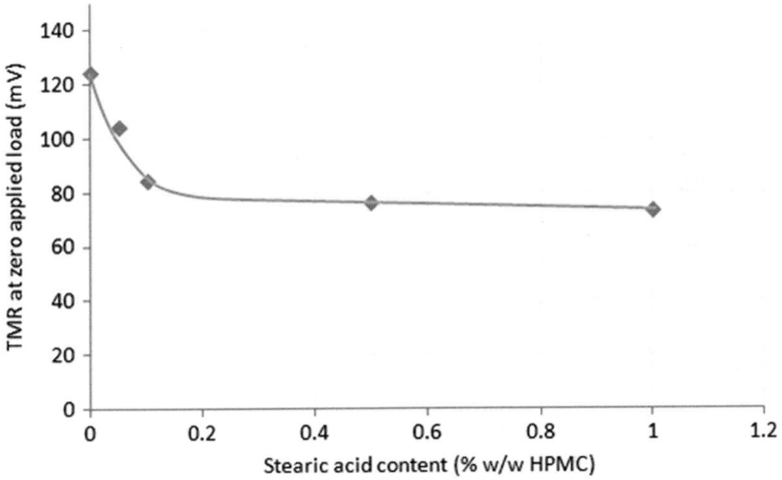

Fig. 16.20 Variation of TMR (at zero normal load) versus stearic acid content (% w/w)

consequence adhesion and friction mechanisms, at a local scale, must be connected. The study of the friction properties of HPMC films at nanoscale underlines clearly that nanoadhesion interplays in friction measurements of HPMC films. The nanoadhesive component of nanofriction force (F_T^0) is in good correlation with the adhesion force obtained by force curve experiments. Figure 16.21 represents the evolution of the nanoadhesion force versus the friction force (TMR) at zero applied load for HPMC–SA films.

The linear relationship obtained in Fig. 16.21 shows that the adhesion properties contribute linearly to friction properties at nanoscale. The dependence scale of

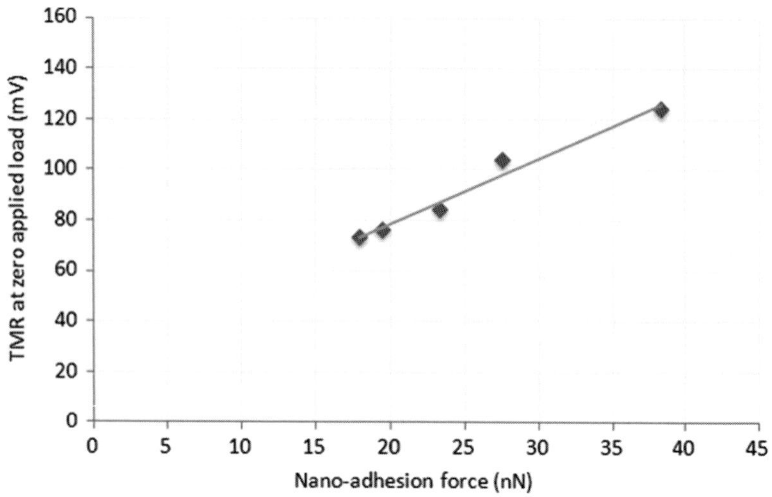

Fig. 16.21 Correlation between nanofriction and nanoadhesion for HPMC–stearic acid films

the adhesion contribution depends on the HPMC film compositions, as well as the environment conditions (relative humidity, temperature, etc.). Stearic acid induces the formation of a weak boundary layer [65] at the film top surface. This is later formed by migration and surface segregation of stearic acid molecules. This weak boundary layer has low cohesion leading to sliding properties during tip–film surface contact and low adhesion (hydrophobic and low molecular weight compared to pure HPMC).

16.7 Conclusions

In this chapter, the surface physicochemical properties of HPMC-formulated films were investigated. The incorporation of additives in formulation may affect the film morphology, nanoadhesion, and/or nanofriction behaviors. Even if the two additives have very different hydrophobic/hydrophilic characters, they may constitute an asset to illustrate what is possible to happen at the film surface in formulated systems.

Incorporation of PEG into HPMC films induces strong changes on surface morphology, hydrophobic/hydrophilic character, and adhesion and friction characteristics. The variations observed suggest an insertion of PEG molecules in HPMC matrix, resulting in an increase in HPMC cluster size (nanodomains). PEG plays the role of a plasticizer by insertion mechanism within HPMC matrix. Concerning the adhesion property, it was found that the hydrophilic character of PEG leads to increase adhesion at nanoscale. The nanofriction experiments on HPMC–PEG films show the same tendencies, indicating that capillary forces play a major role in adhesive and sliding properties at nanoscale due to meniscus formation.

For HPMC films containing stearic acid, the surface topography becomes smoother with stearic acid content, due to the presence of hydrophobic agent on films surface. At nanoscale, using contact mode AFM and friction mode (FFM), we have demonstrated that the addition of stearic acid decreases both nanoadhesion and nanofriction. Stearic acid additive plays the role of a surface lubricant in HPMC-formulated films. Stearic acid surface migration at the top surface is favored by phase separation between HPMC and SA.

For both additives (PEG and SA), the interplay between nanoadhesion and nanofriction was demonstrated.

References

1. E. Doelker, Adv. Polymer Sci. **107**, 199–265 (1993)
2. S. Kamel, N. Ali, K. Jahangir, S.M. Shah, A.A. El-Gendy, Express Polym. Lett. **2**, 758–778 (2008)
3. J.L. Ford, Int. J. Pharm. **179**, 209–228 (1999)
4. M.L. Navarro-Tarazaga, R. Sothornvit, M.B. Pérez-Gago, J. Agric. Food Chem. **56**, 9502–9509 (2008)
5. O.E. Pérez, C. Carrera Sánchez, J.M. Rodríguez Patino, A.M.R. Pilosof, Biomacromolecules **7**, 388–393 (2006)
6. R. Villalobos, J. Chanona, P. Hernandez, G. Gutiérrez, A. Chiralt, Food Hydrocolloids **19**, 53–61 (2005)
7. R. Villalobos, P. Hernandez-Munoz, A. Chiralt, Food Hydrocolloids **20**, 502–509 (2006)
8. S.R. Bechard, O. Quraishi, E. Kwong, Int. J. Pharm. **87**, 133–139 (1992)
9. V. Coma, I. Sebti, P. Pardon, A. Deschamps, H. Pichavant, J. Food Prot. **64**, 470–475 (2001)
10. A. Fahs, M. Brogly, S. Bistac, M. Schmitt, Carbohydr. Polym. **80**, 105–114 (2010)
11. P. Sakellariou R.C. Rowe, Prog. Polym. Sci. **20**, 889–942 (1995)
12. J.W. McGinity, L.A. Felton, *Aqueous Polymeric Coatings for Pharmaceutical Dosage Forms*, 3rd edn. (Informa Healthcare USA Inc., New-York – London, 2008)
13. C.L. Li, L.G. *Martini*, J.L. *Ford*, M. *Roberts*, J. Pharm. Pharmacol. **57**, 533–546 (2005)
14. D.A. Alderman, Int. J. Pharm. Technol. Prod. Manuf. **5**, 1–9 (1984)
15. D.L. Wise, *Handbook of Pharmaceutical Controlled Release Technology* (Marcel Dekker, Inc, New-York – Basel, 2000)
16. F.I. Uchegbu, A.G. Schätzlein, *Polymers in Drug delivery*. (Taylor & Francis, CRC Press, Philadelphia, 2006)
17. J.L. Ford, Int. J. Pharm. **179**, 209–228 (1999)
18. R.C Rowe, P.J. Sheskey, S.C. Owen, *Handbook of Pharmaceutical Excipients*, 5th edn. (Pharmaceutical Press and American Pharmacists Association, Washington DC, 2006)
19. J. Swarbrick, *Encyclopedia of Pharmaceutical Technology*, 3rd edn. (Informa Healthcare, New-York – London, 2007)
20. J.A.Q. Gallo, F. Debeaufort, A. Voilley, J. Agric. Food Chem. **47**, 108–113 (1999)
21. N.E. Suyatma, L. Tighzert, A. Copinet, V. Coma, J. Agric. Food Chem. **53**, 3950–3957 (2005)
22. S. Benita, *Microencapsulation: Methods and Industrial Applications*, 2nd edn. (Taylor & Francis, CRC Press, Philadelphia, 2006)
23. K. Ridgway, *Hard Capsules: Development and Technology*. (The Pharmaceutical Press, London, 1987)
24. G. Samaranayake, W.G. Glasser, Carbohydr. Polym. **22**, 1–7 (1993)
25. J. Siepmanna, N.A. Peppas, Adv. Drug Delivery Rev. **48**, 139–157 (2001)
26. R.C Rowe, P.J. Sheskey, S.C. Owen, *Handbook of Pharmaceutical Excipients*, 5th edn. (Pharmaceutical Press and American Pharmacists Association, Washington DC, 2006)

27. J.W. McGinity, L.A. Felton, *Aqueous Polymeric Coatings for Pharmaceutical Dosage Forms*, 3rd edn. (Informa Healthcare USA Inc., New-York – London, 2008)
28. *Methocel Cellulose Ethers, Technical Handbook*, The Dow Chemical Company, Form No. 198-02075-700 AMS, Midland, 2000
29. N.A. Camino, O.E. Pérez, A.M.R. Pilosof, Food Hydrocolloids **23**, 1089–1095 (2009)
30. R. Daniels A. Barta, Eur. J. Pharm. Biopharm. **40**, 128–133 (1994)
31. C. Wollenweber, A.V. Makievski, R. Miller R. Daniels, Colloids Surf. A **172**, 91–101 (2000)
32. Y.M. Li, G.Y. Xu, X. Xin, X.R. Cao D. Wu, Carbohydr. Polym. **72**, 211–221 (2008)
33. A. Fahs, *Thèse de Doctorat, Université de Haute Alsace*, Mulhouse, France, 2009
34. P. Sakellariou, R.C. Rowe, Prog. Polym. Sci. **20**, 889–942 (1995)
35. M.L.N. Hansen, D. Plackett, Biomacromolecules **9**, 1493–1505 (2008)
36. E. Ayranci, S. Tunc, Food Chem. **72**, 231–236 (2001)
37. V. Coma, I. Sebti, P. Pardon, A. Deschamps, H. Pichavant, J. Food Prot. **64**, 470–475 (2001)
38. R. Villalobos, P. Hernandez-Munoz, A. Albors, A. Chiralt, Food Hydrocolloids **23**, 526–535 (2009)
39. LA. Felton, T. Austin-Forbes, T.A. Moore, Drug Dev. Ind. Pharm. **26**, 205–210 (2000)
40. F. Fuertges, A. Abuchowski, J. Controlled Release **11**, 139 (1990)
41. R.C. Rowe, P.J. Sheskey, S. Owen, *Handbook of Pharmaceutical Excipients*. (Pharmaceutical Press, Washington DC, 2006)
42. P. Gabbott, *Principles and Applications of Thermal Analysis*. (Blackwell, Oxford, 2008)
43. H.A. Lieberman, M.M. Rieger, G.S. Banker, *Pharmaceutical Dosage Dorms – Disperse Systems*, 2nd edn. (Marcel Dekker, New-York – Basel, 1998)
44. F. Debeaufort, A. Voilley, J. Agric. Food Chem. **45**, 685–689 (1997)
45. H.J. Park, C.L. Weller, J. Food Sci. **58**, 1361–1364 (1993)
46. I. Sebti, F. Ham-Pichavant, V. Coma, J. Agric. Food Chem. **50**, 4290–4294 (2002)
47. L. Yang, A.T. Paulson, Food Res. Int. **33**, 571–578 (2000)
48. O.E. Pérez, C.C. Sánchez, J.M. Rodríguez Patino, A. Pilosof, Biomacromolecules **7**, 388–393 (2006)
49. N.A. Camino, O.E. Pérez, A.M.R. Pilosof, Food Hydrocolloids **23**, 1089–1095 (2009)
50. J.K. Sears, J.R. Darby, *Mechanism of Plasticizer Action*. (John Wiley & Sons, Hoboken, 1982)
51. L. Yang, A.T. Paulson, Food Res. Int. **33**, 571–578 (2000)
52. B. Bhushan, *Nanotribology and Nanomechanics: An Introduction*. (Springer, Berlin – Heidelberg, 2005)
53. O. Noel, M. Brogly, G. Castellein, J. Schultz, Langmuir **20**, 2707 (2004)
54. M. Brogly, O. Noel, H. Awada, G. Castelein, J. Schultz, C.R. Chimie **9**, 99–110 (2006)
55. H. Awada, G. Castelein, M. Brogly, Surf. Interface Anal. **37**, 755 (2005)
56. Y. Liu, T. Wu D.F. Evans, Langmuir **10**, 2241–2245 (1994)
57. J.N. Israelachvili, *Surface Forces and Microrheology of Molecularly Thin Liquid Films*. (CRC Boca Raton, FL, Boca Raton, 1995)
58. J.N. Israelachvili, *Intermolecular and Surface Forces*, 2nd edn. (Academic, London, 1991)
59. T. Stifter, O. Marti B. Bhushan, Phys. Rev. B **62**, 13667–13673 (2000)
60. A. Fahs, M. Brogly, S. Bistac, M. Schmitt, Carbohydr. Polym. **80**, 105–114 (2010)
61. S. Bistac, A. Galliano, Tribol. Lett. **18**, 21–25 (2005)
62. S. Song, S. Ren, J. Wang, S. Yang, J. Zhang, Langmuir **22**, 6010–6015 (2006)
63. N.J. Brewer, B.D. Beake G. Leggett, Langmuir **17**, 1970–1974 (2001)
64. N. Garoff, S. Zauscher, Langmuir **18**, 6921–6927 (2002)
65. S.J. O'Shea, M.E. Welland, T. Rayment, Appl. Phys. Lett. **61**, 2240–2242 (1992)

Chapter 17
Surface Growth Processes Induced by AFM Debris Production. A New Observable for Nanowear

Mario D'Acunto

Abstract Loss of material due to abrasion, adhesion, erosion or other types of wear mechanisms is a fundamental phenomenon occurring between two surfaces in relative motion on each other. Generally, in a wide range of length scales, from macroscale down to nanoscale, wear is quantified by measuring the volume loss after a wear test, and the quantification of the wear volume is the main observable to be measured in a wear test. In this chapter, we present some recent results showing that in precise experimental conditions, as ultrahigh vacuum (UHV) environments, surface growth processes induced by atomic force microscopy (AFM) tip sample abrasion can be estimated to have an accurate knowledge of atomic and molecular onset mechanisms involving the occurrence of wear mechanisms, mainly abrasion.

In fact, recent UHV scratching AFM experiments made on ionic crystals showed the formation of small clusters, larger aggregates or regular patterns on the surface being scanned, and a theory capable of capturing the basic mechanisms producing the formation of such structures has been proposed. Such cluster structures, generally self-organised in regular structures, are mainly produced by the flux of adatoms generated by the AFM tip stripping off adatoms during the continuous passage of the probe tip on the surface being analysed. In UHV environments, surface diffusion is the dominant mass transport mechanism, and a non-equilibrium thermodynamic framework for the self-organised growth process has been developed demonstrating that the surface growth processes maintain a sort of coherence with respect to the flux rates of the adatomic debris induced by the AFM tip during the wear test making the wearing and the surface growth specular. As a consequence, the physical nature of the growth processes induced by AFM debris could represent a new observable to be measured for a new and accurate comprehension of wear mechanisms on nanoscale.

17.1 Introduction

Wearing processes are defined as the succession of events whereby atoms, products of chemical conversion, fragments, etc. are induced to leave the system and identified in a manner that embody or suggest solutions [1, 2]. Wear processes include six principal, relatively distinct phenomena that have one thing in common, i.e., the removal of solid material from rubbing surfaces [3, 4]. These are (1) abrasive, (2) adhesive, (3) fatigue, (4) impact by erosion and percussion, (5) corrosive and electrical-arc-induced wear [5]. These are not distinct mechanisms, but rather combinations of the abrasive, adhesive and corrosive wear mechanisms. In many cases, wear is initiated by one mechanism and it can proceed by other wear mechanisms, thereby complicating failure analysis. Anyway, wear by all mechanisms, except by fatigue mechanism, occurs by gradual removal of material. Laboratory investigations of wear are usually carried out either to examine the mechanisms by which wear occurs, or to simulate practical applications. Wear is usually measured either by removing the specimen at intervals and weighing or measuring it, or by continuously measuring its position with an electrical or mechanical transducer and deducing the wear from its change in dimensions. Because the wear process is an interfacial phenomenon, it is difficult or impossible to observe the wear occurrence in real time when it is induced by a loading scanning tip. The removed material can be generally estimated at the end of the wear test. As a consequence, it is made difficult to know the threshold characteristic for the onset wear or the geometry of the worn sample during the test. Wear volume and its quantification is the main paradigm for the analysis of the occurrence of wear mechanisms.

As a consequence, quantitative analysis for a basic knowledge of wear mechanisms has been addressed to write formula or equations capable of capturing the specific role of all the variables entering in the wear mechanisms. Actually, there are few such equations for wearing of materials that have been validated by researchers other than the author of the equations. At the macroscale, wear is a result of a combination of processes including fracture, plasticity and third-body effects, and has generally been described using the Holm–Archard's phenomenological model. In an abrasive wear process, Holm and Archard describe the wear volume of macroscopic plastic contacts with multiple asperities to be proportional to the load force F_{N} and sliding distance d, which can be written as $V = k_{\mathrm{wear}} F_{\mathrm{N}} d$, where k_{wear} is the dimensional wear coefficient [6]. Later, a similar expression was proposed by Bhushan for elastic contacts [7]. Rabinowitz proposed a model for abrasive wear [8], assuming a conical asperity of the harder surface, infinitely rigid, that when submitted to a load indents the softer surface. By knowing the characteristic geometry of the indenting asperity and applied load, after the softer surface becomes fully plasticised, the worn volume can be calculated in a similar manner as made by Holm and Archard. Analogously, all these approaches are based on the same basic idea: the worn volume is directly proportional to the real contact area, where the real area of contact is proportional to the ratio between the applied load and the hardness of the softer material for fully plastic contacts, or to the ratio between the applied load and the elastic local hardness for Greenwood-Williamson Gaussian multiple elastic contacts [6].

Single asperity experiments are possible on nanoscale, a length scale where the surfaces and their properties play a major role with respect to bulk properties. As a consequence, in the last two decades, much knowledge has been addressed to deal the tribological properties at the atomic and molecular scale. Many of such research have been performed using the scanning probe microscopy (SPM) family, and some fundamental properties of the wear mechanisms have been outlined [4, 5]. In a typical nanoscale wear experiment performed using an SPM tool, the wear process is as follows: a probe tip scans repeatedly a small area of the surface of interest. The stressing passage of the atomic force microscopy (AFM) tip involves production of removed debris and other complex phenomena dependent on the sample-tip mechanical contact that, in some cases, can be studied with the same tip. Analogously to macroscale, also on nanoscale, wear volume is the main observable to be quantified after a wear test. For a complete meaning of the occurrence of wear mechanisms, the geometry of the worn sample around the scratched areas can be imaged for having a supplemental knowledge. Nevertheless, the experiments made on ionic crystals showed that, in a scratch test performed using a UHV environment, a secondary effect due to debris produced by the stressing AFM tip is the epitaxial surface adatoms re-crystallisation leading to the formation of some structures as atomic clusters, or graining mounds around the scratched area [9–11]. A new approach for the comprehension of some basic wear mechanism looking at the surface growth patterns induced by the abrasive wearing AFM scratching tests has been recently proposed. It has been demonstrated that on sample made of Al(111), the growing structures are formed via surface diffusion by the adatoms stripped off by the AFM tip are an indirect measure of the basic wear mechanisms, because the system maintains a sort of coherence between the flux rates of debris production due to AFM tip and the surface growth processes dominated by the surface diffusion [12]. The adatoms are nothing but the debris produced during the tip-sample stressing contact, and their ability to re-organise around the surrounding surface being worn is strongly dependent by their rate production, that is, ultimately, by the rapidity of the activated wear mechanisms. The self-organised ordered growth structures are a secondary effect of the wear processes.

The chapter is organised as follows: Sect. 17.2 is devoted to show some relevant experiments on nanowear, focusing the attention on the re-crystallisation processes induced by AFM tip. In Sect. 17.3 an analytical expression for the adatoms flux rates produced during an AFM scratch test is proposed and related to the growth structures and surface diffusion. Finally, in Sect. 17.4, a further theoretical description of the regular self-organised growth structures produced in scratching tests is accurately presented and discussed using a non-equilibrium thermodynamics picture.

17.2 Single Asperity Nanowear Experiments

Wear volume is generally the main observable for the quantification of removed material during a wear test on macroscale (Fig. 17.1a). On nanoscale, where the SPM based techniques give the possibility to perform experiments in the regime of single asperity, the quantification of wear volume can be often the main observable from a wear test. Geometrically, AFM wear experiments are analogous to the traditional pin-on-disk or ball-on-disk wear experiments [3, 4] (Fig. 17.1b).

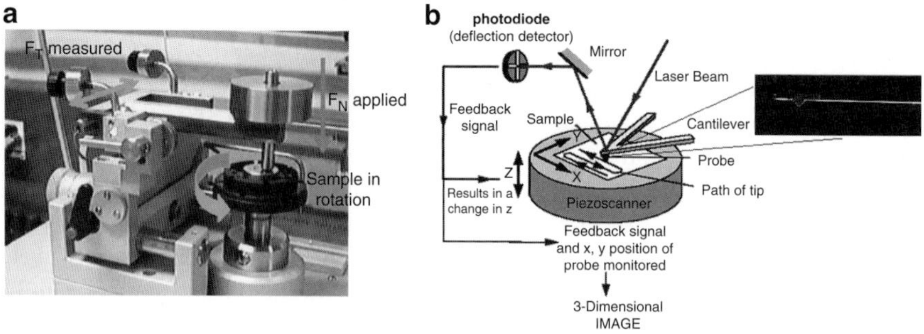

Fig. 17.1 A typical set up for the measurements of wear on macroscale. (**a**) In such machine, a flat or a sphere shaped indenter is loaded on to the test sample with a precisely known weight. A spin is mounted on a stiff lever, designed as frictionless force transducer. As a sample is either rotating or sliding, resulting frictional forces acting between the pin and the sample are measured by very small deflections of the lever. Volume of the material lost is measured after a friction run test and wear coefficients for both the sample and material are calculated (reprinted from CSM instruments, Switzerland, with permission). (**b**) A typical AFM configuration experimental set up. The zoom represents a cantilever with a pyramidal shape tip

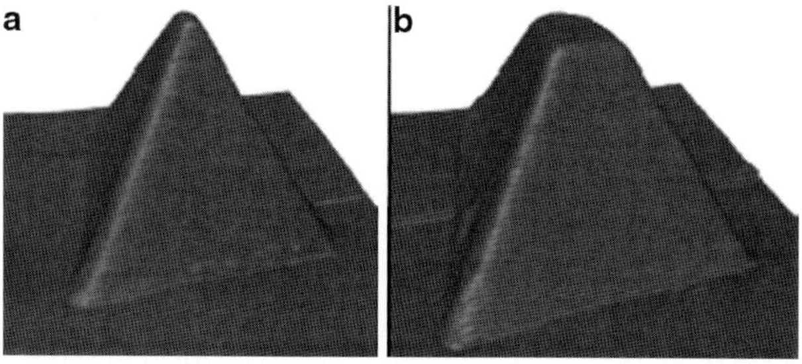

Fig. 17.2 Images of a fresh AFM tip (**a**) and the same tip after a wear test (**b**). The AFM tip was scanned 45 times across a sodium trisilicate glass substrate in ammonium hydroxide solution (pH $\sim$ 11) at an applied force of 125 nN. The vertical axis (z) has the same scale as the x- and y-axes (reprinted from [13], with permission)

Wear of silicon nitride AFM tips has been previously studied and attributed to adhesive wear, plastic deformation, low cycle fatigue fracture and ambient solution [13]. In these experiments, tip wear was induced by scanning small patterns in raster mode at a tip velocity correspondent to several microns per second on a wide variety of substrates. Due to the important role of the substrate in tip wear, the tip shape before and after each wear experiment was characterised by scanning sharp silicon spikes $\sim$600 nm tall with a tip radius of curvature of $<$10 nm. Because the spikes are significantly sharper than the AFM tip, imaging the spikes produces an image of the AFM tip. Typical images before and after a wear experiment are shown in Fig. 17.2, and the height changes as a function of the number of scans or contact force are shown in Fig. 17.3.

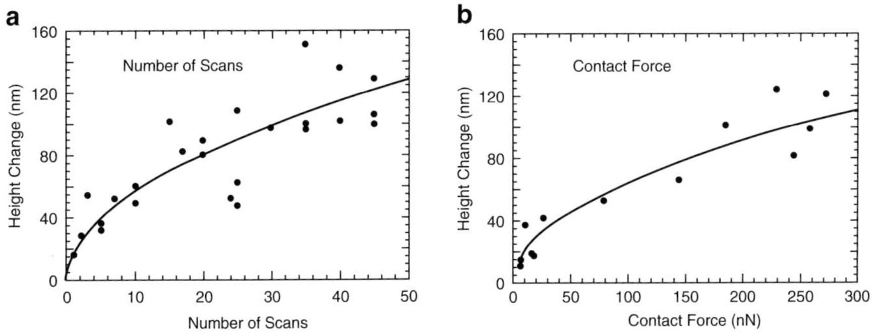

Fig. 17.3 (**a**) Height change in AFM tip as a function of the number of scans on trisilicate glass in ammonium hydroxide environment (pH $\sim$ 11) at a contact force of $\sim$120 nN. (**b**) Height change as a function of contact force over 15 scans. Each scan was $3 \times 3\,\mu\,m^2$ at 5.2 Hz. Each tip wear measurement was undertaken with a fresh AFM tip. The continued line represents least squares fit, while the circles represent the experimental data (reprinted from [13], with permission)

To interpret the observed dependence of tip wear on time and normal force F_{N}, it is useful to consider the rate of tip wear, $\mathrm{d}h_{\mathrm{tip}}(t)/\mathrm{d}t$, which is related to the experimentally measured net change in tip height, $H_{\mathrm{tip}}(F_{\mathrm{N}}, t)$, by:

$$\frac{\mathrm{d}h_{\mathrm{tip}}}{\mathrm{d}t} = \frac{\partial H_{\mathrm{tip}}(F_{\mathrm{N}}, t)}{\partial t}. \tag{17.1}$$

Since one is interested in the volume removal rate, $\mathrm{d}V_{\mathrm{tip}}/\mathrm{d}t$, which equals product of the tip area, A_{tip}, and $\mathrm{d}h_{\mathrm{tip}}/\mathrm{d}t$, the following relation has been obtained [13]:

$$\frac{\mathrm{d}V_{\mathrm{tip}}}{\mathrm{d}t} = A_{\mathrm{tip}}\frac{\mathrm{d}h_{\mathrm{tip}}}{\mathrm{d}t} = \frac{1}{2}kF_{\mathrm{N}}(\alpha A_{\mathrm{tip}})^{1/2}, \tag{17.2}$$

where k is a constant of proportionality and α is a coefficient that depends by the tip geometry (for a square pyramidal tip with a tip angle of 70°, $\alpha = 1.99$). Normally, one expects the local volume removal rate to vary linearly with the contact area ($= A_{\mathrm{tip}}$). The extra factor of $(A_{\mathrm{tip}})^{1/2}$ in (17.2) suggests that the volume wear rate changes proportionally to the length of the tip, $l = (A_{\mathrm{tip}})^{1/2}$. It has been suggested that one possible source of this unusual dependence on A_{tip} could be a gradual stress-activated production of debris on the substrate as the tip passes over [13]. If the time constant for debris formation is small relative to the duration of the applied stress, the debris concentration will be approximately uniform and constant under the AFM tip. If the time constant for debris formation is long relative to the duration of applied stress, the concentration of possible precursors of debris sites under the influence of the tip will increase in a linear fashion from the loading edge of the tip to the trailing edge. Further, the maximum concentration will be proportional to the time required for the tip to pass over any given point on the surface. This time is proportional to the length of the flat portion of the tip, $l = (A_{\mathrm{tip}})^{1/2}$. The stress at a given substrate location under tips with large flat areas is sustained for longer times than tips with small flat areas, producing higher concentrations of precursors states and yielding faster wear.

The measurement of wear volume by observing the wear tip after a wear test has been the main experimental effort of some recent experiment made by Gotsmann and Lantz [14], Lantz et al. [15] and Baskaran et al. [16]. In these experiments, the wearing tip at different velocities and loading conditions has been investigated. The measurements were generally carried out in ambient conditions. The tip shape has been estimated during the wear experiment to gain more insights into the occurrence of the wear process. For this, the adhesion between the tip and sample surface, as measured in a force–distance test, is used as an indicator of in situ wear. In fact, to a first approximation, the adhesion between the tip and a flat surface can be considered proportional to the contact radius R with a constant energy u_{adh}, making possible to write the pull-off force as $F_{\mathrm{adh}} = u_{\mathrm{adh}} R$ [16]. It is important to note that the contact radius, R, is assumed to be the tip radius for a flat surface, as would be the case in worn tips (Fig. 17.3a). The AFM is configured to record adhesion data in automatic manner after a previously specified length, and in the experiments made by Bhaskaran et al., this length was fixed to 0.8 mm of sliding, while keeping the force at the tip almost constant with unwanted variations less than 1% [16] (Fig. 17.4).

Assuming a slow atom-by-atom wear process, the erosion of the conical tip with shape $R(h)$, where h is the position along the vertical axis of the tip, can be described by the relation [14]:

$$\frac{\partial R}{\partial d} = \frac{\omega b}{v} \frac{\partial R}{\partial h} \exp[\{-(E - V_{\mathrm{act}}\tau) + P V_{\mathrm{eff}}\}/k_{\mathrm{B}}T], \qquad (17.3)$$

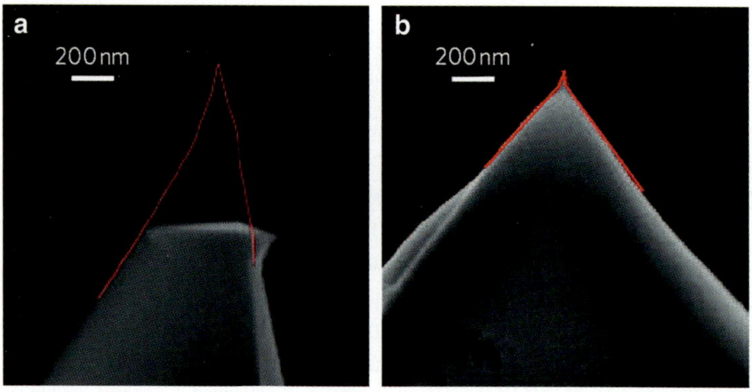

Fig. 17.4 Scanning electronic microscope (SEM) images of silicon tip (**a**) and a comparable ultra-sharp Si-DLC tip (**b**) under similar loading and environmental conditions after 2 m of sliding wear on an SiO_2 surface. The red outline shows the tip shape before the wear experiment (reprinted from [16], with permission)

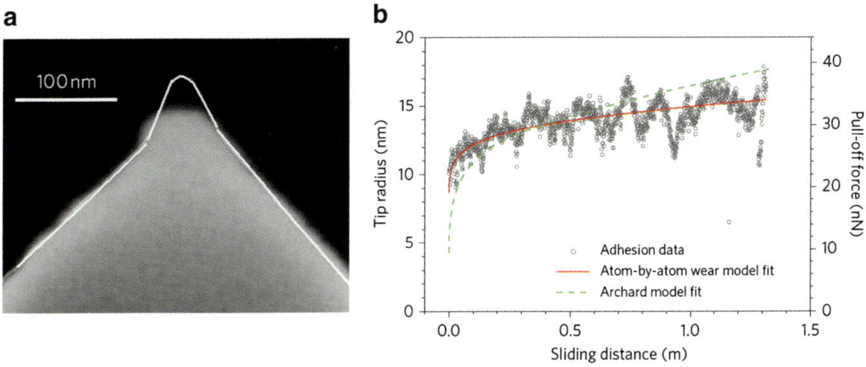

Fig. 17.5 SEM image of an Si-DLC tip after the wear test, with the outline of the unworn tip indicated. (**a** and **b**) Plot of tip radius and pull-off force (y axis) against sliding distance (x axis). The *solid red* line indicates the fit using the atomistic attrition model (17.3) and the *dashed green line* indicates Archard's model fit (**b**) (reprinted from [16] with permissions)

where ω is an attempt frequency due to phononic vibration, E is the activation energy required to break the atomic bond, such value is reduced by the sliding-induced shear stress, V_{act} is an activation volume and $k_B T$ is the thermal energy. Equation (17.3) is used to fit the experimental data obtaining the experimental mean values: $E - V_{\text{act}}\tau = 1.0 \pm 0.1\,\text{eV}$ and $V_{\text{eff}} = 3.4(\pm 2) \times 10^{-28}\,\text{m}^3$, for the activation energy and the effective activation volume, respectively. Archard's law does not always fit the experimental data, whereas a smooth erosive atom-by-atom model does [14–16]. In such experiments, the mean wear observable is still the wear volume addressed to quantify the height change of AFM tips, nevertheless, the great accuracy of the experimental data leaves open many questions, for example, although the duration of the wear test is divided in regular periods for the observation of the tip shape, the nonlinearity of the wear processes and the unmeasured time dependence of the wear activation rates do the some characteristics of the wear mechanisms necessarily underestimated (Fig. 17.5).

17.2.1 Surface Growth Processes Induced by AFM Tip: Experimental Results

Nevertheless, a series of nanowear experiments where the SPM in UHV has been used to realise and detect atomic-scale abrasion on ionic crystal have been performed opening the possibility of a new paradigm for nanowear. In fact, such wear tests showed how the debris extracted from ionic crystal surface by a silicon tip is reorganised on the atomic scale (Fig. 17.6).

The dependence of the mean lateral force as a function of the time was investigated scratching across a single line for a long time at fixed load and velocity (5,120 scans at $F_N = 11.2\,\text{nN}$ and scanning velocity of $2.5\,\mu\,\text{m/s}$). The mean lateral force

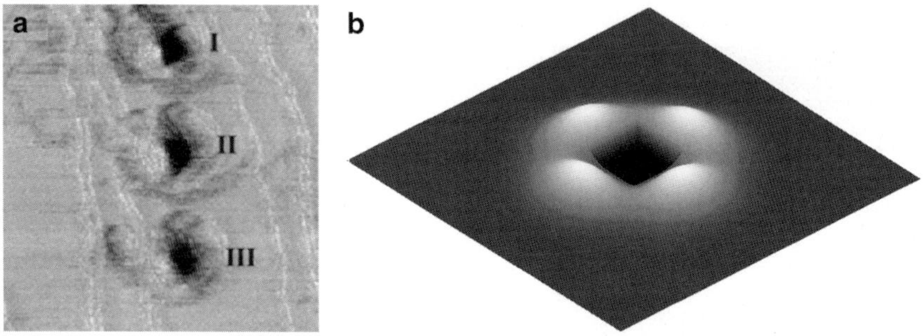

Fig. 17.6 (**a**) Lateral force images of pits and mounds produced by 256 scratches on $5 \times 5\,\mathrm{nm}^2$ areas with $v = 25$, 50 and 100 nm/s (*upper to lower*) and $F_{\mathrm{N}} = 14.1\,\mathrm{nN}$, frame size 74 nm. (**b**) Simulated surface centred around a scratched test area, for further details on the simulation see Sects. 17.3 and 17.4 (reprinted from [9] and [12], respectively, with permissions)

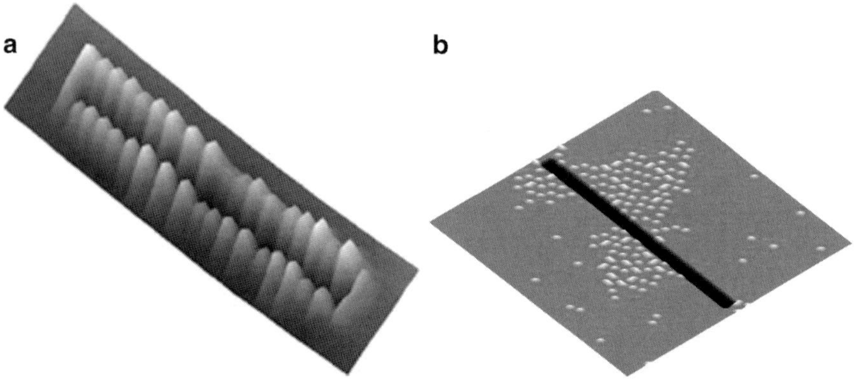

Fig. 17.7 The image (**a**) represents the topography of a groove formed on KBr(001) after 512 scratches along the (100) direction with $F_{\mathrm{N}} = 26.6\,\mathrm{nN}$ and $v = 2.67\mu\,\mathrm{m/s}$ as obtained by Socoliuc et al. [10]. (**b**) A simulated surface composed by a linear scratch and its surrounding area after the wear test, the rippled dots are distributed not only in proximity of the scratched line, for further details on the simulated image see Sects. 17.3 and 17.4 [12] (reprinted from [10] and [12], respectively, with permission)

increased continuously from an initial value as a function of the number of scratches N. The behaviour of the lateral force can be considered a clear demonstration of the abrasive character of the wear process. In fact, the ions seem to be removed and released by pairs or by small clusters of ion pairs, and no crack events were observed in thousand of scans, and the regular arrangement of the deposited material resembles an epitaxial growth of thin ionic films, where the ions are piled up pair by pair. Such experiments introduce a new *observable* on nanowear, where the new observable is the growth process that is directly connected to the production of the ionic pair debris, rather than observing roughly the classical wear volume observable. In fact, the AFM scratch process was used to produce nanoscale regular patterns on KBr(001)in UHV, where the initially flat surface is rearranged in a quasi-periodic pattern of mounds and pits or regular ripple structures (Fig. 17.7).

The growth processes leading to the formation of ripple structures and mounds in a UHV AFM experiment has recently been modelled using a continuum approach for aluminium [12]. Although a molecular approach could be more adequate to describe the experimental results, the continuum approach seems to capture the physics of the whole process, showing that the growth structures obtained (regular ripples or large mounds) are strictly connected to the specific wear rate activated. As a consequence, a new approach to study and quantify nanowear is made possible observing the growth structures around the AFM scratched areas. This new nanowear method is based on a new observable (growth process) rather than the wear volume quantification. In environments different from UHV, the growth processes could be totally different making it difficult to observe them, leading to wear volume quantification to be the main observable. The next two paragraphs describe the continuum approach for the growth processes induced by the AFM tips and their connection with wear debris rates.

17.3 A Model for Wear Debris Production in a UHV AFM Scratching Test

The introduction of a new observable for nanowear experiments is made possible if there is a direct connection between the rates of debris produced by the AFM tip and the surface growth processes. The rates for the debris production are related to the flux of adatoms made available for the surface diffusion after the passage of the stressing tip. The flux intensity is described by a time function, F, that is related to the statistical counting of the atoms stripped off (debris), while, in a UHV environment, surface diffusion, described by the coefficient D, is the main parameter characterising the adatom dynamics. A possible direct connection between F and D implies that a *coherence* is maintained between the wearing process induced by the AFM tip and the subsequent surface growth process induced by the debris. Also, if a coherence between wearing and growth is established, it implies that on changing the flux rates, different growth structures should be observed. This coherence has been demonstrated by D'Acunto in a recent paper focused using the aluminium as a material of reference because its Al/Al(111) homoepitaxial growth [12].

In this paragraph, we focus the attention on the basic atomic wear mechanisms while in the next paragraph, the growth processes are described using a nonequilibrium thermodynamics approach. The growth of structures induced by the formation of debris in a UHV AFM scratching test can be schematised in the following steps: (1) formation of debris (atoms) induced by the AFM tip scratching the substrates; (2) the debris jump and diffuse around the worn area (dynamical adatoms); (3) the diffusion process induces nucleation of adatoms; (4) depending on the intensity and time function of the flux of adatoms, the nucleation process can lead to the formation of self-organised array of dots or mounds (Fig. 17.1).

The AFM imaging recorded at the end of the scratching process suggests that, considering the whole process wear + growth as an *inverse problem*, i.e., it is more useful to look at the recording data of the surface growth process taking information from the abrasive wear process while studying the growing structures. Consequently, it could be possible to describe wear mechanisms (onset wear conditions, mature wear threshold, etc.) considering the growth phenomena. We try to demonstrate that such processes are essentially conditioned by the rate of production of the adatoms made available by the abrasive wear process for the nucleation and subsequent growth of regular patterns.

During a wear test performed using an AFM probe tip, the scanning tip stresses the sample surface, and debris can form during the passage of the tip. Theoretical studies and recent experimental results indicate that on nanoscale, the debris seem to be composed essentially by single atomic dimensions (atom-by-atom loss process) [14–16]. Deviations from an atom-by-atom loss process may be possible; however, fractures induced by AFM tip have not yet been observed experimentally, and the abrasive wear processes can be considered as smooth processes.

During the scratching test, the area continuously stressed by the probe tip changes because of the abrasive test. For simplicity, it is considered that the probe tip is significantly harder than sample, so that tip wearing can be neglected. As matter is removed, the z-component of the worn surface area can be expressed using the following relation:

$$\frac{\partial h(x, y, t)}{\partial t} = -\Omega \nabla \mathbf{j} = F_{\text{outgoing}}, \tag{17.4}$$

where the time evolution of the surface $h(x, y, t)$ is calculated on the scratched area, the debris current $\mathbf{j}$ is due to the stressing role of the probe tip, and Ω is the atomic debris volume and F_{outgoing} is the flux of the adatoms stripped off by the probe tip, diffusing along the sample surface. We are interested in describing a wear progress where the current $\mathbf{j}$ is derivable from effective free energy F

$$\mathbf{j} = -\frac{\delta F}{\delta \nabla z}, \tag{17.5}$$

as a consequence, the progression of the damage is greater where the pit is deeper caused by the scratching test. The evaluation of the surface abrasive erosion, the current of adatoms or the effective free energy change due to abrasion can be complicated and experimentally difficult to test. This is because the imaging of the scratched area and its surroundings is made immediately after the test, as a result, it is difficult to control the time evolution of the erosion previously produced in real time. In the next sub-paragraph, a derivation of the effective free energy as a function of the number of stripped off atoms during the scratch test is given.

17.3.1 Localisation of the Free Energy Changes Due to Stressing AFM Tip

The study of the surface growth processes of the adatomic debris produced during the abrasion is an indirect method for the analysis of the basic wear mechanisms during the scratching. Here it is shown that the growth processes depend essentially on the adatom fluxes, and the different possible fluxes are directly connected to the activated wear mechanisms. All the quantities such as the surface current density **j**, the free energy and the chemical potential μ are unknown quantities, and it is practically impossible to determinate them during the scratch test. On the contrary, it is possible to quantify the progressive incidence of the activated wear mechanisms by acquiring the image of a region surrounding the scratched area with the same scratching unworn AFM tip. As it is demonstrated here, this is because the growth structures (small clusters or large mounds) are strongly dependent on the incidence of the adatomic debris flux induced by the scratching probe tip.

When the activated wear process takes place, the damage can be localised in discrete points very close to initial defects. The eroded surface of the worn area can be used to quantify the free energy variation and its connection with chemical potential. In a 1 + 1 system, the z-component of the surface can be written as:

$$z(x,t) = \langle z \rangle - \delta h(t) f(x), \tag{17.6}$$

where $f(x)$ is a characteristic local function of the wearing rate. The time evolution of the abrasive process involves that the time derivative of z-component is equal to

$$(\mathrm{d}\delta h/\mathrm{d}t) f(x) = -\partial \mathbf{j}\,(x,t)/\partial x. \tag{17.7}$$

The surface current density $\mathbf{j}\,(x,t)$ is usually assumed to be related to the chemical potential μ (the chemical potential is the free energy for atom), by a linear relation as:

$$\mathbf{j}\,(x,t) \cong -\partial \mu(x,t)/\partial x. \tag{17.8}$$

Relations (17.7) and (17.8) imply that the chemical potential μ correspondent to the abrasion wear has the form

$$\mu(x,t) = \mu_1(t) f(x), \tag{17.9}$$

where f describes the profile of the scratched surface, and the function $\mu_1(t)$ is related to $\delta h(t)$ by the following relation which follows from (17.8) to (17.9),

$$(\mathrm{d}\delta h/\mathrm{d}t) f(x) = \left(\frac{\partial^2 f(x)}{\partial x^2}\right) \mu_1(t). \tag{17.10}$$

On the other hand, $\mu(x,\mathrm{t})$ is the free energy per atom, so that during the abrasive process, the free energy, $\mathscr{F}$, variation per unit time is, if Ω is the volume per atom (debris):

$$\dot{\mathscr{F}} = \Omega^{-1} \int \mathrm{d}x\, \mu(x,t) \frac{\partial z(x,t)}{\partial t} = \Omega^{-1} \frac{\mathrm{d}h(t)}{\mathrm{d}t} \mu_1(t) \int \mathrm{d}x\,\mathrm{d}y f^2(x). \tag{17.11}$$

Once that the profile of the worn substrate is known, the kinetic approach permits to calculate the time derivative of the free energy of the worn space and eventually connect its value to the variation of free energy of the growth structures. A computational calculation can be addressed in this direction in order to simulate the wearing and growing processes in real time in their appropriate time scales. In a linear scratch a sinusoidal profile is often found [10], as a consequence (17.11) reduces to

$$\dot{\mathscr{F}} = \Omega^{-1} A \frac{\mathrm{d}h(t)}{\mathrm{d}t} \mu_1(t), \tag{17.12}$$

where A is area of the worn surface. The surface height can be approximated as $h(t) = V/A$ and the volume $V = V_0 - n_\mathrm{w}\Omega$, where V_0 is the initial volume of the region interested to the scratch and $n_\mathrm{w}\Omega$ is the volume removed, where n_w is the number of Al atoms stripped off and Ω their individual volume. After these changes (17.12) becomes:

$$\dot{\mathscr{F}} = \dot{n}_w \mu_1(t) \quad \text{or} \quad \dot{\mathscr{F}} = F\mu_1(t). \tag{17.13}$$

At least, (17.13) states that the time derivative of the total free energy of the wear process considered as granular production of debris is given by the flux and the chemical potential.

17.3.2 Flux of Adatoms Induced by the AFM Stressing Tip

There is a rough experimental evidence that, on nanoscale, debris is essentially composed by single atoms or small clusters of atoms [9, 10, 14–16]. It has been proposed and estimated that basic wear mechanisms can be described in a similar way to solid state rate activated chemical reaction between solid, where the mechanical stresses induced by the AFM tips can be considered as a catalyser [17–21]. As a consequence, the formation of debris can be described following the Kramers theory [17].

The probability that an atom is stripped off during the passage of a probe tip is given by [17–22]:

$$\Gamma = \nu \exp\{-[\varepsilon_\mathrm{b} - \delta\varepsilon_\mathrm{s}]/k_\mathrm{B}T, \tag{17.14}$$

where ε_b is the binding energy, $\delta\varepsilon_\mathrm{s}$ represents the stress-induced deviation which is a function of the applied load and lateral force, ν is the attempt frequency, k_B is Boltzmann's constant and T is the temperature. The stress-induced deviation is a combined effect of the compression and shear stresses and it lowers the energy barrier allowing the atoms to be stripped off from the substrates. Once the atomic debris is stripped off, it moves diffusely along the surface. Under the tip stress, the maximum possible rate of debris production, $\Delta n_i/\Delta t$ at the ith point contact, should be given by the relation for thermal activated detachment:

$$\frac{\Delta n_i}{\Delta t} \cong \frac{A_\mathrm{r}}{a^2} \Gamma_\mathrm{i}, \tag{17.15}$$

where Δn_i is the number of atoms stripped off during the contact with the probe tip, Δt is the residence time of the probe tip on the ith point contact, A_r is the real area of contact between the tip and the sample, a is the van der Waals average radius of a single atom and Γ_{i} is the rate of activation for the creation of AFM induced debris and A_{r}/a^2 gives the number of atoms simultaneously in contact with the probe tip.

It can be assumed that a continuum mechanics could give a good approximate value for the real area of contact even if the contact is discrete. For a paraboloid shape of the probe tip in non-adhesive contact with a flat surface, the real area of contact and penetration depth are respectively

$$A_{\mathrm{r}} = \pi\left(\frac{3F_{\mathrm{N}}R}{4E^*}\right)^{2/3} \quad \text{and} \quad z_{\mathrm{d}} = \frac{1}{2}\left(\frac{3F_{\mathrm{N}}}{4E^*\sqrt{R}}\right)^{2/3}, \tag{17.16}$$

where $E^* = [(1-v_1^2)/E_1 + (1-v_2^2)/E_2]^{-1}$is the reduced elastic modulus, $E_{1,2}$ are the Young moduli and $v_{1,2}$ is the Poisson's ratios for the Al sample ($E_1 \cong 70\,\mathrm{GPa}$, $v_1 = 0.33$) and silicon probe tip ($E_2 = 140\,\mathrm{GPa}$, $v_2 = 0.28$), respectively, F_{N} is the applied load and $R = 15\,\mathrm{nm}$ is the tip radius. Making use of a value $a \cong 0.2\,\mathrm{nm}$ for the Al lattice period and a real area of contact $A_{\mathrm{r}} \cong 8\,\mathrm{nm}^2$ for an applied load of 25 nN (17.16) approximately 10^2 atoms are simultaneously in contact with the probe tip at the ith point of contact. If we consider that when the wear mechanism is active, at the ith point contact, a single atom is stripped off and this corresponds to $\sim$1% of stripped off adatoms. Such percentage corresponds to an activated rate $\Gamma_{\mathrm{I}} \sim 10^{-2}$, for unity of Δt. Analogously, for other different percentage of stripped off adatoms a correspondent activated rate is obtained. The values of the stripping probability, as given in (17.14), permit the calculation of the stressing deviation for the decrease of the activation energy. It can be supposed that the decrease of activation energy, $\delta\varepsilon_{\mathrm{s}} = \delta\varepsilon_1 + \delta\varepsilon_2$, has two major contributions depending on the normal stress, or sharing stress, respectively. The first contribution is essentially due to the ratio between the load and the real area of contact, and it can be written as follows [21]:

$$\frac{\delta\varepsilon_1}{n_{\mathrm{c}}\varepsilon_{\mathrm{b}}} = \frac{1}{n_i}\frac{z_{\mathrm{d}}}{(A_{\mathrm{r}}/\pi)^{1/2}}, \tag{17.17}$$

where n_{c} is the surface number of coordination for Al(111) lattice constant, ε_{b} the Al–Al binding energy, z_{d} the penetration depth, A_{r} the tip-sample real area of contact and $n_i = A_{\mathrm{r}}/a^2$ is the number of Al atoms simultaneously supporting contact with the probe time. The second contribution strongly depends on the ratio between the lateral force and the lateral area of contact [21]:

$$\frac{\delta\varepsilon_2}{\varepsilon_{\mathrm{b}}} = \frac{1}{n_{\mathrm{l}}}\frac{z_{\mathrm{d}}^2\psi}{(A_{\mathrm{l}}/\pi^{1/2})} \cong \frac{z_{\mathrm{d}}\psi}{n_{\mathrm{l}}}, \tag{17.18}$$

where $\psi = \textit{sticking force}/\varepsilon_{\mathrm{b}}$ is calculated at the maximum value of the lateral force in the sticking phase during the typical stick–slip motion of the probe tip sliding on the Al surface, $A_{\mathrm{l}} \cong \pi(z_{\mathrm{d}})^2$ is the lateral area of contact calculated at the edge of the penetration depth and $n_{\mathrm{l}} = A_{\mathrm{l}}/a^2$ is the number of atoms simultaneously in contact with the probe tip on the lateral side towards the sliding tip motion. The role of sticking force can be complicated, because of the difficulty regarding its accurate quantitative determination. In addition, the existence of a connection between penetration depth and sticking force implies that the aluminium atoms set at the edge of the penetration depth should entirely support the shearing stress provoked by the approaching probe tip along the sliding direction. Nevertheless, we can have an estimation of the sticking force considering that the stressing-induced deviation $\delta\varepsilon_s$ is due entirely to $\delta\varepsilon_2$ as can be deduced using (17.17) and (17.18). Indeed, we can have an idea of the magnitude of the stressing energy $\delta\varepsilon_s$ considering that in condition of low wear regime, ~1% of stripped off atoms at ith point of contact, the correspondent value of Γ is $\sim 10^{-2}$. Calculation of the stressing energy gives $\varepsilon_b - \delta\varepsilon_s \cong 0.9$–1 eV using a thermal energy of 2.58×10^{-2} eV, and an attempt frequency $\nu \cong 10^{13}$ Hz. For a binding energy $\varepsilon_b = 3.3$ eV, we deduce that the stressing energy contribution for the activation of the atom debris is $\delta\varepsilon_s \cong \delta\varepsilon_2 - 2.3$ eV. This value for the induced stressing deviation corresponds to a sticking force approximately half applied load.

As a first approximation, the total number density of adatoms produced by the wear abrasion process during the scratch test and freely available for growth surface and patterning formation can be given by [12, 21]:

$$n_w = n\Gamma t N_{eff}, \tag{17.19}$$

where n_w represents the atom number produced during the wear process, $n = \rho_0 V$ is the initial number of atoms located on the whole scratched volume V and ρ_0 is the Al surface density of atoms, t is the scratch test time and N_{eff} is the effective total number of scans producing debris. The number of effective scans producing wear debris, N_{eff}, and the rate of debris production, Γ, are strictly connected. Generally, there is a threshold value for the N scans above in which the wear mechanisms are activated, before the scan value is reached, Γ is negligible. Moreover, the occurrence of surface defects can increase Γ reducing N_{eff}. The knowledge of Γ also gives the efficiency measurement of the wear activated process giving the rate of the atoms leaving their primitive position and diffusing through the surface, but (17.19) does not give the time evolution of the stripped off adatoms during the scratch test. As a consequence, the time dependence of n_w is crucial both for the quantification of the wear incidence or the nucleation of clusters and for the subsequent growth of regular structures.

The flux of debris (adatoms) is a direct measurement of the efficiency of the wear process, and therefore, the structures that we observe at the end of the scratching test are an indirect measurement of the efficiency of the wear process. The problem at this point could be the connection of the activated wear abrasive mechanisms with the function of the flux F. The wear number is connected to the flux via the expression:

$$n_w = \int_0^t F(\tau)\mathrm{d}\tau, \tag{17.20}$$

where the flux F is expressed as atoms per surface unit cell and second, and t is the entire time length of the scratch test. n_{w} can be connected to the free energy produced during the wearing (see Eq. 17.13). If we consider that at ith contact point, almost one adatom debris is produced, a suitable mean value for the flux can be given by the production of one adatom for residence time, i.e., $F \approx 10^{-17}$ (adatoms $\mathrm{m}^{-2}\,s^{-1}$). Two possible time functions for the flux of the atomic debris that are considered in this chapter have been suggested. The choice is between a Poissonian-like production of independent pulse of adatoms or a flux that considers the progressive nucleation of worn areas inside the scratched volume. In the first case, one possible functional form for the rate flux, $F_1(t)$ (expressed for unity of lattice cell), can be the shot-noise-like form as [12]:

$$F_1(t) = \eta \sum_{m=\bar{m}}^{M} \delta(t - m\tau_0), \qquad (17.21)$$

where $\eta \sim \Gamma A_{\mathrm{r}}/a^2$ denotes the numerical density of adatoms produced during the AFM scratching and τ_0 corresponds to the residence time, $\bar{m}$ is a threshold value above which the wear process starts, M is the ratio between the total time of the scratch test and the residence time taken as an entire number and $M\tau_0$ is the duration of the scratching test. At any pulse, a number of η adatoms start to diffuse along the sample surface. Analogously to (17.21) for the flux rate, it is possible to have a continuous flux rate given by the nucleation of wear domains during the progressive AFM scratching [21, 22]:

$$F_2(t) = \eta(1 - \exp(-\Gamma t)), \qquad (17.22)$$

where Γ and η are the same as in (17.21). A time function flux, as described in (17.22), is a typical example of a nucleation mechanism for progressive wear domains where the occurrence of progressive defects is described by a sigma function. It has been previously demonstrated that nucleation of defects provides a realistic mechanism for the growing of worn domains during gross wear when a critical threshold is reached [21].

In the next section, the surface growth induced by abrasive adatoms is calculated making use of continuum methods. The flux of adatoms generated during the probe tip scratch is connected to the parameters of the equation expressing the surface growth. In the next sub-section, the flux functions (17.21) and (17.22) are used for calculating the number of cluster density.

17.3.3 *Evaluation of Number Cluster Density via Nucleation Theory*

The preview analysis is supported by the numerical density time evolution of the surface growing structures. The array of cluster structures or the large coarse-graining mounds strongly depends on the flux rate of the adatoms produced during the scratching test. The quantification of the number of the structures composed on the surface can be performed using a simple model, depending on the diffusion constant and flux time function [23, 24]. Once the atoms are stripped off during the wearing, in the course of deposition, the atoms arriving from the gas phase with a rate F diffuse with rate D until they meet a second adatom and create a dimer. Mean field nucleation theory (MFNT) and nonlocal MFNT are powerful tools for extracting fundamental microscopic parameters such as diffusion barriers and adsorbate–adsorbate binding energies from island density data [24–27].

Let us consider the numerical density of the surface structures formed via surface diffusion by the adatom flux produced during the scratching test. The nucleation process can be easily described for a cluster structures with an average cluster–cluster distance l on a flat substrate in a stationary state. All clusters are assumed to be stable, i.e., neither dissociating nor diffusing. A capture number is defined as the probability per time and diffusivity unit for a random walker to be captured by a certain island [12]. If the capture numbers can be calculated, the cluster density can be predicted with D as the only parameter. Conversely, D and thereby the adatom diffusion barrier, E_{d}, can be extracted from experimental cluster-density data. For freshly deposited adatoms, it takes about a time $t = l^2/D$ to migrate in a region of a linear dimension l. During this time interval, other adatoms are deposited to a density $n = Ft$, and the initial adatom has a probability nl^2 to meet another adatom during its migration. By assuming that the encounter causes the instantaneous nucleation, the encounter probability is the same as the nucleation probability per site. The latter is given by $(a/l^2)^2$ with a being the lattice constant, since there is a nucleus within a region of l^2. By equating the two results and considering irreversible solidification of adatoms, if the nucleation size is i, the mean cluster–cluster separation is given by $l \sim (Da^2/F)^{i/2(i+2)}$. The time evolution density quantification for such surface structures in the presence of a time-dependent rate flux can be written by making use of a system of two relatively simple equations. The time-dependent equations for the $N(t)$ which is the total number of growing clusters (or equivalently the nucleation density, the aggregation of more than two clusters is excluded) and $N_1(t)$ which is the density of clusters of size 1 adatoms. Introducing dimensionless variables [The following dimensionless variables are introduced: $t' = \sqrt{D\eta/T}t, N'(t) = \sqrt{DT/\eta}N(t), N_1'(t) = \sqrt{DT/\eta}N_1'(t)$, the period in (17.23) is rescaled as $\tau_0' = D\eta\tau_0$], and considering the capture numbers equal to unity, the system equations are as follows [28]:

$$\dot{N}_1(t) = F(t) - DNN_1(2N_1 + N); \qquad (17.23)$$

$$\dot{N}(t) = D(N + N_1)N_1^2(t), \qquad (17.24)$$

where the rate flux of adatoms, $F(t)$, can assume the functional form as given by a shot noise-like flux (17.21) or by a sigma-like function (17.22). The fluxes depend on the activated wear processes; therefore, the formation of surface structures due to adatom debris is an indirect measure of the fundamental wear-activated mechanisms. Equations (17.23) and (17.24) have been integrated using the two different fluxes, (17.21) and (17.22) respectively, and the results are illustrated in Fig. 17.8.

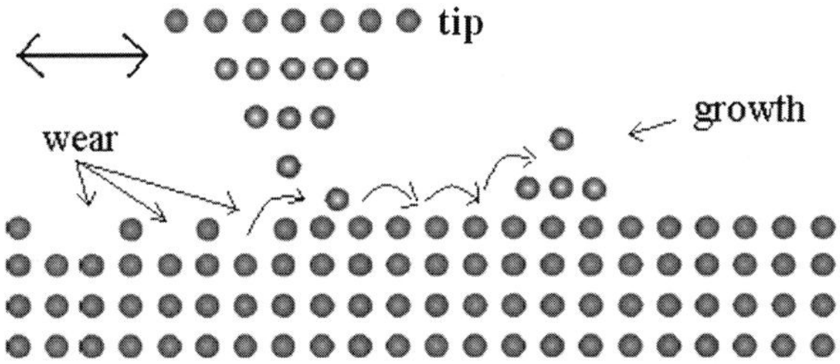

Fig. 17.8 A schematic sketch of the abrasive wear process and the subsequent cluster growth as considered in this paper. The AFM tip scratches the sample surface continuously, the debris is removed atom-by-atom, and hence, the atoms diffuse on the surface growing in clustering structures or coarse-graining in mounds. The growing structures observed depend on the surface energy landscapes that the atoms meet during the surface diffusion. The length of such structures depend on the diffusion constant D and the flux rate F (reprinted from [12] with permission)

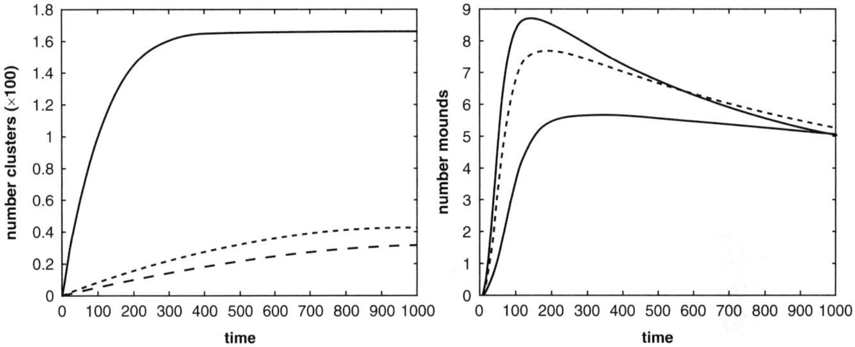

Fig. 17.9 The numerical density of growing structures as a function of time using two type of adatoms fluxes in (17.23) and (17.24). On *left*, the number of cluster density is calculated using the Poissonian shot noise-like flux function as given in (17.21), with $\bar{m} = 0, \eta = 1, \tau_0 = 10^{-3}$ s. The three curves are obtained for different loads, $F_N = 20$ nN (*dotted-dashed line*), 25 nN (*dashed line*), 50 nN (*continue line*). The number density reproduces the growing clusters as in Fig. 17.10a, c, see the next section. On *right*, the number density of growing structures giving coalescent mounds, correspondent to Fig. 17.10b. In this case, such mounds are obtained integrating (17.23) and (17.24) using the sigma-like form for the adatom flux (17.22). The loads used in the numerical integration are $F_N = 20\,nN$ (*dotted-dashed line*), 25 nN (*dashed line*), 50 nN (*continue line*). When the arrival of adatoms is given by a sigma-like function, the mound structures tend to grow driven by instability as in an Ostwald ripening process. The growing process tends to reduce the ratio surface/volume, and a coalescent process suddenly gives growing structures around the scratched area as in Fig. 17.10b. The time period used in the integration of (17.23 and 17.24) was to the residence time $\Delta t \sim 10^{-3}$ s (reprinted from [12], with permission)

Figure 17.8 illustrates the two possible regimes occurring for the surface growing structures and depending by the flux-type produced during the wear abrasion process, as shown in Fig. 17.9. The system equations as (17.23) and (17.24) derived by the nonlocal MFNT theory confirm the growing structures as described by a continuum approach summarised by a partial differential equation (PDE), that is shown in the next section. The adatoms flux generated by nucleating worn islands gives a

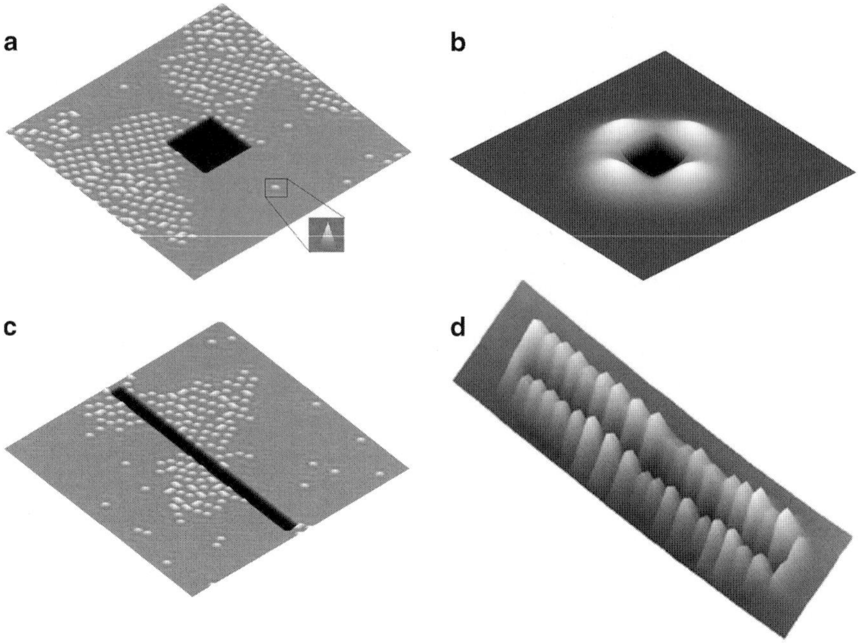

Fig. 17.10 AFM simulated images (**a**–**c**) as observed at the end of the wear test integrating (17.31), to areas (100×100L, corresponding to nearly 200 × 200 nm real area) around the scratched regions. The scratched area can be a square (**a** and **b**) or a line (**c**). The abrasive wear process leads to the formation of regular clusters or coarse-graining mounds due to surface diffusion of adatoms debris. (The parameters used are, respectively: $\alpha = 0.2$, $\chi = 1.5$, $\varphi = 1.0$ in the image (**a**); $\alpha = 0.2$, $\chi = -0.75$, $\varphi = 1.5$ in the image (**b**); $\alpha = 0.23$, $\chi = 1.0$, $\varphi = -2.0$, in the image (**c**)). The image **d** represents the topography of a groove formed on KBr(001) after 512 scratches along the (100) direction with $F_N = 26.6\,\text{nN}$ and $v = 2.67\,\mu\,\text{m/s}$ as obtained by Socoliuc et al. [10], it has been inserted as a comparison with the surface topography, shown in (**c**), see Fig. 17.7 (Reprinted from [10] and [12], with permission)

sigma-like curve where many dots as a function of time are produced (Fig. 17.10b). In contrast, the shot-noise like flux generates a suddenly saturating curve with the production of few islands. This implies that the few islands give coarse-graining producing mounds rather than clusters. Practically, the growth surface regime depends on the nature of the flux, and ultimately on the type of wear process. In the case of Poissonian shot-noise like flux, the number density of the cluster adatoms can be calculated by the relation [12, 24, 27, 28]:

$$n \cong C(\xi) \cdot (D/F)^{-\xi} \exp(E_d/\xi k_B T), \tag{17.25}$$

where C is a prefactor and ξ is an exponent valuable by nonlocal MFNT as a function of the nucleation size i. For $n \sim 150$ clusters, nucleation size $I = 1$ (at hence $\xi = 1$), and $D \cong 1.8 \times 10^{-11}\,\text{m}^2/\text{s}$, mean flux $F \sim 4 \times 10^{-17}\,\text{m}^{-2}\text{s}^{-1}$, and making use of an energy diffusion as $E_d \sim 25\,\text{meV} \sim 10 k_B T$ [29], a prefactor

$C \cong 3 \times 10^3$ is obtained. It is also relevant that the MFNT theory in its nonlocal version is a powerful tool for the quantification of the number density of the clusters structure, giving a confirmation of the results obtained via nonequilibrium thermodynamics approach as in Sect. 17.4.

17.4 Continuum Approach for the Surface Growth Induced by Abrasive Adatoms

The growth of regular patterns induced by a mature wear process can be described by a continuous height function $h(x, y, t)$, where x, y are the surface coordinates, on the average, the surface receives a flux of adatoms, F, produced by the stressing probe tip making the scratch. Such flux depends on the incidence of the activated wear production of single adatoms-by-adatoms debris.

The various processes by which matter is redistributed along the surface are represented by functions of the derivatives of h with respect to space coordinates [30]. Moreover, if the whole stripped off mass goes in the growth process, a mass-conservation condition will be fulfilled:

$$\partial h/\partial t = \Omega \nabla \mathbf{j} = F_{\text{incoming}}, \tag{17.26}$$

where, similar to (17.4), Ω is the atomic volume, $\mathbf{j}$ is the adatom (debris) number current density within the surface and F_{incoming} $(= F_{\text{outgoing}})$ is the total adatom flux produced during the wear process, growing in mounds or clusters. Equation (17.26) connects the time-dependent growth process to the wear production of the debris adatoms. In fact, the current density is referred to the surface diffusion current of adatoms produced during the stressing passage of the AFM probe tip.

Only the mass transport mechanism of surface diffusion is considered. The surface flux of atoms $\mathbf{j}$ can be taken in order to be proportional to the surface gradient of a chemical potential, μ, [30, 31]:

$$\mathbf{j} = -\frac{D\rho}{k_{\mathrm{B}}T} \nabla_{\mathrm{s}}\mu \ \text{on}\, z = h(x, y, t), \tag{17.27}$$

where D is the diffusion constant, ρ is the surface density of the lattice sites, $\nabla_{\mathrm{s}}\mu$, is the surface gradient of the chemical potential and $k_B T$ is the thermal energy. Note that (17.27) which describes the surface current for the growing adatoms is very close to (17.4) used to describe the surface current adatoms produced during the scratching tests. The two main processes, i.e., wearing and growing, may be seen as mirror processes. The relative specular affinity between the two processes gives the possibility of studying the basic wear mechanisms reproducing the growth structures.

Equation (17.27) can be manipulated using the thermodynamics of stressed solids with a reference state for the film of zero strain, and under the additional assumptions of isotropic surface energy, zero surface stress, the gradient of the chemical potential can be written as [32–34]

$$\nabla_s \mu = \Omega \nabla_s (U_e + k\gamma(h) + \Phi), \tag{17.28}$$

where U_e is the elastic energy density, k is the free surface mean curvature, $\gamma(h)$ is the isotropic surface energy and $\Phi(h, |\nabla h|^2) = (\mathrm{d}\gamma/\mathrm{d}h)(1 + |\nabla h|^2)^{-1/2}$ is the surface chemical potential related to wetting interactions with the substrate (wetting potential), and is a function of the height and slope [32, 33]. The morphology of a surface where a scratch test takes place needs to be described with an erosion process, and localised accumulations of atoms because of surface diffusion can be described combining (17.28), (17.27), (17.26) and (17.4) [32–35]:

$$\frac{\partial h}{\partial t} = -\sigma(x, y) + D\sqrt{1 + |\nabla h|^2}\nabla_s^2 (U_e + k\gamma + \Phi), \tag{17.29}$$

where $\sigma(x, y)$ represents the eroded scratched area, and it describes the wear-dependent function that is the localised source of adatom debris production, where the AFM scratch test is performed. Equation (17.29) describes the time evolution of a surface where a localised scratched area produces adatomic debris, via surface diffusion growth in regular clusters or mounds. The last term on the right side of (17.29) has revealed rich dynamics of surface evolution such as self-assembly of quantum dots. Equation (17.29) has been used for describing the nonlinear evolution of epitaxially strained thin solid film subjected to surface diffusion driven Asaro–Tiller–Grinfeld (ATG) instability on solid substrate in the presence of wetting interactions [32–34]. Equation (17.29) captures many physical properties connected to the linear instability of a planar surface of a stressed solid to small perturbations, finding that the planar surface is unstable for wave numbers less than a critical value. This instability is manifested by a mass transport via surface diffusion, where wetting potential plays a key role in the competition between the elastic energy and the surface energy, because the elastic energy in the solid is a destabilising factor, while the surface energy is a stabilising one for the surface growing structures.

In a nanoscale AFM-induced wear process, the adatom density is generally poor, as a result, the film only changes its thickness weakly, locally relating to specific sites that remain under a threshold value above which the film becomes unstable following an ATG instability process. As a consequence, our physical situation is rather different from an epitaxially growing experiment. Nevertheless, it is shown here that (17.29) is still able to capture the physics of the surface growing properties.

In a standard scratching test, where the AFM probe tip scratches a limited region, which can be a square or a line. The surface being worn and undergoing growth process can be divided in three distinct regions. In the first one, the first term on the right side of (17.29) describing the erosion process is predominant in the scratched area. The second region is conditioned by the mismatch lattice due to the stressing around the scratched area, and the third region is the surface area far from the scratched area, this area can be considered as unaffected by the scratching test. Such division implies that we can integrate (17.29) in the third region without considering the erosion term. Otherwise, in the first region, we can consider only the erosion term. Being interested in having a representation of the growth process at the end of the scratch test, we can set the worn region as $h = -h_0$.

Equation (17.29) is an expression considering both the slow erosion dynamics driven by the AFM tip scanning velocity and the fast time growth process dominated by the surface diffusion. The process of erosion and the process of growing surface structures have different time scales. Slow time is for the erosion process that depends on the scanning velocity of the probe tip, and fast time is for the growing structures that depend on the surface diffusion. The different time scales give the possibility to focus the attention on the growth process, which is the target of this section, looking at the growth structures at the end of the scratch test, i.e., when the wear volume has been eroded by the AFM tip. It can be supposed that the mass is conserved, i.e., all the mass produced during the scratch needs to be found in the growth structures. In our simulation, this means recording a surface where the volumetric dimensions of the hole produced during the scratch is equal to the volume of the growth structures, when the volume is calculated with respect to the planar surface not involved in the scratch or the growth debris deposition. It is possible to manage (17.29) in order to simplify it and to draw a numerical quantification. First of all, we need to introduce new lengths, (L/Λ) with $\Lambda \ll 1$, and time scale, τ. The time scale represents how fast the surface perturbations grow on the scale length L. The new set of rescaled variables is selected in order to have $\nabla \sim \Lambda, \partial_\tau \sim \Lambda^2$, for computational reasons, $\Lambda = 1/20$ is opted [12]. The length scale and the characteristic time scale become, respectively [32],

$$L = \frac{\gamma(1-\nu)}{4\mu\varepsilon_z^2(1+\nu)} \quad \text{and} \quad \tau = \frac{L^2}{D}, \tag{17.30}$$

where $\gamma = 1{,}070\,\text{mJ/m}^2$ is the Al surface energy [36], $\mu = 26\,\text{GPa}$ is its shear elastic modulus [36], ν is the Al Poisson ratio (0.33), $\varepsilon_z \cong 3 \times 10^{-2}$ is a lattice misfit value (uniform equibiaxial stress) and $D = 1.8 \times 10^{-11}\,\text{m}^2/\text{s}$ is the diffusion constant. [The diffusion constant is $D = a^2\nu_0 \exp[-E_\text{d}/k_\text{B}T]$, using $E_\text{d} = 25\,\text{meV}$ as a suitable barrier height (i.e., the activation energy for diffusion), $a = 4 \times 10^{-20}\,\text{m}^2$ as the Al lattice constant and $\nu = 10^{13}\,\text{Hz}$ as the attempt frequency, we obtain $D = 1.8 \times 10^{-11}\,\text{m}^2/\text{s}$.] Making use of such parameters we obtain $L \cong 5\,\text{nm}$ and $\tau \cong 10^{-6}\,\text{s}$. After appropriate rescaling, the following equation for h is obtained (details for the derivation can be found in [32, 37]):

$$\partial_\tau h = \nabla^2[\alpha h + \nabla^2 h + \nabla^4 h + (h\nabla^2 h + \chi|\nabla h|^2 + \varphi h^2)], \tag{17.31}$$

where α is a slope parameter, and the parameters $\chi/\alpha \sim \partial\Phi(L,0)/\partial|\nabla h|^2$ and $\varphi/\alpha \sim \partial\Phi(L,0)/\partial h$ are connected to the derivative dependence of the wetting potential with respect to the slope or the height of the surface. Because (17.31) could describe a conserved quantity (in this case h), any term in the right-hand side may be written in the form of a divergence.

The linear analysis of (17.31) can be made assuming a planar surface with a small perturbation $h \sim e^{\lambda t + ikr}$. Inserting this expression for h into (17.31), we obtain a dispersion relation as $\lambda = -\alpha^2 k^2 + k^4 - k^6$ [12]. Therefore, the instability occurs for $\alpha < \alpha_c = 1/4$ at a wave number $k_c = \sqrt{2}/2$, which corresponds to stationary, spatial periodic patterns. The presence of the wetting potential in (17.31) changes significantly the dispersion relation modifying the linear instability range, and almost two bifurcate modes of growth are made possible [12, 32].

Dots or mounds can be obtained with (17.31) making use of a weakly nonlinear analysis of stationary solutions near the instability threshold. Consider the 2 + 1 case of a 3D film with a 2D surface whose evolution is described with (17.31). The weak nonlinear evolution of the surface structure, when the instability is guaranteed taking $\alpha = \alpha_c - \varepsilon$, where ε is a very small quantity, can be expressed as [37]:

$$h = \varepsilon \sum_{n=1}^{3} A_n(\mathbf{X}, T) e^{i\mathbf{k}_n \mathbf{X}} + c.c. + \varepsilon^2 B(\mathbf{X}, T), \tag{17.32}$$

when A_n ($n = 1, 2, 3$) are three complex amplitudes of the unstable modes with wave vectors k_n, respectively, B is a real amplitude, $\mathbf{X} = \varepsilon \mathbf{x}$, and $T = \varepsilon^2 t$. Making use of the multiple-scale analysis near the bifurcation point to obtain the following system of coupled equations at the third and fourth orders of ε for A_1 (similar expressions are obtained for A_2 and A_3) and B, respectively:

$$\begin{aligned}\frac{\partial A_1}{\partial T} &= 2\alpha^2 A_1 + (k_1 \cdot \nabla)^2 A_1 - r_1 A_2^* A_3^* + i r_2 k_1 \nabla (A_2^* A_3^*) \\ &\quad - \left[\lambda_1 |A_1|^2 + \lambda_2 \left(|A_2|^2 + |A_3|^2\right)\right] A_1 - w A_1 B; \qquad (17.33a) \\ \frac{\partial B}{\partial T} &= \frac{1}{4}\nabla^2 B + 2s\nabla^2 \left(|A_1|^2 + |A_2|^2 + |A_3|^2\right), \qquad (17.33b)\end{aligned}$$

where the parameters are as follows:

$$\begin{aligned}s = -1 + \chi + 2\varphi; \quad w = 1/4 - \varphi; \quad r_1 = 1 - \chi - 2\varphi; \quad r_2 = 1 - 2r_1; \\ \lambda_1 = -[(1+\chi)^2 - 3\varphi(1+\chi) + 2\varphi^2]; \quad \lambda_2 = -[1 - \varphi + 3\varphi\chi + 2\varphi^2 - \chi^2].\end{aligned} \tag{17.34}$$

The numerical calculation of (17.33a and 17.33b) in 3D can be made by means of a pseudopotential code with boundary conditions [12, 32, 37]. The computational cell of size 20 × 20L is discretised into a 256 × 256 grid, and the normalised time step $\Delta t = 0.1$ for a total time scale $t = 10^4 \Delta t$. Refining the computational grid and the time step leads to no difference in the simulation results. Results of the numerical simulations are shown in Fig. 17.10. It is possible to observe the formation of regular arrays of dots or mounds for suitable parameters, α, χ, φ, as predicted by the weak nonlinear analysis.

As shown in Fig. 17.10a, the wear volume produced in the scratched area is $20 \times 20 \times 1\,\text{nm}^3$, and using a 0.2 nm lattice constant and an atomic volume $\Omega \cong 1.5 \times 10^{-29}\,\text{m}^3$ for the Al sample, the number of growth clusters is nearly 150 with a narrow size distribution. With such dimensions, a cluster is composed by nearly 300 Al atoms, for an average volume of 3–4 nm^3. The clusters can be represented as pyramids with a squared base, as in the zoom set in Fig. 17.10a, side $\approx$ 2.5–3 nm, with a height of approximately 1–1.5 nm. Such results for the cluster area dimensions are confirmed in Fig. 17.10c, where a different set of χ and φ parameters was used. Note that in the numerical simulation, the wear volume is *manually* added, and the images reproduce the state of the surface at the end of the scratching test, where the mass conservation is imposed, i.e., the wear volume is equal to the summarised cluster volumes. Figure 17.10b shows the formation of large mounds surrounding the scratched area. Such structures can be imputed to Oswald ripening and coarse-graining.

Self-organising structures can be determined looking at steady-state solutions [32,37]. It is well known that regular spatial arrays of clusters can be obtained using (17.33) when $A_1 = A_2 = A_3 = A$, and $B = 0$, where

$$A = \frac{-r_1/2 + |r_1| \sqrt{r_1^2/4 + 2\alpha^2(\lambda_1 + 2\lambda_2)}}{2(\lambda_1 + 2\lambda_2)}. \tag{17.35}$$

The possibility of a derivation of steady-state solutions is particularly useful for the numerical simulation. A relation, as (17.35), can be used as a parameter of control for the stability of the solutions of (17.31). The described stability conditions determine regions in the (χ, φ) plane where self-organisation of hexagonal arrangements of clusters can be observed [32]. Inside these regions, self-organised surface structures with hexagonal symmetry can be stable in a certain range of the α parameter. The complete dependence of such region on the critical parameter α, such as other surface structures is beyond the scope of this study. Here the research is limited to find the parameter regions in the (χ, φ) plane leading to self-organised array of clusters or coarse-graining mounds. The parameter regions in the (χ, φ) plane corresponding to the different surface morphologies can be obtained slightly increasing parameter α. This is because the dispersion curves obtained by (17.30) have a larger range of k values which provide a mound solution for (17.31).

The morphology was obtained after scratch test results by minimising of the total free energy or of the chemical potential [see (17.12) and (17.13)]. Since it is difficult to find an analytical expression connecting the parameters used in (17.31) and the flux forms as given in (17.21) and (17.22), an alternative approach is used here. The growth structures can be found using suitable parameters falling in different regions of the instability range as shown in the linear analysis or using the MFNT (Sect. 17.3.3) that connects the growth structures to the diffusion constant, D, and the flux, F.

The conditions and the corresponding parameters that minimise the chemical potential with the presence of a growing volume V that yield the equilibrium morphology have to be found. The chemical potential (17.28) is composed of three terms, the elastic energy density (connected to the strain energy), the surface energy and the wetting potential. For small volumes V, the elastic energy (proportional to V) is small and the surface free energy ($k\gamma$) is mainly due to the sides of the clusters. The area of these clusters is proportional to $\sqrt{Vh}$, it follows that the clusters should be as small as possible. If the wetting potential dominates in the chemical potential, then the wetting interaction between the film and the substrate can lead to the self-organisation of dots with uniform sizes only if the wetting potential is strongly dependent on the surface slope [32].

For larger cluster sizes, the elastic energy dominates with respect to the other two terms. The energy of a cluster can be written as $F(r) \approx -V(r)\Delta\rho\Delta\mu + \gamma k(r) + \Phi(r)$, where $V(r)$ is the cluster volume as a function of the radius, $\Delta\rho$ is the adatom number density, $\Delta\mu$ is the variation of the chemical potential for the formation of clusters and $k(r)$ is the surface curvature induced by the cluster. The first volume term in the total energy decreases with cluster size while the second surface term increases with cluster size due to increase of surface area, the third term is the wetting potential connected to the cluster. The balance condition requires $\mathrm{d}F(r)/\mathrm{d}r = 0$ that estimates the stable cluster size. The derivative, with respect to surface height of the wetting potential, directly gives the ratio φ/α, with which a positive or a negative value can make the cluster size stable or unstable. When a condition such as $\mathrm{d}F(r)/\mathrm{d}r = 0$ is approached, the new arriving adatoms prefer to grow in new clusters, favourable in an energetic point of view, leaving the old structures stable. The wetting potential drives the self-organisation that gives the narrow shape and the regular array of the cluster structures. The wetting interaction between the growth structure and the substrate also depends on the derivative of the surface energy $\gamma(h)$. We have not yet explicated a form for the surface energy, nevertheless, at the cluster–surface boundary the surface energy generally has a drastic variation that implies a severe change in wetting potential, this can lead to the self-organisation of dots with almost uniform sizes if the wetting potential is strongly dependent on the surface slope. Many authors have investigated the stability of regular dot patterns with respect to patterns having other symmetries [32, 34, 37]. It is shown that, with the increase of parameter α, a regular dot pattern can become unstable leading to stripe patterns.

Mound structures surrounding the scratched area can be obtained for choosing a suitable parameter choice as given in (17.35). The reduced incidence of the role of the wetting potential determines the suitable parameters, therefore a coarse-grained mechanism very close to the Ostwald ripening is activated. Ostwald ripening is a process where large clusters grow at the expense of small clusters through an exchange of monomers. The process is governed by the inverse dependence of the chemical potential on the cluster size. For a system containing both large and small clusters, the chemical potential difference causes material to be transferred from small to large clusters. The reduction of the effects of the wetting potential implies that in the case of homoepitaxy Al/Al the elastic energy tends to remain rather limited, and in absence of the wetting potential, only the surface energy should be considered. The evolution tends to reduce the surface energy, and therefore, the total surface of the clusters. Since the adsorbate volume practically does not change in time, and the incoming flux of adatoms is extremely low and slow time scale, the ratio surface/volume decreases, and therefore the average cluster size increases. Adatoms can eventually escape from smaller clusters and diffuse on the surface to bigger clusters where the free energy per atom, which is the chemical potential, is lower in smaller clusters. Therefore, the coarse-graining of the few mounds is strongly favoured as shown in Fig. 17.10b. Equation (17.31) is a powerful method to capture the physical behaviour. As demonstrated, there are different classes of parameters in (17.35) that lead to self-organised regular cluster arrays or coarse-graining mounds.

The different growth structures, dots or mounds, have been obtained in an empirical way that is for suitable choices of the parameters in (17.31), because the plane (χ, φ) provides the formation of dots or mounds based on the nonequilibrium thermodynamics continuum model that complete the previous evaluation of the density cluster number via molecular nucleation theory as made in Sect. 17.3.3.

17.5 Conclusions and Future Perspectives

In this chapter, recent experimental and theoretical results showing that, in some conditions, such as Ultrahigh Vacuum (UHV) environments, a new observable can be evaluated for an accurate knowledge of atomic and molecular onset mechanisms involving the occurrence of wear mechanisms, mainly abrasion, have been presented. In fact, recent UHV scratching atomic force microscopy (AFM) experiments showed the formation of small clusters, larger aggregates or regular patterns on the surface being scanned, and a theory capable of capturing the basic mechanisms producing the formation of such structures has been recently proposed. Such cluster structures, generally self-organised in regular structures, are mainly produced by the flux of adatoms generated by the AFM tip stripping off adatoms during the continuous passage of the probe tip on the surface being analysed. In UHV environments, surface diffusion is the dominant mass transport mechanism, and a nonequilibrium thermodynamic framework for the self-organised growth process has been developed where the surface growth processes maintain a sort of coherence of the flux rates induced by the AFM tip during the wear test making the wearing and the surface growth specular.

The introduction of a new observable for nanowear experiments is made possible if there is a direct connection between the rates of debris produced by the AFM tip and the surface growth processes. The rates for the debris production are related to the flux of adatoms made available for the surface diffusion after the passage of the stressing tip. The flux intensity is described by a time function, F, that is related to the statistical counting of the atoms stripped off (debris), while, in a UHV environment, surface diffusion, described by the coefficient D, is the main parameter characterising the adatoms dynamics. A possible direct connection between F and D implies that a *coherence* is maintained between the wearing process induced by the AFM tip and the subsequent surface growth process induced by the debris. In addition, if a coherence between wearing and growth is established, it implies that on changing the flux rates, different growth structures should be observed. This coherence has been demonstrated by D'Acunto in a recent paper using a nonequilibrium thermodinamical focused using the aluminium as a material of reference because its Al/Al(111) homoepitaxial growth [12].

The actual theoretical justification of the experimental results has been addressed using a continuum approach, an atomistic simulation reproducing the surface growth processes induced by the AFM tip could be open new insights on the occurrence of basic wear mechanisms on atomic and molecular scale.

Acknowledgements

The author likes to acknowledge E. Ciulli, F. Dinelli and E. Gnecco for useful discussions and suggestions.

References

1. K. Ludema, *Friction, Wear and Lubrication, a Textbook in Tribology* (CRC Press, Boca Raton, 1996)
2. K. Ludema, in *Fundamentals of Tribology and Bridging the Gap between the Macro- and Micro/Nanoscale*, ed. by B. Bhushan (Kluwer, Dordrecht, 2001)
3. I.M. Hutchings, *Tribology: Friction and Wear of Engineering Materials* (CRC Press, Boca Raton, 1992)
4. B. Bhushan, *Springer Handbook of Nanotechnology* (Springer, Berlin, 2004)
5. B. Bhushan, *Principles and Applications of Tribology* (John Wiley, New York, 1999)
6. R. Colaço, in *Fundamentals of Friction and Wear on the Nanoscale*, ed. by E. Gnecco, E. Meyer (Springer, Berlin, 2006)
7. B. Bhushan, *Introduction to Tribology* (Wiley, New York, 2002)
8. E. Rabinowitz, *Friction and Wear of materials* (John Wiley, New York, 1965)
9. E. Gnecco, R. Bennewitz, E. Meyer, Phys. Rev. Lett. **88**, 215501 (2002)
10. A. Socoliuc, E. Gnecco, R. Bennewitz, E. Meyer, Phys. Rev. B **68**, 115416 (2003)
11. T. Filleter, W. Paul, R. Bennewitz, Phys. Rev. B **73**(1–10), 155433 (2006)
12. M. D'Acunto, Phys. B **405**, 793 (2010)
13. W. Maw, F. Stevens, S.C. Langford, J.T. Dickinson, J. Appl: Phys. **92**, 5103 (2002)
14. B. Gotsmann, M.A. Lantz, Phys. Rev. Lett. **101**, 125501 (2008)
15. M.A. Lantz, D. Wiesmann, B. Gotsmann, Nat. Nanotechnol. **4**, 586 (2009)
16. H. Baskaran, B. Gotsmann, A. Sebastian, U. Drechsler, M.A. Lantz, M. Despont, P. Jaroenapibal, R.W. Carpick, Y. Chen, K. Sridharan, Nat. Nanotechnol. **5**, 181 (2010)
17. R. Bassani, M. D'Acunto, Tribol. Int. **33**, 443 (2000)
18. S. Kopta, M. Salmeron, J. Chem. Phys. **113**, 8249 (2000)
19. M. D'Acunto, Tribol. Int. **36**, 553 (2003)
20. M. D'Acunto, Nanotechnology **15**, 793 (2004)
21. M. D'Acunto, Nanotechnology **17**, 2954 (2006)
22. M. D'Acunto, in *Scanning Probe Microscopy in Nanoscience and Nanotechnology*, ed. by B. Bhushan (Springer, Berlin, 2010), p. 647
23. A. Pimpinelli, J. Villain, D.E. Wolf, Phys. Rev. Lett. **69**, 985 (1992)

24. H. Brune, Surf. Sci. Rep. **31**, 121 (1998)
25. J.A. Venables, D.J. Ball, Proc. R. Soc. Lond. A **322**, 331 (1971)
26. J.A. Venables, Philos. Mag. **27**, 697 (1973)
27. S. Ovesson, Phys. Rev. Lett. **88**, 116102 (2002)
28. D.B. Chrisey, G.K. Hubler, *Pulsed Laser Deposition of Thin Films* (John Wiley, New York, 1994)
29. T.R. Mattsson, H. Metiu, Appl. Phys. Lett. **75**, 926 (1999)
30. J. Lapujoulade, Surf. Sci. Rep. **20**, 191 (1994)
31. W.W. Mullins, J. Appl. Phys. **28**, 333 (1957)
32. A.A. Golovin, S.H. Davis, P.W. Voorhees, Phys. Rev. E **68**, 056203 (2003)
33. Y. Pang, R. Huang, Phys. Rev. B **74**, 075413 (2006)
34. J.-N. Aqua, T. Frisch, A. Verga, Phys. Rev. B **76**, 165319 (2007)
35. B.J. Spencer, S.H. Davis, P.W. Voorhees, Phys. Rev. B **47**, 9760 (1993)
36. K.K. Kalazhokov, Z.K. Kalazhokov, K.B. Khokonov, Technic. Phys. **48**, 272 (2003)
37. S.M. Cox, P.C. Matthews, Phys. D **175**, 196 (2003)

Chapter 18
Frictional Stick-Slip Dynamics in a Deformable Potential

Germaine Djuidjé Kenmoé and Timoléon Crépin Kofané

Abstract Frictional stick-slip dynamics is carefully studied theoretically and numerically in a model of one oscillator interacting with a nonsinusoidal substrate potential. We focus our attention on a class of parameterised one-site Remoissenet–Peyrard potential $U_{\mathrm{RP}}(X, r)$, whose shape can be varied as a function of parameter r and which has the sine-wave shape as a particular case. The evolution of the static friction as a function of the deformable parameter r is calculated. Our numerical investigation, which involves bifurcation diagrams, shows a rich spectrum of dynamical behavior including periodic, quasiperiodic and chaotic states. On the other hand, and for a good selection of the parameter systems, the motion of the particle involves periodic stick-slip, erratic and intermittent motions, characterized by force fluctuations, and sliding. This study suggests that the transition between each of the motions strongly depends on the shape parameter r. However, the stick-slip phenomena can be observed for all values of the shape parameter r in the range $|r| < 1$. The analytical analysis of the dry friction reveals that the dynamics depends non-trivially on the shape parameter r. Thus, the variable periodic potential allows us to consider a variety of substrate potential shapes useful in the description of real physical systems.

18.1 Introduction

The study of sliding friction is one of the oldest problems in physics and certainly one of the most important from a practical point of view [3, 20]. In spite of this, the microscopic origin of sliding friction is still not well understood [14, 16, 25, 27]. The interest in elucidating dissipative processes [3, 26] on various spatial and temporal scales, from microscopic to macroscopic, that may occur in all machines and mechanisms is justified by modern experimental technologies that have made it possible to study wearless friction between clean and atomically flat surfaces [2, 19]. In particular, much attention has been recently paid within the field of nanotribology in the understanding of the nature of friction at the microscopic scale [19]. Sheared liquids confined between two atomically smooth solid surfaces provide a good example of a system where a broad range of phenomena and different behaviors have been experimentally observed. The majority of surface force apparatus measurements has been performed with atomically flat mica surface [1,4,10,31,32]. On the other hand, great attention is paid to the dynamics of dry friction, i.e., velocity-independent friction. Moreover, it is now well understood that the dynamic behavior in several fields of science with dry friction is nonlinear because Coulomb's laws [3] make a distinction between static friction force and kinetic friction force. The static friction force is the force necessary to start sliding or a constraining force during sticking (stick regime), while the kinetic friction force is the force necessary to keep sliding at a constant velocity (slip regime). If the static friction force is larger than the kinetic one, or the kinetic friction force drops rapidly at small speeds, the sliding surfaces alternately switch between sticking and slipping in a more regular or irregular fashion. This unstable motion is repeated in rapid succession until the slide reaches a certain velocity called the critical velocity [5].

Tremendous advances in computing hardware and methodology have dramatically increased the ability of theorists to simulate tribological processes. This has led to an explosion in the number of computational studies over the last decade, and allowed increasingly sophisticated modelling of sliding contacts. Although it is not yet possible to treat all the length scales and time scales that enter the friction coefficient of engineering materials, computer simulations have revealed a great deal of information about the microscopic origins of static and kinetic friction, the behaviour of boundary lubricants and the interplay between molecular geometry and tribological properties. These results provide valuable input to more traditional macroscopic calculations. Given the rapid pace of development, simulations can be expected to play an expanding role in tribology.

Considerable progress has been made in recent years in understanding stick-slip phenomena. But we know that almost all studies on stick-slip motion available in the literature involve only quasi-classical potentials with constant shape such as sine-wave potential [9, 12, 24], the Lennard–Jones potential [15] or the Fenkel–Kontorova model [28]. The results of these studies are very encouraging, but in these systems, the shape of the nonlinear one-site potential may deviate considerably from that attributed to the local potential. Thus, the classical systems, which are generally associated with rigid potentials, cannot be satisfactorily used to describe them. Introduction of deformable potentials in the description of those systems is therefore indispensable to obtain general results that can be applied to them [7, 8]. Considerable effort has been made in the last 2 decades to use deformable potentials in the context of the solitary waves [6, 13, 21–23], and recently in the chaotic systems [17, 18, 29, 30]). In spite of these efforts, little is known about stick-slip phenomena, which involve deformable potentials. Therefore, the present chapter is a contribution to the growing field of the concept of friction, and the problem of stick-slip motion in a more complex periodic nonsinusoidal potential, which is of interest to our present investigation.

The remainder of this chapter presents recent results from various aspects of mechanical, frictional and stick-slip phenomena with variable periodic substrate potential. Section 18.2 describes the model, work on static friction and presents the equation of the field displacement. Our analysis starts in Sect. 18.3 by the numerical investigations including computation of bifurcation diagrams. The properties of the pure dry friction force, defined as the dissipated energy per unit length are carried out in Sect. 18.4. Finally, Sect. 18.5 summarizes the results.

18.2 The Model and Equation of motion

18.2.1 Potential and geometry

Consider a model that includes a particle of mass m which moves along a variable periodic substrate. The particle is pulled by a linear spring with force constant K connected to a stage that moves with a velocity V. The surface between the particle and the plate is rough so that the plate exerts a dry friction force on the particle that sticks to it. A wide range of potential has been employed in tribology studies. Many of the studies use simple springs and sine-wave potentials. The Lennard–Jones model, which gives a more realistic representation of typical inter-atomic interactions is also commonly used in studies of general behaviour in tribology (see [15]). In this chapter the interaction between the particle and the plate is described by a nonsinusoidal Remoissenet–Peyrard (RP) potential $U_{\mathrm{RP}}(X,r)$ given by (18.1) and (18.2) (see Fig. 18.1 for a sketch of the model).

The RP potential introduced by Remoissened and Peyrard [21,22] can be derived from the general expression

$$U_{\mathrm{RP}}(X,r) = A(r)\frac{1+e\cos(X)}{[1+r^2+2r\cos(X/m)]^p} \tag{18.1}$$

Here, r is the shape parameter with range $-1 < r < 1$, $A(r)$ is a normalizing amplitude function, m and p are integers and $e = \pm 1$. Figure 18.2 illustrates this deformation in the case where $A(r) = U_0(1-r)^2$, $m = p = 1$ and $e = -1$. The corresponding potential is given by the following equation [20]

$$U_{\mathrm{RP}}(X,r) = U_0(1-r)^2\frac{1-\cos(X)}{1+r^2+2r\cos(X)} \quad |r| < 1, \tag{18.2}$$

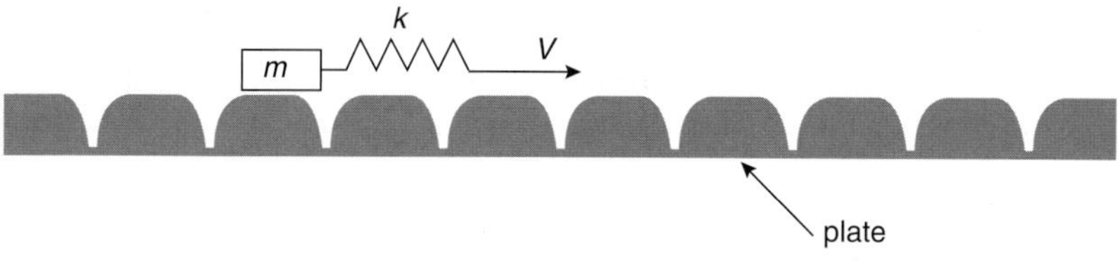

Fig. 18.1 Schematic sketch of the geometric model

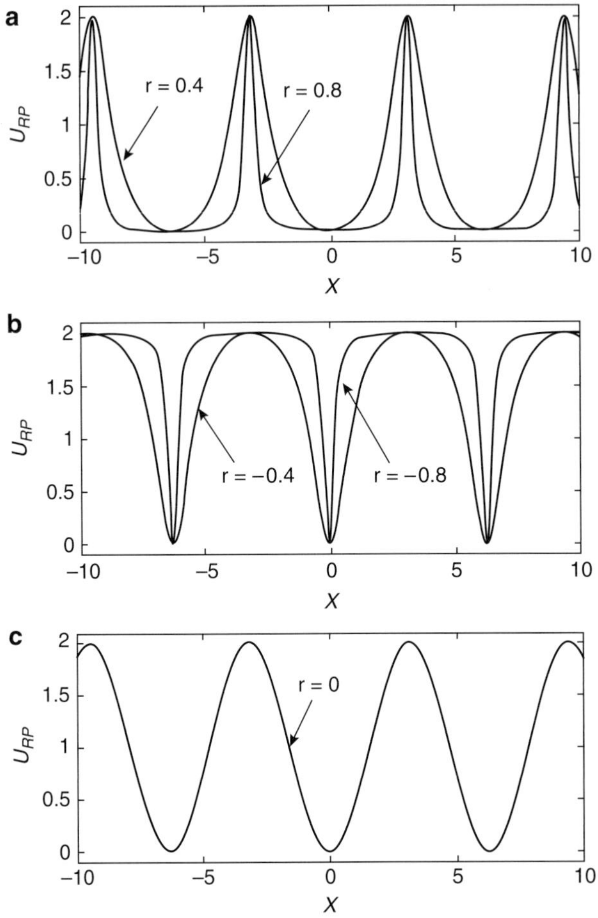

Fig. 18.2 Variability of the shape of the RP potential, (18.2), depending on the parameter r

This potential has constant amplitude and is 2π-periodic in X, where X denotes the displacement field, U_0 is a constant which measures the amplitude of the potential. The shape of the potential is defined by the parameter r. For $r > 0$, the nonsinusoidal RP potential has flat bottoms separated by thin barriers (see Fig. 18.2a, for $r = 0.4$ and $r = 0.8$), while for $r < 0$, it has the shape of sharp wells separated by flat wide barriers (see Fig. 18.2b, for $r = -0.4$ and $r = -0.8$). The potential defined by the (18.2) is reduced to the familiar sine-wave (sw) potential as defined in (18.3) through a continuous deformation when the parameter r goes to zero (see Fig. 18.2c).

$$U_{\mathrm{sw}}(X) = U_0(1 - \cos(X)) \tag{18.3}$$

The shape parameter r could account for the temperature or pressure dependence, or the geometry of the surface of the metallic surface as in our models.

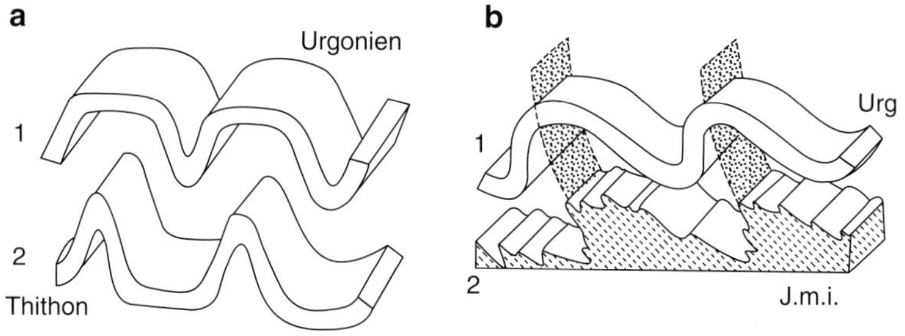

Fig. 18.3 Vertical variations of different types of the folds

In nature, we can observe the probable extension of the RP models to the fold systems. The structural analysis of the terrestrial globe is based on tectonics. Tectonic objects are more or less complex but their study can be reduced to that of elementary objects which are surfaces on the one hand, and lines, on the other. Some are created by distortions; others prior to the distortion display by their attitude, the characteristics of this distortion. In a lot of cases, it is considered that the observed structures result from a progressive distortion spread over a long length. But the slowness of the distortions makes them imperceptible: according to this view, our feet stand on a technically "living" earth. This is confirmed by the discoveries of these last decades in geotectonics, with the moving of the lithospheric plates. The shape can also vary by the change fold mode; one especially observes fluently the passage, from top to bottom, from concentric folds to flattening folds. Figure 18.3 shows a vertical variation of the different fold mode [11].

18.2.2 Frictional Force and Static Friction as a Function of the Shape Parameter

Static and kinetic frictions involve different aspects of the interaction between surfaces. The existence of static friction implies that the particle becomes trapped in a local potential energy minimum. When the bottom surface is held fixed, and an external force is applied to the particle, the system moves away from the minimum until the derivative of the potential energy balances the external force. The static friction f_s is the maximum force the potential can resist, i.e., the maximum slope of the potential. When this force is exceeded, the system begins to slide, and kinetic friction comes in to play. The kinetic frictions $f_k(v)$ is the force required to maintain sliding at a given velocity v. Thus, calculations must not only focus on the nature of potential energy minima to probe the origins of static frictions, but must also address dissipation mechanisms to understand kinetic friction.

Within our model, the frictional force acting on the particle is determined by the particle–plate interaction and is equal to

$$F_{\mathrm{f}} = -\frac{\partial U_{\mathrm{RP}}(X,r)}{\partial X} - \mu \frac{\mathrm{d}X}{\mathrm{d}t} \tag{18.4}$$

The first term is the potential component of the frictional force and the second one describes the dissipation forces between the particle and the plate. In our variable periodic substrate potential, the resistant force is

$$F_r = (2\pi/b)U_0(1-r^2)^2 \frac{\sin(2\pi X/b)}{(1+r^2+2r\cos(2\pi X/b)^2}. \tag{18.5}$$

The maximum value of this force can be interpreted as the static frictional force which also depends on the nature of the potential corrugated. This force is the smallest driving force for which no stationary states exist. When r goes to zero ($r = 0$), the static frictional force equals [15, 16]

$$F_{s0} = (2\pi/b)U_0. \tag{18.6}$$

As noted in the introduction, static friction arises because the system has managed to lock into a local free energy minimum. The total energy needed to activate the system out of this free energy minimum increases with increasing system size. However, for any finite system, thermal fluctuations will eventually lead to activated diffusion in the absence of any lateral force. The static friction corresponds to the maximum derivative of the free energy as the system is driven out of its local minimum. In our model, the results of the calculation of the static friction as a function of the shape parameter r is given in Fig. 18.4. This figure shows the evolution of the static frictional force as a function of the shape parameter r. One can easily see the symmetric property as a function of r. Static friction exists between two objects that are not moving against each other (like a desk on the ground). It will provide just enough force to resist a force you put on the object. The number given for static friction is a limit. It is the highest force you can put on an object before it starts moving. The real reason to distinguish between static and kinetic friction is to provide a simple explanation for *slip-stick oscillations* between two contacting surfaces.

18.2.3 Equation of Motion

From the Lagrangian for the nonconservative systems, the equation of the motion for the particle is

$$m\frac{\mathrm{d}^2X}{\mathrm{d}t^2} + \mu\frac{\mathrm{d}X}{\mathrm{d}t} + K(X - Vt) + \frac{\partial U_{\mathrm{RP}}(X,r)}{\partial X} = 0 \tag{18.7}$$

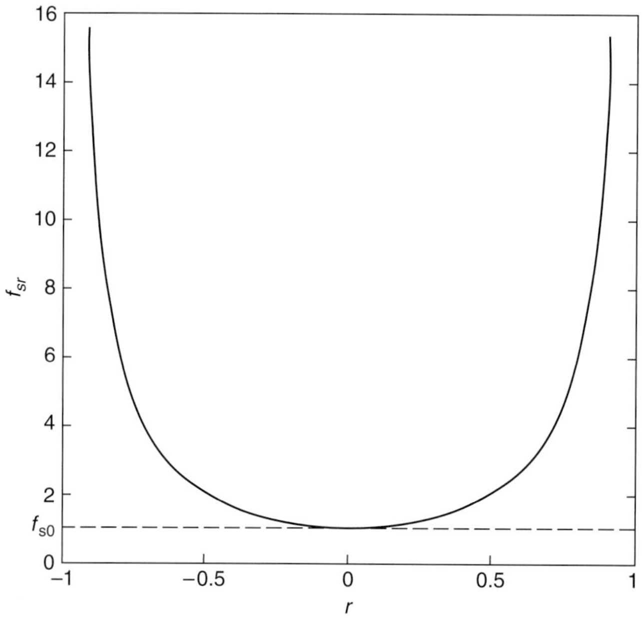

Fig. 18.4 Evolution of the static frictional force as a function of the shape parameter r

As indicated earlier the second term accounts for dissipation due to phonons and/or other excitations. The third term in (18.7) is the driving force due to the stage motion. The remaining term is due to the spatially periodic nonsinusoidal potential interaction between the particle and the plate.

In the next section, we will focus on the dynamical behavior of the particle as the driving velocity v of the stage and the shape parameter r are varied. It is convenient to introduce dimensionless space and time coordinates $x = \frac{2\pi X}{b}$ and $\tau = \Omega t$ where $\Omega = \sqrt{K/m}$ is the frequency of the free oscillations of the particle. Equation (18.7) can be rewritten then in a dimensionless form as

$$\ddot{x} + \alpha\dot{x} + c(1-r^2)^2 \frac{\sin x}{(1+r^2+2r\cos x)^2} + (x - v\tau) = 0 \tag{18.8}$$

where $\dot{x} \equiv \frac{\mathrm{d}x}{\mathrm{d}\tau}$ and $\alpha = \mu/(m\Omega)$, are the dimensionless friction constants which accounts for dissipation due to phonons and/or other oscillations, and $v = V/(b\Omega)$ is the dimensionless stage velocity, $c = (4\pi^2 U_0)/(Kb^2)$ is a dimensionless measure of the strength of the periodic potential compared to that of the spring. Equation (18.8) can describe a dissipative parametrically driven pendulum, in a nonsinusoidal substrate potential. Below, we discuss the dependence of the motion of the particle on the parameters of the system. The main objective is to deduce information on the microscopic properties of the system from the observed dynamics of the particle. For this purpose, one needs to understand the dependence of the dynamics on the mechanical (external) parameters and the parameters of the embedded system (internal).

18.3 Numerical Results

Computer simulations have played an important role in understanding tribological processes. In order to understand dynamical processes of the particle, (18.8) has been integrated numerically with fourth-order Runge–Kutta scheme with the time step equal to 0.055 and the initial conditions $\dot{x}(0) = x(0) = 0$. Depending on the parameters of the system, this model can present regular or chaotic behaviors. Thus, in a particular set of the parameters of the system, and increasing the velocity v, the behavior of the particle presents: stick-slip, intermittent and sliding motions. The transition between the stick-slip and intermittent motions is above a critical velocity $v_c^{(1)}$, while $v_c^{(2)}$ corresponds to the transition between intermittent and sliding motions. These two critical velocities strongly depend on the physical parameters of our model, and particularly on the shape parameter r.

It is well known that the results of tribological experiments can be strongly influenced by the mechanical properties of the entire system that produces sliding. In much the same way, the results from simulations depend on how the relative motion of the surfaces is imposed. However, the extra complexity of the interactions in these systems leads to a richer variety of processes.

18.3.1 Phase Space and Stroboscopic Observation

To present the dynamical behavior of the particle, we use the phase space representation of the expression $z = x - v\tau + \alpha v$ versus $\dot{z} = \dot{x} - v$. With the variation of the parameters c and v, the numerical results are shown in Figs. 18.5a–c, where one can see a periodic structure or limit cycle with period 1 (see Fig. 18.5a), where the curve is obtained for the set of parameters $c = 5$, $\alpha = 0.5$, $v = 3$, and $r = -0.5$. Moreover, with the parameters $c = 5$, $\alpha = 0.5$, $v = 1$ and $r = 0.2$, we obtain a limit cycle with period 2 (see Fig. 18.5b). Taking now the parameters $c = 5$, $\alpha = 0.5$, $v = 1$ and $r = -0.5$, we obtain a chaotic structure (see Fig. 18.5c). Furthermore,

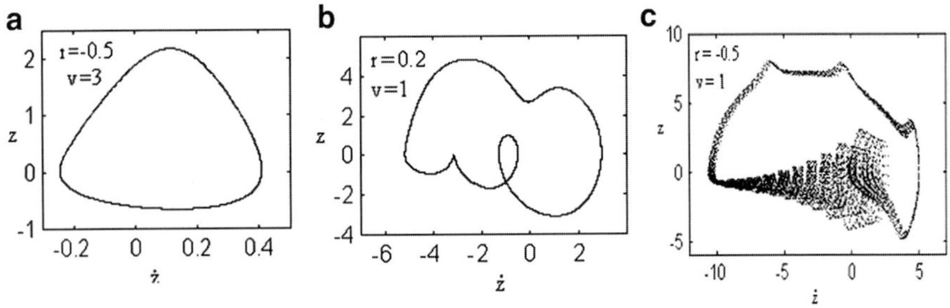

Fig. 18.5 Phase space representation of the motion of the particle for selected values of the parameters v and r, with $c = 5$ and $\alpha = 0.5$. (**a**) Limit cycle for $r = -0.5$ and $v = 3$; (**b**) period 2 for $r = 0.2$ and $v = 1$; (**c**) chaotic behavior for $r = -0.5$ and $v = 1$

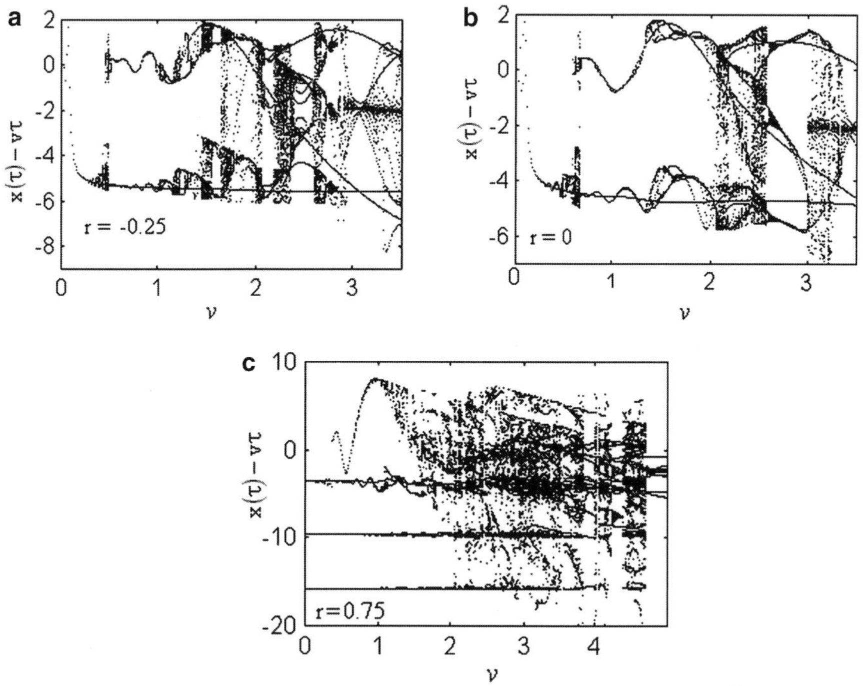

Fig. 18.6 Stroboscopic representation plot $x(\tau) - v\tau$ at time $\tau = 2\pi n/v$ for $n = 1, 2, \ldots, 100$ versus v, for three selected values of parameter r indicated in the graph; with $c = 5$ and $\alpha = 0.5$

numerical solutions show that for a given damping coefficient α, and after a time $\tau \gg 2\pi/v$, the function $x(\tau) - v\tau$ usually becomes periodic with period $2\pi n/v$ (n integer) and $x(\tau + 2\pi n/v) = x(\tau) + 2\pi n$, where the integer n depends on the parameters c and r, and also on the initial conditions. For the parameters $c = 4$, $\alpha = 0.5$ and for $n = 1, 2, 3, \ldots, 1{,}000$, Figs. 18.6a–c and 18.7a–c display the value of $x(\tau) - v\tau$ for integer multiples of periods. One can see in Fig. 18.6a–c, the stroboscopic representation as a function of v for three values of the shape parameter r. While Fig. 18.7a, b shows the stroboscopic representation as a function of the shape parameter r for certain values of the velocity v. In the range $-0.5 < r < 0.5$, the system is regular at the low velocity. This regular behavior disappears with increasing velocity v. These stroboscopic representations display bifurcation and chaotic motions and show the range of velocity v or parameter r where the motion is chaotic or regular. As we will see below, for the other set of the system parameters, the model under consideration exhibits stick-slip phenomena.

18.3.2 Stick-Slip Phenomena

Stick-slip motion refers to the fast movement that occurs between two sides of a fault when the two sides of the fault become unstuck. Stick-slip displacement on

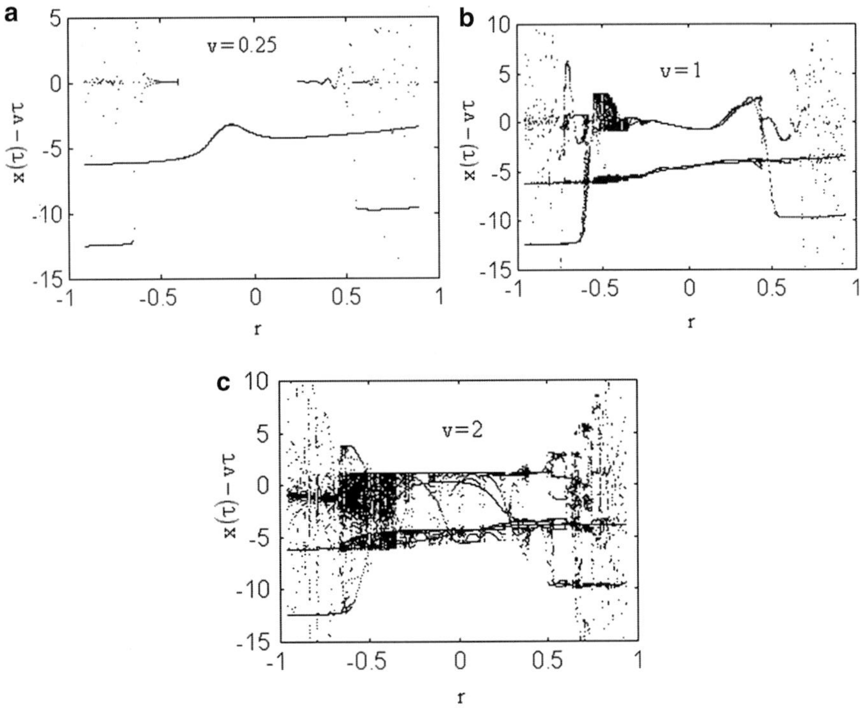

Fig. 18.7 Stroboscopic representation plot $x(\tau) - v\tau$ at time $\tau = 2\pi n/v$ for $n = 1, 2, \ldots, 100$ versus r, for three selected values of parameter $v$indicated in the graph; with $c = 5$ and $\alpha = 0.5$

a fault radiates energy in the form of seismic waves, creating an earthquake. The stick-slip motion is characterized by two regimes: in the first regime, the particle is trapped (creep) by the substrate and makes microscopic oscillations in the minima of the potential

$$U(X, r, t) = U_{\mathrm{RP}}(X, r) + \frac{1}{2}K(X - Vt)^2. \tag{18.9}$$

Figure 18.8a, b shows the form of this potential for two selected values of the shape parameter$r$. Then, a rapid sliding motion takes place and moves the particle in the next minima of the potential. Figure 18.9a, in the phase portrait$(x, \dot{x})$, presents the typical phenomena of stick-slip motion. In this figure, one can then observe positions where the particle is trapped by the substrate and makes microscopic oscillations, stops for an instant when its velocity $\dot{x}$ becomes zero. When this velocity $\dot{x}$ which is zero, increases suddenly, the particle slides rapidly and goes into the next minima of the potential and the same scenario begins. One can observe that the particle goes from one minimum to another in a rapid and regular way, i.e., in a periodic way. Moreover, the form of oscillations, which is present in Fig. 18.9b, is the same in each well of the potential and this characterizes the

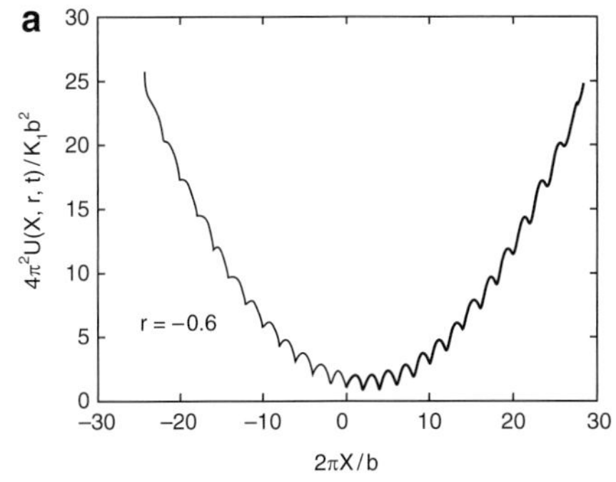

Fig. 18.8 Dimensionless total potential $4\pi^2U(X,r,t)/Kb^2$versus the dimensionless particle coordinate $2\pi X/b$, drawn for $c = 5$ and $\alpha = 4$, (**a**) $r = -0.6$; (**b**) $r = 0.6$

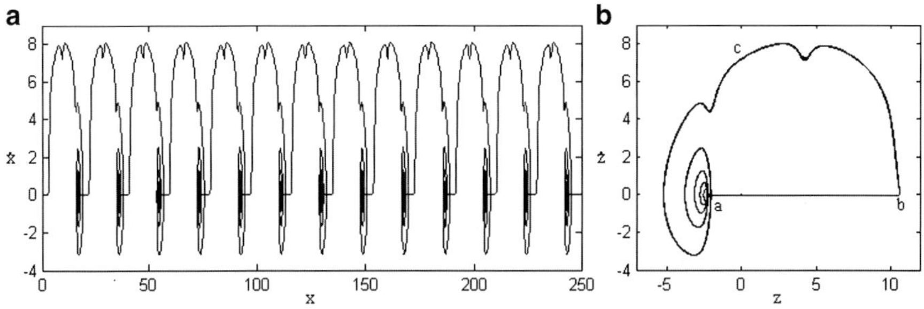

Fig. 18.9 With the set of parameters $c = 4$, $r = 0.6$, $\alpha = 0.4$ and $v = 0.06$. (**a**) Displacement of the particle in the stick-slip regime; (**b**) form of the oscillation in the minima of the potential

periodic stick-slip motion. Note that this form of oscillations strongly depends on the parameters of our physical system. As shown in Fig. 18.9b, one can observe the interval $[a, b]$ which corresponds to a very slow motion (creep) of the particle located in the minima of the potential. The particle starts to slide at the saddle point b, where the instability occurs, approaches the maximal velocity at the point c, and comes again to rest at the point a, but in the next minima of the potential. At the point of instability b, the spring force reaches a maximum value corresponding to the static friction force f_s (see Fig. 18.10a) which has been previously calculated as a function of the shape parameter r(see Fig. 18.4). During the sliding, the spring force $f_r = -(x - v\tau)$ decreases until it reaches a value of the kinetic friction f_k, where the sliding ceases and the particle is trapped again at a potential minimum. The evolution of the spring force is plotted in Fig. 18.10a–c in three regimes of the particle motion corresponding to the stick-slip, intermittent and smooth sliding motions, respectively. This evolution permits us to detect the critical velocities $v_c^{(1)}(r)$ and $v_c^{(2)}(r)$, which indicate the transition between each regime as a function of the shape parameter r.

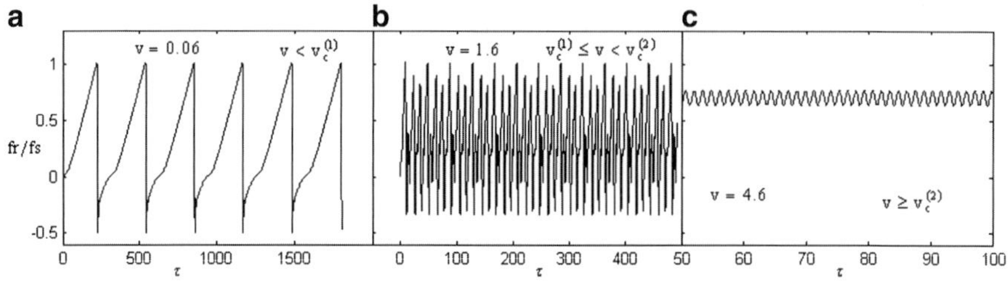

Fig. 18.10 Time series of the spring force corresponding to the three regimes of the particle motion: (**a**) stick-slip regime, for $v = 0.06$; (**b**) intermittent regime, for $v = 1.6$; (**c**) sliding for $v = 4.6$. Parameters are the same as in Fig. 18.8, except for the velocities

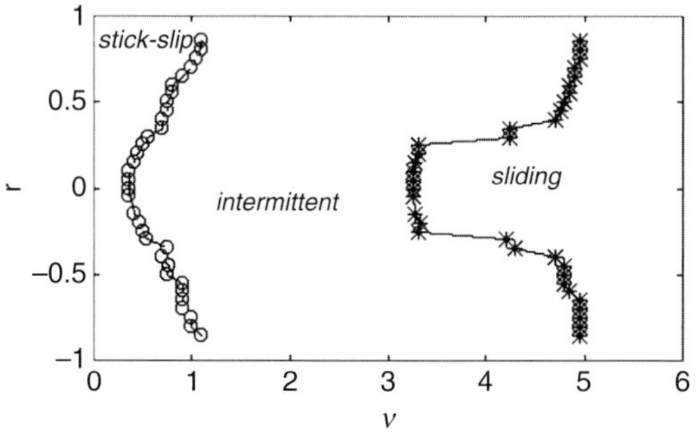

Fig. 18.11 Dynamical phase diagram for the motion of the particle for $c = 4$ and $\alpha = 0.4$. The first curve $v_c^{(1)}(r)$ indicated the boundary between the stick-slip and intermittent motions, and the curve $v_c^{(2)}(r)$ is the velocity boundary of the sliding motion. *Open circles* and *stars* indicate the results of the numerical calculations of $v_c^{(1)}(r)$ and $v_c^{(2)}(r)$, respectively

18.3.3 Influence of the Shape Parameter on the Transition from Stick-Slip Motion to Modulated Sliding State

As the stage velocity increases, the stick-slip motion of the particle becomes more erratic and intermittent, and then changes to periodically modulated sliding state. Figure 18.11 shows the dynamical phase diagram (in the r–v plane), which presents regions of the shape parameter r that correspond to different regimes of the motion of the particle. Stick-slip motion and smooth sliding occur, respectively, to the left at the line $v_c^{(1)}(r)$ and to the right of the line $v_c^{(2)}(r)$. The system exhibits an intermittent motion in the range of parameters between these two curves. The lines $v_c^{(1)}(r)$ and $v_c^{(2)}(r)$ describe the r dependence of the critical velocities corresponding to the transitions between different states of motion. These critical velocities have been found by numerical analysis of the solutions of (18.8). As the driving velocity varies from $v_c^{(1)}(r)$ to $v_c^{(2)}(r)$, the motion of the particle bifurcates from the periodic stick-slip motion to the modulated sliding state. Above$v_c^{(1)}(r)$, the stick-slip motion becomes erratic and intermittent. For a wide range of the system parameters, we find that the motion is chaotic (see Fig. 18.10b). In this figure, it appears that the transition from the periodic stick-slip motion to smooth sliding which occurs in high velocity also depends on the shape parameter r.

18.4 Pure Dry Friction

In this section we consider very small velocities $V \ll 1$, but consider times, which are larger so that $Vt \to L_0$ is finite. In this case, the stage is effectively at rest during the fall of the particle into a minimum potential. At all times, except during the fall, the particle is in a minima of the total potential $U(X, r, t)$. One of the key parameters used to characterize the beginning of the drop is the second derivative $\mathrm{d}^2U(X, r, t)/\mathrm{d}X^2$, which changes sign when instability occurs. After a dropping motion, the particle is still at a minimum potential. When there are several minima of $U(X, r, t)$ the question of where it comes to rest can only be solved using the equation of motion. The position L_0 and X_0 of the stage and the particle at the beginning of the drop, respectively, are now determined at the saddle point given by

$$\frac{\mathrm{d}U(X, r, t)}{\mathrm{d}X} = 0 \tag{18.10a}$$

and

$$\frac{\mathrm{d}^2U(X, r, t)}{\mathrm{d}X^2} = 0 \tag{18.10b}$$

Such a scheme is expected to be valid if we assume that the viscous friction is sufficiently large so that the particle comes at rest in the next minimum. The energy dissipated during the drop is

$$\Delta W = U(X_0, r, L_0) - U(X_1, r, L_0) \tag{18.11}$$

ParametersX_0, L_0 and $U(X_0, r, L_0)$ can be evaluated analytically. $U(X_1, r, L_0)$ is evaluated numerically, where X_1 is the next larger value of X that satisfies (18.10a).

In addition, the notations

$$x_0 = \frac{2\pi X_0}{b}, x_1 = \frac{2\pi X_1}{b}, l_0 = \frac{2\pi L_0}{b} \tag{18.12}$$

are introduced in dimensionless form. In this sense, x_0 is the solution of the cubic equation

$$a_3 \cos^3 x + a_2 \cos^2 x + a_1 \cos x + a_0 = 0 \tag{18.13}$$

where the coefficients $a_i\,(i = 1, 2, 3)$ depend on the shape parameter as

$$\begin{aligned} a_3 &= 8r^3 \\ a_2 &= 12r^2(1+r^2) - 2cr(1-r^2)^2 \\ a_1 &= c(1-r^2)^2(1+r^2) + 6r(1+r^2)^2 \\ a_0 &= (1+r^2)^3 + 4cr(1-r^2)^2 \end{aligned} \tag{18.14}$$

To obtain the solution of (18.13), it is useful to introduce the transformations

$$\alpha = \frac{a_2}{a_3}, \beta = \frac{a_1}{a_3}, \gamma = \frac{a_0}{a_3} \tag{18.15}$$

Applying the transformation of Thirnaus, all solutions of (18.13) is given as follows:

$$x_0 = \arccos\left(\sqrt{\frac{-4}{3p^3}}\cos\left(\frac{\theta}{3} + \frac{2k\pi}{3}\right) - \frac{\alpha}{3}\right),$$

where

$$\theta = \frac{1}{3}\arccos\left(q\sqrt{\frac{-27}{4p^3}}\right), k = 0, 1, 2 \tag{18.16}$$

with

$$p = \beta - \frac{\alpha^2}{3}$$

and

$$q = -\frac{\alpha}{27}(2\alpha^2 - 9\beta) + \gamma \tag{18.17}$$

and

$$l_0 = x_0 + c(1-r^2)^2 \frac{\sin x_0}{(1+r^2+2r\cos x_0)^2} \tag{18.18}$$

In the calculation, the condition $p < 0$ is needed for any value of the shape parameter r.

With the value of l_0, we look for the next value $x = x_1$, which satisfies (18.10a), i.e.,

$$x_1 + c(1-r^2)^2 \frac{\sin x_1}{(1+r^2+2r\cos x_1)^2} = l_0 \tag{18.19}$$

The dry friction force F is defined as the dissipated energy per unit length, i.e., $F = \Delta W/b$, which in dimensionless form is, $f = F/(Kb/4\pi^2)$. This dry friction force f is given by

$$\begin{aligned} f(c) = \frac{1}{2}(x_0 - l_0)^2 + c(1-r)^2 \frac{1-\cos x_0}{1+r^2+2r\cos x_0} - \frac{1}{2}(x_1 - l_0)^2 - c(1-r)^2 \\ \times \frac{1-\cos x_1}{1+r^2+2r\cos x_1} \end{aligned} \tag{18.20}$$

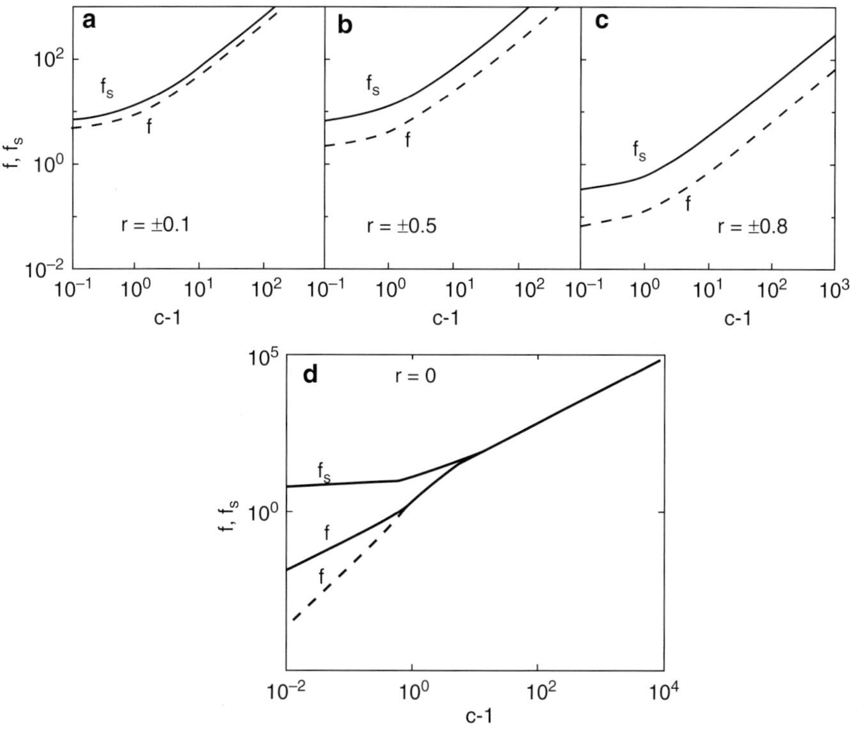

Fig. 18.12 Dimensionless dry friction force as a function of $c - 1$for arbitrary slow motion ($v \to 0$), for selected values of the shape parameter r indicated on the graphs. The parameter c is the dimensionless measure of the amplitude of the potential. *Dashed line* is obtained by the approximated analytical estimates

It appears in Sect. 18.2.2 that the static friction force depends much on the measure of the strength of the periodic nonsinusoidal RP potential and the shape parameter r. It also appears in (18.20) that the dimensionless dry friction depends on the dimensionless measure of the strength of the periodic nonsinusoidal RP potential c and the shape parameter r. Due to its analytical complexity, it has been determined only numerically through the variable c (with all the information about the function determined in the range $10^{-2} < (c-1) < 10^4$ in the log–log plotting). Further, from Fig. 18.12a–c, it should be noted that these two forces present a symmetric behavior as a function of the shape parameter r.

In the particular case where $r = 0$, the average friction force can be calculated analytically for two limiting cases (see [12])

$$f = \frac{9}{2}(c-1)^2 - \frac{18}{5}(c-1)^3 + \frac{531}{175}(c-1)^4 + O((c-1)^5) \text{for } c-1 \ll 1 \quad (18.21)$$

and

$$f = 2\pi c - 2\pi^2 + \frac{8}{3}\pi^{3/2}c^{-1/2} - \pi c^{-1} + O((c)^{-3/2}) \text{ for c} \gg 1 \qquad (18.22)$$

The corresponding dimensionless static frictional force here is $f_s = 2\pi c(r = 0)$. In this case, it is interesting to compare numerical results of this force with the approximated estimates (18.21, 18.22). Thus, we carry out a direct numerical evaluation of dry friction force (18.20) with $a_3 = a_2 = 0$, $a_1 = c$ and $a_0 = 1$. This allows us to obtain $x_0 = \arccos(-1/c)$, $l_0 = \arccos(-1/c) + \sqrt{c^2 - 1}$ and x_1 is evaluated numerically with $x_1 + c \sin x_1 = l_0$. Plotting on the same graph, the approximated (dashed line) dry friction force and the exact one (solid line), the comparison can easily be done. For these calculations, the exact result which is in good agreement with the approximation that was considered has been reproduced in the range $(c - 1) > 1$ as indicated in Fig. 18.12d.

18.5 Conclusion

In this chapter, it has been possible to describe dynamics stick-slip friction in a nonlinear model that contains a periodic nonsinusoidal substrate potential. The nonlinearity of the problem gives rise to a variety of complex features, such as bifurcation, chaotic motion, periodic stick-slip, erratic and intermittent motions, characterized by force fluctuations, and sliding above the critical velocity $v_c^2(r)$. Intensive investigations have been carried out on (18.8) and particular attention is allocated to investigations on dry friction which depends on the parameter r. The substantial impact of the shape parameter on static and dry frictions has been shown. These forces have a symmetric behavior as a function of the parameter r. It should be noted that the critical velocities, which present the transition between the different regimes of the particle motion, strongly depend on the shape parameter r. Calculations also demonstrated that a variation of the shape parameter affects the dynamical phase diagram (leading to a shift of the boundary lines $v_c^1(r)$ and $v_c^2(r)$. This deformable model leads to a variety of phenomena, which may contribute to a better description of the dry friction. In general, the shape variation of the potential is commonly ignored since all studies are limited to the sinusoidal potential, which is the leading term in the nonsinusoidal RP potential. As the periodic sinusoidal substrate potential is of importance, our results have potentially an equally wide range of applications in physics and chemistry.

References

1. Berman, W.A. Ducke, J.N. Israelachvili, Origin and characterization of different stick-slip friction mechanisms. Langmuir **12**, 4559–4563 (1996)
2. Bhushan, J.N. Israelachvili, U. Landman, Nanotribology: friction, wear and lubrication at the atomic scale. Nature (London) **374**, 607–616 (1995)

3. Bowden, D. Tabor, *Friction and Lubrication* (Oxford University Press, Oxford, 1954)
4. Crassous, E. Charlaix, J.L. Loubet, Nanoscale investigation of wetting dynamics with a surface force apparatus. Phys. Rev. Lett. **78**, 2425–2428 (1997)
5. Daikhin, M. Urbakh, Frictional forces in an electrolytic environment. Phys. Rev. E **59**, 1921–1931 (1999)
6. Dikande, T.C. Kofane, Classical-statistical mechanics of kink-bearing deformable systems: continuum study and lattice discreteness corrections. Solid-State Commun. **89**, 283–288 (1994)
7. Djuidjé, J.A. Kenfack, T.C. Kofané, Stick-slip motion in a driven two-nonsinusoidal Remoissenet–Peyrard potential. Physica D **191**, 31–48 (2004)
8. Djuidjé, T.C. Kofané, Frictional stick-slip dynamics in a nonsinusoidal Remoissenet–Peyrard potential. Eur. Phys. J. B **55**, 347–353 (2007)
9. Djuidjé, J.A. Kenfack, T.C. Kofané, Nonlinear spring model for frictional stick-slip motion. Eur. Phys. J. B **70**, 353–361 (2009)
10. Georges, A. Tonck, J.L. Loubet, D. Mazuyer, E. Georges, F. Sidoroff, Rheology and friction of compressed polymer layers adsorbed on solid surfaces. J. Phys. II **6**, 57–76 (1996)
11. Gidon, *Structures Tectoniques*, édition du BRGM, Orléans (1987).
12. Helman, W. Baltensperger, J.A. Holyst, Simple model for dry friction. Phys. Rev. B **49**, 3831–3838 (1994)
13. Kofane, A.M. Dikande, Phonons response to nonlinear excitations in a new parametrized double-well one-site potential lattice. Solid-State Commun. **86**, 749–754 (1993)
14. Landman, W.D. Luedtke, N.A. Burnham, R.J. Colton, Atomistic mechanisms and dynamics of adhesion, nanoindentation and fracture. Science **248**, 454–461 (1990)
15. Mark, M.H. Müser, *Computer Simulations of Friction, Lubrication and Wear*, ed. by B. Bhushan. Handbook of Micro/Nano Tribology, 2nd edn (CRC, Boca Raton, FL 1999)
16. Matsukawa, H. Fukuyama, Theoretical study of friction: one-dimensional clean surfaces. Phys. Rev. **49**, 17286–17292 (1994)
17. Nana, T.C. Kofane, E. Coquet, P. Tchofo-Dinda, Intermittent-type chaos in nonsinusoidal driven oscillators. Phys. Scripta **62**, 225–231 (2000)
18. Nana, T.C. Kofane, E. Coquet, P. Tchofo-Dinda, Chaotic behaviour in deformable models: the asymmetric double periodic oscillators. Chaos Solitons Fractals **12**, 73 (2001)
19. Persson, *Sliding Friction: Physical Properties and Applications* (Springer, Berlin 1998)
20. Rabinowicz, *Friction and Wear of Materials* (Wiley, New York 1965)
21. Remoissenet, M. Peyrard, A new simple model of a kink bearing Hamiltonian. J. Phys. C: Solid State Phys. **14**, L481 (1981)
22. Remoissenet, M. Peyrard, Solitonlike excitations in a one-dimensional atomic chain with a nonlinear deformable substrate potential. Phys. Rev. B **26**, 2886–2899 (1982)
23. Remoissenet, M. Peyrard, Soliton dynamics in new models with parametrized periodic double-well and asymmetric substrate potentials. Phys. Rev. B **29**, 3153–3166 (1984)
24. Rozman, M. Urbakh, J. Klafter, Stick-slip dynamics of interfacial friction. Physica A **249**, 184–189 (1998)
25. Shinjo, M. Hirano, Dynamics of friction: the superlubric state, Surf. Sci. **283**, 473–487 (1993)
26. Singer, H.M. Pollock, *Fundamentals of Friction: Macroscopic and Microscopic Processes* (Kluwer Academic, Dordrecht, 1992)
27. Sokoloff, Theory of the contribution to sliding friction from electronic excitations in the microbalance experiment. Phys. Rev. B **52**, 5318–5322 (1995)
28. Vanossi, A.R. Bisshop, V. Bortolani, Role of substrate geometry in sliding friction. Nanotechnology **15**, 790 (2004)
29. Yamgoue, T.C. Kofane, Chaotic responses of a deformable system under parametric and external excitations. Chaos Solitons Fractals **17**, 155–167 (2003a)
30. Yamgoue, T.C. Kofane, Dynamics of driven coupled oscillators with shape deformable potential. Chaos Solitons Fractals **15**, 119–129 (2003b)
31. Yoshizawa, P. McGuiggan, Israelachvili, Identification of a second dynamic state during stick-slip motion. Science **259**, 1305–1308 (1993)
32. Yoshizawa, Y.L. Chen, J. Israelachvili, Relation between adhesion and friction. J. Phys. Chem. **97**, 4128–4140 (1993)

Chapter 19
Capillary Adhesion and Nanoscale Properties of Water

Michael Nosonovsky and Bharat Bhushan

Abstract Water capillary bridges often condense at contact spots between small particles or asperities. The capillary adhesion force caused by these bridges is a major component of the attractive adhesion force, and thus it significantly affects the nanotribological performance of contacting surfaces. The capillary force is caused by Laplace pressure drop inside the capillary bridge which can be deeply negative (stretched water). In addition, the so-called disjoining pressure also plays a significant role in liquid-mediated nanocontacts. Recent atomic force microscope measurements indicate that phase behavior of water in the tiny capillary bridges may be different from macroscale water behavior. In particular, a metastable state with a deeply negative pressure, boiling at low temperatures, and ice at room temperature has been reported. Understanding these effects can lead to a modification of the traditional water phase diagram by creating a scale-dependent or nanoscale phase diagram.

19.1 Introduction

Recent advances in nanotechnology, including micro/nanoelectromechanical systems (MEMS/NEMS), micro/nanofluidic devices, bio-MEMS, and other tiny devices, have stimulated new areas of nanoscience [1, 2]. Physical properties of many materials at the nanoscale are different from their properties at the macroscale due to the so-called scale effect. For example, it has been reported that the mechanical properties of many materials and interfaces, such as the yield strength, hardness, the Young modulus, and the coefficient of friction, are different at nanoscale compared with their values at macroscale. Significant literature is devoted to the scale effect and scaling laws in mechanics. The reasons for the scale effect are discussed, including the geometrical reasons (such as high surface-to-volume ratios at the nanoscale) and physical reasons (different physical mechanisms acting at the nanoscale compared to those at the macroscale) [3–5].

Besides high surface-to-volume ratios, an important feature of a small mechanical system is that capillary forces are often present [6–8]. When two solid bodies contact air, water tends to condense near the contact spots. This is because a certain amount of vapor is always present in air. Even under low relative humidity (RH), it is not possible to completely eliminate condensation in the form of water capillary bridges forming menisci near the contact spots, such as the tips of the asperities of rough solid surfaces [9–16]. The menisci are usually concave, and therefore, the pressure inside them is reduced compared with the pressure outside in accordance with the Laplace theory. This leads to the attractive capillary force between the contacting bodies that are proportional to the foundation area of the meniscus. For small systems, this capillary force may become very significant, dominating over other forces such as the van der Waals adhesion or electrostatic forces. Therefore, understanding the properties of water in capillary bridges is very important. Normally, the phase state of water is uniquely characterized by its pressure and temperature. The phase diagram of water shows whether water is in its solid, liquid, or vapor state at a given temperature and pressure. However, at the nanoscale, the situation may be different. Both molecular dynamics (MD) simulations and experimental studies with the atomic force microscope (AFM) and the surface force apparatus (SFA) show that the state of nanoscale volumes of water at a given pressure and temperature is not always the same as prescribed by the macroscale water phase diagram [17–20]. In particular, confined small water volumes demonstrate quite unusual properties, such as the melting point depression [21]. Several effects that can be detected by AFM are related to this unusual phase behavior of the nanoscale capillary bridges.

First is the metastability of small volumes of water. When pressure is below the liquid–vapor transition line of the water phase diagram, vapor is the most stable state. However, the liquid–vapor transition involves energy barriers, and thus a metastable liquid state is possible [22]. At the macroscale, random fluctuations are normally large enough to overcome the barriers. The metastable states are very fragile, and thus they are not normally observed, except for special circumstances (e.g., superheated water). However, at the nanoscale, the barriers are large compared with the scale of the system, and metastable states can exist for long intervals of time [22–24].

Second, it was argued that ice could form at room temperature inside the capillary bridges. The water in such a bridge can be neither liquid nor solid, but form a liquid–ice condensate. This explains the slow rearrangement of the bridges between tungsten AFM tips and graphite surfaces reported in recent experiments [25]. The presence of ice-like structured water adsorbed at the silicon oxide surface was suggested to be responsible for the large RH dependence of the adhesion force in the single-asperity contact between silicon oxide surfaces in AFM experiments [26,27].

Third, the pressure inside the capillary bridge is lower than the pressure outside. At the reduced pressure, the water boiling point is lower than 100°C. There is experimental evidence that boiling in the capillary bridges indeed happens in concave water menisci [28]. Investigating nanoscale water phase transitions using AFM data can lead to modifications of the conventional water phase diagram by introducing a scale dependence. Such a diagram should depend on the characteristic size of the system, in addition to the pressure and temperature dependencies. In the present chapter we discuss theoretical and experimental data about phase transitions in capillary bridges.

19.2 Metastable Liquid Capillary Bridges

In this section, we will discuss metastable capillary bridges between AFM tips and samples with water under negative pressure (tensile stress).

19.2.1 Negative Pressure in Water

When a phase transition line in the phase diagram is crossed, it is normally expected that the substance would change its phase state. However, such a change would require additional energy input for nucleation of seeds of the new phase. For example, the liquid–vapor transition requires nucleation of vapor bubbles (this process is called cavitation), while the liquid–solid transition requires nucleation of ice crystals. If special measures are taken to prevent nucleation of the seeds of the new phase, it is possible to postpone the transition to the equilibrium state phase [22, 29, 30]. In this case, water can remain liquid at a temperature below 0°C (supercooled water) or above 100°C (superheated water) at the atmospheric pressure. Such a state is metastable and, therefore, fragile. A metastable equilibrium can be destroyed easily with a small energy input due to a fluctuation. As the stable equilibrium state, a metastable state corresponds to a local energy minimum; however, an energy barrier separating the metastable state from an unstable state is very small (Fig. 19.1).

An interesting and important example of a metastable phase state is "stretched" water, i.e., water under tensile stress or negative pressure. When liquid pressure is reduced below the liquid–vapor equilibrium line for a given temperature, it is expected to transform into the vapor state. However, such a transition requires the formation of a liquid–vapor interface, usually in the form of vapor bubbles, which needs additional energy input and, therefore, creates energy barriers. As a result, the liquid can remain in a metastable liquid state at low pressure, even when the pressure

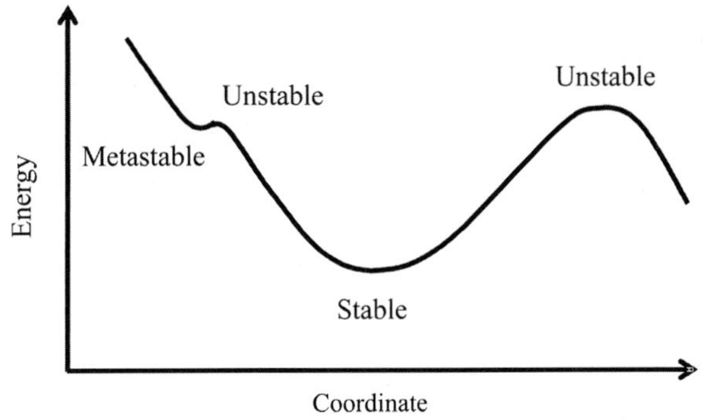

Fig. 19.1 Schematic of an energy profile of a system showing stable, unstable, and metastable equilibriums. The metastable equilibrium is separated by a very small energy barrier

is negative [29]. There are two factors that affect a vapor bubble inside liquid: the pressure and the interfacial tension. While the negative pressure (tensile stress) is acting to expand the size of the bubble, the interfacial tension is acting to collapse it. For a small bubble, the interfacial stress dominates; however, for a large bubble, the pressure dominates [22]. Therefore, there is a critical radius of the bubble, and a bubble with a radius larger than the critical radius would grow, whereas one with a smaller radius would collapse. The value of the critical radius is given by

$$R_C = \frac{2\gamma_{LV}}{P_{sat} - P}, \quad (19.1)$$

where γ_{LV} is the liquid–vapor surface tension, P_{sat} is the saturated vapor pressure and P is the actual liquid pressure [22]. A corresponding energy barrier is given by

$$E_b = \frac{16\pi(\gamma_{LV})^3}{3(P_{sat} - P)^2}. \quad (19.2)$$

Taking $\gamma_{LV} = 0.072\,\text{N/m}$ (water) and $P_{sat} - P = 10^5\,\text{Pa}$ (atmospheric pressure) yields $R_c = 1.4\,\mu\text{m}$ and $E_b = 6.3 \times 10^{-13}$ J. Cavitation (bubble formation) becomes likely when the thermal fluctuations have energy comparable with E_b.

With a further decrease of pressure in the negative region, the so-called spinodal limit can be reached (Fig. 19.2). At that pressure, the critical cavitation radius becomes of the same order as the thickness of the liquid–vapor interface. In this case, there is a lower energy path of nucleation connecting the liquid to the gas phase by

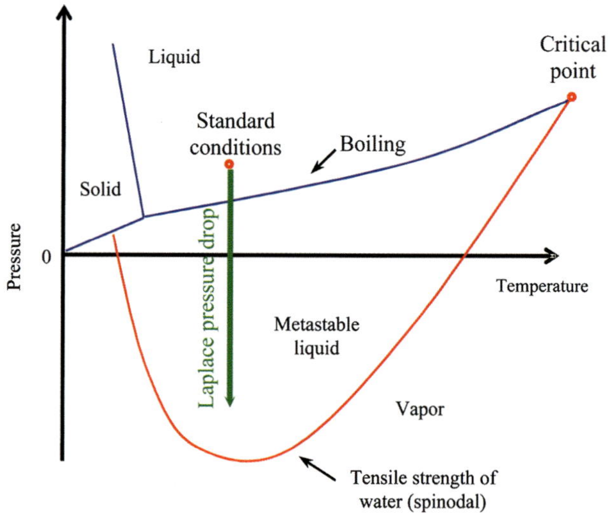

Fig. 19.2 Schematic of a phase diagram of water showing solid, liquid, vapor, and metastable liquid phases. While the energy barriers separating the metastable states are small, at the nanoscale these barriers are very significant

expansion of a smoothly varying density profile [22]. In the phase diagram, the line that corresponds to the spinodal limit is expected to go all the way to the critical point. This critical spinodal pressure at a given temperature effectively constitutes the tensile strength of liquid water. Various theoretical considerations, experimental observations, and MD simulation results have been used to determine the value of the spinodal limit. At room temperature, the spinodal pressure is between −150 and −250 MPa [31].

Experimental observations of stretched water have been known for a long time. For example, water in a cylinder sealed with a piston is subjected to negative pressure when tensile force is applied to the piston [29]. However, it is very difficult to obtain deeply negative pressure at the macroscale because of bubble nucleation. The pressure values that have been reported constitute −19 MPa with the isochoric cooling method [32], −17.6 MPa with the modified centrifugal method [33], and −25 MPa with the acoustic method [34]. The situation is different at the micro- and nanoscale. The pressure of −140 MPa in the microscopic aqueous inclusions in quartz crystals was reported [35]. Tas et al. [23] showed that water plugs in hydrophilic nanochannels can be at a significant absolute negative pressure due to tensile capillary forces. Yang et al. [24] reported the negative pressure of −160 MPa in liquid capillary bridges, as will be discussed below.

An interesting example of negative pressure in nature which has been discussed recently is in tall trees, such as the California redwood (*Sequoia sempervirens*). Water can climb to the top of the tree due to the capillary effect, and if a tree is tall enough (taller than approximately 10 m), the pressure difference between the root and the top of the tree is greater than 1 atm. Thus a negative pressure may be required to supply water to the top [36].

The existence of water at tensions as high as 160 MPa requires an explanation. The absolute pressure is larger than the largest positive pressure at which bulk water can be found on the earth's surface, for example, the pressure of water at the deepest part of the Marianas Trench in the Pacific Ocean (the deepest point of the Ocean) is about 100 MPa [37]. No common van der Waals bonded liquid can sustain such a large tension. The reason for the existence of such large tension limits lies in the strength of the hydrogen bonds which must be broken in order to form a cavity that is large enough to act as a stable nucleus for cavitation [37].

19.2.2 Negative Pressure in Capillary Bridges in AFM Experiments

The first group which suggested investigating stretched water in AFM experiments was Yang et al. [24]. They noticed that very small values of the AFM tip radius (on the order of 10 nm) correspond to a small size of meniscus (meniscus foundation area on the order of 10^{-16} m^2) and thus even a small force of 10 nN corresponds to huge pressure differences inside and outside the meniscus on the order of 100 MPa or 1,000 atm. This contradicted the conventional wisdom that the pressure inside the menisci is reduced compared with the ambient, due to the Laplace pressure drop, however, not more than by 1 atm. An attempt was made to accurately find the values of the pressure inside the bridges. This is a difficult task since the exact geometrical shape of the capillary bridge is not known. Yang et al. [24] conducted the experiment at different levels of relative humidity and in ultrahigh vacuum (UHV). They estimated the shape of the meniscus assuming that its curvature is given by the Kelvin equation which relates the meniscus curvature to the relative humidity of air.

Caupin et al. [38] questioned the results by Yang et al. [24] and paid attention to the fact that the Kelvin equation already assumes a certain pressure drop (dependent on the RH) between the meniscus and the ambient, and it is inconsistent to estimate meniscus curvature with the Kelvin equation when the task is to experimentally measure the pressure drop. In addition, they claimed that the capillary bridges correspond to the most stable state and therefore cannot be called "metastable." Caupin et al. [38] concluded that the results of Yang et al. [24] do not set the "world record" of lowest observed negative pressure in stretched water.

In their response, Yang et al. [39] pointed out that their experiments indicate that for significant relative humidity (RH $> 10\%$) there is a good agreement between the experimentally observed values of the adhesion force and those predicted by the Kelvin equations, whereas for low RH $\leq 10\%$, the Kelvin equation overestimates the pressure drop. Therefore, the calculated values of pressure inside the capillary bridges are within the reasonable error margin (the factor of two) for RH $> 10\%$, whereas for $1 \leq$ RH $\leq 10\%$ the experimental values are 3–5 times lower than those predicted by the Kelvin theory. This is not unexpected, since the Kelvin theory predicts unboundedly growing pressures for RH $\rightarrow 0\%$, and at some low values of RH it is not applicable.

Regarding the metastability of the bridges, Yang et al. [39] noted that "in the AFM context, the bridge is fragile since applying a small external force results in the rupture of the bridge. As soon as the AFM tip starts its motion (but before the bridge fractures), the bridge can become metastable." Their overall conclusion was: "Our objective was not to establish 'the world record' (as Caupin et al. [38] formulated it), but to attract attention of the community of physicists who investigate the properties of stretched water to AFM experimental data, which have been ignored by that community and which correspond to very low pressures. However, one has to admit that the pressure in AFM nanoscale capillary bridges is lower than in other experiments (with quartz inclusions or with direct measurement), and thus the AFM experiments indeed set a record of very low pressures!"

Several additional studies of capillary-induced negative pressure have been conducted later. Boyle et al. [40] investigated the light emission spectrum from a scanning tunneling microscope as a function of RH and showed that it provides a novel and sensitive means for probing the growth and properties of a water meniscus on the nanometer scale. Their modeling indicated a progressive water filling of the tip–sample junction with increasing RH or, more pertinently, of the volume of the localized surface plasmons responsible for light emission; it also accounts for the effect of asymmetry in the structuring of the water molecules with respect to the polarity of the applied bias. Choi et al. [41] performed a molecular dynamic simulation of meniscus growth. Tas et al. [42] measured capillarity-induced negative pressure of several bars for five different liquids (ethanol, acetone, cyclohexane, aniline, and water). They also investigated the probability of the cavitation using the computational fluid dynamics (CFD) approach.

19.2.3 Disjoining Pressure

Another important phenomenon for nanoscale contacts is the so-called disjoining pressure. Atoms or molecules at the surface of a solid or liquid have fewer bonds with neighboring atoms than those in the bulk. The energy is spent for breaking the bonds when a surface is created. As a result, the atoms at the surface have higher energy. This surface energy or surface tension, γ, is measured in N/m, and it is equal to the energy needed to create a surface with unit area. For a curved surface, the energy depends on the radius of curvature, as, at a convex meniscus surface, atoms have fewer bonds on average than at a flat surface, and therefore it is easier for the water molecules to leave the liquid (evaporate). For a concave meniscus, quite oppositely, it is easier for vapor molecules to reach the liquid (condense).

A flat water–air interface is in thermodynamic equilibrium with a certain amount of vapor in air, the partial pressure of which is referred to as "the saturated vapor pressure," P_{sat}, so that evaporation and condensation between the flat interface and vapor occur at the same rate. If the partial pressure of vapor in air, P_{V}, is lower than P_{sat}, evaporation prevails over condensation, whereas in the opposite case ($P_{\mathrm{V}} > P_{\mathrm{sat}}$) condensation prevails over evaporation. The ratio of the two values is referred to as the relative humidity, $\mathrm{RH} = P_{\mathrm{V}}/P_{\mathrm{sat}}$. For a concave interface, the equilibrium occurs at $P_{\mathrm{V}}/P_{\mathrm{sat}} < 1$, whereas for a convex interface it occurs at $P_{\mathrm{V}}/P_{\mathrm{sat}} > 1$. The exact value of $P_{\mathrm{V}}/P_{\mathrm{sat}}$ for a meniscus of a given curvature is given by the Kelvin equation. The pressure drop of water with density ρ risen to the height h in a capillary tube is $\Delta P = \rho g h$. Using the Laplace equation and the hydrostatic formula for vapor pressure $P = P_0 \exp(-gh/RT)$, where R is the gas constant, P_0 is the pressure at the surface ($h = 0$), and T is the temperature, yields the Kelvin equation

$$\gamma_{\mathrm{LV}} \left(\frac{1}{R_1} + \frac{1}{R_2} \right) = \rho R T \ln \frac{P_{\mathrm{V}}}{P_{\mathrm{sat}}}, \tag{19.3}$$

where R_1 and R_2 are the principle radii of curvature of the meniscus ($R_{\mathrm{k}} = (1/R_1 + 1/R_2)^{-1}$ is referred to as the Kelvin radius). Equation (19.3) relates the interface curvature at the thermodynamic equilibrium with the ratio of actual and saturated vapor pressure, $P_{\mathrm{V}}/P_{\mathrm{sat}}$ (relative humidity). According to the Kelvin model, a concave meniscus with a negative curvature given by (19.3) may form at any relative humidity. An important example of such a meniscus is in condensed water capillary bridges at nanocontacts, e.g., between an AFM tip and a sample.

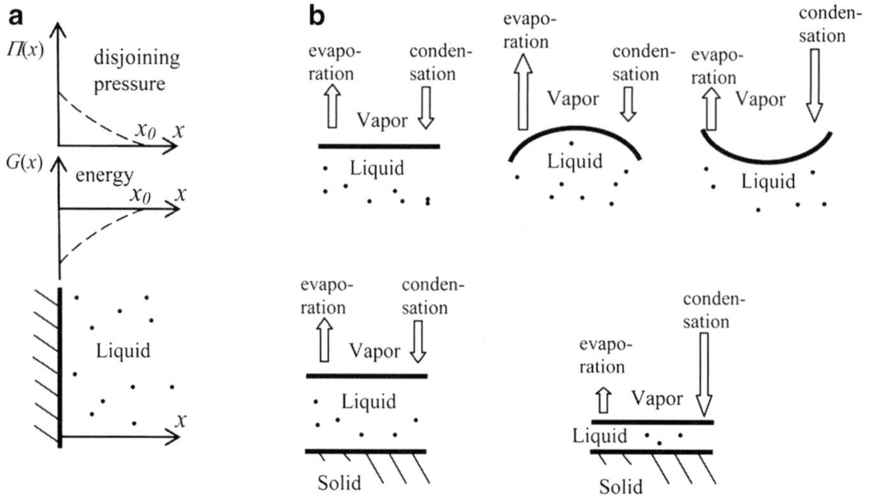

Fig. 19.3 (**a**) Disjoining pressure near a solid surface. (**b**) The effect of meniscus curvature (flat, concave and convex) and liquid layer thickness on the evaporation–condensation equilibrium

Another important effect is the so-called disjoining pressure, first investigated by Derjaguin and Churaev [43]. The adhesion force between the solid and water has a certain range x_0 and decreases with the distance x from the interface (Fig. 19.3a). As a result, water in the layer next to the interface is subjected to higher pressure $P = P_0 + \Pi_d(x)$ than the bulk pressure, where $\Pi_d(x)$ is the so-called disjoining pressure. For a thin liquid layer of thickness $H < x_0$, the evaporation/condensation equilibrium will be shifted towards condensation, since an additional attractive force acts upon water molecules from the solid surface. As a result, a thin water layer can be in equilibrium with undersaturated vapor, $P_V/P_{sat} < 1$; so the effect of the disjoining pressure on the thermodynamic equilibrium is similar to the effect of concave meniscus [44]. The influence of the disjoining pressure and the related wetting films in narrow confinement has to be considered as they may significantly change the meniscus curvature (Fig. 19.3b). A typical effect of nanoscale confinement is the appearance of capillarity-induced negative pressure.

19.2.4 Calculating Pressure in Capillary Bridges

One particularly important example of negative pressure is that in condensed water capillary bridges between very small bodies in contact, such as colloidal particles [45], or small asperities and porous gels [46] and, in particular, between an AFM tip (with a radius on the order of 10 nm) and a flat sample (Fig. 19.4).

There are several ways to estimate the capillary force between an asperity of radius R and a flat substrate. An approximate value, in the case of a small meniscus,

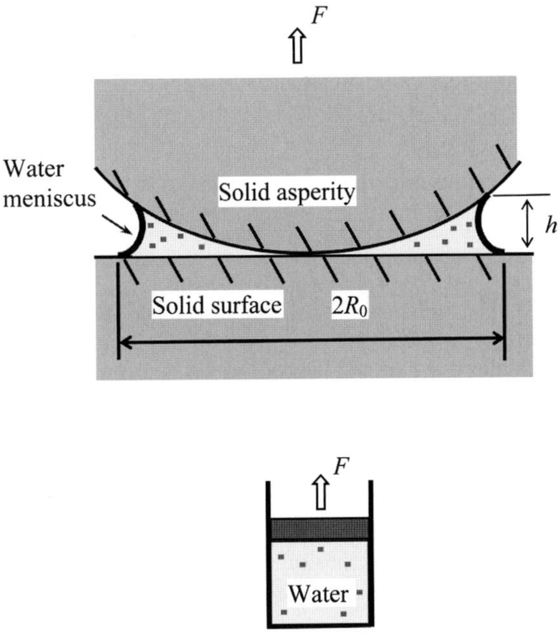

Fig. 19.4 A water capillary bridge between an AFM tip and a flat sample. The Laplace pressure inside the bridge is reduced because the bridge is concave. The bridge is under tension, similar to water in a sealed tube with a piston with the applied tensile force F

comparing with R, is given by

$$F_{\mathrm{cap}} = 4\pi R \gamma_{\mathrm{LV}} \cos\theta, \tag{19.4}$$

where θ is the contact angle between water and the substrate material. Note that the capillary bridges are usually formed between solid asperities only if the material of the asperities is hydrophilic; in other words, if $\theta < 90°$, so that $F_{\mathrm{cap}} > 0$. Equation (19.4) is based on the assumption that the meniscus radius is small compared with R and that the meniscus has an almost cylindrical shape. With the decreasing radius of the asperity, the capillary force is scaled as R, while the area is scaled as R^2. Thus, the pressure difference inside and outside the bridge is scaled as R^{-1}, and it can reach very high values, leading to deeply negative pressure inside the bridges for small asperities. A more accurate calculation of the capillary force should involve the exact calculation of the meniscus shape and pressure inside the capillary bridge, for which the Kelvin model can be used. When the contact takes place in air, a capillary bridge is almost always present due to the condensation of the water vapor from air. The menisci have negative curvature, which can be estimated from the Kelvin equation in the form [47]

$$\frac{1}{R_{\mathrm{k}}} = \left(\frac{1}{R_1} + \frac{1}{R_2}\right) = \frac{\rho R T \ln\left(P_{\mathrm{V}}/P_{\mathrm{sat}}\right)}{\gamma V}, \tag{19.5}$$

where V is the molar volume of water, RT is the temperature times the gas constant, γ is the water surface tension. Equation (19.5) describes the equilibrium between water molecules leaving the meniscus due to evaporation and vapor molecules approaching the meniscus. In particular, for water at $T = 300\,\mathrm{K}$, the Kelvin radius in nanometers is $R_k = 0.53/\ln(P_V/P_{sat})$ nm.

The pressure inside the meniscus is smaller than outside, and the pressure difference ΔP is given by the Laplace equation [48]

$$\frac{1}{R_k} = \left(\frac{1}{R_1} + \frac{1}{R_2}\right) = \frac{\Delta P}{\gamma} \tag{19.6}$$

Note that the total curvature $1/R_k$ is constant at any point of the meniscus, whereas the values of the radii can change. This reduced pressure leads to the attractive meniscus force between the tip and the sample. This meniscus force (or capillary force) is a significant (and often the dominant) part of the adhesion force, and it can be measured with the AFM. Recent results show that the pressure inside these small bridges can be deeply negative, approaching the spinodal pressure [24].

The attractive adhesion pull-off force, measured by an AFM, is given by the sum of the van der Waals component (F_v), the chemical bonding component (F_b), the electrostatic component (F_{el}), and two components of the capillary force (the Laplace force, F_L, and the line tension force F_T) [9]

$$F_{pull-off} = F_v + F_b + F_{el} + F_L + F_T. \tag{19.7}$$

However, the capillary force is absent when the contact takes place in UHV, since no condensation occurs in that case. Assuming that F_v, F_b, and F_{el} are the same in air and in UHV, Yang et al. [24] decoupled the capillary force from the experimental value of the pull-off force by comparing the AFM data, obtained in UHV and air for same samples. After dividing the Laplace force by the force acting area, they found that the Laplace pressure in the capillary bridges is deeply negative down to -160 MPa.

The metastable state of the capillary bridges is observed because the size of the bridges is smaller than the critical cavitation radius given by (19.1) and, therefore, no cavitation occurs. Thus, water in a capillary bridge with the characteristic size smaller than R_c may remain in the metastable liquid state at pressures below zero [24, 49, 50]. The metastable state is fragile, and thus the capillary bridges can collapse. The rupture of the capillary bridges, which corresponds to the pull-off of the tip, may be associated with the collapse of the bridge because the tensile strength limit is reached.

Adhesion force measurements results for silicone SiO_2 AFM tip against the Si(100) substrate at different relative humidities (RH) are presented in Table 19.1 [11] along with the pressure drop through the curved meniscus surface, ΔP, calculated from (19.5) and (19.6). The water pressure inside the meniscus is smaller than the ambient by ΔP. It is observed that for typical RH, the pressures inside the

Table 19.1 Adhesion force F_a and corresponding pressure drop ΔP as a function of relative humidity for SiO_2 AFM tip in contact with the Si sample, based on Ref. [11]

RH	<1%	15%	45%	65%	85%
F_a (nN)	30	35	45	100	100
ΔP(MPa)	>100	171	89	48	18

meniscus is negative down to −25 MPa; however, the capillary bridge is still liquid [24, 49, 50].

Numerous experiments with capillary force measurements using small AFM tips indicate that a significant capillary adhesion force is present (Fig. 19.5a) [24]. Since the foundation area of the capillary bridges is small in these experiments, this force corresponds to huge ΔP and, therefore, to negative water pressure in the bridges (Fig. 19.5b). The data had good reproducibility when measured at the same spot on the sample. The data variation is due to roughness effect when measured at different locations on the sample, since the capillary force is very sensitive to roughness variation [51–53].

This result also follows directly from the Kelvin equation. As the RH(and P_v/P_{sat}) is reduced, the Kelvin radius becomes very small, and thus the pressure drop, calculated from (19.5) and (19.6), $\Delta P = \ln(P_v/P_{sat})RT/V$, exceeds the atmospheric pressure. According to the Kelvin model, it is possible in principle to achieve unlimitedly low negative pressure by reducing p/ps. In practice, however, the metastable region in the phase diagram is limited by the critical cavitation radius (19.1) and by the spinodal limit at a given temperature, which should be taken into consideration for a scale-dependent phase diagram. In addition, small values of RH correspond to small size of the menisci, and as the size approaches the atomic scale, the continuum theory of water, used for the derivation of the Kelvin equation, would break down. In the limit of zero RH there is no meniscus at all as well as in the limit of RH = 100% there is no pressure drop; so the dependence of the capillary force and Laplace pressure upon RH with a distinct maximum is present. We will discuss the RH dependence of the capillary force in more detail in the subsequent sections.

19.3 Capillarity-Induced Low-Temperature Boiling

Another important consequence of the water pressure reduction in the capillary bridges is the possibility of boiling at temperatures lower than 100°C. If a capillary bridge is large enough for cavitation to occur, the pressure inside the bridge is reduced compared with the ambient, and the boiling point is depressed by the amount ΔT. The Clausius–Clapeyron equation [54] provides the slope of the solid–liquid coexistence curve as

$$dP/dT = L/(T\Delta V) \tag{19.8}$$

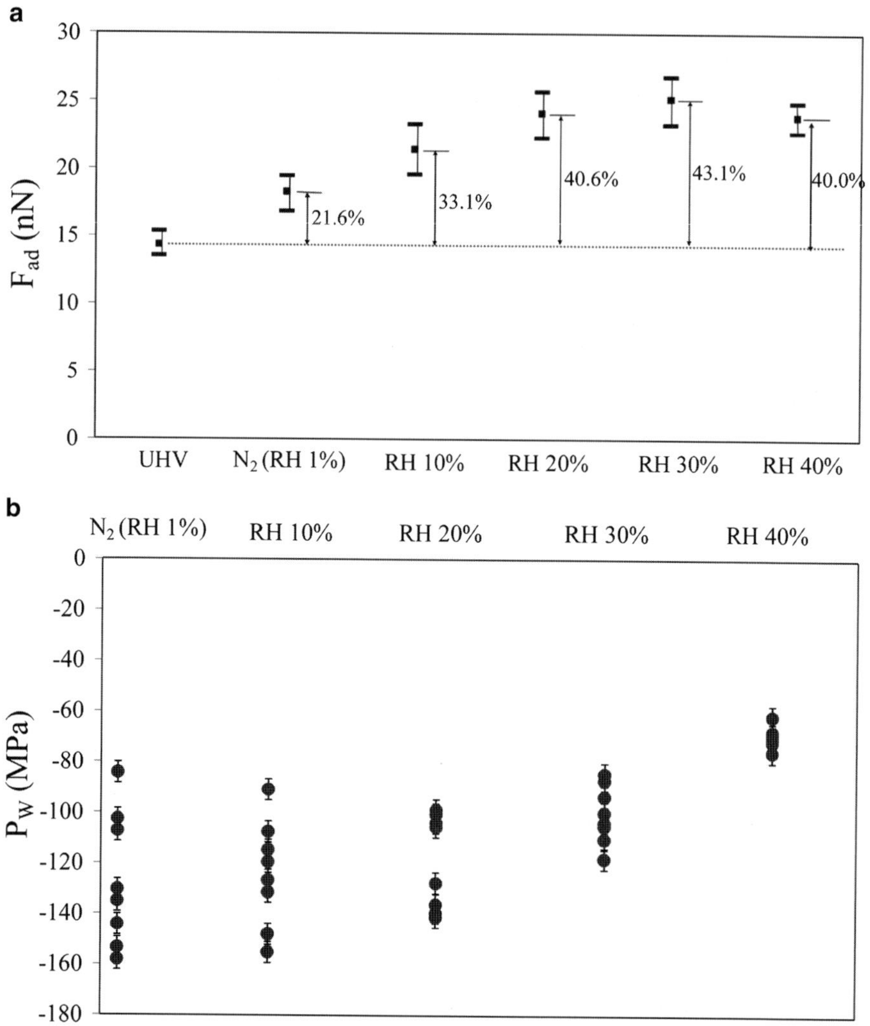

Fig. 19.5 (**a**) Adhesion force as a function of relative humidity for a Si AFM tip with radius of 25 nm in contact with a flat Si sample and (**b**) pressure inside the capillary bridge calculated from the adhesion force data (based on data from Ref. [24])

where L is the latent heat and ΔV is the volume change associated with the phase transition. For the ideal gas approximation, the change of pressure $\Delta P = P_2 - P_1$ is related to the change of the boiling temperature $\Delta T = T_2 - T_1$ as

$$\ln(P_2/P_1) = \Delta h/R(1/T_1 - 1/T_2), \quad (19.9)$$

where Δh is the change of specific enthalpy [54]. Note that ΔP is given by (19.6).

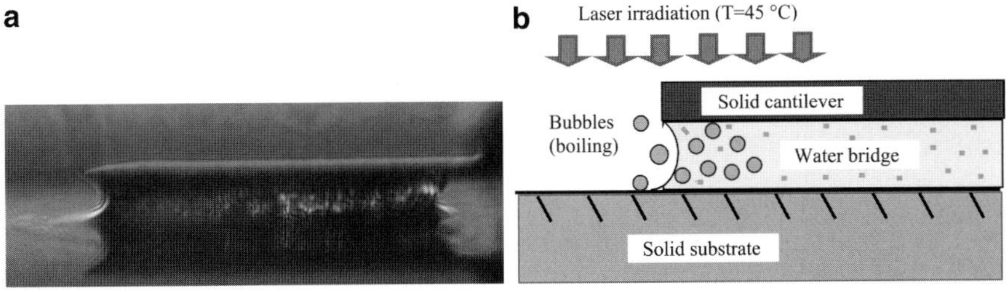

Fig. 19.6 (**a**) Optical image and (**b**) schematic of a water meniscus between a cantilever and a flat surface. The pressure inside the meniscus is reduced and a moderate temperature rise caused by laser irradiation may lead to boiling (credit to S.-H. Yang from NIST) [28]

For example, for a meniscus between an AFM cantilever and a flat surface with a gap between them of 3 μm and the principal radii of curvature of the meniscus $R_1 = R_2 = -2\,\mu\text{m}$ (Fig. 19.6), we find $R_k = (1/R_1 + 1/R_2)^{-1} = -1\,\mu\text{m}$ and the pressure drop, based on (19.6) is $\Delta P = 72.9\,\text{kPa}$ (or 0.72 atm.). Thus, the pressure inside the meniscus is $P = 28.1\,\text{kPa}$ (or 0.28 atm.), which corresponds to the boiling point of about 333 K. Yang et al. [28] reported observation of boiling in a meniscus between the cantilever and a flat Si sample under laser irradiation that provided a temperature rise of only up to 45°C [55].

Szoszkiewicz and Riedo [15] investigated the velocity dependence of the nanoscopic friction forces between an AFM tip and a glass sample changing either the sample temperature or the humidity in the experimental chamber. They found that the water gas–liquid phase diagram is the same at the macroscopic scale as at the nanoscopic tip–sample contact.

19.4 Room Temperature Ice in Capillary Bridges

In this section we will discuss experimental reports of ice in the capillary bridges observed at room temperature. First we will investigate the humidity dependence of the adhesion force and possible explanations of this dependence by the presence of ice. Then we will review the evidence of ice-like water in the confinement.

19.4.1 Humidity Dependence of the Adhesion Force

Numerous AFM experiments indicate that the adhesion force has a strong dependence on RH. Most results show that the capillary force first increases with increasing relative humidity up to approximately 30% and then decreases with a further increase of the relative humidity after that [11, 24, 26, 56] (Fig. 19.7). The reasons

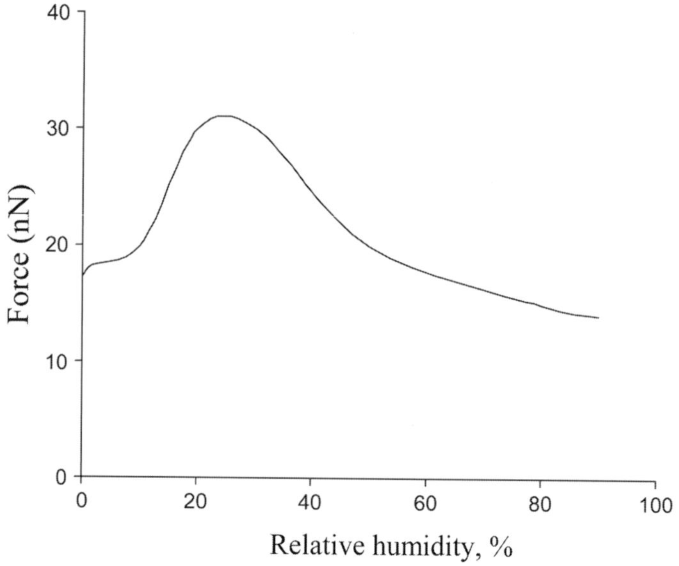

Fig. 19.7 A typical humidity dependence of the adhesion force measured by an AFM (based on data from Ref. [44]). The dependence has a maximum at relative humidity close to 30%

for this trend are still not completely clear. A number of experiments have been conducted also to study the kinetics of the water bridge condensation [15].

Xu et al. [12] investigated water menisci between mica- and Pt-coated AFM tips. They found that at high RH (>20%), capillary condensation caused water to form droplets and two-dimensional islands around the contact point. The droplets evaporated in a short period of time, but the islands remained for much longer periods (hours). At low RH (<20%), the tip contact produced a circular depression in the local polarizability. The pull-off force was constant (2 nN) for RH $< 20\%$, increased rapidly (8 nN) for RH $= 20\%$, and then slowly decreased down to 3 nN for RH $= 95\%$. The dependence is different in hydrophobic and hydrophilic materials [57]. Jang et al. [58] showed, using the Monte-Carlo simulation, that the humidity dependence of the adhesion force is related to the structural properties of the underlying water bridge. A wide bridge whose width does not vary much with tip-surface distance can increase its volume as distance is increased. In this case, the adhesion force decreases as humidity rises. Narrow bridges whose width decreases rapidly with increasing tip-surface distance give the opposite result. Among other reasons suggested for the anomalous RH dependence of the capillary force were an ultrathin water film [59] and various chemical effects [60, 61].

Equation (19.4) states that F_{cap} is independent of the relative humidity, and it provides no information about the pressure inside the capillary bridge. However, both the pressure inside the bridge and the size of the bridge depend upon RH. From simple geometrical considerations, it may be argued that, assuming $cos\, y = 1$, the size of the meniscus foundation is given by $a = (2RR_1)^{1/2}$, where R is the tip

Fig. 19.8 Simplified geometry of a meniscus surface between a spherical tip and a flat sample

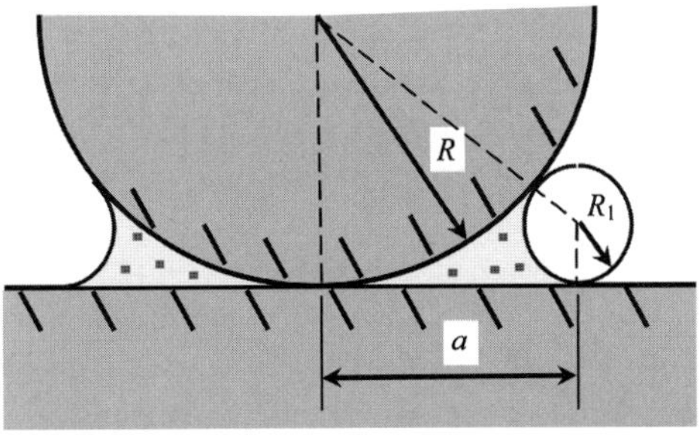

radius (Fig. 19.8). Thus, the force $F_L \sim a^2 \Delta P \sim R$ is independent of RH, keeping in mind that $\Delta P \sim 1/R_k \sim 1/R_1$. For low humidity, the pressure difference inside and outside the bridge is large; however, the size of the bridge is small due to high curvature of the meniscus. For very low humidity, the size of the bridge may be insufficient to generate a capillary force. On the other hand, for high humidity, the pressure difference is small; however, the meniscus is large (but, of course, its size is limited by the radius of the tip). Thus, it is expected that in the limiting cases of RH = 0% and RH = 100% the capillary force would vanish, since at RH = 0% there is no capillary bridge and at RH = 100% the Laplace pressure drop is zero. The capillary force achieves its maximum at a certain value of RH.

19.4.2 Ice in the Capillary Bridges

Another explanation of the humidity dependence of the capillary force was suggested by Asay and Kim [26] who investigated the contact of silicon oxide surfaces. They argued that the capillary force alone cannot explain the observed magnitude of the humidity dependence. The origin of the large RH dependence is due to the presence of ice-like structured water adsorbed at the silicon oxide surface at room temperature. Thus, the force is due to the rupture of an ice–ice bridge at the center of the contact region with the structure, thickness, and viscoelastic behavior of the adsorbed water layer influencing the adhesion force.

Water often forms adsorbed layers at the surface of many materials, and these layers may have an ice-like structure and be connected to the capillary bridge [18,62–65]. Asay and Kim [27] also studied the molecular configuration of adsorbed water as a function of relative humidity using attenuated total reflection (ATR) infrared spectroscopy. A completely hydrogen bonded ice-like network of water grows up to three layers as the relative humidity increases from 0 to 30%. In the relative humidity range of 30–60%, the liquid water structure starts appearing, while the ice-like structure continues growing to saturation. Above 60% relative humidity, the liquid water configuration grows on top of the ice-like layer.

A somewhat similar suggestion was made by Jinesh and Frenken [25], who investigated the contact between a tungsten tip and a graphite surface and found that water in the nanoscale capillary bridges may include ice or be a mixture of liquid water with ice. Such a mixture would have high viscosity and may act like glue, demonstrating elastic response. Assuming Coulombian friction with the coefficient of friction m, the capillary adhesive force, F_{cap}, will lead to the component of the friction force, μF_{cap}, which is present even when no external load is applied to the asperity. Jinesh and Frenken [25] examined the strength and relaxation time of the capillary bridge and found it much more similar to ice than to liquid water. They also observed stick-slip motion with the periodicity of 0.41 nm, which is typical for ice structure. Thus, the condensate solidifies and acts like glue rather than like a lubricant. On the other hand, earlier results obtained with the SFA indicate that water serves as a lubricant [66].

19.4.3 Water Phase Behavior Near a Surface and in Confinement

here is no adequate theoretical explanation why ice can be present in the capillary bridges at room temperature; however, it is clear that this effect is related to the adsorption of water to solid surfaces. The ability of water to form thin layers on hydrophobic surfaces is associated with the disjoining pressure caused by the spatial dependence of the van der Waals adhesion force [43]. It is well known that molecules at a surface have higher energy than those in the bulk, and thus it is usually energetically unprofitable to create a surface. For example, water tends to cover a hydrophilic surface, since the wetted state corresponds to lower energy. Due to the dependence of the adhesion force of the distance (on the scale of nanometers), the energy of the water molecules in the vicinity of the solid surface is lower than the energy of those far away. Thus, it is often not only energetically profitable for water to cover the solid, but also to create a thin layer. It is noted that not only the solid phase can be adsorbed at a surface. Zhang et al. [67] showed that very thin (5–80 nm) gas film could exist for a long time at the interface between a hydrophobic solid and water (Fig. 19.9).

Nanobubbles remained a controversial topic until the early 2000s, because there are theoretical arguments against their existence. First, according to the Laplace equation, pressure in a gas bubble at the thermodynamic equilibrium is greater

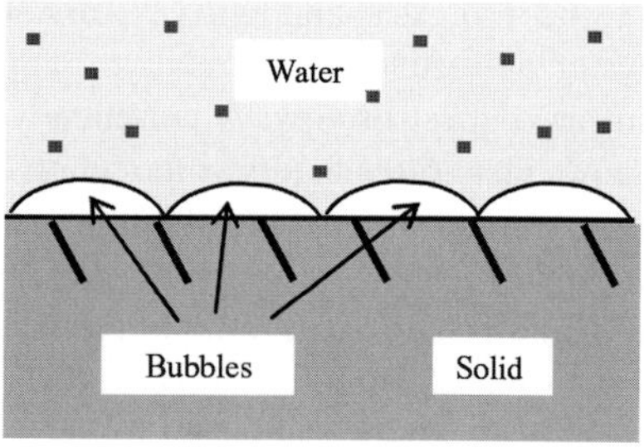

Fig. 19.9 Small gas bubbles form a layer on a solid wall, which can significantly reduce the friction

than that in the liquid outside. For bubbles with a very small radius, the pressure difference through the interface may be very big, and thus the pressure inside is very high. Such high pressure would result in increasing gas solvability, so that the nanobubbles would rapidly dissolve. Second, the total free energy per unit area of a system in which water is in contact with a hydrophobic surface always increases when a gas layer or nanobubbles are formed, unless the surface is extremely rough; so the existence of the nanobubbles is not energetically favorable. Despite these thermodynamic objections, some experimental studies with the AFM and other equipment have demonstrated the existence of nanobubbles at the solid–water interface [67–71].

Nanobubbles can play a significant role in a number of processes, such as the long-range hydrophobic attractive force. Singh et al. [72] showed that long-range hydrophobic interactions can exist underwater due to superhydrophobicity. They found that rough superhydrophobic surfaces experience attractive forces over the separation distance between the tip and sample of up to 3.5 μm, due to the spontaneous evaporation of the intervening, confined water. A gas bubble may form at the contact due to vapor bubble nucleation or due to air dissolved in water. In a sense, such a bubble in water formed due to the cavitation between hydrophobic surfaces is a mirror image of a capillary bridge in air between hydrophilic surfaces formed due to the condensation, and it leads to the attractive force between the tip and sample (Fig. 19.10).

A number of studies are devoted to water phase behavior in confinement [21,73–75]. Mashl et al. [73] used MD simulations to show that water confined to carbon nanotubes of a critical size under ambient conditions (1 bar, 300 K) can undergo a transition into a state having ice-like mobility with an amount of hydrogen bonding similar to that in liquid water. Maeda [74] experimentally investigated phase transitions of capillary-held liquids in a slit-like pore. When the surfaces were placed at a separation very close to the Kelvin length, it was possible to detect a stage in which the system was in an apparent kinetic equilibrium between two physical states: with and without the liquid connecting the two surfaces. Giovambattista et al. [75] used MD simulation to investigate the phase behavior of water confined between two nanoscale hydrophobic or hydrophilic plates at room temperature, varying the

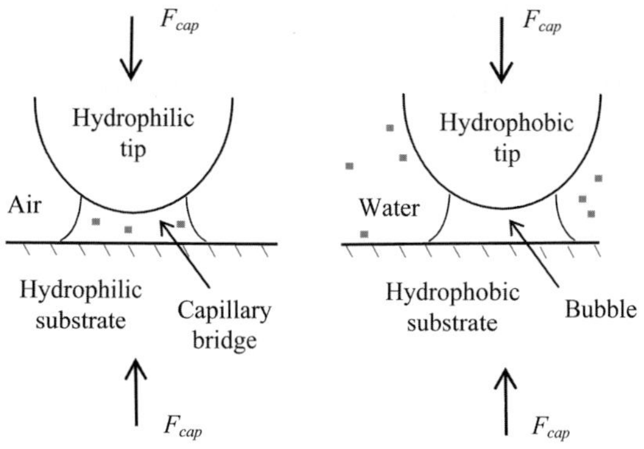

Fig. 19.10 Similarity of the attractive capillary force in the case of a condensed water bridge between hydrophilic surfaces in air and cavitation of a bubble between hydrophobic surfaces in water

pressure ($-0.15 \leq P \leq 0.2$ GPa) and plate separation ($0.4 \leq d \leq 1.6$ nm). With hydrophobic plates, capillary evaporation occurred between the plates at low P. The threshold value of d at which this transition occurs decreases with P (e.g., 0.41 nm at $P = 0.05$ GPa, 0.41 nm at $P = 0.1$ GPa), until, at high P, no capillary evaporation occurred. For $d = 0.41$ nm and $P > 0.1$ GPa, the system crystallizes into a bilayer ice. They developed a P–d phase diagram showing the vapor, liquid, and bilayer ice. With hydrophilic plates, water always remained liquid even at the P at which bulk water cavitates. This indicates that size is important for a phase diagram due to the finite range of the adhesion force in addition to the critical cavitation radius and curvature of the meniscus, which were discussed in the preceding sections.

19.5 Conclusions

We discussed how the size of the system affects the phase diagram. First, if the size is smaller than the critical cavitation radius, a metastable liquid phase can exist. Second, due to the capillary effects, water pressure can be significantly different from the ambient pressure, which affects phase behavior. Third, adhesion can lead to adsorbed layers of another phase at the surface. These factors make the phase behavior of water at the nanoscale much more complicated than at the macroscale and should be taken into consideration in the design of microfluidic and MEMS/NEMS devices. AFM studies of water phase behavior on the nanoscale remain crucial not only because it is the key issue for the reliability of MEMS/NEMS devices, but also because little is known about nanoscale water behavior under tension. For example, it is not known at this point whether water in the capillary bridges is in its liquid or solid state, or a mixture of the two states. As new experimental and simulation data will be obtained, it will be possible to make judgments on these problems with greater confidence.

Acknowledgements

Prof. M. Nosonovsky acknowledges the support of the INSIC TAPE and UWM Research Growth Initiative research grants and would like to specially thank Dr. Seung-Ho Yang from NIST for discussions and demonstrations of negative pressure and room temperature boiling in water capillary bridges.

References

1. B. Bhushan (ed.), *Nanotribology and Nanomechanics – An Introduction*, 2nd edn. (Springer, Heidelberg, 2008)
2. B. Bhushan (ed.), *Springer Handbook of Nanotechnology*, 3rd edn. (Springer, Heidelberg, 2010)

3. B. Bhushan, M. Nosonovsky, Scale effects in friction using strain gradient plasticity and dislocation-assisted sliding (microslip). Acta Mater. **51**, 4331–4345 (2003)
4. A. Carpinteri, N. Pugno, Are scaling laws on strength of solids related to mechanics or to geometry? Nature Mater. **4,** 421–423 (2005)
5. M. Nosonovsky, B. Bhushan, Multiscale friction mechanisms and hierarchical surfaces in nano- and bio-tribology. Mater. Sci. Eng. R **58**, 162–193 (2007)
6. B. Bhushan, J.N. Israelachvili, U. Landman, Nanotribology: friction, wear and lubrication at the atomic scale. Nature **374**, 607–616 (1995)
7. B. Bhushan, Adhesion and stiction: mechanisms, measurement techniques and methods for reduction. J. Vac. Sci. Technol. B **21**, 2262–2296 (2003)
8. B. Bhushan, Nanotribology and nanomechanics. Wear **259**, 1507–1531 (2005)
9. R.L. Fisher, J.N. Israelachvili, Direct measurement of the effect of meniscus forces on adhesion: a study of the applicability of macroscopic thermodynamics to microscopic liquid interfaces. Colloids Surfaces **3**, 303–319 (1981)
10. M. Binggeli, C.M. Mate, Influence of capillary condensation of water on nanotribology studied by force microscopy. Appl. Phys. Lett. **65**, 415–417 (1994)
11. B. Bhushan, S. Sundararajan, Micro/nanoscale friction and wear mechanisms of thin films using atomic force and friction force microscopy. Acta Mater. **46**, 3793–3804 (1998)
12. L. Xu, A. Lio, J. Hu, D.F. Ogletree, M. Salmeron, Wetting and capillary phenomena of water on mica. J. Phys. Chem. B **102**, 540–548 (1998)
13. B. Cappella, G. Dietler, Force–distance curves by atomic force microscopy. Surf. Sci Rep. **34**, 1–104 (1999)
14. B. Bhushan, C. Dandavate, Thin-film friction and adhesion studies using atomic force microscopy. J. Appl. Phys. **87**, 1201–1210 (2000)
15. R. Szoszkiewicz, E. Riedo, Nanoscopic friction as a probe of local phase transitions. Appl. Phys. Lett. **87**, 033105 (2005)
16. R. Szoszkiewicz, E. Riedo, Nucleation time of nanoscale water bridges. Phys. Rev. Lett. **95**, 135502 (2005)
17. P.A. Thompson, M.O. Robbins, Origin of stick-slip motion in boundary lubrication. Science **250**, 792–794 (1990)
18. M. Odelius, M. Bernasconi, M. Parrinello, Two-dimensional ice adsorbed on mica surface. Phys. Rev. Lett. **78**, 2855–2858 (1997)
19. E.A. Jagla, Boundary lubrication properties of materials with expansive freezing. Phys. Rev. Lett. **88**, 245504 (2002)
20. O.M. Braun, M. Peyrard, Dynamics and melting of a thin confined film. Phys. Rev. E **68**, 011506 (2003)
21. H.K. Christenson, Confinement effects on freezing and melting. J. Phys.: Condens. Matt. **13**, R95–R133 (2001)
22. E. Herbert, F. Caupin, The limit of metastability of water under tension: theories and experiments. J. Phys.: Condens. Matter **17**, S3597–S3602 (2005)
23. N.R. Tas, P. Mela, T. Kramer, J.W. Berenschot, A. van der Berg, Capillary induced negative pressure of water plugs in nanochannels. Nanoletters **3**, 1537–1540 (2003)
24. S.-H. Yang, M. Nosonovsky, H. Zhang, K.-H.Chung, Negative pressure in water capillary bridges at nanocontacts. Chem. Phys. Lett. **451**, 88–92 (2008)
25. K.B. Jinesh, J.W.M. Frenken, Capillary condensation in atomic scale friction: how water acts like a glue. Phys. Rev. Lett. **96**, 166103 (2006)
26. D.B. Asay, S.H. Kim, Effects of adsorbed water layer structure on adhesion force of silicon oxide nanoasperity contact with humid ambient. J. Chem. Phys. **124**, 174712 (2006)
27. D.B. Asay, S.H. Kim, Evolution of the adsorbed water layer structure on silicone oxide at room temperature. J. Phys. Chem. B **109**, 16760–16763 (2005)
28. Yang, S.-H., R. Gates, M. Nosonovsky, R. Cook, National Institute of Standards and Technology, private communication (2007)
29. D.H. Trevena, *Cavitation and Tension in Liquids* (Adam Hilger, Bristol, 1987)
30. J. Frenkel, *Kinetic Theory of Liquids* (Dover Publications, New York, 1955)

31. S.B. Kiselev, J.M.H. Levelt-Sengers, Q. Zheng, Physical limit to the stability of superheated and stretched water. in *Proceedings of the 12th International Conference on the Properties of Water and Steam,* ed. by H.J. White et al. (Begell House Inc., 1995), pp. 378–385.
32. S.J. Henderson, R.J. Speedy, A Berthelot–Bourdon tube method for studying water under tension. J. Phys. E **13**, 778–782 (1980)
33. R. Apfel, M. Smith, The tensile strength of di-ethyl ether using Briggs's method. J. Appl. Phys. **48**, 2077–2078 (1977)
34. M. Greenspan, C. Tschiegg, Radiation-induced acoustic cavitation: apparatus and some results. J. Res. Natl. Bur. Stand. Sect. C **71**, 299–312 (1967)
35. Q. Zheng, D.J. Durben, G.H. Wolf, C.A. Angell, Liquids at large negative pressures: water at the homogeneous nucleation limit. Science **254**, 829–832 (1991)
36. M.T. Tyree, Plant hydraulics: the ascent of water. Nature **423**, 923 (2003)
37. Q. Zheng, J. Green, J. Kieffer, P.H. Poole, J. Shao, G.H. Wolf, A.C. Angell, Limiting tensions for liquids and glasses from laboratory and MD studies, in *Liquids Under Negative Pressure*, ed. by A.R. Imre et al. (Kluwer, Dorchester, 2002), pp. 33–46
38. F. Caupin, E. Herbert, S. Balibar, M.W. Cole, Comment on 'Nanoscale water capillary bridges under deeply negative pressure' [Chem. Phys. Lett. **451**, 88 (2008)]. Chem. Phys. Lett. **463**, 283–285 (2008)
39. S.-H. Yang, M. Nosonovsky, H. Zhang, K.-H. Chung, Response to the comment on 'Nanoscale water capillary bridges under deeply negative pressure' by Caupin et al. Chem. Phys. Lett. **463**, 286–287 (2008)
40. M.G. Boyle, J. Mitra, P. Dawson, The tip-sample water bridge and light emission from scanning tunnelling microscopy. Nanotechnology **20**, 335202 (2009)
41. H.J. Choi, J.Y. Kim, S.D. Hong, M.Y. Ha, J. Jang, Molecular simulation of the nanoscale water confined between an atomic force microscope tip and a surface. J. Phys. Chem. B **113**, 4688–4697 (2009)
42. N.R. Tas, M. Escalante, J.W. van Honschoten, Capillary negative pressure measured by nanochannel collapse. Langmuir **26**, 1473–1476 (2010)
43. B.V. Derjaguin, N.V. Churaev, Structural component of disjoining pressure. J. Colloid Interface Sci. **49**, 249–255 (1974)
44. D.B. Asay, M.P. de Boer, S.H. Kim, Equilibrium vapor adsorption and capillary force: exact Laplace-Young equation solution and circular approximation approaches. J. Adhesion Sci Technol. **24**, 2363–2382 (2010)
45. P.A. Kralchevsky, K. Nagayama, Capillary Forces between colloidal particles. Langmuir **10**, 23–36 (1994)
46. W.D. Machin, J.T. Stuckless, Capillary-condensed water in silica gel. J. Chem. Soc. Faraday Trans. **81**, 597–600 (1985)
47. W. Thomson, On the equilibrium of vapour at a curved surface of liquid. Philos. Mag. **42**, 448–453 (1871)
48. J.S. Rowlinson, B. Widom, *Molecular Theory of Capillarity* (Clarendon, Oxford, 1982)
49. M. Nosonovsky, B. Bhushan, Capillary effects and instabilities in nanocontacts. Ultramicroscopy **108**, 1181–1185 (2008)
50. M. Nosonovsky, B. Bhushan, Phase behavior of capillary bridges: towards nanoscale water phase diagram. Phys. Chem. Chem. Phys. **10**, 2137–2144 (2008)
51. S.-H. Yang, H. Zhang, M. Nosonovsky, K.-H. Chung, The effect of contact geometry upon the pull-off force measurement with a colloidal probe. Langmuir **24**, 743–748 (2008)
52. J.L. Streator, R.L. Jackson, A model for the liquid-mediated collapse of 2-D rough surfaces. Wear **267**, 1436–1445 (2009)
53. I. Goryacheva, Y. Makhovskaya, Adhesion effects in contact interaction of solids. C.R. Mecanique **336**, 118–125 (2008)
54. Y.A. Çengel, M.A. Boles, *Thermodynamics – An Engineering Approach*, 3rd edn., (McGraw-Hill, Boston, MA, 1998)
55. O. Marti, A. Ruf, M. Hipp, H. Bielefeldt, J. Colchero, J. Mlynek, Mechanical and thermal effects of laser irradiation on force microscope cantilever. Ultramicroscopy **42–44**, 345–350 (1992)

56. M. He, A.S. Blum, D.E. Aston, C. Buenviaje, R.M. Overney, R. Luginbühl, Critical phenomena of water bridges in nanoasperitycontacts. J. Chem. Phys. **114**, 1355 (2001)1360
57. M. Fujihira, D. Aoki, Y. Okabe, H. Takano, H. Hokari, J. Frommer, Y. Nagatani, E. Sakai, Effect of capillary force on friction force microscopy: a scanning hydrophilicity microscope. Chem. Lett. **7**, 499–500 (1996)
58. J. Jang, J. Jeon, S. Hwang, Monte Carlo study on the water bridge that produces the pull-off force in atomic force microscopy. Colloids Surfaces A **300**, 60–64 (2007)
59. X. Xiao, L. Qian, Investigation of humidity-dependent capillary force. Langmuir **16**, 8153–8158 (2000)
60. D.L. Sedin, K.L. Rowlen, Adhesion forces measured by atomic force microscopy in humid air. Anal. Chem. **72**, 2183–2189 (2000)
61. Jones, R., H.M. Pollock, J.A.S. Cleaver, C.S. Hodges, Adhesion forces between glass and silicon surfaces in air studied by AFM: effects of relative humidity, particle size, roughness, and surface treatment. Langmuir **18**, 8045–8055 (2002)
62. J. Hu, X.-D. Xiao, D.F. Ogletree, M. Salmeron, Imaging the condensation and evaporation of molecularly thin films of water with nanometer resolution. Science **268**, 267–269 (1995)
63. P.B. Miranda, L. Xu, Y.R. Shen, M. Salmeron, Icelike water monolayer adsorbed on mica at room temperature. Phys. Rev. Lett. **81**, 5876–5879 (1998)
64. U. Raviv, J. Klein, Fluidity of bound hydration layers. Science **297**, 1540–1543 (2002)
65. A. Michaelides, K. Morgenstern, Ice nanoclusters at hydrophobic metal surfaces. Nat. Mater. **6**, 597–601 (2007)
66. Y. Zhu, S. Granick, Viscosity of interfacial watery. Phys. Rev. Lett. **87**, 096104 (2001)
67. R. Steitz, H. Gutberlet, T. B. Klosgen, R. Krastev, S. Schemmel, A.C. Simonsen, G.H. Findenegg, Nanobubbles and their precursor layer at the interface of water against a hydrophobic substrate. Langmuir **19**, 2409–2418 (2003)
68. X.H. Zhang, A. Khan, W.A. Ducker, A nanoscale gas state. Phys. Rev. Lett. **98**, 136101 (2007)
69. N. Ishida, T. Inoue, M. Miyahara, K. Higashitani, Nano Bubbles on a hydrophobic surface in water observed by tapping-mode atomic force microscopy. Langmuir **16**, 6377–6380 (2000)
70. A. Poynor, L. Hong, I.K. Robinson, S. Granick, A. Zhang, P.A. Fenter, How water meets a hydrophobic surface. Phys. Rev. Lett. **97**, 266101 (2006)
71. T.R. Jensen, M.O. Jensen, N. Reitzel, K. Balashev, G.H. Peters, K. Kjaer, Th. Bjornholm, Water in contact with extended hydrophobic surfaces: direct evidence of weak dewetting. Phys. Rev. Lett. **90**, 086101 (2003)
72. S. Singh, J. Houston, F. van Swol, C.J. Brinker, Superhydrophobicity: drying transition of confined water. Nature **442**, 526 (2006)
73. R.J. Mashl, S. Joseph, N.R. Aluru, E. Jakobsson, Anomalously immobilized water: a new water phase induced by confinement in nanotubes. Nano Lett. **3**, 589–592 (2003)
74. N. Maeda, Phase transitions of capillary-held liquids in a slit-like pore. J. Phys. Chem. B **110**, 25982–25993 (2006)
75. N. Giovambattista, P.J. Rossky, P.G. Debenedetti, Effect of pressure on the phase behavior and structure of water confined between nanoscale hydrophobic and hydrophilic plates. Phys. Rev. E **73**, 1–14 (2006)

Chapter 20
On the Sensitivity of the Capillary Adhesion Force to the Surface Roughness

Michael Nosonovsky, Seung-Ho Yang, and Huan Zhang

Abstract The adhesion force between rough surfaces in contact is significantly affected by nanoscale asperities. The dominant component of the adhesion force is the capillary force caused by condensed water menisci. The capillary force depends upon the meniscus shape and dimensions. We show, both theoretically and experimentally, that a small change in surface roughness may result in a big change in the meniscus geometry and, consequently, in the adhesion force. This high sensitivity (or instability) of the adhesion force with respect to surface roughness has been overlooked by conventional contact mechanics models that consider adhesion, which assume that the effect of roughness is eliminated with decreasing size of roughness details. The date of atomic force microscopy confirms that the adhesion force between rough surfaces is not a stable function of the roughness.

20.1 Introduction

Modern advances in the contact mechanics are stimulated by the development of nanotechnology. At the nanoscale, forces between surfaces or adhesion forces are significant and often dominate over forces associated with deformation. The two main models of single-asperity mechanical contact that include the adhesion force were developed in the 1970s, namely, the Johnson–Kendall–Roberts (JKR) and Derjaguin–Muller–Toporov (DMT) models [1, 2]. A significant literature is also devoted to the extension of these models [3–7]. Several models have also been proposed for multiple-asperity contacts that combine the adhesion force with a rough surface [8, 9]. These models usually consider the adhesion force with the Lennard–Jones potential or with the so-called Maugis–Dugdale simplified potential [5, 6] and are combined with the models of a rough surface [10]. The emergence of atomic force microscopy (AFM) in the past two decades has allowed the measurement of very small adhesive forces between samples and AFM tips with radii down to the nanometer scale generating new experimental data available [11–21].

The word *adhesion* does not denote a fundamental physical force, but rather it may mean the combination of all attractive forces acting between two bodies. The adhesion force consists of several components, including the relatively weak and long-range van der Waals force, the relatively strong and short-range chemical bonding force, the electrostatic force, and the capillary forces [22]. The capillary force is caused by water menisci or capillary bridges, which are formed at the contact spots due to condensation when the contacting materials are hydrophilic. The dominating component of the capillary force is the so-called Laplace force, caused by the reduced pressure in the meniscus (the Laplace pressure) compared with the pressure outside [22].

Unlike electrostatic, van der Waals, and chemical forces, the capillary force is absent when the contact takes place in vacuum. Therefore, the effect of the capillary force can be estimated by comparing the adhesion force data measured by AFM in air and in ultra-high vacuum (UHV). Such measurements indicate that for chemically non-active surfaces, the capillary force dominates over other components of the adhesion force even at small relative humidity (RH). The capillary force normally constitutes 30–90% of the total adhesion force, depending upon materials combination, RH, and AFM tip radius, [12–20, 23, 24].

Despite the fact that the capillary force dominates adhesion, most contact mechanics models do not consider it and account only for the effect of the van der Waals force and contact force. It has been argued that the Maugis–Dugdale approximation accounts well for the capillary force, because it assumes a constant stress acting up to a certain constant distance h, which can be easily identified with the Laplace pressure and meniscus height (Fig. 20.1a) [4,6]. While this is true in the case of an ideally smooth surface, in a more realistic case of a rough surface, the meniscus height may be different. A two-dimensional asperity with the radius R is shown in contact with a flat substrate. The meniscus surface, which has a constant Kelvin radius R_k [25], comes in contact with the substrate and asperity under contact angles θ_1 and θ_2, respectively, as shown in Fig. 20.1b. The meniscus foundation radius a_0 can be determined from geometrical considerations, and the capillary force is $F_0 = 2a_0 L\Delta P$, where L is the length of the asperity and ΔP is the pressure difference between the meniscus and the ambient. Consider now a very small bump upon the asperity surface (Fig. 20.1c). The meniscus may present a different shape and a different foundation radius a_1 is possible, yielding the force $F_1 = 2a_1 L\Delta P$. Thus, a tiny change in the asperity shape can lead to a significant change of the capillary force.

The above shows that a small variation of the surface roughness may result in a large change of the meniscus. Furthermore, in the case of a rough surface, the meniscus can have multiple configurations. Thus, the capillary force is not a stable function with respect to the surface perturbation or roughness. In addition, a nanoscale meniscus itself can be in a metastable state. This issue has been overlooked by most theories, although due to the instability of the meniscus, conventional contact models that use the Maugis–Dugdale potential do not work well for the capillary force, even though the latter is the most important adhesion force component in applications.

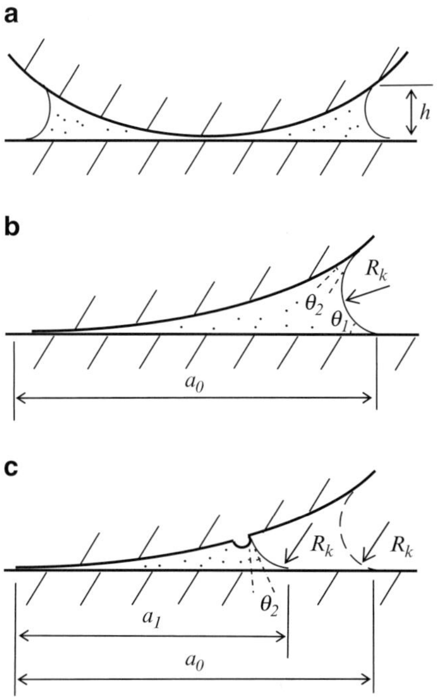

Fig. 20.1 (**a**) Water meniscus with a vertical thickness h between a flat solid and a spherical asperity. (**b**) The contact angle θ between the solids and the water meniscus with radius R_k and foundation radius a_0. (**c**) Small roughness (a bump) upon the asperity may result in a big change of the radius of the meniscus, a_0

The purpose of the present paper is to develop a contact model which accounts for the instability of the capillary force with respect to the roughness. In order to achieve this, we first derive a general formulation for the capillary force based on the minimum free energy principle. Then we illustrate numerically the effects of small bumps on the capillary force with a specific but important two-dimensional case. Following that, similar effects are demonstrated experimentally for the three-dimensional situation with the AFM measurement of the pull-off force between a colloidal probe and a flat substrate.

20.2 Capillary Force Between Rough Surfaces

A liquid-mediated mechanical contact involves two types of processes: thermodynamic processes, such as evaporation and condensation, and the mechanical processes of bringing the surfaces together or separating them. These processes may have very different time constants. It has been argued that breaking a water meniscus bridge between two asperities may be a non-equilibrium thermodynamic process [23, 26]. On the other hand, in many situations it may be assumed that mechanical processes are slow compared to the thermodynamic ones, and, therefore, the meniscus is at a thermodynamic equilibrium [16]. With this assumption, we will formulate a general expression for the free energy of the liquid meniscus between rough solid surfaces as a function of surface geometry and separation distance between the surfaces. Then we will calculate the force between the two solids as a derivative of the energy with respect to the separation distance, assuming a fast and a slow mechanical process.

20.2.1 Shape of the Meniscus

When two elastic bodies are in contact at the nanoscale, they can often be treated as continuum bodies; however, the attractive adhesion force acting between them should be taken into consideration. Two competing theories of elastic contact with adhesion that were developed in the early 1970s are the DMT and JKR models. The DMT model assumes that the adhesive forces act outside of the contact area and yields the pull-off force, needed to separate a sphere of radius r and a flat surface, $P_{\text{pull-off}} = 2\pi wr$, where w is the work of adhesion. The JKR model assumes that the adhesion forces act inside the contact area and yields the pull-off force of $P_{\text{pull-off}} = 1.5\pi wr$. Analytical results for the transition between DMT and JKR were presented by Maugis [3] using a simplified model of adhesion based upon the Dugdale crack model [8]. Both theories and their extensions are widely used in the literature, especially, after the AFM experimental data on adhesion force measurement became available in the 1980s [7]. Figure 20.2 shows the number of reference to the original JKR and DMT papers in the Scopus.com on-line database since 1996 (the database does not provide earlier references). It is seen that the number of references continues to grow and that the two theories have became a standard way of analyzing the single-asperity contact at the nanoscale. Numerous extensions of the DMT and JKR theories, e.g. for the multi-asperity contact of randomly rough surfaces, for elastic plastic materials, and for numerical evaluation of the contact, have been proposed [27].

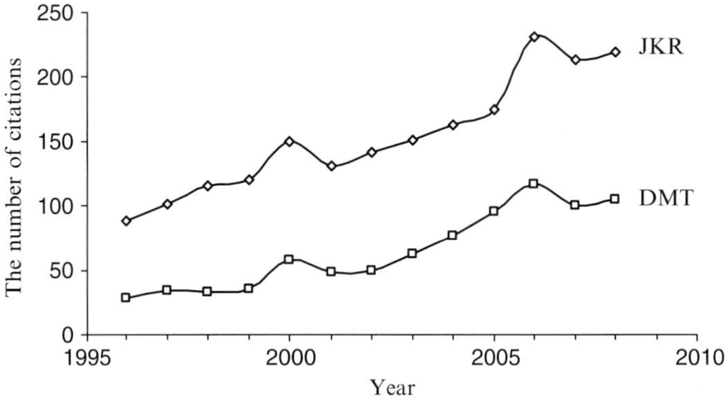

Fig. 20.2 Number of references to the original JKR and DMT publications after 1996 per year in the Scopus database. The data for 2008 (146 and 70 references) are normalized for the 8 month period

The practical motivation for these studies is in their application to the nanotechnology, micro/nanoelectromechanical systems, and other small devices. The small size of these applications leads to large surface-to-volume ratios and, therefore, a correct evaluation of the surface and adhesion forces is crucial for them. Adhesion is a generic term that involves attractive forces of several natures. Among them is the van der Waals adhesion force, chemical bonding, electrostatic, and capillary forces. Many experimental data show that the capillary adhesion force is a very significant and often dominant component of the total adhesion force.

The capillary force is caused by condensed water bridges (menisci) at the contact between the two bodies. The water vapor is always present in air, even at low RH, so the processes of evaporation and condensation take place. If the vapor pressure is equal to the saturation value (RH = 100%), the evaporation and condensation of liquid water occur at the same rate so a liquid layer can be present and it is in thermodynamic equilibrium with vapor. The rate of evaporation and condensation, however, depends on the curvature of the liquid–vapor interface. This is because the surface energy is curvature-dependent. A liquid molecule at a concave surface has in average a larger number of neighboring liquid molecules; therefore, its energy is lower than the energy of a molecule at a flat surface and the rate of evaporation is lower. Thus, the evaporation–condensation thermodynamic equilibrium can be reached even at RH < 100%. The exact relation between the meniscus curvature and RH at equilibrium is given by the Kelvin equation:

$$\frac{\gamma}{R_{\mathrm{m}}} = \rho RT \ln(RH), \tag{20.1}$$

where γ is the water surface tension, R is the universal gas constant, ρ is the water density, and R_{m} is the mean curvature radius of the meniscus related to the principal radii of curvature as $1/R_{\mathrm{m}} = 1/R_1 + 1/R_2$. As nanoscale asperities can have a very small curvature radius, the meniscus can form even at low RH.

For high meniscus curvature, the pressure inside the capillary bridges is significantly reduced comparing to ambient in accordance with the Laplace equation, $\Delta P = \gamma/R_{\mathrm{m}}$, and the pressure can drop deeply into the negative region (negative pressure, of course, is nothing else than the tensile stress) [29]. For example, based on the Kelvin calculation, the value of RH = 30% corresponds to $\Delta P = 162$ MPa (about 1,620 atm.) and RH = 10%. The reduced pressure inside the bridges gives a rise to the attractive capillary force equal the foundation radius of the menisci times ΔP.

Note that the shape of the meniscus should satisfy two types of the boundary condition. First, at the solid–liquid–air contact line the Young equation for the contact angle θ should be satisfied

$$\cos\theta = \frac{\gamma_{\mathrm{SA}} - \gamma_{\mathrm{SL}}}{\gamma}, \tag{20.2}$$

where γ_{SA} and γ_{SL} are the solid–air and solid–liquid surface free energies. Second, at the liquid–air interface, the mean curvature is constant and given by the Kelvin equation. Finding the shape of the meniscus between two solid surfaces with a given topography is a non-trivial and complicated mathematical problem, so no conclusion about the existence or uniqueness of the solution can be made in the general case.

It is an accepted postulate in surface science and tribology that there are no absolutely smooth surfaces. Every nominally flat surface, including even atomically smooth surfaces, has roughness. When a contact model is built, it is often assumed that the contribution of the small-scale roughness can be neglected and surfaces can be modeled as smooth. In most cases this assumption is justified. However, this is not necessarily so when the capillary effects are investigated, as stability issues may be very important in this case [30]. Consider an elastic sphere with the radius r in contact with the flat surface (contact radius a) with the meniscus near the contact spot of the mean radius R_{m} (Fig. 20.3a). If the sphere and the flat are smooth, there is a unique position of the meniscus and the unique value of the adhesion pull-off force, corresponding to the foundation radius a_{f}, as it is assumed by the DMT theory. However, if small-scale roughness with a typical size of a detail $b < R_{\mathrm{m}}$ is introduced, the foundation radius of the meniscus can have a spectrum of values between $b < a < a_{\mathrm{f}}$. While the maximum value of a is close to the foundation radius of the smooth surface (Fig. 20.3b), the minimum radius is comparable with b. Furthermore, with the decreasing size of roughness detail, the average value of b does not reach a_{f} (Fig. 20.3c):

$$\lim_{b \to 0} \langle a \rangle \to \frac{a_{\mathrm{f}}}{2} \neq a_{\mathrm{f}}. \tag{20.3}$$

This illustrates that the small-scale roughness cannot be neglected and therefore the contact models that assume a smooth surface can be misleading.

20.2.2 Capillary Force

Let us now assume that the two bodies can move mechanically and separate from each other. Depending on the velocity of the mechanical process, three regimes are possible (Table 20.1). *First*, for a slow mechanical process, there is enough time for the thermodynamic processes of evaporation and condensation to come into equilibrium, and both Kelvin's and Young's equations are satisfied. *Second*, for a fast mechanical process, there may not be enough time for the meniscus to come into thermodynamic equilibrium, but it may be assumed that the meniscus is in quasi-static mechanical equilibrium. The meniscus then has a constant pressure inside, which, however, may be different from that predicted by the Kelvin equation, which is valid only for thermodynamic equilibrium. In this case, the liquid–vapor interface still has a constant mean curvature, based on the Laplace equation, and Young's equation is satisfied at the triple line. The volume of the meniscus is constant, as there is not enough time to change the volume by condensation or evaporation. *Third*, for a very fast mechanical process, the meniscus may not be in a

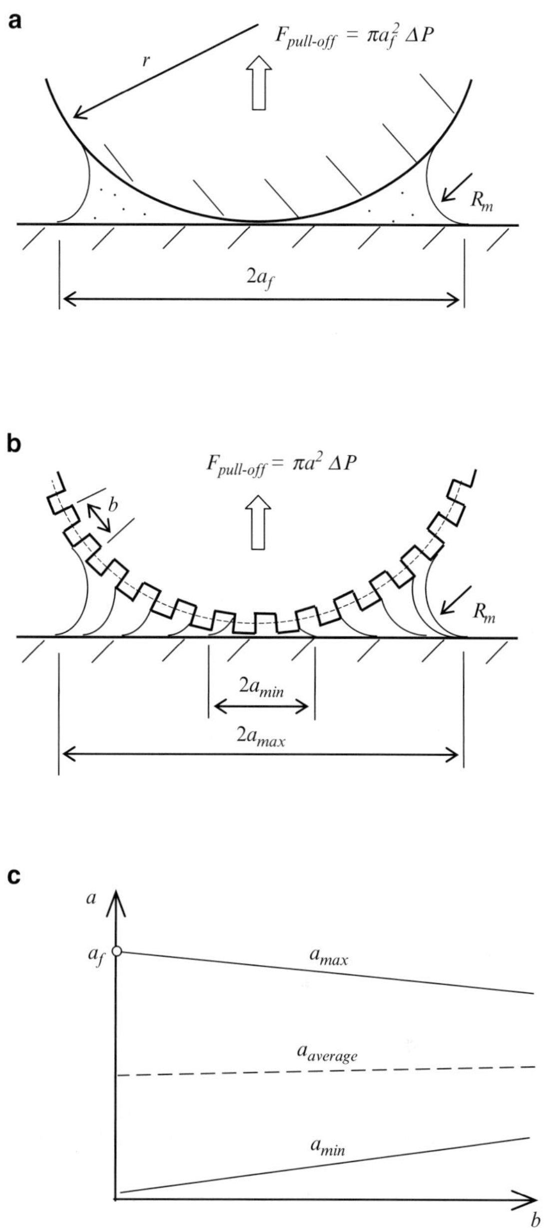

Fig. 20.3 (**a**) Smooth elastic sphere in contact with the flat surface. (**b**) Small-scale roughness detail of typical size b can pin the capillary bridge and change dramatically the size of the meniscus. (**c**) Maximum, minimum, and average radius of the capillary bridge as a function of the size of the meniscus

quasi-static mechanical equilibrium, and in this case dynamic processes, such as interface waves, occur. Then neither Kelvin's nor Young's equations are satisfied, whereas the volume of the meniscus is still conserved.

Which of the three regimes takes place, depends upon the relation of the time constants of the three processes. The meniscus nucleation time, T_n, which obeys

Table 20.1 Volume, surface, and the triple line for mechanical process for different relative speeds of the mechanical, thermodynamic, and hydrodynamic processes (T_m, T_n, and T_d are time constants of the corresponding processes)

	Slow mechanical process, $T_m \gg T_n, T_m \gg T_d$	Fast mechanical process, $T_n \gg T_m$ $T_m \gg T_d$	Very fast mechanical process $T_m \ll T_d, T_m \ll T_n$
	Thermodynamic and mechanical quasi-equilibrium	Mechanical quasi-static equilibrium	Hydrodynamic process
Volume	Not conserved	Conserved	Conserved
Liquid–vapor interface	Kelvin equation	Laplace equation	Dynamic
Triple line	Young's equation	Young's equation	Dynamic

an Arrhenius law [16] governs the thermodynamic process, whereas, $T_m = H/v$, which is separation distance (taken as a typical length of the liquid–vapour interface) divided by separation velocity v, is responsible for the mechanical process, and $T_d = 2\pi(\rho H^3/\gamma_{LV})^{1/2}$ is the period of a capillary wave vibration at a liquid–vapor interface of size H, where ρ is liquid density [28]. In many practically important situations, the first regime occurs, since T_n is on the order of 1 ms [16], T_m is on the order of 0.1 s, and $T_d = 0.75$ ns (calculated for $\rho = 10^3\,\text{kg/m}^3$, $\gamma_{LV} = 70\,\text{mJ/m}^2$, $H = 10$ nm).

The derivative of the energy with respect to the vertical separation distance between the two solids, H, is equal to the normal force. Assuming that the mechanical process is slow and equations (20.1)–(20.2) are satisfied because the meniscus is at a thermodynamic equilibrium, we find from (8)

$$F = \frac{\partial W}{\partial H} = \oint_t \sin\theta\gamma_{LV}\mathrm{d}t + A_{SL}\Delta P. \tag{20.4}$$

The capillary force includes two terms. The first term corresponds to the so-called circumferential force and may be assumed to be small compared to the second term, or the Laplace force [23]. Note that the capillary force is being determined as a force acting upon one of the bodies from the meniscus. Therefore, only the A_{SL}, triple line, t, and contact angle θ, which correspond to one body, should be considered in (20.4).

In the case of a fast quasi-static mechanical process, setting $\delta V = 0$ in (7) yields [31]

$$\delta W = \delta A_{SL}\left[\cos\theta - \frac{\gamma_{SV} - \gamma_{SL}}{\gamma_{LV}}\right]\gamma_{LV} + \delta H\left(\oint_t \sin\theta\gamma_{LV}\mathrm{d}t + A_{SL}\Delta P\right)$$
$$- [\gamma_{LV}(1/R_1 + 1/R_2) + \Delta P]\delta H A_{SL} + \delta(\Delta P)V \tag{20.5}$$

and

$$F = \frac{\partial W}{\partial H} = \oint_t \sin\theta\gamma_{LV}\mathrm{d}t - A_{SL}\gamma_{LV}(1/R_1 + 1/R_2) \tag{20.6}$$

20.3 Case-Study: Two-Tiered Roughness

In the preceding section we discussed the general case of a capillary force between two rough surfaces. Unfortunately, the shape of the meniscus is too complicated in the general three-dimensional case. It is impossible to find a general solution for the capillary force from (20.4). In this section we will consider as an example a simple case of a two-dimensional rigid cylinder of radius R, with a smaller roughness, represented by a bump of radius r, separated by distance H from a solid flat surface (Fig. 20.4). The meniscus of radius R_{k} may form between the cylinder and the flat surface.

It is assumed here that the contact angles with water of both the flat substrate and the cylinder are equal to zero. This assumption does not affect significantly the qualitative nature of the problem; however it simplifies the geometry considerably. The width of the meniscus, as defined in Fig. 20.4, is given by

$$L = L_1 + L_2 = \sqrt{(z - H)(2R - z + H)} + \sqrt{r^2 + 2R_{\mathrm{k}}(r + z) - z^2} \quad (20.7)$$

For the case of small separation between the asperity and the plane, $H \sim 0$, which is a typical case in AFM measurements [12], (20.7) simplifies to

$$L = \sqrt{z(2R - z)} + \sqrt{r^2 + 2R_{\mathrm{k}}(r + z) - z^2} \quad (20.8)$$

If no rough detail is present, the size of the meniscus is given by

$$L = 2\sqrt{R_{\mathrm{k}}R} \quad (20.9)$$

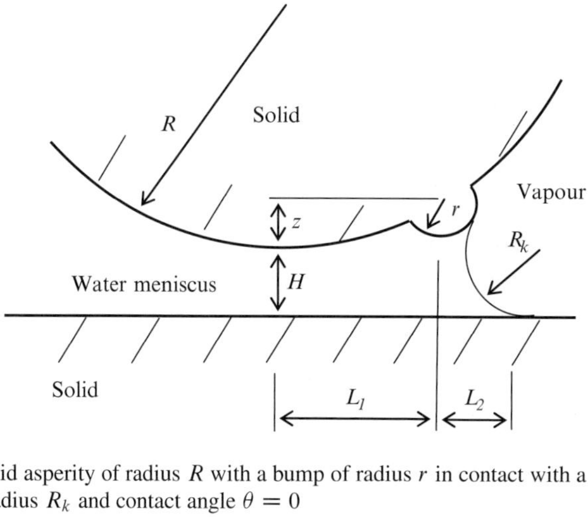

Fig. 20.4 A solid asperity of radius R with a bump of radius r in contact with a flat surface, with a meniscus of radius R_k and contact angle $\theta = 0$

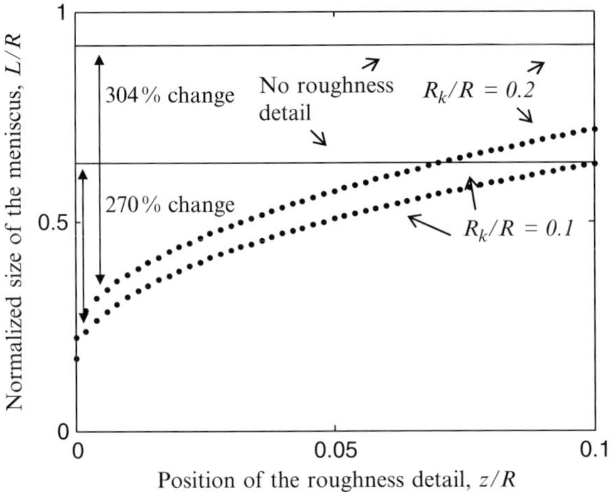

Fig. 20.5 Calculated dependence of the meniscus size upon the position of the small roughness detail. It is observed that introducing the roughness detail and changing its position may lead to a significant change of the meniscus size and, therefore, of the capillary force

Dependencies of L/R upon z/R for different values of $R_k/R = 0.1$ and 0.2 are presented in Fig. 20.5 as dotted curves based on (20.8). Corresponding values of L/R, for the case when no roughness is present, are also shown as solid lines. It is observed that introducing the roughness detail and changing its position, z/R, may lead to a significant change in the meniscus size (from $L/R = 0.17$ to 0.63 and from $L/R = 0.22$ to 0.89) and, therefore, in the capillary force. Thus the meniscus force is very sensitive to the position of the roughness.

In this section, the two-dimensional case has been considered. Theoretically, the effect of a small bump upon the meniscus shape is more pronounced for a two-dimensional profile than for a three-dimensional profile. Nevertheless, a qualitatively similar trend is expected for three-dimensional cases. The following section presents adhesion measurement made for rough surfaces to back up this claim.

20.4 Experimental Data

In this section, we will discuss the stability of the meniscus force between rough surfaces from experimental data. We measured adhesion force between a spherical glass colloidal probe attached to an AFM cantilever and a silicon sample at different tilt angles. The Kelvin radius of the meniscus is estimated as $R_k = 2$ nm and the mean surface roughness of the colloidal probe $\sigma = 10$ nm for the rough probe and $\sigma = 2$ nm for the smooth probe. The results for the force as a function of the tilt angle are presented in Fig. 20.6a, b. It is observed that a small change of the sample stage tilt angle greatly affects the measured adhesion force value. Although both the

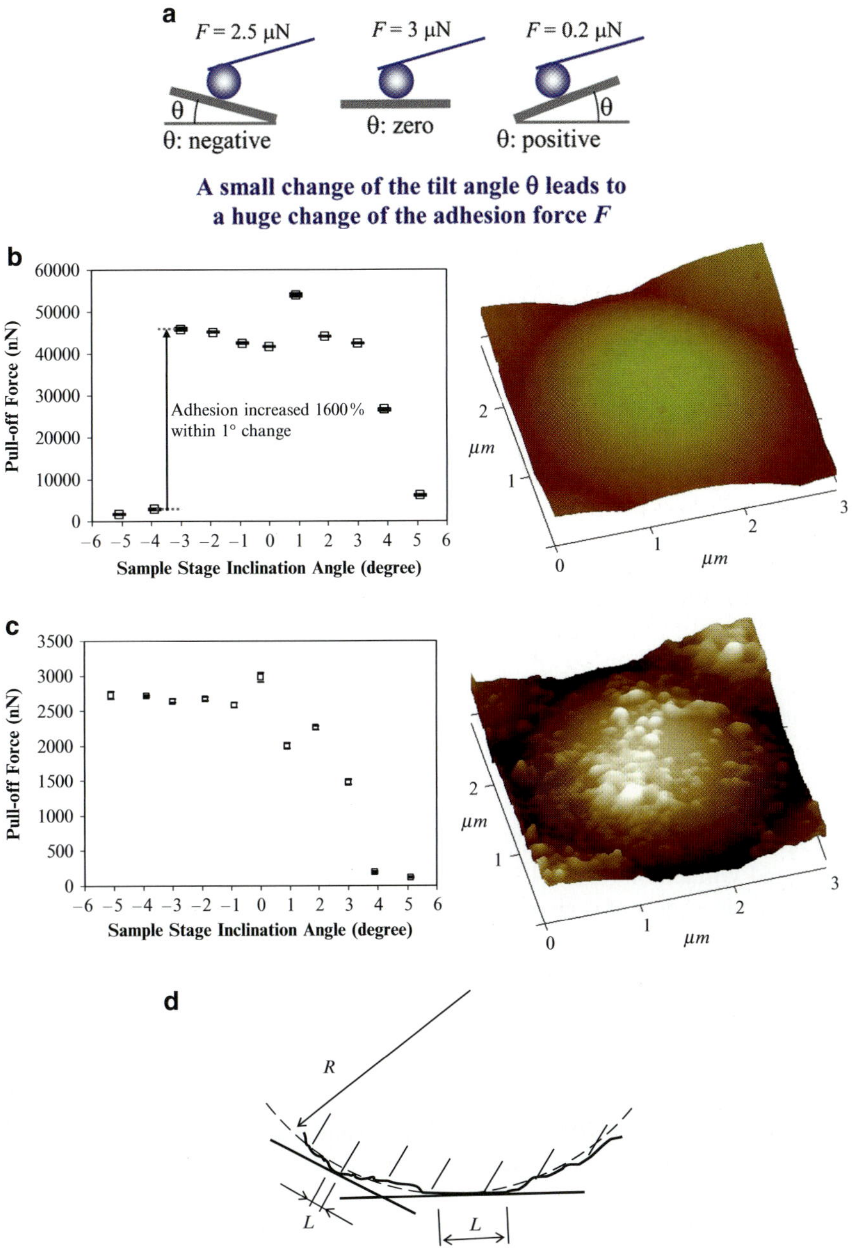

Fig. 20.6 (**a**) Spherical colloidal probe in contact with a flat sample (AFM images represent probe topography). Measured adhesion force as a function of tilt angle of the stage (**b**) for a relatively smooth probe and (**c**) for a rough probe. (**d**) Schematics of a flat surface in contact with a rough sphere. The size of contact L can vary significantly with the change of the tilt angle due to roughness. As a result, the meniscus size and the capillary force would also change significantly

colloidal probe and the sample were not perfectly smooth, statistical fluctuations of their profiles cannot account for such a huge difference in the force. Indeed, as observed from a comparison of Fig. 20.6a, which shows results for a smooth sample, and Fig. 20.6b, which shows results for a much rougher sample, the change of roughness affects the magnitude of the force, but not the relative deviation of the force. Note that the values of the force have high reproducibility (the error bars provided by the AFM are shown in the figure, although they are very sensitive to the tilt angle).

In order to obtain the three-dimensional topography of the colloidal probe, the grating sample (MikroMasch, TGT01[1]) is scanned by the colloidal probes. The grating sample has an array of the sharp silicon peaks (this explains the mirror images in the corners) and the given AFM topographies, in Fig. 20.6a, b, are the negative images of the colloidal probes.

Assuming a single-asperity contact, the traditional analysis based upon the DMT model predicts that the pull-off force is given by

$$F = 2\pi w R \qquad (20.10)$$

where $R = 20\,\mu\text{m}$ is the radius of the colloidal probe and $w = 2\gamma_{\text{LV}} \cos\theta$ [6]. Taking the typical value for water $\gamma_{\text{LV}} = 7\,\text{mJ/m}^2$ [32] and $\cos\theta = 0.5$, $F = 8.8\,\text{mN}$. The JKR model yields $F = 1.5\pi w R = 6.6\,\text{mN}$. The theoretically predicted value exceeds the experimentally observed values of $0.1\,\text{mN} < F < 3\,\text{mN}$ for the rough surface, whereas it is within the range of the observed values for the smooth surface, $2\,\text{mN} < F < 55\,\text{mN}$. The theoretical analysis also does not predict a dependence of the pull-of force upon the tilt angle, which is observed experimentally.

For the multiple-asperity contact, the traditional analysis based upon the Greenwood and Williamson (GW) [10] methodology states that individual asperity contacts are independent of each other. Thus, the net pull-off force is equal to the sum of pull-off forces at individual asperity junctions. The influence of roughness is characterized by the ratio σ/δ_c, where

$$\delta_c = 0.75(\pi^2 w^2 R / E^{*2})^{1/3}, \qquad (20.11)$$

where E^* is the composite elastic modulus [8, 9]. Taking $E^* = 44\,\text{GPa}$, $\delta_c = 0.59\,\text{nm}$ (based on $w = 70\,\text{J/m}^2$, $R = 20\,\mu\text{m}$) and thus $\sigma/\delta_c > 10$. At $\sigma/\delta_c > 2$ the effective adhesion is eliminated [6], so the multiple-asperity model also cannot explain the experimental values.

As the sample and colloidal probe separate, capillary bridges at the lower asperities are broken until only one or a few bridges are left. Thus the pull-off force

[1] Certain instruments and materials are identified to adequately specify the experimental procedure. Such identification does not imply recommendation or endorsement by the National Institute of Standards and Technology, nor does it imply that the materials or instruments identified are necessarily the best available for the purpose.

depends upon the tip radius and shape at a particular asperity, which may vary significantly as shown in Fig. 20.6c. Thus the experimentally measured pull-off force as a function of the tilt angle is not stable, as has been described in the preceding sections. In order to calculate the meniscus force, an exact shape of the contact spot between the colloidal probe and sample should be known.

20.5 Conclusions

Traditional contact models might be inadequate when adhesion is involved, since they do not properly account for the adhesion contribution. We showed that the capillary force, which is the main contributor to the adhesion, is not a stable function of the surface geometry. A small change in the geometry, such as a nanoscale bump or groove, may result in a large change in the capillary meniscus shape and, consequently, in the capillary force. In general, the shape of the meniscus should satisfy Young's equation at the triple line and Kelvin's equation at the liquid–vapor interface, assuming that a thermodynamic state of quasi-equilibrium is achieved. The quasi-equilibrium is achieved if the mechanical process of bringing the surfaces together or separating them is slower than the thermodynamic process of evaporation and condensation. In the opposite case, when the mechanical process is faster, a mechanical quasi-static equilibrium may be assumed, with the Young equation still satisfied at the triple line, and with constant mean curvature of the liquid–vapor interface and a constant liquid volume.

A case study of a two-dimensional cylindrical surface with a small bump, representing a roughness detail, in contact with a flat surface was considered numerically. It was found that a change in the position of the small roughness detail may result in a big change of the capillary force, which is therefore not stable with respect to the roughness distribution.

Experimental data for adhesion force between a spherical colloidal probe and a tilted flat sample support the conclusion that the adhesion force is not stable with respect to the roughness distribution. A small variation in the tilt angle leads to a significant change of the adhesion force. Therefore, the results predicted by the conventional single-asperity and multiple-asperity contact mechanics models do not agree with the experimental data due to the instability of the adhesion force.

Our work shows that the meniscus and capillary force are unstable with respect to the roughness distribution during the adhesion force measurement. This instability should be taken into account during the contact analysis of rigid bodies at the nanoscale.

Acknowledgements

Dr. M. Nosonovsky acknowledges the support of the INSIC TAPE and UWM Research Growth Initiative research grants.

References

1. K.L. Johnson, K. Kendall, A.D. Roberts, Proc. Roy. Soc. Lond. **A324**, 301–313 (1971)
2. B.V. Derjaguin, V.M. Muller, Yu. P. Toporov, J. Colloid. Interface Sci. **53**, 314–326 (1975)
3. D. Maugis, J. Colloid Interface Sci **150**, 243–269 (1992)
4. D. Maugis, *Contact, Adhesion and Rupture of Elastic Solids* (Springer, Berlin, 1999)
5. K.L. Johnson, Proc. Royal Soc. Lond. **A453**, 163–179 (1997)
6. K.L. Johnson, Tribol Int **31**, 413–418 (1998)
7. G.G. Adams, M. Nosonovsky, Tribol Int **33**, 441–442 (2000)
8. K.N.G. Fuller, D. Tabor, Proc. R. Soc. London **A324**, 327–342 (1975)
9. D. Maugis, J. Adhesion Sci. Tech. **10**, 161–175 (1996)
10. J.A. Greenwood, J.B.P. Williamson, Proc. Royal Soc. Lond. **A295**, 300–319 (1966)
11. R.W. Carpick, N. Agrait, D.F. Ogletree, M. Salmeron, Langmuir **12**, 3334–3340 (1996)
12. X. Xiao, L. Qian, Langmuir **16** 8153–8158 (2000)
13. L. Sirghi, N. Nakagiri, K. Sugisaki, H. Sugimura, O. Takai, Langmuir **16**, 7796–7800 (2000)
14. M. He, A.S. Blum, D.E. Aston, C. Buenviaje, R.M. Overney R. Luginbühl, J. Chem. Phys. **114**, 1355–1360 (2001)
15. R. Jones, H.M. Pollock, J.A.S. Cleaver, C.S. Hodges, Langmuir **18**, 8045–8055 (2002)
16. R. Szoszkiewicz, E. Riedo, Phys. Rev. Lett. **95** 135502 (2005)
17. J. Jang, M.A. Ratner, G.C. Schatz, J. Phys. Chem. B. Letters **110**, 659–662 (2006)
18. E.J. Thoreson, J. Martin, N.A. Burnham, J. Collolid. Interface Sci. **298**, 94–101 (2006)
19. H.-J. Butt, M. Farshchi-Tabrizi, M. Kappl, J. Appl. Phys. **100**, 024312 (2006)
20. S.-H. Yang, H. Zhang, S.M. Hsu, Langmuir **23**, 1195–1202 (2007)
21. S.-H. Yang, H. Zhang, M. Nosonovsky, K.H. Chung, Langmuir **24**, 743–748 (2008)
22. J.N. Israelachvili, *Intermolecular and Surface Forces*, 2nd edn. (Academic, London, 1992)
23. L.R. Fisher, J.N. Israelachvili, Colloids Surf **3**, 303–319 (1981)
24. B. Bhushan, S. Sundararajan, Acta Mater. **46**, 3793–3804 (1998)
25. J.S. Rowlinson, B. Widom, *Molecular Theory of Capillarity* (Clarendon, Oxford, 1982)
26. N. Maeda, J.N. Israelachvili, M.M. Kohonen, PNAS **100**, 803–808 (2003)
27. G.G. Adams, S. Müftü, N.M. Azhar, J. Tribol. **125**, 700–708 (2003)
28. L. Landau, E. Lifshitz, *Fluid Mechanics* (Pergamon Press, London, 1959)
29. S.H. Yang, M. Nosonovsky, H. Zhang, and K.H. Chung, "Negative Pressure in Water Capillary Bridges at Nanocontacts," Chem. Phys. Lett. **451** (2008) 88–92
30. M. Nosonovsky, B. Bhushan, Ultramicroscopy, **108**, 1181–1185 (2008)
31. M. Nosonovsky, Langmuir **23**, 3157–3161 (2007)
32. A.V. Adamson, *Physical Chemistry of Surfaces* (Wiley, New York, 1990)

Part III
Industrial Applications

Chapter 21
Nanoimaging, Molecular Interaction, and Nanotemplating of Human Rhinovirus

Markus Kastner, Christian Rankl, Andreas Ebner, Philipp D. Pollheimer, Stefan Howorka, Hermann J. Gruber, Dieter Blaas, and Peter Hinterdorfer

Abstract Capturing and detecting viruses require specialized platforms and analysis techniques. The atomic force microscope (AFM) offers not only tremendously high spatial and topographic resolution, but also allows localizing single molecules and determining their interaction on a molecular level. AFM makes possible (a) generating nanopatterned surfaces, (b) detecting protein and viral pathogen captured in these specific regions over time, and (c) gathering information about binding parameters. A section of this chapter describes, for instance, the measurement of kinetic parameters governing the first step of viral infection – the attachment of a human rhinovirus to a cell – at the single-molecule level. We review the principles of AFM and its applications in the analysis of nanopatterned surfaces and protein arrays, in the detection of proteins and viruses, and in the characterization of the interactions of these viruses with their cognate receptors under physiological conditions.

21.1 Introduction

The large virus family *picornaviridae* includes many animal and human pathogens, among them, three species of human rhinoviruses (HRVs), the causative agents of the common cold – a well-known, widespread, and recurring disease [1]. The non-enveloped icosahedral virions recognize different types of receptors for cell entry and infection; the major group binds intercellular adhesion molecule 1 (ICAM-1) and the minor group binds members of the low-density lipoprotein receptor (LDLR) family. The receptor of recently identified HRVs, tentatively classified as species C, is unknown [2]. The minor group species A virus HRV2 and its cognate LDL receptors are the subject of a series of investigations involving atomic force microscope (AFM). When analyzing interactions such as those between viruses and their receptors, biological samples require special handling to preserve activity and prevent denaturation. Since it allows the sample to be visualized in its native state and immersed in buffer solution, AFM suits these requirements very well. The advent of dynamic force microscopy (DFM) with its lower imaging force modes reduced the physical impact on the analyte and resulted in a further increase in resolution

[3,4]. This sensitive imaging method enabled extensive studies of HRV2 in its native form. In a number of investigations, the virus was attached to a recombinant receptor derivative that, in turn, was immobilized to mica surfaces. In this way, the in vivo situation was replicated almost perfectly [5].

Molecular recognition force microscopy (MRFM) is the method of choice for gaining insights into the interaction behavior of two components, for example, of virus and receptors or antibodies. The MRFM makes evaluating single-molecule interactions possible. As discussed later in this chapter, an extensive analysis yielded information on the time-dependent transition of single to multiple bonds between receptors and virus, and provided real-time statistical data on the distribution of receptors bound to single virus particles [6].

Deriving the above-mentioned structural and kinetic parameters from individual virions requires a stable platform providing the following conditions. (a) The possibility of working under near physiological conditions without the need for staining; (b) use of minute amounts of analyte; (c) high selectivity and specificity for the force parameters to be determined. If these techniques are to be incorporated into a "virus-detection platform of the future," such an instrument should be able to manage environmental samples and body fluids. Thus far, only a few methods have shown the potential to generate readout for site-isolated nano- and micro-scale biological entities at the single particle level [7]. Extreme miniaturization will offer the potential for scanning large and dense libraries of recognition sites with short diffusion times to screen complex chemical and biological systems. In another section of this chapter, we describe the HRV2–VLDR system that was established on a planar platform for extensive characterization of the interaction under physiological conditions, which can be considered a first step in this direction.

Procedures such as nanoshaving, nanografting, and native protein nanolithography (NPNL) were applied to write, read, and erase proteins on a platform, as exemplified in virus-receptor capturing patterns. In addition, non-AFM-related methods, such as electron beam lithography (EBL), nanocontact printing, photolithography, and nanodispensing, also demonstrated the potential of nanopatterning biocompatible surfaces (Table 21.1). Despite promising technological achievements and efforts, these methods have yet to prove their ability to create multiplexed biomolecule nanoarrays in order to become versatile, highly sensitive high-density setups in the evolving field of nanotechnology.

21.2 Contact Mode AFM Imaging

Presently, a wide range of scientific fields from molecular biology, solid state physics, and material science are profiting from the fast evolution of atomic force microscopy. One of its main domains is the visualization and characterization of surface properties and conditions, from the micrometer down to the nanometer scale [36]. A variety of biological samples, membranes, cells, and molecules have been studied extensively [37]. In detail, its ability to resolve nanometer-sized structures

Table 21.1 A variety of promising techniques are available that generate nanopatterns with biological functions

Overview of techniques for generating nanopatterns with biological functions	
Technique	Comments/Refs
Nanopatterning	For review see [8,9]
Soft-lithographic techniques	
Microcontact printing (muCP)	
Nanoimprint lithography (NIL)	10-nm sized dots [10]
Nanosphere lithography (NSL)	
Electron beam lithography (EBL)	100-nm sized features [11]. 200 nm–5,000 mm sized vials [12]
	3D nanopatterning, based on local activation of porous silicon [13]
Focused ion-beam lithography (FIBL)	300 nm sized features [14]
Conductive atomic force microscopy (c-AFM)	[15]
Atomic clusters	<3 nm sized features [16]
Nanopatterning and/or probe deposition	
Dip-pen nanolithography (DPN)	<100 nm sized features [17–20]
Programmable nanobuilding blocks	Based on 28 nm sized features generated by nanobuilding blocks [21]
Native protein nanolithography	Can be used for reading, writing, and erasing protein arrays [22]
Probe deposition	
Nanodispensing	Direct dispensing
Nanopipett	Nanopipette capable of dispensing into Ø 300 nm sized vials [14]
Nano eNabler	Printer, based on microcantilevers, generating >250 nm sized spots [23, 24]
Nanofluidic	[25–29]
Self-addressable probes	The probes can be added in bulk, and they will find the way to their unique spot(s) on the chip on their own via a zipcode–anti-zipcode system [30, 31]
Self-assembling protein arrays	The probes can be expressed in the desired spot directly on the chip by using cell-free expression [32–34]

The depicted techniques are classified according to type of nanopatterning and/or probe deposition. Adapted from [35]

of soft biological samples under near-physiological conditions without the need for labeling is one of its most striking advantages. Furthermore, the high signal-to-noise ratio when detecting single molecules and the exceptionally good vertical resolution must be emphasized.

As a member of the scanning probe microscopy family, the AFM operates by moving a sharp tip on a micro-scaled cantilever horizontally across a surface. The line-by-line movement of the piezo over the surface allows recording and reconstruction of the topographical map of the scanned surface [38]. Interaction between tip and sample causes deflections of the cantilever, which are usually detected by a segmented photodiode. The magnitude of the deflection is registered by the changes in the angle of a laser beam reflected at the end of the cantilever. In order to keep the force applied by the tip to the surface at a constant level, the deflection signal is monitored continuously by an electronic feedback loop. If required, this feedback can be used to adjust the tip–surface distance by varying the voltage applied to the piezos. This imaging mode, termed constant-force mode, is characterized by continuous contact between AFM tip and sample. However, soft biological samples are usually prone to deformation or even destruction due to the strong force induced by the tip. To overcome this, careful force adjustments must be made in advance by employing force–distance measurements. Thereby, the indentation force can be reduced to a minimum (e.g., $<100\,\mathrm{pN}$) [39].

In contrast to conventional topography imaging, however, strong interaction forces can be used to manipulate specimen surfaces. As an example of one of the first controlled nanomanipulations of biomolecules using the AFM, Hansma et al. demonstrated as early as 1992 the dissection of isolated DNA adsorbed on mica [40]. The force of the AFM tip applied to the sample was in the range of 5 nN – more than ten times the normal imaging force. The important role of the indentation force the AFM tip applies was discovered in several other studies. In one of these works, the modulation of polypeptide loops of individual bacteriorhodopsin molecules was analyzed [39, 41]. Bacteriorhodopsin (BR) is a light-driven proton pump packed into highly ordered two-dimensional crystals called purple membranes [42]. These membranes are integrated into the plasma membrane of *Halobacterium salinarium*. A change in the bacteriorhodopsin surface conformation was revealed, depending on the force applied by the stylus of an AFM tip (Fig. 21.1). The bacteriorhodopsin trimer is organized in a trigonal lattice with $6.2 \pm 0.2\,\mathrm{nm}$ side length on the cytoplasmic surface. When the force induced by the tip was kept at low values, around 100 pN, the subunits in the trimer displayed a distinct protrusion (loop connecting the α-helices E&F) from the lipid surface that spanned $0.8 \pm 0.2\,\mathrm{nm}$ (Fig. 21.1a, upper section and Fig. 21.1b). Subsequently, increasing the applied force to 200 pN caused the EF loops to bend, providing a "view" of the shorter loops underneath (Fig. 21.1a, lower section and Fig. 21.1c). This nanomanipulation was shown to be fully reversible when changing back to lower forces, which indicated that loop EF is a flexible element in the membrane (Fig. 21.1d). In summary, this work showed that precise control of the forces applied during the imaging of biological samples is of great importance and can furthermore lead to new biological insights.

Inspired by the possibilities such experiments open up, researchers have applied forces in the range of nanonewtons to a variety of surfaces and materials in order to analyze the influence at the nanometer scale. Nanomanipulations have been used to investigate (a) many different specimens ranging from metal substrates to polymers, (b) macromolecular biological assemblies, and (c) even individual biomolecules [45]. Precise control of spatial dimensions and recently developed applications for substrates created by nanoindentation have advanced to a very high level. The diverse field of surface nanopatterning by means of nanoshaving or nanografting is described in further detail in Sect. 21.5.

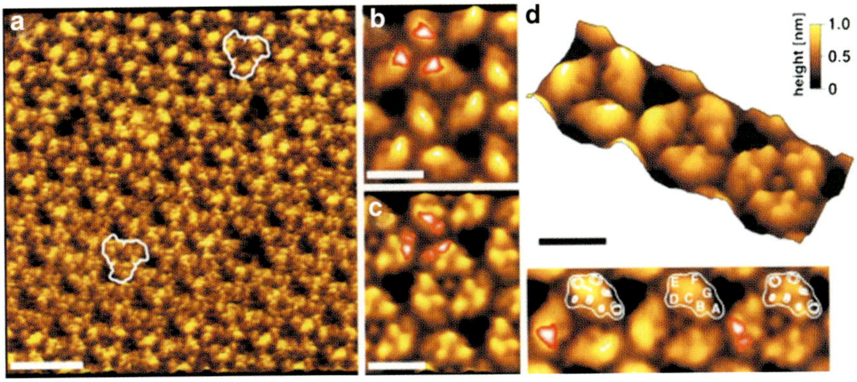

Fig. 21.1 Mechanical modulation of individual polypeptide loops in bacteriorhodopsin (BR). (**a**) The cytoplasmic surface topography of the *purple membrane*. In the region at the top of the image, the force applied to the AFM probe was 100 pN. While scanning the surface line by line, the force was increased by 100 pN (**a**, *center-bottom region*) and the minimized again to 100 pN (**a**, *bottom region*). The *white contour lines* mark BR trimers. The force-induced conformational change of BR is fully reversible [41]. Correlation averages of the cytoplasmic BR surface recorded at 100 pN (**b**) and 200 pN (**c**). Surface regions exhibiting a standard deviation above 0.1 nm are depicted in *red-to-white shades*. (**d**) Interpretation of the force-dependent conformational change of BR. (**d**, *top*) Perspective view of the transitions from low force (100 pN; **d**, *left*) to high force BR trimers (200 pN; **d**, *right*). The central trimer is a graded overlay of the left trimer recorded at 100 pN and the right trimer recorded at 200 pN. (**d**, *bottom*) Corresponding topograph; the helices derived from the atomic model of BR ([43]) are superimposed. The single protrusion evident at low force (**d**, *bottom left*) is identified as the loop connecting helices E and F. The individual protrusions evident at high force (200 pN, **d**, *bottom right*) represent loops connecting helices A and B, loops connecting helices C and D, the C-terminal end, and what remains of the compressed EF loop. Surface regions exhibiting a standard deviation above 0.1 nm are depicted in *red-to-white shades*. The contact mode topograph was recorded in buffer solution (100 mM KCl, 10 mM *Tris*–HCl, pH 7.8). *Scale bars* represent 10 nm (**a**) and 4 nm (**b–d**). The topographs exhibit a vertical brightness range of 1.2 nm (**a**) and 1.0 nm (**b–d**). **a–c** are displayed as reliefs tilted by 5°. Adapted from [44]

21.3 Dynamic Force Microscopy Imaging

The well-known AFM standard imaging mode, the contact mode, is not applicable to all types of biological specimens. Owing to weak attachment to the support, proteins might be pushed away or soft samples could be damaged by the AFM tip during imaging [46]. With the advent of DFM methods such as tapping mode [47, 48] and MAC mode [49, 50], these problems have largely been resolved. These techniques are ideal for imaging weakly attached or soft samples.

The development of DFM modes and their capability of studying biological samples in a liquid environment represent an important step in the application of AFM in the biological sciences. The characteristic feature of DFM methods is that the tip oscillates close to its resonant frequency while it scans the surface, and touches the sample only at the end of its downward movement. In tapping-mode AFM, a piezo actuator is used to generate cantilever oscillation. To apply a homogenous force on the surface, the feedback loop electronics keep the amplitude reduction at a constant value. The amplitude is reduced intermittently, only on contact between tip and sample. These amplitude modulations are used to generate the topographical image of the scanned sample. Since the tip touches the sample surface only at the very end of its downward movement, lateral forces acting on the sample are largely reduced and sample indentation and disruption are minimized [51].

In DFM operation, the parameters important for balancing the feedback are the oscillation amplitude and the amplitude set point (amplitude reduction). Before imaging starts, a suitable setpoint with about 20% amplitude reduction is set in a so-called amplitude–distance cycle. Accordingly, an oscillation amplitude of 80% of the free amplitude is employed during imaging. Usually, oscillation amplitudes are adjusted to small values of about 5 nm to prevent damage to soft biological samples [52]. In DFM modes, precise control of the interaction between tip and sample is a major task. By balancing the involved attractive and repulsive forces, the resolution can be enhanced substantially [53, 54].

21.3.1 Magnetic AC Mode (MAC mode) AFM Imaging

In MAC mode AFM, the magnetically coated cantilever oscillation is excited by an external alternating magnetic field supplied by a coil below the sample stage (Fig. 21.2). The cantilever is excited directly by a magnetic force which features sinusoidal oscillation at a defined frequency and a spectrum that provides one pronounced resonance peak [54, 56]. Moreover, the signal-to-noise ratio is improved

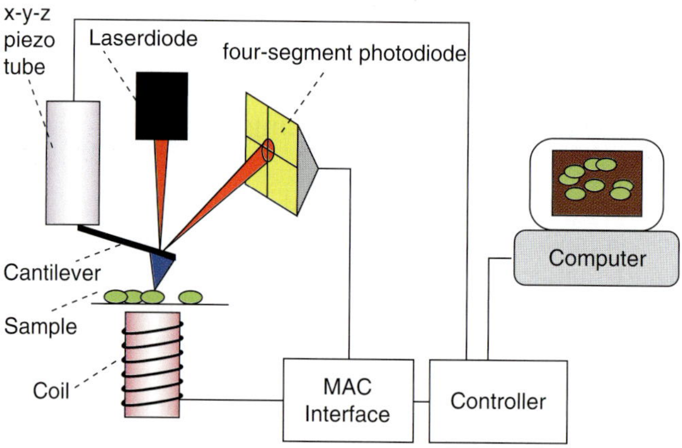

Fig. 21.2 Schematic diagram of a MAC mode AFM. MAC mode is a gentle, nondestructive AFM imaging technique which employs a magnetic field to drive a magnetically coated cantilever, providing precise control of the oscillation amplitude and the applied force. Adapted from [55]

dramatically because the magnetic field drives the cantilever directly [57]. With this type of excitation mode, the background resonances normally visible in tapping mode are absent, and the amplitude–frequency curve follows closely the characteristics of a damped harmonic oscillator [4]. Adjusting both parameters, amplitude and frequency, is thus easily done by regulating the current through the coil and the frequency. This allows sensitive adjustment of both oscillation frequency and phase response, paving the way for high-resolution imaging [58]. Furthermore, the oscillation amplitude is, in contrast to the deflection used in contact-mode AFM, insensitive to thermal drift. Accordingly, MAC mode provides more stable low-force imaging conditions and reduces troubles due to force drift.

To demonstrate the capabilities of the advantageous MAC mode, it was used to image native protein surfaces at a resolution, comparable to contact mode AFM [39, 59]. The topographic images in Fig. 21.3a, b show antibodies bound to the hexagonal surface structure of Sendai purple membrane. The coordinated arrangement of the two antibody F_{ab} fragments on the trimeric lattice of bacteriorhodopsin molecules is depicted. In this specific conformation, the F_c part of the antibody rotates freely and is therefore not observable. The F_{ab} fragments appear far from each other at an angle of approximately 180°. Fourier analysis of the trimeric arrangement of the BR molecules within the hexagonal lattice derived a lateral resolution of 1.5 nm from the magnified image (Fig. 21.3b). The particular BR molecules that are coupled by the two related F_{ab} fragments were marked in black. Contour lines surround hidden BR trimers covered by the antibody. A three-dimensional illustration of the high-resolution image shows the BR structures protruding from the lipid bilayer and the antibody with its two F_{ab} fragments (Fig. 21.3c).

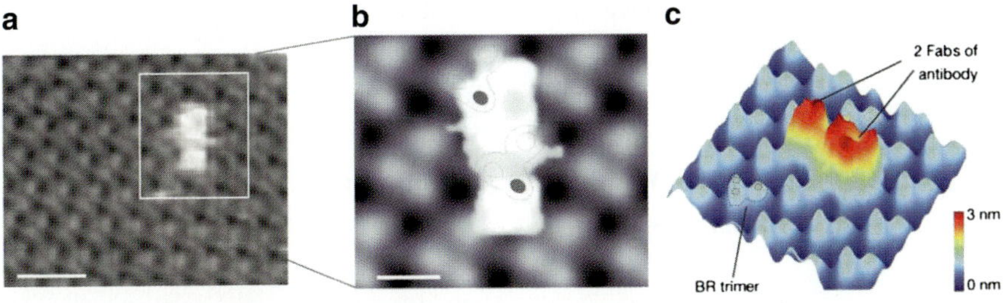

Fig. 21.3 Localization of antibody-binding epitopes on Sendai PM. (**a**) Topographical image obtained by DFM of the cytoplasmic side of Sendai PM to which an antibody is bound. The hexagonal lattice structure of the membrane can be discerned over the whole scan size in this Fourier-filtered image. The *white box* indicates the region for subsequent analysis. *Image size*: 40 nm. *Scale bar*: 10 nm. (**b**) The hexagonal lattice structure of BR trimers and the arrangement of the two F_{ab} fragments of the antibody are clearly visible at higher magnification using MAC mode. The *dotted contour lines* trace BR trimers covered by the antibody. The BR molecules that are engaged by the two F_{ab} fragments of the antibody are marked in *black*. *Image size*: 20 nm; *scale bar*: 4 nm. (**c**) Three-dimensional presentation of (**b**) showing the heights of the structures. The BR trimers (indicated by *dotted lines*) protrude 0.8 nm from the lipid bilayer, whereas the two F_{ab} fragments (see labeling) of the antibody are 2–3 nm in height. Adapted from [59]

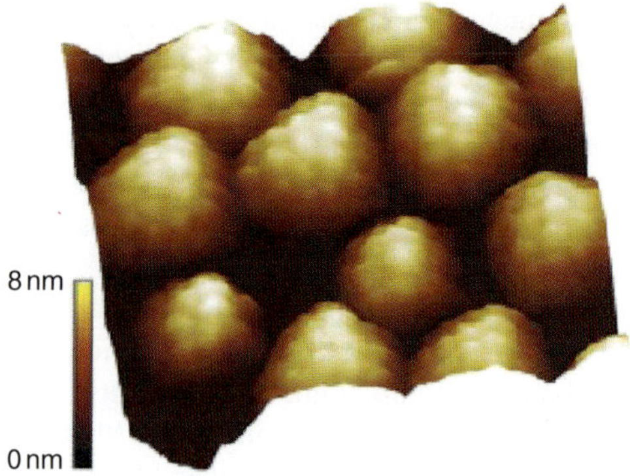

Fig. 21.4 Virus particles adsorbed onto bare mica surfaces. Three-dimensional representation of a densely packed HRV2 layer prepared by physical adsorption from solution onto mica by using bivalent cations. The close apposition of the virions results in some deformation. On most of the virus particles, substructures can be recognized. Roughly 20 protrusions were counted on each virus particle. *Scan size*: 80 nm. Adapted from [5]

Another case in which the advantages of the MAC mode are corroborated is the characterization of pathogens. More precisely, using MAC mode, high resolution images of densely immobilized HRV particles under near-physiological conditions were recorded and published (Fig. 21.4). Moreover, even substructure details on the viral capsids were visualized in buffer solution under physiological conditions [5]. These studies clearly demonstrate that MAC mode is the method of choice when analyzing the topography of soft and weakly attached biological samples at nanometer resolution under physiological conditions while avoiding significant distortions and displacements [60].

21.4 Introduction to Molecular Recognition Force Spectroscopy

Molecular recognition is one of the central principles in numerous biological mechanisms such as the immune response, virus genome replication and transcription, or interactions within cells in general. As a consequence, the emerging field of nanobiotechnology exploits their selectivity and specificity to develop bioanalytical and biomedical devices such as biosensors [61]. A variety of techniques are currently available to investigate inter- and intra-molecular forces between biomolecules. Among these tools, the most popular ones are presently the AFM [62], optical tweezers [63, 64], the biomembrane force probe [65], and the surface force apparatus [66]. They are capable of detecting molecular forces ranging from several piconewtons up to the rupture of covalent bonds in the range of several nanonewtons [67].

The potential of piconewton force sensitivity combined with nanometer positional accuracy makes the AFM a powerful tool for exploring the forces and the dynamics of the interactions between individual ligands and receptors [36, 68].

With this versatile method, many different types of interactions have been studied either on isolated receptor–ligand systems or the complex unfolding patterns of single biomolecules. By conducting many sequential measurements, single-molecule force spectroscopy (SMFS) allows determination of distributions of molecular properties of inhomogeneous systems and enables real-time observation of rarely populated transients, which are difficult or impossible to capture by conventional methods [69]. Accordingly, such single-molecule experiments offer several advantages, for example, for investigating the first step of viral infection – the attachment of a virus to a cell. Section 21.6.2 describes a study in which HRV was immobilized on an AFM tip via a flexible poly(ethylene glycol) (PEG) linker, and the unbinding forces dominating virus–receptor interaction were measured. Hence it was possible to count the number of receptor molecules interacting with the virus and to estimate kinetic rates. In conclusion, the AFM offers one of the most precise force sensors, capable of mapping and studying single receptors at extremely high lateral resolution on the nanometer scale under near-physiological conditions. Several applicable functionalization methods have been developed in recent years to couple the growing number of molecules that are of interest for single-molecule spectroscopy to an AFM tip. Thus, selected chemical approaches for appropriate tethering are described in the following section.

21.4.1 AFM Tip Chemistry

In contrast to AFM imaging modes, SMFS requires an upgrade of the AFM tip to a monomolecular biological sensor. Thus, a single biological ligand (or receptor) must be tethered to the outer apex of the AFM tip in order to detect receptors immobilized on a sample surface via a ligand–receptor complex formation.

Generating tip-linked probe molecules can be facilitated by using different immobilization strategies. The optimal coupling protocol depends mainly on the physical nature of the tip and the ligand to be bound [70]. Nonspecific adsorption of bioligands (e.g., proteins) to the AFM tip can be an adequate approach in some cases [71]. Since this adsorption is based only on noncovalent, weak bonds such as hydrophobic interactions, stacking, hydrogen bonds, or electrostatic interaction, it often fails in stability and reproducibility. However, by chemical crosslinking of tip-adsorbed biomolecules (e.g., using glutaraldehyde), unwanted detachment of adsorbed ligands can be avoided [72]. The most commonly used way of functionalizing AFM tips is covalent attachment of ligands via bifunctional crosslinkers. Two approaches have been used in several studies and are described here in further detail. Using gold-coated tips enables direct attachment of (thiolated) linker molecules by forming a strong covalent-like thiol–gold bond [70], whereas silicon or silicon nitride tips need additional activation steps. The latter kind of tip is most commonly

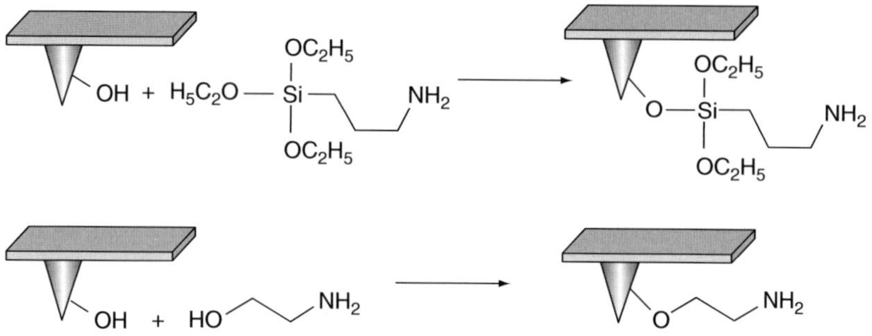

Fig. 21.5 Aminofunctionalization strategies for silicon or silicon–nitride tips. The APTES (3-aminopropyltriethoxysilane) approach is shown in the *upper part* and ethanolamine hydrochloride functionalization in the *lower part*. Adapted from [73]

employed in amino-functionalization using either organosilanes such as APTES (3-aminopropyltriethoxysilane) (Fig. 21.5, upper part) or ethanolamine hydrochloride (Fig. 21.5, lower part). The optimal surface density of approximately 2,000 reactive amino sites per square-micrometer can be achieved using either strategy [73]. In fact, the ethanolamine procedure is easier to perform, but comes with side effects, such as higher chemical stress on the cantilever. In contrast, organosilane coupling must be performed under rigorously controlled conditions, but results in a very high and reproducible coating density. APTES is usually coupled via gas-phase deposition, whereas the ethanolamine procedure is performed in solvent. Subsequently, the amino groups generated on the tip surface are ready to couple the distensible and flexible linker molecule.

Aside from a couple of small ligands such as biotin [74], the linker molecule consists of two different chemical reactive end groups. One end of the heterobifunctional linker molecule is usually coupled to the amino groups of the tip, establishing stable amide bonds. This is done by an activated carboxylic acid (e.g., *N*-hydroxysuccinimide, NHS) binding to the amino groups of the tip, resulting in a stable amide bond. The other linker end binds the bioligand. Itself a linker material, PEG is commonly used at a stretched length of about 5–10 nm (assembled of ~20 ethylene glycol units). The PEG is a perfectly suited polymer due to its nonadsorptive, nontoxic, and inert properties, and its high elasticity. In SMFS experiments, this polymer is responsible for the typical nonlinear stretching behavior observed in the force–distance curve, which allows straightforward discrimination of specific receptor–ligand interaction from unspecific adhesion. Moreover, PEG allows the tethered ligand to move and orientate freely, thereby reducing the risk of crushing the bioligand between tip and surface.

Depending on the chemical nature of the ligand, the second reactive end of the heterobifunctional PEG crosslinker can vary, so several heterobifunctional crosslinkers have been developed for optimal and stable coupling of individual ligands. The 2-pyridyldithiopropionyl-PEG-NHS (PDP-PEG-NHS) linker [75] (Fig. 21.6a) is best suited for bioligands containing a free thiol group (e.g., cysteins). Such molecules can be bound easily and with high efficiency by forming a covalent

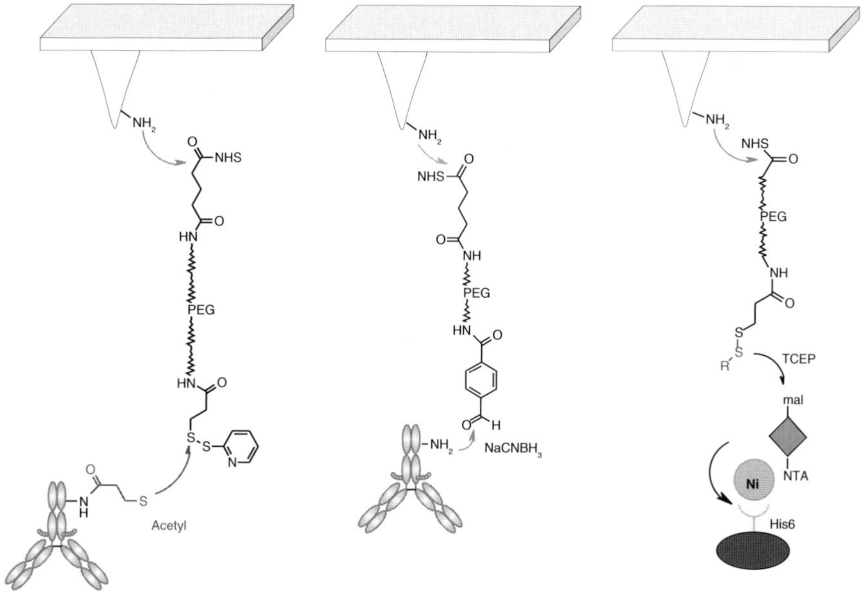

Fig. 21.6 The heterobifunctional crosslinker typically used in SMFS (*left*), PDP-PEG-NHS, allows binding of thiolated ligands or proteins prederivatized with SATP. (*middle*) Ligands with amino groups can be bound directly, using aldehyde-PEG-NHS. (*right*) Hexahistidine-tagged proteins or peptides can be bound noncovalently using the NTA–Ni^{2+}–hexahistidine complex. Adapted from [75]

disulfide bond. In addition, prederivatization of thiol-free proteins with SATP [*N*-succinimidyl 3-(acetylthio)-propionate] allows artificial generation of reactive thiol groups. Since this prederivatization requires a purification step (e.g., via gel filtrations), an adequate amount of protein is necessary. A preferable alternative route of coupling thiol-free proteins as ligands is the aldehyde-PEG-NHS linker [76] (Fig. 21.6b), which allows proteins to be bound directly via their lysine residues (using the primary amine). However, both end groups of this linker are amino reactive, resulting in unwanted loop formation as a side reaction. This can be reduced significantly by using chloroform as a solvent, because the NHS ester is more likely to bind the amine groups. In detail, to suppress unwanted cross-reactions, it turned out to be advantageous to use this linker at a three to sixfold higher concentration. To stabilize the Schiff base formed, a final reduction step using sodium-cyanoborohydride ($NaCNBH_3$) is required.

This heterobifunctional crosslinker turned out to be ideally suited to flexible tethering of single HRV particles to an AFM tip. In detail, the aldehyde-PEG-NHS linker was used for direct viral particle coupling via their endogenous lysine residues. The results of these SMFS measurements taken between viral particles and their associated receptors are described in Sect. 21.6.2 of this chapter. In contrast to the above-mentioned strategies, site-directed coupling is possible when the ligand has a hexa-histidine (His_6)-tag. In this case, a four-step protocol has proven to be suitable for ligand coupling. After amino-functionalization of the tips, the exposed primary amine is converted into a reactive thiol using SATP. After a deprotection step, the nitrilotriacetate-PEG-NHS (NTA-PEG-PDP) linker (Fig. 21.6c) can subsequently be coupled to the tip, resulting in an exposed NTA group. In the last step, a noncovalent but tight and stable nitrilotriacetic acid (NTA)–Ni^{2+} – ion–His_6-protein triple complex is formed. When very high receptor–ligand unbinding forces are expected (e.g., more than 100–200 pN at a loading rate of ~1,000 pN s^{-1}), NTA should be replaced with *tris*–NTA [77], which is known to form an even stronger triple complex. In summary, the above-mentioned coupling schemes provide a wide variety of ways to convert a normal AFM tip into a highly specific molecular biosensor.

21.4.2 Applications of Molecular Recognition Force Spectroscopy

Force spectroscopy experiments allow the determination of interaction forces between a tip-mounted ligand and receptor molecules that are immobilized on a surface. Such interaction forces are measured in force–distance cycles. For this purpose, the tip approaches the surface vertically at a fixed lateral position and is subsequently retracted while monitoring the cantilever deflection (Fig. 21.7a).

This deflection of the cantilever is monitored by back-reflecting a laser beam from the cantilever onto a differential photo detector, which converts the cantilever movements into an electrical signal. This electrical signal V derived from the photodetector can be converted into force using two conversion parameters. First of all, the sensitivity of this detector must be measured. To this end, the cantilever is pressed into a hard surface, and the deflection is monitored during approach. The slope s of this curve is the sensitivity of the optical detection system. The absolute cantilever deflection x, in units of length, is calculated by multiplying the photodiode signal with the inverse of the optical detection system sensitivity: $x = V/s$. It must be mentioned that the determined sensitivity is only correct when the sample acts like a hard, nondeformable material, which is often true for purified molecules attached to a hard surface. In the case of soft cells, the value obtained for the sensitivity may be incorrect due to sample deformation by the tip. In this case, it is necessary to assess the sensitivity of the detector using a hard support in a separate measurement. The second and final conversion is done by applying Hooke's law:

$$F = k\Delta x, \tag{21.1}$$

where k is the cantilever spring constant. Details on how to measure the spring constant are summarized in a recent review [79].

The force resolution of the AFM is, in principle, limited by the thermal noise movement of the cantilever. For small thermal deflections, the cantilever can be approximated by a simple harmonic oscillator with one degree of freedom [80].

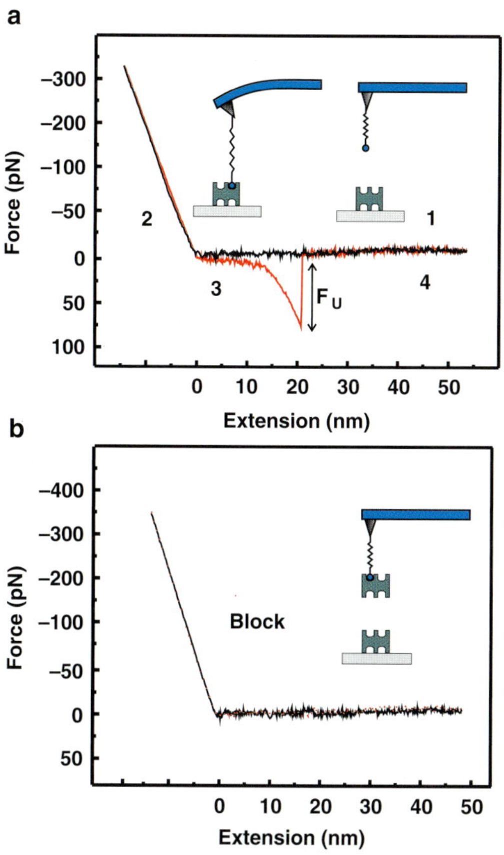

Fig. 21.7 Principles of single-molecule force measurements. Avidin was adsorbed onto mica, and biotin was attached to an AFM tip via a PEG linker. A force–distance cycle was acquired under buffer conditions. The approach curve (*black line*) and the retracting curve (*red line*) are shown. (**a**) The force–distance cycle exhibits an avidin–biotin unbinding event in the retracting curve with an unbinding force of 80 pN (*arrow*). The parabolic retraction force curve reflects the extension of the distensible crosslinker–avidin–biotin connection. (**b**) The specificity of the avidin–biotin unbinding events is shown in blocking experiments in which avidin is injected into the solution to block the biotin sites on the AFM tip. As a consequence, the specific recognition signals disappear and the trace looks like a retrace. Adapted from [78]

According to the equipartition theorem, the cantilever has an average thermal energy of $1/2(k_\mathrm{B}T)$, resulting in

$$\frac{1}{2}k < \Delta x^2 >= \frac{1}{2}k_\mathrm{B}T, \tag{21.2}$$

where k_B is Boltzmann's constant and T the absolute temperature. Inserting this into (21.1) leads to an expression for the force sensitivity ΔF.

$$\Delta F = \sqrt{k_B T k}. \quad (21.3)$$

For single-molecule force measurements, the best results are obtained mainly with cantilevers featuring small spring constants (in the range of 0.01–0.10 N m^{-1}) and short lengths (<50 μm) because they offer lower force noise.

Various features may be distinguished in a force curve. At the beginning of a force–distance cycle (position 1, black trace line, Fig. 21.7a), the cantilever deflection remains zero. As soon as the tip touches the surface, the cantilever is bended upwards (position 2, Fig. 21.7a), reflecting the repulsive force between the apex of the tip and the surface. The cantilever is retracted from the surface (red retrace line, Fig. 21.7a) when an adjusted maximum force value is reached. At first, this retraction leads to reduction of the cantilever bending until the repulsive force drops to zero. In the case of receptor–ligand interaction, further retraction of the tip bends the cantilever downwards, reflecting an attractive force between tip and surface (position 3, Fig. 21.7a). The nonlinear shape of this force–distance is determined by the entropic properties of the flexible PEG crosslinker. At a certain critical force, termed unbinding force, the ligand dissociates from the receptor and the cantilever jumps back to its resting position [81]. If there is no interaction, the retraction and approach curves look similar (position 4, Fig. 21.7a). The specificity of ligand–receptor binding is usually demonstrated by blocking experiments. To this end, free ligands are injected into the solution to block the recognition sites on the sample surface. As a consequence, almost all recognition signals disappear completely (Fig. 21.7b). Another approach uses a tip functionalized with biomolecules resembling the ligands, but without the specificity for the receptors on the surface. In cell studies, mutants lacking receptor expression may provide elegant additional control.

Force–distance cycles are typically repeated more than a thousand times, usually resulting in more than a hundred specific receptor–ligand rupture forces, and finally provide a distribution of unbinding forces. This distribution can be represented by a force histogram or, ideally, by an empirical probability density function (PDF) (Fig. 21.8) [83, 84]. In the latter case, single Gaussian functions of unitary area are calculated from the mean and variances of every value of the unbinding force. The Gaussian functions are added up and finally normalized to arrive at the empirical PDF. The maximum of the distribution indicates the most probable unbinding force at which a single ligand–receptor bond dissociates under the applied force. The advantage of this representation over simple histograms is that the data are weighted by their accuracy, thus yielding a better resolution. In addition, the reliability, reproducibility, and specificity of the measured unbinding forces are confirmed by comparing data obtained using different tips or other samples, and from blocking experiments.

A recent study by Puntheeranurak et al. [85] demonstrated the potential of molecule recognition force spectroscopy. The authors proved the presence, and assessed the dynamics, of plasma membrane transporters in intact cells on the single-molecule level. The topology, conformational changes, and initial substrate–carrier interactions of Na^+–glucose co-transporter (SGLT1) were probed in living cells on a single-molecule level. The AFM tips tethered by an epitope-specific antibody directed against the extramembranous C-terminal loop 13 showed significant

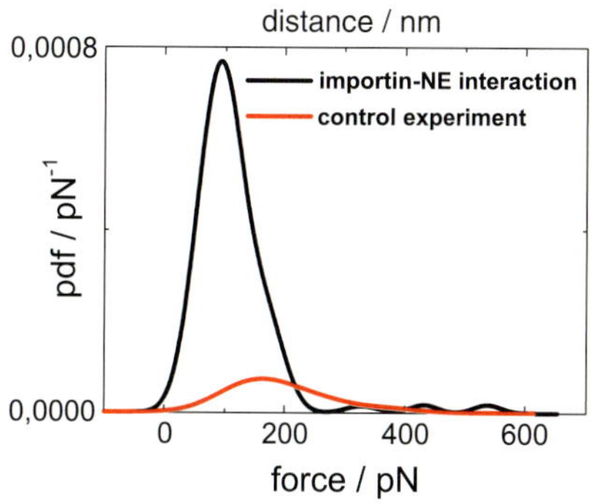

Fig. 21.8 Example of a PDF of the unbinding force observed between a tip-bound transport shuttling factor (importin β) and the related binding sites in the nuclear envelope (*black line*). The *red line* represents the control experiment, in which importin b was removed from the tip by imidazol. Adapted from [82]

recognition events on *SGLT1*-transfected Chinese hamster ovary (CHO) cells. Furthermore, force spectroscopy results, obtained using 1-thio-D-glucose functionalized AFM tips, provided evidence that, in the presence of Na^+, a sugar binding site appears on the transporter surface.

Another application of molecule recognition force spectroscopy was reported by Pfister et al. [86]. They confirmed the presence of heat shock protein 60 (HSP60) on the membrane of stressed cells by detecting, at a high lateral resolution, specific single-molecule binding events between the monoclonal antibody AbII-13 tethered to AFM tips and HSP60 molecules on cells. The interaction force determined was 59 ± 2 pN, which correlated well with the 51 ± 1 pN measured using isolated HSP60 attached to mica surfaces. In summary, with the help of molecule recognition force spectroscopy, they found clear evidence of the occurrence of HSP60 on the surface of stressed living and fixed human umbilical venous endothelial cells (HUVEC).

21.4.3 Topography and Recognition Imaging

Localization and specific detection of receptor binding sites on cells or biological surfaces represents a challenge in the field of biophysics. Very commonly used methods include several optical techniques that involve immunostaining [42,87,88]. However, the data obtained by these techniques are often far from single-molecule resolution and lack additional information, for instance, about the topography of the surfaces analyzed. In this field, AFM has evolved as a key technique for investigating biosurfaces and intermolecular forces in the piconewton range under near-physiological conditions [37]. Moreover, it offers unique possibilities by providing high-resolution imaging with excellent lateral precision to analyze individual biological molecules on surfaces. The recently developed method termed as topography and recognition imaging (TREC) combines the high lateral resolution in imaging and force measurements to localize and identify specific binding sites on various biosurfaces [62, 89, 90].

As in MRFM, the tip must be upgraded to a "smart" molecular sensor by tethering a ligand molecule to its apex. A flexible PEG linker as described in Sect. 21.4 is used to couple the ligand to the tip and facilitates binding to the cognate receptor molecule on the surface. Furthermore, MAC mode is used as the driving force oscillating the chemically functionalized tip to obtain topography information and, simultaneously, support recognition events. The MAC mode is characterized by a magnetically coated cantilever oscillating in an alternating external magnetic field. In conventional MAC mode, the whole oscillation signal amplitude is used to provide the input for the feedback loop of the AFM. By using TREC, the oscillation signal is divided into two parts and is processed in different pathways. The lower part is used to drive a modified feedback loop and to derive the topography image [91]. The recognition image is constructed from the envelope of the lower part of the oscillation signal (Fig. 21.9a). A critical parameter of a successful TREC experiment

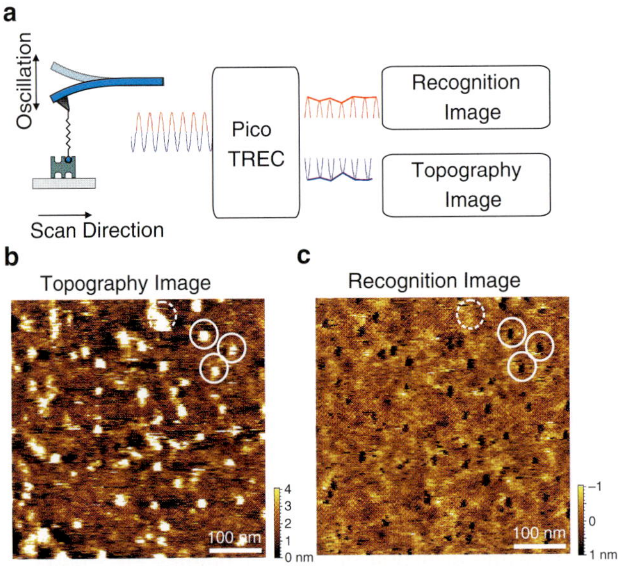

Fig. 21.9 Simultaneous topography and recognition imaging. (**a**) Signal processing for obtaining topography and recognition images simultaneously. The raw cantilever deflection signal of the oscillating cantilever is split via the TREC-box into *upper parts* (corresponding to the recognition image) and *lower parts* (corresponding to the topography image). (**b**) Topographical image of avidin molecules adsorbed to mica acquired with a biotin-tethered tip. (**c**) Simultaneously acquired recognition image. A good correlation between topography and recognition was found (*solid circles*).Topographical spots without recognition denote structures lacking specific interaction (*dashed circle*). Adapted from [91]

is the Q-factor of the cantilever used for measurements. Too high a Q-factor would lead to high levels of mutual influence, i.e., crosstalk, between both parts of the oscillation signal. To ensure, that the upper and lower part of the cantilever oscillation is separated sufficiently, the cantilever should therefore feature a low Q-factor [92].

For a proof-of-principle experiment, single avidin molecules were adsorbed on mica surface and a biotinylated AFM tip was used to study simultaneous TREC on these biomolecules [78]. The topography image on the left side (Fig. 21.9b) shows spatially separated avidin molecules as bright spots adsorbed onto the mica surface. The spots where the biotinylated tip interacted with the avidin molecules on the surface are represented as black dots in the image on the right side (Fig. 21.9c). The position correlation of avidin molecules in the topography image and the recognition spots in the recognition image are definite. In detail, this result indicates high-level accessibility of avidin to the tip-tethered biotin.

With this methodology, topography and recognition images can be obtained simultaneously and definite receptor sites in the recognition image can be correlated to structures in the topography image. With the help of versatile tip-functionalization methods, TREC is applicable to a multitude of ligands. In summary, this approach offers the possibility to recognize many types of proteins, either in isolated form or in complex biological structures such as cells and under near-physiological conditions [93, 94].

21.5 Nanolithography

21.5.1 Applications of Nanolithography

In recent years nanopatterning of biological surfaces has received increasing attention [95–98] in scientific and technological fields. Artificial biological surfaces can be extremely useful in numerous biological applications, ranging from nanobiochips, tissue engineering, nanobiosensors, and drug screening to fundamental studies of molecular biology and cell interaction [95–101]. Smaller biochips with more reaction sites and smaller test sample volumes are further key elements offered by biomolecular nanoarrays. This technology features the potential of higher sensitivity and throughput screening for molecular diagnostics [102–105]. Evolving methods such as DNA hybridization arrays establish rapid and efficient monitoring of the expression levels of thousands of genes for medical diagnosis [95]. Furthermore, the development of protein nanoarrays offers exciting approaches in the field of protein-based diagnostics and therapeutics [95, 96]. This wide range of concepts for the realization of biophysical and biomedical applications is evolving continuously [95, 96, 98–101].

Several requirements have been derived from these applications. The following common important tasks are prominent in most of them: firstly, one crucial issue is the positioning of individual biomolecules within defined regions in or on a substrate. The region of interest is specified by a defined feature size, shape, and spacing on the nanometer scale. However, the native biological functionality and properties should be preserved. Secondly, a point that is often neglected are the properties of the regions surrounding the substrate. High biomolecular resistivity is of great importance for proper discrimination between a functional site and a passive region. Fabricating nanostructured surfaces and functionalization with specific proteins require individual engineering strategies and methods. A variety of promising techniques for generating nanopatterns with biological functions are available. The fields of molecular surface science and surface chemistry have contributed substantially and achieved a remarkable level of sophistication [106–112].

21.5.1.1 Fabrication of Nanoarrays

Scanning probe techniques, such as STM [113] and AFM [36], are standard in microscopy and are important analytical tools capable of imaging surfaces down to the nanometer scale. In addition to their conventional uses in topographical imaging [114], scanning probe microscopes have been employed increasingly in nanolithography. In contrast to AFM, lithographic techniques, such as EBL and focused ion-beam lithography (FIBL), originally developed for the semiconductor industry, have been combined with sophisticated surface chemistry procedures to devise new nanofabrication approaches [115–117]. This interdisciplinary exchange gives rise to a number of methods suitable for fabricating nanoscale features of biomolecules. Further techniques for generating nanopatterns have been published, ranging from bottom-up self-assembly approaches [118–124], Dip-pen nanolithography (DPN) [17–20], conductive AFM [15], scanning near-field photolithography [125] to stamping techniques such as nanoimprinting lithography (NIL) [10, 126, 127] and NPNL [22, 35, 128–130] (Table 21.1).

Among both the novel and the established methods, only a few are potential candidates for future techniques in nanoscale pattern formation and template generation. AFM nanolithography, due to its nanoscale positioning and imaging capability, has proven its ability to create site specific and localized functional structures. Additional morphological, physical, and even functional properties of patterns generated can be characterized immediately by AFM without transferring the patterned surface to another analyzing system. This combined fabrication and characterization function of AFM nanolithography implies time and resource-saving in situ creation and characterization of patterns.

The remainder of this section focuses on nanolithographic approaches, including nanoshaving and nanografting by AFM on planar surfaces. To prepare a planar surface for AFM nanolithography, the basis should include certain key features that facilitate nanoshaving or nanografting. Ideally, the substrate is covered by a homogenous, dense molecular layer to prevent unspecific adsorption to the surface. Furthermore, during the patterning process, the AFM tip should be able to remove the building blocks completely to provide full accessibility to the underlying surface. Molecular layers of bioinert, hydrophilic polymers have been employed successfully in the fabrication of such surfaces [131]. Dense layers of PEG formed by self-assembly have proven efficient in shielding large surfaces against proteins [132, 133]. While relatively long PEGs with a molecular mass of 2,000–5,000 g mol^{-1} yield the most biocompatible surfaces and present an attractive surface for the adsorption of biomolecules, homogenous layers of shorter oligo(ethylene glycol) chains with 3–6 monomer units are suitable to prevent very efficient protein interactions with the surface [134, 135].

Self-assembled monolayers (SAMs), which are formed by adsorption of an active surfactant to a solid surface, exhibit valuable adaptive properties to a number of technologically relevant surface substrates. Widely used systems for modifying the surface properties of metallic and inorganic substrates are, for instance, thiol-SAMs deposited on gold and triethoxysilanes on silicon dioxide [136, 137]. The characteristic properties of a SAM, such as its thickness, structure, surface energy, and stability can be tuned easily. Functional groups for specific interaction can also be inserted into the building blocks and a variety of methods [138–142] are available to characterize SAMs and other functionalized surfaces at resolutions on the nanometer scale. In addition to PEG layers and SAMs, several types of resists, such as proteins, exist which meet the above-defined requirements for coating surfaces as a basis for nanolithography. Below, two main techniques of preparing structures and patterns in surface resists, namely, nanoshaving and nanografting, are described [130].

21.5.1.2 Nanoshaving

In principle, nanoshaving is based on a scanning probe microscope lithographic technique in which a resist material is removed mechanically by an AFM tip to generate patterns down to the nanometer scale. Important factors that influence the resolution and sharpness of the pattern borders are dependent on local displacement by a sharp AFM tip, instantaneous removal of the displaced material, and prevention of re-adsorption to the created structure [143]. In 1994, Wendel et al. [144] were among the first to describe the fabrication of holes and narrow line-shaped trenches down to 50 nm in width in poly(methyl methacrylate) (PMMA) photoresist and in gold layers.

Figure 21.10a, b shows a schematic AFM-based lithography illustration. To create structures or regular patterns in the surface resist, switching to higher tip–sample interaction forces is necessary (e.g., several nN). Fabrication locations are usually selected in regions with flat surface topography, such as plateau areas. To displace the SAM molecules, the AFM tip exerts high local pressure on the contact area. In contrast to normal contact mode imaging (using $\sim$0.01 GPa), the pressure exerted by the tip can reach several GPa in the process of nanoshaving. Preventing re-adsorption of the displaced resist molecules is quite important for achieving sharp pattern borders. Using solvents in which thiols as a resist possess sufficient solubility, such as ethanol or 2-butanol, makes sharp patterns possible, as shown in

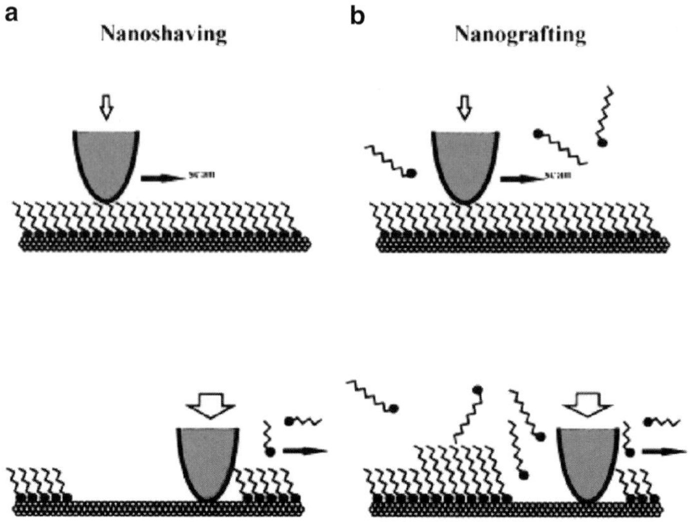

Fig. 21.10 Schematic diagrams of basic manipulation mechanisms using AFM. The imaging and fabrication modes are depicted in the *top and bottom rows*, respectively. Adapted from [143]

Fig. 21.11. In contrast to thiol SAMs on gold, siloxane layers can be patterned using nanoshaving in various media [143, 145, 146]. Nanoshaving of a homogenous PEG film on a glass surface is shown as an example in Fig. 21.12. The PEG layer on a glass surface was characterized by AFM in order to determine the homogeneity and thickness of the film. AFM has proven to be a sensitive tool for probing the quality of PEG films [147, 148]. AFM images of the PEGylated (MW of 500 Da) glass did not reveal any visible surface defects and indicated densely packed surface coverage (Fig. 21.12a). Subsequently, displacing PEG building blocks in a defined 2,500 nm^2 square provided a flat substrate area for determining the thickness of the PEG layer. The derived step size of 2 nm (Fig. 21.12b) matches well with the thickness of PEG films in other studies [131, 149, 149–152].

Several applications exploit the advantages of nanoshaving. It has been used to indirectly immobilize biomolecules on surfaces with SAMs as resists [153–157]. In other studies, patterned surfaces created by nanoshaving were presented as templates for directly accumulating thiolated peptide nanotubes [132] and various proteins on gold substrates [156, 157]. Zhao and coworkers showed that trenches with lengths of up to 1 μm and widths of 150 nm are possible, and further immobilization of nanotubes and IgG proteins onto the patterned gold can be achieved. With their two-step nanoshaving and protein immobilization method, two different antibodies were bound specifically.

Numerous approaches seem promising as means of fabricating high-affinity binding spots towards versatile nanopatterned capturing strategies. In summary, nanoshaving can be employed under ambient conditions that may prove useful when

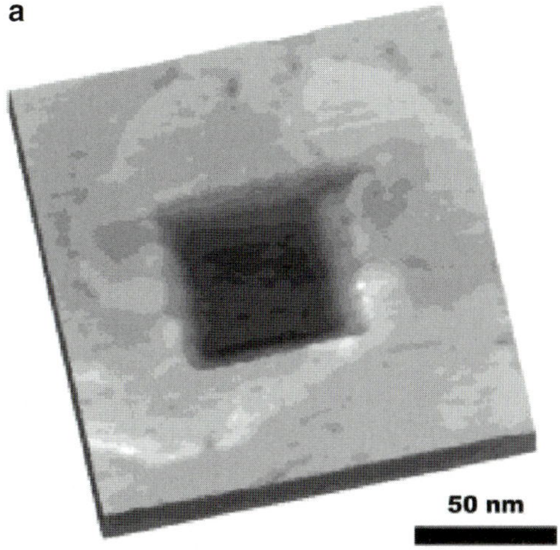

Fig. 21.11 The topograph shows a $160 \times 160\,\text{nm}^2$ AFM scan of a nanoshaved C_{18} S-SAM assembled on a gold surface (111). The thiols were displaced from a central $50 \times 50\,\text{nm}^2$ square in 2-butanol. Adapted from [143]

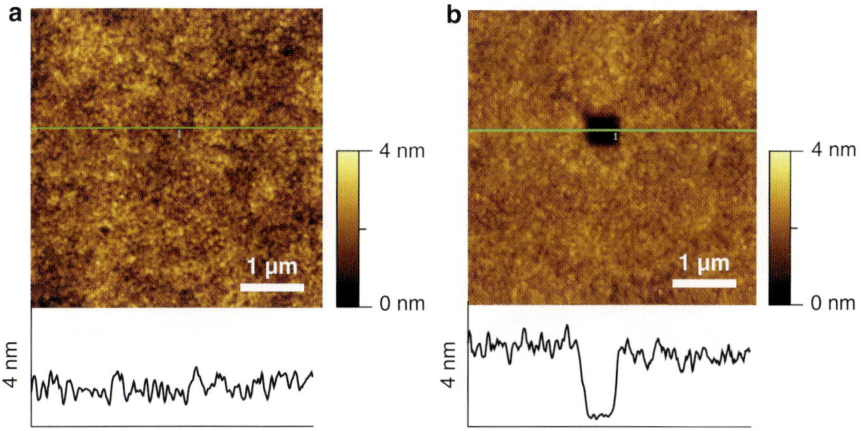

Fig. 21.12 AFM analysis of PEG films on glass. (**a**) Image of a PEG 500 film on a glass surface. (**b**) A $500 \times 500\,\text{nm}^2$ square was scratched into the PEG film. The step size in the height profile is $\sim 2\,\text{nm}$. The scan range is $5 \times 5\,\mu\text{m}^2$ for both images. Scanning mode was contact mode. Adapted from [131]

fabricating multicomponent biomolecular nano structures. An interesting approach in which nanoshaving is used to create defined receptor patterns to capture receptor molecules for detecting viral particles is described later in this chapter in Sect. 21.6.3.2.

21.5.1.3 Nanografting

Nanografting [143, 158] is an enhancement to the method of nanoshaving. In comparison to other nanofabrication techniques, nanografting facilitates precise control of the size and geometry of patterned structures while generating a second type of functionalization in that region. Small trenches and pits as wide as 2–4 nm have been shown [159]. SAMs are the primary aim of surface modification using nanografting.

In principle, the procedure comprises two steps: The nanoshaving process, as described above, is used to remove the layer of molecules on the surface, while a second surfactant molecule with a greater affinity to the surface than the molecule being displaced is present in the solvent (Fig. 21.10b). This second surfactant immediately forms a new SAM in the patterned area and hinders the removed molecules from re-adsorbing. The requirements for successful nanografting are similar to those for nanoshaving and involve easy displacement of an existing surfactant and rapid formation of the SAM in the grafted regions. Thiol-SAMs together with gold surfaces present a favorable combination that fulfills these requirements [130].

The fabrication of two nanografted structures filled with the same SAM is shown in Fig. 21.13. The cross-section in Fig. 21.13b indicates the difference in height between the surrounding $C_{10}S$ matrix monolayer and the grafted $C_{18}S$ nanoislands. Furthermore, the possibility of creating structures that feature two different SAMs separated by only a few nanometers is depicted in Fig. 21.13c. This is confirmed by the height profile in Fig. 21.13d: Height differences amount to 7.5 ± 1.0 and 12.0 ± 1.5 Å between the nanografted areas and the matrix monolayer. In recent years, the progress of nanografting has been adopted by several groups to create patterned substrates as biochips for capturing specific biomolecules [160–167]. The following section describes the development of such a bioactive patterned surface.

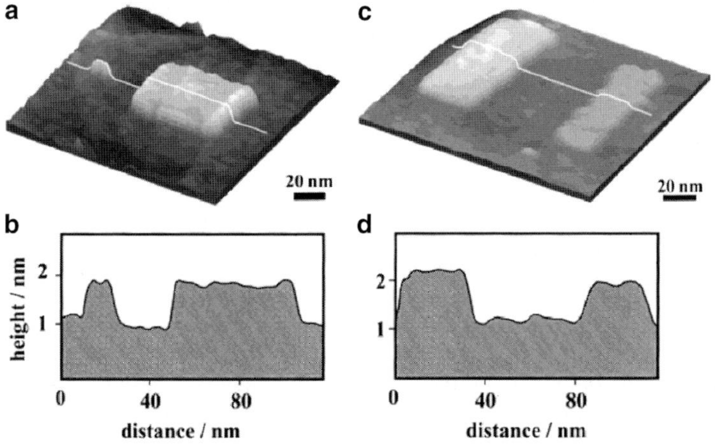

Fig. 21.13 (**a**) Fabrication of two C_{18} S nanoislands (3×5 and $50 \times 50\,nm^2$) in the matrix of a C10S monolayer using nanografting. As shown in the cursor profile (**b**), the C_{18} S islands are 8.8 Å higher than the surrounding C_{18} S monolayer, which is consistent with the theoretical value for crystalline-phase SAMs. (**c**) Fabrication of multicomponent patterns using nanografting. The dimensions of the C_{22} S (*left*) and C18S (*right*) islands are 30×60 and $20\times60\,nm^2$, respectively. (**d**) The corresponding cursor profile shows that the C_{18} S and C_{22} S islands are, respectively, 7.5 ± 1.0 and 12.0 ± 1.5 Å taller than the C_{10} S/Au matrix monolayer. Adapted from [143]

21.5.2 Native Protein Nanolithography

Nanografting as a basis for creating nanopatterned surfaces capable of immobilizing fragile biomolecules seems to be a valuable approach. A successful application of nanografting is herein described using recent work from Tinazli et al. [22]. They showed an elegant design for a biochip based on nanografting histidine-tagged proteins bound to NTA-SAMs-covered substrates. As already mentioned, SAMs are widely used to create monolayer resists on gold substrates in several protein-capturing methods. The NTA–histidine-tag system they used was originally developed for protein purification and analysis in the biological sciences and is a well-known platform. These metal complexes are able to bind proteins tagged with histidines with high specificity. Alkylthiols with functional NTA residues have been synthesized to form metal-chelating SAMs [22]. Proteins containing oligohistidine-peptide tags are able to bind with great affinity, thus forming a dense and uniformly oriented layer on NTA-presenting SAMs [168].

In contrast to conventional nanografting, in which the AFM is operated in contact mode and exerts constant force on the surface, the AFM cantilever is oscillated in this study [22]. The oscillating tip in tapping-mode nanografting enables easier switching between the low-force imaging mode, which avoids deformation of the soft biological surface and the high displacement force mode during nanografting. The forces for imaging in tapping mode are relatively low, reaching maximum values of 1–2 nN, while the forces applied to remove attached proteins range up to 50 nN.

The principle of the oscillating nanografting process is illustrated in Fig. 21.14a. Proteasome complexes were used as test proteins in the depicted example. NTA-SAM functionalized gold surfaces were able to capture this protein complex by means of a site-specifically tethered His-tag fused to the α-subunit. An essential point is that the SAM beneath the protein layer was not affected in its function by nanografting, and subsequently captured His-tagged proteins successfully. To achieve multiplexed lateral organization of proteins, His-tags were fused to different positions on the proteasomes. Nanografting of the monolayer determined a height difference of 15 nm where $\alpha N{-}His_6$ proteasomes had been coupled to the functional SAM chips. The $\beta C - His_6$ proteasome complexes, however, had a homogenously organized side-on orientation with a height of 11 nm after nanografting. Different orientations of His-tags on the target protein combined with the so-called erase-and-write method enable the fabrication of complex proteins assemblies. Figure 21.14b shows how such a multiplexed, laterally organized protein assembly is generated using a series of different proteins in defined orientation. The composition of the vertical stripes was achieved by sequentially erasing and grafting layer by layer of $\alpha N - His_6, \beta C - His_6$, and finally maltose-binding protein-His_{10}(MBP-His10). The gradual height decrease, shown in the cross-section profile in Fig. 21.15, indicates the characteristic assembly of a multiplexed protein structure fabricated using NPNL. This work on writing, reading, and erasing nanopatterns represents a promising step toward future soft protein platforms, which have the potential to become high-density nanoarrays with complex multiplexed surface structures.

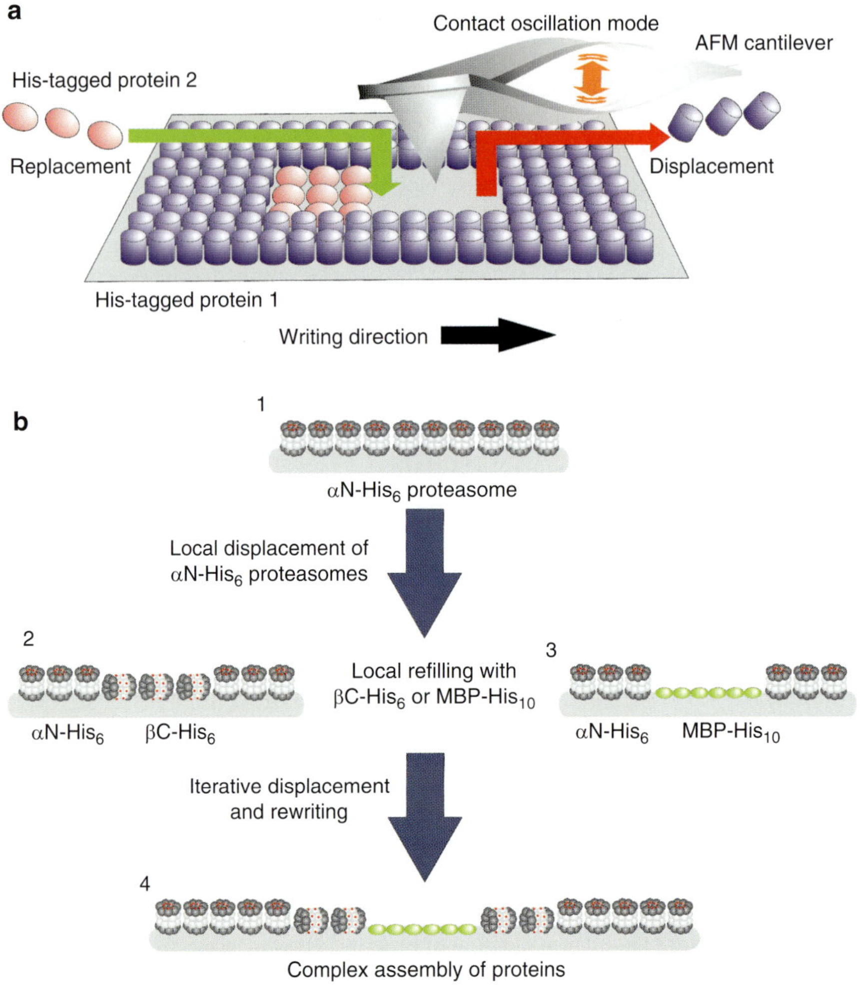

Fig. 21.14 Fabrication of rewritable protein nanoarrays on SAMs by means of NPNL. (**a**) Uniformly oriented His-tagged proteins are removed ("displacement") using an AFM tip in contact oscillation mode and substituted either simultaneously or sequentially with other His-tagged proteins ("replacement"). (**b**) A monolayer of proteasomes is replaced locally by different proteins in subsequent displacement and refilling processes. Proteins and protein complexes in different orientations (here, βN- or bC-His_6 proteasome or maltose-binding protein, MBP-His_{10}) are organized in nanostructured arrays and assemblies (1–4). Adapted from [22]

21.6 Imaging and Force Measurements of Virus–Receptor Interactions

Detecting and characterizing pathogens under physiological conditions is becoming increasingly important for a multitude of scientific, biomedical, and diagnostic applications. The ability to capture pathogens, such as viral particles, from solution

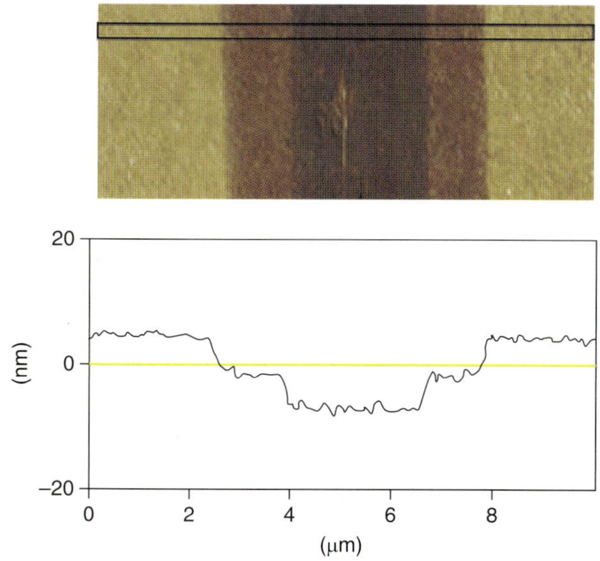

Fig. 21.15 Multiplexed lateral organization of protein assemblies on rewritable biochips. Multiplexed protein assemblies are organized in erase-and-write processes as vertical stripes of sequence $\alpha N - His_6|\beta C\text{-}His_6|MBP\text{-}His_{10}|\beta C\text{-}His_6|\alpha N\text{-}His_6$ (see Fig. 14b.4). The section analysis shows the characteristic height differences of the proteins. AFM images are topographic, and the z-scale represents 40 nm. Adapted from [22]

and to immobilize them specifically at defined sites on a substrate is a great opportunity for AFM to reveal its potential in biological applications. This section gives a short overview of the characterization of surfaces occupied by viral particles and leads on further to the assembly of nanometer-sized virus detection platforms.

Rhinoviruses and their cognate receptors are described in Sect. 21.6.1 of this chapter. Section 21.6.2 summarizes recent studies about the interaction between receptor and virus at the single-molecule level and demonstrates the time-dependent transition of single to multiple virus–receptor bonds. Sect. 21.6.3.1 describes how planar receptor-functionalized substrates were used as model system for site-specific capture of rhinovirus particles from solution with outstanding specificity. Finally, recent work (Sect. 21.6.3.2) illustrates that the AFM can create nanopatterned substrates by nanoshaving, which is employed for capturing, detecting, and characterizing virions, using the very same tip and under near-physiological conditions.

21.6.1 Virus Particle Immobilization and Characterization

In contact mode, the most frequently employed AFM mode, a pointed probe scans a surface horizontally. Therefore, particularly with soft biological samples, the force applied must be carefully adjusted to a minimum; high forces during scanning might result in deformation of the sensitive biomolecules and/or in contamination of the AFM tip [44, 169]. In DFM, these unfavorable lateral forces can indeed be reduced to very low values [56]. While scanning the surface line by line, the cantilever is oscillated close to its resonance frequency at an amplitude of only a few nanometers. Under these conditions, the tip touches the sample only at the end of its downward movement, thereby minimizing dramatically the lateral forces during imaging. For these reasons, DFM is perfectly suited to investigating soft biological samples [60]. For example, DFM was employed successfully in studies of the interactions of pathogens, such as the HRV, with antibodies or their receptors.

HRVs belong to the *picornaviridae* family and are, with a diameter of about 30 nm, amongst the smallest viruses known. As the major cause of the common cold, they are not usually life threatening, but represent an important economic burden in the form of lost working days and huge expenses for medication that generally only alleviates the symptoms. Three-dimensional structures of several HRV serotypes have been solved by cryo-electron microscopy and X-ray crystallography (http://viperdb.scripps.edu/) [170]. The viral capsid surrounding an RNA molecule of about 7,100 bases consists of 60 copies of each of four polypeptide chains assembled on an icosahedral lattice (Fig. 21.16). In addition to phylogenetic classification, HRVs are divided into two receptor groups: those in the major group (89 serotypes) bind human ICAM-1, and those in the minor group (12 serotypes, including HRV2) bind various members of the LDLR family (Fig. 21.17a, b) [172–174]. By using cryo-electron microscopy, the binding site of ICAM-1 on major group viruses was found to be at the base of the canyon, a cleft encircling the vertices at the fivefold axes of the icosahedron [175]. With the same technique, the binding site of a recombinant fragment of the very-low-density lipoprotein receptor (VLDLR) on the minor group virus HRV2 was identified; contact residues were derived from the X-ray structure of the complex [171, 176]. Interestingly, despite great structural similarity between the two HRV groups, the geometry of receptor attachment is very different; up to five ligand-binding modules of VLDLR attach simultaneously to the five

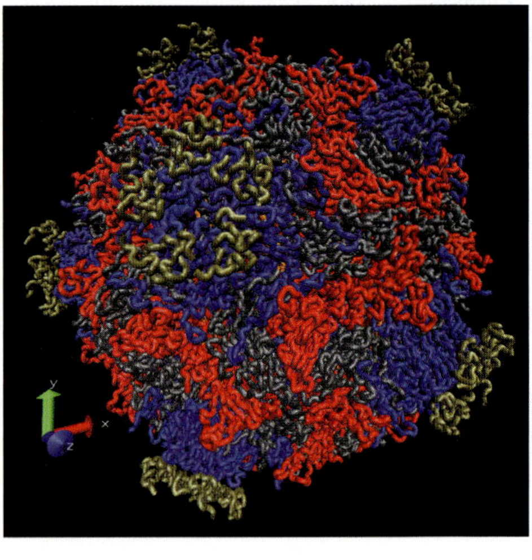

Fig. 21.16 The reconstructed structure of HRV consists of [6] copies of each of four polypeptide chains assembled on an icosahedral lattice. (PDB code: 1V9U) displayed using VMU (Molecular Graphics Viewer). Structural data from http://viperdb.scripps.edu/

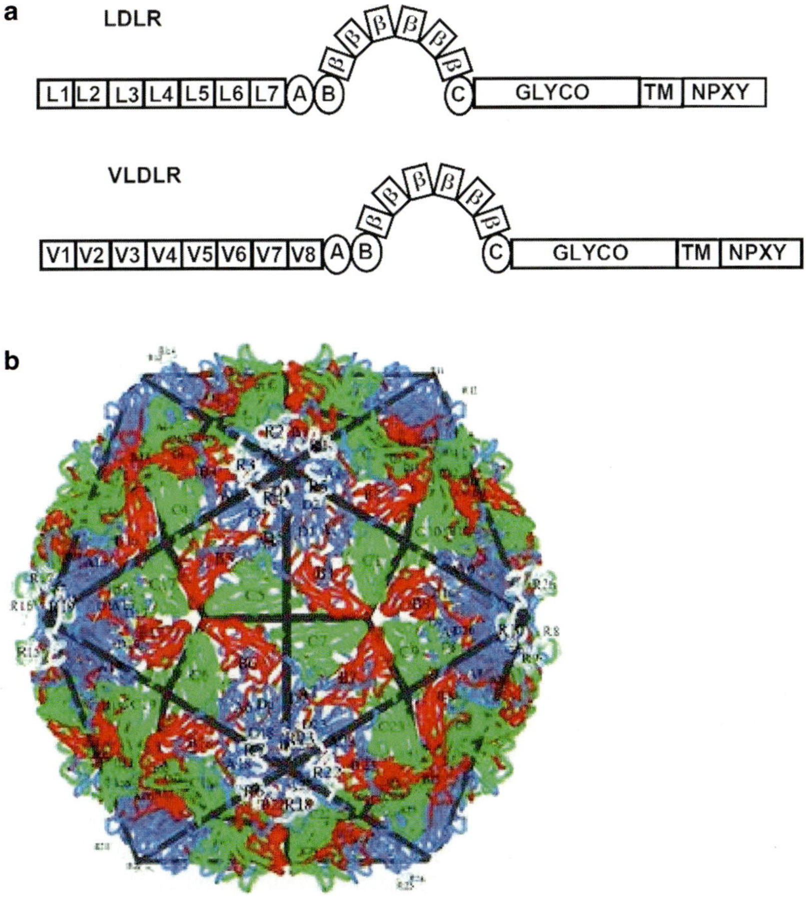

Fig. 21.17 (**a**) Scheme of the modular organization of members of the LDLR family. LDLR and VLDLR have seven and eight terminal ligand-binding repeats, L1–L7 and V1–V8, respectively. Three epidermal growth factor precursor-like regions (A–C) flank a six-bladed beta-propeller (β) with YWTD motifs. The adjacent domain (GLYCO) carries O-linked oligosaccharides. The receptors are anchored in the membrane via a single transmembrane domain (TM). The intracellular C termini carry NPXY clathrin-localization signals. Adapted from [171]. (**b**) Human Rhinovirus (HRV) 2/Receptor Fragment Complex. The visualized structure shows the HRV serotype 2 bound to a fragment of its cellular receptor protein, the very-low-density lipoprotein (VLDL). Structural data from http://viperdb.scripps.edu/

symmetry-related sites in the virus. This binding most probably occurs in a coordinated manner and involves all five modules [177–179]. Because there are 12 fivefold axes per viral particle, 12 copies of VLDLR might bind simultaneously [180].

Kienberger and coworkers achieved high-resolution AFM topographs of virus–receptor complexes (HRV2 complexed with MBP-VLDLR5x3, a recombinant concatemer of five copies of ligand-binding repeat three of human VLDLR fused to maltose binding protein) with various numbers of attached receptor molecules.

The virus–receptor interactions were studied in detail and showed that viral particles attach in a specific and site-directed manner to a model cell membrane. The resolution of AFM topography images of the virus capsid revealed substructures comparable to those seen in cryo-EM studies [5].

In order to analyze these virus–receptor interactions, an artificial membrane interface that mimics the cell surface was employed. A supported lipid bilayer containing Ni^{2+}–NTA lipids as functional groups in the outer leaflet was assembled at high surface density onto mica by using the Langmuir–Blodgett technique. The specific interaction between Ni^{2+}–NTA lipids embedded in the bilayer and His_6-tagged MBP-VLDLR 1–8 receptors (the entire ligand binding domain with all eight repeats) was used to assemble a planar matrix of laterally diffusing receptor proteins. This led to specific and tight immobilization of the virions [181]. Previously, Kubalek et al. had used a similar strategy to obtain two-dimensional crystals of a His_6-taged HIV reverse transcriptase (Fig. 21.18a) [5, 182].

The topographical image in Fig. 21.18b shows a densely packed layer of virus particles covering the surface after 2 h of incubation. The viral particles are associated in a tightly packed single layer order. In previous studies, it was shown that Ca^{2+} is required for specific virus–receptor binding [183, 184]. Accordingly, a densely packed virus layer resulted only when Ca^{2+} was present in the buffer solution. As shown in the cross-section profile (Fig. 21.18b, inset), the height difference between areas that were covered with viral particles and those that were not is about 30 nm. The measured height of about 30 nm is in good agreement with the diameters obtained from X-ray crystallography [185] and electron microscopy [186]. When zooming into higher magnifications, the topographic images feature a homogenously packed layer of viral particles with semicrystalline patches extending over ~500 nm (Fig. 21.18c). The orientation of the patches seemed to differ, and some defect structures were observed. On a smaller scale, a hexagonal arrangement of virus particles on the model cell membrane was visible (Fig. 21.18d). Kienberger and colleagues also showed that superposition and averaging of the two-dimensional virus crystal results in an approximately hexagonal lattice with a mean centre-to-centre distance of 35 nm. In Fig. 21.18d, the average unit cell of the virus is shown in the inset next to the Fourier spectrum, which illustrates the hexagonal arrangement of the spots [187].

This method of immobilizing virus particles specifically via receptor proteins to a fluid supported bilayer turned out to facilitate substantially the detailed topographic imaging of tight, crystalline arrangements of virus particles. To derive high-resolution images of the virus capsid, direct immobilization of the particles to the mica surface by means of divalent cations (i.e., Ni^{2+}) turned out to be more promising, since the soft bilayer compromises high resolution [188]. The Ni^{2+} acts as a bridge between the negatively charged mica and the virus [55]. The tight fixation of the particles and the use of MAC mode allowed acquiring high resolution topography images [52]. By carefully adjusting the imaging parameters, such as low oscillation amplitude and a minimal amplitude reduction, the force exerted by the tip was reduced to a minimum. The result was an image of a densely packed monolayer of viral particles, similar to that obtained from the receptor-containing lipid bilayer

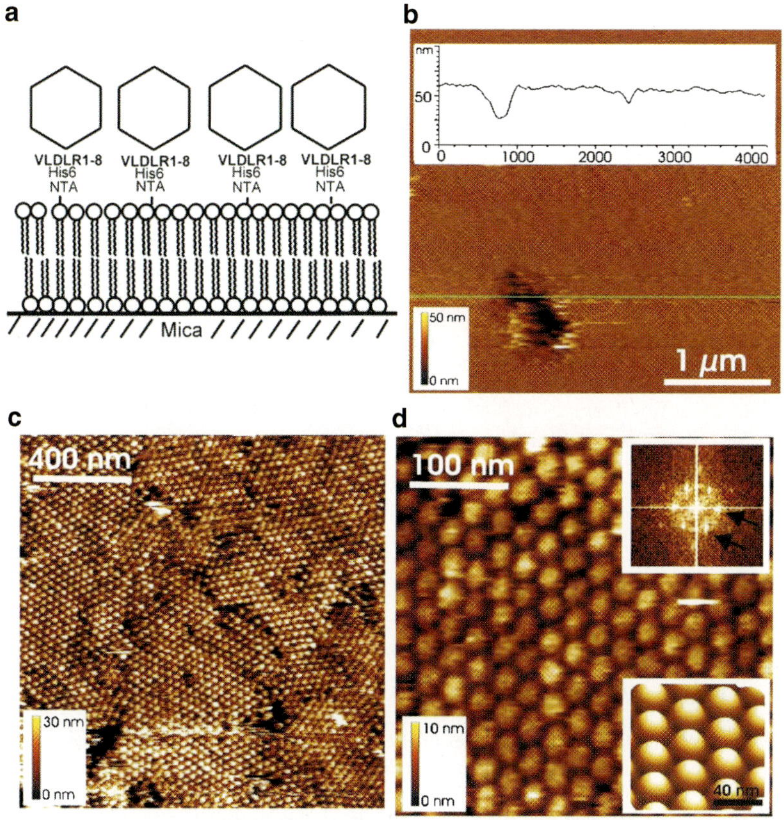

Fig. 21.18 Crystalline arrangement of HRV2 on a receptor-containing lipid bilayer. (**a**) Sketch illustrating the specific binding of HRV2 to receptor molecules (MBP-VLDLR1–8) anchored via His_6-NTA to a lipid bilayer. A high surface density of receptors on the bilayer leads to full coverage with virus. (**b**) Large-scale topographical image of a densely packed virus layer immobilized on a receptor-containing membrane. According to the cross-section profile (*inset*), the height of the layer was determined to be 31 nm. *Scan size* was 4 μm. (**c**) Crystalline patches of virus particles specifically bound to the receptor-containing membrane. The crystals were oriented differently (*scan size* 1.5 μm). (**d**) Hexagonal arrangement of the virions on the model cell membrane at a *scan size* of 400 nm. *Inset* (*upper right*): the Fourier spectrum exhibits hexagonally arranged spots (*arrows*) at frequencies corresponding to the center-to-center distance of HRV2 in the crystal package. *Inset* (*lower right*): averaged three-dimensional representation of the virus arrangement with a center-to-center distance of 35 nm. Adapted from [5]

above (Fig. 21.19a). The height of the layer (~30 nm) was again derived from a small region not covered by viral particles (Fig. 21.19b). A regularly spaced pattern of ~3 nm was observed on almost all viral particles (Fig. 21.19c). This might reflect protruding loops of the capsid proteins on the virion surface known from cryo-EM and X-ray imaging and demonstrates the higher resolution obtained via direct immobilization.

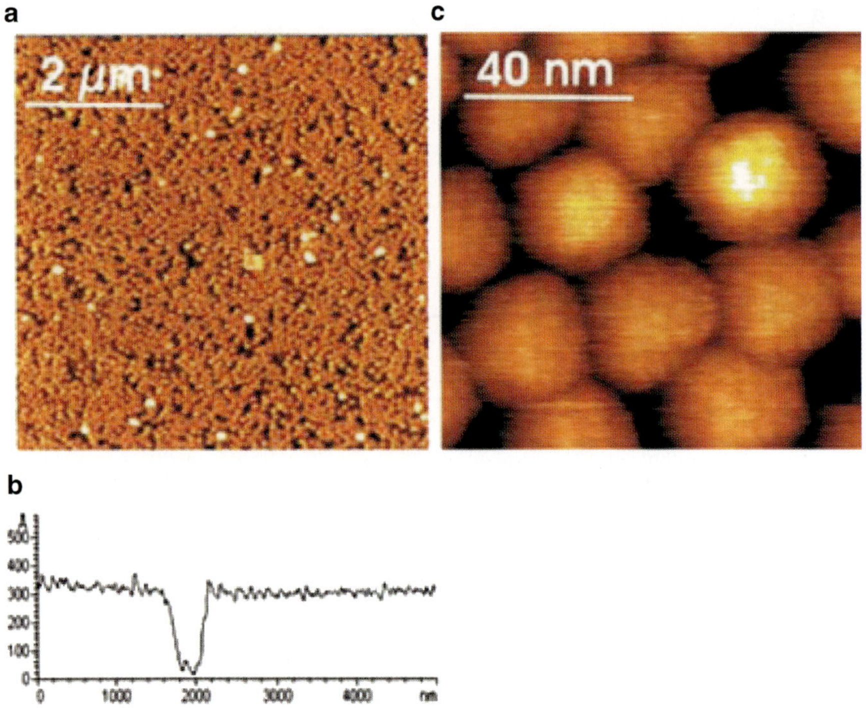

Fig. 21.19 MAC mode topography images of HRV2. (**a**) Densely packed HRV2 monolayer. *Black regions* correspond to holes, whereas *white spots* are viruses bound on top of the first HRV2 layer. The peak-to-peak amplitude of the cantilever oscillation was 5 nm at 8-kHz excitation frequency; the cantilever spring constant was 0.1 N m^{-1}. The image was acquired at a *scan size* of 6 m. (**b**) Cross-section profile of a homogenous virus monolayer containing a large defect hole. The height of the virus layer was determined to be 31 $\pm$ 3 nm. (**c**) Small scan-size image of single HRV2 particles. The close apposition of the virions results in deformation of the capsids. A regular pattern of $\sim$3 nm can be observed on most of the capsids. Image *scan size* was 100 nm. Adapted from [52]

In order to investigate receptor–virus complexes, virus particles were again adsorbed directly onto mica surface via Ni^{2+} in the same way as described before. Subsequently, MBP-VLDLR5x3 was injected into the liquid cell, and the binding of the receptors to the viral particles was analyzed by AFM. The topographic image showed individual distinguishable receptor molecules bound to the vertices of the virus particles (Fig. 21.20). The dimensions of the viral receptors were about $8 \times 1.5\,nm^2$ and are comparable to the dimensions found when immobilizing only VLDLR receptor molecules on a flat substrate. Up to three receptor molecules per viral particle were visible on the upper part of the particles, as far as it was accessible to the AFM tip. Control experiments, which included blocking the virus–receptor interaction by complexing the Ca^{2+} with EGTA and measurements in the absence of Ca^{2+}, showed hardly any receptor on the viral particles.

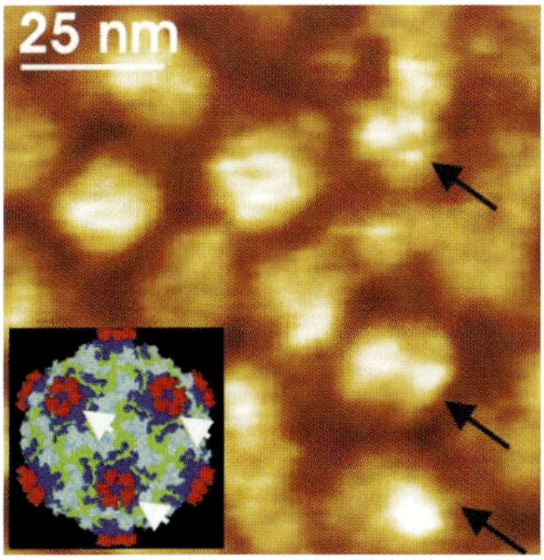

Fig. 21.20 Imaging of virus–receptor complexes. Receptor–virus complexes with one, two, and three receptor molecules (*arrows*). *Scan size* was 100 nm. The *inset* shows the X-ray structure of HRV2 carrying 12 sets of five receptor modules (PDB accession code 1V9U), displayed with RasWin. *Red spots* depict the bound receptor molecules on the fivefold axes (*arrows*). Adapted from [5]

In conclusion, AFM in combination with low-force DFM modes, such as MAC mode, enabled stable imaging in buffer solution over many hours without mechanical distortion of virus particles either bound specifically to the model cell membrane or unspecifically to the mica surface.

21.6.2 Virus–Receptor Interaction Analyzed by Molecular Recognition Force Spectroscopy

As already mentioned in previous sections, HRV2 attaches to members of the LDLR family (e.g., the VLDLR) on the cell membrane to initiate infection. It can be released with EDTA immediately after attachment to the cell, but within some minutes it becomes tightly bound and nondissociable [189]. This was taken to indicate either recruitment of multiple receptors, thus enforcing an initial bond with a single receptor, and/or engulfment within membranes, as the virus enters in clathrin-coated vesicles [190]. Subsequently, it presumably dissociates from its receptors upon arrival in the mildly acidic milieu (pH 6.5–6.0) of early endosomes [191]; finally, the virus is delivered to endosomal carrier vesicles and late endosomes, from where its RNA genome is released into the cytosol.

In this section, we investigate the interaction between HRV2 and the VLDLR on the single-molecule level. We demonstrate a time-dependent transition of single to multiple virus–receptor bonds. By conducting many sequential measurements, our single-molecule force trajectories provide dynamic and statistical real-time information of the distribution of molecular receptor bonds to single virus particles.

21.6.2.1 Theoretical Description

In force spectroscopy, an interaction between two partners is probed by applying an external force. The distribution of measured rupture forces $p(F)$ is given by [192]

$$p(F) = \frac{k(F)}{r} \exp\left(-\int_0^F \frac{k(F')}{r} \mathrm{d}F'\right), \tag{21.4}$$

where $k(F)$ is the dissociation rate under the influence of an external force F and r is the rate of force loading ($r = vk_{\mathrm{eff}}$, with v being the pulling rate and k_{eff} being the spring constant at the moment of rupture). A single bond connection can be described by a single energy barrier, which separates the bound from the unbound state. In this case, the dissociation rate under the influence of an external force is given by $k(F) = k_{\mathrm{off}} \exp(Fx_\beta / k_{\mathrm{B}}T)$. Inserting this into (21.1) shows that the mode of the force distribution F^* exhibits a logarithmic dependence on the loading rate [193, 194]:

$$F^* = \frac{k_{\mathrm{B}}T}{x\beta} \ln\left(\frac{x_\beta r}{k_{\mathrm{B}}T k_{\mathrm{off}}}\right). \tag{21.5}$$

Multiple bond connections may be broken in two modes of loading: (a) in parallel, where each bond shares the force, and (b) in a so-called zipper, where the full force is felt by a single bond until failure, then it is passed on to the next one. Here, we assume parallel loading, since the force is applied to a single tethered virus particle bound to multiple receptors. In addition, the long tethers used in this study do not cause close coupling, therefore all individual bonds can fail independently, and the force is redistributed among the remaining bonds. Thus, the failure rate of the n bonded state to n-1 intact bonds is [195]:

$$k_{n \to n-1}(F) = n k_{\mathrm{off}} \exp\left(\frac{F x_\beta}{n k_{\mathrm{B}} T}\right), \tag{21.6}$$

where k_{off} is the failure rate of a single bond without any force, x_β is the separation of the energy barrier from the equilibrium position, and $k_{\mathrm{B}}T$ is the thermal energy.

The rate of failure of all N identical bonds is therefore:

$$k_{n \to 0}(F) = \left[\sum_{n=1}^{N} \frac{1}{n k_{\mathrm{off}}} \exp\left(-\frac{F x_\beta}{n k_{\mathrm{B}} T}\right)\right]^{-1}. \tag{21.7}$$

The distribution of rupture forces can be derived by inserting (21.4) into (21.1). It can be solved only numerically, but an analytical expression can be found for the most probable rupture force [195]:

$$\frac{1}{r} = \frac{x_\beta}{k_{\mathrm{off}} k_{\mathrm{B}} T} \sum_{n=1}^{N} \frac{1}{n^2} \exp\left(-\frac{F^* x_\beta}{k_{\mathrm{B}} T}\right). \tag{21.8}$$

Note that for $n = 1$, (21.5) is equal to the well-known single barrier case (21.2). A direct consequence of the above result is that the rupture force of N bonds is not N-times the rupture force of a single bond rupture.

21.6.2.2 Unbinding Force Measurements of HRV2–VLDLR Interaction

Molecular recognition force spectroscopy allows studying the interaction between proteins [196, 197] and, as discussed here in detail, between live HRV2 and the VLDLR [198]. For this purpose, HRV2 was bound to AFM tips utilizing polymer linkers. We used an aldehyde-PEG-NHS crosslinker that couples native proteins via their endogenous lysine residues to AFM tips [199]. The VLDLR was immobilized on a mica surface using the same protocol as above [198]. Force–distance cycles of virus-modified tip and receptors bound to mica (Fig. 21.21a) in HEPES-buffered salt solution with 2 mM Ca^{2+} were measured. Interaction events are represented as characteristic nonlinear force signals (Fig. 21.21b). These characteristic shapes originated from the elastic extension of the PEG crosslinker. This made possible distinguishing specific unbinding events from unspecific adhesion.

Within each ligand-binding module of VLDLR, a Ca^{2+} ion is tightly coordinated and is essential for virus binding [200, 201]. Indeed, replacing the Ca^{2+}-containing buffer with a Ca^{2+}-free buffer resulted in a drop in binding probability (percentage of curves displaying an unbinding signal) from $\sim$10% to <1% (see inset in Fig. 21.21c). Re-substitution with a Ca^{2+}-containing buffer restored the binding probability to its original level. This is proof of the measured force signals being a consequence of the virus–receptor interaction.

In most force–distance cycles, unbinding events with a unitary unbinding force value of 82 ± 4 pN at a retraction velocity of 300 nm s^{-1} were found. Occasionally, force signals with multiple unbinding force values were observed, and those were attributed to ruptures of multiple receptor molecules and one virus particle. The unbinding events showed two different profiles in the force–distance cycles: (a) sequential rupture of the receptor–virus bonds (Fig. 21.21c) and (b) simultaneous rupture of all receptor–virus bonds (Fig. 21.21d). The first case (stepwise rupture) was found with a frequency of less than 10%. In this mode, multiple receptors initially bound to the virus. The AFM tip stretches the PEG tethers of the bound receptors until one of the bonds breaks. The remaining slack in the other tethers allows the applied force to drop, followed by a short region corresponding to the loading of the remaining tethers, which finally results in the rupture of the remaining bonds. However, in the majority of cases, single rupture events with a broad range of unbinding forces were observed. This can be explained by taking into account that breaking the first bond leads to a redistribution of force amongst the remaining bonds, and the abrupt increase in the applied load reduces bond lifetime to the nanosecond level [202]. Therefore, the remaining bonds rupture very shortly after the first one is broken, and the finite bandwidth of the AFM causes the multiple bond breakages to be registered as a single rupture event. The distribution of such force signals showed multimodal behavior (Fig. 21.22a, b).

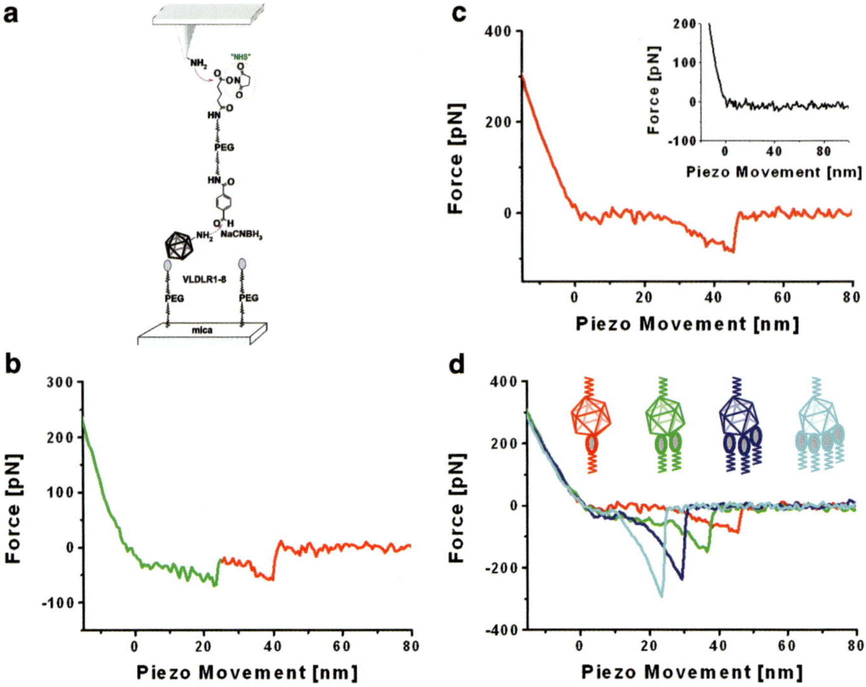

Fig. 21.21 (**a**) Schematic representation of the immobilization scheme used to bind HRV2 to the tip and MBP-V1-8 to mica, and control of attachment. Tip and mica were functionalized with ethanolamine hydrochloride and aldehyde-PEG-NHS prior to incubation with virus and receptor, respectively, and the analytes were fixed by simultaneous application of $NaCNBH_3$. (**b**) A typical force–distance cycle showing the nonlinear stretching of the tether, which allows discriminating between specific and unspecific events. The *inset* shows a force–distance cycle after specifically inactivating the receptor via removal of Ca^{2+} ions. This inactivation is reversible as re-addition of Ca^{2+} restored the specific binding events. (**c**) An example of serial rupturing of two receptors bound to the virus. (**d**) Force–distance cycles showing instantaneous rupture of single (*red*), double (*green*), triple (*blue*), and quadruple (*cyan*) bonds are shown. Note the higher stiffness of the tethers with higher number of bonds (springs in parallel produce higher stiffness). Figure taken from [198]

The simultaneous rupture force distribution was a function of the period of time for which the tip was in contact with the surface (encounter time). Using 17-ms encounter time resulted in a bimodal force distribution with the main peak at 69 pN and a second peak at 123 pN (Fig. 21.22a). Those peaks were interpreted as the simultaneous unbinding of one and two receptors, respectively, from a single virus particle. With increasing encounter time, more multiple receptor–virus unbinding events were observed. Consequently, the force distribution showed a multimodal force spectrum. Figure 21.22b shows the distribution at 32-ms encounter time. Peaks at 149, 203, and 273 pN were observed. Those corresponded to the simultaneous unbinding of two, three, and four receptor molecules, respectively, from a single virus particle. This increase in the number of bonds with encounter time is a clear indication that several single receptors bind sequentially to the virus.

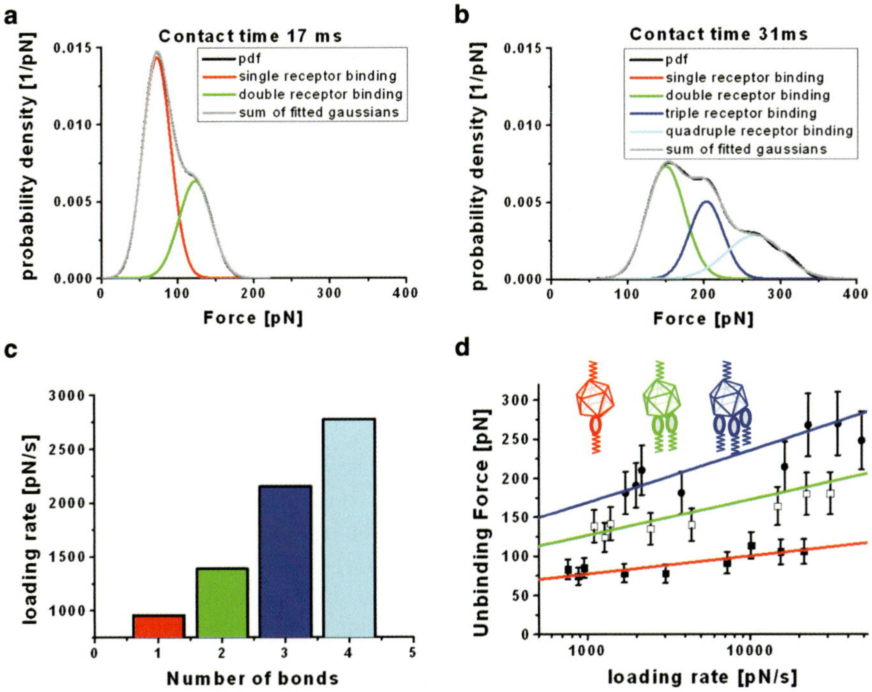

Fig. 21.22 Distribution of rupture forces (*black line*) as a function of the contact time. (**a**, **b**) Extending the contact times shifts the forces towards higher values, indicating binding of more than one receptor molecule. Sums of Gaussians were fitted to the measured curves, yielding an almost perfect reproduction of the experimental data. (**c**) Average of instantaneous loading rate acquired at a certain pulling rate for the different numbers of receptor bound to a virus particle. (**d**) Dynamic force spectra of one (*solid squares*), two (*empty squares*), and three (*solid circles*) MBP-V1–8 molecules bound to HRV2. The results are very well represented by a Markov binding model (21.5) for the double bond (*green*) and triple bond (*blue*) interaction using parameters derived from the single barrier model (21.2) fit (*red*). The error bars account for the uncertainty in the determination of the spring constant and the uncertainty in finding the most probable rupture force. Figure taken from [198]

21.6.2.3 Dynamic Force Spectroscopy

As discussed before, the unbinding force of receptor–virus interactions depends on the pulling rate. By varying the pulling rate, the interaction potential can be probed. Here, we used pulling rates in the range between 150 and 3,000 nm s^{-1}. The most probable unbinding force of single, double, and triple receptor–virus interaction was plotted against the loading rate. Critical to this analysis is that, for a given pulling rate, the effective spring constant k_{eff}, and therefore also the loading rate r, is different for single, double, and triple receptor unbinding (Fig. 21.22c).

The most probable unbinding force of a single receptor–virus interaction showed a logarithmic dependence on the loading rate. Fitting it with the single barrier model (21.2) resulted in the separation of the energy barrier from the minimum $x_\beta = 0.4$ nm and the thermal off-rate constant $k_{\text{off}} = 0.05\,\text{s}^{-1}$. Even though the dependence of the most probable unbinding force on the double and triple receptor–virus interactions seems to follow a logarithmic dependence, fitting them with a single barrier model would give incorrect parameters. Instead, we used the parameters obtained from the single receptor–virus interaction ($x_\beta = 0.4$ nm and $k_{\text{off}} = 0.05\,\text{s}^{-1}$), inserted them into (21.5), and calculated the behavior expected of double ($N = 2$) and triple ($N = 3$) receptor virus interaction. The results were in good agreement with the dynamic force spectra measured for double and triple receptor–virus interaction (Fig. 21.22d).

21.6.2.4 Kinetic On-Rate Constant Obtained from Force Measurements

Assuming that the virus–receptor interaction can be approximated with pseudo first-order kinetics, estimation of the kinetic on-rate constant k_{on} from single-molecule unbinding force measurements requires determining the interaction time τ and the effective concentration c_{eff}, for $k_{\text{on}} = (\tau c_{\text{eff}})^{-1}$ [203]. The interaction time was calculated from the binding probability at different encounter times using $P = A\{1-\exp[-(t - t_0)\tau]\}$ [204], with t_0 being the lag time and A the maximum observable binding probability. A is equal to 1 if unbinding events are found in every force–distance cycle (i.e., binding probability 1). A least square fit resulted in $\tau = 24 \pm 1$ ms. The effective concentration describes the number of binding partners (receptor molecules) within the effective volume V_{eff} available for free equilibrium interaction [203]. V_{eff} can be described as a sphere with radius r_{eff}, with the latter being the sum of the equilibrium crosslinker length (3 nm) and the diameter of HRV2 (30 nm). Assuming four possible binding partners (cf. Fig. 21.22b), k_{on} can be estimated to be $1 \times 10^6\text{M}^{-1}\text{s}^{-1}$. With the kinetic off-rate as determined above, the dissociation constant $K_{\text{D}} = k_{\text{off}}/k_{\text{on}}$ was calculated to be 24 nM.

21.6.3 *Virus Immobilization on Receptor Arrays*

The need to detect viral particles in very small sample volumes requires specific immobilization to target surfaces [128, 205–208]. Label-free readout methods, such as AFM, provide clear advantages over other methods, which require extensive sample preparation and labeling. The capability of high resolution topographical imaging and the possibility to create arrays in the nanometer scale by means of nanolithography are perfectly suited to capturing and analyzing viral particles [7, 209–212]. In summary, several methods have been used in the past to create arrays of receptors with dimensions on the submicrometer or even nanometer scale. For instance, successful application of methods such as photochemical patterning, chemical photolithography, microcontact printing, or microspotting has been reported in this field by several research groups [160].

However, these methods are only applicable if the proteins examined are robust or when only a small part of a more labile protein needs to be active in further experimental steps. To prepare receptor fields, which should act as highly sensitive detectors for virus particles, the proteins must be attached in an active and functional state to a solid substrate. The activity of the immobilized protein should be at the highest possible level. This becomes even more important when the pattern size is reduced towards the nanometer scale. Thus, specific orientation of receptors on the substrate is required to ensure their functionality and prevent sensitivity loss [213–216]. As already noted in Sect. 21.5, the surface surrounding the functional sites on the substrate should feature high resistivity against unspecific attachment of biomolecules [130]. Finally, fabrication and readout of biofunctionalized and micro- or nanopatterned surfaces should be possible under near-physiological conditions. The following Sects. 21.6.3.1 and 21.6.3.2 describe two different examples of constructing such receptor platforms in order to capture HRV particles from an aqueous solution.

21.6.3.1 Receptor Arrays for Selective and Efficient Capturing of Viral Particles

A recently published study [131] reported a successfully applied strategy for selective immobilization of VLDL receptors on micropatterned gold surfaces and subsequent capture of HRV2 particles onto these receptor areas. The ability of NHS-carboxyl and bis-NTA surfaces to bind receptors, and in a next step, viral particles, was analyzed in this work. To achieve a high surface density of viral particles, three different coupling strategies were compared (Fig. 21.23). In a first attempt, viral particles were coupled covalently onto a gold surface via NHS-carboxyl SAMs in strategy A (Fig. 21.23a). In strategy B, the recombinant VLDL receptor derivative was immobilized covalently only onto the NHS-carboxyl SAM (Fig. 21.23b). The

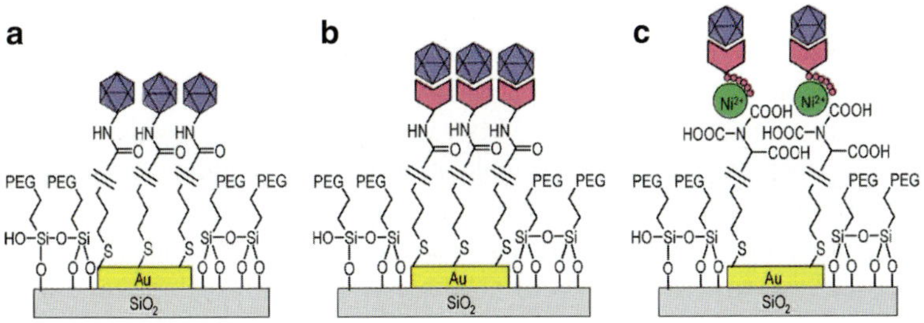

Fig. 21.23 Schematic representation of three coupling strategies used for immobilizing viral particles onto gold surfaces. (**a**) Covalent coupling of viral particles onto NHS-carboxyl surfaces. (**b**) Coupling of viral MBP-VLDLR-His6 receptors onto NHS SAMs, and binding of viral receptors via molecular recognition. (**c**) Immobilization of the His_6-tagged VLDL receptors onto bis-NTA SAMs via Ni(II) chelate complexes, and molecular recognition of viral particles. For visual clarity, simplified structures of the SAM components are displayed. Adapted from [131]

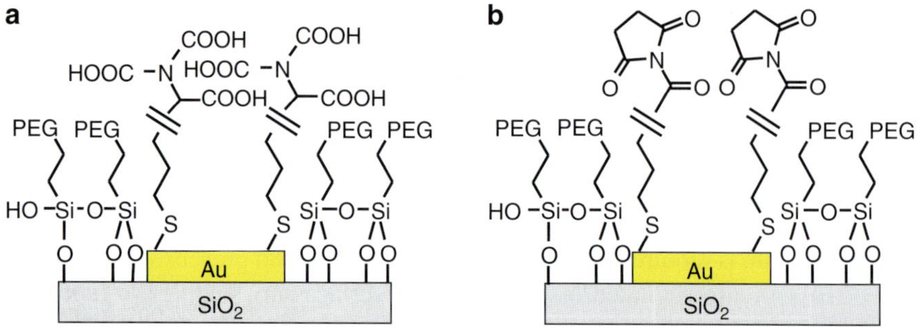

Fig. 21.24 Schematic drawing of microstructured substrates featuring a PEG layer on the glass areas and gold surfaces derivatized with a self-assembled monolayer of alkanethiols terminated with (**a**) bis-NTA groups and (**b**) NHS-activated carboxyl groups. For visual clarity, simplified structures of the SAM components are displayed. Adapted from [131]

viral particles were then specifically captured onto the receptor sites by means of molecular recognition. Finally, in approach C (Fig. 21.23c), a MBP-VLDLR-His$_6$ receptor, which was coupled to a bis-NTA SAM via a nickel chelate, captured the viral particle from solution. These experiments also assessed the ability of the PEG polymer film to prevent unspecific adsorption of viral particles to the surrounding glass surface.

In more detail, glass substrates with microscale gold continents were produced in a photolithographic process and subsequently modified with SAMs of carboxyl or bis-NTA alkanethiols. The surrounding glass support was coated with a dense layer of PEG (Fig. 21.24). The pegylation step turned out to be necessary in order to prevent unspecific adsorption of receptor molecules or viral particles to the glass support. A multistep functionalization procedure enabled coupling of functional SAMs to the gold continents and covering the glass support with a protective layer (Fig. 21.25). In the first step, a protective octylmercaptan SAM (Fig. 21.25, step i) was assembled on the gold sites to protect them against the following modification steps. Subsequently, this protective SAM was removed by heating the substrates (Fig. 21.25, step iii) after the PEG layer had formed on the surrounding glass support in the second step (Fig. 21.25, step ii) [133]. The following step consisted of either the formation of alkanethiol SAMs with carboxyl groups, which must be activated in a further step, or the functionalization with alkanethiols carrying terminal bis-NTA groups (Fig. 21.25, step vi).

As expected, the coupling strategies varied considerably in density of captured particles. The final AFM analysis of all three strategies revealed that approach C featured excellent high densities of captured viral particles (Fig. 21.26b). The MAC mode image showed a densely packed layer of viral particles on the bis-NTA functionalized gold continent. Minute numbers of particles had adsorbed unspecifically onto the pegylated glass support (Fig. 21.26a). The functionalized gold continents, however, were fully covered, which lead to a very high selectivity factor of over 3,30,000.

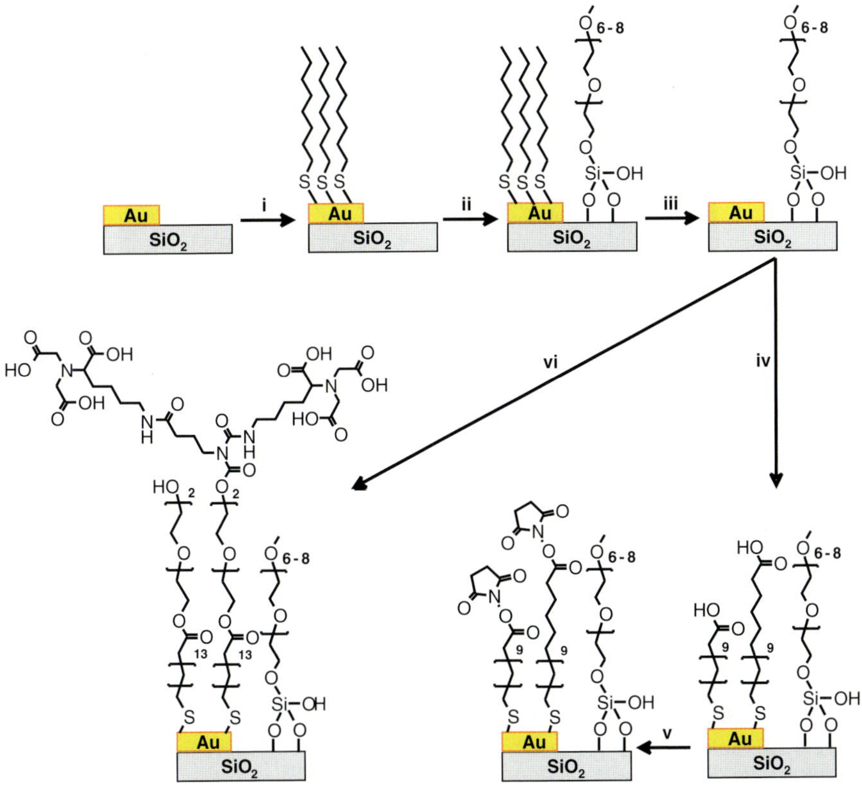

Fig. 21.25 Multistep functionalization of gold-on-glass substrates to obtain NHS-carboxyl and bis-NTA SAMs on gold, and a homogenous protective PEG film on glass. Adapted from [131]

It seems that bis-NTA coupling is superior to NHS coupling because the chelate-forming group preserves its capacity to bind receptors, whereas the activated ester suffers hydrolysis and inactivation [217]. Moreover, when coupled via a hexahistidine tag, the VLDL receptor might be in a more favorable orientation for binding the viral particles than when tethered more randomly as is the case in NHS binding. The study demonstrates that specific and orientated immobilization of receptors plays an important role in micropatterned biofunctionalization [216, 218–220].

21.6.3.2 Atomic Force Microscopy-Derived Nanoscale Chip for Detecting Human Pathogenic Viruses

The study in [128] describes the successful development of a biochip which serves as a platform for detecting HRV particles. NPNL, as described in Sect. 21.5, was used to create nanometer-sized patterns of viral receptor spots to capture viral particles from solution. During nanoshaving, the AFM tip–surface interaction force

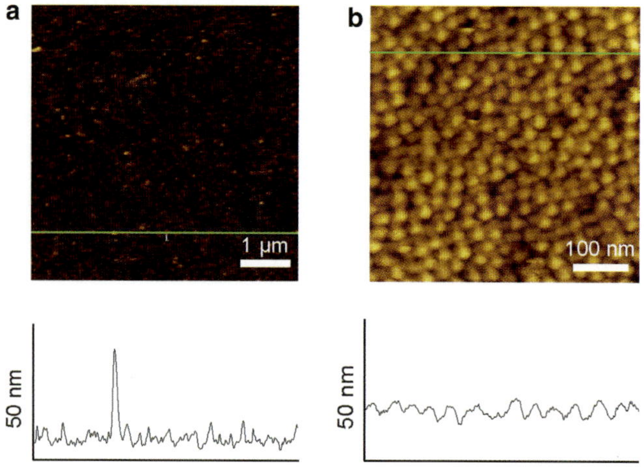

Fig. 21.26 The gold surface was covered with densely spaced clusters of viral particles as displayed in a magnified view (**a**) corresponding to a surface density of 99,200 (3,100 particles per 100 μm^2). In comparison, the very few particles that adsorbed nonspecifically onto PEG areas (**b**) resulted in a low surface density of 0.3 ± 0.2 particles per 100 μm^2). Adapted from [131]

was adjusted carefully such that the His_6-tagged bovine serum albumin (BSA) molecules were removed selectively only from an underlying SAM which was left intact [22]. Subsequently, His_6-tagged very-low-density lipoprotein receptor (MBP-VLDLR-His_6) derivatives were added to fill the patterns by means of specific and oriented binding to the underlying bis-NTA groups of the exposed SAMs. Finally, the receptor arrays were used to capture HRV2 particles and were read out using high-resolution MAC mode under native conditions. Several parameters, such as the kinetic on-rate and the time-dependent binding of the virus to receptor sites, were derived by continuous analysis of the receptor spots using AFM.

Figure 21.27 shows a schematic sketch of the fabrication of the AFM-based label- and amplification-free detection platform for HRVs. NPNL was applied to reveal the functional bis-NTA groups beneath a protective BSA-His_6 layer. Subsequently, added His_6-tagged viral receptors diffused into the nanopatterned structures and bound to the bis-NTA SAM. Finally, the receptor arrays provided highly specific capturing sites for binding HRV2 by means of molecular recognition.

Ultra-flat template-stripped gold provides the ideal solid support for the formation of mixed alkanethiols SAMs. Such gold substrates feature largely homogenous flat gold surfaces with exceptionally low surface roughness [168]. After assembling and activating the bis-NTA SAM, the coverage protein for AFM nanolithography was added. The His_6-tagged BSA formed a homogenous, dense protective layer on the functional SAM. Figure 21.28 depicts the full potential of AFM nanolithography to create various nanopatterns with precise control of pattern size and geometry [143]. The force applied by the tip to the surface was in the range of around 12 nN. Accordingly, the relatively weak force needed to remove BSA proteins confirmed

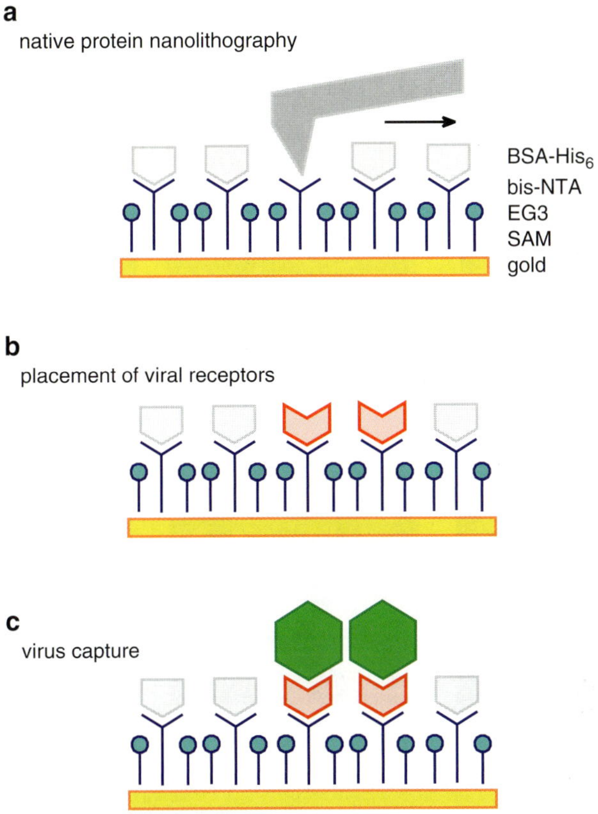

Fig. 21.27 Schematic illustration of AFM-based virus detection. (**a**) NPNL is used to make free areas in a layer of antiadsorptive proteins (BSA). The BSA layer is formed on a SAM consisting of a mixture of thiols terminated with 95% tri(ethylene glycol) (EG3) and 5% bis-NTA groups. (**b**) Injection of hexahistidine (His_6)-tagged viral receptors into the AFM liquid cell results in defined and stable binding to the exposed bis-NTA groups. (**c**) The immobilized viral receptors detect cognate viral particles. Adapted from [128]

that the underlying functional SAM was not affected by the lithography process. The measured height difference of 5–9 nm between the lithographically formed structures and the surrounding protein layer indicates that a single BSA layer had formed. The holes in the BSA layer were then filled with His_6-tagged VLDL receptor molecules. The specific and orientated binding of the receptors kept them highly accessible to the cognate viral particles. As shown in Fig. 21.29a, a 2×2 square shaped array was created in the BSA layer and readily filled with VLDL receptors as depicted in Fig. 21.29b. Injecting a solution containing viral particles and incubating for 2 h resulted in densely occupied receptor sites (Fig. 21.29c). The height increase observed for the recognition spots was about 30 nm. This is in perfect agreement with the diameter of HRVs as determined by other methods [185].

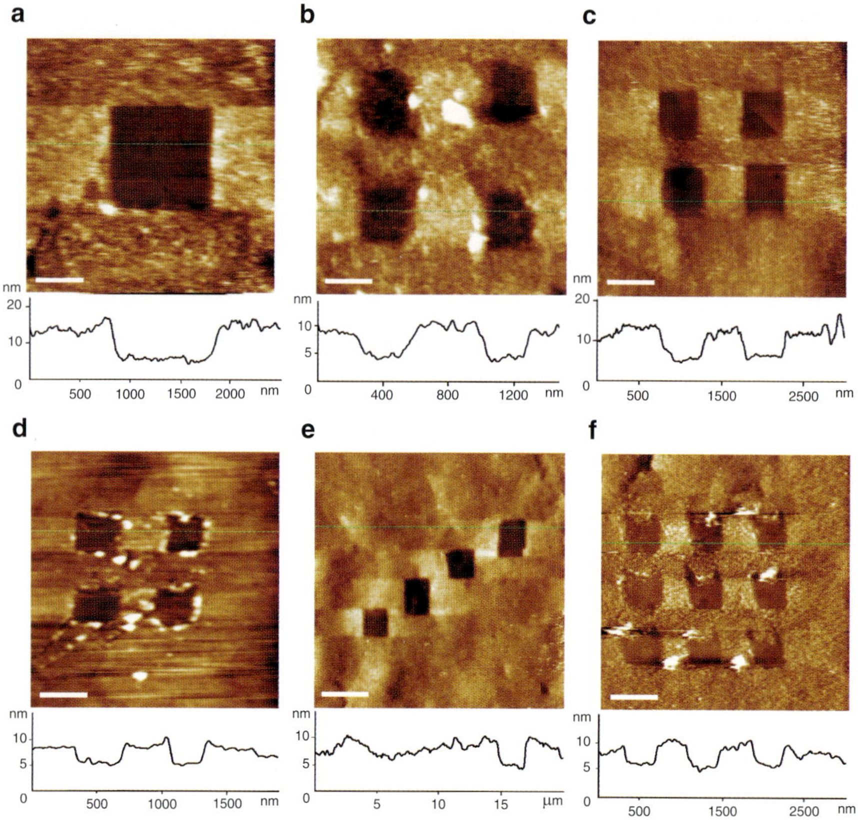

Fig. 21.28 Gallery of nanopatterns within the structuring matrix of BSA. The size of the holes ranged from $200 \times 200\,nm^2$ to $2 \times 2\,\mu m^2$ with a depth of 5–9 nm. (**a**) Single hole with $1 \times 1\,\mu m^2$. *Scale bar*: 400 nm. *z*-scale: 0–15 nm. (**b**) 2×2 array, $300 \times 300\,nm^2$. *Scale bar*: 300 nm. *z*-scale: 0–10 nm. (**c**) 2×2 array, $500 \times 500\,nm^2$. *Scale bar*: 500 nm. *z*-scale: 0–10 nm. (**d**) 2×2 array, $300 \times 300\,nm^2$. *Scale bar*: 400 nm. *z*-scale: 0–10 nm. (**e**) Diagonal arrangement of four holes, $2 \times 2\,\mu m^2$. *Scale bar*: 4 μm. *z*-scale: 0–8 nm. (**f**) 3×3 array, $300 \times 300\,nm^2$. *Scale bar*: 400 nm. *z*-scale: 0–8 nm. Adapted from [128]

Viral particle binding was observed almost exclusively at the preformed receptor spots. Accordingly, these findings emphasize that the receptor was present only in the defined areas, facilitating the binding of viral particles.

To derive the binding kinetics of HRV2 to the receptor spots, the nanopattered region was scanned continuously by the AFM while a solution containing viral particles was injected (Fig. 21.30) [59]. A selected series of topographic images showed the obvious time-dependent binding behavior of the virions (Fig. 21.30a). Monitoring the indicated area for about 2 h revealed an initial lag phase followed by a logarithmic increase in the number of bound viruses (Fig. 21.30b) [221]. The time course of virus binding was described by a pseudo first-order kinetic model,

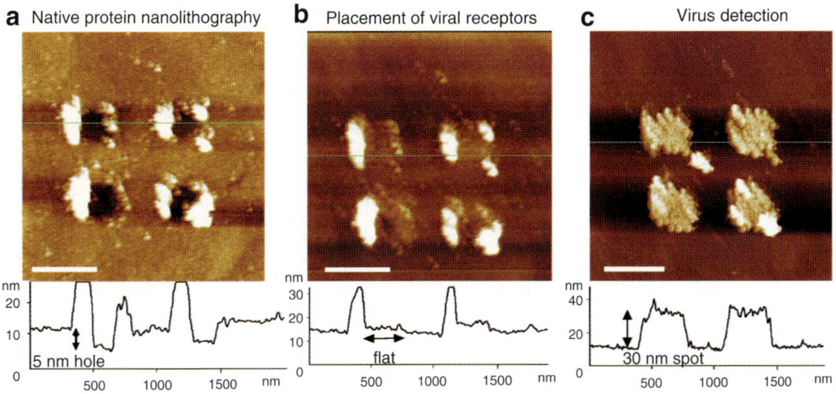

Fig. 21.29 In situ receptor filling and virus detection. (**a**) 2 × 2 array of rectangular shaped holes (5 nm in depth; 250 × 250 nm^2) made in the BSA layer. The *scan size* was 1.8 × 1.8 μm^2. *z*-scale: 0–20 nm. *Scale bar*: 500 nm. (**b**) Injection of VLDLR1–8-His$_6$ (100 mg ml^{-1} final concentration) led to specific binding to the holes. The holes were completely filled, as observed in the cross-section profile (*lower panel*). Incubation time was 30 min. *z*-scale: 0–25 nm. *Scale bar*: 500 nm. (**c**) Human rhinovirus serotype 2 (HRV2; 10 mg ml^{-1} final concentration) was injected, and specific binding to the preformed VLDLR pattern was observed. The height of the virus spots was about 30 nm (cf. cross-section profile showing the height of the sample along the *solid line*). *z*-scale: 0–30 nm. *Scale bar*: 500 nm. Adapted from [128]

because the viral particles in solution were in excess over the receptor molecules tethered to the surface. Accordingly, the resulting kinetic on-rate constant was $k_{on} = 6.2 \times 10^5 M^{-1} s^{-1}$.

Finally, binding specificity was assessed using another, related rhinovirus – HRV14. This virus belongs to the major receptor group and uses ICAM-1 as a receptor. Hence, no binding of HRV14 to the recognition spots was observed. In conclusion, a highly specific platform for detecting HRV particles was successfully developed that shows potential for the usage as a biochip in diagnostic applications and biomedical research.

Acknowledgements

We acknowledge Robert Schlapak, Alma Halilovic, and Friedrich Schaeffler for valuable discussions and Ingrid Abfalter and Martina Rangl for manuscript editing.

This work was supported by the Austrian Research Promotion Agency (FFG, Austrian Nanoscience Initiative VO104-08-BI, project 819703 NSI-NABIOS), the Austrian Science Fund Projects P-14549, P-18693-B09, N-00104, the European Fund for Regional Development (EFRE) and the government of Upper Austria.

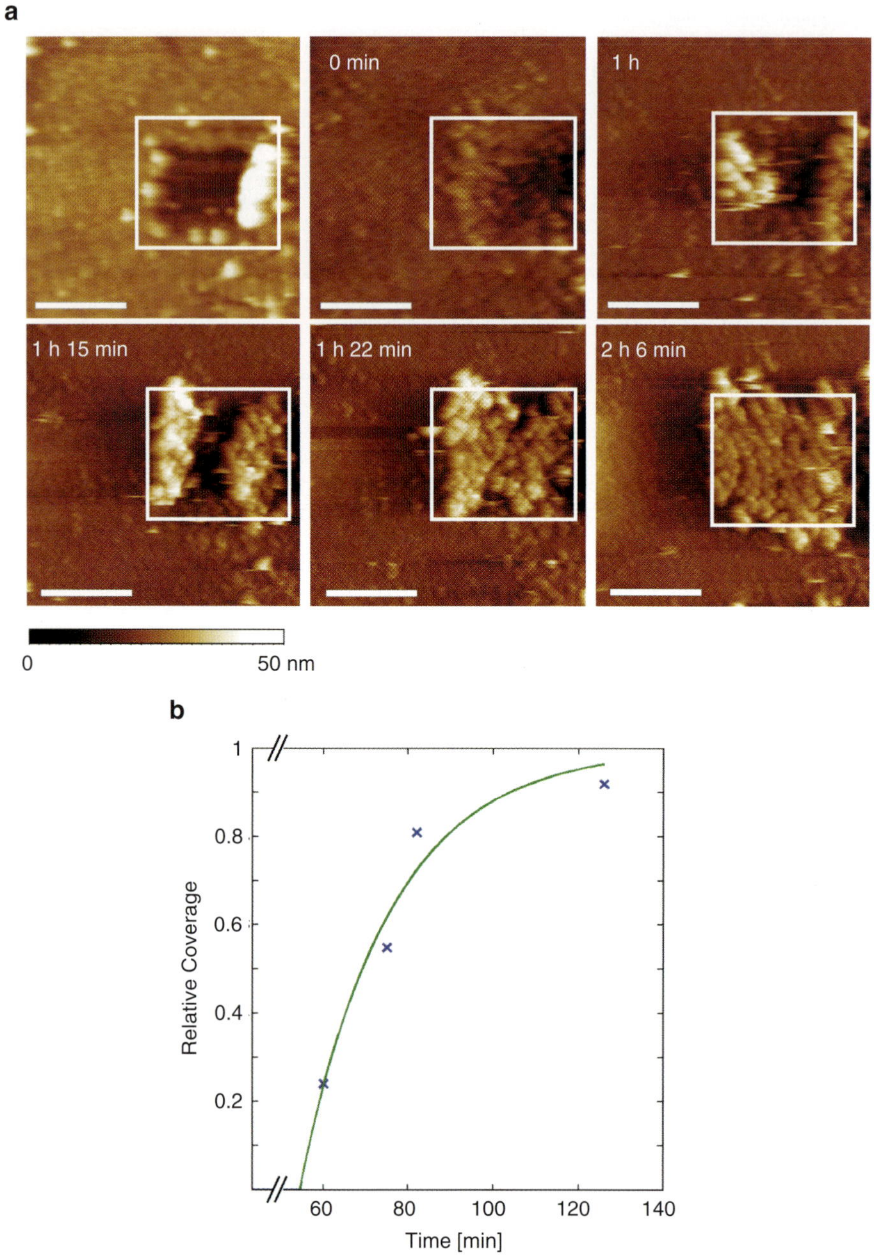

Fig. 21.30 Binding kinetics in real time. (**a**) A single rectangular element ($400 \times 400\,nm^2$) nanostructured into the BSA layer by NPNL. The *scan size* was $1 \times 1\,\mu m^2$. For all images, the z-scale was 0–50 nm (except *upper left*; z-scale: 0–20 nm). *Scale bar*: 300 nm. The hole was filled with His_6-tagged VLDLR1–8 (0 min). After washing and HRV2 injection ($10\,mg\,ml^{-1}$ final concentration), the same surface area was scanned repeatedly after 1 h (*upper right*), 1 h 15 min (*lower left*), 1 h 22 min (*lower middle*), and 2 h 6 min (*lower right*). (**b**) Analysis of virus coverage with respect to incubation time. The area covered by viruses was calculated using a threshold algorithm (relative coverage: crosses) and fit at different incubation times using a pseudo first-order reaction model (*solid curve*). Adapted from [128]

References

1. J.M. Rollinger, M. Schmidtke, The human rhinovirus: human-pathological impact, mechanisms of antirhinoviral agents, and strategies for their discovery. Med. Res. Rev. doi:10.1002/med.20176 (2009)
2. K.E. Arden, C.E. Faux, N.T. O'Neill, P. McErlean, A. Nitsche, S.B. Lambert, M.D. Nissen, T.P. Sloots, I.M. Mackay, Molecular characterization and distinguishing features of a novel human rhinovirus (HRV) C, HRVC-QCE, detected in children with fever, cough and wheeze during 2003. J. Clin. Virol. **47**(3), 219–223 (2010)
3. O.H. Willemsen, M.M. Snel, K.O. van der Werf, B.G. de Grooth, J. Greve, P. Hinterdorfer, H.J. Gruber, H. Schindler, Y. van Kooyk, C.G. Figdor, Simultaneous height and adhesion imaging of antibody-antigen interactions by atomic force microscopy. Biophys. J. **75**(5), 2220–2228 (1998)
4. H. Schindler, D. Badt, P. Hinterdorfer, F. Kienberger, A. Raab, S. Wielert-Badt, V. Pastushenko, optimal sensitivity for molecular recognition MAC-mode AFM. Ultramicroscopy. **82**(1–4), 227–235 (2000)
5. F. Kienberger, C. Rankl, V. Pastushenko, R. Zhu, D. Blaas, P. Hinterdorfer, Visualization of single receptor molecules bound to human rhinovirus under physiological conditions. Structure. **13**(9), 1247–1253 (2005)
6. C. Rankl, L. Wildling, J. Wruss, H.J. Gruber, D. Blaas, P. Hinterdorfer, Multiple receptors involved in human rhinovirus attachment to live cells. Proc. Natl. Acad. Sci. USA. **105**(46), 17778–17783 (2008)
7. R.A. Vega, D. Maspoch, K. Salaita, C.A. Mirkin, Nanoarrays of single virus particles. Angew. Chem. **117**(37), 6167–6169 (2005)
8. L. Montelius, in *Nanotechnology for Biomedical Devices.*, ed. by C.S.R.R. Kumar (Wiley, New York, 2006)
9. T. Blattler, C. Huwiler, M. Ochsner, B. Stadler, H. Solak, S. Janos, H.M. Grandin, Nanopatterns with biological functions. J. Nanosci. Nanotechnol. **6**, 2237–2264 (2006)
10. J.D. Hoff, L.-J. Cheng, E. Meyhöfer, L.J. Guo, A.J. Hunt, Nanoscale protein patterning by imprint lithography. Nano Lett. **4**(5), 853–857 (2004)
11. G.-J. Zhang, T. Tanii, Y. Kanari, I. Ohdomari, Production of nanopatterns by a combination of electron beam lithography and a self-assembled monolayer for an antibody nanoarray. J. Nanosci. Nanotechnol. **7**, 410–417 (2007)
12. S. Ghatnekar-Nilsson, L. Dexlin, C. Wingren, L. Montelius, C.A.K. Borrebaeck, Design of atto-vial based recombinant antibody arrays combined with a planar wave-guide detection system. Proteomics. **7**(4), 540–547 (2007)
13. S. Borini, S. D'Auria, M. Rossi, M.A. Rossi, Writing 3D protein nanopatterns onto a silicon nanosponge. Lab Chip. **5**, 1048–1052 (2005)
14. A. Bruckbauer, D. Zhou, D.-J. Kang, Y.E. Korchev, C. Abell, D. Klenerman, An addressable antibody nanoarray produced on a nanostructured surface. J. Am. Chem. Soc. **126**(21), 6508–6509 (2004)
15. J. Gu, C.M. Yam, S. Li, C. Cai, Nanometric protein arrays on protein-resistant monolayers on silicon surfaces. J. Am. Chem. Soc. **126**(26), 8098–8099 (2004)
16. R.E. Palmer, C. Leung, Immobilisation of proteins by atomic clusters on surfaces. Trends Biotechnol. **25**(2), 48–55 (2007)
17. G. Agarwal, R.R. Naik, M.O. Stone, Immobilization of histidine-tagged proteins on nickel by electrochemical dip pen nanolithography. J. Am. Chem. Soc. **125**(24), 7408–7412 (2003)
18. K.-B. Lee, S.-J. Park, C.A. Mirkin, J.C. Smith, M. Mrksich, Protein nanoarrays generated by dip-pen nanolithography. Science. **295**(5560), 1702–1705 (2002)
19. M. Lee, D.-K. Kang, H.-K. Yang, K.-H. Park, S.Y. Choe, C. Kang, S.-I. Chang, M.H. Han, I.-C. Kang, Protein nanoarray on prolinker surface constructed by atomic force microscopy dip-pen nanolithography for analysis of protein interaction. Proteomics. **6**(4), 1094–1103 (2006)
20. J.-M. Nam, S.W. Han, K.-B. Lee, X. Liu, M.A. Ratner, C.A. Mirkin, Bioactive protein nanoarrays on nickel oxide surfaces formed by dip-pen nanolithography. Angew. Chem. **116**(10), 1266–1269 (2004)

21. N.F. Steinmetz, G. Calder, G.P. Lomonossoff, D.J. Evans, Plant viral capsids as nanobuilding blocks: construction of arrays on solid supports. Langmuir. **22**(24), 10032–10037 (2006)
22. A. Tinazli, J. Piehler, M. Beuttler, R. Guckenberger, R. Tampe, Native protein nanolithography that can write, read and erase. Nat Nano. **2**(4), 220–225 (2007)
23. M. Lynch, C. Mosher, J. Huff, S. Nettikadan, J. Johnson, E. Henderson, Functional protein nanoarrays for biomarker profiling. Proteomics. **4**(6), 1695–1702 (2004)
24. S. Nettikadan, K. Radke, J. Johnson, J. Xu, M. Lynch, C. Mosher, E. Henderson, Detection and quantification of protein biomarkers from fewer than 10 cells. Mol. Cell. Proteomics. **5**(5), 895–901 (2006)
25. A. Bernard, B. Michel, E. Delamarche, Micromosaic immunoassays. Anal. Chem. **73**(1), 8–12 (2000)
26. D. Juncker, H. Schmid, E. Delamarche, Multipurpose microfluidic probe. Nat. Mater. **4**(8), 622–628 (2005)
27. P. Nath, S. Roy, T. Conlisk, A.J. Fleischman, A system for micro/nano fluidic flow diagnostics. Biomed. Microdev. **7**(3), 169–177 (2005)
28. C. Situma, M. Hashimoto, S.A. Soper, Merging microfluidics with microarray-based bioassays. Biomol. Eng. **23**(5), 213–231 (2006)
29. M. Wolf, D. Juncker, B. Michel, P. Hunziker, E. Delamarche, Simultaneous detection of C-reactive protein and other cardiac markers in human plasma using micromosaic immunoassays and self-regulating microfluidic networks. Biosens. Bioelectron. **19**(10), 1193–1202 (2004)
30. C.M. Niemeyer, Functional devices from DNA and proteins. Nano Today. **2**(2), 42–52 (2007)
31. R. Wacker, C.M. Niemeyer, DDI-muFIA – a readily configurable microarray-fluorescence immunoassay based on DNA-directed immobilization of proteins. ChemBioChem. **5**(4), 453–459 (2004)
32. P. Angenendt, L. Nyarsik, W. Szaflarski, J. Glökler, K.H. Nierhaus, H. Lehrach, D.J. Cahill, A. Lueking, Cell-free protein expression and functional assay in nanowell chip format. Anal. Chem. **76**(7), 1844–1849 (2004)
33. A. Buxboim, M. Bar-Dagan, V. Frydman, D. Zbaida, M. Morpurgo, R. Bar-Ziv, A single-step photolithographic interface for cell-free gene expression and active biochips. Small. **3**(3), 500–510 (2007)
34. N. Ramachandran, E. Hainsworth, B. Bhullar, S. Eisenstein, B. Rosen, A.Y. Lau, J.C. Walter, J. LaBaer, Self-assembling protein microarrays. Science. **305**(5680), 86–90 (2004)
35. C. Wingren, C.A.K. Borrebaeck, Progress in miniaturization of protein arrays – a step closer to high-density nanoarrays. Drug Discov. Today. **12**(19–20), 813–819 (2007)
36. G. Binnig, C.F. Quate, C. Gerber, Atomic force microscope. Phys. Rev. Lett. **56**(9), 930–933 (1986)
37. J.K. Horber, M.J. Miles, Scanning probe evolution in biology. Science. **302**(5647), 1002–1005 (2003)
38. D. Sarid, R. Coratger, F. Ajustron, J. Beauvillain, Scanning force microscopy – with applications to electric, magnetic and atomic forces. Microsc. Microanal. Microstruct. **2**(6), 649 (1991)
39. D.J. Muller, D. Fotiadis, S. Scheuring, S.A. Muller, A. Engel, Electrostatically balanced subnanometer imaging of biological specimens by atomic force microscope. Biophys. J. **76**(2), 1101–1111 (1999)
40. H.G. Hansma, J. Vesenka, C. Siegerist, G. Kelderman, H. Morrett, R.L. Sinsheimer, V. Elings, C. Bustamante, P.K. Hansma, Reproducible imaging and dissection of plasmid DNA under liquid with the atomic force microscope. Science. **256**(5060), 1180–1184 (1992)
41. D.J. Müller, G. Büldt, A. Engel, Force-induced conformational change of bacteriorhodopsin. J. Mol. Biol. **249**(2), 239–243 (1995)
42. M. Koopman, A. Cambi, B.I. de Bakker, B. Joosten, C.G. Figdor, N.F. van Hulst, M.F. Garcia-Parajo, Near-field scanning optical microscopy in liquid for high resolution single molecule detection on dendritic cells. FEBS Lett. **573**(1–3), 6–10 (2004)

43. N. Grigorieff, T.A. Ceska, K.H. Downing, J.M. Baldwin, R. Henderson, Electron-crystallographic Refinement of the Structure of Bacteriorhodopsin. J. Mol. Biol. **259**(3), 393–421 (1996)
44. D. Fotiadis, S. Scheuring, S.A. Müller, A. Engel, D.J. Müller, Imaging and manipulation of biological structures with the AFM. Micron. **33**(4), 385–397 (2002)
45. X.N. Xie, H.J. Chung, C.H. Sow, A.T.S. Wee, Nanoscale materials patterning and engineering by atomic force microscopy nanolithography. Mater. Sci. Eng.: R: Rep. **54**(1–2), 1–48 (2006)
46. S. Karrasch, M. Dolder, F. Schabert, J. Ramsden, A. Engel, Covalent binding of biological samples to solid supports for scanning probe microscopy in buffer solution. Biophys. J. **65**(6), 2437–2446 (1993)
47. C.A.J. Putman, K.O. Van der Werf, B.G. De Grooth, N.F. Van Hulst, J. Greve, Tapping mode atomic force microscopy in liquid. Appl. Phys. Lett. **64**(18), 2454–2456 (1994)
48. P.K. Hansma, J.P. Cleveland, M. Radmacher, D.A. Walters, P.E. Hillner, M. Bezanilla, M. Fritz, D. Vie, H.G. Hansma, C.B. Prater, Tapping mode atomic force microscopy in liquids. Appl. Phys. Lett. **64**(13), 1738–1740 (1994)
49. J. Tamayo, A.D.L. Humphris, M.J. Miles, Piconewton regime dynamic force microscopy in liquid. Appl. Phys. Lett. **77**(4), 582–584 (2000)
50. W. Han, S.M. Lindsay, M. Dlakic, R.E. Harrington, Kinked DNA. Nature. **386**(6625), 563 (1997)
51. M. Lantz, Y.Z. Liu, X.D. Cui, H. Tokumoto, S.M. Lindsay, Dynamic force microscopy in fluid. Surf. Interface Anal. **27**, 354–360 (1999)
52. F. Kienberger, R. Zhu, R. Moser, C. Rankl, D. Blaas, P. Hinterdorfer, Dynamic force microscopy for imaging of viruses under physiological conditions. Biol. Proc. Online. **6**(1), 120–128 (2004)
53. V. Vie, M.C. Giocondi, E. Lesniewska, E. Finot, J.P. Goudonnet, C. Le Grimellec, Tapping-mode atomic force microscopy on intact cells: optimal adjustment of tapping conditions by using the deflection signal. Ultramicroscopy. **82**(1–4), 279–288 (2000)
54. J. Tamayo, R. Garcia, Deformation, contact time, and phase contrast in tapping mode scanning force microscopy. Langmuir. **12**(18), 4430–4435 (1996)
55. F. Kienberger, R. Zhu, R. Moser, D. Blaas, P. Hinterdorfer, Monitoring RNA release from human rhinovirus by dynamic force microscopy. J. Virol. **78**(7), 3203–3209 (2004)
56. W.H. Han, S.M. Lindsay, T.W. Jing, A magnetically driven oscillating probe microscope for operation in liquids. Appl. Phys. Lett. **69**(26), 4111–4113 (1996)
57. G. Ge, D. Han, D. Lin, W. Chu, Y. Sun, L. Jiang, W. Ma, C. Wang, MAC mode atomic force microscopy studies of living samples, ranging from cells to fresh tissue. Ultramicroscopy. **107**(4–5), 299–307 (2007)
58. A. Raab, W. Han, D. Badt, S.J. Smith-Gill, S.M. Lindsay, H. Schindler, P. Hinterdorfer, Antibody recognition imaging by force microscopy. Nat. Biotechnol. **17**(9), 901–905 (1999)
59. F. Kienberger, H. Mueller, V. Pastushenko, P. Hinterdorfer, Following single antibody binding to purple membranes in real time. EMBO Rep. **5**(6), 579–583 (2004)
60. F. Kienberger, C. Stroh, G. Kada, R. Moser, W. Baumgartner, V. Pastushenko, C. Rankl, U. Schmidt, H. Muller, E. Orlova, C. LeGrimellec, D. Drenckhahn, D. Blaas, P. Hinterdorfer, Dynamic force microscopy imaging of native membranes. Ultramicroscopy. **97**(1–4), 229–237 (2003)
61. A.P.F. Turner, Biochemistry: biosensors-sense and sensitivity. Science. **290**(5495), 1315–1317 (2000)
62. P. Hinterdorfer, Y.F. Dufrene, Detection and localization of single molecular recognition events using atomic force microscopy. Nat. Meth. **3**(5), 347–355 (2006)
63. K. Svoboda, C.F. Schmidt, B.J. Schnapp, S.M. Block, Direct observation of kinesin stepping by optical trapping interferometry. Nature. **365**(6448), 721–727 (1993)
64. M.S.Z. Kellermayer, Delayed dissociation of in vitro moving actin filaments from heavy meromyosin induced by low concentrations of Triton X-100. Biophys. Chem. **67**(1–3), 199–210 (1997)
65. E. Evans, K. Ritchie, R. Merkel, Sensitive force technique to probe molecular adhesion and structural linkages at biological interfaces. Biophys. J. **68**(6), 2580–2587 (1995)

66. D. Leckband, J. Israelachvili, Molecular basis of protein function as determined by direct force measurements. Enzyme Microb. Technol. **15**(6), 450–459 (1993)
67. H. Clausen-Schaumann, M. Seitz, R. Krautbauer, H.E. Gaub, Force spectroscopy with single bio-molecules. Curr. Opin. Chem. Biol. **4**(5), 524–530 (2000)
68. B.P. Jena, J.K. Hörber (eds.), *Atomic force microscopy in cell biology*. Methods in Cell Biology. Vol. 68 (Academic Press, London, UK, 2002) pp. v–xiv, 1–415
69. R. Nevo, C. Stroh, F. Kienberger, D. Kaftan, V. Brumfeld, M. Elbaum, Z. Reich, P. Hinterdorfer, A molecular switch between alternative conformational states in the complex of Ran and importin beta1. Nat. Struct. Biol. **10**(7), 553–557 (2003)
70. A. Ebner, L. Wildling, R. Zhu, C. Rankl, T. Haselgrubler, P. Hinterdorfer, H.J. Gruber, Functionalization of probe tips and supports for single-molecule recognition force Microscopy. Top. Curr. Chem. **285**, 29–76 (2008)
71. V.T. Moy, E.L. Florin, H.E. Gaub, Intermolecular forces and energies between ligands and receptors. Science. **266**(5183), 257–259 (1994)
72. J. Wong, A. Chilkoti, V.T. Moy, Direct force measurements of the streptavidin-biotin interaction. Biomol. Eng. **16**(1–4), 45–55 (1999)
73. A. Ebner, P. Hinterdorfer, H.J. Gruber, Comparison of different aminofunctionalization strategies for attachment of single antibodies to AFM cantilevers. Ultramicroscopy. **107**(10–11), 922–927 (2007)
74. C.K. Riener, C.M. Stroh, A. Ebner, C. Klampfl, A.A. Gall, C. Romanin, Y.L. Lyubchenko, P. Hinterdorfer, H.J. Gruber, Simple test system for single molecule recognition force microscopy. Anal. Chim. Acta. **479**(1), 59–75 (2003)
75. A.S.M. Kamruzzahan, A. Ebner, L. Wildling, F. Kienberger, C.K. Riener, C.D. Hahn, P.D. Pollheimer, P. Winklehner, M. Holzl, B. Lackner, D.M. Schorkl, P. Hinterdorfer, H.J. Gruber, Antibody linking to atomic force microscope tips via disulfide bond formation. Bioconjug. Chem. **17**(6), 1473–1481 (2006)
76. A. Ebner, L. Wildling, A.S.M. Kamruzzahan, C. Rankl, J. Wruss, C.D. Hahn, M. Holzl, R. Zhu, F. Kienberger, D. Blaas, P. Hinterdorfer, H.J. Gruber, A new, simple method for linking of antibodies to atomic force microscopy tips. Bioconjug. Chem. **18**(4), 1176–1184 (2007)
77. J. Tang, A. Ebner, B. Kraxberger, M. Leitner, A. Hykollari, C. Kepplinger, C. Grunwald, H.J. Gruber, R. Tampe, U.B. Sleytr, N. Ilk, P. Hinterdorfer, Detection of metal binding sites on functional S-layer nanoarrays using single molecule force spectroscopy. J. Struct. Biol. **168**(1), 217–222 (2009)
78. F. Kienberger, A. Ebner, H.J. Gruber, P. Hinterdorfer, Molecular recognition imaging and force spectroscopy of single biomolecules. Acc. Chem. Res. **39**, 29–36 (2006)
79. M.L.B. Palacio, B. Bhushan, Normal and lateral force calibration techniques for AFM cantilevers. Crit. Rev. Solid State Mater. Sci. **35**(2), 73–104 (2010)
80. J.L. Hutter, J. Bechhoefer, Calibration of atomic-force microscope tips. Rev. Sci. Instrum. **64**(7), 1868–1873 (1993)
81. F. Kienberger, V.P. Pastushenko, G. Kada, H.J. Gruber, C. Riener, H. Schindler, P. Hinterdorfer, Static and dynamical properties of single poly(ethylene glycol) molecules investigated by force spectroscopy. Single Mol. **1**(2), 123–128 (2000)
82. M. Rangl, R. Nevo, I. Liashkovich, V. Shahin, Z. Reich, A. Ebner, P. Hinterdorfer, Stable, non-destructive immobilization of native nuclear membranes to micro-structured PDMS for single-molecule force spectroscopy. ChemPhysChem. **10**(9–10), 1553–1558 (2009)
83. P. Hinterdorfer, W. Baumgartner, H.J. Gruber, K. Schilcher, H. Schindler, Detection and localization of individual antibody-antigen recognition events by atomic force microscopy. Proc. Natl. Acad. Sci. USA. **93**(8), 3477–3481 (1996)
84. W. Baumgartner, P. Hinterdorfer, W. Ness, A. Raab, D. Vestweber, H. Schindler, D. Drenckhahn, Cadherin interaction probed by atomic force microscopy. Proc. Natl. Acad. Sci. USA. **97**(8), 4005–4010 (2000)
85. T. Puntheeranurak, L. Wildling, H.J. Gruber, R.K.H. Kinne, P. Hinterdorfer, Ligands on the string: single-molecule AFM studies on the interaction of antibodies and substrates with the Na+-glucose co-transporter SGLT1 in living cells. J. Cell Sci. **119**(14), 2960–2967 (2006)

86. G. Pfister, C.M. Stroh, H. Perschinka, M. Kind, M. Knoflach, P. Hinterdorfer, G. Wick, Detection of HSP60 on the membrane surface of stressed human endothelial cells by atomic force and confocal microscopy. J. Cell Sci. **118**(8), 1587–1594 (2005)
87. T. Schmidt, G.J. Schutz, W. Baumgartner, H.J. Gruber, H. Schindler, Imaging of single molecule diffusion. Proc. Natl. Acad. Sci. USA. **93**(7), 2926–2929 (1996)
88. K.I. Willig, S.O. Rizzoli, V. Westphal, R. Jahn, S.W. Hell, STED microscopy reveals that synaptotagmin remains clustered after synaptic vesicle exocytosis. Nature. **440**(7086), 935–939 (2006)
89. C.M. Stroh, A. Ebner, M. Geretschlager, G. Freudenthaler, F. Kienberger, A.S. Kamruzzahan, S.J. Smith-Gill, H.J. Gruber, P. Hinterdorfer, Simultaneous topography and recognition imaging using force microscopy. Biophys. J. **87**(3), 1981–1990 (2004)
90. C. Stroh, H. Wang, R. Bash, B. Ashcroft, J. Nelson, H. Gruber, D. Lohr, S.M. Lindsay, P. Hinterdorfer, Single-molecule recognition imaging microscopy. Proc. Natl. Acad. Sci. USA. **101**(34), 12503–12507 (2004)
91. A. Ebner, F. Kienberger, G. Kada, C.M. Stroh, M. Geretschlager, A.S.M. Kamruzzahan, L. Wildling, W.T. Johnson, B. Ashcroft, J. Nelson, S.M. Lindsay, H.J. Gruber, P. Hinterdorfer, Localization of single avidin-biotin interactions using simultaneous topography and molecular recognition imaging. ChemPhysChem. **6**(5), 897–900 (2005)
92. J. Preiner, A. Ebner, L. Chtcheglova, R. Zhu, P. Hinterdorfer, Simultaneous topography and recognition imaging: physical aspects and optimal imaging conditions. Nanotechnology. **20**(21), 215103 (2009)
93. L. Chtcheglova, L. Wildling, J. Waschke, D. Drenckhahn, P. Hinterdorfer, Nano-scale dynamic recognition imaging on vascular endothelial cells. Biophys. J. **93**(2), L11–L13 (2007)
94. M. Duman, M. Pfleger, R. Zhu, C. Rankl, L.A. Chtcheglova, I. Neundlinger, B.L. Bozna, B. Mayer, M. Salio, D. Shepherd, P. Polzella, M. Moertelmaier, G. Kada, A. Ebner, M. Dieudonne, G.J. Schutz, V. Cerundolo, F. Kienberger, P. Hinterdorfer, Improved localization of cellular membrane receptors using combined fluorescence microscopy and simultaneous topography and recognition imaging. Nanotechnology. **21**(11), 115504 (2010)
95. C.M. Niemeyer, C.A. Mirkin, *Nanobiotechnology*, vol. 1. (Wiley-VCH, Weinheim, 2004)
96. D. Kambhampati, *Protein Microarray Technology*. (Wiley-VCH, Weinheim, 2004), p. XXXII, 243S
97. G. Zheng, F. Patolsky, Y. Cui, W.U. Wang, C.M. Lieber, Multiplexed electrical detection of cancer markers with nanowire sensor arrays. Nat. Biotechnol. **23**(10), 1294–1301 (2005)
98. G. Shekhawat, S.-H. Tark, V.P. Dravid, MOSFET-embedded microcantilevers for measuring deflection in biomolecular sensors. Science. **311**(5767), 1592–1595 (2006)
99. Y. Cui, Q. Wei, H. Park, C.M. Lieber, Nanowire nanosensors for highly sensitive and selective detection of biological and chemical species. Science. **293**(5533), 1289–1292 (2001)
100. K.L. Christman, V.D. Enriquez-Rios, H.D. Maynard, Nanopatterning proteins and peptides. Soft Matter. **2**, 928–939 (2006)
101. W. Demidov, Nanobiosensors and molecular diagnostics: a promising partnership. Expert Rev. Mol. Diagn. **4**(3), 267–268 (2004)
102. D.P. Malinowski, Multiple biomarkers in molecular oncology. I. Molecular diagnostics applications in cervical cancer detection. Expert Rev. Mol. Diagn. **7**, 117–131 (2007)
103. C. Lin, Y. Liu, H. Yan, Self-assembled combinatorial encoding nanoarrays for multiplexed biosensing. Nano Lett. **7**(2), 507–512 (2007).
104. M.M. Ling, C. Ricks, P. Lea, Multiplexing molecular diagnostics and immunoassays using emerging microarray technologies. Expert Rev. Mol. Diagn. **7**, 87–98 (2007)
105. M. Freemantle, *Downsizing Chemistry: Chemical Analysis and Synthesis on Microchips Promise a Variety of Potential Benefits*, vol. 77 (American Chemical Society, Washington, DC, ETATS-UNIS, 1999), p. 10
106. J. Sagiv, Organized monolayers by adsorption. 1. Formation and structure of oleophobic mixed monolayers on solid surfaces. J. Am. Chem. Soc. **102**(1), 92–98 (1980)

107. J. Gun, R. Iscovici, J. Sagiv, On the formation and structure of self-assembling monolayers: II. A comparative study of Langmuir–Blodgett and adsorbed films using ellipsometry and IR reflection–absorption spectroscopy. J. Colloid Interface Sci. **101**(1), 201–213 (1984)
108. R. Maoz, J. Sagiv, On the formation and structure of self-assembling monolayers. I. A comparative atr-wettability study of Langmuir–Blodgett and adsorbed films on flat substrates and glass microbeads. J. Colloid Interface Sci. **100**(2), 465–496 (1984)
109. L. Netzer, J. Sagiv, A new approach to construction of artificial monolayer assemblies. J. Am. Chem. Soc. **105**(3), 674–676 (1983)
110. L. Netzer, R. Iscovici, J. Sagiv, Adsorbed monolayers versus Langmuir-Blodgett monolayers – why and how? I: From monolayer to multilayer, by adsorption. Thin Solid Films. **99**(1–3), 235–241 (1983)
111. R.G. Nuzzo, D.L. Allara, Adsorption of bifunctional organic disulfides on gold surfaces. J. Am. Chem. Soc. **105**(13), 4481–4483 (1983)
112. M.H. Saavedra, T.J. Mullen, P. Zhang, D.C. Dewey, S.A. Claridge, P.S. Weiss, Hybrid strategies in nanolithography. Rep. Prog. Phys. **73**(3), 03650 (2010)
113. G. Binnig, H. Rohrer, C. Gerber, E. Weibel, Surface studies by scanning tunneling microscopy. Phys. Rev. Lett. **49**(1), 57 (1982)
114. T. Uchihashi, T. Ishida, M. Komiyama, M. Ashino, Y. Sugawara, W. Mizutani, K. Yokoyama, S. Morita, H. Tokumoto, M. Ishikawa, High-resolution imaging of organic monolayers using noncontact AFM. Appl. Surf. Sci. **157**(4), 244–250 (2000)
115. P.M. Mendes, S. Jacke, K. Critchley, J. Plaza, Y. Chen, K. Nikitin, R.E. Palmer, J.A. Preece, S.D. Evans, D. Fitzmaurice, Gold nanoparticle patterning of silicon wafers using chemical e-beam lithography. Langmuir. **20**(9), 3766–3768 (2004)
116. P. Mendes, M. Belloni, M. Ashworth, C. Hardy, K. Nikitin, D. Fitzmaurice, K. Critchley, S. Evans, J. Preece, A novel example of X-ray-radiation-induced chemical reduction of an aromatic nitro-group-containing thin film on SiO^2 to an aromatic amine film. ChemPhysChem. **4**(8), 884–889 (2003)
117. P.M. Mendes, J.A. Preece, Precision chemical engineering: integrating nanolithography and nanoassembly. Curr. Opin. Colloid Interface Sci. **9**(3–4), 236–248 (2004)
118. J. Groll, K. Albrecht, P. Gasteier, S. Riethmueller, U. Ziener, M. Moeller, Nanostructured ordering of fluorescent markers and single proteins on substrates. ChemBioChem. **6**(10), 1782–1787 (2005)
119. A. Valsesia, P. Colpo, T. Meziani, P. Lisboa, M. Lejeune, F. Rossi, Immobilization of antibodies on biosensing devices by nanoarrayed self-assembled monolayers. Langmuir. **22**(4), 1763–1767 (2006)
120. F.A. Denis, P. Hanarp, D.S. Sutherland, Y.F. Dufrêne, Nanoscale chemical patterns fabricated by using colloidal lithography and self-assembled monolayers. Langmuir. **20**(21), 9335–9339 (2004)
121. R. Michel, I. Reviakine, D. Sutherland, C. Fokas, G. Csucs, G. Danuser, N.D. Spencer, M. Textor, A novel approach to produce biologically relevant chemical patterns at the nanometer scale:? selective molecular assembly patterning combined with colloidal lithography. Langmuir. **18**(22), 8580–8586 (2002)
122. Y. Cai, B.M. Ocko, Large-scale fabrication of protein nanoarrays based on nanosphere lithography. Langmuir. **21**(20), 9274–9279 (2005)
123. H. Agheli, J. Malmström, E.M. Larsson, M. Textor, D.S. Sutherland, Large area protein nanopatterning for biological applications. Nano Lett. **6**(6), 1165–1171 (2006)
124. H. Yan, S.H. Park, G. Finkelstein, J.H. Reif, T.H. LaBean, DNA-templated self-assembly of protein arrays and highly conductive nanowires. Science. **301**(5641), 1882–1884 (2003)
125. S. Sun, M. Montague, K. Critchley, M.-S. Chen, W.J. Dressick, S.D. Evans, G.J. Leggett, Fabrication of biological nanostructures by scanning near-field photolithography of chloromethylphenylsiloxane monolayers. Nano Lett. **6**(1), 29–33 (2005)
126. D. Falconnet, D. Pasqui, S. Park, R. Eckert, H. Schift, J. Gobrecht, R. Barbucci, M. Textor, A novel approach to produce protein nanopatterns by combining nanoimprint lithography and molecular self-assembly. Nano Lett. **4**(10), 1909–1914 (2004)

127. V.N. Truskett, M.P.C. Watts, Trends in imprint lithography for biological applications. Trends Biotechnol. **24**(7), 312–317 (2006)
128. H. Artelsmair, F. Kienberger, A. Tinazli, R. Schlapak, R. Zhu, J. Preiner, J. Wruss, M. Kastner, N. Saucedo-Zeni, M. Hoelzl, C. Rankl, W. Baumgartner, S. Howorka, D. Blaas, H.J. Gruber, R. Tampe, P. Hinterdorfer, Atomic force microscopy-derived nanoscale chip for the detection of human pathogenic viruses. Small. **4**(6), 847–854 (2008)
129. R. Zhu, A. Ebner, M. Kastner, J. Preiner, S. Howorka, P. Hinterdorfer, Topography and recognition imaging of protein-patterned surfaces generated by AFM nanolithography. ChemPhysChem. **10**, 1478–1481 (2009)
130. P. Mendes, C. Yeung, J. Preece, Bio-nanopatterning of surfaces. Nanoscale Res. Lett. **2**(8), 373–384 (2007)
131. P.D. Pollheimer, M. Kastner, A. Ebner, D. Blaas, P. Hinterdorfer, H.J. Gruber, S. Howorka, Receptor arrays for the selective and efficient capturing of viral particles. Bioconj. Chem. **20**, 466–475 (2009)
132. J. Piehler, A. Brecht, R. Valiokas, B. Liedberg, G. Gauglitz, A high-density poly(ethylene glycol) polymer brush for immobilization on glass-type surfaces. Biosens. Bioelectron. **15**(9–10), 473–481 (2000)
133. C. Boozer, Q. Yu, S. Chen, C.-Y. Lee, J. Homola, S.S. Yee, S. Jiang, Surface functionalization for self-referencing surface plasmon resonance (SPR) biosensors by multi-step self-assembly. Sens. Actuat. B: Chem. **90**(1–3), 22–30 (2003)
134. K.L. Prime, G.M. Whitesides, *Self-Assembled Organic Monolayers: Model Systems for Studying Adsorption of Proteins at Surfaces*, vol. 252. (American Association for the Advancement of Science, Washington, DC, ETATS-UNIS, 1991), p. 4
135. C. You, M. Bhagawati, A. Brecht, J. Piehler, Affinity capturing for targeting proteins into micro and nanostructures. Anal. Bioanal. Chem. **393**(6), 1563–1570 (2009)
136. F. Schreiber, Self-assembled monolayers: from ‘simple’ model systems to biofunctionalized interfaces. J. Phys.: Condens. Matter. **16**(28), R881 (2004)
137. S. Onclin, B.J. Ravoo, D.N. Reinhoudt, Engineering silicon oxide surfaces using self-assembled monolayers. Angew. Chem. Int. Ed. **44**(39), 6282–6304 (2005)
138. C.S. Fadley, R.J. Baird, W. Siekhaus, T. Novakov, S.Å.L. Bergström, Surface analysis and angular distributions in x-ray photoelectron spectroscopy. J. Electron Spectrosc. Relat. Phenom. **4**(2), 93–137 (1974)
139. D. Briggs, J.T. Grant, (eds.), *Surface Analysis by Auger and X-Ray Photoelectron Spectroscopy* (SurfaceSpectra Ltd. and IM Publications, Manchester, 2003)
140. J. Als-Nielsen, D. Jacquemain, K. Kjaer, F. Leveiller, M. Lahav, L. Leiserowitz, Principles and applications of grazing incidence X-ray and neutron scattering from ordered molecular monolayers at the air-water interface. Phys. Rep. **246**(5), 251–313 (1994)
141. R.J. Good, Contact angle, wetting, and adhesion: a critical review. J. Adhes. Sci. Technol. **6**, 1269–1302 (1992)
142. F.L. McCrackin, E. Passaglia, R.R. Stromberg, H.L. Steinberg, Measurement of the thickness and refractive index of very thin films and the optical properties of surfaces by ellipsometry. J. Res. Natl. Bureau Stand. A.: Phys. Chem. **67A**(4), 363–377 (1963)
143. G.-Y. Liu, S. Xu, Y. Qian, Nanofabrication of self-assembled monolayers using scanning probe lithography. Acc. Chem. Res. **33**(7), 457–466 (2000)
144. M. Wendel, S. Kuhn, H. Lorenz, J.P. Kotthaus, M. Holland, Nanolithography with an atomic force microscope for integrated fabrication of quantum electronic devices. Appl. Phys. Lett. **65**(14), 1775–1777 (1994)
145. X.-D. Xiao, G.-Y. Liu, D.H. Charych, M. Salmeron, Preparation, structure, and mechanical stability of alkylsilane monolayers on mica. Langmuir. **11**(5), 1600–1604 (1995)
146. W. Kiridena, V. Jain, P.K. Kuo, G.-Y. Liu, Nanometer-scale elasticity measurements on organic monolayers using scanning force microscopy. Surf. Interface Anal. **25**(6), 383–389 (1997)
147. R. Schlapak, P. Pammer, D. Armitage, R. Zhu, P. Hinterdorfer, M. Vaupel, T. Fruhwirth, S. Howorka, Glass surfaces grafted with high-density poly(ethylene glycol) as substrates for DNA oligonucleotide microarrays. Langmuir. **22**(1), 277–285 (2006)

148. R. Schlapak, D. Armitage, N. Saucedo-Zeni, M. Hohage, S. Howorka, Dense passivating poly(ethylene glycol) films on indium tin oxide substrates. Langmuir. **23**(20), 10244–10253 (2007)
149. M. Doi, *Introduction to Polymer Physics* (Clarendon Press, Oxford, 1996)
150. A.K. Kenworthy, K. Hristova, D. Needham, T.J. McIntosh, Range and magnitude of the steric pressure between bilayers containing phospholipids with covalently attached poly(ethylene glycol). Biophys. J. **68**(5), 1921–1936 (1995)
151. S. Rex, M.J. Zuckermann, M. Lafleur, J.R. Silvius, Experimental and Monte Carlo simulation studies of the thermodynamics of polyethyleneglycol chains grafted to lipid bilayers. Biophys. J. **75**(6), 2900–2914 (1998)
152. X.Y. Zhu, Y. Jun, D.R. Staarup, R.C. Major, S. Danielson, V. Boiadjiev, W.L. Gladfelter, B.C. Bunker, A. Guo, Grafting of high-density poly(ethylene glycol) monolayers on Si(111). Langmuir. **17**(25), 7798–7803 (2001)
153. M. Kaholek, W.-K. Lee, B. LaMattina, K.C. Caster, S. Zauscher, Fabrication of stimulus-responsive nanopatterned polymer brushes by scanning-probe lithography. Nano Lett. **4**(2), 373–376 (2004)
154. J.E. Headrick, M. Armstrong, J. Cratty, S. Hammond, B.A. Sheriff, C.L. Berrie, Nanoscale patterning of alkyl monolayers on silicon using the atomic force microscope. Langmuir. **21**(9), 4117–4122 (2005)
155. I.A. Banerjee, L. Yu, R.I. MacCuspie, H. Matsui, Thiolated peptide nanotube assembly as arrays on patterned Au substrates. Nano Lett. **4**(12), 2437–2440 (2004)
156. N. Nuraje, I.A. Banerjee, R.I. MacCuspie, L. Yu, H. Matsui, Biological bottom-up assembly of antibody nanotubes on patterned antigen arrays. J. Am. Chem. Soc. **126**(26), 8088–8089 (2004)
157. Z. Zhao, I.A. Banerjee, H. Matsui, Simultaneous targeted immobilization of anti-human IgG-coated nanotubes and anti-mouse IgG-coated nanotubes on the complementary antigen-patterned surfaces via biological molecular recognition. J. Am. Chem. Soc. **127**(25), 8930–8931 (2005)
158. S. Xu, G.-Y. Liu, Nanometer-scale fabrication by simultaneous nanoshaving and molecular self-assembly. Langmuir. **13**(2), 127–129 (1997)
159. S. Xu, S. Miller, P.E. Laibinis, G.-Y. Liu, Fabrication of nanometer scale patterns within self-assembled monolayers by nanografting. Langmuir. **15**(21), 7244–7251 (1999)
160. G.R. Luis, L. Jian, Atomic force microscope nanolithography: dip-pen, nanoshaving, nanografting, tapping mode, electrochemical and thermal nanolithography. J. Phys.: Condens. Matter. **21**(48), 483001 (2009)
161. J.-F. Liu, S. Cruchon-Dupeyrat, J.C. Garno, J. Frommer, G.-Y. Liu, Three-dimensional nanostructure construction via nanografting:? positive and negative pattern transfer. Nano Lett. **2**(9), 937–940 (2002)
162. M. Liu, N.A. Amro, G.-Y. Liu, Nanografting for surface physical chemistry. Annu. Rev. Phys. Chem. **59**(1), 367–386 (2008)
163. C. Staii, D.W. Wood, G. Scoles, Verification of biochemical activity for proteins nanografted on gold surfaces. J. Am. Chem. Soc. **130**(2), 640–646 (2007)
164. J. Liang, R. Luis, G. Scoles, Nanostructuring, imaging and molecular manipulation of dithiol monolayers on Au(111) surfaces by atomic force microscopy. J. Phys. Chem. C. **111**(46), 17275–17284 (2007)
165. J. Liang, G. Scoles, Nanografting of alkanethiols by tapping mode atomic force microscopy. Langmuir. **23**(11), 6142–6147 (2007)
166. Y. Hu, A. Das, M.H. Hecht, G. Scoles, Nanografting de novo proteins onto gold surfaces. Langmuir. **21**(20), 9103–9109 (2005)
167. M.A. Case, G.L. McLendon, Y. Hu, T.K. Vanderlick, G. Scoles, Using nanografting to achieve directed assembly of de novo designed metalloproteins on gold. Nano Lett. **3**(4), 425–429 (2002)
168. A. Tinazli, J. Tang, R. Valiokas, S. Picuric, S. Lata, J. Piehler, B. Liedberg, R. Tampé, High-affinity chelator thiols for switchable and oriented immobilization of histidine-tagged

proteins: a generic platform for protein chip technologies. Chem. – Eur. J. **11**(18), 5249–5259 (2005)
169. H.G. Hansma, J.H. Hoh, Biomolecular imaging with the atomic force microscope. Annu. Rev. Biophys. Biomol. Struct. **23**(1), 115–140 (1994)
170. B.L. Semler, *Molecular Biology of Picornaviruses* (American Society for Microbiology Press, Washington, DC, 2002), p. XIX, 502S
171. N. Verdaguer, I. Fita, M. Reithmayer, R. Moser, D. Blaas, X-ray structure of a minor group human rhinovirus bound to a fragment of its cellular receptor protein. Nat. Struct. Mol. Biol. **11**(5), 429–434 (2004)
172. A.C. Palmenberg, D. Spiro, R. Kuzmickas, S. Wang, A. Djikeng, J.A. Rathe, C.M. Fraser-Liggett, S.B. Liggett, Sequencing and analyses of all known human rhinovirus genomes reveal structure and evolution. Science. **324**(5923), 55–59 (2009)
173. C.R. Uncapher, C.M. Dewitt, R.J. Colonno, The major and minor group receptor families contain all but one human rhinovirus serotype. Virology. **180**(2), 814–817 (1991)
174. M. Vlasak, M. Roivainen, M. Reithmayer, I. Goesler, P. Laine, L. Snyers, T. Hovi, D. Blaas, The minor receptor group of human rhinovirus (HRV) includes HRV23 and HRV25, but the presence of a lysine in the VP1 HI loop is not sufficient for receptor binding. J. Virol. **79**(12), 7389–7395 (2005)
175. N.H. Olson, P.R. Kolatkar, M.A. Oliveira, R.H. Cheng, J.M. Greve, A. McClelland, T.S. Baker, M.G. Rossmann, Structure of a human rhinovirus complexed with its receptor molecule. Proc. Natl. Acad. Sci. USA. **90**(2), 507–511 (1993)
176. E. Neumann, R. Moser, L. Snyers, D. Blaas, E.A. Hewat, A cellular receptor of human rhinovirus type 2, the very-low-density lipoprotein receptor, binds to two neighboring proteins of the viral capsid. J. Virol. **77**(15), 8504–8511 (2003)
177. J. Wruss, D. Rünzler, C. Steiger, P. Chiba, G. Köhler, D. Blaas, Attachment of VLDL receptors to an icosahedral virus along the 5-fold symmetry axis: multiple binding modes evidenced by fluorescence correlation spectroscopy. Biochemistry. **46**(21), 6331–6339 (2007)
178. J. Wruss Jr., P.D. Pollheimer, I. Meindl, A. Reichel, K. Schulze, W. Schöfberger, J. Piehler, R. Tampé, D. Blaas, H.J. Gruber, Conformation of receptor adopted upon interaction with virus revealed by site-specific fluorescence quenchers and FRET analysis. J. Am. Chem. Soc. **131**(15), 5478–5482 (2009)
179. Q.-A. Jordi, K. Tuende, P. Joan, C. Oliviero, F. Ignasi, V. Nuria, B. Dieter, Minor group human rhinovirus–receptor interactions: Geometry of multimodular attachment and basis of recognition. FEBS Lett. **583**(1), 235–240 (2009)
180. T. Konecsni, L. Kremser, L. Snyers, C. Rankl, F. Kilár, E. Kenndler, D. Blaas, Twelve receptor molecules attach per viral particle of human rhinovirus serotype 2 via multiple modules. FEBS Lett. **568**(1–3), 99–104 (2004)
181. F. Kienberger, R. Moser, H. Schindler, D. Blaas, P. Hinterdorfer, Quasi-crystalline arrangement of human rhinovirus 2 on model cell membranes. Single Mol. **2**(2), 99–103 (2001)
182. E.W. Kubalek, S.F.J. Le Grice, P.O. Brown, Two-dimensional crystallization of histidine-tagged, HIV-1 reverse transcriptase promoted by a novel nickel-chelating lipid. J. Struct. Biol. **113**(2), 117–123 (1994)
183. K. Lonberg-Holm, N.M. Whiteley, Physical and metabolic requirements for early interaction of poliovirus and human rhinovirus with HeLa cells. J. Virol. **19**(3), 857–870 (1976)
184. S.K. Basu, J.L. Goldstein, M.S. Brown, Characterization of the low density lipoprotein receptor in membranes prepared from human fibroblasts. J. Biol. Chem. **253**(11), 3852–3856 (1978)
185. N. Verdaguer, D. Blaas, I. Fita, Structure of human rhinovirus serotype 2 (HRV2). J. Mol. Biol. **300**, 1179–1194 (2000)
186. E.A. Hewat, D. Blaas, Structure of a neutralizing antibody bound bivalently to human rhinovirus 2. EMBO J. **15**(7), 1515–1523 (1996)
187. G. Kuznetsov Yu, A.J. Malkin, T.A. Land, J.J. DeYoreo, A.P. Barba, J. Konnert, A. McPherson, Molecular resolution imaging of macromolecular crystals by atomic force microscopy. Biophys. J. **72**(5), 2357–2364 (1997)

188. Y.G. Kuznetsov, A.J. Malkin, R.W. Lucas, M. Plomp, A. McPherson, Imaging of viruses by atomic force microscopy. J. Gen. Virol. **82**(9), 2025–2034 (2001)
189. J. Noble Harvey, K. Lonberg Holm, Sequential steps in attachment of human rhinovirus type 2 to HeLa cells. J. Gen. Virol. **25**(1), 83–91 (1974)
190. L. Snyers, H. Zwickl, D. Blaas, Human rhinovirus type 2 is internalized by clathrin-mediated endocytosis. J. Virol. **77**(9), 5360–5369 (2003)
191. E. Prchla, E. Kuechler, D. Blaas, R. Fuchs, Uncoating of human rhinovirus serotype 2 from late endosomes. J. Virol. **68**(6), 3713–3723 (1994)
192. S. Katletz, C. Stroh, C. Rankl, U.M. Titulaer, P. Hinterdorfer, Force induced lysozyme – 1534 HyHEL5 antibody dissociation and its analysis by means of a cooperative binding model. Biophys. J. **99**(1), 323–332 (2010)
193. E. Evans, K. Ritchie, Dynamic strength of molecular adhesion bonds. Biophys. J. **72**(4), 1541–1555 (1997)
194. E. Evans, K. Ritchie, Strength of a weak bond connecting flexible polymer chains. **76**(5), 2439–2447 (1999)
195. P.M. Williams, Analytical descriptions of dynamic force spectroscopy: behaviour of multiple connections. Anal. Chim. Acta. **479**(1), 107–115 (2003)
196. M.I. Chang, P. Panorchan, T.M. Dobrowsky, Y. Tseng, D. Wirtz, Single-molecule analysis of human immunodeficiency virus type 1 gp120-receptor interactions in living cells. J. Virol. **79**(23), 14748–14755 (2005)
197. T.M. Dobrowsky, Y. Zhou, S.X. Sun, R.F. Siliciano, D. Wirtz, Monitoring early fusion dynamics of human immunodeficiency virus type 1 at single-molecule resolution. J. Virol. **82**(14), 7022–7033 (2008)
198. C. Rankl, F. Kienberger, L. Wildling, J. Wruss, H.J. Gruber, D. Blaas, P. Hinterdorfer, Multiple receptors involved in human rhinovirus attachment to live cells. Proc. Natl. Acad. Sci. USA. **105**(46), 17778–17783 (2008)
199. A. Ebner, L. Wildling, A.S. Kamruzzahan, C. Rankl, J. Wruss, C.D. Hahn, M. Holzl, R. Zhu, F. Kienberger, D. Blaas, P. Hinterdorfer, H.J. Gruber, A new, simple method for linking of antibodies to atomic force microscopy tips. Bioconj. Chem. **18**(4), 1176–1184 (2007)
200. F. Kienberger, C. Rankl, V. Pastushenko, R. Zhu, D. Blaas, P. Hinterdorfer, Visualization of single receptor molecules bound to human rhinovirus under physiological conditions. Structure. **13**(9), 1247–1253 (2005)
201. K. Lonberg-Holm, N.M. Whiteley, Physical and metabolic requirements for early interaction of poliovirus and human rhinovirus with HeLa cells. J. Virol. **19**(3), 857–870 (1976)
202. T.A. Sulchek, R.W. Friddle, K. Langry, E.Y. Lau, H. Albrecht, T.V. Ratto, S.J. Denardo, M.E. Colvin, A. Noy, Dynamic force spectroscopy of parallel individual Mucin1-antibody bonds. Proc. Natl. Acad. Sci. USA. **102**(46), 16638–16643 (2005)
203. W. Baumgartner, H.J. Gruber, P. Hinterdorfer, D. Drenckhahn, Affinity of trans-interacting VE-cadherin determined by atomic force microscopy. Single Mol. **1**(2), 119–122 (2000)
204. P.W. Atkins, *Physical Chemistry* (Oxford University Press, Oxford, 1998)
205. K.-B. Lee, E.-Y. Kim, C.A. Mirkin, S.M. Wolinsky, The use of nanoarrays for highly sensitive and selective detection of human immunodeficiency virus type 1 in plasma. Nano Lett. **4**(10), 1869–1872 (2004)
206. M.A. Cooper, F.N. Dultsev, T. Minson, V.P. Ostanin, C. Abell, D. Klenerman, Direct and sensitive detection of a human virus by rupture event scanning. Nat. Biotechnol. **19**(9), 833–837 (2001)
207. F. Patolsky, G. Zheng, O. Hayden, M. Lakadamyali, X. Zhuang, C.M. Lieber, Electrical detection of single viruses. Proc. Natl. Acad. Sci. USA. **101**(39), 14017–14022 (2004)
208. A.J.-C. Eun, L. Huang, F.-T. Chew, S. Fong-Yau Li, S.-M. Wong, Detection of two orchid viruses using quartz crystal microbalance-based DNA biosensors. Phytopathology. **92**(6), 654–658 (2002)
209. J.C. Smith, K.-B. Lee, Q. Wang, M.G. Finn, J.E. Johnson, M. Mrksich, C.A. Mirkin, Nanopatterning the chemospecific immobilization of cowpea mosaic virus capsid. Nano Lett. **3**(7), 883–886 (2003)

210. Y.G. Kuznetsov, J.G. Victoria, W.E. Robinson, Jr., A. McPherson, Atomic force microscopy investigation of human immunodeficiency virus (HIV) and HIV-infected lymphocytes. J. Virol. **77**(22), 11896–11909 (2003)
211. Y.F. Drygin, O.A. Bordunova, M.O. Gallyamov, I.V. Yaminsky, Atomic force microscopy examination of tobacco mosaic virus and virion RNA. FEBS Lett. **425**(2), 217–221 (1998)
212. M.R. Falvo, S. Washburn, R. Superfine, M. Finch, F.P. Brooks Jr, V. Chi, R.M. Taylor 2nd, Manipulation of individual viruses: friction and mechanical properties. Biophys. J. **72**(3), 1396–1403 (1997)
213. M.J. Dabrowski, J.P. Chen, H. Shi, W.-C. Chin, W.M. Atkins, Strategies for protein-based nanofabrication: Ni2+-NTA as a chemical mask to control biologically imposed symmetry. Chem. Biol. **5**(12), 689–697 (1998)
214. N.-P. Huang, J. Vörös, S.M. De Paul, M. Textor, N.D. Spencer, Biotin-derivatized poly (l-lysine)-g-poly(ethylene glycol): a novel polymeric interface for bioaffinity sensing. Langmuir, **18**(1), 220–230 (2002)
215. A. Breitwieser, S. Kupcu, S. Howorka, S. Weigert, C. Langer, K. Hoffmann-Sommergruber, O. Scheiner, U.B. Sleytr, M. Sara, 2-D protein crystals as an immobilization matrix for producing reaction zones in dipstick-style immunoassays. Biotechniques. **21**(5), 918–925 (1996)
216. X. Chen, R. Ferrigno, J. Yang, G.M. Whitesides, Redox properties of cytochrome c adsorbed on self-assembled monolayers: a probe for protein conformation and orientation. Langmuir. **18**(18), 7009–7015 (2002)
217. G.T. Hermanson, *Bioconjugate Techniques*, 2nd edn. (Academic, London, 2008)
218. M. Veiseh, M.H. Zareie, M. Zhang, Highly selective protein patterning on gold-silicon substrates for biosensor applications. Langmuir. **18**(17), 6671–6678 (2002)
219. H.E. Gaubert, W. Frey, Highly parallel fabrication of nanopatterned surfaces with nanoscale orthogonal biofunctionalization imprint lithography. Nanotechnology. **18**(13), 135101 (2007)
220. F.A. Zoller, C. Padeste, Y. Ekinci, H.H. Solak, A. Engel, Nanostructured substrates for high density protein arrays. Microelectron. Eng. **85**(5–6), 1370–1374
221. H. Nygren, Kinetics of antibody binding to surface-immobilized antigen. Analysis of data and an empiric model. Biophys. Chem. **52**(1), 45–50 (1994)

Chapter 22
Biomimetic Tailoring of the Surface Properties of Polymers at the Nanoscale: Medical Applications

Valeria Chiono, Emiliano Descrovi, Susanna Sartori, Piergiorgio Gentile, Mirko Ballarini, Fabrizio Giorgis, and Gianluca Ciardelli

Abstract New-generation biomaterials are information-rich and incorporate biologically active components derived from Nature. The chapter reviews current approaches for the realization of biomimetic coatings for tissue engineering applications, via functionalization with bioactive molecules such as native long chains of extracellular matrix (ECM) proteins as well as short peptide sequences derived from intact ECM proteins. Biomimetic materials provide cues for cell–matrix interactions, favouring cell adhesion, proliferation, spreading, and differentiation. The recently developed scanning probe techniques for optical and spectroscopic characterization of surfaces are also described, providing advanced topographical imaging and sensitive force measurements. The spectroscopic imaging of surfaces at the nanoscale provides insight into the function and structure of biomimetic molecules and represents a tool for tailoring the surface design.

22.1 Introduction

22.1.1 Biomimetic Material Design Criteria for Biomedical Applications

Tissue engineering, an emerging field in the area of human health care, combines biology and materials science and engineering to generate products with suitable biochemical and physical–chemical performance to repair or replace portions of whole tissues (i.e. bone, cartilage, blood vessels, bladder, etc.) [1]. One of the challenges in tissue engineering applications is to preserve cells normal physical activities on synthetic scaffolds and maintain tissue-specific function. Tissue engineering is aimed at the realization of artificial substrates (scaffolds) able to support cell adhesion, proliferation, migration, and differentiation through suitable bulk and surface physico-chemical and mechanical characteristics. Cells are supported in vivo by a three-dimensional extracellular matrix (ECM) consisting of nanofibres containing different chemical ligands that interact with cell–surface receptors.

The ECM is a complex mixture of matrix molecules, including the glycoproteins fibronectin, collagens, laminins, proteoglycans, and non-matrix proteins including growth factors. Thus, the ECM contains a variety of peptide epitopes that promote integrin-mediated cell adhesion and other receptor-linked functions. These epitopes are organized in a specific pattern at the nanometer scale within each protein molecule and at the micrometer scale in fibrillar and other structures [2]. Receptors transmit stress from the ECM through the membrane to the cytoskeleton within the cell in a dynamic and concerted manner [3]. The ECM may also sequester and present a number of morpho-regulatory molecules, such as growth factors, that control cell division, differentiation and multicellular morphogenesis [2,3]. Growth factors are released due to dissociation or enzymatic degradation, and then bind to cell–surface receptors and trigger signalling. As a consequence, natural ECM provides cell signalling molecules, including cell adhesion molecules and growth factors. Cells interact with their nanofibrous ECM via cell–surface proteins, such as integrins, to activate intracellular signalling pathways regulating cellular processes, e.g. cell shape, mobility, and proliferation. The three-dimensionality of ECM nanofibres is a key factor in regulating cell adhesion, proliferation, migration and differentiation: the molecular events associated with cells interacting with three-dimensional nanofibres are different to those on two-dimensional substrates such as Petri dishes. Cells also respond to a variety of chemical ligands [4] and mechanical cues in the ECM environment [5]. The interplay between cells and ECM is dynamic and complex, with physical and chemical properties of ECM eliciting different cellular responses. A biomimetic substrate is one which seeks to emulate the functions of native ECM by reproducing its three-dimensionality, nanofibrous topography and plethora of chemical motifs. Thus, the realization of advanced substrates in the field of tissue engineering requires an intensive investigation of the mechanisms guiding cell–matrix interactions in order to control biomaterial properties to elicit desirable cellular responses. Particularly, the surface of scaffolding materials should be properly designed in order to direct cell response and tissue regeneration.

Biomolecular recognition of materials by cells may be achieved by two major design strategies. One approach consists of the incorporation and release of soluble bioactive molecules, such as growth factors through the use of suitable biomaterial carriers. The local presence of growth factors plays an important role in controlling survival, migration, proliferation, and differentiation of the various cell types. However, growth factor therapy is a difficult task because of the high biological activity (in pico- to nanomolar range), pleiotrophic effects (acting on variety of targets), and short biological half-life (few minutes to hours) [6] of these protein drugs. Thus, growth factors should be administered locally to achieve an adequate therapeutic effect with little adverse reactions. The short biological half-life of growth factors demands for a delivery system that protects and slowly releases locally the protein over a prolonged period of time. In this review chapter, this biomolecular recognition strategy will not be treated. Many review articles are available on this topic [7–10].

The other approach deals with the incorporation of cell-binding peptides into biomaterials. Figure 22.1 shows a schematic representation of a biomaterial surface

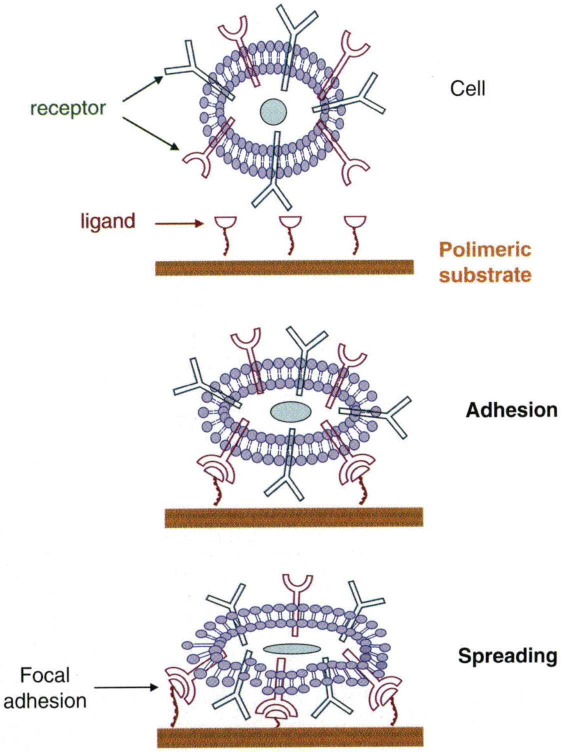

Fig. 22.1 Schematic representation of cell adhesion and spreading promoted by a biomaterial surface functionalized with cell adhesive ligands

functionalized with cell adhesion ligands, promoting initial cell attachment and further cell spreading. Focal adhesions are sites where integrins and proteoglycans mediate adhesion links to the actin cytoskeleton.

Cell-binding peptides include native long chains of ECM proteins (fibronectin, vitronectin, and laminin) and short peptide sequences derived from ECM proteins. The incorporation of peptide sequences may render the biomaterial degradable by specific protease enzymes or induce specific cellular responses. Native long chains of ECM proteins tend to be randomly folded upon adsorption to the biomaterial surface such that the receptor binding domains are not always sterically available. However, the short peptide sequences are relatively more stable during the modification process than long chain proteins such that nearly all modified peptides are available for cell binding. In addition, short peptide sequences can be massively synthesized in laboratories more economically. The most common cell-binding domain is the Arg-Gly-Asp (RGD) sequence [11]. Other non-RGD containing cell-binding domains exist, such as Arg-Lys-Arg-Leu-Glu-Val-Glu-Leu-Ser-Ile-Arg-Thr (RKRLQVQLSIRT) [12], Tyr-Ile-Gly-Ser-Arg (YIGSR) and Ile-Lys-Val-Ala-Val (IKVAV) in laminin [13]; Arg-Glu-Asp-Arg-Val (REDRV), Arg-Glu-Asp-Val (REDV), and Leu-Asp-Val (LDV) in fibronectin [14]; Asp-Gly-Glu-Ala (DGEA) in collagen I [15]. Whereas RGD is a promiscuous cell adhesion recognition site, different peptides are selective for specific cell lines:

for instance, the laminin-derived peptides and REDV sequences promote neurite outgrowth and endothelial cell adhesion and spreading, respectively. Moreover, the heparin-binding domain of adhesive ECM proteins, such as fibronectin, has been found to be a cofactor in promoting cell adhesion and spreading [16–18]. A more complete cell response was obtained by providing the cells with both the cell-binding and heparin-binding domains of fibronectin [17, 18]. Changes in the ratio between the RGD and the heparin-binding domains have been found to affect cell response and to enhance cell attachment and differentiation [19, 20]. Surface modification may be performed after scaffold production and has the advantage not to change the bulk physical–chemical and mechanical properties of the material. As a disadvantage, the functionalities are only available at the exposed surface. In this book chapter, approaches for the realization of biomimetic materials will be reviewed, including the spectroscopy and microscopy methods for their tailoring and characterization.

22.1.2 Techniques for the Characterization of Surfaces at the Nanoscale

In this section, the broad family of scanning probe-based characterization techniques (SPM) is introduced. More specifically, Atomic Force Microscopy (AFM), Scanning Near-field Optical Microscopy (SNOM) and Tip-Enhanced Raman Spectroscopy (TERS) are addressed. In addition, it might be helpful to provide a brief presentation of Confocal Laser Scanning Microscopy (CLSM), although it does not strictly belong to the SPM family. These techniques have been extensively used in a wide range of disciplines such as life science, solid-state physics, and materials science. The underlying working principle of characterization techniques based on a scanning probe relies on a serial interrogation of the sample over an area (or a volume, in the case of CLSM) raster scanned by the probe. Consequently, data collection is sequentially performed, in which the physical parameters involved in the interaction of the probe with the sample (e.g. light intensity, mechanical/electrostatic forces, spectra, current) are measured and stored in the memory of a computer controlling the overall scanning process. Then, data are arranged in a matrix form allowing data rendering in the form of false-colour images, in which any value of the measured parameter(s) is represented by a colour. Given the aforementioned general working principle, SPM techniques do use a material tip, while CLSM makes use of a focused laser light to probe the surface of the sample. With respect to the latter technique, it is worth to underline that CLSM allows an in-depth scan of transparent volumes of the sample, according to the so-called "optical slicing" mode. Furthermore, scanning techniques are spatially limited by the finite size of the probe interacting with the sample. As a consequence, the lateral resolution that can be attained by means of a SPM microscope is defined by the actual size of the very end of the tip, which ultimately interacts with the surface. Typical sizes of AFM-based tips are of the orders of few nanometres. On the other hand, both axial and lateral resolutions are limited by diffraction of light in the case of CLSM, while the diffraction limit is overcome in SNOM, perhaps attaining nanometric optical resolution. AFM has been invented in 1985 by Binnig, Quate and Gerber [21], exploiting a sharp tip scanning across the sample surface and monitoring the interactions probe-sample. The technological core of the AFM consists of a silicon-based beam, e.g. a cantilever, with an integrated tip at one end. The cantilever is typically made of silicon or silicon nitride and its size depends on the particular operating mode of AFM (e.g. length ranging from few microns to 500 μm and thickness from 0.5 to 5 μm). At the end of the cantilever, a sharp tip is hosted. When the probe is brought in close contact to the sample surface, the electrostatic forces between the very apex of the tip and the portion of the surface interacting with the tip make the cantilever to be deflected or otherwise mechanically perturbed. Once such a perturbation is quantitatively determined, a feedback loop guarantees the strength of the mechanical perturbation on the cantilever remains constant during the scan, thus moving up or down the sample (or the probe) in order to compensate changes in the interaction sample-probe. In this basic configuration, the tip is forced to preserve the same interaction strength between the sample and the tip, that ultimately leads to a map of the point-by-point vertical displacements of the piezo-stage holding the sample or the probe. As described above, the cantilever experiences a force from the sample surface. Several methods exist for detecting such a force acting on the cantilever. Although the very first AFM setup was based on the detection of a tunnelling current between a conductive tungsten tip and the backside of the cantilever [21], one of the most widely used relies on the amplification principle implemented by the so-called optical lever in the beam deflection method [22]. A laser beam is focused on the back of the cantilever (typically metal-coated for increasing reflectivity) and reflected onto a segmented photodetector. The photodetector can be segmented into four quadrants, thus allowing the possibility of monitoring the mechanical perturbations acting on the cantilever in terms of bending, torsion, amplitude, and phase of oscillation. Nevertheless, other methods for detecting tiny forces on the cantilever can be used, such as piezo-resistive or interferometric [23, 24]. AFM can work in both static and dynamic mode. In static mode, the probe experiences a repulsive, short-range force (Fig. 22.2) and the cantilever bends over. According to this operation mode, the cantilever bending is directly detected by measuring the vertical displacement of the laser reflected onto the four quadrant detector, that is ultimately proportional to the force acting on the tip. In order to detect very faint forces ($\sim$10–8 N), a low spring constant associated to the cantilever (typically, from 0.01 to 1 N/m) is needed for producing detectable displacement of the reflected spot.

When long-range attractive forces ($\sim$10–12 N) are aimed to be detected, the static operation mode is not sensitive enough to provide a reliable measurement of the cantilever deflection. For this reason, the cantilever is made vibrating and dynamic operation modes are used. Tiny amounts of piezoelectric material integrated within the probe holder are suitable for exciting the cantilever oscillation (dithering). In this case, variations in amplitude or phase of the cantilever oscillation are detected upon a long-range interaction tip/sample. Within the limit of small oscillations, the cantilever can be approximated by a harmonic oscillator having a

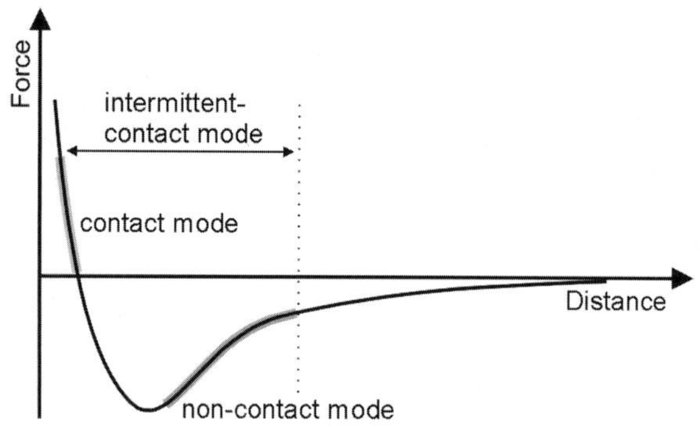

Fig. 22.2 Schematic of electrostatic forces acting on the AFM tip as a function of the tip/sample distance. Long-range forces probed in non-contact mode are attractive, while short-range forces probed in contact-mode are highly repulsive

resonance frequency $\omega_0 = \sqrt{k_0/m}$, where k_0 is the cantilever spring constant and m the mass of the oscillator. When immersed in a force field possessing a vertical gradient, the resonance frequency is shifted according to the following formula $\omega = \sqrt{(k_0 - (\partial F/\partial z))/m}$ where F represents the force field and z the distance from the sample surface. Depending on the force gradient sign, the resonance frequency can be shifted towards higher or lower values, thus allowing to probe and discriminate attractive forces having positive or negative gradients. In the former case, the AFM is operating in non-contact mode, whilst in the latter case, the AFM is operating in intermittent-contact mode, since the amplitude of oscillation of the cantilever basically equals the tip/surface distance of interaction.

The main advantage of the AFM dynamic operation modes is the fact that the sample surface is not in direct contact with the probe. Such a feature allows highly non-invasive topographic/morphological mapping of soft material surfaces while being not invasive at all. When optical characterization on a local basis is considered, SNOM is an invaluable tool at providing optical information with a lateral resolution of a few nanometers. Nowadays, a variety of different implementations of SNOM exists. Starting from the visionary work of Synge in 1929 [25] in which the basic idea of apertured SNOM was presented, the concept of a microscopy technique based on the local interaction of light mediated by a nanoprobe has been extended including, but not limited at, Scattering-type SNOM (s-SNOM) [26], collection mode SNOM [27], illumination-mode SNOM [28], PSTM (Photon Scanning Tunnelling Microscopy) [29], coupled to several detection methods ranging from interferometric homodyne [30], heterodyne [31, 32], to spectrometric (fluorescence [33, 34] and Raman [35]). A comprehensive description of SNOM-related techniques is outside the scope of this work. However, in the following, the two main operational modes of SNOM will be addressed, namely, collection/illumination mode with both aperture and apertureless probes (tapered optical fibres and microfabricated, cantilever-based tips) and scattering mode with AFM/STM dielectric or metallic tips. Strictly connected to the latter operational

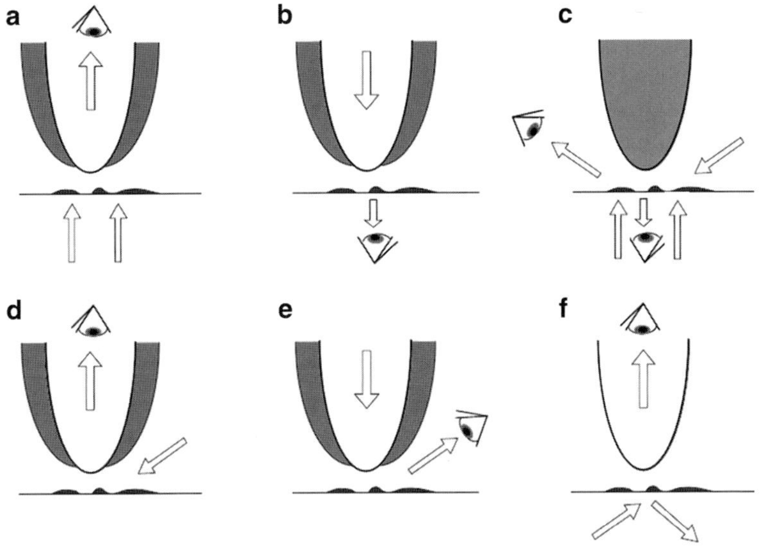

Fig. 22.3 Illustrative summary of the most widely used SNOM configurations: (**a**) collection mode, (**b**) illumination mode, (**c**) scattering mode, (**d**) oblique illumination, (**e**) oblique collection, (**f**) PSTM. Configuration (**c**) typically makes use of AFM/STM probes, eventually metal-coated, configuration (**f**) utilizes bare transparent probes, while configurations (**a**, **b**, **d**, **e**) make use of apertured metal-coated transparent probes

mode, the TERS microscopy will be also presented and briefly discussed. An illustrative summary of the most widely used SNOM configurations is provided in Fig. 22.3.

The original idea behind SNOM is to scan the sample surface by means of a nanosized aperture in an opaque screen. The aperture should be placed in close proximity to the surface, in such a way that the sample/probe interaction takes place in the near-field region of illuminating light. The difficulty of approaching the aperture to the sample surface can be circumvented by opening a nanometer-sized transparent aperture at the apex of a sharp tip. The probe/sample distance can be accurately feedback-controlled by exploiting the shear-force mechanism based on electrostatic perturbations of tip oscillations [36], thus allowing topographical data to be collected along the scan. In collection mode, the aperture acts as a local collector of light. The sample is illuminated either at normal incidence (Fig. 22.3a) or oblique incidence (Fig. 22.3d) and the light locally coupled within the probe through the nanoaperture is finally detected. An image is collected after a complete scan is completed. The lateral resolution of the image is ultimately defined by the size of the aperture.

In illumination mode, light injected into the transparent probe is guided till its apex and then it leaks out through the nanoaperture. Once brought close to the sample, the probe acts as a nanosource of light, illuminating the sample over small areas at each scanning step. The lateral size of the illuminating spot is defined by the diameter of the aperture. Light detection can be done in transmission (Fig. 22.3b) or in scattering configuration (Fig. 22.3e). In order to prevent light diffraction effects through the nanoaperture to be dominant, the aperture at the tip apex is maintained at the same nanometric distance to the surface with the help of the aforementioned feedbacked shear-force mechanism. Historically, the most widely used probes for SNOM are based on monomode optical fibres. After the cladding is removed, the transparent core of a glass fibre is made tapered, then metalized and patterned with an aperture at the apex. The metal coating prevents the light either to leak out of the fibre before reaching the tip end (in illumination mode) or to be coupled and transmitted within the fibre otherwise than the very apex (in collection mode). Two main methods are used for the fabrication of optical tapers. In the heating and pulling method, the optical fibre is locally heated and simultaneously stretched [37]. During the stretching, the fibre becomes thinner and thinner until it breaks up. By accurately controlling the overall process, it is possible to obtain probes with very small tip radii at low cost. Another method involves an hydrofluoric acid (HF) etching of monomode fibres, obtaining larger tip angles and larger optical power throughput in transmission [38]. The bare transparent fibre is then metal-coated. Typically, aluminium is used, because of its small skin depth (12.5 nm). In order to leave an optical aperture at the probe apex, the metal evaporation is made by rotating the tip maintained at a given angle with respect to the evaporation direction, in such a way that the apex remains uncovered because of a shadowing effect [39]. In alternative, Focused Ion Beam (FIB) milling can be used, thus providing much more regular shapes and controlled size of the aperture. It must be noted that in the PSTM configuration (Fig. 22.3f), the probe is not required to be metal-coated since the detected field leaking out of the sample are mostly evanescent. The major drawbacks of fibre-based SNOM are the fragility of the probes and the mechanical damping of the shear-force mechanism mainly due to the elasticity of the fibre. These two problems very often cause tip crashes against the sample surface. In addition, the aforementioned fabrication processes are serially done by hand, and are not suitable for a parallel batch implementation. Beside fibre-based probes, micromachined, cantilever-based probes are also available. In apertureless (or scattering mode) SNOM, an oscillating sharp probe is far-field illuminated, along with the sample [40, 41]. Light scattered by the tip is collected back, providing the local optical information suited. In order to distinguish the tiny amount of light scattered by the tip apex out of the huge amount of light power reflected and/or scattered by the sample, the tip is made vibrating in a direction perpendicular to the sample surface. Then, an efficient background subtraction is obtained by means of a lock-in-based detection tuned at the sampling frequency of the oscillating probe. In this configuration, no transparent probes are required. For this reason, the most widely used probes in scattering SNOM consist of AFM silicon (or silicon nitride) tips. Nevertheless, it might be worth to recall that a new kind of SNOM probe has been recently developed with the aim to obtain a cantilever-based transparent tip suitable for illumination/collection mode as well as for scattering mode [42, 43]. In addition of being batch microfabricated, this new probe design benefits of the AFM-based control of the probe/sample distance instead that the inefficient shear-force

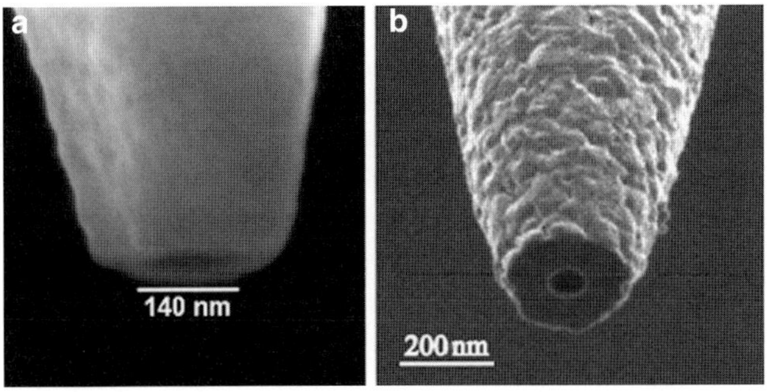

Fig. 22.4 Apertured apex of (**a**) a metal-coated microfabricated cantilever-based SNOM probe and (**b**) a metal-coated tapered monomode fibre

mechanism. In Fig. 22.4, an exemplary illustration of several SNOM probes is presented. A fruitful extension of the scattering mode SNOM is represented by the combination of near-field microscopy with Raman spectroscopy. As it is well known, Raman scattering is an anelastic interaction radiation-matter occurring with a very low cross-section. Accordingly, it is very difficult to perform Raman imaging by exploiting the few amount of light leaking out of a SNOM aperture and a Raman amplification effect would be desirable. Using Surface Enhanced Raman Scattering (SERS), Raman emission is greatly increased by exploiting a coupling mechanism between Raman emission and Localized Surface Plasmons (LSP) on metallic (silver or gold) nanostructures. In SERS, Raman signal can benefit of amplification factors up to 15 orders of magnitude [44]. However, SERS is not intended to be an imaging technique. Rather, it is possible to produce a strong (plasmonic) field enhancement at the end of sharp metallic tips raster scanning the sample surface. By employing suitable illumination conditions, LSP are coupled at the tip apex. Using this technique, called TERS [45, 46] Raman emission is locally amplified during the scan, and detected point-by-point according to an arrangement similar to Fig. 22.3f. As a final remark, it is worth to recall that recent advances in probe engineering are demonstrating the possibility of producing an enhanced "hot spot" at the apex of microfabricated, fully metal-coated transparent SNOM probes by injecting radially polarized laser light from the back side of the cantilever hosting the tip [47].

22.2 Realization of Biomimetic Surfaces by Coating Strategies

22.2.1 Generalities

Biomaterial surface plays an important role in biological systems due to the biological reactions occurring at material/biological environment interface. The understanding of the interaction between the material surface and biological system

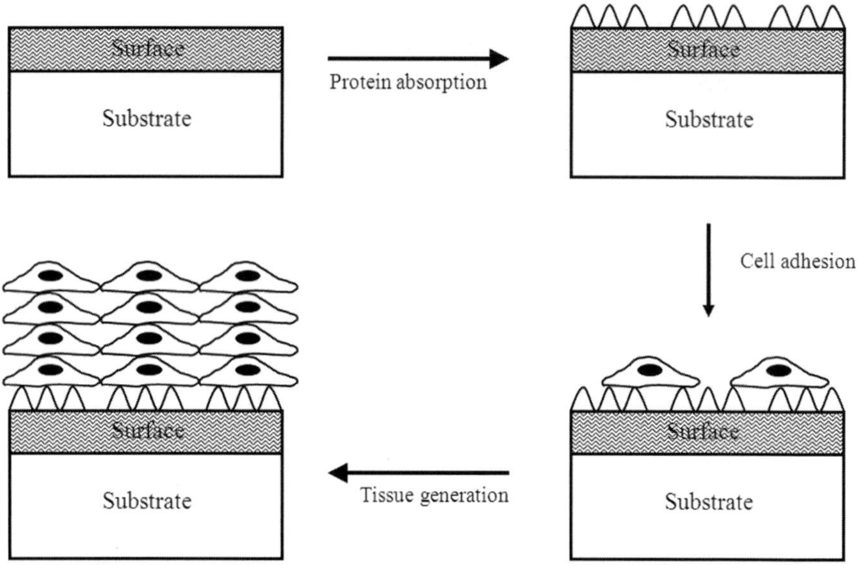

Fig. 22.5 Schematic representation of the biological events evoked by the surface properties of the biomaterials implanted in the host

is an important requirement for the development of biomaterial interfaces with tailored properties to the application. When a material is implanted in vivo, the surface chemical composition, energy, roughness and topography determine the reactions occurring at the interface between the host and biomaterial surfaces [48]. These reactions include initial water adsorption, subsequent bio-molecule (mainly proteins) adsorption and later cell attachment to the material surface [49] (Fig. 22.5).

The surface properties of biomaterials determine the binding strength and structural arrangement of water molecules, which may affect later protein–surface interactions [50]. The interaction between the surface water shell and the biomolecular hydration water influences the fundamental kinetic processes and the thermodynamics at the interface. Thus, the composition of the adsorbed layer is a key mediator to cell behaviour on the biomaterial surface. For instance, it may determine if proteins denature or not, their orientation and degree of coverage. As a consequence, cell–surface interaction is mediated by interaction between cells and surface bound proteins. The orientation of the adsorbed proteins can stimulate a constructive cell response, favouring wound repair and tissue regeneration, while the proteins in an unrecognizable state may indicate that the foreign materials should be removed or isolated [49].

In living tissues, cells adhere to and interact with their extracellular environment via specialized cell–cell and cell–ECM contacts [51], therefore the maintenance of tissue-specific function of artificial tissues depends on cell/scaffold and cell/cell interactions [52]. There are two major categories of cell–biomaterials interactions: specific and unspecific. Specific interactions are controlled by the adsorbed protein layer containing specific ligands for certain cell types [49], whereas unspecific interactions are usually depending on general surface properties and affect the adhesion of multiple cell types. For instance, unspecific adhesion to material surfaces is mediated by common cell characteristics including cell–surface properties, such as the negative charge of cell membrane, ubiquitous lipophilic membrane proteins, or lipophilic proteins of ECMs [53]. In Tissue Engineering, cell adhesion to implant surface anticipates other biological events such as cell spreading, migration and differentiation and cell function. As many conventional materials do not possess the adequate surface characteristics to be used as scaffolds for tissue engineering, surface modification methods may be applied to properly direct biological interactions without affecting bulk properties. Surface modification may tailor the substrate surface morphological, chemical and biological characteristics and may also be applied to provide accessible chemical functional groups for the immobilization of biologically active species.

In the past, research has been addressed to understand the effects of surface chemistry and topography on the adsorption of ECM proteins and the presentation of cell adhesion ligands [54]. However, it is not possible to develop a biomaterial surface able to adsorb a precise mixture and arrangement of ECM proteins; moreover, the initially adsorbed protein molecules may be subsequently denatured or replaced. The aim of current research is the realization of biomaterial surfaces which resist unspecific protein adsorption and are functionalized with molecules regulating host biology in a controlled manner [55]. Biomimetic surfaces may be designed taking inspiration from the structure-function analysis of the ECM components that cells use to organize themselves into tissues. The structural and physico-chemical characteristics of natural ECM are a model for the development of advanced biomaterials mimicking ECM key features: the presence of adhesion molecules and the release of growth factors.

Coating of a surface with biomimetic macromolecules is one of the simplest surface modification approaches. Various coating techniques have been applied including Langmuir–Blodgett (LB) deposition, self-assembled monolayers (SAMs), layer-by-layer (LbL) self-assembly, and others.

22.2.2 Coating Methods

22.2.2.1 Langmuir–Blodgett Films

Since its introduction in the 1930s by Langmuir and his colleague Blodgett [56], LB transfer has been developed as bottom-up technique for the surface modification of materials. This technology is based on the physical–chemical properties of organic molecules such as lipids, phospholipids, or glycolipids, to orient themselves at air/water interface, to minimize their free energy and form an insoluble monolayer (the Langmuir film). The classical materials forming monolayers are insoluble amphiphiles, composed of two distinct molecular regions: a hydrophilic (water-loving) headgroup, easily soluble in water, and a hydrophobic (water-hating) tail, soluble in non-polar solvents. For biological components, the tail groups are most often composed of one or two long alkyl chains. In detail, drops of solutions of an amphiphilic molecule in a volatile and not-water miscible solvent are deposited and spread on pure water surface. As the organic solvent evaporates, a monolayer interfacial film forms having headgroups in contact with water and tail groups in contact with air. If the concentration of the amphiphilic molecules is low, they do not interact strongly with each other and the layer can be regarded as a *two-dimensional gas*.

If the available surface area of the monolayer is reduced by a compression system, the intermolecular distance decreases and the surface tension diminishes. The hydrocarbon chains of the molecules tend to exert a repulsive force on each other. The force exerted by the film for unit length is called surface pressure (π) and is equal to the reduction of surface tension due to the presence of the film ($\pi = \gamma_0 - \gamma$, with γ_0 being the surface tension of water and γ the surface tension of water in the presence of the monolayer). During the compression, the monolayer undergoes a self-organization process with several phase transformations (from gas to liquid to solid states with increasing the compression degree and decreasing the monolayer surface area), leading to an insoluble and organized monolayer at air/water interface. The behaviour of the π surface pressure as a function of the surface area per molecule has been reported in several specific books and reviews dedicated to LB films [57–59]. When the surface pressure is sufficiently high, the monolayer can be transferred on a solid support called substrate. The universally known LB technique is based on the vertical movement of a solid substrate through a monolayer/air interface (Fig. 22.6). The deposition will be obtained by raising or lowering the substrate through the interfacial film depending on the hydrophilic (Fig. 22.3) or hydrophobic nature of the solid substrate surface.

Multilayers may be obtained by successive deposition sequences, with a different structure called X-type, Y-type and Z-type (Fig. 22.7). In particular, the Y-type

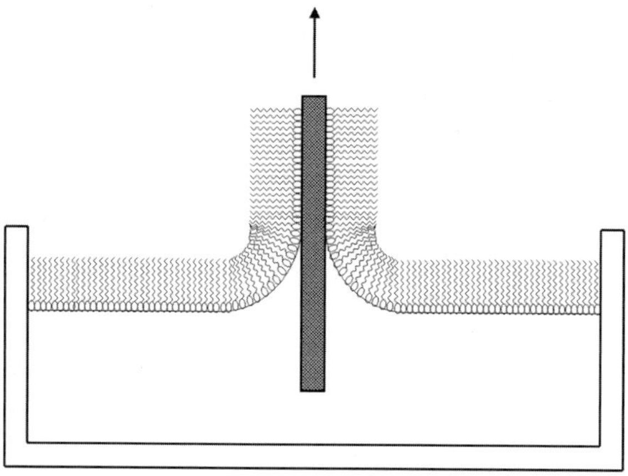

Fig. 22.6 Deposition of a floating monolayer on a solid hydrophilic substrate

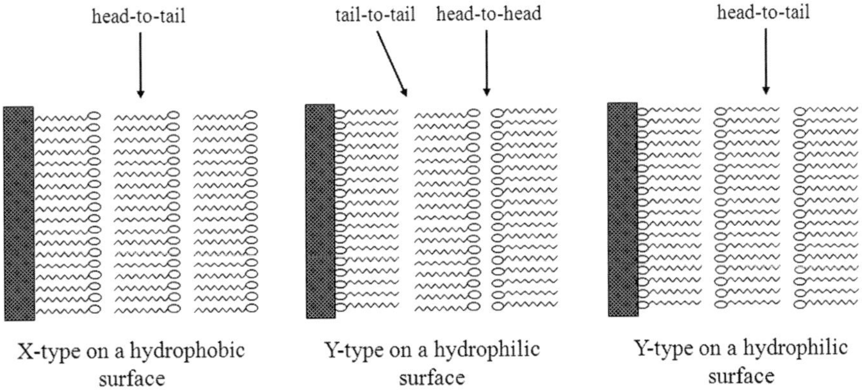

Fig. 22.7 Different types of deposited multilayer LB films

deposition leads to head-to-head and tail-to tail configurations which resemble the natural lamellar staking of the biological membranes, therefore Y-stacked LB films are candidate for developing models of natural membranes.

Briefly, the LB-technique enables: (1) the precise control of the monolayer thickness, (2) homogeneous deposition of the monolayer over large areas, and (3) the possibility to make multilayer structures with varying layer composition. An additional advantage of the LB technique is that monolayers can be deposited on almost any kind of solid substrate.

However, the application of LB transfer is mainly limited to scientific models at small scales and commercial success has not been achieved to date [60, 61]. The speed of this fabrication procedure is controlled by the transfer rate of the film from the water–air interface to the solid substrate, which needs to occur at a certain rate (usually relatively slow) to maintain film architecture. Also, the substrate must possess a regular shape (e.g. a planar solid substrate), which limits the ability to make LB architectures using economic process technology.

Typical applications of LB technique include the realization of biomimetic coatings from peptide amphiphiles. The well-known minimal cell recognition sequence Arg-Gly-Asp (RGD) has the ability to bind a variety of cells through ligand–receptor interactions. Although RGD peptide sequence may be immobilized on surfaces through chemical methods, the linking of synthetic lipid tails to the peptide sequence allows them to self-assemble into biomimetic films with organized structure and to assume well-defined secondary and tertiary structure conformations which promote cell adhesion [62–66]. Pakalns et al. have covalently linked synthetic RGD-containing peptides to dialkyl lipid tails to form novel amphiphiles, capable of self-assembling into stable LB monolayers with biologically active interfaces [67]. These LB monolayers, once deposited on a solid surface, function as biomimetic coatings presenting uniformly oriented peptide sequences to cells and having a flexible loop conformation [67].

In the contest of biomimetic studies, LB films have been assembled from phospholipids or glycolipids, which are essential components of biological membranes.

These molecules are complex lipids, containing two hydrocarbon chains and a polar head per molecule. Various LB films based on lipid bilayers, unilamellar/bilamellar lipids vesicles, detergent micelles, and other biologically important molecules and structures which possess appropriate amphipathic properties have been used as biomembranes [68]. The functionalization of LB biomimetic membranes by association or incorporation of macromolecules with specific biological activities (enzymes, antibodies, or specific ligands), the possibility to deposit LB films on any type of substrate and the structural organization (highly ordered molecular array) and ultrathin thickness of LB films open new perspectives in the field of nanobiotechnology.

For instance, new biospecific membranes made of organized ultrathin films at the nanoscale and containing biomolecules in a functionalized and orientated position, may be directly interfaced with a transducer to obtain miniaturized bio-electronic devices and other nanoanalytical tools.

In this context, the development of functional proteolipidic nanostructures corresponding to highly organized molecular assemblies with orientated biological recognition sites is of great interest. For instance, acetylcholinesterase (AChE) has been immobilized at the surface of a miniaturized biomimetic nanostructure, obtained by interfacial liposome fusion and LB technique [69]. By inserting a non-inhibitor monoclonal antibody in a functional position, this nanostructure is able to sequester AChE in a suitable orientation and to maintain the enzyme activity for several months. This molecular assembly has been intimately associated with a sensor [69].

In conclusion, LB technique is a powerful method to elaborate functionalized biomimetic membranes, preserving specific characteristics of the biological membranes such as fluidity and asymmetry, and allowing the orientation of functional macromolecules in the bilayer structure. Moreover, the association of LB biomembranes with proper transducers, respectively allowing recognition and signal transduction in a single device, opens new perspectives for the design of biomimetic nanosensors investigating the biological processes at the molecular level.

22.2.2.2 Self-Assembled Monolayers

SAMs technique was introduced by Nuzzo and Allara [70] and Whitesides and coworkers [71]. SAMs are ordered molecular assemblies that form spontaneously by the adsorption of a surfactant with a specific affinity of its headgroup to a specific substrate. Most studies of SAMs to date have been carried out on monolayers on gold surfaces because this system offers a combination of high structural order, flexibility in the structure of groups exposed at the extreme surface, and ease of preparation and analysis. Typically, thiols, sulfides, or disulfides are adsorbed on gold, as gold displays a strong specific interaction with sulphur.

To form an alkanethiol monolayer, a clean gold substrate, freed from organic impurities, is simply immersed into a dilute ($\sim 10^{-3}$ M) solution of the thiol molecule in an organic solvent at room temperature. The immersion time varies from a few minutes to several hours, or even days, depending on the system. The chemisorption of thiols (RSH) on gold implies the formation of a RS–Au bond.

SAM films are assembled in two distinct phases: a very fast step (adsorption of the molecules onto the substrate), which takes a few minutes, and a slower one (organization and structuration), which lasts several hours. The hydrocarbon chains are trans-extended and tilted approximately 25–30° from the normal to the surface, as a result of reestablishment of van der Waals contact between the chains in an assembly with ~5 Å S–S distance, larger than the distance of ~4.6 Å for perpendicular alkyl chains in a close packed layer [72].

SAMs provide well-defined structures and chemistries; spatially defined arrays of SAMs can be prepared by combining self-assembly with patterning methods such as micro-contact printing and photolithography [73]. In addition, SAMs can be used to immobilize peptides, proteins, and other biomolecules. Complex systems can be constructed by co-adsorbing several thiols with different terminal functional groups or different chain length. A mixed thiol/oligo-ethylene glycol thiol monolayer has been assembled on a gold surface [74]: a thiol molecule that contains the protein binding group (biotin) has been diluted with a thiol resisting protein binding (oligo-ethylene glycol). The streptavidin protein has then been bound to biotin: streptavidin has binding pockets on opposite sides of the molecule and it can be used as a linker to bind other biotinylated molecules (Fig. 22.8).

The ability to control interfacial processes has important implications from the point of view of both fundamental and technological advances.

As compared to LB films, SAMs based on organothiolates offer an essential advantage as the monolayer is more strongly bound to the substrate. In the case of LB films, the grafting is achieved essentially by electrostatic interactions, whereas

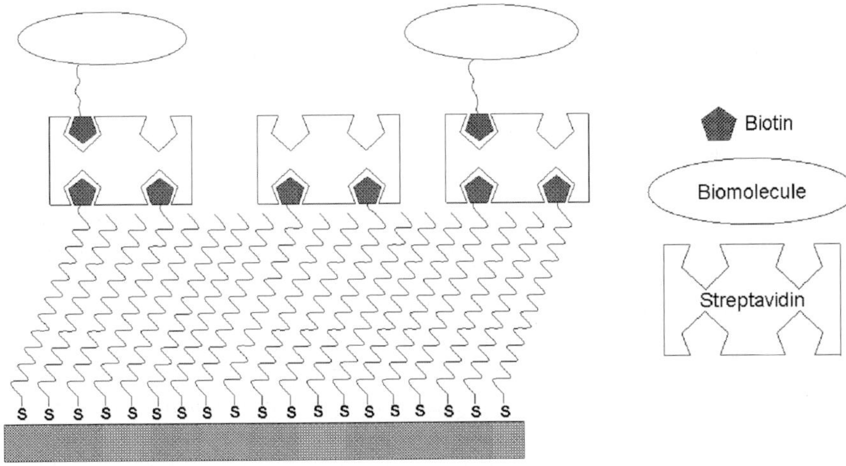

Fig. 22.8 Schematic representation evidencing how biomolecules functionalized with biotin groups can be selectively immobilized onto a gold surface using a streptavidin linker bound to a mixed biotinylated thiol/ethylene glycol thiol self-assembled monolayer

the monomers that form a SAM are anchored to the substrate through stable, covalent Au-thiolate bonds.

General disadvantages of SAM are the low loading of biological components due to the monolayer nature, the limited number of substrate types (mainly, noble metal surfaces adsorbing thiols and silica surfaces adsorbing silanes) and the limited stability of the films under physiological conditions.

22.2.2.3 Layer-by-Layer Coating

An alternative of LB deposition and SAMs is LbL assembly, introduced by Decher and Hong for preparing structured controlled thin films for biological applications [75]. The LbL process, commonly driven by electrostatic interactions, involves sequentially dipping of charged substrate into solutions of oppositely charged polyelectrolytes and allowing the polymer to absorb and reverse the charge of the substrate surface. A rinsing step is included between one adsorption process and another, to remove excess and to prevent cross-contamination of the polyelectrolyte solutions. The LbL technique is versatile and offers the benefit of solvent-free processing as well as the ability to coat all available surfaces of virtually any material with uniform ultrathin films of precisely controlled thickness. By this cyclic deposition, it is possible to obtain multilayered films with the desired structure (Fig. 22.9).

The thickness of a single bilayer is generally below 1 nm, therefore LbL allows a control of film/coating properties at the nanoscale [60]. LbL is not only applicable to polyelectrolyte/polyelectrolyte systems: any kind of charged species can be used to prepare LbL films, including nucleic acids, DNA, nanoparticles, nanowires, polypeptides, etc. [75]. LbL assembly can be driven by electrostatic attractions or other forces, such as hydrogen bonding, charge transfer, covalent bonding, biological recognition, and hydrophobic interactions [75]. Charged organic materials are typically used as substrates to deposit LbL films; however, in some case hydrophobic surfaces have also been applied for LbL assembly driven by hydrophobic interactions [76]. Due to the wide spectrum of available substrates and deposition materials as well as the variety of the assembly methods, LbL provides a versatile platform for the design of advanced nanostructured materials at the molecular level. Moreover, multiple interactions in the multilayers may increase their stability in physiological conditions.

LbL coatings/films are advantageous over SAMs as they allow a higher loading of biological components due to their multilayered nature. Moreover, LbL technique is much faster and simpler than LB method and results in much more stable films.

LbL technique offers an excellent platform for the development of biomimetic films/coatings. The simplicity of coating with LbL films on geometrically complex substrates provides a method for the control of protein and cell adhesion on biomaterial substrates. Proteins have been found to interact with LbL substrates by electrostatic forces, hydrophobic interactions, hydrophilic repulsion, and hydrogen bonds [77]. In detail, when the outside charge of LbL coating is the same as

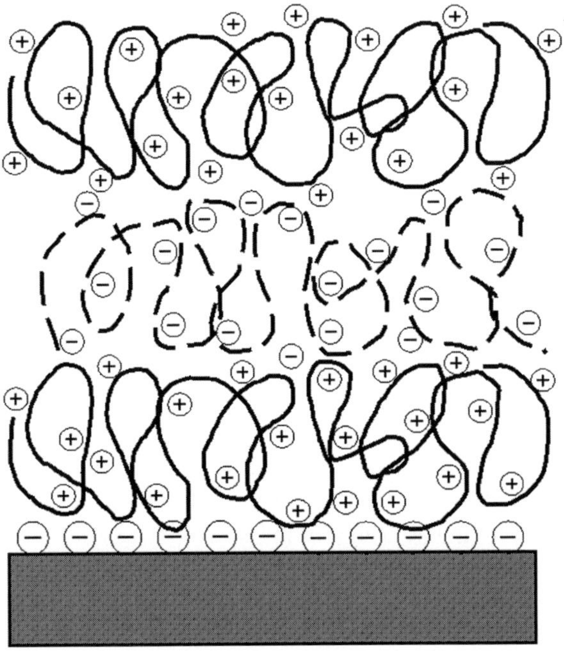

Fig. 22.9 Schematic representation of LbL coating on a negatively charged substrate (three deposited layers)

the protein of interest, a thin protein layer is adsorbed by hydrogen bonding and hydrophobic interactions [77]. On the other hand, when the charges are opposite, electrostatic interactions dominate leading to the deposition of a thick protein layer on the multilayer. The interactions between LbL coating and proteins may also cause a change in the protein secondary structure, depending on the degree of protein/multilayer interaction [77].

The ease of control of multilayer structure allows to direct and manipulate cell adhesion, differentiation, proliferation, and function. Cell adhesion and spreading on LbL films/coatings may be modulated by multilayer film biofunctionality, hydration, charge, and rigidity. In this context, Berg et al. have chemically grafted the RGD amino acid sequence to the terminal layer of a LbL film and have found an increased murine fibroblast adhesion with increasing RGD density [78]. On the other hand, Richert et al. have improved chondrosarcoma and smooth muscle cell adhesion on LbL films by increasing the substrate rigidity through cross-linking [79, 80]. Mendelsohn et al. have found that the adhesion of murine fibroblasts on LbL films may be increased with increasing the degree of ionic cross-linking due to a reduced hydration degree [81]. Hydrophobicity has been found to be an additional factor to control cell attachment: Salloum et al. have investigated the influence of the multilayer hydrophobicity as controlled through layer chemistry on the adhesion and spreading of smooth muscle cells. In this work, the weakest adhesion and highest cellular mobility have been reported for zwitterionic films [82]. Witter et al. assembled LbL films from biological polymers (poly(L-lysine)/dextran sulphate, PLL/DS) and evaluated the adsorption of fibronectin (Fn) on the multilayers as a function of the layer number and charge of the terminal layer and the time-dependent morphological changes of a model cell system (human umbilical vein endothelial cells, HUVEC) as a function of the layer number, charge of the terminal layer and amount of Fn [83]. Fn is an ECM protein containing the specific RGD amino acid sequence, which modulates cell adhesion and spreading on substrates. In the reported work, Fn has been found to adsorb preferentially on PLL-terminated layers, due to their positive charge and low hydration degree and to remain principally at the film surface. HUVEC spreading increased and their circularity decreased in the presence of Fn; Fn-coated PLL-terminated multilayers provided the best substrates for HUVEC spreading.

As an alternative to the functionalization of the outer layer, some authors have grafted peptide sequences on the polyelectrolytes for LbL assembly. For instance, elastin-like polypeptides with $(VPGXG)_n$ peptide sequence have been conjugated to polyethyleneimine and polyacrylic acid polyelectrolytes for further LbL assembly and modulation of in vitro cell adhesion [84].

Polyelectrolyte multilayers have also been assembled on electrospun nanofibres [85] as substrates mimicking the structural features of ECM matrix, a highly hydrated network hosting three major components: fibrous elements (e.g. collagen, elastin, and reticulin), space filling molecules (e.g. glycosaminoglycans covalently linked in the form of proteoglycans), and adhesive glycoproteins (e.g. fibronectin, vitronectin, and laminin).

LbL method also allows the incorporation of biologically active molecules such as bioactive drugs [86], specific nucleic acids [87, 88], and growth factors [89], through chemical bonding, adsorption or embedding in polyelectrolyte films without loss of biological activity.

In conclusion, the extreme versatility of LbL technique makes it fundamental for the incorporation of the achievements of nanotechnology in biomaterials and biomedical devices.

22.2.2.4 Surface Biomineralization

Natural bone is an inorganic/organic composite material consisting of collagen and mineral (apatites), leading to an excellent balance of strength and toughness. Composite scaffolds have been frequently produced to mimic the natural bone structure, based on hydroxyapatite or calcium phosphate (providing osteoconductivity) and a polymer phase (providing the continuous structure and desired flexibility to achieve high porosity and high surface area) [90]. For instance, composite bilayered films based on gelatine and hydroxyapatite/bioactive glasses with a biomimetic composition (30 wt% gelatine) [91] and three-dimensional porous hydroxyapatite/collagen composites with a random pore structure obtained by freeze-drying and cross-linked by microbial transglutaminase (mTGase) have been produced and characterized [92].

In some cases, synthetic biomaterials mimic nature not by influencing cells directly through receptor-binding epitopes but indirectly, for instance, by initiating and regulating the formation of bioactive inorganic structures (for example, mineralized bone or shell). Biomimetic coating strategies have been frequently applied to confer osteoinductive and osteoconductive properties to scaffolds aimed at bone repair by the release of growth factors and the application of a calcium phosphate layer, respectively [93]. The biomimetic coating technique involves the nucleation and growth of bone-like crystals on a substrate by immersing it in a supersaturated solution of calcium phosphate under physiological conditions of temperature (37°C) and pH (7.4). The method, originally developed by Kokubo [94], has been refined by several groups of investigators [95–101]. It is simple to perform, cost-effective and can be applied even to heat-sensitive, non-conductive, and porous devices of large dimensions and with complex surface geometries. The apatite generated via the biomimetic process in SBF is a partially carbonated hydroxyapatite, more similar to the natural bone apatite (calcium deficient $Ca/P{\sim}1.5$) than the stoichiometric hydroxyapatite crystals ($Ca/P = 1.67$). For instance, titanium alloy implants bearing a fine, dense, amorphous layer of calcium phosphate (to promote crystal growth) has been immersed in a supersaturated solution of calcium phosphate containing growth factors such as BMP-2 for 48 h under physiological conditions, to promote the coating with a crystalline latticework of the inorganic components into which BMP-2 had been incorporated [93]. The growth of apatite crystals is significantly affected by the substrate type (material composition and porosity), the ionic concentration of the SBF, and the pH value [102].

A highly innovative biomimetic method for the photo-activated deposition of apatite crystals on substrates containing photosensitive polymers (photozymes) have been recently developed [103]. Photozymes are amphiphilic photoactive polymers, having fluorophores such as xanthene dyes (fluorescein, rose bengal, eosin, eritrosin), naphthalene, carbazole, and hydrophilic units, containing strong electrolytes such as sulphonate groups or weak electrolytes such as the primary amino groups of chitosan chains [104–106]. Photozymes may be statistical, block, or graft copolymers and they are generally water soluble. Aromatic chromophores undergo strong interactions, which induce the coiling of the macromolecules and the formation of hydrophobic domains surrounded by charged fragments of the chains. The hydrophobic packed domains may solubilize hydrophobic guest molecules and play the role of nanoreactors with photochemical activity [106]. Among photozymes, chitosan-fluorescein (CH-FL) has been demonstrated to be photo-bioactive: the classical biomimetic approach has been found to be successful only by proper excitation of its chromophoric groups [103]. Figure 22.10 shows the morphological characteristics of the surface of a CH-FL cast film after incubation in SBF under irradiation. The same film incubated in SBF under dark has been found not to be covered with apatite crystals.

On the basis of this innovative approach, a not bioactive substrate of interest may be coated with photozymes by a conventional technique such as LbL, and then immersed in SBF under appropriate irradiation exciting the chromophoric groups

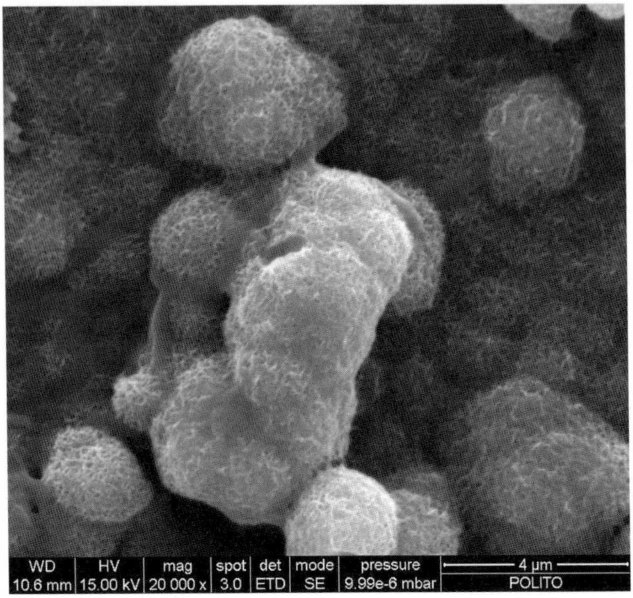

Fig. 22.10 Scanning electron microscopy image of the surface of CH-FL cast films after 7d incubation in SBF under irradiation

of the photozymes. LbL coatings based on photozymes have been successfully obtained [107]. Nanostructured multilayered photoactive films have been produced on cross-linked gelatin substrates via the LbL method, using alternative layers of a stat-copolymer photozyme (forming pseudomicellar structures) and a graft-copolymer photozyme (forming compact layers) [107]. These innovative nanostructured materials may have important applications as multifunctional coatings for scaffolds for bone repair: the photosensitizing properties of the different types of photozymes are combined, leading to osteointegration by hydroxyapatite precipitation induced by the graft-copolymer photozyme and anti-microbial action promoted by the stat-copolymer photozyme.

22.3 Realization of Biomimetic Surfaces by Chemical Modification

It is well known that both very hydrophilic and hydrophobic surfaces are unfavourable to cell attachment, whereas surfaces with moderate wettability are able to adsorb a proper amount of proteins, and can preserve their natural conformation, resulting in a positive cell response [108]. The relationship between wettability and cell attachment differs for different cell lines and cell culture conditions [109] but it can be generally stated that cell adhesion is maximized on surfaces with contact angles ranging from 60° to 80°. Most synthetic biodegradable polymers are hydrophobic, therefore several studies have been carried out in order to increase their surface hydrophilicity. One convenient method is plasma treatment, which can introduce hydroxyl, carboxyl, amino, and sulfate groups on polymer surfaces using different carrier reactive gases such as air, NH_3, SO_2, and CO_2. Another widely used method is the functionalization and coating with hydrophilic layers through grafting polymerization (Fig. 22.11). The method involves the introduction of initiators (radicals or peroxide groups) on the polymer surface by plasma treatment, ozone oxidation, γ-ray or UV irradiation, electron beam, and laser treatment [110–112]. As stated in the previous section, polymer surfaces may be functionalized with biomimetic molecules to make them more conducive to cell attachment and spreading. Biomimetic molecules are generally ECM components (polysaccharides, proteins or their fragment [3, 113]). Due to the inert nature of most of commercial polymers, a proper surface functionalization has to be performed before the covalent bonding of bioactive molecules. In the following sections, various chemical approaches for the realization of biomimetic surfaces are treated.

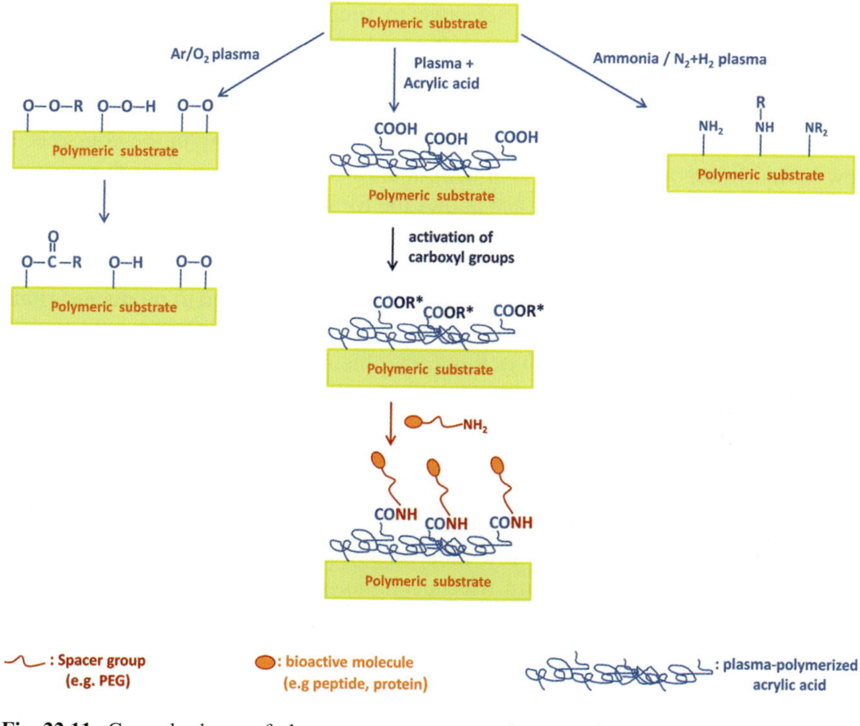

Fig. 22.11 General scheme of plasma treatments on a polymer surface and biomimetic functionalization

22.3.1 *Introduction of Functional Groups on Polymer Surfaces by Irradiation and Chemical Techniques*

22.3.1.1 Plasma-Surface Modification of Polymers

Plasma can be regarded as the fourth state of matter and it is composed of highly excited atomic, molecular, ionic, and radical species. It is typically obtained when gases are excited into energetic states by radio frequency, microwave, or electrons from a hot filament discharge [114]. Plasma surface modification involves the interaction of the plasma-generated excited species with a polymeric interface. The plasma process results in a physical and/or chemical modification of the first few molecular layers of the polymer surface. The effectiveness of the treatment is determined by the plasma source gases, the configuration of the plasma system and the plasma operating parameters.

Plasma surface activation employs gases, such as argon, oxygen, nitrogen, hydrogen, and ammonia. When these gases are exposed to the plasma, they dissociate and react with the surface, creating different chemical functional groups on the surface (Fig. 22.11).

Surface modification is often sensitive to time and environmental exposure, where the surface may lose its plasma-induced physical and chemical properties. For instance, hydrophilized surfaces do not remain hydrophilic for a long time due to mobile, low-molecular weight monomers that migrate from the bulk to the air–surface interface. This makes the surface revert to hydrophobic in the course of a few hours. In argon/air plasma treatment, the transfer of energy from the reactive species towards the polymer surface allows the formation of radicals on the macromolecule surface, followed by the formation of unstable hydroperoxides through rapid binding with oxygen, when samples are exposed to air afterwards. In a subsequent phase, the unstable hydroperoxides are easily decomposed to produce various oxygen-based polar functionalities (hydroxyl groups, esters, ketones, aldehydes, carboxylic acids, carboxylic esters). Similar problems have been also reported when using ammonia or nitrogen plasma treatment. Ammonia has been extensively used for the amination of surfaces. Alternatively, a mixture of hydrogen and nitrogen can be used for the same purpose. However, it has been shown that the treated surfaces often undergo rapid restructuring during the post-plasma period, with a significant degradation of the introduced functionalities [115]. In order to increase the stability of plasma-treated materials, the low-molecular weight oxidation products on surface can be removed by washing with alcoholic solution [116]. Otherwise some investigators have suggested that storage under water or alcohol reduces the rate of hydrophobicity recovery [117]. One additional difficulty in the case of the use of ammonia gas is that primary as well as secondary and tertiary amines can form on the surface. However, only primary amines can be useful for the attachment of proteins or other biomolecules to the modified surface [118]. To minimize the rearrangement and/or to functionalize the surface with a specific molecule, the generated functional groups can be reacted to form covalent linkages.

22.3.1.2 Plasma-Grafting Polymerization

In plasma-grafting polymerization, polymers are first exposed to the plasma to create radicals on the surface in the polymer chains. Afterwards, the polymers are exposed to a vapour or an aqueous or organic solution of the monomer (e.g. acrylic acid, allylamine) and pre-formed radicals initiate polymerization reactions. Radicals are produced only close to the surface of the polymers, which means that the modification is restricted to a depth of a few nanometers. Plasma polymerization is chemically different from conventional radical polymerization: the resulting polymer layers have different chemical composition and physicochemical properties. In plasma environment, the polymerization process involves the activation of monomers to radicals, recombination of the formed radicals, and reactivation of the recombined molecules, resulting in the formation of a grafted brush layer, containing cross-linked, fragmented, and rearranged units from the monomers. Generally plasma polymers show a high elastic modulus and do not exhibit a distinct glass transition temperature due to their partially cross-linked nature [119].

Plasma-grafting polymerization is often employed to alter the surface hydrophilicity of polymers or to introduce functional groups, in order to subsequently immobilize bioactive molecules. A commonly used approach is to graft poly(acrylic acid) or poly(methacrylic acid) to introduce carboxyl groups, followed by the activation of the carboxyl groups with carbodiimide; finally, activated carboxyl groups react with the amino groups of the target proteins or peptides [120] (Fig. 22.11). Alternatively, amino groups can be yielded by grafting allylamine, and subsequently they can react with activated carboxylic groups of bioactive molecules. It has been reported that plasma-polymerized allylamine may be applied for the immobilization of biomolecules such as polysaccharides [121] and DNA [122]. Several researches report that the grafted layer is not stable in aqueous environment because of the dissociation of low-molecular weight materials from the deposited cross-linked coating [112]. In biomedical applications, these dissolved small molecules may induce inflammatory reaction and initiate undesirable side effects, leading to the failure of a biomedical device. Therefore, it is of fundamental importance that the plasma-polymerized film is made stable by specific post-plasma treatments.

22.3.1.3 UV Irradiation

UV irradiation of polymer surfaces generates reactive sites which can be converted to functional groups upon exposure to gas, or can be exploited to initiate polymerization (UV-induced graft polymerization). Differently from other techniques, such as ionized gas treatments, in UV irradiation the depth of surface reactivity can be tailored selecting the wavelength and thus the absorption coefficient. Photo-induced grafting is a useful technique to achieve permanent alteration of the surface chemistry without affecting the bulk polymer, and it is performed with low cost of operation and mild reaction conditions. UV irradiation has been used to introduce carboxylic acid functionalities to poly(methyl methacrylate) (PMMA) [123], to activate polystyrene surfaces for protein immobilization and tissue engineering applications [124]. UV irradiation has also been used to initiate radical graft polymerization of bioactive compounds. For instance, *N*-vinylpyrrolidone has been photografted to the surface of polypropylene films to generate antimicrobial materials [125]. Moreover, UV treatment can affect the optical properties of the polymer, and UV light can be blocked by particles, which may impair treatment uniformity [126].

22.3.1.4 Hydrolysis and Aminolysis

For polyesters, hydrolysis and aminolysis can be used to produce the reactive groups. Aliphatic polyesters, such as poly(glycolic acid) (PGA) and poly(lactic acid) (PLA), are degraded by water via autocatalytic cleavage of main chain ester bonds. The hydrolysis process can be catalyzed by specific enzymes (hydrolases). The cleavage of the ester bonds of the polymers results in the generation of carboxylic acids and hydroxyl groups at the newly exposed chain ends, resulting in a hydrophilic and rough surface, suitable for cell attachment [127]. Moreover Yamaoka et al. have directly grafted gelatin on PLA surface using a gelatin/basic solution [128]. An aminolysis method was developed by simply treating ester containing polymers and copolymers, such as PCL, PLA, and PU, in solution of diamino compounds, followed by grafting of proteins using glutaraldehyde as coupling reagent [129]. In the hydrolysis/aminolysis mediated surface modification, the molecular weight of the biodegradable materials is decreased, thus the reaction conditions should be well controlled [130].

22.3.2 Immobilization of Bioactive and Biomimetic Compounds

22.3.2.1 Biomimetic Surfaces by Chemical Modification

Biomimetic surfaces may be obtained by immobilization of peptide sequences on biomaterial surface using various chemical approaches. The peptides conformation, density, and spatial distribution at the nanometer scale may all influence cell adhesion (e.g. focal adhesion formation), cellular proliferation, apoptosis, migration, differentiation [131, 132], and tissue formation [15, 16]. The RGD sequence was identified as a minimal essential cell adhesion in several ECM proteins but many other bioactive sequences exist, such as REDV, PHSRN, and KNEED, from fibronectin, YIGSR in laminin and these peptides have been used *separately* or in combination in order to effectively mimic the natural tissue.

In a work of Santiago and colleagues [133] peptides containing IKVAV, RGD, and YIGSR sequences have been covalently attached to the surface of polycaprolactone, using a two-step procedure that employs a treatment with 1,6-hexanediamine followed by the use of 1-ethyl-3-(dimethylaminopropyl) carbodiimide (EDC). Over a period of 48 h, adipose-derived stem cells preferentially attached to IKVAV-functionalized PCL as compared to YIGSR- and RGD-treated PCL disks. However, IKVAV-treated surfaces were more hydrophilic than RGD- and YIGSR-functionalized surfaces and this physical parameter also influences cell adhesion results.

A spacer group between the polymer chain and the peptide segment is often introduced, in order to increase peptide conformational mobility, enhancing the interactions with the integrin receptors on the cell surface [134]. Bae and colleagues have reported a surface modification of polyurethanes (PU), designed to promote endothelization using peptides and a poly(ethylene glycol) (PEG) spacer. PU surface was firstly modified with PEG as a spacer for an effective endothelial cell activity, and then two kinds of endothelial cells-adhesive peptides (GRGDS and YIGSR) were chemically immobilized. These bioactive PU surfaces revealed higher HUVEC activities than unmodified PUs in vitro. In addition, the combination of GRGDS plus YIGSR was found to synergistically increase human umbilical vein endothelial cell proliferation rates, compared to single peptide immobilized surfaces.

A different biomimetic approach is applied in blood-contacting applications, where materials must resist to protein and cell adhesion. In order to improve the biocompatibility of blood-contacting materials and to avoid thrombus formation, phosphorylcholine and its derivatives are used. This strategy is suitable to mimic the cell membrane structure, which consists primarily of a thin layer of amphipathic phospholipids. It was observed that modification of the polyurethane surface with 2-methacryloyloxyethyl phosphorylcholine units, effectively reduces protein adsorption and platelet adhesion [135].

22.3.3 Not-Conventional Approaches Towards Nanoscale Tailoring of Biomimetic Surfaces

The success of biomimetic functionalization strategies may be limited in vivo by the adsorption of serum proteins to the implanted materials, that will be then responsible for the cell-biomaterial interaction [136]. Therefore, alternative approaches to biomimetic materials, not involving a direct, covalent immobilization of specific functional groups to the polymer surface should be considered to circumvent this problem.

These alternative strategies should be designed starting with a thorough understanding of the mechanisms with which Nature constructs its own materials. The first pioneering attempts in this direction can be dated back to the early 1990s when bioartificial materials were developed. Bioartificial polymeric materials were initially conceived to produce new systems combining the characteristics of synthetic polymers, such as good mechanical properties, good processability, low production and transformation costs, with the specific properties of biodegradability and compatibility with cells and tissues associated to biopolymers [137]. Although materials showing interesting properties were obtained, with several applications in the biomedical field especially as membranes for bioseparations [138] or in Tissue Engineering [139, 140], these products were never demonstrated to direct cell adhesion or proliferation processes through appropriate molecular signals stemming from the specific peptidic or saccharidic sequences of the natural component.

A step forward in this direction was attempted by constructing bioartificial materials through template polymerization. The very general concept of this process, inspired by the natural routes to nucleic acids and proteins as in the basics of the Central Dogma of Synthetic Biology is to use a template directed synthesis of macromolecules from their monomers [141–144] involving the molecular recognition, ligation, and product dissociation steps [145]. By using template polymerization of synthetic monomers onto biological templates, bioartificial polymeric blends with a higher degree of miscibility between the two components than by simple mixing of pre-formed polymers are obtained [146, 147]. However, a true biomimetic behaviour of these bioartificial materials has been not fully demonstrated, yet. In the framework of the topic of this chapter, the concept of template directed polymerization and the realization of biomimetic surfaces through functionalization combine ideally in the so-called molecular imprinting approach. Molecular imprinting is a technique to create template-shaped cavities in polymer matrices with a memory of the template molecules. The best established method to achieve this goal is through the polymerization of a monomer in the presence of the template [148–150] (Figs. 22.12 and 22.13).

In cross-linked polymers, the molecular recognition can be obtained through either the set up of covalent interactions between print molecules and monomers [151] or the development of reversible non-covalent interactions (mainly hydrogen

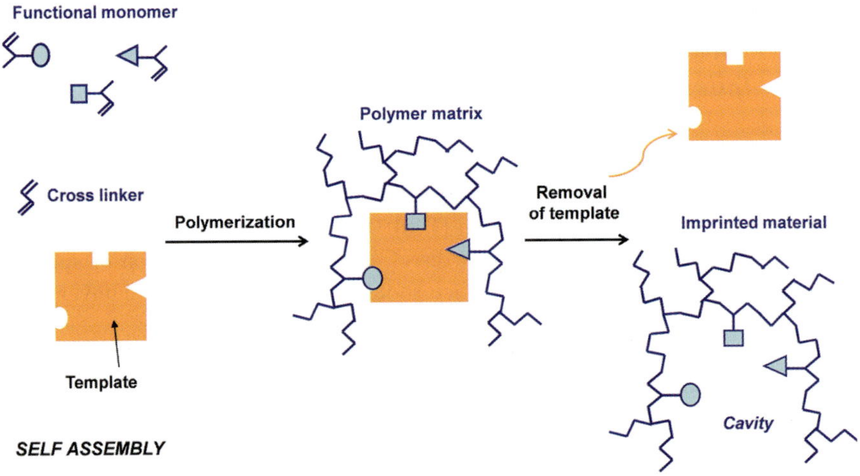

Fig. 22.12 Schematic representation of the molecular imprinting process: self-assembling of functional monomers complementary to the template; and monomer(s)–template complex formation; polymerization in the presence of a cross-linking agent; extraction of the template and formation of the free recognition cavities

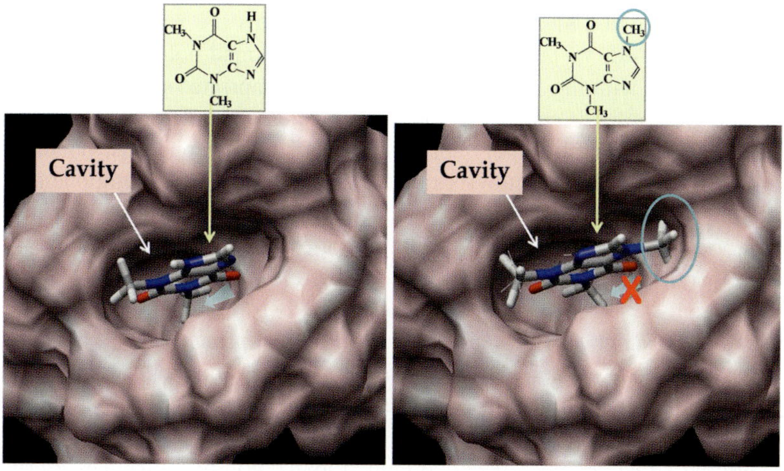

Fig. 22.13 Computer simulation of a Theophylline molecule (the template) inside the binding pocket of a molecularly imprinted polymer. The figure below shows how a similar competitor molecule such as caffeine is fitting less efficiently inside the imprinted cavity due to the presence of a methyl group (*circled*) (Courtesy of Susanna Monti and Chiara Cappelli)

bonding) between them [152]. In both cases, after the establishment of the interactions between functional monomers and templates, a polymerization process follows, usually in the presence of polyfunctional monomers (cross-linkers), which "freezes" the macromolecule network in the configuration capable of interacting with the imprinting molecule. When aiming at imprinting the material to construct a biologically active surface, two main problems arise: (1) recognition in aqueous environment may be impaired by the fact that water may interfere with the selective formation of hydrogen bonding; (2) biomolecules are usually variable in conformation depending on the surrounding medium, reducing their complementarity towards the imprinting site.

The imprinting of proteins is a typical example: it has been attempted by several different approaches including surface grafting of the molecularly imprinted polymer, immobilization of the template on the surface, shape recognition with polymeric gels, imprinting of silica-based materials, recognition through the placement of few constrains on the polymer [153, 154]. Despite the moderate success of these solutions in some cases, it is quite evident that creating an imprint with an array of different functionalities complementary to a protein template is a challenging task. For this reason one of the most smart solutions to tackle the imprinting of proteins remains the so-called epitope approach: if a short peptide which represents only a small exposed fragment of a protein structure is used as template, the resultant macroporous MIP recognizing the imprinted peptide should also be able to recognize the whole protein molecule. This system is present in Nature where, in the recognition mechanism, an antibody interacts with only a small part of an antigen (the epitope, the antigenic site of a macromolecule). This strategy has been proved to be successful only with relatively simple polypeptides [155]. However, since only small fragments of ECM and extramembranal proteins are involved in processes such as cell adhesion or targeting, the imprinting of small polypeptidic molecular domains can be a winning solutions when designing imprinted surfaces for biomedical application in fields such as regenerative medicine or selective drug targeting. Two molecularly imprinted polymer formulations (a macroporous monolith and nanospheres) were obtained towards epitopes of fibronectin, a component of the ECM, showing good specificity and selectivity towards similar peptide sequences. The epitope effect was observed, being the particles able to recognize and rebind not only the specific peptide sequence, but also a larger fibronectin fragment. Moreover, the cytotoxicity tests showed normal adhesion and proliferation of fibroblasts cultured in the medium that was put in contact with the imprinted polymers. Imprinted particles deposited on the surface of polymeric films, maintained their specific recognition and rebinding behaviour. The functionalized materials showed to promote cell adhesion and proliferation [156]. It has to be noted here that with respect to traditional strategies, the "negative" functionalization of molecular imprinting, since it produces recognition sites on the material surface, gives access to a material, which is "per se" non-immunogenic, since it does not present biologically active moieties to the immune system, but it is interacting with the living environment only by means of the imprinting cavities that are complementary to the spatial arrangement of the functional groups of the biomolecule epitope (mimicking recognition properties in Nature such as antibody/antigen or enzyme/substrate-inhibitor). An interesting application in tissue engineering is the combination of molecular imprinting with scaffold microfabrication technologies (e.g. soft-lithography) to enhance cellular adhesion and to guide cell growth. Vozzi et al. prepared imprinted PMMA scaffolds by using poly(dimethylsiloxane) (PDMS) microstructured molds functionalized with gelatin, an adhesion protein. In this way cell adhesion and proliferation on a typically non-cell-adhesive polymer was achieved [157].

22.4 Scanning Probe Techniques for Optical and Spectroscopic Characterization of Surfaces at High Resolution

In the following section, several examples of organic surface characterization by means of the aforementioned techniques are provided. The variety of the applications considered demonstrates the flexibility of scanning probe techniques and their importance over a wide research area.

22.4.1 Dynamic-Mode AFM for the Characterization of Organosilane Self-Assembled Monolayers

AFM is a powerful investigation tool used for biomedical applications, as it provides high resolution imaging of proteins, nucleotides, membranes, and living cells [158].

AFM allows the visualization of biological samples also in buffers, providing topographical maps with high vertical resolution (<1 nm). The study of bio-molecule assembly in physiological conditions is important in order to understand the function of biological molecules. Particularly, as a result of continuous developments in sample preparation, imaging techniques, and instrumentation, AFM is used as a complement of X-ray crystallography and electron microscopy (EC) for the determination of protein structures [159]. Dynamic force microscopy methods have been developed to analyze soft biological surfaces as they are less invasive imaging techniques than contact mode AFM, due to reduced lateral forces during scanning.

In the examples presented below, it is presented the use of non-contact AFM for the morphological characterization of ultra-thin organic layers deposited on flat surfaces. When dealing with soft samples, it is essential to avoid any possible indentation, scratching and/or alteration of the surface due to the scanning tip. For this reason, contact-mode AFM should be avoided and dynamic operation mode has to be preferred. As described in Sect. 22.1.2, non-contact mode AFM is based on the detection of amplitude and phase changes in the oscillation of the cantilever upon electrostatic perturbations by the surface. It is commonly acknowledged to refer amplitude variations to changes in topography, and phase variations (also) to inhomogeneities in composition, adhesion, friction, viscoelasticity, and other properties of the sample. Organosilane SAMs have been widely used to tailor the surface properties (both physical and chemical) of glass, quartz, oxidized Si wafers, and silica particles [160, 161]. This kind of functionalization (Sect. 22.2.2.2) has been extensively applied also for the immobilization of molecules of biological interest, such as enzymes, immunosensors, and DNA microarrays [162–165]. The final goal was the development of biosensing systems, which are widely used in the fields of medical care and drug manufacturing. The formation of organosilane monolayers in the liquid phase has been widely studied, although the dependence on temperature, solvents, and silane concentration is still not completely clear. To overcome the limitations of liquid-phase processes, deposition from vapour phase has been proposed because the lower molecular density in the vapour phase should decrease the extent of self-polymerization of the silane precursor. Here, a gas-phase procedure was used for deposition on silicon surfaces. In particular, a vacuum system equipped with a plasma source was employed for the deposition of a 3-aminopropyltriethoxysilane (3-APTES) SAM on a silicon surface. The surface activation step, using the air plasma treatment prior to the monolayer deposition, was essential to clean any organic contamination and to ensure hydroxylation of the surface. With this particular procedure there is no need of an extra oxidation step, resulting in a much quicker and cheaper process.

Here we show how AFM can be used to ivestigate the properties of organosilane SAM when formed using different procedures, such as wet or dry salinization. Topographic and phase images (2−μ m scan scale) of SiO2/Si substrate modified via wet-based silanization are presented in Fig. 22.14a, b. Particles and aggregates with diameter sizes betwwen 100 and 500 nm and heights of a few nm are clearly recognizable. A flat topography (Fig. 22.15c) is revealed when AFM is carried out onto a portion of sample free from aggregates. However, phase imaging AFM carried

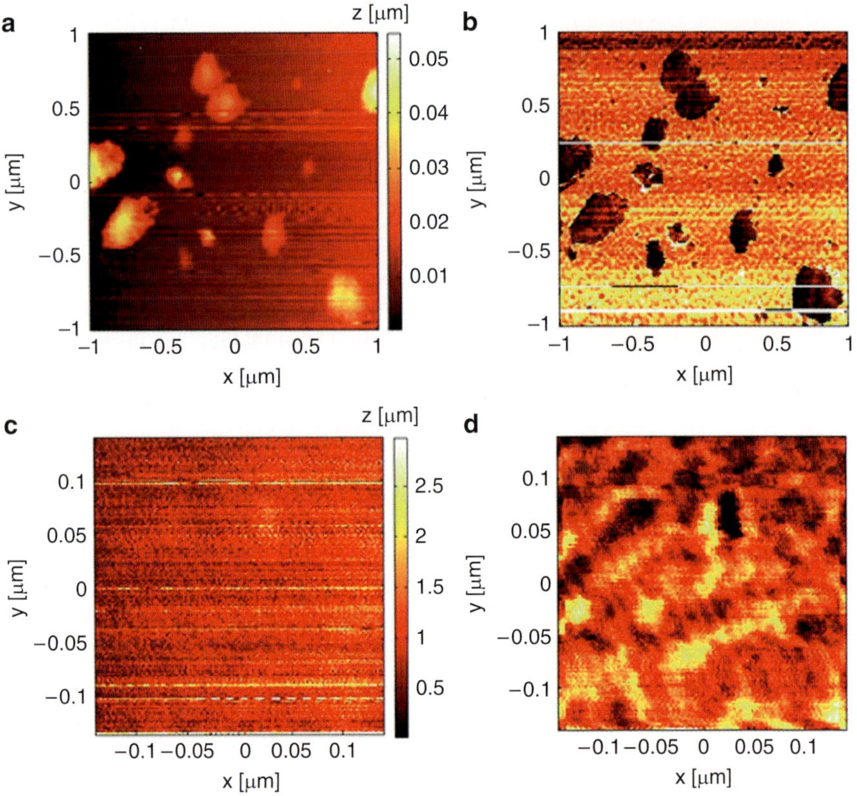

Fig. 22.14 Non-contact AFM (**a**) topography and (**b**) phase images (scanning area 2 μm) of SiO2/Si_APTES-*wet*. Non-contact AFM (**c**) topography and (**d**) phase images (scanning area 0.3 μm) of SiO2/Si_APTES-*wet*

out onto the same area (Fig. 22.15d) indicates the presence of linearly-structrured nanometer-sized features. This can be explained as closely packed silane molecules covering the silicon surface. In particular, figs. 22.14c-d represent a fine example of the superior sensitivity of phase imaging with respect to topographic imaging.

The vapour-phase method produces a much cleaner surface: topographic and phase images of Si_APTES-*dry*, shown in Fig. 22.15a-b, respectively, reveal the presence of smaller (diameter less than 30 nm) and sparser particles as compared to surfaces silanized with the already mentioned solution-phase treatment.

AFM has been further exploited to verify the efficient binding of protein A on APTES-dry functionalized substrates. Results shown in Fig. 22.16 show the presence of closely packed proteins bounded to the surface. It is worthwhile to note that AFM does not alter the soft surface of the sample, and that the phase imaging technique allows a clear visualization of the boundaries of small aggregates even when the height is barely detectable by topographic imaging.

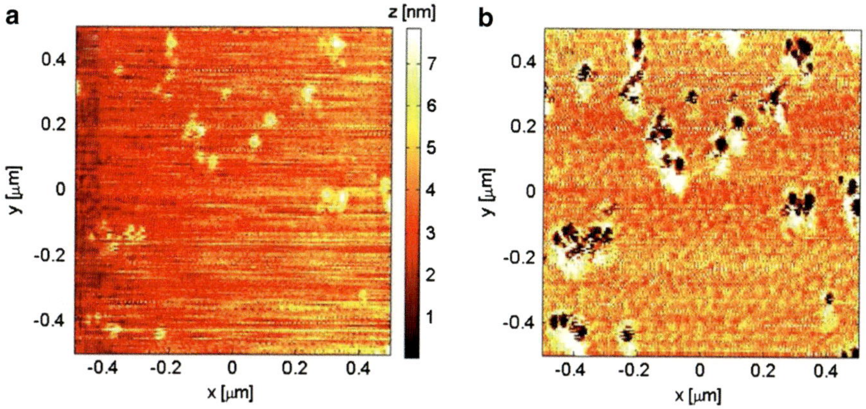

Fig. 22.15 Non-contact AFM (**a**) topography and (**b**) phase images (scanning area 1μm) of SiO2/Si_APTES-*dry*

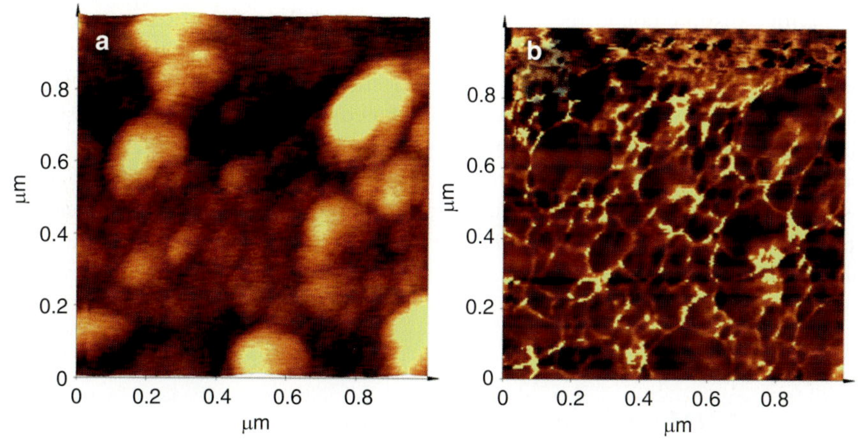

Fig. 22.16 Non-contact AFM images of protein A bound to a silicon-oxide surface APTES functionalized: (**a**) topography (z-scale 2 nm), (**b**) phase. Scan area is 1 × 1 μm

One more advantage of AFM is the possibility to investigate the function–structure relationship of the biomolecular systems [166], meaning that inter- and intra-molecular interactions of biomolecules can be studied at the molecular level. These interaction forces are of fundamental importance, as they determine the structure, dynamics, and function of biomolecules. For example, the binding strength of receptor–ligand complexes may influence many biological processes, such as receptor signalling and immune system protection. This kind of analysis foresees the AFM tip functionalization, for instance, via non-specific adsorption of sensor proteins or of a layer of derivatized bovine serum albumin (BSA) molecules. On the other hand, the adsorption forces may be lower than the interaction forces between ligand molecules that are bound to the tip and target molecules that are bound to the support, leading to desorption. Moreover, the adsorbed molecules may undergo denaturation or degradation during bonding, or squeezing during the scanning process. To overcome these problems, covalent functionalization of AFM tip may be performed. Interaction forces of tip-bound ligands and surface-immobilized receptors are measured in force-distance cycles (Force Spectroscopy) [167]. Force measurements by AFM allow the analysis of molecular interactions in terms of kinetic rate constants, structural parameters of binding pockets, molecular dynamics of the recognition process, and the energy range of interactions [168]. As compared to conventional methods, single-molecule AFM experiments present a few advantages: (1) determination of the distribution of the molecular properties of non-homogeneous systems, by means of many sequential measures; (2) determination of dynamic and statistical surface properties, not detectable by conventional ensemble methods; and (3) real-time monitoring of population transients.

Dynamic force microscopy can be combined with single-molecule force measurements for the localization of specific binding sites at a high resolution. If the functionalized tip oscillates while scanning a surface, binding of tip-bound ligand with substrate-bound receptor sites is recorded. In recognition imaging, the surface topography affects the minima [169, 170], whereas the ligand–receptor recognition affects the maxima of cantilever sinusoidal oscillations. Topographic and recognition information can thus be separated in space and time. For instance, the distribution of VE-cadherin adhesion proteins for vascular endothelial cells has been identified on endothelial cell membranes by this approach [171]. VE-cadherins are Ca-dependent proteins responsible for cell-to-cell adhesion. A VE-cadherin-modified tip has been used to probe the area of endothelial cell membrane. The adhesion molecules have been found to assemble into 20–300 nm sized aggregates [162].

22.4.2 SNOM for Fluorescence Imaging

Since a few decades, SNOM has found applications in a variety of fields in the biological area [172–175]. The study of the interactions between biomolecules, their structure–function relationship, the mechanism of cellular drug delivery, the localization of proteins in cell membranes are just a few examples of the widespread potential of near-field imaging in biomedicine. In order to exploit the detecting capability of SNOM for the optical properties of a target molecule, fluorescent materials are typically conjugated to the target biomolecules. The biomolecules conjugate with the fluorescent materials, which facilitates the detection of the fluorescence properties and the topography of the target via SNOM. This method has been proven to be very useful for investigating molecular interactions at the nanoscopic scale [176–178]. Park et al. have demonstrated that SNOM is a powerful tool for the detection of protein interactions at the nanoscopic level when the protein is conjugated with gold nanoparticles (AuNPs) [179]. In this work, a glass surface was first patterned with a micro-scale line of (3-aminopropyl) trimethoxysilane

(APTMS) by micro-contact printing (mCP) and was subsequently reacted with N-hydroxysuccinimide (NHS)-biotin. The AuNP-conjugated streptavidin was then applied to the biotin-modified glass surface and SNOM was employed to detect the resulting specific interaction between streptavidin and biotin on the glass surface [179]. Investigation of cell surface at nanometer scale is a fundamental issue of cell biology, since the morphology of the cell membrane and its molecules distribution may provide important clues to cell physiology and pathology. SNOM imaging is advantageous in the biomedical field as it offers a unique technology to simultaneously map cell–surface topography and investigate the local optical properties (near-field) of the inner cell structures. SNOM allows to overcome the drawbacks of the traditionally applied techniques for investigating cell surface, such as scanning electron microscopy (SEM) and confocal scanning microscopy, including SEM invasive protocols (like strong dehydration methods and metal coating of biological specimens) which may disrupt fragile structures and nanoscale membrane details, and low resolution power of confocal microscopy, which is diffraction limited. Moreover, SEM analysis does not allow detection and localization of specific structures using immunolabelling approaches, and cannot provide information on the cell interior.

Enderle and co-workers [172] have tackled the problem of accurate localization of proteins within the substructure of cells. In particular, they studied the colocalization of the malaria *Plasmodium falciparum* parasite proteins with an erythrocyte skeletal protein. SNOM and confocal fluorescence maps of infected erythrocytes are shown in Fig. 22.17. The superior spatial resolution of SNOM, estimated to be about 100 nm, made it feasible to excite the fluorescence only from the outermost layer of the cell, and thus to image proteins found only inside the membrane cell. The distribution of the antibody on the membrane could not be revealed through the confocal image (spatial resolution of about 300 nm). An example of characterization of molecules at the level of the subcellular compartments inside single living human cells is given in [174]. The authors show near-field fluorescence spectra acquired in proximity of the mitochondrial membrane inside human breast adenocarcinoma cells, which have been previously treated with a fluorescence dye. In Fig. 22.18 SNOM fluorescence spectra taken at different distances from the mitochondria are displayed. The fluorescence intensity increases from position 1 (optical probe on the membrane, far from the mitochondria) up to a maximum found in position 3 (the probe is in its closest position to the mitochondria).

Trevisan et al. have explored the cell membrane of oligodentrocytes (ODC) and their underlying microtubules by SNOM [180]. The authors have set up a new method to visualize peculiar elements of the ODC cytoskeleton, using α-tubulin antibodies and DAB (3–3,3-diaminobenzidine tetrahydrochloride)-based immunocytochemistry technique. The developed method has allowed the detection of data about the 30–100 nm region below the cell surface, with particular reference to the presence (or absence) of cytoskeleton structures.

As microfabricated cantilevered SNOM probes are now available, some efforts have been spent in investigating the possibility of obtaining super-resolved optical imaging without creating an optical aperture at the tip apex. Working with sharp

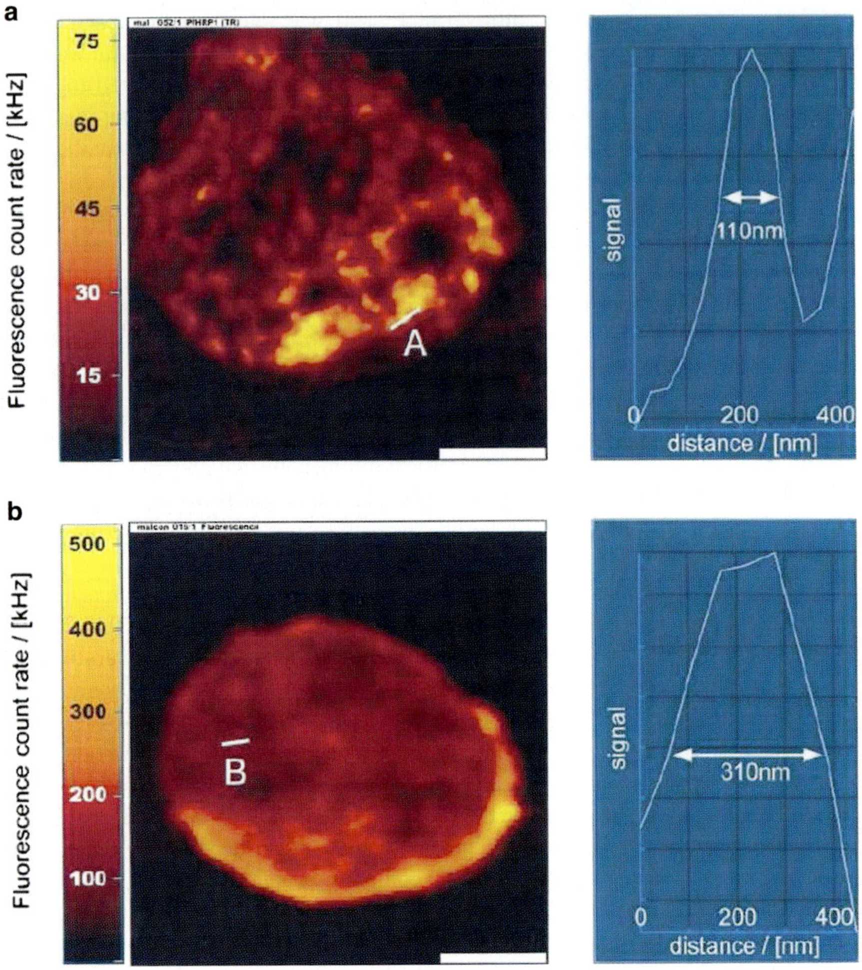

Fig. 22.17 On the left, fluorescence images obtained through (**a**) SNOM and (**b**) confocal microscopy. On the *right*, cross-section of the smallest feature found in the image, showing the estimated spatial resolution of the two techniques (reprinted with permission from [172])

microfabricated tips with no aperture would lead to obvious benefits in terms of fabrication costs and AFM-topographic resolution. This latter issue is, in fact, a limiting factor of SNOM apertures, in which the overall tip plateau containing the 50–100 nm aperture is often more than 150 nm wide: more than 15 times the average diameter of a standard AFM tip.

On the other hand, although narrower tip apex would represent an advantage in terms of spatial resolution, it is not clear how light confinement and coupling would work in the absence of an optical transparent nanoaperture. After the work of Aeschimann et al. [181, 182], it has been demonstrated that super-resolving images

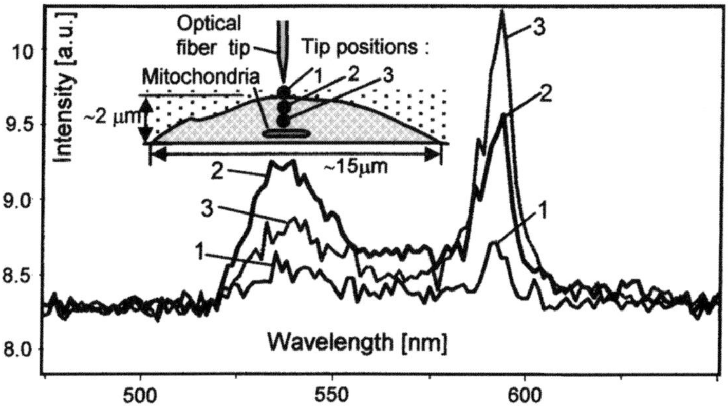

Fig. 22.18 Near-field fluorescence spectra measured inside a living breast carcinoma cell as a function of the position of the optical probe inside the cell. The spectrum belonging to position 1 was measured on the membrane cell, while positions 2 and 3 indicate spectra taken inside the cell and right on the mitochondria, respectively (reprinted with permission from [174])

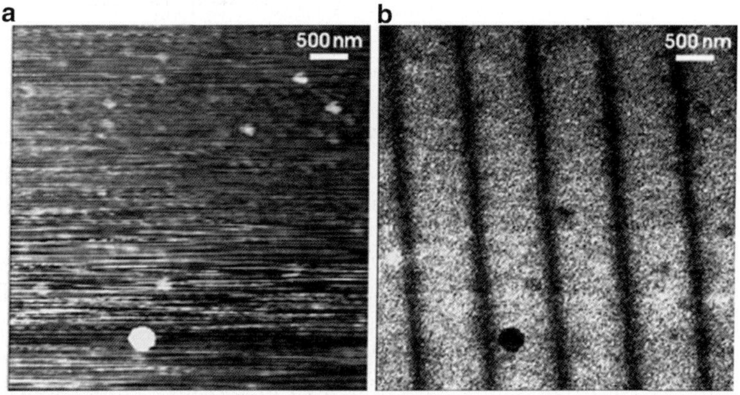

Fig. 22.19 (**a**) Topography and (**b**) near-field excited fluorescence image of a homogeneous fluorescent layer locally bleached by electron beam. Bleached dark stripes are 100 nm wide and separated by 1 mm. Illumination-mode apertureless SNOM clearly resolves the subwavelength features of the sample (reprinted with permission from [181, 182])

can be obtained even in the case of fully coated quartz probes. For this purpose, a test sample has been fabricated, showing almost undetectable topographical features, but strong optical contrast when used in fluorescence measurements. The idea is to create a nanometric pattern forming a homogeneous fluorescent layer by means of an electron beam. As an example, in Fig. 22.19, a sample showing dark (bleached) stripes (100 nm wide and separated by 1 μ m) is illustrated. Fig. 22.19a shows the tipography image of a homogeneous fluorescent layer bleached by means of an electron beam. It can be seen that no features emerge and the surface is determined to be flat. Conversely, Fig. 22.19b shows the fluorescent SNOM image taken with an apertuless tip. Here the features are clearly resolved with subwavelength resolution. As the topographical features are not correlated to the actual pattern created by electron beam, it can be stated that the periodic arrays of dark and bright regions detected upon SNOM excitation are not artefacts induced by the topography. Accordingly, it is possible to provide an estimation of the optical super resolution capabilities of the probes. In this case, a lateral resolution down to 70 nm is demonstrated [181, 182].

In conclusion, the SNOM technique applied to biomimetics allows to: (1) observe the tailoring of the biomimetic functionalization at the molecular level and (2) study the cell membrane characteristics at the molecular level as well as 30–100 nm under the cell membrane, when cells have been in contact with a properly designed substrate.

22.4.3 TERS for Chemical Mapping at the Nanoscale

Among the family of vibrational spectroscopic techniques, Raman Spectroscopy (RS) has been shown as an invaluable tool for the chemical characterization of surfaces as well as for biological systems, where a deep understanding of the composition and the dynamics occurring at an interface is required. The major drawback of RS is the extremely low intensity of the Raman signal, which makes it difficult to detect. Methods for the enhancement of the Raman scattering have been proposed, being the SERS the most remarkable example. Although proved to be capable of enhancing the Raman intensity up to a factor of 10^{14} [183, 184], SERS is not suited for imaging. As the sample has to be deposited as a thin layer on a rough noble-metal film, the enhancement may greatly vary over the surface and will depend on the substrate preparation. On the other hand, aperture Scanning Near-field Optical Microscopy (a-SNOM) overcomes this issue and provides a robust tool for imaging the fluorescence of single molecules with unprecedented spatial resolution [185–187]. However, a-SNOM still suffers from poor illumination power, which limits the intensity of the Raman scattering, and from spatial resolution above 50 nm. As an alternative for the existing techniques, TERS has been proposed [188]. In its simplest configuration, a TERS setup consists of an AFM in combination with a Raman spectrometer. If a suitable metalized tip is brought close to the sample surface, inside the laser focal spot, the incident electric field is locally enhanced. In most cases, a common AFM tip may be used, where Ag or Au is preferably used for the coating. The optical field enhancement that eventually generates the amplified Raman signal, originates from the so-called lightning-rod effect. If the incident radiation has a polarization component along the tip axis, a strong surface charge density is induced at the tip end, giving rise to the field enhancement effect. This concept is clearly demonstrated in Fig. 22.20. The scattered light intensity of gold-coated tip is plotted versus the incident wavelength. A resonant peak centred at around 580 nm is observed for P-polarized incident radiation (the electric field oscillates in a plane parallel to the tip axis), which is the typical resonant behaviour expected for nanometer-sized gold tips. Thus TERS is combining the imaging capabilities of an

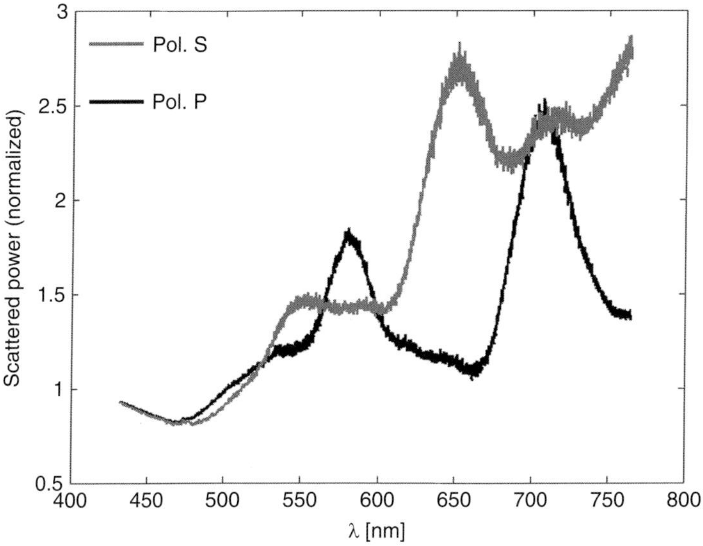

Fig. 22.20 Scattered light intensity from a gold-coated tip for the S- and P-polarizations as a function of the incident wavelength

AFM setup, though with higher lateral resolution (spacial resolutions down to 15 nm have been reported [189]), with the exceptional field enhancement typical of SERS. In biology, the feasibility of TERS at the nanometer scale has been demonstrated on DNA-stand sequencing [190, 191], bacteria and biofilms chemical characterization [192, 193].

An application of TERS that is of great interest in the field of biological science concerns the study of biofilms. In natural systems bacteria live as colonies over a solid surface, progressively aggregate and eventually form biofilms. As biofilms can provide a habitat for pathogenic microorganisms, they may cause persistent infections. A detailed knowledge of the structure and composition of a biofilm is necessary for the optimization of biocides and wastewater treatment, but the role and arrangement of the different biopolymers are largely unknown due to the lack of chemical information or insufficient spatial resolution provided by the microscopy techniques used so far. A method for the chemical characterization capable of spatial resolution well beyond the diffraction limit is thus required. Schmid and co-workers [192] have shown the potential of TERS on the characterization of calcium alginate fibres (Fig. 22.21). It can be appreciated how the carbon-corrected Raman peaks (bottom figure) are only visible in the TERS spectra, while no signal could be detected in the far field by conventional confocal techniques.

The detailed study of bacterial metabolism may also shed some light on the processes at the basis of host-bacteria and drug-bacteria interactions. Depending on the chemical composition, antibiotics function on different parts of the bacterial cell wall. To enhance the efficiency of antibiotics, it is crucial to know the chemical

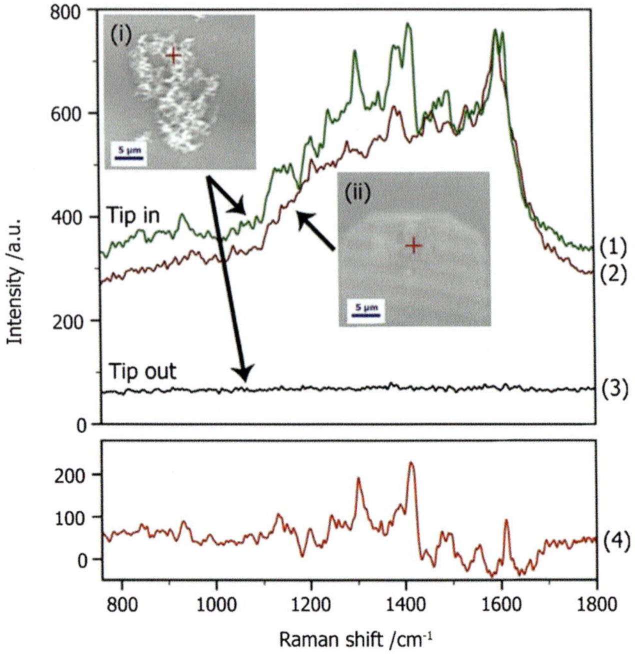

Fig. 22.21 TERS spectra of calcium alginate fibres. *Displayed curves* indicate (1) raw TERS data; (2) carbon-contaminating spectrum; (3) retracted-tip Raman spectrum; and (4) carbon-corrected TERS spectrum (reprinted with permission from [192])

characteristics of the cell wall. Neugebauer et al. have investigated the molecular composition of the outer layer of *Staphylococcus Epidermidis* cells embedded into an ECM by means of TERS [193]. In this case, a rich TER spectrum emerges (Fig. 22.22), showing a series of peaks that could not be resolved in the far field (Fig. 22.22, bottom).

A fine example of the advantage, in terms of spatial resolution, of using tip-enhanced against conventional surface-enhanced Raman spectroscopy is given by Yeo and co-workers [194]. They analyzed the heme membrane protein (cytochrome c, Cc) and found that both the heme and amino acid vibrational bands of Cc could be clearly identified by means of TERS. Conversely, in SERS spectra the weakest peaks were obscured by the strongest ones, leading to convoluted signal. Due to its superior sensitivity and spatial resolution, TERS has been suggested as the method of choice for the analysis of proteins, for which complete band assignment was possible, and for structural characterizations of large biomolecules.

The goal of further studies is to improve TERS towards a robust tool for the analysis of biological samples, which allows for example the understanding of the distribution of different biopolymers inside the biofilm matrix or the visualization of proteins in lipid membranes. Surely, one of the major steps forward for TERS will be the investigations of biosystems in physiological conditions.

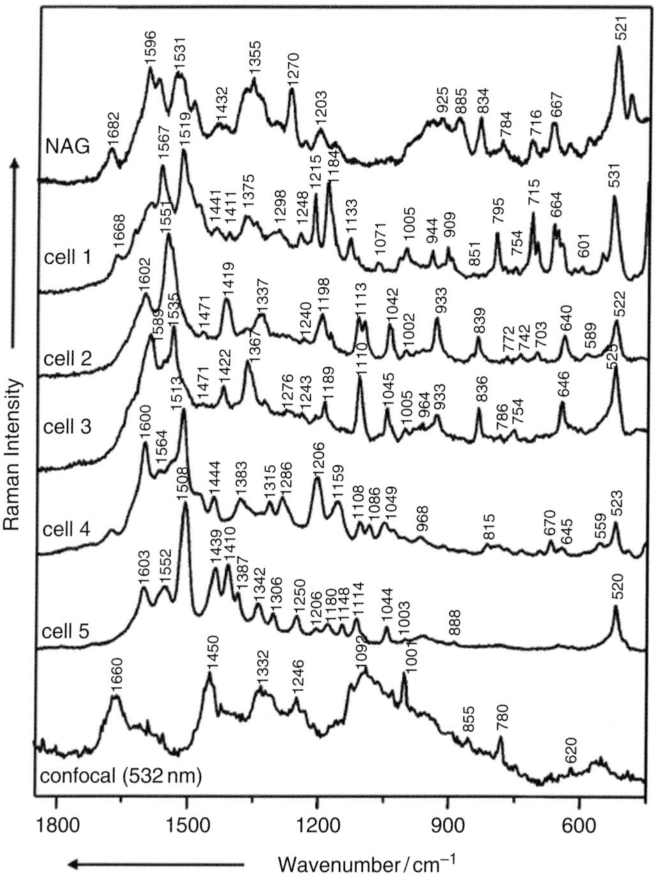

Fig. 22.22 TERS spectra of *S. Epidermidis* cells. The spectrum at very bottom was acquired through conventional confocal Raman spectroscopy (reprinted with permission from [193])

In conclusion, the most exciting prospects of Raman spectroscopy of surfaces are connected with the SERS single-molecule measurements and with the high spatial resolution TERS characterization of surfaces. The single-molecule SERS gives the possibility to track structural and chemical transformations of single molecules, which provides useful insights into reaction mechanisms. SERS measurements have a molecular resolution (practically a few molecules) and therefore they may be interesting for investigating extremely small amounts of biologically relevant molecules. The main target of biochemical TERS investigations are peptides, DNA, or RNA, as TERS measurements allow sequencing of such systems. The first, indirect approach is to selectively label DNA or RNA with Raman-active dyes and then measure TER spectra along the analyzed strand. Direct (without labelling) sequencing by means of TERS can also be possible, although more complicated experimentally.

Recently, Bailo and Deckert have demonstrated this concept by measuring reliable TERS spectra along no-labelled single RNA strand [195, 196].

22.5 Conclusions

In conclusion, the present chapter, although far from being exhaustive over the existing literature on the strategies to functionalize materials and surfaces, to impart them a biomimetic behaviour, is nevertheless giving a contribution to select the more appealing strategies in the field. The common philosophy of these approaches is to develop materials by design (i.e. tailor a material, starting with its chemical composition, constituent phases, and microstructure, in order to obtain a desired set of properties suitable for a given application), to obtain materials or surfaces that are seeking to replicate or mimic biological processes and materials. The success of these engineered materials in biomedical applications is necessarily related to the contribution of biological and medical concepts, not only in the validation of the cell and tissue compatibility and/or the clinical efficacy of the developed product, but also in defining the biological mechanisms involved in the damage/illness and in the healing/repair process as a support in the design of the material. It is, however, likely that the development of a material with a more precise design of structural and functional properties will give a new burst to the biomedical industry and allow biomedical engineers to gain a central role in this arena which is currently dominated by medical and biologically related strategies. A fundamental support in this direction will be given by the advanced methods of surface characterization at the nanoscale, with scanning probe methods having a leading role with their ability to identify functionalization at the molecular level and their effect of developing cell–material interaction identifying mechanical forces at the nanoscale.

In the future, novel functional medical materials are expected to be developed with wider fields of applications. The challenge is also to re-think from scratch the way in which inspiration is drawn from nature: advanced biomimetic materials should not only replicate the active motifs of biological molecules but rather their functional behaviour.

References

1. R. Langer, J.P. Vacanti, Science **260**, 920 (1993)
2. N. Huebsch, D.J. Mooney, Nature **26**, 426 (2009)
3. B.J. Hinz, Biomech **43**, 146 (2009)
4. F. Rosso, A. Giordano, M. Barbarisi, A. Barbarisi, J. Cell. Physiology **199**, 174 (2004)
5. S. Li, J.L. Guan, S. Chien, Annu. Rev. Biomed. Eng. **7**, 105 (2005)
6. M.A. Tria, M. Fusco, G. Vantini, R. Mariot, Exp. Neurol. **127**, 178 (1994)
7. L.A. Pfister, M. Papaloïzos, H.P. Merkle, B. Gander, J. Peripher. Nerv. Syst. **12**, 65 (2007)
8. M.J. Whitaker, R.A. Quirk, S.M. Howdle, K.M. Shakesheff, J. Pharm. Pharmacol. **53**, 1427 (2001)

9. T.P. Richardson, W.L. Murphy, D.J. Mooney, Crit. Rev. Eukaryot. Gene Expr. **11**, 47 (2001)
10. J.E. Babensee, L.V. McIntire, A.G. Mikos, Pharm. Res. **17**, 497 (2000)
11. X. Niu, Y. Wang, Y. Luo, J. Xin, Y. Li, J. Mater. Sci. Technol. **21**, 571 (2005)
12. M. Nomizu, Y. Kuratomi, K.M. Malinda, S.Y. Song, K. Miyoshi, A. Otaka, S.K. Powell, M.P. Hoffman, H.K. Kleinman, Y. Yamada, J. Biol. Chem. **273**, 32491 (1998)
13. S. Vukicevic, F.P. Luyten, H.K. Kleinman, A.H. Reddi, Cell **63**, 437 (1990)
14. S. Aota, T. Nagai, K.M. Yamada, J. Biol. Chem. **266**, 15938 (1991)
15. B.A. Dalton, J. Cell. Sci. **108**, 2083 (1995)
16. A. Woods, J.R. Couchman, S. Johansson, M. Hook, EMBO J. **5**, 665 (1986)
17. D.A. Puleo, R. Bizios, Bone **12**, 271 (1991)
18. J. Laterra, J.E. Silbert, L.A. Culp, J. Cell. Biol. **96**, 112 (1983)
19. A. Rezania, K.E. Healy, Biotechnol. Prog. **15**, 19 (1999)
20. K.E. Healy, A. Rezania, R.A. Stile, Ann. N. Y. Acad. Sci. **875**, 24 (1999)
21. G. Binnig, C.F. Quate, C. Gerber, Phys. Rev. Lett. **56**, 930 (1986)
22. G. Meyer, N.M. Amer, Appl. Phys. Lett. **53**, 1045 (1988)
23. P.F. Indermühle, G. Schürmann, G.A. Racine, N.F. de Rooij, J. Micromech. Microeng. **7**, 218 (1997)
24. T. Akiyama, L. Aeschimann, L. Chantada, N.F. de Rooij, H. Heinzelmann, H.P. Herzig, O. Manzardo, A. Meister, J. Polesel-Maris, R. Pugin, U. Staufer, P.J. Vettiger, Appl. Phys. **46**, 6458 (2007)
25. E.H. Synge, Phil. Mag. **6**, 356 (1928)
26. R. Hillenbrand, F. Keilmann, Phys. Rev. Lett. **85**, 3029 (2000)
27. E. Betzig, M. Isaacson, A. Lewis, Appl. Phys. Lett. **51**, 2088 (1987)
28. J. Guerra, Appl. Opt. **29**, 3741 (1990)
29. D. Barchiesi, D. Van Labeke, Microsc. Microanal. Microstruct. **5**, 435 (1994)
30. L. Gomez, R. Bachelot, A. Bouhelier, G. Wiederrecht, S. Chang, S. Gray, F. Hua, S. Jeon, J. Rogers, M. Castro, S. Blaize, I. Stefanon, G. Lerondel, P.J. Royer, Opt. Soc. Am. B **23**, 823 (2006)
31. A. Nesci, R. Dändliker, H.P. Herzig, Opt. Lett. **26**, 208 (2001)
32. A. Bek, R. Vogelgesang, K. Kern, Rev. Sci. Instrum. **77**, 043703 (2006)
33. S.K. Sekatskii, V.S. Letokhov, Appl. Phys. B.: Opt. Lasers **63,** 525 (1996)
34. S. Hell, M. Booth, S. Wilms, C. Schnetter, A. Kirsch, D. Arndt-Jovin, T. Jovin, Opt. Lett. **23**, 1238 (1998)
35. M. Futamata, A. Bruckbauer, Chem. Phys. Lett. **341**, 425 (2001)
36. H. Heinzelmann, D.W. Pohl, Appl. Phys. A **59**, 89 (1994)
37. G.A. Valaskovic, M. Holton, G.H. Morrison, Appl. Opt. **34**, 1215 (1995)
38. A. Sayah, C. Philipona, P. Lambelet, M. Pfeffer, F. Marquis-Weible, Ultramicroscopy **71**, 59 (1998)
39. E. Betzig, J.K. Trautmn, T.D. Harris, J.S. Weiner, R.L. Kostelk, Science **251**, 1468 (1991)
40. U.C. Fischer, D.W. Phol, Phys. Res. Lett. **62**, 458 (1989)
41. D. Courjon, J.M. Vigoureux, M. Spajer, K. Sarayeddine, S. Leblanc, Appl. Opt. **9**, 3734 (1990)
42. L. Aeschimann, T. Akiyama, U. Staufer, N.F. deRoij, L. Thiery, R. Eckert, H. Heinzelman, J. Microscopy **209**, 182 (2003)
43. E. Descrovi, L. Aeschimann, I. Soboleva, F. De Angelis, F. Giorgis, E. Di Fabrizio, J. Nanosci. Nanotech. **9**, 6460 (2009)
44. S. Nie, S.R. Emory, Science **275**, 1102 (1997)
45. R.M. Stockle, Y.D. Suh, V. Deckert, R. Zenobi, Chem. Phys. Lett. **318**, 131 (2000)
46. N. Hartschuh Anderson, L. Novotny, J. Microscopy **210**, 234 (2003)
47. P. Tortora, E. Descrovi, L. Aeschimann, L. Vaccaro, H.P. Herzig, R. Dändliker, Ultramicroscopy **107**, 158 (2007)
48. H.J. Mathieu, Surf. Interface Anal. **32**, 39 (2001)
49. C.J. Wilson, R.E. Clegg, D.I. Leavesley, M.J. Pearcy, Tissue Eng. **11**, 1 (2005)
50. B. Kasemo, Surf. Sci. **500**, 656 (2002)
51. H.K. Kleinman, D. Philp, M.P. Hoffman, Curr. Opin. Biotechnol. **14**, 526 (2003)

52. M. Li, M.J. Mondrinos, M.R. Gandhi, F.K. Ko, A.S. Weiss, P. Lelkes, Biomaterials **26**, 5999 (2005)
53. S. Drotleff, U. Lungwitz, M. Breunig, A. Dennis, T. Blunk, J. Tessmar, A. Gopferich, Eur. J. Pharm. Biopharm. **58**, 385 (2004)
54. J.M. Anderson, A. Rodriguez, D.T. Chang, Semin. Immunol. **20**, 86 (2008)
55. B.D. Ratner, S.J. Bryant, Annu. Rev. Biomed. Eng. **6**, 41 (2004)
56. K.B. Blodgett, J. Am. Chem. Soc. **57**, 1007 (1935)
57. G.G. Roberts, *Langmuir-Blodgett Films* (Plenum, NY, 1990)
58. A. Ulman, *An Introduction to Ultrathin Organic Films* vol. 101 (Academic, NY, 1991)
59. G.L. Gaines Jr, *Insoluble Monolayers at Liquid-Gas Interfaces* (Wiley-Interscience, NY, 1966)
60. G. Decher, Science **277**, 1232 (1997)
61. M. Tirrell, in *Unsolved Problems in Nanotechnology: Chemical Processing by Self-assembly*, ed. by L.S. Fan, M. Feinberg, G. Hulse, T.L. Sweeney, J.L. Zakin (Ohio State University, 2003), 24–25 April, pp. 63–70
62. P. Berndt, G.B. Fields, M. Tirrell, J. Am. Chem. Soc. **117**, 9515 (1995)
63. Y.C. Yu, P. Berndt, M. Tirrell, G.B Fields, J. Am. Chem. Soc. **118**, 12515 (1996)
64. Y.C. Yu, T. Pakalns, Y. Dori, J.B. McCarthy, M. Tirrell, G.B. Fields, Methods Enzymol. **289**, 571 (1997)
65. G.B. Fields, J.L. Lauer, Y. Dori, P. Forns, Y.C. Yu, M. Tirrell, Biopolymers **47**, 143 (1998)
66. Y.C. Yu, M. Tirrell, G.B. Fields, J. Am. Chem. Soc. **120**, 9979 (1998)
67. T. Pakalns, K.L. Haverstick, G.B. Fields, J.B. McCarthy, D.L. Mooradian, M. Tirrell, Biomaterials **20**, 2265 (1999)
68. A.P. Girard-Egrot, S. Godoy, L.J. Blum, Adv. Coll. Interf. Sci. **116**, 205 (2005)
69. S. Godoy, B. Leca-Bouvier, P. Boullanger, L.J. Blum, A.P. Girard-Egrot, Sens Actuators B **107**, 82 (2005)
70. R.G. Nuzzo, D.L. Allara, J. Am. Chem. Soc. **105**, 4481 (1983)
71. C.D. Bain, G.M. Whitesides, J. Am. Chem. Soc. **111**, 7164 (1989)
72. F. Schreiber, Prog. Surf. Sci. **65**, 251 (2000)
73. N.L. Jeon, R.G. Nuzzo, Y. Xia, M. Mrksich, G.M. Whitesides, Langmuir **11**, 3024 (1995)
74. L. Häussling, H. Ringsdorf, F.J. Schmitt, W. Knoll, Langmuir **7**, 1837 (1991)
75. G. Decher, J.B. Schlenoff (eds.), *Multilayer Thin Films-Sequential Assembly of Nanocomposite Materials* (Wiley-VCH, 2002)
76. A. Delcorte, P. Bertrand, E. Wischerhoff, A. Laschewsky, Langmuir **13**, 5125 (1997)
77. Z. Tang, Y. Wang, P. Podsiadlo, N.A. Kotov, Adv. Mater. **18**, 3203 (2006)
78. M.G. Berg, S.Y. Yang, P.T. Hammond, M.F. Rubner, Langmuir **20**, 1362 (2004)
79. L. Richert, F. Boulmedais, P. Lavalle, J. Mutterer, E. Ferreux, G. Decher, Biomacromolecules **5**, 284 (2004)
80. L. Richert, A.J. Engler, D.E. Discher, C. Picart, Biomacromolecules **5**, 1908 (2004)
81. J.D. Mendelsohn, S.Y. Yang, J. Hiller, A.I. Hochbaum, M.F. Rubner, Biomacromolecules **4**, 96 (2003)
82. W. Tan, T.A. Desai, J. Biomed. Mater. Res. A **72A**, 146 (2005)
83. C.R. Wittmer, J.A. Phelps, W.M. Saltzman, P.R. Van Tassel, Biomaterials **28**, 851 (2007)
84. M. Swierczewska, C.S. Hajicharalambous, A.V. Janorkar, Z. Megeed, M.L. Yarmush, P. Rajagopalan, Acta Biomaterialia **4**, 827 (2008)
85. D. Grafahrend, J.L. Calvet, K. Klinkhammer, J. Salber, P.D. Dalton, M. Moller, D. Klee, Fibers. Biotechnol. Bioeng. **101**, 609 (2008)
86. K.C. Wood, J.Q. Boedicker, D.M. Lynn, P.T. Hammond, Langmuir **21**, 1603 (2005)
87. M. Dimitrova, Y. Arntz, P. Lavalle, F. Meyer, M. Wolf, C. Schuster, Y. Haikel, J.C. Voegel, J. Ogier, Adv. Funct. Mat. **17**, 233 (2007)
88. F. Meyer, V. Ball, P. Schaaf, J.C. Voegel, J. Ogier, Biochim. Biophys. Acta **1758**, 419 (2006)
89. S. Müller, G. Koenig, A. Charpiot, C. Debry, J.C. Voegel, P. Lavalle, D. Vautier, Adv. Funct. Mater. **18**, 1767 (2008)
90. X. Liu, P.X. Ma, Ann. Biomed. Eng. **32**, 477 (2004)

91. P. Gentile, V. Chiono, F. Boccafoschi, F. Baino, C. Vitale-Brovarone, E. Verne', N. Barbani, G. Ciardelli, J. Biomater. Sci. Polym. Edn. **21**, 1207 (2009).
92. G. Ciardelli, P. Gentile, V. Chiono, M. Mattioli-Belmonte, G. Vozzi, N. Barbani, P. Giusti, J. Biomed. Mater. Res. A **92A**, 137 (2009)
93. Y. Liu, J.P. Li, E.B. Hunziker, K. de Groot, Phil. Trans. R. Soc. A. **364**, 233 (2006)
94. T. Kokubo, H. Kushitani, S. Sakka, T. Kitsugi, T. Yamamuro, J. Biomed. Mater. Res. **24**, 721 (1990)
95. B. Gondolph-Zink, Orthopade **27**, 96 (1998)
96. F. Barrere, P. Layrolle, C.A. van Blitterswijk, K. de Groot, Bone **25**, 107S (1999)
97. F. Barrere, P. Layrolle, C.A. van Blitterswijk, K. de Groot, J. Mater. Sci. Mater. Med. **12**, 529 (2001)
98. F. Barrere, B.C. Van, K. de Groot, P. Layrolle, Biomaterials **23**, 2211 (2002)
99. F. Barrere, C.A. van Blitterswijk, K. de Groot , P. Layrolle, Biomaterials **23**, 1921 (2002)
100. Y. Liu, P. Layrolle, J. de Bruijn, C. van Blitterswijk, K. de Groot, J. Biomed. Mater. Res. **57**, 327 (2001)
101. P. Habibovic, F. Barrère, C.A. van Blitterswijk, K. de Groot, P. Layrolle, J. Am. Ceram. Soc. **85**, 517 (2002)
102. R. Zhang, P.X. Ma, Macromol. Biosci. **4**, 100 (2004)
103. V. Chiono, P. Gentile, F. Boccafoschi, I. Carmagnola, M. Ninov, G. Georgieva, G. Ciardelli, Biomacromolecules **11**, 309 (2010)
104. M. Nowakowska, E. Sustar, J.E. Guillet, J. Photochem. Photobiol. A: Chem. **80**, 369 (1994)
105. M. Nowakowska, L. Moczek, K. Szczubialka, Photoactive modified chitosan, Biomacromolecules **9**, 1631 (2008).
106. J.E. Guillet, N.A.D. Burke, M. Nowakowska, S. Paone, Macromol. Symp. **134**, 41 (1998)
107. V. Chiono, I. Carmagnola, P. Gentile, F. Boccafoschi, C. Tonda-Turo, G. Ciardelli, G. Georgiev, M. Ninov, V. Georgieva, I. Pashkuleva, R. Reis, in *Proceedings of 9th IEEE Conference on Nanotechnology* (2009), p. 798
108. Z. Ma, Z. Mao, C. Gao, Colloids and Surfaces B: Biointerfaces **60**, 137 (2007)
109. J.H. Lee, G. Khang, J.W. Lee, H.B. Lee, J. Colloid Interface Sci. **205**, 323 (1998)
110. J.K. Shim, H.S. Na, Y.M. Lee, H. Huh, Y.C. Nho, J. Membr. Sci. **190**, 215 (2001)
111. F. Poncinepaillard, B. Chevet, J.C. Brosse, J. Appl. Polym. Sci. **53**, 1291 (1994)
112. C. Situma, Y. Wang, M. Hupert, F. Barany, R.L. McCarley, S.A. Soper, Anal. Biochem. **340**, 123 (2005)
113. P.X. Ma, Adv. Drug Del. Rev. **60**, 184 (2008)
114. J.Y. Chu, L.P. Chen, N.H. Wang, Mat. Sci. Eng. **R36**, 143 (2002)
115. F. Arefi-Khonsari, M. Tatoulian, G. Placinta, J. Kurdi, J. Amouroux, in *Plasma Processing of Polymers*, ed. by R. d'Agostino, P. Favia (Kluwer Academic Publishers, Dordrecht, 1997)
116. T. Murakami, S. Kuroda, Z. Osawa, J. Coll. Interf. Sci. **202**, 37 (1998)
117. I.J. Chen, E. Lindner, Langmuir **23**, 3118 (2007)
118. M. Tatoulian, F. Bretagnol, F. Arefi-Khonsari, J. Amouroux, O. Bouloussa, F. Rondelez, A.J. Paul, R. Mitchel, Plasma Process. Polym. **2**, 38 (2005)
119. N. Inagaki, in *Plasma Surface Modification and Plasma Polymerization* (Technomic Publishing Company Inc., Lancaster, USA, 1996)
120. S. Sartori, A. Rechichi, G. Vozzi, M. D'Acunto, E. Heine, P. Giusti, G. Ciardelli, React Funct Polym **68**, 809 (2008)
121. L. Dai, H.A. St. John, J. Bi, P. Zientek, R.C. Chatelier, K. Griesser, J. Surf. Interface Anal. **29**, 46 (2000)
122. Q. Chen, R. Forch, W. Knoll, Chem. Mater. **16**, 614 (2004)
123. C. Situma, Y. Wang, M. Hupert, F. Barany, R.L. McCarley, S.A. Soper, Anal. Biochem. **340**, 123 (2005)
124. A. Welle, S. Horn, J. Schimmelpfeng, D. Kalka, J. Neurosci. Meth. **142**, 243 (2005)
125. C.M. Xing, J.P. Deng, W.T. Yang, J. Appl. Polym. Sci. **97**, 2026 (2005)
126. J.M. Goddard, J.H. Hotchkiss, Prog. Polym. Sci. **32**, 698 (2007)
127. Y.S. Nam, J.J. Yoon, J.G. Lee, T.G. Park, J. Biomater. Sci. E **10**, 1145 (1999)

128. T. Yamaoka, Y. Takebe, Y. Kimura, Kobunshi Ronbunshu **55**, 328 (1998)
129. Y.B. Zhu, C.Y. Gao, X.Y. Liu, T. He, J.C. Shen, Tissue Eng. **10**, 53 (2004)
130. Z. Ma, Z. Mao, C. Gao, Colloids Surf B Biointerfaces **60**, 137 (2007)
131. N.D. Gallant, K.E. Michael, A.J. Garcia, Mol. Biol. Cell. **16**, 4329 (2005)
132. G.A. Silva, C. Czeisler, K.L. Niece, E. Beniash, D.A. Harrington, J.A. Kessler, S.I. Stupp, Science **303**, 1352 (2004)
133. L.Y. Santiago, R.W. Nowak, J.P. Rubin, K.G. Marra, Biomaterials **27**, 2962 (2006)
134. B. Li, J. Chen, J.H.C. Wang, J. Biomed. Mat. Res. Part A **79A**, 989 (2006)
135. K. Kobayashi, K. Ohuchi, H. Hoshi, N. Morimoto, Y. Iwasaki, S. Takatani, J. Art. Org. **8,** 237 (2005)
136. M.S. Shoichet, Macromolecules **43**, 581 (2010)
137. P. Giusti, L. Lazzeri, M.G. Cascone, in *Bioartificial Polymeric Materials, in the Polymeric Materials Encyclopedia* (CRC Press, Boca Raton, 1996)
138. D. Silvestri, N. Barbani, G. Ciardelli, C. Cristallini, P. Giusti, J. Appl. Biomat. Biomech. **3**, 83 (2005)
139. V. Chiono, G. Vozzi, M. D'Acunto, S. Brinzi, C. Domenici, F. Vozzi, A. Ahluwalia, N. Barbani, P. Giusti, G. Ciardelli, Mater. Sci. Eng. **29**, 37 (2009)
140. G. Ciardelli, V. Chiono, G. Vozzi, A. Ahluwalia, N. Barbani, C. Cristallini, M. Pracella, P. Giusti, Biomacromolecules **6**, 1961 (2005)
141. V.A. Kabanov, Makromol. Chem. Suppl. **3**, 41 (1979)
142. A.K. Srivastava, S.K. Nigam, A. Shukla, S. Saini, P. Kumar, N. Tewari, J. Macromol. Sci. Rev. Macromol. Chem. Phys. **C27**, 171 (1987)
143. C.H. Bamford, Makromol. Chem. **1**, 52 (1980)
144. G. Challa, Y.Y. Tan, Pure Appl. Chem. **53**, 627 (1981)
145. S.H. Gellmann, B. Clare, in *Templated Organic Synthesis*, ed. by F. Diederich (ETH-Zürich), P.J. Stang (University of Utah) (Wiley-VCH, Weinheim, 2000)
146. C. Cristallini, N. Barbani, P. Giusti, L. Lazzeri, M.G. Cascone, G. Ciardelli, Macromol. Chem. Phys. **202**, 2104 (2001)
147. P. Cerrai, G.D. Guerra, M. Tricoli, S. Maltinti, N. Barbani, L. Petarca, Macromol. Chem. Phys. **197**, 1 (1996)
148. G. Wulff, Angew. Chem. Int. Ed. Engl. **34**, 1812 (1995)
149. L.I. Andersson, I.A. Nicholls, K. Mosbach, Adv. Mol. Cells Biol. **15B**, 651 (1996)
150. P.A.G. Cormack, K. Mosbach, React. Funct. Polym. **41**, 115 (1999)
151. G. Wulff, A. Sarhan, Angew. Chem. Int. Ed. Engl. **11**, 341 (1972)
152. K. Mosbach, Trends Biochem. **19**, 9 (1994)
153. A. Bossi, F. Bonini, A.P.F. Turner, S.A. Piletsky, Biosens Bioelectron **22**, 1131 (2007)
154. D.E. Hansen, Biomaterials **28**, 4178 (2007)
155. A. Rachkov, N. Minoura, Biochim. et Biophys. Acta **1544**, 255 (2001)
156. A. Rechichi, C. Cristallini, U. Vitale, G. Ciardelli, N. Barbani, G. Vozzi, P. Giusti, J. Cell. Mol. Med. **11**, 1367 (2007)
157. G. Vozzi, I. Morelli, F. Vozzi, C. Andreoni, E. Salsedo, A. Morachioli, P. Giusti, G. Ciardelli, Biotechnol. Bioeng. **106**, 804 (2010)
158. A. Engel, D.J. Müller, Nat. Struct. Mol. Biol. **7**, 715 (2000)
159. D.J. Müller, H. Janovjak, T. Lehto, L. Kuerschner, K. Anderson, Prog. Biophys. Mol. Biol. **79**, 1 (2002)
160. J.D. Swalen, D.L. Allara, J.D. Andrade, E.A. Chandross, S. Garoff, J. Israelachvili, T.J. McCarthy, R. Murray, R.F. Pease, J.F. Rabolt, K.J. Wayne, H. Yu, Langmuir **3**, 932 (1987)
161. S. Fiorilli, P. Rivolo, E. Descrovi, C. Ricciardi, L. Pasquardini, L. Lunelli, L. Vanzetti, C. Pederzolli, B. Onida, E. Garrone, J. Colloid Interf. Sci. **321**, 235 (2008)
162. D. Niwa, Y. Yamada, T. Homma, T. Osaka, J. Phys. Chem. B **108**, 3240 (2004)
163. G. Zhang, T. Tanii, T. Zako, T. Funatsu, I. Ohdomari, Sens. Actuat. B **97**, 243 (2004)
164. T. Strother, W. Cai, X. Zhao, R.J. Hamers, L.M. Smith, J. Am. Chem. Soc. **122**, 1205 (2000)
165. L. Zhang, T. Strother, W. Cai, X.P. Cao, L.M. Smith, R.J. Hamers, Langmuir **18**, 788 (2002)
166. D.P. Allison, P. Hinterdorfer, W.H. Han, Curr. Opin. Biotechnol. **13**, 47 (2002)

167. M. D'Acunto, G. Ciardelli, A. Rechichi, F.M. Montevecchi, P. Giusti, in *Fundamentals of Friction and Wear at Atomic Scale*, ed. by E. Gnecco, E. Meyer (Springer–Verlag, Berlin Heidelberg, 2007)
168. E. Evans, Annu. Rev. Biophys. Biomol. Struct. **30**, 105 (2001)
169. M. Lantz, Y.Z. Liu, X.D. Cui, H. Tokumoto, S.M. Lindsay, Surf. Interface Anal. **27**, 354 (1999)
170. C.M. Stroh, A. Ebner, M. Geretschlager, G. Freudenthaler, F. Kienberger, A.S.M. Kamruzzahan, S.J. Smith-Gill, H.J. Gruber, P. Hinterdorfer, Biophys. J. **87,** 1981 (2004)
171. L.A. Chtcheglova, J. Waschke, L. Wildling, D. Drenckhahn, P. Hinterdorfer, Biophys J. **93**, L11 (2007)
172. T.H. Enderle, T. Ha, D.F. Ogletree, D.S. Chemla, C. Magowan, S. Weiss, Proc. Natl. Acad. Sci. **94**, 520 (1997)
173. A. Jauss, J. Koenen, K. Weishaupt, O. Holliricher, Single Mol. **3**, 232 (2002)
174. F.H. Lei, G.Y. Shang, M. Troyon, M. Spajer, M. Morjani, J.F. Angiboust, M. Manfait, Appl. Phys. Lett. **79**, 2489 (2001)
175. J. Hwang, L.A. Gheber, L. Margolis, M. Edidin, Biophys. J. **74**, 2184 (1998)
176. T. Yoshino, S. Sugiyama, S. Hagiwara, D. Fukushi, M. Shichiri, H. Nakao, J.-M. Kim, T. Hirose, H. Muramatsu, T. Ohtani, Ultramicroscopy **97**, 81 (2003)
177. C. Szymanski, C. Wu, J. Hooper, M.A. Salazar, A. Perdomo, A. Dukes, J.D. McNeill, J. Phys. Chem. B **109**, 8543 (2005)
178. J. Kim, J. Kim, K.I.B. Song, S.Q. Lee, E.U.N.K. Kim, S.E.U.L. Choi, Y. Lee, K.H.O. Park, J. Microsc. **209,** 236 (2002)
179. H.K. Park, Y.T. Lim, J.K. Kim, H.G. Park, B.H. Chung, Ultramicroscopy **108**, 1115 (2008)
180. E. Trevisan, E. Fabbretti, N. Medic, B. Troian, S. Prato, F. Vita, G. Zabucchi, M. Zweyer, NeuroImage **49**, 517 (2010)
181. L. Aeschimann, PhD thesis, University of Neuchatel, 2004
182. L. Aeschimann, T. Akiyama, U. Staufer, N.F. deRoij, L. Thiery, R. Eckert, H. Heinzelman, J. Microscopy **209**, 182 (2003)
183. M. Fleischmann, P.J. Hendra, A.J. McQuillan, Chem. Phys. Lett. **26**, 163 (1974)
184. S. Nie, S.R. Emory, Science **275**, 1102 (1997)
185. E. Betzig, J.K. Trautman, Science **257**, 189 (1992)
186. H.F. Hess, E. Betzig, T.D. Harris, L.N. Pfeiffer, K.W. West, Science **264**, 1740 (1994)
187. E. Betzig, J.K. Trautman, T.D. Harris, J.S. Weiner, R.L. Kostelak, Science **251**, 1468 (1991)
188. J. Wessel, J. Opt. Soc. Am. B **2**, 1538 (1985)
189. J. Steidtner, B. Pettinger, Phys. Rev. Lett. **100**, 236101 (2008)
190. K.F. Domke, D. Zhang, B. Pettinger, J. Am. Chem. Soc. **129**, 6708 (2007)
191. E. Bailo, V. Deckert, W. Ange, Chem. Int. Ed. **47**, 1658 (2008)
192. T. Schmid, A. Messmer, B.S. Yeo, W.H. Zhang, R. Zenobi, Anal. Bioanal. Chem. **391**, 1907 (2008)
193. U. Neugebauer, P. Rosch, M. Schmitt, J. Popp, C. Julien, A. Rasmussen, C. Budich, V. Deckert, Chem. Phys. Chem. **7**, 1428 (2006)
194. B.S. Yeo, S. Madler, T. Schimd, W.H. Zhang, R. Zenobi, J. Phys. Chem. **112**, 4867 (2008)
195. E. Bailo, V. Deckert, Chem. Soc. Rev. **37**, 921 (2008)
196. E. Bailo, V. Deckert, Angew. Chem. Int. Ed. **47**, 1658 (2008)

Chapter 23
Conductive Atomic-Force Microscopy Investigation of Nanostructures in Microelectronics

Christian Teichert and Igor Beinik

Abstract Conductive atomic-force microscopy (C-AFM), where a conductive, biased probe is scanned in contact mode across the surface under investigation is one of the most prominent scanning probe microscopy based techniques to study electrical properties of dielectric and semiconducting thin films on the nanometer scale. The technique, originally developed to evaluate the homogeneity in gate dielectrics is also successfully applied to study electrical and electronic properties of semiconductor nanostructures. The chapter starts with the discussion of the technical implementation of the technique (both under ambient conditions and in ultra-high vacuum) and the experimental peculiarities due to contact mode. The concepts of two-dimensional current maps acquired at constant tip-to-sample bias and local current voltage maps will be introduced for the example of thin silicon gate oxide and high-k dielectric thin films. Applicability of C-AFM to semiconductor nanostructures is demonstrated for supported semiconductor nanowires and free standing nanorods. Characterization of antiphase defects in ternary alloys and ZnO based multilayer varistor films show the technique's potential for device evaluation. An outlook is devoted to the so-called photoconductive AFM where photocurrents are detected under simultaneous illumination with monochromatic light.

23.1 Introduction

The trend of miniaturization in semiconductor technology pushes conventional test methods for materials and devices to their limits and demands for characterization techniques operating on the nanometer scale. Since the invention of atomic-force microscopy (AFM) in the mid 1980s, a lot of progress has been made concerning the development of derivatives of the AFM method, especially focused on the electrical characterization of materials used in semiconductor processing [1, 2]. Among these techniques are scanning capacitance microscopy (SCM) [3, 4] and scanning spreading resistance microscopy (SSRM) [5, 6] which are primarily used to image the distribution of dopant concentrations within semiconductors. The development of Kelvin probe force microscopy (KPFM) [7, 8] which is also called scanning

Kelvin force microscopy [1] allows the detection of tip-surface contact potential differences. This can be used to study devices under applied bias [9], grain boundaries in ferroelectric thin films [10], or even for failure analysis in dynamic random access memories [11]. Investigation of electromechanical responses in ferro- and piezoelectric films resulted in the development of piezoresponse force microscopy (PFM) [12, 13]. Another derivate of AFM is electrostatic force microscopy (EFM), which allows the detection of charges within dielectric films [14] or to study charging processes in metal or semiconductor nanoclusters embedded in an isolating matrix [15].

A further technique, being the topic of this book chapter, is conductive atomic-force microscopy (C-AFM) which has been developed in the mid 1990s in order to measure currents through thin silicon gate dielectrics on the nanometer scale [16–20]. The technique – also known under the term tunneling atomic-force microscopy (TUNA) – operates in contact mode employing conductive, biased probes. Simultaneously to the topography the current between tip and sample is recorded as a function of the applied bias between sample and tip. C-AFM is a superior tool to study the quality of dielectric thin films with respect to their dielectric parameters and thickness homogeneity on the nanometer scale. (An important application of C-AFM – but not being topic of this chapter – is its utilization to modify surfaces and create structures on the nanometer scale by local oxidation [21–24] allowing even to fabricate nanoscale devices [25–27].) Originally exclusively applied to characterize SiO_2 films C-AFM has been more recently successfully applied to study the so-called high-k dielectric thin films [28–31]. Further recent applications include conductivity measurements on phase-separated semiconductor films [32, 33] and free standing nanorods (NRs) [34] as well as electrical characterization of electroceramic devices [35].

Here, we will first consider the experimental aspects of C-AFM with special focus on current amplification [36,37], tip issues [38,39], and the implementation of the technique under ultra-high vacuum conditions [20,40]. Second, using the investigation of gate dielectrics, the common measurement routines as two-dimensional (2D) current maps recorded at constant tip–sample bias and the sampling of local current–voltage (I–V) characteristics by ramping the bias will be introduced. For the latter, a so-called pseudo-breakdown voltage (PBDV) will be defined [28]. Third, conductivity measurements of phase-separated semiconductor nanostructures will be demonstrated. Then, first results of C-AFM based exploration of free standing ZnO NRs and ZnO based electroceramics will be presented. Afterwards, we will present a rather detailed outlook to a recently developed derivative of C-AFM, namely photoconductive AFM (PC-AFM) [41,42] allowing to detect currents under sample illumination, i.e., measuring local photocurrents. This dedicates the technique especially to the characterization of materials for photovoltaic applications. The chapter will finish with an overall summary and a perspective to further fields of potential applications of C-AFM.

23.2 Technical Implementation of C-AFM

The C-AFM is a conventional AFM operating in a contact regime utilizing a conductive cantilever and tip. Figure 23.1 presents a principle scheme of the C-AFM setup. Besides conventional AFM scanner and cantilever detection technique, the setup employs a conductive AFM probe, an external voltage source, and a current amplifier connected to the AFM tip. The voltage is applied between the conductive AFM tip and the sample to obtain the desired electrical information. The resulting current is measured using a highly sensitive amplifier as a current-to-voltage converter. The amplification is especially important for samples with high resistance or even isolating behavior, where the measured currents can be as low as several femto- to picoamperes. The C-AFM technique bears a striking resemblance to scanning tunneling microscopy (STM) using similar technical solutions for the current signal acquisition. However, there is a fundamental difference in the operational principle. In contrast to STM, the conductive AFM measures the current signal completely independent from the topography which is simultaneously recorded via the cantilever deflection.

In order to obtain a potential drop and current across the sample, an external bias is applied to the sample. (In principle, tip biasing is also possible but rarely used.) It is also worth to notice that the contact to the sample should be preferentially ohmic, in other case the influence of the additional barrier has to be considered carefully. The commonly used range of the applied voltage is $\pm 10\,\mathrm{V}$. The application of higher voltages is possible using external voltage sources, though additional circuit protection should be implemented.

The current signal is amplified and converted to a voltage signal either by electrometer or logarithmic amplifiers, depending on the desired range of magnitude. The main difference between these two amplifiers is a feedback loop. In the case of the electrometer, the feedback loop is performed using a feedback resistor R_f (see Fig. 23.1), whereas a logarithmic amplifier utilizes a nonlinear element (diode) for this purpose. The voltage signal is more convenient to deal with, and it can be

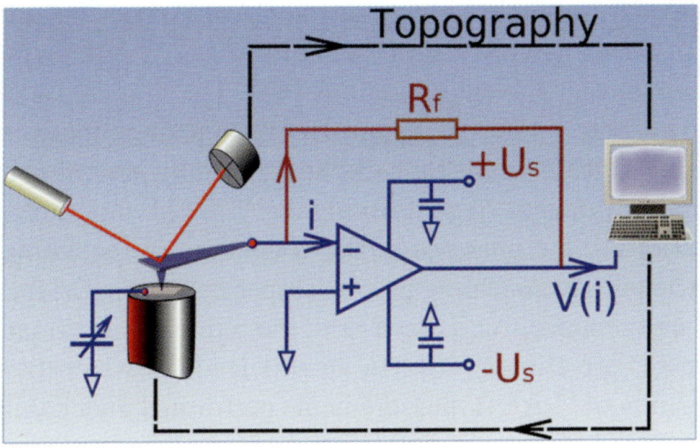

Fig. 23.1 Schematic drawing of the conductive AFM setup

easily converted to a digital form using an analog-to-digital converter (ADC). Usually, the AFM electronics provides several voltage input/output channels equipped with the high-resolution ADC's, which are appropriate for both the current signal acquisition and sample biasing. For the materials under investigation, the currents to be detected will cover a range from fA (tunneling current throughout oxides) to mA for semiconducting or conducting material. Over the last decade, several commercial solutions have been elaborated which exploit one of the above-mentioned approaches. A system based on operational amplifiers (OA) is usually able to detect lower currents (down to tens of fA) with small noise levels, while systems based on logarithmic amplifiers overlap a wider range of currents (1 pA–1 mA) [43, 44]. However, the use of logarithmic amplifiers restricts the operational mode to a single voltage operation in most of the cases.

Low-noise, high-precision OA with inputs based on isolated field effect transistors are widely used in commercial and home-built systems as a main component. Figure 23.1 presents a simple electrometer scheme built on the basis of such an amplifier. This setup is usually able to measure currents down to a few tenth of femtoamperes when properly assembled. In principle, this scheme is a current-to-voltage converter with the simple conversion law $U_{\mathrm{out}} = I_{\mathrm{inp}} R_{\mathrm{f}}$. Although having superior sensitivity as well as low inherent noise, the scheme requires a solution of several problems related to the design and assembly of current-to-voltage converters for direct current (DC) amplification. The noise induction and electrical leaks should be carefully analyzed and suppressed if possible. Among possible solutions are the isolation of the inputs of the OA from the printed circuit board and minimization of the current paths. It has to be mentioned that the amplifier should be placed in close proximity to the AFM tip in order to avoid picking up noise.

The input offset voltage of the utilized OA has also to be taken into account during the design of the amplifier. The presence of small voltages at the input of the OA is unfortunately practically almost unavoidable. Optimally, the value of the offset voltage should not exceed 1–2 mV. For the best performance, it should be 0.5 mV or less, since even a small offset voltage may account for the appearance of significant currents at the output even when the sample is grounded. Parameters, like voltage noise, drift, and power-supply rejection should also be considered. There are several commercially available high-precision OA that meet the above-mentioned demands. One of those, namely, OPA129, is utilized in a home-built C-AFM setup for measurements under ambient conditions [37]. The noise level of the amplifier finally achieved is about 70 fA, the input offset voltage does not exceed 1 mV.

Commercially available setups for C-AFM are usually operating in air. However, one gains several advantages of performing the C-AFM experiments under UHV conditions. First of all, the presence of the water film in the ambient can be easily avoided in vacuum by annealing at low temperatures prior to the measurement. Second, when applying positive sample bias, the oxide layer formation and anodic oxidation processes are eluded [20]. (In air this leads to water dissociation [24].) A comparative study of C-AFM measurements performed under ambient and UHV conditions is given in [44].

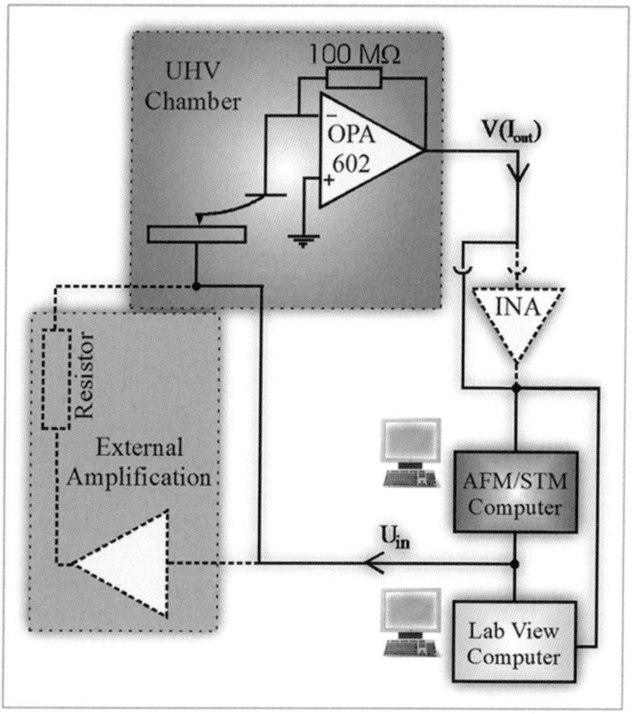

Fig. 23.2 C-AFM setup for the measurements under UHV (Adapted from [44])

A UHV implementation of the C-AFM setup is in fact only little different from the one used in the ambient with the one difficulty to place the OA close to the AFM tip, i.e., into UHV. Figure 23.2 shows a UHV C-AFM setup that has been used since early 2000 [20]. This setup is based on a combined Omicron AFM/STM system operating at room temperature. The system has also been equipped with a high voltage unit which extends the applicable voltage range from ± 10 V up to ± 150 V. The amplification factor was set to 10^8 V/A for the input amplification stage and can be additionally increased by a factor of 2 or 20 at the second amplification stage. The resulting noise level in the system is approximately 0.2 pA. In order to have more flexibility in data acquisition and analysis, the system has been supplemented with a data acquisition module (National Instruments) and LabView software.

Particular attention during C-AFM experiments has to be paid to the contact properties which depend to a large extent on the tip conditions. Although any conductive AFM probe, e.g., conventional silicon probes, can be used in C-AFM, the probes with conductive, wear resistive coatings are favored. The reason is the high stability of the tip geometry and its electrical properties. There are several types of commercially available conductive probes designed especially for C-AFM. These are conventional AFM probes for contact mode with one of the following coatings: heavily doped diamond, PtIr, TiN, W_2C, and Au. The best results are

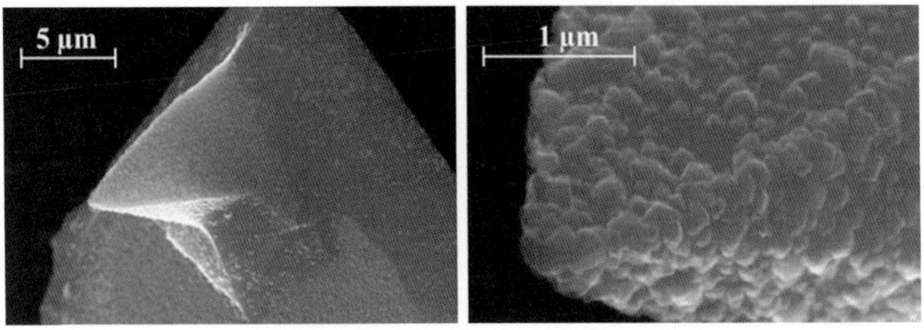

Fig. 23.3 SEM micrographs of a diamond coated AFM tip. In the *right image*, the granular structure of the diamond coating is well resolved

usually obtained with diamond coated tips. Figure 23.3 shows SEM micrographs of a diamond coated AFM tip used for 2D current imaging. There are two types of diamond coated C-AFM probes on the market, one is heavily boron doped (p-type) and another one is heavily nitrogen doped CVD (n-type) [45]. The doping is so high that the Fermi level is shifted into the valance or the conduction band of the diamond, respectively. Thus the tip can be assumed as metallic. The resulting resistivity of the diamond coating is normally about 0.5–1 Ω cm. Despite the fact that the metal coatings like Au or PtIr have lower resistivity (0.01–0.025 Ω cm), their lifetime is low compared to the diamond coated ones since the soft coating is worn much faster. The typical curvature radius of probes with metal coating is about 35 nm, whereas for the diamond coated ones it can be more than 100 nm, resulting in lower resolution on rough surfaces. However, the diamond coatings exhibit a nanoroughness in the 10 nm regime (see Fig. 23.3) which leads to excellent resolution on flat surfaces. Recently, conductive probes fabricated entirely of diamond have become commercially available (NaDiaProbe from nanoScience Instruments). They are manufactured entirely from electrically conductive ultrananocrystalline diamond (UNCD®) [46] with a resistivity less than 1 Ω cm, therefore problems caused by coating damage are no longer existing. One additional advantage of the coated and entirely diamond probes is the low adhesion and low surface energy of diamond, which helps to avoid tip contamination and thus yields reproducible contrast during current imaging. Another important parameter is the work function of the coating which has influence on the type of the tip-to-sample electrical contact and barrier height in the case of barrier formation. The work functions for several materials used as C-AFM tip coatings are summarized in Table 23.1.

An evaluation of probes for AFM electrical modes can be found in [47]. The search for an appropriate probe is a kind of compromise between desired resolution and probe life time. From a practical point of view it is always useful to check several probes and choose the one with the best performance since optimal conditions may vary significantly for different materials and surface morphologies.

Table 23.1 Work functions of materials used as conductive coatings in C-AFM probes

Coating	Pt [48]	Au [48]	W_2C [49]	PtIr [50]	Diamond [51]	TiN [52]
Work function [eV]	5.65	5.10	3.8	5.6	5.7	4.8–5.3

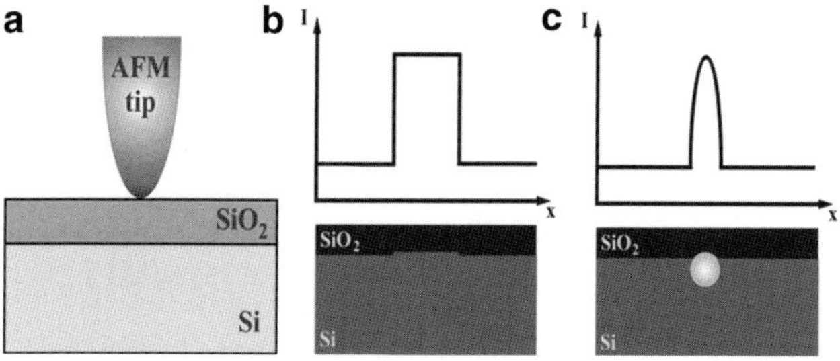

Fig. 23.4 Information which can be gained from C-AFM investigations of gate dielectrics. (**a**) The arrangement of conductive tip and sample corresponds in principle to a metal oxide semiconductor (MOS) structure. Leakage currents may result from a reduction in the oxide thickness (**b**) or from electrical defects (**c**)

23.3 C-AFM to Study Gate Dielectrics

As has been mentioned in the introduction, C-AFM has been developed to detect the tunneling currents through silicon gate dielectrics on the nanometer scale. With the shrinking dimensions of modern complementary metal oxide silicon (MOS) transistors, also the thickness of the gate oxide is further and further decreasing approaching the limit where direct tunneling is dominating the desired Fowler–Nordheim (FN) tunneling mechanism [53]. For silicon oxide, this thickness is reached at about 3 nm. Therefore, there are currently strong efforts to replace silicon oxide by high dielectric constant gate oxides, the so-called high-k dielectrics [54]. In any case, thickness fluctuations and dielectric quality become more and more crucial for the device performance. When applying the C-AFM setup to a gate oxide on silicon, an MOS structure arises. This is illustrated in Fig. 23.4 together with situations that result in a current increase with respect to the nominal current. Such leakage currents can result from a decrease in oxide thickness due to thickness fluctuations in the gate oxide fabrication and/or interfacial roughness (Fig. 23.4b), from electrical defects in the oxide (Fig. 23.4c) or from an accumulation of trapped charges over time (the so-called electrical stress). In the following paragraph, the common experimental procedures to obtain these information are illustrated for the C-AFM investigation of silicon gate oxide.

23.3.1 Local Current–Voltage Characteristics, Dielectric Breakdown, and Two-Dimensional Current Maps

For acquiring a local current–voltage (I–V) curve, the C-AFM tip will be located at a certain sample position and a sufficient force has to be applied. Then, the tip–sample voltage will be ramped from 0 V till the preselected final value as illustrated in Fig. 23.5a. For each data point of the I–V curve, the voltage is increased during a certain delay time, which is necessary due to the capacitive parts within the circuit. After the delay time, the current is measured with 50 kHz of sampling rate and averaged over the acquisition time. The curve ends at the preselected voltage or once a preselected maximum current (e.g., 10 pA to avoid dielectric breakdown) has been reached. For a typical I–V curve with 200 data points and typical delay and acquisition times of 20 ms and 50 ms, respectively, a measurement takes about 14 s. Assuming Fowler–Nordheim tunneling [53], the recorded I–V curve can be fitted by

$$I = A_{\text{eff}} \frac{q^2 m_0}{8\pi h m_{\text{eff}}} \frac{1}{t(E)^2} \frac{\beta^2 V^2}{\phi d_{\text{ox}}^2} \exp\left(\frac{8\pi\sqrt{2 m_{\text{eff}} q}}{3h} v(E) \frac{d_{\text{ox}}}{\beta V} \phi^{3/2}\right), \qquad (23.1)$$

where $\underline{A}_{\text{eff}}$ is the effective tunneling area, q is the electron charge, m_0 is the free electron mass, m_{eff} is the effective mass of the electron in the oxide ($m_{\text{eff}}/m_0 = 0.4\ldots0.5$) [7, 8], h is the Planck's constant, d_{ox} is the oxide thickness, V is the applied voltage and ϕ is the barrier height which is 3.25 eV for silicon oxide on silicon. $t(E)$ and $v(E)$ take into account that image charge lowering [5] occurs during the FN tunneling mechanism. $t(E)$ and $v(E)$ do not vary in a wide range. At electric fields between 10 and 25 MV/cm, $t(E)$ varies between 1.011 and 1.013 and $v(E)$ is between 0.925 and 0.937 [17, 19]. β is the field enhancement factor and arises from the nonplanar geometry of the tip. This factor has to be taken into account when injection from the tip occurs (positive sample voltage). For injection from the silicon substrate, however, this factor can be assumed as 1. The effective tunneling area depends on the actual tip geometry as well as on the oxide thickness [19]. In Fig. 23.5b, a FN fit for a local I–V curve of a 4.5 nm thermal silicon gate oxide is presented and the influence of the parameter A_{eff} is demonstrated. In order to obtain statistical information on the thickness fluctuations, I–V-curves for a large array of measurement points have to be recorded as is demonstrated in Figs. 23.5c, d. Besides determining the local oxide thickness by FN fits, it is also useful just to extract a so-called PBDV [28]. The value is defined as the voltage for which a certain small measurable current (e.g., 1 or 2 pA) is detected. The distribution of these values is also a measure for thickness fluctuations. As has been demonstrated in [20], one can further determine the voltage where true dielectric breakdown occurs which results in significantly higher currents (e.g., 20 pA) and possible modification of the gate oxide [24]. However, in this case care has to be taken that the lateral separation of the individual measurement spots is sufficient to exclude an influence during consecutive measurements by charge injection from adjacent spots [24]. For

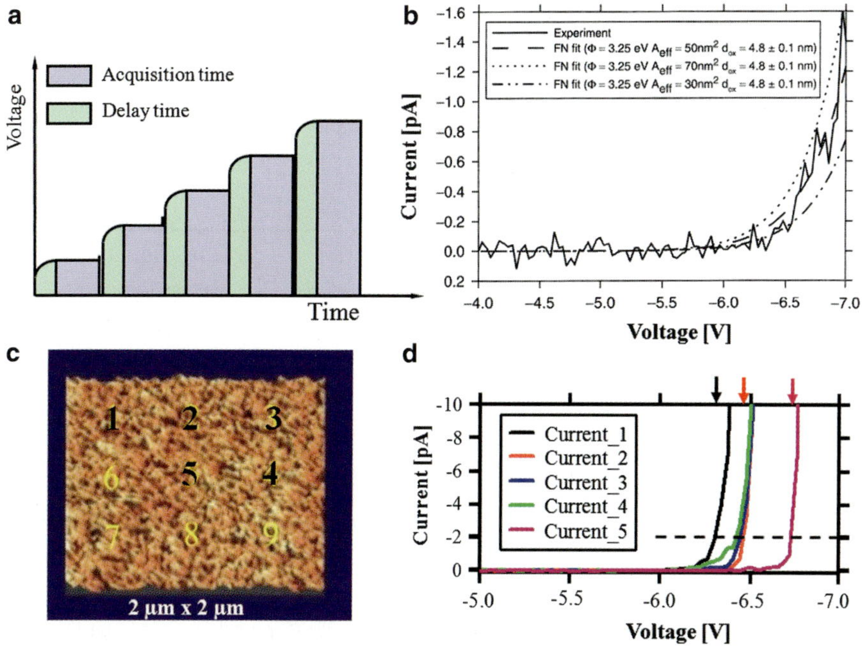

Fig. 23.5 Procedure to obtain local I–V characteristics on gate oxides. (**a**) Voltage ramp applied to the sample. (**b**) Part of an I–V curve measured in UHV on a nominal 4.5 nm thick silicon oxide film with corresponding FN fits [20]. (**c**) Array of measurement locations on a 2 µm × 2 µm area of the 4.5 nm SiO_2 film and (**d**) corresponding I–V curves. The *arrows* in (**d**) mark the pseudobreakdown voltages at which a current of 2 pA (*dashed line*) is reached

device quality silicon gate oxides the histograms of the breakdowns can be fitted by Gaussian distributions as is presented in Fig. 23.6a–c) for different gate oxide thicknesses. The width of these distributions is a measure for the thickness homogeneity. It has to be noted that the peak breakdown voltage for a given thickness does also depend on the measurement speed [20]. Extrinsic defects as caused by intentional contamination of the gate oxide, e.g., with aluminum, cause low breakdown voltages which are clearly outside the Gaussian distribution (see Fig. 23.6d).

In the case that one is interested in the local arrangement of either the above-mentioned extrinsic defects or of the lateral distribution of thickness fluctuations, 2D current maps recorded at a constant voltage are the method of choice. Such a 2D-CAFM image is presented in Fig. 23.7 again for a 4.5 nm thermally grown SiO_2 film measured in UHV for a sample voltage of −3.8 V. The resulting currents are in the range of 1 pA. There is no general decrease in current due to passivation as it is frequently observed for C-AFM measurements in air or if a water film is present on the surface [24]. The results presented in Figs. 23.5–23.7 have all been obtained employing Si tips with boron doped diamond coating. It has to be noted here that the electrical oxide thickness determined by C-AFM is not fully independent of the tip coating [55].

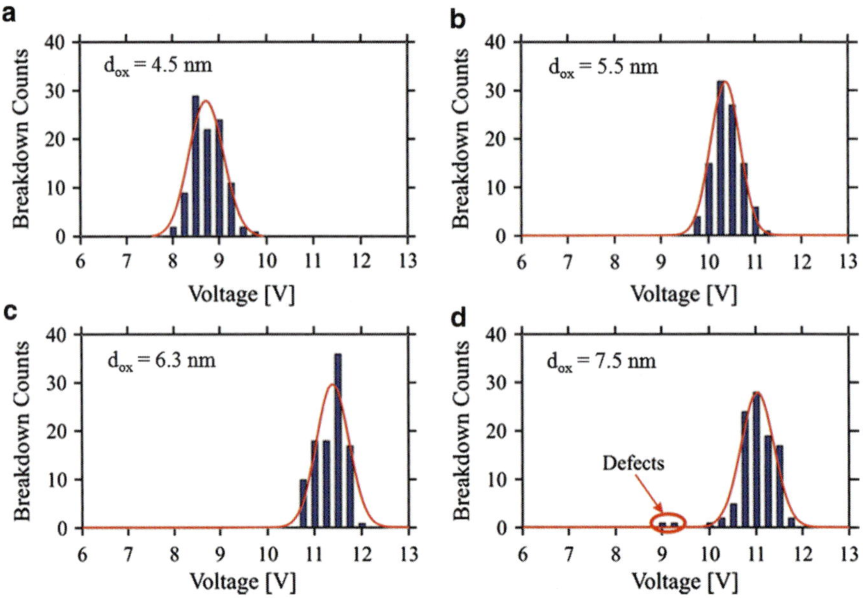

Fig. 23.6 (**a–c**) Histograms of dielectric breakdown voltages (for reaching 20 pA) on thermally grown silicon oxides with 4.5, 5.5, and 6.3 nm, respectively. (**d**) Histogram of a dielectric breakdown on a 7.5 nm silicon oxide intentionally contaminated with aluminum. In all cases, 100 I–V curves were analyzed (Adapted from [20])

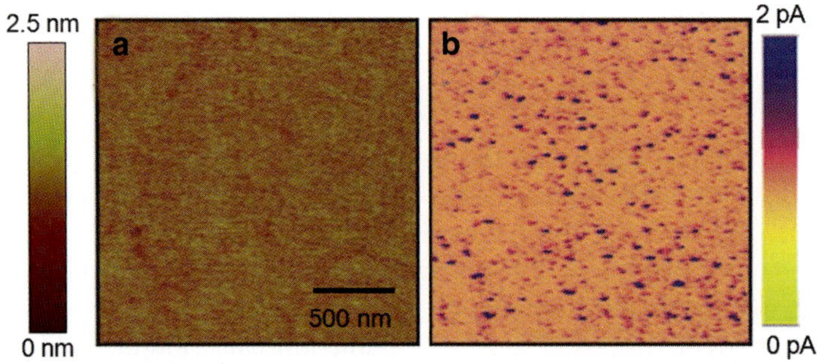

Fig. 23.7 (**a**) 2 μm × 2μm topography image and (**b**) corresponding 2D current image from a C-AFM measurement of 4.5 nm SiO_2 film recorded at −3.8 V sample bias in UHV using diamond coated conductive probes

A third procedure for C-AFM is the so-called electrical stress measurement, where – instead of a voltage ramp for acquiring I–V-curves – a constant voltage is applied for a certain period of time without scanning the tip. With this approach, the current evolution can be recorded vs. time and as a function of the applied voltage on different spots of the surface. Thus, it is possible to obtain information about the electrical stress influence on various material systems [56]. However, it has to be considered that possible drift effects, which cause an undesired movement of the AFM tip on the surface can influence the experimental results. Therefore, especially in the case of long-term experiments it is necessary to be very careful with the interpretation.

23.3.2 *Investigation of High-k Dielectrics*

The C-AFM procedures developed in the late 1990s and described above are fully applicable to study high-k materials [58]. However, the first use of C-AFM in the characterization of the most common high-k dielectrics like hafnium oxide has been reported – to our knowledge – only in 2004. There, the electrical homogeneity of HfO_2 and $HfAlO_x$ films on silicon has been studied in air as a function of sample annealing [59, 60]. Because of the hydrophilicity of most of the high-k dielectrics which are under consideration to replace silicon oxide [61], it is certainly advisable to perform the C-AFM measurements for this kind of research in vacuum.

In the following, ultra-high vacuum C-AFM measurements of hafnium oxide and zirconium oxide films are presented. The films have been prepared by atomic layer deposition on n-type Si wafers [28]. Again, Si tips with boron doped diamond coatings have been used. In Fig. 23.8, the surface morphology for three different HfO_2 film thicknesses is presented together with histograms of the PBDVs. The surface roughness increases from 0.15 nm for the 5 nm film to 0.5 nm for the 19 nm thick film which has been attributed to the onset of crystallite formation in the initially amorphous matrix. In fact, two different crystallographic phases, namely an orthorhombic and a monoclinic one have been detected by X-ray diffraction [28]. This situation is reflected in the corresponding histograms of the PBDVs for the 19 nm film which is significantly broadened with respect to that of the 5 nm amorphous film. In fact, already at 10 nm film thickness, outliers on both sides of the narrow distribution of PBDVs are observed. Figure 23.9 summarizes the results obtained by recording 2D current maps. Let us first focus on the occurrence of leakage currents in the thin amorphous films, e.g., for an 8 nm ZrO_2 film (Fig. 23.9a, b). Increased currents (appearing bright in the 2D current map) are mainly occurring at depressions in the topography. Thus, they can mainly be attributed to a reduction in the thickness of the amorphous film. Such situations are labeled with (1) in the images as well as in the corresponding sketch presented in Fig. 23.9e. Occasionally, increased currents are also observed although there is no deviation from the mean surface level (2). Such situations may result from the interfacial roughness between dielectric film and Si substrate as is also indicated in Fig. 23.9e. Analyzing the 2D C-AFM results of the 10.5 nm HfO_2 film (Fig. 23.9c, d), we can observe four additional situations labeled (3)–(6). Label (3) marks protrusions which exhibit unexpectedly an increased current, they can be attributed to crystallites showing higher conductance (lower-k value) than the amorphous matrix (Fig. 23.9f). Contrary to this there exist crystallites [marked with (4)] exhibiting very low tunneling current. These crystallites have a higher-k value and are responsible for the high PBDVs observed

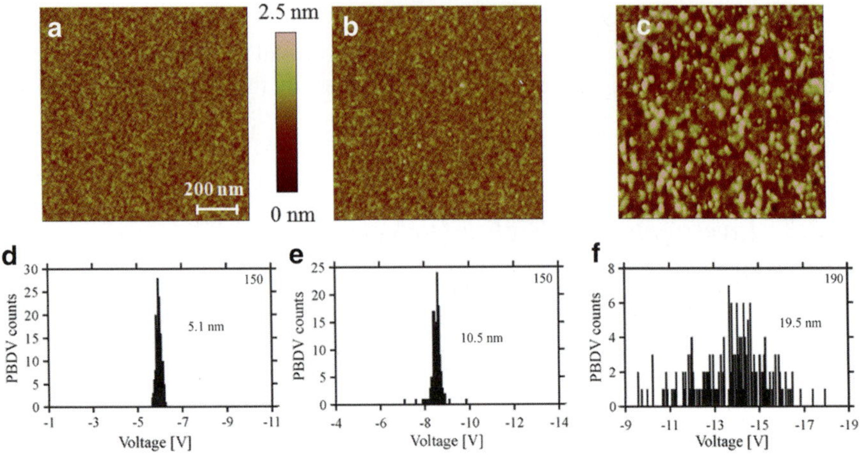

Fig. 23.8 (**a–c**) 1 μm × 1 μm AFM topography images of HfO_2 films with 5.1, 10.5, and 19.5 nm thicknesses, respectively. The vertical scale is 2.5 nm in all images. (**d–f**) Corresponding histogram of PBDVs (for reaching 1 pA) as a function of the applied sample voltage. The numbers in the upper right corner of each graph indicate the total number of I–V curves analyzed (Adapted from [28])

in the corresponding histogram (Fig. 23.8e). Label (6) marks increased currents just between adjacent crystallites indicating leakage currents at grain boundaries.

The combination of local I–V-curves, analysis of PBDVs, and 2D current maps allows to locate increased leakage currents due to thickness fluctuations and/or defects in high-k dielectric thin films. The lateral resolution in these C-AFM investigations is about 10–15 nm as can clearly be seen in Fig. 23.9b, d. The defects can be identified at least qualitatively. Quantitative interpretation is, however, not easy since also the native oxide layer – present between the high-k dielectric layer and the silicon substrate – has to be taken into account. Finally, we have to point out that performing C-AFM investigations of the same samples and employing the same type of probes under ambient conditions resulted in poor lateral resolution and we faced problems with tip passivation and sample modification due to the water film present on the surface.

C-AFM has been applied in the last years for a variety of single high-k dielectrics and high-k stacks including also other materials like Al_2O_3 [31], Pr_2O_3 [57], and Ta_2O_5 [62]. However, most of the studies are still performed under ambient conditions. As we demonstrated above and as has been investigated recently by a detailed comparative study [44], improved lateral resolution and reliability of the data is only obtained by measurements performed under vacuum. To our opinion, the situation will be different when applying the technique to organic dielectric thin films used in the growing field of organic or molecular electronics [63]. However, here the problem of damaging the soft film due to the relatively large lateral forces in contact mode arises. This might be overcome by applying the so-called jumping mode scanning force microscopy [64].

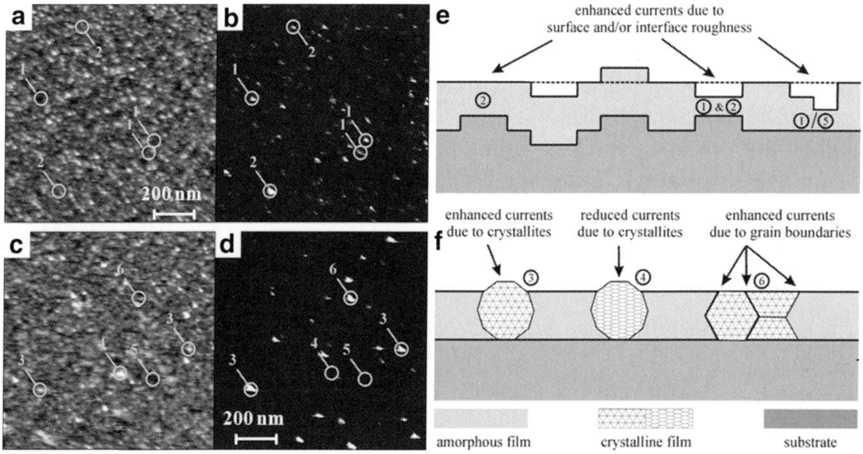

Fig. 23.9 Interpretation of 2D current maps acquired in UHV for atomic layer deposited HfO_2 and ZrO_2 films: (**a**) 1 μm × 1 μm topography and (**b**) 2D current images of an 8.2-nm ZrO_2 film recorded at −5.7 V. The *z* scale is 4 nm in the topography image and 5 pA in the current image. (**c**) 1 μm × 1 μm topography and (**d**) 2D current images for the 10.5-nm HfO_2 film measured at −8.8 V. The *z* scale is 2.5 nm in the topography image and 5 pA in the current image. The *circles* with the corresponding numbers mark areas with distinct topographical and current characteristics. (**e**, **f**) Schematic illustration of thickness fluctuations (**e**) and crystallite formation (**d**) having an influence on current characteristics in the high-k dielectric thin films (Adapted from [28])

23.4 Conductivity Measurements of Phase-Separated Semiconductor Nanostructures

In the fabrication of functional semiconductor thin films for lasers, solar cells, and photodetector applications, phase-separated nanostructures like supported nanowires (NWs) nanorods (NRs), or antiphase defects are frequently evolving. These structures may be advantageous or disadvantageous for the device performance. In any case, their electrical properties have to be studied on the nanometer scale. Besides KPFM, C-AFM is a technique to reveal the electrical characteristics of such phase-separated semiconductor nanostructures as we will demonstrate in two exemplifying cases. Prior to this we will consider in the following several aspects of using C-AFM to study semiconductor nanostructures.

C-AFM enables one to detect currents as low as several hundreds of femtoamperes as described in the experimental section. Therefore this technique is sufficiently sensitive to detect small variations in the local conductivity of the sample surface. In most of the cases, one deals with the formation of Schottky contacts. This point becomes crucial in the exploration of self-organized structures on a sample surface. In the case of high loads, above 100 nN, the measured resistance (or current) is dominated by the spreading resistance rather than by the contact resistance. Despite the success achieved in the field of the SSRM for doping profiling with excellent concentration sensitivity and determination of junction position with high spatial resolution [70], it remains difficult to measure structures with the size of the order of several nanometers in one of their dimensions. Another issue is the lateral resolution which drops when the contact area is increasing. Reducing the loading force below 10–20 nN – using soft cantilevers – improves the lateral resolution and one can get access to the local variation of the contact resistance and Schottky barrier height. Thus, the difference in material properties can be resolved at low loading forces in C-AFM.

23.4.1 Exploration of Supported Nanowires and Nanodots

Applying C-AFM, we examine here InAs nanostructures, namely, NWs and nanodots (ND) grown by molecular beam epitaxy (MBE) on GaAs. Two different self-organized GaAs substrates have been used as stepped templates, the details of their preparation can be found in [32]. Here, we demonstrate that C-AFM is suitable for the characterization of low-dimensional structures, like supported NWs and ND beyond the limits of conventional AFM.

23.4.1.1 C-AFM of InAs NW on GaAs(110)

In preliminary measurements, a sample containing InAs NW was measured under ambient conditions using DCP11 probes from NT-MDT with a force constant in the range of 5–12 N/m. The applied loading forces ranged from 30 to 100 nN. In Fig. 23.10, the topography (a) and current map (b) recorded at a sample bias of −5 V are presented. The topography image Fig. 23.10a reveals a characteristic multiatomic step array with step edges running along [1–10], [1–12] and [−112] separated by (110) terraces. The current absolute value ranged from 0 to 15 pA. The current map reveals an increase of conductivity at the step edges of the GaAs substrate indicated by current peaks occurring along the [1–10] direction but almost no current along the [1–12] and [−112] step edges.

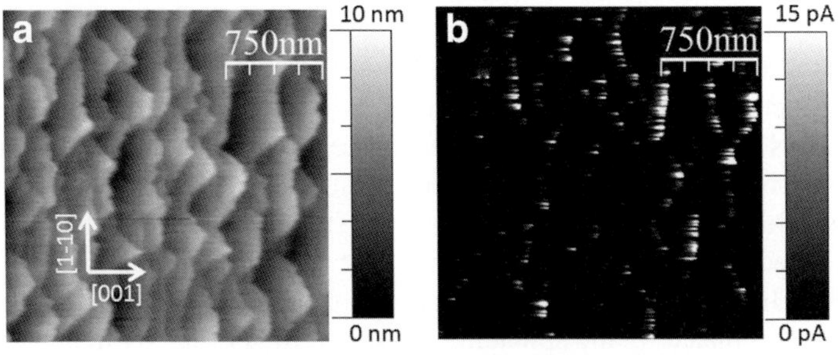

Fig. 23.10 (**a**) 2.5 μm × 2.5 μm topography image of InAs/GaAs NW measured under ambient conditions. (**b**) 2D current map recorded at −5 V sample bias (**b**)

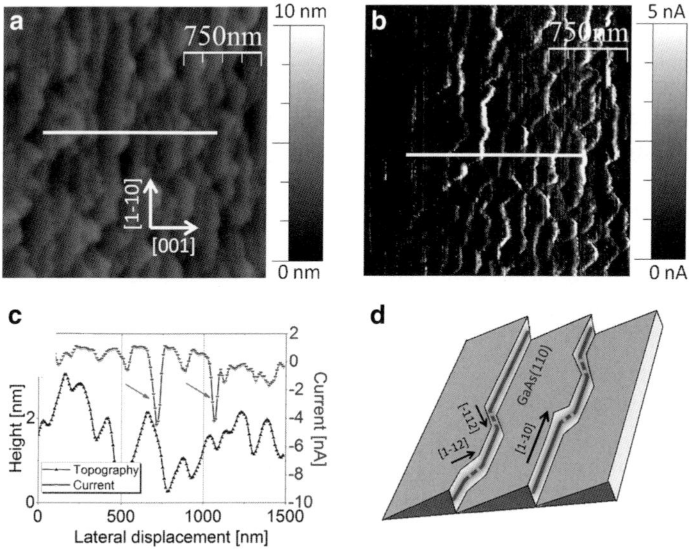

Fig. 23.11 C-AFM results of InAs NW on GaAs recorded in UHV. (**a**) 2.5 μm × 2.5 μm topography image, (**b**) 2D current map recorded at −9.1 V sample bias, (**c**) cross-sections along the line marked in (**a**, **b**), (**d**) schematic drawing of InAs NW growth on the giant steps of the GaAs surface, comparatively small amounts of InAs on the [1–12] and [−112] edges depicted by the *dashed line* (Adapted from [33])

The topography and current map recorded under UHV conditions for the same sample are depicted in Fig. 23.11a, b. The applied sample bias was −9.1 V and the tip load was about 100 nN. In this case the current image clearly shows stripes of high conductivity running along the [1–10] direction. In contrast to the images recorded at ambient conditions, weak but detectable conductivity is also observed along the [1–12] and [−112] directions. The overlap of the corresponding cross-sections of the topography and current images taken along the [001] direction which is depicted in Fig. 23.11c clarifies that the regions of maximum conductivity coincide with the center of the step bunches [32].

When considering the tip-to-sample contact we assume that the Schottky barrier height for InAs is lower than the height for the GaAs (the electron affinity for the GaAs is 4.07 and 4.9 eV for InAs). Therefore the areas of higher conductivity can be attributed to the presence of InAs.

The difference between the current maps measured in UHV and under ambient conditions can be attributed to instabilities caused by the presence of adsorbed water on the surface exposed to air. Since the current maps were acquired at negative bias voltages, anodic oxidation of the surface can be ruled out. The instabilities can be explained via contact abruption during the imaging process. When the conductive tip comes into contact with the adsorbed water film on the biased surface water dissociation takes place and contact abruption occurs. This appears in the current maps as a current termination and lateral spreading of the current peaks. Therefore

UHV C-AFM current maps demonstrate much higher stability than those taken in the ambient.

Referring to the current map measured in UHV (Fig. 23.11b) InAs preferentially decorates step edges along the [1–10] direction. The InAs decoration is less pronounced along the [1–12] and [−112] step edges. This result is illustrated in Fig. 23.11d.

The latter finding is in good agreement with earlier cross-sectional transmission electron microscopy (TEM) investigations [71], which revealed that the formation of InAs NW is a result of step decoration of the self-organized GaAs layer. The TEM measurements showed that InAs preferentially decorated the [1–10] step edges rather than those along [1–12] and [−112]. In contrast to cross-sectional TEM investigations with little statistics, the 2D current maps reveal the uniformity in lateral arrangement of the NW.

23.4.1.2 C-AFM of InAs ND on GaAs(110)

The current maps of the surface containing InAs ND measured under UHV conditions were acquired under similar conditions to those used for InAs NW imaging. The morphology of the surface under investigation is shown in Fig. 23.12a, b and is schematically depicted in the Fig. 23.12d. As it was shown in [32], the specific morphology of the template along with the H-assisted surface cleaning preliminary to MBE growth offers preferred nucleation sites resulting in lateral confinement of InAs. This leads to the transition from 2D growth to 3D mound formation which is energetically and kinetically most favorable. The analysis of the correlation between topography (Fig. 23.12a) and current map (Fig. 23.12b) simultaneously with the corresponding profiles (Fig. 23.12c) taken along the tilt direction on the terraces bounded by {1–15} facets, revealed several current peaks originating from the terrace. The current from the terrace apex is most pronounced and has a magnitude of ~4 nA. The comparison with data obtained from the surface containing InAs NW enables us to conclude that InAs deposits form three-dimensional (3D) clusters at the apexes of the triangular terraces with an average lateral size of ~65 nm. The estimated density of the clusters was $\sim 4 \times 10^9\,\text{cm}^{-2}$. Additionally, it was found that the entire surface of the terraces is covered by small and dense 2D islands which are about 10 nm in the lateral dimension [32]. The currents measured from these 2D islands were smaller than those measured for 3D islands at the same bias voltage. This observation is hard to interpret, since the effective contact area between the tip and the surface can be smaller than the size of the 2D islands, causing therefore smaller currents at the same current density. However, this still may serve as an evidence of some chemical exchange which takes place between In atoms in the 2D nuclei and the substrate Ga atoms underneath and leads to the formation of InGaAs [32]. As a result we would expect a smaller concentration of In atoms in 2D islands and therefore an increase in the Schottky barrier height.

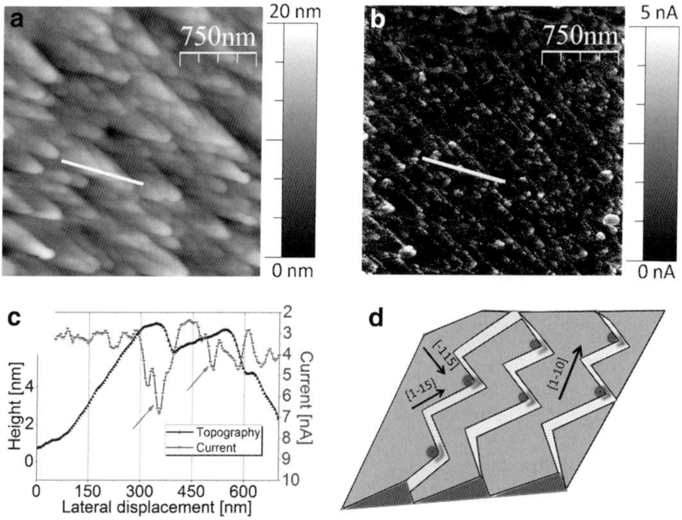

Fig. 23.12 C-AFM results of InAs/GaAs ND measured under UHV conditions. (**a**) 2.5 μm × 2.5 μm topography image, (**b**) 2D current map recorded at −9.1 V sample bias. Current peaks are marked by *arrows*, (**c**) corresponding cross-section profiles, schematic drawing of InAs ND growth on the terrace apexes of GaAs surface (**d**)

23.4.2 Investigation of Defects in Ternary Semiconductor Alloys

Arrowhead defects (AD) refer to a special kind of defects appearing during growth of ternary epitaxial III–V semiconductor films with zincblende structure [74]. Films such as GaInP grown on misoriented Ge substrates, serve as a relatively simple modeling system for solar cell applications where the GaInP thin layer is usually an integral part of more sophisticated stack structures [75]. The appearance of different kinds of defects like antiphase domain boundaries and ADs in these stacked epilayers correlates with the solar cell efficiency. Therefore, the investigation of the AD's electrical properties is of practical interest. Here, we performed C-AFM measurements of GaInP grown on a misoriented Ge substrate, for details of the sample preparation see [33, 74].

Figure 23.13a shows a tapping mode image in 3D presentation of the surface containing several AD's. It has been recorded in air using super sharp SSS-NCHR AFM probes from Nanosensors. As marked in Fig. 23.13a, we denote the left and right side planes as "L" and "R," respectively. The cross-sections taken along [010] and [001] (Fig. 23.13b) reveal that the terminating planes of the AD are composed of two alternating subplanes inclined 12° and 6° with respect to (100) plane. The comparison of the cross-section reveals, the terminating planes are morphologically equivalent. Surprisingly, the C-AFM investigations reveal a difference in the electrical behavior between the adjacent terminating planes and also with respect to the surrounding crystalline GaInP film. Figure 23.14 presents the morphology and

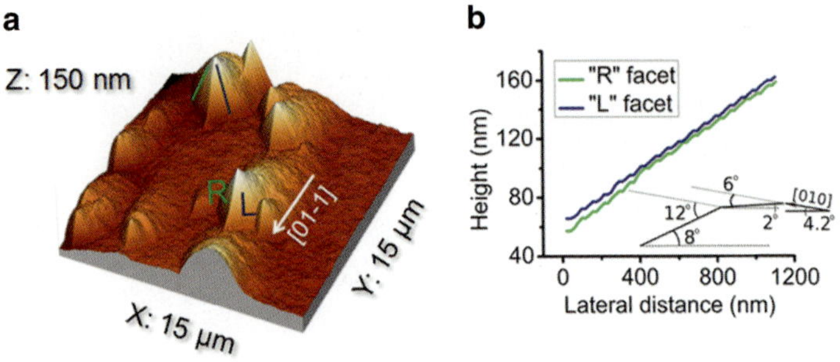

Fig. 23.13 C-AFM results of arrowhead defects (AD) in GaInP/Ge(100). (**a**) 15 μm × 15 μm AFM image in 3D presentation, (**b**) corresponding cross-section that reveals the step-structure of the AD's terminating planes which consist of two alternative subplanes inclined 6° and 12° towards the (100) plane, as is schematically depicted in the *inset* of the cross-section

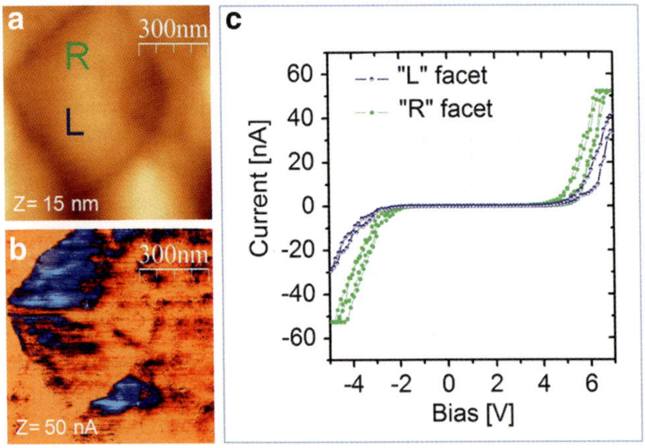

Fig. 23.14 C-AFM results of AD in GaInP/Ge(100): (**a**) 3 μm × 3 μm topography image, "L" and "R" denote the left and right facets, respectively, (**b**) corresponding current map obtained under UHV conditions at −3 V bias. The regions with the higher current have *cyan color*, (**c**) *I*–*V* curves recorded from the two adjacent planes of the AD

corresponding current map obtained by C-AFM under UHV conditions using conductive diamond coated DCP11 cantilevers from NT-MDT with force constants in the range from 3 to 10 N/m. In order to extract quantitative information, the current density for each plane and the surrounding area has been calculated. The current density for the "L" and "R" planes amounts to 278 ± 35 and 445 ± 35 $\mathrm{mA/m^2}$, respectively, averaged for 0.1 $\mathrm{mm^2}$ of the projected area close to the AD apex and a bias of −3 V. The difference in current densities between the two terminating planes was found to be ∼170 ± 35 $\mu\mathrm{A/m^2}$ at −3 V. The *I*–*V* characteristics obtained from both plane types (Fig. 23.14c) of the AD exhibit a rectifying behavior. The characteristic for the "L" plane differs from the one for the "R" plane and the defect-free surface (not shown). The absolute value of the current varies from 0 to 50 nA when the bias is swept in the range of ± 10 V. Despite the skewness of the I–V curves, as an evidence of the Schottky-like behavior, the backward current branch (positive voltage region) indicates significant current appearing already at 5 V. In our opinion, this behavior is caused by a strong field emission due to high electric field at the tip apex. It was found that field emission partially occurs also in the forward current region. A more detailed discussion on the current-to-voltage behavior of the AD can be found in [33]. As demonstrated here, the sensitivity to work function differences makes C-AFM suitable for electrical characterization of different materials with nanometer lateral resolution.

23.5 C-AFM Investigations of Nanorods

Arrays of free standing, semiconducting NRs are an emerging field of nanostructure research. Among them are Si [76], III–V [77], and ZnO [78–80]. In particular, ZnO bears the potential of future development in a broad range of electronic applications. ZnO NRs are of particular interest for applications in UV photosensors [81], solar cells [82], and energy harvesting devices [83].

For the imaging of semiconductor NWs, so far mainly scanning and TEM have been used. However, it was demonstrated recently that AFM in tapping mode also yields reliable information on the NR morphology [34, 84], including information on transition facets between the (0001) top and {10–10} side facets. To gain – in addition to morphology – information on electrical properties of individual NR it seems to be natural to apply C-AFM which is one of the most appropriate techniques for this purpose. However, the main problem of C-AFM measurements is the proper adjustment of the feedback loop parameters for the contact mode, which is a severe problem on rough surfaces.

Here, we present C-AFM results of ZnO nanorod arrays grown by thermal evaporation (TE) and hydrothermal (HT) techniques in terms of their electrical properties [34]. As substrates, silicon and ITO were used. For an extended discussion of the differences between the HT and TE grown sample see [80,85]. The measurements have been performed under ambient conditions using CDT-CONTR (Nanosensors^{TM}) silicon cantilevers with a boron doped diamond tip coating with force constants in the range from 0.02 to 0.77 N/m. The sample morphology was initially characterized by tapping mode AFM (TM-AFM) using PPP-NCHR cantilevers with force constants varying from 10 to 130 N/m. In Fig. 23.15, topographies of the TE and HT samples with corresponding height distribution plots are presented. ZnO NR diameters vary in a wide range from 150 to 1,200 nm for the TE grown sample whereas the rods of the HT grown sample are significantly smaller in diameter (20–80 nm). Rod lengths are $\sim$1.5 μm as determined from SEM.

As is seen in Fig. 23.16, 2D current maps of the upright free-standing TE NR can be measured indeed. The loading force during the C-AFM characterization was of the order of 5–10 nN. We have managed to obtain current maps on NRs with diameters as small as 300–500 nm, measuring "as grown" samples without any

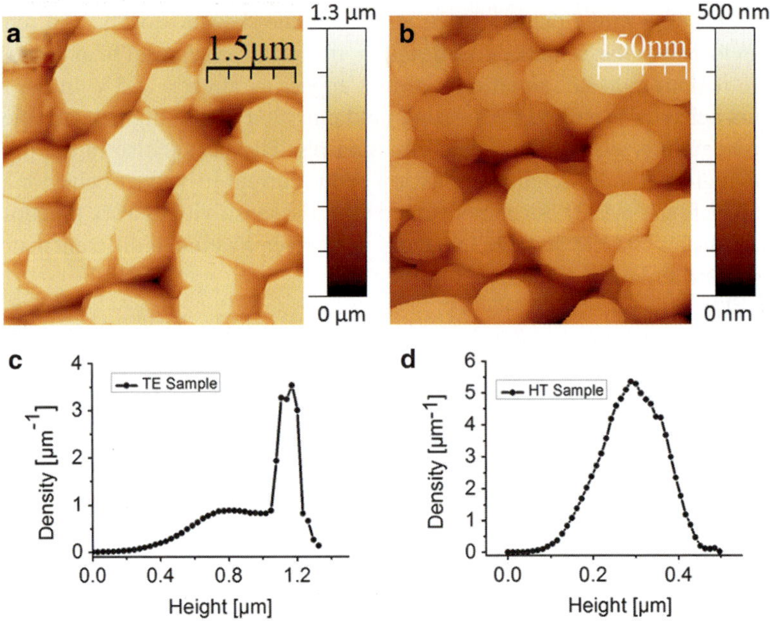

Fig. 23.15 (**a**) 5 µm × 5 µm tapping mode image of ZnO nanorods grown by TE and (**b**) 5 µm × 5 µm tapping mode image of HT grown samples, (**c**), (**d**) corresponding height distribution plots taken from the larger 10 µm images

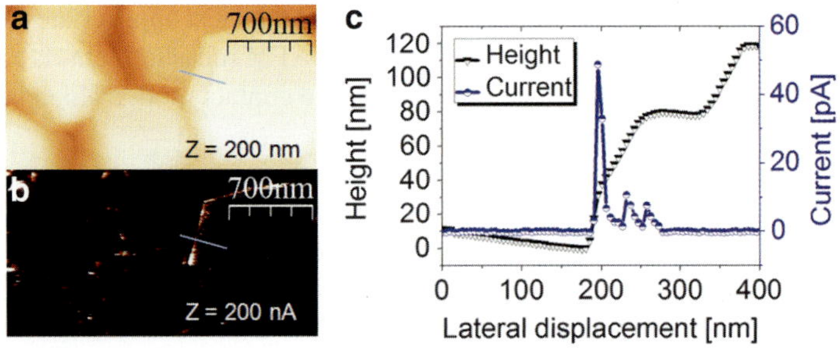

Fig. 23.16 2.5 µm × 1.25 µm C-AFM images of ZnO nanorods grown by TE on Si: (**a**) topography, (**b**) current map at −0.5 V, (**c**) topography and current cross-sections taken across the edge of the NR as marked in (**a**) and (**b**)

additional treatment, like filling the NR array with a polymer matrix, etc. The topography image (Fig. 23.16a) reveals clearly both the top and side facets of the NR. The current map has been recorded at −0.5 V which was a practical limit in the negative voltage region because of the large current peaks originating from the side facet of the NR resulting in amplifier saturation. We did not apply positive voltages during the scans across the surface, since application of positive voltages would lead to modification of the electrical properties of the surface. For the conditions used here we did not notice any significant variation of the current originating from the top facets. It is worth mentioning that the topography and current map depicted in Fig. 23.16 are both recorded for the trace scan direction (left-to-right), the current spikes appear in this case preferentially on the left side of the NRs. The current varied in the range from 0 to 200 pA (which is the amplifier limit). A comparison of the topography and current profiles (Fig. 23.16c) reveals that the maximum current peaks appear simultaneously with the moments of rapid height increase. According to [83–86] a possible explanation for this observation is piezoelectric charge carrier generation in the ZnO NR due to bending of the NR. However, the observation of this phenomenon by C-AFM technique is currently under strong discussion [87–89].

Here, the observation of currents on the edges of ZnO NRs may not be solely related to the piezoelectric response. In order to address this problem we performed C-AFM experiments on a ZnO NR containing step-like structures which are presented in Fig. 23.17. The sample morphology is represented in 3D and 2D presentation in Fig. 23.17a, b, respectively, and excludes a possible bending due to the applied stress. The sample was grown on an ITO conductive substrate. C-AFM measurements were performed at the grounded sample with PtIr5 coated PPP-EFM cantilevers from Nanosensors™. The nominal value of the spring constant for this type of cantilever is 2.8 N/m. A special feature of this sample is that the side facet is "replaced" by a stepped plane. Therefore, the obtained structure is an intermediate in morphology between nanorod and nanopyramid, which can provide an insight into the electrical transport at the step edges. The current maps recorded in both trace and retrace scan directions are represented in the Fig. 23.17c, d respectively. They demonstrate basically the same behavior as the current maps of ZnO NR grown on a Si substrate (Fig. 23.16), the currents appear preferentially on the left side of the NR during the trace scan and vice versa during the retrace scan. The current spikes originating from the step edges reach an absolute value of $\sim$100 pA. From the comparison of the current maps measured on different ZnO NR samples we draw the following conclusions. First, the edge effect appears only in the case if the contact between the tip and surface leads to the formation of a Schottky barrier. Second, the appearance of the current spikes is related to instabilities in the feedback loop. Third, the charge carrier transport is strongly enhanced at the edges. The latter is not consistent with the fact that the contact resistance should increase when the contact area is getting smaller. As mentioned above, the C-AFM measurements have been performed at low tip-to-surface forces. According to [90], this is the region where the abruption–formation of Schottky nano-contacts to ZnO takes place. Additionally, an investigation of the electrical transport through Schottky nano-diodes [91] indicates that the barrier is lowered significantly if the contact size is smaller than the characteristic length in the material. Therefore, the current spikes might be caused by feedback instabilities which led to a decrease of the contact area resulting in a lowering of the barrier height.

The small diameter and high aspect ratio of the HT grown samples hamper 2D measurements in the contact mode. Though, the characterization of their electrical properties can be done via $I-V$ curve measurements as presented in Fig. 23.18. The $I-V$ characteristics for both samples have been recorded with a diamond coated

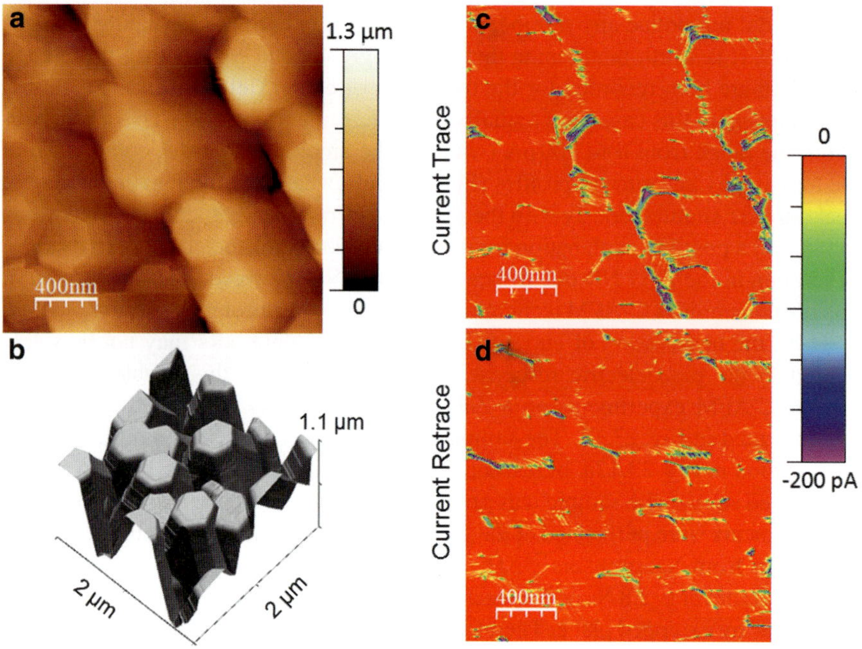

Fig. 23.17 2 μm × 2 μm C-AFM images of ZnO nanorods grown by TE on ITO: (**a**, **b**) morphology, (**c**) trace and (**d**) retrace current maps recorded with grounded sample and PtIr coated probe

DCP11 cantilever from NT-MDT company. Since the morphology of the HT grown NRs is not achievable via the measurements in contact mode, the positioning of the conductive probe should be done in tapping mode. In order to obtain I–V characteristics, the system is switched to contact mode after probe positioning in tapping mode. Small lateral displacements (up to 20–30 nm) are unavoidable. Therefore, the I–V characterization of HT grown samples presented in Fig. 23.18a has been performed in "hit-or-miss" fashion and ohmic curves which occurred due to the contact with the ITO substrate were removed from the plot. The rest of the I–V curves demonstrated distorted rectifying behavior, perhaps due to additional field emission during the experiment. The characteristic feature of the I–V curves is the presence of a jump at around +1.8 V sample bias (reverse direction). The characterization of the TE grown NR has been performed entirely in contact mode, to increase the accuracy during the probe positioning. The low-speed scan direction was turned off at the moment when the NR position and geometry was defined. The scheme of the experiment is depicted in Fig. 23.18b, the diameter of the measured NR was ~700 nm and the length was ~1.3 μm. The I–V curves taken from the side and top facets are presented in Fig. 23.18c, d, respectively. They clearly demonstrate rectifying behavior with rather significant reverse current and a characteristic jump at +4.5 V in the case of the measurements from the side facet. During the measurements of the I–V

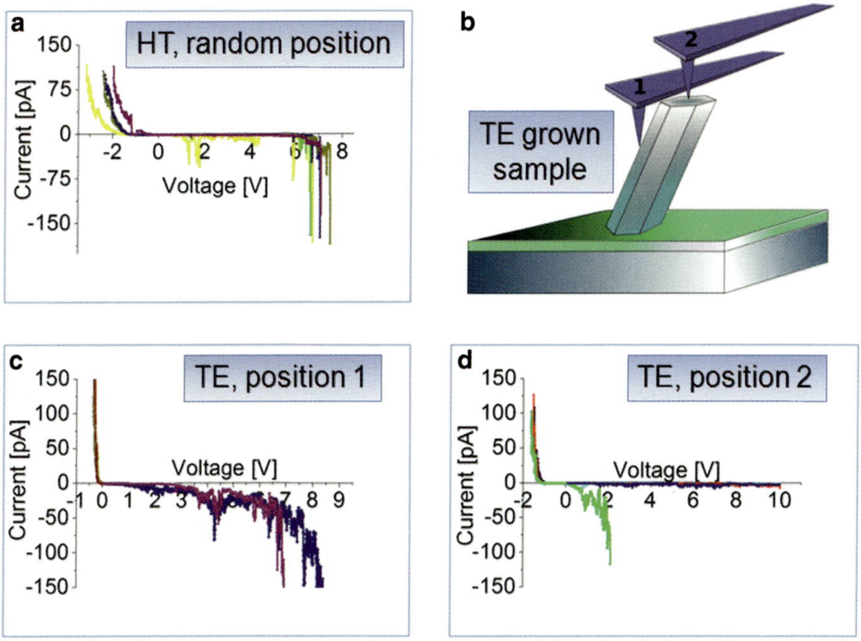

Fig. 23.18 (**a**) I–V curves recorded from HT grown NR on ITO, (**b**) scheme of the experiment for the case of a TE grown sample, (**c**, **d**) I–V curves measured from the side facet and from the top terrace of the TE grown NR, respectively

characteristics, a single breakdown occurred at +1.8 V (Fig. 23.18d) during the first voltage ramping which recovered and did not appear afterwards again during several subsequent I–V cycles.

As is well known, piezoelectric properties of ZnO NRs can be accessed via the well-established PFM technique nevertheless C-AFM can contribute additional information to this subject. Since conductivity and free charge carrier concentration are interrelated quantities, the influence of the latter on the piezoelectric response can be studied by combined C-AFM and PFM. The correlation between C-AFM and PFM data for single upright standing ZnO NR was shown recently. It was found that the resistivity of a single ZnO NR grown on a silver substrate varied from 0.1 to 155 Ω cm, whereas the piezoelectric coefficient ranged from 0.4 to 9.5 pm/V [92]. It was shown recently that the metal–insulator transition in a single VO_2 NW can be detected by force-to-distance AFM characterization [93]. Here, C-AFM measurements could allow a simultaneous access to the changes both in mechanical and electrical behavior.

Finally we want to mention that the C-AFM technique has already been frequently applied for investigations of lying NRs or nanotubes of various materials, like carbon nanotubes (CNT) [94, 95], WO_3 [96], CuO [97], and others. As it was shown, the AFM tip can be used as a local electrostatic gate, and the gating action can be correlated with the structure of the single wall CNTs bundles [94, 98]. One

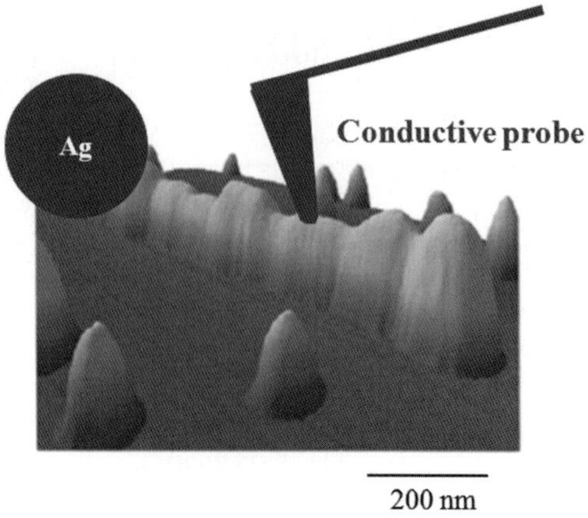

Fig. 23.19 The sketch of an experiment for the electrical transport investigations of lying NR, in this case an organic semiconductor fiber [99]

can use this approach also for electrical characterization of organic NWs as is illustrated in Fig. 23.19. Arrays of several tens of micrometers long organic nano-fibers can be grown on highly resistive substrates like mica and are easily accessible for quality AFM analysis [99]. For the carrier transport investigation the nano-fibers should be contacted from one side (e.g., using conductive silver paint) and a conductive AFM probe can be used as the second contact.

23.6 Application of C-AFM to Electroceramics

In electroceramic devices like $PbZr_xTi_{1-x}O_3$ (PZT) actuators, ZnO varistors, positive temperature coefficient resistors, and low temperature cofired ceramics, local leakage currents may finally determine life-time and reliability of the components [65]. It has been found that the degradation mainly occurs at specific locations near electrode ends or in the vicinity of cracks [66]. Conventional device testing techniques, measuring currents on a macroscopic scale, however, fail in order to reveal the current paths and the underlying mechanisms on the microscopic and nanometer scale. Here, scanning probe based techniques come into play again [58,67]. Interestingly, C-AFM has been introduced just recently into the research fields of positive temperature coefficient resistors [68] and varistors [35]. The results of the latter will be presented in the following.

ZnO based metal-oxide varistor ceramics with certain additives exhibit excellent nonlinear current voltage characteristics. The grain boundaries, acting with typical switching voltages of about 3.5 V are responsible for that extraordinary macroscopic material's behavior [69]. Here, we studied multilayer varistor components (Bi_2O_3 doped ZnO with silver/palladium inner electrodes) which were polished in a way to enable electrical inspection close to an inner electrode layer (see Fig. 23.20c) [35]. Measurements have been performed in air with a positive voltage applied to the inner Ag/Pd electrode and employing the approved Si tips with boron doped

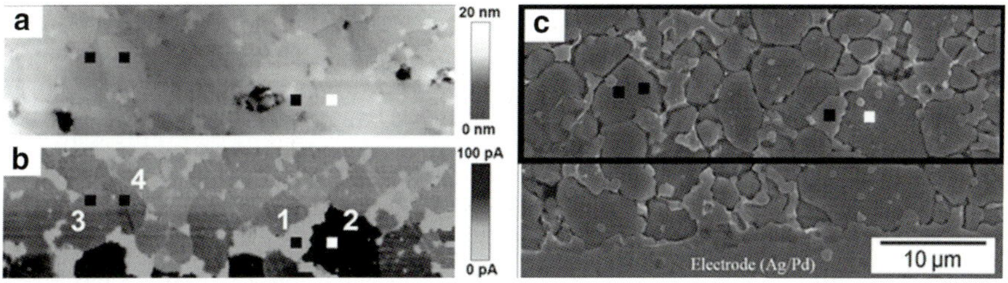

Fig. 23.20 C-AFM investigation of a polished ZnO multilayer varistor: (**a**) 50 μm × 12.5 μm topographic image and (**b**) corresponding 2D current image recorded at +10 V. The SEM image of the etched surface (**c**) shows a slightly larger area to visualize the distance to the inner metal electrode. The *numbered squares* mark the spots for acquiring IV characteristics (Adapted from [35])

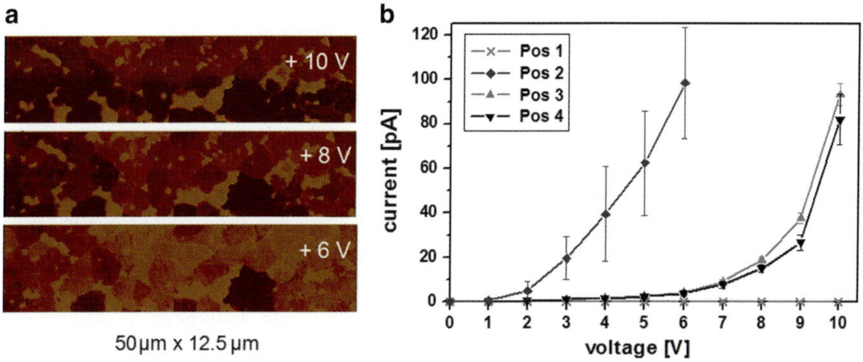

Fig. 23.21 Presentation of a few C-AFM measurements recorded at different applied voltages from the sample area presented in Fig. 23.20. (The current scale has been cut at 50 pA.) (**b**) I–V curves acquired for the positions (1)–(4) marked in Fig. 23.20 (Adapted from [35])

diamond coatings. Figure 23.20 shows a C-AFM overview image of a typical region of the doped polycrystalline ZnO layer about 5 μm away from the inner electrode. The individual crystallites are slightly visible in the topography image (Fig. 23.20a). Further relevant information on the grain structure can be obtained from the 2D current map recorded at the highest voltage applicable in our setup (Fig. 23.21b). The crystallite marked with (1) exhibits no detectable current although being close to the electrode. This behavior is indicating an insulating spinel or Bi–O phase. The ZnO grains show different, but homogeneous current levels allowing to distinguish individual grains. Even parts of twin crystals can be discriminated [e.g., positions (3) and (4)].

Figure 23.21 demonstrates how quantitative information on the electrical characteristics of individual grains is obtained. Since conventional recording of local I–V curves was strongly influenced by the local surface conditions, we applied here the following, rather time consuming but reliable acquisition procedure. As indicated in Fig. 23.21a, a series of current images was recorded at voltages from 0 to +10 V with increments of 1 V. (In Fig. 23.21a, just the 2D current maps for 6, 8, and 10 V are presented.) Then, the current values are averaged at the desired positions on an area of 1 μm × 1 μm. Reassembling the resulting data leads to a discrete I–V curve which is representative for a given crystal or even domain. For the positions marked in Fig. 23.20, the I–V curves – collected in the described way – are presented in Fig. 23.21b. At position (2) in a mono-domain ZnO grain, a typical nonlinear I–V curve has been measured with a switching voltage below the ideal value of 3.5 V [35]. Between the twin half's [marked by (3) and (4)] and the electrode, there are two grain boundaries (see Fig. 23.20c) resulting in a switching voltage in the range between 6 and 7 V. These characteristics can also be related to the crystal orientation of the grains as revealed by electron backscattering diffraction [35]. The slight differences in the electrical characteristics for positions (3) and (4), however, are probably due to different work functions of the polished surface plane in contact to the tip. These results for ZnO multilayer varistors demonstrate the applicability of C-AFM to local electrical characterization of electroceramics.

23.7 Outlook to Photoconductive AFM

One of the biggest advantages of scanning probe microscopy (SPM) based techniques is their high versatility and applicability to different fields of research. It was shown that the C-AFM technique is an efficient tool for the local electrical characterization of semiconductor materials on the nanometer scale. However, in some cases information obtained by C-AFM can be significantly enhanced via supplementary sample illumination. This modification of C-AFM is referred to as PC-AFM. The technique was originally developed for 2D mapping of the photoelectric properties of organic thin films [41, 100]. The PC-AFM setup is schematically depicted in Fig. 23.22. The sample surface is biased and stays under illumination while current passing throughout the AFM tip is measured as a function of illumination intensity and/or wavelength. Basically, there are in principle two possible geometries of sample illumination, either the illumination from the back side of the sample [42, 100] or from the top [101] as depicted in Fig. 23.22. The first procedure holds much favor because of the easier focusing and higher illumination intensity at the measurable region. Whereas the latter has to be done at grazing incidence making light focusing rather problematic. Since the majority of materials demonstrate weak photoresponse, a high intensity light source is preferable. The simplest way is to use a laser. The difference in currents measured by PC-AFM between illuminated and dark situation carries important information like the optical absorption edge and the presence of electrically active defects.

PC-AFM is also used to track processes on the surface of a working solar cell. Recently local photocurrents in blend polymer/fullerene solar cells were mapped by C-AFM with a resolution down to 20 nm [42, 100, 102]. Its efficiency was demonstrated also for direct imaging of photocurrent pathways in ZnO based solar cells [103], WO_3, TiO_2, and composite WO_3/TiO_2 thin films [104].

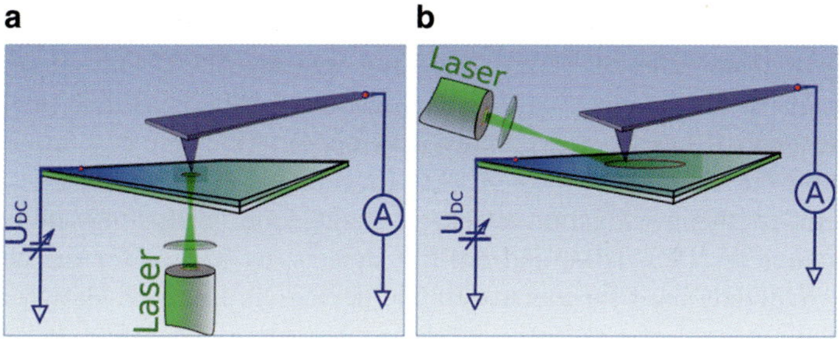

Fig. 23.22 Schematic drawings of the photoconductive AFM (PC-AFM) setups: (**a**) illumination from the back side of the sample and (**b**) illumination from the top

In principle, the majority of semiconductors demonstrates photoresponse to a greater or lesser extent. Microscopically this effect is relatively easy to investigate since the integral photocurrent is sufficiently large in most of the cases. This is in contrast with the situation realized in PC-AFM where the amount of "working" material is limited by the volume under the tip-to-sample contact. The fraction of the photocurrent becomes vanishingly small comparable to the dark current. Therefore photocurrent detection by using lock-in technique seems suitable. This approach has been already successfully implemented in STM [72] to perform illumination-dependent studies, but was not applied to AFM so far. Since most of the commercially available AFMs use a laser in their feedback loop for recording the topography signal, this causes an additional problem for measurements under illumination. Staying under the illumination from two lasers at the same moment, the conductivity of the sample may be influenced by each of them, therefore, making the obtained data hard to interpret. In order to avoid the above-mentioned problem one can use additional modulation of the feedback loop laser. When coupled with the lock-in technique this will allow to detect the photocurrent from the illumination laser separately, if the light from the illumination source is modulated as well [101]. It is also possible to overcome this problem by using a piezoresistive cantilever with the conductive tip as it was demonstrated in [73]. To the best of our knowledge, all the present PC-AFM systems are operating in air. UHV would also give several additional advantages, like higher stability, better noise characteristics and easier data interpretation.

23.8 Overall Summary and Perspectives

This chapter has been devoted to review the potential of C-AFM for the nanometer scale characterization of materials used in micro- and nanoelectronics. The experimental peculiarities of the technique like operation in contact mode and detection of sub-pA currents have been discussed. Special emphasize was paid to the implementation of the method under ultra-high vacuum. The measurement strategies like acquisition of current–voltage curves, determination of breakdown voltages, obtaining of histograms of PBDVs, and the analysis of 2D current maps have been introduced for the electrical characterization of gate dielectrics (the field for which the method has been originally developed). Although this research topic still remains the main target of the C-AFM technique, C-AFM possesses increasing potential for the investigation of phase-separated low conducting semiconductor nanostructures as NWs, NRs, and dot-like defects, as well as for the characterization of materials used for electroceramic devices as has been shown in detail. Finally, the potential of a recently developed derivative of C-AFM, PC-AFM has been discussed.

However, there are further potential application topics of C-AFM to be mentioned here: The electrical properties of superconducting thin films are accessible [105] as well as the electrical properties of supported graphene layers [106, 107]. A growing field of applications, especially in conjunction with PC-AFM, is the characterization of organic semiconductors [108]. Here, for instance photovoltaic blends [109], thin multilayered composite films with embedded CNTs [110], and growth-related electrical properties [111] can be studied by C-AFM. It will be only a matter of time that C-AFM investigations will also enter the field of piezotronics and piezophototronics [112].

Last but not least one has certainly to emphasize that C-AFM is just one suitable SPM based technique to study electrical properties of functional films and nanostructures. Comprehensive conclusions on the physical mechanisms investigated will always require its application in conjunction with other electrical characterization techniques like KPFM, SSRM, SCM, EFM, and PFM.

Acknowledgments

The authors wish to thank S. Kremmer, A. Benkitsch, F. Kuchar, E. Pischler, S. Peissl, H. Wurmbauer, A. Andreev, Y. Hou, M. Schloffer, L. Wang, and M. Kratzer for their contributions to C-AFM investigations. Samples have been kindly provided by G. Tallarida and M. Fanciulli (Milan), P. Tejedor and B. Galiana (Madrid), A.B. Djurišić (Hongkong), G. Brauer (Rossendorf), and P. Supancic (Leoben). Research has been partially funded by the Austrian Science Fund (FWF) under Project # P19636-N20 and Austrian Exchange Service (ÖAD) under Projects # IT 9/2004 and # ES 17/2007.

References

1. A. Avila, B. Bhushan, Crit. Rev. Solid State Mater. Sci. **35**, 38 (2010)
2. S.V. Kalinin, A. Gruverman (eds.), *Scanning Probe Microscopy, Electrical and Electromechanical Phenomena at the Nanoscale*, vol. 1 and 2, 1st edn. (Springer, Heidelberg, Germany, 2007)

3. D.W. Abraham, C. Williams, J. Slinkman, H.K. Wickramasinghe, J. Vac. Sci. Tech. B **9**, 703 (1991)
4. M. Dreyer, R. Wiesendanger, Appl. Phys. A**61** 357 (1995)
5. P. DeWolf, J. Snauwaert, T. Clarysse, W. Vandervorst, L. Hellemans, Appl. Phys. Lett. **66**, 1530 (1995)
6. P. De Wolf, F. Geva, T. Hantschel, W. Vandervorst, R.B. Bylsma, Appl. Phys. Lett. **73**, 2155 (1998)
7. M. Nonnemacher, M.P. O'Boyle, H.K. Wickramasinghe, Appl. Phys. Lett. **58**, 2921 (1991)
8. J.M.R. Weaver, D.W. Abraham, J. Vac. Sci. B **9**, 1559 (1991)
9. G.H. Buh, H.J. Chung, J.H. Yi, I.T. Yoon, Y. Kuk, J. Appl. Phys. **90**, 443 (2001)
10. X.F. Chen, W.G. Zhu, W.W. Liu, O.K. Tan, X. Yao, Ferroelectrics **252**, 201 (2001)
11. M. Ruprecht, G. Benstetter, D. Hunt, Microelectron. Reliab. **43**, 17 (2003)
12. P. Guethner, K. Dransfeld, Appl. Phys. Lett. **61**, 1137 (1992)
13. M. Alexe, A. Gruverman (eds.), *Nanoscale Characterization of Ferroelectric Materials*, 1st edn. (Springer, Berlin, 2004)
14. S. Morita, Y. Sugawara, Thin Solid Films **393**, 310 (2001)
15. D.M. Schaadt, E.T. Yu, S. Sankar, A.E. Berkowitz, Appl. Phys. Lett. **74**, 472 (1999)
16. M.P. Murrell, M.E. Welland, S.J. O'Shea, T.M.H. Wong, J.R. Barnes, A.W. McKinnon, M. Heyns, S. Verhaverbeke, Appl. Phys. Lett. **62**, 786 (1993)
17. S.J. O'Shea, R.M. Atta, M.P. Murrell, M.E. Welland, J. Vac. Sci. Technol. B**13**, 1945 (1995)
18. T.G. Ruskell, R.K. Workman, D. Chen, D. Sarid, S. Dahl, S. Gilbert, Appl. Phys. Lett. **68**, 93 (1996)
19. A. Olbrich, B. Ebersberger, C. Boit, Appl. Phys. Lett. **73**, 3114 (1998)
20. S. Kremmer, C. Teichert, E. Pischler, H. Gold, F. Kuchar, M. Schatzmayr, Surf. Interface Anal. **33**, 168 (2002)
21. J.A. Dagata, J. Schneir, H.H. Harary, C.J. Evans, M.T. Postek, J. Bennet, Appl. Phys. Lett. **56**, 2001 (1990)
22. H.C. Day, D.R. Allee, Appl. Phys. Lett. **62**, 2691 (1993)
23. P. Avouris, R. Martel, T. Hertel, R. Sandstrom, Appl. Phys. A**66**, 659 (1998)
24. S. Kremmer, S. Peissl, C. Teichert, F. Kuchar, H. Hofer, Mat. Sci. Eng. B**102**, 88 (2003)
25. P.M. Campbell, E.S. Snow, P.J. McMarr, Appl. Phys. Lett. **66**, 1388 (1995)
26. R. Held, T. Heinzel, P. Studerus, K. Ensslin, M. Holland, Appl. Phys. Lett. **71**, 2689 (1997)
27. J.T. Sheu, K.S. You, C.H. Wu, K.M. Chang, J. Vac. Sci. Technol. B **20**, 2824 (2002)
28. S. Kremmer, H. Wurmbauer, C. Teichert, G. Tallarida, S. Spiga, C. Wiemer, M. Fanciulli, J. Appl. Phys. **67**, 074315-1-7 (2005)
29. O. Bierwagen, L. Geelhaar, X. Gay, M. Piešinš, H. Riechert, B. Jobst, A. Rucki, Appl. Phys. Lett. **90**, 232901-1-3 (2007)
30. M. Lanza, M. Porti, M. Nafria, G. Benstetter, W. Frammelsberger, H. Ranzinger, E. Lodermeier, G. Jaschke, Microelectron. Reliab. **47**, 1424 (2007)
31. D. Martin, M. Grube, W.M. Weber, J. Rüstig, O. Bierwagen, L. Geelhaar, H. Riechert, Appl. Phys. Lett. **95**, 142906-1-3 (2009)
32. P. Tejedor, L. Díez-Merino, I. Beinik, C. Teichert, Appl. Phys. Lett. **95**, 123103 (2009)
33. I. Beinik, B. Galiana, M. Kratzer, C. Teichert, I. Rey-Stolle, C. Algora, P. Tejedor, J. Vac. Sci. Technol. B **28**, C5G5 (2010)
34. G. Brauer, W. Anwand, D. Grambole, W. Egger, P. Sperr, I. Beinik, L. Wang, C. Teichert, J. Kuriplach, J. Lang, S. Zviagin, E. Cizmar, C.C. Ling, Y.F. Hsu, Y.Y. Xi, X. Chen, A.B. Djurišić, W. Skorupa, Phys. Status Solidi C **6**, 2556 (2009)
35. M. Schloffer, C. Teichert, P. Supancic, A. Andreev, Y. Hou, Z. Wang, J. Eur. Ceram. Soc. **30**, 1761 (2010)
36. S. Kremmer, E. Pischler, C. Teichert, F. Kuchar, in *Proceedings of IEEE-NANO 2001 – 1st IEEE Conference on Nanotechnology*, Maui, Hawaii (IEEE, Piscataway, NJ, 2001), p. 162
37. S. Kremmer, S. Peissl, C. Teichert, F. Kuchar, in *Proceedings 28th International Symposium for Testing and Failure Analysis*, Phoenix, AZ, 3–7 Nov 2002 (ASM International, Materials Park, OH, 2002), p. 473

38. S.J. O'Shea, R.M. Atta, M.E. Welland, Rev. Sci. Instrum. **66**, 2508 (1995)
39. M.A. Lantz, S.J. O'Shea, M.E. Welland, Rev. Sci. Intrum. **69**, 1757 (1998)
40. C.A. Peterson, R.K. Workman, D. Sarid, B. Vermeire, H.G. Parks, D. Adderton, P. Maivald, J. Vac. Sci. Technol. A **17**, 2753 (1999)
41. H. Sakaguchi, F. Iwata, A. Hirai, A. Sasaki, T. Nagamura, Jpn. J. Appl. Phys. **38**, 3908 (1999)
42. D.C. Coffey, O.G. Reid, D.B. Rodovsky, G.P. Batholomew, D.S. Ginger, Nano Lett. **7**, 738 (2007)
43. L. Aguilera, M. Lanza, M. Porti, J. Grifoll, M. Nafría, X. Aymerich, Rev. Sci. Instrum. **79**, 073701 (2008)
44. L. Aguilera, W. Polspoel, A. Volodin, C. Van Haesendonck, M. Porti, W. Vandervorst, M. Nafria, X. Aymerich, J. Vac. Sci. Technol. B **26**, 1445 (2008)
45. F.A.M. Koeck, R.J. Nemanich, Diamond Relat. Mater. **15**, 217 (2006)
46. K.-H. Kim, N. Moldovan, C. Ke, H.D. Espinosa, X. Xiao, J.A. Carlisle, O. Auciello, Small **1**, 866 (2005)
47. T. Trenkler, T. Hantschel, R. Stephenson, P. De Wolf, W. Vandervorst, L. Hellemans, A. Malavé, D. Büchel, E. Oesterschulze, W. Kulisch, P. Niedermann, T. Sulzbach, O. Ohlsson, J. Vac. Sci. Technol. B **18**, 418 (2000)
48. R.C. Weast (ed.), *CRC Handbook of Chemistry and Physics*, 67th edn. (CRC Press, Boca Raton, 1986)
49. P. Sun-Jian, X. Wang-Zhao, M. Hou-Shi, M. Zhang-Geng, Y. Zhao-Xing, M. Liu-Wei, Q. Xue-Zeng, Acta Electron. Sin. **30**, 655 (2002)
50. M. Klaua, D. Ullmann, J. Barthel, W. Wulfhekel, J. Kirschner, R. Urban, T.L. Monchesky, A. Enders, J.F. Cochran, B. Heinrich, Phys. Rev. B **64**, 134411 (2001)
51. O. Groning, O. M. Kuttel, P. Groning, L. Schlapbach, J. Vac. Sci. Technol. B **17**, 1970 (1999)
52. G.S. Lujan, T. Schram, L. Pantisano, J.C. Hooker, S. Kubicek, E. Rohr, J. Schuhmacher, O. Kilpelä, H. Sprey, S.D. Gendt, K.D. Meyer, in *Proceedings of the ESSDERC 35th European Solid-State Device Research Conference* (2002), p. 583
53. R.H. Fowler, L.W. Nordheim, Proc. R. Soc. London Ser. A **119**, 173 (1928)
54. J. Robertson, Rep. Prog. Phys. **69**, 327 (2006)
55. W. Frammelsberger, G. Benstetter, J. Kiely, R. Stamp, Appl. Surf. Sci. **253**, 3615 (2007)
56. M. Porti, M. Nafría, X. Aymerich, A. Olbrich, B. Ebersberger, J. Appl. Phys. **91**, 2071 (2002)
57. P. Fiorenza, R. Lo Nigro, V. Raineri, S. Lombardo, R.G. Toro, G. Malandrino, I.L. Fragal̀, Appl. Phys. Lett. **87**, 231913-1-3 (2005)
58. S.A. Landau, N. Junghans, P.-A. Weiß, B.O. Kolbesen, A. Olbrich, G. Schindler, W. Hartner, F. Hintermaier, C. Dehm, C. Mazuré, Appl. Surf. Sci. **157**, 387 (2000)
59. X. Blasco, J. Pétry, M. Nafria, X. Aymerich, O. Richard, W. Vandervorst, Microelectron. Eng. **72**, 191 (2004)
60. J. Pétry, W. Vandervorst, X. Blasco, Microelectron. Eng. **72**, 174 (2004)
61. J. Robertson, Rep. Prog. Phys. **69**, 327 (2006)
62. E. Atanassova, P. Lytvyn, R. V. Konakova, V.F. Mitin, D. Spassov, J. Phys. D **42**, 145301 (2009)
63. W. Brütting (ed.), *Physics of Organic Semiconductors* (Wiley-VCH, Weinheim, 2005)
64. P.J. de Pablo, J. Colchero, J. Gómez-Herrero, A.M. Baró, Appl. Phys. Lett. **73**, 3300 (1998)
65. Q. Feng, C.J. McConville, J. Electroceram. **14**, 213 (2005)
66. W.L. Warren, D. Dimos, R. Waser, Mat. Res. Soc. Bull. **21**, 40 (1996)
67. B.D. Huey, D.A. Bonnell, Solid State Ionics **131**, 51 (2000)
68. P. Fiorenza, R. Lo Nigro, P. Delugas, V. Raineri, A.G. Mould, D.C Sinclair, Appl. Phys. Lett. **95**, 142904-1-3 (2009)
69. D.R. Clarke, J. Am. Ceram. Soc. **82**, 485 (1999)
70. P. Eyben, T. Janssens, W. Vandervorst, Mat. Sci. Eng. B **124–125**, 45 (2005)
71. X.M. Zhang, D.W. Pashley, I. Kamiya, J.H. Neave, B.A. Joyce, J. Cryst. Growth **147**, 234 (1995)
72. T. Takahashi, K. Takada, M. Takeuchi, Ultramicroscopy **97**, 1–6 (2003)
73. H. Masuda, M. Takeuchi, T. Takahashi, Ultramicroscopy **105**, 137–142 (2005)

74. B. Galiana, E. Barrigón, I. Rey-Stolle, V. Corregidor, P. Espinet, C. Algora, E. Alves, Superlatt. and Microstr. **45**, 277 (2009)
75. R.R. King, D.C. Law, C.M. Fetzer, R.A. Sherif, K.M. Edmondson, S. Kurtz, G.S. Kinsey, H.L. Cotal, J.H. Ermer, N.H. Karam, in *Proceedings of the 20th Europ. Photovolt. Solar Energy Conf. and Exhibition*, Barcelona, 2005
76. Y. Cui, L.J. Lauhon, M.S. Gudiksen, J.F. Wang, C.M. Lieber, Appl. Phys. Lett **78**, 2214 (2001)
77. T. Martensson, C.P.T. Svensson, B.A. Wacaser, M.W. Larsson, W. Seifert, K. Deppert, A. Gustafsson, L.R. Wallenberg, L. Samuelson, Nano Lett. **4**, 1987 (2004)
78. L.E. Greene, M. Law, D.H. Tan, M. Montano, J. Goldberger, G. Somorjai, P. Yang, Nano Lett. **5**, 1231 (2005)
79. K.H. Tam, C.K. Cheung, Y.H. Leung, A.B. Djurišić, C.C. Ling, C.D. Beling, S. Fung, W.M. Kwok, Y.H. Leung, W.K. Chan, D.L. Phillips, L. Ding, W.K. Ge, J. Phys. Chem. B **110**, 20865 (2006)
80. Y.F. Hsu, Y.Y. Xi, A.B. Djurišić, W.K. Chan, Appl. Phys. Lett. **92**, 133507 (2008)
81. L. Luo, Y. Zhang, S.S. Mao, L. Lin, Sens. Actuators A **127**, 201 (2006)
82. I. Gonzalez-Valls, M. Lira-Cantu, Energy Environ. Sci. **2**, 19 (2009)
83. Z.L. Wang, J. Song, Science **312**, 242 (2006)
84. G. Brauer, W. Anwand, D. Grambole, W. Skorupa, Y. Hou, A. Andreev, C. Teichert, K.H. Tam, A.B. Djurišić, Nanotechnology **18**, 195301 (2007)
85. Y.F. Hsu, Y.Y. Xi, A.B. Djurišić, W.K. Chan, Appl. Phys. Lett. **92**, 133507 (2008)
86. J. Zhou, P. Fei, Y. Gao, Y. Gu, J. Liu, G. Bao, Z.L. Wang, Nano Lett. **8**(9), 2725 (2008)
87. J. Liu, P. Fei, J. Song, X. Wang, C. Lao, R. Tummala, Z.L. Wang, Nano Lett. **8**(1), 328 (2008)
88. M. Alexe, S. Senz, M.A. Schubert, D. Hesse, U. Gösele, Adv. Mater. **20**, 4021 (2008)
89. Z.L. Wang, Adv. Mater. **20**, 1 (2008)
90. B. Pérez-García, J. Zúñiga-Pérez, V. Muñoz-Sanjosé, J. Colchero E. Palacios-Lidón, Nanoletters **7**(6), 1505 (2007)
91. G.D.J. Smit, S. Rogge, T.M. Klapwijk, Appl. Phys. Lett. **81**, 3852 (2002)
92. D.A. Scrymgeour, J.W.P. Hsu, Nano lett. **8**, 2204 (2008)
93. W. Fan, S. Huang, J. Cao, E. Ertekin, C. Barrett, D.R. Khanal, J.C. Grossman, J. Wu, Phys. Rev. B **80**, 241105(R) (2009)
94. M. Ishikawa, M. Yoshimura, K. Ueda, Jpn. J. Appl. Phys. **41**, 4908 (2001)
95. L. Rispal, Y. Stefanov, F. Wessely, U. Schwalke, Jpn. J. Appl. Phys. **45**, 3672 (2006)
96. R. Delamare, M. Gillet, E. Gillet, P. Guaino, Surf. Sci. **601**, 2675 (2007)
97. G. Cheng, S. Wang, K. Cheng, X. Jiang, L. Wang, L. Li, Z. Du, G. Zou, Appl. Phys. Lett. **92**, 223116 (2008)
98. M. Freitag, M. Radosavljevic, W. Clauss, A. T. Johnson, Phys. Rev. B **62**, 2307 (2000)
99. C. Teichert, G. Hlawacek, A. Yu. Andreev, H. Sitter, P. Frank, A. Winkler, N.S. Saricifci, Appl. Phys. A **82**, 665 (2006)
100. J. Lee, I. Choi, S. Hong, S. Lee, Y.I. Yang, Y. Kim, J. Yi, Ultramicroscopy **108**, 1090 (2008)
101. W. Brezna, G. Strasser, J. Smoliner, Physica E **40**, 1229 (2008)
102. L.S.C. Pingree, O.G. Reid, D.S. Ginger, Nano Lett. **9**, 2946 (2009)
103. L. Heng, D. Tian, L. Chen, J. Su, J. Zhai, D. Han, L. Jiang, Chem. Commun. **46**, 1162 (2010)
104. S. Wang, X. Zhang, G. Cheng, X. Jiang, Y. Li, Y. Huang, Z. Du, Chem. Phys. Lett. **405**, 63 (2005)
105. J.D. Pedarnig, K. Siraj, M.A. Bodea, I. Puica, W. Lang, R. Kolarova, P. Bauer, K. Haselgrübler, C. Hasenfuss, I. Beinik, C. Teichert, Thin Solid Films **518**, 7075 (2010)
106. S. Banerjee, M. Sardar, N. Gayathri, A.K.B. Raj, Appl. Phys. Lett. **88**, 062111 (2006)
107. J.A. Kellar, J.M.P. Alaboson, Q.H. Wang, M.C. Hersam, Appl. Phys. Lett. **96**, 143103 (2010)
108. L.S.C. Pingree, O.G. Reid, D.S. Ginger, Adv. Mater. **21**, 19 (2009)
109. O. Douhéret, L. Lutsen, A. Swinnen, M. Breselge, K. Vandewal, L. Goris, J. Manca, Appl. Phys. Lett. **89**, 032107 (2006)
110. H.J. Park, K.A. Oh, M. Park, H. Lee, J. Phys. Chem. C **113**, 13070 (2009)
111. N. Kumaran, P.A. Veneman, B.A. Minch, A. Mudalige, J.E. Pemberton, D.F. O'Brien, N.R. Armstrong, Chem. Mater. **22**, 2491 (2010)
112. Z.L. Wang, J. Phys. Chem. Lett. **1**, 1388 (2010)

Chapter 24
Microscopic Electrical Characterization of Inorganic Semiconductor-Based Solar Cell Materials and Devices Using AFM-Based Techniques

Chun-Sheng Jiang

Abstract Atomic force microscopy (AFM)-based microscopic electrical techniques provide new insights into the characterization of electronic devices, which is useful for understanding device physics and improving device performance. This chapter introduces the fundamentals of some major AFM-based electrical techniques, including scanning Kelvin probe force microscopy (SKPFM), scanning capacitance microscopy (SCM), and conductive AFM (C-AFM), and reviews some recent characterization studies on inorganic semiconductor solar cell materials and devices. A potential measurement on junctions of solar cells presents SKPFM as a powerful tool for two-dimensional junction location identification by direct electrical measurement. The potential measurement further reveals electrical field distributions under bias voltages that relate to the defects on the junctions. Grain boundaries (GBs) of polycrystalline thin-film materials are characterized by measuring carrier depletion or band bending using SCM and SKPFM, which provide the measurements of individual GBs and are thus able to correlate the GB electrical properties directly to the GB structure. Structural and electrical properties of a-Si:H and nc-Si:H thin films and devices are also characterized through a prototype structure of a-Si:H and nc-Si:H mixed phase. Localized electrical properties on the nc-Si:H phase, and phosphorus and boron doping effects on the film structure are reviewed.

24.1 Introduction

Solar cells are among the most attractive and practical solutions for clean and renewable energy. Development of the first-generation solar cells, crystalline silicon (Si)-based solar cells (For a textbook, see [1]), started several decades ago, and this technology currently occupies more than 80% of the terrestrial solar cell market. In recent years, the second-generation solar cells, thin film-based cells, have progressed aggressively in their light-electricity conversion efficiencies in both laboratory and manufactory environments (For a review, see [2, 3]). The critical advantage of the thin-film cell is the low cost afforded by its several μm thin

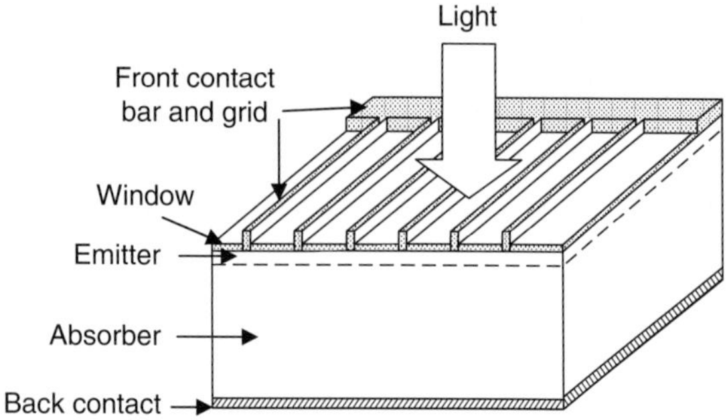

Fig. 24.1 A schematic illustrating a typical solar cell structure with multiple components

absorber material, in contrast to the 200–300 μm thickness of the Si wafer of crystalline Si-based cells. The thin-film cell is steadily increasing its market share.

An inorganic semiconductor solar cell (Fig. 24.1) is typically constructed of multiple components, including front-contact grid, antireflection window, emitter, absorber, and back-contact layers [1, 2]. The window layer serves to reduce light reflection on the surface of the cells. The emitter layer partners with the absorber layer to form a p–n or p–i–n junction, where photogenerated carriers are collected. The absorber layer is the structure where light is mostly absorbed and the minority carrier is generated. The front and back contacts are electrodes ohmically connected between the device and the external circuit. Since solar cells consist of multiple components with respective electronic properties, and their layer dimensions range widely from nanometers to submillimeters, characterization of their electronic properties with fine spatial resolutions is highly desirable for understanding device physics and improving device performance. State-of-the art solar cells such as III–V- and crystalline Si-based cells are made of single crystalline or large crystalline grains ranging in size from millimeters to centimeters. This type of cell exhibits uniformity within the plane of the cell, and the cell is essentially one-dimensional [1]. However, second-generation thin-film solar cells are characterized by severe in-plane nonuniformity. This makes their electrical characterization with nanometer resolutions especially desirable.

Atomic force microscopy (AFM)-based electrical characterizations [4–8] were developed shortly after AFM was invented [9–11]. Using proximate probes, these characterizations can sense not only the structural but also the electrical properties of a sample in the vicinity of the probe, and thus provide unique local electronic information with fine resolutions. This chapter provides an introduction to some of major AFM-based electrical characterization techniques and reviews some of the recent characterization studies of inorganic semiconductor solar cell materials and devices conducted at National Renewable Energy Laboratory (NREL).

24.2 AFM-Based Nanoelectrical Characterization Techniques

AFM measures a sample surface corrugation by probing the atomic force between the tip and the sample [9–11]. When a sharp tip is brought close to the surface, the small force between the tip and the sample (~nN) is detected by a laser-reflection on an extremely soft cantilever (~N/m), on which a sharp tip with an apex radius of ~10 nm is attached. In addition to the atomic force, variable electrical signals between the tip and the sample can be detected and further used for the characterization of local electrical properties. Examples of such electrical signals are Coulomb force, capacitance, and electric current, which are used to measure electrostatic potential, carrier concentration, and local conduction path or conductivity on a sample surface, in the applications of scanning Kelvin probe force microscopy (SKPFM) [4–6], scanning capacitance microscopy (SCM) [7, 8], and conductive AFM (C-AFM), respectively. In this section, the fundamentals of these techniques are introduced.

24.2.1 Scanning Probe Force Microscopy

SKPFM measures a surface electrical potential distribution with fine resolutions of tens of nanometers, by probing the Coulomb force between the tip and the sample. If the work function of the AFM tip is different from that of the sample, when the tip and the sample are electrically connected together or both connected to the ground level, there is a contact potential difference (CPD), and this CPD makes a charge transfer between the tip and the sample. This charge transfer makes the Fermi level aligned and generates a Coulomb force between the tip and the sample. The charging energy and Coulomb force are:

$$E = \frac{1}{2} C V_{\mathrm{CPD}}^2,$$

$$F = \frac{\partial E}{\partial Z},$$

where C, V_{CPD}, and Z are the capacitance, CPD, and separation distance between the tip and the sample [6]. However, the Coulomb force is very difficult to detect because of the small capacitance (~aF) due to the small AFM tip size (~10 nm). It is also difficult to separate the Coulomb force from the atomic force that dominates during topographic imaging. To enhance the Coulomb force and separate it from the atomic force, AC voltage with amplitude of ~10–1,000 mV is usually applied to the tip. In one-dimensional approximation, the Coulomb force can be written as follows:

$$
\begin{aligned}
F &= F_0 + F_1 + F_2, \\
F_0 &= \frac{1}{2}\frac{\partial C}{\partial Z}(V_{\mathrm{CPD}}^2 + \frac{1}{2}V_{\mathrm{ac}}^2), \\
F_1 &= \frac{\partial C}{\partial Z}V_{\mathrm{CPD}}V_{\mathrm{ac}}\sin(\omega t), \\
F_2 &= -\frac{1}{4}\frac{\partial C}{\partial Z}V_{\mathrm{ac}}^2\cos(2\omega t),
\end{aligned}
$$

where V_{CPD} equals the work function difference ($W_{\mathrm{tip}} - W_{\mathrm{s}}$) of the tip and the sample in the thermal equilibrium state, and equals ($W_{\mathrm{tip}} - W_{\mathrm{s}} - V_{\mathrm{tip}} + V_{\mathrm{s}}$) if the tip and the sample were applied with bias voltages of V_{tip} and V_{s} [12]. The first term, F_0, is a constant with time and applies a constant bending on the cantilever. The second term, F_1, oscillates in the same frequency as the AC voltage, and its amplitude is proportional to V_{CPD}. The third term, F_2, oscillates in a double frequency of the AC voltage, and the amplitude is dominated by the capacitance characteristics and the amplitude of AC voltage. F_1 can be detected using a lock-in amplifier, and its amplitude is proportional to V_{CPD}. Therefore, the experimental setup for probing the term F_1 is called electrostatic force microscopy (EFM). EFM gives a two-dimensional mapping of the Coulomb force; it is a qualitative measurement of the sample surface potential.

Using $F_1 \propto V_{\mathrm{CPD}}$ and $V_{\mathrm{CPD}} = W_{\mathrm{tip}} - W_{\mathrm{s}} - V_{\mathrm{tip}} + V_{\mathrm{s}}$, if a proper voltage $V_{\mathrm{tip}} = V_{\mathrm{s}} + W_{\mathrm{tip}} - W_{\mathrm{s}}$ is applied to the tip, V_{CPD} and F_1 are minimized, and the tip voltage V_{tip} in this case is the Kelvin probe signal, which is essentially similar to the classical Kelvin probe signal, with fine resolutions concentrating on the local sample area [12]. In the experimental setup shown in Fig. 24.2, AFM uses noncontact mode for tracing the topography because probing Coulomb force term F_1 requires that the tip be oscillating. The cantilever oscillation signal is filtered to two components with frequencies corresponding to topographic (AFM) and surface potential (SKPFM) measurements. The electrical component is then detected by a lock-in amplifier, and the output of the amplifier is further sent to a negative feedback circuit. The output of the feedback circuit is the Kelvin probe V_{KP}. The qualitative description would be: If V_{tip} deviates from V_{KP}, V_{CPD} is large. Consequently, F_1 is large, and then V_{tip} changes and tries to reduce the V_{CPD} until V_{tip} reaches V_{CDP}. In this sense, F_1 resembles the error signal in AFM, V_{tip} to the voltage applied to Z piezo, and the Kelvin probe signal to the Z height signal.

A typical AFM tip used in SKPFM has three local oscillation maxima (Fig. 24.3): The first, in the ~0–20 kHz range, is the low frequency; the second in the ~50–90 kHz range, is the first resonant oscillation, and the third, in the ~300–500 kHz range, is the second resonant oscillation. The first resonant oscillation is mostly used for noncontact-mode topographic measurement. Either a low frequency in the ~5–20 kHz range or the second resonant oscillation is used for the SKPFM measurement. For the majority of tips, the second resonant oscillation has a stronger oscillation amplitude and thus a better signal sensitivity or energy resolution (<10 mV) than in the low-frequency range (~50 mV) [5]. However, it has been found that the potential image obtained by using the second resonant frequency

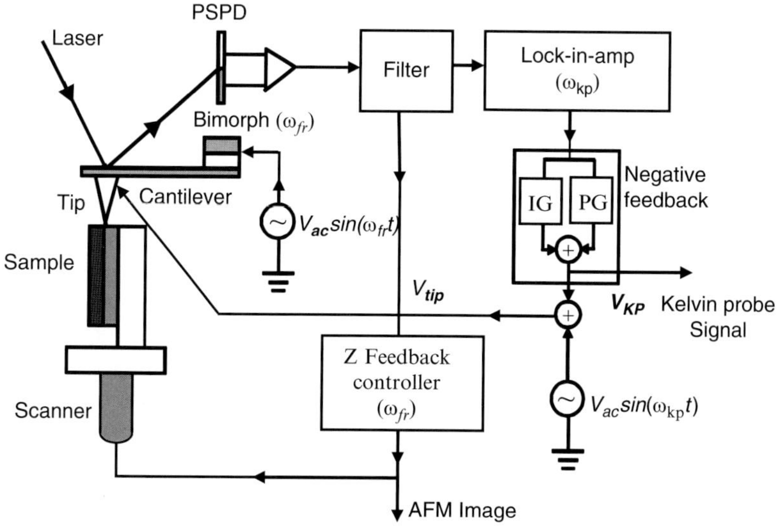

Fig. 24.2 A schematic of SKPFM experimental setup based on the noncontact mode AFM. The cantilever oscillation signal is filtered to two components for topographic and potential measurements

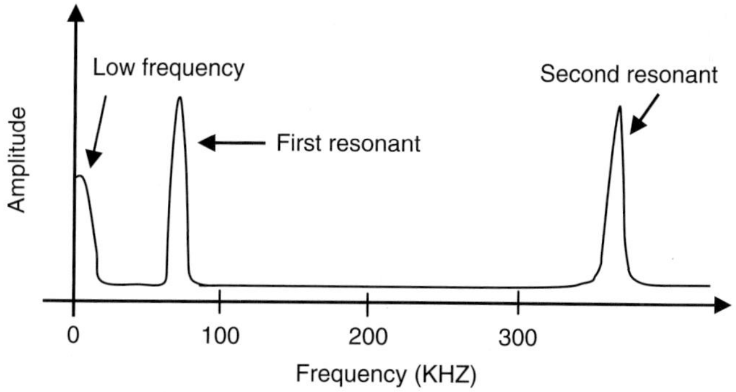

Fig. 24.3 Oscillation characteristics of a typical cantilever used for SKPFM. Three local oscillation maxima in the low frequency, first resonance, and second resonance are shown

includes more topographic effect than one obtained by using the low frequency, which is not suitable for imaging on a rough surface with corrugations larger than ~50 nm. During the tip scanning, topographic and electrical images were obtained simultaneously by using atomic force and Coulomb force at the corresponding frequencies; thus the oscillations cannot be completely separated by the filters and lock-in amplifiers due to the nonlinear relations of the total force and the respective atomic and Coulomb forces. Because the oscillations between first and second resonant frequencies are more closely related to each other than those between the first and the off-peak low frequency, the potential image obtained by using the second resonant frequency includes more topographic effect than one obtained by using the low frequency when imaging on a rough surface.

24.2.2 Scanning Capacitance Microscopy

SCM measures the two-dimensional distribution of carrier concentration in resolutions of tens of nanometers or equivalent to the size of the AFM tip (For a review, see [13]). In the contact mode, the AFM tip, a thin insulator layer on the sample surface such as an oxide layer on Si, and the semiconductor sample form a metal–insulator–semiconductor (MIS) structure (Fig. 24.4). Capacitance of the MIS structure directly relates to the local carrier concentration in the semiconductor sample. The MIS capacitance is the sum of serially connected oxide capacitance (C_{O}) and semiconductor capacitance (C_{S}) [14]:

$$C_{\mathrm{MIS}} = \frac{C_{\mathrm{O}}C_{\mathrm{S}}}{C_{\mathrm{O}} + C_{\mathrm{S}}}.$$

When the gate voltage is varied from the tip, typical capacitance–voltage curves (Fig. 24.5) at a high frequency (>1 kHz) can be separated to three regions: carrier accumulation, depletion, and inversion (inversion region is not shown in Fig. 24.5) [14]. When the gate voltage makes the underlying region a carrier accumulation, capacitance of the semiconductor increases exponentially with the voltage, and the limiting factor for the MIS capacitance C_{MIS} is the oxide capacitance C_{O}, and C_{MIS} saturates to the value of C_{O}. When the gate voltage makes the semiconductor a depletion region, the n-type semiconductor capacitance decreases as the voltage is reduced (Fig. 24.5a), due to an increase in the depletion width. The capacitance, $C_{\mathrm{S}}(V) = \partial Q/\partial V \approx \varepsilon e/\lambda$ (Q,ε, e, and λ are the charge amount, dielectric constant, single charge, and depletion width, respectively), is determined by the carrier modulation on the depletion edge. The semiconductor capacitance is the limiting factor for the total capacitance in this case. When the gate voltage makes the semiconductor an inversion layer, although there is a dense layer of minority carriers under the oxide/semiconductor interface, the minority carriers do not response an AC voltage of high frequency (>1 kHz) because of the slow thermal-exciting rate of minority carriers [14]. The C–V curves show a deep depletion behavior, instead of a low-frequency inversion capacitance.

If the semiconductor is p-type with majority carriers of holes, the C–V curves show similar behaviors, with the gate voltage polarity flipped opposite (Fig. 24.5b). The $\mathrm{d}C/\mathrm{d}V$ vs. V are maximized at a voltage around the flat band condition [13], slightly deviating from the flat band to a weak accumulation for a low carrier concentration ($\sim 10^{16}/\mathrm{cm}^3$), and to a weak depletion for high carrier concentration ($\sim 10^{18}/\mathrm{cm}^3$). The maximum $\mathrm{d}C/\mathrm{d}V$ voltage is critical in understanding the SCM measurements [15, 16].

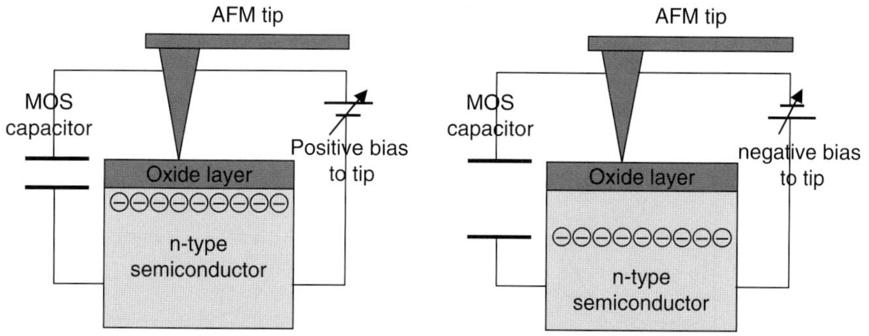

Fig. 24.4 A schematic illustrating that the AFM tip, the oxide layer on the sample, and the semiconductor sample form an MOS structure. Capacitance of the structure relates to the carrier concentration and increases (decreases) with application of a positive (negative) bias voltage to the gate for an n-type semiconductor

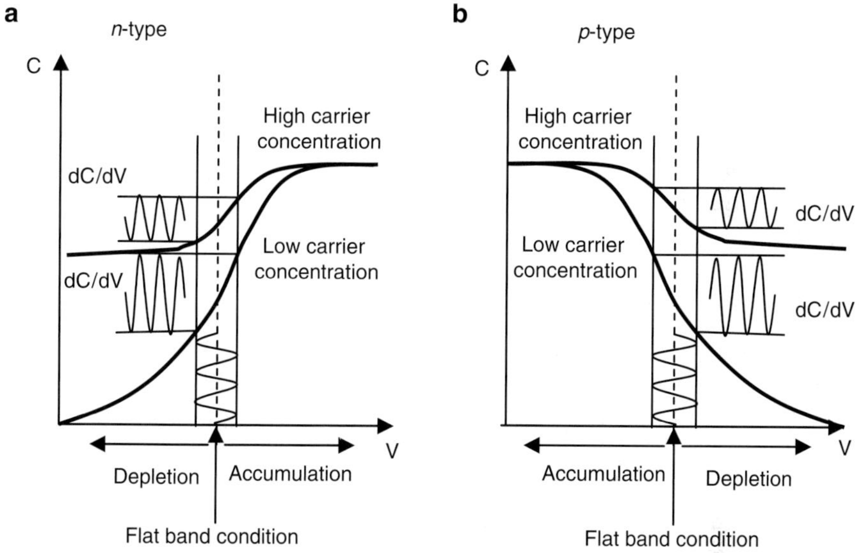

Fig. 24.5 Schematics of typical capacitance–voltage curves of MIS structures with high and low carrier concentrations and both n- and p-doping. The SCM $\mathrm{d}C/\mathrm{d}V$ signal with its amplitudes and signs relates to carrier concentrations and doping types in the semiconductor and has the maximum amplitude around the flat band voltage

With a low (high) carrier concentration, the increasing rate of the capacitance in the depletion region or the slope of the C–V curve is large (small) because of the large (small) screening or Debye lengths (see Fig. 24.5). In the experimental setup, the carrier concentration is determined by measuring $\mathrm{d}C/\mathrm{d}V$ as the SCM signal, while scanning the AFM tip over the sample surface. The magnitude of $\mathrm{d}C/\mathrm{d}V$ is qualitatively related to the doping concentration. The carrier concentration can

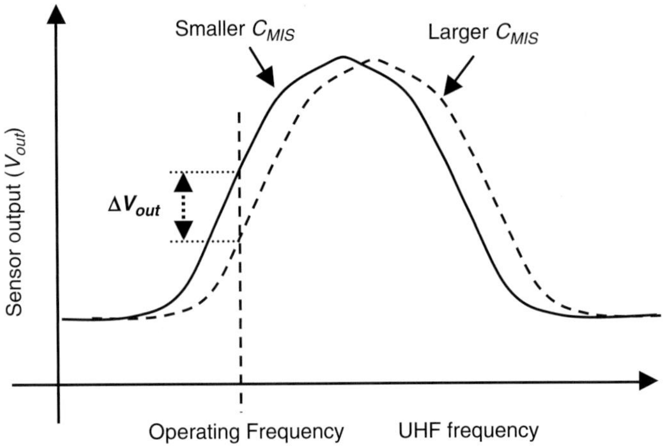

Fig. 24.6 A schematic illustrating how an SCM sensor works. A small change in the MIS capacitance induces the resonance curve shift, and makes the sensor output voltage at a fixed operating frequency change $\Delta V_{\rm out}$. $\Delta V_{\rm out}$ is linear to the capacitance change $\Delta C_{\rm MIS}$, if $\Delta C_{\rm MIS}$ is small

be quantitatively derived by molding and by measuring on a reference sample as calibration [13].

Experimentally, the capacitance between the sharp tip and the sample is very small (∼aF). It is detected by an ultrahigh frequency (UHF) resonant circuit [17]. A UHF resonator with a frequency of 0.9–1.0 GHz is coupled to a detection circuit through a resonant circuit that contains tip/oxide/sample MIS structure. The resonant frequency is determined by the total stray capacitance that is several orders (∼pF) of magnitude larger than the targeted MIS capacitance. However, the sensor is even capable of detecting a capacitance change of 10^{-21} F with a resonant bandwidth of 1 Hz [17]. The amplitude of the UHF signal coupled to the detection circuit is decided by the resonant frequency of the resonant circuit and the actual operating frequency (Fig. 24.6). The coupled UHF signal is further converted to a DC signal proportional to the UHF amplitude.

The change of DC output voltage $\Delta V_{\rm out}$ is linear to the change of MIS capacitance ΔC if ΔC is relatively small [13]. Therefore, an SCM image composed directly from $\Delta V_{\rm out}$ represents all the ΔC at the corresponding pixels, which is in turn determined by the local carrier concentration of semiconductor samples. In fact, the $V_{\rm out}$ signal from the detection circuit is amplified by a preamplifier, and $\Delta V_{\rm out}$ was subsequently detected by a lock-in amplifier. An AC voltage $V_{\rm ac}$ with amplitude from tens of millivolts to several volts and a frequency of 10–100 kHz is applied to the sample to modulate the MIS capacitance or generate ΔC. A DC voltage $V_{\rm dc}$ is also added to the sample to drive the MIS to the maximum $\mathrm{d}C/\mathrm{d}V$ condition, which is close to the flat band. Charges trapped in the oxide and at the oxide/semiconductor interface can significantly shift the flat band condition. Taking an SCM image under a proper V_{dc} is critical due to the fact that the dC/dV contrast can be even reversed under different V_{dc} values [15, 16].

A high-quality thin oxide layer with thickness of a few nanometers is critical for a reliable SCM measurement [13], which requires a uniform thickness with minimum structural and electronic defects. Indeed, a contrast in an SCM image can result solely from nonuniform oxide thickness [15]. Structural defects can significantly affect the SCM image. Mobile charges in the oxide layer cause hysteresis when measuring dC/dV–V_{dc} curves. Charges trapped at the oxide/semiconductor interface can significantly shift the flat band voltage and further change the amplitude of the maximum dC/dV [15]. To achieve the high-quality oxide, a sample is often chemical-mechanically polished (CMP) by diamond-coated pads with grains from rough (~10 μm) to fine (0.1 μm) sizes [13]. A final polishing step is to use silica colloid with particle size of 0.05 μm. After the CMP, annealing at ~300°C under an ultraviolet illumination for 30 min has been found to effectively improve the oxide quality, which is expected to eliminate the electronic defects [13, 18]. As for the AFM probes, Si tips that are suitable for contact mode and coated with metals are used because of the requirement of good electrical conduction. The tip is electrically connected to the nearby capacitance sensor by a short wire to avoid increasing the tray capacitance.

24.2.3 Conductive AFM

In the contact mode of AFM, if a bias voltage is applied between the sample and the tip, the current through the tip can be measured (Fig. 24.7), which is related to the local electrical conduction path or local conductivity of the sample. To achieve the measured current fully determined by the electrical properties of the sample, an ohmic contact between the tip and the sample is desired but is difficult to obtain in most cases for semiconductor samples. Therefore, comparison of the current images measured in opposite polarities and under various bias voltages is useful to identify whether the current images are dominated by the sample property or by the tip/sample contact. To get good conduction, a metal-coated tip or a highly doped diamond-coated tip is used. The diamond-coated tip has an obvious advantage of stiff wear-resistance. A relatively larger contact force than in SCM and normal contact mode AFM is often used to enhance the electrical contact of the tip and the sample, which wears out the tip quickly if using a metal-coated tip. There are several versions of C-AFM, including tunneling AFM (TUNA), C-AFM, and scanning spreading resistance microscopy (SSRM); all setups of the C-AFM modes are similar except for the current sensor range, TUNA in ~10 fA–100 pA, C-AFM in ~10 pA–1 μA, and SSRM in ~10 pA–100 μA in a logarithmic scale. Design of the current range is due to specific detection of the current for the specific samples and purposes.

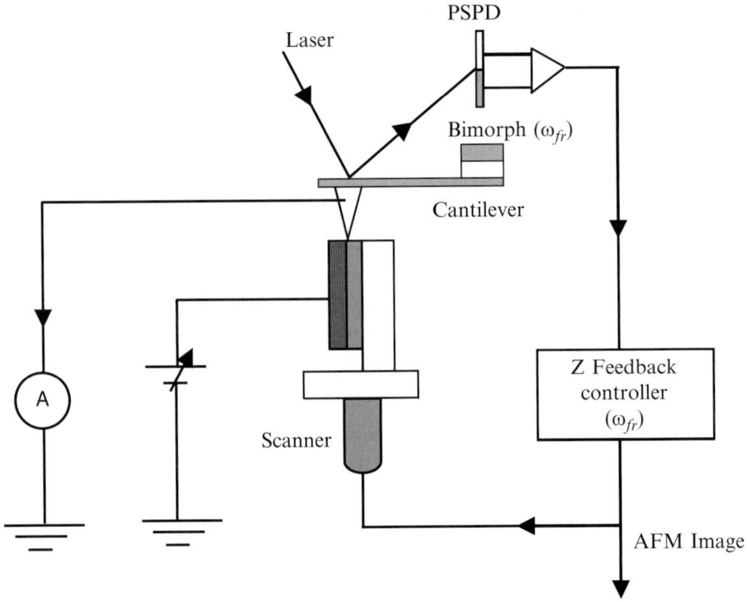

Fig. 24.7 A schematic of C-AFM. With a V_b applied between the tip and the sample, the electric current through the tip is measured by a current sensor

24.3 Characterization of Junctions of Solar Cells

The core structure of a solar cell device has an emitter (n-type) on the absorber (p-type) to form the junction of the photodiode or solar cell where minority carrier collection happens. Typically, the emitter is a thin ∼10 nm to submicrometer layer, and the junction depletion is of similar width. Therefore, characterization techniques with nanometer resolutions are useful for the electrical characterizations of the junctions. This section describes junction location identifications using SKPFM and discusses measurements of electric potential and field on the junctions.

24.3.1 Junction Location Determination

The junction depth or emitter layer thickness is one of the main device parameters that determine device performance. There is an optimal emitter layer thickness [1]; thicker than this would cause a poor short-circuit current (J_{sc}) due to carrier recombination in the highly doped/defected emitter layer. On the other hand, if the layer is thinner than optimal, poor junction quality and/or high sheet resistance would degrade the efficiency. Secondary ion mass spectrometry (SIMS) is a common technique for measuring the junction depth by measuring both donor and acceptor concentrations throughout the junction region [19, 20]. However, the conventional

SIMS is a one-dimensional technique that measures the depth of individual points on a solar cell. The analysis spot size is in tens of micrometers, which limits the lateral spatial resolutions. On the other hand, scanning electron microscopy (SEM) exhibits a secondary electron contrast of the emitter layer and depletion region, but is difficult to accurately determine the p–n junction location. Admittance spectroscopy is a powerful tool for measuring carrier concentration and depletion width (For a review, see [21]) but cannot provide the junction location or depth.

SKPFM measures the two-dimensional (2D) electrical potential distributions with nanometer resolutions. Electric field distribution can be further derived from the measured potential, and the maximum electric field corresponds to the junction location. Therefore, SKPFM can provide 2D junction location identifications. The following sections describe junction identifications in three solar cells: multicrystalline Si, GaInNAs, and Cu(In,Ga)Se_2 [22–24].

24.3.1.1 Junction Identification in Multicrystalline Si Solar Cells [22]

Solar cell devices are often made with surfaces exhibiting micrometer-scale nonuniformity. Examples of such devices are polycrystalline thin-film cells and multicrystalline Si (mc-Si) cells [25, 26]. Textured surface in mc-Si cells is naturally formed after slicing a cast ingot and chemical etching, and it is useful for light trapping. For these devices, it is difficult to provide precise junction depth using SIMS measurement due to the fact that the analysis spot size is often lager than the texture features. Therefore, a 2D junction location determination made by using SKPFM is a valuable characterization for the mc-Si cells.

In sample preparation for the SKPFM measurement, an mc-Si device was cleaved, and the junction was exposed on the cross sections [22]. The two cleaved pieces were glued together in a sandwich structure by using a conductive epoxy. In this configuration, the front sides of the two-piece devices face each other, with a conductive epoxy layer between them. Because the surface of as-cleaved cross sections is rough for SKPFM measurement, the cross section was polished by CMP. The sandwich structure is for protecting the cross section from edge damage during the polishing, because the junction of solar cells is located close to the front edge of the device. Because a bias voltage (V_b) must be applied to the device in SKPFM measurement, the polishing was performed carefully to avoid shunting the device.

Potential profiles along the junction at various V_b values are shown in Fig. 24.8a. Each profile in the figure was an average of multiple measurements to enhance the signal/noise ratio. On the potential profile at $V_b = 0$, the small potential variations of less than 100 mV were due to the surface potentials on different components of the device, but not to the built-in potential along the junction. The effect of Fermi level (E_F) pinning at an interface state between the oxide and Si is the reason the surface-potential contrast is much smaller than the built-in potential in the bulk [27–29]. However, when a V_b is applied to the device, the change of the potential across the junction can be measured by SKPFM, as shown in Fig. 24.8a. This is because the E_F position relative to the band configurations should not change with the application

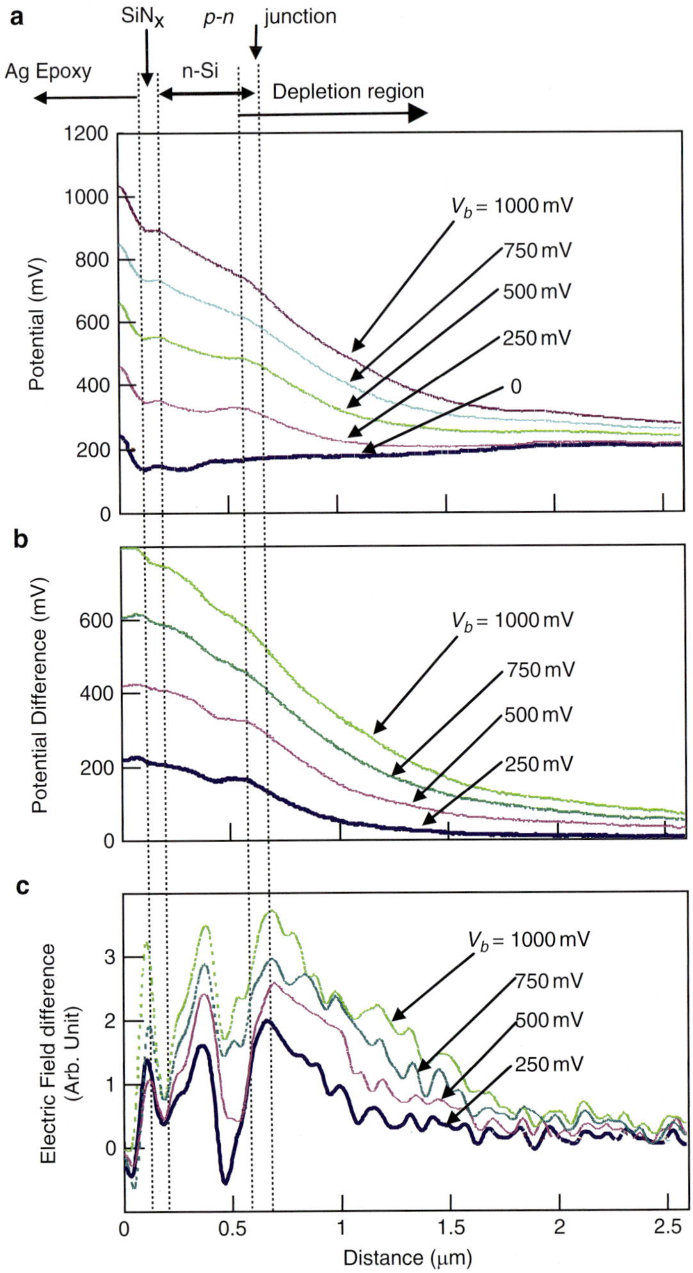

Fig. 24.8 (**a**) SKPFM potential profiles taken on a mc-Si solar cell at various reverse V_b values; each profile is an average of multiple line scans from an SKPFM image. (**b**) V_b-induced changes in the potential as deduced from the profiles at given V_b values and $V_b = 0$ in (**a**). (**c**) V_b-induced changes in the electric field as deduced by taking the first derivative of the profiles in (**b**); the maximum field corresponds to the p–n junction. The structure of the device is shown at the top of the illustration

of a small V_b; the V_b-induced potential change in the bulk is thus about identical to that on the surface [28, 29].

The V_b-induced changes in the surface potential can be deduced by subtracting the potential profile at $V_b = 0$ from the profiles at the various V_b values (Fig. 24.8b), and the potential changes in the bulk are about identical to those on the surface. By this procedure, the surface effect on the potential measurement can be avoided [28, 29]. This is a crucial procedure for the application of SKPFM to photovoltaic characterizations, because SKPFM measures on the surface that is affected by the subsurface region within a depletion depth from the surface. However, the characterization for photovoltaic properties is requested to reflect the properties in the bulk. The potential variation on each curve in Fig. 24.8b is slightly smaller than the corresponding V_b. This potential loss is considered to occur at the silver (Ag)-epoxy and other connections to the device. Another reason for the potential loss is that, in the potential measurement, the charging energy and electrostatic force between the tip and the sample dominate the Kelvin probe signal measurement. However, considering the much larger area of the cantilever versus the tip apex, the charging energy and electrostatic force between the cantilever and the sample, other than between the tip and the sample, also cause a loss of potential or a smoothing in the potential contrast, depending on the area of the cantilever and the tip length.

The location of the p–n junction was deduced by taking the first derivative of the potential profile, which is a profile of V_b-induced change in electric field, as shown in Fig. 24.8c. The maxima on the curves correspond to the location of the p–n junction. The interface of the Ag-epoxy/Si-device could be identified from AFM images, where a sharp change in topography was observed. The junction location and depth (~490 nm) were identified from the profile of V_b-induced electric-field change. The starting point of the depletion region in the n-layer was observed on the potential profiles at $V_b = 250$ mV and $V_b = 500$ mV, where the potential began to decrease in the depletion region.

The potential profiles under various V_b values were simulated (Fig. 24.9) using a one-dimensional solar cell simulation software PC1D.[1] The electric field at all V_b values is maximal at the n/p interface. However, the V_b-induced changes in the electric field show plateaus in the depletion region, and the width of the plateau increases with increasing reverse V_b. This is because carrier concentration is very small in the depletion region so that V_b is applied uniformly to the region. Thus, the additional electric field to the built-in field is uniformly applied. The depletion width increases with increasing V_b, and so does the electric-field plateau. However, the plateau was not observed in the measurement where the electric field exhibits a broad peak. If the peak corresponds to the center of the plateau, the junction location as deduced from the measurement has a ~40-nm deviation from its true location, as indicated by the dashed lines in Fig. 24.9c. The tail of the electric field from the plateau to zero fields is longer on the p-type than on the n-type regions, which

[1] The software is issued by ARC Photovoltaics Center of Excellence at the University of New South Wales, Australia.

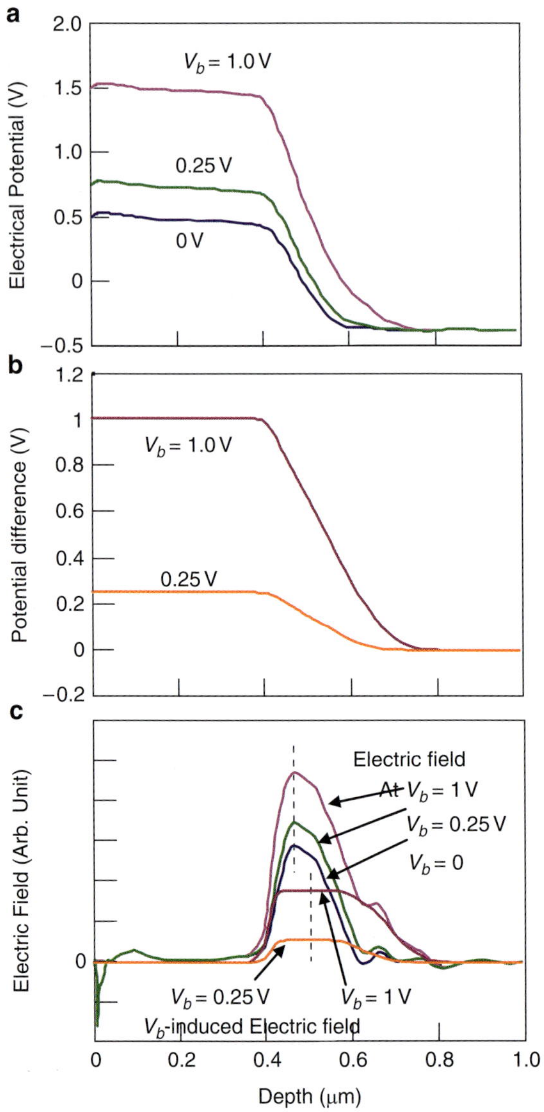

Fig. 24.9 (**a**) A simulation of the electric potential in the mc-Si device bulk at various V_b values. (**b**) V_b-induced potential changes as deduced from the potential profiles in (**a**). (**c**) The electric field at various V_b values as deduced from the potential profiles in (**a**) and V_b-induced changes in the electric field as deduced from the potential changes in (**b**). *Dashed lines* in (**c**) indicate the true junction location and the center of the electric field plateau that should correspond to the junction location measured by SKPFM, which has a ∼40 nm deviation

show on both the measurement and simulation curves, because of the lower doping concentration in the p-type than in the n-type regions.

Another difference between the measurement and simulation is that the simulation exhibits a much smaller width of the potential change (~300 nm at $V_b = 0.25$ V) than that of the measurement (~840 nm at $V_b = 0.25$ V). Two reasons for this discrepancy may exist: (1) The oxide layer on top of the device may make the distribution of the surface potential wider than in the device bulk. Indeed, the depletion width measured on an as-cleaved III–V device is much smaller than on the polished Si device. (2) The SKPFM measurement may image the depletion area at a considerably greater-than-actual width. Although both measurement and calculation of the SKPFM on a charged line (~1 nm in width) and a charged point show spatial resolutions of several tens of nanometers [30–32], with tip sizes of ~10 nm, a recent calculation of the SKPFM on a square shows considerable broadening of the square edge in the simulated SKPFM image, due to the long-range nature of Coulomb force [33].

The most valuable capability of SKPFM used in the junction identification is the 2D identification. Figure 24.10a, b shows an AFM and the corresponding SKPFM images taken under a reverse $V_b = 1$ V. The distinction between the Ag epoxy and Si device is clear on both images. The electric-field image (Fig. 24.10c) was deduced by taking a 2D first derivative of the SKPFM image:

$$|\vec{E}| \propto |\nabla\varphi| = \left|\frac{\partial\varphi}{\partial x}\vec{i} + \frac{\partial\varphi}{\partial y}\vec{j} + \frac{\partial\varphi}{\partial z}\vec{k}\right|,$$

where $\vec{E}$ and φ denote the electric field and potential, respectively. The bright curve in Fig. 24.10c corresponds to the sharp change in surface potential at the epoxy/device interface. Contrast of the electric field image was subsequently enhanced to show the p–n junction as the bright band in Fig. 24.10d. The shape of the junction is clearly shown following the topography of the device, i.e., the junction depth is identical everywhere over the measured area of the device. This clearly demonstrates the uniqueness of the SKPFM capability in 2D p–n junction identification.

24.3.1.2 Junction Backshift in a GaInNAs Cell [23]

It is possible to achieve n- and p-doping with either foreign specimens such as zinc (Zn) and Si in GaAs or electronic defects formed in the material growth. The electronic defect doping widely exists in solar cell materials, such as GaInNAs epitaxial layer, and CdTe and Cu(In, Ga)Se_2 thin films. SIMS identifies the junction location by measuring the concentrations of foreign specimens but cannot apply to a junction formed by the electronic defect doping. Such junction identification needs a direct electrical characterization. As shown in the following, an unintentional backshift of the junction in a GaInNAs cell is observed by the SKPFM measurement.

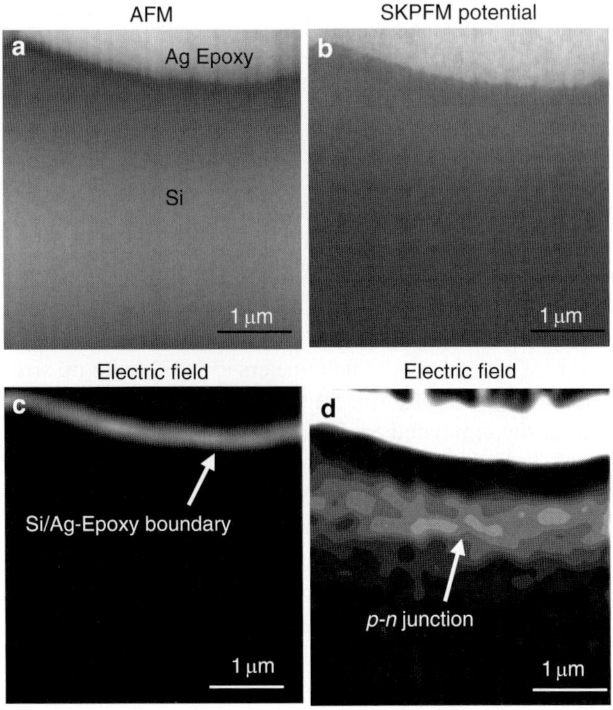

Fig. 24.10 (**a**, **b**) AFM and the corresponding SKPFM images taken on a mc-Si solar cell at a reverse $V_b = 1$V, with gray scales of 160 nm and 0.8 V, respectively. (**c**) An image of the electric field as deduced from (**b**). (**d**) The same image as (**c**) with an enhanced contrast to present the p–n junction, which conformably follows the topographic feature of the cell

GaInNAs with a few percent of nitrogen (N) and indium (In) is considered promising for a lattice-matched four-junction solar cell, as the third cell [34], but roughness caused by In segregation leads to poor epitaxial qualities [35, 36]. To circumvent this problem, a small amount of bismuth (Bi) surfactant was used during the molecular beam epitaxy (MBE) growth, which effectively improved the epitaxial layer smoothness [34–36]. However, the Bi incorporation induced serious device degradation, especially in the short-wavelength range of quantum efficiency measurements, regardless of its improvement upon the epitaxial quality [34]. SKPFM measurement on the devices with various Bi incorporation amounts showed backshifts of the junction location, which could induce device degradation [23].

The GaInNAs(Bi) solar cell grown by MBE was cleaved along the [110] direction, and the junction was exposed onto the [110] cross-sectional surface. The cleaved surfaces are atomically flat with a few steps in heights of several atomic layers. Most III–V devices can be cleaved atomically flat in the [110] direction and thus are suitable for SKPFM measurement with no further polish needed. Figure 24.11 shows the surface root-mean-square (RMS) roughness measured by AFM. Without a Bi flux present, the surface roughness is ~2 nm, which is the In-segregation-related

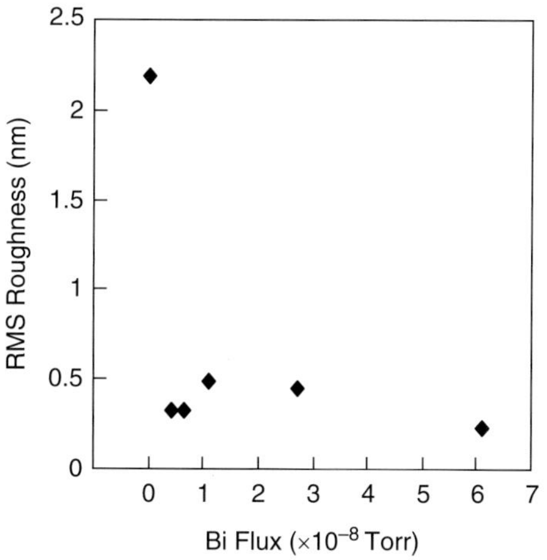

Fig. 24.11 Changes in the RMS roughness on the surface of GaInNAs (Bi) epitaxial layers with BEP of Bi evaporations as a surfactant. The roughness was measured by AFM and shows a drop with a little Bi evaporation, but stays around with further increases in the Bi evaporation

surface roughening [35, 36]. With a Bi evaporation in a beam equivalent pressure (BEP) of 0.4×10^{-8} torr, the surface roughness dropped about one order of magnitude, to ~0.3 nm. However, further increase of BEP in the 10^{-8} torr range did not change the roughness significantly. Bi is considered to act as surfactant during the MBE growth, through increasing the diffusion length relative to the arsenic (As)-terminated surfaces and lowering the step energy by reducing adatom bond strength at the step edges [36]. Although increasing the Bi BEP in the 10^{-8} torr range does not make the epitaxial layer smoother, it induces significant device degradation. Each potential profile in Fig. 24.12 was measured on the sample with BEP of 0.44×10^{-8} torr. Measurement of the potential profiles under various V_b values is for avoiding the surface E_F pinning effect. The profile at $V_b = 0$ shows a 0.35 V potential drop at the p–n junction, which is much smaller than the bandgap of the absorber layer and is due to surface E_F pinning [27–29].

The potential measurements on the devices with various BEPs show the direct observation of Bi-induced junction shifts (Fig. 24.13). With increasing BEP to 0.67×10^{-8} torr, location of the potential drop stays close to the front contact of the device (Fig. 24.13a). However, the potential profile shape is significantly broadened in the absorber layer, and the maximum electric field location (Fig. 24.3b) is slightly shifted toward the back side. With further increasing BEP, both potential and electric field profiles exhibit clear shifts of the p–n junction to the back side of the absorber layer. The larger the BEP, the closer the junction is to the interface between the absorber and the substrate.

A conversion from p- to n-types occurred in the absorber layer due to the evaporation of Bi. Bi acts as a surfactant, and only a small amount of it is incorporated into the absorber layer. In fact, no Bi was detected by SIMS above the background level ($\sim 10^{16}/\text{cm}^3$) in all the samples in Fig. 24.13. A small amount of Bi incorporation may be enough to change the doping configuration in the small background doping concentration of $1.5 \times 10^{14}/\text{cm}^3$. Admittance spectroscopy measurement has found a defect level associated with the Bi incorporation [34]. However, I–V

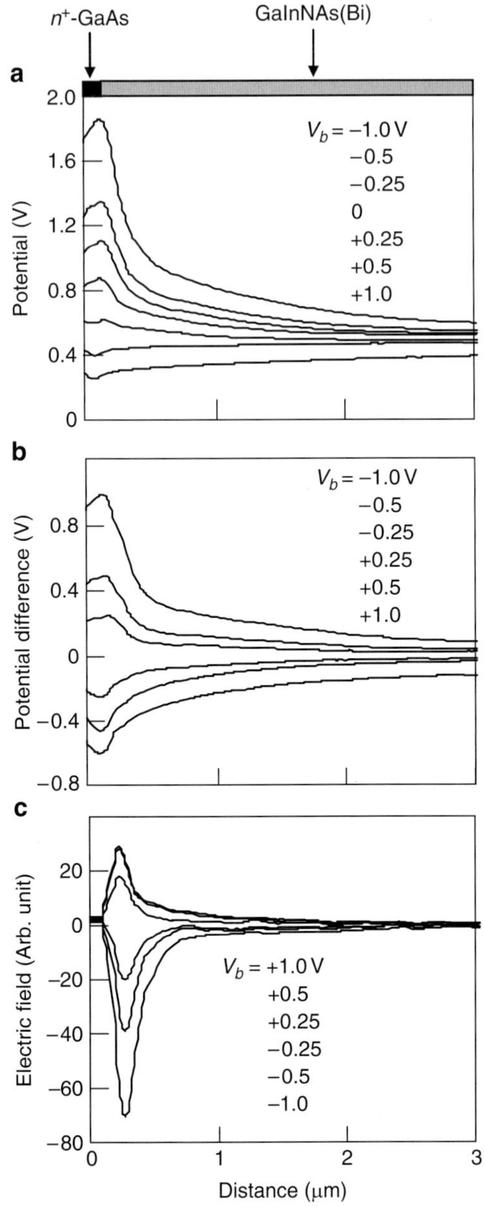

Fig. 24.12 (**a**) SKPFM electrical potential at various V_b values. (**b**) The V_b-induced potential changes as deduced from the profiles at given V_b values and $V_b = 0$ in (**a**). (**c**) V_b-induced changes in the electric field as deduced from (**b**), taken on a cross section of the GaInNAs (Bi) device with Bi BEP of 0.41×10^{-8} torr. Positive and negative V_b values represent, respectively, the forward and reverse bias voltages

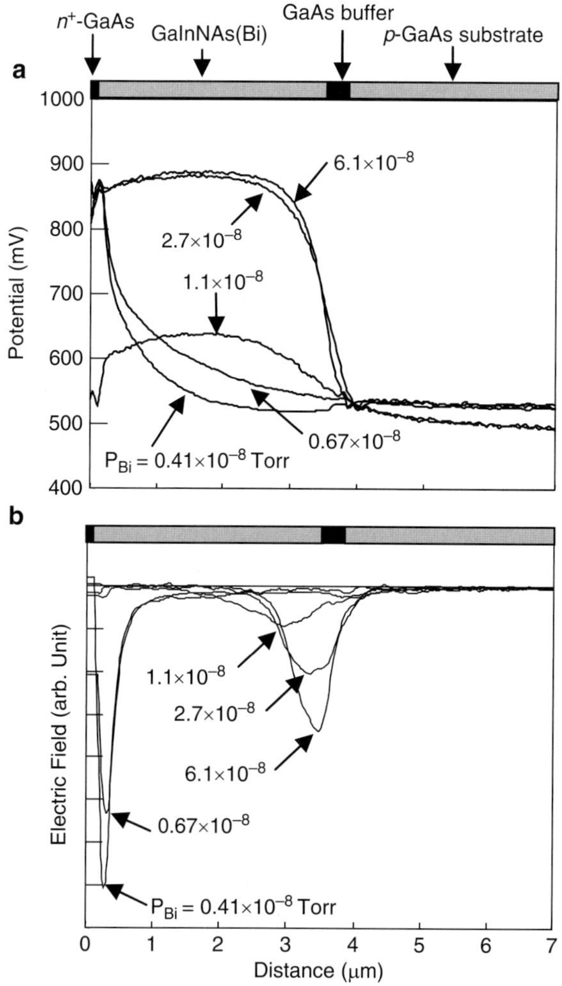

Fig. 24.13 (**a**) SKPFM potential profiles. (**b**) The electric field profiles as deduced from the potentials in (**a**), taken on the cross sections of the GaInNAs(Bi) devices with various Bi BEP during MBE growth of the devices. Both potential and electric field profiles exhibit backshift of the junction location with the Bi evaporations

measurements indicate an independency of dark saturation current J_0 on the Bi incorporation, possibly due to a much larger nitrogen-related background defect concentration that dominates the carrier transport across the junction [34]. The fact that the p–n junction location is closer to the interface of absorber/substrate with a higher Bi BEP suggests a buildup effect of the Bi incorporation and a gradual change in the donor concentration during MBE growth. It can be a buildup of Bi incorporation, defects, or impurities.

24.3.1.3 Junction Location in $Cu(In, Ga)Se_2$ Cells [24]

Thin-film solar cells based on the $Cu(In, Ga)Se_2$ (CIGS) material system have reached a high level of performance, both on a laboratory scale and in large-area modules [37, 38], and it is one of the most promising materials for low-cost photovoltaic applications. Typical structure of CIGS solar cells is constructed of metal grid, MgF_2 antireflection coating, ZnO transparent conductive oxide (TCO), CdS emitter, CIGS absorber, and molybdenum (Mo) back-contact layers, and soda lime glass substrate [37, 39, 40]. The highest conversion efficiency has been achieved by depositing a CIGS film using "three-stage co-evaporation" and forming a CdS buffer layer using chemical bath deposition (CBD) [37, 39, 40]. In contrast to this great improvement in efficiencies, however, understanding of the basic electronic properties of this material is relatively poor. One of the most important issues, determination of the p–n junction location, is not yet settled, although it should be considered one of the most critical factors in solar cell performance. Based on measurements of photoluminescence (PL) and electron-beam induced current (EBIC) [41, 42], it has been postulated that the p–n junction is located on the CIGS side, close to the CIGS/CdS interface. However, many researchers involved in investigation of this material system explain their experimental data and build theoretical models based on the natural assumption that the junction is a heterojunction, abrupt on the p-GIGS/n-CdS interface (For examples, see [43–45]). Therefore, a solid determination of the location of the p–n junction is highly desirable. SKPFM was used to investigate the electrical potential on the p–n junction of CIGS cells, and the results indicate that the junction is indeed located in the CIGS absorber close to the CIGS/CdS interface (30–80 nm) [24].

To get a relatively flat cross-sectional surface that is necessary for the SKPFM measurement, the device structure was grown on a Mo-coated GaAs(001) wafer instead of the soda lime glass that is normally used. A $\sim$2 μm CIGS film was evaporated on the substrate using "three-stage co-evaporation," and a $\sim$50 nm CdS emitter layer was subsequently grown on the film by CBD. An undoped ($\sim$50 nm thickness) and an aluminum (Al)-doped ($\sim$500 nm thickness) ZnO window layer was then deposited using radio frequency sputtering. Sodium (Na) was also incorporated using an NaF source during the growth of the CIGS film, to ensure identical properties to the devices grown on Mo-coated soda lime glass substrates. The sample was cleaved along the [110] direction of the substrate. The cross-sectional surface is atomically flat on the GaAs substrate, and flatness adequate for high-quality SKPFM measurement on the device structure could be found in some areas [24]. The SKPFM measurement was impossible if the device was grown on the glass substrates. CMP on cross sections of devices on soda lime glass substrate was also tried; however, it was found that the polishing often makes the devices shorted and causes their polarity to be reversed, possibly due to a mix-up of the layer components of the device by the polish.

AFM images provide information on device structures. As shown in Fig. 24.14a, the CIGS film was recognized as the large grains, in contrast to the small grains in the CdS/ZnO region. This difference in grain size reveals a boundary between the

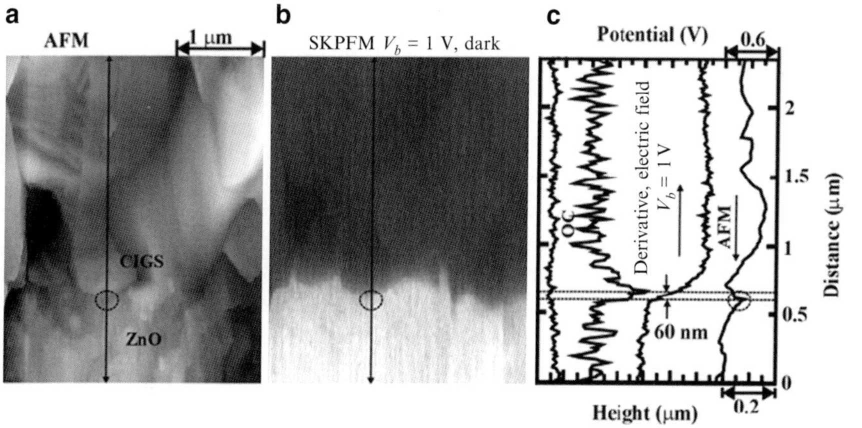

Fig. 24.14 (**a**) AFM and (**b**) corresponding SKPFM images of a CIGS solar cell taken under a reverse V_b of 1 V. Example line profiles along the lines indicated in (**a**) and (**b**) are shown in (**c**). The dashed circles show the location of the CIGS/CdS interface, as identified from the AFM line profile. A profile of electric field in (**c**), as derived from the potential profile, reveals the p–n junction location in CIGS and ~60 nm away from the CIGS/CdS interface. A potential profile taken under OC at the same line position is also shown in (**c**)

CIGS and the CdS/ZnO regions. However, the boundary between the CdS and ZnO layers is not discernible from the topographic image, because the topography of the CdS and ZnO layers is similar. Complementary SEM images taken in the secondary electron mode and the back-scattering mode show a CdS thin layer between the CdS and ZnO regions, as well as similar grain structures on both regions of CdS and ZnO, which is consistent with the AFM observations [46]. This demonstrates that the boundary seen in the AFM image is the boundary between CIGS and CdS layers. The corresponding SKPFM image (Fig. 24.14b) taken under a 1 V reverse V_b shows a boundary of potential contrast that is consistent with the boundary shape of CIGS/CdS in the AFM image (Fig. 24.14a).

From the AFM line profile (Fig. 24.14c), the location of the CIGS/CdS interface can be determined, as indicated by the dashed circle in the figure. Because the V_b is a reverse bias, the potential in the CdS and ZnO regions is higher than in the CIGS region. The potential starts to drop from the interface and extends ~250 nm into the CIGS film. This indicates that a built-in electric field exists on the CIGS side close to the interface, and the p–n junction is located 0–250 nm from the interface.

The maximum electric field location as derived from the potential profile indicates that p–n junction in the present line scan in Fig. 24.14c is located ~60 nm from the CIGS/CdS interface. In all the effective line scans that have good SKPFM signal/noise levels, and in which the location of the GIGS/CdS interface can be determined from the AFM profile, the p–n junction is located 30–80 nm from the CIGS/CdS interface. This is an important characterization for the CIGS device that measures the junction location with high resolutions. Also shown in Fig. 24.14c is an SKPFM line profile taken in open circuit at the same line position. On the p–n junction, the potential profile is relatively flat over the CIGS, CdS, and ZnO layers, showing a slight increase ($\sim$110 mV) of the potential in the CIGS layer relative to that in the ZnO layer. Again, the small potential contrast is due to surface E_F pinning.

The band diagram (Fig. 24.15) of the device bulk in the thermal equilibrium state was calculated by solving Poisson's equation [46] and using the device structure and the parameters in the literature [47–49]. SKPFM measures the work function in the thermal equilibrium state. Because E_F is constant throughout the sample, the work function follows the profile of the vacuum level. Because the diagram is for electrons that have a negative charge, the measured potential should be compared with the mirror image of the vacuum level, as indicated by "Potential" in Fig. 24.15. On the depletion region around the p–n junction, there is a big potential increase ($\sim$1 V) close to the bandgap of CIGS. However, this potential increase does not exist on the measured potential profile due to the surface E_F pinning. From the location of the CIGS/CdS interface, the potential starts to decrease because of the different doping concentrations in the materials. The potential decreases throughout the CdS layer because of the thin layer thickness ($\sim$50 nm) and low doping concentration ($\sim 10^{16}/\mathrm{cm}^3$). At the CdS/ZnO interface, the conduction band minimum has a discontinuity due to the band offset [49]. Accordingly, the potential decreases slightly. From the interface between the intrinsic and the n-doped ZnO layers, the potential increases a little due to the different doping levels. The difference in the potential between n-type CIGS and n-type ZnO regions is $\sim$100 mV, which is consistent with the measurement, $\sim$110 mV lower potential on the ZnO region than that on the CIGS region.

With a V_b applied to the p–n junction, SKPFM can measure the potential drop because the voltage should not unpin the surface E_F (Fig. 24.16b). The "intrinsic" potential height in the bulk without V_b is slightly smaller than the energy bandgap, and the bandgap of CIGS is 1.12 eV. To make the potential height on the surface similar to that of the bulk in the "intrinsic" case without a V_b, a reverse V_b of 1 V was applied to the device, and it should reveal the p–n junction properly. With the V_b, the potential height in the bulk should be $\sim$2 V. The measured potential height on the surface across the p–n junction ($\sim$0.7 V, Fig. 24.14c) is smaller than the applied V_b (1 V). Depletion widths of 60 nm in the n-region and 190 nm in the p-region correspond to a doping level of $\sim 2 \times 10^{17}$ and $\sim 6 \times 10^{16}/\mathrm{cm}^3$, respectively, by taking $\varepsilon = 12$ [50] and a bandgap of 1.12 eV and assuming that the carriers are completely depleted in the depletion region. If the effect of nonuniform depletion of the carriers is included, the doping levels should be a little larger. The bulk of CIGS film fabricated by the three-stage co-evaporation is p-type by unintentional electronic doping. The n-type region close to the CIGS/CdS interface may be doped by the diffusion of Cd^{2+} during the formation of the CdS layer and postanneals [41]. These Cd^{2+} ions would occupy copper (Cu) sites and act as donors.

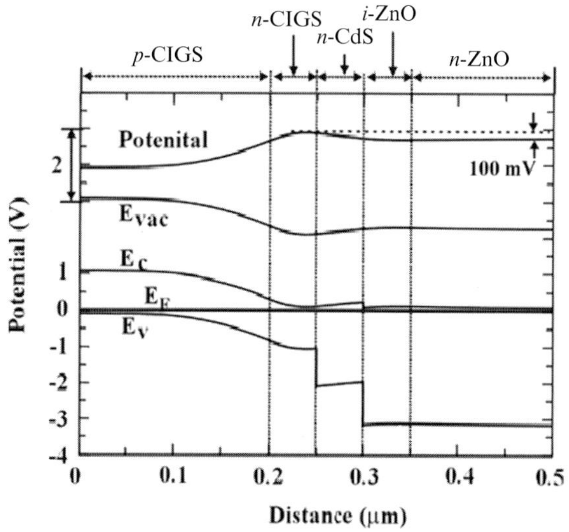

Fig. 24.15 A calculated band diagram of the CIGS device in the bulk containing p-CIGS, n-CIGS, n-CdS, i-ZnO, and n-ZnO layers. The SKPFM potential corresponds to the profile "Potential" indicated in the figure, which is a mirror image of the vacuum level in the diagram

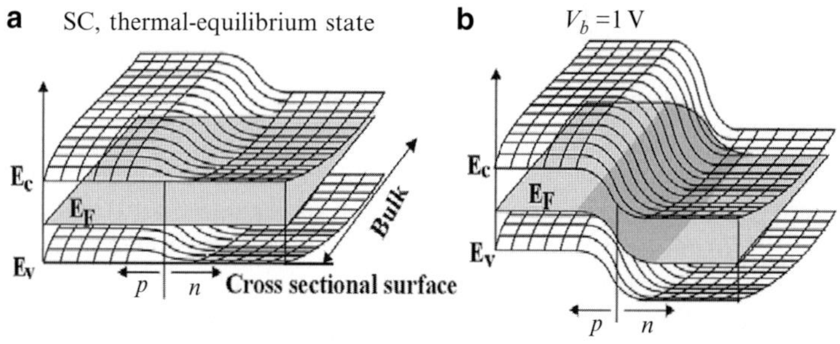

Fig. 24.16 Schematics illustrating two-dimensional band diagrams on the p–n junction of a CIGS device: (**a**) is in the thermal equilibrium state showing flat bands on the surface and a band bending on the p–n junction in the bulk; (**b**) is under a reverse V_b showing that the band bending on the p–n junction is visible on the surface by applying a V_b even with a surface band bending

24.3.2 *Electrical Potential and Field on Junctions*

SKPFM provides an innovative technique for the real spatial potential measurements with nanometer resolutions. The electrical potential on junctions of solar cell devices is critical for the device performance, as it collects minority carriers. The potential profile is affected by defects in the junction area, so it can also be a characterization of the defects that may be mobile under a V_b. In the following, characterization of electric field uniformity in hydrogenated amorphous Si (a-Si:H) and hydrogenated amorphous Si and germanium (Ge) alloy (a-SiGe:H) cells is described [51], and the potential measurement of III–V single- and multijunction cells is presented [27, 29].

24.3.2.1 Electric Field Uniformity in a-Si:H and a-SiGe:H Cells [51]

a-Si:H and hydrogenated nanocrystalline Si (nc-Si:H)-based thin-film solar cells have been greatly improved by optimizing the device structure and fabrication processes [52–54]. However, the device physics are not thoroughly understood due to the complexity of material properties and device structure. For example, characterization of the spatial distribution of the electric potential across the n–i–p structures is not well achieved in many situations. The electric potential plays a major role in photovoltaic actions, especially in the case of a-Si:H and a-SiGe:H cells, because of the short diffusion length of carriers due to low carrier mobility. In contrast to theoretical modeling of the potential distribution [55,56], experimental characterizations have been limited to indirect measurements such as I–V, C–V, and electroabsorption measurements. Therefore, the potential measurement of the n–i–p junction of the devices under various V_b values is expected to be helpful to understanding device operation [23, 57].

a-Si:H and a-SiGe:H n–i–p structures were deposited by using plasma-enhanced chemical vapor deposition (PECVD) on chromium (Cr)-coated GaAs(001) wafers. Using the Cr-coated GaAs wafer substrate, instead of the glass or stainless-steel substrates that are normally used, allows for flat cross sections to be cleaved, as in the case of CIGS device described in Sect. 24.3.1. The thin Cr layer (~100 nm) served as the back contact of the device. A thin a-Si:H n-layer and a thin nc-Si:H p-layer were used to provide a built-in potential in the n–i–p junction. Indium tin oxide (ITO, ~100 nm thick) was deposited on the p-layer as front contact. SKPFM measurements were performed on four samples in this study. Samples A and B were two a-Si:H solar cells made under the same conditions except that sample A had two buffer layers (~10 nm) between the n and i and the i and p layers. Samples C and D were two a-SiGe:H alloy solar cells made under the same conditions except that sample C had a flat bandgap obtained by using a constant GeH_4/SiH_4 ratio through the i-layer deposition, whereas Sample D had a profiled bandgap achieved by varying the GeH_4/SiH_4 ratio during the i layer deposition. The Ge content was higher at the p side than at the n side of the i layer to enhance the hole collection. Samples C and D had buffer layers at the *n/i* and *i/p* interfaces. Optimized buffer layers have improved cell performance for a-Si:H and a-SiGe:H solar cells [58], and a proper bandgap profile is a key factor in achieving good performance for a-SiGe:H [59].

An AFM and the corresponding SKPM images taken on a cross section of sample A under $V_b = 0$ are shown in Fig. 24.17. Device structure of the ITO, Si, and Cr layers, as well as the GaAs substrate, can be identified from the AFM image, as indicated in the illustration. However, the n, i, p, and buffer layers cannot be identified from the AFM images. The GaAs [110] substrate is atomically flat. The

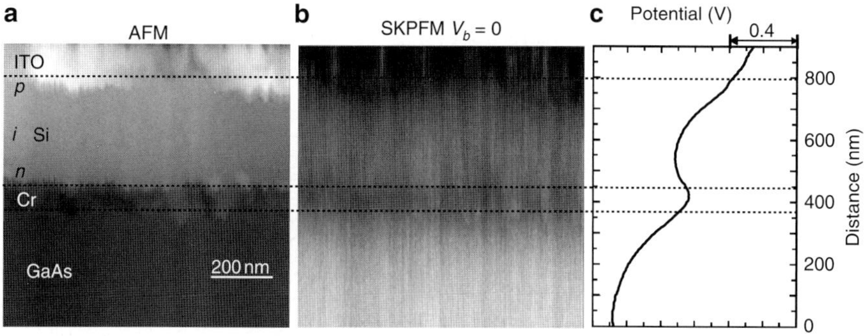

Fig. 24.17 (**a**) AFM and (**b**) the corresponding SKPFM images taken on a cross section of the a-Si:H cell with buffer layers (Sample A) deposited on Cr-coated GaAs substrate; (**c**) a potential profile obtained from an average of the potential signal in (**b**) along the lateral direction. ITO, Si:H, and Cr layers, and the GaAs substrate can be identified from the AFM image. The potential profile is different from that in the bulk, affected by surface charges

corrugation on cross sections of the device is less than 50 nm, which is adequately flat for the SKPFM measurement. The potential profile is displayed in Fig. 24.17c. The potential shows a valley around the Cr layer and a decrease (~450 mV) along the n–i–p direction in the junction region. A higher potential means a smaller work function and vice versa. Because work function of Cr is larger than those of n-type a-Si:H and n-type GaAs, the potential valley in Fig. 24.17c reflects the Schottky barriers at the Cr/a-Si:H and Cr/GaAs interfaces. The potential decrease in the junction region is qualitatively consistent with the built-in potential in the device. However, the measured potential cannot be quantitatively explained by the bulk properties of the materials. Because the conduction band is nearly aligned between GaAs and a-Si:H [60, 61], the potential should be about the same at both sides of the potential valley. However, the potential distribution shows a large difference between the n-type a-Si:H and n-type GaAs (~380 mV). Also, the potential decrease in the junction region is much smaller than the built-in potential (~1.2 V) [62]. The differences between the measured surface potential and the potential in the bulk originate from the nonuniformly distributed surface charges, or the surface E_F pinning. Therefore, as described before, measuring the change in the potential induced by a V_b is a way to avoid the effect of surface charges.

The V_b-induced potential changes on sample A are shown in Fig. 24.18a. As shown in Fig. 24.18b, the measured potential changes between a given V_b and $V_b = 0$ agree with the values of applied V_b, indicating that the potential change on the surface is the same as in the bulk. The V_b-induced change of the electric field distribution is deduced in Fig. 24.18c. In an ideal a-Si:H n–i–p junction, the potential should be linear, and thus the electric field should be constant through the i-region of the junction. However, the electric field does not distribute uniformly in an actual device. It shows two peaks close to the n- and p-layers. The result is consistent with

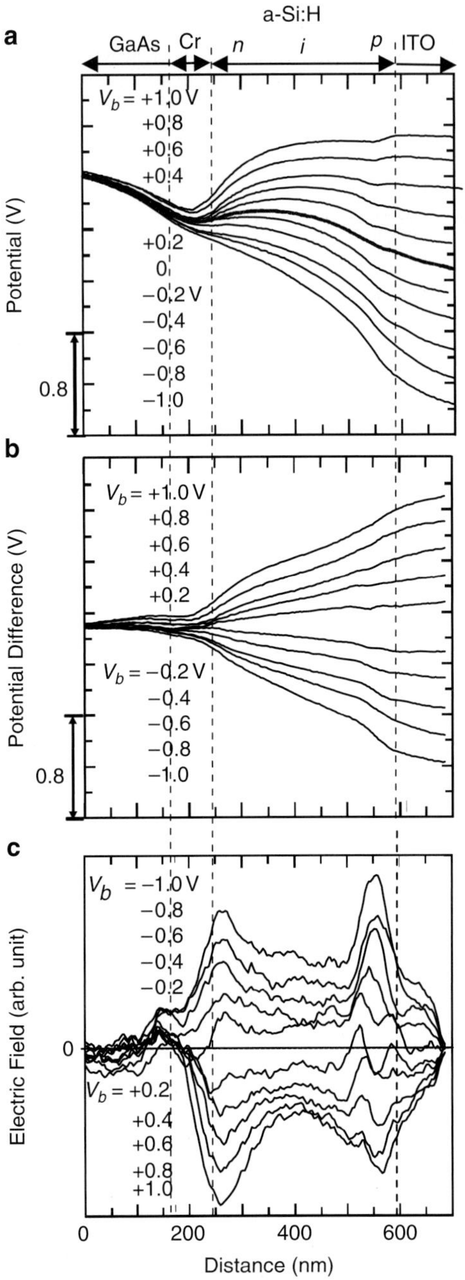

Fig. 24.18 (**a**) Potential profiles of sample A measured under various V_b values; (**b**) V_b-induced potential changes as deduced from profiles at the given V_b values and $V_b = 0$ in (**a**); and (**c**) profiles of V_b-induced electric field deduced from the potentials in (**b**). The electric field exhibits significant nonuniformity on the n–i–p junction, stronger on the n and p sides than on the middle of the i-layer

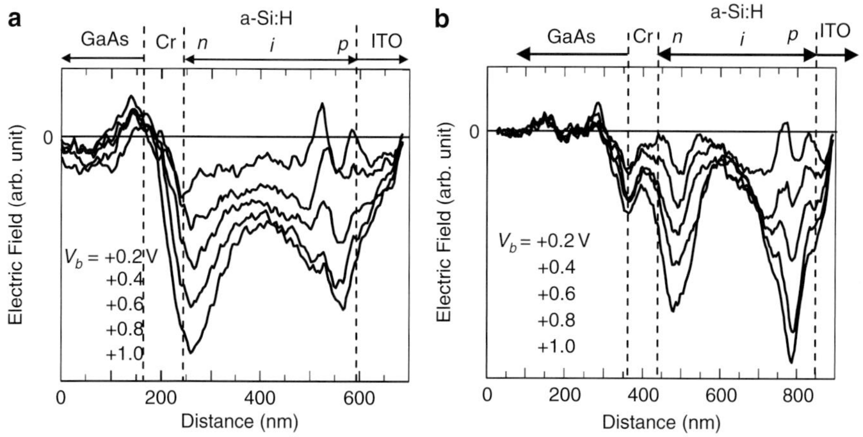

Fig. 24.19 V_b-induced changes in the electric field on the junction of an a-Si:H device (**a**) with (sample A) and (**b**) without (Sample B) the buffer layers at the n/i and i/p interfaces. The electric field is even more nonuniform in the device without the buffer layer. The nonuniformity relates to the charged defects at the interfaces

simulations in the literature [55, 56]. The electric field extends into both front and back-contact layers, which most likely resulted from the measurement resolution.

The potential distribution in the a-Si:H cell without the buffer layers (sample B) was compared with that of the sample A (Fig. 24.19). Only the profiles under forward biases are shown because they represent the actual situation when solar cells are in operation. These profiles show that the electric field is more nonuniform in the cell without the buffer layers (sample B) than in the one with the buffer layers (sample A). This result implies that there is a higher charge density (traps or defects) at the *n/i* and *i/p* interfaces in sample B than in sample A. The incorporation of buffer layers at the *n/i* and *i/p* interfaces reduces the nonuniformity of the electric field distribution by reducing the trapped charges at the interfaces. Indeed, the buffer layers lead to improvements of V_{oc}, FF, and E_{ff} [51, 58].

Electric field distributions in the a-SiGe:H cell with a flat bandgap (sample C) and the one with a profiled bandgap (sample D) are shown in Fig. 24.20. In sample C, the V_b-induced electric field is relatively weak at the n side as compared to the p side of the i-layer, indicative of a high charge density at the i/p interface. It is known that the incorporation of Ge increases the defect density in the i layer and reduces hole collection/diffusion length. Under a forward bias in the dark, a significant amount of holes diffuse into the i layer. Because the diffusion length is much shorter for holes than electrons, it is speculated that a lot of charge accumulates in the i/p interface and causes a high electric field at that point. This phenomenon is even more severe on the cell with the profiled bandgap (sample D) due to a high defect density resulting from the high Ge content at the region near the i/p interface. The purpose of the profiled bandgap is to enhance light absorption in the p-side of the i-layer to overcome the drawback of short hole collection length. By this method,

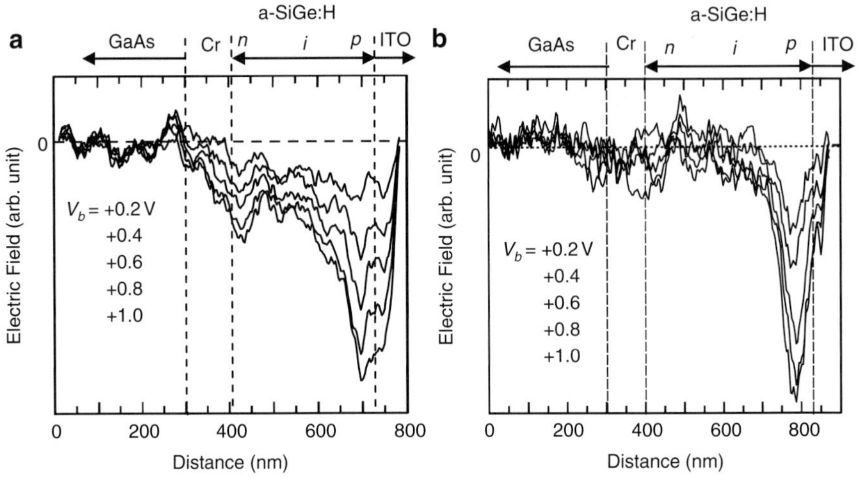

Fig. 24.20 V_b-induced changes in the electric field on the junction of an a-SiGe:H alloy cell with (**a**) a flat (Sample C) and (**b**) a profiled (Sample D) bandgap. The electric field exhibits severe nonuniformity, stronger on the p-side than on the n-side. The nonuniformity is even more severe on the device with the profiled bandgap, showing more charged defects induced by the higher Ge contents

FF has been improved significantly, but the high defect density at the *i*/*p* interface caused by the Ge profiling has a negative effect on the *FF* under short wavelength illumination. Therefore, an optimized bandgap profile is important for improving the cell efficiency [59].

24.3.2.2 Potential Profiles in III–V Single- and Multiple-Junction Cells [27, 29]

An electron in the valence band is excited to the conduction band and leaves a hole in the valence band by absorption of a photon. The minority carrier, electron or hole depending on whether the absorber semiconductor is p- or n-type, is then collected by the junction. Even if an electron is excited to a higher energy state in the conduction band and/or leaves a higher energy state for the hole, the electron (hole) will relax to a state at the bottom (top) of the conduction (valence) band, and the hot electron (hole) energy will lose. Multijunction solar cells are a natural extension of single-junction cells, because they preserve more energy from the photons at the high-energy end of the solar spectrum in terms of increased cell V_{oc} [63, 64]. In fact, III–V multijunction cells have for years held records for energy-conversion efficiencies [65–67] and are unlikely to ever lose their leadership position. In multijunction cell designs, serially connected two-junction – the tandem-junction configuration – has been considered the most practical because of its simple design and low manufacturing cost in comparison with four-terminal devices. The basic structure of the

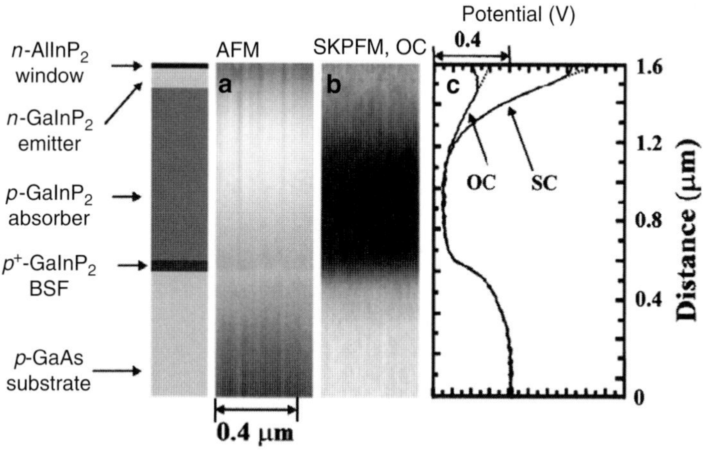

Fig. 24.21 (**a**) An AFM image and (**b**) the corresponding SKPFM image taken on the cross section of a $GaInP_2$ single-junction cell under OC. Potential profile that is an average of the SKPM image along the lateral direction is given in (**c**). The potential profile measured in SC is also shown in (**c**). The device structure is shown on left side of the figure

tandem-junction cell is that two subcells responsible for absorbing the respective ranges of solar spectra are serially connected by a highly electrical conductive tunneling junction that is formed by a very thin ($<\sim$10 nm) *n*/*p* stack spatially opposite the p–n junction direction of the subcells. In the following, a potential measurement on a $GaInP_2$ single-junction cell is presented [27], followed by a characterization of potential distribution between two subcells of a $GaInP_2$/GaAs tandem-junction cell [29].

A $GaInP_2$ solar cell (Fig. 24.21) containing an n-$AlInP_2$ window, an n-$GaInP_2$ emitter, a p-$GaInP_2$ absorber, and a P^+-$GaInP_2$ back surface field (BSF) layer was grown on a p-type GaAs substrate by metal-organic chemical vapor deposition [68]. Figure 24.21 shows an AFM and the SKPFM images taken in open circuit (OC), as well as potential profiles taken under OC and SC conditions. In the topographic image (Fig. 24.21a), there is a single atomic step, which was created upon cleaving due to stress at the $GaInP_2$/GaAs interface. On the profiles, the potential shows identical decreases ($\sim$350 mV) near the $GaInP_2$/GaAs interface, and it increases on the p–n junction differently in OC and SC – slowly in OC and steeply in SC.

Both OC and SC are in thermal equilibrium state and should give identical potential profiles if there is no illumination on the sample. However, they show a big difference in the potential difference on the p–n junction: $\sim$250 mV for OC and $\sim$780 mV for SC. This is because AFM uses a laser to probe the deflections of the cantilever, and illumination by laser light on the samples is difficult to avoid. This unintentional illumination produces a V_{oc}. In SC, however, the potential value at the front contact must be the same as that of the thermal equilibrium state, because the potentials at the front and the back contact are both on the ground level. The

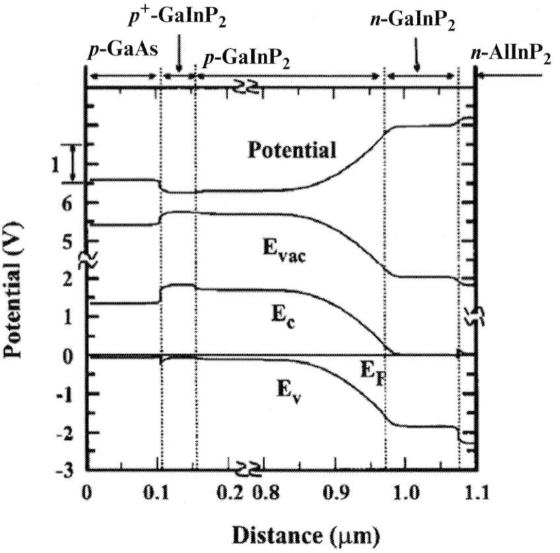

Fig. 24.22 A calculated band diagram of the $GaInP_2$ device bulk containing a n-$AlInP_2$ window, n-$GaInP_2$ emitter, p-$GaInP_2$ absorber and p^+-$GaInP_2$ BSF layers, and the GaAs substrate, as indicated by the *dotted lines*. The profile "Potential," which is a mirror image of the vacuum level, corresponds to the measured SKPFM potential

potential difference between OC and SC according to the SKPFM measurement is ~530 mV.

The band diagram (Fig. 24.22) of the device bulk in thermal equilibrium state was calculated using the parametersin the literature [61,63,69–74]. A portion of the diagram in the absorber from 0.2 to 0.8 μm, which shows flat bands, is not shown in order to display the band features in detail. At the interface between the GaAs substrate and the p^+-$GaInP_2$ BSF layer, there is a potential decrease (~330 meV) because of the bandgap difference (~470 meV) [61,63,71] and the conduction band offset (~140 meV) [61,72–74]. At the interface between the p^+-$GaInP_2$ BSF and the p-$GaInP_2$ base layers, the potential increases slightly (~60 meV) due to the different doping levels. Although there is a bandgap difference between the ordered absorber (1.81 eV) and disordered p^+-BSF(1.89 eV) layers [63,71], the band offset occurred mainly in the conduction bands [73,74], so it does not affect the potential in the thermal equilibrium state. On the p–n junction, there is a large potential increase close to the value of the $GaInP_2$ bandgap (~1.8 eV). Finally, at the interface between the window and emitter layers, there is a discontinuity in the conduction band because of the band offset [69,70,72]. Accordingly, the potential increases slightly at the interface. Although there is a difference of bandgap (~0.5 eV) between the two materials [63,69,71], this difference does not significantly affect the potential, because the bandgap difference of the n-type materials mainly affects the valance band alignment, not the conduction band alignment.

The calculated diagram is compared with the potential measurement in SC, which is close to the thermal equilibrium state. The decrease in the potential (Fig. 24.21c) can be assigned to the potential drop at the GaAs/$GaInP_2$ interface. The small potential dip on the BSF layer was not observed (Fig. 24.22). This is due to inadequate spatial and/or energy resolutions of the SKPFM performed at room temperature. If the measurement cannot probe the small dip, the measured potential decrease will correspond to the potential difference between the GaAs substrate and the $GaInP_2$ absorber layer, and the value in the band diagram is $330 - 60 = 270$ meV. The literature gives a range (60–210 meV) of the conduction band offset between $GaInP_2$ and GaAs [72, 75, 76]. We used the value of 60 meV in the calculation of Fig. 24.22, which gives the largest potential decrease of 270 meV. The difference (∼80 meV) between the measurement (∼350 meV) and the diagram (270 meV) is associated mainly with the different doping levels of the absorber and the substrate (∼60 meV). That implies that the E_F on the cleaved surface is pinned at similar positions above the valance band maximum for the $GaInP_2$ absorber layer and the GaAs substrate. From the measured potential (∼350 meV), the E_F pining position of the p-type GaAs(110) surface (∼0.88 eV below the conduction band minimum) [77], and the conduction band offset between $GaInP_2$ and GaAs (∼60 meV) [72], it is suggested that the surface E_F position of the p-type $GaInP_2$ material is at ∼1.29 eV below the conduction band minimum (Fig. 24.23). The fact that the increase in potential at the p–n junction is much smaller than that in the band diagram is also due to the surface E_F pinning. In this case, however, surface E_F may be pinned at different energy locations between n- and p-type materials due to electron- or hole-trap-like surface states [77–79]. The different pinning position makes the nonzero surface potential between the n- and p-type materials. From the potential measurement (∼0.78 eV) and the surface E_F position of the p-type

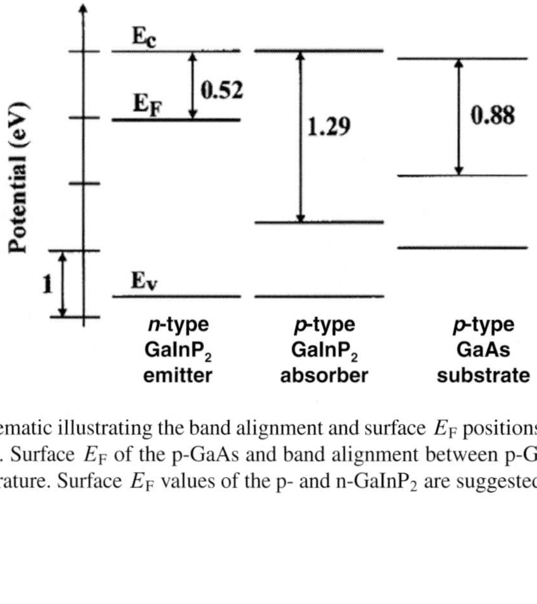

Fig. 24.23 A schematic illustrating the band alignment and surface E_F positions of the multilayers in the $GaInP_2$ cell. Surface E_F of the p-GaAs and band alignment between p-GaAs and p-$GaInP_2$ are cited from literature. Surface E_F values of the p- and n-$GaInP_2$ are suggested from the SKPFM measurements

material, the surface E_F for n-type $GaInP_2$ is pinned at $\sim$0.5 eV below the conduction band minimum.

Next, characterization of potential distributions between the top $GaInP_2$ and bottom GaAs cells of a $GaInP_2$/GaAs tandem junction cell is discussed [29]. Structure of the top cell is similar to the single-junction cell described above (Fig. 24.24a). The bottom cell contains an n-$GaInP_2$ emitter and a p-GaAs absorber layer. The top and bottom cells are serially connected by a heavily doped GaAs tunneling junction. Figure 24.24b shows the potential profiles taken on the cross sections of the cell under SC, both with and without illumination. Halogen lamps filtered with glass filters were used to illuminate from the front side of the device. The glass filters pass either long (<1.72 eV) or short wavelengths (>2.13 eV), matching the photon energy to the bandgaps of the top (I_{top}) or bottom (I_{bot}) cells.

Without illumination ($I_{top} = 0$, $I_{bot} = 0$), the potential shows a decrease around the bottom and tunneling junctions and an increase on the top junction. On illumination of I_{top}, even at a small intensity of 0.32 mW/cm^2 (Fig. 24.24b), the potential distribution changes dramatically; the potential increase on the top p–n junction is flattened, and the potential on the bottom and tunneling junctions becomes a peak. Since the photons are mostly absorbed by the top cell, redistribution of the charge density with the photo-excited carriers makes the top junction flattened and the bottom junction accumulated with extra charges [27]. The extra charges raise the potential in the bottom junction and induce the potential peak together with the potential decrease in the tunneling junction. Because the carrier flow is blocked on the bottom junction, the peak position of the profile corresponds to the boundary between the bottom and tunneling junctions, shown by a vertical dotted line in Fig. 24.24b. Therefore, the potential heights on the bottom (V_{bot}), tunneling (V_{tun}), and top (V_{top}) junctions can be determined, as shown in Fig. 24.24b. Since the surface E_F positions are not determined on the multilayers of the device structure and the measured potential heights depend on the E_F pinning positions, the absolute potential heights of V_{top}, V_{bot}, and V_{tun} cannot be explained. However, rather than the absolute potential heights and the potential shape, changes in the potentials with illumination and bias voltages V_b values can be well determined and should not relate to the surface charges.

With increasing intensity of I_{top} (I_{bot}), the potential height on the bottom (top) junction increases and that on the top (bottom) junction decreases (Fig. 24.24b). If two channels of illumination (I_{bot} and I_{top}) are conducted, keeping the intensity of I_{bot} (I_{top}) constant and increasing I_{top} (I_{bot}) gradually, the potential (Fig. 24.24c) changes with a similar tendency to the case of one channel illumination (Fig. 24.24b). This means that there are dual states of the potential distributions: One is negative V_{bot} and positive V_{top}, and the other is positive V_{bot} and V_{top} around zero. This also means that the distribution of the potential relies on the competition between I_{top} and I_{bot}, but not solely on the illumination of I_{top} or I_{bot}. Therefore, the potential distribution depends on the competition of the minority-carrier density excited on the top and bottom junctions. This relationship is better represented by plots between the potential heights of V_{top}, V_{bot}, and V_{tun}, as well as the total potential height (V_{total}) that is the potential difference between the inner bulk and

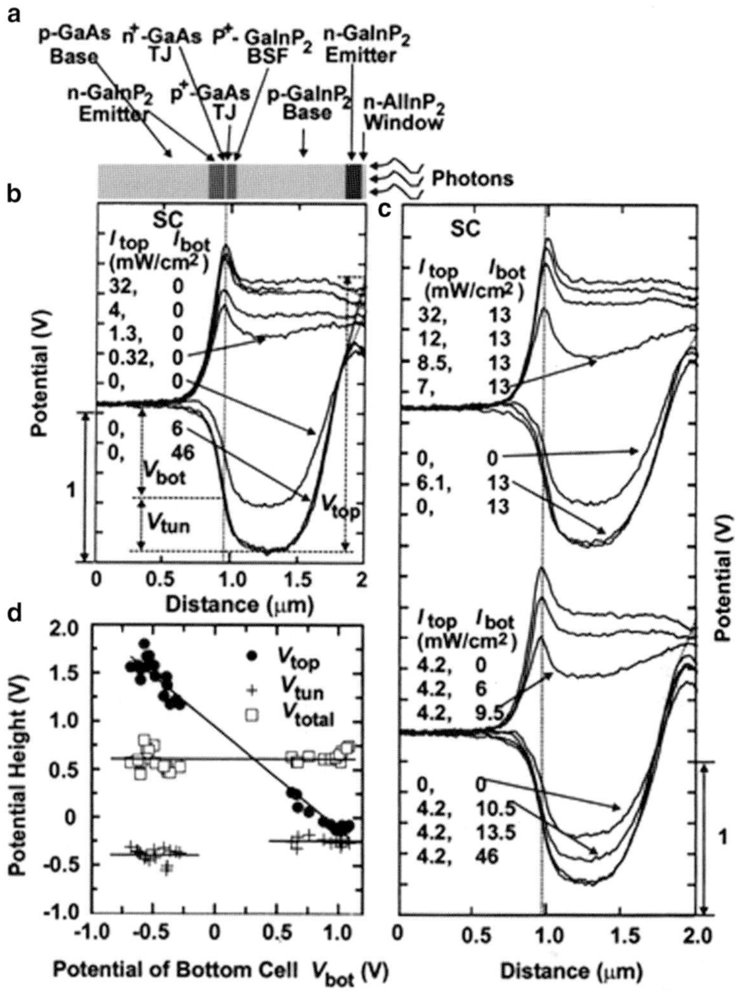

Fig. 24.24 Potential profiles measured on the cross section of a $GaInP_2$/GaAs tandem-junction cell under SC: (**a**) shows the structure of the device; (**b**) and (**c**) were taken under one and two channels of illumination, respectively; (**d**) plots the relations between the potentials on the top (V_{top}), bottom (V_{bot}), and tunneling (V_{tun}) junctions, and the total potential between the front contact and inner bulk (V_{total}). The dotted lines in (**b**) indicate how the potentials are measured from the potential profiles. The potential distributions can be separated to two groups of V_{top} and V_{bot}, i.e., the potentials are in the dual stable states, as shown in (**d**)

the front contact (Fig. 24.24d). V_{total} is approximately constant because the measurement is under SC. V_{tun} at positive V_{bot} (with I_{top}) is slightly higher than that at negative V_{bot} (with I_{bot}). This is due to the incorporation of V_{bot} to V_{tun}, because the bottom and tunneling junctions are spatially close ($\sim$100 nm) to each other. V_{top} decreases approximately linearly with V_{bot}: $V_{top} = aV_{bot} + b$, $a = -0.91 \pm 0.02$,

$b = 0.82 \pm 0.02$. This linear relationship is because voltages on the top and bottom junctions have the same absolute value and opposite signs under SC. The offset b reflects the difference of work function in the multiple layers. The deviation of the linear slope (a = –0.91 $\pm$ 0.02) from negative one is due to the incorporation of V_{tun} to V_{bot}.The potential heights of V_{bot} and V_{top} are separated into two data groups with many intensities of I_{top} and I_{bot} (Fig. 24.24d). This means that the charge accumulation and band flattening happen on either of the junctions, and the potential distributions are in dual states.

Under SC, the current of the top cell [1,2] ($J_{top} = J_0[\exp(eV_{top}/\alpha KT)-1]-J^L_{top}$) equals that of the bottom cell ($J_{bot} = J_0[\exp(eV_{bot}/\alpha KT) - 1] - J^L_{bot}$) and their voltages have relations of ($V_{top} = -V_{bot}$), where e, α, K, and T are single charge, diode ideality factor, Boltzmann constant, and temperature, respectively. From the relationships of the currents and voltages, the voltages of V_{top} and V_{bot} change with ($J^L_{top} - J^L_{bot}$), and change into the same absolute values with opposite signs when ($J^L_{top} - J^L_{bot}$) changes from J_0 to $-J_0$. Because J_0 is the dark saturate current with a small value, a small change of ($J^L_{top} - J^L_{bot}$) around the value of zero would induce a large change of V_{top} and V_{bot}. Therefore, the voltages V_{top} and V_{bot} separate to two groups, i.e., the dual states of potential distribution, under illuminations even with a small change of light intensity ($I_{top} - I_{bot}$).

Next, the potential distribution under various V_b values (Fig. 24.25) is discussed, which is a more practical situation because it takes power out of the solar cell. Figure 24.25a shows the potential profiles measured under V_b and illuminations of I_{bot} or I_{top}. The potential heights of V_{bot}, V_{top}, V_{tun}, and V_{total} are plotted vs. V_b in Fig. 24.25b. The sign of V_b was taken according to the polarity at the front contact relative to the grounded back contact. Negative V_b represents forwarded bias and taking power out of the device. With decreasing V_b, in the case of illumination of I_{bot}, the decrease in the potential profile happens on the top junction V_{top}, whereas V_{bot} and V_{tun} do not significantly change until V_b approaches V_{oc} under the specific illumination. The potential heights measured in OC are also plotted in Fig. 24.25b, and they should be consistent with the data measured under the V_b values. When V_b is decreased to $V_b < V_{oc}$, V_{top}, V_{bot}, and V_{tun} all decrease slightly with V_b (upper panel of Fig. 24.25b).

With photon absorption on the top cell ($I_{top} > 0$, $I_{bot} = 0$), however, the decrease of potential with V_b happens on the bottom junction V_{bot}. V_{top} does not show significant change. V_{tun} decreases slightly with V_{bias}, most likely due to the incorporation of the decreasing V_{bot}. When the external illumination is turned off, the potential changes are similar to the case of illumination of I_{bot}. This is due to the unintentional illumination of the laser light used in AFM. In all cases of I_{bot}, I_{top}, and no intentional illumination, V_{total} changes in the whole range linearly with V_b, and the slope is close to one, $0.95 \pm 0.02, 0.95 \pm 0.01$, and 0.94 ± 0.02 for I_{bot}, I_{top}, and no intentional illumination, respectively. From the measurement of the potential change, it is concluded that the potential decrease with V_b happened always on the junction with the charge accumulations; the potential did not significantly change or changed only a little on the other junction in which the flattening of band bending happened.

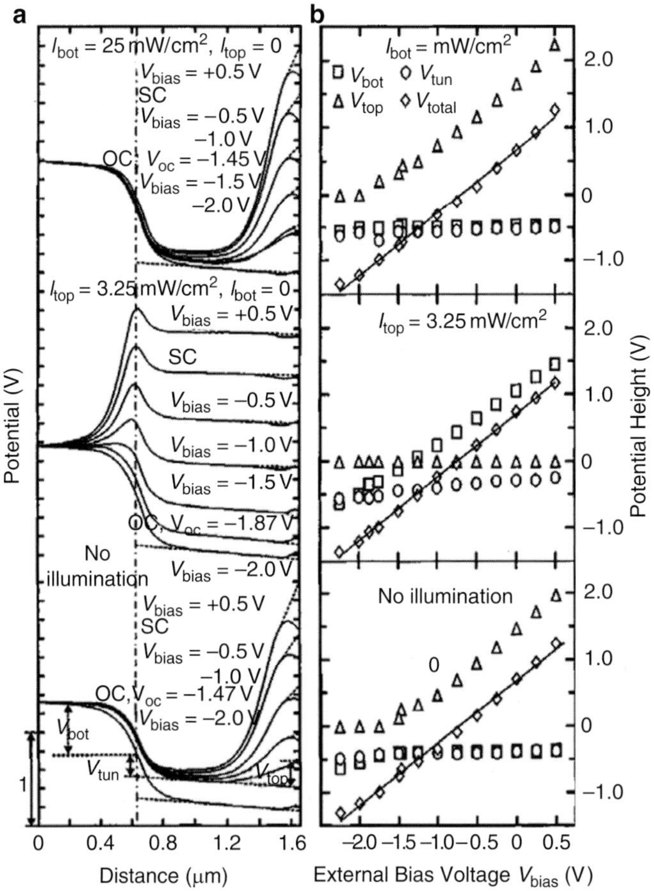

Fig. 24.25 (**a**) Potential profiles measured on the cross section of a $GaInP_2$/GaAs tandem-junction cell with various V_{bias} under illuminations of I_{top}, I_{bot}, and no intentional illumination; (**b**) shows the plots of potential heights V_{top}, V_{bot}, V_{tun}, and V_{total} vs. V_{bias} under the respective illumination conditions, showing that the potential change with V_{bias} always happens in the junction with charge accumulation or with a reversed bias voltage

Schematic I–V curves of a tandem cell combined from the I–V curves of the top and bottom cells are shown to understand the change of the potentials with V_b (Fig. 24.26). At each current point, the voltage of the tandem cell is the sum of the corresponding voltages of the top and bottom cells. In the case of $I_{top} > I_{bot}$ (Fig. 24.26a, c), with change of V_b, V_{bot} shows a big change, whereas V_{top} changes only a little (see the dotted lines between Fig. 24.26a, c). In the case of $I_{top} < I_{bot}$ (Fig. 24.26b, d), however, the big change happens on the top-junction V_{top}. This is because a small current change would induce a big potential change in the reverse-biased junction (i.e., the charge-accumulated junction) and a small change in the forward-biased junction, due to the exponential shape of the I–V curves. In the case

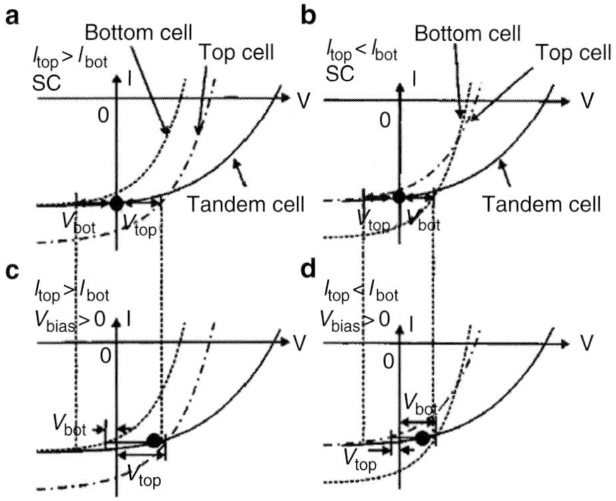

Fig. 24.26 Schematics of I–V curves of a tandem-junction cell in the conditions of (**a**) and (**b**) under SC, (**c**) and (**d**) under a forward bias voltage V_{bias}, (**a**) and (**c**) with photocurrent of $I_{top} > I_{bot}$, and (**b**) and (**d**) $I_{top} < I_{bot}$. Each I–V curve is composed of the I–V curves of serially connected top and bottom subcells, and a bias voltage on the tandem-junction cell can be assigned to the respective top and bottom cells

of SC, if we change the illumination from $I_{top} > I_{bot}$ to $I_{top} < I_{bot}$ (Fig. 24.26a, b), both V_{bot} and V_{top} change dramatically, changing their signs from negative to positive or from positive to negative. This corresponds to the switching back and forth between the dual states discussed above.

24.4 Characterization of Grain Boundaries of Polycrystalline Materials

The AFM-based microelectrical characterization techniques introduced in this chapter, SKPFM, SCM, and C-AFM, all have their corresponding techniques without the fine resolutions. Classical scanning Kelvin probe, which uses a submillimeter-sized probe and monitors an AC current generated by oscillating the probe, measures work function on a solid surface with resolutions in similar sizes to the probe. Admittance spectroscopy, which measures capacitance vs. DC voltage under various frequencies, is widely used in photovoltaic characterizations to measure the depletion width, carrier concentration, and defect levels in the bandgap [21]. Typical sample size for the admittance spectroscopy measurement is mm–cm, and thus has no fine spatial resolutions. The counterpart of C-AFM in macroscopic scale is I–V measurement with measurement area in mm–cm sizes. Therefore, the advantages of the AFM-based techniques are their fine spatial resolutions in nanometer ranges. The actual sizes of GBs in polycrystalline materials are in one to a few atomic layers. The distance electrically impacted by charges trapped at a GB is ~10 nm to submicrometers. There are numerous and extensive characterization studies on the structural and electrical properties of the GBs [80–83], however, most of the characterizations have been carried out on a macroscopic sample size with a huge amount of GBs; the characterizations result from collective effects of the GBs, in contrast to the fact that the electronic properties of individual GBs depends on specific GB structures. Therefore, the characterizations are expected to perform on individual GBs. In this section, carrier depletion on the GBs of polycrystalline Si thin films is described [84, 85]. In addition, electrical potential barrier measurement on GBs of CIGS thin films is presented [86–89].

24.4.1 Carrier Depletion and Grain Misorientation on Individual Grain Boundaries of Polycrystalline Si Thin Films

Polycrystalline Si thin film is a promising candidate in next-generation solar cells because of its advantages in material saving, low cost, and potential high performance. In recent years, efficiency of the devices improved steadily [90, 91]. Enlargement of grain size and passivation of the GBs are considered effective ways to stimulate research and further improve efficiency [91–94]. Most of the literature reports that GBs in Si films are detrimental to the efficiency because GBs in the absorber act as recombination centers for photo-excited carriers, and GBs in the junction area increase dark saturate current J_0 [91–93,95–97]. The GBs could negatively impact all device performance parameters of E_{ff}, J_{sc},V_{oc}, and FF [92,95–97]. On the other hand, positive impacts of GBs were proposed by a 2D device simulation and by a quantum efficiency analysis, suggesting that the efficiency would be enhanced if the GBs were doped to a considerable length-scale by diffusion of impurities in the emitter [98, 99]. In this case, the minority carriers become majority carriers at the type-inversed GB, and the GBs could help collections of minority carriers. Such minority-carrier collections can suppress the negative impact, such as potential fluctuation on the GBs.

SCM measures two-dimensional carrier concentrations with nanometer resolutions and is expected to probe the carrier depletion on the GBs. Carrier depletion on the GBs means either a charged deep level in the bandgap or a charged impurity ion, which would deteriorate the photovoltaic performance of the device [84, 85].

24.4.1.1 Probing Carrier Depletion on Grain Boundaries of Polycrystalline Si Thin Films Using SCM [85]

Si thin films with ~500 nm thickness were deposited by hot-wire chemical vapor deposition (HWCVD) from pure SiH_4 gas onto ~100-nm-thick silicon seed layers formed by aluminum-induced crystallization on glass substrates. Description of the deposition process and epitaxial alignment of the HWCVD growth can be found in the literature [100, 101]. Typically, the individual grain sizes are ~10 μm.

The AFM and SCM measurements were carried out on as-grown Si films, and the AFM image is shown in Fig. 24.27a. The image reveals that the surface morphology varies from grain to grain, depending on the grain orientation. There are several types of surface morphologies, which may correspond to the low-index orientations in the film growth direction. However, the SCM measurement (image is not shown) does not show clear features due to the low quality of the as-grown oxide layer on the film surface. CMP polishing of the film significantly improved the quality of SCM images. An AFM image and the corresponding SCM image on a polished sample are shown in Fig. 24.27b, c. From the AFM image, it is evident that the corrugation on the film surface was reduced from ~50 nm before polishing to <2 nm after polishing. In the SCM image, carrier depletion on the GBs is clearly observed as dark lines. Because the film surface is flat, the GBs cannot be identified from the AFM image. However, the GBs can be recognized from the high-quality SCM image. They can be identified from the depletion on the GBs or from the subtle topography-like features on the grains. An identification of the grain structure is shown in Fig. 24.27d, based on the data from Fig. 24.27c.

The SCM measurement shows that the GBs indeed exhibit carrier depletions and that the carrier depletion behavior is not uniform among the GBs. For example, the GBs between the grains A/C, B/C, and A/B in Fig. 24.27c, d exhibit clear carrier depletions, and the depletion on the GBs between grains A/C and B/C is deeper than that of A/B. However, the GBs between grains C/D and C/E do not exhibit significant depletions. A deeper depletion represents a larger GB charge density and a greater depletion width. This nonuniformity of the GB depletions indicates that the electronic properties of the GBs vary, due to specific GB structures that are decided by grain orientations and the facets adjacent to the GBs. A zoomed-in SCM image is shown in Fig. 24.28a, and a profile across the line in Fig. 24.28a is shown in Fig. 24.28b. From the line profile, the depletion width at each side of the GB is ~100 nm.

Most researches indicate that the electrically active GBs in Si thin-film solar cells are among the main causes of losses in efficiency because they negatively impact all the J_{sc}, V_{oc}, and FF [92, 95–97]. Recombination of photo-excited carriers with significantly large recombination velocity ($>\sim 10^4$ cm/s) at the GBs makes the minority-carrier diffusion length ($<\sim$μm) smaller than the film thickness [92,93,95–97]. On the other hand, it was reported that GB passivation by posthydrogenation effectively improved J_{sc} [92, 93]. In addition, control of grain orientations and subsequently the GB structures, as well as increasing the grain size, is expected to relieve negative impacts of the GBs. Small-angle or low-index GBs with fewer defects and dangling bonds are expected to reduce the recombination at the GBs. The SCM measurement on the GBs suggests that the electronic properties depend on detailed GB structures. Smaller recombination velocities are expected in GBs without detectable carrier depletions.

V_{oc} of the devices is negatively impacted by the GBs through the increase in dark saturation current density J_0 [92, 93]. Most studies believe that the carrier transport

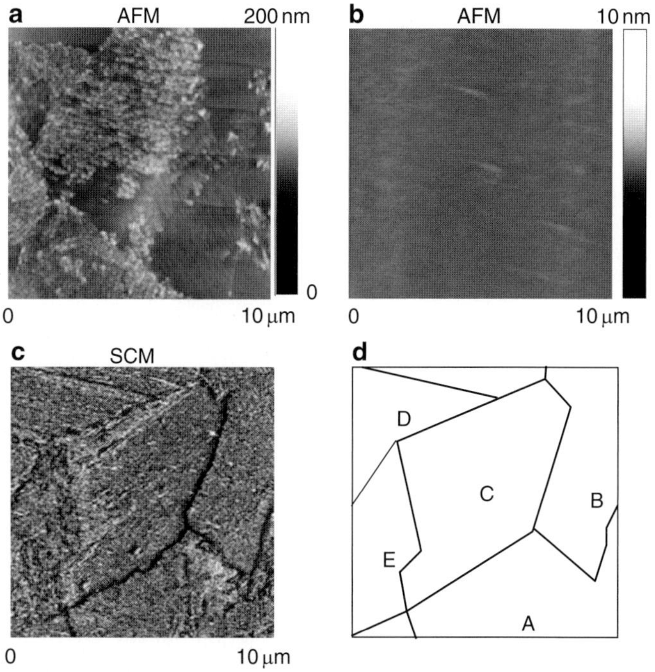

Fig. 24.27 (**a**) An AFM image taken on an as-grown polycrystalline Si thin film showing the grain orientation-dependent topographic features; (**b**) and (**c**) are a pair of AFM and the corresponding SCM images taken on a polished Si film; (**d**) shows a map of the grain structure as identified from the carrier depletion on the GBs and from the textures shown in (**c**). The SCM image shows highly nonuniform carrier depletion among the GBs, depending on specific GB structures

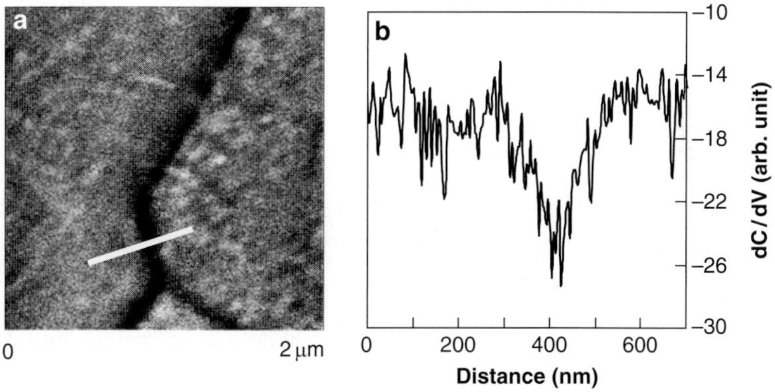

Fig. 24.28 (**a**) A zoomed-in SCM image taken on a polished Si film, and (**b**) a dC/dV line profile along the line indicated in (**a**) showing the amplitude of the SCM signal and the depletion width (∼100 nm) of the GB potential barrier

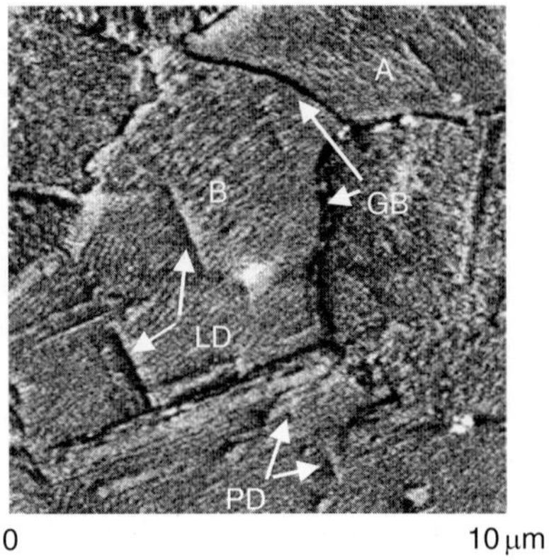

Fig. 24.29 An SCM image taken on a polished Si film showing intragrain line defects (LD) and point defects (PD), as well as the surface topography-like texture on the grains. These intragrain defects are considered detrimental to PV performance

mechanism in the p–n junction is governed by carrier recombination at the GBs, where the defect states act as recombination centers [93]. A few studies reported that it is governed by the recombination at intragrain defects [102]. In addition, band-edge fluctuations at both conduction and valence bands increase J_0 by reducing the excitation energy [103]. Both the electrostatic potential fluctuation induced by the charges at the GBs and the bandgap fluctuation at the GBs due to local defect configurations result in the band-edge fluctuations at conduction and valence bands [103]. Similar to the effects on J_{sc}, hydrogenation passivation of the GBs significantly improved V_{oc} [91, 92], and the SCM measurement also suggests a possible improvement of V_{oc} by controlling the GB structure.

In addition to the depletion on the GBs, the SCM images also exhibit intragrain defects and texture-like features on grains. An example of the SCM images is shown in Fig. 24.29. Four examples of the intragrain defects are indicated in the image: Two of them are line defects, labeled LD, and two others are point defects, labeled PD. The intragrain defects are not observed on every SCM image because only the defects at the interface of oxide-layer/Si-film or in the region close to the interface with a distance less than the deletion width can be detected by SCM. If the defects are deeper in the film bulk than the depletion width, they are screened by carriers and cannot be detected. SCM measures the carrier depletion on the charged defects. However, the nature of the defects is not clear; they could be vacancy, interstitial, substitutional, or agglomeration of the point defects, as well as line defects of dislocations and 2D defects of stacking faults. Similar to the GBs, these charged intragrain defects can be active recombination centers for minority carriers and can result in potential fluctuations. Therefore, the intragrain defects are considered detrimental to photovoltaic performance.

The texture-like features on the grains (Fig. 24.29) originate from nonuniformity of the oxide layer on top of the film [15, 104]. These features are similar to the topographic features on the film surface (Fig. 24.27a) before the surface was polished. For example, fine stripes on grain A in Fig. 24.29 trend in the direction from top left to bottom right, and they trend from top right to bottom left on grain B. Although the surface was polished flat during the polishing, the oxide layer may have been grown nonuniformly, following the facet features of the grain orientations. This nonuniformity of the oxide layer may include layer thickness and charged defects at the interface between the oxide layer and the film, which are sensitive to the SCM measurements [104].

24.4.1.2 Comparison of Carrier Depletion and Grain Misorientation on Individual Grain Boundaries of Polycrystalline Si Thin Films [84]

Electron backscattering diffraction (EBSD) [105] measures grain orientations on individual grains with submicrometer resolutions, and thus gives misorientation direction and angle adjacent to a GB. Then the GB structure provided by the EBSD measurement is compared with the GB carrier depletion behavior measured on the same sample area. It is thus expected that the structural and electronic properties of the GBs are correlated.

Using an electron gun of SEM, EBSD scans an electron beam on a sample and collects Kikuchi patterns from every pixel. Subsequently, a map of the grain orientation is obtained by analyzing the Kikuchi patterns. A common requirement for high-quality EBSD measurement is a flat sample surface to avoid shadowing the electron diffraction beams from a rough surface. The polished Si films provide good sample preparation for the high-quality EBSD measurement.

Two pairs of SCM and EBSD images are shown in Fig. 24.30. From the grain structure in both the SCM and EBSD images, it is apparent that Fig. 24.30a, b were taken on the same area, and Fig. 24.30c, d on the same area. In addition to the color scheme representing the grain orientations, thin and bright lines in the EBSD maps (Fig. 24.30b, d) represent GBs with small misorientation angles, and thick and dark lines with large angles. The misorientation directions are labeled on each of the GBs. Also, the colored lines represent twin GBs, red lines the $\Sigma 3$ GBs and pink lines the $\Sigma 9$ GBs. For random GBs, the GB structure is determined not only by the grain misorientation, but also by the grain facets adjacent to the GBs. However, for the twin GBs like Σ 3 and $\Sigma 9$, because the symmetry and the small formation energy at these GBs are obvious, most of GBs can be determined only from the grain misorientation.

Comparing the SCM image and the EBSD map, the following remarks were obtained: (1) The $\Sigma 3$ boundaries do not exhibit carrier depletion. There are many $\Sigma 3$ boundaries because of the small formation energy. However, the $\Sigma 3$ boundaries cannot be found in the SCM images. (2) Some of the $\Sigma 9$ GBs exhibit carrier depletion, and some do not. For example, carrier depletion shows on boundaries A and C in Fig. 24.30d, but does not show on boundaries B and D. (3) For random GBs, a rule or a trend was not found so far that describes the relationship between the carrier depletion behaviors and the grain misorientation. Interestingly, it was found

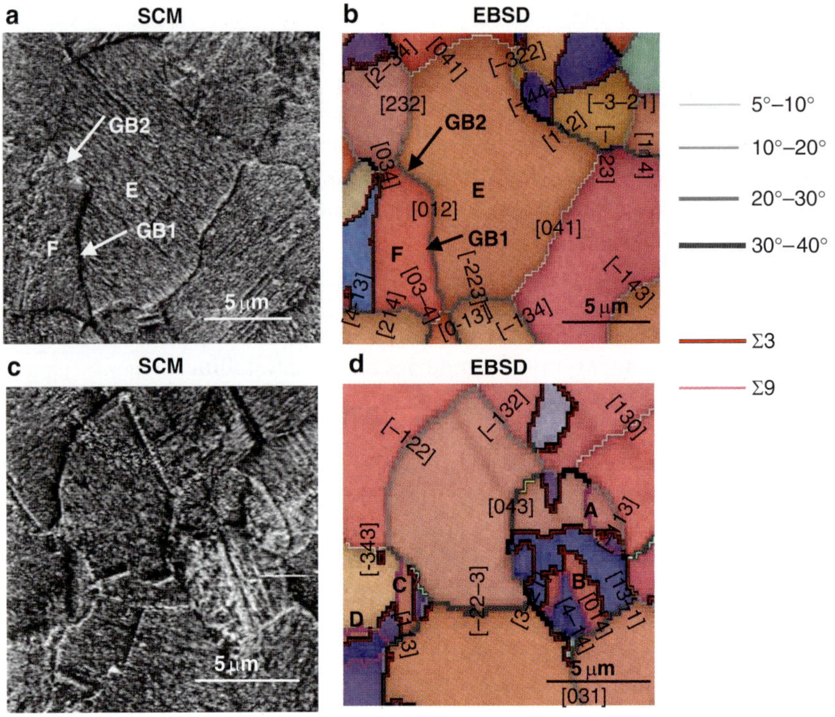

Fig. 24.30 Two pairs of SCM images (**a**) and (**c**) and EBSD maps (**b**) and (**d**) taken on one area (**a**) and (**b**) and another area (**c**) and (**d**). In addition to the color-scheme presenting the grain orientations, *dark* and *thick* (*bright* and *thin*) *lines* in (**b**) and (**d**) and shown on the *right side* of the illustrations represent GBs with large (small) misorientation angles. *Colored lines* present twin GBs. Misorientation directions are labeled on each GB in (**b**) and (**d**)

that the GBs can behave differently even for GBs that are between the same two grains. For example, GB1 between the lower parts of grains E and F in Fig. 24.30b shows clear carrier depletion. However, GB2 between their higher parts does not show significant carrier depletion. This observation clearly demonstrated that the grain misorientation by itself cannot determine the GB structure and subsequently the electronic properties.

The SCM measurements show no significant carrier depletion on the $\Sigma3$ GBs. The $\Sigma3$ GBs are considered electrically inactive and intentionally introduced in mc-Si ingots [106, 107]. $\Sigma3\{111\}$ GB is a coherent first-order twin-boundary; its atomic structure contains no dangling and strained bonds [80]. This atomic structure is believed to have a very minor effect on the electronic structure, and it does not create charged deep levels. Also, because there are no dangling bonds and strained lattice spacing, the energy is high to accommodate impurity ions in the GB. Therefore, there is no significant carrier depletion on $\Sigma3\{111\}$ GBs. On the other hand, the structure model of $\Sigma3\{112\}$ GB suggests that the GB contains strained bonds but no dangling bonds [82]. Chen et al. probed weak, if any, recombination at both the $\Sigma3\{111\}$ and $\Sigma3\{112\}$ GBs in clean mc-Si samples [83]. However, the carrier recombination was much enhanced in contaminated samples, and the recombination strength is much stronger on the $\Sigma3\{112\}$ than on the $\Sigma3\{111\}$ GBs. This corresponds to the stronger impurity gettering property of the $\Sigma3\{112\}$ GB [83], due to the strained lattice spacing [82].

Two possible alternative GB structures exist for the $\Sigma9\{221\}$ GBs; both contain strained bonds, but no dangling bonds [80, 81]. The incoherent structure of the GB is essentially similar to $\Sigma3\{112\}$, containing five- and seven-membered rings at the core of the structure unit [80, 81]. The SCM measurements show that some of the $\Sigma9\{221\}$ GBs are charged and some are not, suggesting that the intrinsic $\Sigma9\{221\}$ GB does not create charged deep levels, which is supported by the GB structures containing no dangling bonds. Impurity gettering can be responsible for the measured carrier depletion, as the strained lattice spacing could enhance the impurity gettering at the GB and result in increased carrier depletion. The SCM measurements also indicate a stronger gettering strength on the $\Sigma9\{221\}$ GBs than on either the $\Sigma3\{111\}$ or $\Sigma3\{112\}$ GBs, which is consistent with the literature [83].

For random GBs, the structure is rather complicated; it depends not only on the grain misorientation, but also on the GB planes. The electronic properties can be even more complicated due to their strong impurity gettering. For specific GBs, the structural and electronic properties depend on both misorientations and GB planes. For a large number of GBs, the GB properties may also have a statistical relationship with misorientations, if considering that a GB with a large misorientation angle about a high-index direction has, in general, high densities of dangling and strained bonds. A large number of comparisons between the SCM and EBSD measurements and a proper quantitative measurement of the carrier depletion are necessary for this study.

24.4.2 Electrical Potential Barrier on Grain Boundaries of Chalcopyrite Thin Films

One of the most surprising achievements for CIGS thin-film solar cells is that their ~19.9% efficiency is higher than that demonstrated for single-crystal CIGS cells (~13%) [37, 108, 109]. Two controversial arguments on the role of recombination in GBs suggest that the GB is either electrically active or inactive as a recombination center for photogenerated carriers, possibly due to the different GB structures generated by the different fabrication processes of the films and the measurement techniques used [110–112]. However, even if the GB is electrically inactive, this property alone is not sufficient to explain the high performance of the polycrystalline cells compared to their single-crystal counterparts. In terms of electrical potential, three fundamental questions on the GB of CIGS cells arise: (1) Is there band bending or trapped charges at the GB? (2) If yes, does the band bending or trapped charges originate from deep levels in the bandgap or from impurity ions gettered at the GBs?

(3) Are the trapped charges detrimental to the photovoltaic (PV) performance of this material?

SKPFM is a tool capable of probing band bending on the GBs. Oxide layers on Si thin films or Si wafers with a few nanometers thick prevent the band bending or electrical potential barrier on GBs from being measured by SKPFM, due to the fact that SKPFM measures the surface potential. However, this is not the case with CIGS. The GB potential in CIGS thin films is exposed to the surface and has been measured by SKPFM [86–89, 113, 114]. The potential has been further revealed originating from Na segregation to the GBs [88].

24.4.2.1 Measurement of Electrical Potential on the Grain Boundaries [86, 87]

The CIGS films presented in this section were deposited by three-stage co-evaporation process on Mo-coated soda lime glass substrates. The film was rinsed in high-purity water before the potential measurement. Water-rinsing was used to remove Na residing on the film surface, which diffused from the soda lime glass substrate onto the surface [115]. The surface Na makes the measured potential peak on the GB broad due to the Na-induced surface dipoles [86]. Figure 24.31 shows an AFM and the corresponding potential images measured by the SKPFM in air. Comparing the AFM (Fig. 24.31a) and SKPFM (Fig. 24.31b) images and the example line profiles (Fig. 24.31c), the electrical potential is higher on the GBs than on the grain surface. Both topographic and potential images measured in ultrahigh vacuum (UHV; see Fig. 24.32) are similar to those measured in air (Fig. 24.31) [89]. However, the noise level or the energy resolution is improved in the UHV-SKPFM (Fig. 24.32c). The primary purpose for measuring the potential in UHV is to avoid the effect of air-molecule absorption and surface contamination. However, annealing the sample at 110°C in UHV for 30 min to desorb water molecules from the sample surface did not significantly change the surface potential. Further argon (Ar) ion sputtering at 600 V for several seconds likely damaged the surface structure and caused disappearance of the potential contrast on the GB.

The higher potential on the GB indicates that there is a downward band bending around the GB in the p-type material (Fig. 24.33a) and that the GB is positively charged. Such local built-in potential makes the local work function smaller on the GB than on the grain surface. The measured potential height ($\sim$150 mV) is small compared to the bandgap ($\sim$1.15 eV) of CIGS. Most likely, the depletion on the grain surface reduces the contrast of potential between the GB and the grain surface and makes the measured potential peak on the GB smaller than the potential barrier on the GB in the bulk (Fig. 24.33b). Such surface depletion exists widely on semiconductor surfaces. Therefore, the local built-in potential in the bulk should be between the measured height ($\sim$150 mV) and the bandgap ($\sim$1.15 V).

The potential barrier on the GBs was further examined with varying Ga content in the film [87]. The bandgap of CIGS films can be tuned from 1.01 to 1.68 eV by adjusting Ga content of Ga/(In+Ga) from 0 to 100% [116]. Although a Ga content

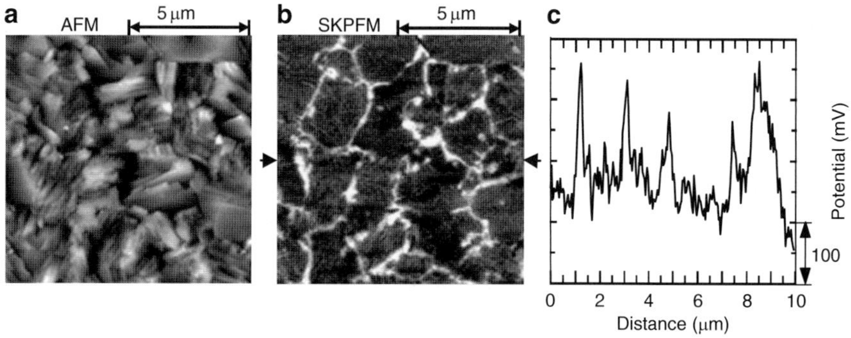

Fig. 24.31 (**a**) AFM and (**b**) the corresponding SKPFM images of the CIGS film taken by air-SKPFM; (**c**) a line profile along the arrows in (**b**), showing the higher potential on the GBs than on the grain surface, i.e., a potential barrier on the GBs

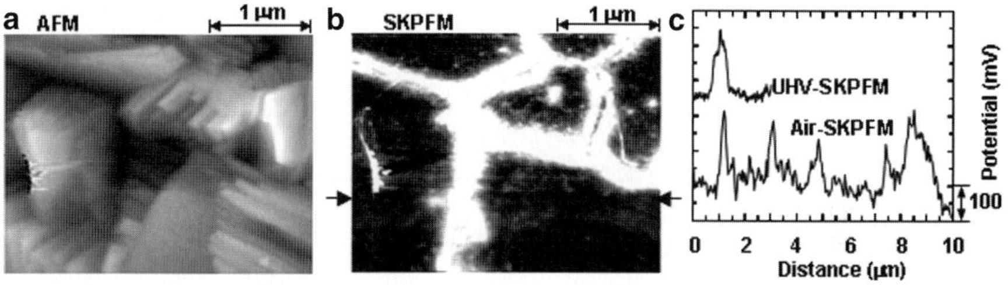

Fig. 24.32 (**a**) AFM and (**b**) the corresponding SKPFM images of the CIGS film taken by UHV-SKPFM; (**c**) a line profile along the arrows in (**b**) together with the line profile in Fig. 24.31c for comparison. The UHV-SKPFM shows similar potential measurement to the air-SKPFM, with improved signal/noise ratio or energy resolution

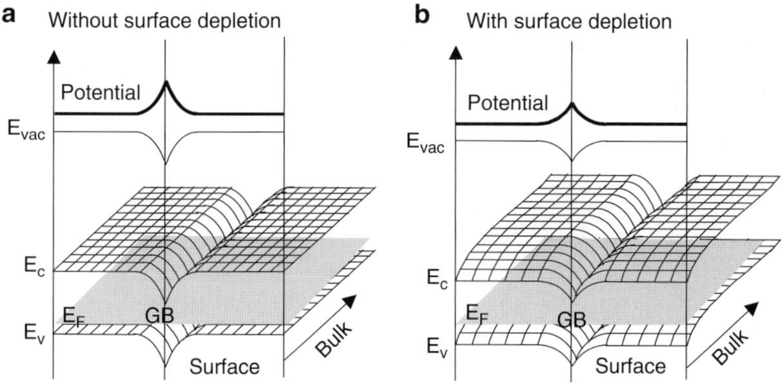

Fig. 24.33 Schematics of two-dimensional band diagrams around the GB, both (**a**) without and (**b**) with consideration of surface band bending. The surface band bending considerably reduces the GB potential barrier and makes the barrier height on the surface smaller than in the bulk

of ~65% provides the optimal bandgap value for optimal conversion efficiency, as theoretically predicted from the solar spectra [1], the actual Ga content in the current record-efficiency device is ~28% [37]. This difference between the theoretical optimal bandgap and the experimental Ga content of the highest-efficiency device implies that, in addition to the bandgap, other factors also play major roles in conversion efficiency.

With varying the Ga content in the film, it was found that the potential on the GB drops sharply in a range of ~28–38% Ga (Fig. 24.34a), within the energy resolution of ~50 mV. The theoretical curve for efficiency in Fig. 24.34b is predicted solely from the bandgap consideration [1]. By comparing the plot of potential in Fig. 24.34a to the theoretical and experimental plots of the conversion efficiency in Fig. 24.34b, one sees that the potential on the GB correlates nominally with the measured device efficiency. The increase of efficiency in the range of 0–28% Ga could relate to the increase in bandgap, and the potential on the GB in this Ga range stays strong. However, at the higher Ga content, the absence of the potential on the GB seems to be significant compared to the effect of the bandgap widening. Thus, the drop in the measured efficiency is slower compared to the precipitous drop in

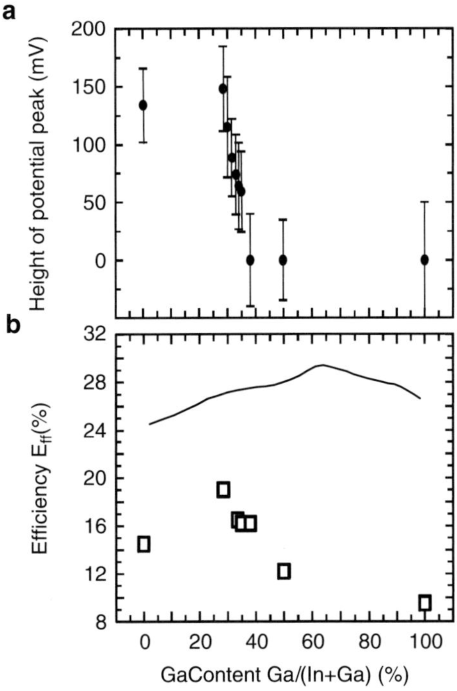

Fig. 24.34 (**a**) Measured potential barrier height on the GBs vs. Ga content of Ga/(In + Ga) in the CIGS film; (**b**) theoretical (*solid line*) and experimental (*squares*) conversion efficiencies of CIGS devices vs. Ga content. The potential barrier height and device efficiency show a nominal correlation with changing the Ga content

the potential at the GB. There could be many factors that affect the conversion efficiency. Ga content could significantly impact grain size and defect densities, and thus change the diode ideality factor and dark saturate current J_0, and subsequently the V_{oc} and FF of the device [117]. Illumination-induced change in work function or surface photovoltage was also measured by SKPFM in the dark and under illumination [114]. It was found that the GB potential barrier either reduces or does not change under illumination, which demonstrates positively charged GB in the p-type semiconductor.

24.4.2.2 Na Impurity in the Grain Boundaries [88]

The SKPFM measurement probed the potential barrier or band bending around the GB [86, 114], but it could not determine the root cause for the potential barriers. The GB potential can originate from charged deep levels and/or impurity ions. It is known that a moderate amount of Na incorporation in CIGS films significantly improves solar cell performance [118, 119]. The devices were grown on Mo-coated soda lime glass substrates, and Na in the soda lime glass is known to diffuse into the CIGS film during high-temperature growth and room-temperature storage. It is believed that Na diffuses largely through GBs, resulting in a high concentration of Na at the GBs [120]. The Na incorporation was reported to improve the structure of the film [118]. However, the effect of Na on the electrical properties of GBs was not adequately studied. By comparing the potential measurements on CIGS films with and without Na, it was found that Na is responsible for the potential barrier rather than the other possibility of intrinsic gap states.

For comparison purposes, two CIGS films were grown in the same deposition run on both Mo-coated soda lime and boron silicate glasses that either did or did not contain Na [88]. This procedure ensured the same growth conditions but with different substrates, resulting in CIGS films with and without Na. Both films had the [220]/[204]-preferred film orientations. The film grown on the soda lime glass shows a higher electrical potential on the GB (150–200 meV; see Fig. 24.35) than on the grain surface. In contrast, the film grown on the boron silicate glass appears featureless in the SKPFM image, except for small potential variations ($<\sim$30 meV) that may be caused by topographic inclusions (Fig. 24.36).

Comparison of the SKPFM images leads to the conclusion that Na is critical for the GB potential barrier. It is believed that Na is concentrated more at the GBs than in the grain bulks. Positive Na^+ ions at the GBs may be responsible for the GB potential. Na at the GBs may lead to the formation of point defects [121], which makes the GBs less p-type or even n-type. Taking the carrier concentration of $\sim 10^{16}/cm^3$ in the CIGS material and the depletion width of $\sim$500 nm from the SKPFM measurement, sheet charge density at the GB is $\sim 10^{12}/cm^2$. This is a possible defect concentration at the GB, in an approximate ratio of one charge per 100 CIGS unit cells. A local built-in potential (150 meV) around the GB was proposed by assuming a similar property of the GB (internal surface) compared to the film surface [122], which is consistent with the measurement ($\sim$150–200 meV). This

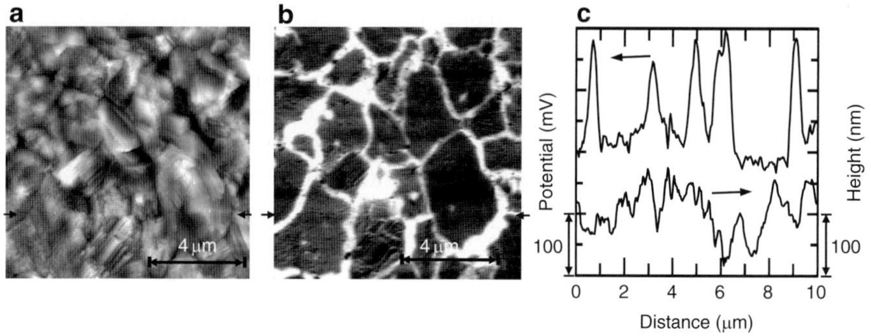

Fig. 24.35 (**a**) AFM and (**b**) the corresponding SKPFM images of a CIGS film grown on soda lime glass substrate. The film contains Na diffused from the glass substrate. *Gray scales* in (**a**) and (**b**) are 200 nm and 200 meV; (**c**) shows example line profiles of the AFM and SKPFM along the *arrows* in (**a**) and (**b**)

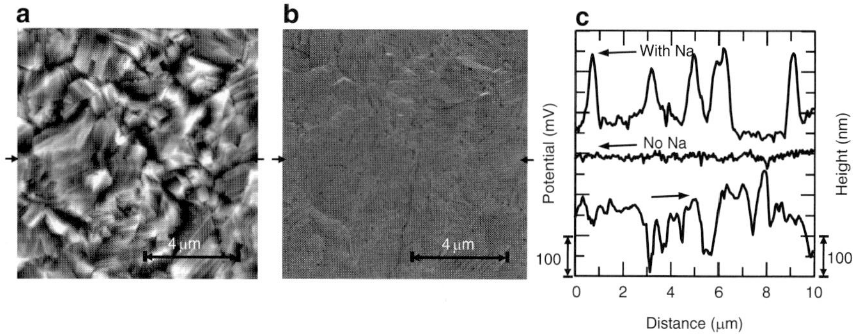

Fig. 24.36 (**a**) AFM and (**b**) the corresponding SKPFM images of a CIGS film grown on boron silicate glass substrate. The film does not contain Na. *Gray scales* in (**a**) and (**b**) are 200 nm and 200 meV; (**c**) shows example line profiles of the AFM and SKPFM along the *arrows* in (**a**) and (**b**), together with the potential line profile in Fig. 24.35c for comparison. The potential image is featureless, showing no potential barrier or charged deep levels at the GBs, which indicates inactive GBs

long-range potential covering the depletion width is induced by a small amount of Na segregation (0.05 Å). Further segregation of Na can induce charge dipoles, which further lower the local work function right at the GB that may be in a short range of a few atomic layers [122]. However, this short-range local work function reduction cannot be probed by the SKPFM measurement, because the resolution of the measurement is not enough to resolve the potential on the small width of the GB.

Regarding the structure of the GB, it is proposed that there may be a second Na-containing phase of Na_2Se_x, Na_2SeO_3, or $Na_xCu_{1-x}In_yGa_{1-y}Se_2$ (etc.) at the GB [120, 121]. The Na-containing phase involves a relatively large amount of Na, depending on the volume of the phase. The doping type and amount in the Na-containing phase, as well as the band offset between the Na-containing phase and the adjacent CIGS material, can be responsible for the measured GB potential.

The fact that the film containing no Na does not exhibit a potential barrier on the GBs indicates that the intrinsic GB with no foreign impurities does not create band bending around the GB and does not have charged deep levels. In general, charged deep levels exist widely on semiconductor GBs, such as Si and CdTe [123]. A theoretical calculation based on a $\Sigma3\{112\}$ GB atomic arrangement indicates that the GB of CIGS does not have gap states [88]; all the gap states are pushed out of the bandgap by relaxations of the atoms at the GB. The large relaxation of atoms is primarily caused by the strong repulsion between Se–Se atoms because of the large electronegativity.

Na is a shallow donor in most semiconductors, so it would not be very detrimental to PV performance. Further, if Na donor concentration is high enough, it will create large band bending around the GB and even possibly cause the depletion region to be inverted. In the case of a carrier inversion, the GB could form a conduction channel for minority carriers, and the GB could help minority carrier collections.

24.5 Localized Structural and Electrical Properties of nc-Si:H and a-Si:H Thin Films and Devices

nc-Si:H is one of the most promising solar cell materials to be actively studied in recent years [52,53]. Unlike a-Si:H, nc-Si:H shows very little light-soaking-induced degradation [124]. The material is proposed as an absorber in a single-junction cell or in middle/bottom cells of multijunction cells [52, 53]. However, due to the structural complexity that contains nc-Si:H grains, GBs, a-Si:H tissues, and voids [125], a high-level understanding of material and device physics has not yet been achieved. In PECVD, a-Si:H and nc-Si:H, as well as their mixed-phase materials, can be fabricated by optimizing processing parameters, mainly hydrogen dilution ratio $R = H_2/SiH_4$ [126–128]. Both structural and electrical characterizations using AFM-based techniques are difficult on either nc-Si:H or a-Si:H. This is because the fully nc-Si:H material contains too much nc-Si:H phase, making it difficult to recognize the phase separation and phase boundaries, whereas highly a-Si:H contains too little nc-Si:H phase that is characterized. Therefore, a-Si:H and nc-Si:H mixed-phase films and devices with $\sim$10% of nc-Si:H volume fraction were fabricated as a prototype model material, which provided suitable characterizations [129–133]. On the other hand, the best a-Si:H solar cells are deposited under the conditions close to the a-Si:H/nc-Si:H transition, but still in the a-Si:H regime [134], whereas the best nc-Si:H solar cells are deposited under the conditions close to the transition, but in the nc-Si:H regime [135]. Therefore, characterization of a-Si:H and nc-Si:H mixed-phase material is assumed to be useful for optimizing both a-Si:H and nc-Si:H. In this section, localized electrical properties of photovoltage and conductivity of the mixed-phase devices is presented [130, 131], and the doping effect on the structure of the mixed-phase film is reviewed [132, 133].

24.5.1 *Localized Electrical Properties of a-Si:H and nc-Si:H Mixed-Phase Devices*

The mixed-phase solar cells show an V_{oc} ranging from 0.5 to 1.0 V, which is between the V_{oc} values of typical a-Si:H and nc-Si:H cells. One interesting phenomenon for the mixed-phase solar cells is that V_{oc} increases after light soaking [127, 128], the opposite of what is commonly observed for a-Si:H and nc-Si:H solar cells where the V_{oc} decreases after light soaking due to the Staebler–Wronski effect [124]. The original explanation for the light-induced V_{oc} increase in the mixed-phase solar cell was based on light-induced structural changes from crystalline to amorphous phases [127, 128]. Subsequently, a complementary model of two parallel-connected diodes (two-diode model) was proposed [136]. There are two key points made in the two-diode model. First, the a-Si:H phase and nc-Si:H phase can be considered as two separate diodes with significantly different characteristics. Second, I–V characteristics of the nc-Si:H diode in the mixed-phase cells degrade with light soaking. The degradation of nc-Si:H causes a decrease in the electric current, when a forward voltage larger than V_{oc} of the nc-Si:H cell is applied, resulting in an increase in V_{oc} of the mixed-phase solar cells [136]. However, the size of the nanocrystallites observed by X-ray diffraction and Raman spectroscopy is very small, ranging from a few nanometers to 30 nm [137, 138]. It is difficult to believe that such small grains can form complete diodes through the entire thickness of the intrinsic layer. C-AFM was used to measure the local current flow and found that the nanocrystallites aggregated, and the aggregation regions showed a significantly higher forward current than the amorphous phase [129]. It was further found that photovoltage of the mixed-phase solar cell showed localized features on the p-layer of the n–i–p structure with significantly smaller values on the nc-Si:H aggregates than on the surrounding a-Si:H areas [130]. Furthermore, degradation in current flowing through the nc-Si:H aggregation with light soaking was observed, consistent with the two-diode model [131]. In the following, the localized photovoltage is described [130], and the light-soaking-induced degradation in the nc-Si:H aggregations is discussed [131].

24.5.1.1 Localized Photovoltage on a-Si:H and nc-Si:H Mixed-Phase Devices [130]

Mixed-phase Si:H and SiGe:H n–i–p structures were deposited [129–131] onto a stainless-steel substrate. The n layer is an a-Si:H layer doped with P, and the p layer is an nc-Si:H layer doped with B. The i layer was deposited under the condition of the a-Si:H and nc-Si:H transition regime. Due to the high sensitivity of the transition to the plasma properties, the mixed-phase solar cell shows different characteristics at different locations even on the same substrate, namely, an amorphous phase in the center, a mixed-phase signature in the outer region, and substantially nanocrystalline characteristics in the corner of the sample [129–131].

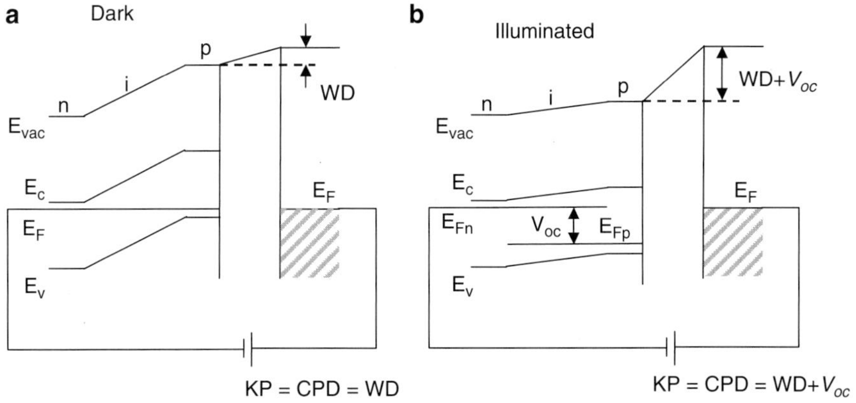

Fig. 24.37 Schematic band diagrams in the SKPFM measurements. In (**a**), the SKPFM measures the WD between the tip and the sample surface in the thermal equilibrium state; in (**b**), it measures the sum of WD and V_{oc} when the sample is illuminated

The SKPFM measurements were performed on the three regions. During the SKPFM measurements, the laser used for the AFM operation with 1.85-eV photon energy illuminated the sample. In the thermal equilibrium state in the dark, the CPD measured by SKPFM is the work function difference (WD) between the sample and the tip, as illustrated by Fig. 24.37a. However, when the solar cell device is illuminated, the CPD is the sum of WD and the local photovoltage, as shown in Fig. 24.37b. To obtain the photovoltage, the WD of the p layer must be measured and subtracted from the measured CPD. A p-type nc-Si:H layer was directly deposited on stainless-steel substrates, and the CPD was measured. Because the p layer is heavily doped and there is no p–n junction in the sample, the measured CPD should be close to the WD, even though the sample is illuminated by the laser light during the measurement.

Figure 24.38 shows SKPFM and the corresponding AFM images taken on the a-Si:H, substantially nc-Si:H, and mixed-phase regions, as well as on the p-type nc-Si:H single layer. The potential distributions are relatively uniform on the a-Si:H and nc-Si:H single-phase, as shown in Fig. 24.38a, c, as well as on the p layer in Fig. 24.38g. The microstructures on the morphology of the a-Si:H phase (Fig. 24.38b) show a small surface roughness, while the cauliflower-like structures in the nc-Si:H phase (Fig. 24.38d) are aggregates of nanometer-size grains, which results in a large surface roughness. On the mixed-phase area, a few large cauliflower-like structures appear on the topographic image (Fig. 24.38f), which corresponds to the valleys in the SKPFM image (Fig. 24.38e). The sizes of the potential valleys are larger than the corresponding nc-Si:H aggregates that appeared in the AFM image.

Figure 24.38i shows line profiles of the SKPFM images of the a-Si:H, nc-Si:H, and their mixed phase along the lines indicated by the arrows in Fig. 24.38a, c, e,

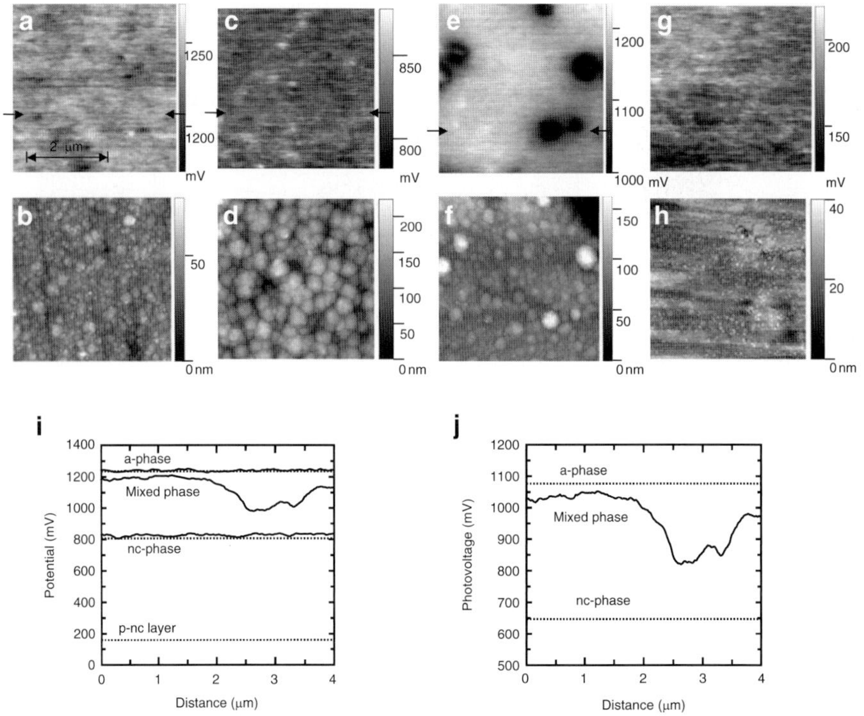

Fig. 24.38 An SKPFM potential and the corresponding AFM images taken on (**a**, **b**) the a-Si:H, (**c**, **d**) the nc-Si:H, and (**e**, **f**) their mixed-phase regions of an n–i–p solar cell. (**g**) and (**h**) were taken on a p-type nc-Si:H layer directly deposited on the stainless-steel substrate. (**i**) Potential profiles: *solid lines* are the potentials along *arrows* in (**a**), (**c**), and (**e**); *dotted straight lines* are averaged potential values over several SKPFM images. (**j**) The photovoltage line profile in the mixed-phase region deduced from the potentials in (**i**)

respectively. The three dotted straight lines are the averaged values through several SKPFM images of the a-Si:H and nc-Si:H phases and the p layer. The difference between a-Si:H and nc-Si:H is 0.45 V, which is consistent with the V_{oc} difference between a-Si:H and nc-Si:H solar cells measured by I–V characteristics. Figure 24.38j shows the photovoltage line profile of the mixed phase obtained by subtracting the p-layer potential from the potential profiles in Fig. 24.38i. Two important features are observed from this plot. First, the minimum photovoltage value in the nc-Si:H valleys is much larger than the average value in the fully nc-Si:H phase, as indicated by the bottom straight line. Second, the transition from the low to high photovoltages is a gradual change within a distance of about 1 µm. The two phenomena should be related and caused by the lateral charge redistribution. The electrical interaction between the two phases by the relatively conductive p layer and the charge redistribution near the a-Si:H/nc-Si:H boundary in the bulk of the i layer could be a mechanism.

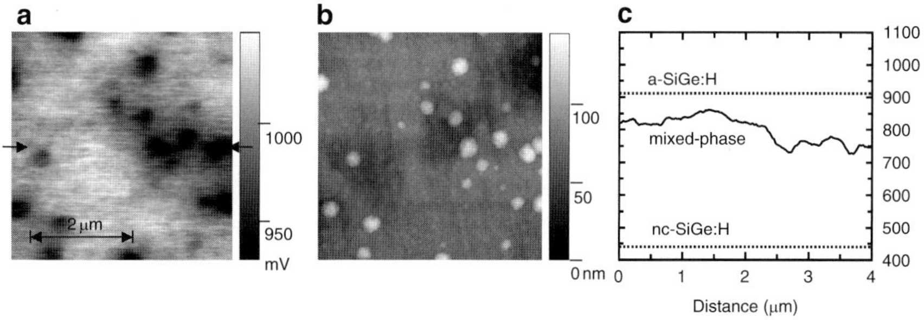

Fig. 24.39 (**a**) An SKPFM potential and (**b**) the corresponding AFM images taken on a mixed-phase a-SiGe:H solar cell. The *solid curve* in (**c**) shows a potential line profile along the *arrows* in (**a**), and the two *dotted straight lines* show averaged potential values taken on the a-SiGe:H and nc-SiGe:H regions

Although there is a lateral interaction between the a-Si:H phase and nc-Si:H aggregation areas, the photovoltage characteristic shows a clear distinction between the two phases. This is consistent with the C-AFM measurement on the mixed-phase cells [129]. The two-dimensional electrical current images show much larger local current flow through the nc-Si:H aggregates than through the surrounding a-Si:H matrix. However, unlike the gradual potential transition that is related to the charge redistribution around the a-Si:H/nc-Si:H boundary, the transition of local current flow between the two phases is sharp at the boundary, because of the much smaller bandgap and higher conductivity in the nc-Si:H aggregates than in the surrounding a-Si:H phase.

Local photovoltage distributions in mixed-phase SiGe:H alloy solar cells are shown in Fig. 24.39. In general, the results are similar to the mixed-phase Si:H cells. However, some differences are observed. First, the photovoltage values are lower than the corresponding values in the mixed-phase Si:H cells. Second, the size of the nc-SiGe:H aggregates is also smaller than that of the mixed-phase Si:H cells. Third, the photovoltage of the nc-SiGe:H aggregate is more heavily influenced by the surrounding a-SiGe:H matrix, showing smaller depth in the valley than the mixed-phase Si:H cells. Fourth, the transition from the low to high photovoltage regions becomes even broader.

The phenomena above may result from the small size of nc-SiGe:H aggregates and/or heavy electrical interaction between the a-SiGe:H phase and nc-SiGe:H aggregates. The electrical interaction may relate to the electronic structure in the i layer, such as band alignment at the a-SiGe:H/nc-SiGe:H boundary. In addition, the conductivity of the SiGe:H alloy materials is higher than that of the Si:H materials. The heavier photovoltage smooth-out in the mixed-phase SiGe:H cell could be partially caused by the high conductivity in the SiGe:H materials.

24.5.1.2 Effects of Light-Soaking and Thermal Annealing on Local Conductivity of nc-Si:H [131]

As mentioned above, the a-Si:H and nc-Si:H mixed-phase cell shows an interesting V_{oc} increase with light-soaking [127, 128], opposite to a fully a-Si:H and nc-Si:H cell where a V_{oc} decrease happened [124]. Subsequently, the V_{oc} increase in the mixed-phase and V_{oc} decrease in the fully a-Si:H and nc-Si:H devices recovered with thermal annealing. To investigate the light-soaking and thermal annealing effects, the mixed-phase device sample was cut into four identical pieces. Two pieces were subjected to light soaking, and the other two were kept in the initial state. The light soaking was carried out under white-light illumination with an intensity of $100\,\mathrm{mW/cm^2}$ for more than 1,000 h at 50°C. After light soaking, one piece of the light-soaked samples was annealed at 150°C for 2 h in vacuum, and the V_{oc} values were recovered to values similar to those in the initial state. Then local current maps of the three pieces at the initial, light-soaked, and annealed states were measured using C-AFM.

A forward V_b during the C-AFM measurements was fixed at 0.75 V, which was between the V_{oc} of a-Si:H and nc-Si:H cells. The local current on the nc-Si:H aggregates is much larger than on the surrounding a-Si:H phase, as shown in Fig. 24.40a–f, where the high-current regions are nc-Si:H aggregates, corresponding to the hill-like structures on the AFM image shown in Fig. 24.40g, h. As the AFM tip scans multiply over the same area on the sample surface, the current image changes, as compared among the three pairs of C-AFM images in Fig. 24.40, which results probably from the surface modifications by the AFM tip. In the first scan as shown in Fig. 24.40a, b, the current flow is very nonuniform in each individual aggregate. The nonuniformity was reduced in the second scan as shown in Fig. 24.40c, d. The improvement in the current uniformity could be caused by a mechanical modification of the p layer when the AFM tip contacted the sample surface, because it does not significantly depend on V_b. However, in the following scans, the current value got smaller and smaller with continuing scans. The reduction of current from the previous scans could result from both reoxidation of the sample surface caused by high local current and wearing off the tip. In this case, the change in the current image depends on V_b. The tip-induced oxidation process was also reported in the literature [139, 140].

The changes in C-AFM image from different scans resulted in an unsuccessful in situ light-soaking experiment because it could not be distinguished whether the changes in the images were caused by the light soaking or by an artifact of sample surface modification during the measurements. To ensure a fair comparison, the same measurement condition must be used. In reality, the samples were pretreated to reach the initial, light-soaked, and annealed states, as described above. Then, the C-AFM images were taken on the fresh sample surface areas and use relatively fresh tips. In most cases, the C-AFM images in the second scan were the best, where the current values were relatively uniform in an individual aggregate and the average value was the largest among the multiple scans. Therefore, the data presented below were from the second scans.

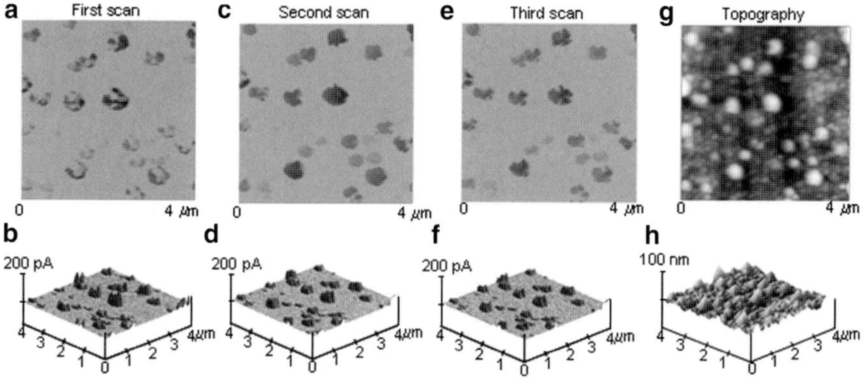

Fig. 24.40 2D and corresponding 3D current images on a mixed-phase Si:H cell from (**a**, **b**) the first C-AFM scan, (**c**, **d**) the second scan, and (**e**, **f**) the third scan. AFM topographic images of the same area are shown in (**g**) and (**h**). The 3D images are for better presentation of current values. Current images change considerably in different scans due to the surface modification by the AFM tip

A large number of C-AFM images were taken at different areas of the samples in different states for statistical comparisons. Although the current values showed a large scattering, a larger aggregate exhibits a higher average current value. Figure 24.41 shows an example of the C-AFM images in the three states, with slightly larger average current values in the annealed state than in the initial states, but significantly smaller average current values in the light-soaked state. Figure 24.42 plots the average current value versus aggregate area. Although the data are very scattered, the average current values with the same size of aggregates at the initial and annealed states are similar and significantly larger than that at the light-soaked state. Linear fitting of the current data versus aggregate area gives current densities of $(9.4 \pm 0.7) \times 10^{-1}$, $(2.8 \pm 0.3) \times 10^{-1}$, and $(10 \pm 1) \times 10^{-1}$ $(\mathrm{nA}/\mu\mathrm{m}^2)$ for the initial, light-soaked, and annealed states, respectively. The similar current density on aggregates with different sizes reveals that the current flow in individual aggregates is limited by the size of aggregates, but not by the tip–sample contact. This means that the current route spread out in the nc-Si:H aggregates, but was not confined in small areas under the tip. The large scattering in the current data reflects structure fluctuations in the aggregates. Each aggregate contains a large number of small grains. The GBs and amorphous component between the nanometer-size grains significantly affected the electron transport and resulted in the fluctuations in current density.

The local current-density and local photovoltage measurements by C-AFM and SKPFM reveal that the a-Si:H and nc-Si:H phases in the mixed-phase solar cells can be distinguished, and thus [129, 130] they support the model that the V_{oc} of the mixed-phase solar cells can be considered from the two parallel-connected diodes [136]. The light-induced decrease in average current density in the nc-Si:H aggregates further supports this model [131]. The V_{oc} of a mixed-phase cell is the forward

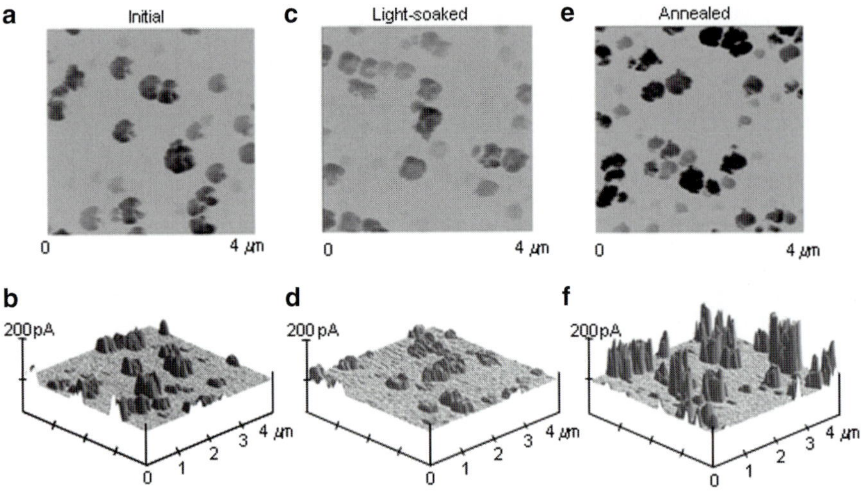

Fig. 24.41 Examples of 2D and 3D current images measured on an a-Si:H and nc-Si:H mixed-phase cell in (**a**, **b**) the initial state, (**c**, **d**) the light-soaked-state, and (**e**, **f**) the thermal annealed state. The images show similar current values between the initial and annealed states, and significantly smaller current for the light-soaked state

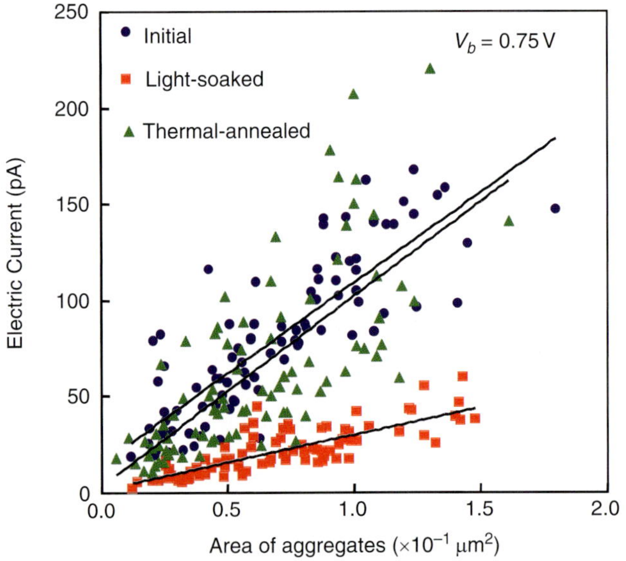

Fig. 24.42 Electric current as a function of the aggregation area in the initial, light-soaked, and annealed states. Straight lines are linear fitting of the data and represent current densities. Current density in the light-soaked state is significantly smaller than in the initial and annealed states

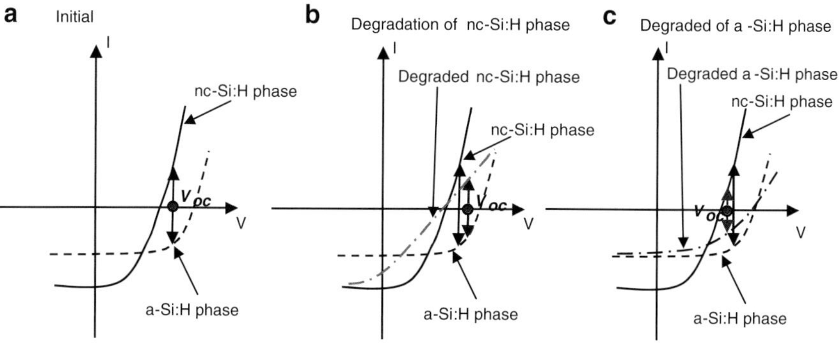

Fig. 24.43 Schematics of I–V curves and V_{oc} positions in (**a**) the initial state and (**b**, **c**) nc-Si:H phase degradation and a-Si:H phase degradation, respectively. A V_{oc} of a mixed-phase device is the voltage, at which a forward current density through the nc-Si:H phase equals the reverse current through the a-Si:H phase

bias voltage, at which the reverse photocurrent flowing through the amorphous phase equals the forward current through the nc-Si:H aggregates, as schematically shown in Fig. 24.43a. The V_{oc} values of mixed-phase solar cells should be between those of a-Si:H cells (~1.0 V) and nc-Si:H cells (~0.5 V). If the current through the nc-Si:H phase around this bias voltage decreases, the compensating reverse current through the a-Si:H phase also decreases, resulting in an increase in V_{oc} of the mixed-phase cell, as shown in Fig. 24.43b. It is well known that a-Si:H solar cells degrade after light soaking [124], as shown in Fig. 24.43c, correspondingly decreasing the reverse current under a given bias voltage. As a result, the V_{oc} of the mixed-phase cell would decrease, which is opposite to what was observed. If the currents through the a-Si:H and nc-Si:H phases both decrease after light soaking, the V_{oc} could increase or decrease, depending on which is the dominant factor. The fact is that the light-induced V_{oc} increase appeared in most mixed-phase cells. Therefore, the current-density decrease in the nc-Si:H aggregates is the dominant factor for the light-induced V_{oc} increase [136]. The nc-Si:H aggregates in the mixed phase contain many nanometer-size grains and GBs. Although the mechanism of the light-induced current decrease in the nc-Si:H aggregate is not clear, defect generation in the GBs inside the aggregates is a logical speculation.

24.5.2 Doping Effects on nc-Si:H Phase Formation

PECVD fabrication of a-Si:H and nc-Si:H, which decomposes SiH_4 or Si_2H_6 molecules and deposits Si radicals on foreign substrates, has been used in industrial massive manufacturing of solar modules [52,53]. One of the advantages in fabricating Si:H-based thin-film devices is that the n- and p-doped layers can be produced during film growth by adding P- or B-containing gases. High-performance solar cell devices have been successfully manufactured by adding PH_3 in the n-layer and BF_3, $(CH)_3B$, or B_2H_6 in the p-layer. Therefore, P- and B-doping effects on the structural and electronic properties of the films are of primary importance in fundamental material study and industrial applications. In the following, characterizations using AFM-based techniques, along with other compositional and structural characterization techniques, are described: SIMS was used to measure the P- and B-doping concentrations, the Raman spectroscopy the crystallinity of the films, and the cross-sectional transmission electron microscopy (X-TEM) the film structure.

24.5.2.1 Phosphorus and Boron Doping Effects [132, 133]

PH_3 or BF_3 gases were added [132, 133] in H_2 and Si_2H_6 gas mixtures during the deposition of a-Si:H and nc-Si:H mixed-phase films. It was found that as the doping gases are added, the hydrogen dilution threshold to reach the a-Si:H/nc-Si:H transition changes [132]. The characterization results from three typical samples of intrinsic, P-doped, and B-doped films are presented here [132, 133]. The samples were made with different hydrogen dilution ratios, but with similar crystalline volume fractions as measured by Raman spectroscopy. The two doped films have similar dopant concentrations as measured by SIMS, which allow a fair comparison of the doping effects.

To reach similar crystallinity, the hydrogen dilution ratio was 2.3 times higher for the P-doped film than for the intrinsic film. This demonstrates that a few percent P-doping significantly impacts the structure of the film, which makes the material change toward to amorphous structure. However, increasing the hydrogen dilution allowed the crystallinity to change back toward nanocrystalline. Because the doping efficiency is similar for P- and B-doping [141, 142], the similar P and B concentrations in the films give a good comparison of doping effects on the film properties. To reach similar dopant concentrations, the PH_3 and BF_3 flow rates were adjusted. The SIMS results show high doping concentrations of $\sim 1.5 \times 10^{21}/cm^3$ in both P- and B-doped films. The dopant concentrations are uniform throughout the entire film thickness.

X-TEM images of the three samples are shown in Fig. 24.44. The intrinsic Si:H film in Fig. 24.44a contains cone-shaped structures [132, 143]. The high-resolution X-TEM image in Fig. 24.44b on the cone-area exhibits a high density of nanograins with sizes of several nanometers. The cones are embedded in the a-Si:H matrix. No clear nanograins were found outside the cones, which demonstrates that the intrinsic mixed-phase Si:H film has a clear a-Si:H and nc-Si:H phase separation. The formation of the cone structure has been explained by a faster growth rate in the nc-Si:H phase than in the a-Si:H phase. From the cone angle of $\sim 70°$, it is estimated that the nc-Si:H phase growth rate is ~ 1.2 times that of the a-Si:H phase [144]. This faster nc-Si:H growth rate causes the nc-Si:H volume fraction to increase with the film growth, which has not yet reached the steady state. With the heavy doping of PH_3, however, the aggregations of nanograins are no longer observed, as shown in Fig. 24.44c, although the Raman spectra show a similar crystalline volume

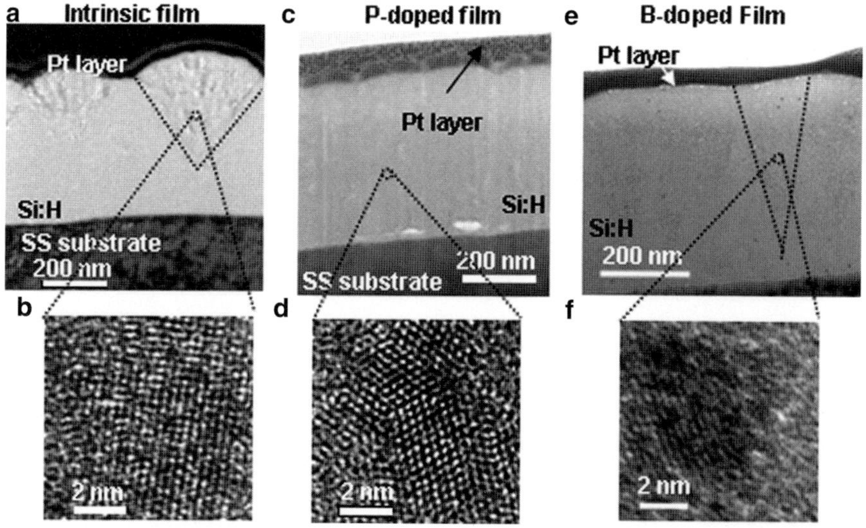

Fig. 24.44 X-TEM images taken on (**a**, **b**) intrinsic film, (**c**, **d**) P-doped film, and (**e**, **f**) B-doped film. (**a**), (**c**), and (**e**) are low-resolution images; (**b**), (**d**), and (**f**) are high-resolution images. The images show cone-shaped aggregation of nc-Si:H phase, isolated nc-Si:H grains, and cone-shaped aggregation with smaller cone angle and dilute grain density, in the intrinsic, P-doped, and B-doped films, respectively

fraction. The high-resolution X-TEM image of Fig. 24.44d still shows many nanograins with similar sizes of several nanometers. These nanograins are isolated and randomly distributed in the matrix of the amorphous phase. Therefore, the PH_3 doping significantly impacted the mixed-phase Si:H film structure.

With the heavy B-doping, the cone structures are still observed [133], as shown in Fig. 24.44e. The high-resolution X-TEM image taken in the cone area, as shown in Fig. 24.44f, exhibits nanograins with similar sizes to those observed in the intrinsic and P-doped films. However, the structure of the film is different from both the intrinsic and P-doped films although the Raman spectra show a similar nc-Si:H volume fraction to that of the intrinsic and P-doped films. First, the cone angle ($\sim 25^o$) is significantly smaller than that of the intrinsic film ($\sim 70^o$), indicating a growth rate in the nc-Si:H phase of $\sim$1.03 times that in the a-Si:H phase [144]. Second, the cone density is significantly higher than that of the intrinsic film. The cone tops coalesce at the top of the film. Third, the nanograin density inside the cones is significantly lower than that in the cones of the intrinsic film. Therefore, the crystalline volume fraction in the cones is low, which makes the X-TEM contrast of the cones much lower than the contrast in the intrinsic film. Inside the cone, besides the nc-Si:H grains, a-Si:H tissue also occupies a considerable volume fraction.

C-AFM was used to investigate the local electrical conduction [132, 133]. In C-AFM measurements, there are two possible contact barriers of tip/film and film/substrate. Because the contacts are similar to Schottky barriers, the barrier directions of the two contacts are opposite each other. From the current images

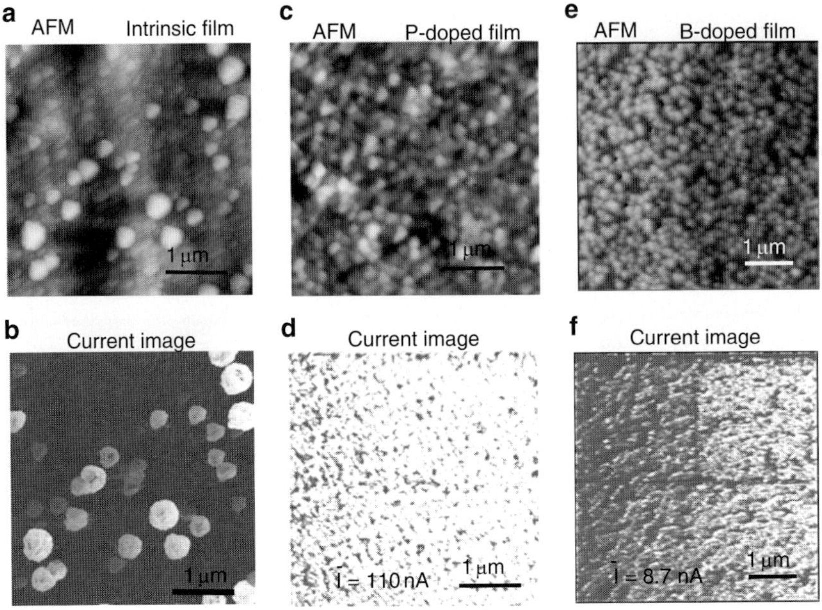

Fig. 24.45 (**a**, **c**, **e**) AFM images and (**b**, **d**, **f**) the corresponding current images taken on (**a**, **b**) the intrinsic, (**c**, **d**) the P-doped, and (**e**, **f**) the B-doped films, under a forward V_b of 0.75 V. The current mainly flows through the nc-Si:H aggregations on the intrinsic film, and everywhere on the P- and B-doped films, due to the respective structures of the films

under either a positive or negative voltage between the tip and substrate, it was recognized that the tip/film is the resistance-dominant contact. Therefore, the current images taken under a forward bias voltage of 0.75 V to the tip/film contact are shown here. In this case, the current images mainly measure the conduction in the film but are not limited by the tip/film contact.

The AFM image in Fig. 24.45a shows the cone tops of the nc-Si:H aggregations on the intrinsic film. The cone tops disappear on the P-doped film, as shown in Fig. 24.45c, which is consistent with the X-TEM observations. On the B-doped films, there are high-density particle-like features of 100–200 nm, as shown in Fig. 24.45e, where the coalescent small cone tops are consistent with the X-TEM observations as well. On the intrinsic film, the clear correspondence between the nc-Si:H aggregations in Fig. 24.45a and the high current areas in Fig. 24.45b reveals that the carrier transport paths are through the nc-Si:H aggregations.

The current image on the P-doped film does not exhibit localized features. However, the high currents are not uniform, which is possibly caused by the complexity of the film structure. The current may depend closely on the local structure of the contact area with the probe. If the probe is just on a nanograin, the current will be higher than on an amorphous tissue area. The average current value over the image is $\overline{I} = 110\,\text{nA}$, which is three orders of magnitude larger than that of the intrinsic film. On the B-doped film, no localized features are observed, although the X-TEM image shows cone-shaped nc-Si:H aggregations. This could be because of the high density of cones and the fact that the cone tops coalesce to each other on the film surface. The reasons for the smaller average current value ($\overline{I} = 8.7\,\mathrm{nA}$) in the B-doped film as compared to the P-doped film could be a lower mobility of the majority carriers (holes) and differences in the film structures.

24.5.2.2 Film Growth Mechanisms [132, 133]

Based on the atomic hydrogen-enhanced surface diffusion model [145], the formation of cone-shaped aggregations in the intrinsic film and the isolated nanograins in the P-doped film can be explained as follows. In the intrinsic film, once a nucleation is formed, the surface of the grain may have a higher sticking coefficient or a lower surface energy for Si radicals such as SiH_3. This leads to a continuing growth of the nc-Si:H cone. The growth rate of the nc-Si:H phase is determined by the net diffusion-in of Si radicals into the cones balanced by the diffusion-in and diffusion-out of mass [146]. Therefore, the "long-range" diffusion is the main factor determining the shape of cone structures and the nanocrystalline volume. For P-doped film, PH_3 molecules are adsorbed on the growth surface with a large probability because PH_3 can adsorb on the surface without dissociating to form a stable layer with pentavalent phosphorus bridging adjoining silicon atoms (For a review, see [147]). This absorption of PH_3 prevents the "long-range" diffusion of Si radicals and thus prevents the nanocrystalline phase growth on existing Si grains and the nanocrystalline phase clustering. However, the increased hydrogen dilution provides more hydrogen termination of Si bonds on the growth surface. This makes the diffusion of Si radicals better in small, local areas, which are not disturbed by PH_3 molecules. This enhanced local diffusion rate is beneficial for new nucleation in amorphous tissue, and such a nucleation rate is the main factor that determines the nanocrystalline volume fraction in the P-doped films.

In the case of B-doped film, the cone-shaped nc-Si:H aggregations indicate that the long-range diffusion of Si radicals is still the main mechanism for nucleation. The mechanisms for the smaller cone-angle and less dense nanograins inside cones can be explained as follows. First, the cone density in the B-doped film is much larger than that of the intrinsic film. B-related radicals may promote nc-Si:H nucleation [148]. If there are dense cones, the net-diffusion mass into each cone may be reduced. Consequently, the growth rate and cone angle are reduced, because the net-diffusion-in mass would be the dominant factor in the nc-Si:H growth rate, the nc-Si:H/a-Si:H volume ratio inside the cones, and the cone angle. Second, similar to PH_3 molecules, B radicals on the surface may also reduce the long-range diffusion rates of Si radicals. However, this effect should be much weaker than PH_3 molecules because the dominant factor of the nc-Si:H aggregation formation should still be the long-range diffusion. This reduced long-range diffusion rate can reduce the net-diffusion mass into the cones and slow the cone growth rate. Third, the B radicals may also reduce the H terminations on Si bonds in areas around the B radicals [142] and thus reduce the nucleation rate locally in amorphous tissue. The explanation is consistent with the fact that no clear nanograins were found outside the cones. Fourth, B radicals may have a catalytic effect for SiH_4 or Si_2H_6 to decompose, and to promote the growth rate of the film [142, 149]. The enhanced rate of decomposition inside and outside the cones promotes the growth of local amorphous phases and suppresses the growth of the nanocrystalline phase. This effect makes the crystalline volume fraction inside the cones of the B-doped material smaller than that inside the cones of the intrinsic film.

24.6 Summary

Some characterizations of solar cell materials and devices can fall to several categories. Basic I–V and spectra response measurements give solar cell performance parameters of E_{ff},V_{oc}, J_{sc},FF, and quantum efficiency. Electro-optical techniques in macroscopic sampling sizes of μm–mm characterize electronic/electrical and optical properties of devices that directly relate to or determine the solar cell performances. Examples of such electro-optical techniques are admittance spectroscopy, minority carrier lifetime and diffusion length, sheet and bulk resistance, light reflection and absorption measurements, etc. Microscopy tools such as TEM, SEM, and AFM are structural tools that investigate microscopic structures of materials down to atomic resolutions.

To understand the roles of microscopic structure on the macroscopic electro-optical properties and photovoltaic performances, investigations of microscopic electronic/electrical properties of the materials and devices with resolutions of nm–submicrometer are often highly desirable and helpful. SEM-based techniques, cathode luminescence and EBIC provide such microelectrical characterizations. In this chapter, some of the relatively new AFM-based microelectrical characterization techniques were introduced for measuring different electronic/electrical properties from the SEM-based techniques. Theses techniques, SKPFM, SCM and C-AFM, measure the two-dimensional distributions of electrical potential, carrier concentration, and electrical conduction, on a sample surface or subsurface region.

Some of recent characterization studies on inorganic semiconductor solar cell materials and devices conducted by using the AFM-based techniques were reviewed. In Sect. 24.3, studies on junctions of solar cell devices were reviewed. SKPFM presents a powerful tool for identification of junction or determination of junction depth, in the light of two-dimensional capability and the fact that the identification is from the electrical measurements and thus not limited to junctions formed by foreign specimen doping. Three identifications of the junctions were presented: the one on an mc-Si cell demonstrated that the shape of the junction conformably follows the surface topography of the device; the one on a GaInNAs cell demonstrated that a backshift of the junction with incorporation of Bi is responsible for the device degradation; and the one on CIGS revealed that the junction is in the absorber close to the CdS/CIGS heterointerface. None of these junction identifications are reached by conventional SIMS analysis due to the two-dimensional requirement in the first identification and the electronic doping in the second and third identifications.

SKPFM is also powerful for the real-space potential and field measurements. The potential and field relate directly to the mobile charges and defects in the materials. The measurement of a-Si:H and a-SiGe:H revealed a nonuniform field distribution on the junctions. Buffer layers effectively improved, and Ge incorporation significantly degraded the field uniformity. The measurement on a $GaInP_2/GaAs$ tandem-junction cell exhibits dual stable states of the potential distribution under SC and provided the first real-space experimental evidence for the $I-V$ model of tandem-junction cells.

In Sect. 24.4, carrier depletion, charge trapping, or potential barrier on GBs of polycrystalline materials were reviewed. The tens-nanometer resolution is one of the most attractive merits of AFM-based techniques, which provide the electrical characterization of individual GBs of thin-film solar cells with grain sizes in micrometers. SCM measurements revealed the highly nonuniform electronic properties in the polycrystalline Si thin films due to specific GB structures: The low-order $\Sigma 3$ twin GBs do not, the $\Sigma 9$ twin GBs in some cases do, and most random GBs do exhibit carrier depletion or charge trapping, which should be detrimental to device performance. SKPFM enables direct measurement of the potential barrier or band bending on GBs of CIGS films and shows that the barrier results from Na segregation to the GBs. Without the Na incorporation, SKPFM shows no band bending on the GB and thus no charged deep level, which indicates that the CIGS GBs are inactive for carrier recombination. This might be the reason for the superior performance of the thin-film device.

In Sect. 24.5, localized structural and electrical properties on a-Si:H and nc-Si:H mixed-phase films and devices were reviewed. The mixed-phase material is a prototype of nc-Si:H for practical characterizations using the microscopic tools. Local photovoltage and conductivity measurements of the nc-Si:H phase proved the two-diode model for explanation of light-soaking-induced V_{oc} increase. Further, doping effect on the structural and electrical properties of the mixed-phase films were reviewed: The nc-Si:H grains aggregated to cone-shaped clusters in the intrinsic film; P-doping made the nc-Si:H grains isolated and distributed throughout the film; B-doping made the nc-Si:H cone-angle smaller and the nc-Si:H grain in the cones dilute.

In addition to the techniques described above, there are newly emerging AFM-based methods that characterize the microelectrical properties in nanometer scales. Scanning microwave microscopy,[2] which measures reflection of microwaves at the tip/sample interface, provides simultaneously both local carrier concentration and local conductivity with fine resolutions equivalent to the AFM tip. This technique is promising if the real and imaginary parts of the complex dielectric constant can be successfully separated, which correspond to the local capacitance and conductance

[2] A description is available on Website of Agilent Technology: http://cp.literature.agilent.com/litweb/pdf/5989–8817EN.pdf.

of the sample. Another interesting SPM-based microelectrical technique sets up an STM/AFM in an SEM chamber and uses two probes of an electron beam and an STM/AFM tip [150, 151]. The interaction of the two probes provides microelectrical characterizations such as electron transport barriers on GBs of polycrystalline materials.

References

1. For a textbook, see M.A. Green, *Solar Cells* (Prentice-Hall Inc., Englewood Cliffs, NJ, 1982)
2. For a review, see R.H. Bube, *Photovoltaic Materials* (Imperial College Press, London, 1998)
3. M.A. Green, K. Emery, Y. Hishikawa, W. Warta, Prog. Photovolt. Res. Appl. **18**, 144 (2010)
4. M. Nonnenmacher, M.P. O'Boyle, H.K. Wickramasinghe, Appl. Phys. Lett. **58**, 2921 (1991)
5. A. Kikukawa, S. Hosaka, R. Imura, Appl. Phys. Lett. **66**, 3510 (1995)
6. Y. Martin, D.W. Abraham, H.K. Wickramasinghe, Appl. Phys. Lett. **52**, 1103 (1988)
7. J.R Matey, J. Blanc, Appl. Phys. **57**, 1437 (1985)
8. C.C. William, W.P. Hough, S.A. Rishton, Appl. Phys. Lett. **55**, 203 (1989)
9. G. Binning, C.F. Quate, Ch. Gerber, Phys. Rev. Lett. **56**, 930 (1986)
10. T.R. Albrecht, S. Akamine, T.E. Carver, C.F. Quate, J. Vac. Sci. Technol. **A8**, 3386 (1990)
11. G. Meyer, N.M. Amer, Appl. Phys. Lett. **56**, 2100 (1988)
12. J.M.R. Weaver, D.W. Abraham, J. Vac. Sci. Technol. **B9**, 1559 (1991)
13. For a review, see C.C. Williams, Annu. Rev. Mater. Sci. **29**, 471 (1999)
14. E.H. Nicollian, J.R. Brews, *MOS (Metal Oxide Semiconductor) Physics and Technology* (John Wiley & Sons Inc., New York, 1982)
15. D. Goghero, V. Raineri, F. Giannazzo, Appl. Phys. Lett. **81**, 1824 (2002)
16. K.M. Wong, W.K. Chim, Appl. Phys. Lett. **88**, 083510 (2006)
17. R.C. Palmer, E.J. Denlinger, H. Kawamoto, RCA Rev. **43**, 194 (1982)
18. V.V. Zavyalov, J.S. McMurray, C.C. Williams, Rev. Sci. Instrum. **70**, 158 (1999)
19. A. Bentzen, B.G. Svensson, E.S. Marstein, A. Holt, Sol. Energy Mater. Sol. Cells **90**, 3193 (2006)
20. S.E. Asher, K. Ramanathan, D.W. Niles, H. Moutinho, in *Proceedings of the 15th NCPV Photovoltaic Program Review Meeting*, Denver, CO, Sept 1998 (unpublished), p. 126
21. For a review, see S.S. Hegedus, W.N. Shafarman, Prog. Photovolt. Res. Appl. **12**, 155 (2004)
22. C.-S. Jiang, H.R. Moutinho, R. Reedy, M.M. Al-Jassim, A. Blosse, J. Appl. Phys. **104**, 104501 (2008)
23. C.-S. Jiang, A. Ptak, B. Yan, H.R. Moutinho, J.V. Li, M.M. Al-Jassim, Ultramicroscopy **109**, 952 (2009)
24. C.-S. Jiang, F.S. Hasoon, H.R. Moutinho, H.A. Al-Thani, M.J. Romero, M.M. Al-Jassim, Appl. Phys. Lett. **82**, 127 (2003)
25. A.G. Aberle, in *Proceedings of the Fourth World Conference Photovoltaic Energy Conversion*, Waikoloa, HI, 2006, p. 1481
26. T. Yamazaki, Y. Uraoka, T. Fuyuki, Jpn. J. Appl. Phys. Part 1 **45**, 2441 (2006)
27. C.-S. Jiang, H.R. Moutinho, D.J. Friedman, J.F. Geisz, M.M. Al-Jassim, J. Appl. Phys. **93**, 10035 (2003)
28. S. Saraf, A. Schwarzman, Y. Dvash, S. Cohen, D. Ritter, Y. Rosenwaks, Phys. Rev. B **73**, 035336 (2006)
29. C.-S. Jiang, D.J. Friedman, J.F. Geisz, H.R. Moutinho, M.J. Romero, M.M. Al-Jassim, Appl. Phys. Lett. **83**, 1572 (2003)
30. C.-S. Jiang, H.R. Moutinho, M.J. Romero, M.M. Al-Jassim, L.L. Kazmerski, Appl. Phys. Lett. **88**, 061909 (2006)
31. E. Strassburg, A. Boag, Y. Rosenwaks, Rev. Sci. Instrum. **76**, 083705 (2005)
32. Y. Rosenwaks, R. Shikler, Th. Glatzel, S. Sadewasser, Phys. Rev. **B70**, 085320 (2004)

33. Y. Shen, D.M. Barnett, P.M. Pinsky, Rev. Sci. Instrum. **79**, 023711 (2008)
34. A.J. Ptak, R. France, C.-S. Jiang, R.C. Reedy, J. Vac. Sci. Technol. **B26**, 1053 (2008)
35. M.R. Pillai, S.-S. Kim, S.T. Ho, S.A. Barnett, J. Vac. Sci. Technol. **B18**, 1232 (2000)
36. S. Tixier, M. Adamcyk, E.C. Young, J.H. Schmid, T. Tiedje, J. Cryst. Growth **251**, 449 (2003)
37. I. Repinds, M.A. Contreras, B. Egaas, C. DeHart, J. Scharf, C.L. Perkins, B. To, R. Noufi, Prog. Photovolt: Res. Appl. **16**, 235 (2008)
38. Y. Tanaka, N. Akema, T. Morishita, 17th EC Photovoltaic Solar Energy Conference, Munich, October 2001, p. 989
39. U.S. Patent No. 5,441,897, 15 Aug 1995 and U.S. Patent No. 5,436,204, 25 July 1995
40. M.A. Contreras, B. Egaas, K. Ramanathan, J. Hiltner, A. Swartzlander, F. Hasoon, R. Noufi, Prog. Photovolt. Res. Appl. **7**, 311 (1999)
41. K. Ramanathan, R. Noufi, J. Granata, J. Webb, J. Keane, Sol. Energy Mater. Sol. Cells **55**, 15 (1998)
42. M.A. Contreras, H. Wiesner, D. Niles, K. Ramanathan, R. Matson, J. Tuttle, J. Keane, R. Noufi, in *Proceedings of the 25th IEEE PVSC Conference*, Washington, D.C., 13–17 May 1996, p. 809
43. U. Rau, M. Schmidt, Thin Solid Films **387**, 141 (2001)
44. T. Minemoto, T. Matsui, H. Takakura, Y. Hamakawa, T. Negami, Y. Hashimoto, T. Uenoyama, M. Kitagawa, Sol. Energy Mater. Sol. Cells **67**, 83 (2001)
45. R. Klenk, Thin Solid Films **387**, 135 (2001)
46. C.-S. Jiang, F.S. Hasoon, H.R. Moutinho, H.A. Al-Thani, M.J. Romero, M.M. Al-Jassim, Mat. Res. Soc. Symp. Proc. **763**, 309 (2003)
47. O. Madelung, *Semiconductor: Other Than Group IV Elements and III–V Compounds* (Springer-Verlag, New York, 1992)
48. M. Morkel, L. Weinhardt, B. Lohmuller, C. Heske, E. Umbach, W. Riedl, S. Zweigart, F. Karg, Appl. Phys. Lett. **79**, 4482 (2001)
49. T. Minemoto, Y. Hashimoto, T. Negami, H. Takamura, Y. Hamakawa, J. Appl. Phys. **89**, 8327 (2001)
50. N.N. Syrbu, M. Bogdanash, V.E. Tezlevan, I. Mushcutariu, Physica B **229**, 199 (1997)
51. C.-S. Jiang, H.R. Moutinho, Q. Wang, M.M. Al-Jassim, B. Yan, J. Yang, S. Guha, Mat. Res. Soc. Symp. Proc. **808**, 587–592 (2004)
52. B. Yan, G. Yue, G. Guha, Mater. Res. Soc. Symp. Proc. **989**, 335 (2007)
53. K. Yamamoto, M. Toshimi, T. Suzuki, Y. Tawada, T. Okamoto, A. Nakajima, Mater. Res. Soc. Symp. Proc. **507** (1998)
54. J. Meier, J. Sitznagel, U. Kroll, C. Bucher, S. Fay, T. Moriaty, A. Shah, Thin Solid Films **451–452**, 518 (2004)
55. M. Hack, M. Shur, J. Appl. Phys. **58**, 997 (1985)
56. A.H. Pawlikiewicz, S. Guha, Mater. Res. Soc. Symp. Proc. **118**, 599 (1988)
57. C.-S. Jiang, H.R. Moutinho, M.J. Romero, M.M. Al-Jassim, Y.Q. Xu, Q. Wang, Thin Solid Films **472**, 203 (2005)
58. A. Banerjee, X. Xu, J. Yang, S. Guha, Mater. Res. Soc. Symp. Proc. **377**, 675 (1995)
59. S. Guha, J. Yang, A. Pawlikiew, T. Glatfeilter, R. Ross, S.R. Ovshinsky, Appl. Phys. Lett. **54**, 2330–2332 (1989)
60. X. Xu, J. Yang, A. Banerjee, S. Guha, K. Vasanth, S. Wagner, Appl. Phys. Lett. **67**, 2323 (1995)
61. M.E. Levinshtein, S.L. Rumyantsev, in *Handbook Series on Semiconductor Parameters*, vol. 1, ed. by M. Levinshtein, S. Rumyantsev, M. Shur (World Scientific, River Edge, NJ, 1996)
62. L. Jiang, J.H. Lyou, S. Rane, E.A. Schiff, Q. Wang, Q. Yuan, Mater. Res. Soc. Symp. Proc. **609**, A18.3 (2000)
63. S.R. Kurtz, P. Faine, J.M. Olson, J. Appl. Phys. **68**, 1890–1895 (1990)
64. S.R. Kurtz, D. Myers, J.M. Olson, in *Proceedings of the 26th IEEE PVSC Conference*, Anaheim, CA, 29 Sept 1997, p. 875
65. T. Takamoto, K. Sasaki, T. Agui, H. Juso, A. Yoshida, K. Nakaido, III–V compound solar cells. SHARP Tech. J. **100** (2010)

66. R.R. King, A. Boca, W. Hong, X-Q Liu, D. Bhusari, D. Larrabee, K.M. Edmondson, D.C. Law, C.M. Fetzer, S. Mesropian, N.H. Karam, 24th European Photovoltaic Solar Energy Conference, Hamburg, Germany, 21–25 Sept 2009
67. X. Wang, N. Waite, P. Murcia, K. Emery, M. Steiner, F. Kiamilev, K. Goossen, C. Honsberg, A. Barnett, 24th European Photovoltaic Solar Energy Conference, Hamburg, Germany, 21–25 Sept 2009
68. K.A. Bertness, S.R. Kurtz, D.J. Friedman, A.E. Kibbler, C. Kramer, J.M. Olson, Appl. Phys. Lett. **65**, 989 (1994)
69. D.P. Bour, J.R. Shealy, G.W. Wicks, W.J. Schaff, Appl. Phys. Lett. **50**, 615 (1987)
70. S.-H. Wei, A. Zunger, Appl. Phys. Lett. **72**, 2011 (1998)
71. A. Gomyo, T. Suzuki, S. Iijima, Phys. Rev. Lett. **60**, 2645 (1988)
72. J. Chen, J.R. Sites, I.L. Spain, M.J. Hafich, G.Y. Robinson, Appl. Phys. Lett. **58**, 744 (1991)
73. S. Froyen, A. Zunger, A. Mascarenhas, Appl. Phys. Lett. **68**, 2852 (1996)
74. J.J. O'shea, C.M. Reaves, S.P. Debaars, M.A. Chin, V. Narayanamurti, Appl. Phys. Lett. **69**, 3022 (1996)
75. M.A. Haase, M.J. Hafich, G.Y. Robinson, Appl. Phys. Lett. **58**, 616 (1991)
76. T.W. Lee, P.A. Houston, R. Kumar, X.F. Yang, G. Hill, M. Hopkinson, P.A. Claxton, Appl. Phys. Lett. **60**, 474 (1992)
77. K.J. Choi, J.-L. Lee, Appl. Phys. Lett. **74**, 1108 (1999)
78. H.O. Jacobs, H.F. Knapp, S. Muller, A. Stemmer, Ultramicroscopy **69**, 19 (1997)
79. Ch. Sommerhalter, Th.W. Matthes, Th. Glatzel, A. Jager-Waldau, M. Ch. Lux-Steiner, Appl. Phys. Lett. **75**, 286 (1999)
80. For a review, see D.B. Holt, B.G. Yacobi, *Extended Defects in Semiconductors* (Cambridge University Press, Cambridge, 2007), pp. 292–331
81. For a review, see C. Grovenor, J. Phys. **C18**, 4079 (1985)
82. H. Sawada, H. Ichinose, Scripta Mater. **44**, 2327 (2001)
83. J. Chen, T. Sekiguchi, D. Yang, F. Yin, K. Kido, S. Tsurekawa, J. Appl. Phys. **96**, 5490 (2004)
84. C.-S. Jiang, H.R. Moutinho, M.J. Romero, F. Liu, M.M. Al-Jassim, in *Proceedings of the 34th IEEE PVSC Conference*, Philadelphia, PA, June 2009
85. C.-S. Jiang, H.R. Moutinho, B. To, P. Dippo, M.J. Romero, M.M. Al-Jassim, Mater. Res. Soc. Symp. Proc. **1066**, 87 (2008)
86. C.-S Jiang, R. Noufi, J.A. AbuShama, K. Ramanathan, H.R. Moutinho, J. Pankow, M.M. Al-Jassim, Appl. Phy. Lett. **84**, 3477 (2004)
87. C.-S. Jiang, R. Noufi, K. Ramanathan, J.A. AbuShama, H.R. Moutinho, M.M. Al-Jassim, Appl. Phys. Lett. **85**, 2625 (2004)
88. Y. Yan, C.-S. Jiang, R. Noufi, S.-H. Wei, H.R. Moutinho, M.M. Al-Jassim, Phys. Rev. Lett. **99**, 235504 (2007)
89. C.-S Jiang, R. Noufi, K. Ramanathan, J.A. AbuShama, H.R. Moutinho, M.M. Al-Jassim, in *Proceedings of the 31st IEEE PVSC Conference*, Orlando FL, Jan 2005, p. 251
90. M.J. Keevers, T.L. Young, U. Schubert, M.A. Green, 22nd European Photovoltaic Solar Energy Conference, Milan, Sept 2007
91. P.A. Basore, in *Proceedings of the 29th IEEE PVSC Conference*, New Orleans, LA, May 2002, p. 49
92. Yamazaki, Y. Uraoka, T. Fuyuki, Thin Solid Films **487**, 26 (2005)
93. G. Beaucarne, S. Bourdais, A. Slaoui, J. Poortmans, Appl. Phys. **A79**, 469 (2004)
94. N. Kawamoto, A. Matsuda, N. Matsuo, Y. Seri, T. Nishimori, Y. Kitamon, H. Matsumura, H. Hamada, T. Miyoshi, Jpn. J. Appl. Phys. **45**, 2726 (2006)
95. K. Kurobe, Y. Ishikawa, Y. Yamamoto, T. Fuyuki, H. Matsunami, Sol. Energy Mater. Sol. Cells **65**, 201 (2001)
96. T. Matsui, T. Yamazaki, A. Nagatani, K. Kino, H. Takakura, Y. Hamakawa, Sol. Energy Mater. Sol. Cells **65**, 87 (2001)
97. Y. Ishikawa, Y. Yamamoto, T. Hatayama, Y. Uraoka, T. Fuyuki, Jpn. J. Appl. Phys. 40, 6783 (2001)
98. G. Beaucarne, S. Bourdais, A. Slaoui, J. Poortmans, in *Proceedings of the 28th IEEE PVSC, Conference*, Anchorage, Alaska, 2000, p. 128

99. E. Christoffel, M. Rusu, A. Zerga, S. Bourdais, S. Noël, A. Slaoui, Thin Solid Films 403–404, 258 (2002)
100. Q. Wang, C.W. Teplin, P. Stradins, B. To, K.M. Jones, H.M. Branz, J. Appl. Phys. **100**, 093520 (2006)
101. C.W. Teplin, H.M. Branz, K.M. Jones, M.J. Romero, P. Stradins, S. Gall, Mater. Res. Soc. Symp. Proc. **989**, 133 (2006)
102. R. Brendel, R.B. Bergmann, B. Fischer, J. Krinke, R. Plieninger, U. Rau, J. Reiss, H.P. Strunk, H.Wanka , J. Werner, in *Proceedings of the 26th IEEE PVSC Conference*, Anaheim, California, 1997, p. 635
103. J.H. Werner, J. Mattheis, U. Rau, Thin Solid Films **480–481**, 399 (2005)
104. J.T. Heath, C.-S. Jiang, H.R. Moutinho, M.M. Al-Jassim, Mater. Res. Soc. Symp. Proc. **1268** (2010)
105. A.J. Schwartz, M. Kumar, B.L. Adams, *Electron Backscatter Diffraction in Material Science* (Kluwer Academic/Plenum Publishers, New York, 2000)
106. A.L. Endrös, Sol. Energy Mater. Sol. Cells **72**, 109 (2002)
107. M. Kitamura, N. Usami, T. Sugawara, K. Kutsukake, K. Fujiwara, Y. Nose, T. Shishido, K. Nakajima, J. Cryst. Growth **280**, 419 (2005)
108. C.H. Champness, in *Proceedings of the 29th IEEE PVSC Conference*, 2002, p. 732
109. L.S. Yip, I. Shih, in *Proceedings of the 1st World Conference on Photovoltaic Energy Conversion*, 1994, p. 210
110. K. Bothe, G.H. Bauer, T. Unold, Thin Solid Films **403–404**, 453 (2002)
111. B.M. Keyes, P. Dippo, W.K. Metzger, J. AbuShama, R. Noufi, J. Appl. Phys. **94**, 5584 (2003)
112. O. Lundberg, M. Bodegard, L. Stolt, Thin Solid Films **431–432**, 26 (2003)
113. S. Sadewasser, Thin Solid Films **515**, 6136 (2007)
114. D.F. Marron, S. Sadewasser, A. Meeder, Th. Glatzel, M.Ch. Lux-Steiner, Phys. Rev. **B71**, 033306 (2005)
115. C. Heske, R. Fink, E. Umbach, W. Riedl, F. Karg, Appl. Phys. Lett. **68**, 3431 (1996)
116. D.S. Albin, J.J. Carapella, J.R. Tuttle, R. Noufi, Mater. Res. Soc. Symp. Proc. **228**, 267 (1992)
117. M.A. Contreras, K. Ramanathan, J. AbuShama, F. Hasoon, D.L. Young, B. Egaas, R. Noufi, Prog. Photovolt. Res. Appl. **13**, 209 (2005)
118. K. Granath, M. Bodegard, L. Stolt, Sol. Energy Mater. Sol. Cells **60**, 279 (2000)
119. N. Kohara, T. Negami, M. Nishitani, Y. Hashimoto, T. Wada, Appl. Phys. Lett. **71**, 835 (1997)
120. D. Braunger, D. Hariskos, G. Bilger, U. Rau, H.W. Schock, Thin Solid Films **361–362**, 161 (2000)
121. S.-H. Wei, S.B. Zhang, A. Zunger, J. Appl. Phys. **85**, 7214 (1999)
122. C. Heske, G. Richter, Z. Chen, R. Fink, E. Umbach, J. Appl. Phys. **82**, 2411 (1997)
123. Y. Yan, K.M. Jones, C.-S. Jiang, X.Z. Wu, R. Noufi, M.M. Al-Jassim, Physica B **401–402**, 25 (2007)
124. D.L. Staebler, C.R. Wronski, Appl. Phys. Lett. **31**, 292 (1977)
125. D. Azulay, I. Balberg, V. Chu, J.P. Conde, O. Millo, Phys. Rev. B **71**, 113304 (2005)
126. For a review, see A.V. Shah, J. Meier, E. Vallat-Sauvain, N. Wyrsch, U. Kroll, C. Droz, U. Graf, Sol. Energy Mat. Sol. Cells **78**, 469 (2003)
127. K. Lord, B. Yan, J. Yang, S. Guha, Appl. Phys. Lett. **79**, 3800 (2001)
128. J. Yang, K. Lord, B. Yan, A. Banerjee, S. Guha, D. Han, K. Wang, Mater. Res. Soc. Symp. Proc. **715**, 601 (2002)
129. B. Yan, C.-S. Jiang, C.W. Teplin, H.R. Moutinho, M.M. Al-Jassim, J. Yang, S. Guha, J. Appl. Phys. **101**, 033712 (2007)
130. C.-S. Jiang, H.R. Moutinho, M.M. Al-Jassim, L.L. Kazmerski, B. Yan, J.M. Owens, J. Yang, S. Guha, in *Proceedings of the 4th World Conference on Photovoltaic Energy Conversion*, Waikoloa, HI, 2006, p. 1552
131. C.-S. Jiang, B. Yan, H.R. Moutinho, M.M. Al-Jassim, J. Yang, S. Guha, Mater. Res. Soc. Symp. Proc. **989**, 15 (2007)
132. C.-S. Jiang, B. Yan, Y. Yan, C.W. Teplin, R. Reedy, H.R. Moutinho, M.M. Al-Jassim, J. Yang, S. Guha, J. Appl. Phys. **103**, 063515 (2008)

133. C.-S. Jiang, Y. Yan, H.R. Moutinho, M.M. Al-Jassim, B. Yan, L. Sivec, J. Yang, S. Guha, Mater. Rec. Soc. Symp. Proc. **1153**, 345 (2009)
134. J. Yang, A. Banerjee, K. Lord, S. Guha, in *Proceedings of the 28th IEEE PVSC Conference*, Anchorage, AK, 15–22 Sept 2000, p. 742
135. T. Roschek, T. Repmann, J. Müller, B. Rech, H. Wagner, in *Proceedings of the 28th IEEE PVSC Conference*, Anchorage, AK, 15–22 Sept 2000, p.150
136. B. Yan, J. Yang, S. Guha, in *Proceedings of the 3rd World Conference on Photovoltaic Energy Conversion*, Osaka, Japan, 2003, p. 1627
137. D.L. Williamson, Sol. Energy Mater. Sol. Cells **78**, 41 (2003)
138. J.M. Owens, D. Han, B. Yan, J. Yang, K. Lord, S. Guha, Mater. Res. Soc. Symp. Proc. **762**, 339 (2003)
139. X.-Z Bo, L.P. Rokhinson, Y. Haizhou, D.C. Tsui, J.C. Sturm, Appl. Phys. Lett. **81**, 3263 (2002)
140. S. Myhra, Appl. Phys. A **76** 63 (2003)
141. P. Kumar, B. Schroeder, Thin Solid Films **516**, 580 (2008)
142. T. Matsui, M. Kondo, A. Matsuda, J. Non-Cryst. Solids **338–340**, 646 (2004)
143. R.W. Collins, A.S. Ferlauto, G.M. Ferreira, C. Chen, J. Koh, R.J. Koval, Y. Lee, J.M. Pearce, C.R. Wronski, Sol. Energy Mater. Sol. Cells **78**, 143 (2003)
144. C.W. Teplin, E. Iwaniczko, B. To, H. Moutinho, P. Stradins, H.M. Branz, Phys. Rev. **B74**, 235428 (2006)
145. A. Matsuda, Thin Solid Films **337**, 1 (1999)
146. C.W. Teplin, C.-S. Jiang, P. Stradins, H.M. Branz, Appl. Phys. Lett. **92**, 093114 (2008)
147. T.I. Kamins, *Polycrystalline Silicon for Integrated Circuit Applications* (Kluwer Academic, Boston, 1997), p. 32
148. T. Toyama, W. Yoshida, Y. Sobajima, H. Okamoto, J. Non-Cryst. Solids **354**, 2204 (2008)
149. S. Nakayama, I. Kawashima, J. Murota, J. Electrochem. Soc. Solid State Sci. Technol. **133**, 1721 (1986)
150. M.J. Romero, C.-S. Jiang, R. Noufi, M.M. Al-Jassim, Appl. Phys. Lett. **87**, 172106 (2005)
151. M.J. Romero, F. Liu, O. Kunz, J. Wong, C.-S. Jiang, M.M. Al-Jassim, A.G. Aberle, Mater. Res. Soc. Symp. Proc. **1153**, 287 (2009)

Chapter 25
Micro and Nanodevices for Thermoelectric Converters

J.P. Carmo, L.M. Gonçalves, and J.H. Correia

Abstract This focus of this chapter is the presentation of micro and nanodevices for thermoelectric converters. Examples of applications for these converters are (1) in the conversion of electrical energy from temperature gradients and (2) in cooling devices. These converters are of solid-state type and use pairs of thermoelectric p- and n-type materials, which were obtained by thin-films depositions. In this context, issues such the fabrication and characterization details of materials are discussed. The materials selected to serve as p- and n-type structures were the antimony telluride (Sb_2Te_3) and the bismuth telluride (Bi_2Te_3). The thin-films depositions of both Bi_2Te_3 and Sb_2Te_3 materials require a precise controlled process to achieve the highest possible thermoelectric figure-of-merit, ZT, and at the same time to achieve the desire composition. This goal is achieved with the co-evaporation of antimony (Sb)/bismuth(Bi) and telluride (Te) technique. It is also analyzed the influence of the parameters involved in the depositions (e.g., the temperature which the substrate is subjected, and the evaporation rates of Sb/Bi and Te), and their impacts in the final composition of Sb_2Te_3 and Bi_2Te_3 thin-films. Advanced issues and trends related to the fabrication of super-lattices for use in thermoelectric converters are also presented in this chapter.

25.1 Introduction

The thermoelectricity consists in the conversion of temperature differences on electric energy and the contrary. Three effects are present in the thermoelectricity: the Seebeck effect, the Peltier effect, and the Thomson effect. Thomas Seebeck (1770–1831) observed the arise of a voltage, ΔV [V], after subjecting the junction of two different materials to a temperature gradient, ΔT [K], and an electric current flowing in a load [1]. This effect can be quantified by the way of the Seebeck coefficient, $\alpha[\mu\text{V K}^{-1}]$:

$$\alpha = \Delta V/\Delta T \tag{25.1}$$

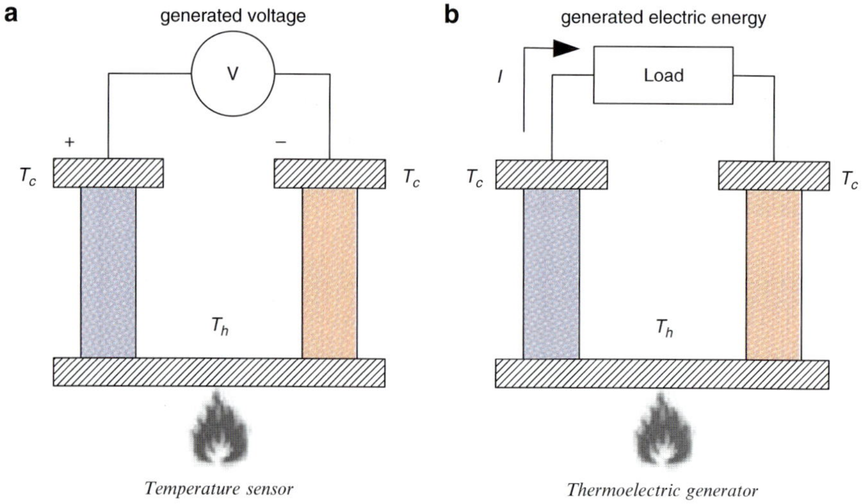

Fig. 25.1 Seebeck effect, used as temperature sensor (**a**) or energy generator (**b**)

The Seebeck effect can be used in two types of applications: for temperature sensors and thermoelectric generators [1]. Figure 25.1 shows the two latter examples, where T_h [K] and T_c [K] ($> T_\mathrm{h}$) are the temperatures in the hottest and in the coolest sides, respectively. In these examples, the temperature gradient ΔT is equal to $T_\mathrm{h} - T_\mathrm{c}$ (and it is always equal for temperatures either in K or in °C).

Jean Peltier (1785–1845) observed in 1834 that the flow of an electric current through the junction of two different materials had the consequence to release (heating) or to absorb (cooling) thermal energy. Figure 25.2 illustrates these two examples, where the temperature in the junction rises or decreases, respectively [2]. Moreover, he observed that the released or the absorbed energy was proportional to the electric current. This effect is quantified by the Peltier coefficient, π. Some years later, William Thomson (later known as Lord Kelvin) established the relation between the Peltier and the Seebeck coefficients:

$$\pi = \alpha T, \tag{25.2}$$

where T [K] is the room temperatures.

25.1.1 Macrodevices

In the 1960s, the thermoelectric phenomena were of major concern by researchers, when it was generally accepted that one of those days, the Peltier elements would be used in all cooling applications. However, in the four past decades, there was not a significant progress in this area. This was mainly due to the coefficient of performance (COP) of the Peltier elements being 4–5 times below to those found

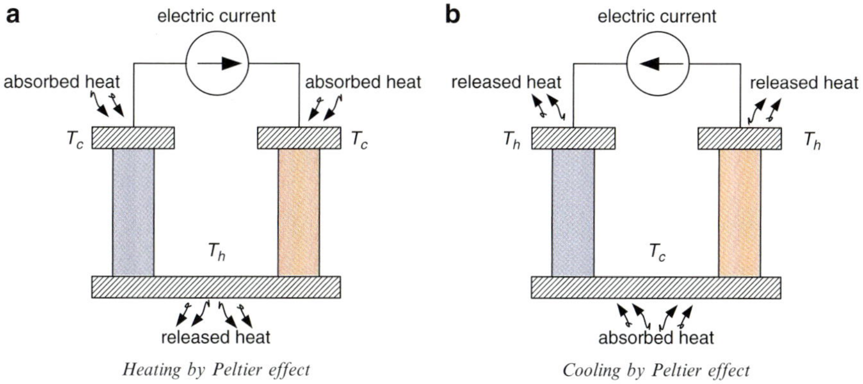

Fig. 25.2 Using the Peltier effect for heating or cooling

in conventional coolers (based on the Carnot cycle). Additionally, the unitary limit of the figure-of-merit, Z [K^{-1}] (a performance measure of thermoelectric materials), it was seemed as an impossible barrier to pass, but also unexplainable. The chalcogenide compounds, the bismuth telluride (Bi_2Te_3) and the antimony telluride (Sb_2Te_3) were known for decades as the best thermoelectric materials at room temperature. The figure-of-merit of a material is calculated by the following equation:

$$Z = \frac{\alpha^2}{\rho\kappa}, \tag{25.3}$$

where α[$\mu V\,K^{-1}$] is the Seebeck coefficient, ρ[Ω m] is the electric resistivity and κ [$W\ m^{-1}K^{-1}$] is the thermal conductivity. The figure-of-merit can also be calculated for a specific room temperature, T [K], and it results in the following dimensional quantity:

$$ZT = \frac{\alpha^2}{\rho\kappa}T. \tag{25.4}$$

In spite of its inferior performance, the cooling applications based on Peltier devices do not require maintenance because there are no movable parts. Figure 25.3 shows a conventional Peltier module, constituted by thermoelectric elements between the two ceramic foils. The ceramic foils must be good thermal conductors to minimize their effect on the performance of the Peltier device. The smallest device (between the wires) is a miniaturized Peltier module made with thermoelectric micropairs is also depicted.

25.1.2 Microdevices

The use of thin-films deposition techniques with microsystems technologies renewed interest in the thermoelectricity. Integration of efficient solid-state thermoelectric microdevices with microelectronics is desirable for local cooling and

Fig. 25.3 A photograph showing conventional and miniaturized Peltier modules

thermoelectric microgeneration, since they can be used to stabilize the temperature of devices, decrease noise levels and increase operation speed. An array of such devices can also be used for lab-on-chip applications, where the precise temperature control is required. Despite the range of exciting applications, only few approaches to manufacture thermoelectric devices with small dimensions were reported up to now [3–6]. Also, the Peltier devices can convert thermal energy to electrical energy, because they are reversible. Electronic devices with low-power consumptions can be supplied with these energy converters. Examples of electronic devices include electronics embedded in wearables, wristwatches, among others. The last application has been widely explored by Seiko [7] and by Citizen [8]. The Citizen's Eco-Drive Thermo watch contains a generator composed by 1,242 thermoelectric elements capable to produce a voltage of 640 mV with a power of 13.8 μW, for each degree temperature difference. The Seiko's Thermic watch can generate 200 mV of voltage with 1,040 Peltier elements under the same conditions.

Micropelt was the first company to deploy ready-to-use microcoolers in the market. This company was created under a joint-venture formed by the Infinion Technologies and the Fraunhofer Institute for Physical Measurement Techniques (IPM), and it presents a fabrication process that allows the fabrication of thermoelectric elements with reduced dimensions. This fabrication process uses two wafers (with the n-type and p-type elements) that are glued together in order to construct submillimeter-sized Peltier elements. These Peltier present cooling capabilities above 100 W cm^{-2} [9] for gradient temperatures up to 30 K at the room temperature.

Efforts to use MEMS (Micro-Electro-Mechanical Systems) techniques in the fabrication of thermoelectric devices had been made by the Jet Propulsion Laboratory (JPL, NASA) and by the IPM. An electrodeposition process is used by the JPL [5], while at the same time a sputtering and dry-etching process was developed by the IPM to fabricate thermoelectric structures. Additionally, Völklein [10] explores the use of tellurides for fabricating thermoelectric sensors. Mr. Völklein also developed a technique to measure the thermal conductivity on thin-films.

Concerning the use of patterning techniques on thermoelectric films, few approaches were reported in the literature. Da Silva et al. [11] uses lift-off with SU-8 photoresist to create vertical columns of thermoelectric materials. But due to the maximum working temperature of SU-8 photoresist, the thermoelectric properties of the thin-films incorporated in the devices are worse than those obtained in bulk materials, since 200–300°C of substrate temperature is required to fabricate high-quality thermoelectric thin-films. Böttner et al. [3] uses dry etching to pattern thermoelectric devices. Shafai [12] reported the patterning of thin-films by wet etching but no details were given about the process.

Doctor Wijngaards [13] fabricated Peltier microdevices in membranes by the way of a process compatible with CMOS fabrication processes. However, the thermoelectric properties of the polycrystalline silicon material used in his work created temperature differences of only 2 K. Schavaevitz [14] presented at the Massachusetts Institute of Technology (MIT), a MEMS device to generate electricity from a thermal combustion microcamera. Such a device is based on a silicon–germanium thermoelectric converter. In this case, an electrical power of 450 mW was obtained from a temperature of 650°C inside the combustion camera.

The ThermoLife® Energy Corp. (USA, California) imported technology from the former D.T.S. (Germany) to fabricate thermoelectric microgenerators [15]. This company reported the ability to produce 30 μW – corresponding to 3 V, 10 μA – from temperature gradients of 5 K in an area less than 1 cm^2. Stordeur and Stark were one of the first to fabricate thermoelectric microgenerators in flexible membranes [16].

New developments also took place at TU-Dresden in order to use flexible membranes to fabricate thermoelectric devices to provide autonomous energy sources for application in microsystems [17]. In this case, the electrodeposition and photolithography techniques were employed for fabricating the thermoelectric devices with thin-films of n- and p-types. However, these devices have the drawback to produce voltages of only 250 mV with temperature differences of 30 K.

To finish, it must be told that the works developed by Gao Min [18], Völklein [19] and Jacquot [20] are important contributions in the modeling field, and these models of cooling microdevices still are references in the thermoelectric area.

25.1.3 Nanodevices and Superlattices

Bismuth and antimony tellurides are the materials with the best known thermoelectric performances. However, the silicon process compatibility of polycrystalline silicon germanium alloys makes them good thermoelectric material candidates. Devices, based on bismuth (Bi), antimony (Sb) and tellurium (Te) can support up to 250°C. For higher temperatures, silicon–germanium (Si–Ge) compounds can be used, despite having poorer thermoelectric properties.

In today's best commercial thermoelectric modules (made of Bi, Sb, and Te compounds), the figure-of-merit, ZT, is about 1, despite many attempts to find compounds with higher performance. In conventional tridimensional crystalline systems, the quantities S, σ, and κ are interrelated and are difficult if not impossible to control them independently to improve the ZT [21, 22]. An increase of S

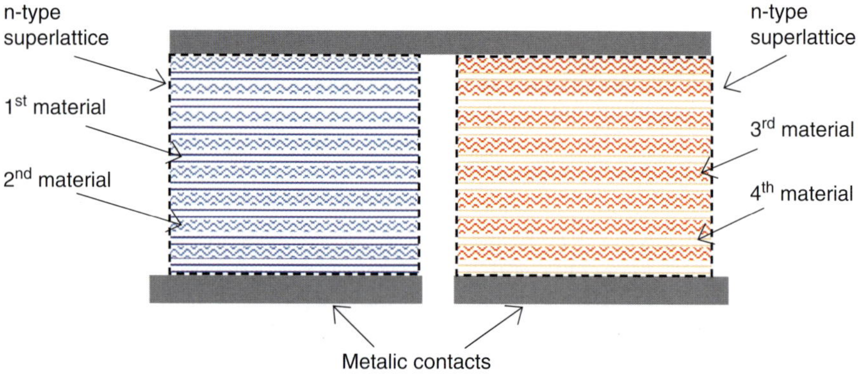

Fig. 25.4 A thermoelectric pair made of superlattice materials. Each material is composed of alternating layers of two different materials, with thickness in the range of tens of nanometers

usually results in a decrease of σ. A decrease of σ leads to a decrease of the electronic contribution to κ. However, if the dimensionality of the material is decreased, the new variable of length scale becomes available to control the properties of the material [23], due to differences in the density of electronic states. New phenomena are introduced when reducing the tridimensional solid crystalline structures to bidimensional superlattices (see Fig. 25.4), one-dimensional nanowires, or quantum dots, with new opportunities to control S, σ, or κ. The introduction of many interfaces in the structure can scatter phonons more effectively than electrons and allows enhanced ZT in such (new) materials. Recent work with PbTe [24], SiGe [25], and BiSbTe [26, 27] superlattices demonstrated an enhancement of ZT. Whereas $\mathrm{ZT} = 2.4$ and $\mathrm{ZT} = 1.4$ were measured in p-type and n-type Bi–Sb–Te superlattices [28], respectively, no equivalent improvement was found yet in Si–Ge superlattices, when compared to bulk the material.

The current efforts concerning the research of thermoelectric materials are focused in the use of nanostructures in thermoelectric applications. In fact, the thin-film technology is helping to increase the efficiency of these devices. In the last decade, new horizons are opening in the thermoelectricity, in the development of microsystems, and in new techniques to make possible to have figures-of-merit, ZT, behind the unitary bound ($\mathrm{ZT} > 1$) [29, 30]. The biggest challenge in the current research of thermoelectric materials is the reduction of thermal conductivity, and the use of thin-films techniques is one way to achieve that reduction. Molecular Beam Epitaxy (MBE) and Chemical Vapor Deposition (CVD) are two possible thin-films deposition techniques to achieve thermoelectric materials based on quantum confinement [26, 31].

The use of thin-film processes in thermoelectric structures limits the thickness of the deposited films to few micrometers. Using this thickness, the achieved heat-flow density has higher value, compared with traditional large-scale devices. If $10\,\mathrm{W\,cm^{-2}}$ can be found in typical large-scale device, then $500\,\mathrm{W\,cm^{-2}}$ can be supported in a thin-film device. However, this density could not be attended with

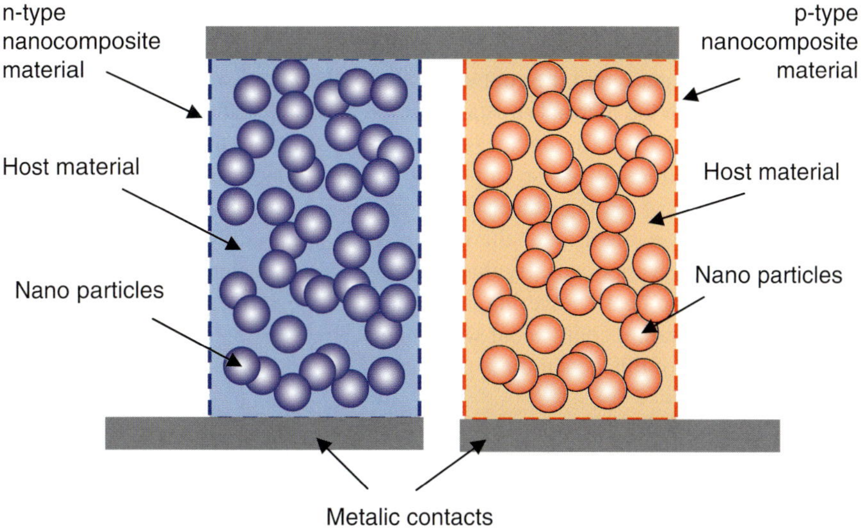

Fig. 25.5 A thermoelectric pair made of nanocomposite materials

conventional heatsinks. For lower densities applications, efforts are also being done to achieve bulk materials (rather than films) with increased figures-of-merit. A periodic structure is the major mechanism to reduce the thermal conductivity and to support the enhanced figure-of-merit in superlattices. However, nanocomposites become a natural step for extending the success in superlattices to more scalable materials. Randomly distributed nanostructures in nanocomposite materials (see Fig. 25.5) can lead to a reduction in the thermal conductivity below that of an alloy of the same overall chemical stoichiometry [32]. These materials can be prepared by either wet-chemistry, ball-milling or by inert-gas condensation methods. The nanometer or micrometer-sized particles are then hotpressed to obtain dense and mechanically strong, bulk nanocomposites.

25.2 Thermoelectric Converters Models

Since each thermoelectric pair can produce a voltage near $1.46\,\mu\mathrm{V\,K^{-1}}$, many pairs connected in series are necessary to generate a usable voltage. The maximum power in a thermoelectric generator, calculated with (25.5), is obtained when the load resistance equals the internal resistance, $R[\Omega]$. Such a power is

$$P_{\mathrm{MAX}} = \frac{V_{\mathrm{OUT}}^2}{4R} = \frac{[n(\alpha_\mathrm{p}\alpha_\mathrm{n})\Delta T]^2}{4n(R_\mathrm{n} + R_\mathrm{p} + R_\mathrm{j} + 4R_\mathrm{c})}, \tag{25.5}$$

where n is the number of thermoelectric elements (pairs of thermoelectric p–n junctions), $\alpha[\mu\mathrm{V\,K^{-1}}]$ is the Seebeck coefficient, ΔT [K] is the temperature difference between the hot side and the cold side of thermoelectric elements, and $R[\Omega]$ is the internal electric resistance of the thermoelectric device. The indexes p and n refer to p-type and n-type materials, respectively, and the indexes j and c refer, respectively, to materials of contacts and the contact itself. To obtain maximum power, it is also important to match the thermal resistance of the generator with the heatsink (on the cold side) and hot object (in the hot side), not represented in the previous equation. In several applications, it is also important to analyze the impact of the generator in the temperature of the hot object. If a human body generator is designed, then it will not suit comfortable if much thermal power is absorbed from the skin (the sensation of cold will be noticed). By the other hand, when designing a thermoelectric generator for waste heat recovering (e.g., recovering heat from a laptop CPU), an increase of temperature could occur where the heat is generated.

The efficiency of this generator is given by [33]

$$\varphi_{\max} = \frac{q_c}{W} = \frac{T_1}{T_2 - T_1} \times \frac{\sqrt{1 + ZT_M} - \frac{T_2}{T_1}}{\sqrt{1 + ZT_M} + 1}, \tag{25.6}$$

where Z represents the figure-of-merit of the device, calculated with the help of (25.7), T_1 [K] and T_2 [K] are the temperatures at hot and cold sides and T_M [K] is the mean temperature given by $(T_1 + T_2)/2$. Despite the use of the figure-of-merit to quantify the quality of a thermoelectric material, this parameter is also used to quantify the performance of a thermoelectric device, and it is given by

$$Z = \frac{(\alpha_p - \alpha_n)^2}{4\kappa\rho + \frac{8\kappa\rho_c H}{LL_c} + \frac{4\kappa_m \rho H_m}{H} + \frac{8\kappa_m \rho_c H_m}{LL_c} + \frac{4\rho\gamma Ll}{H} + \frac{8\rho_c \gamma l}{L_c}}, \tag{25.7}$$

where L[m] and H[m] are, respectively, the length and the height of thermoelectric materials, L_c [m] is the length of contacts between the thermoelectric material and the metal pads, l [m] is the length of cold area, α[μV m^{-1}] is the Seebeck coefficient, ρ[Ω m] is the electrical resistivity, κ[W m^{-1}K^{-1}] is the thermal conductivity, and ρ_c[Ω m] is the contact electrical resistivity. The device is supported by an isolating membrane with thickness H_m [m] and thermal conductivity κ_m [W m^{-1}K^{-1}]. The quantity γ[W m^{-2}K^{-1}] represents a coefficient to the include radiation and convection losses, and it is normally in the range $5 < \gamma < 10\,\mathrm{W\,m^{-2}\,K^{-1}}$.

The maximum temperature difference, $\Delta T_{\max} = (T_h - T_c)_{\max}$ [K], between the hot and the cold sides without external load ($Q_L = 0$) can be obtained:

$$\Delta T_{\max} = \frac{ZT_c^2}{2}. \tag{25.8}$$

This maximum temperature difference is obtained when the current, I[A], has a specific value I_{opt} [A], such that $\partial T_c/\partial I = 0$. Under this condition, such a current is

$$\frac{\partial T_c}{\partial I} = 0 \leftrightarrow I_{opt} = \frac{(\alpha_p - \alpha_n)T_c}{R}. \tag{25.9}$$

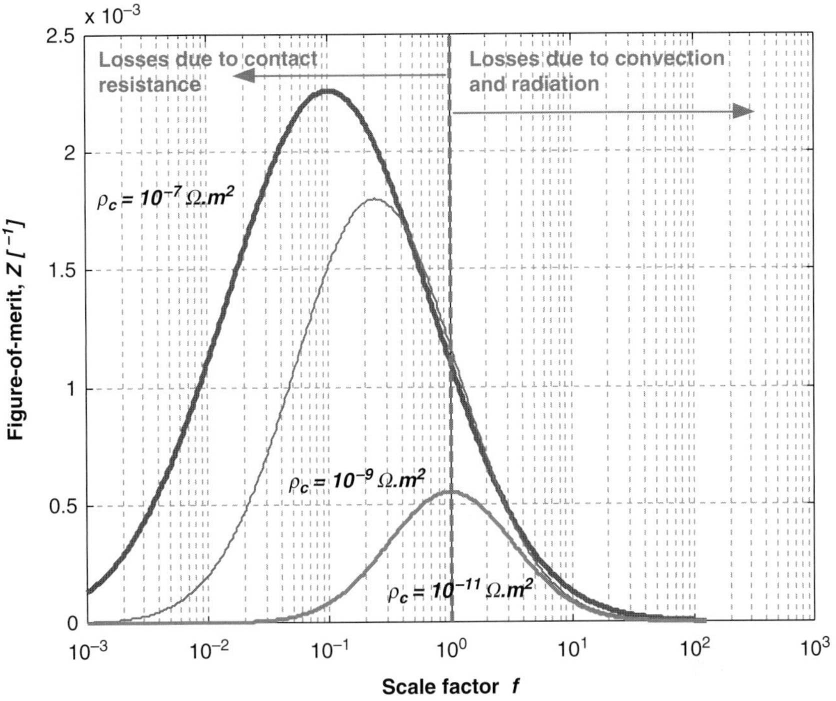

Fig. 25.6 The effect of scaling the device in the *XY* plane. The heights of the device and of the support membrane are constant ($H = 10\,\mu\text{m}$, $H_\text{m} = 10\,\mu\text{m}$), while the other dimensions are scaled by the value on the horizontal axis

The next step is the minimization of the temperature on the cold side, whose value is:

$$T_\text{c} = \frac{\sqrt{1 + 2ZT_\text{h}} - 1}{Z}, \quad I = I_\text{opt}. \tag{25.10}$$

The previous equations help to predict the effect of reducing the dimensions of a planar thermoelectric device, when the scaling from millimeters to micrometers is done. The graph of Fig. 25.6 presents the effect of scaling down the dimensions of devices in the *XY* plane. The scale factor $f = 1$ (in the horizontal axis) represents a device with dimensions $L = W = 1$ mm, $H = 10\,\mu\text{m}$ and $H_\text{m} = 10\,\mu\text{m}$. Lower values of f represent devices where the dimensions (L and W) were reduced by a factor f.

Since the losses by radiation and convection are less relevant in lower dimension devices, the figure-of-merit increases when the device is smaller. However, a low-contact resistivity must be ensured, to keep this higher figure-of-merit. Considering both effects, an optimum dimension exists where the figure-of-merit is maximized.

Figure 25.7 presents the complete model of a thermoelectric device [14], including the thermal and the electric domains. This thermoelectric model is for micro-cooling applications (where the Peltier effect is present). The electrical model (on

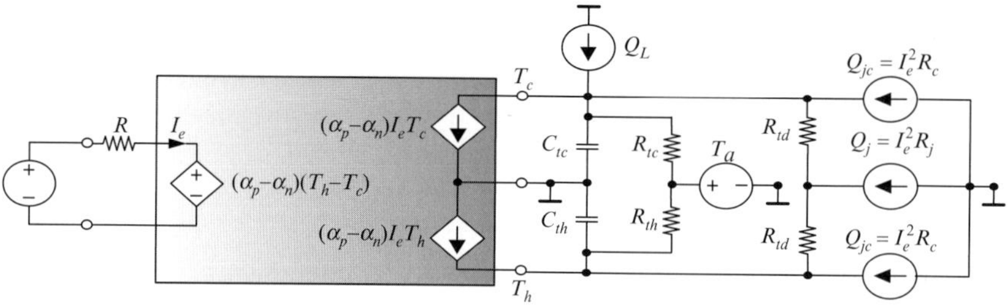

Fig. 25.7 The complete model of a Peltier microcooler, including electrical and thermal domains

the left of Fig. 25.7) includes the electrical equivalent resistance of the device, $R[\Omega]$, a voltage source that provides power to the device and a voltage source modeling the Seebeck effect of junctions – given by $(\alpha_p - \alpha_n)(T_h - T_c)$. The resultant current is I_e [A]. On the right side of Fig. 25.7, the thermal model is presented. The two current sources, $(\alpha_p - \alpha_n)I_eT_c$ and $(\alpha_p - \alpha_n)I_eT_h$, represent, respectively, the cooling and the heating by the Peltier effect. The capacitors $C_{t,c}$ [F] and $C_{t,h}$ [F] are the heat capacity on the cold and the hot sides. The resistances $R_{t,c}[\Omega]$ and $R_{t,h}[\Omega]$ are the losses by convection and by radiation to the ambient temperature (T_a [K]). The resistance $R_{t,h}[\Omega]$ is usually very small. The quantity $T_h \approx T_aR_{t,d}$ represents half of the thermal resistance between the hot side and the cold side, including support membrane effects. The variables Q_j [J] and Q_{jc} [J] represent the heating by joule, respectively, in the thermoelectric materials and in the contacts. The load applied on the cold side of the device is represented by Q_L [J].

A thermoelectric cooler can be analyzed using finite element modeling. After designing the structure, two types of loads are considered: the first one is due to the Peltier effect on the hot and cold sides, whereas the second is due to the joule heating.

25.2.1 Peltier Effect on Hot and Cold Sides

The cooling by Peltier effect is calculated with (25.11) and the heating with (25.12) (T_C represents the temperature on the cold side and T_H the temperature on the hot side):

$$Q_c = Q_{nc} + Q_{pc} = (\pi_n + \pi_p)I = (\alpha_p - \alpha_n)T_cI \quad [\text{J}], \tag{25.11}$$

$$Q_h = Q_{nh} + Q_{ph} = \left(\pi_n + \pi_p\right)I = \left(\alpha_p - \alpha_n\right)T_hI \quad [\text{J}]. \tag{25.12}$$

This load is applied on p–n junction of the device. Q_C should be negative, since heat is removed from the cold side of the device.

25.2.2 Joule Heating

The joule heating should be considered in all electrical resistances of the device. This includes the p- and the n-type materials, the metallic contact between them and the interface contact resistances. The metallic contact resistance is usually neglected, since this value is very small compared with those found on p- and n-type materials, due to its low resistivity. W is the width of elements and the other symbols are as defined in (25.7). The total equivalent resistance of a thermoelectric n–p pair is

$$R_{\mathrm{eq}} = R_{\mathrm{e}} + 2R_{\mathrm{c}} = \frac{\rho_{\mathrm{n}} L_{\mathrm{n}}}{W_{\mathrm{n}} H_{\mathrm{n}}} + \frac{\rho_{\mathrm{p}} L_{\mathrm{p}}}{W_{\mathrm{p}} H_{\mathrm{p}}} + 2\left(\frac{\rho_{\mathrm{cn}}}{L_{\mathrm{c}} W_{\mathrm{n}}} + \frac{\rho_{\mathrm{cp}}}{L_{\mathrm{c}} W_{\mathrm{p}}}\right). \tag{25.13}$$

The joule heating should be calculated and applied in these volumes. The convection and the radiation should be considered in all surfaces. Figure 25.8 shows the surface temperature of a planar Peltier microcooler, on top of a Kapton polyimide substrate. A temperature difference of 16 K (between the hot side and the cold side) was obttained in simulations [35].

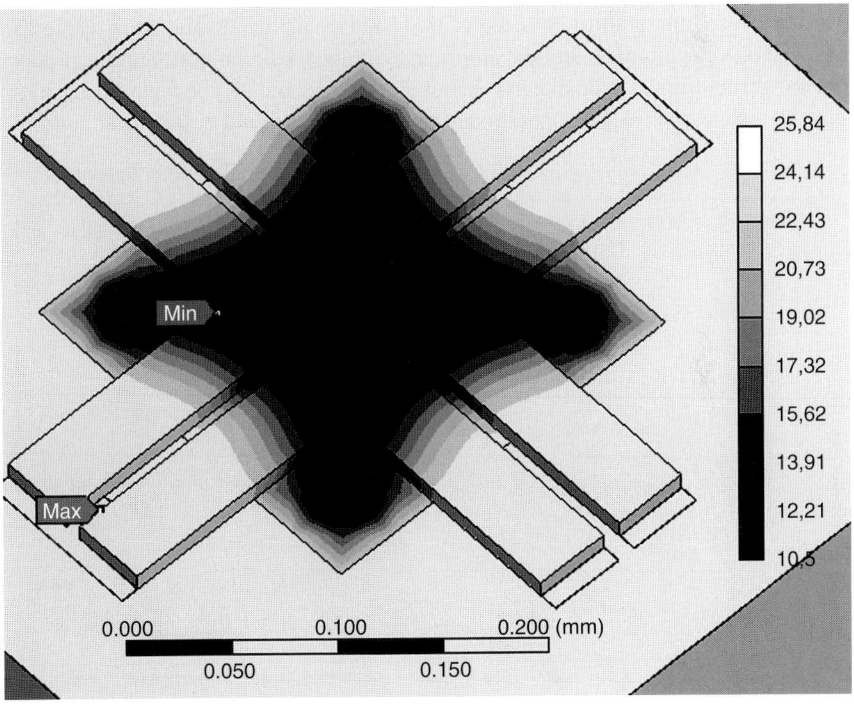

Fig. 25.8 FEM simulation of a microdevice shows the possibility of achieving a temperature difference between hot and cold sides of 16°C (or 16 K)

25.3 Thin-Films Technology for Thermoelectric Materials

The way how materials are deposited is not a pacific issue, and the following example shows exactly this problem. Considering a single junction of a n–p thermocouple having a Seebeck voltage of only 400 μVK^{-1}, thus, to achieve a usable voltage in generator devices, more than 4,000 thermocouples must be connected in series. If these 4,000 thermocouples are to be fitted in a 1 cm^2 device then, the size of each thermocouple is about 100 $\mu m \times 200$ μm. This means that the fabrication methods used in macrosized thermoelectric devices cannot be used in these microdevices. The materials can be deposited by thin-film deposition processes (physical and chemical vapor deposition or electrochemical deposition).

Another controversial issue is the material type to fabricate the thermoelectric structures. Figure 25.9 shows the figure-of-merit, Z[K^{-1}], calculated for different materials at different temperatures [36] and helps to understand this problem.

Of the great number of materials investigated, those based on bismuth telluride, lead telluride and silicon–germanium alloys emerged as the best for operating at temperatures near 300 K, 900 K and 1,400 K, respectively. Near the room temperature (250–350 K), tellurium (Te), bismuth (Bi), antimony (Sb), and selenium (Se) composites show the highest figure-of-merit values. For this reason, they are used in many of the commercial Peltier devices. The typical thermoelectric properties at room temperature of some of these materials are displayed in Table 25.1 (the nickel, chromium, bismuth, antimony, n-doped silicon–germanium, p-doped silicon–germanium, n-type material bismuth telluride, and p-type material antimony telluride, respectively). For operation at temperatures around 800 K, lead antimony

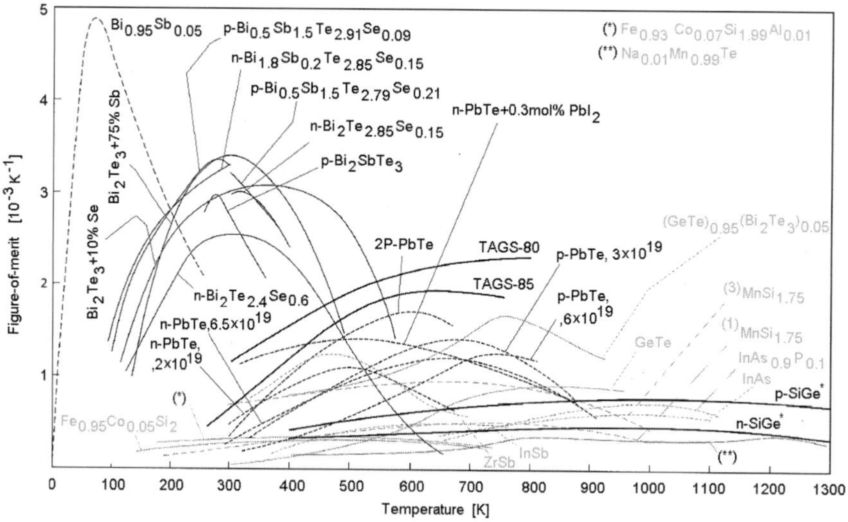

Fig. 25.9 The figure-of-merit, $Z[K^{-1}]$, calculated for different materials at different temperatures [36]

Table 25.1 The thermoelectric properties of some materials

Material	α [$\mu V\,K^{-1}$]	[$\mu\Omega$ m]	κ[$Wm^{-1}K^{-1}$]	ZT	Temp. [K]
Ni	−18	0.07	91	0.015	300
Cr	18	0.13	94	0.008	300
Bi	−60	1.15	8.4	0.11	300
Sb	40	0.42	18.5	0.062	300
Si–Ge (n)	−242	17.8	4.2	0.94	1,200
Si–Ge (p)	−240	31.9	4.38	0.5	1,200
Bi_2Te_3 (n)	−240	10	2.02	0.86	300
Sb_2Te_3 (p)	92	3.23	1.63	0.48	300

telluride shows the highest figure-of-merit. A ZT value around 1 was reported at 800 K [37]. However, there are environmental restrictions to the use of lead. Silicon–germanium is a candidate material for operation at temperatures above 1,000 K. A figure-of-merit around unity was achieved at 1,200 K [38]. These materials also have the advantage of easy integration with microelectronics.

Due to its compatibility with IC technology, polycrystalline SiGe alloys and polycrystalline Si are commonly used in thermopile applications. Their use in microcoolers has been investigated; however, their performance is very low compared to that of tellurium compounds, which have been used for many years in conventional large-area Peltier devices [14]. Tellurium compounds (n-type bismuth telluride, Bi_2Te_3 and p-type antimony telluride, Sb_2Te_3) are well-established room temperature thermoelectric materials and are widely employed by the industry, in conventional thermoelectric generators and coolers. Several deposition techniques have been investigated for their suitability for fabricating thin-films materials. The direct evaporation of the bulk materials for the deposition of Bi_2Te_3 films was demonstrated by Da Silva to be non-suitable. The large differences in vapor pressure of bismuth and tellurium result in a compositional gradient along the thin-film thickness [32]. Other techniques explored for the deposition of Bi_2Te_3 thin-films are thermal co-evaporation [39], electrochemical deposition [40], co-sputtering [3], flash evaporation [41] and metal-organic chemical vapor deposition (MOCVD) [42]. Although all these approaches are in principle suitable, the co-evaporation was used in this work to obtain both n-type Bi_2Te_3 and p-type Sb_2Te_3 thin-films, because it allows to precisely control the estequiometry of the deposited thin-film with the lowest costs. Only MOCVD is better than the co-evaporation to obtain thin-films with good uniformity and with the desired estequiometry. However, in order to do the deposition of thin-films by MOCVD, a reactor chamber is needed. Thus, this process is too expensive and it requires additional security procedures, when compared with co-evaporation [43].

The influence of deposition parameters on the thin-film performance was analyzed in detail. For obtaining data that were statistically significant, more than 100 samples were fabricated. One essential parameter in the fabrication of thermoelectric microdevices is the film-to-substrate adhesion. To investigate this issue three different substrate materials have been used: glass, silicon, and polyimide (kapton). However, for actual thermoelectric applications, a kapton film was chosen as substrate, because of the low-thermal conductivity (0.12 $Wm^{-1}K^{-1}$). Moreover, the thermal expansion coefficient ($12 \times 10^{-6}K^{-1}$) closely matches the thermal expansion coefficient of the telluride thin-films, thus reducing residual stress and increasing the adhesion. The information obtained on silicon and glass is important for MEMS-based thermoelectric devices, where micromachining is applied for thermal definition of the microstructure. Flexible substrates enable the integration with many novel types of devices, however, also introduce complications, such as the uncommon mechanical properties of the composite film–substrate.

25.3.1 Bismuth and Antimony Tellurides Depositions

Two different approaches can be used for on-chip integration of thermoelectric (TE) devices: transversal (off-plane or vertical) and lateral (in-plane), depending on the direction in which heat is transported, relative to the surface of the device. These fabrication examples felt on the lateral heat-flow approach, due to its easier fabrication process and compliance with planar technology [44]. Figure 25.10 shows a process flow used for fabricating thermoelectric converters. The fabrication steps are as follows: a thin-layer of metal (aluminum – Al) is deposited by a direct current (DC) sputtering on a polyimide substrate. Then, the patterning of the metal layer is done, and the contacts are obtained (a). Next, the n-type Bi_2Te_3 thin-film is deposited by thermal co-evaporation (b). The next step starts with the deposition of a negative photoresist (PR) layer, followed by an expose to UV (ultra-violet) light with a mask made of glass and nickel (to block the UV light) placed between the PR and the UV light source. After the exposition to the UV light, the UV-protected areas will be removed during the PR development (c). The n-type elements are patterned by photolithography (d). The n-type thin-film is etched in HNO_3 and p-type Sb_2Te_3 thin-film is deposited by thermal co-evaporation (e). A new layer of PR is deposited, exposed to a UV light source and developed (f). The thermoelectric layer made of Sb_2Te_3 is patterned by wet etching in $HNO_3 : HCl$ bath (g) and finally, the PR is removed (h). A protective layer of silicon nitride (Si_3N_4) can also be deposited by low-temperature hotwire chemical vapor deposition (HW-CVD) and patterned if required by the application.

The thermoelectric thin-film structures were deposited by thermal co-evaporation (see the deposition setup in Fig. 25.11) with the help of a high-vacuum chamber (with the base pressure settled to about 1.33×10^{-4} Pa). Two large molybdenum boats (baffled boxes, with a volume of 4cm^3) were used at the same time, one for each of the elementary materials required to produce the desired compound. The power applied to each boat is controlled independently, using two computed proportional-integral derivative (PID) controllers [45] to maintain the deposition rate at user-defined constant values, during the deposition process. Two thickness monitors (quartz crystal oscillators) are carefully placed inside the chamber in such a way that each of them receives material only from the boat it is monitoring. A metal sheet is placed between the two boats to ensure deposition of a material at the

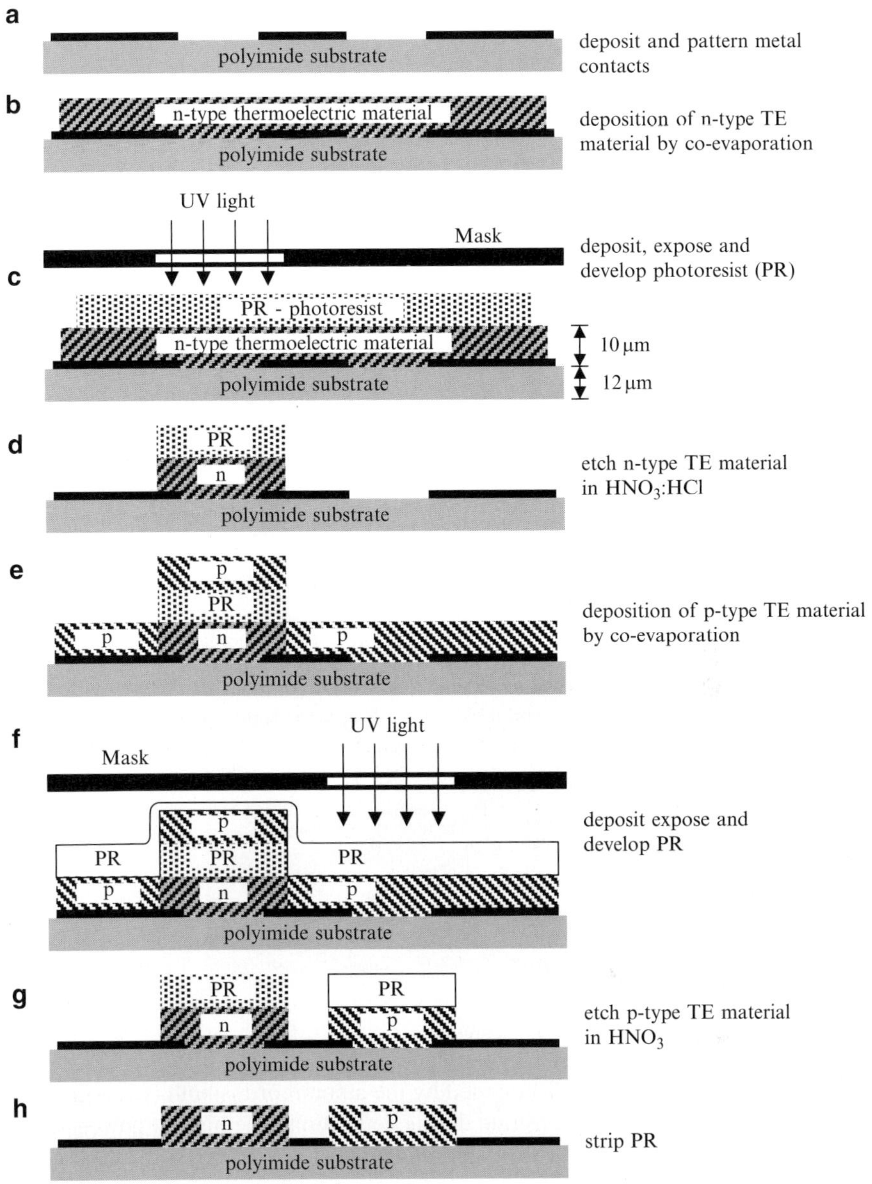

Fig. 25.10 A sequence of fabrication steps of thermoelectric devices

respective quartz crystal sensor only. Substrates are heated to the temperature set point, $T_{\mathrm{sub}}[°\mathrm{C}]$, in the range 150–270° C.

The graphs showing the influence of the evaporation rate, R, of each material on thermoelectric properties of the compounds are presented in Fig. 25.12. The evaporation flow rate ratio, $R = \mathrm{Fr}_{\mathrm{Te}}/\mathrm{Fr}_{\mathrm{Bi,Sb}}$, is defined as the amount (in volume of the

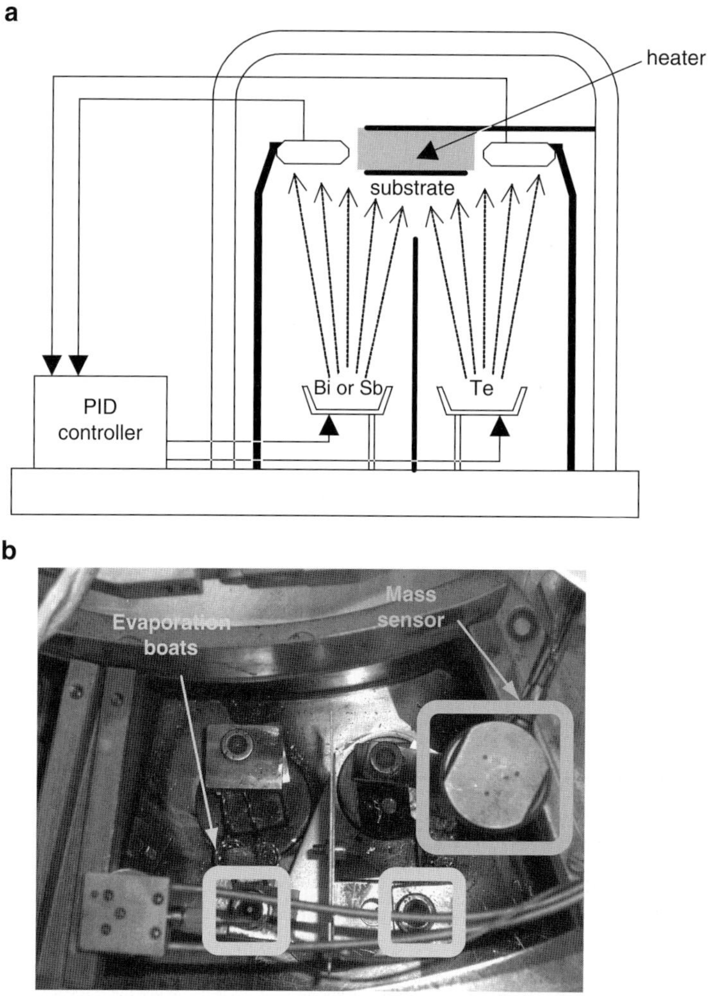

Fig. 25.11 The system used in the co-evaporation

deposited film) of tellurium (Te) divided by the amount of bismuth (Bi) – or antimony (Sb) – that arrives the substrate during deposition. The highest power-factor (PF) was obtained with a Bi (or Sb) evaporation rate of 2 $\text{Å}s^{-1}$ and a Te evaporation rate of 6–7$\text{Å}s^{-1}$, which corresponds to an evaporation flow rate ratio in the range 3–3.5.

The best values of T_{sub} for Bi_2Te_3 and for Sb_2Te_3 thin-films were about 270°C and 220°C, respectively. Finally, it must be noted that all thin-films were deposited on a polyimide (kapton) foil with a thickness of 25 μm. Figure 25.13 shows two SEM cross-section and surface images of both Bi_2Te_3 and Sb_2Te_3 thin-films, where their polycrystalline structure can be confirmed. Also, the former temperatures and the optimal evaporation flow rate ratios, R, were those which resulted in thin-films with larger grain size. This is of major concern, because a crystalline structure with

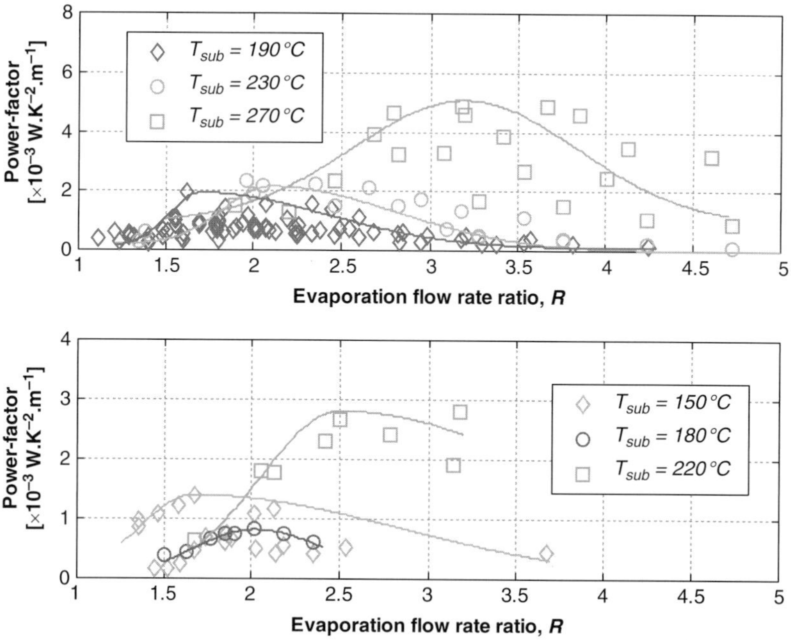

Fig. 25.12 The power-factor of Bi_2Te_3 (*top*) and Sb_2Te_3 (*bottom*) thin-films as a function of the Te/Bi evaporation flow rate ratio, R, and the respective curve fittings (*solid lines*)

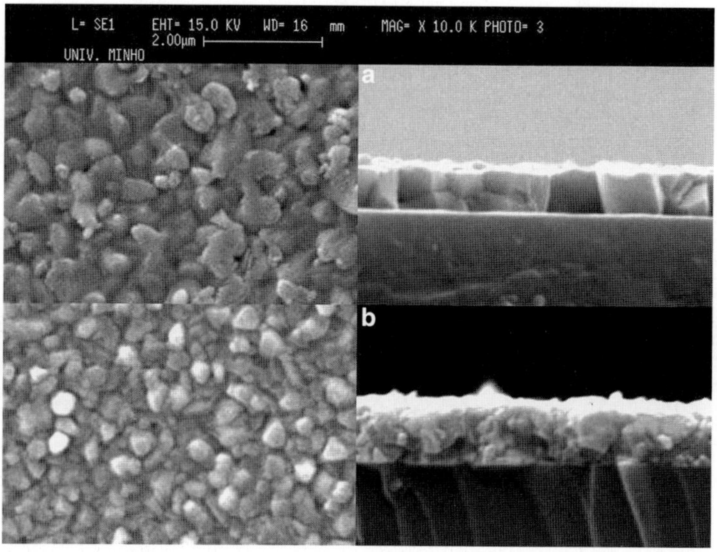

Fig. 25.13 SEM top view (on left) and cross-sectional (on right) images of (a) Bi_2Te_3 (on top) (b) Sb_2Te_3 (on bottom) thin-films

an increased grain size is less resistive, whose consequence is a thin-film with an increased thermoelectric figures-of-merit, ZT.

25.3.2 *Optimization of Thermoelectric Properties*

The in-plane thin-film electrical resistance was measured using the conventional four probe van der Pauw method, at room temperature. The thermal conductivity [$W\,m^{-1}K^{-1}$] was measured using the method proposed by Völklein [46], and values of 1.3$W\,m^{-1}K^{-1}$ and 1.8 $Wm^{-1}K^{-1}$ were obtained for the Bi_2Te_3 and Sb_2Te_3 thin-films (which were deposited as conditions to obtain the maximum power-factor), respectively. The measurements of the Seebeck coefficient were made, by connecting one side of the thin-film to a fixed temperature (heated metal block) and the other side to a heat sink at room temperature.

Figure 25.14 shows a X-ray diffraction (XRD) spectrum of the best Bi_2Te_3 (top plot) and Sb_2Te_3 (bottom plot) thin-film, which reveal their polycrystalline structures. The peaks agree with the diffractograms for polycrystalline Bi_2Te_3. The peaks also agree with the power diffraction spectra for polycrystalline Sb_2Te_3. Tables 25.2

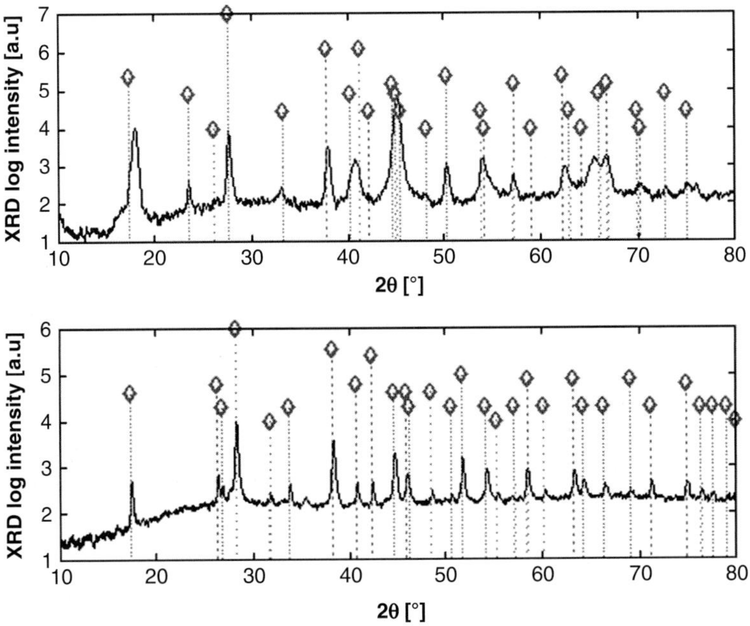

Fig. 25.14 XRD analysis of Bi_2Te_3 (*top plot*) and Sb_2Te_3 (*bottom plot*) thin-film analysis. In the both plots, the peaks agree well with the respective power diffraction spectrums (*dotted lines* and *diamonds*)

Table 25.2 Properties of the best samples of Bi_2Te_3 thin-films

TF	T_{sub}[°C]	R	%Te by EDX	α [μVK^{-1}]	ρ [μΩ m]	PF [WK^{-2}m^{-1}]	ZT at 300 K
#1	190	1.70	–	−180	16.6	1.95×10^{-3}	0.39
#2	230	2.10	62.8	−156	11.3	2.16×10^{-3}	0.43
#3		3.10	62.2	−152	13.4	1.72×10^{-3}	0.34
#4	240	3.20	59.1	−180	16.6	1.95×10^{-3}	0.40
#5	270	3.20	62.0	−248	12.6	4.87×10^{-3}	0.97
#6		3.90	–	−220	10.6	4.57×10^{-3}	0.91

Table 25.3 Properties of the best samples of Sb_2Te_3 thin-films

TF	T_{sub}[°C]	R	%Te by EDX	α [μVK^{-1}]	ρ [μΩ m]	PF [W K^{-2}m^{-1}]	ZT at 300 K
#1	150	1.47	54.5	91	7.6	1.09×10^{-3}	0.22
#2		1.67	61.4	140	14.0	1.40×10^{-3}	0.28
#3	180	2.02	59.1	158	30.3	0.82×10^{-3}	0.16
#4		2.35	62.4	156	39.1	0.62×10^{-3}	0.12
#5	220	2.50	67.3	156	9.2	2.66×10^{-3}	0.53
#6		3.18	73.5	188	12.6	2.81×10^{-3}	0.56

and 25.3 show the most important thermoelectric characteristics obtained from the measurements applied on the best samples of Bi_2Te_3 and Sb_2Te_3 thin-films. The measurements done in the best samples show an absolute value of the Seebeck coefficient in the range of 91–248 μ VK^{-1}. An in-plane electrical resistivity of 7.6–39.1μΩ m was obtained. The measurements for the Bi_2Te_3 and Sb_2Te_3 thin-films also reveal figures-of-merit, ZT, of 0.97 and 0.56, and power-factors, *PF*, of 4.87×10^{-3}W K^{-1}m^{-2} and 2.81×10^{-3} WK^{-1}m^{-2}, respectively.

25.4 Superlattices for Fabrication of Thermoelectric Converters

25.4.1 Why Superlattices?

The dimensionless thermoelectric figure-of-merit, ZT, in bismuth/antimony/telluride bulk alloys has remained around the unity for more than 50 years. The best results up to now showed that a peak ZT of 1.4 at 100°C could be achieved in a p-type nanocrystalline BiSbTe bulk alloy [47]. Such nanocrystalline bulk materials were made by hot pressing nanopowders that were ball-milled from crystalline ingots under inert conditions. Also it was observed that the ZT improvement was the result of low thermal conductivity caused by the increased phonon scattering by grain boundaries and defects. But the most important result was a ZT of about 1.2 at the room temperature and 0.8 at 250°C, making these materials useful for cooling and power generation [47]. However, the obtainance of these materials is not an easy task, because nanocrystalline structures must be obtained (see Fig. 25.5).

The superlattices are an alternative but simpler way to improve the thermoelectric characteristics of the structures, because it is only need to alternate layers of two different materials with thickness in the range of tens of nanometers (see Fig. 25.4). Previous work [27] demonstrates that superlattices can be fabricated by CVD deposition of alternating layers. However, the cluster formation during the deposition of few nanometers can avoid a homogeneous film thickness. A low-substrate temperature is used to improve the thickness homogeneity of the thin-films.

25.4.2 Materials and Properties

In order to enhance the ZT, a high Seebeck coefficient, an increased electrical conductivity, and a decreased thermal conductivity are needed, but the conventional bulk materials properties often counter each other [48]. Because the performance of thermoelectric devices is determined by the figure-of-merit, ZT, of the thermoelectric materials. In order to obtain the optimal performance of a superlattice thermoelectric device, both the superlattice structure and the device must be optimized in the design and the fabrication process [49]. There are available thermoelectric structures made of InGaAs/InGaAsP superlattices and such study had proven that the superlattice at a certain periodic length has minimal thermal conductivity, which, in general, corresponds to an enhanced ZT [49].

Superlattices with improved thermoelectric figures-of-merit, ZT, were also fabricated by several research teams around the world. These thermoelectric structures can be made with materials such as Bi_2Te_3/Sb_2Te_3, Bi_2Te_3/Bi_2Se_3 [27,28,34,50–56] and PbSeTe/PbTe [57]. Despite the small thermoelectric performance of isolated Bi_2Te_3 and Sb_2Te_3 alloys, the thermoelectric figure-of-merit, ZT, can be significantly increased to about 2.4 in Bi_2Te_3/Sb_2Te_3 superlattices [27].

25.4.3 Fabrication

A huge number of conventional techniques already used by the microelectronics industry can be successfully applied for fabricating thermoelectric devices. Such techniques include the photolithography, electroplating, wafer dicing, and pick-and-place systems. This allows the fabrication of a wide spectrum of thermoelectric devices that range from the simple modules for handling a few microwatts [58] (in cooling or energy conversion) to more complex module arrays (to handle higher power levels) [59].

The quality of the deposited thin-films (either both constitutive layers or the entire superlattice) must be high in order to successfully fine-tuning the thin-film thicknesses in a superlattice so as to hinder lattice vibrations (thus decreasing heat flow) and transmit electrons (increasing electrical conductivity), using materials that intrinsically have a large thermopower; this generates a thermoelectric material with a particularly high figure-of-merit [60]. The conventional imaging techniques based on X-rays diffractions do not offer satisfactory resolution to make a quality assessment of the surface of the deposited thin-films. In this context, the Scanning Probe Microscopy (SPM) technique fills such a gap, because high-resolutions of a few picometers can be obtained. This is due to the resolution to be only dependent of the size of the probe–sample interaction volume [61].

References

1. T.I. Seebeck, Abhandlungen der Deutschen Akademie der Wissenschaften zu Berlin, 265–373 (1822)
2. J.C. Peltier, Annales de Chimie et the Physique, LVI **56**, 371–386 (1834)
3. H. Böttner, J. Nurnus et al., J. Microelectromech. Syst. **13**, 414–420 (2004)
4. L.W. da Silva, M. Kaviany, J. Mem. Science **14**, 1110 (2005)
5. G.J. Snyder et al., Nat. Mater. Lett. **2**, 528–531 (2003)
6. H. Böttner, in *Proceedings of ICT'05*, Clemson, SC, USA, 2005
7. Seiko electronics, http://www.sii.co.jp/info/eg/thermic_main.html
8. Citizen electronics, http://www.citizen.co.jp/english/release/01//03basel/thermo.htm
9. Micropelt, http://www.micropelt.com
10. Nanocoolers, http://www.nanocoolers.com
11. F. Völklein et al., in *Proceedings of 18th International Conference on Thermoelectrics – ICT '99*, Baltimore, MD, USA, 1999, pp. 285–293
12. L.W. da Silva, M. Kaviany, J. Microelectromech. Syst. **14**, 1110 (2005)
13. C. Shafai, Natl. Libr. Canada **76** (1998)
14. D.D.L. Wijngaards et al., Sens Actuators A Phys Sens **85**, 316–323 (2000)
15. S. Schaevitz, *A MEMS Thermoelectric Generator* (Massachusetts Institute of Technology, Cambridge, MA, 2000)
16. G. Min, D.M. Rowe, Solid-State Electron. **43**, 923–929 (1999)
17. F. Volklein et al., Sens Actuators A Phys Sens **75**, 95–101 (1999)
18. A. Jacquot et al., Sens Actuators A Phys Sens **116**, 501–508 (2004)
19. I. Stark, M. Stordeur, in *Proceedings of 18th International Conference on Thermoelectrics – ICT'99*, Baltimore, MD, USA, 1999, pp. 465–472
20. M. Stordeur, I. Stark, in *Proceedings of 16th International Conference on Thermoelectrics – ICT'97*, 1997
21. W. Qu et al., J. Micromech. Microeng. **11**, 146–152 (2001)
22. U. Ghoshal, R. Schmidt, in *Proceedings of IEEE International Solid-State Circuits Conference – ISSCC'00*, San Francisco, CA, USA, 2000
23. R. Venkatasubramanian, Semiconduct. Semimet. **71**, 175–201 (2001)
24. G. Chen et al., Int. Mater. Rev. **48** (2003)
25. M.S. Dresselhaus, in *Semiconductors and Semimetals:Recent Trends in Thermoelectric Materials Research III*, vol. 71, ed. by T.M. Tritt (Academic, London, 2001), pp. 1–121
26. R. Venkatasubramanian et al., in *Proceedings of 1st Natl Thermogenic Cooler Workshop*, Center for Night Vision and Electro-Optics, Fort Belvoir, VA, 1992, pp. 196–231
27. R. Venkatasubramanian et al., Nature **413**, 597 (2001)
28. L.M. Gonçalves et al., J. Micromech. Microeng. S168–S173 (2007)
29. H. Zou et al., Thin Solid Films **408**, 270–274 (2002)
30. L. da Silva et al., J. Appl. Phys. **97**, 114903 (2005)
31. D. Kim et al., Thin Solid Films **510**, 148–153 (2006)
32. L. da Silva, M. Kaviany, in *Proceedings of IMECE 2002–2002 ASME International Mechanical Engineering Congress & Exposition*, New Orleans, LA, USA, 2002. Paper No. 2-8-1-6

33. D.M. Rowe (ed.), *Handbook of Thermoelectric* (CRC, Boca Raton, FL, 1987), pp. 211–237
34. L.M. Goncalves et al., Sens. Actuators A Phys. Sens. **145–146**, 75–80 (2008)
35. L. Bell, Science **321**, 1457–1461 (2008)
36. D.D.L. Wijngaards, Phd Thesis, Tu Delft, 2003
37. V. Fano, in *CRC Handbook of Thermoelectrics*, ed. by D.M. Rowe (CRC, Boca Raton, FL, 1987), p. 261
38. C.B. Vining, in *CRC Handbook of Thermoelectrics*, ed. by D.M. Rowe (CRC, London, 1987), p. 329
39. Z. Helin et al., Thin Solid Films **408**, 270–274 (2002)
40. J. Lim et al., in *Proceedings of ICT2002*, Long Beach, 2002, pp. 535–539
41. A. Foucaran, Mater. Sci. Eng. B **52**, 154–161 (1998)
42. A. Giani et al., Mater. Sci. Eng. B **64**, 19–24 (1999)
43. Z.M. Lia et al., J. Cryst. Growth **311**, 4679–4684 (2009)
44. D.D.L. Wijngaards, R.F. Wolffenbuttel, IEEE Trans. Electron. Devices **52** (2005)
45. Y.F. Chan et al., IEEE Trans. Ind. Electron. **54**, 1898–1906 (2007)
46. F. Völklein, in *Proceedings of Symposium on Microtechnology in Metrology and Metrology in Microsystems*, Delft, The Netherlands, 2000
47. B. Poudel et al., Science **320**, 634–638 (2008)
48. T.C. Harman et al., Science **297**, 2229–2232 (2002)
49. Y. Zhang et al., J. Microelectromech. Syst. **16**, 1113–1119 (2007)
50. L.M. Gonçalves et al., Thin Solid Films **518**, 2816–2821 (2010)
51. L.M. Gonçalves et al., Vaccum **82**, 1499–1502 (2008)
52. L.M. Gonçalves et al., Sens. Actuators A Phys. Sens. **130–131**, 346–351 (2006)
53. L.M. Gonçalves et al., in *Proceedings of IECON 2009*, Porto, Portugal, 2009, pp. 4076–4080
54. L.M. Gonçalves et al., in *Proceedings of MME 2008*, Aachen, Germany, 2008, pp. 49–52
55. L.M. Gonçalves et al., in *Proceedings of Transducers 2005*, Seoul, Korea, 2005, pp. 904–907
56. J.P. Carmo et al., Electron. Lett. **45**, 803–805 (2009)
57. H. Böttner et al., MRS Bull. **31**, 211–217 (2006)
58. J.R. Heath, Nature **445**, 492–493 (2007)
59. J.P. Carmo et al., IEEE Trans. Ind. Electron. **57**, 861–867 (2010)
60. G. Min et al., IEEE Trans. Energy Convers. **32**, 528–534 (2007)
61. M. Miles, Science **277**,1845–1847 (1997)

Index

Printing and Binding: Stürtz GmbH, Würzburg